August 22–26, 2012
Istanbul, Turkey

Association for Computing Machinery

Advancing Computing as a Science & Profession

MobiCom'12

Proceedings of the 18th Annual International Conference on Mobile Computing and Networking

Sponsored by:

ACM SIGMOBILE

Supported by:

AveaLabs, Koc University, HP, Microsoft Research, and Tübitak

Association for
Computing Machinery

Advancing Computing as a Science & Profession

The Association for Computing Machinery
2 Penn Plaza, Suite 701
New York, New York 10121-0701

Notice to Past Authors of ACM-Published Articles
ACM intends to create a complete electronic archive of all articles and/or other material previously published by ACM. If you have written a work that has been previously published by ACM in any journal or conference proceedings prior to 1978, or any SIG Newsletter at any time, and you do NOT want this work to appear in the ACM Digital Library, please inform permissions@acm.org, stating the title of the work, the author(s), and where and when published.

ISBN: 978-1-4503-1159-5 (Digital)

ISBN: 978-1-4503-1749-8 (Print)

Additional copies may be ordered prepaid from:

ACM Order Department
PO Box 30777
New York, NY 10087-0777, USA

Phone: 1-800-342-6626 (USA and Canada)
+1-212-626-0500 (Global)
Fax: +1-212-944-1318
E-mail: acmhelp@acm.org
Hours of Operation: 8:30 am – 4:30 pm ET

Printed in the USA

ACM MobiCom 2012 General Co-Chairs' Welcome

It is our distinct pleasure to welcome you to the *18th Annual International Conference on Mobile Computing and Networking (MobiCom)* in Istanbul, Turkey, a unique locale straddling two continents. This year's conference continues its tradition of being the premier forum for presentation of research results addressing emerging issues in networks, systems, algorithms, and applications that support the symbiosis of mobile computers and wireless networks.

The three-day conference also includes two keynote addresses by Professors Ian F. Akyildiz and P. R. Kumar. Professor Akyildiz is well known for his contributions to various aspects of wireless networking. His insights into new research directions are of high interest to the MobiCom community. In his keynote address, he will introduce nanonetworking, a new networking paradigm with a direct impact on many key aspects of our future daily lives. This year, we are also pleased to host Professor P.R. Kumar, who will receive the 2012 SIGMOBILE Outstanding Contribution Award in Istanbul. With his wide range of interests and, indeed, outstanding contributions, Professor Kumar is a keystone of our research community. In his talk, he will share his vision on tackling three very important problems in wireless networking.

The main symposium is also preceded and followed by a total of six full-day workshops, each focusing on specific areas of interest of the broader MobiCom community. There are also Posters, Demonstrations, and Student Research Competitions, which we believe will bring together researchers with a wide range of interests.

As always, this year's conference is a collaborative effort of a large number of volunteers. Program Chairs Alex Snoeren and Lili Qiu, and the Program Committee members carried out the hard task of forming an excellent technical program from a large number of submissions. Workshop Chairs and Program Committees they led organized the technical programs of their events. Finance Chair (Tracy Camp), Registration Chair (Mehmet Can Vuran), and Local Arrangements Chairs (Tuna Tuğcu and Berk Canberk) were responsible for attendee relations. Giovanni Pau and Jiang Xie led the Publicity and Web related activities. Workshops Chair Tommaso Melodia headed the coordination of the six workshops. Falko Dressler organized the demo and poster session. Wenye Wang and Matteo Cesana secured generous financial supports from National Science Foundation and European Network of Excellence TREND for student travel grants. They also organized the Student Research Competition. We thank the entire organization committee for their impeccable efforts.

We would like to thank ACM staff, Adrienne Griscti, April Mosqus, and Stephanie Sabal as well as Lisa Tolles of Sheridan Printing Company.

Finally, we would like to thank ACM SIGMOBILE for sponsoring the conference; and our Gold Supporter AVEA Labs; our Silver Supporters Microsoft Research, Hewlett Packard Labs, and Koç University; and our Bronze Supporter The Scientific and Technological Research Council of Turkey (TÜBİTAK) for their generous financial support. We also extend our special thanks to the National Science Foundation and European Network of Excellence TREND for providing funds for student travel grants.

Özgür B. Akan
MobiCom 2012 General Co-Chair
Koç University, Turkey

Eylem Ekici
MobiCom 2012 General Co-Chair
The Ohio State University, USA

ACM MobiCom 2012 Program Chairs' Welcome

Welcome to Istanbul and to *ACM MobiCom 2012,* the 18th Annual Conference on Mobile Computing and Networking. MobiCom is known for top-quality research in mobile computing and networking, and this year is no different. The technical program represents the center of gravity of the conference, and is the result of a seven-month-long effort on behalf of the organizers—and even longer than that on behalf of the submitting authors. We are delighted with this year's program and hope you have a stimulating, enjoyable, and fruitful experience at the conference!

The Call for Papers attracted 212 papers submissions from 26 countries, out of which 32 were selected for inclusion in the program. All papers (except for a small handful that were rejected without review due to format violations) received three initial reviews, and 110 manuscripts were selected for further consideration. These papers received an additional set of reviews, and an on-line discussion identified 70 manuscripts for further discussion at the in-person PC meeting. The TPC assembled in Austin, Texas, where we crafted the final program over the course of two days. Each paper was reviewed in a double-blind fashion, and TPC members with conflicts of interest were excused from the room during discussion. A senior TPC member managed the review process for papers authored by either of the TPC co-chairs.

We would like to thank the TPC members for their hard work and dedication to the review process. This year marked a return to a smaller PC, which we believe led to increased consistency and a higher energy level in the TPC meeting, but also resulted in a larger reviewing load for the individual members: The average TPC member reviewed 25 papers. We also required attendance at the TPC meeting, which involved significant travel for many international members. Despite the jet lag, the discussions were heated and productive, and we hope you enjoy the results.

Due to logistical constraints, this year's publication schedule was more compact than most. We appreciate the willingness of the authors and shepherds to meet the resulting challenges, including a last-minute change in camera-ready formatting, which led to a slightly more varied paper length in this year's proceedings than is typical.

Finally, we are indebted to Eylem Ekici and Ozgur Akan for arranging the conference, allowing us to focus entirely on selecting the technical program. As always, Eddie Kohler and Geoff Voelker gave generously of their time to support HotCRP and Banal, respectively, both of which were instrumental in managing the submission process. Peter Steenkiste graciously and seamlessly managed the out-of-band review process for papers with which we had a conflict. Thank you all! We look forward to seeing everyone at the conference.

Lili Qiu
ACM MobiCom'12 Program Co-Chair
University of Texas, Austin, USA

Alex C. Snoeren
ACM MobiCom'12 Program Co-Chair
University of California, San Diego, USA

Table of Contents

Session 4: Cellular Networks

Session Chair: Ranveer Chandra *(Microsoft Research)*

Session 5: Potpourri

Session Chair: Ben Greenstein *(Google)*

Session 6: Sensing

Session Chair: Moustafa Youssef *(Egypt-Japan University of Science and Technology)*

Session 7: Wireless Video

Session Chair: Sunghyun Choi *(Seoul National University)*

Session 8: Leveraging MIMO

Session Chair: Alex Snoeren *(University of California, San Diego)*

Session 9: Localization

Session Chair: Victor Bahl *(Microsoft Research)*

Session 10: Low Power

Session Chair: Robin Kravets *(University of Illinois at Urbana-Champaign)*,

Session 11: Algorithms

Session Chair: Xinbing Wang *(Shanghai Jiao Tong University)*

Posters

Demos

ACM MobiCom 2012 Conference Organization

General Co-Chairs: Özgür B. Akan, *Koç University, Turkey*
Eylem Ekici, *The Ohio State University, USA*

Program Co-Chairs: Lili Qiu, *University of Texas at Austin, USA*
Alex C. Snoeren, *University of California, San Diego, USA*

Workshops Chair: Tommaso Melodia, *State University of New York, SUNY at Buffalo, USA*

Local Arrangements Co-Chairs: Tuna Tuğcu, *Bogazici University, Turkey*
Berk Canberk, *Istanbul Technical University, Turkey*

Publicity Co-Chairs: Giovanni Pau, *University of California, Los Angeles, USA*
Jiang (Linda) Xie, *University of North Carolina at Charlotte, USA*

Finance Chair: Tracy Camp, *Colorado School of Mines, USA*

Registration Chair: Mehmet Can Vuran, *University of Nebraska-Lincoln, USA*

Student Grants & SRC Co-Chairs: Wenye Wang, *North Carolina State University, USA*
Matteo Cesana, *Politecnico di Milano, Italy*

Demos and Posters Chair: Falko Dressler, *University of Innsbruck, Austria*

Steering Committee: Victor Bahl, *Microsoft Research, USA*
David B. Johnson, *Rice University, USA*

Program Committee: Victor Bahl, *Microsoft Research, Redmond, USA*
Suman Banerjee, *University of Wisconsin-Madison, USA*
Elizabeth Belding, *University of California, Santa Barbara, USA*
Srdjan Capkun, *ETH Zurich, Switzerland*
Augustin Chaintreau, *Columbia University, USA*
Ranveer Chandra, *Microsoft Research, Redmond, USA*
Sunghyun Choi, *Seoul National University, Korea*
Romit Roy Choudhury, *Duke University, USA*
Samir Das, *SUNY Stony Brook, USA*
Deepak Ganesan, *University of Massachusetts, Amherst, USA*
Ben Greenstein, *Google, USA*
Dirk Grunwald, *University of Colorado, Boulder, USA*
Marco Gruteser, *Rutgers University, USA*
Kyle Jamieson, *University College London, UK*

Program Committee (continued):

Sneha Kasera, *University of Utah, USA*
Sachin Katti, *Stanford University, USA*
Edward Knightly, *Rice University, USA*
Ulas Kozat, *DOCOMO Innovations, USA*
Robin Kravets, *University of Illinois, Urbana-Champaign, USA*
Baochun Li, *University of Toronto, Canada*
Yunhao Liu, *Tsinghua University, China*
Songwu Lu, *University of California, Los Angeles, USA*
John C. S. Lui, *City University of Hong Kong, China*
Patrick McDaniel, *Pennsylvania State University, USA*
Thyaga Nandagopal, *National Science Foundation, USA*
Dina Papagiannaki, *Telefonica I+D, Spain*
Ram Ramjee, *Microsoft Research, India*
Peter Steenkiste, *Carnegie Mellon University, USA*
Karthik Sundaresan, *NEC Labs, USA*
Kun Tan, *Microsoft Research, China*
Xinbing Wang, *Shanghai Jiao Tong University, China*
Moustafa Youssef, *Egypt-Japan University of Science & Technology, Egypt*
Heather Zheng, *University of California, Santa Barbara, USA*

ACM MobiCom 2012 Sponsor & Supporters

Sponsor: sigmobile

Gold Supporter: avealabs

Silver Supporters: KOÇ UNIVERSITY, hp, Microsoft Research

Bronze Supporter: TÜBİTAK

Keynote Talk

Nanonetworks: A New Frontier in Communications

Ian F. Akyildiz
Broadband Wireless Networking Lab
School of Electrical and Computer Engineering
Georgia Institute of Technology
Atlanta, GA 30332, USA
ian@ece.gatech.edu

Abstract
Nanotechnology is enabling the development of devices in a scale ranging from one to a few one hundred nanometers. Nanonetworks, i.e., the interconnection of nano-scale devices, are expected to expand the capabilities of single nano-machines by allowing them to cooperate and share information. Traditional communication technologies are not directly suitable for nanonetworks mainly due to the size and power consumption of existing transmitters, receivers and additional processing components. All these define a new communication paradigm that demands novel solutions such as nano-transceivers, channel models for the nano-scale, and protocols and architectures for nanonetworks. In this talk, first the state-of-the-art in nano-machines, including architectural aspects, expected features of future nano-machines, and current developments are presented for a better understanding of the nanonetwork scenarios. Moreover, nanonetworks features and components are explained and compared with traditional communication networks. Novel nano-antennas based on nano-materials as well as the terahertz band are investigated for electromagnetic communication in nanonetworks. Furthermore, molecular communication mechanisms are presented for short-range networking based on ion signaling and molecular motors, for medium-range networking based on flagellated bacteria and nanorods, as well as for long-range networking based on pheromones and capillaries. Finally, open research challenges such as the development of network components, molecular communication theory, and new architectures and protocols, which need to be solved in order to pave the way for the development and deployment of nanonetworks within the next couple of decades are presented.

Categories & Subject Descriptors: C.2.m.
Computer Communication Networks: Miscellaneous.

Keywords: nanonetworks, nano-scale devices, nano-machines

Bio
I. F. AKYILDIZ received his BS, MS, and PhD degrees in Computer Engineering from the University of Erlangen-Nuernberg, Germany, in 1978, 1981 and 1984, respectively. Currently, he is the Ken Byers Distinguished Chair Professor with the School of Electrical and Computer Engineering, Georgia Institute of Technology, Director of the Broadband Wireless Networking Laboratory and Chair of the Telecommunications Group at Georgia Tech. Dr. Akyildiz is an Honorary Professor with School of Electrical Engineering at the Universitat Politecnica de Catalunya, and Director of N3Cat (NaNoNetworking Center in Catalunya) in Barcelona, Spain, since June 2008. He is the Editor-in-Chief of Computer Networks (Elsevier) Journal since 2000, the founding Editor-in-Chief of the Ad Hoc Networks Journal (Elsevier) in 2003, the founding Editor-in-Chief of the Physical Communication (PHYCOM) Journal (Elsevier) in 2008, and the founding Editor-in-Chief of the Nano Communication Networks (NANOCOMNET) Journal (Elsevier) in 2010.

Dr. Akyildiz is an IEEE FELLOW (1996) and an ACM FELLOW (1997). Dr. Akyildiz received the 1997 IEEE Leonard G. Abraham Prize award (IEEE Communications Society) for his paper entitled "Multimedia Group Synchronization Protocols for Integrated Services Architectures" published in the IEEE Journal of Selected Areas in Communications (JSAC) in January 1996. Dr. Akyildiz received the 2003 Best Tutorial Paper Award (IEEE Communications Society) for this paper entitled "A Survey on Sensor Networks" published in the IEEE Communications Magazine, August 2002. Dr. Akyildiz received the Best Paper Award for "Interferer Classification, Channel Selection and Transmission Adaptation for Wireless Sensor Networks" in the Ad Hoc and Sensor Networks (AHSN) symposium at IEEE ICC, June 2009.

He received the "Don Federico Santa Maria Medal" for his services to the Universidad of Federico Santa Maria in Chile in 1986. He served as a National Lecturer for ACM from 1989 until 1998 and received the ACM Outstanding Distinguished Lecturer Award for 1994. Dr. Akyildiz received the 2002 IEEE Harry M. Goode Memorial award (IEEE Computer Society) with the citation "for significant and pioneering contributions to advanced architectures and protocols for wireless and satellite networking". He also received the 2003 ACM SIGMOBILE Outstanding Contribution Award for his "pioneering contributions in the area of mobility and resource management for wireless communication networks," September 2003.

Dr. Akyildiz received the 2004 Georgia Tech Faculty Research Author Award for his "outstanding record of publications of papers between 1999-2003," April 2004. He also received the 2005 Distinguished Faculty Achievement Award from School of ECE, Georgia Tech, April 2005. Dr. Akyildiz received the Georgia Tech Outstanding Doctoral Thesis Advisor Award for his 20+ years service and dedication to Georgia Tech and producing outstanding PhD students. He also received the 2009 ECE Distinguished Mentor Award by the Georgia Tech School of Electrical and Computer Engineering Faculty Honors Committee. Dr. Akyildiz received the Best Paper Award for "Interferer Classification, Channel Selection and Transmission Adaptation for Wireless Sensor Networks" in the Ad Hoc and Sensor Networks (AHSN) symposium at IEEE ICC, in June 2009. He also received the Best Paper Award for "Deployment

MobiCom'12, August 22–26, 2012, Istanbul, Turkey.
ACM 978-1-4503-1159-5/12/08.

Algorithms for Wireless Underground Sensor Networks using Magnetic Induction" in the IEEE Global Communications Conference (Globecom), in December 2010. Dr. Akyildiz received the 2010 IEEE Communications Society Ad Hoc and Sensor Networks Technical Committee (AHSN TC) Technical Recognition Award with the citation: "For pioneering contributions to wireless sensor networks and wireless mesh networks," in December 2010. He received the 2011 IEEE Computer Society W. Wallace McDowell Award for pioneering contributions to wireless sensor network architectures and communication protocols, in May 2011. He also received the 2011 TUBITAK (Turkish National Science Foundation) Exclusive Award for outstanding contributions to the advancement of scholarship/research at international level. Dr. Akyildiz is the author of an advanced textbook on "Wireless Sensor Networks" published by John Wiley and Sons in June 2010. Dr. Akyildiz is the author of an advanced textbook on "Wireless Mesh Networks" published by John Wiley and Sons in February 2009. Dr. Akyildiz serves on the advisory boards of several research centers, journals, high tech companies, conferences and publication companies. His current research interests are in Nanonetworks, Cognitive Radio Networks and Wireless Sensor Networks.

Keynote Talk

Three Theories for Delays, Clocks and Security in Wireless Networks

P.R. Kumar
Department of Electrical and Computer Engineering
Texas A&M University
College Station, TX 77843, USA
prk@tamu.edu

Abstract

We propose three theories, which can be regarded as attempts to characterize and establish guaranteed properties of wireless networks:

(i) How and to what extent can we deliver packets with hard delay bounds?

(ii) How and to what extent can we synchronize clocks in wireless networks?

(iii) Can we develop provably secure protocols for the entire life-cycle of wireless networks that also optimize a utility measure while operating in a hostile environment?

For the first problem, consider an access point serving several clients over unreliable wireless links. Suppose packets arrive for/from the clients, with each packet having a hard deadline, after which it is dropped. We characterize precisely the mix of delivery ratios, channel unreliabilities and hard deadline that the access point can guarantee, under some models.

For the second problem, consider a wireless network where clocks at the nodes are linear, though with different rates (skews) and offsets. Nodes can exchange packets with their neighbors, with direction dependent delays. We characterize precisely to what extent clocks can and cannot be synchronized and delays determined. Under a random model the end-to-end error can be kept bounded irrespective of network size.

Concerning the third problem, traditionally, wireless protocols have been developed to provide performance. As attacks are identified, the protocols are fortified against the identified vulnerabilities. However, holistic guarantees are not provided against other attacks. We seek to reverse this paradigm. We propose a provable approach that guarantees the protocol suite is secure when the nodes are subject to certain assumptions. The protocols take a set of good nodes mingled with unknown malicious nodes from primordial birth to an operating network, while attaining min-max of a utility function. The maximization is over protocols announced and followed by the good nodes, and the minimization is over all behaviors of the malicious nodes. Further, the malicious nodes are reduced to either cooperating or jamming [Joint work with Vivek Borkar, Nikolaos Freris, Scott Graham, I-Hong Hou, Yih-Chun Hu, Jonathan Ponniah and Roberto Solis].

Categories & Subject Descriptors: C.2.1.

Computer Communication Networks: Network Architecture and Design: Wireless Communication.

Keywords: wireless networking; delay bounds; clock synchronization; security

Bio

P. R. Kumar obtained his B. Tech. in Electrical Engineering (Electronics) from I.I.T. Madras in 1973, and D.Sc. in Systems Science and Mathematics from Washington University, St. Louis in 1977. From 1977-84 he was in the Department of Mathematics at the University of Maryland Baltimore County, and from 1985-2011 he was in the Department of Electrical and Computer Engineering, and the Coordinated Science Laboratory, at the University of Illinois. Currently he is at Texas A&M University, where he holds the College of Engineering Chair in Computer Engineering. Kumar has worked on problems in game theory, adaptive control, stochastic systems, simulated annealing, neural networks, machine learning, queueing networks, manufacturing systems, scheduling, wafer fabrication plants, information theory, wireless networks, sensor networks, and cyberphysical systems. He received the Donald P. Eckman Award of the American Automatic Control Council, the Fred W. Ellersick Prize of the IEEE Communications Society, and the IEEE Field Award for Control Systems. He was a Guest Chair Professor and Leader of the Guest Chair Professor Group on Wireless Communication and Networking at Tsinghua University, Beijing, and is an Honorary Professor at IIT Hyderabad. He is a Fellow of IEEE, the Academy of Sciences of the Developing World, and a member of the National Academy of Engineering, USA. He received the Distinguished Alumnus Award from IIT Madras, the Alumni Achievement Award from Washington University in St. Louis, and the Daniel C. Drucker Eminent Faculty Award from the College of Engineering, University of Illinois. He was awarded an honorary doctorate by ETH, Zurich.

MobiCom'12, August 22–26, 2012, Istanbul, Turkey.
ACM 978-1-4503-1159-5/12/08.

AutoMAC: Rateless Wireless Concurrent Medium Access

Aditya Gudipati
adityag1@stanford.edu
Stanford University
Stanford, CA, USA

Stephanie Pereira
sfp@stanford.edu
Stanford University
Stanford, CA, USA

Sachin Katti
skatti@stanford.edu
Stanford University
Stanford, CA, USA

Abstract

Current wireless network design is built on the ethos of avoiding interference. In this paper we question this long-held design principle. We show that with appropriate design, successful concurrent transmissions can be enabled and exploited on both the uplink and downlink. We show that this counter-intuitive approach of encouraging interference can be exploited to increase network capacity significantly and simplify network design. We design and implement AutoMAC, a novel MAC and PHY protocol that exploits recently proposed rateless coding techniques to provide such concurrency. We show via a prototype implementation and experimental evaluation that AutoMAC can provide a 60% increase in network capacity on the uplink compared to traditional Wifi that does omniscient rate adaptation and a 35% median throughput gain on the downlink PHY layer as compared to an omniscient scheme that picks the best conventional bitrate.

Categories and Subject Descriptors

C.2 [**Computer Systems Organization**]: Computer - Communication Networks

General Terms

Algorithms, Performance, Reliability, Design

Keywords

Wireless Communication, MAC Protocol, Interference exploitation, Successive Interference Cancellation

1. INTRODUCTION

A basic precept of current wireless design is that at any instant only one transmission should be happening on any given frequency. For example, WiFi networks attempt to ensure that at any time a single client transmits to the AP via a random access MAC protocol. Similarly, cellular networks use synchronous scheduling techniques such as OFDMA, and the AP ensures that on any given subcarrier, only one node is transmitting at any instant. Correspondingly, when the AP is transmitting to clients, it ensures that only one client's packet is transmitted on any given frequency at any time [1].

The key reason for this design approach is of course to avoid the harmful effects of interference. If concurrent transmissions are allowed on the same frequency, packets interfere and the receiver cannot decode the constituent packets. Avoiding such interference when the basestation is transmitting is simple since the basestation controls to whom it transmits. But, when client nodes are transmitting, current network designs spend significant overhead (in the form of control traffic such as RTS/CTS, backoffs) to sense and avoid interfering with each other. Consequently, such an interference avoidance approach limits throughput, both fundamentally since only one node can transmit at any time, as well as due to protocol overheads such as RTS/CTS that are needed to avoid interference.

In this paper, we take the opposite approach. Instead of avoiding interference, we show that with the appropriate design, networks can systematically exploit interference to significantly increase throughput. We design and implement AutoMAC, a new PHY/MAC design that encourages multiple clients to transmit simultaneously to the AP on the uplink and similarly allows the AP to transmit packets to multiple clients concurrently on the downlink. Inevitably inerference results and the current design would be forced to throw away the receptions since they are undecodeable. Instead in AutoMAC, we design a novel *rateless PHY layer encoding and decoding algorithm* that allows the receivers to decode all the constituent packets from interfered receptions. The rateless code provides another critical benefit: it allows successful concurrent transmissions without requiring that the sender know either their channel to the receiver nor the identities of the other concurrent senders, each node simply sends a continous stream of rateless encoded symbols until the receiver decodes. As we will show in Sec. 2, traditional PHY layer modulation and coding techniques would incur enormous MAC layer coordination overhead to realize the benefits of exploiting interference. In AutoMAC, the rateless property of its PHY ensures that concurrency benefits can be realized with a simple, low overhead MAC.

We provide detailed intuition on how AutoMAC exploits interference in the following section 2, but we first summarize our main contributions. Theoretically, we show that AutoMAC is asymptotically optimal, i.e. it achieves the capac-

[1]The above discussion is for a single antenna. MIMO allows multiple transmissions on the same frequency, but separates them on the spatial dimension, and our approach naturally generalizes on the spatial dimension.

ity of the Gaussian Multiple Access channel on the uplink, and the Gaussian broadcast channel on the downlink. For experimental evaluation, we have prototyped AutoMAC in the GNURadio [3] SDR platform and evaluated it in an indoor testbed via experiments using USRP2s, as well as trace drive simulations. We compare AutoMAC with the standard 802.11 MAC which performs omniscient rate adaptation. This scheme has perfect channel knowledge, and always picks the optimal bitrate. However, it suffers from the inefficiencies of the 802.11 MAC protocol. Our evaluation shows that:

- On the uplink, AutoMAC's decoding algorithm can handle and decode an interfered signal that consists of up to 3 concurrent transmissions. AutoMAC thus completely eliminates hidden terminals. Similarly, on the downlink, AutoMAC can send 3 packets concurrently to different nodes in the same transmission.
- AutoMAC achieves a median throughput that is 35% *better than the omniscient scheme* on the downlink due to its ability to multiplex concurrent packets.
- In comparison with 802.11 carrier sense with omniscient rate adaptation, we show that AutoMAC achieves nearly 60% throughput improvement on the uplink in dense networks. The source of the gains come from both being able to exploit interference, as well as having an efficient MAC protocol.

Finally, we note that there has been considerable interest in recent work in combating collisions and interference [4]. However, all of them still consider interference as a necessary evil that can be dealt with in smarter ways when it inevitably happens, but the main goal is still to avoid it. AutoMAC makes a fundamental leap: it shows that interference should be encouraged and exploited to increase capacity. We believe AutoMAC thus represents the next step in the evolution of our understanding of how to handle interference, from avoiding it in current designs with CSMA [23, 7], to dealing with it [11, 4, 9, 8], to exploiting it.

2. INTUITION

To give intuition about AutoMAC, we use the classic example of two nodes Alice and Bob connected to an AP. Throughout this section, we assume that the channel attenuation between Alice and Bob to the AP is h_1 and h_2 respectively. For exposition simplicity, we assume channel reciprocity and hence the channels from the AP to the two nodes have the same parameters. We assume that all nodes transmit with power P, and the noise at all nodes is N.

2.1 Exploiting Interference on the Uplink

To avoid interference on the uplink, the optimal current system will schedule Alice and Bob to transmit one after the other in separate time slots. Assuming no MAC overhead, a fair (equal) time allocation to both Alice and Bob, and capacity achieving codes, information theory tells us that the throughput of the system is given by:

$$R_{nint} = \frac{1}{2}\left(\log_2\left(1+\frac{|h_1|^2P}{N}\right)+\log_2\left(1+\frac{|h_2|^2P}{N}\right)\right). \quad (1)$$

In our approach, Alice and Bob are explicitly asked to interfere. In other words, Alice and Bob transmit concurrently and the AP receives

$$y = h_1x_1 + h_2x_2 + n \quad (2)$$

How might the AP decode Alice or Bob's signals? One possible approach is for the AP to try and decode one of the packets while treating everything else (including the other interfering transmission) as noise, and if successful, subtract the decoded packet out and then decode the second packet interference free. This approach is referred to in the literature as Successive Interference Cancellation (SIC).

Assuming Alice's transmission is decoded first, the optimal throughput that Alice can achieve is given by:

$$R_A = \log_2\left(1+\frac{|h_1|^2P}{|h_2|^2P+N}\right) \quad (3)$$

Once Alice's transmission is decoded, the AP can subtract her contribution to the received signal h_1x_1 (feasible now since the AP knows both h_1 and x_1), to obtain $y' = h_2x_2 + N$. The AP can now decode Bob's signal at a rate which is exactly the same as when Bob alone was transmitting and there was no interference from Alice:

$$R_B = \log_2\left(1+\frac{|h_2|^2P}{N}\right). \quad (4)$$

The benefits of exploiting interference are now apparent. Specifically, Bob is able to decode his signal interference-free, as if Alice had not transmitted at all! Hence the extra throughput that Alice obtained is over and above the throughput Bob would have obtained if there was no interference, and thus strictly better than the interference-free case. If the decoding order is reversed (i.e. if Bob's packet is decoded first and then subtracted out to decode Alice's), Alice's transmission would see an interference free channel and thus any throughput Bob would achieve would be essentially for free compared to the non interference case. Assuming the AP equally alternates between these two decoding orders, the sum throughput of the system can be shown to be:

$$R_{sic} = \log_2\left(1+\frac{(|h_1|^2+|h_2|^2)P}{N}\right) \quad (5)$$

As we can see $R_{sic} > R_{nint}$, i.e. the system throughput achieved by SIC is strictly greater than that achieved by the optimal interference-avoidance scheme.

Said another way, for every transmission that the optimal interference-avoidance system makes (either from Alice or Bob), the other node can also transmit concurrently and get extra throughput without hurting the ongoing transmission which can be decoded interference free after SIC. The above intuition is not our contribution, information theory calls this the multiple access channel and its well known that exploiting interference in this manner provides significantly higher performance than avoiding interference. Fig. 1 plots the achievable throughput combinations for Alice and Bob when they are exploiting interference compared to the conventional interference avoidance scheme.

2.2 Exploiting Interference on the Downlink

In the downlink, the AP is transmitting to Alice and Bob. The current practice is to send to Alice and Bob one after the other using power P for each transmission. In that case, the downlink system throughput of the optimal interference

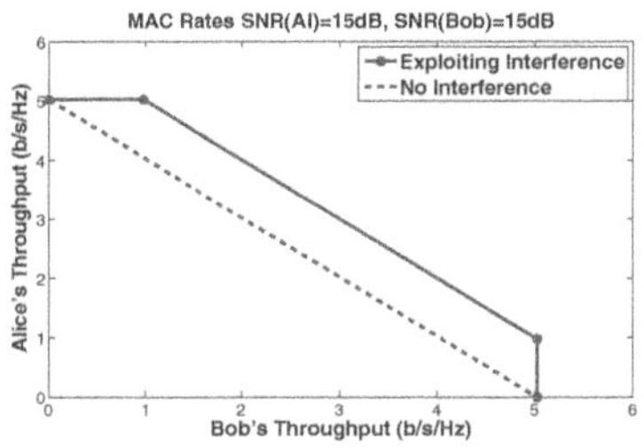

Figure 1: Throughputs Alice and Bob can achieve by exploiting interference compared to the optimal interference avoidance scheme in the uplink.

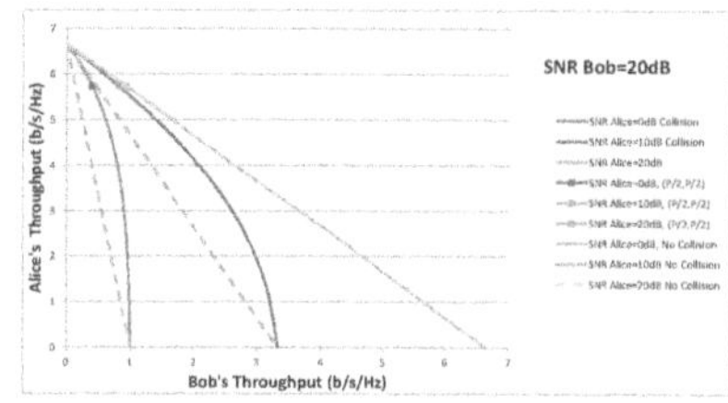

Figure 2: Throughputs Alice and Bob can achieve by exploiting interference compared to the optimal interference avoidance scheme in the uplink.

avoidance scheme is

$$R_{nint} = \frac{1}{2}\left(\log_2\left(1+\frac{P|h_1|^2}{N}\right)+\log_2\left(1+\frac{P|h_2|^2}{N}\right)\right). \quad (6)$$

We show that we can do better using ideas from information theory [1] where the problem of downlink transmission is termed the broadcast channel and the best possible rates for each of the users has been characterized analytically. Wireless is a broadcast medium, so every transmission from the AP reaches both Alice and Bob, even if it is not meant for them. AutoMAC takes advantage of this property to send packets concurrently for both Alice and Bob in the same transmission.

Suppose the AP needs to send the symbols x_1 and x_2 for Alice and Bob respectively. It has a power budget of P per transmission that must be allocated between x_1 and x_2. Since the AP has no channel knowledge, we decide to split the power equally among Alice and Bob. Thus, the AP sends $(x_1 + x_2)/\sqrt{2}$ so that each symbol has power $P/2$.

How can Alice and Bob decode? Bob receives the signal $h_2x_1/\sqrt{2} + h_2x_2/\sqrt{2} + N$. Notice that this is similar to the uplink interference case, except that both symbols experience the same channel and the channel depends on the receiver, i.e. Alice or Bob. One can imagine using the same successive interference cancellation (SIC) approach to decode. However, there is one critical difference. In the downlink, Alice or Bob require only the packet intended for them, while in the uplink, the AP needed to decode all the packets. This implies that the node with the weaker channel from the AP should simply try to decode his packet treating everything else as interference. The reason is that it will take him the same amount of time to decode either his or the other packet, since both of them are allocated equal power and experience the same channel. Hence there is no reason to try decoding anything else, the node with the weak channel just decodes the packet intended for it.

However, the node with the stronger channel doesn't have to do the same. Instead, since the AP has to transmit enough to ensure that the weaker channel node can decode, the number of transmissions will be greater than that required by the stronger channel node to decode the packet of the weak node. Hence, the strong node first decodes the weak node's packet, and then applies SIC to cancel the interference, and decode his own packet interference free.

To understand rigorously why allowing users to interfere with each other produces higher rates, we refer the reader to [21]. The intuitive (but less precise) reason is that the Shannon capacity function is concave with increasing power: increasing the power of a transmission provides diminishing returns beyond a certain level. Hence instead of using all the transmission power for one node, an AP is best served by dividing it and sending to multiple nodes concurrently. To show the effect visually, Fig. 2 plots the optimal rates achieved using the above scheme by Alice and Bob for various realistic SNR combinations. The scheme achieves higher sum throughput when Alice and Bob have different SNRs and the same sum throuhgput as traditional interference-free schemes when Alice and Bob have equal SNRs.

2.3 What's the Catch?

While the above intuitions on how exploiting interference can help improve throughput in both the uplink and the downlink have been well understood, realizing them in practice has been challenging due to one basic challenge: rate adaptation. Modern wireless networks perform rate adaptation, i.e. the senders estimate the channel quality to the receiver and pick the best bitrate that maximizes throughput while ensuring that the receiver can decode. Even assuming that nodes can do perfect rate adaptation (i.e. pick the optimal bitrate) given the channel to their receiver, the above scheme fails in the presence of interference. For example in the uplink, if multiple clients transmit concurrently, the effective SINR each of their transmissions will see will be lower than the channel SNR the transmitters had estimated earlier because the receiver has to treat the other interfering transmissions as noise when decoding. Consequently, none of the transmissions will be decoded since bitrates exhibit a threshold behavior [5], if the actual SINR is lower than the minimum required to decode that bitrate, decoding fails. This is consistent with the observations in [18] which doubts the future of SIC based protocols. The other design choice is of course to coordinate, i.e. each node tries to learn in advance the identity of the concurrent senders, their channels to the AP, and then appropriately encode its own transmission for the SINR his transmission will experience at the receiver. However, that approach adds enormous coordination overhead and is at odds with the random access nature of WiFi.

The same effect holds on the downlink since clients are using the same SIC approach to decode their respective packets, and the SINR a packet experiences is a function of both the channels as well as the interference. Remember that the strong channel node is using SIC to decode its packet, while the weak channel node decodes assuming everything else is noise. To ensure that they can decode, the AP has to encode their respective packets such that they are decodeable at the SINR of their respective transmissions. To do so, the AP would need to have accurate knowledge of the channels to both clients, else once again decoding will fail due to the threshold effect of current encoding techniques. Such channel estimation adds overhead and might be impractical in dense or mobile networks.

2.4 AutoMAC

The basic challenge in exploiting interference is that the sender does not know the SINR his packet will experience at the receiver in advance, yet it has to be able to encode its data such that the receiver can decode even when there are collisions and bootstrap SIC. In other words, we want the sender's encoding to have a *rateless* property, i.e. the sender should be able to generate a stream of rateless transmissions which enable the receiver to decode the original data after a certain number of rateless transmissions commensurate with the actual channel SINR are received, and in effect achieve the same throughput as one would achieve if the sender had known the SINR in advance and had encoded its data at the correct rate.

AutoMAC leverages recent work on rateless codes [6, 15] to solve the above problem. These codes allow the receiver to bootstrap SIC, since the senders transmit a stream of rateless coded transmissions until the receiver can decode. Hence eventually one of the colliding packets will be decoded and SIC kickstarts. So a simple MAC protocol would be to let both senders keep transmitting with the rateless code until their transmissions get ACKed. The receiver would employ SIC and ACK the senders as soon as their respective packets are decoded.

However, the above simple MAC protocol fails. Consider the case when the channels from Alice and Bob to the AP are equally strong, e.g. 10dB. Using the above MAC protocol they would keep transmitting concurrently until both their packets could be decoded. For simplicity, lets assume the AP tries to decode Alice's packet first. Since it is treating Bob's packet as noise when decoding Alice's, the effective SINR for Alice's transmission is 0dbB. Consequently Alice has to transmit quite a few rateless coded transmissions before the AP can decode. For example, with the Strider rateless code [6], Alice has to send 15 rateless transmissions before the AP can decode her packet. However, once the AP decodes, it can completely cancel Alice's contribution and decode Bob's signal interference free. In effect, this means Bob's transmissions see a SINR of 10dB, and can be decoded with far fewer transmissions compared to Alice. Again with Strider's rateless code, with a SNR of 10dB, Bob would need to only send 5 rateless transmissions. But since the AP does not ACK until it can decode Bob's packets fully, which it cannot until Alice's packet is decoded, Bob ends up sending 10 wasteful transmissions or nearly 3× the number he actually needs to transmit. Thus the naive MAC protocol leads to tremendous waste.

AutoMAC's second main contribution is a novel PHY/MAC protocol that solves the above challenges and allows one to exploit collisions through rateless codes. The key technique is *speculative ACKing*. Intuitively, in the above example, if the AP could speculatively ACK Bob's transmissions after it has received a sufficient number (around 5) to ensure it can decode Bob's packet after eventually decoding Alice's packet, then Bob can move on to his next packet and ensure that channel time is not wasted. The basic problem is that the AP has to estimate even before it has decoded either Alice or Bob's packets, when to speculatively ACK. The decision is dependent on the SNR of the channel from Bob which the AP now has to estimate in the presence of interference. AutoMAC designs a novel detection and prediction algorithm that allows it to accurately estimate channels under interference, and speculatively ACK to avoid waste.

3. DESIGN

In this section we describe the design of AutoMAC.

3.1 MAC Protocol

AutoMAC uses a simple AP driven MAC protocol. In AutoMAC the AP has an opportunity to transmit after every packet transmission in the network. In the uplink, after receiving the concurrent transmissions, the AP sends an ACK/NACK packet to inform the clients if any of their packets have been decoded. In the downlink, the AP is of course transmitting a packet. The basic idea is that whenever the AP is transmitting a packet (either data or ACK), if it wants to accept transmissions from the clients in the next slot, the AP appends a short contention advertisement at the end of its transmission. The contention advertisement informs nodes of the number, n of users that may transmit concurrently in the next slot. The advertisement may also include requests from the AP for retransmission of specific packets from particuar nodes, which as we will see in Sec. 3.2.1 is needed to ensure robustness when decoding fails after speculative ACKing.

After hearing the contention advertisement, nodes that have outstanding traffic enter into a contention period to determine which n nodes may transmit concurrently. We build on the recently proposed frequency domain backoff technique [17] that allows nodes to pick a random subcarrier out of the 64 available ones, and send a single tone on that subcarrier. The contention period lasts for 4 time slots, and the clients randomly pick one slot out of the four and one subcarrier out of the 64 to transmit the tone. Next, the AP just retransmits the information about the subcarriers and time slots on which it received tones in the contention period. The n nodes with the n smallest subcarriers and time slots win the contention period, and transmit concurrently after the contention period ends. Since there are a total of 256 time-frequency slots, the probability of two clients picking the same slot and subcarrier in a reasonably sized network is low. However, if network size grows, the contention period can always be increased. Note that the contention period is only needed when the AP wants to accept transmissions from new clients. If not, the existing set of clients continue to transmit concurrently after receiving the ACK/NACK packet.

Note that the AP can choose to not advertise any contention spots. In this case, two events can occur. One the AP itself may chose to transmit in the next slot. Or, the AP might ask the nodes which were concurrently transmitting in the previous slot to continue. The contention advertisement contains a special field to inform all nodes about these cases. The above MAC protocol provides each node with an equal amount of channel transmission time. However that by itself does not ensure fairness, as the discussion in Sec. 2 shows, the throughput a node achieves is also a function of the order in which it is decoded by SIC and the SNRs of the other colliding transmissions. In Sec. 3.6 we describe a novel algorithm that ensures fairness even under collisions.

3.2 Exploiting Interference on the Uplink

The n nodes that win the contention period transmit concurrently after the contention period ends. Each node uses the standard OFDM PHY with 64 subcarriers spread over

20MHz. As with normal WiFi, data symbols are sent only on 48 subcarriers, the remaining 16 are used for pilots and padding. The only change we make is replacing convolutional coding with the rateless code. We first describe how each node generates its packet.

3.2.1 Packet Encoding

Each node uses a rateless code to encode its data. AutoMAC is *orthogonal to the choice of the rateless code*, and can use either Strider [6] or spinal codes [15]. Since Strider has linear decoding complexity and is simpler to implement, we choose to use it in our current design.

Strider encodes data using the following steps:

1. Packet data is divided into chunks of size S KB. In each block we have D data blocks of length L bytes each. Current default values are $D = 30$, $L = 200$ and $S = 6$.
2. Each of the D data blocks is passed through the *base code* (Strider's current implementation uses a 1/5 rate tubo code and a QPSK constellation as the base code), to produce D blocks with $5L/2$ complex symbols each.
3. The D blocks are passed through Strider's encoding matrix R, which is a $M \times D$ matrix, where M stands for the number of transmissions. To generate the i'th transmission, the sender uses the i'th row of the encoding matrix and so on. The sender can generate as many transmissions as it needs since Strider is a rateless code.

The output of the Strider encoder is a frame of $5L/2$ complex symbols. These are divided into 48 equal sized sets which are mapped to the 48 OFDM subcarriers typical in WiFi. Next the standard OFDM PHY is used to generate the actual transmitted wireless signal, which includes doing a IFFT, adding a cyclic prefix, adding pilots on four subcarriers and adding a preamble at the start of the frame.

3.2.2 Decoding All the Constituent Packets with SIC

The concurrently transmitted frames go through the wireless channel and are received at the AP. Assuming h_{ij} is the channel for the j'th rateless transmissions from the i'th node. The signal at the AP for the j'th transmission is

$$y_j = \sum_{i=1}^{K} h_{ij} x_{ij} + N \tag{7}$$

where x_{ij} is the rateless symbol transmitted by the i'th node in the j'th slot and N is the noise.

AutoMAC uses successive interference cancellation to decode the transmissions from the K nodes. To kickstart the process, the AP attempts to decode the first node's packet treating everything else as interference. Hence the first node's transmissions at the AP can be modeled as

$$y_j = h_{1j} x_{1j} + \sum_{i=2}^{K} h_{ij} x_{ij} + N \tag{8}$$

Lets assume we need M rateless transmissions from the first node to decode its packet using Strider's rateless code. To decode the original data, Strider first needs to estimate the channel for each of the M rateless transmissions.

Per Frame Channel Estimation: AutoMAC uses known training symbols at the start of the transmitted OFDM frames. The estimation uses the least square algorithm due to its low complexity. Since the training symbols are defined in the frequency domain – each OFDM subband is narrow enough to have a flat frequency response, AutoMAC estimates the channel in the frequency domain as a complex scalar value at each subcarrier. Specifically, let $\mathbf{X} = (P_{ij}[0], \cdots, P_{ij}[64])$ be the vector of the training symbols used across the 64 subcarriers for a single OFDM symbol for the j'th transmission from node i, and C be the number of such OFDM training symbols. Let $\mathbf{Y_j}^{(c)}$, $c = 1, \cdots, C$, be the corresponding values at the receiver after going through the channel. The least squares algorithm estimates the channel frequency response of each subcarrier k for the j'th transmission as, $\hat{H}_{ij}[k]$, as follows:

$$\hat{H}_{ij}[n] = \frac{1}{C}\left[\frac{1}{P_{ij}[n]}\left(\sum_{c=1}^{C} Y_j^{(c)}[n]\right)\right]$$

Decoding the Packet using Strider: Next, AutoMAC decodes the data from the first node using the Strider rateless code. Since the AP has received M transmissions from node 1, it estimates the channel for all M transmissions as above. Since we are using OFDM, the received symbol on the n'th subcarrier can be written as

$$Y_j[n] = H_{1j}[n] X_{ij}[n] + N \tag{9}$$

It then uses Strider's rateless decoding technique to decode the original packet from node 1. We skip the details of the rateless decoding algorithm and refer the reader to [6], but for our discussion it suffices to assume that the packet from node 1 is now decoded.

Subtracting Node 1's contributions from the collision: Next, the AP subtracts node 1's transmissions from the received collisions. Note that the AP has to do this for all the M rateless transmissions from node 1. However, since the actual interference happens in the time domain, we have to subtract the decoded signal in the time domain even though the rateless decoding happens after the FFT in the frequency domain.

To do so, we first compute the time domain channel response by applying the inverse fast Fourier transform (IFFT) to the frequency response of the channel estimated above. This is done for all M transmissions. Next, the AP re-encodes the decoded data using Strider's rateless code and OFDM to re-generate the M rateless transmissions. Each rateless transmission is then passed through the time-domain channel response computed above to obtain the exact interference contribution in the received signal.

The time domain channel response is represented using a standard finite impulse response (FIR) filter in the digital domain. To generate the digital samples for cancellation, AutoMAC convolves the known signal with the FIR filter representing the channel. Let $s_{1j}[n]$ be the known transmitted digital sample at time n by node 1 in the j'th rateless transmission. These digital samples are fed into the FIR filter. The output $x_{1j}[n]$ of the filter is the linear convolution of $\hat{h}_{1j}[n]$ (the time domain FIR representation of the channel response) $s_{1j}[n]$:

$$x_{1j}[n] = \sum_{k=0}^{F-1} \hat{h}_{1j}[k] s[n-k]$$

After this step, the radio subtracts the estimates of the transmit signal from the received samples $y_j[n]$:

$$\hat{y}_j[n] = y_j[n] - x_{1j}[n]$$

The above process is repeated recursively until all n packets are decoded. As soon as a particular node's packet is decoded, a spot opens up in the next slot for another node to transmit concurrently. The AP signals this open slot via the contention advertisement discussed in Sec. 3.1.

3.3 Accurate Channel Estimation Under Interference

To apply the above SIC technique, the AP needs to now estimate channels from all the n nodes. The above Least Squares algorithm works very well when there is a single node transmitting, but can be inaccurate if the training symbols used for channel estimation collide with other training symbols. To ensure that we obtain accurate channel estimates, AutoMAC leverages the AP driven MAC protocol.

Specifically, after the contention period ends, the succesful nodes transmit training blocks in separate time slots in order of their contention winning sequence. In other words, the node with the smallest contention subcarrier and time slot index transmits 4 OFDM symbols, the first one is a symmetric pseudorandom sequence for time and frequency synchronization via Schmidl-Cox [16], and the next two symbols are the preamble used for channel estimation. The fourth symbol is used for identifying the node and the packet. Next, there is a short guard interval of $2\mu s$, and then the node with the next highest contention subcarrier index transmits the training symbols and so on. Such clean training symbol transmission one after the other is possible because the nodes can synchronize on the reception of the contention advertisement, which essentially acts like a common clock for the network. AutoMAC is now free to use the standard least squares algorithm for all nodes since pilots and preambles are obtained cleanly for all of them. After n such training block transmissions, all the nodes start transmitting their rateless encoded frames concurrently.

Estimating Wireless Hardware Distortions: Apart from the channel, the training symbols help estimate two other practical hardware effects that distort the signal. The first one is *carrier frequency offset* (CFO) which exists because the oscillators at the sender and receiver are never perfectly synchronized to the same frequency. Consequently, there is always a small frequency offset δf_i, between the i'th transmitter and receiver. The CFO causes a linear displacement in the phase of the received signal that increases over time, i.e.

$$y_j[n] = \sum_{i=1}^{n} h_{ij} e^{j2\pi n \delta f_i T} x_{ij} + N \qquad (10)$$

In AutoMAC, the receiver calculates the CFO for every transmitter thats part of a collision. We use the standard Schmidl-Cox algorithm [16] on training symbols. Remember that the first symbol is pseudorandom repeating sequence that is used by the Schmidl-Cox algorithm for packet detection. By correlating shifting halves of samples, the receiver can figure out the start of the packet where the peak occurs. The phase of the correlation peak also provides an estimate of the fractional CFO, defined as a fraction of the OFDM subcarrier width.

When decoding a packet, CFO correction is applied on all the received samples. However, once decoded AutoMAC has to cancel the decoded packet's contribution from the received signal. Hence it has to apply back the CFO distortion to ensure that it cancels exactly the signal that interfered in the time domain. Applying CFO back is a matter of multiplying the time domain samples by a factor of $e^{j2\pi\delta f_i n}$.

3.4 Speculative ACKing

AutoMAC sends multiple packets in every time-slot from different clients to the AP. The channel between each client and the AP will be different, i.e. some will be stronger than the others. Consequently, packets from different clients can be decoded at different times. The exact time is a function of two factors: the order of decoding, and the relative SNRs of the different client-AP channels that are involved in the concurrent transmission.

However, as described, the above PHY and MAC protocol will keep sending rateless transmissions for packets even though they could have been decoded with far fewer transmissions. The reason is of course that the AP does not ACK a packet until it has actually decoded that packet. For example the packet that is decoded last after interference contributions from all other transmissions are subtracted out is ACKed only after the sender has sent at least as many transmissions as the first packet that has to be decoded *while considering all other packets as interference*. Clearly decoding the first packet requires far more rateless transmissions since its effective SINR is low due to interference from the other packets, while the last decoded packet requires far fewer since it sees a much better SNR.

AutoMAC designs a novel speculative ACKing technique to combat the above waste. The key insight is an algorithm that accurately estimates how many rateless transmissions any packet will take to be decoded. Prior work on Strider [6] has shown that there is a fixed relationship between the number of rateless transmission required to decode a packets and the SINR at which the transmisisons are received. Hence, if a node can estimate the effective SINR for any packet, it can accurately estimate how many rateless transmissions it will take to decode that packet.

The challenge then is to estimate the effective SINR for any packet. This is a function of both the raw channel from the specific client, the position at which that client's packet is decoded, and the raw channels from the other clients. To keep track, the receiver keeps track of the total SINR we have accumulated for a specific packet p as follows

1. For every transmission, the node estimates the channel for all the packets that are part of the transmission. Note that this is possible because of the MAC protocol that allows nodes to send training blocks before each transmission one after the other.
2. For packet p, the receiver keeps track of total power received as $T_p = \sum_i^K h_i P$ where h_i is the channel gain for p's i'th rateless transmission and P is the transmit power. It does the same for all the other packets that are received in any transmission.
3. The total interference I_p for decoding p is computed as the sum of the powers for all the other packets that are concurrently transmitted with p and act as interference for p's decoding. In other words they get decoded after p does.
4. The effective SINR for p after receiving M transmissions that contain it is then computed as $T_p/(I_p + M * N)$.

To estimate if M transmissions are enough, the receiver consults the SINR vs number of transmissions graph for Strider. If M is equal to or greater than the number required, then the receiver speculatively ACKs packet p.

Note that when the receiver speculatively ACKs, p will likely not actually be decoded yet. That happens when all the other packets that are supposed to be decoded before p get decoded. Of course, there will be cases where the speculative ACK might be optimistic, i.e we may need more rateless transmissions to decode p. In that case, the receiver can send an ACK that specifically requests more rateless transmissions for packet p from the appropriate sender. Fig. 3 demonstrates an example of AutoMAC's MAC protocol including speculative ACKing for a three node scenario with Alice, Bob and Charlie transmitting to the AP.

3.5 Exploiting Broadcast on the Downlink

AutoMAC uniquely allows the AP to use the same encoding technique on the downlink. Assuming the AP wishes to transmit n packets on the downlink to n different nodes, it uses the following procedure

1. Encode each packet using Strider's rateless code to create a stream of rateless frames, without yet passing them through the OFDM PHY.
2. Take the n rateless encoded frames, add them up in the complex domain to generate one frame.
3. Next it prepends the appropriate headers which include information about all the packets being sent together. Finally, the frame is passed through the OFDM PHY and transmitted.

3.5.1 Decoding at the Clients

To decode the above rateless encoded frames, the client uses a similar procedure as the uplink. Lets say the client has accumulated m such rateless frames each having a packet intended for it among the n packets that are encoded together. In effect, except for its own packet, everything else acts as interference. The client can use a technique similar to successive interference cancellation to decode its own packet.

Remember from the discussion in Sec. 2 that in the downlink case, the client with the strongest channel should decode its packet after having decoded and cancelled all the other packets. The next strongest channel node should decode his second-last and so on. How can each client know the relative strengths of the SNRs of all the other clients? Assuming such channel knowledge is untenable for AutoMAC since its main goal is to eliminate the need for such coordination.

AutoMAC's key insight is to let every client pretend that it is decoding its own packet last, i.e. every client believes that it has the best channel initially. Eventually, the node with the actual strongest channel will accumulate enough transmissions to ensure that the effective SINR is sufficient to decode the packet intended for it. As soon as that happens, it will speculatively ACK just like in the collision case. The next speculative ACK will come from the node with the second strongest channel and so on. Thus, in effect the AP will get an ordering of the channel strengths. The AP uses this information to signal to all the clients the order in which they should decode the transmissions to recover their own packets. Each client now applies SIC in the specified decoding order to decode their own packet.

Finally, note that speculative ACKs serve the same purpose on the downlink as the uplink. After a node speculatively ACKs, a new slot opens up for the AP to transmit a packet to a different node in the downlink. The AP picks one of the outstanding packets, according to the fairness policy discussed next.

3.6 Fairness

The MAC protocol as described does not ensure fairness. To see why consider the uplink case where nodes are concurrently transmitting. The node whose packet is decoded while treating all other packets as interference experiences very low SINR, while the node whose packet is decoded last achieves the best SNR. Thus even though both nodes use the same power and get the same channel time, the node decoded last will achieve much better performance than the node decoded first. The same effect exists in the downlink since the exact same SIC algorithm is used in there too.

AutoMAC attempts to ensure the following notion of fairness. A node should not achieve a lower throughput with AutoMAC than that it would achieve with a conventional MAC protocol that allocates equal channel time to all nodes. The reasoning is that AutoMAC should never be less fair than what a well designed conventional MAC protocol would have achieved in current networks.

Ensuring Uplink Fairness: In order to ensure uplink fairness as described above, AutoMAC requires the AP to estimate the maximum possible throughput that each contending node could achieve using a conventional PHY/MAC scheme. This estimate is arrived at and subsequently updated based on the signal strength observed whenever the corresponding node transmits a packet. The actual throughput that the node would achieve in a conventional setting with time based fairness can then be calculated by simply dividing this estimate by the number of contending nodes. This throughput value is deemed to be the node's *fair share*.

The key idea is whenever a node that has been under-served, wins the contention, the AP should allow it to transmit multiple packets without contending again. By under-served, we mean the node's actual throughput is significantly lower than its fair share. To achieve this, all nodes are required to set one bit in their transmitted packets based on whether or not they have more data to send. When the AP decodes an under-served node's packet and observes this bit to be set, it asks the corresponding node to continue transmitting the next batch without going into contention.

Ensuring Downlink Fairness: AutoMAC aims to provide the same time-based notion of fairness on the downlink, i.e. traffic to each client gets an equal amount of channel time on the downlink. Similar to the uplink, the client decoding his packet last experiences an interference free channel, as if the AP was transmitting only to that client. Thus as long as the AP ensures that every node gets equal opportunity to decode its packet last, it will ensure the required fairness.

However, the challenge is that in the downlink the AP does not have the same luxury of choosing random packet decoding orders, the order is fixed by the relative channel strengths of the clients. Hence, the probability of a weak channel node being the last to decode is lower than that of a strong channel node. Consequently, the AP uses a slightly different algorithm. For each client, it keeps track of how many times it has been asked to decode last. When the next opportunity arises to pack multiple packets in a downlink transmission, it picks the client whose packets have been decoded last the least number of times. It then searches for a client with outstanding packets which it knows in the past has had a weaker channel than the first client (because it was asked to decode before the first client). That client's transmission is added to the broadcast transmission. The process is repeated re-

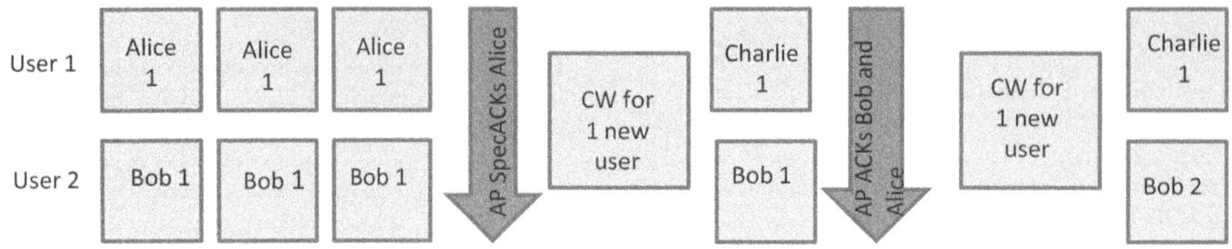

Figure 3: AutoMAC: MAC protocol with Speculative ACKing

cursively until no other clients can be found to fit the above condition. Note that with the above algorithm, sometimes we may find no other client's transmission to send along with the first client's transmission, and thus lose the throughput gains we could get from multiplexing concurrent transmissions. However, AutoMAC consciously makes this tradeoff to ensure fairness. Over the long run, the above algorithm ensures that each client's packets gets equal amount of time being decoded last in the downlink broadcast transmissions, ensuring the desired fairness.

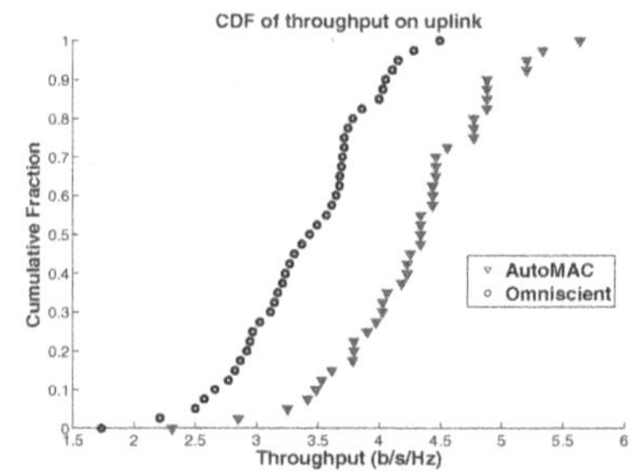

Figure 4: AutoMAC: CDF of throughput gains on uplink

4. EVALUATION

We evaluate AutoMAC on an indoor testbed of 15 USRP2s and trace driven simulations. For evaluating the AutoMAC's PHY layer gains we compare with **omniscient scheme**. This scheme has perfect advance knowledge of the channel strength, and picks the maximum possible bitrate that can be decoded error free. The bitrate choices are from the 8 different bitrates available in the 802.11 standard. For the evaluation of AutoMAC's MAC protocol, we compare AutoMAC with the **802.11 protocol doing omniscient rate adaptation**. This scheme also has perfect advance channel knowledge, and picks the best possible Wifi bitrate. However, it suffers from the inefficiencies of the 802.11 MAC protocol. This is done in order to exclude the benefits of Rate Adaptation, achieved due to Strider's implementation [6] in our analysis.

As observed in [6] we achieve a throughput that is roughly 30% higher than the omniscient scheme on the uplink. The rest of our findings are summarized below:

- In our testbed experiments, AutoMAC outperforms the omniscient scheme by 35% on the downlink too!
- AutoMAC's decoding algorithm can decode all constituent packets within collisions with upto 3 packets. AutoMAC thus eliminates hidden terminals in our testbed. Further, on the downlink it can multiplex upto 3 packets together.
- AutoMAC's design accurately estimates channels, frequency and sampling offsets even under collisions.
- In networks with contention, AutoMAC provides a throughput gain of 60% over 802.11 MAC doing omniscient rate adaptation on the uplink and 50% over an 802.11 style MAC ensuring time based fairness.

5. INDOOR TESTBED EXPERIMENTS

In this set of experiments, we evaluate AutoMAC using experiments in our indoor testbed of USRP2s. We compare with the omniscient scheme which has perfect advance channel knowledge and picks the best bitrate which can be decoded. This is done in order to exclude the benefits of rate adaptation via Strider and focus on the gains due to exploiting interference.

5.1 AutoMAC: Exploiting Interference on the Uplink

Method: In this experiment, we statically place one USRP node which acts as the AP. Next, we randomly place 10 client nodes in the testbed so that they have different channels to the AP. For each client-AP link, we transmit 1000 packets between the two nodes using all the different bitrates, and pick the one which achieves the maximum throughput. Next, we connect all the client nodes together to a common clock as described in prior work [13] to ensure that they transmit one after the other according to an omniscient schedule that allocates equal channel time to all nodes and avoids collisions. For AutoMAC, we use the same clock setup, but allow a random set of two nodes to concurrently transmit.

Recall from Sec. 3.3 that in order to accurately estimate the channel under interference, we require the training blocks to arrive in different time slots. In a commerical deployment we envision the clients would synchronize based on the reception of the conntention advertisement from the AP which essentially acts like a clock. However, in our setup we need to rely on an external common clock for the synchronization. This is the distinguishing factor from the collision experiments done in [6], which were asynchronous.We repeat this experiment 10 times for the same location of the nodes and take the average throughput for either scheme, expressed in terms of bits/second/Hz. We then change the locations of the client nodes to get a different SNR and repeat the above procedure. We plot the CDF of throughputs achieved by the two schemes in Fig. 4.

5.1.1 Accuracy of Channel and CFO Estimation

In this experiment, we evaluate the accuracy of AutoMAC's channel and CFO estimation techniques. We select 3 client nodes and allow them to transmit to the AP using AutoMAC. Next, we immediately make them transmit to the AP separately one after the other without collisions using the standard WiFi OFDM PHY . Note that while doing Successive Interference Cancellation, any error in channel estimates of the stronger client, leads to increased interference for all the

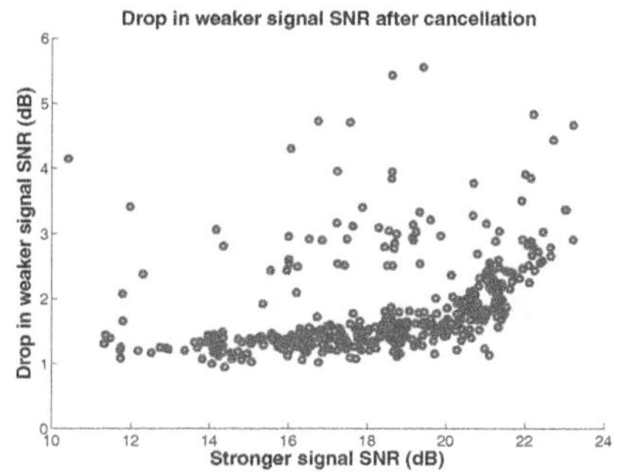

Figure 5: Reduction in Channel SINR post cancellation due to errors in Channel and CFO Estimation

clients that need to be decoded subsequently. We plot the corresponding reduction in channel SINR for the weaker client as a function of the signal strength of the stronger client in Fig. 5.

5.1.2 Accuracy of Speculative ACKing

In this experiment, we evaluate the accuracy of AutoMAC's speculative ACKing technique. We check when the AP speculatively ACKs in anticipation of being able to decode a packet after SIC, what fraction of the time it actually is succesful. Using the results of the experiment, we observe that the AP is totally accurate in atleast 80% of the scenarios. In 10% of the scenarios, the AP sends a speculative ACK before it is actually capable of decoding. We believe, these errors do not cost much in terms of the system throughput, since the AP can request more rateless transmissions from the appropriate sender. These additional transmissions contribute constructively towards the packet decoding and hence are not wasteful. However, in 10% of the scenarios, the AP fails to send the speculative ACK on time. This results in wasteful transmissions (typically just 1 extra transmission) which could have been avoided. Overall, the inaccuracies of speculative ACKing have negligible effect on the system throughput $(2-3\%)$.

5.2 AutoMAC: Exploiting Interference on the Downlink

Method: In this experiment, we statically place one USRP node which acts as the AP. Next, we randomly place client nodes in the testbed so that they have different channels to the AP. For each client-AP link, we transmit 1000 packets between the two nodes using all the different bitrates, and pick the one which achieves the maximum throughput. Finally, we repeat the experiment with AutoMAC's downlink technique. We plot the CDF of throughputs achieved by AutoMAC and the conventional PHY in Fig. 6.

Analysis: AutoMAC provides a median throughput gain of 35% over the optimal conventional downlink scheme. The reason is of course higher concurrency, a AutoMAC AP manages to deliver packets to multiple destinations in every transmission. The gains vary and as in the uplink, are a function of the relative SNRs between the clients. We plot the relative gains of AutoMAC as compared to the omniscient scheme as a function of the relative SNRs in Fig. 7. The figure shows that the throughput of the omniscient scheme is almost as good as AutoMAC when the SNRs of the clients are similar, but the gains increase as the relative SNR increases. This is consistent from the theoretical insight about broadcast channels as discussed in Sec. 2 and in [1].

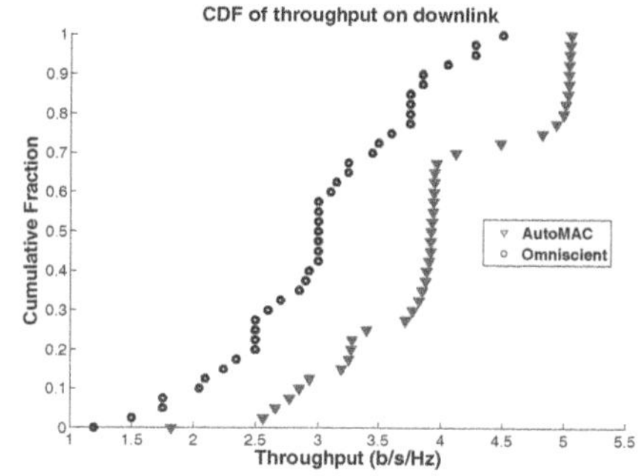

Figure 6: AutoMAC: CDF of throughput gains on downlink

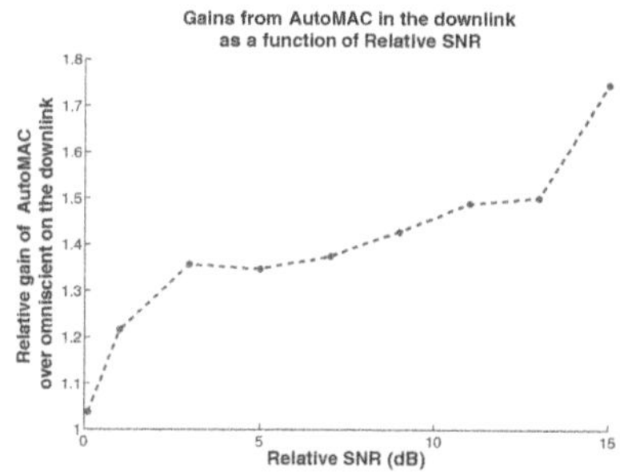

Figure 7: Gains from AutoMAC in the downlink as a function of Relative SNR

6. TRACE DRIVEN EMULATION

Similar to prior work [22, 10] we turn to trace driven emulation to evaluate AutoMAC's performance in a larger network setting with varying traffic. We compare against a state-of-the-art 802.11 MAC protocol which performs optimal rate adaptation. We also compare against an 802.11 style MAC protocol with CSMA and exponential backoffs which also ensures equal channel time fairness. However, since these techniques require fast turnaround times to send ACKs and synchronized feedback etc, which the USRP2s are not equipped to do. Additionally, it is hard to generate controllable high-contention and varying traffic in experimental settings.

Trace: We collect real channel information for the simulations via a high precision channel sounder [2]. The channel sounder is an equipment designed for high precision channel measurement, and provides almost continuous channel state information over the entire measurement period, and can measure channel SNRs as low as -3dB. Our experiments are conducted at night on the band between 2.426 and 2.448GHz which corresponds to WiFi channel 6, and include some interference from the building's WiFi infrastructure which operates on the same channel. The transmit node of the channel sounder is placed at ten different locations in our testbed, and its channel to the receive node is measured over a period of 100 seconds where the receive node records and estimates detailed channel state information for all frequencies in the 20Mhz channel. We collect around 100000 measurements over a 100 second period for 10 different locations of the transmit node.

Emulator: We feed this trace to a custom emulator written in MATLAB with the AutoMAC's implementation. AutoMAC allows for 2 concurrent transmissions on the uplink in each time slot. The emulator implements a 802.11 style MAC

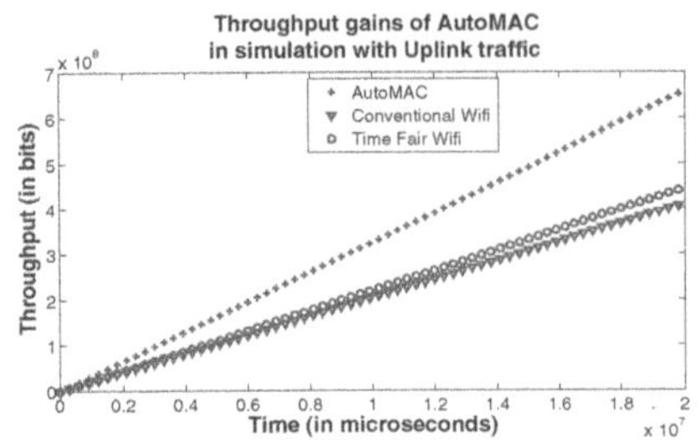

Figure 8: AutoMAC: Throughput gains of AutoMAC in simulations with Uplink Traffic

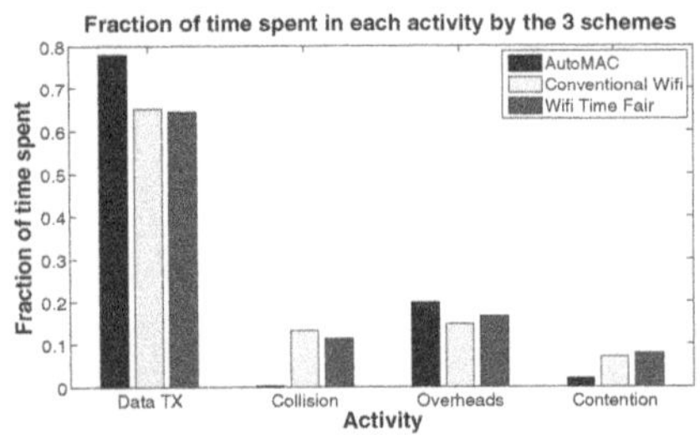

Figure 9: Fraction of time spent by clients in different states across the 3 schemes

with ACKs, CSMA and exponential backoff with the default parameters. This scheme is allowed perfect knowledge of the channel strength and can pick the optimal conventional bitrate. This is done in order to ensure that the gains of rate adaptation [6] do not show up in our analysis.

To better judge the source of the gains, the emulator also implements an 802.11 style MAC with CSMA and exponential backoffs, which ensures a notion of time based fairness. That is, it ensures all contending nodes get roughly the same channel time. This is done by allocating a fixed time window to each node whenever it wins the contention. The user can only transmit on the channel for this fixed period of time irrespective of its channel strength.

6.1 Performance

Method: We compare AutoMAC with the 802.11 MAC protocol which does omniscient rate adaptation and with an 802.11 style MAC which ensures time based fairness. In these experiments, we simulate the network with 8 nodes connected to an AP. We plot the evolution of throughput on the uplink with time in the three scenarios in Fig. 8.

Analysis: AutoMAC provides a 60% gain on the uplink over the conventional 802.11 MAC protocol with omniscient rate adaptation, and a 50% gain over the 802.11 style MAC protocol ensuring time based fairness. The gain due to imposing time based fairness on 802.11 MAC is consistent with the observations in [20].

In Fig. 9, we plot the fraction of time spent by the network in different activities (i.e. useful data transmission, collisions, Overheads (SIFS, ACK, packet header) and contention) for the three different MAC protocols.

A close observation of Fig. 9 leads us to a better understanding of the source of the gains.We observe the following:

- The breakup of times spent looks similar for the Conventional Wifi and the Time fair Wifi. The main difference between the two protocols shows up only when we analyse the individual nodes seperately.

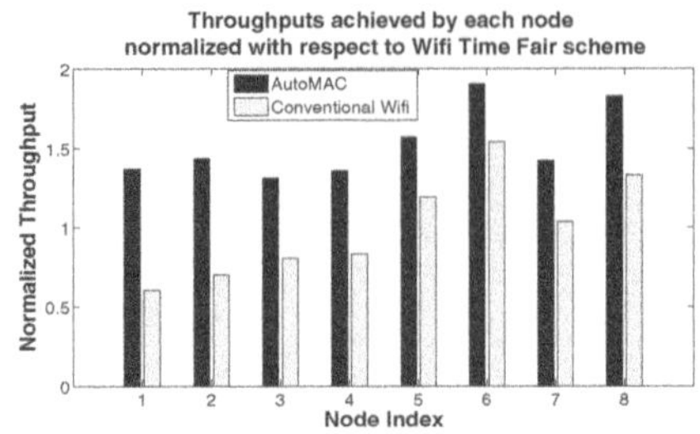

Figure 10: AutoMAC guarantees atleast as much fairness as the optimal conventional scheme

- Note that the nodes in other MAC protocols waste a significant portion of their channel access oppurtunities due to lack of coordination and inability to decode from collisions. The centralized MAC protocol of AutoMAC reduces collisions to a minimum. Here, collision is defined as observing more than 2 simultaneous transmissions on the channel.
- The contention time is also comparatively lower in AutoMAC, due to the frequency domain backoff technique [17].
- The comparison of overheads is slightly more involved. Overheads here include the time spent by nodes in interframe spaces,(i.e. SIFS, DIFS), while receiving ACK and while transmitting packet headers and training symbols. The time spent in receiving ACK is comparatively lower in AutoMAC, since, AutoMAC requires an ACK to be sent only once every batch. However, the interframe spaces and packet headers are significantly higher in AutoMAC due to the smaller packet size. Overall, the time spent in overheads is comparable across the 3 schemes.
- Due to AutoMAC's efficient MAC protocol, the fraction of time spent in sending useful data is 20% higher as compared to the conventional MAC protocols.

Keep in mind that beyond these gains, the protocol also gains from being able to exploit the uplink interference in the PHY layer.

6.2 Fairness

Fig. 10 plots the node-wise throughputs of AutoMAC and traditional Wifi normalized by the node-wise throughputs of the time fair Wifi scheme. Note that AutoMAC achieves atleast as much throughput for all the individual nodes as a Wifi scheme with time based fairness. Conventional Wifi is worse for certain nodes since the MAC cannot perfectly schedule all the nodes to ensure fairness. As the figure shows, AutoMAC provides fairness and in fact higher throughput due to factors discussed earlier.

7. RELATED WORK

AutoMAC is related to prior work in rateless codes and hybrid ARQ. Rateless codes such as LT [14] and Raptor codes [19] allow one to automatically achieve the capacity of an erasure channel without knowing the packet loss probability in advance. However, these techniques require whatever packets are received to be correctly decoded, and do not work in wireless channels where packets are corrupted. AutoMAC builds on recent work on rateless coding for Gaussian channels [6, 15]. AutoMAC's key contribution is to design a MAC protocol that exploits these rateless codes to systematically encourage and take adavntage of interference.

AutoMAC is related to prior work on interference cancellation [9, 4, 11, 12, 8]. However, all prior techniques require that the colliding packets be encoded at the correct bitrate to enable them to decode collisions. For example in SIC, if the colliding packets have been encoded at a bitrate corresponding to the idle channel (which will happen because the colliding hidden terminal senders cannot know in advance that they will collide), SIC will fail to work [4, 18]. Zigzag has a similar but less acute problem, since it also needs correct decoding of its interference free chunks, which requires the packets to be encoded at the correct bitrate. Further Zigzag needs the same set of packets to collide across successive collisions. AutoMAC does not have any of these problems, since its rateless property automatically adjusts the effective bitrate to enable it to decode collisions, and it can decode even if collisions are between different sets of packets.

8. CONCLUSION

AutoMAC provides a rateless MAC design that systematically exploits interference and consistently achieves very good performance across both uplink and downlink scenarios. AutoMAC suggests a number of avenues for future work including extending it to 802.11n MIMO scenarios as well as OFDMA.

9. ACKNOWLEDGEMENT

The first author was supported by Stanford Graduate Fellowship during the course of this research. The authors acknowledge the financial support from NSF POMI grant. We are also indebted to the reviewers of this paper for their detailed feedback and to our colleagues at Stanford University whose inputs have been invaluable to this paper.

10. REFERENCES

[1] T. Cover and J. Thomas. *Elements of Information Theory.* Wiley, 2006.

[2] G. Czink, B. Bandemer, and A. Paulraj. Stanford July 2008 radio channel measurement campaign. In *COST 2010*, October 2008.

[3] Free Software Foundation. Gnuradio. http://gnuradio.org.

[4] S. Gollakota and D. Katabi. ZigZag decoding: combating hidden terminals in wireless networks. In *SIGCOMM '08: Proceedings of the ACM SIGCOMM 2008 conference on Data communication*, pages 159–170, New York, NY, USA, 2008. ACM.

[5] A. Gudipati and S. Katti. Automatic rate adaptation. In *ACM HotNets*, 2010.

[6] A. Gudipati and S. Katti. Strider: Automatic rate adaptation and collision handling. In *ACM SIGCOMM*, 2011.

[7] R. Gummadi, R. Patra, H. Balakrishnan, and E. Brewer. Interference Avoidance and Control. In *7th ACM Workshop on Hot Topics in Networks (Hotnets-VII)*, Calgary, Canada, October 2008.

[8] D. Halperin, J. Ammer, T. Anderson, and D. Wetherall. Interference cancellation: Better receivers for a new wireless mac. *Hotnets*, 2007.

[9] D. Halperin, T. Anderson, and D. Wetherall. Taking the sting out of carrier sense: interference cancellation for wireless lans. In *MobiCom '08: Proceedings of the 14th ACM international conference on Mobile computing and networking*, pages 339–350, New York, NY, USA, 2008. ACM.

[10] D. Halperin, A. Sheth, W. Hu, and D. Wetherall. Predictable 802.11 packet delivery from wireless channel measurements. In *ACM SIGCOMM*, 2010.

[11] S. Katti, S. Gollakota, and D. Katabi. Embracing wireless interference: analog network coding. In *SIGCOMM '07: Proceedings of the 2007 conference on Applications, technologies, architectures, and protocols for computer communications*, pages 397–408, New York, NY, USA, 2007. ACM.

[12] L. Li, K. Tan, Y. Xu, H. Vishwanathan, and Y. Yang. Remap decoding: Simple retransmission permutation can resolve channel collisions. In *ACM MOBICOM*, Sept. 2010.

[13] S. K. Lin and D. Katabi. Random access heterogenous mimo networks. In *ACM SIGCOMM*, 2011.

[14] M. Luby. Lt codes. In *Proc. of FOCS 2002*, 2002.

[15] D. J. Perry and H. Balakrishnan. Rateless spinal codes. In *ACM HotNets*, 2011.

[16] T. Schmidl and D. Cox. Robust frequency and timing synchronization for ofdm. *IEEE Transactions on Communications*, Dec. 1997.

[17] R. S. Sen and S. Nelakuditi. No time to countdown: Backing off in the frequency domain. In *ACM MobiCom*, 2011.

[18] S. Sen, N. Santhapuri, R. R. Choudhury, and S. Nelakuditi. Successive interference cancellation: a back-of-the-envelope perspective. In *Proceedings of the 9th ACM SIGCOMM Workshop on Hot Topics in Networks*, Hotnets-IX, pages 17:1–17:6, New York, NY, USA, 2010. ACM.

[19] A. Shokrollahi. Raptor codes. *IEEE/ACM Trans. Netw.*, 14(SI):2551–2567, 2006.

[20] G. Tan and J. Guttag. Time-based fairness improves performance in multi-rate wlans. In *Usenix Annual Technical Conference*, 2004.

[21] D. Tse and P. Vishwanath. *Fundamentals of Wireless Communications.* Cambridge University Press, 2005.

[22] M. Vutukuru, H. Balakrishnan, and K. Jamieson. Cross-layer wireless bit rate adaptation. In *ACM SIGCOMM, Barcelona, Spain*, August 2009.

[23] M. Vutukuru, K. Jamieson, and H. Balakrishnan. Harnessing Exposed Terminals in Wireless Networks. In *5th USENIX Symposium on Networked Systems Design and Implementation*, San Francisco, CA, April 2008.

No Symbol Left Behind: A Link-Layer Protocol for Rateless Codes

Peter Iannucci, Jonathan Perry, Hari Balakrishnan, and Devavrat Shah
Massachusetts Institute of Technology
Cambridge, Mass., USA
{iannucci,yonch,hari,devavrat}@mit.edu

ABSTRACT

Recently, rateless codes have introduced a promising approach to obtaining wireless throughput higher than what is achieved by fixed-rate codes, especially over time-varying channels. Rateless codes like Raptor [26, 22], Strider [9], and spinal codes [23, 24] naturally process all the information available at the receiver corresponding to a packet, whether from one or many frame transmissions. However, a profitable deployment of rateless codes in a wireless network requires a link-layer protocol to coordinate between sender and receiver. This protocol needs to determine how much coded data should be sent before the sender pauses for feedback from the receiver. Without such feedback, an "open-loop" sender would not know when the packet has been decoded, but sending this feedback is not free and consumes a significant fraction of the packet transmission time. This paper develops RateMore, a protocol that learns the probability distribution of the number of symbols required to decode a packet (the *decoding CDF*), and uses the learned distribution in a dynamic programming strategy to produce an optimal *transmission schedule*. Our experiments show that RateMore reduces overhead by between 2.6× and 3.9× compared to 802.11-style ARQ and between 2.8× and 5.4× compared to 3GPP-style "Try-after-n" HARQ.

Categories and Subject Descriptors: C.2.1 [Network Architecture and Design]: Wireless communication

General Terms: Algorithms, Design, Performance

Keywords: Wireless, rateless, protocol, link-layer, HARQ

1. INTRODUCTION

For network applications on mobile devices to perform well, network protocols must cope with significant variations in wireless channel conditions caused by user and device movement, as well as interference from sources internal and external to the wireless network. These variations can occur even over short (sub-packet) time scales, posing a fundamental challenge for wireless network protocols.

Recent work on *rateless codes* over wireless networks has shown significant promise in achieving high performance over time-varying channels [26, 22, 6, 9, 24]. These codes are "regret-free" in the sense that the receiver never wishes that a different modulation or code rate had been chosen by the transmitter – all received symbols are equally useful and none are discarded. Under a traditional bit rate adaptation approach, the sender decides in advance on one of a pre-determined set of fixed-rate codes and modulation schemes (see §2). If the chosen rate is too aggressive, the resulting received symbols are nearly useless.

Rateless codes offer a natural way to adapt to channel variations via the *prefix property*: in a rateless code, an encoding with a higher rate is a prefix of any lower-rate encoding. Rather than discarding symbols which fail to decode, the receiver uses them again to form part of a lower-rate encoding once additional symbols arrive. No symbol is left unused in decoding a rateless message. The greater the number of symbols received, the higher the probability of a successful decoding.

Ratelessness does not excuse the sender and receiver from adapting to channel conditions. Instead, it unifies the processes of sensing conditions and reacting to them. For a good rateless code, the number of symbols required for decoding closely tracks changes in the prevalent channel conditions. This means that a rateless sender and receiver can skip the customary estimation of channel quality and focus on the central problem of estimating, based on feedback, the number of symbols to be transmitted.

A comprehensive rateless link protocol would include a mechanism for reliable feedback and a mechanism for dealing with channel contention. It would allow application-level end-to-end latency constraints to impose limits on packet aggregation, and it would provide theoretical guarantees that a suitable performance metric is optimized. It would circumscribe what the sender needs to know about the channel in order to optimize performance, and it would include the cost of conveying this information to the sender in the optimization.

In this paper, we identify efficiency as the right performance metric and show how the sender can maximize it given the feedback delay and a cumulative probability distribution function called the decoding CDF. Efficiency, given by the fraction of channel occupation time which is strictly necessary for reliable delivery, provides a code-neutral measure of protocol overhead. Our paper develops an efficiency-maximizing protocol, RateMore, and presents its implementation and evaluation over spinal codes [23], Strider [9], and Raptor codes [26, 22] on stationary and fading channels. RateMore accommodates soft single-hop latency constraints and provides reliable delivery. Multi-hop latency constraints and unreliable service classes are interesting unexplored directions.

We assume common half-duplex radios with 802.11-style framing and acknowledgments, and we consider operation above a minimum signal-to-noise ratio (SNR) such that acknowledgement packets sent at the lowest 802.11 convolutional code rate are reliable. We also assume that once a sender has contended for the medium, it uses

MobiCom'12, August 22–26, 2012, Istanbul, Turkey.

802.11's short inter-frame space (SIFS) mechanism to bypass contention during feedback hand-offs until the transaction is complete. Under these conditions, the feedback delay is known to the transmitter a priori via a calculation we review. Our analysis shows that the decoding CDF and the feedback delay are sufficient knowledge for the sender to maximize the average throughput of the link over all possible strategies of transmitting data and pausing for feedback.

The decoding CDF is not only sufficient, but it is practical to obtain. We demonstrate how the sender can obtain approximate yet satisfactory knowledge of the decoding CDF (e.g. performing within 1.57% of full knowledge) from just a handful of acknowledgments.

Our evaluation compares RateMore with two alternatives borrowed from fixed-rate coding. The first is an analogue of automated repeat requests (ARQ) fashioned after 802.11, and the second is incremental redundancy after the pattern of the 3GPP cellular standard, sending a fixed number of additional coded symbols between pauses for feedback. RateMore proceeds from a sound mathematical framework and outperforms both of these ad hoc approaches.

Our results show that RateMore achieves an efficiency of over 90% across all our experiments. Our experiments also show that RateMore reduces overhead by 2.6× to 3.9× compared to 802.11-style classical ARQ and by 2.8× to 5.4× compared to 3GPP-style "Try-after-n" hybrid ARQ. These translate to throughput improvements of up to 26% even under "static" channel conditions. Over fluctuating Rayleigh-fading channels, we show that RateMore performs within 1.57% of the ideal adaptation.

2. RELATED WORK

2.1 Type-I HARQ

Hybrid ARQ (HARQ) systems [19] marry forward error correction with ARQ to raise the probability of successful reception. Type-I HARQ systems transmit coded messages and retransmit on failures, and are widely used in wireless standards such as Wi-Fi [14]. A common design technique is to specify several modes, or "bit rates", each suitable for a small range of channel conditions. Bit rate adaptation algorithms then choose what mode to use by picking a combination of modulation (symbol set) and code.

2.2 Type-II HARQ

Type-II HARQ allows the exchange of coded data over several round-trip interactions between the transmitter and receiver. Using more interactions has been shown to allow improvement over ARQ's performance under poor conditions, while not sacrificing performance under favorable conditions . Two approaches exist for combining transmissions. In HARQ with Chase Combining [5], the retransmission repeats parts of the original packet, which are then combined at the receiver. HARQ with Incremental Redundancy (IR) sends different coded bits in every transmission. This approach has been used successfully in the 3GPP LTE protocol [15], and a protocol has been built to achieve IR-HARQ over 802.11 networks in ZipTx [18].

Soljanin et al. discuss the design of HARQ with incremental redundancy [27]. The develop such a protocol for an ensemble of LDPC codes as well as Raptor codes. The basic idea is to transmit only as many coded symbols as required for the receiver to decode the message with high probability using the best possible maximum-likelihood (ML) decoding rule, presupposing a very high SNR. If the receiver is successful, the transmitter moves to the next message. If not, the transmitter sends the next sets of coded symbols in such a way that given what receiver has received (the transmitter knows the past channel condition at the receiver), there is a high probability of successful ML decoding if the SNR is only moderately high. This is repeated until successful decoding is achieved.

Another related work is by Anastasopoulos [1]. The primary difference arises in the fact that their work does not take feedback delays into account. They model the decoding CDF in terms of an error exponent, and assume that the underlying code cannot decode with more than some maximum number of symbols. They model the channel as a Markov chain and try to infer its state. Our approach is different both in the learning strategy (learning the decoding CDF) and in applying a dynamic programming strategy to compute the transmission schedule.

Most proposed HARQ schemes have a small number of effective rates to choose from. This is achieved using rate-compatible puncturing [11, 17, 10], where the mother code is partitioned into a series of punctured words, such that the decoder has good probability of successfully processing any prefix of the series. In comparison, rateless codes do not require such gentle constructions, and natively offer a free choice of rates.

Systems with either Type-I or Type-II HARQ schemes use bit rate adaptation algorithms. These algorithms take a *reactive* approach to combating temporal channel variations: systems passively monitor channel conditions in the form of the signal-to-noise ratio [4, 13], frame loss rate [29, 3], or bit-error rate [28, 25].

2.3 Partial Packet Recovery

Partial packet recovery schemes are protocols designed to recover packets that were received with errors. Data is divided into small fragments, and some method is used to detect which fragments were received correctly. The transmitter then only retransmits erroneous fragments. Seda [8] adds an overhead of a CRC-8 and an identifier to each block, and sends correction blocks alongside fresh blocks. PPR [16] uses confidence information from the PHY layer instead of a CRC to determine which bits need retransmission, and incorporates a dynamic algorithm to determine what feedback should be sent by the receiver. Maranello [12] again uses CRC to protect each block, but reduces overhead by only transmitting CRCs when the packet contains errors.

Partial packet recovery can be viewed as a finer-grained Type-I HARQ: the PPR system sends smaller blocks, retransmitting a block on error. The difference to non-PPR HARQ is the aggregation of several blocks onto a single transmission, eliminating the fate-sharing of bits in the non-PPR case.

2.4 Rateless Codes

Our experiments use three recently developed rateless codes—Raptor codes, Strider [9], and spinal codes [23]—so we start by summarizing the key ideas in these codes and the salient features of the implementations of these codes used in our experiments.

Raptor code. Raptor codes [26, 7], which are built on LT codes [20], achieve capacity for the Binary Erasure Channel where packets are lost with some probability. Not much is known about how close Raptor codes come to capacity for additive Gaussian noise (AWGN) channels and binary symmetric channels (BSC). However, there have been several attempts made to extend Raptor codes for the AWGN channel [22, 27, 2]. We adopt a similar construction to [22] in this paper, with an inner LT code generated using the degree distribution in the Raptor RFC [21], and an outer LDPC code as suggested by Shokrollahi [26].

Strider. Strider realizes the layered approach to rateless codes of Erez et al [6]. This approach combines existing fixed-rate base codes to produce symbols in a rateless manner. By carefully selecting linear combinations of symbols generated by the base codes, they show that the resulting rateless code can achieve capacity as the number of layers increases, provided the fixed-rate base code achieves capacity at some fixed SNR. The Strider implementation in our experiments was built using reference code from Gudipati [9], with recommended parameters.

Spinal codes. Spinal codes [23] use a hash function to generate transmitted symbols. The rich structure obtained using the hash function can be exploited to achieve rates very close to the channel capacity, with a practical encoder and decoder. In addition to their ability to approach capacity, spinal codes also work well on short messages (256-1024 bits), making them appealing from a latency perspective.

3. OPTIMAL TRANSMISSION SCHEDULE

We have yet to define the decoding CDF or to show how to coordinate the sender and receiver via a *transmission schedule* derived from this CDF. This section introduces the decoding CDF and applies it to the schedule optimization problem, treating the CDF as a known quantity. We also show how to compute the feedback delay. The solution of the optimization problem proceeds from these two inputs via a dynamic programming algorithm, which we demonstrate with an example. The next section will show how to learn the CDF using feedback from the receiver.

3.1 Decoding CDF

We abstract a rateless code as an encoder which produces an infinite sequence of encoded symbols (bits or constellation points/channel usages) from a finite message, and a decoder which takes any prefix of this infinite sequence and returns either nothing or the correct message. Supposing that all input messages are protected equally, and that the channel parameters drawn from some unknown distribution, the behavior of the code on this channel is characterized by a single function giving the probability with which a message can be decoded correctly after a certain number of symbols have been received by the decoder.

We assume that this probability increases monotonically with the number of symbols received; otherwise, a better (and admittedly more expensive) decoder could attempt to decode with only the first symbol, then with the first two, and so on, guaranteeing monotonicity. If every message is eventually decoded, then the function can be viewed as the cumulative distribution for a random variable that we denote n. Changes to the code parameters, channel conditions, or code block length will affect this distribution.

The decoding CDF has three desirable features from the perspective of the protocol:

1. It succinctly captures all of the uncertainties in the system, including those due to fluctuating outcomes on a stationary channel, time-varying channel parameters, and uncertainty about these parameters. Moreover, a schedule determination algorithm that relies only on the decoding CDF is insulated from the details of the code.

2. It enables the protocol to explicitly compute, and thus maximize, the expected throughput of a transmission schedule.

3. It can be learned from receiver feedback. §4 shows how beliefs about the decoding CDF can be updated from the number of symbols needed to decode each packet. Alternatively, the CDF can be estimated using offline simulations of the behavior of the code. Either way, the sender and receiver have common knowledge of the decoding CDF.

3.2 Optimization Problem

Given a decoding CDF, we would like to obtain a rule for the sender that determines when it should transmit and when it should pause for feedback. This rule takes the form of a sequence of positive integers n_i, $i \in \{1,2,\ldots\}$. Each n_i indicates cumulatively how many symbols should be transmitted before the sender begins its ith pause.

If feedback were free (i.e., took 0 time), then a throughput-maximizing sender would pause after each symbol transmission. In reality, it takes many microseconds to turn the radios from transmit to receive mode, re-synchronize, complete any ongoing decoding operations to determine whether a positive ACK should be sent, encode the ACK, and turn the radios around again to return control to the sender. The *feedback delay* is the time between the sender's last symbol prior to a pause for feedback and the same sender's first useful symbol following the pause for feedback. With 802.11a/n timings and the most reliable ACK coding rate,

$$T_{\text{feedback}} = \text{SIFS} + \text{preamble} + \text{ACK payload} + \text{SIFS} + \text{preamble}$$
$$= 64\ \mu\text{s} + 4\ \mu\text{s} \cdot \left\lceil \frac{\text{ACK bits} + 6}{24} \right\rceil$$

On a 20 MHz channel with the standard guard interval, 802.11 sends 12 million symbols per second. Thus, if the sender and receiver have only one coded transmission in play at a time, the cost of a single bit of feedback could equal the cost of 816 symbols. This number can be reduced by aggregating many packets to divide the large constant cost of feedback across more useful bits. We explore the details of such aggregation in §5. Even for Strider, the code with the largest packet size of those we considered, transmitted frames are only on the order of 3750 symbols each, so that a pause for feedback after each frame would occupy 18% of the total time spent on transmission plus feedback.

Let

$$n_f = \frac{T_{\text{feedback}}}{\#\text{ aggregated packets}} \cdot 12 \text{ million } \frac{\text{symbols}}{\text{sec}}$$

This is the amortized cost of feedback in units of foregone symbols. Note that higher layers cannot tolerate an arbitrary increase in latency caused by excessive aggregation, so one can never drive the cost of feedback to zero. Suppose that n_f already takes into account as much amortization as is possible subject to higher-layer latency constraints.

We wish to send a packet whose length before any coding is b bits, after coding it using a rateless code. The decoding CDF for the rateless code specifies $\mathbb{P}(n > \cdot)$, the probability that decoding will be successful after receiving n symbols.

Consider a general transmission schedule for reliable delivery. The sender first transmits n_1 symbols, then pauses for feedback. If decoding fails, the sender transmits $n_2 - n_1$ symbols before pausing a second time. In sum, before the ith pause, the sender transmits n_i symbols. We seek an assignment of values to n_1, n_2, ... that minimizes the average time spent delivering each b-bit packet.

Let

$$p_i = \mathbb{P}(\text{first success after } i \text{ feedback rounds})$$
$$q_i = \mathbb{P}(\text{stop after } i \text{ feedback rounds}).$$

These two quantities differ if the sender gives up on this packet without success and moves on to the next one. The sender will spend an average of $\sum_i q_i(n_i + i \cdot n_f)$ symbol-times on each message, including time spent transmitting and time spent waiting for feedback.

The *efficiency* of a transmission strategy is the fraction of this time *strictly necessary* for reliable delivery. Its formal definition is motivated by two observations.

1. For any feedback-based link-layer protocol, it is essential that the sender solicit feedback at least once for each network-layer packet to ensure that it has been received, so the sender can then proceed to the next packet.

2. It is necessary for the transmitter to transmit, on average, at least $\mathbb{E}[n]$ symbols. If the transmitter sends less than this number of symbols, then it follows that some fraction of packets are not decoded correctly.

Combining these two observations, we define efficiency as

$$\eta = \frac{\mathbb{E}[n] + n_f}{\sum_i q_i(n_i + i n_f)} \tag{1}$$

Note that if n_f is nonzero, the only way to achieve perfect efficiency will be for the sender to guess n correctly every time. If n has positive entropy arising from unpredictable channel variations, as in practice, we will be unable to achieve 100% efficiency, but our goal is to come as close as possible to the ideal efficiency.

Minimizing the time spent delivering a message is equivalent to maximizing η. In principle, this problem seems like a difficult multi-dimensional search, but it turns out that the optimal n_i assignments can be obtained by dynamic programming using only the decoding CDF as input. The optimal substructure in this problem is revealed by interpreting the transmitter's decisions in the framework of a dynamic game, as we explain next.

3.3 Dynamic Game Formulation

Suppose that we are playing a game against nature, and that nature chooses n from the known decoding CDF distribution but does not tell us the value. Our first move is to transmit n_1 symbols; nature's behavior is to use its hidden knowledge of n to determine whether the game ends or continues. We then transmit $n_2 - n_1$ symbols, and nature once more determines whether the game ends, and so on.

Our score at the end of the game is the negative of the total time we spent transmitting symbols or waiting for feedback (we measure the time in units of "symbol-time"). The negative is because we want the game to end as soon as possible. Because nature only acts to end the game, the optimal strategy depends only on n_f and the distribution of n (the decoding CDF), but not on any information arising during gameplay.

To develop an optimal strategy for the game, we apply backward induction. Supposing we find ourselves at some node of the decision tree where i symbols have already been transmitted, we must decide how many additional symbols j_i^* we will transmit before pausing to minimize our expected time-to-completion. For some choice of j, the corresponding expected time is

$$t_{ij} = j + n_f + \mathbb{P}(n > i + j \mid n > i) t_{(i+j)j^*_{i+j}}$$

That is, we pay an immediate cost of $j + n_f$, and with some probability given by the decoding CDF and expressed in the third summand above, the game will continue and we will incur additional costs according to our optimal strategy. The third term is somewhat counter-intuitive: it says that with probability $\mathbb{P}(n > i + j \mid n > i)$, the additional time required is $t_{(i+j)j^*_{i+j}}$; the tricky part is that the expression of t_{ij} now depends on $t_{\ell j'}$, where the index $\ell = i + j$ is *greater than* i.

We address this issue below, but for now observe that choosing

$$j_i^* = \arg\min_{j>0} t_{ij}$$

$$t_i^* = t_{ij_i^*} \text{ so that we can write}$$

$$t_i^* = j_i^* + n_f + \mathbb{P}(n > i + j_i^* \mid n > i) t^*_{i+j_i^*} \tag{2}$$

produces the optimal strategy[1].

The reason is that if we know the optimal strategy and the corresponding expected time for all $i' > i$, then we can compute the strategy and expected time for i. So, if n were bounded above by some finite value, one can use dynamic programming to find the optimal strategy to minimize the expected completion time for all i. We would then choose $n_1 = j_0^*$, $n_{i+1} = n_i + j^*_{n_i}$.

[1]The strategy is a subgame perfect equilibrium.

3.4 Finite Termination by Tail Fitting

Some finesse is required to terminate the infinite recursion onto larger and larger i. The problem is that one may easily encounter a point where there is no information available from the decoding CDF for some number of symbols, n^*, required to make progress; i.e., we have no information about $\mathbb{P}(n > n^* = i + j^*)$. What RateMore does in this situation is to revert to the best possible periodic-rate schedule given all the information known. It produces a schedule where the sender pauses after sending j^* symbols, obtaining the feedback, and then continuing, until either the packet is decoded or the sender gives up. If the packet gets decoded, the sender will have obtained information about the decoding CDF for this point in the state space, which will improve the subsequent operation of the protocol.

To determine j^*, we replace the tail of our distribution for n with an *analytic form*, which produces a stationary optimal strategy. Rearranging the terms in Equation (2) and replacing the t's with a fixed (stationary) value on both sides, we find

$$\begin{aligned} t^* &= j^* + n_f + \mathbb{P}(n > i + j^* \mid n > i) t^* \\ 1 - \frac{j^* + n_f}{t^*} &= \mathbb{P}(n > i + j^* \mid n > i) \\ &= \frac{\mathbb{P}(n > i + j^*)}{\mathbb{P}(n > i)} \end{aligned}$$

The memoryless (geometric) distribution satisfies this property. We fit a geometric tail onto our distribution and compute j^* and t^* for it to bootstrap the recursion onto finite i.

We could instead have set the complementary CDF to be zero above some finite n, but this introduces undesirable oscillations into the computed expected time and strategy variables, and does not offer any insight into what the transmitter should do if it does eventually find itself in the position of choosing a j for some i larger than this limit. With the geometric tail, the transmitter simply falls back to sending the stationary number j^* of symbols before each pause. This distinction is especially important during the process of learning the empirical CCDF, because the sender will have no data for the probability of $n > n_{\max}$, where $n_{\max}$ is the largest n we have observed so far.

For the memoryless distribution with geometric parameter $0 < \beta < 1$, we solve to obtain

$$\begin{aligned} \beta^{-j} + j\log\beta &= 1 - n_f \log\beta \\ &\Rightarrow k = \ln(k + \gamma), \text{ where} \\ &k \triangleq -j\log\beta \\ &\gamma \triangleq 1 - n_f\log\beta \end{aligned}$$

Solving iteratively,

$$k_1 = \gamma \qquad k_{i+1} = \ln(k_i + \gamma)$$

$$j^*_{\text{tail}} = -\frac{k_\infty}{\log\beta} \qquad t^*_{\text{tail}} = \frac{j^* + n_f}{1 - \beta^{j^*}}$$

In practice, the iteration converges rapidly.

Another useful special case of the dynamic programming algorithm is for $n = c + g$ where c is a constant and g is geometrically distributed with parameter β. In this case, we reuse the above calculation for j^*_{tail}, but take $n_i = c + i \cdot j^*_{\text{tail}}$. That is, the first step is to send $c + j^*_{\text{tail}}$ symbols, and each subsequent step is to send

j^*_{tail} symbols. This form for the distribution of n turns out to be a reasonable practical approximation to the empirical behaviors of the spinal code and Strider.

3.5 An Example

This section will use Figure 1 to illustrate the behavior of the dynamic programming strategy using a synthetic decoding CDF for purposes of illustration. We proceed top-to-bottom, then left-to-right. The top-left plot shows a CDF and when we would pause for feedback with various delays, n_f. The mapping of colors[2] to feedback costs is shown on the right axis. The middle-left and bottom-left plots show the estimated remaining time and optimal forward strategy for the same CDF after some number of symbols have been transmitted. The circles show the best division into feedback intervals, starting at 0 and stepping forward according to the bottom-left plot.

The top-right plot shows how the optimal strategy chooses a variable level of pre-feedback confidence of decoding. For the constant-plus-geometric distribution, for example, the lines would be flat. The other two plots on the right column show how tail-fitting gives us reasonable behavior even after dozens of round-trips (which might be reasonable if the cost of feedback is low enough).

4. LEARNING THE DECODING CDF

The decoding CDF is the primary input to the dynamic programming algorithm, specifying the probability that a given number of symbols will be required by the receiver for successful decoding. These probabilities depend on current channel conditions. When the channel can be characterized by a single parameter, we can obtain CDFs for different values of that parameter using off-line simulations or experiments. For example, in the case of the additive white Gaussian noise channel (AWGN) we could obtain CDFs for a range of signal-to-noise ratio (SNR) values. In practice, however, wireless radios operate in complex environments, and we expect the channel to have too many parameters for such an approach to be practical; we expect these parameters to vary unpredictably over time; and we expect that off-line simulations will differ significantly from the actual implemented system.

A robust alternative to this approach is to *learn* the CDF on the fly: that is, to estimate the CDF based on the recent history of number of symbols required for successful decoding at the receiver. The sender always consults the strategy derived from the most recent CDF estimate. This form of online learning directly accommodates variations in channel conditions due to fading and mobility.

The most general empirical distribution for the probability of successful decoding after any number number of symbols is the multinomial distribution. Thus, a very general Bayesian approach would be to learn this multinomial distribution beginning from a Dirichlet prior. This entails maintaining a histogram over the number of symbols required for decoding so far. While straightforward, a model with such a large state-space leads to slow learning and slow adaptation to variations in the channel.

Ideally, we would like to maintain a parametric distribution with a minimal number of parameters as our surrogate for the actual CDF. With this in mind, we propose two approaches: (a) Gaussian approximation, and (b) Constant-plus-Geometric approximation. From these, we pick the Gaussian approximation, for the reasons explained below.

Gaussian approximation. Our inspection of the decoding CDFs indicates that the Gaussian distribution should be a reasonable approximation at low SNR. Maximum-likelihood (ML) estimation for the Gaussian distribution requires nothing more than computing the empirical mean and variance of the number of symbols needed for successful decoding, which can be accomplished with a few accumulators. In the face of time-varying conditions, the empirical mean and variance can be *filtered* using a moving average or a similar scheme.

We chose to use Algorithm 1 for learning a Gaussian CDF. The algorithm keeps exponentially-weighted accumulators to estimate the mean, μ, and variance, σ^2. The parameter α ranges from 0 (no memory) to 1 (unlimited memory). The averages track the input with a time constant of $1/\ln(1/\alpha)$. This scheme has two advantages over a traditional exponentially-weighted moving average of the form $y \leftarrow (1-\alpha)x + \alpha y$. First, the start-up transient dies out more quickly because we weight our initial conditions less heavily. Second, the estimator's behavior is well-defined for $\alpha = 1$: inputs are retained forever, and we recover the ML estimator.

Algorithm 1 Gaussian learning with exponentially weighted moving average. Long-term performance is not sensitive to initial values.

```
function init():
    samples ← 1
    sum ← 100 (for instance)
    sumsq ← sum^2 + 10^2 (for instance)
function learn(sample, α):
    samples ← samples · α + 1
    sum ← sum · α + sample
    sumsq ← sumsq · α + sample^2
function get_ccdf(x):
    μ = sum/samples
    σ^2 = sumsq/samples − μ^2
    return normal_ccdf(μ, σ^2, x)
```

We present results in §7 showing that under the Gaussian approximation, performance of RateMore when the true decoding CDF is not available differs by only a few percent from performance when the true CDF is available. At low SNR, the Gaussian approximation with a memory parameter of $\alpha = 0.99$ performs within 1% of the known-CDF case. We also show that very aggressive learning rates of $\alpha = 0.80$ perform very well on simulated fading channels.

Constant-plus-Geometric approximation. The Constant-plus-Geometric approximation treats the random variable n as $n = c + g$, where c is a constant and g is a geometric random variable with parameter β. Like the Gaussian, this family of distributions has two parameters, c and β. We considered this approximation because of the analogy with the presence of *error exponents* in most good codes. When a code has an error exponent, the probability of a decoding error after transmitting n symbols scales as $\exp(-\gamma n)$ for some constant γ as long as n is large enough to push the rate below the channel capacity. This minimal n corresponds to the constant summand c, and the exponential drop in error probability with increasing n corresponds to the geometric drop in probability of unsuccessful decoding with increasing g.

Additionally, if this approximation is good, the the resulting dynamic programming strategy has a simple form: for the first transmission, send a number of symbols $c + j^*_{\text{tail}}$, then pause for feedback. If the receiver has not decoded the transmission, send j^*_{tail} more symbols before each subsequent request for feedback.

Unfortunately, despite this elegance, this distribution does not have a convenient maximum-likelihood estimator. The most likely value of the constant c is upper bounded by the min of n observed

[2]Unfortunately, this chart is best viewed in color; we recognize that it is a problem when not viewed online or on a color print-out; note, however, that the different "levels" correspond to different feedback costs.

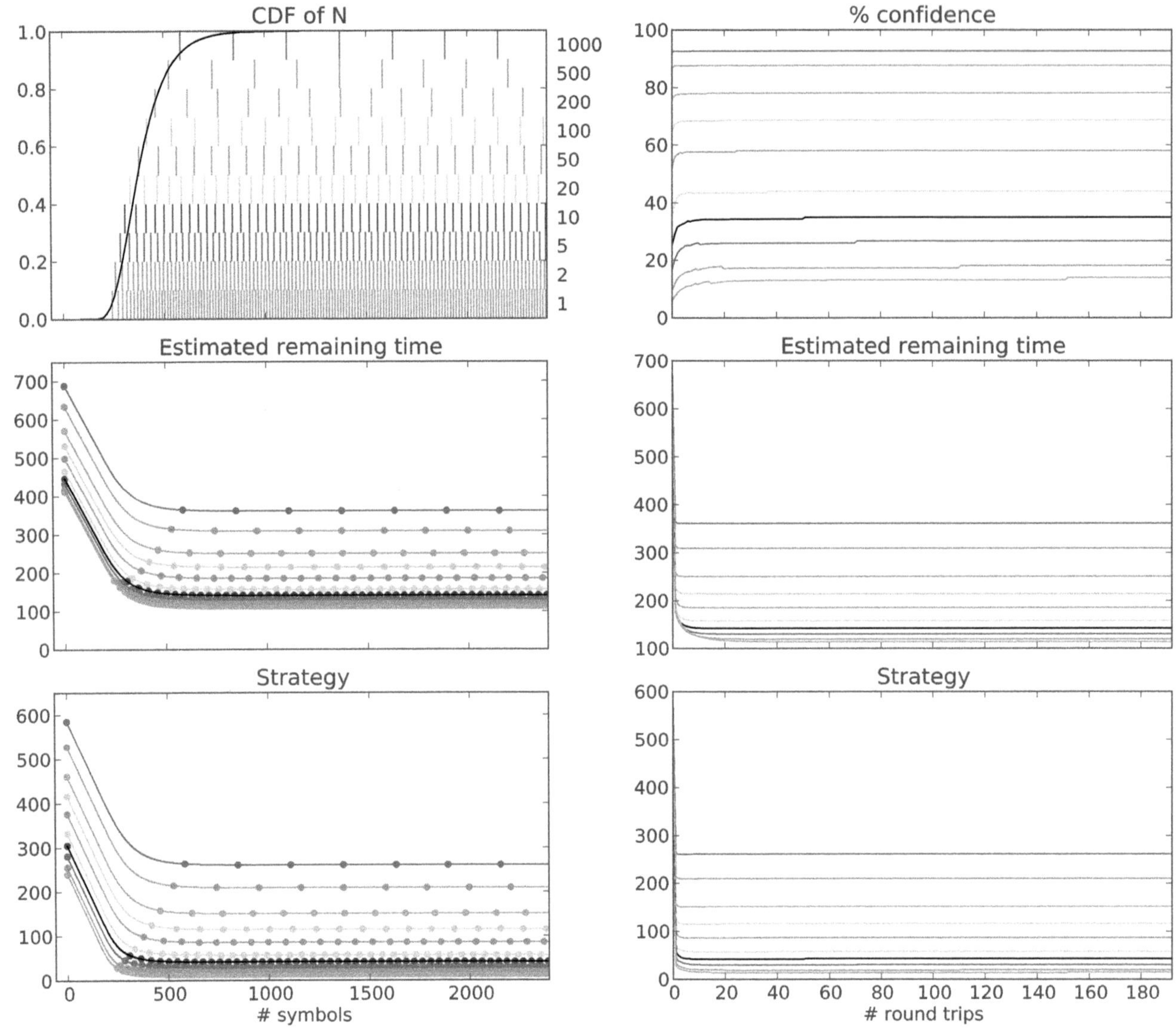

Figure 1: An illustration of the output of the dynamic programming algorithm

over all decoding attempts. In practice, this will lead the dynamic programming algorithm to under-estimate the number of symbols necessary for decoding. For this reason, we performed our experiments using the Gaussian approximation.

5. PACKING PACKETS AND BLOCK ACKS

Rateless codes, particularly in high bandwidth situations, benefit from a method to pack symbols belonging to multiple distinct network-layer packets into the same link-layer frame, and using a "block ACK" scheme (as in 802.11e/n) to amortize the cost of sending feedback about the decoding state of each packet. Packing packets into a single frame is a useful amortization not only when the packets are small in size, but also when only a few more symbols are required to be transmitted to successfully decode the packet. As link rates increase, it is likely that any code (whether rateless or using incremental redundancy) that decodes using symbols from previous frame transmissions will benefit from packing multiple distinct packets into the same frame and using block ACKs.

If per-packet latency were not a concern, we could pack as many packets as we wish into one frame, making n_f as small as we wish. In practice, however, we care about this metric: for example, if we consider a high-definition video/audio teleconference, the maximum wireless per-packet latency might be set to 50 ms, which at a bit rate of 1.5 Mbits/s works out to $n_f =$. RateMore's dynamic programming method assumes that n_f is exogenously given by such latency requirements, and that it has already been amortized over multiple in-flight packets packed into one frame.

To efficiently pack and unpack symbols from multiple packets into a single transmitted frame, we observe that because the sender's decoding CDF is learned from the receiver's feedback, both the sender and receiver can agree on the exact CDF and hence the exact feedback schedule currently in use. The sender encodes the length of the packet (in symbols) in the Signal field as in 802.11, which is transmitted inside a 4 μs OFDM symbol. From this field, the receiver of a packet knows immediately how many symbols it will receive in all. The problem is to allocate a certain number of symbols for

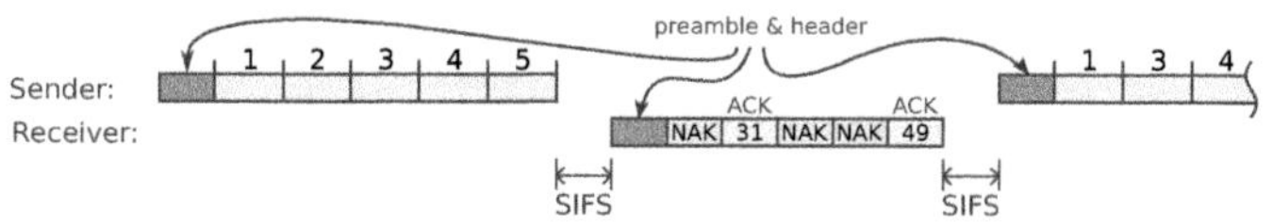

Figure 2: Symbols from many packets are aggregated into a frame in order to amortize the cost of the short inter-frame spaces (SIFS, 16µs) and preambles/headers.

packet 1, then the symbols for packet 2, and so on, as shown in Figure 2.

The solution to this problem is as follows. When the sender transmits the first frame, the receiver consults its copy of the feedback schedule to see how many symbols should have been included for a packet for which no symbols have previously been sent (obviously, this number would be the same for every packet). For example, the schedule may recommend sending 100 symbols before pausing for feedback. If a frame is received with length 500 symbols, the receiver infers that the sender wants to deliver 5 packets (as in Figure 2).

After the frame ends and the receiver finishes attempting to decode each packet, it sends an acknowledgment (ACK) frame to the sender including a short field (e.g., 6 bits) for each such packet. If the field for packet i is all ones (NAK), the packet has *not* been decoded successfully, so the sender should consult its feedback schedule (also known to the receiver) and include more symbols for that packet in the next frame. Otherwise, decoding was successful; in this case, the ACK field encodes the *fraction of this last frame's symbols for packet i that was needed to achieve a successful decoding*. Using this feedback, the sender can calculate the number of symbols that were needed to decode any given packet to estimate the decoding CDF using the learning algorithm (§4).

In general, because the sender and the receiver both agree on the updated CDF, the receiver will now know how many symbols for each unfinished packet will be in the next frame. For example, in Figure 2, three packets out of five were not decoded successfully. Each of those packets will have the *same* number of additional symbols sent in the next frame (for example, 20 symbols, according to the feedback schedule). In addition, the frame would include symbols from new packets; each of those new packets would send the same number of symbols given by the feedback schedule for new packets.

The key insight in this scheme is that we do not need to explicitly communicate the number of symbols per packet being transmitted because both parties know the schedule. This scheme works correctly if frames and ACKs are not lost, which can be engineered by ensuring that the preamble and headers are sent at low rates (one can extend this protocol to handle frame and ACK loss with some effort).

6. SOFTWARE IMPLEMENTATION

Our implementation comprises several interconnected simulations written in C++ and Python.

Codes and Channel Models. To obtain raw decoding CDFs, we performed thousands of experiments with the Spinal, Strider, and Raptor codes using the parameters shown in Table 1. The experiments produced decoding CDFs for SNRs from −5 to 35 dB on both stationary additive white Gaussian noise (AWGN) and fast-fading Rayleigh channels. The decoders had access to channel information. The slow-fading experiments used the decoding CDFs from the stationary channel, because in this regime the channel coefficients fluctuate over a time scale much longer than a packet (e.g., 45 ms for a receiver moving at 10 km/hr).

Code	Parameter	Value
Spinal	Message size	256 bits
	k	4 bits/half symbol
	B	256 candidates/beam
	Constellation	QAM-2^{20}
Strider	K	33 layers
	M	1500 bits/layer
	Message size	50490 bits
	Outer code	1/5-rate turbo
Raptor	Message size	9500 bits
	LT degree dist.	As per RFC [21]
	LDPC design	As per Shokrollahi [26]
	LDPC rate	0.95
	Constellation	QAM-64

Table 1: Parameters for the underlying rateless codes.

To simulate slow Rayleigh fading, we directly synthesized a Gaussian process with the appropriate Doppler spectrum and used the result as a time series of channel coefficients. Noise power was maintained at a constant fraction of the average signal power. Given the Rayleigh channel coefficient and the ratio of average signal power to noise power, we computed an appropriate instantaneous SNR. For the "ideal adaptation" results in the next section, we then selected the corresponding raw decoding CDF from our empirical data by interpolating between CDFs at the nearest SNRs. RateMore itself learns the CDFs online using the Gaussian approximation. Interestingly, despite the approximation, the system's performance is as good as using the raw CDFs.

Timing. To obtain accurate values for n_f as well as throughput and latency, we built a timing model for transmission and feedback on a 20 MHz half-duplex OFDM channel designed to mimic 802.11a/n. We included the overhead associated with preamble synchronization, and assumed that all feedback was transmitted at the lowest 802.11a/n rate of 6 Mbps for reliable delivery. We did not include a simulation of contention, but we used the SIFS interval ($16\mu s$) to determine how long the parties would have to wait for the ACK stage.

Because the nature of our transmission schedule is to periodically incur negative acknowledgements, we wanted to ensure that in such cases the communication proceeds to successful completion before the channel is relinquished. This avoids receivers having to maintain large buffers for several ongoing conversations. Therefore, in our simulation the transmitter resumes transmitting after SIFS elapses from the receipt of an ACK frame.

Dynamic programming algorithm. Our implementation allows for the possibility of "sparse" CDFs whose values are known only for some isolated values of the number of symbols n. For the Spinal code, we found that choosing the values of n corresponding to $1/8^{\text{th}}$-passes sacrificed virtually no performance while reducing the search space for the algorithm by a factor of roughly 8. For Strider, n should be a multiple of 3840 symbols.

The sender runs Algorithm 2 to determine how many symbols should be sent before each feedback. The points at which the CDF is known are given by x_i. The steady-state values j^*_{tail} and t^*_{tail} are computed as described in §3.4, and i_{tail} denotes the index into the x_i where recursion is terminated and the steady-state behavior takes over.

A value j_i returned from Algorithm 2 specifies that after a failed decode attempt with x_i symbols, the next decode attempt should

Algorithm 2 Given decoding CDF (sampled at x_i) and n_f, compute the transmission schedule by minimizing Expected Time To Completion (ETTC).

$j_i, t_i \leftarrow j^*_{\text{tail}}, t^*_{\text{tail}}$ for all $i \geq i_0$
for $i = i_{\text{tail}} - 1$ **to** 0 **do**
  # compute strategy after x_i transmitted symbols:
  for $j = 1$ **to** $i_{\text{tail}} - i$ **do**
    # when transmitting $x_{i+j} - x_i$ more symbols:
    $\text{ETTC}_j \leftarrow (x_{i+j} - x_i) + n_f + t_{i+j} \cdot \mathbb{P}(n > x_{i+j} | n > x_i)$
  # choose strategy that minimizes expected time:
  $j_i, t_i \leftarrow \arg\min\{\text{ETTC}_j\}, \min\{\text{ETTC}_j\}$
return $j_0, j_1, \cdots$

occur with x_{i+j_i} symbols. A sender that gets a NAK after x_i symbols should transmit $x_{i+j_i} - x_i$ more symbols before the next feedback.

Given a feedback schedule, we use Equation (1) to compute the fraction η of transmission time that is well-spent, and the overhead $1 - \eta$ that is wasted.

ARQ and Try-after-n HARQ. We implemented two alternative feedback schedules for comparison, which we refer to as ARQ and Try-after-n HARQ. With ARQ, the sender picks some number of symbols to transmit, then waits for feedback in the form of an ACK. If it receives no ACK, it drops the packet and starts over. This scheme effectively chooses a rated version of the rateless code, and does not incorporate incremental redundancy. In principle, ARQ could choose a different rate for each SNR value. However, existing ARQ systems with rate adaptation, like 802.11, do not have such a large number of rates. 802.11n with one spatial stream has only eight. We therefore require ARQ to pick eight rates to cover the SNR range from −5 to 35 dB. These eight rates are selected to maximize efficiency over all SNRs. For the Spinal code, this maximization led us to select rates of 3.4, 6.2, 11, 19, 30, 43, 56, and 72 Mbits/s.

For Try-after-n HARQ, we picked a family of eight parameter values, as before, and sent n additional symbols of incremental redundancy before each solicitation of feedback.

7. EVALUATION

We evaluate RateMore in simulation on stationary, slow-, and fast-fading AWGN channels. Our principal results are summarized in Table 2.

7.1 Baseline comparison to ARQ, Try-after-n HARQ

Figure 3 shows the overhead of RateMore compared to ARQ and "Try-after-n" HARQ running atop spinal codes with a feedback value $n_f = 20$. These experiments are using a stationary channel simulator that introduces Gaussian noise of different variances. We run 20 trials of 100 packets at each SNR.

Across a range of SNRs, the reduction in overhead is between 2.6× and 3.8× relative to ARQ, and between 2.8× and 5.4× relative to HARQ. These reductions are significant; they translate to link-layer throughput improvements up to 26%.

7.2 Slow-fading Rayleigh channel

Using our simulated Rayleigh fading coefficients for a Doppler velocity of 10 meters per second, we compared the throughput of RateMore running with known CDFs and SNR against the throughput of RateMore with Gaussian CDF learning. Figure 4 shows a typical trace of 100 milliseconds of adaptation on a fading channel. The average SNR is 15 dB. We found that long integration times (e.g. learning parameter α close to 1) were not necessary to obtain good performance. In fact, with $\alpha = 0.8$, corresponding to an exponential

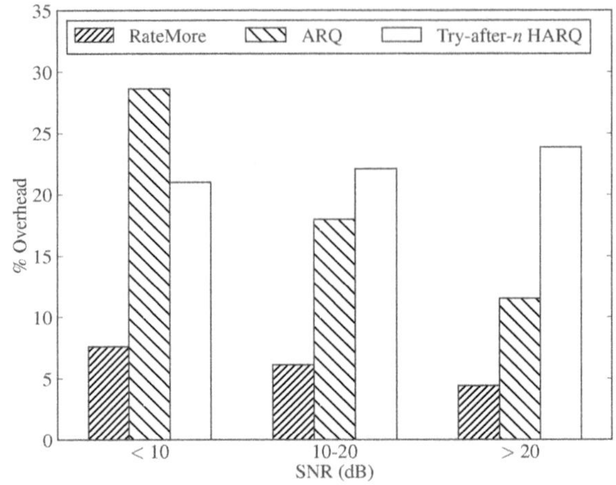

Figure 3: With spinal codes and $n_f = 20$. At low, medium, and high SNR, respectively, RateMore reduces overhead by 3.8×, 2.9×, and 2.6× relative to ARQ and 2.8×, 3.6×, and 5.4× relative to Try-after-n HARQ.

time constant of only 4.5 packets, we found aggregate throughput with learning to be within 1.57% of throughput under known CDFs and SNR. Figure 5 is a detail of Figure 4.

7.3 Efficiency

Figure 6 shows the efficiency of RateMore (we defined this metric in Equation (1)) across a range of SNRs for different values of the feedback cost, n_f. These graphs are for spinal codes (the other codes show similar results). We have grouped the data according to "low", "medium", and "high" SNRs for illustration. In general, efficiency is extremely high, always over 95% at medium and high SNRs, and over 88% for low SNR values, even when n_f is as high as 100. The conclusion is that RateMore produces a good transmission schedule across different channel conditions.

7.4 How well does learning work?

Figure 7 shows the results of experiments that evaluate how much we lose in our *learning* the decoding CDF compared to *knowing* the true CDF. We include the comparison for spinal codes, but the results are similar for other codes. At low and medium SNRs, the cost of learning, i.e., how much we lose, is a negligible 2%. At higher SNR values, the cost is always less than 6%, which is still tolerable.

The conclusion is that our Gaussian approximation of the decoding CDF, which is a simple two-parameter fit (mean and variance) works extremely well, as does the simple filtering method to estimate the mean and variance.

7.5 Streaming with a low-latency requirement

For spinal codes, Strider, and Raptor, we compared overhead and channel occupation fraction using RateMore for streaming multimedia designed to emulate a Skype voice and HD video call (Figure 8). The main constraint here is latency: we require that the number of packets in flight, times the size of a packet, should represent no more than 100 ms of audio or audio and video. The Strider and Raptor codes we used have single-packet sizes larger than this limit for the low-rate voice call. In order to meet the latency requirements, these codes would be forced to transmit partly-empty packets. Because RateMore is agnostic to the details of a specific rateless code, it enables a direct comparison of different rateless coding schemes with

Fig., §	Experiment	Result
3, §7.1	RateMore vs. ARQ, Try-after-n HARQ	Overhead reduced by up to 5.4×; throughput increased by up to 26.6%.
4, 5, §7.2	Learning under slow Rayleigh fading	Throughput only 1.57% below case of known CDFs and SNR.
6, §7.3	RateMore efficiency for various n_f	Better than 88% efficiency even when $n_f = 100$: time is well spent.
7, §7.4	Impact of learning on throughput	RateMore's learning has an impact of only 0.25%-6%.
8, §7.5	Streaming with a latency requirement	Block ACK scheduling and code agnosticism enables efficient streaming.
9, §7.6	Streaming on a fast-fading channel	Less than 15% channel occupancy even at low bandwidth.

Table 2: Summary of principal results.

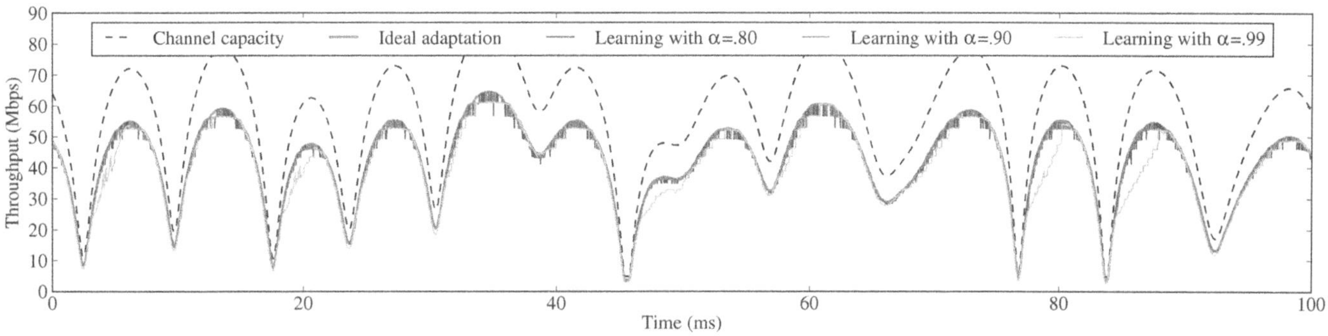

Figure 4: Learning performance of RateMore with spinal codes on a Rayleigh fading AWGN channel. Noise power is 15 dB below average faded signal power, giving an equivalent (in terms of capacity) stationary SNR of 12.8 dB. Doppler velocity is 10 m/s. The learning parameter α can be set to a very aggressive value of .8 for an aggregate throughput within 1.57% of the known-SNR "ideal adaptation" throughput. With $n_f = 10$.

different packet sizes, including for instance the effects of relatively smaller packet sizes in spinal codes.

7.6 Streaming on a fast-fading channel

When run on CDFs generated for a fast-fading Rayleigh channel, RateMore still maintains low overhead and channel occupation (Figure 9).

8. CONCLUSION

The recent innovations in practical rateless codes for wireless networks are an exciting development because they promise a better way to deal with the fundamental problem of time-varying channel conditions caused by mobility and interference. This paper presented the case for a different approach to link-layer protocols than the traditional methods, motivated by the property that with a rateless code, there is no obvious pause point to ask for feedback. We showed how using the decoding CDF provides a code-independent way to encapsulate the essential information of the underlying code. We developed RateMore, which incorporates a dynamic programming strategy to compute the transmission schedule, a simple learning method to estimate the decoding CDF, and a block ACK protocol to amortize feedback.

Our results show that RateMore reduces overhead by between 2.6× and 3.9× compared to 802.11-style ARQ and between 2.8× and 5.4× compared to 3GPP-style "Try-after-n" HARQ, which are the best existing deployed approaches. We demonstrated the reduced overhead of RateMore in experiments using three different rateless codes (Raptor, Strider, and Spinal codes). These significant reductions in overhead translate to good throughput improvements. For example, for spinal codes, the throughput improvement is as high as 26% compared to both ARQ and "Try-after-n" HARQ.

In conclusion, we believe that RateMore provides a practically useful link-layer protocol to coordinate between sender and receiver in wireless networks using rateless codes, improving performance when channel conditions are variable.

ACKNOWLEDGMENTS

We thank the reviewers and Bob Iannucci for helpful comments. Two generous graduate fellowships supported this work: the Irwin and Joan Jacobs Presidential Fellowship (Perry and Iannucci), and the Claude E. Shannon Assistantship (Perry). We thank the members of the MIT Center for Wireless Networks and Mobile Computing, including Amazon.com, Cisco, Intel, Mediatek, Microsoft, and ST Microelectronics, for their interest and support.

REFERENCES

[1] A. Anastasopoulos. Delay-optimal hybrid ARQ protocol design for channels and receivers with memory as a stochastic control problem. In *Communications, 2008. ICC'08. IEEE International Conference on*, pages 3536–3541, 2008.

[2] R. Barron, C. Lo, and J. Shapiro. Global design methods for raptor codes using binary and higher-order modulations. In *IEEE MILCOM*, 2009.

[3] J. Bicket. Bit-Rate Selection in Wireless Networks. Master's thesis, M.I.T., Feb. 2005.

[4] J. Camp and E. Knightly. Modulation Rate Adaptation in Urban and Vehicular Environments: Cross-Layer Implementation and Experimental Evaluation. In *MobiCom*, 2008.

[5] D. Chase. Code Combining: A Maximum-Likelihood Decoding Approach for Combining an Arbitrary Number of Noisy Packets. *IEEE Trans. on Comm.*, 33(5):385–393, May 1985.

[6] Erez, U. and Trott, M. and Wornell, G. Rateless Coding for Gaussian Channels. *IEEE Trans. Info. Theory*, 58(2):530–547, 2012.

[7] O. Etesami, M. Molkaraie, and A. Shokrollahi. Raptor codes on symmetric channels. In *ISIT*, 2005.

[8] R. Ganti, P. Jayachandran, H. Luo, and T. Abdelzaher. Datalink streaming in wireless sensor networks. In *SenSys*, pages 209–222, 2006.

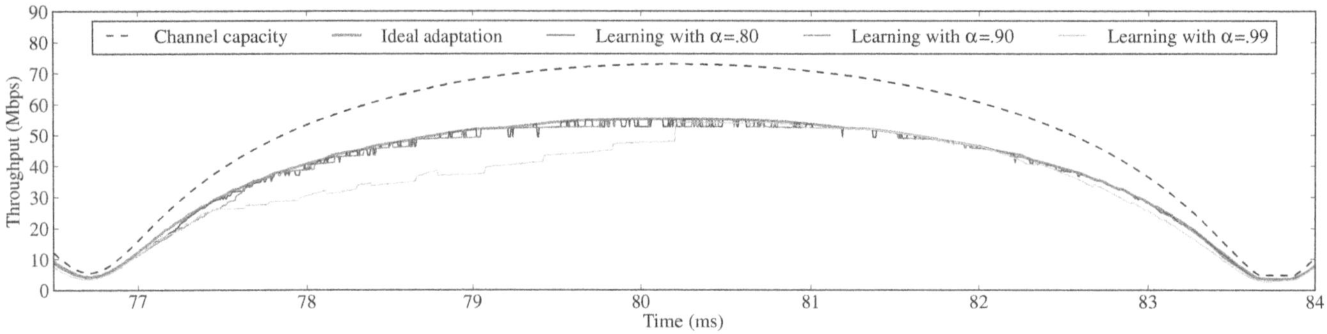

Figure 5: Detail of Figure 4.

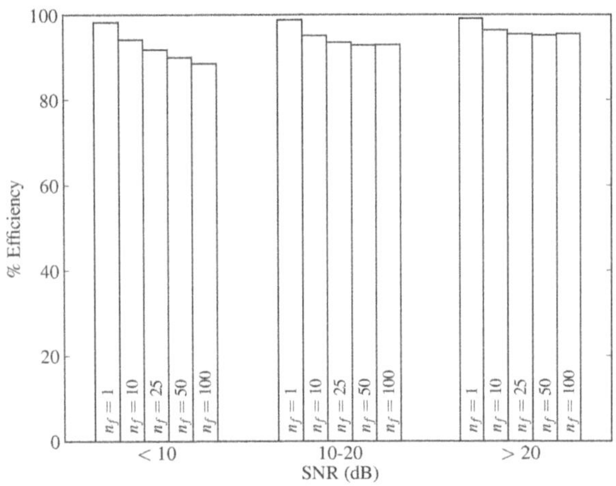

Figure 6: Even though n is random, RateMore makes good guesses and thus wastes very little time sending unnecessary symbols or waiting for unnecessary feedback.

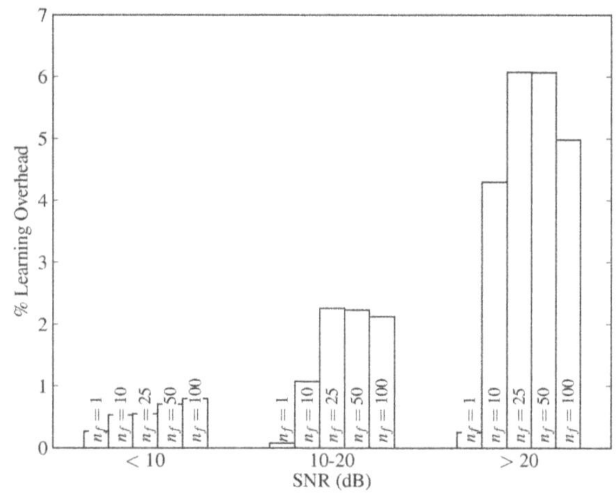

Figure 7: Gaussian learning performance on a stationary channel with spinal codes. Mean and variance are smoothed by an exponentially-weighted moving average with parameter $\alpha = .99$. Steady-state performance is comparable to performance when CDFs are fully known.

[9] A. Gudipati and S. Katti. Strider: Automatic rate adaptation and collision handling. In *SIGCOMM*, 2011.

[10] J. Ha, J. Kim, and S. McLaughlin. Rate-compatible puncturing of low-density parity-check codes. *IEEE Trans. Info. Theory*, 2004.

[11] J. Hagenauer. Rate-compatible punctured convolutional codes (rcpc codes) and their applications. *IEEE Trans. on Comm.*, 1988.

[12] B. Han, A. Schulman, F. Gringoli, N. Spring, B. Bhattacharjee, L. Nava, L. Ji, S. Lee, and R. Miller. Maranello: practical partial packet recovery for 802.11. In *NSDI*, pages 14–14, 2010.

[13] G. Holland, N. Vaidya, and P. Bahl. A Rate-Adaptive MAC Protocol for Multihop Wireless Networks. In *MobiCom*, 2001.

[14] IEEE Standard 802.11: Wireless LAN Medium Access Control and Physical Layer Specifications, Aug. 1999.

[15] J. Ikuno, M. Wrulich, and M. Rupp. Performance and modeling of LTE H-ARQ. In *Proc. International ITG Workshop on Smart Antennas (WSA 2009), Berlin, Germany*, 2009.

[16] K. Jamieson and H. Balakrishnan. PPR: Partial Packet Recovery for Wireless Networks. In *SIGCOMM*, 2007.

[17] J. Li and K. Narayanan. Rate-compatible low density parity check codes for capacity-approaching ARQ scheme in packet data communications. In *Int. Conf. on Comm., Internet, and Info. Tech.*, 2002.

[18] K. C. Lin, N. Kushman, and D. Katabi. ZipTx: Exploiting the Gap Between Bit Errors and Packet Loss. In *MobiCom*, 2008.

[19] C. Lott, O. Milenkovic, and E. Soljanin. Hybrid ARQ: theory, state of the art and future directions. In *2007 IEEE Information Theory Workshop on Information Theory for Wireless Networks*, pages 1–5. IEEE, July 2007.

[20] M. Luby. LT codes. In *FOCS*, 2003.

[21] M. Luby, A. Shokrollahi, M. Watson, and T. Stockhammer. Raptor Forward Error Correction Scheme for Object Delivery. RFC 5053 (Proposed Standard), Oct. 2007.

[22] R. Palanki and J. Yedidia. Rateless codes on noisy channels. In *ISIT*, 2005.

[23] J. Perry, H. Balakrishnan, and D. Shah. Rateless spinal codes. In *HotNets-X*, Oct. 2011.

[24] J. Perry, P. A. Iannucci, K. E. Fleming, H. Balakrishnan, and D. Shah. A rateless wireless communication system using spinal codes. In *SIGCOMM*, Aug. 2012. To appear.

[25] S. Sen, N. Santhapuri, R. Choudhury, and S. Nelakuditi. AccuRate: Constellation-based rate estimation in wireless networks. *NSDI*, 2010.

[26] A. Shokrollahi. Raptor codes. *IEEE Trans. Info. Theory*, 52(6), 2006.

[27] E. Soljanin, N. Varnica, and P. Whiting. Incremental redundancy hybrid ARQ with LDPC and Raptor codes. *IEEE Trans. Info. Theory*, 2005.

[28] M. Vutukuru, H. Balakrishnan, and K. Jamieson. Cross-Layer Wireless Bit Rate Adaptation. In *SIGCOMM*, 2009.

[29] S. Wong, H. Yang, S. Lu, and V. Bharghavan. Robust Rate Adaptation for 802.11 Wireless Networks. In *MobiCom*, 2006.

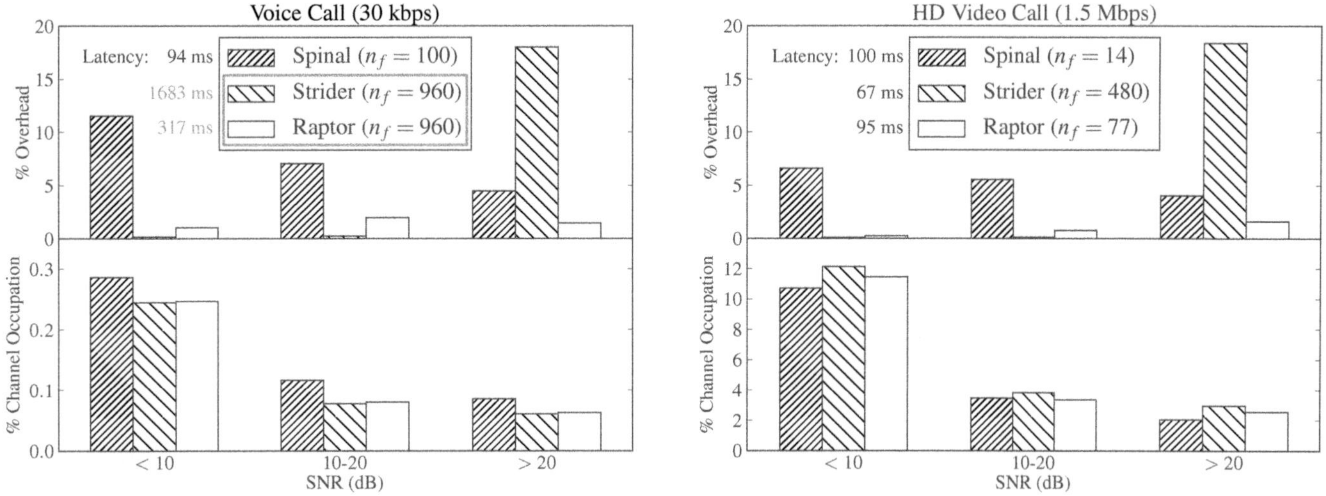

Figure 8: RateMore is agnostic to the details of a specific rateless code, and achieves low overhead in latency-critical environments. n_f has been normalized according to the largest number of parallel streams compatible with a given bandwidth-latency product. Cases are shown for a minimum-quality Skype voice call and for an HD video call. For an ideal user experience, latency should be less than 100 ms. Strider and Raptor require relatively large packets, which prevents the use of aggregation on the voice call – in fact, a single packet contains more than 100 ms of audio, as shown in red. In order to meet the latency target at the expense of channel occupation, packets could be fired off partly empty.

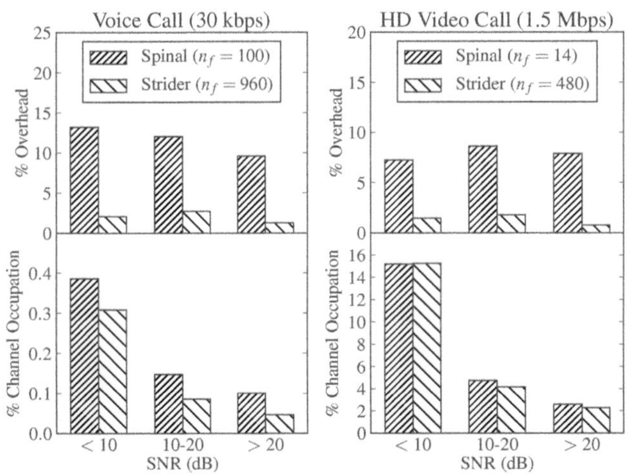

Figure 9: RateMore overhead and channel occupation on fast-fading Rayleigh channels. n_f has been normalized as in Figure 8

Rate Adaptation for 802.11 Multiuser MIMO Networks

Wei-Liang Shen[†‡], Yu-Chih Tung[†], Kuang-Che Lee[†], Kate Ching-Ju Lin[†],
Shyamnath Gollakota[°], Dina Katabi[♯] and Ming-Syan Chen[†‡]
[†]Research Center for IT Innovation, Academia Sinica
[‡]Department of Electrical Engineering, National Taiwan University
[°]CSE, University of Washington
[♯]CSAIL, Massachusetts Institute of Technology

ABSTRACT

In multiuser MIMO (MU-MIMO) networks, the optimal bit rate of a user is highly dynamic and changes from one packet to the next. This breaks traditional bit rate adaptation algorithms, which rely on recent history to predict the best bit rate for the next packet. To address this problem, we introduce TurboRate, a rate adaptation scheme for MU-MIMO LANs. TurboRate shows that clients in a MU-MIMO LAN can adapt their bit rate on a per-packet basis if each client learns two variables: its SNR when it transmits alone to the access point, and the direction along which its signal is received at the AP. TurboRate also shows that each client can compute these two variables passively without exchanging control frames with the access point. A TurboRate client then annotates its packets with these variables to enable other clients to pick the optimal bit rate and transmit concurrently to the AP. A prototype implementation in USRP-N200 shows that traditional rate adaptation does not deliver the gains of MU-MIMO WLANs, and can interact negatively with MU-MIMO, leading to low throughput. In contrast, enabling MU-MIMO with TurboRate provides a mean throughput gain of 1.7x and 2.3x, for 2-antenna and 3-antenna APs respectively.

Categories and Subject Descriptors C.2.2 [**Computer Systems Organization**]: Computer-Communications Networks

General Terms Algorithms, Design, Performance, Theory

Keywords Multiuser MIMO networks, Rate adaptation

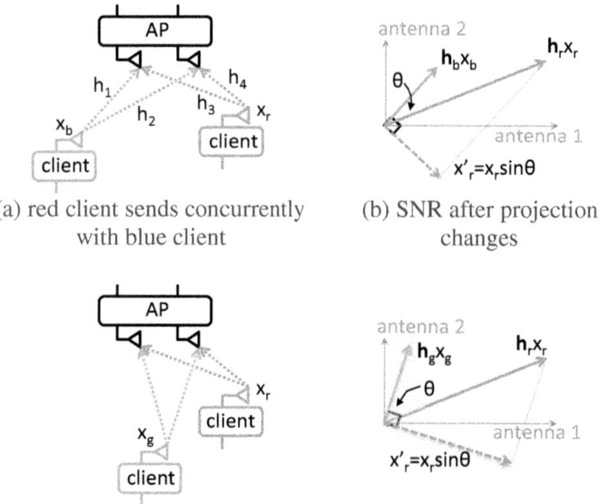

(a) red client sends concurrently with blue client

(b) SNR after projection changes

(c) red client sends concurrently with green client

(d) SNR after projection is larger than in (b)

Figure 1—Optimal bitrate changes after projection. The optimal bitrate depends on the client that is transmitting concurrently. The smaller angle between the concurrent clients leads to a larger amount of SNR reduction after projection.

1. INTRODUCTION

Wireless LANs are facing two trends: First, the number of antennas on an access point is increasing steadily, with typical APs today having two or three antennas [1]. Second, there is a proliferation of small WiFi devices, e.g., sensors, smart phones, and game consoles, which have a small form factor and strict power limitations, and hence typically use a single antenna. These trends cause a multi-antenna access point to spend a significant fraction of its time communicating with a single antenna client. As a result, wireless LANs will not deliver the maximum number of concurrent transmissions enabled by their infrastructure. To address this problem, researchers have advocated the use of multiuser MIMO (MU-MIMO) LANs, where multiple single-antenna clients communicate concurrently with a multi-antenna AP. They demonstrated that decoding such concurrent transmissions is feasible both on the uplink and downlink [31, 7, 37]. They also developed a MAC protocol that allows clients to contend for concurrent transmissions to a multi-antenna AP [31]. So far, however, research on MU-MIMO WLANs has not addressed the bit rate selection problem, and simply assumed that the transmitters know the best bit rate [31, 7, 37]. This assumption is valid on the downlink where there is only one transmitter, the AP, and hence the problem can be reduced to standard 802.11n rate adaptation. The scenario on the uplink, however, is quite different: it has multiple concurrently transmitting clients that collectively have to pick the best bit rates to their AP. The decisions made by these clients are not independent; they interact in a complex manner that intrinsically differs from existing 802.11 networks.

To see the problem, consider the scenario in Fig. 1(a) where two single-antenna clients transmit concurrently to a 2-antenna access point. Recall that a 2-antenna AP receives signals in a 2-dimensional space defined by its two antennas, as shown in Fig. 1(b). The basic approach for decoding the concurrent packets is as follows [33]: The AP first projects the incoming signal on a direction orthogonal to one of the clients, say the blue client. This eliminates the signal of the blue client and allows the AP to decode the red client. The AP then uses interference cancellation to subtract the red client's signal and decode the blue client. Note that the success of this decoding process depends on the AP being able to decode the red client after projecting its signal on a direction orthogonal to the blue client. This projection however reduces the SNR of the red client, as evident from the reduction in the length of the projected red vector in Fig. 1(b). This means that the red client should transmit at a bit rate supported by its SNR after projection; otherwise the AP becomes unable to decode its signal. Note also that the SNR after projection and hence the optimal bit rate depends on the angle between the signals of the two clients, i.e., θ. For example, if the red client transmits its next packet with the green client, as in Fig. 1(c), then its SNR after projecting on a direction orthogonal to the green client will be different, as in Fig. 1(d), and hence the red client's optimal bit rate for the next packet will change.

Thus, in a MU-MIMO LAN, the optimal bit rate of a client changes depending on the set of clients that transmit with it. Since this set may vary from one packet to the next, the optimal bit rate changes on a per packet basis. This breaks the basic assumption underlying existing 802.11 bit rate adaptation algorithms, which use the bit rate that fits recent packets as a predictor for the best bit rate for the next packet [17, 27, 9, 24, 8, 35].

This paper presents TurboRate, a bit rate adaptation protocol suitable for concurrent MU-MIMO 802.11 clients. TurboRate enables a MU-MIMO client to pick the optimal bit rate for each packet it transmits, even when the bit rate changes from one packet to the next.

At a high level, TurboRate works as follows. Each client listens to the AP's transmissions (including its beacons) to learn the channel coefficients from the AP to itself. The client uses this information to *passively* compute two variables: 1) the direction along which the AP receives its signal, and 2) its SNR if it were to transmit to the AP alone (i.e., its SNR without projection). For example, in the case of a 2-antenna AP, the direction along which the AP receives a client's signal can be identified by the direction of its channel vector, e.g., $\mathbf{h}_b = (h_1, h_2)$ for the blue client as shown in Fig. 1(b), and the client's SNR can be computed as $\|\mathbf{h}\|^2 P/N_0$, where $\mathbf{h}$ is the vector of channels that the client passively measures from the AP's transmissions, P is the client's transmission power, and N_0 is the noise level at the AP, which we include in the beacons.[1] When clients contend for concurrent transmissions, the client that wins the contention first starts its packet with a special header that includes the direction along which the AP receives its signal. A client that wants to transmit concurrently with the first client uses this information to project its signal orthogonal to the first client and compute the reduction in its SNR. It then maps its SNR after projection to the optimal bit rates using standard SNR-bitrate tables [17, 27]. Additional concurrent clients can join the transmission and compute their optimal bit rate using the same process.

A notable feature of TurboRate is that it works in a distributed random access manner. Specifically, a client, e.g., the blue client in Fig. 1(a), can win the contention and transmit, picking its bit rate as usual without knowing whether other clients have packets and may transmit concurrently. A client, like the red client, that decides to transmit concurrently with the first client does not have to confer with it; it simply picks a bit rate that does not interfere with the first client's reception.

We built a prototype of TurboRate using the USRP-N200 radio platform and evaluated it over a 10 MHz channel. Our implementation uses an OFDM PHY-layer and supports the various modulations (BPSK, 4-64 QAM) and coding options used in 802.11. Our results are as follows:

- Activating MU-MIMO with existing bit rate selection fails to deliver its gains and can lead to a significant throughput reduction. In particular, we experimented with different client positions that span the range of inter-client reception angle, i.e., $\theta \in [0, \pi/2]$. The results show that, in 90% of the studied cases, enabling MU-MIMO without addressing its special needs for per-packet bit rate adaptation reduces the throughput below that achieved with a single client. Further, in about 50% of the cases, the network throughput reduces to zero because the clients' rates overshoot the capacity of the network.
- TurboRate's bit rate selection enables MU-MIMO to deliver its gains. With TurboRate, MU-MIMO produces an average throughput gain of 1.7x in the case of 2-antenna AP and 2.3x in the case of 3-antenna AP.

TurboRate enables distributed bitrate adaptation for MU-MIMO LANs. The closest to our work is 802.11n$^+$ [23], which supports per packet bit rates, but addresses a different problem in which concurrent clients communicate with different APs. It also assumes that concurrent clients have a different and increasing number of antennas (i.e., one client has a single antenna, the second has two antennas, and the third has three antennas). In contrast, TurboRate can support clients, with the same or different numbers of antennas, transmitting to an AP in a MU-MIMO LAN.

2. UNDERSTANDING RATE SELECTION IN MU-MIMO

Before describing our proposed rate adaptation protocol, we conduct theoretical analysis and testbed measurements to understand how MU-MIMO concurrent transmission changes a client's optimal bit rate and the implications of picking the wrong bit rate. We focus on the scenario in Fig. 1(a), where two single-antenna clients communicate with a 2-antenna AP. The maximum bit rate of both clients is limited by the need to ensure that the access point can still decode the signals. Let x_b be the symbol transmitted by the blue client and x_r be the symbol transmitted by the red client, concurrently. The 2-antenna AP receives the combined signals in a 2-dimensional antenna space, as shown in Fig. 1(b),

$$\begin{pmatrix} y_1 \\ y_2 \end{pmatrix} = \begin{pmatrix} h_1 \\ h_2 \end{pmatrix} x_b + \begin{pmatrix} h_3 \\ h_4 \end{pmatrix} x_r + \begin{pmatrix} n_1 \\ n_2 \end{pmatrix},$$

where the vector $\mathbf{h}_b = (h_1, h_2)$ is the channels of the blue client and the vector $\mathbf{h}_r = (h_3, h_4)$ is the channels of the red client in the antenna space, as shown in Fig. 1(b), and n_1 and n_2 are the noise observed at the AP's two antennas. For simplicity, we assume that n_1 and n_2 are independent and follow the same Gaussian distribution $n_1, n_2 \sim \mathcal{CN}(0, N_0)$, where N_0 is the average noise power at the AP.

Say the AP is interested in decoding the red client, x_r. To null out the interfering signal, x_b, the AP uses a technique called *zero-forcing (ZF)* [33] to project the received signal on a direction orthogonal to x_b, i.e., $(h_2, -h_1)$, which can be formalized as follows:

$$y_{proj} = h_2 y_1 - h_1 y_2 = (h_2 h_3 - h_1 h_4) x_r + (h_2 n_1 - h_1 n_2).$$

[1] Note that the direction along which an AP receives a client's signal stays stable with the channels, despite that the signal rotates in the complex I-Q plane. This is because this direction is expressed in the AP antenna space, not in the I-Q plane [15, 13].

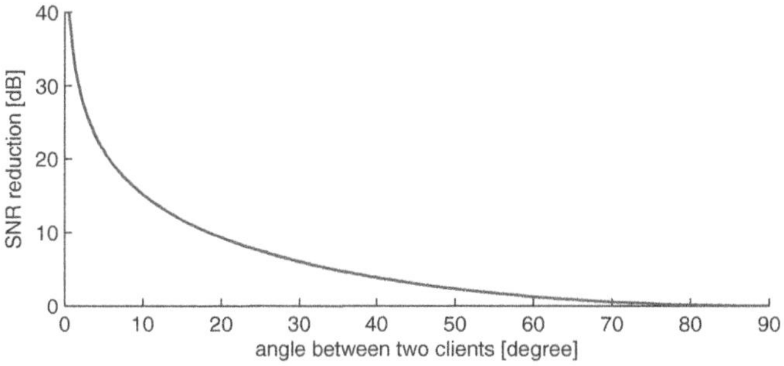

Figure 2—SNR reduction in dB due to projection, as a function of the angle between the two clients at the AP.

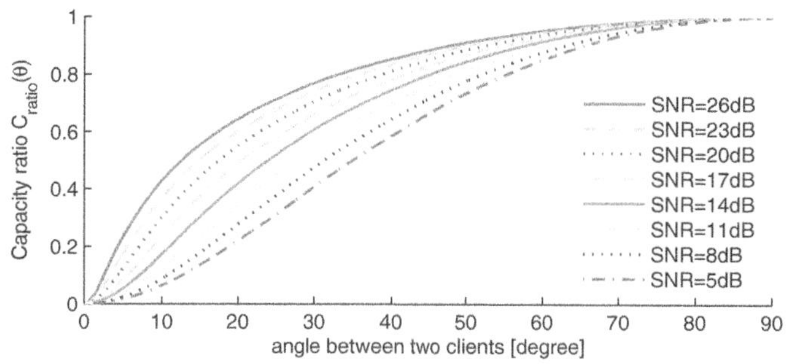

Figure 3—Capacity ratio after projection.

It then decodes the projected signal as:

$$x_r' = \frac{y_{proj}}{h_2h_3 - h_1h_4} = x_r + n_r = x_r + \frac{h_2n_1 - h_1n_2}{h_2h_3 - h_1h_4}. \quad (1)$$

We can observe from the above equation that the noise after projection, n_r, is scaled up. The SNR hence decreases after projection, and becomes

$$\begin{aligned} SNR_{proj} &= \frac{\mathbb{E}[|x_r|^2]}{\mathbb{E}[|n_r|^2]} \\ &= \frac{|h_2h_3 - h_1h_4|^2}{\|(h_2, -h_1)\|^2 N_0}\mathbb{E}[|x_r|^2] \\ &= \frac{|(h_2,-h_1)\cdot(h_3,h_4)|^2}{\|(h_2,-h_1)\|^2\|(h_3,h_4)\|^2}\frac{\mathbb{E}[\|(h_3,h_4)x_r\|^2]}{N_0} \\ &= \cos^2(\pi/2 - \theta)SNR_{orig} \\ &= \sin^2(\theta)SNR_{orig}, \end{aligned} \quad (2)$$

where $(\cdot)$ denotes the inner product, θ is the angle between the channels of two clients, (h_1,h_2) and (h_3,h_4), as in Fig. 1(b), and $SNR_{orig} = \mathbb{E}[\|(h_3,h_4)x_r\|^2]/N_0$ is the SNR of x_r when the red client transmits alone, i.e., without projection. Geometrically, we can see from Fig. 1(b) that the amplitude of x_r after projection is reduced to $sin(\theta)x_r$, matching the above derivation that SNR_{proj} equals $\sin^2(\theta)SNR_{orig}$. The amount of SNR reduction for the red client in dB due to projection orthogonal to the blue client can be expressed as:

$$\begin{aligned} \Delta SNR &= 10\log_{10}(SNR_{orig}) - 10\log_{10}(SNR_{proj}) \\ &= -20\log_{10}\sin(\theta). \end{aligned} \quad (3)$$

We note two important points:

- First, the direction along which a client is received is defined by its channel vector at the AP. In our example, the blue client is received along the direction (h_1,h_2), and the red client is received along the direction (h_3,h_4). Thus, the angle between two clients, θ, is in the antenna space, not the I-Q plane [15, 13]. Hence, this angle does not change with signal rotation in the complex I-Q plane.
- For general scenarios where a client communicates with an M-antenna AP in the presence of k concurrent transmissions ($k < M$), we can still compute the SNR reduction of this client based on Eq. (3). The only difference is that the AP needs to decode by projecting along the direction orthogonal to all the k concurrent transmissions. In this general case, θ hence becomes the angle between the client and the k-dimensional subspace S spanned by the k concurrent transmissions in the AP's M-dimensional antenna space. The value of $\sin\theta$ can be computed by

$$\sin\theta = \frac{|\mathbf{h}_\perp \cdot \mathbf{h}|}{\|\mathbf{h}_\perp\|\|\mathbf{h}\|}, \quad (4)$$

where $\mathbf{h}$ is the channel vector of the client that we want to decode and $\mathbf{h}_\perp$ is the vector that is orthogonal to the subspace spanned by the k concurrent transmissions[2].

(a) How does zero-forcing affect the SNR of the signal? We can see from Eq. (3) that the reduction in SNR due to zero-forcing is independent of the original SNR of the client, and solely depends on the angle between the clients. Fig. 2 plots the reduction in SNR as a function of the angle between the two clients. It shows that, when the angle is smaller than 45 degree, the SNR reduction exceeds 3 dB. A reduction in SNR larger than 3 dB requires an 802.11 node to reduce the transmission bit rate at least one bit-rate lower [27]. Depending on the actual value of the SNR reduction, it might be insufficient to just go down one bit rate lower. In fact, if the reduction in SNR is such that the SNR after projection is less than 4 dB, a client will be unable to use even the lowest 802.11 bit rate and hence should not transmit concurrently with the ongoing client.

(b) How does the SNR reduction impact the optimal bit rate? Even though the SNR reduction is independent of the original SNR, the change in the optimal bit rate depends on the original SNR. Since the optimal bit rate tends asymptotically to the capacity, we estimate the change in the optimal bit rate as the change in the capacity before and after projection. The ratio of the capacity after projection to the original capacity can be formulated as a function of the angle between the two clients' signals at the AP as follows:

$$C_{ratio}(\theta) = \frac{B\log_2(1+SNR_{proj})}{B\log_2(1+SNR_{orig})} = \frac{log(1+\sin^2(\theta)SNR_{orig})}{\log(1+SNR_{orig})},$$

where B is the bandwidth of the channel, and θ is the angle between the two clients at the AP.

Fig. 3 plots the capacity ratio in different SNR_{orig} regimes. The figure shows that, for a particular angle, e.g., $\theta = 30$ degree, a link with a low original SNR experiences a larger capacity drop than that with a high original SNR. It means that the low SNR regime is more sensitive to SNR reduction, and will likely require decreasing the bit rate to support concurrent transmissions. The figure also shows that the median capacity reduction, i.e., the reduction corresponding to an angle of 45 degree, is about 30%. This means that, assuming the distribution of the angle between two clients is uniform over all angles, one would expect the throughput of two concurrent clients in a MU-MIMO to be about 1.7x the throughput

[2]The orthogonal vector $\mathbf{h}_\perp$ can be computed by $\mathbf{h}_\perp = \mathbf{h} - proj_S(\mathbf{h})$, where $proj_S(\mathbf{h})$ is the projection of $\mathbf{h}$ onto the subspace S.

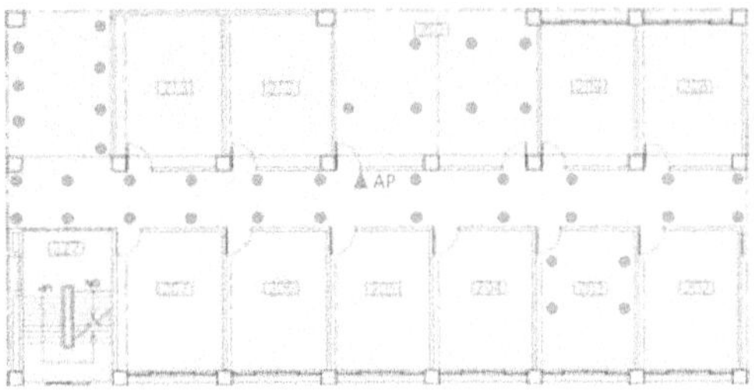

Figure 4—The testbed. Orange dots refer to client locations. Blue triangle refers to the location of the AP.

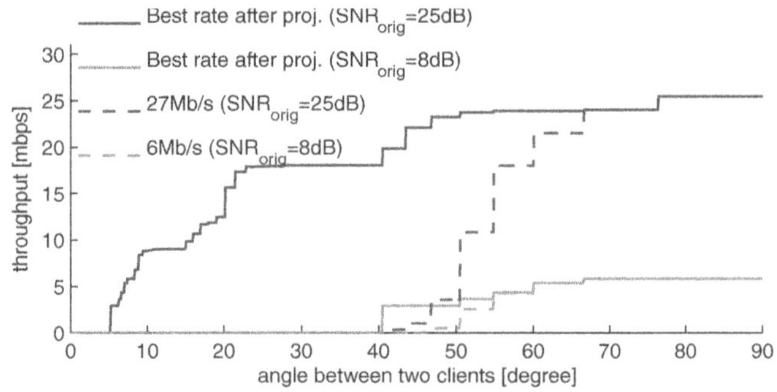

Figure 5—Throughput gain in TurboRate.

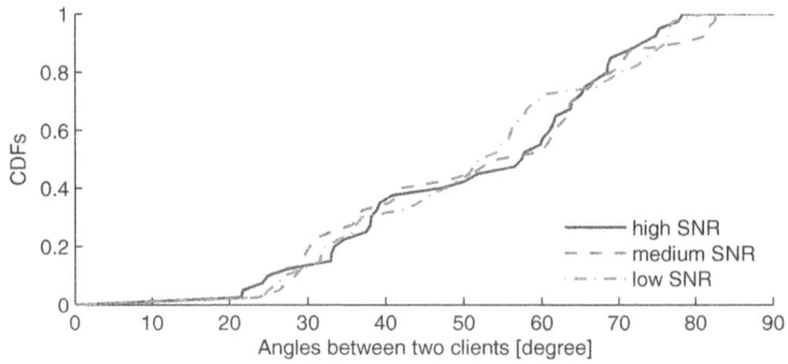

Figure 6—Angles between two clients in real channels.

of a single client transmitting to the same 2-antenna AP. We will see in §6.2 that the median throughput gain in TurboRate is 1.7x for 2-antenna AP scenarios, which shows that TurboRate matches the expected theoretical performance of MU-MIMO.

(c) What are the implications of ignoring MU-MIMO in rate adaptation? The above argument shows that the channel capacity of a client changes when it joins a concurrent transmission because the channels of the two clients interact together. The client should react to that change in capacity by adopting a different bit rate than it would adopt if it were transmitting alone. If the client does not react then it might exceed the capacity of its channel leading to its packets becoming undecodable. This also impacts all other clients that are transmitting concurrently, because the aggregate rate of all clients exceeds the combined channel capacity. As we argued earlier, these client channels are not independent from each other; they are related by the angle between the directions along which the AP receives them.

To illustrate this point, we collect empirical measurements using USRP-N200 [2] on a 10 MHz channel. We use the testbed in Fig. 4. We fix the location of the 2-antenna AP, and vary the locations of the two clients. We empirically measure the packet delivery ratio for different bit rates in the entire 802.11 operational range, and compute the throughput by multiplying the rate by the packet delivery ratio corresponding to the SNR after projection.

We plot in Fig. 5 the throughput of a client whose original optimal bit rate is 27 Mb/s (i.e., 64-QAM, 3/4 coding rate on a 10 MHz channel) if it were to transmit alone. The dotted blue line shows the throughput of this client if it does not change its rate as a function of the angle between its signal at the AP and that of the concurrent client. The solid blue line is the throughput of the client if it reacts by changing its rate to take into account the angle between its channel and that of the concurrent client, and the resulting SNR reduction. The figure shows that if the client does not change its bit rate, then for any angle smaller than 38 degree, it will get zero throughput. This is because the original bit rate significantly exceeds the capacity of its channel after projection. In contrast, if it does adapt, then it can continue enjoying a significantly higher throughput even for small angles.

For comparison, we also plot in red the same graphs for a low SNR client whose original optimal bit rate if transmitting alone is 6 Mb/s (QPSK, 1/2 coding rate). Note that this client will get a zero throughput for any angle smaller than 40 degree, even if it reduces its bit rate to the lowest rate (i.e., the solid red line) for concurrent transmissions. Thus, a client whose optimal bit rate when transmitting alone is 6 Mb/s should check the angle it has with the other client who has proceeded it to transmit, and if the angle is smaller than 40 degree, it should abstain from contending for the channel.

(d) What are the practical values for the angle between the signals of two clients at a shared AP? The analysis so far assumes that the angle between the two clients ranges from 0 to 90 degrees. We next use empirical measurements to check the distribution of the angle between the channels of two clients. Again the measurements are conducted using USRP-N200 [2] in the testbed shown in Fig. 4. We fix the location of the 2-antenna AP, and vary the locations of the two clients. We collect measurements for 100 different choices of clients' locations, picked at random from Fig. 4.

Fig. 6 plots the CDFs of the angle between the directions along which the two clients are received. The CDFs are taken over different client locations. The figure shows that the angles are uniformly distributed between 20 and 80 degree in all SNR regimes. Note that an angle of 90 degree shows that the two clients are received orthogonal to each other at the AP and hence their channels do not interact. In contrast, a small angle means that the signals of the two clients interfere significantly and the total capacity is far from the sum of the two capacities. Since the empirical results show that the angle can take a wide range of values, the client has to measure this angle and react appropriately.

3. TURBORATE

TurboRate addresses rate selection on MU-MIMO uplinks. We consider a MU-MIMO MAC protocol similar to SAM [31], where clients contend for concurrent transmissions and join the ongoing transmissions one after another (see [31] for details). In such an MU-MIMO MAC, a client that wins the contention needs to select its best bit rate immediately before data exchange. It however has no idea who and how many other clients will win the contention after it, and transmit concurrently with it. For example, say the AP has three antennas; the first client that wins the contention does not know whether other clients might contend and win the second and third concurrent transmission opportunities. Further, the second client that wins the contention knows only about the first client, but does not know whether there will be a third concurrent client.

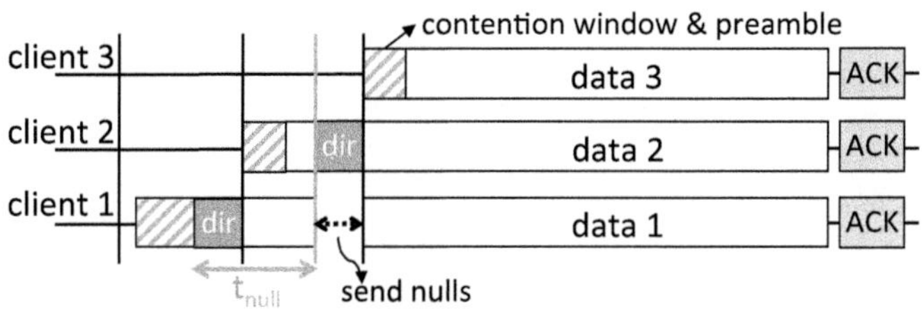

Figure 7—Rate Adaptation Protocol. Each client annotates its packets with the direction of its signal at the AP. This information enables potential concurrent clients to select their bit rates. To ensure single-antenna clients can decode these annotations, we force the ongoing transmissions to pause their streams when the contention winner sends the annotated header.

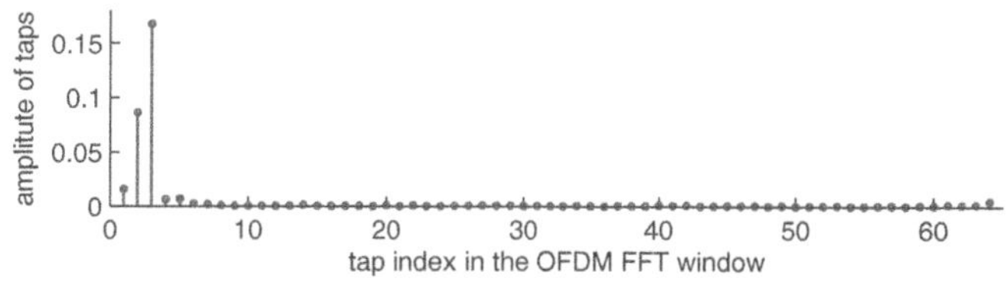

Figure 8—Amplitude of time-domain channels. Only first few taps of the 20 MHz 802.11 channel have a noticeable amplitude. To reduce the overhead, a client only announces those taps.

We would like a bit rate adaptation protocol that enables each client to select its bit rate by considering only those clients that won the contention before it, and without worrying about the clients that may win the contention after it.

TurboRate realizes the above goal. At a high level, TurboRate works as follows: Each client passively learns the direction along which it is received at the AP and its SNR if it transmits alone, i.e., SNR_{orig}. During contention, the client learns the direction of other clients that won the contention before it, and uses this information to compute its SNR after projection, SNR_{proj}, and the corresponding optimal bit rate. The AP decodes the concurrent clients using zero-forcing with successive interference cancellation (ZF-SIC) [33].

The next few subsections describe the protocol in detail.

3.1 Learning a Client's Direction and SNR Passively

TurboRate requires the client to know its own SNR to the AP and the direction along which its signal is received at the AP. Both parameters can be directly derived from the client's channels to the AP. The client's SNR can be easily computed using the preamble [27][29]. As for the direction, a client is received along the direction of its channel vector, i.e., $\mathbf{h}$, where the elements of $\mathbf{h}$ are the channels from the client to the AP's antennas. So, to estimate these variables the client needs to learn its channels to the AP.

One naive mechanism to learn the channels is to have the AP explicitly tell each client its channel values. This solution, however, has a high overhead that increases with the number of clients. In contrast, TurboRate enables the clients to learn their channels to the AP passively by listening to the AP's transmissions including its beacons. Specifically, the clients leverage *channel reciprocity* [16]. Reciprocity refers to the property that the channels in the forward and reverse directions are the transpose of each other because electromagnetic waves travel forward and backward the same way. The feasibility of reciprocity has been verified empirically in [16, 4]. Using reciprocity, every client can exploit the beacons to learn the channels from the AP and estimate the reverse channels. Updating the channels using periodic beacon frames is sufficient because the coherence time of indoor channels is typically between 0.2 second to multiple seconds [33], which is longer than the beacon interval 0.1s. Clients can further refine the estimation opportunistically by overhearing the downlink packets from the AP.

TurboRate also makes the AP measure its noise level and include it in its beacons. Given the channel vector and the AP noise power, each client can estimate its original SNR and the direction along which it is received at the AP.

3.2 Exchanging the Channel Directions

To compute the best rate, a TurboRate client has to consider the SNR reduction after projecting along the direction orthogonal to all the ongoing transmissions. The SNR reduction after projection depends on the angle between its signal and all the ongoing transmissions. To compute this angle, a TurboRate client not only needs to know its own channels, but also the directions of all the ongoing transmissions. A client however can only learn its own channels from the beacons. To enable the client that joins later to know the direction of the ongoing transmissions, each TurboRate client that wins contention announces its channel direction by annotating the PLCP header. Clients that later contend for concurrent transmissions use this information to select their rates.

This simple solution addresses the problem in a 2-antenna AP scenario because all clients can hear the information sent by the first contention winner. This solution, however, does not generalize to more than two antennas. To see this, lets consider a 3-antenna AP that can support up to three concurrent transmissions. While all clients can hear the PLCP header of the first winner, single-antenna clients are unlikely to successfully decode the second client's header in the presence of the first client's ongoing transmission. Thus, to decode this header information, TurboRate forces all the clients to stop transmitting when a client broadcasts its direction. In particular, TurboRate forces clients with ongoing transmissions to pause their streams and send null samples for a period of time that is long enough for the new client to broadcast its direction information.

The issue, however, is that ongoing transmitters do not know when will a new client win contention and broadcast its direction. For example, the client that first wins contention and starts transmitting does not know when the second client wins contention and broadcasts. To avoid this uncertainty, as shown in Fig. 7, in TurboRate the first winner always pauses its transmission at a pre-defined timeslot t_{null} after it wins the contention. In particular, as soon as the second client wins contention, it transmits its preamble[3], stays idle, and then broadcasts its direction in the clear when the first winner keeps silent at t_{null}. More generally, in a network support M concurrent transmissions, TurboRate forces all the ongoing clients to pause their transmissions at times $k * t_{null}$, for all $k = 1, \cdots, M-2$, after the first client wins.

Enabling this protocol however requires the client that wants to join the concurrent transmission to win the contention before t_{null} because the information has to be sent by the winner at t_{null} exactly. To satisfy this constraint, the client must give up the transmission opportunity if it wins the contention after t_{null}. The efficiency of the above protocol hence depends on the value of t_{null}. A large t_{null} defers the information exchange and the data packets of later contention winners, while a small t_{null} decreases the opportunity of concurrent transmissions. We will verify in §6.4 that setting $t_{null} = CW_{min}/2$ balances out the above tradeoff and produces a relatively low overhead.

[3] As we describe in §3.5, other contending clients use this preamble to count the number of ongoing transmissions.

Finally, we perform the following optimizations to minimize the overhead of the channel state information.

- Since 802.11 typically operates on a 20MHz OFDM channel, each client has to learn the channels over the 48 occupied OFDM subcarriers. It is however a high overhead to broadcast the direction of each subcarrier. We observe that after transforming the channels across all the 64 subcarriers to the time domain, there are only few taps with a noticeable amplitude. In particular, we empirically measure the amplitude of time-domain taps in the OFDM FFT window in our testbed. The result we plot in Fig. 8 shows that only about 5 taps have a relatively large amplitude. This is expected because there are only a limited number of distinguishable paths between a transmitter and a receiver in an indoor environment. Thus, a TurboRate client only announces the first few significant taps, e.g., 5 taps, of the time-domain channels. We will demonstrate in §6.1 that discarding the other taps results in a negligible error. The other clients can recover the channel information by transforming them back to the frequency-domain channels.

- The channel of each subcarrier in an M-antenna AP scenario is an M-dimensional vector, in which each element is the channel between the client's antenna to one of the AP's M antennas. Instead of sending the M-dimensional channel vector $\mathbf{h} = [h_1, h_2, \cdots, h_M]$, the client only requires to inform the direction of that vector, which is equivalent to the direction of a scaled vector $\mathbf{h}^d = [1, h_2/h_1, \cdots, h_M/h_1]$. Scaling the vector reduces the size of representing the direction to $M-1$ complex numbers for each subcarrier.

After the above optimizations, the overhead of transmitting the direction information is 5 and 10 complex numbers for the 2-antenna and 3-antenna AP scenarios respectively, which are about 3 and 6 BPSK symbols.

3.3 Estimating the Best Bit Rate

Next we focus on how each client uses its SNR and the channel directions of ongoing transmissions to select its best bit rate. Lets consider a general scenario where a client wins the $(k+1)^{th}$ contention and transmits a concurrent stream to an M-antenna AP in the presence of k ongoing transmissions. Let $\mathbf{h}_{k+1}$ denote the vector of the client's channels to the AP and $\mathbf{h}_i^d$, $i = 1, \cdots, k$, denote the directions, i.e., scaled channel vectors, of the k ongoing transmissions. To estimate its SNR after projection, SNR_{proj}, the client first estimates its own SNR to the AP, SNR_{orig}. Then the client subtracts the SNR reduction caused by the projection, ΔSNR, which as explained in §2, can be estimated using $\mathbf{h}_{k+1}$ and $\mathbf{h}_i^d$. Note that the subspace spanned by the k ongoing transmissions is the same as that spanned by their directions. We can therefore use the directions of the ongoing signals to compute the SNR reduction.

After estimating the SNR after projection in each OFDM subcarrier, the client computes the effective SNR (ESNR) [17], which can then be mapped to the best bit rate. ESNR considers the impact of frequency selectivity across multiple OFDM subcarriers, and hence can more accurately predict the bit rate.

3.4 Decoding at the AP

A simple way for an M-antenna AP to decode M concurrent streams is to use zero forcing for each of the streams. In particular, the AP decodes each stream by projecting the signal along the direction orthogonal to all the other concurrent streams. This decoder however might make the bit rates selected by the clients undecodable. To see why this is the case, lets consider a 3-antenna AP scenario where three clients communicate with the AP and join the concurrent transmissions one after another. Say the AP is interested in decoding the second stream. Recall that the second client estimates its SNR after projection, SNR_{proj}, according to the angle between its signal and the direction of the first client. If the AP ignores this fact and simply decodes the second stream by projecting along the direction orthogonal to both the first and the third clients, it will produce a SNR different from SNR_{proj}. This is because it projects on the orthogonal direction of a different subspace and leads to a different amount of SNR reduction after projection. Thus, the AP is unlikely to successfully decode the second client.

To ensure that the rate selected by each client can be decoded correctly, a TurboRate AP instead uses *zero-forcing with successive interference cancellation* (ZF-SIC) [33]. Using ZF-SIC, the AP decodes the kth stream after removing all the interfering streams that join after the kth stream. For example, lets again consider the 3-antenna AP scenario. The AP first decodes the third stream by projecting along the direction orthogonal to the plane of the first and the second clients. It then re-encodes the third stream and subtracts it from the received signals. The AP then decodes the second stream by projecting the resulting signal along the direction orthogonal to the first client. It then subtracts the second client and decodes the first client using a standard decoder.

Using ZF-SIC, the AP decodes the kth client after cancelling the interfering clients that joined after it. This property allows the AP to decode the packet sent at a rate chosen by the kth client, which the client computes using only its SNR and the channel directions of the $k-1$ clients that won contention before it.

3.5 TurboRate's Medium Access Protocol

Fig. 7 shows an instantiation of our MAC in a network where the AP has three antennas and can support three concurrent transmissions. Similar to the MAC protocol in SAM [31], each client listens to the medium and counts the number of concurrent streams by cross-correlating with the known preamble. If the number of existing streams is less than the number of antennas supported by the AP, the clients contend for the medium using 802.11's contention window and random backoff. Clients can continue contending for transmission opportunities until they detect that the number of concurrent streams equals the number of antennas at the AP. Unlike the MAC protocol in SAM, TurboRate allows clients to contend for concurrent transmissions, only if their SNR after projection is higher than the 802.11 operational SNR range, i.e., 4 dB. In addition, the contention winner selects its best rate before data exchange, and annotates the direction of its channels in the header. To ensure that the clients can overhear the information annotated by the contention winners, TurboRate makes the ongoing transmissions pause their streams as mentioned in §3.2.

3.6 Supportung Clients with Multiple Antennas

TurboRate's design can be easily generalized to support clients with multiple antennas. For example, a 2-antenna client and a single antennas client can transmit concurrently to a 3-antenna AP. TurboRate can be generalized to such scenarios by having a multi-antenna client contend for each of its antenna independently. The only difference, however, is that because radios typically cannot transmit and receive at the same time, a multi-antenna client contends for all the antennas it can transmit on, before starting the concurrent data transmission.

4. ADDITIONAL ISSUES

This section addresses the following additional issues.

Acknowledgements: Since the AP has multiple antennas, it can send the acknowledgements to all the clients concurrently on the downlink using beamforming [7].

Fragmentation and Aggregation: In TurboRate, all concurrent clients end their transmissions at about the same time. To do so, nodes may need to fragment or aggregate packets. TurboRate leverages the methods used in existing link layer protocols, e.g., packet fragmentation [18] and packet aggregation [6].

Retransmissions: A TurboRate client needs to re-transmit the packet if it is not ack-ed. The retransmission can be with a different subset of clients. The client thus must select a different rate and fragment or aggregate the packet differently.

Time Synchronization: To avoid inter-symbol interference (ISI), the concurrent clients need to synchronize their transmissions within a cyclic prefix of an OFDM symbol [31]. TurboRate applies the method proposed in [31], which allows concurrent clients to estimate the OFDM symbol boundary of the first stream and synchronize their transmission to it. To cope with the small delays due to hardware turn-around time and channel propagation, both the cyclic prefix and the OFDM FFT window are scaled up by the same factor. Such scaling does not increase the overhead, but allows the system to tolerate synchronization error [30].

Frequency Offset: To avoid inter-carrier interference, concurrent clients should have the same carrier frequency offset (CFO) at the AP. In TurboRate, all clients compensate their offset using a mechanism proposed in [28, 30]. Specifically, all the clients overhear the PLCP header sent by the first contention winner and estimate the frequency offset with respect to the first client. All the concurrent clients synchronize their frequency-domain signals by compensating this offset.

Fairness: In TurboRate, the first client that wins the contention for a concurrent transmission is likely to have a higher rate than the others since it computes its rate using its original SNR without projection. TurboRate however is still fair because every client wins the first contention with an equal probability, as in 802.11. In TurboRate, a client has the opportunity to transmit the first stream without lowering its rate. It can transmit concurrently and benefit from the throughput gain of MU-MIMO if it loses the first contention.

Imperfect Cancellation: In practice, Successive Interference Cancellation (SIC) cannot cancel the interfering signals perfectly, resulting in some residual noise. The magnitude of the residual noise depends on the power at which the interfering signal is received. A high residual noise might hinder the AP from decoding the existing signals after SIC. To avoid this, a TurboRate client determines whether it can join ongoing transmissions by applying a technique similar to the one proposed in [23]. In particular, a TurboRate client is only allowed to transmit concurrently if its interfering power at the AP is below a threshold.[4] This limits the amount of residual noise and allows the AP to successfully decode the existing signals after SIC.

Hidden Terminals: TurboRate provides a rate adaptation algorithm for MU-MIMO and is orthogonal to the issue of hidden terminals. In general, MU-MIMO enables concurrent transmissions and hence may increase the probability of hidden terminals. However, one can address the hidden terminal problem in MU-MIMO networks by using mechanisms like retransmissions [21], RTS-CTS [23, 22], and ZigZag [14].

[4]This threshold in practice is set to about 25–27dB [23].

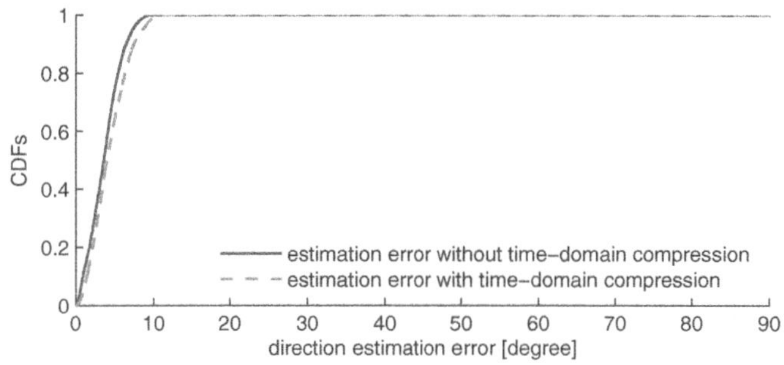

Figure 9—Accuracy of direction estimation. Leveraging channel reciprocity allows clients to measure the direction of the channels accurately in a passive way. The additional estimation error caused by compressing the time-domain channels is negligible.

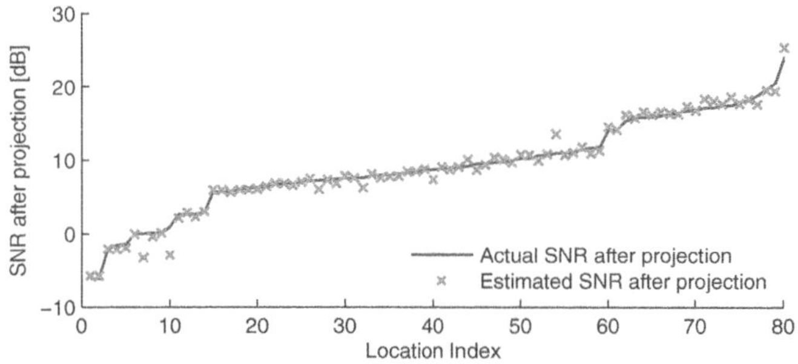

Figure 10—Accuracy of SNR estimation. The estimated SNR after projection closely matches the actual SNR after projection in the 802.11 operational SNR range.

5. IMPLEMENTATION

We build a prototype of TurboRate using the USRP-N200 radio platform [2] and the UHD software package. Each USRP-N200 is equipped with an RFX2400 daughterboard, and operates on a 10 MHz channel. We build a multi-antenna AP by combining multiple USRP-N200 boards using an external clock [3] and making them act as a MIMO node. Each node runs a PHY layer similar to that in 802.11a, i.e., including OFDM subcarriers and using modulations (BPSK, 4-64QAM) and standard 802.11 code rates [5]. Since we operate at the bandwidth of 10MHz, the possible bit rates range from 3 to 27 Mb/s.

Due to the timing constraints limited by software radio, we implement all the components of our design except contention and ACK. To allow multiple clients to transmit concurrently, we leverage USRP-N200 timestamps to synchronize the clients within a cyclic prefix as follows. We make the AP broadcast a trigger signal. Each client records the timestamp of detecting the trigger, $t_{trigger}$, waits a pre-defined period of time, $t_{\triangle}$, and sets the timestamp of the beginning of its transmission to $t_{start} = t_{trigger} + t_{\triangle}$. In our testbed, $t_{\triangle}$ is set to 0.1s, which is long enough to tackle the delays introduced by software.

6. RESULTS

We evaluate the performance of TurboRate in the testbed environment shown in Fig. 4. Our evaluation focuses on answering the following questions:

- Are the estimate of the direction of the channels and the SNR after projection accurate enough for a client to select its best bit rate?

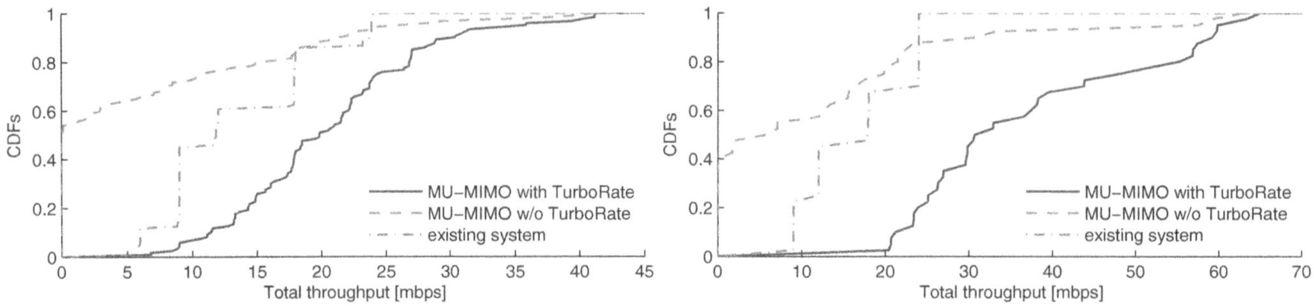

(a) Total throughput in the 2-antenna AP scenario

(b) Total throughput in the 3-antenna AP scenario

Figure 11—Throughput Comparison. The figure compares the throughput with and without TurboRate for the 2-antenna and 3-antenna AP scenarios. TurboRate delivers a total throughput gain of 1.7x and 2.3x as compared to existing 802.11. Without considering the effect of projection, the concurrent client in many cases selects a rate that makes its stream undecodable after projection and leads to zero throughput.

- What is the throughput gain achieved by TurboRate?
- Where does the throughput gain come from?
- How much extra overhead is introduced by TurboRate?

6.1 Micro Benchmark

The performance of bit rate selection in TurboRate relies on the accuracy of two estimates: the directions of the concurrent clients, which the client learns from the annotation in their packets, and the client's computation of its SNR after projection. We empirically measure the accuracy of these two variables.

(a) Accuracy of Signal Direction Estimate: The errors of the signal direction estimate come from two sources: 1) the estimation error due to learning the channels using reciprocity, and 2) the information loss due to compression, i.e., due to expressing the channel using only 5 time-domain taps, as mentioned in §3.2. We check how these two errors impact the accuracy of the estimate.

Experiment: We consider a 2-antenna AP scenario where a single-antenna client communicates with the AP. The client and the AP are randomly assigned to the locations in Fig. 4. We measure their uplink and downlink channels, and calibrate the tx and rx chains using the method proposed in [4]. Since our protocol makes each client send only five taps to reduce the overhead, we further compare the accuracy of the direction after performing the following compression: convert the direction of the downlink channels after calibration to the time domain, keep only 5 taps and reset the rest to zero, and convert them back to the frequency domain. We define the estimation error as the angle between the direction of the actual uplink channel and the estimated direction of the channel, i.e., the direction of the downlink channel after calibration, with and without compression.

Results: Fig. 9 plots the CDFs of the estimation error across all experiments. The figure shows that the medium estimation error is only 4 degree, which means that the estimated direction is close to the actual direction. The additional estimation error caused by compression in time-domain information is negligible. The results show that clients can exploit the channel reciprocity property to estimate this information accurately in a passive way. Exchanging only 5 taps of time-domain information introduces a minimum estimation error, but decreases the overhead significantly.

(b) Accuracy of SNR Estimation: We next check how accurate a client can estimate SNR_{proj} using the method mentioned in §3.3.

Experiment: We focus on the scenario where two single-antenna clients communicate with a 2-antenna AP. In each experiment, the AP transmits 10 known symbols for the clients to learn their channels using reciprocity, followed by both clients transmitting a 1500B data packet concurrently. We compress the direction of the channel as mentioned in the previous experiment, and use the channels learned from the known symbols and the noise at the AP to estimate the SNR after projection. We compare the estimated SNR after projection to the actual projected SNR, which is computed *at the AP* by using ZF to decode the received concurrent packets. We repeat the same experiment with different random locations of nodes in Fig. 4.

Results: Fig. 10 compares the estimated SNR to the actual SNR after decoding. The results show that the estimation is accurate when the SNR after projection is larger the 802.11 operational SNR, i.e., 4 dB, as shown in Fig. 10 along the y-axis. We however note that the estimation error in the extremely low SNR regime (i.e., lower than 4 dB) does not harm our system because OFDM does not work properly in this critical regime, and hence we do not allow the client to transmit concurrently. For the operational SNR regime, the average estimation error is about 0.5 dB, which has little impact on bit-rate selection.

6.2 Throughput Gain of TurboRate

We next investigate the throughput gain delivered by enabling TurboRate in MU-MIMO. We compare the throughput of three systems: 1) MU-MIMO with TurboRate, which is our proposed protocol, 2) MU-MIMO without TurboRate, in which clients transmit concurrently, but select their rates only according to their own SNRs to the AP without considering the interaction between the concurrent transmissions, and 3) the existing system, in which only a single client is allowed to transmit to a multi-antenna AP using diversity gain [33]. We compare their performance in 2-antenna AP and 3-antenna AP scenarios respectively.

Experiment: We first focus on the scenario in Fig. 1(a), where two single-antenna clients transmit concurrently to a 2-antenna AP. We repeat the experiment with random assignment of node locations in Fig. 4. Each experiment consists of three phases: First, two clients transmit concurrently at the rates selected by TurboRate. Second, both clients transmit concurrently at the bit rates selected based on their own SNRs to the AP without projection. Third, one of the two clients is picked randomly and made to transmit alone at the best bit rate supported by its own SNR without projection. In each phase, each concurrent client transmits a 1500 byte packet, and uses the ESNR [17] to lookup the optimal rate.

Results: Fig. 11(a) plots the CDFs of the total throughput of three different systems. The figure shows that enabling TurboRate in

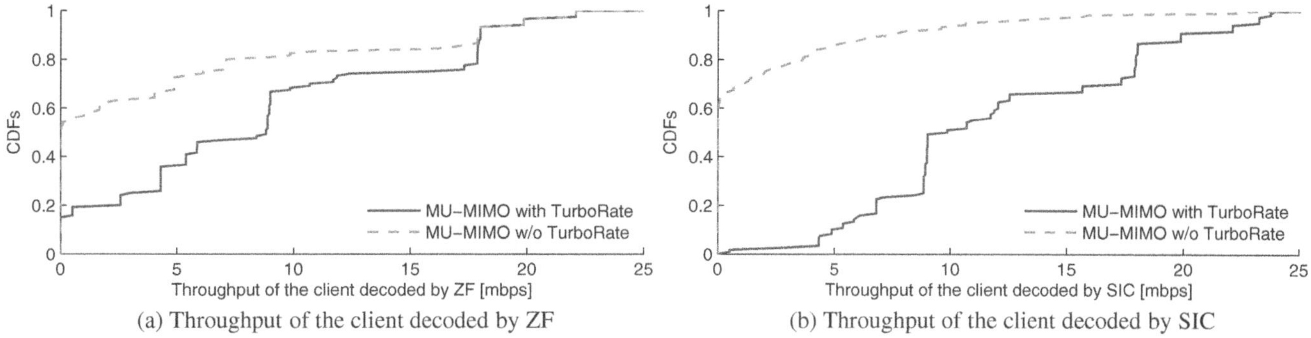

Figure 12—Throughput of individual clients. Without TurboRate, the client has a high probability to pick a wrong rate, which is undecodable by zero forcing and leads to zero throughput. With TurboRate, it can not only select the optimal rate, but also determine whether it should refrain from transmitting concurrently due to an extremely low SNR after projection. This is why the other client can be decoded correctly after interference cancellation and still achieve high throughput in TurboRate.

a MU-MIMO network ensures decodability and allows clients to achieve high throughput. Compared to existing 802.11 where only one client is allowed to transmit, the throughput gain from enabling concurrent transmissions with TurboRate's bit rate selection is about 1.7x, matching the analysis in §2. In contrast, concurrent transmissions with traditional bit rate adaptation could cause one client to be decoded incorrectly and leave residual interference, as a result harming the other client. The results show that traditional bit rate selection hampers the gain of MU-MIMO, and leads to large throughput reductions compared to existing 802.11 (about 50% of the cases are reduced to zero throughput).

Experiment: We next check the performance in a 3-antenna AP scenario where three single-antenna clients transmit concurrently to the AP. Each experiment consists of three phases: In phases 1 and 2, three clients transmit 1500 byte packets concurrently at the bit rate selected by TurboRate and selected only based on their own SNRs, respectively. In phase 3, we randomly pick one of the three clients and make it transmit alone at its best rate. We repeat the experiment with random assignment of nodes locations in Fig 4.

Results: Fig. 11(b) plots the CDFs of the total throughput of the three compared systems. The total network throughput of TurboRate in the 3-antenna AP scenario increases by 2.3x over existing 802.11 where only one stream is allowed. Further, without TurboRate, the gain of MU-MIMO cannot be achieved. Note that the throughput of three concurrent MU-MIMO clients is *not* three times as high as a single client. The reason is that the second and third concurrent clients lose some of their SNRs due to projection. This is a natural limitation of MU-MIMO (not a limitation of TurboRate.)

6.3 Implications of Not Using MU-MIMO Rate Adaptation

To better understand TurboRate's throughput gains, we zoom in on the throughput that the individual clients can achieve in the 2-antenna AP experiment mentioned in the last section. TurboRate decodes one client using zero-forcing (ZF), projecting the received signal along the direction orthogonal to the other client. It decodes the other client using successive interference cancellation (SIC), i.e., it is decoded after removing the interfering signal decoded by ZF.

Results: We first plot in Fig. 12(a) the throughput of the client decoded by ZF. Our findings are:

- Without considering the effect of projection, the client is very likely to choose a bit rate that exceeds its capacity after projection. This results in 54% of experiments with zero throughput.
- For 15% of the experiments, the SNR after projection is lower than the 802.11 operational SNR range. Most of these cases occur when the client is in the low original SNR regime and thus more sensitive to SNR reduction after projection. For this critical regime, TurboRate plays an important role to enable the client to detect these situations and prevent interfering with the ongoing transmission by refraining from transmitting concurrently.

We next plot in Fig. 12(b) the throughput of the client decoded by SIC. The figure shows the following:

- Without TurboRate, the AP cannot remove the interference from the other client because it did not adapt to SNR reduction after projection and picked a wrong rate. In this case, the AP cannot decode the other client and subtract its signal and hence also fails to decode this client correctly using SIC. This reduces the throughput of this client to zero as well. The client decoded by SIC can only obtain positive throughput if the angle between the two clients is by chance large enough such that the AP can still decode the other client correctly even after projection.
- In TurboRate, the client decoded by SIC can achieve a throughput comparable to that when it transmits alone because the AP can correctly decode and remove the interfering client. There might be a small residual interference left after interference cancellation due to imperfect hardware linearity. The results however show that this small interference does not hinder the AP from decoding the client after interference cancellation.

The above results verify that TurboRate is not only necessary for the client that joins the ongoing transmission and is decoded by projection, but also essential for the client that wins the earlier contention.

6.4 Overhead

We finally check how much extra overhead is introduced by TurboRate due to direction announcement. The overhead includes two parts: 1) the transmission time required for sending the annotated information, and 2) the idle period for ensuring correct reception of the information. To analyze the overhead, we have to consider the dynamics of node contention in a large scale network. This is hard to do in a USRP testbed because of the long delay and the difficulty of experimenting with realtime contention. We hence use Matlab to simulate the dynamics of 802.11 contention

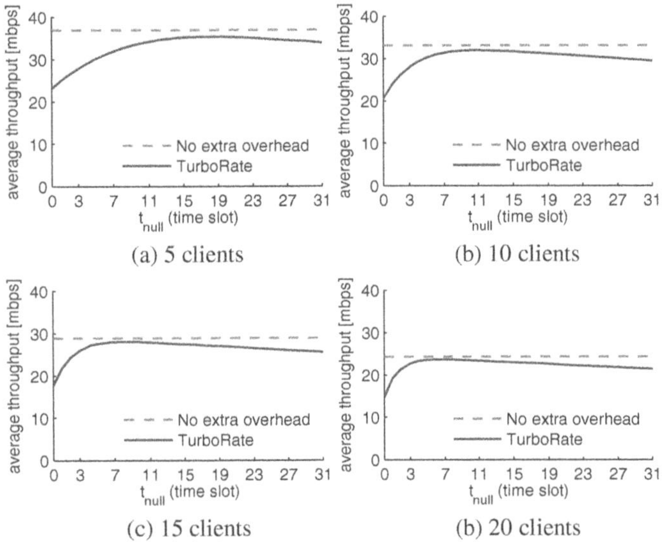

Figure 13—Overhead in a 3-antenna AP scenario. By setting $t_{null} = CW_{min}/2$, the extra overhead introduced by TurboRate results in 4% throughput reduction, which is fairly small.

in a large network. We use a 3-antenna AP scenario with many single-antenna clients that contend for concurrent transmissions.

In the simulations, we implement all the components of our protocol, including contention, backoff, interframe timing (i.e., SIFS and DIFS), PLCP header, MAC header and ACK. We randomly assign a channel vector to each client, and assume that all clients always have packets to send. The first contention winner transmits a 1500-byte packet, and the second and third winners end their transmissions at the same time as the first client. Each simulation compares the average total throughput of 10,000 transmissions [5] of the following schemes: 1) no extra overhead: each client starts sending its data packet immediately after it wins the contention without any overhead of information exchange, and 2) TurboRate: the first client pauses its transmission at a pre-defined time-slot t_{null} such that the second client can broadcast its direction to the rest of clients. We test different values of t_{null} and force all clients to give up the opportunity of the second transmission if no one rolls a random number smaller than t_{null} in the second contention. Clients can however start the third contention after t_{null}, regardless of the outcome of the second contention. This is because the AP has only 3 antennas and hence the third client does not need to announce its direction to other potential contenders.

Results: Fig. 13 plots the throughput of two schemes for varying numbers of clients. The figures show that the throughput decreases with increasing number of clients due to increased probability of collisions. A small t_{null} decreases the concurrent transmission opportunity, while a large t_{null} forces the client who wins the third contention to wait for the information and postpone its transmission. The maximum throughput can be achieved by balancing the above tradeoff and picking the optimum t_{null}. The optimum choice of t_{null} however changes with the number of contending clients. To avoid the complexity, we can simply set $t_{null} = 15$, i.e., half of CW_{min} defined in 802.11. After that, as compared to the throughput without extra overhead, the average throughput loss due to TurboRate's overhead, including the loss of concurrent transmission opportunities and the time used for exchanging information, is 4%, which is fairly small.

[5]We compute the throughput as the total number of bits in the correctly-decoded packets divided by the channel time occupied by 10,000 transmissions, where each transmission includes up to three concurrent packets.

7. RELATED WORK

Related work falls in the following two areas:

(a) Multi-user MIMO WLANs: MU-MIMO advocates having multiple clients concurrently communicate with a single AP or multiple receivers. The gain of MU-MIMO WLANs has been verified theoretically [11, 12, 34] and realized empirically [31, 7, 37, 15, 23]. SAM [31] allows multiple single-antenna clients to communicate concurrently with a multi-antenna AP. Beamforming [7, 37] deals with the downlink, and allows an AP to communicate concurrently with multiple single-antenna clients. IAC [15] makes multiple APs connect to each other and act as a virtual MIMO node to communicate with multiple clients concurrently. All these practical MU-MIMO systems leave rate adaptation an open issue. 802.11n$^+$ [23] enables random access for concurrent MIMO transmissions and takes bit rate selection into account. 802.11n$^+$ however works only when nodes have different numbers of antennas. It also uses an handshake to exchange information which introduces additional overhead. In contrast, TurboRate enables distributed bit rate adaptation for MU-MIMO LANs without explicit coordination and can support clients, with the same or different numbers of antennas, to communicate concurrently with an AP.

(a) Bit rate adaptation: There is a rich literature on rate adaptation for legacy 802.11a/b/g. They assume that the channels do not change for a short period of time, and hence can exploit historical performance, like loss rate [8, 36], SNR [17, 27, 9], BER [24], soft values [35], to predict the optimal bit rate for the next packet. Historical-based rate adaptation is then extended to 802.11n MIMO networks, where two multi-antenna nodes communicate with each other [26, 25, 10]. Such assumption holds for a single pair of MIMO nodes, but not in MU-MIMO, in which the best bit rate of a client changes with its concurrent transmitters. Prior work on downlink MU-MIMO rate adaptation [20, 32] enables a single transmitter to select the best rates for all its clients. Uplink rate adaptation is however much more difficult because, in a random access network, a client could transmit concurrently with a different subset of clients and hence requires to adapt its rate on a per-packet basis depending on who are the current transmitters. Compared to prior uplink MU-MIMO rate selection algorithms [38, 19], which require the AP to explicitly coordinate between the clients for each packet and tell each client the bit rate it has to use, TurboRate enables distributed rate adaptation and hence can be applied in a random access network.

8. CONCLUSION

This paper introduces TurboRate, a distributed rate adaptation protocol for MU-MIMO LANs. It decomposes rate adaptation in dynamic MU-MIMO LANs to the estimation of two variables: the SNR when a client transmits alone to the AP and the direction of the client's signal received at the AP. The short-term stability of these two parameters allows each TurboRate client to measure them in a passive way, but still be able to adapt its optimal rate on a per-packet basis, depending on who are the concurrent clients. Our prototype implementation shows that enabling MU-MIMO with traditional rate adaptation reduces the throughput in most cases, while enabling TurboRate in MU-MIMO increases the network throughput by 1.7x and 2.3x over existing 802.11 for 2- and 3-antenna AP scenarios respectively.

9. ACKNOWLEDGMENTS

This work is partially supported by the National Science Council of R.O.C. under contract No. NSC 100-2221-E-001-005-MY2 and NSF Grant CNS-1117194. We thank the members of the MIT Center for Wireless Networks and Mobile Computing, including Amazon.com, Cisco, Intel, Mediatek, Microsoft, and ST Microelectronics, for their interest and support.

10. REFERENCES

[1] Cisco Aironet 3600 Series. http://www.cisco.com/en/US/products/ps11983/index.html.

[2] Ettus Inc., Universal Software Radio Peripheral. http://ettus.com.

[3] Jackson Labs Technologies Inc., Fury GPSDO. http://www.jackson-labs.com/index.php/products/fury.

[4] System Description and Operating Principles for High Throughput Enhancements to 802.11. IEEE 802.11-04/0870r, 2004.

[5] Local and Metropolitan Area Networks Specific Requirements Part 11: Wireless LAN Medium Access Control (MAC) and Physical Layer (PHY) Specifications. *IEEE Std 802.11*, 2007.

[6] *IEEE Std 802.11n-2009*, pages c1 –502, 2009.

[7] E. Aryafar, N. Anand, T. Salonidis, and E. W. Knightly. Design and Experimental Evaluation of Multi-User Beamforming in Wireless LANs. In *ACM MobiCom*, 2010.

[8] J. Bicket. *Bit-rate Selection in Wireless Networks*. PhD thesis, Massachusetts Institute of Technology, 2005.

[9] J. Camp and E. Knightly. Modulation Rate Adaptation in Urban and Vehicular Environments: Cross-layer Implementation and Experimental Evaluation. In *ACM MobiCom*, 2008.

[10] C.-B. Chae, A. Forenza, R. Heath, M. McKay, and I. Collings. Adaptive MIMO Transmission Techniques for Broadband Wireless Communication Systems. *IEEE Communications Magazine*, 48(5):112 –118, may 2010.

[11] A. El Gamal and T. Cover. Multiple User Information Theory. *Proceedings of the IEEE*, 68(12):1466 – 1483, dec. 1980.

[12] D. Gesbert, M. Kountouris, R. Heath, C.-B. Chae, and T. Salzer. Shifting the MIMO Paradigm. *IEEE Signal Processing Magazine*, 24(5):36 –46, sept. 2007.

[13] S. Gollakota, F. Adib, D. Katabi, and S. Seshan. Clearing the RF Smog: Making 802.11n Robust to Cross-technology Interference. In *ACM SIGCOMM*, 2011.

[14] S. Gollakota and D. Katabi. Zigzag Decoding: Combating Hidden Terminals in Wireless Networks. In *ACM SIGCOMM*, 2008.

[15] S. Gollakota, S. D. Perli, and D. Katabi. Interference Alignment and Cancellation. In *ACM SIGCOMM*, 2009.

[16] M. Guillaud, D. Slock, and R. Knopp. A Practical Method for Wireless Channel Reciprocity Exploitation through Relative Calibration. In *Signal Processing and Its Applications*, 2005.

[17] D. Halperin, W. Hu, A. Sheth, and D. Wetherall. Predictable 802.11 Packet Delivery from Wireless Channel Measurements. In *ACM SIGCOMM*, 2010.

[18] R. Handel and M. Huber. *Integrated Broadband Networks; An Introduction to ATM-Based Networks*. Addison-Wesley Longman Publishing Co., Inc., 1991.

[19] Y. Kim, S. Cho, and D. K. Kim. Low Complexity Antenna Selection based MIMO Scheduling Algorithms for Uplink Multiuser MIMO/FDD System. In *IEEE VTC Spring*, 2007.

[20] V. Lau and M. Jiang. Downlink Scheduling and Rate Adaptation Design of Multi-User, Multiple-Antenna Base Station with Imperfect CSIT. In *IEEE GLOBECOM*, 2005.

[21] L. E. Li, K. Tan, H. Viswanathan, Y. Xu, and Y. R. Yang. Retransmission $\neq$ Repeat: Simple Retransmission Permutation Can Resolve Overlapping Channel Collisions. In *ACM MOBICOM*, 2010.

[22] K. C.-J. Lin, Y.-J. Chuang, and D. Katabi. A Light-Weight Wireless Handshake. *ACM SIGCOMM Comput. Commun. Rev.*, 42(2):28–34.

[23] K. C.-J. Lin, S. Gollakota, and D. Katabi. Random Access Heterogeneous MIMO Networks. In *ACM SIGCOMM*, 2011.

[24] K. C.-J. Lin, N. Kushman, and D. Katabi. ZipTx: Harnessing Partial Packets in 802.11 Networks. In *ACM MobiCom*, 2008.

[25] I. Pefkianakis, Y. Hu, S. H. Wong, H. Yang, and S. Lu. MIMO Rate Adaptation in 802.11n Wireless Networks. In *ACM MobiCom*, 2010.

[26] F. Peng, J. Zhang, and W. Ryan. Adaptive Modulation and Coding for IEEE 802.11n. In *IEEE WCNC*, 2007.

[27] H. Rahul, F. Edalat, D. Katabi, and C. Sodini. Frequency-Aware Rate Adaptation and MAC Protocols. In *ACM MOBICOM*, 2009.

[28] H. Rahul, H. Hassanieh, and D. Katabi. SourceSync: a Distributed Wireless Architecture for Exploiting Sender Diversity. In *ACM SIGCOMM*, 2010.

[29] G. Ren, H. Zhang, and Y. Chang. SNR Estimation Algorithm based on the Preamble for OFDM Systems in Frequency Selective Channels. *IEEE Transactions on Communications*, 57(8):2230 –2234, aug. 2009.

[30] K. Tan, J. Fang, Y. Zhang, S. Chen, L. Shi, J. Zhang, and Y. Zhang. Fine-Grained Channel Access in Wireless LAN. In *ACM SIGCOMM*, 2010.

[31] K. Tan, H. Liu, J. Fang, W. Wang, J. Zhang, M. Chen, and G. M. Voelker. SAM: Enabling Practical Spatial Multiple Access in Wireless LAN. In *ACM MobiCom*, 2009.

[32] M. Torabi, W. Ajib, and D. Haccoun. Discrete-Rate Adaptive Multiuser Scheduling for MIMO-OFDM Systems. In *IEEE VTC 2008-Fall*, 2008.

[33] D. Tse and P. Vishwanath. *Fundamentals of Wireless Communications*. Cambridge University Press, 2005.

[34] P. Viswanath and D. Tse. Sum Capacity of The Vector Gaussian Broadcast Channel and Uplink-downlink Duality. *IEEE Transactions on Information Theory*, 49(8):1912 – 1921, aug. 2003.

[35] M. Vutukuru, H. Balakrishnan, and K. Jamieson. Cross-Layer Wireless Bit Rate Adaptation. In *ACM SIGCOMM*, 2009.

[36] S. H. Y. Wong, H. Yang, S. Lu, and V. Bharghavan. Robust Rate Adaptation for 802.11 Wireless Networks. In *ACM MobiCom*, 2006.

[37] H. Yu, L. Zhong, A. Sabharwal, and D. Kao. Beamforming on Mobile Devices: a First Study. In *ACM MobiCom*, 2011.

[38] Y. Zhang, C. Ji, Y. Liu, W. Malik, D. O'Brien, and D. Edwards. A Low Complexity User Scheduling Algorithm for Uplink Multiuser MIMO Systems. *IEEE Transactions on Wireless Communications*, 7(7):2486 –2491, july 2008.

Achieving High Data Rates in a Distributed MIMO System

Horia Vlad Balan Ryan Rogalin Antonios Michaloliakos Konstantinos Psounis
Giuseppe Caire
University of Southern California
{hbalan,rogalin,michalol,kpsounis,caire}@usc.edu

A distributed MIMO system consists of several access points connected to a central server and operating as a large distributed multi-antenna access point. In theory, such a system enjoys all the significant performance gains of a traditional MIMO system, and it may be deployed in an enterprise WiFi like setup. In this paper, we investigate the efficiency of such a system in practice. Specifically, we build upon our prior work on developing a distributed MIMO testbed, and study the performance of such a system when both full channel state information is available to the transmitters and when no channel state information is available. In the full channel state information scenario, we implement Zero-Forcing Beamforming (ZFBF) and Tomlinson-Harashima Precoding (THP) which is provably near-optimal in high SNR conditions. In the scenario where no channel information is available, we implement Blind Interference Alignment (BIA), which achieves a higher multiplexing gain (degrees of freedom) than conventional TDMA. Our experimental results show that the performance of our implementation is very close to the theoretically predicted performance and offers significant gains over optimal TDMA. We also discuss medium access layer issues in detail for both scenarios. To the best of our knowledge, this is the first time that the theoretical high data rates of multiuser MIMO systems have been showcased in a real world distributed MIMO testbed.

Categories and Subject Descriptors

C.2.2 [**Computer System Organization**]: Computer Communication Networks

General Terms

Design, Experimentation, Performance

Keywords

Software Radio, Synchronization, Virtual MIMO, Wireless

1. INTRODUCTION

One of the most fundamental challenges in wireless networking today is the increasing demand for higher data rates. In theory, denser deployments, which suppress interference and achieve high spectral efficiency via spatial multiplexing, may be realized in a cost effective manner via a distributed multiuser MIMO system. More specifically, in an enterprise environment, several access points (APs) can be connected to a central server and operate as a large distributed multi-antenna AP, ensuring that all transmitted signal power serves the purpose of data transmission, rather than creating "interference."

In this work, we investigate the efficiency of such a system in practice by focusing on the downlink, that is, the transmissions from the APs to the clients. The first challenge that needs to be addressed is how to achieve tight phase and timing synchronization of the different APs in order to allow for distributed multiuser MIMO precoding. We address this challenge by building upon our prior work on developing a distributed MIMO testbed (see Section 3). We then proceed to study the performance of such a system both when full channel state information is available to the transmitters (full CSIT) and when no channel state information is available (no CSIT).

In the full CSIT scenario we implement two schemes (see Section 4): Zero-Forcing Beamforming (ZFBF) and Tomlinson-Harashima Precoding (THP) [30]. ZFBF is known for its simplicity and high performance (see [6] and references therein), and has been implemented and studied recently in the case of a centralized MIMO system [2]. THP is provably near-optimal in high SNR conditions, the typical use case for WiFi. In the no CSIT scenario, we implement the so called Blind Interference Alignment (BIA) scheme [16], which achieves higher multiplexing gain than orthogonal access (TDMA) by exploiting antenna switching at the receivers in conjunction with a special form of linear precoding at the transmitters (see Section 5). It is worth noting that to the best of our knowledge, neither THP nor BIA have been implemented in the past even in the simpler case of a centralized MIMO setup.

We study the performance gains of the three schemes in our distributed MIMO testbed, comprised of two APs and two clients with one antenna each (see Section 6). Our results are particularly promising. Our implementation of both ZFBF and THP achieve close to 85% of the theoretically achievable gain, an 85% improvement in absolute terms in comparison to TDMA (this gain grows linearly with the number of clients when enough antennas and a sufficiently rich environment is present). Also, our implementation of BIA achieves up to 60% of the theoretically achievable gain, i.e. a 22% improvement in absolute terms in comparison to optimal TDMA. We also study the performance of these schemes through simulations in a variety of scenarios, and we compare it to that of Dirty Paper Coding (DPC), an information theoretic coding scheme that is known to be capacity achieving in the full CSIT case, but is too complex to be implemented in practice. As a matter of fact, THP can be regarded as the practical approximation of the DPC idea. Finally, we discuss medium access layer issues for both the full CSIT and the no CSIT scenarios (see Section 7). To the best of our knowledge, this is the first time that the theoretical high data rates of a MIMO system have been showcased in a real world distributed MIMO testbed.

2. A MULTIUSER MIMO PRIMER

2.1 Precoding schemes

Since the introduction of multiuser MIMO [7] and the solution of its information theoretic capacity region [29], there has been extensive research on low-complexity practical precoding schemes able to achieve a large fraction of capacity with low complexity. Under the assumption of full CSIT, the capacity achieving scheme consists of Dirty-Paper Coding (DPC) [10] combined with optimal power allocation.

While DPC is essentially an information theoretic tool for realizing perfect interference pre-cancellation with no power penalty, its practical implementation is notoriously difficult [4,13]. THP is a well-known low-complexity scheme for the pre-cancellation of known interference which was proposed and widely studied in the context of precoding for Inter-Symbol Interference channels [14]. At high SNR, THP achieves the same rate as DPC minus a fixed penalty called the "shaping loss", which is an asymptotic power loss (see Section 4.2). For this reason, THP was proposed as a low-complexity alternative to DPC for multiuser MIMO in [30].

An even simpler alternative to non-linear precoding (DPC and THP) is represented by *linear* precoding, i.e., beamforming. Among linear beamforming schemes, the simplest and best known is Zero Forcing Beamforming (ZFBF), consisting of a column-normalized version of the right pseudo-inverse of the channel matrix. ZFBF may outperform THP at low SNR. However, when ZFBF is used in conjunction with practical QAM coded modulation [22] it is affected by the same shaping loss as THP at high SNR. In addition, it has been shown [31] that when the number of users is larger than the one of the transmit antennas, ZFBF may use user selection to approach the system sum capacity.

The conclusion of the above discussion is that, in general, there is no clear a priori ranking of the achievable performance between THP and ZFBF: this depends on the actual channel statistics (scattering environment), operating SNR, number of clients K and jointly coordinated AP antennas M. However, when ZFBF is used with practical codes based on QAM constellations, THP is generally superior to ZFBF unless the system works in low SNR (typically not the case for WLAN broadband access, targeting high spectral efficiency and operating in high SNR) or the number of clients is significantly larger than the number of AP antennas, and greedy selection of the clients is performed. The reason why ZFBF has been the preferred choice in testbed implementations (see Section 2.3) is that it is considered to be easier and lower complexity than THP. Our work, reported in this paper, shows that this is not the case.

The need for CSIT imposes certain constraints on the system hardware, and PHY/MAC protocols. For example, in a Time-Division Duplexing (TDD) scenario (common in wireless LANs), CSIT can be acquired at the APs from incoming uplink pilot symbols, provided that the RF hardware is designed in order to preserve the reciprocity of the uplink and downlink baseband equivalent channels. In order to be able to schedule the downlink clients, it is necessary that these clients send their uplink pilots right before the downlink slot, such that their channel state information is not outdated. This requires the design of special uplink pilots and downlink data frames, which are different from legacy 802.11 WLANs (see the discussion in Section 7). When the hardware is not specifically designed to achieve uplink-downlink reciprocity and/or when legacy MAC protocols must be used, it is impossible to have fresh CSIT of the scheduled clients for each downlink slot and precoding schemes based on CSIT cannot be used. In this case, it is known that if the clients are *statistically equivalent* (i.e., they have identical marginal channel statistics of their channel vectors), then TDMA is optimal. In order to achieve a performance gain over conventional TDMA it is therefore necessary to induce artificially *different* channel statistics for the clients.

Following this idea, a scheme known as Blind Interference Alignment (BIA) was proposed in [16]. The scheme assumes that each client terminal has M antennas, but only one RF chain. At each point in time, only the output of one antenna is demodulated and sampled. Therefore, the complexity and power consumption of the client terminals is the same as in a conventional single antenna terminal, with the addition of a switch that connects the different antennas to the RF demodulation front-end. The client channel statistics are differentiated by assigning to each client a specific antenna switching pattern. In this way, each client "sees" a different time-varying sequence of channel vectors. In [16] it is shown that BIA can send MK independent information streams over blocks of $M + K - 1$ channel uses, and, in idealized conditions, each client is able to perfectly remove the interference from streams destined to the other clients, even though the APs have no CSIT at all.

2.2 Performance

Different schemes can be evaluated in terms of their achievable rates, implicitly assuming ideal coding. Studying achievable rates is much easier than implementing a whole coding scheme and evaluating throughput and packet error rate, since the achievable rates can be calculated on the basis of closed-form information theoretic formulas (see Sections 4 and 5). In general, a coarse performance measure is provided by the so-called "system degrees of freedom" (DoF), also known as "spatial multiplexing gain". This is the number of non-interfering data streams that we can simultaneously send to the clients, per time-frequency slot. Mathematically, the systems DoFs are given by the limit:

$$d_{\text{sum}} = \lim_{\text{SNR}\to\infty} \frac{\text{sum\_rate(SNR)}}{\log \text{SNR}}, \quad (1)$$

where SNR denotes the ratio between total transmit energy per channel use and noise power spectral density. A system achieving a total DoFs count equal to d_{sum} has high-SNR sum rate of $\text{sum\_rate(SNR)} = d_{\text{sum}} \log(\text{SNR}) + o(\log(\text{SNR}))$.

It turns out that for all schemes considered in this work, the $o(\log(\text{SNR}))$ terms are effectively a constant, i.e. a $O(1)$ term. It is well-known that THP and ZFBF achieve under full CSIT $d_{\text{sum}} = \min\{M, K\}$ DoFs (recall that K is the number of clients and M the number of antennas), which is the maximum achievable. BIA achieves $d_{\text{sum}} = \frac{MK}{M+K-1}$, which is close to $\min\{M, K\}$ when either $M \gg K$ or $K \gg M$. Finally, TDMA (or, equivalently, any orthogonal multiple access scheme) achieves $d_{\text{sum}} = 1$.

The notion of DoFs becomes particularly relevant when fixed QAM constellations are used (e.g., determined by some standard coded modulation scheme), such that the maximum number of information bits per symbol per stream is capped by $\log_2$(constellation size). With fixed size QAM constellations, the achieved spectral efficiency at high SNR approaches $d_{\text{sum}} \times \log_2$(constellation size). We shall see in Section 6 that our testbed, comprising two coordinated APs $M = 2$ and two clients $K = 2$, and transmitting data streams formed by 16QAM symbols (carrying a maximum of 4 information bits per symbol per stream), achieves sum rates close to 8 bps/Hz for the CSIT schemes (THP and ZFBF), sum rates close to $\frac{16}{3}$ bps/Hz for the BIA scheme, and sum rate close to 4 bps/Hz for TDMA, reflecting the DoFs that we expect from the theory.

2.3 Related work with software radio testbeds

The performance of multiuser precoding schemes in real world scenarios has been investigated recently using software radio testbeds. A related study [2] focuses on ZFBF performance when using a single access point with multiple antennas. Achieving multiplexing gains with *distributed* transmit antennas (i.e., using different APs jointly coordinated by a central server through a wired backhaul network) is discussed in [15] for systems that do not provide phase synchronization between the different APs. This precludes the use of the known full-multiplexing achieving precoding schemes. Therefore, in [15] an alternative based on interference alignment and cancellation is proposed. Other streams of work related to spatial multiplexing in wireless systems have considered the use of directional antennas to limit interference [19], the use of additional antennas to transmit concurrently with other nodes without harming the ongoing transmissions [18], and applying multiuser detection techniques to the WiFi uplink [28].

Other related implementations are SourceSync [23], achieving diversity gains through the use of distributed space time coding, and Fine Grained Channel Access [27], which implements practical uplink bandwidth sharing. These systems rely on frame alignment but do not achieve tight phase synchronization. Our solution (see Section 3) achieves sufficiently accurate phase synchronization across the different jointly coordinated APs such that full distributed precoding is possible. This condition is *necessary* to implement full CSIT precoding schemes (in particular THP and ZFBF) for distributed multiuser MIMO. It should be remarked, however, that BIA only requires frame synchronization and sufficient carrier phase stability, and therefore BIA might be supported by system solutions such as SourceSync and Fine Grained Channel Access.

3. DISTRIBUTED MIMO

In this section we briefly describe the distributed MIMO testbed over which we implement THP, ZFBF, and BIA. We pay special attention to the requirement to synchronize the clocks of the APs which transmit simultaneously.

3.1 The Synchronization Requirement

Most precoding schemes designed for multiuser MIMO transmission assume that multiple transmit antennas are hosted on the same AP and share the baseband circuitry and the clocking circuitry which produces the passband signals. In fact, for schemes exploiting full CSIT it is essential that the signal amplitudes and the relative phase and timing offsets of the signals received from different antennas remain unchanged between channel estimation and the actual transmission.

In contrast, schemes such as Blind Interference Alignment [16] and distributed space-time coding through Alamouti encoding [1] and its generalizations do not require any CSIT and can be implemented under relaxed synchronization requirements, limited to frame synchronization and sufficient phase stability over the number of packets spanning a precoding block. For example, for the BIA scheme that we have implemented we need phase stability over three consecutive packets, forming a BIA precoding block, as detailed in Section 5.

Since full CSIT schemes achieve the full sum-DoFs of the distributed MIMO system, and in high SNR the advantage in spectral efficiency can be very significant, it is interesting to explore the possibility of achieving this level of synchronization in practice. We have implemented a testbed that achieves phase synchronization and enables full CSIT precoding for distributed multiuser MIMO. It is also evident that the schemes requiring less strict synchronization can be implemented in our testbed as well.

3.2 Platform Description

We have implemented our synchronization method in the FPGA of the WARP [24] software radio platform. The radio platform is comprised of a central board containing the programmable logic circuitry and up to four daughterboards serving as radio frequency front-ends. The FPGA and the digital-to-analog and analog-to-digital circuitry present on the daughterboards are clocked from a single oscillator called the sampling clock, while the transceivers on the RF front-ends derive their carrier frequency signals used in modulation and demodulation from a second oscillator, called the carrier clock.

All transmitters are connected to a central server through individual gigabit Ethernet links. The server is responsible for the joint encoding of the transmitted signals and for passing the resulting waveforms, in the form of frequency domain soft symbols, to the transmitter radios.

3.3 The Synchronization Method

Phase drifts and oscillators. Any discussion on phase synchronization of distributed wireless transmitters must necessarily start with the mechanisms through which phase errors occur. Digital wireless transmission systems are constructed using a number of clock sources, among which the two most important ones are the sampling clock and the carrier clock. In a typical system, signals are created in a digital form in baseband at a sampling rate on the order of megahertz, then passed through a digital-to-analog converter (DAC). Through the use of interpolators and filters, the DAC creates a smooth analog waveform signal which is

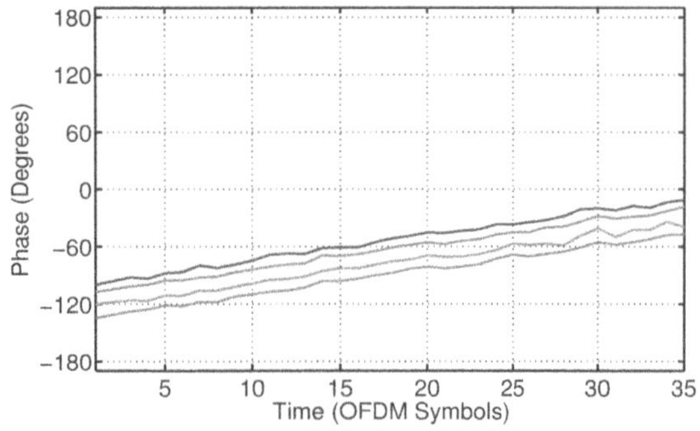

Figure 1: Pilot phases

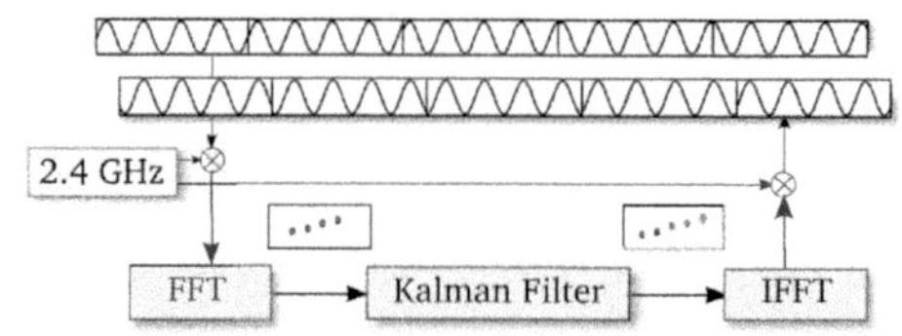

Figure 2: AirSync Schematic. The baseband signals are processed through an FFT which feeds phase estimates into a Kalman Filter. The IFFT produces a phase-adjusted data signal, with the same phase drift as the main transmitter. The modulation and demodulation use the same carrier clock.

then multiplied by a sinusoidal carrier produced by the carrier clock. The result is a passband signal which is then sent over the antenna.

Wireless receivers, in turn, use a chain of signal multiplications and filters to create a baseband version of the passband signal received over the antenna. Some designs, such as the common superheterodyne receivers, use multiple high frequency clocks and convert a signal first to an intermediate frequency before bringing it back to baseband. Other designs simply use a carrier clock operating at the same nominal frequency as the carrier clock of the transmitter and perform the passage from passband to baseband in a single step. We will be focusing on such designs in the ensuing discussion. After baseband conversion, the signal is sampled and the resulting digital waveform is decoded.

There are four clocks in the signal path: the transmitter's sampling clock and carrier clock and the receiver's carrier clock and sampling clock. All four clocks manifest phase drift and jitter. The drift effect, when linear in time and happening at a relatively stable rate, can be assimilated to the presence of a carrier frequency offset.

Denote by $\omega_n = \frac{2\pi n}{NT_s}$ the subcarrier frequency and with ω_c the carrier frequency. Let the timing error of the sampling clock be Δt_s and the timing error of the carrier clock be Δt_c and assume that they are on the same order of magnitude. The phase error due to the sampling clock will be $\omega_n \Delta t_s$ while the phase error due to the carrier clock will be $\phi_i = \omega_c \Delta t_c$. Since ω_c is much greater than ω_n, the dominant phase rotation is due to the carrier clock and does not depend on the subcarrier frequency. Moreover, since time errors are additive, if the time error is approximately linear in time (linear clock drift) then the phase error will also be linear in time and almost equal for all subcarriers.

The assumptions behind the above statement are verified by the results presented in Figure 1. We have constructed an experiment in which a transmitter sends several tone signals, i.e. simple unmodulated sine waves, on several different subcarrier frequencies. In the absence of phase drift each of the tone signals would exhibit a constant phase when measured over several OFDM frames. In reality, the phase is not constant and the frame to frame phase shifts of the signals can be measured and recorded. In the figure, phase drifts have been plotted over the duration of a few tens of frames, a time length corresponding to that of a packet transmission according to the WiFi standards. Our experiment confirms what was anticipated above: the drift is indeed linear and does not depend on the subcarrier frequency. This allows us to design a scheme through which the drift can be tracked and predicted.

Design of a synchronization method. In a nutshell, our method consists of performing frame alignment similarly to prior work [23,27], and then broadcasting a reference signal from a master AP, on to which the rest of the APs transmitters will lock. Thanks to the high clock precision of the WARP platform, the carrier offset is sufficiently low to preclude the need for further frequency offset compensation at baseband frequency before decoding the signal through a Fourier transform. In general, when it is needed, such additional carrier estimation and compensation can be performed using standard methods [25]. No matter how small the residual carrier frequency offset is, it will lead to phase changes in the signal received from a transmitter, from one symbol to the next. We chose to track these phase shifts in real-time and compensate them separately for every frame. We base the tracking of the phase drift on pilot tone signals broadcasted by a master transmitter.

In order to reduce self-interference at the secondary transmitters, the tone signals are placed outside the data band, from which they are separated by a guard interval. The secondary transmitters place an analog baseband filter around their data band further limiting their interference with the pilots. Self-interference could have been avoided using a number of other techniques such as nulling by antenna placement [9], digital compensation [11], or simply relying on the OFDMA-like property of a frame aligned system [27] and preventing the secondary transmitters from using the pilot subcarriers.

In addition to tracking the common drift, the secondary APs measure the initial phase of each tone of the master AP in order to obtain absolute estimates, that is the intercepts of the lines describing the phase drifts in Figure 1. For this, the master AP transmits an initial synchronization header containing a set of known channel estimation symbols. The initial phase estimate, combined with the phase drift measured using the pilot signals, suffices to predict the absolute phase of any particular tone.

Figure 2 illustrates the process of creating a phase synchronous signal at the secondary transmitter. The secondary transmitter overhears a packet sent by the primary transmitter and uses the initial pseudo-noise sequence in order to determine the block boundary timing of this packet. Using a discrete Fourier transform the secondary transmitter decodes the successive frames of the incoming packet. It then employs the CORDIC algorithm on the complex-valued received soft symbols in order to obtain their phases in radians. The phases of the out-of-band pilot signals are tracked throughout the entire packet transmission in order to estimate the phase drift from the primary sender. The measurements from the four different pilots are averaged and passed through a simplified Kalman filter which maintains an accurate estimate and predicts, based on the current drift estimate, the phase drift after the passage of a few further frames.

The phase estimates are used in synthesizing a synchronized signal. The secondary transmitter uses an inverse discrete Fourier transform, whose output frames are timed such that they align with the frames of the main sender's signal. For every subcarrier the secondary transmitter rotates the soft symbol to be sent by an angle corresponding to the subcarrier's estimated phase offset. The result is a tone that, while not having the same phase as the corresponding tone from the main transmitter, follows that tone at a fixed, preknown phase difference.

Since the subcarrier tones of all transmitters are now synchronized, the transmitters act together as a commonly hosted set of MIMO radio-frequency front-ends.

3.4 Centralized joint encoding

By transmitting phase synchronous signals from multiple access points we have created the equivalent of a distributed MIMO transmitter, capable of employing multiuser MIMO precoding strategies in order to transmit to multiple users at the same time. However, the use of multiple access points complicates the design of the transmitter system. For most of the precoding schemes available, the encoding of the waveforms to be transmitted over the antennas must be done jointly, since reaching a single user usually involves transmitting over multiple antennas. While in theory the joint encoding process could be duplicated at each access point given the binary information destined to each user, we chose to do the encoding only once, at a central server and send the resulting waveforms to each access point for transmission[1].

Our central server has an individual gigabit Ethernet connection to each of the WARP radios serving as access points. We divide the downlink time into slots and in each slot schedule for transmission a number of packets destined to various users. For each of the access points, the server computes the waveform of the signal to be transmitted in the next downlink slot. However, it does not perform any phase correction at this point. The only information used in the precoding is the data to be transmitted and the channel state information between each access point antenna and each user antenna. The server assumes that all access points are phase synchronous, like in a normal MIMO system. The server transmits their corresponding waveforms to all secondary transmitters and finishes by sending the last waveform to the primary transmitter. The primary transmitter starts transmitting right away and the secondary transmitters follow.

The design of the phase alignment method ensures its scalability. There is no added overhead for synchronizing a larger number of secondary transmitters.

[1]This approach is practical in enterprise networks where a number of access points are already connected to a common server.

4. PRECODING USING TRANSMITTER CSI

Consider a system with M single-antenna jointly coordinated access points and K single-antenna clients. When using OFDM, the time and frequency selective fading channel is decomposed into subcarriers, where for each OFDM symbol and subcarrier the channel is characterized by a single frequency-domain complex coefficient for each transmit/receive antenna pair. The resulting baseband channel model for a single subcarrier over any given OFDM symbol is described by

$$\mathbf{y} = \mathbf{H}^{\mathsf{H}}\mathbf{x} + \mathbf{z}, \tag{2}$$

where the vectors $\mathbf{y}$, $\mathbf{x}$ and $\mathbf{z}$ and the matrix matrix $\mathbf{H}$ have complex entries, and ^{H} denotes Hermitian transpose. For the sake of notation simplicity, we neglect in Equation(2) both the OFDM symbol and the subcarrier (time and frequency indices), since they are irrelevant at this point. The received signal sample at client receiver k is the k-th entry of the vector $\mathbf{y}$, the symbol transmitted by access point j is the j-th entry of the vector $\mathbf{x}$, and the (j, k)-th element of the matrix $\mathbf{H}$ indicates the channel coefficient between the j-th access point transmitter and the k-th client receiver. Finally, $\mathbf{z}$ denotes a K-component vector of independent, identically distributed, additive white circularly symmetric Gaussian noise samples (AWGN).

Let $\mathbf{u}$ be a K-vector containing the symbols destined to each user. The transmitted vector $\mathbf{x}$ is obtained as a function of $\mathbf{u}$. The mapping $\mathbf{u} \mapsto \mathbf{x}$ is called *precoding*. When the mapping function depends on the matrix $\mathbf{H}$, assumed known to the transmitters, we say that the scheme makes use of full CSIT. If the mapping is independent of $\mathbf{H}$, then the scheme is "blind". Next, we review the two full CSIT schemes that we have implemented in our testbed: ZFBF (a linear mapping) and THP (a non-linear mapping).

4.1 Zero-Forcing Beamforming

The precoding mapping in ZFBF is given by

$$\mathbf{x} = \mathbf{V}\mathbf{u} \tag{3}$$

where $\mathbf{V}$ is a scaled version of the Moore-Penrose right pseudo-inverse of $\mathbf{H}^{\mathsf{H}}$, normalized in order to have unit-norm columns. Assuming $M \geq K$, [2] the matrix $\mathbf{H}$ has rank K with probability 1 for any practically relevant scattering environment. The pseudo-inverse takes on the form $\mathbf{H}^{\dagger} = \mathbf{H}(\mathbf{H}^{\mathsf{H}}\mathbf{H})^{-1}$ and the precoding matrix $\mathbf{V}$ is given by

$$\mathbf{V} = \mathbf{H}^{\dagger}\mathbf{\Lambda}^{1/2} \tag{4}$$

where $\mathbf{\Lambda} = \text{diag}(\lambda_1, \ldots, \lambda_K)$ and

$$\lambda_k = \frac{1}{[(\mathbf{H}^{\mathsf{H}}\mathbf{H})^{-1}]_{k,k}}$$

(inverse of the (k, k) diagonal element of $(\mathbf{H}^{\mathsf{H}}\mathbf{H})^{-1}$). It is immediate to check that $\mathbf{V}$ in (4) has unit norm columns.

With ZFBF precoding, the original channel (Equation 2) reduces to $\mathbf{y} = \mathbf{\Lambda}^{1/2}\mathbf{u} + \mathbf{z}$. Hence, each k-th client sees a separate (spatially decoupled) channel of the form $y_k = \sqrt{\lambda_k}u_k + z_k$. The total transmit power is given by $\mathbb{E}[\|\mathbf{x}\|^2] = \mathbb{E}[\|\mathbf{u}\|^2]\sum_{k=1}^{K} q_k$, where $q_k = \mathbb{E}[|u_k|^2]$ is the power allocated to the k-th data stream. The corresponding achievable sum

[2]If $K > M$, greedy user selection is used to serve a number of clients not larger than M.

rate assuming a Gaussian random coding ensemble is given by

$$R^{\text{zfbf}}_{\text{sum}}(\text{SNR}) = \sum_{k=1}^{K} \log(1 + \lambda_k q_k), \quad (5)$$

where, without loss of generality, we normalize the noise variance per component to 1, and let $\sum_k q_k = \text{SNR}$. The above rate can be maximized with respect to the power allocation to the data streams, i.e., with respect to $q_1, \dots, q_K$, subject to the constraint that the total power must be not larger than SNR.

In the case of a distributed MIMO system, both inaccuracies in channel state information and phase synchronization errors between the transmitter antennas can lead to power leakage from each client's channel to the others, creating an error floor for downlink transmission. In previous work, during the development of our system, we have measured the amount of leakage arising from this cumulated effects. We have discovered that the residual self-interference power due to imperfect zero-forcing is, on average, about 24 dB below the useful power.

4.2 Tomlinson-Harashima Precoding

In THP the mapping from the data symbol vector $\mathbf{u}$ to the transmitted symbol vector $\mathbf{x}$ is *non-linear*. Consider again the channel model (2). THP imposes a given precoding ordering, and it pre-cancels sequentially the interference of already precoded signals. Without loss of generality, consider the natural precoding ordering to be from 1 to K, and let $\mathbf{h}_k$ denote the k-th client's $M \times 1$ channel vector. Let $\mathbf{H} = \mathbf{QR}$ be the QR factorization of $\mathbf{H}$, such that $\mathbf{R}$ is $K \times K$ upper triangular with real non-negative diagonal coefficients, and $\mathbf{Q}$ is such that $\mathbf{Q}^{\mathsf{H}}\mathbf{Q} = \mathbf{I}$. THP precoding is formed by the concatenation of a linear mapping, defined by the unitary matrix $\mathbf{Q}$, with a non-linear mapping that does the interference pre-cancellation. Let $\hat{\mathbf{u}} = \text{THP}(\mathbf{u})$ denote the non-linear mapping of the data vector $\mathbf{u}$ into an intermediate vector $\hat{\mathbf{u}}$, that will be defined later. The linear mapping component of THP is then given by

$$\mathbf{x} = \mathbf{Q}\mathbf{\Sigma}^{1/2}\hat{\mathbf{u}}, \quad (6)$$

where $\mathbf{\Sigma} = \text{diag}(q_1, \dots, q_K)$, and q_k is, as before, the power allocated to the k-th stream. It follows that the channel reduces to $\mathbf{y} = \mathbf{L}\mathbf{\Sigma}^{1/2}\hat{\mathbf{u}} + \mathbf{z}$, where $\mathbf{L} = \mathbf{R}^{\mathsf{H}}$ is lower triangular. The signal seen at client k receiver is given by

$$y_k = [\mathbf{L}]_{k,k}\sqrt{q_k}\hat{u}_k + \underbrace{\sum_{j<k}[\mathbf{L}]_{k,j}\sqrt{q_j}\hat{u}_j}_{\text{interference}} + z_k \quad (7)$$

Next, we look at the non-linear mapping $\mathbf{u} \mapsto \hat{\mathbf{u}}$. The goal is to pre-cancel the term indicated by "interference" in Equation 7. Notice that this term depends only on symbols $\hat{u}_j$ with $j < k$. Therefore, the elements $\hat{u}_1, \dots, \hat{u}_K$ can be calculated sequentially. A simple pre-subtraction of the interference term at each step would increase the effective transmit power.

The key idea of THP is to introduce a modulo operation that limits the transmit power of each precoded stream $\hat{u}_k$. This is defined as follows. Assume that the data symbols u_k are points from a QAM constellation uniformly spaced in the squared region of the complex plane bounded by the interval $[-\tau/2, \tau/2]$ on both the real axis and the imaginary axis. Then, for a complex number s, let s modulo τ be given by $[s]_{\text{mod } \tau} = s - Q_\tau(s)$, where $Q_\tau(s)$ is the point $(n + jm)\tau$ with integers n, m closest to s. In short, $Q_\tau(s)$ is the quantization of s with respect to a square grid with minimum distance τ on the complex plane, and $[s]_{\text{mod } \tau}$ is the quantization error. We let

$$\hat{u}_k = \left[u_k - \frac{\sum_{j<k}[\mathbf{L}]_{k,j}\sqrt{q_j}\hat{u}_j}{[\mathbf{L}]_{k,k}\sqrt{q_k}} \right]_{\text{mod } \tau}. \quad (8)$$

In this way, the symbol $\hat{u}_k$ is necessarily bounded into the squared region of side τ, and its variance (assuming a uniform distribution over the squared region, which is approximately true when we use a QAM constellation inscribed in the square) is given by $\mathbb{E}[|\hat{u}_k|^2] = \tau^2/6$. Letting $\tau = \sqrt{6}$ we have that the precoded symbols have unit energy and that the transmit power for stream k is exactly given by q_k.

Let's focus now on receiver k and see how the modulo precoding can be undone. The receiver scales the received symbol y_k by $[\mathbf{L}]_{k,k}\sqrt{q_k}$ and applies again the same the modulo τ non-linear mapping. Simple algebra then shows that

$$\widehat{y}_k = \left[u_k + \frac{z_k}{[\mathbf{L}]_{k,k}\sqrt{q_k}} \right]_{\text{mod } \tau}. \quad (9)$$

It follows that the interference term is perfectly removed, but we have introduced a distortion in the noise term. Namely, while u_k is unchanged by the modulo operation, since by construction it is a point inside the square, the noise term $\frac{z_k}{[\mathbf{L}]_{k,k}\sqrt{q_k}}$ is "folded" by the modulo operation, i.e., the tails of the Gaussian noise distribution are folded on the squared region. Noise folding is a well-known effect of THP [14].

As far as the achievable rate is concerned, it is possible to show (see [5,12]) that this is given by

$$R^{\text{thp}}_{\text{sum}}(\text{SNR}) = \sum_{k=1}^{K} \left[\log(1 + |[\mathbf{L}]_{k,k}|^2 q_k) - \log(\pi e/6)\right]_+, \quad (10)$$

where $[\cdot]_+$ indicates the positive part. Again, this sum rate can be optimized with respect to the power allocation $q_1, \dots, q_K$, subject to the sum power constraint $\sum_{k=1}^{K} q_k \leq \text{SNR}$. The rate penalty term $\log(\pi e/6)$ is the shaping loss, due to the fact that THP produces a signal which is uniformly distributed in the square region (therefore, a codeword of n signal components is uniformly distributed in an n-dimensional complex hypercube). [3]

5. PRECODING WITHOUT TRANSMITTER CSI

In the case when CSIT cannot be reliably acquired, when tight phase synchronization cannot be achieved, or when channel reciprocity does not hold because of hardware limitations, we need to resort to "blind" approaches. The classical choice, already implemented in software radio [23], consists of distributed space time coding (e.g., distributed Alamouti [1]). This, however, improves reliability through diversity, but achieves the single DoF of TDMA. Here, we implemented the BIA scheme proposed in [16], able to achieve $d_{\text{sum}} = \frac{MK}{M+K-1}$. We briefly outline the case of $M = K = 2$,

[3]It should be noticed that the same shaping loss is incurred by any other scheme such as ZFBF, BIA, as well as TDMA if practical QAM constellations are used instead of the theoretical Gaussian coding ensemble.

Slot	$t=t_1$	$t=t_2$	$t=t_3$
$\begin{bmatrix}\text{Tx1 Sends}\\ \text{Tx2 Sends}\end{bmatrix}$	$\mathbf{x}^{[1]}+\mathbf{x}^{[2]}=\begin{bmatrix}u_1^{[1]}+u_1^{[2]}\\ u_2^{[1]}+u_2^{[2]}\end{bmatrix}$	$\mathbf{x}^{[1]}=\begin{bmatrix}u_1^{[1]}\\ u_2^{[1]}\end{bmatrix}$	$\mathbf{x}^{[2]}=\begin{bmatrix}u_1^{[2]}\\ u_2^{[2]}\end{bmatrix}$
User 1 Antenna	A	B	A
User 2 Antenna	A	A	B
User 1 Receives	$y_1(t_1)=\mathbf{h}_{1A}^{\mathsf{H}}(\mathbf{x}^{[1]}+\mathbf{x}^{[2]})+z_1(t_1)$	$y_1(t_2)=\mathbf{h}_{1B}^{\mathsf{H}}\mathbf{x}^{[1]}+z_1(t_2)$	$y_1(t_3)=\mathbf{h}_{1A}^{\mathsf{H}}\mathbf{x}^{[2]}+z_1(t_3)$
User 2 Receives	$y_2(t_1)=\mathbf{h}_{2A}^{\mathsf{H}}(\mathbf{x}^{[1]}+\mathbf{x}^{[2]})+z_2(t_1)$	$y_2(t_2)=\mathbf{h}_{2A}^{\mathsf{H}}\mathbf{x}^{[1]}+z_2(t_2)$	$y_2(t_3)=\mathbf{h}_{2B}^{\mathsf{H}}\mathbf{x}^{[2]}+z_2(t_3)$
User 1 Decodes	$\left.\begin{aligned}\widetilde{y_1}(1)&=y_1(t_1)-y_1(t_3)\\ &=\mathbf{h}_{1A}^{\mathsf{H}}\mathbf{x}^{[1]}+z_1(t_1)-z_1(t_3)\\ \widetilde{y_1}(2)&=y_1(t_2)\\ &=\mathbf{h}_{1B}^{\mathsf{H}}\mathbf{x}^{[1]}+z_1(2)\end{aligned}\right\}\Rightarrow\hat{\mathbf{x}}^{[1]}=\begin{bmatrix}\mathbf{h}_{1A}^{\mathsf{H}}\\ \mathbf{h}_{1B}^{\mathsf{H}}\end{bmatrix}^{-1}\begin{bmatrix}\widetilde{y_1}(1)\\ \widetilde{y_1}(2)\end{bmatrix}$		
User 2 Decodes	$\left.\begin{aligned}\widetilde{y_2}(1)&=y_2(t_1)-y_2(t_2)\\ &=\mathbf{h}_{2A}^{\mathsf{H}}\mathbf{x}^{[2]}+z_2(t1)-z_2(t2)\\ \widetilde{y_2}(2)&=y_2(t_3)\\ &=\mathbf{h}_{2B}^{\mathsf{H}}\mathbf{x}^{[2]}+z_2(3)\end{aligned}\right\}\Rightarrow\hat{\mathbf{x}}^{[2]}=\begin{bmatrix}\mathbf{h}_{2A}^{\mathsf{H}}\\ \mathbf{h}_{2B}^{\mathsf{H}}\end{bmatrix}^{-1}\begin{bmatrix}\widetilde{y_2}(1)\\ \widetilde{y_2}(2)\end{bmatrix}$		

Table 1: Blind Interference Alignment for the 2×2 scenario

when each client has two antennas connected through a switch to a single RF front-end.

The fundamental idea of BIA is to differentiate the users by inducing special signatures in their channel temporal variations. This is obtained by allocating to each user an antenna switching sequence, according to which they demodulate the signal from one of their antennas. Only one antenna in every given slot is used, so that a single RF front-end and demodulation chain are needed.

The scheme that we have implemented sends 2 independent streams per client to two clients, over 3 time slots. Figure 4 contains a sketch of the testbed. Receiver 1 uses the switching sequence A,B,A, indicating that it uses antenna A in slots 1 and 3 and antenna B in slot 2 of a precoding frame formed by 3 slots. Receiver 2 uses the switching sequence A,A,B, with analogous meaning. Denoting by $u_i^{[k]}$ the i-th data symbol of user k, with $i=1,2$ and $k=1,2$, the BIA scheme transmits $\mathbf{x}^{[1]}+\mathbf{x}^{[2]}$ in the first slot, $\mathbf{x}^{[1]}$ in the second slot, and $\mathbf{x}^{[2]}$ in the third slot, where $\mathbf{x}^{[1]}$ and $\mathbf{x}^{[2]}$ are formed out of the symbol streams as illustrated in Table 4.2. Letting $\mathbf{h}_{kA}$ and $\mathbf{h}_{kB}$ denote the 2×1 channel vectors seen at antennas A and B of user k, we observe that the 2×2 matrix with columns $[\mathbf{h}_{kA},\mathbf{h}_{kB}]$ has rank 2, and that the channels remain constant over the precoding block spanning 3 slots.

After linear interference cancellation at each client, the achievable sum rate with Gaussian random coding ensembles is given by [16]:

$$R_{sum}=\sum_{k=1}^{K}\frac{\mathbb{E}\left[\log\det\left(\mathbf{I}+\frac{(K+M-1)P}{M^2K}\mathbf{H}_k^H\mathbf{H}_k\right)\right]}{M+K-1} \tag{11}$$

where for the 2×2 case:

$$\mathbf{H}_k=\left[\frac{1}{\sqrt{2}}\mathbf{h}_{kA},\mathbf{h}_{kB}\right] \tag{12}$$

6. PERFORMANCE EVALUATION

6.1 Experimental results

In this section we characterize the performance of the precoding schemes. The first series of experiments studies how the achieved rates vary in respect to variations in the transmitted power, for a single channel realization. We then study the performance of BIA when the channel experiences phase variations. We have focused on the THP and BIA precoding schemes. While we have implemented and tested ZFBF as well, we do not plot the results since the resulting performance values in our setup are quite close to those of THP.

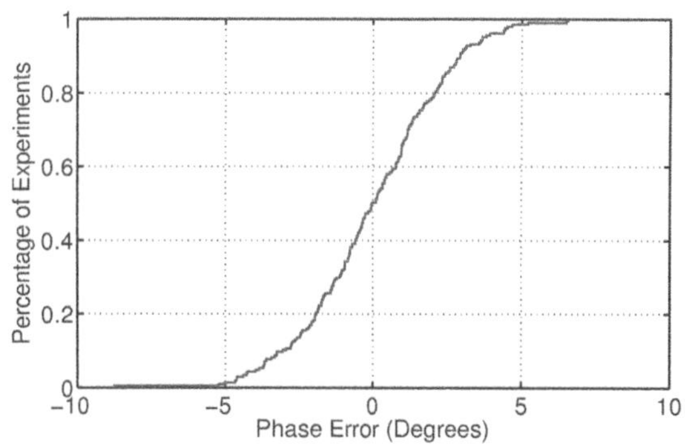

Figure 3: The Precision of the Phase Synchronization. AirSync achieves phase synchronization within a few degrees of the source signal.

Synchronization Accuracy. In this particular experiment we have placed the two transmitters and the two receivers at random locations. We placed a third RF front-end on the secondary sender and configured it in receive mode.

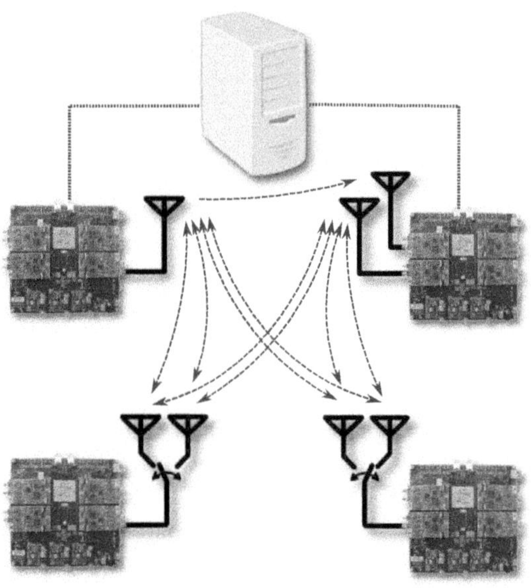

Figure 4: BIA Testbed. When usng Blind Interference Alignment each receiver switches between two antenna modes.

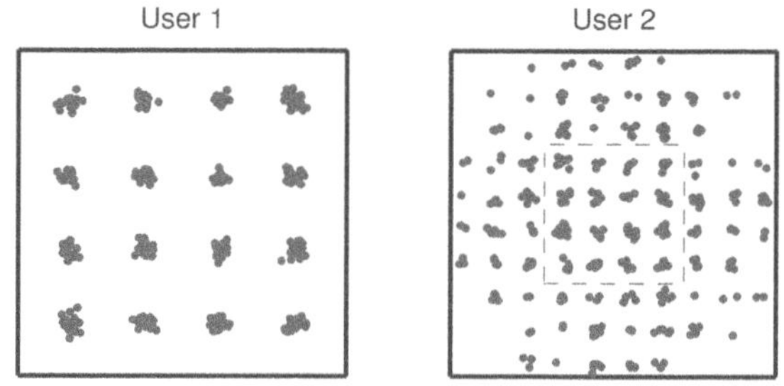

Figure 5: The scattering diagram for Tomlinson-Harashima Precoding.

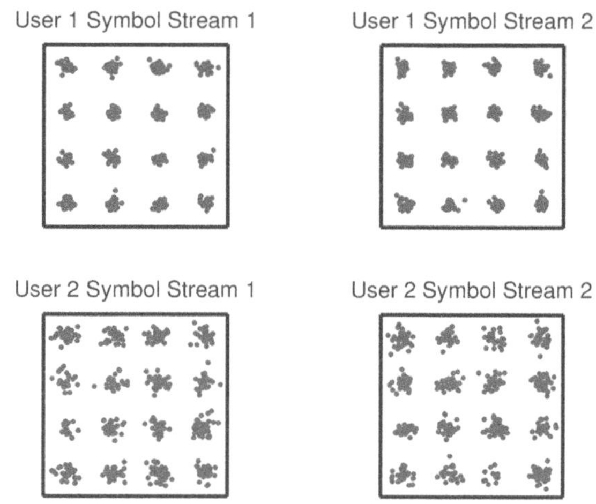

Figure 6: The scattering diagram for Blind Interference Alignment.

The secondary transmitter samples its own synthesized signal over a wired feedback loop and compares it with the main transmitter's signal. The synchronization circuit measures and records the phase differences between these two signals. Since we use the primary transmission as a reference, in this experiment we do not broadcast the signal synthesized by the secondary transmitter in order to protect the primary transmission from unintended interference. We note that the use of a third RF front-end is not needed in the general case.

We have modified the synchronization circuit to produce a signal that is not only phase synchronous with that of the primary transmitter but has the exact same phase when observed from the secondary transmitter. To achieve this, the circuit estimates the phase rotation that is induced between the DAC of the secondary transmitter and the ADC through which the synthesized signal is resampled. It then compensates for this rotation by subtracting this value from the initial phase estimate. It is worth noting that this rotation corresponds to the propagation delay through the feedback circuit and is constant for different packet transmissions, as determined through measurements.

Figure 3 illustrates the CDF of the synchronization error between the secondary transmitter and the primary transmitter. The error is measured on a frame-to-frame basis using the feedback circuit. In decimal degree values, the standard deviation is 2.37 degrees. The 95th percentile of the synchronization error is at most 4.5 degrees.

Achievable rates. We have used the testbed topology illustrated in Figure 4 throughout our experiments, placing the receivers in arbitrary locations in a closed environment. In order to compare the performance of THP and BIA to the one of a typical TDMA system, we introduced a third transmission scheme, in which instead of multiuser precoding we transmit to one user at a time from the closest access point. In this scheme, transmissions to different users happen in a time-shared manner, just like in 802.11. As opposed to 802.11, we assume that different access points do not collide when doing channel access, i.e. they perform perfect downlink scheduling. We investigate the sum rates achievable during downlink transmission. The unit of measure is the number of bits per second per Hertz (bps/Hz) transferred by each scheme, where the comparison was done looking only at the portion of the bandwidth used for data transmission (i.e. we considered only the data carriers and ignored the overhead of null carriers, pilots and cyclic prefix). Since the OFDM framing for all three schemes is identical and similar to the one of 802.11, we obtain a fair comparison of their throughputs.

We have varied the transmitters' signal powers in a proportional way, trying to obtain a typical range of SNRs at the receivers. The receive-side SNR values span the typical high range encountered in WiFi signal transmission, from 15 dB to 30 dB. The received SNR values (or carrier to noise ratios) in our figures were estimated using non-precoded and non-synchronized isotropic broadcasts, measuring the raw received power and comparing it to the receiver noise. The same levels of total transmit power were used in the precoded synchronous transmissions.

We evaluate the SINR (Signal to Noise plus Interference Power Ratio) values of the different symbols streams decoded by the receivers. Determining the symbol SINR values requires more effort in our scenario than in classic point-to-point transmission. Since our system is susceptible to power leakage from one stream to another, we would like to con-

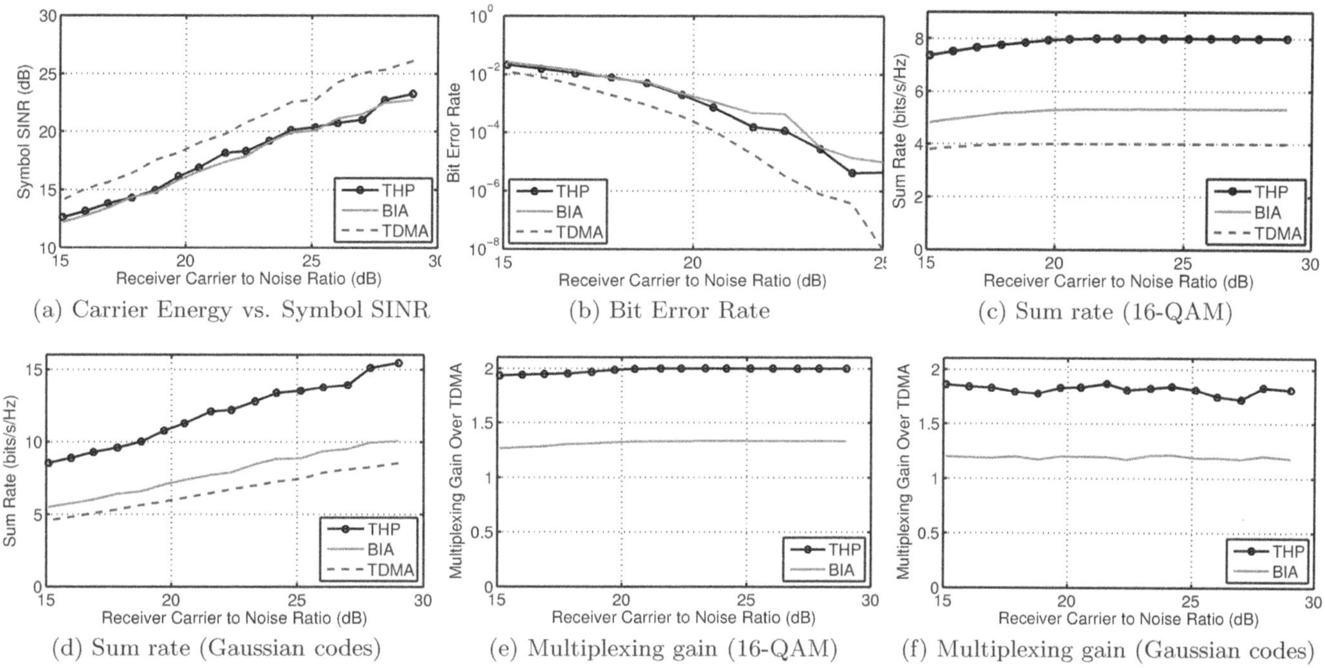

(a) Carrier Energy vs. Symbol SINR (b) Bit Error Rate (c) Sum rate (16-QAM)

(d) Sum rate (Gaussian codes) (e) Multiplexing gain (16-QAM) (f) Multiplexing gain (Gaussian codes)

Figure 7: Experimental Results

tinuously transmit over all channels in order to assess the impact of interference.

To this end we sampled each symbol stream using symbols chosen from a relatively sparse QAM-16 constellation. We measured the variance of the constellation points on the receiver side in order to determine the sum of the noise and interference powers. The amplitude of the constellation reflects the received signal power. At the high SNR values present in our system, the clusters of constellation points are spaced sufficiently to allow for an accurate mapping of the received symbols to constellation points. In order to assess the effects of interference produced by streams that follow other encodings, we have, in some experiments, fixed a QAM-16 constellation on one symbol stream while employing symbols chosen according to a Gaussian or uniform distribution on the other stream. Our results have shown that at the low interference levels measured, none of the statistics collected shows considerable variance depending on the type of interference.

Figure 7a presents the SINR values for symbols received when using each of the three precoding schemes. Figure 7b illustrates the inferred symbol error rates for the QAM-16 constellation transmitted. It can be easily seen that the THP and BIA curves closely follow the TDMA curve, with only a few dB difference.

Figure 7c presents the sum rate achievable by the three different schemes (THP, BIA and plain TDMA) for different levels of the total transmit power, when employing a capacity achieving code on top of the transmitted QAM-16 constellation. Figure 7e presents the relative gains of THP and BIA over TDMA. It can be easily seen that each scheme quickly saturates at the maximum rate of 4 bits/DoF. Since THP and BIA provide extra degrees of freedom, they achieve their theoretical multiplexing gain over TDMA.

We would like to know how the quality of the resulting symbol streams affects the achievable rates. To this end we have estimated the rates achievable when using capacity-achieving codes instead of the QAM-16 modulation. Figure 7d presents the resulting sum rates and Figure 7f presents the multiplexing gains. THP achieves an average increase in sum rate of 85%. While this may seem shy of the theoretical achievable multiplexing gain of 2, we must remember that THP allocates power among two degrees of freedom, while TDMA allocates its whole transmitted power to a single transmitter. The second reason for this discrepancy is the shaping loss present in the rate calculation in the case of THP, which was indicated in Equation 10.

The average gain for BIA is 22%. Again, the transmitted power is distributed between the two transmitters. Additionally, as mentioned in Section 5, BIA suffers from noise enhancement, which affects the received symbols.

In the case of a distributed MIMO system, we would expect that phase synchronization error could lead to random rotations of the received soft symbols. We investigated this effect by comparing the variance of soft symbols corresponding to constellation points of different amplitudes. We would expect that due to random rotations, the variance of the outer constellation points would be higher. However, our measurements could not identify such an effect for any of the transmission schemes.

Since BIA does not provide the transmitter with channel state information, to allow it to guess an appropriate transmission rate, it is interesting to find out by how much the received symbol quality is affected by small variations in the positioning of the antennas. Such an effect is analogous to fast fading, where small phase changes affect the channel amplitude at different frequencies. We have conducted an experiment in which we have varied the transmitter antenna positions within one wavelength of their initial position and measured the channel SINR for the two user symbols. The

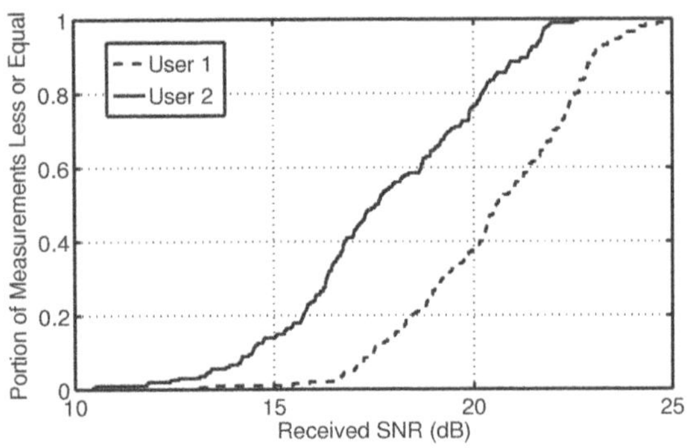

Figure 8: The cumulative distribution function of received SNRs under the Blind Interference Alignment Scheme.

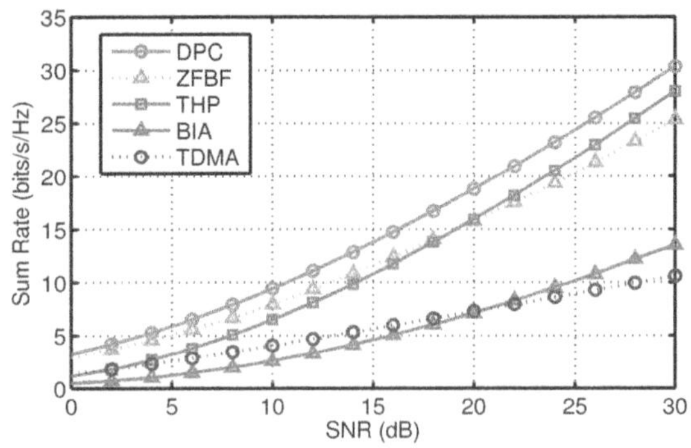

Figure 9: Simulation Results. Sum rate for four users and four access points.

CDFs of the resulting SINR distributions are shown in Figure 8. The high variance of the distribution has profound implications on the design of a coding and medium access scheme for BIA, as will be discussed in Section 7. The higher SNR present in one of the CDFs can be easily explained by the fact that the two symbols are transmitted by antennas placed on different transmitters. The placement of the users relative to the corresponding transmitter determines each symbol's average power.

6.2 Simulation results

To further evaluate the different schemes that were implemented on our testbed, we have also performed off-line simulations to compare them with respect to information theoretic benchmarks. We collected a large set of channel matrices measured at the receivers of our testbed, based on the downlink pilot symbols, and we calculated the achievable sum rate of the various precoding schemes assuming ideal Gaussian random coding, and averaging over the ensemble of measured channel matrices. Specifically, we compare the sum rates between the capacity achieving theoretic DPC, THP and ZFBF, BIA and finally TDMA. In Fig. 9 we can see the sum rates plotted against the SNR (in dBs) at the transmitter for $K = 4$ users and $M = 4$ antennas.

For the full CSIT schemes, we observe that both THP and ZFBF approach the optimal theoretic capacity and show a constant gap of $2.5 - 5$ dB with respect to the system sum capacity (achieved by DPC), at sufficiently large SNR. Notice that THP suffers from the shaping loss even in the case of ideal coding, while ZFBF with ideal coding does not. This makes THP outperform ZFBF only in very high SNR. However, in a real system implementation, coded modulation schemes based on QAM constellations would be used also with ZFBF, and these suffer from the same shaping loss of THP. Therefore, in an actual implementation, the performance of ZFBF would decrease also by roughly 0.5 bit/s/Hz per stream. Thus, the effective gain of THP over ZFBF in an actual implementation can be significant, even at moderate SNR. This is consistent with the testbed experiments shown in Section 6.1, where symbols from a 16QAM constellation were used.

Next, we compare the two schemes that require no CSIT. First, notice that in order for BIA to achieve the theoretic DoFs, we have to be in the high SNR region. It is therefore expected that BIA's sumrate will suffer from the noise enhancement until we reach a high SNR. However, we again remark that these conclusions driven by the use of ideal Gaussian coding may be misleading. For example, when the system is constrained to use constellations of fixed size, as observed before, the advantage of BIA over TDMA can be significant. For example, in the results of Section 6.1, we see that BIA achieves almost a rate 4/3 times larger than TDMA.

7. MEDIUM ACCESS

Our performance evaluation has shown that multiuser transmission at high data rates is possible and that we can approach the theoretical multiplexing gains in a real-world scenario. In the following we discuss issues pertaining to the MAC layer design for the full CSIT and no CSIT cases.

MAC design for full CSIT schemes. As opposed to standard 802.11, distributed multiuser MIMO transmission requires a high degree of coordination between the APs. Part of that coordination, necessary for obtaining frame alignment and phase synchronization, has been detailed in Section 3. The precoding schemes of Sections 4 and 5 require joint precoding of the packets to be transmitted to the different users. It results that the system design of a distributed MIMO system is necessarily more centralized than the current enterprise WLAN design.

We envision a system that, at the MAC layer, resembles more current cellular deployments than the standard CSMA/CA-based large wireless deployment. The downlink and uplink phases of the system should be completely separated in a Time Division Duplexing fashion. In the downlink phase, there will be almost no random behavior, since downlink transmissions are centrally planned by the server and then executed by the APs, see [3] for a very recent study of this matter. With full CSIT, the clients can be selected and jointly precoded on the basis of the knowledge of the system channel matrix $\mathbf{H}$ (see Equation (2)). In particular, the coding rate allocated to each data stream can be dynamically adapted depending on the effective SINR of the stream. This is in line with what is currently done in today's WLAN standards [20], that make use of adaptive coding and modulation based on a family of coded-modulation schemes.

MAC design for schemes with no CSIT. When CSIT is not available, the AP adapts its transmission rate (i.e., its coded-modulation scheme) on the basis of some average signal quality level, reported from the clients through a Channel Quality Indicator (CQI) feedback message, or simply obtained by measuring the quality of the uplink packets. The CQI depends on the *average* SINR, rather than on the instantaneous realization of the channel matrix $\mathbf{H}$, and allows to adapt to coarse and slow signal strength variations (e.g., due to distance and walls attenuation). We assume that for practical reasons, obtaining a quality measure in a timely manner is not possible. For example, in the case of a moving receiver, the symbol SNR figures for BIA vary widely due to phase changes (see Section 6). It follows that the channel quality is, as far as the transmitter is concerned, a random variable with quite high variance and that the rates supported may be very different from one packet transmission to the next. We may choose one of two strategies in order to deal with the rate uncertainty: a conservative rate adaptation approach that chooses a small rate for every transmission or an incremental redundancy scheme.

In incremental redundancy schemes, transmission occurs in blocks. An encoder produces several consecutive encoded versions of a block which are sent over the channel to the decoder. Each successive pass increases the amount of mutual information transfered. At some point the decoder accumulates sufficient information to decode the block and verifies its correctness through a checksum check. At this point the decoder informs the transmitter, through an acknowledgment, that it can proceed to the following information block.

Rateless codes have been proposed as efficient building blocks for incremental redundancy systems. Raptor codes [26], for example, utilize a belief propagation-based decoder and implement the process described above. However, Raptor codes, along with a large class of rateless codes, must be tuned depending on the channel SNR figure in order to effectively approach capacity. More recent approaches do not suffer from this shortcoming. Strider [17] is able to operate in the presence of unknown amounts of interference and when paired with a capacity achieving code comes very close to the actual channel rate. Spinal codes operate with a very dense constellation suitable to many channel conditions [21]. The spinal code decoder makes use of the channel SNR figure, however this quantity can be estimated efficiently [8] for each block transfer and used appropriately in symbol likelihood estimation.

8. DISCUSSION AND FUTURE WORK

In this work we have implemented for the first time ZFBF, THP, and BIA in a distributed MIMO setting. All schemes were shown to achieve almost all of the multiplexing gain predicted by theory in our testbed, consisting of two access points and two clients. In the rest of this section we discuss scalability and implementation overhead issues.

Scalability. For full CSIT schemes, the main issues affecting the SINR of distributed multiuser MIMO are the accuracy of channel estimation and the ability to maintain accurate phase synchronization between the APs. Both issues are not affected by the number of clients, but depend on the number and geometry (distances) of the jointly coordinated APs. In future work, we seek to understand how many APs can be jointly coordinated through the synchronization scheme that we have designed. In general, we may envisage that "clusters" of jointly processed APs can operate according to a distributed multiuser MIMO scheme.

Implementation Overhead. All schemes discussed in this paper, namely ZFBF, THP, BIA, as well as TDMA, require downlink pilots, similarly to 802.11. The main difference when it comes to overhead is the need for CSIT. Specifically, ZFBF and THP need to estimate the channel from the uplink in special slots, and rely on reciprocity to get an estimate for downlink. In contrast, BIA and TDMA do not have this need. An additional benefit of BIA is that it requires a single RF chain and very low power, making it ideal for small devices like smartphones. That said, the performance gain from CSI aware schemes is quite significant, as shown in Section 6. When it comes to implementing the MAC, all schemes require scheduling to materialize high rates. It is part of future work to implement in our testbed MAC protocols that utilize efficient schedulers, as well as to implement several incremental redundancy schemes.

9. ACKNOWLEDGMENTS

The authors would like to thank the reviewers and our shepherd, Xinbing Wang, for their assistance in improving the manuscript of this paper.

This work was supported by the Ming Hsieh Institute, the Army Research Laboratory (ARL) Collaborative Technology Alliance (CTA) number W911NF-09-2-0053, Marie Curie grants 256416 and 274523, Cisco Systems under a CRC grant and NSF grant CIF-0917343.

10. REFERENCES

[1] S. Alamouti. A simple transmit diversity technique for wireless communications. *IEEE J. Sel. Areas Commun.*, 16(8):1451–1458, 1998.

[2] E. Aryafar, N. Anand, T. Salonidis, and E. W. Knightly. Design and experimental evaluation of multi-user beamforming in wireless LANs. In *ACM MobiCom*, Chicago, IL, 2010.

[3] H. Balan, A. Michaloliakos, R. Rogalin, G. Caire, and K. Psounis. Efficient MAC for distributed multiuser mimo systems. In *Information Theory and Applications Workshop*, La Jolla, CA, Feb. 2012.

[4] A. Bennatan, D. Burshtein, G. Caire, and S. Shamai. Superposition coding for side-information channels. *Information Theory, IEEE Transactions on*, 52(5):1872 – 1889, may 2006.

[5] F. Boccardi, F. Tosato, and G. Caire. Precoding schemes for the mimo-gbc. In *Communications, 2006 International Zurich Seminar on*, pages 10 –13, 0-0 2006.

[6] G. Caire, N. Jindal, M. Kobayashi, and N. Ravindran. Multiuser MIMO achievable rates with downlink training and channel state feedback. *IEEE Trans. Inf. Theory*, 56(6):2845–2866, 2010.

[7] G. Caire and S. Shamai. On the achievable throughput of a multiantenna gaussian broadcast channel. *IEEE Trans. Inf. Theory*, 49(7):1691 – 1706, Jul. 2003.

[8] B. Chen, Z. Zhou, Y. Zhao, and H. Yu. Efficient error estimating coding: feasibility and applications. In *ACM SIGCOMM*, pages 3–14, New Delhi, India, 2010.

[9] J. I. Choi, M. Jain, K. Srinivasan, P. Levis, and S. Katti. Achieving single channel, full duplex wireless communication. In *IEEE MobiCom*, Chicago, IL, 2010.

[10] M. Costa. Writing on dirty paper (corresp.). *IEEE Trans. Inf. Theory*, 29(3):439–441, May 1983.

[11] M. Duarte, C. Dick, and A. Sabharwal. Experiment-driven characterization of full-duplex wireless systems. *CoRR*, abs/1107.1276, 2011.

[12] U. Erez, S. Shamai, and R. Zamir. Capacity and lattice strategies for canceling known interference. *Information Theory, IEEE Transactions on*, 51(11):3820 – 3833, nov. 2005.

[13] U. Erez and S. ten Brink. A close-to-capacity dirty paper coding scheme. *Information Theory, IEEE Transactions on*, 51(10):3417 –3432, oct. 2005.

[14] J. Forney, G.D. and M. Eyuboglu. Combined equalization and coding using precoding. *Communications Magazine, IEEE*, 29(12):25 –34, dec. 1991.

[15] S. Gollakota, S. D. Perli, and D. Katabi. Interference alignment and cancellation. In *ACM SIGCOMM*, Barcelona, Spain, 2009.

[16] T. Gou, C. Wang, and S. Jafar. Aiming perfectly in the dark - blind interference alignment through staggered antenna switching. In *IEEE GLOBECOM*, Dec. 2010.

[17] A. Gudipati and S. Katti. Strider: automatic rate adaptation and collision handling. In *ACM SIGCOMM*, Toronto, Ontario, Canada, 2011.

[18] K. C.-J. Lin, S. Gollakota, and D. Katabi. Random access heterogeneous mimo networks. In *ACM SIGCOMM*, pages 146–157, Toronto, Ontario, Canada, 2011. ACM.

[19] X. Liu, A. Sheth, M. Kaminsky, K. Papagiannaki, S. Seshan, and P. Steenkiste. Pushing the envelope of indoor wireless spatial reuse using directional access points and clients. In *IEEE MobiCom*, Chicago, Illinois, USA, 2010.

[20] A. Molisch. *Wireless communications*. Wiley-IEEE, 2005.

[21] J. Perry, H. Balakrishnan, and D. Shah. Rateless spinal codes. In *ACM HotNets*, Cambridge, Massachusetts, 2011.

[22] J. Proakis and M. Salehi. *Digital communications*. McGraw-Hill, New York, NY, 2007.

[23] H. Rahul, H. Hassanieh, and D. Katabi. SourceSync: a distributed wireless architecture for exploiting sender diversity. In *ACM SIGCOMM*, New Delhi, India, 2010.

[24] Rice University. Rice university WARP project.

[25] T. Schmidl and D. Cox. Robust frequency and timing synchronization for ofdm. *Communications, IEEE Transactions on*, 45(12):1613 –1621, dec 1997.

[26] A. Shokrollahi. Raptor codes. *IEEE Trans. Inf. Theory*, 52(6):2551 –2567, Jun. 2006.

[27] K. Tan, J. Fang, Y. Zhang, S. Chen, L. Shi, J. Zhang, and Y. Zhang. Fine-grained channel access in wireless LAN. In *ACM SIGCOMM*, New Delhi, India, 2010.

[28] K. Tan, H. Liu, J. Fang, W. Wang, J. Zhang, M. Chen, and G. M. Voelker. Sam: enabling practical spatial multiple access in wireless lan. In *IEEE MobiCom*, pages 49–60, Beijing, China, 2009.

[29] S. Vishwanath, N. Jindal, and A. Goldsmith. Duality, achievable rates, and sum-rate capacity of gaussian mimo broadcast channels. *Information Theory, IEEE Transactions on*, 49(10):2658 – 2668, oct. 2003.

[30] C. Windpassinger, R. Fischer, T. Vencel, and J. Huber. Precoding in multiantenna and multiuser communications. *Wireless Communications, IEEE Transactions on*, 3(4):1305 – 1316, Jul. 2004.

[31] T. Yoo and A. Goldsmith. On the optimality of multiantenna broadcast scheduling using zero-forcing beamforming. *IEEE J. Sel. Areas Commun.*, 24(3):528 – 541, Mar. 2006.

Argos: Practical Many-Antenna Base Stations

Clayton Shepard[1], Hang Yu[1], Narendra Anand[1], Li Erran Li[2], Thomas Marzetta[2], Richard Yang[3], and Lin Zhong[1]

[1]Rice University, Houston, TX {cws, hang.yu, nanand, lzhong}@rice.edu
[2]Bell Labs, Murray Hill, NJ {erranlli, tlm}@research.bell-labs.com
[3]Yale University, New Haven, CT yry@cs.yale.edu

ABSTRACT

Multi-user multiple-input multiple-output theory predicts manyfold capacity gains by leveraging many antennas on wireless base stations to serve multiple clients simultaneously through multi-user beamforming (MUBF). However, realizing a base station with a large number antennas is non-trivial, and has yet to be achieved in the real-world.

We present the design, realization, and evaluation of *Argos*, the first reported base station architecture that is capable of serving many terminals simultaneously through MUBF with a large number of antennas ($M \gg 10$). Designed for extreme flexibility and scalability, Argos exploits hierarchical and modular design principles, properly partitions baseband processing, and holistically considers real-time requirements of MUBF. Argos employs a novel, completely distributed, beamforming technique, as well as an internal calibration procedure to enable implicit beamforming with channel estimation cost independent of the number of base station antennas. We report an Argos prototype with 64 antennas and capable of serving 15 clients simultaneously. We experimentally demonstrate that by scaling from 1 to 64 antennas the prototype can achieve up to 6.7 fold capacity gains while using a mere 1/64th of the transmission power.

Categories and Subject Descriptors

C.2.1 [**Computer-Communication Networks**]: Network Architecture and Design—*Wireless Communication*

Keywords

Large-scale Antenna Systems (LSAS), Many-Antenna, Massive MIMO, Multi-User MIMO, Beamforming, Conjugate, MRT, Zero-forcing

1. INTRODUCTION

Due to the popularization of smartphones, tablets and data-hungry applications, mobile data traffic is growing exponentially, with the expectation that it will increase 18-fold within 5 years [7]. In response, wireless operators are scrambling to acquire more spectrum resources and deploy more base stations to increase spatial reuse. However, there is a fundamental spectrum efficiency limit to existing cellular network architectures: they are *single-user* systems. That is, a base station serves only one terminal in a given resource block (i.e., time slot, spectrum channel, or code sequence). Information theory shows that this limit can be overcome through multi-user multiple-input multiple-output (MU-MIMO) [9]; one promising form of MU-MIMO is called multi-user beamforming (MUBF). With MUBF, a base station employs multiple antennas to send independent data streams to multiple terminals in the same resource block, effectively improving spatial reuse. As theory shows, the more antennas a base station has, the more terminals it can serve simultaneously, resulting in higher spectral capacity. Not surprisingly, the theory community is envisioning MUBF base stations with hundreds of antennas.

However, building a MUBF base station with many antennas is non-trivial. Scaling up baseband processing, clock distribution, transmission synchronization, and channel estimation raises serious system challenges. As a result, only testbeds with a few antennas have been reported in the literature, e.g., [5, 14]. Emerging wireless standards are similarly restricted to a small number of antennas and terminals. The key question to the proposal of MUBF base stations with many antennas remains: *is it practical at all?*

In this work, we answer this question affirmatively with *Argos*[1], a flexible base station architecture that is scalable up to thousands of antennas and able to serve tens of terminals simultaneously through MUBF. Using commercial off-the-self radio modules, i.e., the WARP platform [4], we have realized an Argos prototype with 64 antennas that is capable of serving 15 terminals through zero-forcing and conjugate MUBF. Extensive experimental characterization using this prototype shows that the spectral capacity increases from 12.7 bps/Hz when using a single-antenna to 85 bps/Hz for Argos employing zero-forcing MUBF, and to 38 bps/Hz for Argos employing the less computationally intensive conjugate MUBF, while using a mere 1/64th of the single-antenna transmission power. We show that the spectral capacity grows nearly linearly with the number of base station antennas and the number of simultaneously served terminals, as suggested by theory. The scale of our prototype and experimentation are only limited by the number of WARP boards that are available to us. To the best of our knowledge, Argos is the first publicly reported many-antenna MUBF base station design and realization ($M \gg 10$). Our work demonstrates the feasibility of the MUBF theory community's proposal, and presents key design principles for a scalable, flexible, and cost-effective realization.

Argos achieves its scalability and flexibility with four novel design principles. (*i*) First, Argos adopts a hierarchical and modular design. This allows it to scale up easily by incrementally adding modules, e.g., WARP boards in the

[1]Argos is a giant with 100 eyes in Greek mythology. The great vision of Argos is analogous to the improved capacity of our many-antenna base station.

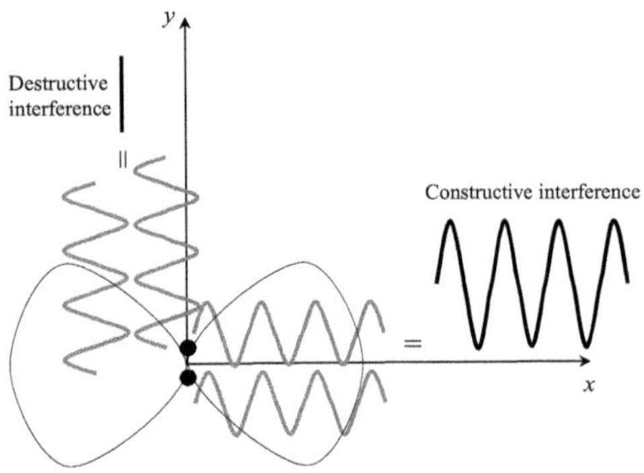

Figure 1: Aerial view of two antennas, represented by two dark dots, emitting identical sine waves at the same frequency. The two waves perfectly reinforce each other along the x axis (constructive interference) but completely cancel each other out along the y axis (destructive interference). Between the two axes the interference gradually varies, producing conical radiation know as a beam pattern.

reported prototype. As Argos scales up it can select the optimal beamforming algorithm by thoroughly analyzing the performance factors and data dependencies of various MUBF techniques. (*ii*) Second, Argos intelligently partitions computation tasks among the different modules in the hierarchy. In the downlink, data to multiple terminals is broadcast to all antennas. Each antenna locally applies its beamforming weights and transmits the combined signal to all terminals simultaneously. In the uplink, I and Q samples from each antenna are combined in upstream modules along the hierarchy. (*iii*) For very large scale operation, Argos leverages a modified version of conjugate beamforming that allows localized weight computation at each antenna. Specifically, traditional conjugate beamforming requires centralized transmission power normalization, while Argos conducts the normalization *locally* at each antenna. This modification allows Argos to scale almost indefinitely with regard to baseband complexity. (*iv*) Finally, Argos employs a novel *internal* calibration procedure that allows implicit beamforming across a large number of base station antennas without explicit channel state information (CSI) estimation, enabling the real-time CSI estimation overhead to become independent of the number of base station antennas. Notably, implicit beamforming requires time division duplex (TDD) operation, which is a substantial modification to the frequency division duplex (FDD) systems primarily used in cellular networks currently.

In summary, we make the following contributions to advance the state of the art of MUBF with many antennas:

- We design and realize Argos, a first-of-its-kind base station architecture that can scale up to thousands of antennas serving tens of terminals with either conjugate or zero-forcing MUBF. We report an Argos prototype with 64 antennas simultaneously serving 15 terminals;

- Using the Argos prototype, we experimentally demonstrate the real-world feasibility of base stations of many-antenna MUBF and its capability to significantly improve capacity;

- The design of Argos contributes multiple novel techniques to address key challenges toward realizing base stations with a large number of antennas, including clock distribution, transmission synchronization, localized weight computation, and channel calibration.

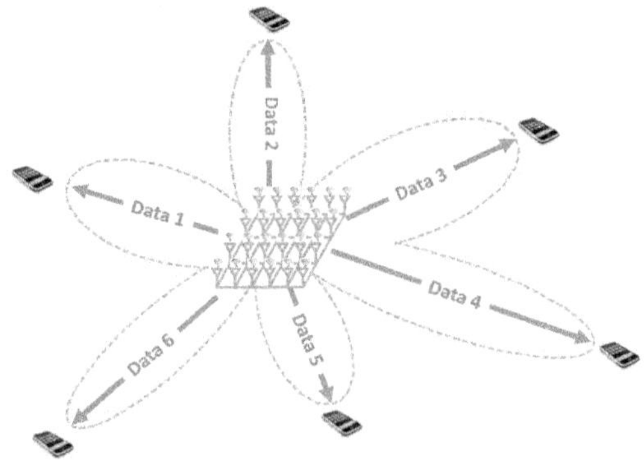

Figure 2: Multi-user beamforming employs baseband precoding and multiple antennas to send independent data streams to multiple terminals at the same time.

In the rest of this paper, we provide the background in Section 2. We present the design and implementation of Argos in Sections 3 and 4, respectively. In Section 5 we evaluate the real-world performance of Argos. In Sections 6 and 7 we discuss related and future work, respectively, and then conclude in Section 8.

2. BACKGROUND

We first provide some background on multi-user beamforming (MUBF) and highlight the key benefits and challenges of using a large number of antennas on base stations.

2.1 Beamforming Basics

Beamforming utilizes multiple antennas transmitting at the same frequency to realize directional transmission. Due to constructive and destructive interference of signals from multiple transmission antennas, the signal strength received at different directions varies spatially, leading to a *beam pattern*, as shown in Figure 1. This beam pattern can be altered by changing the beamforming *weights* applied to each antenna, effectively altering the amplitude and phase of the signal sent from that antenna. *Closed-loop* beamforming employs CSI to calculate the beamforming weights that maximize the signal strength at intended receivers and minimize the interference at unintended ones.

2.2 Single and Multi-user Beamforming

There are two major categories of closed-loop beamforming: *Single-user beamforming* (SUBF) and *Multi-user beamforming* (MUBF). SUBF maximizes the signal strength at a single intended receiver by using beamforming weights that are the complex conjugate of the CSI, which is also known as maximum ratio transmission [15]. MUBF concurrently transmits multiple data streams, each to a different intended receiver as shown in Figure 2. Not surprisingly, information theoretical studies have shown that MUBF can improve spectral capacity manyfold due to its spatial multiplexing gain.

There are many baseband techniques to realize MUBF. We focus on *linear precoding* since other methods are computationally infeasible for practical systems. Let **s** denote a $K \times 1$ vector representing the data-bearing symbols to K users. Linear precoding creates a transmission vector $\mathbf{s}'$ for M antennas, by multiplying the original data vector $\mathbf{s}$ by a $M \times K$ matrix $\mathbf{W}$: $\mathbf{s}' = \mathbf{W} \cdot \mathbf{s}$. Where $\mathbf{W}$ consists of the beamforming weights.

In this work, we study two important forms of linear-precoding for MUBF: *conjugate* and *zero-forcing*. Let $\mathbf{H}$ denote the $M \times K$ channel matrix between the M base station antennas and K concurrent terminals. Let c denote a constant chosen to satisfy a transmission power constraint.

Conjugate: $\mathbf{W} = \mathbf{W}_{conj} = c \cdot \mathbf{H}^*$, where $\mathbf{H}^*$ is the complex conjugate of $\mathbf{H}$. In other words, conjugate beamforming simply takes the complex conjugate of each channel coefficient in $\mathbf{H}$ as the beamforming weight, normalized by c. Indeed, it can be viewed as simultaneous single-user beamforming to K terminals by aggregating the signals intended for these terminals. Conjugate MUBF is sub-optimal and may not perform well with a small M due to inter-terminal interference. This method has only been recently proposed for MUBF with a large number of antennas in [17].

Zero-forcing: $\mathbf{W} = c \cdot \mathbf{W}_{zf} = \mathbf{H}^* \left(\mathbf{H}^{\mathrm{T}}\mathbf{H}^*\right)^{-1}$. Zero-forcing beamforming employs the CSI to precode the data-bearing symbols so that they sum to zero, or a 'null', at unintended receivers. The effectiveness of zero-forcing has been experimentally demonstrated recently [5] with a small number of antennas (four) and terminals (four). Zero-forcing MUBF can keep inter-terminal interference to zero if $K \leq M$. However, due to the required matrix inversion the computational overhead quickly becomes infeasible for real-time applications, as will be discussed in Section 2.4.2.

2.3 Benefits of Many-Antenna MUBF

It is well known in information theory that MUBF with many antennas provide the following key benefits:

First, MUBF can greatly improve spectral capacity through spatial reuse. Roughly speaking, the spectral capacity gain from MUBF is $\min(M, K)$ [9]. A large M allows the base station to serve more terminals concurrently and therefore achieve higher spectral capacity.

Second, a very large M allows a more power-efficient and cost-effective base station. The directional gain from using a large M can be used to compensate for reduced transmission power; that is, a base station can achieve the same capacity with a much lower total transmission power. Under all conceivable propagation conditions doubling the number of base station antennas permits the total radiated power to be reduced by a factor-of-two with no degradation of performance. Only when the number of antennas grows so large that it begins to envelope the terminals or intervening scatterers will this effect cease. Moreover, multi-user beamforming distributes the total transmission power across M antennas, leading to a much lower transmission power per antenna. The base station can therefore leverage cheaper power amplifiers and simpler RF filters. This eliminates the need for active cooling, further reducing power consumption and total cost.

Finally, since power gains are reciprocal, the preceding benefit also applies to terminals. Specifically, it allows battery-constrained terminals to use much lower transmission power to achieve higher capacity.

In Section 5, we will experimentally demonstrate these benefits using the Argos design.

2.4 Challenges to Many-Antenna MUBF

Realizing the key benefits outlined above is, however, non-trivial. Any implementation of MUBF with many antennas faces fundamental timing constraints imposed by the coherence time of the physical wireless channel. MUBF must collect CSI for each terminal then use it to calculate the beamforming weights within a small fraction of the coherence time. Additionally, the computational complexity of MUBF weight calculation grows with the number of antennas, M, and the number of simultaneously served terminals, K. The Argos design has to address both challenges.

2.4.1 CSI Estimation

Acquisition of CSI fundamentally limits the capacity of MUBF with many antennas. MUBF with M antennas to serve K terminals requires CSI between every base station antenna and terminal, or $M{\cdot}K$ channels. Importantly, all $M{\cdot}K$ physical channels must be assessed within a period much shorter than the channel coherence time in order to be useful. The coherence time of a wireless channel depends on how quickly the terminal and environment move. In cellular systems this is typically on the order of a few milliseconds, but can drop below 500 μs [1] with vehicular mobility at or near the terminal. This results in a fundamental tradeoff between the time spent collecting CSI, which dictates how many users can be served simultaneously, and the time allocated to sending beamformed data to those users. This tradeoff is explored theoretically in [16].

Traditionally, CSI is estimated *explicitly*. That is, each base station antenna broadcasts a pilot to the terminal, where the latter then uses this pilot to estimate its channel to each of the base station antennas. In order for this channel estimation to be useful, it has to be fed back to the base station in order to perform downlink beamforming. The reverse of this procedure is then used to find uplink CSI, though feedback is unnecessary for maximum ratio combining at the base station. This method thus requires $O(M{+}K)$ time to send pilots (one pilot from each base station antenna and terminal) and $O(M{\cdot}K)$ estimates that need to be sent back over-the-air (M estimates from K terminals). This overhead is unavoidable in frequency division duplex (FDD) systems, since the physical channel is not reciprocal at different frequencies.

In time division duplex (TDD) systems the physical channel is reciprocal, and thus, theoretically, CSI could be estimated *implicitly*. That is, uplink pilots could be used to perform downlink beamforming, reducing channel estimation overhead to $O(K)$ and eliminating the required feedback. This is often called *implicit beamforming*. However, in practice, the uplink and downlink channels consist of not only the physical channel, but also the channels introduced by the active RF components in the transmit and receive hardware, as will be further discussed in Section 3.3.

2.4.2 Real-time Beamforming Weight Calculation

The computational complexity of MUBF weight calculation also grows with the number of base station antennas, M, and the number of terminals, K. For conjugate MUBF, the beam weight computation is trivial. In hardware, taking the complex conjugate of a signal only needs a bit-flip and an adder. Therefore, the delay introduced by weight calculation is negligible. However, zero-forcing requires the computation of a matrix inverse, a calculation with the complexity of $O(M \cdot K^2)$. Moreover, the inverse algorithm has internal data-dependencies that limit its ability to be parallelized. While the incurred latency is acceptable at small scales, the polynomial time nature of the inverse makes it very challenging for MUBF systems with a large number of antennas. For example, we estimate that a *single* 15 by 15 matrix inverse would require approximately 150 μs using a specialized high performance FPGA implementation reported in [8]. 150 μs is already 30% of the 500 μs coherence time specified by the LTE channel model. Moreover, in a wideband system such as LTE, this inversion has to be performed for every 14 subcarriers [17]. Thus, while these computations may be pipelined, the true overall inversion time incurred will be far greater than 150 μs.

Additionally, existing beamforming techniques incur a high data transmission overhead because the channel estimates and beam weights have to be exchanged between a central controller and each of the antennas. Even using state-of-the art hardware, e.g., InfiniBand, such exchange incurs a sizable latency cost. The fastest InfiniBand bus has 1 μs overhead per hop and 40 Gbps transfer rate [3]; it will incur approximately 5 μs delay per subcarrier group in a 15 by 15 system. For a 20 MHz bandwidth this amounts to over a 700 μs delay. Zero-forcing cannot avoid this data exchange because the inverse calculation requires the full CSI matrix, $\mathbf{H}$. More subtly, even the simplest beamforming algorithm, conjugate, requires full knowledge of $\mathbf{H}$ in order to appropriately scale the power of the steering weights. In Section 3.4, we present a novel localized conjugate beamforming method that eliminates the overhead due to data transfer between the central controller and antennas.

3. DESIGN

The key question we ask in this section is: *how do we design a MUBF base station that can flexibly optimize its architecture over a wide range of M and K?* Before proceeding to answer it, we want to highlight its practical interest: realistic wireless networks often have large variations in many of their properties, including the financial budget for the base stations, the terminal population within the coverage, and the data traffic volume from terminals. Traditional base stations can only scale their transmission power or, equivalently, their cell size, to cope with such variations. In contrast, Argos base stations can also scale the number of antennas to accommodate various deployment needs.

We argue that in order to meet these demands our many-antenna base station must: (*i*) be economically affordable with cost proportional with M, (*ii*) scale as both M and K become very large, and (*iii*) select the optimal beamforming technique given deployment requirements. We next present how our design of Argos accomplishes these attributes.

3.1 Scalability

The first question is: *can MUBF scale up with M, the number of base station antennas?* MUBF entails three distinct phases: channel estimation, weight calculation, and linear precoding. We explore the feasibility and design implications of these as M scales up.

3.1.1 Channel Estimation

Explicit channel estimation *does not* scale well with M or K. As discussed in Section 2.4.1, explicit channel estimation typically requires $M + K$ pilots to be sent, and $M \cdot K$ estimates to be fed back to the base station. This is clearly an unacceptable overhead for large-scale systems, and suggests that Argos *must* employ TDD reciprocity and implicit beamforming to reduce this overhead to K pilots and eliminate the feedback. In order to enable this, however, we must first overcome the asymmetries introduced by the RF hardware. To accomplish this we devise a novel *internal* calibration scheme, which we present in Section 3.3.

3.1.2 Beamforming Methods

Unfortunately, existing beamforming methods are distinctly unscalable, as they all have centralized data requirements and typically have polynomial time complexity, as discussed in Section 2.4.2.

In light of this, we propose a novel beamforming method that allows weights to be computed completely locally, at each base station radio, as described in Section 3.4. Leveraging this method allows additional radios to be added *without* requiring additional bandwidth, enabling Argos to easily scale up to an unprecedented number of base station antennas, e.g., 1000s.

However, while this beamforming method performs well with a very large number, e.g., 100s, of base station antennas serving 10s of terminals simultaneously, it is well known to be sub-optimal for smaller scale systems, e.g., $M = 30, K = 10$. We demonstrate this empirically in our results, Section 5, where we find that zero-forcing results in up to a 4 fold capacity increase over our method. However, this does not account for the data transport and computational overhead of zero-forcing, which becomes prohibitive with a large number of users or high mobility, as described in Section 2.4.2. Thus we conclude that in order to scale optimally Argos must support traditional, centralized beamforming techniques for smaller scale deployments.

3.1.3 Linear Precoding

Linear precoding requires each antenna to transmit a data stream that is the linear combination of K data streams with K beamforming weights. One design option is to apply these weights centrally. Since each antenna transmits a distinct data stream, this would require the central controller to deliver M I and Q sample streams to each of the individual radios. This approach, obviously, does not scale well, since it requires the central controller to have an output bandwidth proportional to M. As M increases to hundreds or even thousands, this becomes exorbitantly expensive and eventually intractable. Thus we conclude that in any efficient scalable design, the beamforming weights should be applied at the radio. This design choice conveniently allows all of the radios to share a common databus for downlink transmission. In contrast, for uplink transmission, the radio leverages the same linear precoding to apply K beamforming weights to the incoming I and Q samples. Since each radio has unique weights, this again results in M unique data streams (that are K wide)! Fortunately, linear precoding requires these streams to simply be added together; conveniently, this can be done anytime two streams merge in the architecture, thus, again, enabling a constant bandwidth databus. Indeed, we see that with careful design decisions linear precoding *can* scale up with constant data rate requirements. Notably, there is still a need for some form of central controller to demodulate the data once it has been

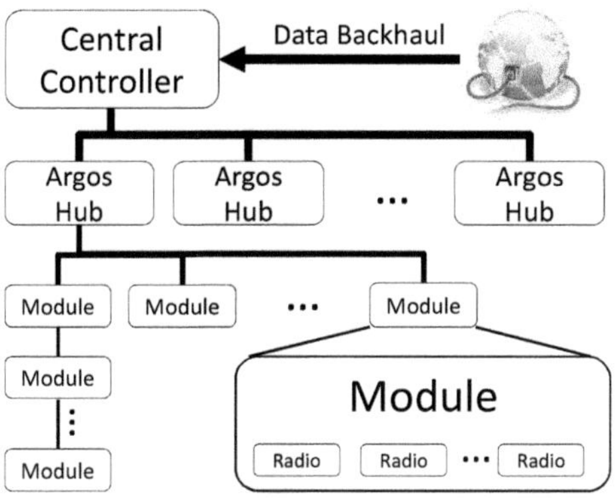

Figure 3: Argos architecture: fat tree structure with daisy-chained leaf nodes.

completely recombined; however this operation is latency insensitive, and computationally trivial.

Thus we find that, yes, MUBF *can* scale up with M, but only with careful design choices and new methods for weight calculation and channel estimation.

3.2 Architecture and Topology

The design choices to enable scalability presented above result in two distinct components: (*i*) a central controller, which handles modulation and demodulation, and (*ii*) the M radio front-ends that locally calculate beam weights and apply linear precoding. The immediate question we need to answer is: how do we interconnect the controller and the radios? On one hand, we can connect all the radios directly to the controller. This requires the controller to have at least M ports. Since M can be dynamic and very large, this would be a unscalable and inefficient design choice. On the other hand, we can daisy-chain all the radios serially. While scalability seems to be maximized, reliability and delay of the system are severely compromised.

Our solution is to add hierarchies to the base station to improve flexibility, and simultaneously achieve a balance between scalability, reliability, and delay. But, what type of hierarchical structure should we adopt? First we note that deploying M separate radios and antennas would be unwieldy, and cost ineffective to manufacture; thus we create our first level hierarchy: a module that contains one or more radio front-ends. Next, in order to achieve flexible, cost-effective, scaling we allow these modules to be connected serially; enabling additional modules to be added atomically with low overhead. Finally, in order to increase reliability and reduce end-to-end latency, we introduce the Argos hub, which allows multiple modules to be connected in parallel. Figure 3 depicts the Argos architecture.

The Argos base station enables unprecedented scalability and deployability, while fulfilling performance and cost constraints. This architecture enables the Argos base station to scale in three directions: by adding more Argos hubs, by increasing the length of the module chains, and by increasing the number of antennas on a module. The hierarchal architecture facilitates deployments of base stations with many antennas to be flexibly distributed geographically by using a single link to an Argos hub, as well as deployments of base stations with a small number of antennas where the hub can be omitted completely, and modules are simply chained together in series. Additionally, if chains become too long to meet latency requirements, Argos hub can simply be added to parallelize connections and reduce latency.

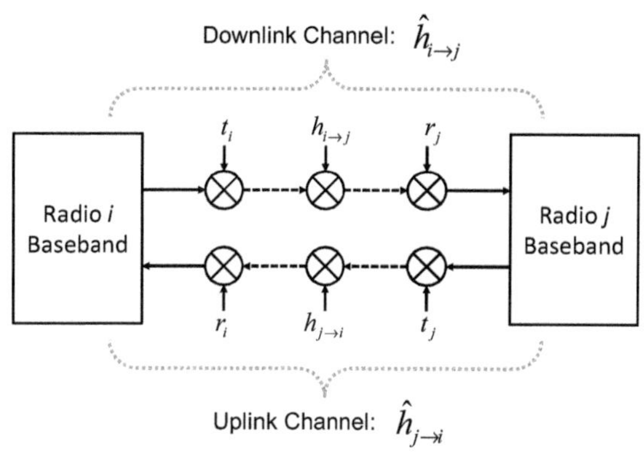

Figure 4: Real channels are not reciprocal due to the differences in transmit and receive hardware. Note that channel reciprocity indicates that within the channel coherence time the physical channel *is* reciprocal: $h_{i\to j} = h_{j\to i}$.

3.3 Channel Calibration

We devise a novel, completely internal, calibration procedure to enable implicit beamforming on many-antenna base stations through TDD channel reciprocity in order to collect CSI data in constant time with respect to M.

For an M antenna base station to multi-user beamform to K terminals, the base station must acquire the downlink channel state information, $\hat{h}_{m\to k}$, for all $m = 1, 2, ..., M$ and $k = 1, 2, ..., K$. The key challenge is to estimate the effective downlink CSI $\hat{h}_{m\to k}$ from the uplink CSI, $\hat{h}_{k\to m}$, acquired from the uplink pilot signals. However, as shown by Figure 4, the uplink and downlink channels are *not* reciprocal due to the random phase and amplitude differences in the RF hardware. This is caused by a combination of dynamic effects from internal clocking structures, such as dividers, multipliers, and PLLs, as well as static effects from manufacturing deviations. Indeed, we verify that simply resetting a given radio i, or even tuning to a different frequency, randomizes the phase effects of t_i and r_i.

The uplink and downlink channels between any two transceivers is a product of (*i*) the frequency response of the transmit hardware, (*ii*) the physical wireless channel, and (*iii*) the frequency response of the receive hardware:

$$\hat{h}_{i\to j} = t_i \cdot h_{i\to j} \cdot r_j \tag{1}$$

In order to estimate the reciprocal channel, $\hat{h}_{j\to i}$, we define a *calibration coefficient*, $b_{i\to j}$, between radios i and j as:

$$b_{i\to j} = \frac{\hat{h}_{i\to j}}{\hat{h}_{j\to i}} = \frac{t_i \cdot h_{i\to j} \cdot r_j}{r_i \cdot h_{j\to i} \cdot t_j} = \frac{t_i \cdot r_j}{r_i \cdot t_j} = \frac{1}{b_{j\to i}} \tag{2}$$

Notably, if both channels are measured within the coherence time then $h_{j\to i} = h_{i\to j}$ due to physical channel reciprocity. Clearly, if we know the calibration coefficient between two radios and one channel estimate, we can find the reciprocal channel:

$$\hat{h}_{i\to j} = b_{i\to j} \cdot \hat{h}_{j\to i} \quad \text{or} \quad \hat{h}_{j\to i} = \frac{\hat{h}_{i\to j}}{b_{i\to j}} \tag{3}$$

Now let's apply this to our scenario where we would like to estimate the downlink CSI from base station antenna m to terminal k ($\hat{h}_{m\to k}$) from the uplink CSI ($\hat{h}_{k\to m}$). To do this we must know the M calibration coefficients between each base station antenna and the terminal, that is, all $b_{m\to k}$. These would be impractical to find in a real-system, as estimating $b_{m\to k}$ requires pilots to be sent between every base station antenna and terminal pair, as well as feedback from each terminal. Moreover, unless the terminal and base station share clocks, which is impossible in a wireless system, their hardware transmit and receive channels drift relatively over time, thus requiring this calibration to happen frequently. This approach would be counter-productive, since estimating $b_{m\to k}$ requires downlink pilots, which could be used to directly estimate $\hat{h}_{m\to k}$.

3.3.1 Internal Calibration

We find that it is possible to *internally* calibrate the base station relative to one of it's antennas, e.g., antenna 1. That is, we find all calibration coefficients $b_{m\to 1}$ (for $m = 2, 3, ...M$) using Equation 2. Note that these coefficients are in fact stable over long periods of time, as we show in Section 5.4, since all base station antennas share clocks. We also find that if we know the calibration coefficient between any two radios and a reference radio, then we can derive the direct calibration coefficient between them:

$$\frac{b_{i\to j}}{b_{i\to y}} = \frac{\frac{t_i \cdot r_j}{r_i \cdot t_j}}{\frac{t_i \cdot r_y}{r_i \cdot t_y}} = \frac{t_y \cdot r_j}{r_y \cdot t_j} = b_{y\to j} \tag{4}$$

Thus if we know the calibration coefficient between our reference antenna and terminal k, $b_{1\to k}$, we can find the downlink CSI:

$$\hat{h}_{k\to m} \cdot \frac{b_{1\to k}}{b_{1\to m}} = \hat{h}_{k\to m} \cdot b_{m\to k} = \hat{h}_{m\to k} \tag{5}$$

This suggests that full CSI can be found by simply sending one pilot from each of the terminals, then just one pilot from the base station's reference antenna! Unfortunately, however, to find $b_{1\to k}$ we must feedback the reference antenna's downlink channel estimate, $\hat{h}_{1\to k}$, from each of the k terminals. This significantly reduces the channel capacity, and quickly becomes infeasible for even a moderate K.

3.3.2 Key Idea: Relative Calibration

Our key idea in solving the calibration problem is that an *absolutely* accurate estimation of downlink CSI, $\hat{h}_{m\to k}$, is unnecessary. For all multi-user beamforming techniques using linear precoding, it is sufficient for beamforming antennas to have a *relatively* accurate estimation. That is, as long as each base station antenna's CSI estimation deviates from the real CSI by the same multiplicative factor, multi-user beamforming will still result in the same beam pattern. To visualize this, refer back to Figure 1; if both antennas were to experience the same phase offset, the resulting spatial beam pattern would remain the same. Thus, we can assume $b_{1\to k} = 1$:

$$\hat{h}_{m\to k} = \hat{h}_{k\to m} \cdot \frac{b_{1\to k}}{b_{1\to m}} \quad \Rightarrow \quad \hat{h}'_{m\to k} = \frac{\hat{h}_{k\to m}}{b_{1\to m}} = \hat{h}_{k\to m} \cdot b_{m\to 1} \tag{6}$$

This means that we estimate *relative* downlink CSI, $\hat{h}'_{m\to k}$, by using *only* uplink pilots, without any feedback! To recapitulate, the entire CSI collection process involves 4 steps:

1. Find all internal calibration coefficients, $b_{1\to m}$, offline by sending pilots to and from every base station antenna m and reference antenna 1.

2. Send K orthogonal pilots from each terminal and determine $\hat{h}_{k\to m}$.

3. Derive all $\hat{h}'_{m\to k}$ from 6.

4. Use $\hat{h}'_{m\to k}$ to calculate the beam weights, then send the beamformed data.

Using this process we can efficiently collect full channel state information at the base station by sending only K terminal pilots, without any feedback from the terminals. This enables us to scale M up without any additional channel estimation overhead, which is a critical feature in order to realize a MUBF system with many antennas.

Note that the measurements of downlink and uplink have to be done within the channel coherence time in order for $h_{m\to 1} = h_{1\to m}$. Since base station antennas do not move, the channel coherence time is much larger than typical base station to terminal coherence times. However, as we show in Section 4.5, this calibration can easily be done well within even highly mobile timing constraints; our prototype completes a single antenna pair calibration within 300 μs.

3.4 Decentralized Beamforming

In order to achieve scalable real-time beamforming weight calculation, Argos employs a novel method that allows weights to be calculated locally at each antenna, and therefore avoid the unscalable data-transport overhead required by existing beamforming techniques. As discussed in Section 2.4.2, to perform traditional conjugate beamforming, the weights must be globally normalized so that no base station radio exceeds its maximum power output. For example, assuming a maximum radio transmit amplitude of 1, and in order to ensure at least one radio transmits at maximum power:

$$c = \left(\max \left(\sum_{k=1}^{K} \|\hat{\mathbf{h}}_{m\to k}\| \right) \right)^{-1} \quad (m = 1, 2, ...M) \tag{7}$$

where c is the scaling factor used in the beamforming weight calculation ($\mathbf{W} = c \cdot \mathbf{H}^*$). Global power scaling is characterized by using a single constant to scale all of the weights. This global scaling is necessary to maintain the ratio between each base station antenna's weight for a given terminal, which ensures per-terminal transmission energy optimality, as proven in [15]. However, each base station antenna must know either c (or $\mathbf{H}$) to properly scale its own beamforming weights. This requires full CSI to be transferred from each module to the central controller, nullifying the benefit from the aforementioned decentralization. To tackle this, we propose a local power scaling approach that closely approximates global normalization.

Argos leverages a key observation that for the different terminals in multiuser beamforming, *the channels corresponding to different terminals are uncorrelated and experience independent fading*. Therefore, statistically speaking, when the number of terminals is large, the actual transmission power at each antenna is very similar. Our solution simply normalizes the total transmission power locally at each base

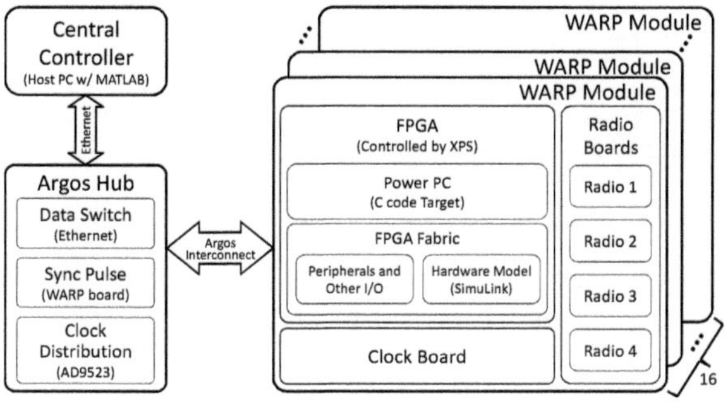

Figure 5: The implementation of Argos using WARP boards, a laptop, an ethernet switch, and an AD9523 based clock distribution board.

station antenna using only the CSI it measures:

$$c_m = \left(\sum_{k=1}^{K} \|\hat{\mathbf{h}}_{m \to k}\| \right)^{-1} \quad (m = 1, 2, ...M) \tag{8}$$

The conjugate beamforming weights are then scaled via:

$$\mathbf{W} = \mathbf{H}^* \cdot \text{diag}(C) \tag{9}$$

Where C is the scaling vector given by $C_{local} = [c_1, c_2, ...c_M]$, from Equation 8; notably the globally scaled conjugate can also be found in this form, using $C_{global} = [c, c, c, c...]$, from Equation 7.

We have experimentally verified the effectiveness of such local power scaling and observed that its performance is almost indistinguishable from the optimal global power scaling method (see Section 5), using equal transmit power for both methods. Moreover, in real deployments, since local power scaling ensures that each radio can utilize its full hardware power capacity, it can *always* achieve equal or greater SNR than global power scaling (since it can send with greater total transmit power), as proven in [22, p. 24]. Notably, if terminals are not approximately equidistant from the base station, then per-terminal power scaling is required to ensure fairness (preventing terminals closer to the base station from being allocated all of the transmission power), but this can be done at a much coarser time scale (i.e., seconds), thus not creating additional overhead or affecting performance.

4. IMPLEMENTATION

In this section we provide a detailed report of our implementation of Argos, which leverages WARP [4], commercially available clock distribution boards, a commodity PC, and an ethernet switch. Figure 5 shows an abstract representation of our implementation. As the first proof-of-concept prototype, our system includes a central controller, an Argos hub and 16 modules, each with 4 radios. The central controller consists of a single host PC, which uses MATLAB to send data, weights, and control commands to the radio modules. The Argos hub is comprised of a 24-port ethernet switch, a clock distribution board, and a WARP board, which uses its GPIO pins to provide transmission synchronization splitting/replication. Due to the limited availability of WARP boards, this board also serves as a radio module, however these roles are functionally separate, and in future generations of the Argos prototype they will be physically separated as well. Each radio module is a single WARP board with 4 radio daughter cards and 4 antennas. Figure 6 depicts the real system: the base station includes 16 WARP boards with 64 antennas that are compactly placed on a custom rack-mount platform. We note that the number of terminals supported by each module is fundamentally limited by its hardware capabilities. In the WARP boards we are using, this bottleneck is the number of multipliers (328 on the Virtex 2 Pro xc2vp70) [26]. We are able to use 240 of these multipliers to provide linear precoding for 15 terminals on the 4 antennas, which requires 60 complex multipliers. The remaining multipliers are used by other functions, and 4 are unusable due to routing constraints. However, the recently released Virtex 7 supports up to 3600 multipliers clocked at a rate of 741 MHz; with multiplexing this would enable 16,672 complex multiplies per 40 MHz sample (neglecting routing overhead and other functions that require multipliers), which would, obviously, alleviate this bottleneck [25].

To the best of our knowledge, our Argos prototype is the first publicly reported many-antenna MUBF system with real-world feasibility. We next elaborate on our implementation.

4.1 Hardware and Software Platform

WARP is a scalable and programmable wireless platform for prototyping advanced wireless systems. Each WARP board allows up to four radio daughter cards to be connected and therefore can contribute up to four active antennas simultaneously to Argos. Each radio board includes a Maxim 2829 transceiver chip [18], which operates at the 2.4 or 5 GHz ISM bands with a 20 MHz bandwidth. WARP conveniently provides a MATLAB-based framework, WARPLab, which allows MATLAB to control the WARP boards and process the transmit and receive data samples. As shown in Figure 5, WARPLab consists of four layers: (*i*) The underlying Simulink model that implements the custom hardware for controlling the FPGA board and radio boards; (*ii*) The Xilinx Platform Studio (XPS) project that integrates and connects all of the hardware components, including the Simulink model, the I/O cores for the serial port, Ethernet port, clocking, etc.; (*iii*) The C code that runs on the PowerPC microprocessor, controls the hardware through memory mapped I/O, and acts as an interface to the Ethernet port; (*iv*) The MATLAB interface that configures the boards, generates the transmit samples, and processes the receive samples.

We have extensively customized the WARPLab framework to enable hardware MUBF, transmission synchronization, clock synchronization, and indirect calibration among base station antennas. These functionalities are essential to for Argos to enable MUBF with many antennas.

4.2 Hardware MUBF

A straightforward, and much easier approach to realize MUBF in WARPLab is to implement it in software within the MATLAB interface; this, in fact, was our first implementation. In this approach the beamformed baseband signal can be directly delivered to the WARP boards without the need of linear-precoding in hardware. However, this method introduces major latency between the CSI collection and data transmission, which increases linearly with the number of base station antennas, and severely degrades performance. This is a result of the same scaling problem discussed in Section 3.1. Therefore, we modified the WARPLab hardware to enable hardware MUBF.

At each base station antenna, applying the beamform-

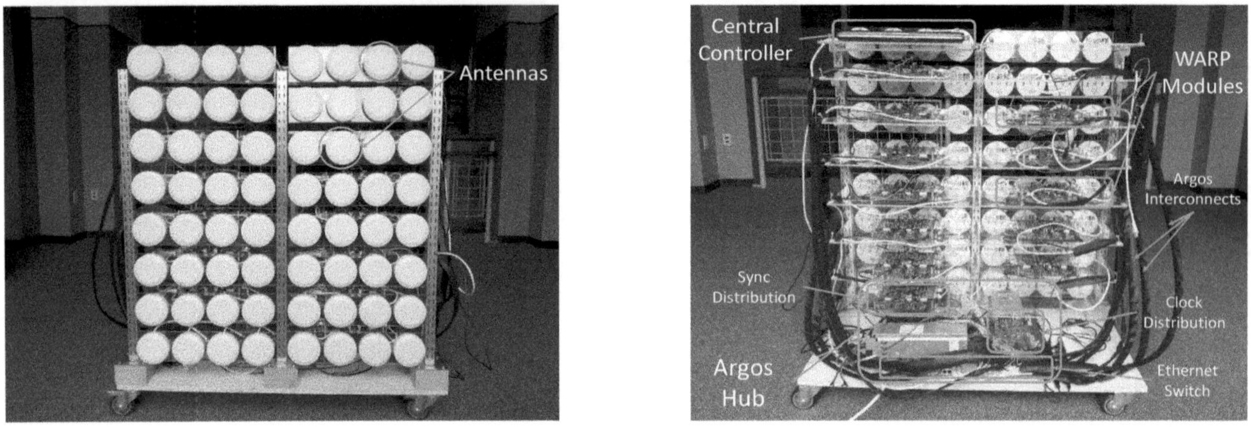

Figure 6: The prototype of Argos with 16 modules and 64 antennas. Left: front side. Right: back side.

ing weights consists of multiplying the baseband symbol intended for each terminal, s_k, by its corresponding beamforming weight, w_k, and then adding them together: $s'_m = \sum_{k=1}^{K} w_k \cdot s_k$ where s'_m is the resultant beamformed signal transmitted by antenna m. Multiplying the signal by a complex number is equivalent to rotating the phase and scaling the amplitude. In hardware, this requires K registers and K parallel complex multipliers (each complex multiplier needs 4 multipliers and 2 adders) in series with 2 K input adders. We store the beamforming weights, $w_k(k = 1, 2, ...K)$, in memory mapped registers. This is important since it enables the PowerPC, and in turn, the MATLAB interface to directly control them.

4.3 Transmission Synchronization

WARPLab has a default function to enable transmission synchronization between multiple WARP boards. It is achieved by using the built-in API command "sendsync()" in the MATLAB interface. However, due to the jitter introduced by the ethernet stack, switch, and cables, such synchronization can lead to a timing offset on the order of 20 samples, depending on the ethernet switch and cable lengths, which makes accurate CSI collection and beamforming impossible.

To address this challenge, we employ a WARP board to distribute the central controller's transmission synchronization signal. As part of the Argos hub, this WARP node leverages directly connected, registered, GPIO to reliably send the sync pulse to the radio modules. Notably, to ensure the modules receive the pulse within 1 clock cycle, the cable lengths should all be within one wavelength, λ. With a channel bandwidth of 20 MHz, λ is 7.5 meters (40 MHz sampling clock), which is a very easy constraint to meet. As stated above, this WARP node serves the dual role of sync distribution and module, thus it "distributes" the sync to itself with an effective cable length of 0. This means the other cables must be less than 7.5 meters, which is not a problem; in our current setup the length is 2 meters. While each board may have a slightly different clock phase, this phase offset is constant (due to the clock synchronization), and explicitly compensated for by the beamforming algorithm.

We have modified the Simulink model, the XPS project, and the C code to enable GPIO-based transmission synchronization. Specifically, we inserted appropriate gateways and registers into the Simulink model, re-mapped the GPIO pins to the appropriate signals in the XPS project, and disabled the traditional ethernet sync in the C code.

4.4 Clock Synchronization

Precise inter-board clock synchronization is critical for Argos, due to its distributed modular architecture. The WARP board requires two reference clocks: a 20 MHz RF clock and a 40 MHz logic/sampling clock. Both clocks can be either forwarded or driven by an external source. In addition, we discovered that the Maxim 2829 transceiver chip on the radio board can use a 40 MHz clock. Therefore, we were able to use a single external source to drive the logic clock, then forward the logic clock to the reference input for the RF clock. This way, inter-board clock synchronization can be achieved in an easily manageable and scalable way.

We leverage a commercial clock distribution evaluation board designed for LTE, the AD9523/PCBZ, to accomplish this. The AD9523 provides 18 clock outputs, which we leverage to drive all of the radio modules. Although we haven't exceeded the capacity of the AD9523, an additional clock distribution board could be connected (as part of an additional Argos hub), which would provide 17 more outputs. Alternatively, the existing modules can forward their clocks to additional modules, through Argos' multi-hop extension.

4.5 Indirect Calibration

For indirect calibration, we need to estimate $b_{m\to1} = \frac{t_m \cdot r_1}{r_m \cdot t_1}$ for each antenna m with respect to the "reference antenna," as described in Section 3.3. Due to buffer constraints, we implement this in a per-module iterative fashion. First, the module containing the reference antenna calibrates internally; that is, the reference antenna sends a pilot while other antennas on the module listen, then each of those antennas sends a pilot, in turn, while the reference antenna listens. These channel estimates are then reported to the central controller so that the reference antenna's buffer can be overwritten. Next, the reference antenna sends a pilot sequence while all the antennas on another module listen, then each of those antennas transmits a pilot, in turn, while the reference antenna listens. Again, the channel estimates are reported to the central controller. The process is then repeated for each module. The calibration procedure is very latency sensitive, as the physical channel should not change between transmission and reception of pilots for any antenna pair. To address this, we implement the calibration locally on the PowerPC in C code and leverage Argos' transmission synchronization to coordinate the send and receive phases. The resulting calibration happens within 300 μs for each antenna pair, which is well within the channel coherence time.

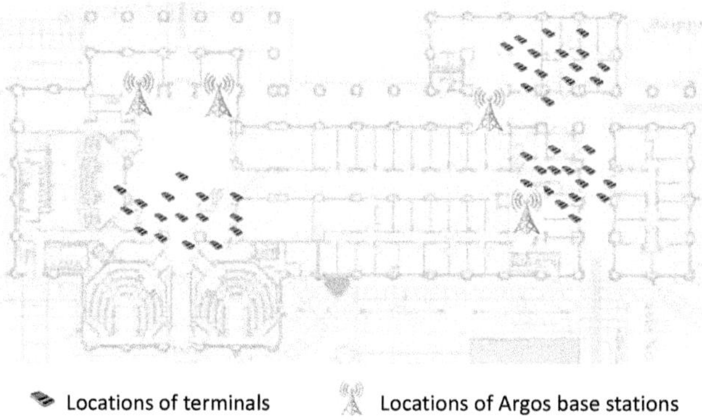

Figure 7: Environments and the locations of the base station and terminals for for the reported experiments. Note that the base station leverages directional antennas in order to serve one sector. Terminals have vertical separation as well, spanning up to three floors.

Another challenge we encountered while performing our indirect calibration approach is the significant amplitude variation for the channels between the reference antenna 1 and other antennas. This is due to the grid-like configuration of our antenna array where different pairs of antennas can have very different antenna spacings. According to our measurement, the SNR difference can be as high as 40 dB, leading to a dilemma for us to properly choose the transmission power for the reference signal. To address this, we isolate the reference antenna from the others, and place it in a position so that its horizontal distance to the other antennas are approximately identical. Such placement of the reference antenna does not affect the calibration performance due to our calibration procedure's isolation of the radio hardware channel from the physical channel.

5. EVALUATION

Leveraging our prototype, we experimentally evaluate the feasibility of Argos in realistic environments. We have the following impressive observation: compared to using a single antenna, Argos can improve spectral capacity over *12 fold* leveraging MUBF with many antennas, using *equal total transmission power.* With 64 antennas and 15 terminals, the spectral capacity can be boosted from 12.7 bps/Hz to 85 bps/Hz (6.7x) for zero-forcing MUBF, and 38 bps/Hz (3x) for conjugate MUBF, while *using a mere 1/64th of the total transmission power.* We find that Argos easily scales from 1 to 64 base station antennas serving 1 to 15 terminals, and that, in general, performance scales linearly with M and K. Finally, we experimentally validate the performance of our localized conjugate beamforming method, as well as our internal calibration procedure.

5.1 Experimental Setup

We employ all 64 antennas at the base station to perform MUBF to 15 concurrent terminals. We use the 2.4 GHz band with a 625 kHz carrier width to avoid frequency fading effects. Since it is relatively easy to move our platform (see Figure 6), we tested various indoor locations (see Figure 7) in order to collect data from diverse environments. There are both LOS and NLOS channels between the base station and terminals. We repeat our experiments multiple times, typically collecting over 3000 measurements at each location, to reliably average out performance.

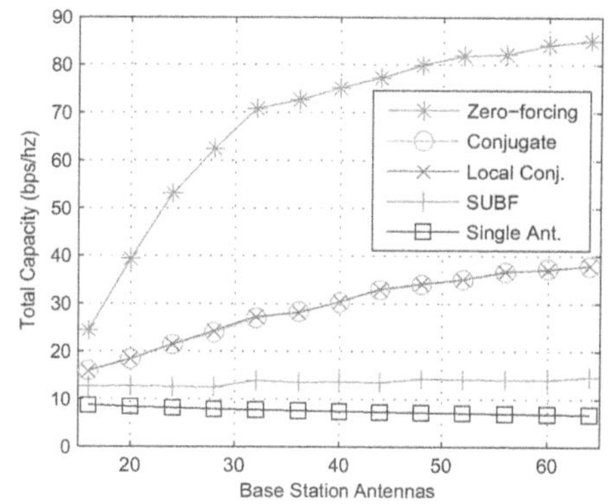

Figure 8: Cell capacity as the number of base station antennas (M) increases from 16 to 64, by 4, serving 15 terminals. In order to compensate for the beamforming gain, total transmission power is $1/M$, implying average power per-antenna for multi-antenna schemes is $1/M^2$.

To obtain the cell capacity, we aggregate the Shannon capacity for each terminal, or $C_{Cell} = \sum_{k=1}^{K} \log(1+SINR_k)$ where $SINR_k$ is the measured SINR at terminal k. We let the base station transmit dummy QPSK-modulated frames to the terminals, which is sufficient to validate the real-world feasibility of Argos since MUBF is a physical layer technique that is orthogonal to the MAC layer and above.

To accurately measure the terminal SINR, we use the RSSI indicator from the Maxim 2829 transceiver on the radio board to report the received signal strength for each transmission, as well as the noise floor after the transmission completes. Since the radio is unable to distinguish signal and interference strength, we slightly stagger the transmission to the intended terminal and that to the unintended terminals. This way we can separately measure the signal power and interference power, and acquire the SINR accordingly. To make sure the channel remains constant during the transmissions we conduct our experiments in an ultra-stable environment, i.e., late at night, without moving people and wireless traffic.

5.2 Improvement of Cell Capacity

The primary purpose of our experiments is to determine the capacity improvement of Argos, in order to ultimately answer the practicality of the many-antenna MUBF base station proposal from the theory community. We report two sets of experiments, which evaluate the scalability with regards to the number of base station antennas, M, and the number of terminals, K, respectively.

5.2.1 Scaling up with M

In the first set of experiments, we vary the number of base station antennas, M, assuming a fixed number of terminals, $K = 15$. Figure 8 shows C_{Cell} as a function of M for a base station with a single antenna (*Single Ant.*), SUBF (*SUBF*), conjugate MUBF (*Conjugate*), our modified localized conjugate MUBF (*Local Conj.*), and zero-forcing MUBF (*Zero-forcing*). In order to compensate for the beamforming gain, total transmission power is scaled by a factor of $1/M$ for all five cases. This enables our experiments to separate the orthogonality gain of scaling up from the well-known,

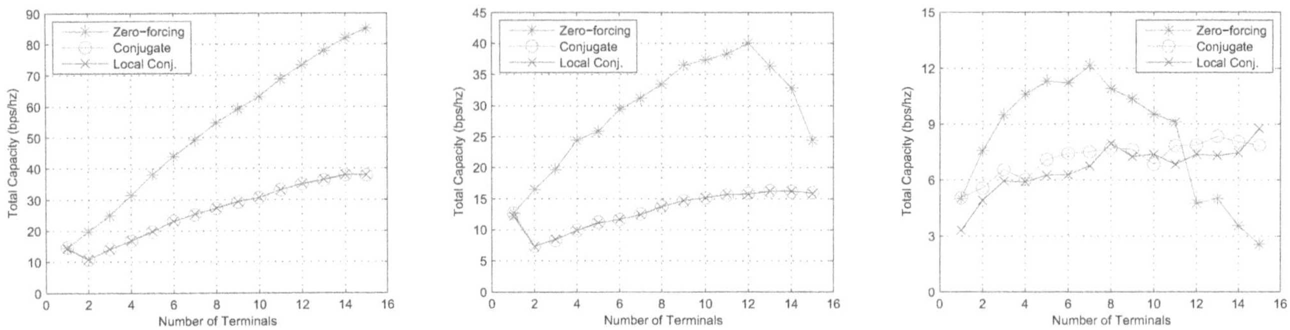

Figure 9: Cell capacity as the number of terminals K increases. Total transmission power is held constant within each plot. Left: $M = 64$; Middle: $M = 15$; Right: $M = 15$ with reduced transmission power.

predictable, beamforming gain. We have the following key observations:

First, when M is much larger than K, both conjugate and zero-forcing MUBF increase the cell capacity as M scales up, *despite reducing the total transmission power proportionally with M.* The beamforming gain from the additional antennas compensates for the power reduction, as demonstrated by the flat performance of SUBF, while simultaneously increasing the natural orthogonality of the terminals. This reduces the inter-terminal interference of conjugate MUBF, and reduces the amount of power wasted to create nulls for zero-forcing MUBF. With $M = 64$ the improvement for conjugate and zero-forcing MUBF over a single antenna are 5.7x and 12.7x for equal power, or 3x and 6.7x for 1/64 power, respectively.

Second, as M drops to K, i.e., $M \approx K = 15$, the performance of zero-forcing drops steeply. This is due to the tightness of the degrees of freedom at the base station; zero-forcing inevitably wastes the majority of transmission power for interference cancelation, leading to a much reduced signal power at the intended terminals. Later, we will show that when $M = K$ this inefficiency can even result in conjugate MUBF out-performing zero-forcing.

5.2.2 Scaling up with K

We next fix M and vary the number of terminals to see how capacity scales with K. In the experiments reported by Figures 9 Left and Middle, the total transmission power is scaled by $1/M$, similar to that in Figure 8, and is held constant regardless of K. Because the total power is split among the terminals, the power per terminal is therefore scaled by $1/K$. In the experiment shown by Figure 9 Right, we reduce the transmission power to the minimum WARP setting in order to demonstrate how the capacity of the three forms of MUBF are affected by power. We have the following observations:

First, when $M \gg K$, as shown in Figure 9 Left, capacity increases approximately linearly with the number of terminals for both conjugate and zero-forcing MUBF; this is attributable to the multiplexing gains from simultaneously serving K terminals.

Second, conjugate beamforming initially loses capacity as the number of terminals increases from 1 (SUBF) to 2 due to the addition of interference from the other terminal, and thus the overwhelming drop in SINR. This loss, however, is quickly compensated for by the multiplexing gains.

Third, as K approaches M, the performance of zero-forcing drops sharply as shown in Figure 9 Middle. This corroborates the second observation made in Section 5.2.1. Additionally, the performance of conjugate flattens, and even starts to decline, as the additional interference from more terminals causes the average SINR to approach 0 dB.

Finally, when the transmission power is reduced, conjugate MUBF performs relatively better than zero-forcing, as shown in Figure 9 Right. This is because the performance of conjugate is inherently limited by interference from other terminals, while the performance of zero-forcing is instead limited by noise, since the interference is explicitly canceled. It is not until the transmission power is reduced to a point where interference has the same magnitude as noise that there is a significant effect on the capacity improvement for conjugate.

5.3 Near-optimality of Localized Conjugate

In order to verify the viability of our localized method for conjugate MUBF, we implement it in Argos and compare it to standard conjugate MUBF with global power scaling. As shown in Figure 10, we see that our local power scaling method (*Local Conj.*) results in a signal power within 1.2 dB of global power scaling (*Conjugate*), but quickly approaches equivalent power as the number of terminals increases. For a fair comparison we ensure that both methods send with the same transmission power, however in a practical deployment our method will always transmit equal or more power. While local power scaling is less efficient for a given transmission power, it ensures that each base station radio is being fully utilized, thus more intelligently adapting to the constraints of real-world hardware. Furthermore, we see in Figures 8 and 9 that the performance difference between global power scaling (*Conjugate*) and local power scaling (*Local Conj.*) is almost indistinguishable.

5.4 Stability of Indirect Calibration

As described in the previous section, we implemented a novel reciprocal calibration method to enable implicit beamforming and efficient TDD operation. Figure 11 shows that this calibration deviates from the mean angle an average of less than 2.6% (maximum 6.7%), and from the mean amplitude less than 0.7% (maximum 1.4%), over a period of 4 hours. Notably, these measurements were taken during the day with normal movement around the base station, indicating the calibration procedure is stable in real-world environments. Angle deviation is calculated by difference in angle from average angle over π, i.e., 2.6% error is equivalent to 0.08 radians. This indicates that our internal calibration scheme can performed very infrequently, i.e., once a day, and thus has negligible performance overhead.

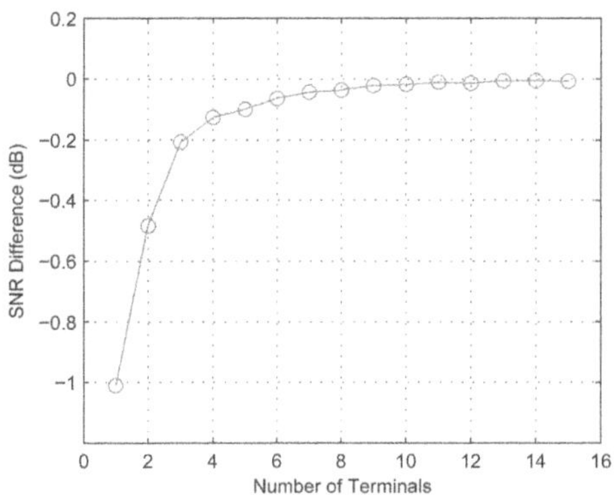

Figure 10: Relative signal power between conjugate and our conjugate with local power scaling, sent at the same transmit power. Local power scaling performs within 1.5 dB of global power scaling, and the difference quickly converges to 0 dB as K increases.

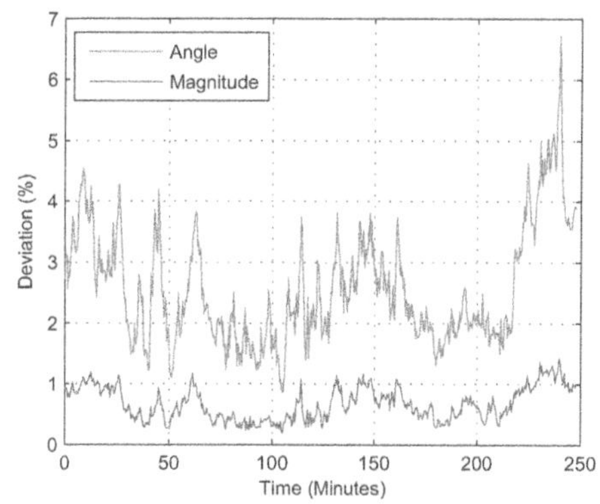

Figure 11: Our calibration procedure exhibits an average instantaneous noise of less than 7% and remains stable indefinitely.

6. RELATED WORK

Argos is directly motivated by multi-user beamforming (MUBF) theory. Recent theoretical works have demonstrated the exciting benefits of large-scale MUBF, or its general form, massive MIMO [12, 13, 20]. In [17], the realization of conjugate beamforming with an infinite number of antennas in a TDD multi-cell system is discussed. Leveraging conjugate beamforming for a distributed architecture is proposed in [20], however our paper explicitly addresses power-normalization, and shows that, in a statistical sense, each base station antenna can perform normalization independently of the others. Single-cell analyses of the performance of both conjugate and zero-forcing beamforming for finite numbers of antennas are obtained in [19]. However, these works are theoretical in nature, and do not address the system and implementation issues such as decentralized power scaling and TDD calibration. Argos is motivated by and built on top of this prior work, and complementarily addresses the architectural and system challenges to realize a many-antenna base station in the real-world.

Endeavors to push MUBF into practice have been observed recently. State-of-the-art wireless standards in cellular networks and WLAN, such as LTE, WiMAX, and 802.11, have all considered incorporating downlink MUBF in their emerging releases, e.g., 3GPP release 9 [1] and 802.11ac [2]. However, MUBF in these standards is restricted to a much smaller scale, e.g., up to eight antennas in 802.11ac. The most recent research efforts towards practical MUBF [5, 14] are limited to a small number of antennas. Argos, comparatively, is the most ambitious MUBF prototype, featuring a much greater number of antennas. As a result, our contribution differs from existing works in that we have identified and addressed a set of unique challenges regarding practicality and scalability. In addition, Argos is programmable and therefore can be used for any large-scale MU-MIMO technique, although we have focused on MUBF in this paper.

There are orthogonal techniques to improve the spatial reuse of spectrum such as sectorization and small cells. First, sectorization uses multiple antennas to form directional beams, each of which covers a range of directions as a *sector*. Terminals in different sectors can be served simultaneously. Therefore, sectorization can be treated as a special case of MUBF to physically separate terminals. Second, small cell techniques, such as femto-cells and pico-cells deploy base stations more compactly, with limited coverage to improve spatial reuse. In each sector or small cell, one can further apply large-scale MUBF to achieve even more spatial reuse and better energy efficiency.

Argos improves spatial reuse efficiency through by employing a large number of antennas on base stations. One can also add antennas at the terminal to further improve spectral efficiency and reduce inter-cell interference. The authors in [21] reported a system with multiple directional antennas on mobile terminals where only one antenna is active at a time to realize uplink directionality. The authors in [27] studied the feasibility of SUBF on terminals, and demonstrated its power efficiency and capacity benefit. Argos is completely complementary to these terminal-based solutions and provides orthogonal benefits.

A few prior works have offered solutions for efficient channel calibration in TDD systems, e.g., [6, 10, 11, 23, 24]. All prior solutions require terminal involvement and feedback in the calibration process, an unacceptable overhead in a large-scale MUBF system. In contrast, the relative calibration in Argos is done internally at the base station without such overhead.

7. DISCUSSION

This paper presents a scalable architecture for many-antenna base stations, a real-world implementation of this architecture, as well as compelling early experimental results. These results motivate further research in the area of many-antenna systems, and raise many practical challenges to be surmounted.

First, the size, cost, and power consumption of a many-antenna base station are significant impediments to adoption. However, we believe that advances in manufacturing combined with the use of specialized hardware in both the analog and digital domains can overcome these barriers.

While Argos supports multiple precoding techniques, an obvious question this work raises is: which precoding technique is optimal under a given scenario? This is a deceivingly hard question to answer, as there are many variables to optimize, such as power, fairness, and spectral efficiency, as well as many factors that impact this optimality, such as transmission power, propagation environment, terminal mobility, hardware capabilities, number of terminals, and number of base station antennas. Moreover, we conjecture that it will be advantageous to dynamically select the precoding technique, as many of these factors continuously change.

Finally, this many-antenna architecture also presents numerous network level challenges, including terminal paging, optimal grouping and scheduling of terminals, and handover between cells. This architecture also raises many opportunities for improving total network capacity; for example, as predicted by [17], the reduced transmit power will significantly improve inter-cell interference. Addressing these challenges and opportunities is the subject of our future work.

8. CONCLUDING REMARKS

We present the design, realization, and evaluation of Argos, a base station architecture that can potentially employ thousands of antennas to serve tens of terminals simultaneously through MUBF. In order to enable this unprecedented scaling in a practical environment Argos employs a hierarchal modular design that facilitates flexible, scalable deployments while simultaneously constraining latency and providing fault tolerance. It also features a novel beamforming algorithm that is completely decentralized and a new calibration method that allows CSI to be collected in constant time with regard to the number of base station antennas.

Our experimental characterization of an Argos prototype with 64-antennas clearly shows the practical benefits of MUBF base stations with many antennas, improving spectral and energy efficiency manyfold simultaneously. Our results are the first publicly reported evidence that many-antenna MIMO systems can produce significant benefits under real-world settings. The scale of our experiments is only limited by the number of Argos modules (WARP boards) currently available to us. The architecture of Argos, however, can easily accommodate many times more modules, each with more radios, potentially allowing thousands of antennas to serve tens of terminals through MUBF.

Acknowlegements

This work was supported in part by NSF grants CRI 0751173, MRI 0923479, NetSE 101283, MRI 1126478 and CNS 1018292. Clayton Shepard was supported by an NDSEG fellowship and grant CNS 1018292. We would like to thank Ashutosh Sabharwal, Edward Knightly, Patrick Murphy, Reinaldo Valenzuela, Cuong Tran, our reviewers, and our shepherd, Sunghyu Choi, for their input and support.

References

[1] 3GPP Release 9. `www.3gpp.org/Release-9`.

[2] IEEE 802.11ac. `mentor.ieee.org`.

[3] InfiniBand. `www.infinibandta.org`.

[4] Rice University Wireless Open Access Research Platform. `warp.rice.edu`.

[5] E. Aryafar, N. Anand, T. Salonidis, and E. Knightly. Design and experimental evaluation of multi-user beamforming in Wireless LANs. In *Proc. ACM MobiCom*, Chicago, Illinois, Sept. 2010.

[6] A. Bourdoux, B. Come, and N. Khaled. Non-reciprocal transceivers in OFDM/SDMA systems: impact and mitigation. In *Proc. IEEE RAWCON*, 2003.

[7] Cisco Inc. Cisco visual networking index: Global mobile data traffic forecast update, 2011-2016. `cisco.com/en/US/solutions/collateral/ns341/ns525/ns537/ns705/ns827/white_paper_c11-520862.html`.

[8] C. Dick, F. Harris, M. Pajic, and D. Vuletic. Implementing a real-time beamformer on an FPGA platform. *Xcell J.*, 2007.

[9] D. Gesbert, M. Kountouris, R.W. Heath, C. Chae, and T. Salzer. Shifting the MIMO paradigm. *IEEE Signal Processing Magazine*, 2007.

[10] M. Guillaud, D.T.M. Slock, and R. Knopp. A practical method for wireless channel reciprocity exploitation through relative calibration. In *Proc. IEEE ISSPA*, 2005.

[11] Y. Hara, Y. Yano, and H. Kubo. Antenna array calibration using frequency selection in OFDMA/TDD systems. In *Proc. IEEE GLOBECOM*, 2008.

[12] J. Hoydis, S. ten Brink, and M. Debbah. Massive MIMO: How many antennas do we need? *arXiv:1107.1709v2 [cs.IT]*, 2011.

[13] H. Huh, G. Caire, H.C. Papadopoulos, and S.A. Ramprashad. Achieving large spectral efficiency with TDD and not-so-many base-station antennas. *IEEE-APS Topical Conf. on APWC*, 2011.

[14] J. Koppenborg, H. Halbauer, S. Saur, and C. Hoek. 3D beamforming trials with an active antenna array. In *Int. Workshop on Smart Antennas*, 2012.

[15] T.K.Y. Lo. Maximum ratio transmission. *IEEE Trans. on Communications*, 1999.

[16] T.L. Marzetta. How much training is required for multiuser MIMO? In *Proc. IEEE ACSSC*, 2006.

[17] T.L. Marzetta. Noncooperative cellular wireless with unlimited numbers of base station antennas. *IEEE Trans. on Wireless Communications*, 2010.

[18] Maxim. Single-/dual-band 802.11a/b/g world-band transceiver ICs. `datasheets.maxim-ic.com/en/ds/MAX2828-MAX2829.pdf`.

[19] H. Ngo, E. Larsson, and T.L. Marzetta. Energy and spectral efficiency of very large multiuser MIMO systems. *IEEE Trans. on Communications*, 2011.

[20] F. Rusek, D. Persson, B. Lau, E. Larsson, T.L. Marzetta, O. Edfors, and F. Tufvesson. Scaling up MIMO: Opportunities and challenges with very large arrays. *arXiv:1201.3210v1 [cs.IT]*, 2011.

[21] A. Amiri Sani, L. Zhong, and A. Sabharwal. Directional antenna diversity for mobile devices: Characterizations and solutions. In *Proc. ACM MobiCom*, 2010.

[22] C. Shepard. Argos: Practical base stations with large-scale multi-user beamforming. Master's thesis, Rice University, April 2012. Available at: `clay.rice.edu/pubs/MasterThesis.pdf`.

[23] J. Shi, Q. Luo, and M. You. An efficient method for enhancing TDD over the air reciprocity calibration. In *Proc. IEEE WCNC*, 2011.

[24] D. Tse and P. Viswanath. *Fundamentals of Wireless Communication*. Cambridge University Press, 2005.

[25] Xilinx Inc. 7 Series FPGAs Overview. `xilinx.com/support/documentation/data_sheets/ds180_7Series_Overview.pdf`.

[26] Xilinx Inc. Virtex-II Pro and Virtex-II Pro X Platform FPGAs: Introduction and Overview. `xilinx.com/support/documentation/data_sheets/ds083.pdf`.

[27] H. Yu, L. Zhong, A. Sabharwal, and D. Kao. Beamforming on mobile devices: A first study. In *Proc. ACM MobiCom*, 2011.

802.11ec: Collision Avoidance without Control Messages*

Eugenio Magistretti
ECE Dept.
Rice University
Houston, TX
emagistretti@rice.edu

Omer Gurewitz
CSE Dept.
Ben Gurion University
Beer Sheva, Israel
gurewitz@cse.bgu.ac.il

Edward W. Knightly
ECE Dept.
Rice University
Houston, TX
knightly@rice.edu

ABSTRACT

In this paper, we design, implement and evaluate 802.11ec (Encoded Control), an 802.11-based protocol *without* control messages: instead, 802.11ec employs correlatable symbol sequences, which together with the timing the codes are transmitted, encode all control information and change the fundamental design properties of the MAC. The use of correlatable symbol sequences provides two key advantages: *(i)* efficiency, as it permits a near order of magnitude reduction of the control time; *(ii)* robustness, because codes are short and easily detectable even at low SINR and even while a neighbor is transmitting data. We implement 802.11ec on an FPGA-based software defined radio. We perform a large number of experiments and show that, compared to 802.11 (with and without RTS/CTS), 802.11ec achieves a vast efficiency gain in conveying control information and resolves key throughput and fairness problems in the presence of hidden terminals, asymmetric topologies, and general multi-hop topologies.

Categories and Subject Descriptors

C.2.1 [**Computer-Communication Networks**]: Network Architecture and Design- Wireless Communication

General Terms

Design, Experimentation, Measurement, Performance

Keywords

WLANs, 802.11, Channel Access, Collision Avoidance, Control Messages, RTS/CTS, Signal Correlation

*While we employ the name 802.11ec, our work does not represent an IEEE standard. We use this name for reasons that will be apparent in the body of the paper.

MobiCom'12, August 22–26, 2012, Istanbul, Turkey.

1. INTRODUCTION

MAC control messages are essential: for example, ACKs convey correctly received data and RTS/CTS exchange can significantly mitigate hidden terminal collisions. However, even though the information conveyed in MAC control messages is small, their duration can be quite long, as in addition to the control information, they also need to include source/destination address, message type, etc., all of which are transmitted at base rate to improve the likelihood that they can be successfully decoded. For example, an ACK message is 14 bytes plus physical layer encapsulation, but contains only one bit of relevant information (that DATA was successfully received). Likewise, RTS/CTS is rarely used in practice precisely due to excessive overhead despite its important role in mitigating collisions.

In this paper, we design, implement, and evaluate 802.11ec (Encoded Control) as a control message free MAC. Instead of control messages, 11ec employs correlatable symbol sequences (CSS's), which together with their transmission timing, convey all control information, and change the fundamental design properties of the MAC. For example, 11ec replaces an 802.11 ACK *message* with a predefined ACK CSS that can be correlated instead of decoded, thereby vastly reducing its duration and dramatically improving its robustness by enabling its reception at low SINR.

Control information can be classified along two dimensions: first, as to whether or not the information in the message can be represented from a small dictionary or codebook. For example, a small dictionary can encode the three different control messages used in 802.11 for data exchange (RTS, CTS, and ACK). Likewise, while the space of all MAC addresses is large (seemingly precluding a small dictionary), each node communicates with only a limited number of addresses at a time. Thus, both MAC addresses and control message type can be encoded from a small dictionary. Second, control information can further be classified according to whether they are necessarily public or can be private. For example, for correctness of the protocol, all nodes must know that a CTS should cause them to defer, i.e., this message must be public; on the other hand, only a data sender need know that its data was correctly received, i.e., its ACK may be private.

802.11ec's key techniques are two fold: first, we use a dictionary of correlatable symbol sequences to convey control information that can be represented by a limited dictionary. For example, instead of CTS that contains physical layer preamble, frame control sequence, type field, frame checksum, destination address, duration field (as well as it incurs a T_{SIFS} delay), we transmit a short (e.g., 127 symbols) CSS from a small dictionary to convey that it is a CTS. For an 802.11a physical layer, we show that this reduces the time to convey the control information by nearly an order of magnitude, from 60 μs to 6.35 μs. Second, we show that the information that cannot be represented by a limited dictionary can be conveyed via CSS timing. For example, nodes overhearing the 802.11 CTS message need to defer for an amount of time as specified by the CTS duration field contained in the message. We show that 11ec nodes can instead simply defer until a channel-clear CSS is transmitted by the receiver (or until a timeout).

802.11ec's second technique is to distinguish between public and private information. Namely, 11ec only uses public CSS's for information that is required to be public, such as conveying channel reservation and channel clear. On the other hand, address fields need not be public, as the identity of the sender and receiver need not be known by other nodes. 11ec ensures that private control information, including addresses and ACKs, is not correlated by other nodes. This has the potential to thwart eavesdroppers not only from decoding data (as data can be encrypted), but even from knowing which nodes are communicating with each other; we show how all private control information, including addresses can only be correlated by the intended receiver.[1]

802.11ec enhances robustness in two ways. First, control information is more likely to be received in 11ec because control information is conveyed in short CSS's that are correlatable even at low SINR. For example, because 802.11ec replaces an ACK message with a CSS, 11ec ACKs are more robust and can be received even in the presence of transmitting interferers. Second, 802.11 is "fragile" to topological factors in that while 802.11 DCF without RTS/CTS yields high performance in fully connected wireless LANs [5], hidden terminals, asymmetric topologies, and general multi-hop topologies can yield severe throughput degradation and unfairness [6]. These latter topologies are becoming increasingly common because of device power asymmetries, e.g., between APs, laptops, and popular smart-phones, and of the wider coverage achievable with the adoption of sub-GHz frequencies, including TV white spaces [16, 19]. While use of RTS/CTS can significantly improve throughput in such challenged topologies, the additional overhead of RTS/CTS can sometimes overwhelm this improvement. Moreover, in fully connected topologies RTS/CTS degrades throughput due to its unnecessary overhead. In contrast, 11ec overcomes these limitations through robust and short-duration control signals, i.e., 11ec minimally penalizes station throughput thus allowing to enable channel reservation independently of the network topology. Consequently, 11ec stations have vastly increased opportunities to obtain channel access thereby dramatically improving the network's fairness in throughput distribution.

We implement correlatable symbol sequences in a software defined radio, perform a large set of experiments and study issues that have not been experimentally investigated previously. We find a correlatable symbol sequence *length* that simultaneously: *(i)* provides sufficient physical-layer robustness, *(ii)* limits communication overhead, and *(iii)* supports large networks. Specifically, we first investigate the trade-offs between sequence length and physical-layer robustness and show that even short sequences, e.g., 127-symbol or 6.35 μs long, can be detected at -6 dB SINR with only 5% false negatives. We demonstrate that our encoded sequences can be detected at an SINR 10 dB lower than 802.11 control messages. Second, we show that 127-symbol code lengths can support more than 50 co-located nodes, with minimal penalty on detection errors.

Finally, we implement 11ec in a measurement-driven emulator, whose inputs are channel measurements collected in a real deployment and real card performance parameters (e.g., BER and multiple supported modulations). We compare 11ec's performance to 802.11 with and without RTS/CTS. We examine a wide set of basic topologies that are at the origin of throughput losses and/or imbalances in 802.11-based networks in order to provide an insight in understanding the performance of larger networks. Our finding is that 11ec can dramatically reduce throughput imbalances by improving the Jain index [12] by up to 88%. Moreover, while such a fairness improvement can often decrease total utilization, 11ec increases channel utilization by more than 10% via the use of short encoded control that simultaneously decreases vulnerability intervals and control overhead. We also study a larger topology and show that 802.11ec can improve the throughput of an under-served flow by a factor of 1255%. Over all flows, we improve Jain index by up to 217% while also improving the channel utilization by up to 44%.

The remainder of the paper is organized as follows. Section 2 discusses coded control CSS's and the design of 802.11ec. Section 3 includes a thorough experimental evaluation of CSS's using a software defined radio platform. Section 4 investigates the benefits of 802.11ec in a measurement-driven emulator. Finally, Sections 5 and Section 6 overview related works and conclude the paper.

2. MAC PROTOCOL DESIGN

11ec collision avoidance realizes and improves on the collision avoidance mechanism of IEEE 802.11 DCF with RTS/CTS, reduces the overhead by nearly an order of magnitude, and practically eliminates collisions, even in hidden terminal topologies. Specifically, 11ec retains the four-way handshake suggested by 802.11, where control messages are replaced by very short correlatable symbol sequences (see Figure 1). It is important to note that: *(i)* the duration of each correlatable symbol sequence is nearly zero; *(ii)* the duration between the start of the reservation signal until the data transmission is negligible, hence practically invulnerable to collisions.

In this section, we define correlatable symbol sequences (CSS's) and explain how control messages can be turned into CSS's. Furthermore, we show that CSS's are a key element for the realization of 11ec efficiency and robustness. The second part of the section provides a detailed description of 11ec including protocol primitives, and an analysis of hidden terminal vulnerability leading to novel collision reduction opportunities.

2.1 Coded Control Information versus Message Control Information

CSS's. Correlatable symbol sequences are predefined pseudo-noise binary codewords; namely, while codewords are deterministically generated, they retain the statistical properties of a sampled white noise. For this reason, the cross-correlation of any such sequence with a matching copy ob-

[1]While development of a complete privacy protocol is beyond the scope of this work, 11ec provides important mechanisms to design such a protocol.

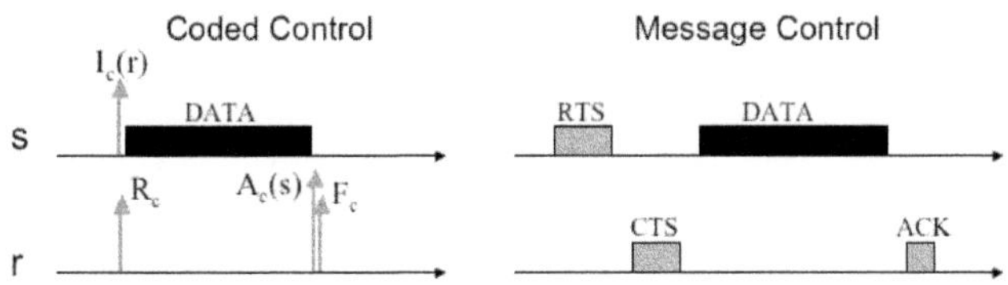

Figure 1: Timeline of a packet exchange with Coded Control versus Message Control.

tains spike values, while it appears random to a listener without prior knowledge of the codeword. An example of a CSS is the 802.11 preamble used for packet detection, symbol synchronization, and radio parameter tuning.[2]

The CSS detection process via cross-correlation enjoys three key advantages over data decoding. First, cross-correlation obtains a large processing gain even for small codewords (e.g., the 802.11 preamble used for detection is 64 symbols), which permits reliable detection even at low SINR. Second, differently from decoding, detection is highly robust to imperfect radio parameter tuning and thus a codeword does not need to be preceded by a preamble. For these reasons, a CSS can be short; for example, in our implementation, 11ec utilizes 127-symbol codewords that can be transmitted in 6.35 μs. Third, detection is almost instantaneous as no decoding is needed. For example, 802.11 inter-control message time is at least $T_{SIFS} = 16$ μs, including about 14 μs of data processing, while in 11ec no control message processing is required; hence, 11ec reduces substantially the short inter-CSS time.

Encoded Control. While consuming a significant amount of airtime, 802.11 control messages usually convey little information. For example, an ACK occupies the medium for up to 60 μs, i.e., an airtime sufficient to transmit 3240 bits at 54 Mbps, while it contains a single bit of relevant information. 11ec replaces control messages with CSS's, which permit to shorten the transmission duration of nearly an order of magnitude to 6.35 μs, while retaining the information content.

According to the 802.11 standard [1], RTS, CTS, and ACK control messages may include up to four information fields: destination address, sender address, duration, and frame control (a fifth field is the frame check-sum that protects the other four). In particular, the frame control field is a 2-byte long sequence of bits representing specific control parameters. The values of most control bits are fixed for control messages; only frame subtype (4 bits) and station power management flag (1 bit) can assume different values. However, the latter does not convey novel information when used in control messages.

In order to represent the information content of the control messages as described above, 11ec considers the size of the dictionary needed to represent such information. Information that can be expressed by a small dictionary is conveyed using CSS's, while information that needs large dictionaries is conveyed with timing codes. First, in 802.11 data exchange, the type field that distinguishes the control messages may assume only three values, i.e., RTS, CTS, and ACK; 11ec conveys the type by associating each message with distinct CSS's. Second, 802.11 control messages include addresses (sender and/or receiver). Since the number of nodes a station communicates with at a time is generally small, i.e., can be represented by a small dictionary, 11ec integrates the addresses in CSS's, i.e., a single CSS may represent the combination of a control type and a specific address. For example, in 802.11 the RTS includes the address of the intended receiver; accordingly, in 11ec, RTS addressed to different receivers are represented by distinct CSS's. Third, some control messages include a duration field that cannot be represented by a small dictionary. For this information, 11ec utilizes a combination of time codes and new control types. For example, 11ec nodes reserve the channel for the duration of a data reception, by transmitting two CSS's corresponding to channel reservation (immediately before the reception) and release (immediately after). The potential interferers do not access the channel during the interval between the two CSS's (or before a timeout expires), i.e., effectively implement a form of virtual carrier sensing.[3]

Public CSS's versus Private CSS's. A second dimension of control messages, and of the corresponding CSS's, is whether they can be private, i.e., carry information relevant only to a specific destination (e.g., acknowledgements), or are necessarily public, i.e., meant to be heard by all neighbors of a node (e.g., channel reservation/release). Accordingly, in the case of a private CSS, only the intended receiver possesses a copy of the correlatable symbol sequence, and thus can correctly detect it; conversely, all nodes possess copies of the public CSS's and can detect them. For example, only the data sender needs to cross-correlate its private acknowledgement CSS, while all nodes must cross-correlate public channel reservation/release CSS's.

Control during Data. Because correlatable symbol sequences can be detected even at sub-noise SINR (e.g., 11ec 127-symbol CSS's can be detected at -6 dB with high reliability), 11ec nodes attempt detection even while receiving data. This technique provides a signaling mechanism effective even in cases in which the receiver is subject to long periods of noisy channel, or undesired data overhearing. In 802.11, if a node receives an RTS while also overhearing data, the node cannot decode the RTS and therefore cannot respond. In contrast, 11ec uniquely enables a node to receive a CSS signaling that another node is requesting to communicate. Therefore, the receiver can send a Request for RTS (RRTS) CSS to reserve the medium when it becomes free and initiate a data exchange [4]. Likewise, because ACKs are also encoded, they can be correctly received even if an interfering terminal is simultaneously transmitting data.

2.2 11ec Channel Reservation Primitives

Wireless MAC protocols perform collision avoidance by silencing the medium in the vicinity of a transmitting link via channel reservation. Channel reservation fundamentally hinges on three key mechanisms: *(i) initiation*, performed by the node that initiates the exchange to request the cooperation of the other endpoint to reserve the channel; *(ii) reservation*, performed to inform nodes potentially hindering the exchange; and *(iii) deferral*, performed by the surrounding terminals in order to avoid disturbing ongoing transmissions. 802.11 implements the three mechanisms via *(i)* RTS; *(ii)*

[2]A detailed discussion of signal correlation can be found in literature [10, 14, 22]. A short introduction is in Appendix A.

[3]Another alternative is to discretize the duration and associate different CSS's to each discrete value.

CTS and data packet - the latter realizes channel reservation in the vicinity of the sender; and *(iii)* NAV and carrier sensing. When RTS/CTS is disabled, during the data transmission the medium is reserved exclusively in the vicinity of the sender. In the following, we show how 11ec implements these three mechanisms via CSS's and timing codes.

A key concept of 11ec channel reservation is very short channel reservation negotiation for near immunity to interruptions, e.g., collisions and capturing by other nodes. Specifically, 11ec channel reservation is based on three basic primitives (the subscript c indicates CSS's).

Initiation: $\mathbf{I_c(r)}$. In 11ec, a sender wishing to start a data exchange performs virtual, and optionally physical, carrier sensing. If the medium is free, the sender waits for a back-off interval similar to 802.11 and then transmits a sender side channel request primitive, in short $I_c(r)$, to request the receiver r to reserve the channel. $I_c(r)$ need only be detected by r and not necessarily by the neighboring nodes; thus, we implement it as a private CSS. In order to convey the identity of the receiver r (as 11ec does not transmit the traditional MAC address), 11ec implements $I_c(r)$ via several CSS's, and associates a distinct CSS with each receiver; i.e., when a sender needs to contact a receiver, it uses the receiver's $I_c(r)$. Nodes in the vicinity of the sender do not detect the initiation $I_c(r)$.

Reservation: $\mathbf{R_c}$. A node r receiving an $I_c(r)$ checks if other nodes are communicating in its vicinity and may hinder its reception. If that is not the case, r immediately transmits a channel reservation primitive R_c to notify potential interferers. In order to realize the reservation, R_c should be detected by all nodes in the vicinity of the receiver r and it is therefore transmitted via a public CSS.

In addition to channel reservation that forces neighbors to defer, R_c implicitly communicates to the transmitter that the channel is available and that data can be transmitted. Instead of providing a distinct CSS to convey the sender address, as in the previous case of the $I_c(r)$, we employ a simple temporal code: Since a receiver r transmits R_c immediately after $I_c(r)$, the sender, in contrast to the other neighboring nodes, interprets the reception of a R_c as an authorization to begin a data transmission.

Deferral: $\mathbf{R_c \rightarrow F_c}$. 11ec implements the deferral and conveys its duration via a combination of CSS's and a simple time code. Specifically, after data reception and acknowledgement, the receiver explicitly releases the channel with a channel free primitive F_c. Thus, nodes receiving R_c need to wait to receive an F_c (or wait a predefined timeout) before accessing the channel; practically, this procedure represents a form of virtual carrier sensing. Because all neighboring nodes need to receive F_c, 11ec implements F_c as a public CSS.

11ec further increases the robustness of R_c/F_c messages by pairing them. Our technique is based on associating a small number of CSS pairs to distinct R_c^i/F_c^i pairs; a receiver randomly picks and transmits any such pair to reserve and free the channel. This feature is particularly useful for a node located in the neighborhood of several receivers that may be active simultaneously, i.e., the receivers may send R_c and F_c that denote overlapping intervals. In that case, the node can still correctly determine the state of the medium in each moment, by associating each overheard reservation R_c^i to its corresponding F_c^i.

Finally, we define an acknowledgement primitive, $A_c(s)$ (see Figure 1), and we implement it as a private CSS associated to each sender s. Few recently proposed packet forwarding schemes leverage the ACK exchange for additional purposes other than data acknowledgement, e.g., network coding [7] and routing [8]. In such cases, 11ec can revert to 802.11 ACKs with small overhead penalty (the major gain of 11ec both in throughput and robustness derives from the novel channel reservation scheme). Note that CSMA/CN [22] uses signatures for acknowledgement, which are similar to our CSS's.

CSS Association. The Initiation and Acknowledgement primitives $I_c(r)$ and $A_c(s)$ require association of unique CSS's to specific nodes. While a detailed investigation of the issue is beyond the scope of this paper, we suggest two simple mechanisms that can be used for this purpose. The first leverages the solution already existing in the 802.11 standard, in which the AP assigns an identifier (AID) upon station association [1]; AIDs can be easily mapped to CSS's. In order to initiate the association, the stations may use a reserved CSS. The second mechanism utilizes multiple hash functions based on the station MAC address. In case of assignment conflicts, which can be easily detected using the address fields in the header of the data frames, the stations can switch the hash function utilized.

Primitive Extensions. As briefly mentioned above, 11ec can support a signaling mechanism to alleviate starvation similar to RRTS [4] via control during data, and can highly reduce the occurrence of exposed terminals via robust CSS acknowledgement. For reason of space, these cases are not covered in this paper.

2.3 Contending Flows and Vulnerability Interval

The shortness and robustness of correlatable symbol sequences dramatically reduces vulnerability to collisions. The vulnerability interval of a packet exchange is twice the time delay from the beginning of a transmission until all potential interferers are prevented from corrupting the exchange, i.e., it includes the whole interval, before and after the beginning of the intended exchange, during which interfering transmissions may start and corrupt the exchange. In the case of hidden terminals in **802.11** with RTS/CTS, the vulnerability interval is twice the delay from the moment a node sends an RTS to the detection of CTS by the hidden terminals. Considering the case of 802.11a/g and 6 Mbps control packets, the total duration of the vulnerability interval can be computed as follows. First, node s transmits the RTS, for a total duration of 52 μs including preambles; second, the RTS propagates to the receiver for up to 1 μs; third, the receiving node r waits $T_{SIFS} = 16$ μs before sending the CTS, due to practical communication issues such as RTS decoding; fourth, the CTS propagates to hidden terminals for up to 1 μs; fifth, the hidden terminals need up to 4 μs to detect the packet; last, additional 2 μs account for potential radio turn-around (all temporal indications are taken from section 17 of the 2007 version of the standard, for 20 MHz bandwidth [1]). The total amounts to 152 μs, i.e., twice 76 μs. In 802.11 without RTS/CTS, the vulnerability interval can be considerably larger, spanning twice the data transmission duration and as large as 4 ms.

802.11ec's CSS's shorten the vulnerability interval and practically reduce the set of potential hidden terminals. The vulnerability interval of 11ec can be computed as follows (see

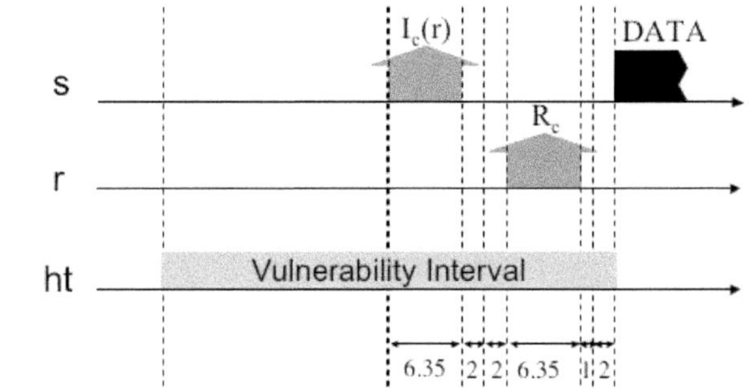

Figure 2: Timeline of the vulnerability interval of 802.11ec (time indications are in μs).

Figure 2, where the intervals below are denoted by numerical time indications). First, node s transmits $I_c(r)$ for 6.35 μs; second, the receiving node r waits for up to 2 μs, in order to detect potential overlapping hidden terminal transmissions as explained below (this interval is a design choice and includes the propagation delay from the transmitter); third, r needs a turn-around time to prepare its radio for transmission, i.e., 2 μs (according to 802.11 standard); fourth, r transmits R_c for 6.35 μs; finally, the hidden terminals need a propagation delay of up to 1 μs to receive the R_c and an additional 2 μs to account for potential radio turn-around. 11ec vulnerable interval is twice 19.7 μs, i.e., 39.4 μs or about 25.9% of the vulnerability interval of 802.11.

CSS's also nearly eliminate the collisions of nodes starting in the same slot when the SINR allows it.[4] In fact, the receiver r can detect simultaneous and overlapping transmissions of multiple $I_c(r)$ because of the CSS processing gain, and request retransmission. Specifically, r waits for a round-trip propagation time in order to detect potentially overlapping transmissions and, in that case, sends a negative acknowledgement which prompts the contending nodes to undergo a quick backoff repetition. However, in order to take advantage of this technique, we need to enlarge the slot size to encompass the half vulnerability duration, i.e., 20 μs (note that this is sufficient, as the F_c of the receiver synchronizes all of its transmitters). Also in cases of no hidden terminals, this choice induces only minor throughput penalties at high data rates, as we show in Section 4.2.

Co-existence with legacy 802.11. For backward compatibility with 802.11, 11ec exactly follows the standard [1] except for the techniques described in this section. For example, a key element for co-existence is the arbitration of the medium, which leverages carrier sensing based on the correlation of the data preambles and backoff mechanism. Accordingly, 11ec uses the same data preamble format as 802.11, and sets the contention window size following the same binary exponential backoff scheme. A more complete discussion of co-existence is beyond the scope of this paper.

[4]Nodes may use power adaptation techniques exclusively to transmit control CSS's while transmitting data at full power, i.e., without modulation rate adaptation requirement or throughput penalty. While this may improve the performance of 11ec, we defer its investigation to future work.

3. EXPERIMENTAL EVALUATION OF CSS

In this section we present an experimental evaluation of correlatable symbol sequences using software defined radios. Specifically, our evaluation covers the following issues, *none of which has been previously experimentally studied in the literature.*

(i) We explore the trade-off between length of the sequences, i.e., overhead, and processing gain, i.e., robustness. **Our finding is that 127-symbol sequences provide a good trade-off between overhead (6.35 μs) and robustness (5% false negatives at -6 dB SINR).**

(ii) We contrast the performance of CSS detection with control message decoding. **Our finding is that 127-symbol CSS's can be reliably detected at about 10 dB lower SINR than 802.11 6 Mbps OFDM control packets.**

(iii) We determine the codebook size that 11ec can support, i.e., the number of distinct CSS's that can be practically used, by studying the cross-correlation between different CSS's and its effect on the probability of false positives. **Our finding is that the design of 127-symbol CSS's via Gold codes can support more than 50 co-located nodes (a total of 127 CSS's), without any penalty on false positives.**

3.1 Experimental Setup

3.1.1 Tools

WARP and WARPLab. Our reference software defined radio is the WARP platform [3]. WARP is an FPGA-based platform, including custom designed radios based on the MAX2829 chipset. WARPLab is a programming environment that permits to drive WARP from a host computer. Relevantly to our experiments, WARPLab supports the execution of micro experiments, each one of approximately 400 μs duration (2^{14} samples at the 40 MHz frequency of DAC/ADC), and access analog sample send/receive buffers and RSSI recordings collected during each experiment. RSSI is measured by the MAX2829 circuit, and digitized by a dedicated 10-bit ADC.

Azimuth Channel Emulator. In order to perform experiments under controllable and repeatable conditions, we used an Azimuth ACE MX channel emulator.[5] The channel emulator permits creation of different network topologies, by tuning the attenuation along each path independently and predictably.

3.1.2 Implementation

We implement CSS transmission/detection and OFDM packet transmission/decoding on WARPLab. Specifically, CSS's are BPSK sequences filtered, upsampled, and transmitted via standard wideband methods. This solution enjoys a practical advantage over alternative solutions, e.g., OFDM-modulated BPSK sequences, due to the lower peak to average power ratio [26]. Finally, in order to reproduce 802.11 as closely as possible, we implement all types of OFDM 802.11a/g modulation and convolutional code pairs in WARPLab.

3.2 Channel Emulator Validation

The results in this section are obtained using a channel emulator. In order to validate the emulator setting, our

[5]Azimuth Systems Inc., `http://www.azimuthsystems.com/`

methodology includes a preliminary validation contrasting results of a cross-correlation experiment performed over the air, with an identical experiment conducted with the emulator. Specifically, our experiment consists of exchanging CSS's between two WARP nodes a and b, under the interference generated by random OFDM transmissions of a third interfering node c. In the first part of the experiment, we deploy the three nodes inside an office building. In an over the air setting, it is difficult to control SINR given interference from 802.11 networks operating in the building. For this reason, we perform the experiment late at night and measure the SINR on links $a-b$ and $c-b$ a few seconds before and after the experiment. Then, we repeat the experiment under controllable and repeatable conditions using the channel emulator, where we can control the SINR with high accuracy. We repeat both experiments several times and for different SINR, and show a representative sample result.

In both experiments, node a transmits 7 repetitions of a 127-symbol CSS. For each newly acquired sample (i.e., at 40 MHz frequency of the WARP platform ADC), node b computes the signal correlation with a local copy of the transmitted CSS. Figures 3(a) and 3(b) show a representative outcome of a realization in which the SINR on link $a-b$ carrying the CSS is -6 dB. The x-axis is the temporal progression of collected samples, while the y-axis is the correlated value. The thick crossing line in the plots represents a possible choice of the detection threshold; such a threshold is strategic in determining the robustness of CSS's, by balancing false positives and false negatives. In the experiments we conduct in this section, the threshold is chosen in order to obtain a false positive probability of 10^{-8}. While we defer more details on how to tune the threshold to Section 3.6, here we observe that because of the threshold design the correlation value on the y-axis is normalized according to the magnitude of correlated I/Q samples. In the figures, correlation spikes are clearly identifiable in coincidence with the reception of each single CSS, as all and only marks exceeding the threshold. Thus, the detection of 127-symbol CSS's is possible at -6 dB with few errors. *By comparing the two plots, we conclude that controllable emulator experiments and over the air experiments provide similar results for identical SINR values.* However, because of the difficulty to constantly control the SINR in over the air settings, we perform the remaining experiments in this section using the channel emulator, thereby also ensuring their repeatability.

3.3 CSS Length versus Robustness Trade-off

The first issue is the trade-off between the CSS length L and robustness under different SINR; we quantify robustness in terms of the probability of false positives and false negatives. The outcome of this assessment is important, as robustness to SINR is one of the two key elements in the choice of CSS length (the other element guiding this choice is the number of CSS's, discussed in Section 3.5). In this subsection, we determine a length that can tolerate significant interference without high communication overhead. In this experiment, we deploy the three node topology above and use the channel emulator to vary the SINR on link $a-b$. Specifically, the link between node a transmitting the CSS and the receiving node b is maintained fixed to -82 dBm, while the attenuation on the interfering link $c-b$ is set in order to obtain the desired SINR. We iterate the experiment

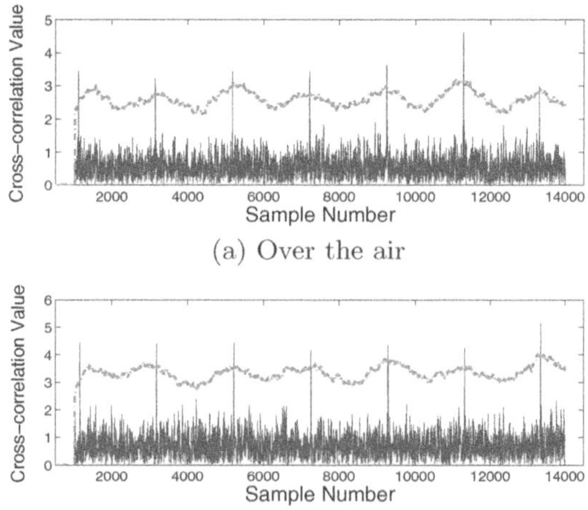

(a) Over the air

(b) Channel emulation

Figure 3: Example of 127-symbol CSS correlation at -6 dB.

for different combinations of SINR and CSS lengths. Each experiment consists in the detection of at least 100 CSS's of lengths L ranging from 63 to 511 symbols. We vary the SINR between 0 dB and -10 dB.

Figure 4 shows the probability of false negatives as a function of SINR and CSS length. Specifically, the x-axis denotes the SINR on link $a-b$, while the y-axis denotes the probability of false negatives. The different curves correspond to different CSS lengths. The figure shows that longer CSS's are more robust due to the processing gain, e.g., at -8 dB, CSS's of length 63 can be detected only 4% of the time (96% of false negatives in the figure), while CSS's of lengths 127, 255, 511 can be detected approximately 30%, 99%, and 100% of the time respectively. However, increasing the CSS length involves an overhead penalty; in fact, while a 63-symbol CSS can be delivered in about 3.15 μs, 127, 255, 511-symbol CSS's require 6.35, 12.75, 25.55 μs respectively. With regard to the probability of false positives, we never obtained more than a single occurrence (out of hundreds of thousands of tests performed) for all the experiments related to a fixed SINR and length combination. Finally, it is relevant to notice that our results show only minor degradation with respect to theoretical performance in AWGN channel. For instance, considering the probability of false positives and false negatives at -8 dB, the length of sequences with similar performance in AWGN would be 47, 81, 198 for the cases of 63, 127, 255 actual lengths. *We conclude that 127-symbol CSS's provide a good compromise between overhead (6.35 μs), and resilience as they can be detected at -6 dB with 5.7% false negatives and no false positives.*

3.4 CSS Detection versus Control Message Decoding

Our second experiment aims to show that control CSS's are more robust to noise than 802.11 control messages, i.e., that CSS's can be reliably detected at considerably lower SINR than control messages. The metrics we use are false positives for the case of CSS detection, and packet error rate for control message decoding. We consider 127-symbol CSS's vs. 160-bit messages transmitted via BPSK modula-

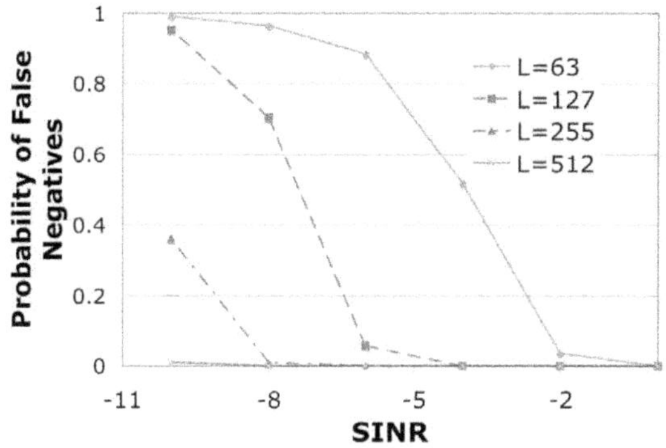

Figure 4: Robustness vs length tradeoff for different CSS lengths.

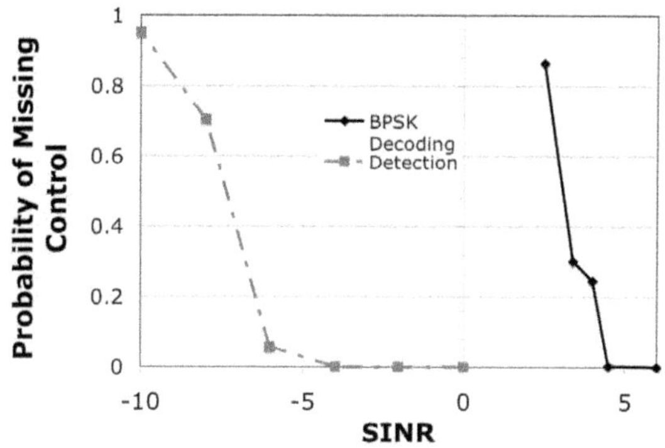

Figure 5: Probability of missing CSS detection vs missing message decoding.

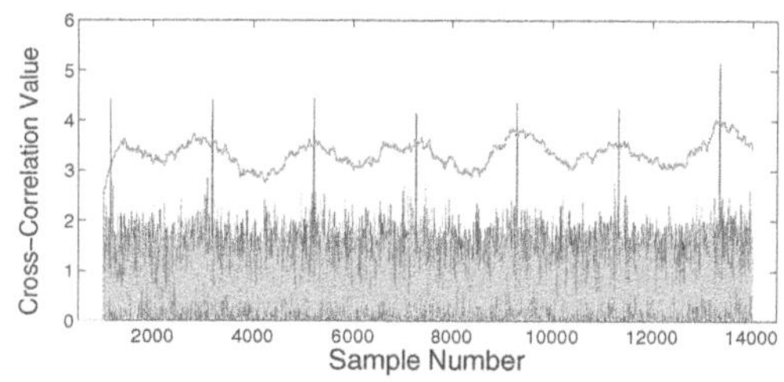

Figure 6: Low cross-correlation of CSS's different from the one transmitted.

tion, with 1/2 rate convolutional coding, corresponding to an RTS packet transmitted at base rate of 6 Mbps in 802.11 OFDM. In order to create the interference scenarios, we follow a methodology identical to the previous experiment. For the case of BPSK modulation, our experiment directly measures the BER out of at least 100000 bit transmissions. Then, we convert the obtained value to packet error rate, by considering a random and independent distribution of the bit errors among the packets (i.e., $1-(1-BER)^{PL}$, where PL is the packet length in bits). Note that the adoption of burst error models, such as Gilbert-Elliot [9], with expected burst length of 6 bits [27] may vary the results by about 1 to 1.5 dB.

Figure 5 shows two curves corresponding to CSS detection and control messages decoding. The x-axis denotes the SINR, while the y-axis denotes the *probability of missing control*, i.e., the probability of false negatives (resp. of packet decoding error) for CSS's (resp. for control messages). The plot shows that control CSS's are substantially more robust than control messages, since their probability of false negatives is much less than the error probability of control packets for any SINR. Furthermore, similar *probabilities of missing control* are obtained for the two control mechanisms, for SINR values separated by about 10 dB. For example, CSS's obtain probability of false detection of 5.7% at -6 dB, while control messages achieve 24% packet error rate at +4 dB, and 0.2% at +4.5 dB. *We conclude that due to the improved robustness of CSS detection with respect to packet decoding, control CSS's are about 10 dB more resilient to noise than 802.11 control messages.*

3.5 11ec Codebook Size

In 11ec, nodes use multiple CSS's and need to be able to reliably detect and discern all of them. In this experiment, we investigate whether cross-correlation between CSS's affects detection accuracy, and we explore the number of distinct CSS's that can be practically used. Specifically, we assess the probability of falsely detecting CSS A when CSS B is transmitted instead. For a given CSS length, a trade-off exists between the number of CSS's that 11ec uses, and the magnitude of the cross-correlation between any CSS pair, which in turn influences the probability of false positives. For this error probability to be small, we use well known sparse binary sequences, with optimal cross-correlation properties. Instances of such sequences have been studied for lengths corresponding to powers of 2, e.g., in cellular communications [21]. Different families of sequences provide larger (resp. smaller) sets of codewords, with larger (resp. smaller) cross-correlation between any codeword pair. Our design implements Gold codes, which provide 127 CSS's for our 127-symbol length, with a theoretical cross-correlation on the order of 12%. The choice of Gold codes permits us to support more than 50 co-located nodes by assigning distinct CSS's pairs to each node representing $I_c(r)$ and $A_c(s)$, while saving several CSS's for F_c^i/R_c^i pairs. In case a larger number of nodes needs to be supported, 11ec can switch to 255-symbol Kasami large codes for example, which allow more than 2000 nodes with 4011 CSS's.

To verify our choice, we emulated a situation in which a CSS A is sent, and 10 nodes try to detect CSS's different from A within the same samples. We repeated the experiment for 100 detection attempts, for 127-symbol Gold codes and SINR from 0 dB to -10 dB. The goal of this experiment is to assess the probability that the other nodes obtain false positives of their own CSS when the signature A is sent. The number of false negatives is immaterial in this experiment. For each SINR experiment, we obtained at most one false positive more than the case of a single CSS detection. Figure 6 shows an example outcome for -6 dB SINR (where the axes have the same meaning as in the experiment in Figure 3). The overlapping plots show the cross-correlation values obtained by 11 different nodes (including the one that expects to detect the transmitted CSS). While the number of spikes is unchanged, the noise looks visually denser due to overlapping plots. *We conclude that by using Gold codes, 11ec can support more than 50 co-located nodes without significant incidence of false positives.*

3.6 Discussion on Signal Correlation

Practical Detection Threshold Selection. The choice of the detection threshold is strategic in balancing the trade-off between false positives and false negatives. For example, as mentioned above, in Figures 3(a) and 3(b) the threshold is denoted by the thick crossing line; corresponding to the chosen threshold, the figures show 0 false positives and 0 false negatives. In general, a higher value of the threshold decreases the probabilty of false positives at the expense of a large probability of false negatives; viceversa, a lower threshold increases the occurrence of false positives. The theory of correlation in Gaussian noise provides an optimal threshold for a target detection SINR [14]. Specifically,

$$T = \sqrt{\frac{L \cdot \mathcal{E} \cdot \mathcal{N}}{2}} * Q^{-1}(P_{FA}) \qquad (1)$$

where Q is the tail probability of the standard normal function. The formula shows that the optimal T depends on noise power $\mathcal{N}$, CSS power $\mathcal{E}$, CSS length L, and on the target probability of false positives P_{FA} that we fix to 10^{-8}. We remark that: *(i)* Generally, the power of the noise (which may be due to interfering transmissions) varies in time, thus the detection threshold should also change; *(ii)* Practically, it is difficult to estimate in advance the power of the noise and the power of the signal. In order to address the latter concern, we establish a lower bound on the SINR of the signal that we aim to detect (in our experiments -6 dB), and we tune T correspondingly (i.e., $T = \sqrt{\frac{L \cdot SINR \cdot \mathcal{N}^2}{2}} * Q^{-1}(P_{FA})$). Unfortunately, this solution is not sufficient because of the difficulty to estimate $\mathcal{N}$. In fact, the receiver can only measure the total power of the incoming signal, which may or may not contain the target CSS. Thus, we conservatively choose to replace $\mathcal{N}$ with the total signal power received, as if the incoming signal did not contain the CSS; practically, when the CSS is actually present, this choice has the effect of tuning T to a higher value than desired, i.e., it increases the occurrence of false negatives. Figures 3 and 6 show that the value of the threshold increases when the signal is present, and decreases otherwise.

Wireless Communications Issues. It is important to note that two issues may affect the performance of correlation, both due to the fact that the transmitter and receiver radio generate independent clocks [10, 22]. First, the clock phases at the transmitter and receiver are in general not aligned; this produces a *phase offset* between the two radios, which causes a fixed rotation of the received symbols of an angle γ. In order to compensate for this effect, we compute the magnitude of the correlation, with a theoretical penalty on the processing gain of about 0.5 dB [20]. Second, while the nominal frequencies of transmitter and receiver clocks are identical, they practically differ by a small Δf; this problem is known as *carrier frequency offset.* Carrier frequency offset produces a continuous rotation of the received symbols. Practically, Δf is sufficiently small (e.g., $\sim$ 1-4 KHz [22]), so that its effect is negligible over the CSS lengths/durations considered in this paper.

Hardware Implementation. The hardware implementation of CSS transmission and detection only requires the replication of components which are already present in off-the-shelf 802.11 chipsets, and specifically of filters and correlators. The basic implementation of 11ec needs four correlators (additional correlators may help increase channel reservations robustness as per Section 2.2). Because the correlated BPSK sequences at most require a sign flip on the received I/Q samples and several summations (with respect to expensive multiplications required to implement 802.11 floating-point correlators), CSS correlators occupy a very limited amount of resources.

4. EXPERIMENTAL EVALUATION OF 11EC

In this section, we present an experimental validation of 11ec using a measurement-based emulator that we design and implement. We perform the following experiments.

(i) We investigate the performance of 11ec in basic topologies that typically incur loss or imbalance for 802.11. **Our finding is that 11ec increases the throughput of underserved flows compared to 802.11 with or without RTS/CTS up to 10-fold (resp. 200-fold), with a benefit of almost 60% (resp. 55.6%) in fairness according to the Jain index.**

(ii) We investigate the performance of 11ec in a larger 5-flow network topology. **Our finding is that compared to 802.11 with or without RTS/CTS, 11ec achieves a gain of 30% (resp. 44%) in airtime utilization, and improves a severely underserved flow's throughput from 160 kbps (resp. 0 kbps) to 2.168 Mbps, for a gain of 1255%.**

The results in this section show that 11ec dramatically *improves network fairness*; furthermore, while such improvement can often decrease total utilization, 11ec remarkably *increases channel utilization.* Unlike 802.11, 11ec gives equal opportunities to weak links characterized by low data rates and strong high data rate links; for this reason, 11ec may sometimes achieve lower cumulative network throughput.

4.1 Measurement-driven Network Emulator

Our measurement-driven emulator is based on the GloMoSim simulator [28].[6] For the sake of realism, we modify GloMoSim in two ways: *(i)* we implement the support for multiple 802.11a/g modulations, i.e., BPSK 1/2, QPSK 1/2, 16QAM 1/2 and 64QAM 3/4 (corresponding to 6, 12, 24, 54 Mbps respectively); *(ii)* we implement a new propagation model that calculates links attenuation using our channel measurements. Specifically, with regard to the former, we perform a set of measurements at the channel emulator using the same transmitter/receiver/interferer setup described in the previous section, and we measure the BER as a function of the SINR. With regard to the second issue, we deploy up to 8 WARP nodes simultaneously in an office building (see Figure 7) and measure the signal strength between any pair, i.e., for any run of the experiment a single node transmits 400 μs packets and all others record the received power. As a result of these two measurements, we manually select for each link in the network emulator the highest data rate that its channel SINR can support with negligible packet error probability. Finally, we integrate all results into our measurement-driven emulator.

CSS Implementation. We implement CSS reception and detection as an autonomous physical layer component,

[6]Despite that the last version of GloMoSim dates to late 2001, the basic operations of the 802.11 MAC layer are consistent with the latest standard. The physical layer includes features such as noise accumulation, which make it preferable to alternative simulators.

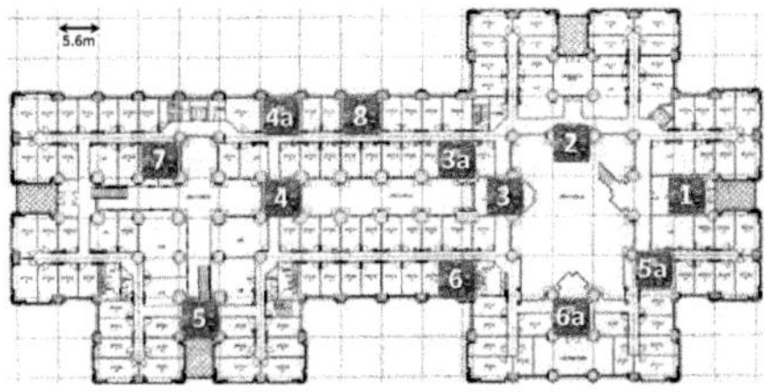

Figure 7: Layout of our office building deployment.

independent of the packet detection architecture of GloMoSim, i.e., CSS's are not simulated via small packets. Specifically, nodes store incoming CSS's, and schedule their evaluation after a delay corresponding to CSS's length, i.e., 6.35 μs for a 127-symbol CSS. For any stored CSS, the emulator keeps track of the variation of the background interference. At the moment of the evaluation the average SINR of the CSS is computed, and CSS detection is triggered if the SINR exceeds a threshold tuned to -6 dB for 127-symbol CSS (see Section 3.3). Our implementation permits each node to simultaneously store, evaluate and potentially detect multiple CSS's overlapping with other CSS's or incoming packets. Note that the original GloMoSim implementation of packet decoding does not support any of the features above, i.e., delayed evaluation and simultaneous multi-signal reception.

Finally, we implemented 11ec's MAC layer state machine by building on GloMoSim's 802.11. In particular, the design integrates the novel procedures corresponding, e.g., to deferral and timeout management.

4.2 Basic Topologies

In this set of experiments, we evaluate the performance of 11ec in a few basic topologies (mostly including two flows) that are characterized by symmetries or asymmetries in link signal strength differences and carrier sensing relationships. This study is important because these topologies are at the origin of throughput losses and/or imbalances in 802.11-based networks [6]. We show that 11ec largely overcomes the problems of 802.11 with and without RTS/CTS.

In our study, we classify the basic topologies into 4 main groups according to the prevalence of one of the two links with respect to the other (e.g., due to higher SINR), and to the carrier sensing relationships between the transmitters. Specifically, the 4 topologies are the following: *(i) Symmetric Hidden Terminals* include the case where two links are formed by transmitters that do not carrier sense each other, and share a common receiver; moreover, the links have similar reception power at the receiver. Specifically, these topologies include links whose signal strengths at the receiver are not different by more than 4 dB; we choose this threshold to exclude capture at any 802.11 modulation. *(ii) Asymmetric Hidden Terminals* include topologies where two non carrier sensing transmitters share a common receiver, but one of the formed links has a significant power advantage. Specifically, these topologies includes links whose signal strengths at the receiver differ by more than 5 dB; the choice of the threshold is to permit to one of the two links to capture over the other at BPSK modulation. *(iii) Information Asymmetries* include link pairs *a*-*b* and *c*-*d* whose transmitters *a* and *c* do not carrier sense each other, and whose receivers differ; moreover, one of the two links *c*-*d* interferes with the other link *a*-*b*, but not viceversa. *(iv) Fully Connected WLANs* include topologies where all nodes carrier sense each other and transmit to a common receiver.

Most of the experiments in this section are performed by reproducing the topologies in our measurement-driven network emulator, with fully backlogged UDP/CBR traffic formed by 1024-byte packets. Each figure includes the bar graph of the throughputs of the flows for the three protocols we compare, namely 11ec, 802.11 with RTS/CTS, and 802.11 without RTS/CTS. Where utilized, RTS/CTS are transmitted at the OFDM 6 Mbps base rate. 11ec implementation includes CSS acknowledgements, but does not support RRTS mechanisms. The experiments in Figures 8(a) to 8(c) involve flow pairs; accordingly, the figures contain groups of six bars that correspond to the throughput of each flow, as achieved by the three protocols. The x-axes of the graphs indicate the data rates of the flows involved in the denoted experiment (when two different values are used, they orderly match the bar pairs), while the y-axes are in Mbps. In Figure 9 we introduce a metric termed *total airtime utilization*, which denotes the time share during which successful data packets are transmitted. Finally, we evaluate fairness according to two well known indicators, namely the Jain index [12], and proportional fairness [15]. The Jain index assumes values in the interval [0, 1]; for both indicators, higher values correspond to higher fairness.

Symmetric Hidden Terminals. We consider four instances of symmetric hidden terminals: three of them are selected from our deployed network, and use modulation rates corresponding to 6/12/24 Mbps, respectively. The last case is artificially generated in order to explore the effect of the use of higher modulation schemes. Figure 8(a) shows that all solutions assign similar throughputs to both flows. However, 11ec and 802.11 with RTS/CTS achieve considerably higher total throughput than 802.11 for most rates. Furthermore, 11ec outperforms 802.11 with RTS/CTS due to the smaller duration of CSS's with respect to control messages (11ec control CSS exchange lasts 19.7 μs, with respect to the 128 μs of 802.11 control messages). This entails a reduction of the probability of collisions (see Section 2.3), and lower control overhead. Both effects become more and more evident as the packet data-rate increases; at 54 Mbps the total throughput gain of 11ec over 802.11 with RTS/CTS is about 25%. Finally, at 54 Mbps data packets are sufficiently short to permit low collision probability to 802.11 without RTS/CTS; nonetheless, 11ec still shows 5% throughput gain.

Asymmetric Hidden Terminals. We consider four instances of asymmetric hidden terminals, all based on actual link power measurements. In these topologies, packets sent at base rate (e.g., control packets) by the sender with high SNR capture over packets sent by the sender with low SNR. However, because of our choice of the data modulation rate as the highest that can be supported by the link in the absence of interference, data packets always collide for both senders. Figure 8(b) shows that the capture effect has disastrous consequences for the flow with low SNR in 802.11 with and without RTS/CTS, while it has no effect on 11ec. For example, in the first instance represented (i.e., the left-most six bars in the figure) 11ec improves the throughput of the under-served flow by 11-fold (resp. by 2-fold) with respect to 802.11 with (resp. without) RTS/CTS. In 802.11 with RTS/CTS the imbalance is due to the fact that in case of overlapping RTS, the RTS of the stronger link is correctly decoded, while the other is ignored. Since 802.11

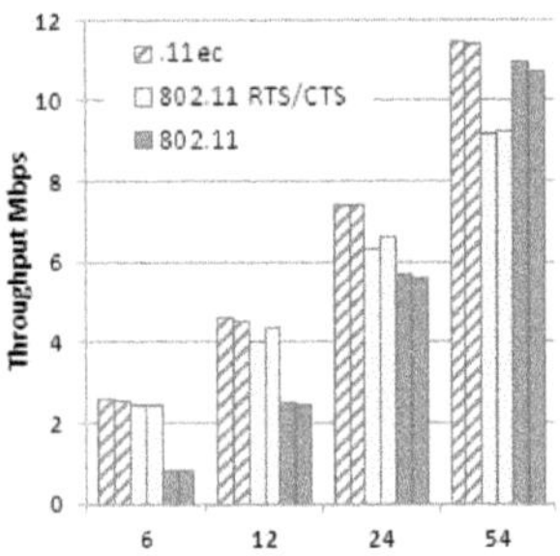

(a) Sym Hidden Terminals

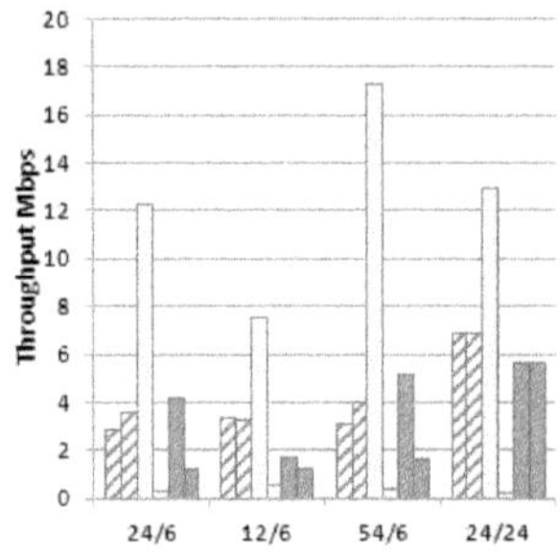

(b) Asym Hidden Terminals

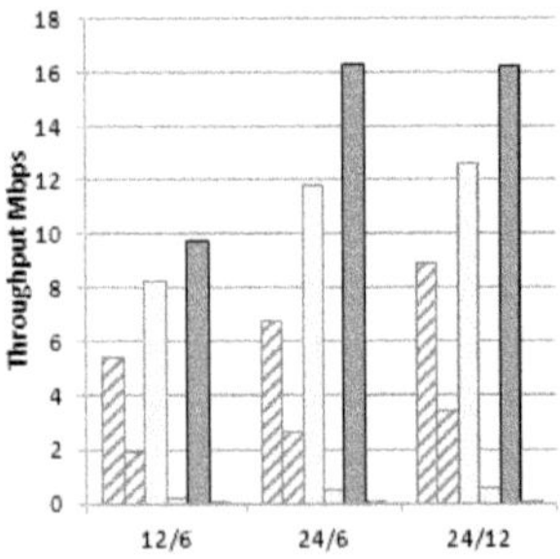

(c) Information Asymmetry

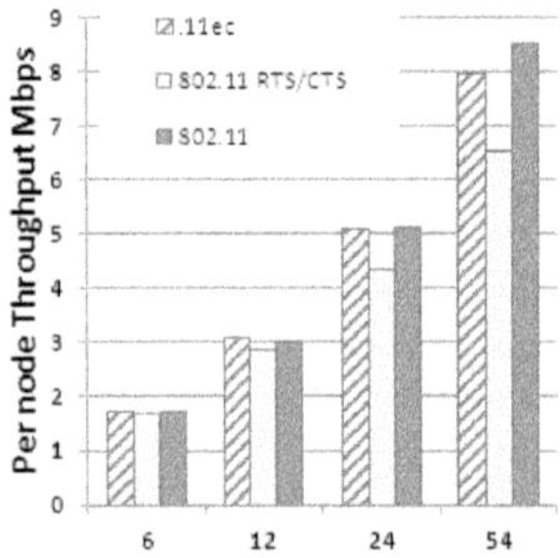

(d) Fully Connected WLANs

Figure 8: Throughput of 11ec, 802.11 with/without RTS/CTS in basic topologies

without RTS/CTS does not use the base rate, but transmits all packets at data rate, the throughput imbalance is originated only by the shorter duration of the data packet of the dominant link, which permits it to enjoy higher success probability. The result for the last instance (i.e., last two bars in the graph) supports this claim: when both links transmit at 24 Mbps, their throughputs do not depend on any SNR imbalance. Finally, in general, because of the larger number of collisions, 802.11 without RTS/CTS has a lower total throughput. In contrast to 802.11, 11ec reduces the imbalance by reducing the number of collisions, and actually reverses the imbalance, i.e., the link with the lower SNR obtains higher throughput. In brief, this is due to the fact that the link with higher SNR has the possibility to collide multiple times with a single packet sent at lower data rate by the link with lower SNR. This deduction is corroborated by the last instance (i.e., the right-most group of six bars) in Figure 8(b); for two links with different SNR, but both using 16QAM 1/2 modulation, the throughputs are identical.

In terms of fairness, considering for example the first instance, 11ec improves the Jain index from approximately 0.52 and 0.76 in 802.11 with and without RTS/CTS to 0.98; similarly, in terms of proportional fairness, 11ec improves the sum of the logs of the rates from 4.69 and 4.82 of 802.11 with and without RTS/CTS to 5.13. The total throughput of 11ec is lower than the competitors; however, this is misleading, and is due to the fact that 11ec improves the throughput of flows with lower data rate. Figure 9 clarifies this aspect, by showing the total airtime utilization for all instances represented in Figure 8(b). 11ec obtains up to 30% higher airtime utilization than the other 802.11 versions.

Information Asymmetry. The three instances we consider are again based on our channel measurements. In these

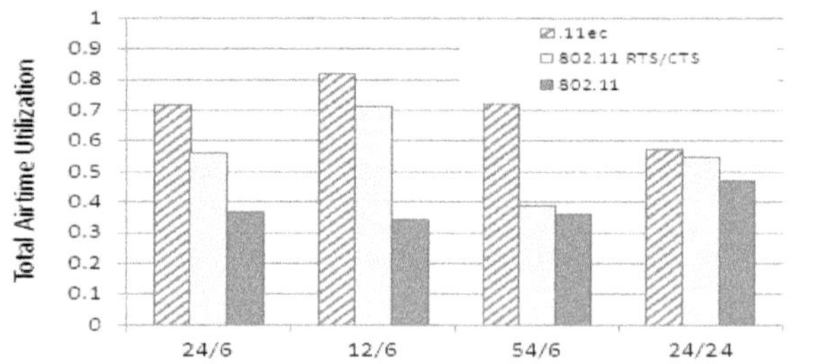

Figure 9: Total airtime utilization in the case of Asymmetric Hidden Terminals.

topologies, the link interfering always succeeds, while the interfered may become severely under-served due to high number of collisions. In fact, the sender of the under-served link cannot perceive when the channel is free at its receiver, and randomly selects transmission instants; in the likely case of collision, the sender of the under-served link decreases its sending rate due to backoff. Figure 8(c) shows that the information asymmetry completely starves the under-served link in 802.11. 802.11 with RTS/CTS slightly outperforms the version without, due to the larger probability of the under-served flow to correctly transmit an entire RTS without being interrupted by the interferer. It is important to notice that several of the bars corresponding to 802.11 flows in this and the next figures are almost or completely invisible. In contrast, even without the feature of control during data, 11ec manages to assign a significant throughput (about 40% of the interfering link) to the under-served link. For example, in the first instance represented (i.e., the left-most six bars in the figure) 11ec improves the throughput of the under-served link by 20-fold (resp. by 200-fold) with respect to 802.11 with (resp. without) RTS/CTS. In terms of fairness, 11ec improves the Jain index from approximately 0.5 in 802.11 with and without RTS/CTS to 0.8, with a 60% gain, and proportional fairness from 4.30 and 3.03 of 802.11 with and without RTS/CTS to 5.14. Similarly to the case of asymmetric hidden terminals, 11ec achieves a cumulative throughput lower than 802.11 but gains up to 26% and 5% in airtime utilization with respect to 802.11 with and without RTS/CTS. These results are due to the small size of control CSS's that have a high probability of being received during free channel intervals.

Fully Connected WLANs. Figure 8(d) shows the average throughput obtained for three carrier-sensing links, and groups of three bars correspond to the three protocols compared. In this scenario, 802.11 without RTS/CTS performs the best due to low overhead and rare collisions. In the worst case of 54 Mbps, 11ec obtains 6% less throughput than 802.11 without RTS/CTS, while 802.11 with RTS/CTS achieves 18% less throughput than 11ec due the control message overhead. At low data rates all protocols perform similarly due to the long packet durations. This result clearly shows that even in the absence of hidden terminals, control CSS's and larger slot size of 11ec do not involve significant throughput penalties.

Final Remarks. First, in the experiments above, we consider fixed (large) packet size and UDP traffic. We notice that: *(i)* as the packet size decreases, the overhead of RTS/CTS becomes more and more relevant, i.e., 11ec is expected to have larger and larger relative gain; *(ii)* throughput imbalances highly affect TCP in most 802.11 cases. By inspecting the traces, we also notice that even in cases of balanced throughput (such as *symmetric hidden terminals*), 802.11 (with and without RTS/CTS) alternately serves for long periods of time one of the two links, by almost starving the other [4]; this is extremely detrimental for TCP performance. Second, our experiments do not implement rate-adaptation, but manually select the best rate achievable based on the links SNR. We observe that rate adaptation does not produce any benefit to 802.11 with RTS/CTS, since control messages need still be transmitted at base rate, and data packets rarely collide. On the other hand, 802.11 without RTS/CTS may benefit in case of hidden terminals, but typically at the price of higher unfairness even in fully connected WLANs due to capture effect [18]. Third, we notice that in the asymmetric topologies, the feature of receiver-side signaling, briefly mentioned at the end of Section 2 and which we do not implement in the experiments in this paper, improves the throughput of links with low SNR due to the capability of a receiver to contend for channel access. Finally, it is remarkable to notice that in contrast to common tenets of related literature, RTS/CTS at 6 Mbps does produce a large performance improvement vs. without RTS/CTS, and only slightly penalizes the throughput in the absence of hidden terminals.

4.3 Network Wide Experiment

Here we investigate larger topologies in order to demonstrate the fairness gains of 11ec in case of multiple flow interactions. We consider a 5-flow topology based on the channel measurements we performed; the flows operate at 24, 24, 24, 54, and 6 Mbps, respectively. Figure 10 shows the detailed bar graph of the throughput of all flows for 11ec and 802.11 with/without RTS/CTS; the flows referred on the x-axis match the node positions in Figure 7. Notice that for need of representation, we do not entirely reproduce the throughput of the flow $5a \rightarrow 1$ in the case of 802.11 without RTS/CTS; that flow's throughput is 21.23 Mbps because of high data rate and strong capture effect. The figure shows that 802.11 with RTS/CTS and 11ec achieve higher fairness than 802.11 without RTS/CTS. In addition, while 802.11 with RTS/CTS almost starves flow $4a \rightarrow 3a$ by assigning 160 Kbps, 11ec manages to assign it 2.168 Mbps for a gain of 1255%. Because the flow operates at 6 Mbps, this has a large effect on the airtime utilization, which increases from 0.5 and 0.45 of 802.11 with and without RTS/CTS to 0.65 of 11ec, for a gain of 30% and 44% respectively. 802.11 without RTS/CTS completely starves that flow. Finally, 11ec significantly increases throughput fairness; specifically, the Jain index is 0.65, 0.23, 0.73 for 802.11 with and without RTS/CTS, and 11ec respectively. 11ec shows about 10% higher proportional fairness with respect to 802.11 with RTS/CTS.

5. RELATED WORK

Many random access MAC protocols have been proposed to mitigate collisions and address hidden terminals, e.g., [4, 13, 25]. In contrast to tones and messages used by such protocols, we design a new primitive of correlatable symbol sequences which, by virtue of being short and robust, increases throughput and fairness with minimal overhead. More recently, while several papers have addressed 802.11 throughput overhead reduction [18, 23], they completely neglect the case of hidden terminals and, because of their more aggressive contention mechanisms, suffer severe throughput penalties in their presence. Ongoing standardization efforts, namely 802.11ah [2], target overhead reduction for sub-GHz communications, with application to smart grids, surveillance systems, etc. The strategy adopted in [2] consists in removing or compressing some information fields of the control messages, for a total of few bytes; compared to 11ec, this has a minor effect on the control message overhead, e.g., due to the preservation of the cumbersome preamble structure. Other techniques have been proposed to improve 802.11 throughput, e.g., [10, 17, 22], but address collision resolution rather than collision avoidance. For this reason, they are complementary to (and can be used in combination with) 11ec. Furthermore, our physical layer model is significantly more simple than [10, 18, 22, 23], since it relies only on the replication of components (correlators) that are already present in common 802.11 cards. Finally, other applications of signal correlation have been recently shown in [11, 24, 29].

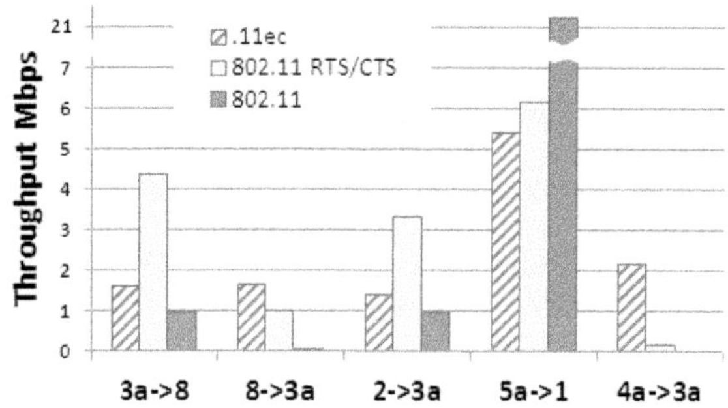

Figure 10: Throughput distribution for a 5-flow topology.

6. CONCLUSIONS

In this paper we introduce 802.11ec, an 802.11-based protocol without control messages. 802.11ec introduces control correlatable symbol sequences which provide robustness and efficiency. Through a wide set of experiments on a software defined radio we show that 6.35 μs correlatable symbol sequences can be detected at an SINR of -6 dB, i.e., 10 dB lower than 802.11 control messages. Finally, we implement 802.11ec on a measurement-based network emulator and show that it improves network fairness by up to 217%, channel utilization by up to 44%, and throughput of under-served flows by 1255%, with respect to 802.11.

7. ACKNOWLEDGEMENTS

This research was supported by NSF grants CNS-1126478 and CNS-1012831 and by a grant from Cisco Systems Inc. O. Gurewitz is in part supported by EU FP7-FLAVIA project, contract number 257263.

8. REFERENCES

[1] IEEE Std 802.11-2007 - Part 11: Wireless LAN Medium Access Control (MAC) and Physical Layer (PHY) Specifications. *IEEE*, 2007. `http://standards.ieee.org/about/get/802/802.11.html`

[2] IEEE P802.11 Sub 1GHz Study Group. http://www.ieee802.org/11/Reports/tgah_update.htm
[3] Rice University WARP project. Available at: http://warp.rice.edu.
[4] V. Barghavan, A. Demers, S. Shenker, L. Zhang. MACAW: a Media Access Protocol for Wireless LAN's. In *Proc. of ACM SIGCOMM*, 1994.
[5] G. Bianchi. Performance analysis of the IEEE 802.11 distributed coordination function. *IEEE Journal on Selected Areas in Communications*, 18(3): 535–547, March 2000.
[6] J. Camp, E. Aryafar, E. Knightly. Coupled 802.11 Flows in Urban Channels: Model and Experimental Evaluation. In *Proc. of IEEE INFOCOM*, 2010.
[7] T. Cui, L. Chen, T. Ho. Energy Efficient Opportunistic Network Coding for Wireless Networks. In *Proc. of IEEE INFOCOM*, 2008.
[8] D.S.J. De Couto, D. Aguayo, J.C. Bicket, R. Morris. A high-throughput path metric for multi-hop wireless routing. In *Proc. of ACM MobiCom*, 2003.
[9] E.O. Elliot. Estimates of Error Rates for Codes on Burst-Noise Channels. *Bell Syst. Tech. J.*, 42(5): 1977–1997, September 1963.
[10] S. Gollakota, D. Katabi. Zig-zag Decoding: Combating Hidden Terminal in Wireless Networks. In *Proc. of ACM SIGCOMM*, 2008.
[11] S. Hong, S. Katti. DOF: A Local Wireless Information Plane. In *Proc. of ACM SIGCOMM*, 2011.
[12] R.K. Jain, D.M.W. Chiu, W. Hawe. A Quantitative Measure of Fairness and Discrimination for Resource Allocation in Shared Computer Systems. *Technical Report TR-301,* DEC Research, 1984.
[13] P. Karn. MACA- A New Channel Access Method for Packet Radio. In *Proc. of ARRL Computer Networking Conference*, 1990.
[14] S.M. Kay. *Fundamentals of Statistical Signal Processing - Vol. 2.* Prentice Hall, 1998.
[15] F.P. Kelly, A.K. Maulloo, D.K.H. Tan. Rate control for communication networks: shadow prices, proportional fairness and stability. *Journal of the Operational Research Society*, 49(3): 237–252, March 1998.
[16] K. LaCurts, H. Balakrishnan. Measurement and Analysis of Real-World 802.11 Mesh Networks. In *Proc. of ACM IMC*, 2010.
[17] T. Li, M.K. Han, A. Bhartia, L. Qiu, E. Rozner, Y. Zhang. CRMA: Collision-Resistant Multiple Access. In *Proc. of ACM MobiCom*, 2011.
[18] E. Magistretti, K.K. Chintalapudi, B. Radunovic, R. Ramjee. WiFi-Nano: Reclaiming WiFi Efficiency via 800 ns Slots. In *Proc. of ACM MobiCom*, 2011.
[19] B. Radunovic, D. Gunawardena, P. Key, A. Proutiere, N. Singh, V. Balan, G. Dejean. Rethinking Indoor Wireless Mesh Design: Low Power, Low Frequency, Full-duplex. In *Proc. of IEEE WiMesh*, 2010.
[20] M. Richards. *Fundamentals of Radar Signal Processing.* McGraw-Hill, 2005.
[21] D.V. Sarwate, M.B. Pursley. Crosscorrelation properties of pseudorandom and related sequences. *Proceedings of the IEEE*, 68(5): 593–619, May 1980.
[22] S. Sen, R.R. Choudury, S. Nelakuditi. CSMA/CN: Carrier Sense Multiple Access with Collision Notification. *Proc. of ACM MobiCom*, 2010.
[23] S. Sen, R.R. Choudury, S. Nelakuditi. No Time to Countdown: Moving Backoff to the Frequency Domain. In *Proc. of ACM MobiCom*, 2011.
[24] K. Tan, H. Liu, J. Fang, W. Wang, J. Zhang, M. Chen, G. Voelker. SAM: Enabling Practical Spatial Multiple Access in Wireless LAN. In *Proc. of ACM MobiCom*, 2009.
[25] F. Tobagi, L. Kleinrock. Packet Switching in Radio Channels: Part II–The Hidden Terminal Problem in Carrier Sense Multiple-Access and the Busy-Tone Solution. *IEEE Transactions on Communications*, 23(12): 1417–1433, December 1975.
[26] R. Van Nee, R. Prasad. *OFDM for Wireless Multimedia Communications.* Artech House, 2000.
[27] A. Willig, M. Kubish, C. Hoene, A. Wolisz. Measurements of a Wireless Link in an Industrial Environment Using an IEEE 802.11-Compliant Physical Layer. *IEEE Trans. on Industrial Electronics*, 49(6): 1265–1282, December 2002.
[28] X. Zeng, R. Bagrodia, M. Gerla. GloMoSim: a library for parallel simulation of large-scale wireless networks. In *Proc. of IEEE PADS*, 1998.
[29] X. Zhang, K. Shin. E-MiLi: Energy-Minimizing Idle Listening in Wireless Networks. In *Proc. of ACM MobiCom*, 2011.

APPENDIX

A. SIGNAL CORRELATION

A CSS s is detected via cross-correlation with a local copy, i.e., at the reception of a complex signal y that may contain s, y is cross-correlated with the complex conjugate of the target CSS $s*$. Formally, for a CSS of length L,

$$\mathcal{C}(\Delta) = \sum_{0}^{L-1} s^*(k)y(k+\Delta) \qquad (2)$$

where Δ represents the position of the correlation with respect to the input signal, i.e., the sample for which we perform the correlation. Note that: *(i)* if $[y(\Delta) \dots y(\Delta+L-1)]$ does not contain exactly s, the value of $\mathcal{C}(\Delta)$ is nearly 0; *(ii)* if $[y(\Delta) \dots y(\Delta+L-1)]$ contains a copy of s with sufficient SINR, the $\mathcal{C}(\Delta)$ obtains a large value proportional to the energy of the signal.

We use cross-correlation as a test statistic for the detection of a target CSS, and repeat its computation at each new sample of the incoming signal. Specifically, detection is performed by setting a threshold T (see Section 3.6); if $\mathcal{C}(\Delta) \geq T$ (resp. $\mathcal{C}(\Delta) < T$), the presence (resp. absence) of the CSS in $y(k+\Delta)$ is declared. The performance of cross-correlation can be quantified in terms of false positives and false negatives. Specifically, after deciding on a threshold T, a false positive is declared when $C(\Delta) < T$ even though the CSS is present within $[y(\Delta) \dots y(\Delta+L-1)]$, and a false negative is declared when $C(\Delta) \geq T$ even though the CSS is absent. Cross-correlation detection provides a processing gain, which is linear in the length of the correlated sequence [14]. This means that a sequence of length $2L$ obtains a value of $C(\Delta)$ twice as large a sequence of length L, i.e., it is considerably more likely to exceed T.

SimpleMAC: A Jamming-Resilient MAC-Layer Protocol for Wireless Channel Coordination

Sang-Yoon Chang
Department of Electrical and Computer Engineering
University of Illinois at Urbana-Champaign
Urbana, IL 61801
chang6@illinois.edu

Yih-Chun Hu
Department of Electrical and Computer Engineering
University of Illinois at Urbana-Champaign
Urbana, IL 61801
yihchun@illinois.edu

Nicola Laurenti*
Dipartimento di Ingegneria dell Informazione
Universitá di Padova
Padova, Italy
nil@dei.unipd.it

ABSTRACT

In wireless networks, users share a transmission medium. To increase the efficiency of channel usage, wireless systems often use a Medium Access Control (MAC) protocol to perform channel coordination by having each node announce its usage intentions; other nodes avoid making conflicting transmissions minimizing interference both to the node that has announced its intentions and to a node that cooperates by avoiding transmissions during the reserved slot. Traditionally, in a multi-channel environment, such announcements are made on a common control channel. However, this control channel is vulnerable to jamming because its location is pre-assigned and known to attackers. Furthermore, the announcements themselves provide information useful for jamming. In this paper, we focus on a situation where multiple wireless transmitters share spectrum in the presence of intelligent and possibly insider jammers capable of dynamically and adaptively changing their jamming patterns.

We develop a framework for effectively countering MAC-aware jamming attacks and then propose SimpleMAC, a protocol resilient to these attacks. SimpleMAC consists of two schemes (the *Simple Transmitter Strategy* and the *Simple Signaling Scheme*) that are easily analyzed using game theory, and show the optimal adversarial behavior under these protocols. We evaluate our schemes mathematically, through Monte Carlo simulations, and by implementation on the WARP software-defined radio platform. SimpleMAC provides very rapid improvement over the alternative of not using any MAC protocol, and eventually converges to optimal performance (over six-fold improvement in SINR, 50% gains in Shannon capacity in a realistic mobile scenario).

Categories and Subject Descriptors

C.2.0 [**Computer-Communication Networks**]: General

*Part of this work was done while Prof. Laurenti was a visiting scholar at the Coordinated Science Laboratory, University of Illinois at Urbana-Champaign.

Keywords

MAC-Layer, Jamming, Wireless

1 Introduction

As wireless features are introduced into more and more electronic devices, it is becoming increasingly important to use scarce radio spectrum as efficiently as possible. An important part of efficient usage is effective coordination of user transmissions. Traditional protocols aim to avoid overlapping transmissions; typical channel access schemes separate users' usage in some combination of time, frequency, and code. As networks increasingly carry data traffic, which is characterized by bursty arrivals, fixed channelization is being replaced by dynamic Medium Access Control (MAC) protocols that change user allocations from frame to frame. In a distributed MAC, each node announces its usage intentions, both to link a transmitter-receiver pair for communication and to help other transmitters minimize interference. In this paper, we refer to this task of exchanging channel usage information as *channel coordination.* In channel coordination, a network user reserves a channel by sending one or more control packets (that contain its channel usage intentions) on a *control channel*, and then uses the reserved *data channel* to send its data traffic.

Channel coordination is only useful when other nodes respect reservations; we call such environments *collaborative environments.* In such environments, channel coordination protocols can provide substantial performance gains. In particular, if we characterize the channel capacity using *Shannon capacity* (as given by the Shannon-Hartley Theorem), we see that when two nodes with equal power levels share a band, coordinating nodes get capacity

$$\frac{B}{2}\log_2\left(1+\frac{S}{N}\right),$$

whereas when they do not cooperate, they get capacity

$$B\log_2\left(1+\frac{S}{S+N}\right),$$

since each node's transmission is interference to the every other node. Whenever the band's signal-to-noise ratio exceeds about 2 dB, coordination provides substantial gains. However, when an attacking node can receive channel coordination information, such as when the attacker is a compromised network node, coordination information can also be used to jam more effectively, since jammers know exactly on which channel to focus their jamming power to disrupt data communication. Also, since the location of the control channel is pre-assigned and known to network users, attackers can jam the control channel, thereby eliminating any benefit legitimate users might gain from channel coordination. In contrast to previous work, we consider attackers that are insiders, intelligent, and adversarial. In particular, such adversaries are capable of reacting to legitimate user strategy ("intelligent"), they have the keys of one or more legitimate nodes ("insider"), and their goal is to minimize the throughput of legitimate users ("adversarial"). We construct a game-theoretic framework to analyze strategies for both adversaries and legitimate users. Our adversarial model represents a worst-case scenario; an attacker with equal power that chooses any other strategy cannot result in worse legitimate user performance. (It is possible that against any particular attacker strategy, another algorithm may provide better performance; optimal performance in such environments is beyond the scope of this paper.)

Our goal is to provide a channel coordination mechanism that maximizes legitimate node throughput in the presence of intelligent jammers. The fundamental difficulty in building adversary-resistant channel coordination is that channel coordination information does good in the hands of cooperative users, because such users will avoid our transmissions, but does harm in the hands of attackers, because attackers will attempt to coincide with our transmissions. We devise SimpleMAC, a scheme in which channel coordination is shared with a specific group of users. SimpleMAC includes the *Simple Transmitter Strategy* (STS), which uses a feedback-based trial-and-error approach to select the group of users that will receive the control packet, and the *Simple Signaling Scheme* (SSS), which establishes a secure communication channel for exchanging control messages that ensures confidentiality against users other than intended recipients and ensures availability against Denial-of-Service (DoS) attacks. With the SSS, a control packet can be received only by intended recipients, and can be received even when the adversary attempts to jam the message. Against an intelligent adversary, SimpleMAC produces desirable results despite its simplicity; our protocol limits the optimal adversarial behavior, quickly outperforms schemes that do not perform channel coordination (the current state-of-the-art channel coordination strategy, which is generally adopted in most security-concerned frameworks [11, 13]), and eventually converges to optimal performance.

Our work applies to both single-channel TDMA systems (with an *energy-limited* attacker, since a power-limited attacker would jam at all times and gain no advantage from channel coordination information) and multi-channel systems (with a *power-limited* attacker). For clarity of presentation, we present our system as applied in multi-channel systems. Because we model legitimate users as cooperative and attackers as malicious, our protocol must simultaneously allow legitimate users to avoid our transmissions and yet prevent attackers from coinciding with them. For this reason, each transmission in our protocol transfer is sent using Frequency Hopping Spread Spectrum (FHSS), in which each transmission is sent while the transmitter hops from one frequency band to another according to a pseudorandom hopping pattern. Channel coordination information thus consists of the time at which a sender plans to send and the frequency hopping pattern the sender plans to use.

The rest of the paper is organized as follows. Section 2 presents security vulnerabilities in current MAC-layer protocols and a brief overview of our approach to resolve these vulnerabilities. After presenting the model to be used in our investigation in Section 3 and setting up our system's theoretical framework in Section 4, we analyze the general jammer behavior in Section 5. We then introduce our scheme in Section 6, and the jammer reaction to our scheme in Section 6.3. Next, we mathematically analyze the performance of our scheme in Section 7 and evaluate it using MATLAB simulations and WARP implementation in Section 8. Lastly, we present conclusions and open problems in Section 10.

2 Problem Statement & Our Contribution

A wide variety of MAC-layer protocols have been proposed for various environments and applications. In this section, we outline a security vulnerability of existing wireless MAC protocols, where the attacker can jam control messages and can use control messages to jam more effectively. We then give a brief overview of our protocol, which is the first to perform channel coordination in a manner that addresses these vulnerabilities.

2.1 Threat Overview & Related Work

To reduce the inefficiencies inherent in simultaneous channel usage, most wireless MAC-layer protocols (with few exceptions, such as ALOHA [22]) attempt to reserve a channel by exchanging channel coordination information. Traditionally, a common control channel is used to exchange channel coordination information among users. There are two important jamming attacks against a control channel: first, the attacker can jam the channel itself, and second, the attacker can use *jamming-relevant information* transmitted on the control channel (such as when and where data transmissions will take place) to facilitate effective jamming.

Our work focuses on *adversarial* entities. Extensive prior work (including [8, 15]) has observed that an attacker can send excessive reservation messages to prevent legitimate nodes from using the channel. Our work is orthogonal, but we suggest limiting each legitimate node to one reservation at a time, and using a central authority to prevent Sybil attacks. Another form of adversarial behavior is channel jamming. Awerbuch et al [5] propose a fair single-channel MAC protocol against a power-limited jammer that does not jam all of the time. Other papers propose mechanisms to avoid jamming [4, 14, 27] but, unlike our work, these approaches are not secure against insider attacks; that is, when jammers are compromised network participants and thus have access to some of the keys of the network nodes, jamming avoidance cannot be assured by this prior work.

2.2 Vulnerabilities in Current Protocols

The use of a common control channel, which is typical in currently available protocols, is vulnerable to jamming attacks. For example, in the IEEE 802.11 WiFi standard [2], nodes use *virtual carrier sense* in which they reserve the channel by exchanging Request to Send (RTS) and Clear to Send (CTS) messages; these messages can be jammed to reduce network performance.

Though virtual carrier sense provides an effective way for a legitimate potential transmitter to avoid collisions with another transmitter, the very mechanism that allows them to mitigate interference also allows an attacker to jam every transmission. In particular, whenever an attacker senses another user's transmission, either through carrier sense or virtual carrier sense, the attacker can jam the corresponding data packet.

In addition to carrier sense and virtual carrier sense, WiFi also uses a Collision Avoidance mechanism in which a node transmitting a frame chooses a backoff interval. The node counts down the backoff interval whenever the channel is idle; this mechanism reduces the probability that two nodes will transmit simultaneously. Several researchers have investigated the attack wherein the attacker chooses incorrect backoff intervals [9,17,21].

In the Out-of-Band signaling scheme [12], each receiver sends a very narrowband busy tone whenever it receives data to indicate the channel is in use. A powerful adversary may be able to jam the busy tone, and even when the jammer is unable to remove the busy tone, a jammer that hears a busy tone knows that a receiver is active within its wireless transmission range. A jammer that jams the data channel whenever it hears a busy tone can effectively deny service to receivers within its interference range. An attacker can also falsely reserve the channel by continuously sending a busy tone.

IEEE 802.16 [3], commonly called WiMAX, uses a centralized scheduling algorithm in which the base station assigns time slots to each user. Since the base station broadcasts control messages, a jammer that knows the location of the control channel can either jam the control channel to disrupt the exchange of control messages or use the received channel scheduling information to jam data transmissions at the assigned time slots (and frequency channels). In WiMAX, the control channel location is a published part of the standard, but even if it were not, an attacker that has compromised a legitimate node must know the location of the control channel. Furthermore, a node can request and be scheduled for time and frequency slots that it does not need, thus wasting time and bandwidth.

Another centralized protocol is Bluetooth [1], in which a master device sends control messages to each slave device in the network (called a *piconet*). Since an attacker who knows the frequency hopping pattern of a scheduled transmission can easily jam that transmission. In Bluetooth, these frequency hopping patterns are a public part of the standard, but even if it were not, an attacker that has compromised a legitimate node must know the location of the control channel. Furthermore, an attacker can become the master and have significant control over other legitimate users.

MAC protocols that do not perform channel coordination suffer from higher probability of collisions between simultaneous transmitters, resulting in more interference. Thus, a protocol that lacks channel coordination functionality yields lower SINR and therefore lower capacity.

In conclusion, currently implemented protocols either mitigate interference from legitimate nodes by regulating their channel usage, in which case jammers can effectively jam during legitimate node usage, or provide no channel coordination and suffer from increased interference. In this paper, we present the first protocol that mitigates interference from both legitimate users and jammers.

2.3 Our Approach

Current protocols, when faced with just one intelligent, insider jammer, will at best reach the Nash equilibrium, in which channel coordination is completely disabled and each message is spread across the entire band [11, 13]; at the Nash equilibrium, there is no reduction in collisions, which reflects a non-cooperative environment. In this paper, we construct a theoretical framework to analyze the dynamics between the adversaries and legitimate users, then propose SimpleMAC, a MAC-layer protocol that performs channel coordination while mitigating the effects of jamming.

In Section 4.1, we describe our MAC-layer framework. A MAC protocol provides reduced probability of collision by exchanging a channel usage plan with other network users; this channel usage plan is *jamming-relevant information* because the plan allows legitimate users to avoid the transmitter but also allows jammers to intentionally collide with the transmitter. We divide our scheme into two components. The *transmitter strategy* selects the set of network nodes with which to share the relevant control message; we call this set, which may vary for each packet, the *recipient list* and denote it S. The *signaling scheme* delivers a control message to each node in the recipient list, and ensures that no other nodes are able to receive the control message. When the adversary is malicious, the ideal recipient list includes all legitimate users and no attackers.

SimpleMAC consists of the *Simple Signaling Scheme* (SSS) and the *Simple Transmitter Strategy* (STS). We develop a jamming-resistant signaling scheme to deliver jamming-relevant control message to exactly the set of nodes in the recipient list S, and a transmitter strategy that decides on the recipient list based on prior receiver feedback. In our transmitter strategy, the transmitter-receiver pair measures the performance of S after sending each packet and uses this information to adapt S for future packet transmissions. The long-term goal for the transmitter is to search for a set S that provides the optimal performance. As a general rule, the choice of $S = \emptyset$ (equivalent to disabling channel coordination) represents baseline performance; after a sufficient number of independent trials, any set that performs significantly worse must contain a jammer. Therefore, the transmitter can determine whether channel coordination has been compromised by comparing the performance of the recipient list with performance when $S = \emptyset$. However, since attackers are intelligent, and thus capable of dynamically changing their jamming strategy, a recipient list S with better performance than when $S = \emptyset$ does not necessarily mean that S excludes all attackers.

SimpleMAC quickly outperforms the case where channel coordination is disabled, eventually converges to the recipient list offering optimal performance, and forces the optimal jammer strategy to be jamming at full power all the time (even though jamming alerts the user and prompts it to stop sharing information with the compromised recipient lists).

3 System Model & Assumptions

We consider an environment with $T + 1$ non-idle transmitters (each transmitter has T potential interference sources), each identified by an index $i \in \mathcal{T} = \{1, \ldots, T + 1\}$, a subset $\mathcal{N} = \{i_1, \ldots, i_N\}$ of which are N jammers. All non-jammers are protocol-compliant. We assume a shared secret key between each pair of nodes, and that all nodes operate in shared spectrum divided into C channels, each with bandwidth W Hz. No online authority governs users. We consider a repeated game with infinite horizon; either the transmission never ends or the users do not know when the transmissions will end. We index the rounds of the game $r \in \{1, 2, 3, ...\}$.

We assume that each user has technical means to transmit on the spectrum, that the attacker ignores any legal prohibitions against interference, and that there is no way to *a priori* determine which node is trustworthy. Also, because of the possibility of jamming, we make the standard assumption [20, 23] that each transmitter sends data using fast frequency hopping on randomly generated hopping patterns chosen independently for each packet. Traditionally, fast frequency hopping is characterized by a hopping time of more than one hop per symbol; here, we only require that the hopping time be faster than the jammer's reaction time.

At the physical layer, we assume there exists a known spreading gain at which any pair of neighbors can communicate with a suitably low bit error rate. Alternatively, we define a *neighbor* as a node that can be reached using a specific spreading gain. In our SimpleMAC, described in Section 6, we use Direct Sequence Spread Spectrum (DSSS) for control communications and Frequency Hopping Spread Spectrum (FHSS) for data communications, and we communicate the frequency hopping pattern in the control message.

The communication between any transmitter-receiver pair is single-hop; that is, the transmitter does not rely on a third node to relay the message to the final destination. For simplicity, in our model, each user is a neighbor of every other user. Thus, when two nodes transmit on the same frequency at the same time, those transmissions interfere with each other, resulting in reduced capacity. Our protocol can be extended to hidden-terminal environments by having both sender and receiver repeat each channel coordination transmission, although the details of this approach are beyond the scope of this paper.

Our protocol is designed for unicast data transmissions, where performance affects only the receiver. Therefore, when a sender transmits to multiple receivers, the feedback of a malicious receiver does not affect the performance of other receivers. However, a malicious receiver may be able to induce a sender to choose $S = \emptyset$ for all transmissions to that receiver. The impact of this selection depends on the transmission priority scheme, so we discuss this attack further in Section 9.

All users, including attackers, share the same power constraint P_c. The case where each attacker is more powerful than a normal user can be modeled by increasing the fraction of nodes which are attackers.

3.1 Performance Metric

When user i transmits to user j, it does so on a frequency channel that varies with time according to a frequency hopping pattern known to user i and user j. At any point in time, the user transmits on frequency channel $c \in \{1, \dots, C\}$. Assuming a flat fading channel with additive white Gaussian noise and Gaussian signals, the Shannon capacity of the link $i \to j$ is:

$$\mathcal{R} = \int_{f_c - W/2}^{f_c + W/2} \log_2 \left[1 + \text{SINR}_{i,j}(f)\right] df \qquad (1)$$

where f_c is user i's carrier frequency, and SINR is the effective signal-to-interference-and-noise ratio at the receiver

$$\text{SINR}_{i,j} = \frac{\gamma_{i,j}\widetilde{P_i}(f)}{\widetilde{N_0} + \sum_{\ell \neq i, \ell \in \mathcal{N}^c}[\gamma_{\ell,j}\widetilde{P_\ell}(f)] + \sum_{k \in \mathcal{N}}[\gamma_{k,j}\widetilde{J_k}(f)]} \qquad (2)$$

In Equation 2, $\gamma_{a,b}$ is the channel gain between transmitter a and receiver b, $\widetilde{N_0}$ is the power spectral density of the noise, $\mathcal{N}^c$ are the indices of legitimate users, $\mathcal{N}$ are the indices of jammers, $\widetilde{P_\alpha}$ is transmitter α's power spectral density for some α, and $\widetilde{J_k}(f)$ is the jammer k's power spectral density.

Shannon capacity ($\mathcal{R}$) is an upper bound on communication rate performance. Shannon capacity is a mathematically simple formulation and is tight in many practical environments; existing codes very nearly achieve Shannon capacity [18]. In order to separate MAC-layer issues from physical-layer decisions such as modulation and coding, we use both SINR and Shannon capacity as representative performance metrics in our mathematical analysis. We observe that Equation 1 exhibits two properties that we use in our analysis: it is decreasing and convex with respect to jamming power and monotonically increasing with respect to the user's signal power. Though we use SINR and capacity as representative measures of performance, our approach generalizes to any utility function that is convex in interference power and monotonically increasing in SINR. (In Section 8, our implementation testbed simulation results show the effective SINR at the receiver, because achieving Shannon channel capacity involves sophisticated coding and modulation, and because the instantaneous capacity is strictly monotonic in the instantaneous SINR.)

Channel capacity $\mathcal{R}$ (Equation 1) serves as our utility function for the legitimate transmitter i. The transmitter's aim is to maximize its capacity $\mathcal{R}$. As $\mathcal{R}$ is a monotonically increasing function of P_i, the transmitter will emit full power. To aggregate capacity (which is an instantaneous metric) over time, we compute its time-average. At time t, given $\{\mathcal{R}_{t'} \mid t' < t\}$, the utility function is:

$$U = \frac{1}{N_t} \sum_{t'=t-N_t+1}^{t} \text{E}[\mathcal{R}_{t'}] \qquad (3)$$

where $\mathcal{R}_\gamma$ is the capacity measured at time γ for some γ. In an infinite-horizon game, Equation 3 is replaced by its limit as $N_t \to \infty$, where N_t represents the time duration of transmission.

3.2 Attacker Model

We consider a jammer that intends to minimize the utility function, Equation 3, subject to its power constraint:

$$\text{minimize } U \text{ subject to } \int_f \widetilde{J_k}(f)\, df \leq P_c,\ \forall k \in \mathcal{N} \qquad (4)$$

We assume that jammers collude. Thus, if one jammer knows a user's frequency hopping pattern, then all jammers can make use of that information. Also, attackers know the protocol and can adaptively change their strategies according to the legitimate users' strategies.

We also consider reactive jammers that jam according to their observations on the target signal, in case they do not receive the user's channel coordination information. To counteract reactive jammers, the user can shorten the frequency hopping time so that the jammers do not have enough time to observe the spectrum and jam the used channel.

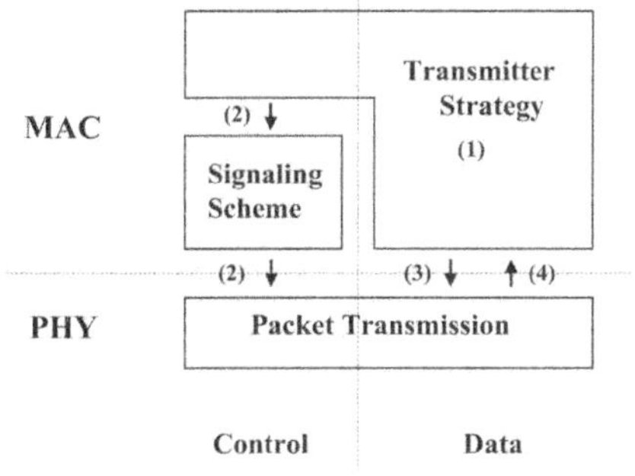

Figure 1: Our Protocol Framework

We do not consider the very strong attack of correlated jamming, where an adversary mimics the target signal with a phase offset of π at equal amplitude, canceling the target signal. Under this attack, assuming the attacker has at least as much power as the legitimate node at a target receiver, no physical layer can provide any throughput [7].

Attackers can choose between narrowband jamming (concentrating its power on one or a subset of frequency channels at a time) or wideband jamming (emitting power across the spectrum at a time) and can freely switch between these strategies. Because the jammer has so much flexibility, we do not consider legitimate user attempts to infer information about a jammer; however, our approach still converges to the optimal performance.

We also consider the possibility of non-Gaussian jamming, since Equation 1 holds only when all received signals are Gaussian signals. Given a Gaussian fading channel with Gaussian signals, where the overall signal power is much greater than the combined power of the jammer network, the optimal jammer strategy is to jam with Gaussian noise [7, 16]. Also, the transmitter can make any received jamming signal appear Gaussian by using a sufficiently long Direct Sequence Spread Spectrum (DSSS) code shared only with the receiver. When jammers do not know the code used by the transmitter, the received jamming signal looks like a Gaussian signal by the Central Limit Theorem.

4 Theoretical Framework

4.1 Overview

We design SimpleMAC from the ground up without making any MAC-layer assumptions. Our MAC-layer framework contains two parts: a transmitter strategy and a signaling scheme. For each packet, a transmitter strategy determines the set of users S that will receive the channel coordination information for that packet; this set is called the *recipient list*. The optimal transmitter strategy will prevent the attacker from gaining any advantage from its knowledge of insider network keys while minimizing interference from legitimate users. A signaling scheme delivers the control message to the recipient list and provides availability (that is, messages are not easily jammed) and confidentiality (that is, nodes not on the recipient list will not receive the control message).

Figure 1 depicts our MAC-layer framework. When sending a packet, the transmitter (1) chooses a subset S of network users, (2) transmits its frequency hopping information to S, (3) transmits the data packet using the previously reserved hopping pattern, and (4) determines the effectiveness of S based on the feedback that it receives from the receiver.

4.2 Collision Between Benign Users

Even when the spectrum is used very sparsely, a randomly selected frequency hopping pattern is likely to collide with the hopping patterns of other nodes. Channel coordination schemes are designed to reduce this inter-transmitter interference. When two nodes wish to use the same channel during the same time slot, they each determine which of the transmitters has priority, for example, based on the time at which each node claimed the channel. The node that does not have priority will not transmit at all, so the corresponding receiver will decode random data in this position. However, since the positions for these lost bits are known *a priori*, a sender mitigates the loss by using channel coding and forward error or erasure correction.

Two nodes may collide when neither node informed the other about its frequency hopping patterns, and, depending on the priority scheme, when only one of the two nodes disclosed its transmission intentions to the other. In our analysis, we assume that transmitter i has the highest priority for transmission, and is targeted by all the jammers. In other words, all nodes in the transmitter's recipient list will avoid interfering with the transmitter (we revisit this assumption in Section 9 and consider the case when all nodes have equal priority). Thus, increasing the number of benign transmitters in S reduces the number of potential interferers, increasing capacity.

4.3 Capacity Expression for Our Framework

In this section, we mathematically derive the capacity of our system when all channels have equal gains and all users emit power uniformly across their chosen channel. Since legitimate users that receive the transmitter's channel coordination information will not interfere, the transmitter's capacity depends on its selection of recipient list S. Equation 1 can be simplified to:

$$\mathcal{R}(S) = W \cdot \log_2 \left[1 + \frac{P}{N_0 + \sum_{\ell \neq i, \ell \in (\mathcal{N}^c \cap S^c)} P_\ell + \sum_{k \in \mathcal{N}} J_k(S) \cdot P_c}\right]$$

where N_0 is the noise power in the channel, P is the transmitter's signal power, P_ℓ is the amount of user ℓ's power that interferes with the transmitter's signal, J_k is the jammer k's power normalized with respect to the power constraint P_c, and

$$J_k(S) \leq \begin{cases} \frac{1}{C}, & \text{if } (S \cap \mathcal{N}) = \emptyset \\ 1, & \text{otherwise} \end{cases}$$

In the $(S \cap \mathcal{N}) = \emptyset$ case, the jammer does not receive the user's channel coordination information, and therefore can at best conduct wideband jamming across C channels, as described in Section 5.

If we further assume that legitimate users not in S emit at full power to maximize their own performance, then $E[P_\ell] = \frac{P_c}{C}, \forall \ell$, since there is a $\frac{1}{C}$ chance that any legitimate user not in S will interfere with the transmitter. Then, using Jensen's inequality, the expected capacity is bounded from below by:

$$\mathrm{E}[\mathcal{R}(S)] \geq W \cdot \log_2 \left[1 + \frac{P}{N_0 + |\mathcal{N}^c \cap S^c| \frac{P_c}{C} + \sum_{k \in \mathcal{N}} J_k(S) \cdot P_c}\right] \quad (5)$$

We use this expression in our analysis.

4.4 Transmitter Strategy

To make future recipient list decisions, each network user records historical performance data for each packet, including the recipient list used for that packet and the resulting performance. One natural choice for the recipient list is the set that has yielded the best average performance in the past. We call this the *Best so far* set and denote it with S_B: $S_B(t) = \underset{\sigma \in \{S(t'), \forall t' < t\}}{\text{argmax}} \bar{\mathcal{R}}(\sigma)$, where $\bar{\mathcal{R}}$ is the time-average performance. When jammers jam at full power, the optimal set is the set that includes all legitimate nodes and excludes all jammer nodes; we denote this optimal set S^*. However, an attacker might choose not to jam when certain nodes are in the recipient list, so S^* may not have the maximum performance for a particular jammer strategy; however, S^*'s performance is optimal in the worst case. Our scheme will converge to at least the performance of S^*, but if the attacker concedes better performance, our scheme can take advantage of the better-performing set.

In order to improve the Best so far recipient list, a sender must explore possible sets from time to time. To reach optimal performance, a transmitter strategy must eventually explore the optimal set. When we do not know the jammer's strategy or the distribution for the number of jammers, the optimal set may be any set, because the attacker may choose to cease all jamming activities when a particular recipient list is chosen. Thus, to provide optimality against arbitrary attackers, a sender must be willing to explore all possible sets. In our framework, as well as our protocol, SimpleMAC, convergence to the optimal set takes exponential time in the average case; however, we will show in Section 8 that we improve over the state-of-the-art within a single round in many cases, and fast convergence is not a goal of our design.

5 Jammer Strategy Analysis

In this section, we assume that the attacker is purely adversarial, as described in Section 3.2. Attackers are capable of using a potentially non-deterministic, time-varying strategy to meet their goal of minimizing capacity. Since an attacker's strategy depends on whether or not it receives the channel coordination information, we study both cases.

5.1 Recipient List With No Jammer

If the recipient list S contains no jammers, then jammers do not learn any jamming-relevant information, and thus do not know which channel will be used for the user's transmission. This limits jammers to a much weaker attack, since they cannot use their compromised keys and gain no advantage from collusion. The only decision to be made in this case is whether to choose narrowband jamming or wideband jamming. We assume that a legitimate user i will uniformly choose any of the C channels, and we observe that $\mathcal{R}$ is a decreasing and convex function of $\widetilde{J_k}(f)$. By Jensen's inequality, the expected capacity $\mathrm{E}[\mathcal{R}]$, under the constraint of Equation 4, is minimized by choosing $\widetilde{J_k}(f) = \frac{P_c}{C \cdot W}$ for each jammer k. Thus, to minimize capacity, jammers will conduct wideband jamming when they do not know the frequency hopping pattern, but conduct narrowband jamming when they do have the information. In our analysis, we assume the jammer uses this strategy when it does not know the frequency hopping pattern.

5.2 Compromised Recipient List

We now analyze the jammer strategy when a jammer does receive the user's channel coordination information. In this scenario, jammers know where to concentrate their power to minimize transmitter capacity. However, in an infinite-horizon repeated game, jammers must also consider how their current action will affect future capacity. Equation 1 shows that jamming with higher power causes more interference and lowers capacity. However, since a user will avoid any set S that appears to contain jammers, jammers may not wish to strongly jam the transmission, hoping to reduce the user's suspicions that S contains a jammer. If the user converges on a new S_B that contains no jammers, then jammers can no longer influence capacity except by wideband jamming. A jammer may then want the *Best so far* set to include a jammer by abstaining from excessive jamming.

In the long run, the jammer knows that the transmitter will explore S^*, and will choose the best set. If the jammer allows another set S' to have better performance than S^*, then the transmitter will pick S', otherwise it will pick S^*. If the jammer's goal is to minimize capacity, they should not concede any additional long-run performance to the sender. Thus the sender will choose $S_B = S^*$, and the optimal jammer strategy will converge to full-power jamming.

CLAIM 1. *Given the general transmitter strategy in Section 4.4, jammer strategy converges to full power over time:*

$$\forall k \in \mathcal{N}, J_k(t) \to 1 \text{ as } t \to \infty$$

PROOF. Omitted due to space constraints. □

6 Our Protocol, SimpleMAC

Our SimpleMAC protocol has two components: the Simple Transmitter Strategy (STS) and the Simple Signaling Scheme (SSS). Despite the simplicity of the schemes, from which SimpleMAC derives its name, SimpleMAC effectively combats intelligent attackers: it quickly outperforms the case where MAC protocol is disabled (which is the standard approach for securing MAC protocols) and has an easily analyzed optimal jammer strategy.

When selecting a recipient list, we determine the effectiveness of recipient list S by comparing the capacity when S is chosen as the recipient list to the capacity when no one knows the recipient list. In the latter case (i.e., when $S = \emptyset$), there is neither gain in capacity from legitimate nodes avoiding the transmitter nor loss in capacity from the jammers using the jamming-relevant information. Whenever the capacity is less than or equal to (with some error margin) the capacity when $S = \emptyset$, the transmitter chooses a new set S before the next transmission, because the current set S provides no advantage over $S = \emptyset$.

SimpleMAC does not try to infer which nodes are jammers and which ones are not; rather, it directly uses channel feedback to determine which recipient lists result in good performance. For example, when node A shares its information with a jammer, any recipient list with node A in it will have decreased performance, so the STS will avoid such list. Similarly, if node A jams only when node B is also in the recipient list, the STS will avoid lists that contain both A and B. Because we make all of our decisions based on actual performance and not behavior, SimpleMAC is immune to collusion.

6.1 Simple Transmitter Strategy

In the STS, for each transmission t, a legitimate user has three options when choosing a recipient list S:

1. *Best so far* (S_B): the set with best average performance among explored sets, as described in Section 4.4.

2. *Randomly explore* (S_R): chosen uniformly at random among all possible sets.

3. *Empty set* ($\emptyset$): $S(t) = \emptyset$.

The transmitter always chooses one of these three strategies. The *Best so far* action, $S(t) = S_B$ corresponds to choosing the set that yielded the highest average capacity among all the recipient lists that have been tried through time $t-1$, which guarantees performance at least as good as $S = \emptyset$, since $S = \emptyset$ has been tried earlier. If jammers jam with sufficient power ($\sum_{k \in \mathcal{N}} J > \frac{T}{C}$), then the set S that yields the highest capacity is S^*. In this case, when the user explores sets occasionally (so that the user eventually visits all possible sets with probability one), the *Best so far* set S_B converges to S^*, since the probability that S^* has been previously chosen approaches one. The user chooses the *Randomly explore* action to search for a set that yields higher capacity than the previous *Best so far* set. Once such a set is found, the set S_R becomes the new *Best so far* until the node discovers another set that yields even higher capacity. The more often the user chooses to explore a random set, the more quickly S_B converges to S^*. The *Empty set* action establishes baseline performance during each time interval, so that slow time-variance in channel conditions do not bias set selection.

The STS operates in rounds. For each transmission t within round r, the user makes an independent random choice among the three options. The probabilities may vary with r, so that in expectation, round r contains $B(r)$ transmissions with the Best so far recipient list, $R(r)$ transmissions on a randomly chosen recipient list, and $E(r)$ transmissions using an empty recipient list. Round r lasts for $B(r)+R(r)+E(r)$ transmissions, and we do not rely on the secrecy of $B(r)$, $E(r)$, $R(r)$.

In order to converge to the optimal performance, we explore a user strategy where the user uses the *Best so far* set more and more often, while occasionally using *Randomly explore* and *Empty set*. One such user strategy is:

$$B(r) = r^{\delta}, R(r) = 1, E(r) = 1, \forall r \qquad (6)$$

In order to converge to the optimal performance for S^*, δ needs to be positive. A higher δ corresponds to more aggressive search for a better *Best so far* set and thus quicker convergence to S^*.

6.2 Simple Signaling Scheme

In order to send a control message to exactly those nodes in a recipient list, we need a signaling scheme that provides confidentiality against malicious entities and reliability in the presence of jamming. We make no attempt to design an efficient signaling scheme; because the overhead of a control message is amortized over the data frame, and because we can choose arbitrarily long data frames, we can reach near-optimal overall protocol performance even with an extremely inefficient signaling scheme. Thus, we simply unicast the control messages to each recipient in the recipient list. We provide confidentiality by encrypting messages with a symmetric key, and availability by using direct sequence spread spectrum (DSSS) using a chip sequence known only to the sender and receiver. In a 50 node network with 20 byte reservation messages (consisting of source address, destination address, and a seed for the hopping pattern), if each reservation covers 100 kB of data (for example, 66 packets each 1500-bytes long), SSS incurs an overhead of not more than 1%, and average overhead of 0.5%. Though the data rate may be higher than the control rate due to our use of DSSS for the control message, we can continue to keep our overhead low by covering more data with each control message, or by replacing repeated-unicast with a jamming-resilient broadcast protocol, of which several have been proposed [6, 10, 24]. SSS simply requires that each node have a pairwise shared key with every other node. Such keys can be established through Diffie-Hellman exchanges over a jamming-resilient broadcast protocol.

6.3 Jammer Reaction to SimpleMAC

In Section 5, we studied the attacker strategy under our general framework and showed that the optimal attacker strategy converges to full power, even though an attacker may wish to avoid detection so that the legitimate user will use a compromised recipient list. In this section, we claim that against the STS, optimal jammers *jam at full power all the time*. The claim holds because B, R, E are independent of the jammers' strategy; unlike the user's selection of recipient lists, the user's choice of action (B, R, and E) does not adapt to jammer strategy. Intuitively, the sender forms a partial order on recipient lists based on their past performance. An attacker that does not jam at full power can jam at a higher power level and yet maintain the same partial order of recipient lists (or a functional equivalent), which shows that any strategy that does not jam at full power cannot be optimal.

CLAIM 2. *Against the STS, the best jammer strategy is to emit at full power all the time, i.e.,*

$$\forall t, \forall k \in \mathcal{N}, J_k(S) = \begin{cases} \frac{1}{C}, & \textit{if } (S \cap \mathcal{N}) = \emptyset \\ 1, & \textit{else} \end{cases}$$

PROOF. Proof is by contradiction. Suppose that an optimal jammer strategy $J = (J_1, J_2, ..., J_N)$ does not jam at full power at some time; we let Γ be the set of times at which J does not use full power. We now show that there exists a different jammer strategy J' that yields less capacity than J while preserving the legitimate user strategy. To find such J', we assume perfect knowledge about the recipient list S. (This does *not* mean that a jammer needs perfect information; rather, it shows that even a jammer with perfect information will still choose the simple strategy of full-power jamming, and therefore *any* attacker should do the same.) J' will only diverge from J when J does not emit at full power,

At time $t' \in \Gamma$, let the two best previously measured recipient lists be A and C, where A is the best and C is the second-best. Then either $S = A$ yields higher capacity than $S = C$ or both sets $S = A$ and $S = C$ yield the same performance. We study the two cases separately:

i) If $\mathcal{R}(A) > \mathcal{R}(C)$, then pick $J'(t')$ such that $\mathcal{R}_{J'}(A) = \frac{\mathcal{R}_J(A)+\mathcal{R}_J(C)}{2}$. This choice preserves the performance order of recipient lists and thus do not change the user's choice of recipient list.

ii) If $\mathcal{R}(A) = \mathcal{R}(C)$, then pick $J'(t')$ such that its corresponding performance is ϵ smaller than that of $J(t')$ for small ϵ. This breaks the tie between A and C since $A \neq C$.

$$\begin{aligned}
\mathrm{E}[\mathcal{R}]_{\mathrm{ss}} &= \lim_{r\to\infty} \mathrm{E}[\mathcal{R}(r)] \\
&= \frac{B}{B+R+E} \cdot W \log_2\left(1+\frac{P}{N_0+\frac{N}{C}}\right) \\
&+ \frac{R}{B+R+E} \sum_{\alpha=0}^{T-N} \left\{ \frac{\binom{T-N}{\alpha}}{2^{T-N}} \right. \\
&\cdot \left[\frac{1}{2^N} \cdot W \log_2\left(1+\frac{P}{N_0+\frac{N}{C}+\frac{\alpha}{C}}\right) \right. \\
&\left.\left. + \frac{2^N-1}{2^N} \cdot W \log_2\left(1+\frac{P}{N_0+N+\frac{\alpha}{C}}\right)\right]\right\} \\
&+ \frac{E}{B+R+E} \cdot W \log_2\left(1+\frac{P}{N_0+\frac{T}{C}}\right) \qquad (9)
\end{aligned}$$

Though this changes the legitimate user's choice of recipient list (because the legitimate user will choose A over C for the *Best so far* set), the legitimate user strategy when jammer picks J' is functionally equivalent to the legitimate user strategy when jammer picks J (because when the jammer picks J, it does not matter whether the user chooses A or C). Therefore, J' yields smaller capacity than J.

Since, in both cases J' yields lower capacity than J while preserving the order of recipient lists (and thus preserving the legitimate user strategy or its equivalent), J is not optimal and there is a contradiction. □

7 Performance Analysis

Under the STS, a legitimate user chooses the recipient list from among three options: *Best so far* (B), *Randomly explore* (R), and *Empty set* (E). We use Equation 5 to determine the expected capacity. Since attackers in S jam at full power, as shown in Section 6.3, $S = S^*$ (including no jammers but all other legitimate users) yields the optimal performance. Thus, the expected capacity varies with time (in units of rounds r) until it reaches the steady state where $S_B = S^*$. The steady-state expected capacity is shown in Equation 9, where α denotes the number of legitimate users outside S (who could potentially cause interference to the transmitter) and $\binom{a}{b} = \frac{a!}{b!(b-a)!}$.

For the transient expected capacity, $\mathrm{E}[\mathcal{R}|\mathit{Random}]$ and $\mathrm{E}[\mathcal{R}|\mathit{Empty}]$ are constant in time, whereas the expected capacity for *Best so far* varies with time. The user chooses $S_B = \emptyset$ at round r if all the previously explored sets contain jammers; otherwise, he chooses the set that contains no jammer and the most nodes (minimizing α). The term $\mathrm{E}[\mathcal{R}(r)|\mathrm{Best}]$ for the rth round is expressed in Equation 10 where β corresponds to the number of times that the user found a jammer-free set, and $\mathcal{B}_i$ are independent Binomial random variables with probability 0.5 ($p = 0.5$) and $T - N$ trials ($n = T - N$), since $T - N$ is the number of protocol-compliant users.

8 Evaluation

In earlier sections, we have analyzed SimpleMAC theoretically, in a manner that is general and not limited to any particular system design. In this section, we evaluate our scheme in practice both using MATLAB simulations and a testbed implementation on the WARP software radio platform [19]. As described in Section 3.1, we use SINR as our metric in this section both because capacity is strictly monotone in SINR and because we can evaluate SINR improvements without needing to make perfect modulation and coding choices that are necessary to achieve channel capacity.

8.1 Methodology and Metric

We built our implementation on the WARP (Wireless Open-Access Research) software-defined radio platform. We used four WARP boards: one acting as the source, one acting as the receiver, and the other two acting as co-existing transmitters. By using the MIMO capabilities of the boards, we built an environment consisting of one source, one receiver, and four other transmitters ($T = 4$), one of which is a jammer ($N = 1$). We divided the spectrum into five channels of equal bandwidth[1] ($C = 5$). Also, we manually calibrated the antenna locations so that the receiver observes approximately the same power from each transmitter.

For the purposes of our evaluation, we filled the queues at each node so that each transmitter transmits packets all the time. This is not a requirement of SimpleMAC; because the recipient list performance estimates will not be updated during periods without traffic, traffic is always present from the perspective of the protocol. In fact, SimpleMAC works even in dynamic environments where the jammer and competing transmitters are sometimes present and sometimes absent; in Section 8.6, we show that SimpleMAC works even better in mobile environments.

At the physical layer, we modulate data using differential quadrature phase-shift keying (DQPSK), and synchronize using a preamble which is a Barker sequence modulated using binary phase-shift keying (BPSK). We divide the entire 12 MHz-wide spectrum (centered at 2.452 GHz) into 300 OFDM subcarriers, so each channel contains 60 subcarriers. We send our control communication across the entire band (300 subcarriers). Our frequency hopping scheme is to split each data message into frames of 60 symbols, which we simultaneously send on each of 60 subcarriers in the chosen channel. We hop from channel to channel between frames.

Our transmitter sends random symbols to the receiver, and we observe the decoded symbols at the receiver. We compare these symbols to determine the error rate, and use that error rate to estimate the signal-to-interference-and-noise ratio (SINR) at the receiver. The expected bit error rate ($\overline{\mathrm{BER}}$) and the expected SINR at the receiver ($\overline{\mathrm{SINR}}$) have the following relationship for DQPSK modulation [25, 26]: $\overline{\mathrm{BER}} = \frac{1}{2}\left(1 - \frac{\sqrt{2}\cdot\overline{\mathrm{SINR}}}{\sqrt{1+4\cdot\overline{\mathrm{SINR}}+2\cdot\overline{\mathrm{SINR}}^2}}\right)$.

We also validated our results using a MATLAB-based simulation. Our simulator works on a per-packet basis: for each time slot, each transmitter chooses a recipient list according to the STS, and the channel selection according to a uniform random distribution. Our channel model is an independent, identically distributed Rayleigh fading channel with AWGN

[1]Our evaluations focus on scenarios with relatively few channels; this is not a limitation of SimpleMAC, but is a performance optimization. SimpleMAC can improve performance regardless of the number of channels, but the optimal number of channels tends to be small relative to the number of transmitters, because from a capacity perspective, it is much better to have a legitimate-to-legitimate node collision than to let spectrum go unused. We have a mathematical formulation for the optimal number of channels, but omit it due to space constraints.

$$
\begin{aligned}
\mathrm{E}[\mathcal{R}(r)|\text{Best}] &= \Pr[\forall r' < r, S_R(r') \cap \mathcal{N} \neq \emptyset] \cdot \mathrm{E}[\mathcal{R}]_{S=\emptyset} + \Pr[\exists r' < r, S_R(r') \cap \mathcal{N} = \emptyset] \cdot \sum_{\alpha=0}^{T-N} \Pr[|S| = \alpha] \cdot \mathrm{E}[\mathcal{R}]_{|S|=\alpha} \\
&= (1-2^{-N})^r \cdot W \cdot \log_2\left(1 + \frac{P}{N_0 + \frac{T}{C}P}\right) + \sum_{\beta=1}^{r} \binom{r}{\beta} (1-2^{-N})^{r-\beta} (2^{-N})^{\beta} \\
&\quad \cdot \left[\sum_{\alpha=0}^{T-N} \Pr[\min_{i=1\ldots\beta} \mathcal{B}_i = \alpha] \cdot W \cdot \log_2\left(1 + \frac{P}{N_0 + \frac{N+\alpha}{C}P}\right)\right] \\
&= (1-2^{-N})^r \cdot W \cdot \log_2\left(1 + \frac{P}{N_0 + \frac{T}{C}P}\right) + \sum_{\beta=1}^{r} \left\{ \binom{r}{\beta} (1-2^{-N})^{(r-\beta)} 2^{-T\beta} \right. \\
&\quad \left. \cdot \sum_{\alpha=0}^{T-N} \left[\left\{ \left(\sum_{\sigma=\alpha}^{T-N} \binom{T-N}{\sigma} \right)^{\beta} - \left(\sum_{\sigma=\alpha+1}^{T-N} \binom{T-N}{\sigma} \right)^{\beta} \right\} \cdot W \cdot \log_2\left(1 + \frac{P}{N_0 + \frac{N+\alpha}{C}P}\right) \right] \right\}
\end{aligned} \tag{10}
$$

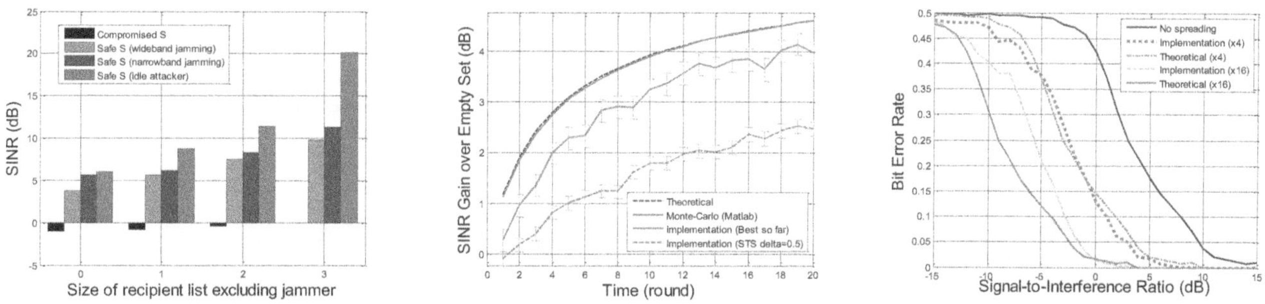

(a) Performance for S

(b) STS performance with time

(c) The SSS performance

Figure 2: Implementation and simulation results

noise. We then compute the number of interfering users (legitimate and jammer) and calculate the resultant SINR, which we then use as feedback for the next round.

To analyze the performance of our STS in our implementation and simulation environments, we use the $S = \emptyset$ performance (corresponding to the no channel coordination) as our reference and study the performance gain over $S = \emptyset$. This gain represents the improvement over a protocol that does not reserve a channel prior to data communication. We study the *SINR gain* which is the SINR observed by the STS divided by the SINR when $S = \emptyset$. As shown in Equation 1, the instantaneous channel capacity (which we use in our theoretical analysis) is strictly monotone in instantaneous SINR, and assuming the optimal fixed strategy for jammers, this relationship extends to the time-average SINR and the time-average capacity in Equation 3.

8.2 Without attack

We first consider the performance of our protocol when all transmitters are protocol-compliant. In this scenario, our protocol minimizes unintentional interference, and the more nodes we include in our recipient list, the better our performance. Figure 2(a) shows the estimated SINR and reflects a 14.1 dB SINR increase between $|S| = 0$ (no coordination) and $|S| = 3$ (full coordination). The performance under full coordination gives us an estimate of the SNR without interference (when one channel is used): $\overline{\text{SNR}} \approx 20\text{dB} = 100$.

8.3 Under attack

We now consider protocol performance under jamming. If the recipient list is compromised and contains the jammer, the jammer can effectively jam the transmitter by following its hopping pattern. Otherwise, the jammer can either choose to jam across all five channels at reduced power per channel, or on a random channel at full power. Figure 2(a) displays the expected SINR at the receiver in each of the three cases. Including more legitimate users in the recipient list yields better SINR, as described in Section 8.2. We also observe a drastic drop in performance when S is compromised; whenever a set contains a jammer, its SINR is below 0 dB, since the jamming power is equal to the signal power, and other legitimate nodes may accidentally interfere. (When there is perfect coordination among legitimate nodes, the only additional noise is the receiver's thermal noise, so the SINR is very close to 0 dB in this case.) These cases with compromised recipient lists thus all perform worse than when disabling channel coordination ($S = \emptyset$), which provides 3.72 dB SINR. Furthermore, wideband jamming is more effective and yields lower SINR for the target transmitter than narrowband jamming, verifying our theoretical analysis in Section 5.1.

Despite the risk of possibly choosing the jammer, channel coordination is still potentially advantageous. Choosing a random recipient list $S = S_R$ has expected performance better than $S = \emptyset$ in expectation (our computation shows an SINR gain of about 1 dB assuming wideband jamming for uncompromised S). Furthermore, once the STS converges to the best possible set, we can reach an SINR of about 9.69 dB in spite of the wideband jamming, which reflects an SINR gain of 5.97 dB over the baseline performance of $S = \emptyset$.

8.4 Data Communication Using the STS

Now that we have established the performance of known-good and known-bad sets, we study the performance of the STS and explore its convergence behavior. For each round, the STS performs three actions (B, R, E) as described in Section 6. In our evaluation, the jammer uses the optimal strategy (full-power jamming using all available information).

Because our metric is SINR gain, and our baseline performance is the empty set, the Empty set has performance of 0 dB. The Randomly explore action chooses a recipient list at random with uniform probability. Therefore, the performance of Randomly explore is independent and has constant expectation across time. Assuming that the user randomly explores at least once per round, the Best so far performance is increasing in time, and converges to the optimal steady-state performance where $S = S^*$, as more sets are explored and the user has more sets from which to choose S_B.

In Figure 2(b), we plot the performance of the Best so far strategy under three evaluation environments: the theoretical analysis corresponding to Equation 10, our simulation, and our testbed implementation. For our implementation, we also plot 95% confidence intervals, which are not shown for our simulation results because our simulation results included enough runs that the confidence intervals would not be visible. Our results show that the performance predicted by our theoretical analysis coincides with our simulations. Our implementation performance is worse than the theoretical and simulation results because our simulation assumes a perfect measurement of SNR, whereas our implementation infers it from the bit error rate; early in the run, when the number of observed bits is small, the BER measurement can deviate from the expected BER, and the STS may as a result make suboptimal choices. However, in later rounds, this performance difference decreases as the implementation gains better information. As a result, the maximum performance difference of 23% (of implementation performance) occurs at round one and decreases to 16% at round 20 in Figure 2(b). We also show the overall (as opposed to Best so far only) performance of STS under our implementation for $\delta = 0.5$. The performance of STS overall (including Empty and Random transmission sets) is monotonically increasing in time and begins outperforming the no-channel-coordination option after round one. We will later show that the STS converges to the Best so far performance in Section 8.6.

8.5 Control Communication Using the SSS

SimpleMAC relies on a robust signaling scheme that can reach each recipient on the recipient list. We implemented our Simple Signaling Scheme in WARP using wideband communication and Direct Sequence Spread Spectrum (DSSS), as described in Section 6.2; our decoder is based on an analog correlator. In order to study the effect of spreading gain, we sent messages using three different code lengths: 1 (no spreading and thus, no redundancy), 4, and 16. We sent each chip on a separate, adjacent OFDM subcarrier. To get various signal-to-interference ratios, we fixed the signal power and varied the interference power to obtain signal-to-interference ratios ranging from -15 dB to 15 dB. Figure 2(c) shows the relationship between the BER performance and the signal-to-interference ratio.

To show the effectiveness of spreading to avoid interference in SSS, we generated theoretical curves by shifting the "no spreading" result by the expected spreading gain. As expected, the implementation result aligned well with the theoretical. When the code length is small, the performance can be slightly better than the theoretical because the x-axis considers only interference but not noise; the processing gain filters both interference and noise, which means that the implementation slightly outperforms the shifted curve. For example, when the code length is 4, the implementation has better performance than the theoretical except between -5 and -1 dB. However, with increasing code length, we use an increasing number of adjacent subcarriers, increasing inter-carrier interference and degrading performance.

8.6 Simulations with Mobility

We have previously shown that without mobility, STS is effective. Using our simulator, we now show that introducing mobility makes STS *more effective*; that is, transmitter performance increases when it is surrounded with mobile users. In our mobile environment, at the beginning of each round, each mobile user is placed at a random position, and we compute the channel gains based on the positions and a path loss model (with path loss exponent 3) and Rayleigh fading. We fix the receiver's location and choose the position of each mobile user with a distribution such that the expected received power corresponds to unit channel gain; that is, the expected received power is the same in the mobile and static cases. We use the same parameter values as we did previously: $C = 5$, $T = 4$ (other than the source transmitter), $N = 1$, SNR (across the entire spectrum)=13 dB. To allow for comparison between scenarios, we normalize all performance to the SINR achieved using a $S = \emptyset$ in the *static case*. We also ran for an increased number of rounds (10^4). We plot performance on a semi-log scale to better show the dynamics of convergence.

In each case, we consider *convergence to improvement*, which shows how quickly the performance exceeds the no-channel-coordination case, i.e., $S = \emptyset$, and *approximate convergence*, which shows when the system reaches within 10% of steady-state performance.

For comparison, the static case, where no user moves, is shown in Figure 3(a) (and reflects the data shown in Figure 2(b)). The scheme converges to improvement instantly at round one, since the expected SINR for Random set exceeds one (as we discussed in Section 8.3), and achieves approximate convergence in about 32 rounds and 316 rounds, respectively, for $\delta = 1$ and $\delta = 0.5$, for an eventual performance gain of 5.31 dB or 3.4.

We expect that mobility will improve performance *for any set S* because of Jensen's Inequality. Using Equation 2, we observe that SINR is convex in channel gain γ of other users, so static channel gains are worse than random channel gains when the expected channel gain is equal. Adding mobility thus improves both SINR for any set S, *including* $S = \emptyset$. As more nodes become mobile (0 in Figure 3(a), 1 in Figure 3(c), 4 in Figure 3(b), and 5 in Figure 3(d)), the Empty set $\overline{\text{SINR}}$ performance monotonically increases. The values for $\overline{\text{SINR}}_{\text{Empty}}$ are shown in Figure 3.

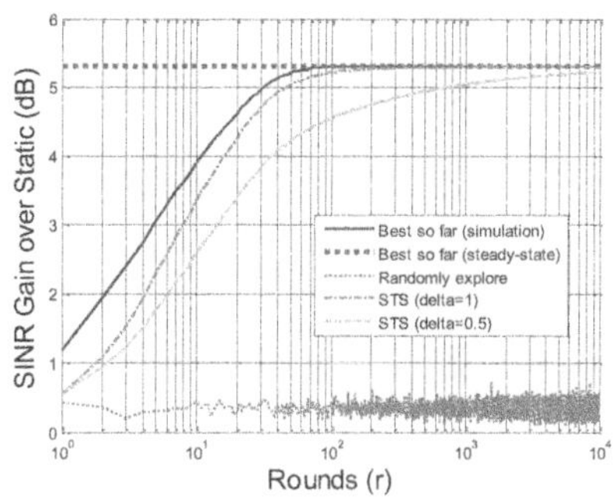

(a) All users are static. $\overline{\text{SINR}}_{\text{Empty}} = 1.26$

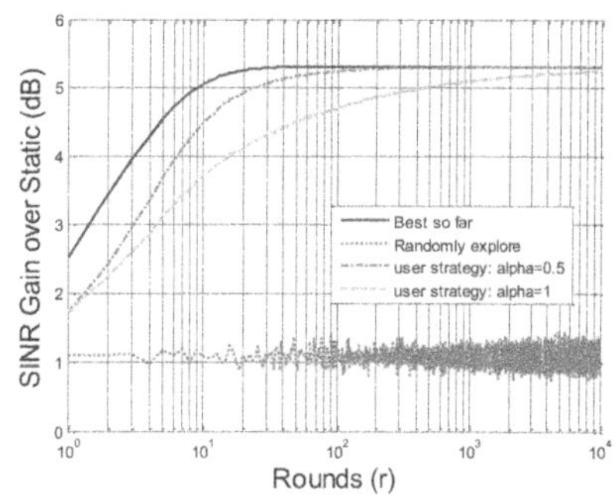

(b) Only legitimate users are mobile. $\overline{\text{SINR}}_{\text{Empty}} = 2.20$

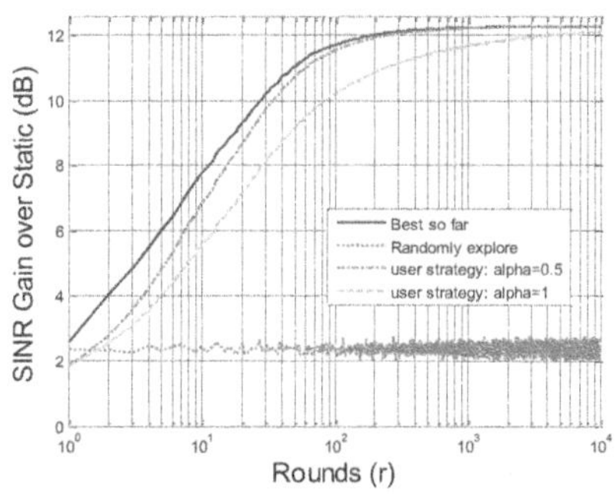

(c) Only jammers are mobile. $\overline{\text{SINR}}_{\text{Empty}} = 1.45$

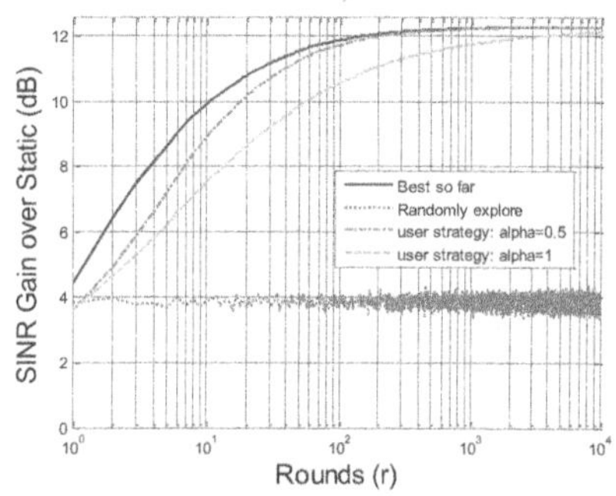

(d) All users are mobile. $\overline{\text{SINR}}_{\text{Empty}} = 3.33$

Figure 3: Simulations with Mobility. Gain is relative to static-case $S = \emptyset$ performance

To show the source of the additional SINR gain, we performed two additional set of experiments, one in which only legitimate users were mobile (Figure 3(b)) and one in which only jammers were mobile (Figure 3(c)). Adding mobility for legitimate users, as shown in Figure 3(b), has minimal impact on the steady-state performance $\overline{\text{SINR}}_{\text{Best}}$, because they avoid collision and create no interference in steady-state where $S_B = S^*$. However, when legitimate users move, convergence is much faster. In particular, *approximate convergence* occurs around 16 rounds and 188 rounds, compared to the static case of 32 rounds and 316 rounds, for $\delta = 1$ and $\delta = 0.5$, respectively. This increased convergence speed is because the jammers' noise contribution is more consistent and therefore has a larger impact on performance, making jammers more easily identifiable.

On the other hand, jammer mobility, as shown in Figure 3(c), results in a substantial improvement of $\overline{\text{SINR}}_{\text{Best}}$. When $S = \emptyset$, the majority of noise comes from legitimate nodes, so jammer has limited impact. However, after many rounds, S_B approaches S^*, so jammers' interference comprises most of the noise, so $\overline{\text{SINR}}_{\text{Best}}$ improves more than $\overline{\text{SINR}}_{\text{Empty}}$.

When all nodes other than the source transmitter are mobile (Figure 3(d)), we get benefits from both user mobility and jammer mobility. leading to faster convergence and better SINR performance for all STS action choices of B,R,E as compared to the stationary scenario. The steady-state $\overline{\text{SINR}}_{\text{Best}}$ values show the mobile case having 7 *dB* improvement, or about five times as much improvement as, compared the static case, which yields 95% capacity improvement over the static case. Thus SimpleMAC performs even better in mobile environments than in stationary ones.

9 Alternative Transmission Priorities

In Section 4.2, we considered a single user that has the highest priority for transmission on the channel, so all other legitimate users will avoid that node's transmissions, and we calculate the performance improvement of that node. In this section, we consider nodes that have equal priority. At the steady-state where $S_B = S^*$ for all legitimate users, and where all users have equal priority, each user will defer a $1 - \sigma$ fraction of its slots, where

$$\sigma = \sum_{k=0}^{T-N} \Pr_k \cdot \frac{1}{k+1}$$

where $\Pr_k$ is the probability that k other users are transmitting on the same channel and has a binomial distribution with parameters of $T - N$ trials and $\frac{1}{C}$ probability. To evaluate our scheme's performance in this egalitarian regime, we multiply the previous performance measurements by a fraction of σ. Since our simulations with mobility showed a capacity gain of 112% for a single priority node (Section 8), we observe a capacity gain of 56% when all nodes have equal priority. Similarly, without mobility, the capacity gain over baseline strategy is 51% with equal priority among all nodes.

Transmission priorities also affect how a node performs with a malicious receiver. If a malicious receiver forces a legitimate transmitter A to send with $S = \emptyset$, then the probability that any other node collides with transmitter A increases. In the worst case scenario, if transmitter A has absolute priority and is always transmitting, then A operates as a narrowband jammer. However, we can also consider the class of transmission priority schemes in which reserved channels always take priority over unreserved channels; under such schemes, in the steady-state, transmitter A will always defer to transmissions of other nodes, actually *increasing* the performance of other nodes. A more complete analysis of transmission priority schemes and malicious receivers is beyond the scope of this paper.

10 Conclusion

This paper introduces SimpleMAC, a MAC protocol that provides effective channel coordination to minimize interference among coexisting transmitters while simultaneously resisting jammers that use channel coordination information to jam more effectively. SimpleMAC avoids control channel jamming and limits jamming-relevant information to a recipient list, adjusting the recipient list to optimize performance. SimpleMAC converges to the optimal performance and forces an optimal jammer to always jam at full power. We used a game-theoretical approach to counter intelligent attackers, and analyzed the effectiveness of our scheme through theory, simulation, and implementation, and observed over 570% increases in SINR and over 50% increases in Shannon capacity gains in a realistic mobile environment.

11 Acknowledgments

This material is based upon work partially supported by USARO under Contract No. W-911-NF-0710287 and by NSF under Contract No. NSF CNS-0953600. We would also like to thank the anonymous reviewers, and our shepherd, Kun Tan, for their feedback.

12 References

[1] *IEEE Std 802.15.1-2005*, 2005.
[2] *IEEE Std 802.11-2007*, 2007.
[3] *IEEE Std 802.16-2009*, 2009.
[4] G. Alnifie and R. Simon. A multi-channel defense against jamming attacks in wireless sensor networks. In *Q2SWinet*, pages 95–104, Oct. 2007.
[5] B. Awerbuch, A. Richa, and C. Scheideler. A jamming-resistant MAC protocol for single-hop wireless networks. In *PODC*, pages 45–54, Aug. 2008.
[6] L. Baird, W. Bahn, M. Collins, M. Carlisle, and S. Butler. Keyless jam resistance. In *Information Assurance and Security Workshop*, pages 143–150, June 2007.
[7] T. Basar. The Gaussian test channel with an intelligent jammer. *IEEE Trans. Info. Theory*, 29(1):152–157, Jan. 1983.
[8] J. Bellardo and S. Savage. 802.11 denial-of-service attacks: real vulnerabilities and practical solutions. In *USENIX Security Symposium*, pages 15–28, Aug. 2003.
[9] A. Cardenas, S. Radosavac, and J. Baras. Performance comparison of detection schemes for MAC layer misbehavior. In *IEEE INFOCOM*, pages 1496–1504, May 2007.
[10] J. Chiang and Y. Hu. Dynamic jamming mitigation for wireless broadcast networks. In *IEEE INFOCOM*, pages 1211–1219, Apr. 2008.
[11] S. T. Chung, S. J. Kim, J. Lee, and J. Cioffi. A game-theoretic approach to power allocation in frequency-selective gaussian interference channels. In *IEEE ISIT*, pages 316–316, June 2003.
[12] L. Dryburgh and J. Hewitt. Signalling system no. 7 (SS7/C7): protocol, architecture, and services. *Cisco Press*, Aug. 2004.
[13] R. Etkin, A. Parekh, and D. Tse. Spectrum sharing for unlicensed bands. *IEEE JSAC*, 25(3):517, Apr. 2007.
[14] K. Firouzbakht, G. Noubir, and M. Salehi. On the capacity of rate-adaptive packetized wireless communication links under jamming. In *ACM WiSec*, pages 3–14, Apr. 2012.
[15] V. Gupta, S. Krishnamurthy, and M. Faloutsos. Denial of service attacks at the MAC layer in wireless ad hoc networks. In *MILCOM*, volume 2, pages 1118–1123, Oct. 2002.
[16] A. Kashyap, T. Basar, and R. Srikant. Correlated jamming on MIMO Gaussian fading channels. In *IEEE ICC*, volume 1, pages 458–462, June 2004.
[17] P. Kyasanur and N. Vaidya. Selfish MAC layer misbehavior in wireless networks. *IEEE Trans. Mobile Comput.*, 4(5):502–516, Sept. 2005.
[18] H. Li and I. Marsland. A comparison of rateless codes at short block lengths. In *IEEE ICC*, pages 4483–4488, May 2008.
[19] P. Murphy, A. Sabharwal, and B. Aazhang. Design of WARP: a flexible wireless open-access research platform. In *Proceedings of EUSIPCO*, pages 53–54, Sept. 2006.
[20] R. Pickholtz, D. Schilling, and L. Milstein. Theory of spread-spectrum communications–a tutorial. *IEEE Transactions on Communications*, pages 855–884, May 1982.
[21] M. Raya, J.-P. Hubaux, and I. Aad. Domino: a system to detect greedy behavior in ieee 802.11 hotspots. In *MobiSys*, pages 84–97, June 2004.
[22] L. G. Roberts. ALOHA packet system with and without slots and capture. *SIGCOMM CCR*, 5(2):28–42, Apr. 1975.
[23] M. Simon, J. Omura, R. Scholtz, and B. Levitt. *Spread spectrum communications handbook.* McGraw-Hill: New York, Mar. 1994.
[24] M. Strasser, S. Capkun, C. Popper, and M. Cagalj. Jamming-resistant key establishment using uncoordinated frequency hopping. In *IEEE SSP*, pages 64–78, May 2008.
[25] C. Tellambura and V. Bhargava. Unified error analysis of DQPSK in fading channels. *Electronics Letters*, 30(25):2110–2111, Dec. 1994.
[26] T. Tjhung, C. Loo, and N. Secord. BER performance of DQPSK in slow Rician fading. *Electronics Letters*, 28(18):1763 –1765, Aug. 1992.
[27] W. Xu, W. Trappe, and Y. Zhang. Channel surfing: defending wireless sensor networks from interference. In *ACM IPSN*, pages 499–508, Apr. 2007.

Frame Retransmissions Considered Harmful: Improving Spectrum Efficiency Using Micro-ACKs

Jiansong Zhang *
Microsoft Research Asia and HKUST
Beijing, China
jiazhang@microsoft.com

Haichen Shen *
Microsoft Research Asia and Tsinghua University
Beijing, China
v-hashen @microsoft.com

Kun Tan
Microsoft Research Asia
Beijing, China
kuntan@microsoft.com

Ranveer Chandra
Microsoft Research
Redmond, WA, USA
ranveer@microsoft.com

Yongguang Zhang
Microsoft Research Asia
Beijing, China
ygz@microsoft.com

Qian Zhang
HKUST
Hong Kong, China
qianzh@cse.ust.hk

ABSTRACT

Retransmissions reduce the efficiency of data communication in wireless networks because of: (i) per-retransmission packet headers, (ii) contention overhead on every retransmission, and (iii) redundant bits in every retransmission. In fact, every retransmission nearly doubles the time to successfully deliver the packet. To improve spectrum efficiency in a lossy environment, we propose a new in-frame retransmission scheme using μACKs. Instead of waiting for the entire transmission to end before sending the ACK, the receiver sends smaller μACKs for every few symbols, on a separate narrow feedback channel. Based on these μACKs, the sender only retransmits the lost symbols after the last data symbol in the frame, thereby adaptively changing the frame size to ensure it is successfully delivered. We have implemented μACK on the Sora platform. Experiments with our prototype validate the feasibility of symbol-level μACK. By significantly reducing the retransmistion overhead, the sender is able to aggressively use higher data rate for a lossy link. Both improve the overall network efficiency. Our experimental results from a controlled environment and an 9-node software radio testbed show that μACK can have up to 140% throughput gain over 802.11g and up to 60% gain over the best known retransmission scheme.

Categories and Subject Descriptors

C.2.1 [**COMPUTER-COMMUNICATION NETWORKS**]: Network Architecture and Design—*Wireless communication*

General Terms

Algorithms, Design, Experimentation, Performance

*This work is done when Jiansong Zhang is a researcher and Haichen Shen is a research intern in Microsoft Research Asia.

MobiCom'12, August 22–26, 2012, Istanbul, Turkey.

Keywords

Retransmission, Error Recovery, Feedback, μACK, WLAN

1. INTRODUCTION

Packets transmissions over a wireless medium are lossy. Losses typically occur because of interference, fading or noise, which in turn causes poor signal to interference and noise ratio (SINR) at the receiver. A low SINR reduces the probability of successfully decoding all bits in the packet.

In existing data networks, such as Wi-Fi or cellular, the onus of recovering a lost packet is with the link layer of the OSI stack. The physical layer (PHY) may include some redundancy, *i.e.*, channel coding, in a frame to correct erroneous bits. But when the errors cannot be recovered, the frame will be retransmitted by the MAC layer. The sender usually relies on an acknowledgment (ACK) packet from the receiver to detect transmission failures. For example, in IEEE 802.11, if the sender does not receive the ACK packet within a fixed timeout period, it retransmits the entire frame.

This frame retransmission is costly. First, each retransmitted frame requires both PHY and MAC headers, which can consume up to 100 bytes. While MAC header can be transmitted using high modulation, the PHY header is sent at the lowest rate and usually includes a sequence of training symbols (preamble) for synchronization and channel estimation. In 802.11g, this header overhead can add up to 52 μs. Second, the sender needs to re-contend for the medium to retransmit the packet. Depending on the number of additional contending nodes, the sender has to perform backoff, possibly several times, before the frame can be successfully delivered to the receiver. For each unsuccessful retransmission, the sender needs to wait for an ACK timeout before it can detect the loss. All reduce the link efficiency in a noisy wireless channel. Finally, retransmitting the entire frame may unnecessarily send redundant bits that may have already been received correctly by the receiver. In previous work, Jamieson, *et. al.*, tried to address the last issue by selectively requesting only erroneous bits in retransmission [12]. But the first two issues remain unexplored.

In this paper, we propose a system that does away with frame-level retransmissions. After a sender gets access to a medium, we dynamically adjust the length of the packet (up to a maximum limit) to ensure that it is reliably delivered. In a zero-loss network the packet length is unchanged; while in a lossy network we pad the frame, on the fly, with the bits that are received erroneously at the receiver. Therefore, the receiver can recover the errors inside a frame, instead of waiting for another frame retransmission. Since the retransmitted symbols do not have additional PHY/MAC headers or contention overhead, they significantly reduce the cost of error recovery. As we show in Section 6, our system can be 140% more efficient than 802.11, and up to 60% better than the existing best known retransmission scheme [12].

We achieve this *in-frame error recovery* by introducing a narrow-band feedback channel, similar to the control channels proposed in RI-BTMA [21] and other schemes [16]. However, instead of transmitting a simple tone (or signature signal), in our system, the receiver uses the control channel to send modulated acknowledgments to the sender for received symbols. We call these tightly synchronized symbol-level acknowledgments as *micro-ACKs* (μACKs).

Implementing the μACK system imposes several challenges. First, the forward and feedback radios should be tightly synchronized. The receiver needs to dynamically generate μACKs based on the decoding results of the data symbols. The sender needs to re-encode erroneous bit based on μACK feedbacks. This needs to occur in real-time in the order of a few symbol durations (tens of μs). Second, the receiver needs to reliably determine the data symbols that are correct or in error. Previous work [12] uses PHY hints to identify erroneous symbols. Although useful in several scenarios, this scheme is not reliable when the link is operating at the modulation's threshold SNR, *i.e.*, when retransmissions are more likely to occur.

This paper presents the design, implementation, and evaluation of μACK. μACK uses a multi-radio architecture with multiple RF front-ends tightly integrated onto one control board. We further exploit a new side-channel inside 802.11 OFDM PHY to transmit a CRC-checksum along with a group of data symbols to facilitate error detection without adding additional overhead. Finally, we design and evaluate the PHY schemes for both, the side control channel and the μACK feedbacks. We show the μACK PHY design is simple, yet reliable for their purpose in our system.

We implement μACK using a high speed software radio platform [18]. Experiments with our prototype validate that symbol-level μACK is practically feasible. We also show that with μACK, the retransmission overhead can be significantly reduced, thereby improving spectrum efficiency. Furthermore, we believe that the idea of μACK has wider applicability, beyond error recovery. For example, using μACK, a sender can detect collisions before the entire transmission is complete, and therefore it can abort earlier to save channel time, similar to [16]. Also, the μACK feedback channel can be used as an extended busy-tone [6], and therefore can mitigate hidden and expose terminal problems.

The rest of paper is organized as follows. Section 2 motivates our work with an analysis of the retransmission overhead. Section 3 presents the detailed design of μACK. We further analyze μACK in Section 4. After describing our implementation of μACK using a high-speed SDR platform in Section 5, we evaluate the performance of μACK in Section 6. Section 7 discusses related work and Section 8 concludes.

2. OVERHEAD OF RETRANSMISSIONS

Although retransmissions help recover a packet, they add significant redundancy and overhead, thereby reducing spectrum efficiency. In this section we present a simple model to quantify this overhead, which motivates the need for the μACK mechanism.

Model: We consider the impact of retransmissions on IEEE 802.11g networks, although the results can similarly be extrapolated for 802.11a/n/b networks. In 802.11g, the timing parameters are, $t_{slot} = 9\mu s, t_{SIFS} = 10\mu s$, and $t_{DIFS} = 2t_{slot} + t_{SIFS} = 28\mu s$. The contention overhead, determined empirically from Atheros cards is $t_{CW} = t_{slot} \cdot \frac{CW_{min}}{2} = 8t_{slot}$ per packet. Each OFDM data symbol is 4 μs.

Using the model presented in [9], at modulation rate R, $4 \cdot R$ data bits are encoded in a symbol. The frame is broken down and encoded in symbols of duration $t_{symb} = 4\mu s$, where each frame is preceded by a 20 μspreamble, and a 6 μs signal extension. Using all these values, the time to transmit a packet of size s_{data} bits at data rate R Mbps, without any retransmissions is:

$$\begin{aligned} T_{data}(R) &= t_{CW} + t_{DIFS} + t_{DATA} + t_{SIFS} + t_{ACK} \\ &= 72\mu s + 28\mu s \\ &\quad +(20\mu s + t_{symb}\lceil s_{data}/(4R)\rceil + 6\mu s) + 10\mu s \\ &\quad +(20\mu s + t_{symb}\lceil s_{ack}/(4R_{ack})\rceil + 6\mu s) \\ &= 162 + 4\cdot(\lceil s_{data}/(4R)\rceil + \lceil s_{ack}/(4R_{ack})\rceil)\mu s \end{aligned}$$

When there are r retransmissions, the time to transmit a packet is:

$$\begin{aligned} T_{ReTx} &= \sum_{i=0}^{r} T_{data}(R_i) \\ &\geq 162(r+1) + 4(r+1)\cdot(\lceil s_{data}/(4R)\rceil \\ &\quad +\lceil s_{ack}/(4R_{ack})\rceil)\mu s \end{aligned}$$

As we see, retransmissions delay the total time to transmit the frame due to the following:

- **Redundant bits ($T_{Redundancy}$):** These are bits that are correctly received by the receiver, yet are part of the retransmission. Depending on the number of bits in error, this adds anywhere from 0 (all bits are lost) to $(4r\cdot(\lceil(s_{data}-1)/(4R)\rceil)$ when only 1 bit is decoded error.

- **Contention ($T_{Contention}$):** A retransmission has to contend for the medium with all other nodes in the medium, just like a fresh transmission. This adds the DIFS and contention overhead, for a total of: $(100\cdot(r-1))\ \mu s$.

- **Header overhead (T_{Header}):** Since the retransmission is just like a new frame, it has to include all training symbols, the PHY and MAC header, as well as an ACK frame. This adds another $((r-1)\cdot(62 + \lceil s_{ack}/(4R_{ack})\rceil))\ \mu s$.

- **Lower data rate ($T_{DataRate}$):** Bits in the retransmission are usually sent using a lower data rate. In most common implementations, the retransmission data rate is one rate lower than the original data rate [2]. For example, when the original packet is sent at 18 Mbps, and the retransmission at 12 Mbps, the extra overhead is: $(r-1) * (T_{data}(12) - T_{data}(18))$.

To get a quantitative feel of these numbers, we present the overhead for a single retransmission in Table 2. We change the packet size and data rate, and note that the retransmission consumes more time than the original transmission because retransmissions are sent at a lower data rate. The contention overhead is assumed to be 100 μs. It will be much larger in congested environments. Also, $T_{Redundancy}$ is assumed to be the worst case, *i.e.*, all but one bit is erroneously decoded. As expected, the contention and header overheads are a much larger fraction of the retransmission overhead when the packets are small in size, or the data rates are higher. This fraction is likely to dominate in IEEE 802.11n MIMO networks with much higher data rates. To summarize, even one retransmission more than doubles the delivery time of a packet, and also reduces the spectrum efficiency because of the factors highlighted above.

(Pkt size, Data Rate)	$T_{data}(\mu s)$	$T_{ReTx} - T_{data}(\mu s)$	$\frac{T_{Redundancy}}{T_{ReTx}}$	$\frac{T_{Contention}}{T_{ReTx}}$	$\frac{T_{Headers}}{T_{ReTx}}$	$\frac{T_{DataRate}}{T_{ReTx}}$
(1500, 54)	389	417	0.53	0.24	0.16	0.07
(1500, 9)	1500	2166	0.61	0.05	0.03	0.31
(500, 54)	241	250	0.30	0.40	0.26	0.04
(500, 9)	611	833	0.53	0.12	0.09	0.26

Table 1: The overhead introduced by one retransmission on changing packet size (in bytes) and the data rate (in Mbps) of the original transmission, and the fraction of overhead introduced by each of the four factors. Contention and packet headers dominate for smaller packets, while redundancy is the largest overhead otherwise.

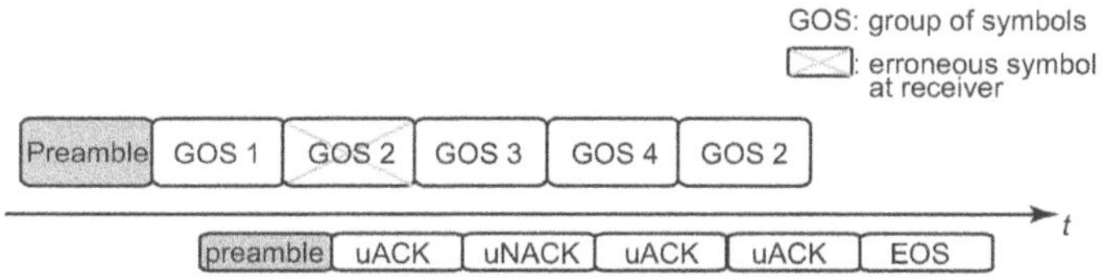

Figure 1: Illustration of a frame transmission in μACK system.

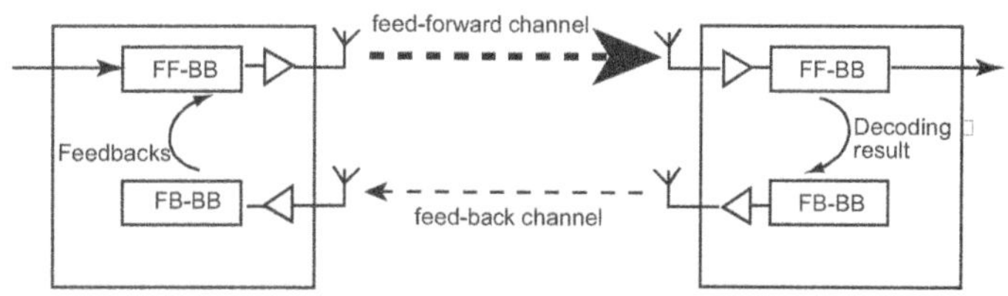

Figure 2: Architecture of μACK. Two tightly synchronized radios are deployed at both sender and receiver. The receiver sends real-time feedback symbols to the sender over a narrow-band channel.

3. μACK DESIGN

In this section, we present our system, that eliminates packet retransmission and thereby mitigates the overhead in recovering errors. The receiver decodes each incoming data symbol and dynamically determines if the symbol is correct or in error. For a group of successfully decoded symbols, the receiver will send an acknowledgment (ACK) to the sender; otherwise, if erroneous symbols are detected, an negative acknowledgment (NACK) is sent. These ACK/NACK are transmitted back to the sender using a separate narrow-band feedback channel, which is tightly synchronized to the feed-forward data communication channel. Since the ACK/-NACKs are at symbol level, we call them micro-ACKs/NACKs (or μACKs/ μNACKs). The sender monitors the feedback channel and marks the symbols that get NACKs. Then, when all data symbols have been sent, the sender re-encodes these symbols and appends them after the last data symbol. The sender continually re-encodes the lost symbols until it receives an acknowledgment of the entire frame from the receiver or a predefined maximum limit has reached. In this way, an μACK sender recovers all transmission errors inside a frame, instead of relying on different retransmission frames.

Figure 1 illustrates a frame transmission in μACK system. In the wide-band feed-forward (FF) channel, the sender sends data symbols to the receiver after a preamble. The receiver, after synchronizing to the preamble, starts immediately a feedback frame to the sender in a narrow-band feedback (FB) channel. Each symbol of the feedback frame acknowledges (or negatively acknowledges) a group of data symbols (GOS) in the feed-forward channel. If the sender gets a NACK, it will mark the corresponding GOS as *lost*. Lost GOSes are re-encoded and appended after the last GOS. As shown in Figure 1, the second GOS contains errors. Then, it is re-encoded after GOS 4, the last group of symbols of the frame. When the entire frame is correctly received, *i.e.*, passed CRC check, the receiver will send back an end-of-stream (EOS) symbol to the sender, which, on receiving EOS, terminates the frame transmission. If the sender does not yet receive an EOS, but also does not have any GOS marked *lost*, it will simply re-encode a GOS from the very beginning (see details in Section3.3).

We now present the μACK feedback design. Based on this design, in Section 3.2 we describe the error and collision detection at the sender. We then describe in details the μACK in-frame error recovery protocols in Section 3.3. We discuss other applications of μACK in Section 3.4. Finally, we finish by discussing several related design issues in Section 3.5.

3.1 μACK feedback

μACK relies on a multi-radio architecture for a receiver to send fine-grained feedback when simultaneously receiving data symbols from the sender. With the increased popularity of wireless communication, multi-radio structure has extensively studied and exploited in previous systems [3]. However, μACK differs from these previous systems in that we integrate both radios into a single board and thus they can be tightly synchronized at micro-second level. Figure 2 shows the system architecture of μACK. Two tightly synchronized radios are deployed at both sender and receiver. The receiver receives data symbols in wide-band FF channel, and based on the decoding results, dynamically modulates and sends feedback symbols in real-time using a narrow-band FB channel. The sender also, in real-time, re-encodes and sends the lost symbols based on received feedbacks.

The key design question here is what shall be the granularity of μACK and how much bandwidth should be allocated to the feedback channel. Ideally, we want to get μACK as fine as possible (*i.e.*, μACK for each received byte) to minimize the overhead of redundant bits. However, too fine granularity μACK will require more bandwidth at the feedback channel to convey this information, which will add overhead to the system. Second, too fine μACK might not be necessary at all since wireless errors are essentially bursty [19]. Bits transmitted within the coherent time usually share the same channel state and are lost in one burst, and therefore may be acknowledged with one μACK. The coherence time depends on the speed of multi-path fading in the wireless channel. In an indoor environment, it varies from around $100\mu s$ (fast fading) to tens of milliseconds (slow fading) [19, 20].

In this paper, we propose to generate one μACK every $20\mu s$, which is much shorter than the coherence time in common indoor wireless channel. This period of time contains five 802.11 OFDM symbols. In theory, the receiver may need to convey only one bit information (*i.e.*, correct or wrong) for every group of symbols it receives. Thus, the feedback channel can be very narrow. For example, the required baud rate the feedback is merely $1/20 = 50KHz$.

Table 2: Symbol encoding in feedback PHY.

Symbol name	Symbol binary ($b_3b_2b_1b_0$)	Chip values
ACK	1100	0111100010
NACK	1001	0011001101
EOS	0110	1100110110

Even with 100% guard-band, it takes around 0.5% overhead of a WiFi $20MHz$ channel. However in practice, we may want more bits in μACK to perform proper coding for reliability. In this paper, we choose the bandwidth of the feedback channel to be $1MHz$. The feedback signal has a width of $500KHz$ and another $500KHz$ spectrum serves as guard-band. We choose $500KHz$ guard-band by referencing to the DECT standard [15], which deploys similar guardband of $576KHz$ with a channel width of $1.728MHz$.

Compared to 802.11, the feedback channel of μACK adds around 5% overhead to the system. But it obviates the need for the original ACK frame and thereby reduces the time-domain overhead. Recall in Section 2, when the frame size is 1500 bytes, the ACK frame normally takes about 8% overhead when the modulation rate is $24Mbps$, which is comparable to our feedback channel overhead. It should also be noted that at higher data rates that involve wider channels (*e.g.*, $40MHz$ channel in 802.11n), the feedback channel overhead decreases reciprocally; while the overhead of ACK frame increases (*e.g.*, 20% with 300Mbps rate) due to shorter transmission time of data symbols [17].

μACK uses very robust code and modulation scheme to ensure high reliability of the feedback. The symbol time of the feedback is equal to the duration of a GOS, which is $20\mu s$. Given 500KHz signal bandwidth, each feedback symbol can be coded into 10 chips. In current design, we map four bits to one feedback symbol and these bits are encoded into 10 chips, each of which is modulated using basic differential binary phase-shift keying (DBPSK). Table 2 summarizes the three feedback symbols currently defined in our system. The remaining symbol values are reserved for future extensions. The entire feedback frame (consisting of smaller ACKs for the data symbols) is preceded with a $20\mu s$ synchronization symbol that allows the sender to detect and synchronize to the feedback frame. The SYNC symbol value is 0x2FF.

3.2 Error detection

μACK requires the receiver to reliably detect erroneous symbol groups to generate real-time fine-grained feedbacks. This partial packet detection problem has been well studied in previous work [8, 12]. In [8], Ganti *et. al.* proposed to split the frame into fragments and insert a CRC check sequence for each fragment. The overhead to transmit these check sequences increase proportionally with the number of fragments in a frame. Alternatively, Jamieson, *et. al.* [12] proposed to use physical layer hints to classify correct or erroneous symbols. These PHY hints usually use the soft-output of the channel decoder. While PHY hints save the overhead of checksums, in our experiments, we find the PHY hint based classifier becomes less reliable when the SNR is just enough to support the modulation rate on a wireless link. Figure 3 shows such an example. The PHY hint in this experiment is the soft-output of Viterbi decoder [11] [1] and the modulation rate is 24Mbps. Figure 3 plots the cumulative fraction of soft-outputs for both correct and erroneous symbols. We see that when the link SNR is 12dB high, using

[1]The output is quantized log-likelihood (LLR) of a correct trellis path in Viterbi decoder.

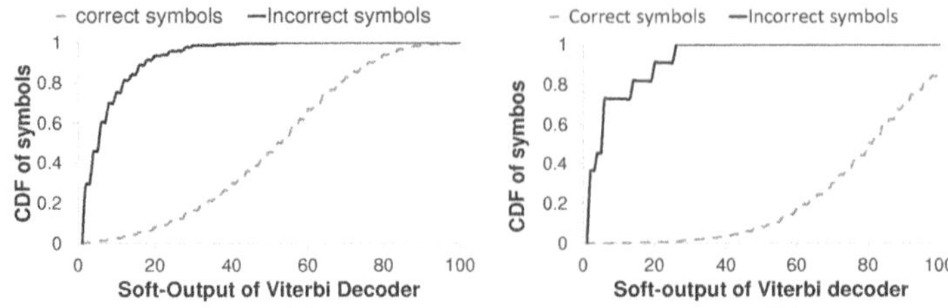

Figure 3: CDF of Viterbi soft-outputs in two SNR settings. The modulation rate is 24Mbps. (a) SNR = 10dB; (b) SNR = 12dB.

PHY hints can detect both correct and erroneous symbols reliably (Figure 3(b)). But when the link SNR reduces slightly to 10dB (just enough to support 24Mbps), the PHY hint cannot easily separate erroneous symbols from the correct ones, and it becomes a difficult tradeoff between higher false negative (choosing a lower threshold) or higher false positive (by choosing a higher threshold). This is expected, since the PHY hints are generally good statistical estimations of error probability for a large amount of symbols [20]. However, they may fall short in reliably predicting correctness of specific set of symbols, especially when the probability of correct and erroneous symbols are comparable at marginal SNR.

Unfortunately, it is not uncommon for a wireless link to work at this marginal SNR regime, as modern rate adaptation mechanisms tend to choose the highest modulation rates for better efficiency. Consequently, in this paper, we allow the sender explicitly embed a CRC checksum along each GOS to facilitate reliable error detection at the receiver.

However, embedding an extra checksum would add extra packet overhead. To avoid this overhead, we exploit a side-channel that has not been used by existing 802.11 OFDM PHY. Current 802.11 PHY deploys four pilot subcarriers in each OFDM symbol, each of which transmits dummy bits. These dummy bits carry no information and are used by the receiver to track channel changes [13]. We argue that such a design, although simple, is not efficient. In this paper, we propose to modulate one bit information on each pilot subcarrier without significantly reducing the channel tracking performance. Specifically, let $P_{i,j}$ denotes the known pilot sequence for symbol i on pilot subcarrier j. Then, instead of inserting $P_{i,j}$, the sender sends $P'_{i,k} = c_k P_{i,j}$, where $c_k = \{1, -1\}$ is a differentially coded binary data. The receiver, however, decodes c_k first, before performing normal pilot tracking. To decode c_k, the receiver uses the pilot value of the previous symbol, $P'_{(i-1),j}$, as a reference, just like a differential demodulation. A group of c_ks can further be protected with error-correction codes, and small number of bit errors may not affect the correctness of c_k. Once c_k is decoded, the receiver can recover $P_{i,j}$ and feed it to normal pilot tracking algorithms. Since in this scheme, we need to detect c_k first, we name our approach as *decision-directed pilot tracking* (DDPT).

We briefly present some reasons on why DDPT does not degrade tracking performance compared to original dummy pilots. We evaluate DDPT in detail in Section 6. First, we note that once c_k is successfully decoded, DDPT is just equivalent to original dummy-bit pilot tracking (DBPT) scheme. This is easy to see, as once c_k is decided, the original pilot value $P_{i,j}$ can be directly recovered. Second, it is reasonable to use a previous received pilot value as a reference to demodulate c_k, since the wireless channel holds stable during the coherence time, which is from $100\mu s$ to several milliseconds for indoor environment like WiFi. The symbol period is about 4 μs which is much shorter than the channel coherence time in our setting. Third, even if there is a sudden interference causing erroneous detection of c_k, which is BPSK modulated with proper channel coding, we note that in this case the original pilot tracking algorithm may also not perform well, since the interference may already corrupt the tracking results. Therefore, the symbols, in either DDPT or DBPT, suffer a strong interference and fail to decode anyway. Finally, to prevent the decision errors from propagating, we perform DDPT only within a GOS. For the first symbol in a GOS, we always insert normal pilot bits, while modulating information only on the remaining four symbols.

With DDPT, μACK can embed up to 16 bits on the pilot subcarriers of a GOS. In this design, we use a simple Hamming (16,11) code to protect this side-channel. There, we have eleven information bits. Ten bits are used to encode a CRC-10 checksum, which is used in B-ISDN and ATM networks. The other bit is used to indicate whether or not the GOS contains retransmission metadata (detailed in next section). We note that the Hamming code is slightly weaker than the 1/2 code used for 6Mbps rate in 802.11, and may have higher bit error rate (Section 6.1.2). In our future work, we will investigate a better coding scheme for the pilot side channel.

Upon decoding a GOS, the receiver computes a CRC checksum of the decoded bits and compares it to the one embedded in the pilot subcarriers. If they match, the receiver sends out a positive μACK; otherwise it sends a negative μACK.

3.3 In-frame recovery protocol

Based on the error detection and feedback mechanisms discussed earlier, μACK can perform in-frame recovery of erroneous symbols. The in-frame recovery protocol works in the following way.

After detecting a preamble, the receiver starts to transmit the synchronization symbol on the feedback channel. Then, for each GOS it receives, the receiver will send a μACK or μNACK based on the correctness of the GOS. All correctly received data are kept in an assembly buffer. If all received data pass the frame CRC-check, the receiver sends an EOS symbol to notify the sender. If after that, the receiver continues receiving GOS, possibly due to the corruption of the EOS symbols, it simply returns another EOS.

Based on the μACKs received, the sender puts the negatively acknowledged bytes into a retransmission queue. After the last GOS is sent, the sender fetches all data from the retransmission queue and starts re-coding them into *retransmission GOS*(RGOS). RGOSs are sent directly after the last GOS of the frame. Since the data in the retransmission queue are no longer continuous, the first RGOS should contain retransmission metadata to identify the corresponding symbols at the receiver. The RGOS containing the retransmission metadata is marked by the *meta-bit* on its pilot side-channel. The format of the retransmission meta-bit is simple. It contains at least 6 bytes, as shown in Figure 4. The first field, *number of symbols*(10-bit), indicates how many of symbols are re-encoded. The field *size* (6-bit) indicates the number of retransmission block entries in the metadata. Each block entry is 32-bit, including 16-bit start position and 16-bit length. In our current implementation, the size of this metadata header cannot exceed one GOS. For example, if the frame is modulated with 6Mbps, the metadata may contains only 2 entries; while for 54Mbps, the metadata header may hold up to 26 entries. If there are more retransmission blocks that cannot be fitted into one metadata header, multiple metadata headers may need to be added.

All acknowledged data are then removed from the retransmission queue. After one round of RGOS, if there is still any data remaining in the queue, the sender will start a second recovery round. A rare, but possible, scenario is that there is no data remaining, and the sender has not yet received an EOS. For example, there might be a hash collision in CRC-10, such that the receiver mistakes an erroneous GOS as a correct one, and returns a positive μACK. Therefore, to handle this corner case, the sender has to mark all data as lost and retransmit every byte in following symbols.

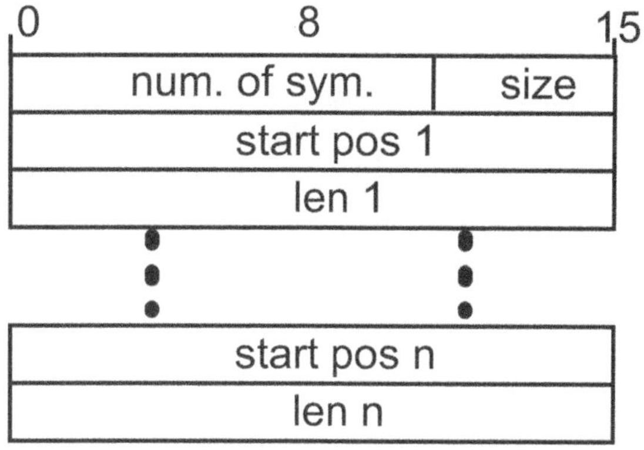

Figure 4: The format of retransmission metadata.

The sender will terminate transmission if it receives an EOS from the feedback channel, or an upper bound of symbols have been transmitted. We set this upper limit to prevent the sender from occupying the channel for an unreasonably long period (Section 3.5). In this paper, we set this upper limit as three times the frame size. The sender will also abort transmission and retry after a backoff time if a significant number of erroneous GOS have been detected ($\geq 50\%$), which usually indicates a persistent collision has occurred.

3.4 Other applications of μACK feedback

Beside the error recovery, the tightly synchronized μACK feedback can also improve the wireless network in following two ways.

Collision detection and early backoff. Using μACK, a sender can detect collisions before the entire transmission is complete. If the sender does not receive the first few μACK symbols, this means the preamble was not likely received by the receiver, which usually implies a collision may occur. Therefore, the sender can immediately stop its transmission, and retry the frame after a backoff time.

Hidden terminal mitigation. μACK can also mitigate the hidden terminal problem without relying on explicit RTS/CTS handshakes, which adds significant overhead. In μACK system, a sender can simply monitor both FF and FB channels. Since the receiver constantly continually sends the μACKs during the data transmission, the feedback can be viewed as a busy tone channel (similar to DBTMA [6]). The contender will defer its transmission once detecting the μACK feedback. Finally, μACK can also help to reduce exposed terminals following the similar heuristic in [6].

3.5 Discussion

We now briefly touch upon some factors in the design of the μACK mechanism.

When to send μACK sync symbol? The receiver should send back synchronization symbol in feedback channel as soon as it detects a frame for it in the data channel. However, in current 802.11 design, the destination address is embedded in the MAC header. That means the receiver can send μACK only after the first GOS has been decoded, adding a large additional delay. Therefore, we propose the extend current PLCP header by an OFDM symbol to store a physical layer address of the receiver. In this way, the receiver can start μACK feedback right after the preamble of the data frame. We note that PHY layer addressing has been previously exploited in [16,23], and we can use similar approach to dynamically allocate them inside a wireless network.

Range Mismatch: An important requirement is to ensure that the FB channel has a similar range as the FF channel. Otherwise, the hidden/exposed terminal problem will become worse. To ensure the same range, both the channels are in the same band, i.e. either 2.4 GHz or 5 GHz. Furthermore, we leverage prior work on channel widths [4] to adjust the transmit power of the FB channel such that its range matches the FF channel.

Rate Anomaly: Packet fairness of IEEE 802.11 hurts the performance of high data rate nodes in the presence of low data rate nodes [22]. μACKs can make the situation worse since a transmission with large number of bit errors will effectively increase its packet size and occupy the medium for a longer period of time, thereby hurting the performance of transmissions over completely reliable links. To solve this problem, we (i) limit the maximum time a node can occupy the medium including retransmissions, and (ii) reduce the probability of a node that just occupied the medium for a long time from immediately regaining access, similar to the technologies proposed in [22].

Rate Adaptation: Using μACK, a sender can get a good estimate of the BER on the link at any time. Therefore, it is possible to dynamically pick the best rate for each symbol inside a frame. We leave this as our future work.

FB channel allocation: In current two-radio implementation of μACK, we statically assign a narrow FB channel for every FF channel. However, in the future, the frequency of the FB channel in a wireless network may be dynamically selected to avoid potential noisy channels. All FB channel may be allocated to a specific portion of spectrum band. For example, in US, the 11 MHz of FB spectrum (for the 11 channels) can be allocated in the unused channels 12 and 13 of IEEE 802.11b in 2.4GHz band. This spectrum is only available for low-power operation, but given that our ACK is low bandwidth, and hence lower power, we expect to be within the FCC regulations for these channels. Alternatively, we may split a portion of existing 802.11 channel for μACK FB use. For example, we may allocate the upper (or lower) 1MHz of 20MHz WiFi channel to feedback, while the remaining 19MHz spectrum is used for FF data communication. Finally, we note that in the future, with full-duplex technology [5], we expect μACK may have single radio designs and the feedback can be sent using the same frequency as the FF channel.

Frame duration field setting: In 802.11 standard, the MAC header contains a duration field that records the expected frame transmission time plus the ACK. A contender after decoding the field will defer according to the value in this field to avoid possible collision to the ACK frame. For μACK, since the sender may dynamically pad the frame with retransmission symbols, the transmission duration may not be known before hand and cannot be accurately set. However, we note that μACK does not need this duration field for correct protocol behavior. This is because of the following two reasons. First, μACKs are sent in a different feedback channel. Therefore, it is safe for a contending sender to pick up the medium immediately after the FF data channel is sensed idle. Second, as discussed in Section 3.4, μACK can effectively detect collisions and mitigate hidden terminals using the feedback. Therefore, it does not need the traditional RTS/CTS handshake, which further relies on this frame duration field to reserve channel time.

4. ANALYTICAL STUDY

We build on the model presented in Section 2 and analyze the performance of the μACK technique. We also compare its performance to Wi-Fi and the most closely related work, PPR [12].

Model: Suppose the symbol error rate is e_s and a frame contains N_s symbols. Assuming the OFDM symbols are independent, the frame error probability is $e_f = 1-(1-e_s)^{N_s}$. To compare different techniques, we define *retransmission overhead* (RO) as the fraction of additional time to successfully transmit a frame.

$$\begin{aligned} RO &= (T_{ReTx} - T_{data})/T_{data} \\ &= T_{ReTx}/T_{data} - 1, \end{aligned}$$

where T_{ReTx} is the time taken to successfully deliver the packet at the receiver including retransmissions. So, the RO is zero if there is no loss, and greater than zero if there is any retransmission. In our model, we assume the sender will persistently retransmit the lost packet until it is successfully received. We note this is only an approximation to practical systems that may only retry up to a a maximum number (*e.g.*, four times in 802.11). A complete model that considers this maximum retransmission number will be our future work.

Wi-Fi: The expected number of frame retransmissions for Wi-Fi is:

$$\begin{aligned} K_{wifi} &= 1*(1-e_f) + 2*e_f(1-e_f) + \ldots \\ &\quad +k*e_f^{(k-1)}(1-e_f) + \ldots \\ &= 1/(1-e_f) \end{aligned}$$

Therefore, extending the formulation of Section 2:

$$T_{ReTx} = T_{data}(R) + \sum_{i=1}^{K_{wifi}} T_{data}(R_i).$$

where R_i is the data rate used for the ith retransmission packet.

PPR: We assume that there is no retansmission aggregation in PPR for latency reduction. Also, we do not model the dynamic programming algorithm that PPR uses to compute the optimal trunks to retransmit. Modeling such a dynamic algorithm is non-trivial. So we leave it as our future work. Here, we just try to build an approximate model: We assume PPR retransmits only erroneous bits and we don't consider the overhead of CRC-checksum for the runs of good bits. Clearly, our approximate model captures an performance upper bound of PPR.

Based on this simplification, we can compute the size of the kth retransmission frame, s_k, in PPR as

$$s_k = e_s s_{(k-1)}.$$

The frame error rate of the kth retransmission is:

$$e_f(k) = 1-(1-e_s)^{(s_k)}.$$

Therefore,

$$T_{ReTx} = T_{data}(R) + \sum_{k=1}^{\infty}\left(\prod_{i=0}^{k-1} e_f(i)\right) T_{data}^{k}(R_k),$$

where $T_{data}^{k}(R_k)$ is the air-time to send the kth retransmission (s_k symbols) with rate R_k.

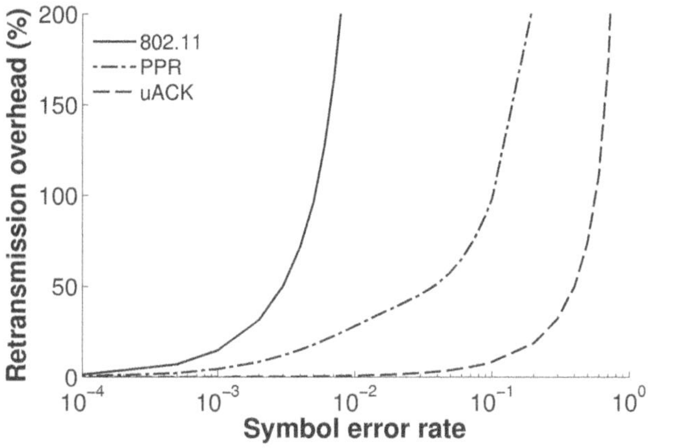

Figure 5: Retransmission overhead versus the symbol loss rate on the wireless link. The packet size is 1500B and the data rate is 54Mbps.

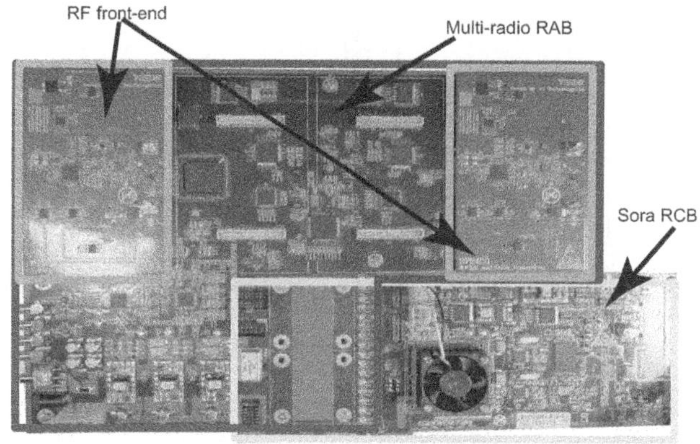

Figure 6: Multiradio RAB for Sora. It can connect to up to four RF front-ends. Two RF boards have been plugged on the board, and there are two slots empty.

μACK: In our system, the erroneous symbols are directly appended at the end of the frame. The number of retransmitted symbols is

$$N_{rx} = N_s \cdot \sum_{i=1}^{\infty}(e_s^i) = N_s e_s/(1-e_s)$$

Let l is the average burst length of erros. Then, the expected frame size in μACK is

$$N_{data} = N_s + N_{rx} + \lambda N_{rx}/l,$$

where λ is the metadata overhead to describe a burst loss in μACK (Section 3.3). Finally, we can derive the expected air-time to send a frame using μACK,

$$\begin{aligned} T_{ReTx}(R) &= t_{CW} + t_{DIFS} + t_{DATA} + t_{\mu ACK\_delay} \\ &= 72\mu s + 28\mu s \\ &\quad +(20\mu s + t_{symb}\lceil N_{data}/(4R)\rceil + 6\mu s) \\ &= T_{data}(R) + t_{symb}(\lceil N_{rx}(1+\lambda/l)/(4R)\rceil) \\ &\quad -T_{ack} + t_{\mu ACK\_delay}, \end{aligned}$$

where $T_{ack} = t_{SIFS} + t_{ACK}$ is the ACK overhead and $t_{\mu ACK\_delay}$ is the latency of μACK feedback.

Figure 5 presents the quantitative results of 802.11, PPR, and μACK under different symbol error rates. When the symbol error rates are low, the link is very reliable and the retransmission overhead is also low. As the link becomes lossy, the retransmission overhead of 802.11 increases exponentially and soon hits a wall at symbol error rate of 10^{-2}, meaning that no matter how many times it retransmits, the packet cannot be delivered. PPR reduces the retransmission overhead by resending only the lost bits. Therefore, the retransmission overhead increases slower than 802.11, but the header and contention overhead is still high. μACK, however, maintains a very low retransmission overhead even in very lossy environments, *i.e.*, the symbol error rate is as high as $1/2$.

5. IMPLEMENTATION

5.1 Platform

We have implemented μACK based on the Sora software radio platform [18]. To fully support μACK, we extended Sora in following two ways:

Multi-radio RAB. We built a new Sora RAB[2] that is able to connect to four RF front-ends as shown in Figure 6. These four RF front-ends are synchronized using the same oscillator, but can be independently programmed. For example, they can be configured in different spectrum bands with different channel widths. Or they can be in different transmitting or receiving modes. Each RF front-end exposes an abstract radio object in Sora system and the software radio applications can separately control each of them.

Streaming Tx interface. Current Sora SDK supports only frame-based transmission control interface. A frame of modulated signal should be entirely transferred into the RCB memory first, before it can be transmitted over the RF front-end [18]. This two-phase Tx method has a latency that is unacceptable for μACK. In this work, we extend the Sora system to support a new streaming Tx interface. Instead of downloading the entire modulated signal, the system can start transmitting a block of samples, while at same time fetching the next sample block from PC memory. Therefore, the modulation process and the signal transmission can be pipelined on Sora. In current implementation, the sample block contains 1.4 μs samples. This way, we can achieve dynamic modulation in real-time at the granularity of OFDM symbols.

5.2 μACK implementation

μACK is implemented as a user-mode SDR application based on Sora UMX API [1]. The program contains two *exclusive threads (ethread)*. One *ethread* runs the standard 802.11 OFDM PHY (SoftWiFi); the other one runs the feedback PHY as described in Section 3.1. We modified about 250 lines of code to SoftWiFi to support the streaming Tx interface and the pilot side-channel. The feedback channel occupies 1MHz frequency, within which 500KHz is used for communication and the rest serves as guard-band. Since the feedback PHY is very simple and contains limited number of symbols (three valid symbols plus the sync symbol), we build a lookup table to modulate μACK feedback and we apply the maximum-likelihood decoding to a received feedback symbol. In other words, the feedback receiver simply compares the demodulated symbol to each valid symbols and pick the one with the minimal Hamming distance.

[2]Radio adaptive board, which is used to connect a 3rd party RF board to Sora RCB.

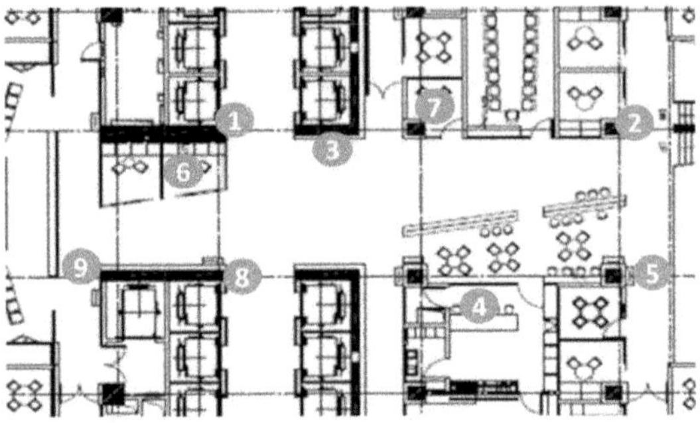

Figure 7: The software radio testbed layout: there are total 9 Sora nodes located in offices and lobby in an indoor environment.

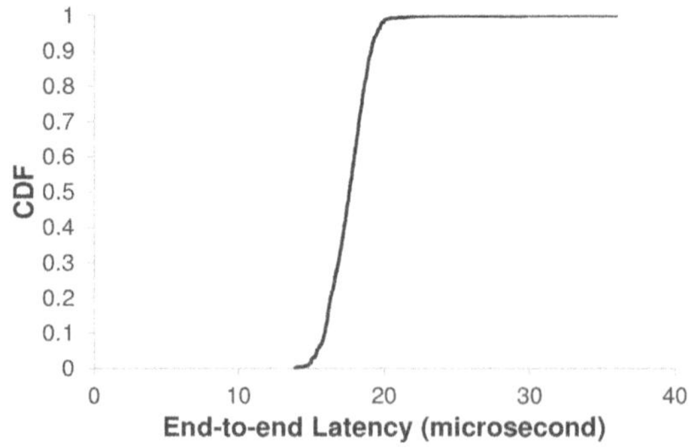

Figure 8: End-to-end delay of μACK feedback.

6. PERFORMANCE EVALUATION

We evaluate μACK in this section. We begin in Section 6.1 the micro-benchmark of our system implementation on the software radio platform. Then, we perform trace-driven simulation to evaluate μACK in a relatively large networking setting and compare μACK to original 802.11 as well as PPR [12].

Method. The micro-benchmarks are mainly conducted in a controlled environment. We connect two nodes' RF front-end using cables and manually add attenuation to create different SNR on the link.

We have built a test-bed contains 9 Sora nodes. Figure 7 shows the layout of our testbed. The Sora nodes are placed in offices and lobby in an indoor environment. The channel on which we collect data is centered at 2422MHz. This is the least busy channel in our office building, but still has uncontrollable interference from neighboring APs. Since we use software radio, we can log the decoding status of each symbol as well as the exact time a frame is received or transmitted. The large PC memory can easily hold a few minutes of frame logs without loss before writing to hard disk. We verify the links between nodes by selecting one node in turn as the sender to broadcast to all other nodes. We find the most links have a packet loss rate between $10 \sim 40\%$. Few links have very good quality to support 48Mbps data rate with few losses. There are also hidden terminals in our testing network. In average, each node may find one hidden node in the network.

We try all data rates on each link and identify the best modulation rate that has the best throughput on the link. We use this best modulation rate for both original frames and retransmissions on the link. Then, we pick up one node as a receiver and all other nodes send a saturate unicast traffic to it. All senders use the best modulation rate measured before. All sending and receiving frames are logged on senders and the receiver. We measure 10 runs for each receiver; each run last 30 seconds. Then, we change the receiver and repeat the measure until all nodes in the network have been covered.

6.1 Micro-benchmark

We first evaluate the end-to-end latency of μACK. This latency defined here is the period between the time when the last sample of a GOS is transmitted and first sample of a corresponding μACK that is received by the sender. This latency includes the demodulation/decoding of the GOS, modulation the acknowledgment, as well as the delay incurred by the hardware platform. Figure 8 shows the distribution of the end-to-end latency measured at the sender. We can see that our system can quickly generate and return μACK in a short delay. The mean value is only about 17.5 μs.

Table 3: Breakdown latency of μACK feedback (μs).

Viterbi decoding	μACK modulation	Hardware
7.5 (5.82, 8.06)	1.96 (1.4, 2.23)	9.103 (8.22, 10.5)

To better understand the latency, we break down the latency of each component in Table 3. The Viterbi decoder introduces about $8\mu s$ delay. This is reasonable. Since the Viterbi algorithm requires certain trellis depth before it can trace-back the results [3], the decoder needs to finish processing symbol $(i+1)$ to output symbol i. Consequently, it adds the processing delay of 2 symbol time. The modulation of μACK is trivial, less than 2 μs. This is because we have optimized the modulation process using lookup tables. Thus, this latency is only the time needed to copy samples of one feedback symbol. The hardware platform contributes the most latency in our system. Since we use a software radio platform, both sending and receiving a symbol will introduce a delay about $1.4\mu s$ to pass the PCIe bus. Therefore, there will add up to 5.4 μs in the end-to-end latency as the samples need to pass the PCIe bus for four times - sending and receiving data/μACK symbols at both sender and receiver. The Sora Fast Radio Link (FRL) used to connect the RCB and the RF front-end contributes the rest latency. Together, the total hardware latency is about 9 μs. We note this hardware latency can be significantly reduced if we implement μACK in silicon.

6.1.1 μACK feedback

Second, we evaluate the reliability of our feedback channel. We perform this measurement in a controlled environment. We connect the two nodes with wires and manually adds attenuation to control the SNR value. Figure 9(a) shows the μACK feedback error rate versus SNR curve of our feedback PHY as discussed in Section 3.1. When SNR is as low as 2.5dB, our μACK feedback has a low error rate of 10^{-2}. The error rate quickly drops to 10^{-4} when SNR is about 6.5dB. Figure 9(b) compares the reliability of μACK feedback channel with the lowest data rate (6Mbps) of 802.11. Each data point presents the BER of μACK feedback and 802.11 6Mbps rate in the same wireless link condition. From the figure, we can conclude that the feedback PHY is much more reliable than the lowest data rate used in 802.11, and adding such a fine-grained feedback will not significantly increase the end-to-end symbol error ratio.

[3](The depth is usually no less than $5K$, K is the constraint size of the convolutional encoder.

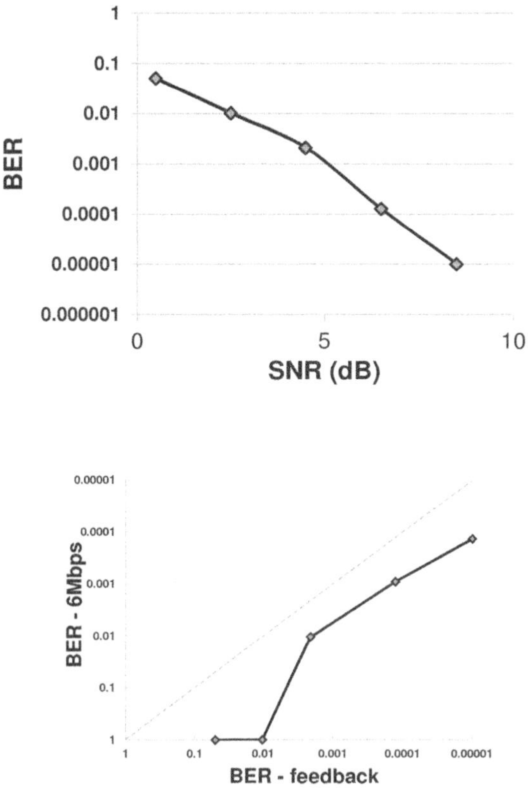

(a) (b)

Figure 9: The reliability of μACK feedback channel. (a) BER of feedback PHY versus SNR. (b) Comparision of μACK feedback with 802.11 OFDM 6Mbps rate.

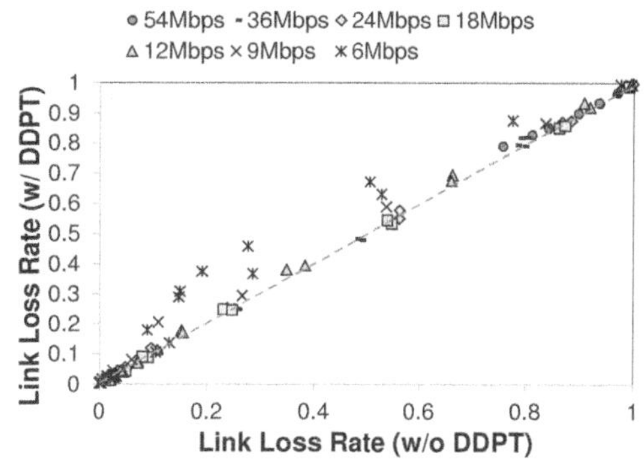

Figure 10: Comparision between DDPT and DBPT.

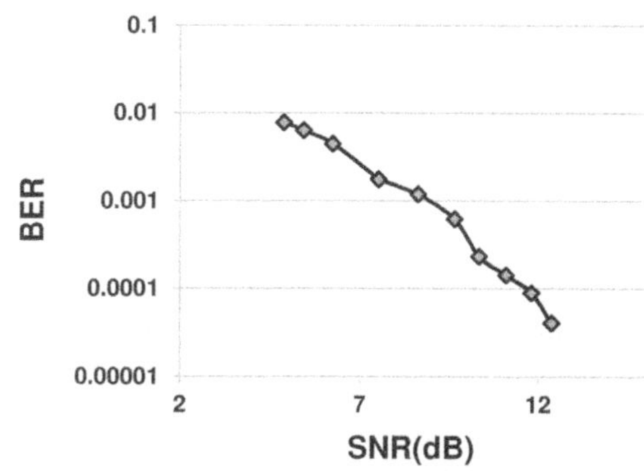

Figure 11: Performance of DDPT side channel.

6.1.2 *DDPT performance*

In this experiment, we evaluate the performance of DDPT (Section 3.2). Again, we perform the experiment in the controlled environment where the channel coherent time is relatively large. The sender broadcasts frames to the receiver over the controlled link. The broadcast traffic is 2Mbps with a frame size of 1000 byte. We configure various SNR values on the link. For each data rate, we first send frames without DDPT (normal 802.11 rates), and then we turn on DDPT. We record the frame loss rate at the receiver. Each measure takes 20 second. Figure 10 plots the results. Each data point in the figure represents the frame loss rate we get for a data rate with and without DDPT. In general, we can see the difference is quite small, especially when the data rates are high. With low rates (*i.e.*, 6Mbps and 9Mbps), using DDPT may slightly increase the frame error rate. This is reasonable. Since 6 and 9Mbps rates are already using BPSK modulation, their bit error rate is comparable to DDPT side channel. Therefore, the error rate added by DDPT modulation may be observable. But with high data rates, the data subcarriers are using high modulation/coding mode, and may already become very erroneous while DDPT is still able to track the channel reliably.

Figure 11 quantifies the BER of the DDPT side channel. The performance is comparable to that of 6Mbps modulation rate of 802.11. For example, to achieve BER of 10^{-3}, DDPT requires 8dB SNR, which is slightly higher (around 1dB) than 6Mbps rate of 802.11.

6.2 Single link performance

We compare the wireless link throughput of three schemes:

1. **802.11g.** This is normal 802.11g that always retransmits the entire frame.

2. **PPR.** We implement PPR [12]. As mentioned earlier in Section 3.2, the PHY hits may work unreliably when the link SNR marginally supports the chosen data rate. For a fair comparison, we also use the pilot side-channel to transmit CRC checksum for PPR as well. After a partial frame is received, the receiver will compute a feedback in an ACK frame and the sender, upon receiving the ACK, send the retransmission in a new frame.

3. **μACK.** μACK exploits the fine-grained feedback from the receiver in a separate channel to re-encode the lost symbols and append them right after the frame. In current implementation, μACK uses more bandwidth – we use total 21MHz spectrum, *i.e.*, 20MHz for forward data channel and 1MHz for μACK feedback. For a fair comparison, we reduce the measured throughput by 5% before we plot it in the figure.

For every SNR setting, we try all data rates for each scheme and we present the result of each scheme with the best modulation rate that yields the highest throughput on the link. For all schemes, we use this best modulation rate for both original frames and retransmissions.

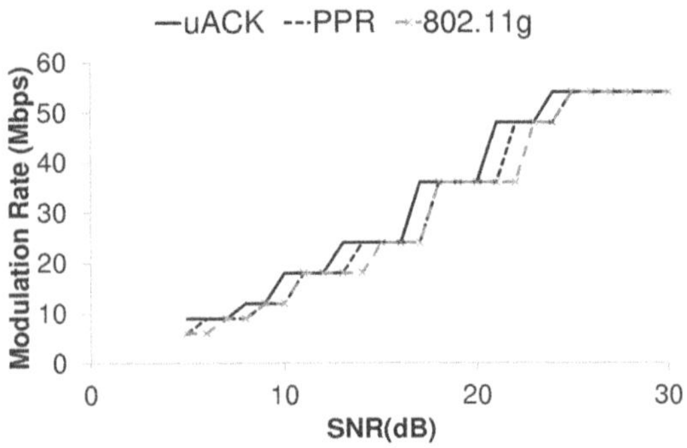

Figure 13: Best rate selected of different schemes at different SNR value.

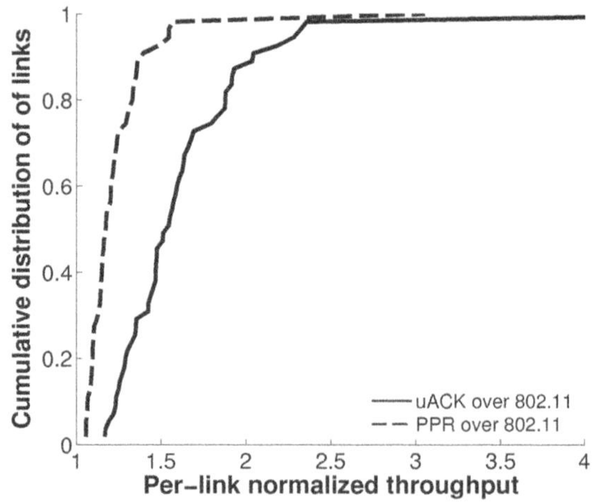

Figure 14: Per-link throughput distribution in the testbed.

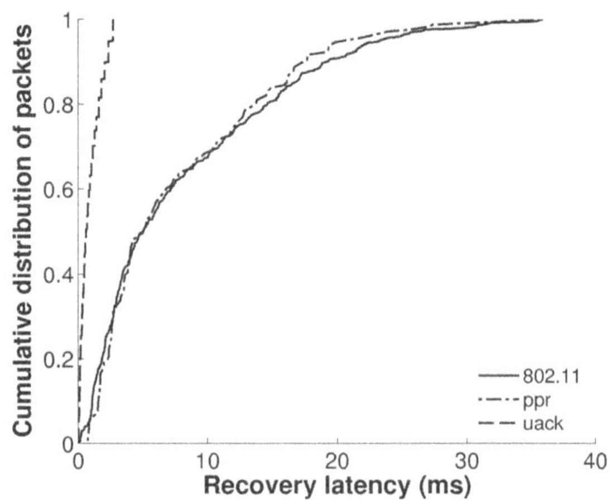

Figure 15: Error recovery latency.

Results. Figure 12 shows the results. When SNR is low, the link is very lossy even though the most reliable rates are applied. Therefore, the retransmission overhead is high for 802.11g. By avoiding retransmission of the redundant data, PPR and μACK can significantly improve the link throughput. With the increase of SNR, the link becomes less lossy, and higher modulation rates can be applied. Therefore, the benefit of PPR also becomes less, since in this case the backoff, headers and ACK dominates the retransmission overhead. They become more severe as the frame size decreases (The bottom row of Figure 12). μACK, however, removes such overheads using a narrow-band feedback channel and thus always gains over other two schemes. Further, as μACK decreases the retransmission overhead, it has better chance to use a higher modulation rate, which gives more throughput gain. Figure 13 shows this effect. We can see that μACK can utilize a higher rate one or two dB earlier than other two schemes. In summary, μACK improves the link throughput by $8 \sim 220\%$ over 802.11g, and $5 \sim 30\%$ over PPR.

During our experiments, we also find PPR sometimes may deliver less throughput than original 802.11g (as shown in Figure 12. The reason is due to PPR's postample. In this situation, the postample adds a fairly amount of overhead, but gains nothing – there is no hidden terminal here. So, we remove the postample and rerun the experiments, listed as *PPR-no post* in Figure 12. PPR-no post always performs better than 802.11g, but still below μACK.

6.3 Testbed experiments

In this section, we present the trace-driven emulation to evaluate μACK performance in a busy network. The trace is collection from a 9-node software radio network as described at the beginning of Section 6. The question we are answering here is what performance gain we can get if we apply μACK in such a network comparing to other schemes, *i.e.*, 802.11g and PPR. The trace-driven emulation is performed as follows: We examine each receiver trace one-by-one. We parse the decoding results of each symbol of each frame. For PPR, once a partial frame is detected, we compute the feedback message that should be embedded inside an ACK. Then, we adjust the ACK size accordingly, and treat the next frame as the retransmission. We further adjust the size of the frame according to the PPR protocol and determine the delivery status based on this new frame size. For μACK, we re-encode lost symbols after the frame. We also refer to the recorded results of the next frame to decide the decoding status of these retransmitted symbols. The backoff window are carefully adjusted according to the decoding results of the previous packet.

Figure 14 shows the per-link throughput distribution of our testbed. The throughput is normalized to 802.11g. We see that PPR has up to 60% (median 20%) throughput gain compared to 802.11g. This gain is much larger compared to a controlled environment in Figure 12. This is because in our test-bed network, the loss is much more prevalent due to the interference from other transmitters. μACK can achieve up to 140% (median 60%) throughput gain over 802.11g and up to 60% (median 30%) gain compared to PPR.

Figure 15 illustrates the delay distribution to recover a lost packet. The recovery delay is defined as the time between the first transmission and the finally delivery of packet. We assume the sender uses up to 4 times retransmission for both PPR and 802.11. We can see that both PPR and 802.11 has large recovery delay (up to 40ms!). This is because each retransmission packet should go entirely contention process to gain the access to the channel. In a busy network, this contending period can be very long. μACK, however, does not give up the channel and all errors are recovered inside the frame. So the recovery delay is very low (up to 4ms). This property is especially beneficial to real-time applications, like real-time conferencing or gaming, that requires very short delay in packet delivery.

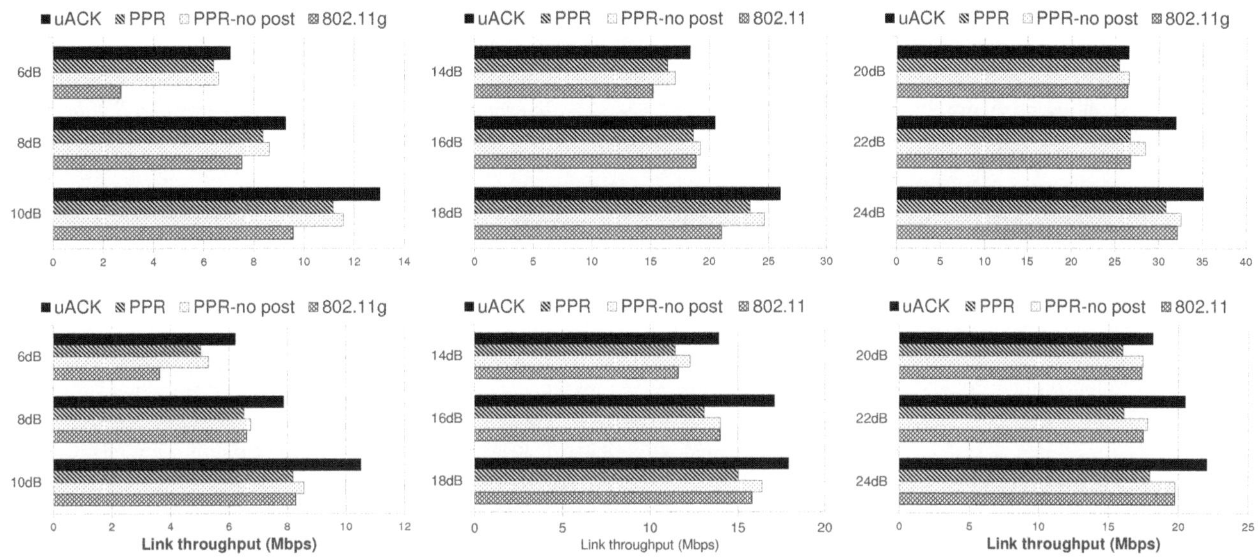

Figure 12: Link throughput of three schemes under different SNR. Top row: packet size is 1500B. Bottom row: packet size is 500B. Left: Low SNR regime. Middle: Middle SNR regime. Right: High SNR regime. X-axisY-axis shows the SNR (dB).

7. RELATED WORK

There is a large body of work on improving the reliability of wireless communication. Due to lack of space we only touch upon the most closely related work in this section, and refer the reader to [19] for more background.

ARQ is a well-known technique to recover from losses, where packets without ACKs are retransmitted. Hybrid ARQ [14] reduces the retransmission overhead using FEC, in addition to ARQ. Type I Hybrid ARQ resends the entire packet (along with FEC and error detection (ED)) when the packet is lost. Type II Hybrid ARQ reduces the ARQ overhead when the medium is mostly loss-free. Packets include ED but no FEC the first time they are sent. Retransmissions include both FEC and ED. Our μACK mechanism is complementary to the 802.11 Hybrid ARQ. Once the mapping from the data including FEC and ED is done to symbols, we only modify the ACK mechanism to be on a finer granularity, i.e. per symbol, instead of per-packet.

The most closely related work to μACKs is PPR [12], which we have referred and compared to in various sections in this paper. PPR reduces retransmission overhead by eliminating redundant bits. We reduce the overhead further, by not requiring separate packet transmissions (and the consequent overhead) for retransmissions.

CSMA/CN [16] utilizes the feedback to detect collision. It uses soft-phy hint to detect collisions. Once a collision is identified, a notification is sent back in the same frequency band, assuming the sender has full-duplex radio. Then, the sender may abort the transmission earlier. μACK has significant different design compared to CSMA/CN. First, μACK utilizes a narrow-band, dedicated feedback channel with little overhead; while the notification in CSMA/CN is a pseudo-random wide-band signal. Second, CSMA/CN notification conveys only one bit information (collision or not); while μACK feedback can contain multiple bits information. As a consequence, μACK design is much more powerful. It can not only detect collisions (similar to CSMA/CN), but also be able to perform in-frame retransmission to improve the error recovery efficiency. Further, the μACK feedback is continuous during the transmission of the data frame, and therefore can serve as an extended busy-tone to mitigate hidden and expose terminal problems.

Using separate channel to send acknowledgments is common in FDD wireless systems, *e.g.*, FDD LTE [7]. However, in such systems, the ACK is still at the frame level. For example, LTE NodeB will send an ACK/NACK subframe in the downlink channel after it receives a sub-frame in the uplink channel from a mobile node. In contrast, μACK sends back feedback symbols during the transmission of the data frame that allows in frame adaption (*e.g.*, symbol retransmission). Further, LTE is a tightly synchronized wireless system and only applied in licensed spectrum bands. However, μACK is designed for a random access wireless network, like WiFi, and is suitable for unlicensed spectrum bands.

Rate control algorithms try to make transmissions more reliable without losing capacity by appropriately adjusting the rate. For example, SoftRate [20] uses the SoftPHY hints to determine the best rate to send the packet. Using μACKs, a sender can get more immediate feedback about symbol error rate, and can in fact adapt the rate in between the transmission, i.e. as and when it receives the μACK for the previous GOS. Another recent rate adaptation algorithm is Strider [10], which proposes using rateless codes, which can significantly improve the reliability of transmissions. Similar to SoftRate, μACKs can improve the performance of Strider, as it can adjust the codes on a GOS time scale instead of the entire packet.

Another body of related work uses control channels. For example, RI-BTMA [21] and DBTMA [6] use busy tones on the control channels to resolve contention and hidden terminals, respectively. Although other work has also looked at offloading control traffic to the control channel, we believe μACK is the first system that integrates the control channel for in-line acknowledgment of symbols as the packet is being transmitted.

8. CONCLUSION

In this paper, we quantify the overhead of frame retransmissions, and show that the overhead is significant in a lossy wireless network. We breakdown the overhead from duplicate headers, contention as well as redundant bits, and propose a new in-frame retransmission scheme using μACKs. Instead of waiting for the entire transmission to end before sending the ACK, the receiver sends smaller μACKs on every few symbols, on a separate narrow feedback channel. Based on these μACKs, the sender only retransmits the lost symbols after the last data symbol in the frame, thereby adaptively changing the frame size to ensure it is successfully delivered.

We have implemented μACK on the Sora platform. Experiments with our prototype demonstrate that μACK can significantly reduce the retransmission overhead. Therefore, the sender can aggressively use higher data rate on a lossy link, which further improves the overall network efficiency. We believe that μACK is a powerful idea and has other applications beyond error-recovery. The feedback mechanism developed in the context of μACK can be used by a variety of future wireless protocols.

9. ACKNOWLEDGEMENTS

The authors would like to thank Dongliang He for his initial work on the frame delivery prediction based on soft-viterbi decoder in MSRA. His work has inspired the error detection scheme in μACK. The authors also thank Prof. Dirk Grunwald and anonymous reviewers for their valuable comments and suggestions to improve the paper.

10. REFERENCES

[1] The sora manual version 1.5. *http://research.microsoft.com/apps/pubs/?id=144847.*

[2] *ANSI/IEEE Std 802.11, Part 11: Wireless LAN Medium Access Control (MAC) and Physical Layer (PHY) Specification.* IEEE Press, 1999.

[3] P. Bahl, A. Adya, J. Padhye, and A. Walman. Reconsidering wireless systems with multiple radios. *SIGCOMM Comput. Commun. Rev.*, 34:39–46, October 2004.

[4] R. Chandra, R. Mahajan, T. Moscibroda, R. Raghavendra, and P. Bahl. A case for adapting channel width in wireless networks. In *Proceedings of the ACM SIGCOMM 2008 conference on Data communication*, SIGCOMM '08, pages 135–146, New York, NY, USA, 2008. ACM.

[5] J. I. Choi, M. Jain, K. Srinivasan, P. Levis, and S. Katti. Achieving single channel, full duplex wireless communication. In *Proceedings of the sixteenth annual international conference on Mobile computing and networking*, MobiCom '10, pages 1–12, New York, NY, USA, 2010. ACM.

[6] J. Deng and Z. Haas. Dual busy tone multiple access (dbtma): a new medium access control for packet radio networks. In *Universal Personal Communications, 1998. ICUPC '98. IEEE 1998 International Conference on*, volume 2, pages 973 –977 vol.2, oct 1998.

[7] J. S. Erik Dahlman, Stefan Parkvall. *4G: LTE/LTE-Advanced for Mobile Broadband.* Academic Press; 1 edition, 2011.

[8] R. K. Ganti, P. Jayachandran, H. Luo, and T. F. Abdelzaher. Datalink streaming in wireless sensor networks. In *Proceedings of the 4th international conference on Embedded networked sensor systems*, SenSys '06, pages 209–222, New York, NY, USA, 2006. ACM.

[9] M. Gast. 802.11 Wireless Networks: The Definitive Guide. In *O'Reilly*, 2005.

[10] A. Gudipati and S. Katti. Strider: automatic rate adaptation and collision handling. In *Proceedings of the ACM SIGCOMM 2011 conference*, SIGCOMM '11, pages 158–169, New York, NY, USA, 2011. ACM.

[11] J. Hagenauer and P. Hoeher. A viterbi algorithm with soft-decision outputs and its applications. In *Global Telecommunications Conference, 1989, and Exhibition. Communications Technology for the 1990s and Beyond. GLOBECOM '89., IEEE*, pages 1680 –1686 vol.3, nov 1989.

[12] K. Jamieson and H. Balakrishnan. Ppr: partial packet recovery for wireless networks. In *Proceedings of the 2007 conference on Applications, technologies, architectures, and protocols for computer communications*, SIGCOMM '07, pages 409–420, New York, NY, USA, 2007. ACM.

[13] H. Meyr, M. Moeneclaey, and S. A. Fechtel. *Digital Communication Receivers: Synchronization, Channel Estimation and Signal Processing.* Cambridge University Press, 1997.

[14] T. K. Moon. *Error Correction Coding.* New Jersey: John Wiley & Sons, 2005.

[15] K. J. Saldanha. *Performance Evaluation of DECT in Different Radio Environments.* Ph.D Thesis, Virginia Polytechnic Institute and State University.

[16] S. Sen, R. Roy Choudhury, and S. Nelakuditi. Csma/cn: carrier sense multiple access with collision notification. In *Proceedings of the sixteenth annual international conference on Mobile computing and networking*, MobiCom '10, pages 25–36, New York, NY, USA, 2010. ACM.

[17] K. Tan, J. Fang, Y. Zhang, S. Chen, L. Shi, J. Zhang, and Y. Zhang. Fine-grained channel access in wireless lan. In *Proceedings of the ACM SIGCOMM 2010 conference*, SIGCOMM '10, pages 147–158, New York, NY, USA, 2010. ACM.

[18] K. Tan, J. Zhang, J. Fang, H. Liu, Y. Ye, S. Wang, Y. Zhang, H. Wu, W. Wang, and G. M. Voelker. Sora: High performance software radio using general purpose multi-core processors. In *6th USENIX Symposium on Networked Systems Design and Implementation (NSDI 2009).*

[19] D. Tse and P. Vishwanath. *Fundamentals of Wireless Communications.* Plenum Press New York and London, 2005.

[20] M. Vutukuru, H. Balakrishnan, and K. Jamieson. Cross-layer wireless bit rate adaptation. In *Proceedings of the ACM SIGCOMM 2009 conference on Data communication*, SIGCOMM '09, pages 3–14, New York, NY, USA, 2009. ACM.

[21] C. Wu and V. Li. Receiver-initiated busy tone multiple access in packet radio networks. In *Proceedings of the ACM SIGCOMM 1987 conference*, SIGCOMM '87, pages 336–342. ACM, 1987.

[22] D.-Y. Yang, T.-J. Lee, K. Jang, J.-B. Chang, and S. Choi. Performance Enhancement of Multi-Rate IEEE 802.11 WLANs with Geographically-Scattered Stations. *IEEE Transactions on Mobile Computing*, pages 906–919.

[23] X. Zhang and K. G. Shin. E-mili: energy-minimizing idle listening in wireless networks. In *Proceedings of the 17th annual international conference on Mobile computing and networking*, MobiCom '11, pages 205–216, New York, NY, USA, 2011. ACM.

RadioJockey: Mining Program Execution to Optimize Cellular Radio Usage

Pavan K. Athivarapu[1], Ranjita Bhagwan[1], Saikat Guha[1],
Vishnu Navda[1], Ramachandran Ramjee[1],
Dushyant Arora[2]*, Venkata N. Padmanabhan[1], George Varghese[3]*
[1]Microsoft Research India [2]Princeton University [3]UC San Diego

ABSTRACT

Many networked applications that run in the background on a mobile device incur significant energy drains when using the cellular radio interface for communication. This is mainly due to the radio-tail, where the cellular radio remaining in a high energy state for up to 20s after each communication spurt. In order to cut down energy consumption, many recent devices employ fast dormancy, a feature that forces the client radio to quickly go into a low energy state after a fixed short idle period. However, aggressive idle timer values for fast dormancy can increase signaling overhead due to frequent state transitions, which negatively impacts the network. In this work, we have designed and implemented RadioJockey, a system that uses program execution traces to predict the end of communication spurts, thereby accurately invoking fast dormancy without increasing network signaling load. We evaluate RadioJockey on a broad range of background applications and show that it achieves 20-40% energy savings with negligible increase in signaling overhead compared to fixed idle timer-based approaches.

Categories and Subject Descriptors

C.2.1 [**Computer Communications Networks**]: Network Architecture and Design—*Wireless communication*; C.4 [**Performance of Systems**]: Design Studies, Performance Attributes

General Terms

Algorithms, Design, Measurement, Experimentation, Performance

Keywords

Cellular, 3G, LTE, 4G, Energy Saving, Signaling Overhead, Fast Dormancy

1. INTRODUCTION

Several applications running on smartphones, tablets and laptops perform network activity while running in the background for tasks such as syncing state (e.g., emails, presence, newsfeeds, etc.). Often these tasks involve only a few seconds of communication interspersed between longer periods of inactivity. Nevertheless, several platforms (e.g., iPhone and Windows Phone) prohibit background applications or severely restrict them, primarily because of the significant energy costs of supporting such workloads over a cellular radio interface.

Even short network bursts have significant energy costs due to the intrinsic nature of cellular radio operation. At the beginning of each communication spurt, the cellular radio wakes up from a low power idle state and transitions to a high energy active state (DCH) to transfer packets. This DCH setup operation takes up to 30 signaling messages and about 2 seconds [9]. After the communication spurt ends, the radio continues to remain in DCH state for a certain inactive duration (e.g., 10-20s) before transitioning back to idle state. This inactive duration is determined by the network operator in both the 3G and the 4G LTE standards [15] and is commonly referred to as the *radio-tail* [2, 18]. A long radio-tail cuts down on repeated DCH setups and the attendant (unmonetized) signaling bandwidth costs. However, while long radio-tails help keep signaling costs low, they incur an energy penalty of up to 16J per tail and severely impact the energy consumption of background applications.

Many recent smartphone models cut down the energy cost of radio-tails by implementing a feature called *Fast Dormancy* (FD), which forces the radio to quickly transition from DCH to a low energy state (Idle or PCH). Fast dormancy is typically invoked with a fixed short inactivity timer (e.g., 3-5s), thereby reducing the energy consumed during the radio tail. However, as we show in Section 2, aggressive fixed idle timer values face two fundamental issues: i) some complex applications such as Outlook (email) have an inherently large variance in their inter-packet arrival distribution; and ii) even for simple applications, network conditions may change as the device moves, significantly altering the inter-packet arrival distribution. Since there is no fixed sweet-spot in inter-packet arrivals, the use of aggressive fixed idle timers can result in unpredictable increases in signaling load as the device repeatedly transitions between idle and DCH states. Indeed, in the early days of fast dormancy, popular devices employing aggressive timers lead to episodes of severe signaling channel congestion [1].

Fundamentally, there is an inherent trade-off between energy saved using fast dormancy and the signaling overhead generated by the device. To date, this trade-off has mostly been explored in an ad-hoc manner through the use of idle timers. Instead, in this paper, we have designed a system called *RadioJockey that uses program analysis to predict end of communication spurts and thereby accurately invokes fast dormancy without increasing signaling costs.*

*Work done while at Microsoft Research India.

MobiCom'12, August 22–26, 2012, Istanbul, Turkey.

The hypothesis behind our program analysis-based approach is that application developers typically invoke a unique set of functions in their code just before/after the end of communication spurts (e.g., cleanup activity such as close socket or flush buffers, update UI elements, etc.) If this hypothesis is true, by mining program execution traces offline, we should be able to automatically identify a set of program features (e.g., system or function calls) that only occur (or never occur) near the end of communication spurts. Since these features are inherent to the way the program operates, they should not be affected by the network dynamics, and closely track dynamic application communication behavior, the two issues that adversely affect a fixed idle timer-based approach.

The advantage of the above program analysis-based approach over a timer-based approach is two-fold. First, during the execution of the program, we can simply invoke fast dormancy as soon as the respective features are detected, thereby being even more aggressive in energy savings than the 3–5s idle timers. Second, despite being more aggressive in invoking fast dormancy, we are still assured that the application will not be communicating imminently (i.e., the idle-timer would anyway have fired, albeit after a 3–5s delay). Thus we avoid adversely impacting the network signaling load as compared to today.

RadioJockey operates in two modes, offline training and online application (Section 3). For offline training, we intercept system calls using either DETOUR [5] for windows applications or systrace for Android applications and collect program execution traces while the application is running in the background with no user interactions. We also simultaneously capture packet traces. From these traces, we extract a variety of features such as function calls, additional attributes like return values, arguments, history, stack depth, call stack, at the time of invoking the call. We mine these traces using the C5.0 decision tree-based classifier and classify periods as active (communication on-going) or inactive (end of communication spurt). A unique decision tree is learned for each application which is then used in the online phase for dynamically invoking fast dormancy during the execution of the application in the background. We have implemented the online runtime component of the RadioJockey system on Windows and show that it results in negligible runtime overhead (Section 4).

We evaluate RadioJockey using a variety of mobile smartphone as well as tablet and laptop applications for the android and windows platforms. The applications evaluated range from simple ones like gnotify, to complex applications like the Outlook email client (Section 5). We show that the rules mined have a high prediction accuracy and negligible false positive rates. A *surprising finding is that system call features alone are sufficient to identify high quality rules*, thereby allowing a simple, generic implementation to support a broad class of legacy, native, and managed applications. For over a dozen background applications, we show that RadioJockey is able to save 20-40% energy consumed by the cellular radio compared to a 3s idle timer while resulting in less than 2% increase in signaling costs for the network operator.

While application developers may be able to modify their code to invoke fast dormancy accurately at the end of each communication spurt [15], there are several advantages to using an automated approach like RadioJockey. First, most application developers are simply unaware of the energy characteristics of the cellular radio; RadioJockey allows these developers to focus on the design of their application while providing the energy savings of fast dormancy automatically. Second, based on the complexity of some of the rules that RadioJockey learns, we suspect that it is non-trivial for a developer to manually identify the end of all communication spurts for large complex applications such as Outlook. Finally, we believe that automated solutions like RadioJockey would allow platforms such as iPhone/Windows Phone to relax constraints imposed on background applications, secure in the knowledge that energy cost of *all background applications are automatically optimized.*

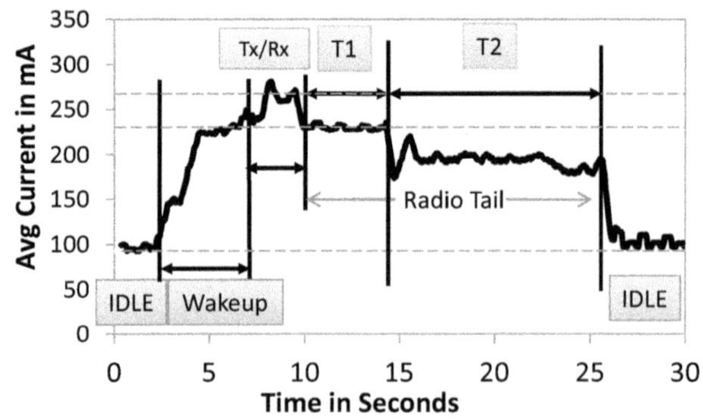

Figure 1: Energy consumption of the cellular interface

	Duration (s)	Energy(mJ)	#Signals
Ramp up IDLE to DCH	2	2	30
Ramp up PCH to DCH	0.5	0.5	12
Default Tail	20	15	2
5-sec Timer + FD	6	6	2
3-sec Timer + FD	4	4	2

Table 1: Cellular Radio Characteristics

In summary, this paper makes the following contributions.

- We present a novel approach for accurately invoking fast dormancy based on mining program execution traces.
- We find that system call features alone are sufficient in identifying high quality rules, allowing generic fast dormancy support for broad class of legacy, native and managed applications.
- We present the design and implementation of RadioJockey that delivers 20-40% energy savings with negligible increase in operator signaling costs.

2. BACKGROUND AND MOTIVATION

2.1 Energy and Signaling Overhead

During normal usage, a cellular radio switches between different internal states called Radio Resource Control (RRC) states depending on the volume of network traffic. These state transitions are not instantaneous, and they incur both energy and signaling costs. This has a major implication on the energy consumed by background applications that typically perform short bursts of network activity for operations such as sending periodic heart beat messages, updating buddy status, checking for new emails, and pulling news feeds, etc. The energy consumption is mainly dominated by the switching overhead for short network transactions.

Figure 1 illustrates the power consumption characteristics of different states of a typical cellular radio during a short network session. Initially, when there is no network activity, the radio is in IDLE state, consuming very low power. It transitions to an active DCH state when there are one or more packets to be transmitted or received. The state transition (ramp up) usually takes up to 2 seconds since it requires authentication and a layer-2 connection to be established, involving exchange of several signaling messages with different entities in the operator's network. The radio remains in DCH state while the network session is active. The radio is transitioned to a lower power state (IDLE or PCH), sometimes via another intermediate state (FACH), using an inactivity timer that is set by the operator. However, the timers are typically very large (for example, 20 seconds) resulting in significant energy cost for the radio-tail. Table 1 summarizes the energy and signaling overheads associated with different cellular states transitions.

To address the high energy drain due to long radio tails, many phone vendors use a feature called fast dormancy that enables quick transition to IDLE. The client device invokes fast dormancy using a shorter inactivity timer to cut down the radio tail. Fast dormancy is achieved by sending a specific signaling message called Signaling Connection Release Indication message (SCRI). In the early days, the mobiles simply used SCRI with cause value set to *unknown-cause* which resulted in releasing the RRC connection and moving the mobile to IDLE state. In some newer networks with support for 3GPP Release 8 Fast Dormancy (HSPA and LTE), the client sets the cause value to *PS Data session end* which allows the network to move the mobile to PCH state. PCH has slightly higher consumption than IDLE but requires fewer signaling messages to ramp-up from from PCH to DCH state (12 instead of 30 for IDLE) [9]. In either case, the signaling cost to transition to DCH is significant.

Note that *current best practices limit fast dormancy usage to background applications* [9]. This is because, when the user is interacting with the application, it is difficult to predict when the user may cause an interaction that results in new network traffic. In fact, iPhone OS v4.2 disabled fast dormancy usage when the screen backlight is on primarily because usage of fast dormancy for foreground applications in v4.1 resulted in significant increase in operator signaling costs [9]. Thus, in this paper, we restrict ourselves to applications that are running in the background while the screen is off, which typically constitutes a significant portion of mobile device usage.

While fast dormancy has been helpful in reducing energy consumption of background applications on mobile devices, existing approaches of employing aggressive timers can cause significant increase in the signaling load in the operator's network, even for networks that support PCH state.

2.2 Drawbacks of Timer based approach

We now describe two fundamental issues with fixed idle timer-based fast dormancy approaches, which can cause an unpredictable and significant increase in signaling load.

First, some applications are very chatty in the background, e.g., email clients such as Outlook and instant messaging clients like Lync and Skype. Figure 2 shows the CDF of inter-packet times when the client is running in the background. Notice that the inter-packet times span a large range of values with no distinctive knee. No matter what value is chosen for a fixed inactivity timer, packets can still arrive immediately after the timer has expired, causing the radio to be woken up immediately, thereby increasing the signaling overhead. RadioJockey operates at the granularity of sessions rather than packets. As we show in Section 5, our program analysis-based approach is able to distinguish between active and inactive sessions since it identifies/predicts the *e*nd of communication spurts.

Secondly, cellular network characteristics can change dramatically due to user mobility or changes in load characteristics. Packets may be delayed when there are sudden latency spikes. If aggressive idle timers are used, such scenarios can cause the signaling load to increase tremendously.

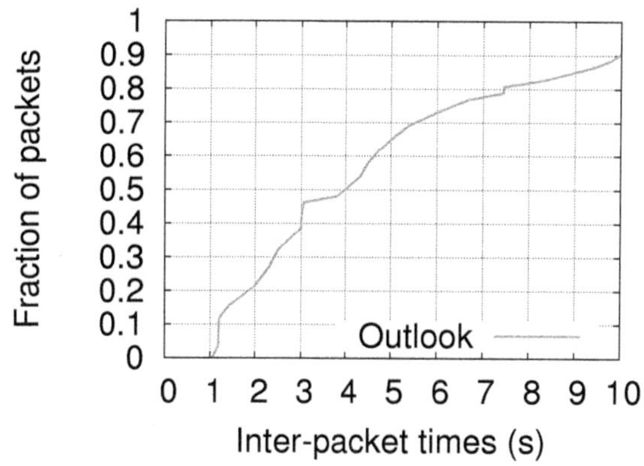

Figure 2: CDF of inter-packet times for Outlook application

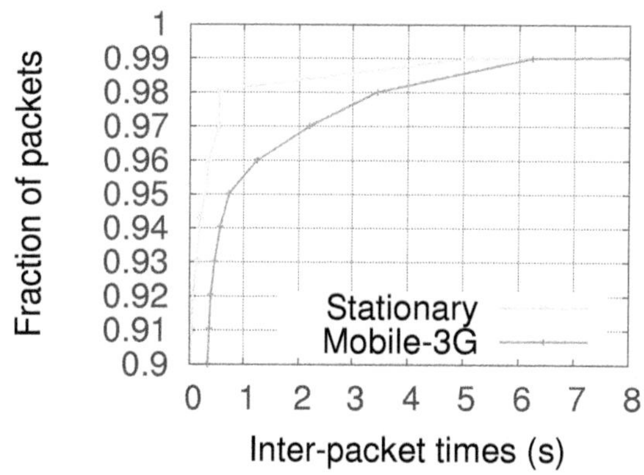

Figure 3: CDF of inter-packet times for Lync application on static and mobile scenarios

Figure 3 shows the CDF of packet inter-arrival distribution for Lync, an instant messenger application running in the background, for two cases: 1) when the node is stationary, and 2) when the node is mobile while attached to a 3G network. The 98 percentile of packet inter-arrival times increases from around 500ms for the stationary node to over 3s for the mobile node. Again, as we show in Section 5, since our program analysis-based approach is looking for rules identifying end of sessions, learning rules from the stationary node trace and applying the rules to the mobile scenario works well, which accords with the intuition that the program execution would remain unaffected by mobility.

In summary, inactivity timer based approaches are oblivious to whether there will a packet transmission in the near future. It is based on ad-hoc timer values, which is prone to errors. We take a different approach. We predict inactivity based on program behavior. Our prediction-based technique has two advantages: 1) it is able to shut down the radio as soon as the communication spurt ends, without waiting for an inactivity timer, and 2) makes it unlikely that there will be an immediate packet transmission, which would otherwise increase signaling costs.

3. RADIOJOCKEY DESIGN

The objective of RadioJockey is to learn *application-level* signatures that predict *network-level* behavior. In this section, we first describe our intuition behind this approach. We then provide rele-

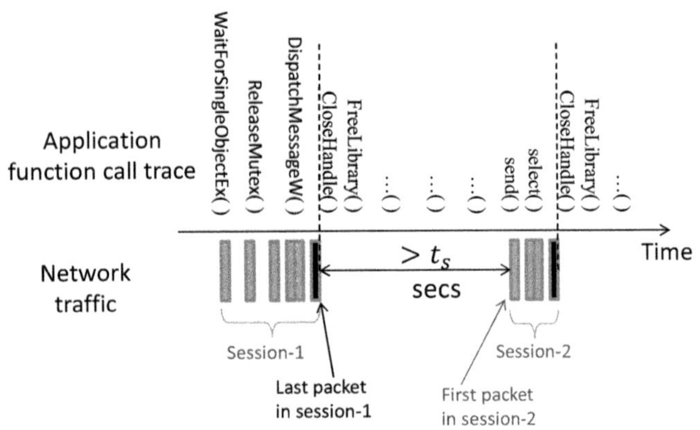

Figure 4: Defining end of sessions traces

vant definitions and describe RadioJockey's learning engine. Next, we describe the design of the run-time engine that implements rule-matching and the invocation of fast-dormancy. Finally, we describe several design decisions we made in the course of determining the specifics of the learning strategy.

3.1 Intuition

A program consists of one or more code paths (sequence of function calls) where each path is invoked in response to application logic and data. The various code paths can be identified by the function calls, stack traces, and system calls invoked as the program executes.

Often, for programs which perform sporadic background network tasks, a certain set of code paths lead to the application actively using the network, and a different set of code paths execute when the application stops using the network. The code paths include operations such as message processing, updating internal state, GUI window updates, memory allocation/deallocation, and closing of sockets/object handles. By identifying code paths that are correlated with the state of the program at the end of network sessions, and also during active network usage, we should be able to predict when the program is likely to stop communicating for a prolonged duration of time. Predicting such inactivity in advance allows us to determine when to invoke fast-dormancy and achieve energy savings while limiting the signaling overhead.

3.2 Definitions

We define an *active network session* as a time period over which the application sends or receives packets such that no two consecutive packets occur more than t_s seconds apart. t_s is the *end-of-session window* parameter. An *end-of-session* (EOS) is an event that occurs when the last packet of an active network session is seen. Therefore, t_s defines when EOS events occur in a given network trace that an application generates. Figure 4 depicts a traffic trace of an application, where packets are grouped into two active network sessions separated by more than t_s seconds.

The period t_s is chosen such that by invoking fast-dormancy immediately after the last packet in a session, there will be energy savings compared to a scenario where the radio remains in the active state between the two sessions. The goal of RadioJockey is therefore to predict whether the most recent packet transferred over the cellular interface happens to be the last packet in the current active network session, and if so invoke fast-dormancy.

If this prediction is correct (i.e., an idle timer based FD would have anyway fired for the same interval, albeit after a 3–5 second delay), we achieve the energy savings without any increase in signaling costs.

To check the feasibility of this idea, we collected function call traces that are executed at runtime for over a dozen background applications. We look at whether there are any unique function calls that are being called near an end of session. Visually inspecting did not seem promising as there were no obvious patterns in function calls called at an EOS event which we could manually discern. Consequently, we decided to use a machine learning approach towards discovering patterns that predict EOS.

3.3 Rule Learning

RadioJockey uses the C5.0 decision tree classifier as its principal learning tool. The algorithm builds trees using the concept of information gain. Starting at the root, it chooses the feature that splits the input data and reduces the entropy of the data set by the maximum amount. The final result is a tree in which branch points at higher levels of the tree correspond to attributes with greater predictive power. To avoid over-fitting, the algorithm includes a pruning step, wherein some branches in the tree are discarded while keeping the error with respect to the training data within a certain confidence threshold.

The input to the decision tree classifier consists of a set of labeled *data-items*. Each data-item consists of a set of key-value pairs, and the label assigned to it can either be "ACTIVE" or "EOS". The decision tree then outputs a set of boolean expressions, or *rules*, over these key-value pairs that indicate when a session is in an ACTIVE state, and when it has reached EOS. We now describe the method by which we generate the input data-items for training Radio-Jockey's classifier.

To profile an application using the classifier, we simultaneously collect raw system call traces and network traces while the application performs background tasks using the network. The required size of the traces varies depending on the complexity of the application. For a simple application, such as gnotify, whose sole purpose is to periodically poll an email server and determine if there is any new mail, the classifier may require only thirty minutes to an hour of data to learn accurate rules. On the other hand for a significantly more complex application such as the Outlook email client that performs multiple kinds of background network-related tasks, the classifier may need several hours of data to learn a potentially larger set of rules that characterize the application.

Figure 5 illustrates how we extract a set of training data-items from the system call traces and the packet traces. We use the network packet trace to divide the function call traces into active and inactive *segments* based on periods of network activity and inactivity. The period between any two consecutive packets is a segment. We create data-items from these segments using the following procedure.

Case 1 ($segment > t_s$)**:** If a segment is longer than t_s seconds, the system calls made from the time of the start of the segment up to a duration of t_w seconds form a data-item labeled EOS (middle data-item in Figure 5). The parameter t_w is called the *shutdown window* parameter and is less than t_s. We truncate every data-item to at most t_w seconds in length because we want to restrict our learning to features that lie within a short time of observing a packet. Without the truncation, the classifier may learn rules for EOS using features that occur well after observing a packet. Such a rule would

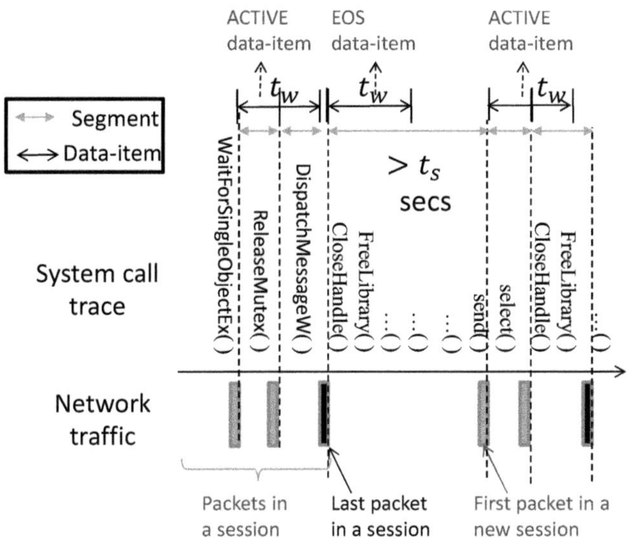

Figure 5: Generating data-items for the decision tree classifier.

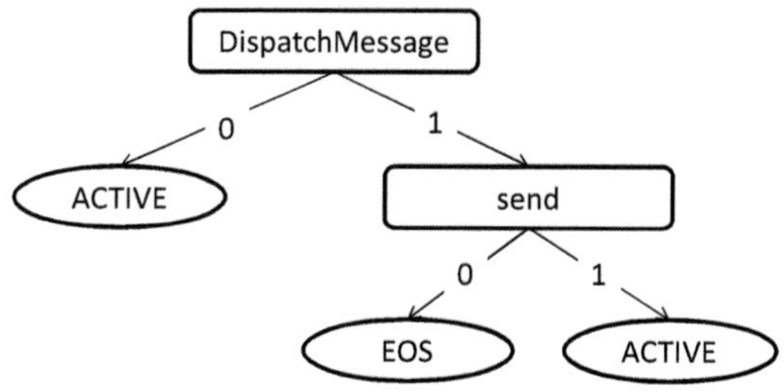

(a) The decision tree learned for Gnotify

DISPATCHMESSAGE = 0 ⇒ ACTIVE (35)
DISPATCHMESSAGE = 1 AND SEND = 0 ⇒ EOS (24)
DISPATCHMESSAGE = 1 AND SEND = 1 ⇒ ACTIVE (1)

(b) The rules learned for Gnotify

Figure 6: Classifying Gnotify Behavior

be matched much after the actual end-of-session, and so the radio would be unnecessarily on from the end-of-session to the time of the match.

Case 2 ($t_w < segment < t_s$)**:** If a segment is shorter than t_s seconds but longer than t_w seconds apart, the system calls made from the first packet up to t_w seconds form an ACTIVE data-item (left data-item in Figure 5). This truncation, ensures that we solely learn rules involving system calls close to a network event.

Case 3 ($segment < t_w$)**:** Often, a large number of segments are formed by observing packets within very short time intervals. Due to inaccuracies in timestamping packets at capture-time, data-items created over very small durations may introduce a large amount of noise into the learning engine. Consequently, we concatenate as many such short consecutive segments as needed to create one ACTIVE data-item that is t_w seconds in duration (right data-item in Figure 5).

Thus every data-item contains system calls made over a period of t_w seconds. For each data-item, we extract binary features determining whether a certain system call existed in it or not. In other words, we use one binary variable for every system call. Its value for the data-item is 1 if the application made the call in this data-item, and 0 otherwise. In addition to these binary features, we also use a feature to indicate the previous data-item's state, that is, whether the data-item just prior to the current one was labeled ACTIVE or EOS. This parameter, which we call PREV_STATE, helps us capture temporal characteristics in the call trace. We found this particularly useful in capturing rules in complex applications, as we explain in Section 3.5.

Using this procedure, from the traces, we therefore generate a set of data-items labeled ACTIVE or EOS. From these, the classifier learns rules in the form of boolean expressions. We provide two *costs* to the learning algorithm. The cost of misclassifying an ACTIVE data-item as EOS has two components: first, we specify the energy expended in shutting-down the radio and then bringing it up again. Second, we specify the corresponding signaling overhead of mis-predicting an EOS. The cost of misclassifying an EOS data-item as ACTIVE involves purely the energy cost of keeping the radio up when it could have been put into fast-dormancy.

Figure 6a shows the tree learned for the gnotify application running on a Windows 7 system. The gnotify application periodically polls gmail and pops up a message when there is a new email (more elaborate results are shown in Section 5). The system call trace used to learn this tree was 5 minutes long. Every path from root to leaf yields one boolean expression. Consequently, the rules that the classifier learned for gnotify are listed in Figure 6b.

The numbers in parenthesis next to each rule are the number of data-items that contribute to learning that rule. They indicate that the first two rules have much higher confidence than the third rule. The DISPATCHMESSAGE system call dispatches an inter-thread message to a specified window procedure in the GUI thread, while the SEND call is used to send data out on a socket. Hence this set of rules indicate that, at run-time, in a data-item, if no DISPATCHMESSAGE is seen, the session is ACTIVE and therefore we should not invoke fast-dormancy. However, if DISPATCHMESSAGE is seen *and* SEND is *not* seen, the rules predict an EOS, and therefore we should invoke fast dormancy. If DISPATCHMESSAGE is seen *and* SEND is seen, then the session is still ACTIVE, and the radio should stay on.

It is easy to see why the SEND call appears in the rules since it directly results in network activity. Learning the DISPATCHMESSAGE call is, on the other hand, not as obvious. We believe that the classifier learns this call because gnotify, after polling the email server (i.e. at EOS), updates certain GUI artifacts (such as setting the last-checked-at message in the tool-tip of the application's icon in the system tray), which registers as a causality between EOS and DISPATCHMESSAGE.

Figure 7 shows a part of the tree learned for the Outlook email application running on a Windows 7 system. Outlook is a significantly more complex application than gnotify, and as a result, our learning engine learned approximately thirty rules of which the figure shows four prominent ones. An interesting observation is that the top-level feature in this tree is the previous state, or PREV_STATE: in Section 3.5 we explain why this is an important feature for complex applications. When PREV_STATE is EOS, the next session is predicted to be ACTIVE. On the other hand, if PREV_STATE is ACTIVE, if the LOCALFREE call is seen, the session remains ACTIVE. If LOCALFREE is not seen, but WSARECV is seen, the session is still ACTIVE. If WSARECV is not seen, the learning predicts that an end-of-session has been reached.

This tree shows the importance of learning different rules for different kinds of ACTIVE sessions that an application can have. Each path ending in an ACTIVE leaf node potentially represents a distinct background network behavior. It is important that our

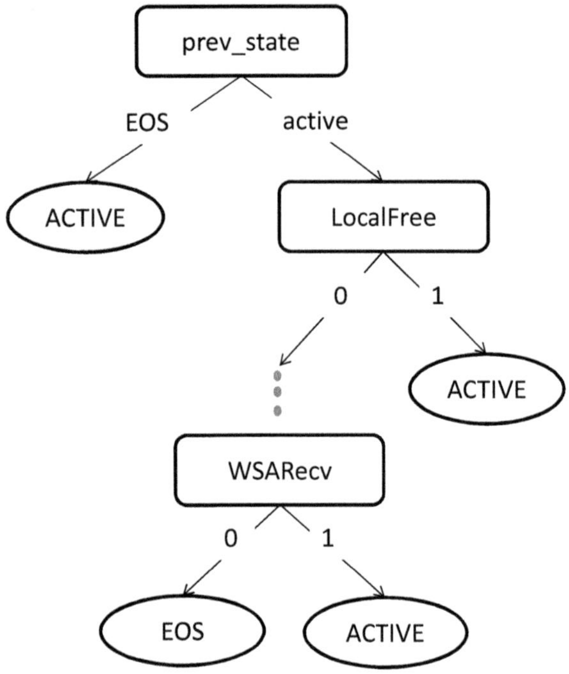

Figure 7: The decision tree learned for Outlook

learning technique capture each of these different behaviors automatically and completely.

3.4 Run-time

We now describe how our run-time engine uses the learned rules to decide when to shut down the network radio.

Whenever an application sends or receives a packet, say at time t, the run-time engine resets state and starts collecting system call traces as part of a new *session*. If the application sends or receives the next packet within the shutdown window t_w, we automatically determine that the current session (started by the first packet) is ACTIVE, and start collecting traces as part of the session started by the second packet. If, on the other hand, we see no network activity for the time window t_w since the first packet, the engine needs to decide whether the system calls in the current session predict EOS or not. If the set of system calls in the current session matches a rule for EOS learned in the rule-learning phase, thereby predicting that the network session has ended, the engine invokes fast dormancy. We elaborate further on this process while describing our implementation in Section 4. We repeat this process every time we see the application send or receive a packet.

3.5 Design Decisions

We now describe various decisions we made in the design of our data collection, learning and run-time engines.

Feature Selection: Initially, our learning engine used a significantly richer feature set that included application-level stack traces. We treated each call-site in the stack trace as a binary feature, just as we did with the system calls. We also included the return values to every system call and the system call as key-value pairs, as opposed to using just a binary variable to indicate whether a system call is present or absent in the data-item. We also tracked nested system calls (where a system call implementation calls another system call before returning to the caller). Surprisingly, we found that using only binary features for system calls was sufficient in learning accurate rules: adding the extra call stack information and the return values did little to increase the power of our learning. Tracking nested system calls made things worse because it introduced non-application-specific features (i.e., OS/library implementation features), which overwhelmed the application-specific features that are of primary interest. While counter-intuitive, this is a welcome result since tracking only the system calls that an application makes is a low-overhead task.

Use of previous state: While learning with only system calls worked well for simple applications, we found that the learning was noisy for more complex applications such as Outlook and Lync, even with stack traces. To address this, we added the binary feature PREV_STATE to every data-item, which tracks whether the previous data-item's label was ACTIVE or EOS. With this addition, our learning for complex applications such as Outlook improved dramatically. On closer observation, we found that these applications had different types of background network activity: while some sessions were long-lived and involved significant traffic, many others were very short and extremely periodic, such as application-level keep-alive messages being sent every 10 seconds. This behavior results in alternating data-items being in the ACTIVE and EOS states. Introducing the PREV_STATE parameter helped the classifier learn a different signature for such short and periodic sessions as compared to the longer, more heavyweight sessions.

Static vs Dynamic Learning: Our results indicate that using off-line, static learning on a per-user basis is very effective at learning accurate rules. We also found that our learning engine has very good turnaround time. For a complex application such as Outlook, using 6500 data-items involving 269 binary features, the classifier ran within 0.5 seconds on a desktop-grade machine. Since this is a fairly quick turnaround time in learning rules, we decided to implement a dynamic learning engine (as described in Section 4) in addition to the static learning engine. In case applications change their behavior over time, the dynamic learner running alongside the application will automatically generate new and updated rules.

4. IMPLEMENTATION

We have implemented the RadioJockey learning engine and run-time engine. Figure 8 describes the RadioJockey components at a high-level.

Learning engine. The learning engine comprises a call-tracing module, a packet-timings module, and the standard C5.0 decision tree classifier. The call-tracing module collects system call traces (on Android phones and tablets), and library call traces (on Windows tablets). System call traces are collected using STRACE. Library call traces are collected by injecting our RadioJockey DLL into the unmodified application's address space. The DLL intercepts calls to the 1634 core Win32 library functions (using DETOUR) and logs them (in-memory) before calling the intended function. Note that we do not need application source-code, nor do we need to re-write the application binary in either case. Using our STRACE and DETOUR approach we can support native as well as legacy applications.

The packet-timings module captures timing information for each packet sent or received by the application. We use LIBPCAP (on Android) and NETMON (on Windows). The key challenge we faced was associating a network packet to the running application process (without modifying the kernel). This was hard because LIBPCAP and NETMON do not support filtering packets by process ID. Our code performs the filtering in user-space. We link a packet to process based on the TCP port number. We get the port numbers opened by the application by periodically (every few milliseconds) snapshotting NETSTAT output. Since establishing a new TCP con-

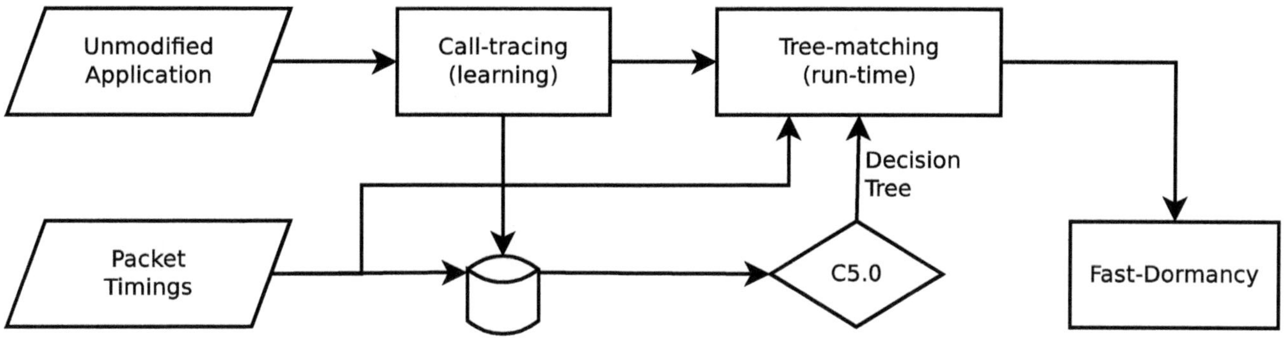

Figure 8: RadioJockey architecture

nection takes several round-trip-times (which are on the order of tens of milliseconds), we are still able to track short-lived sessions, although an in-kernel implementation would be more efficient.

The output of the call-tracing and packet-timing module are processed as described in Section 3.3 and fed into C5.0. The decision tree output by C5.0 is then fed into the run-time engine. Note that our implementation supports both offline and online learning. In offline learning the data used for learning may come from an earlier execution of the application. In online learning the data used for learning comes from the current application instance.

Run-time engine. The RadioJockey run-time engine comprises a tree-matching module, and a fast-dormancy module. The tree-matching module accepts a decision tree over a control channel, and classifies sessions as ACTIVE or EOS based on the logic described in Section 3.4. It shares much of its implementation with the call-tracing module since both use the same data. The decision tree can be updated at any time (and any number of times) without having to restart the application. When the tree-matcher predicts an EOS it informs the fast-dormancy module over a control channel.

The fast-dormancy module uses private APIs to invoke fast-dormancy on the particular platform (Android or Windows).

Overall our implementation consists of 2580 lines of C and C++ code (most of which is for the Windows DLL implementation), and 350 lines of Python scripts to process data-items and C5.0 output. Our RadioJockey DLL, which is injected into the running application's address space, is 1.4 MB in size and includes both the call-tracing and the tree-matching modules.

Avoiding mis-prediction cascades. We designed the RadioJockey run-time to avoid mis-prediction cascades. Specifically, we do not base the current prediction on the previous predicted value; PREV_STATE is always the *actual outcome* of the previous data-item. This also allows us to track the accuracy of our past predictions without incurring any additional overheads. For example, if we predict EOS and we see a packet before t_s has elapsed, we set PREV_STATE to ACTIVE to reflect the actual outcome, and log the false positive. If we don't see a packet within t_s of an ACTIVE prediction, we similarly set PREV_STATE to EOS and log the false negative. If the application behavior changes, or the application enters a phase that was not well-represented in the training data, this is manifested in our log as a string of mis-predictions, which the run-time engine detects. At that point the run-time (temporarily) stops making predictions, which gracefully falls back to existing behavior (of network operator defined idle timer based FD) with no additional signaling overheads.

The run-time engine resumes making predictions after some time. When resuming, it simply needs to wait for two consecutive packets (to set PREV_STATE); all other inputs come from the current data-item. Exponential back-off, where the run-time stops making predictions for exponentially longer intervals when the misprediction rate does not improve, ensures that the run-time closely tracks slight unexpected deviations in application behavior while minimizing overheads if application behavior changes significantly. When the exponential back-off timer exceeds a threshold, the run-time (optionally) switches to dynamic learning (Section 3.5) to relearn the new application behavior.

App	Platform	Avg session(s)	Avg inter-session(s)	Complexity
Litestock	Win	0.51	14.83	Simple
Yahoo Messenger	Win	0.54	28.30	Simple
Seesmic	Win	0.69	35.66	Simple
Miranda IM	Win	0.31	49.38	Simple
Destroy Twitter	Win	1.99	48.77	Simple
Gmail Notify	Win	0.34	69.48	Simple
Desktop Changer	Win	6.99	63.46	Simple
GoChat Facebook	Android	0.18	26.22	Moderate
Google Talk	Win	0.28	27.39	Moderate
Tweetdeck	Win	0.37	24.71	Moderate
Twitdroid	Android	1.81	14.81	Moderate
Live mail	Win	11.42	71.11	Moderate
Lync	Win	1.03	55.73	Complex
Outlook	Win	0.86	18.53	Complex

Table 2: Applications and networks we evaluated RadioJockey on

5. EVALUATION

We evaluate RadioJockey using both real-world experiments and trace-driven simulations across 14 applications on multiple platforms and on multiple network types. We find that our approach is robust to a wide range of application-level and network-level factors.

5.1 Macrobenchmarks

5.1.1 Energy Savings

We ran the 14 applications listed in Table 2. Twitdroid and GoChat are Android applications, and the remaining are Windows applications. We captured call-traces and network-traces for each application. We performed a 10-fold cross-validation. That is, we partitioned the data into 10 parts, trained using 9 of them, and tested the classifier on the 10th; we repeated this 10 times for a different choice of the test partition. To compare with existing approaches, we simulate a *timeout-based* fast-dormancy approach for common timer values used today. The timeout-based approach invokes fast-dormancy when it sees a period of contiguous network inactivity for some fixed time.

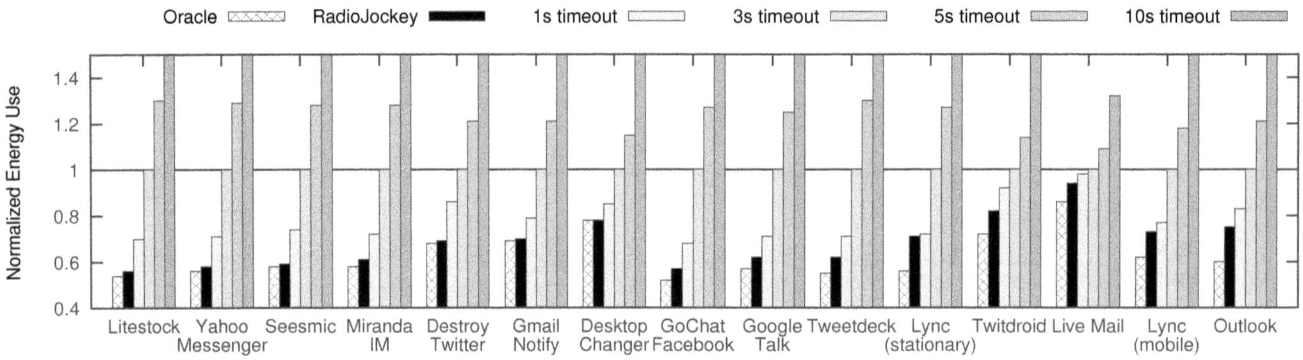

(a) Energy savings

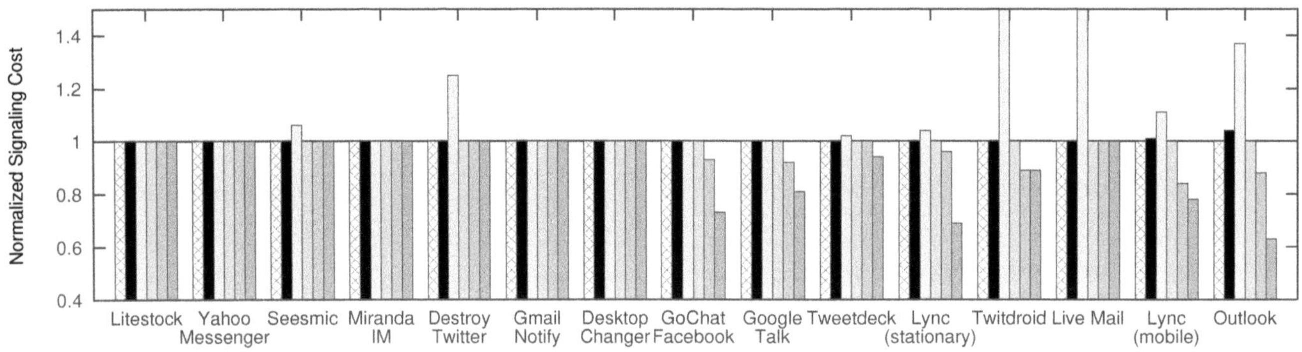

(b) Signaling overhead

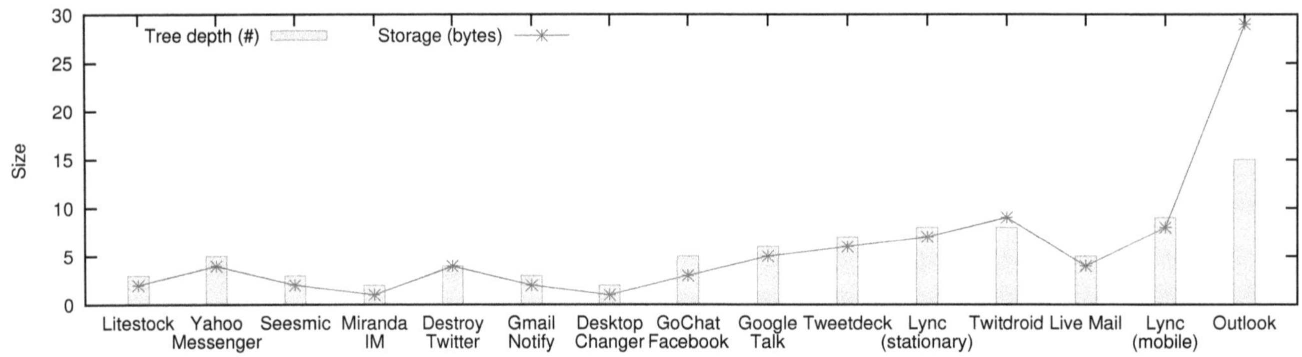

(c) Tree complexity

Figure 9: RadioJockey energy savings and signaling overhead as compared to today, and complexity of the decision tree learned.

We additionally compare against an aggressive timeout-based approach that uses the same shutdown window (t_w) we use. Finally, to determine the maximal savings possible, we simulate an oracle that has perfect future knowledge about when a session ends so it can invoke fast-dormancy at the earliest possible time. We use energy, power, and signaling numbers measured on the Samsung Focus device mentioned earlier (Table 1) for computing these results. We disable RadioJockey's dynamic learning to evaluate the quality of rules we learn statically; enabling dynamic learning would serve only to improve our results.

Figure 9a plots the (normalized) median energy consumed for each approach and Figure 9b plots the (normalized) median signaling messages for each approach. All numbers are normalized to that of the 3 second *timeout-based approach*, which is the most aggressive (in terms of energy-savings) fast-dormancy strategy used by current smartphones [4]. The oracle, of course, consumes the least energy and has no added signaling overhead as compared to the 3s idle timer — this represents the optimal operating point.

For the simple applications, RadioJockey performance is nearly identical to that of the oracle, saving between 44%–22% of energy as compared to today without any additional signaling overheads. For these applications, RadioJockey correctly predicted every EOS (i.e., 0% false-negative) and did not mis-predict any ACTIVE session (i.e., 0% false-positive); the slight energy difference compared to the oracle is due to the small shutdown window that RadioJockey needs to wait before entering fast-dormancy.

While an aggressive timeout-based approach achieves only slightly lower energy savings, it can have significantly higher signaling cost (between 6%–56% more than today across all applications). The reason for this, as mentioned, is that such an approach will be highly sensitive to application and network factors, often mis-predicting an EOS if an application stalls briefly or packets are delayed. As mentioned, each mis-prediction (false-positive) incurs an additional signaling cost of 32 messages (for 3G or for LTE) as compared to today. More conservative timer based approaches have lower signaling overhead (since they enter fast dormancy less often than today), but correspondingly consume a lot more energy than today.

For the moderate complexity applications, RadioJockey still outperforms all timeout based approaches, and achieves 42%–90% of the oracle's energy savings (i.e., 6%–43% savings compared to today) without any added signaling overhead. This is attributed to RadioJockey opting to minimize false-positives (i.e., not incurring added signaling overhead) at the expense of false-negatives (i.e., not saving as much energy as optimally possible) as mentioned earlier. The performance of timeout based approaches is qualitatively similar to before — higher signaling overheads for aggressive timeouts, and lower energy savings for conservative timeouts.

For complex applications, RadioJockey manages to achieve around 65% of the oracle's energy savings (25% compared to today) with only 1%–4% signaling overhead. The timeout based approaches, as before, do not achieve comparable energy savings and signaling overheads.

Next we focus on the complexity of the decision tree learned by RadioJockey (Figure 9c). The figure plots the depth of the tree and the storage overhead (in bytes). The tree depth tracks the number of boolean conditions that need to be checked at run-time. The storage tracks the memory required to store these booleans. As is evident from the graph, as application complexity grows, so does the complexity of the tree. At an absolute level though, run-time processing and storage requirements are minimal (requiring only 15 comparisons and 29 bytes of storage even for applications as complex as Outlook).

5.1.2 Run-time Engine Evaluation

We now focus on experimentally evaluating the RadioJockey run-time engine. The run-time operation needs to be highly performant in order to issue fast dormancy commands to the cellular interface in a timely manner. We ran 4 separate applications of different complexity. Each application ran in isolation for around 2 hours in the background on a Windows 7 based tablet device. The decision tree used for classification is built offline for each application. We measured how often the RadioJockey run-time mis-predicted EOS (false positives), and what fraction of EOS events are predicted accurately (compliment of false negatives). Energy spent by the network interface and signaling overhead is estimated offline using network capture trace and timestamps of EOS predictions when fast dormancy commands are invoked. Each mis-prediction by the run-time engine leads to an unnecessary state transition thereby spending one additional ramp-up and tail energy cost as well as increased signaling messages. Lower accuracy implies reduced energy savings since RadioJockey run-time falls back to using timeouts to invoke fast dormancy for EOS events that are not predicted.

Figure 10 shows percentage values for false negatives, false positives, energy savings, and signaling overhead compared to a default 3 second timer based approach. We find that false negatives to be under 3% for the four applications we tested, and as a result we obtain energy savings of 20%–30%. However, we find that signaling overhead is marginally higher (3%–6%) compared to offline analysis owing to increased mis-predictions in run-time. This is due to

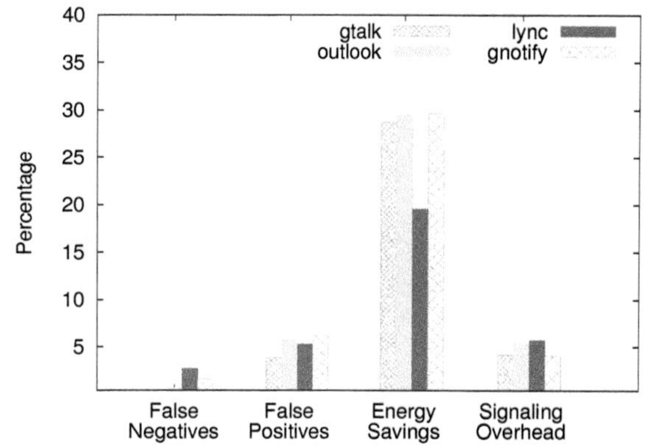

Figure 10: Runtime performance with respect to 3s timeout scheme for four applications.

inaccuracies in our the packet-timings module, which in our current prototype, as mentioned, is implemented in user-space and needs to make heavy use of netstat to link packets to processes. As a result, it experiences scheduler-related delays and jitters. An in-kernel implementation that has direct access to kernel data structures and is not affected by the process scheduler would significantly improve the accuracy of packet timestamps. That being said, our current user-space approach still delivers most of the energy savings with low signaling overheads in practice.

5.1.3 Phone experiments

We next measure the energy saved by RadioJockey on a Samsung Focus Windows phone connected to a 3G network. In order to measure only radio related energy costs, we turn off the screen and measure the base power consumed by the phone using a Monsoon power monitor. Next with the screen still off, we run a representative application (gnotify) on a device tethered to the phone (so all network traffic uses the phone's radio) while measuring the power consumed. We then run the application with the RadioJockey run-time engine enabled, which sends fast-dormancy commands to the fast-dormancy module running on the phone. Finally, for comparison, we measure the power for a simple timeout based fast-dormancy approach where our fast-dormancy module on the phone fires after the configured period of inactivity. We integrate the power over time to compute the energy consumption.

For the 3 second timeout-based approach, the phone consumed 94.86 J of energy, whereas using the RadioJockey engine, the phone consumed only 72.2 J. RadioJockey therefore saves 24% energy over the aggressive 3 second timeout-based approach, which is in line with our simulation results. This demonstrates the expected energy savings promised by RadioJockey can actually be attained in real-life scenarios.

5.1.4 Robustness

Network. To demonstrate the robustness of the decision tree learned by RadioJockey, we experiment by running the learning-engine on a static device, and use the tree to classify the application on mobile device. Figure 11 plots the numbers for the Lync application. Even though the network characteristics (latency, packet loss, bandwidth) for the two scenarios are very different, RadioJockey manages to extract 28% energy savings. This was identical to training and testing on the same network condition. The ability to use traces collected across different network types demonstrates the robustness of RadioJockey to network characteristics, allowing

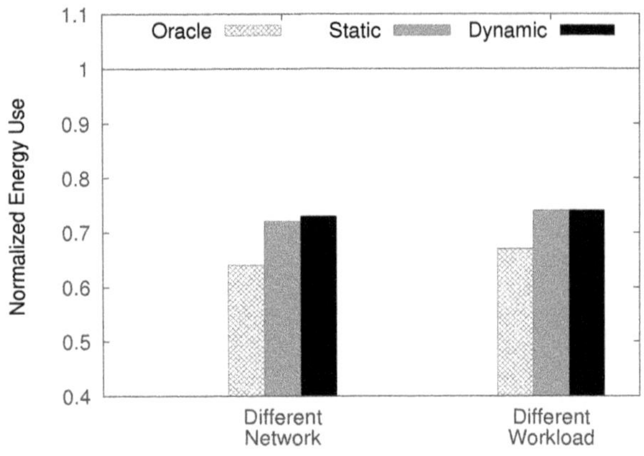

Figure 11: Robustness of the decision tree learned by RadioJockey to changing network and workload conditions.

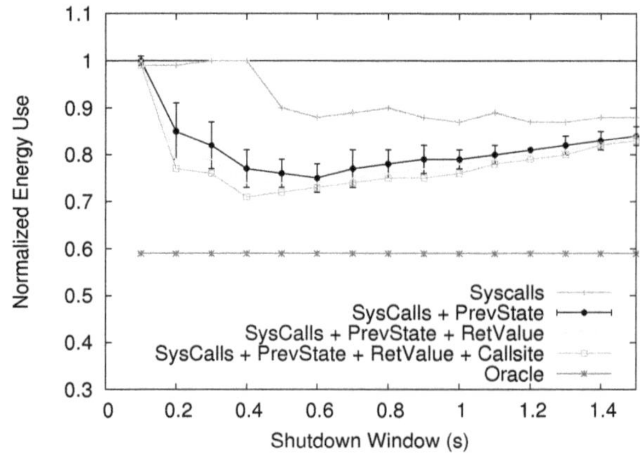

Figure 12: Selecting features and shutdown window size.

Applications	Energy Savings (%)	Signaling Overhead (%)
Outlook	24.03	4.47
GTalk	24.07	4.57
Lync	24.14	0
All	22.8	6.96

Table 3: RadioJockey with multiple applications.

for scenarios where the phone or tablet platform provider may perform the RadioJockey offline learning and provide prediction rules to the users, so that users get some baseline energy savings regardless of their operating scenario.

Workload. To further explore the robustness of the decision tree, we experiment by learning on one workload, and testing on a different workload. We use the Lync application where user one has a small number (9) of friends, while user two has many more (25). We find that RadioJockey saves 26% energy in both cases, further demonstrating the robust nature of the decision trees with respect to workloads. This suggests that RadioJockey traces can additionally be shared across users further reducing the learning burden for individual users.

5.1.5 Multiple applications

So far we have measured each application in isolation. In practice, multiple background applications may be running at any given time, and the radio can go into fast-dormancy only if none of the applications is actively communicating. This naturally reduces the energy savings. We simulate different interleaving of three applications – Outlook, Lync and GTalk. For this we use the decision trees learned earlier, and enter fast dormancy when neither application is classified to be in an ACTIVE session.

Where RadioJockey saves 24% energy for applications in isolation compared to the 3 second timeout approach, for the interleaved trace, RadioJockey saves on average 23%. The signaling overhead only increases marginally to 7%, whereas individually the signaling overhead is below 5% for each application in isolation. The small increase in signaling overhead is due to the interference between different applications. Although we are predicting EOS events accurately, RadioJockey does not predict the start of a session (SOS). As a result, when EOS is predicted for one application leading to fast dormancy, an SOS event for another application can potentially result in waking up the radio immediately resulting in a false positive. However, the data suggests that active sessions of multiple background applications do not overlap often. Thus RadioJockey still manages significant energy savings even while running multiple background applications together with marginal increase in signaling overhead.

5.2 Parameter Selection

We next focus on our choice of features to base RadioJockey's learning algorithm on. Figure 12 performs a parameter sweep for t_w (shutdown window) on the x-axis for Outlook, and plots the energy savings for the choice of application features. Signaling overheads (not shown) are low throughout.

Note first that using only system call information without PREV_STATE performs quite poorly, saving at best only 13% energy (as compared to the oracle saving 41%). The reason, as we have noted, is that the tree did not distinguish between long- and short- ACTIVE session. To better handle this temporal state, we add the PREV_STATE feature.

Adding previous state significantly improves energy savings (increasing it to 25% in the best case). Beyond this, small gains are to be had by including return values, and callsite and stack trace information (adding a further 1% and 4% respectively), however given the significant overheads in acquiring a stack trace at every intercepted function call, we decide to forgo this small boost.

Interestingly we found that for all the applications we tested (and all sets of features we tested), the energy savings curve follows the same pattern — energy savings linearly increase monotonically as the shutdown window size decreases, but only up to a point. Beyond that the energy savings decrease monotonically as the window size decreases further still. This is explained by the fact that when the shutdown window size is large, all EOS are predicted correctly and false-positives are rare, and energy savings is a linear function of how long RadioJockey must wait to make a prediction (i.e., shutdown window size t_w). Beyond the critical shutdown window size, false negatives increase as there is too little data to predict every EOS, which results in lower energy savings.

Based on empirical measurements across different networks, and different workloads, this critical point appears to be stable for a given application (but different for different applications). This suggests that this is an application-dependent quantity that can be learned by the platform during application evaluation time, and used by all users.

6. DISCUSSION

6.1 Network Features

RadioJockey employed features from application execution traces to infer rules that predict an EOS events effectively. We also evaluated whether network features alone can be used for predicting the same. We used features such as TCP flags, HTTP content length, packet payload length, and hash of first few bytes of the packet. The intuition for using hash value is that many applications use custom protocols for communication and most often the first few bytes of the packet carry header related information.

For applications which tear-down network connections at the end of a session, we are able to find good rules that predict EOS based on TCP flags alone. These included simple background applications such as Desktop Changer, FeedReeder, and GNotify. However, for many applications that keep long standing active TCP connections for communicating with the server, we could not find high accuracy rules. Another major drawback of this approach is that it cannot be employed on applications that use encryption.

6.2 Temporal Rule Mining

Prior to using decision tree based classification technique in RadioJockey, we employed temporal rule mining to learn application specific rules from system call traces and network traces. The approach inferred rules of the form $A \rightarrow B \rightarrow ... \rightarrow C$ where A, B and C are predicates that frequently occur in a sequence in the traces, however not necessarily next to each other. A predicate can either be a function name with optional return value parameter or an EOS event (obtained from network trace). We inferred rules which have EOS as one of the predicates. To limit the search space, we only find rules where the first and last predicate occur within a certain time duration. We selected a subset of rules that are able to capture most of the EOS packets in the training set with minimal false positives. In addition, we preferred rules which end with an EOS predicate. At runtime, whenever a series of predicates in a rule are observed, fast dormancy command is issued immediately after the last predicate. As a result temporal rule based approach can potentially save more energy compared to classification based approach that uses system calls captured after the occurrence of a packet, thus taking marginally longer time to invoke fast dormancy.

For most simple applications a small set of temporal rules with at most 4 predicates is sufficient to predict EOS events with high accuracy. However, inferring temporal rules for complex applications like Lync and Outlook, which have a lot of system calls, turned out to be time consuming as the search space blows up exponentially with number of predicates. This was one of the reasons for using classification technique in RadioJockey.

6.3 Foreground Applications

RadioJockey is currently designed to invoke fast dormancy only for background applications during periods when there are no active user interaction. Predicting EOS events for foreground applications turns out to be challenging since user interactions can trigger network communications at any point in time. Accordingly, RadioJockey run-time employs a simple heuristic — enable run-time only when the screen is idle. This ensures that fast dormancy calls for background applications do not interfere with foreground applications. Although this approach may seem very conservative, mobile phone usage is typically dominated by long periods of user inactivity during which background syncing activities are carried out. We plan to focus on foreground applications as a part of future work.

7. RELATED WORK

Understanding cellular radio energy characteristics has received increasing attention in recent years. In particular, the energy cost of the "tail" [2, 18] of the 2G/3G cellular radio, where the radio remains in a high power state for a large inactive duration (determined by the operator) after the end of each data communication spurt, has received significant attention. We describe below related work that has focused on this aspect of the cellular radio.

Measurements and Modeling. Several papers [12, 11, 14] have used active measurements to characterize the cellular radio state transition parameters of different operators. These measurements indicate that default parameter values for the cellular radio energy tail can vary across operators and be as high as 20 seconds, resulting in significant energy drain on smartphones. Falaki et al. [3] analyze smartphone traffic from a large number of users and show that 95% of inter-packet arrival times lie within 4.5 seconds, thereby arguing for a shorter idle timer value for 3G radios.

Analytic models have been proposed for computing the energy consumption of 3G radios [10, 19]. Authors in [19] compute the energy cost of using different inactive timers for 3G radios by modeling traffic and 3G radio characteristics. Pathak et al. [10] use system call tracing to perform fine-grained modeling of energy consumption, including modeling the various states of the 3G radio, for accurate energy accounting of applications on mobile smart phones.

Optimizing for a given tail. Several papers [2, 6, 8, 17, 18], have leveraged pre-fetching and/or delayed scheduling in order to amortize the energy cost of the cellular radio tail. Tailender [2] uses a combination of delayed transfer and prefetching of search query results in order to reduce energy. Cool-Tether [18] takes the energy cost of the tail into account for deciding the number of phones to be combined for tethering. Bartendr [17] prefetches data at good signal strengths for streaming and background transfer applications in order to save energy.

Optimizing the tail. Modern smart phones use idle timer values that range from 3 to 10s for invoking fast dormancy [4], a considerably smaller duration than the typical 12-20s of default tail durations configured by operators [14]. However, these idle timer values are currently chosen in an ad hoc manner, balancing operator sensitivities to signaling costs of low idle timer values with the needs of reducing smartphone energy consumption. Several papers have also proposed to dynamically choose idle timer values based on traffic characteristics instead of static idle timer values used currently [7, 13, 19]. However, as we show in Section 2, it is not always possible to choose an aggressive idle timer value that accurately predicts the end of a communication spurt, given network characteristics and user mobility can cause significant variation in inter-packet arrival times. Thus, these approaches can result in large increases in signaling overhead.

ARO [16] is a tool that collects various profiling data when an application is running (e.g., packet traces, user interacation, etc.) and then analyzes the data offline in order to shed light on the impact of various smartphone application features and their respective radio energy costs. Thus, ARO helps provide developer with sufficient cross-layer information (TCP, radio states, etc.) and the energy cost of different functions of the application, thereby allowing the developer to redesign his code in a more energy efficient manner.

Perhaps closest to RadioJockey is TOP [15] where applications are modified to leverage fast dormancy in order to save energy. Applications in TOP actively inform the network when their communication spurt is finished. Thus, the cellular radio is able to quickly go into a low energy state. However, TOP requires application developers to be aware of the cellular energy tail characteristics and incorporate this into their code design. RadioJockey has the same goal of invoking fast dormancy after the end of a communication spurt but without requiring any support from the application developer. By mining program execution traces, RadioJockey is able to infer rules for end of communication spurts and can thus automatically invoke fast dormancy at the correct times without any change to application code. The complexity of some of the rules that we learn for identifying the end of a communication spurt argues for an approach such as RadioJockey since an application developer may not always be able to identify these rules, especially for large and complex applications.

8. CONCLUSION

Many applications such as email clients, instant messenger, news readers, etc. run in the background on a mobile device and perform network activity for synchronizing state information. However, the energy cost of running these applications in the background on a cellular radio interface can be significant enough that many smartphone platforms prohibit background application or severely restrict them. Recent smartphone models cut down the energy cost by implementing a feature called Fast Dormancy (FD), which forces the radio to go to a low energy state based on a short inactivity timer. However, such idle timer-based approach face two drawbacks: some applications have a large variance in their packet inter-arrival distributioin and second variation in network conditions due to mobility can also change the packet inter-arrival distribution.

In contrast, our system called RadioJockey analyzes program execution traces and mines rules for identifying end of communication spurts. We show that our approach is able to save 20-40% energy savings compared to an idle timer approach for a large class of applications, is much more robust to variations in network conditions, and achieves savings with negligible increase in signaling load for the network operator.

9. REFERENCES

[1] How smartphones are bogging down some wireless carriers. http://bit.ly/Nc0XgL.

[2] BALASUBRAMANIAN, N., BALASUBRAMANIAN, A., AND VENKATARAMANI, A. Energy consumption in mobile phones: a measurement study and implications for network applications. In *Proceedings of the Internet Measurement Conference (IMC)* (Chicago, IL, Nov. 2009), ACM.

[3] FALAKI, H., LYMBEROPOULOS, D., MAHAJAN, R., KANDULA, S., AND ESTRIN, D. A first look at traffic on smartphones. In *Proceedings of the Internet Measurement Conference (IMC)* (Melbourne, Australia, Nov. 2010), ACM.

[4] HUAWEI. Behavior analysis of smartphones. http://bit.ly/Nc0US7.

[5] HUNT, G., AND BRUBACHER, D. Detours: Binary Interception of Win32 Functions. In *Proceedings of the 3rd USENIX Windows NT Symposium* (Seattle, WA, July 1999).

[6] LAGAR-CAVILLA, H. A., JOSHI, K., VARSHAVSKY, A., BICKFORD, J., AND PARRA, D. Traffic backfilling: subsidizing lunch for delay-tolerant applications in UMTS networks. In *Proceedings of the 3rd Workshop on Networking, Systems, and Applications on Mobile Handhelds (MobiHeld)* (Cascais, Portugal, Oct. 2011).

[7] LIERS, F., AND MITSCHELE-THIEL, A. UMTS data capacity improvements employing dynamic RRC timeouts. In *Proceedings of the 16th IEEE International Symposium on Personal, Indoor and Mobile Radio Communications (PIMRC)* (Berlin, Germany, Sept. 2005).

[8] LIU, H., ZHANG, Y., AND ZHOU, Y. TailTheft: Leveraging the wasted time for saving energy in cellular communications. In *Proceedings of the 6th ACM International Workshop on Mobility in the Evolving Internet Architecture (MobiArch)* (Washington, D.C., June 2011).

[9] NOKIA SIEMENS NETWORKS. Understanding Smartphone Behavior in the Network, Nov. 2010. http://bit.ly/OUEkl1.

[10] PATHAK, A., HU, Y., ZHANG, M., BAHL, P., AND WANG, Y. Fine-grained power modeling for smartphones using system call tracing. In *Proceedings of the 6th European Conference on Computer Systems (EuroSys)* (Salzburg, Austria, Apr. 2011), ACM.

[11] PERALA, P., BARBUZZI, A., BOGGIA, G., AND PENTIKOUSIS, K. Theory and practice of RRC state transitions in UMTS networks. In *Proceedings of the 5th IEEE Broadband Wireless Access Workshop (BWA)* (Anaheim, CA, 2009).

[12] PERALA, P., RICCCIATO, F., AND BOGGIA, G. Discovering parameter settings in 3G networks via active measurements. *IEEE communication letters 12*, 10 (2008).

[13] PUUSTINEN, I., AND NURMINEN, J. The effect of unwanted Internet traffic on cellular phone energy consumption. In *Proceedings of the 4th IFIP International Conference on New Technologies, Mobility and Security (NTMS)* (Paris, France, Feb. 2011).

[14] QIAN, F., WANG, Z., GERBER, A., MAO, Z., SEN, S., AND SPATSCHECK, O. Characterizing Radio Resourcde Allocation for 3G Networks. In *Proceedings of the Internet Measurement Conference (IMC)* (Melbourne, Australia, Nov. 2010).

[15] QIAN, F., WANG, Z., GERBER, A., MAO, Z., SEN, S., AND SPATSCHECK, O. TOP: Tail Optimization Protocol for Cellular Radio Resource Alocation. In *Proceedings of the 18th International Conference on Network Protocols (ICNP)* (Kyoto, Japan, Oct. 2010).

[16] QIAN, F., WANG, Z., GERBER, A., MAO, Z., SEN, S., AND SPATSCHECK, O. Profiling resource usage for mobile applications: a cross-layer approach. In *Proceedings of the 9th International Conference on Mobile Systems, Applications, and Services (MobiSys)* (Washington, D.C., June 2011), ACM.

[17] SCHULMAN, A., NAVDA, V., RAMJEE, R., SPRING, N., DESHPANDE, P., GRUNEWALD, C., JAIN, K., AND PADMANABHAN, V. N. Bartendr: A Practical Approach to Energy-aware Cellular Data Scheduling. In *Proceedings of the 16th International Conference on Mobile Computing and Networking (MobiCom)* (Chicago, IL, Sept. 2010).

[18] SHARMA, A., NAVDA, V., RAMJEE, R., PADMANABHAN, V., AND BELDING, E. Cool-Tether: Energy Efficient On-the-fly WiFi Hot-spots using Mobile Phones. In *Proceedings of the 5th International Conference on emerging Networking EXperiments and Technologies (CoNEXT)* (Rome, Italy, Dec. 2009).

[19] YEH, J.-H., CHEN, J.-C., AND LEE., C.-C. Comparative Analysis of Energy-Saving Techniques in 3GPP and 3GPP2 Systems. *IEEE Transactions on Vehicular Technology (TVT) 58*, 1 (2009).

Can We Pay for What We Get in 3G Data Access?

Chunyi Peng* Guan-Hua Tu* Chi-Yu Li Songwu Lu

University of California, Los Angeles, CA 90095, USA

{chunyip,ghtu,lichiyu, slu}@cs.ucla.edu

ABSTRACT

Data-plan subscribers are charged based on the used traffic volume in 3G/4G cellular networks. This usage-based charging system has been operational and received general success. In this work, we conduct experiments to critically assess both this usage-based accounting architecture and application-specific charging policies by operators. Our evaluation compares the network-recorded volume with the delivered traffic at the end device. We have found that, both generally work in common scenarios but may go wrong in the extreme cases: We are charged for what we never get, and we can get what we want for free. In one extreme case, we are charged for at least three hours and 450MB or more data despite receiving no single bit. In another extreme case, we are able to transfer 200MB or any amount we specify for free. The root causes lie in lack of both coordination between the charging system and the end device, and prudent policy enforcement by certain operators. We propose immediate fixes and discuss possible future directions.

Categories and Subject Descriptors

C.2.1 [**Computer-Communication Networks**]: Network Architecture and Design—*Wireless Network*; C.4 [**Performance of Systems**]: *Design Studies*

General Terms

Measurement, Experimentation, Performance

Keywords

Cellular Networks, Mobile Data Services, Charging, Accounting

1. INTRODUCTION

Wireless access to data services is gaining increasing popularity in recent years, thanks to the rapid deployment of 3G/4G cellular networks. Statistics from OECD [25] shows that, 62% of broadband users in the US have subscribed to wireless data plans, with 137M subscribers by June 2010. On the front of mobile data applications, there are 1.2B mobile web users [23], and Facebook claimed more than 425M mobile monthly active users in December 2011 in its IPO statement. The explosive growth of smartphones (e.g., iPhone and Android phone) and tablets (e.g., iPad) will further accelerate this usage trend in the coming years.

While users enjoy wireless data access, it does not come for free. Most operators will charge the user a monthly bill based on the used data volume. The price for this usage-based charging ranges from 1s to 200s of cents for 1MB data in the US depending on the chosen data plans. Different from the flat charging scheme over the Internet, 3G/4G operators do not offer unlimited data usage for smartphone users. Both AT&T and Verizon effectively ended such data plans for new customers in 2011, and T-mobile limits the high-speed data volume in its so-called unlimited data plan.

The 3G charging system has been operational for a few years, and the practice has been generally successful. On one hand, the usage-based charging is not implemented without rationale. The radio spectrum is scarce and mostly licensed, and the offered access speed is bounded by the fundamental limits on channel capacity. On the other hand, the system mostly works as a black box for users, and users do have questions and concerns. Consider Alice, a typical 3G user, as an example. Alice just received a monthly bill of $25.8 for 387.4MB data usage, with a portion of the bill being shown in Figure 1. Even with this itemized data usage, Alice may still have lots of doubts and questions in her mind: (1) Does my iPhone really use 4385KB (but not 2.3MB less) since I remembered I only downloaded a 2MB app from the App Store? (2) How can I find out if the operator made a mistake and overcharged me? Anyway, I heard that up to 20M Americans using their iPhones/iPads are over-charged by 7-14% on average and up to 300% in some cases [26]. (3) For the 31KB item, I remembered I clicked an invalid web page link that did not show me any real content. Why should I be charged? (4) Is there any chance I can somewhat evade the charging system and pay less? The list goes on and long. On the technology side, answers to all these questions reside on the accounting[1] system used by 3G/4G networks.

Date	Time	To/From	Type	Direction	Msg/Kb/Min
03/04	09:21 PM	[illegible]	Internet/MEdia Net	Sent	31KB
03/04	08:07 PM	[illegible]	Internet/MEdia Net	Sent	4385KB
03/04	06:16 PM	[illegible]	Internet/MEdia Net	Sent	65KB
03/04	02:45 PM	[illegible]	MEdia Messaging	Sent	12KB

Figure 1: Example of itemized data usage.

In this paper, we present the arguably *first* work that assesses the 3G accounting system. Our evaluation criterion is *user centric*:

*The first two authors contribute equally to this work.

[1]We do not differentiate accounting from charging in this work by a slight abuse of wording definition.

We pay for what we get. We examine the usage gap between the operator-recorded data volume and the user-logged data amount. We conduct experiments with smartphones on two major US operational 3G networks, while also running similar tests in the third US carrier and two carriers in China and Taiwan. Using the phone-logged traces and the data volume recorded by the 3G accounting system, we analyze accounting behaviors in various scenarios. We focus on two aspects: (1) *How* is data access charged? It concerns the accounting architecture and its implementation within the 3GPP standards; and (2) *What* is charged? It attests the charging policy practice by operators.

Our study yields two main findings. First, we observe that *we be charged for what we never receive in certain scenarios.* The difference is generally small in the normal cases, resulting in about 10s to 100s of KB in typical applications. However, it can go up to 10s of MB for certain applications (e.g., video streaming). Moreover, the gap can grow quite large in extreme conditions. In one extreme case when a UDP session has no control loop, a mobile user receives data from this UDP session and roams into a no-signal zone. His ongoing UDP session continues to be charged by the 3G accounting system, despite no single bit ever received by the user. Our experiments show that this charging proceeds for more than three hours and results in 450MB or more in the charged volume! We also identify its root cause. It turns out that, current cellular accounting standards do not explicitly take feedback from end devices. Charging action is taken at the core components (e.g., GGSN/SGSN in 3G UMTS) inside the cellular infrastructure. The core components simply record the data volume traversing them to/from the given user for charging purpose. Consequently, whenever packet drops occur after traversing these components in the no-signal scenario, the charging system does not know the device status and overcharging may arise. The solution fix is to take feedback directly from the end device or access the device status information already collected within the infrastructure (e.g., at RNC) when making charging decisions. Second, we discover that *we can obtain what we want in data access free of charge.* The key is to exploit the free-of-charge service (e.g., DNS) by both operators and construct a "DNS tunnel" for other data transfer. Using a simple prototype, we are able to transmit 200MB or any amount we specify for free. The root cause is that, operators use application-specific charging policy, and loopholes exist in such policy practice. Operators do not enforce the full flow-based charging scheme by the 3GPP standard, but using only one or two fields in the five-tuple flow ID. Our work also shows that the policy enforcement is indeed operator specific. While all three US operators offer free DNS, the Chinese operator and the Taiwan carrier still charge it as usual. Though our experiments are mainly conducted in 3G networks, the observations are still applicable to 4G LTE networks since they follow almost identical accounting architecture.

The rest of the paper is organized as follows. Section 2 introduces the accounting and data charging process of 3G/4G networks. Section 3 describes the problem statement and study methodology. Sections 4 and 5 report cases where users are charged for data they never receive, and how to build toll-free data services, respectively. Section 6 describes other gray-area cases in charging. Section 7 further discusses architectural and policy issues. Section 8 compares with the related work, and Section 9 concludes the paper.

2. BACKGROUND ON DATA CHARGING

We first introduce 3G charging architecture for data services in context of UMTS, the most widely deployed 3G cellular network technology [17]. We then describe how charging actions proceed in the UMTS network core. Note that, the mechanisms and issues are

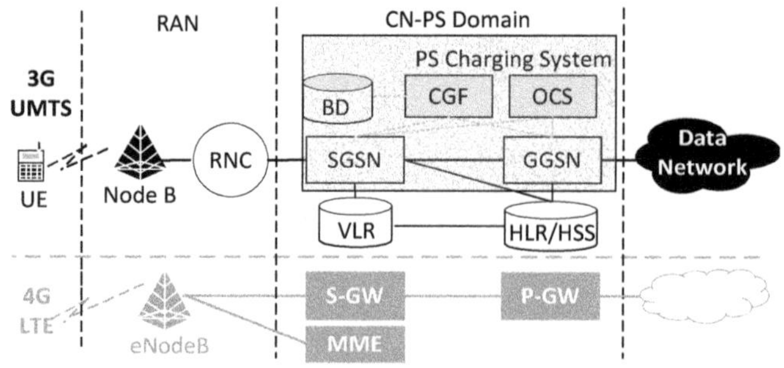

Figure 2: 3G/4G network architecture and charging components in PS domain.

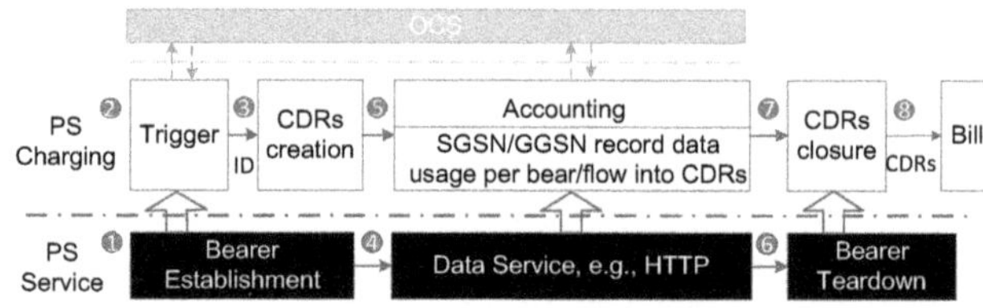

Figure 3: Charging process for a data service flow.

also applicable in 4G Long Term Evolution (LTE) and 3G High-Speed Packet Access (HSPA) networks.

2.1 Data Charging Architecture

Figure 2 shows the overall 3G UMTS network architecture and charging system for data services. The UMTS network consists of the Terrestrial Radio Access Network (RAN) and the core network (CN). Its RAN includes the User Equipment (UE), the Node B, and the Radio Network Controller (RNC). RAN provides wireless access to UEs, and exchanges data session provisioning with the Packet-Switched (PS) core networks.

The major components of the PS core network are the *Serving GPRS Support Node* (SGSN) and *Gateway GPRS Support Node* (GGSN). SGSN is responsible for the delivery of data packets from and to the UEs within its geographical service area. GGSN serves as the hub between the SGSN and the external data networks, e.g., the wired Internet. It 'hides' the 3G UMTS infrastructure from the external network and acts as a router to a subnetwork. GGSN also performs charging, user authentication, and other functions.

In addition to SGSN and GGSN, three more charging components work in the PS charging system: the *Billing Domain* (BD), the *Charging Gateway Function* (CGF), and the *Online Charging System* (OCS). Current cellular networks support both offline and online charging modes [13]. In offline charging, data usage is collected during service provisioning in the form of *Charging Data Records* (CDRs), which are sent to the BD to generate data bills offline. SGSN and GGSN are responsible for collecting data usage and generating CDRs. CGF is used to validate CDRs from SGSNs/GGSNs and transfer CDRs to the BD. In online charging, mobile users have to pre-pay to obtain credits for data services in advance. The OCS authorizes whether or not users have enough credits. GGSN/SGSN deducts data usage from the available credits and stops data services upon zero credit.

2.2 Data Charging Process

We next describe how mobile users are charged for data services. Consider offline charging, and Alice is about to upload one photo to her Facebook, thus starting a PS service (say, HTTP). Figure 3 illustrates the charging procedures during the data service process.

Initially, Alice has no available bearer service connection (which may carry one or multiple PS services). She thus establishes a bearer via Packet Data Protocol (PDP) Context[2] Activation [11] (Step 1). Upon this activation, the UE device is allowed to connect with the external data network through the SGSN and GGSN. This activation also triggers the charging procedure, and GGSN assigns a unique charging ID to the activated PDP context (Step 2). SGSN and GGSN then start to create CDRs using the charging ID (Step 3), and are ready to record the upcoming data volume. In addition to charging per PDP context, 3G also supports charging per data flow, called as *Flow Based Charging* (FBC). FBC separates charging for different services (e.g., web or VoIP) within the same PDP context [14]. One data flow is typically identified by the five-tuple: (1) source IP address or mask, (2) source port number, (3) destination address or mask, (4) destination port number, and (5) protocol ID of the protocol above IP, e.g., TCP or UDP [12]. For example, a HTTP data flow can be represented by *(*, *, *, 80, TCP)*[3].

Now Alice can upload her photo to Facebook. Both SGSN and GGSN route the UE's packets to/from the external data network during the data service session (Step 4). In the meantime, SGSN and GGSN record the traffic volume that arrives at them into corresponding CDRs (Step 5). Both SGSN and GGSN count the payload of GTP-U (GPRS Tunneling Protocol- User Plane) packets as data volume; GTP-U delivers data within cellular networks and runs below the IP protocol. Therefore, the data volume counts all packet headers above IP, including IP, TCP, and HTTP headers, but the MAC header is not counted.

The accounting procedure (Step 5) lasts until this data service completes. It occurs when the UE tears down this bearer (Step 6) in bearer-based charging, or when Alice closes her HTTP session in flow-based charging. CDRs are subsequently closed and transferred to the BD (Step 7). Finally, BD generates a billing item for the proper user based on the charging ID.

The online charging process is similar, though OCS participates in the triggering and accounting steps (Steps 2 and 5) by authenticating the GGSN/SGSN to use user credits. There is also no need to send CDRs to generate a bill since the consumed credits have been deducted (see Figure 3). In the paper, we focus on the offline charging, and the same issues also arise for the online charging.

2.3 On LTE

LTE is a 4G cellular network standard, offering even higher speed. Its architecture is similar to 3G UMTS. The major difference (also shown in Figure 2) is that the functionalities of RAN are performed by eNodeBs, whereas the functionalities of SGSN and GGSN are performed by *Mobility Management Entity* (MME), *Serving Gateway* (S-GW), and *Packet Data Network Gateway* (P-GW) [21]. Moreover, bearer establishment is supported by two procedures, i.e., *Evolved Packet System Bearer Activation* and *Public Data Network Connectivity Procedure* [16]. The charging system for LTE is almost identical to 3G UMTS, with S-GW and P-GW (replacing SGSN and GGSN) in charge of collecting data usage and generating CDRs.

3. PROBLEM AND METHODOLOGY

In this section, we identify the issues to address, and describe the experimental methodology.

[2]PDP contexts provide all the required information for IP packet data connections in cellular networks.

[3]Each of the five tuples can be a wildcard.

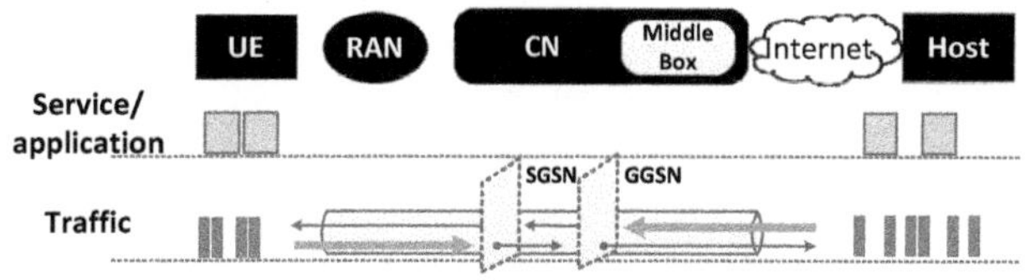

Figure 4: Current mobile data charging

3.1 Issues in 3G Accounting

In this paper, we examine the 3G accounting architecture, as well as the operators' policy practice. The research focuses on accounting, which records the usage volume over time for each user, rather than pricing that sets the unit price for usage and is driven by marketing and cost factors. We study the implication of such architectures and policies on user-perceived charges.

Our evaluation is to compare the user-recorded data volume with the network-recorded usage. Our study is user centric overall, in that *users pay for what they actually get at the end systems*. Anyway, the end systems are where users obtain their data service. This user-centric guideline may not necessarily concur with the infrastructure-oriented view taken by operators. When conflicts happen, one camp may suffer. In this work, our main goal is not to take specific position on what side to stand with, but to illustrate and quantify the discrepancy in different scenarios. To this end, we have built a simple tool, *BillAudit*, to conduct experiments between mobile phones and the Internet via the 3G network sitting in between. We collect the actual usage at the two end systems, and compare with the volume given by the 3G accounting system. The goal is to identify possible limitations and existing loopholes though such cases may occur rarely in reality, and demonstrate their effect on end users. We explore two dimensions of the problem:

- *How* to charge: How does 3G accounting handle various cases of end-to-end data delivery?
- *What* to charge: What is the difference in charging for different types of data traffic?

The first issue concerns the accounting architecture. It is about *how* to charge the mobile user in data services. As described in Section 2, the current 3G architecture takes the SGSN/GGSN based charging approach. It records how much data volume has traversed the intermediate SGSN/GGSN inside the 3G infrastructure (see Figures 2 and 3 for an illustration). This element-based charging takes the local view inside the 3G infrastructure, without coordinating with end devices when making accounting decisions. On the other hand, the data delivery path is always end to end. The end systems (e.g., the UE device or the server) may record usage volume different from what is logged inside the network. Note that the 3G accounting system does not explicitly collaborate with the end systems in its charging decisions. The effect is hence visible when failure or misbehavior occurs over the full delivery path.

In general, the end-to-end data delivery path consists of all six components (shown in Figure 4): the UE, the RAN, the 3G core network (CN), the middlebox, the wired Internet, and the host or server. Note that, the 3G network may deploy middle boxes (e.g., proxy servers, NAT boxes) over the delivery path, as shown in [30], and SGSN/GGSN resides at the CN. Assume that SGSN/GGSN and CN are always functioning. Any other component may fail. Specifically, we consider four cases in Sections 4 and 6: (1) The path segment between UE and RAN (i.e., the wireless delivery between Node B and UE) experiences problems in delivery; (2) The path segment between the Internet and CN has packet drops; (3) The path segment between the middlebox and the host breaks; and (4) The host or server is not accessible.

The second issue concerns on *what* to be charged. It depends on the charging policy. Each 3G operator can define its own application-specific policy on charging. Along this line, we are particularly interested in studying two cases:

1. Given certain type of free data services (e.g., DNS service discovered by our study) offered by operators, is it possible to exploit it to evade charges for other data services?
2. What is the current charging policy for application-level signaling or commands, which do not contribute to real content? Cases include FTP signaling over port 20, invalid HTTP links, HTTP redirects, and Email/IM signaling, etc.

Note that in the second instance, these signaling messages are not the actual content. Operators have every reason to charge it or not to charge it; it is not a right/wrong issue to address. In recent years, 3G/4G operators have been making effort to evolve from "dumb-bit-pipe owners" to "content/service providers". In the role-switching process, the charging policy may also evolve towards more content based; this is an interesting topic for future study.

3.2 Methodology

We conduct a series of experiments to examine the difference between the data volume recorded by operators and the one logged at the end device. In each experiment, we establish end-to-end data sessions from mobile phones to popular Internet services or our deployed server. We then record the data volume charged by operators and the ones observed at mobile phones or servers. We run main tests with two major mobile operators in the US, denoted as Operator-I and Operator-II for privacy concerns. They together offer nationwide coverage for 102.3M users, thus claiming about 50% of US market. For verification purpose, we also run similar tests with the third major US carrier that claims to support 4G LTE, a major carrier in China and Taiwan each. Our mobile devices use several Android phone models: HTC Desire, Samsung Galaxy S1/S2, and Samsung Stratosphere (that supports 4G LTE), running on Android 2.2, 2.3.4, 2.3.6 and 2.3.5, respectively. Our experiments show that all the findings are phone platform independent; this is not hard to understand. In our deployed servers, we use an Apache web server, a FTP server using Wing FTP software [10], and TCP/UDP servers written in Java.

We run experiments for both cases of extreme scenarios and normal settings. The extreme scenarios are carefully created in experiments, and seek to stretch out the charging system in worst-case settings. The normal settings capture users' common usage patterns, including popular protocols such as TCP, typical applications, and daily-life usage.

We use two methods to obtain data usage logged by operators. The first one is to dial a special number to retrieve the remaining monthly data usage in a near real-time mode. Most operators support this Dial-In feature, e.g., via dialing #DATA for Verizon, *DATA# for AT&T, and #932# for T-Mobile in the US. The data usage will then be delivered via a text message after this Dial-In. By logging data usage volume before and after our experiment, we obtain the usage volume observed by the operator during the experiment. The second method is to log onto the mobile operator website and access online data usage records (as shown in Figure 1). Operator-I only supports DIAL-IN method, while Operator-II supports both. We thus use the first method for Operator-I and the second method for Operator-II. In terms of report latency, we found that Operator-I may report data usage in five to ten minutes while Operator-II may take up to six hours to update their records. However, Operator-II provides an itemized data charging volume associated with the timestamp; a new item will be generated when a new PDP context (bearer) is established. We thus conduct experiments with proper time window (> 10 minutues) and establish a new PDP context for each experiment to avoid confusions and cope with latency. Both operators support 1 KB accuracy in data usage report. Since data usage logged by the operators only has timestamps, we need extra mechanisms to ensure that data usage belongs to the specific application or data flows in our experiments. For this purpose, we clean up the runtime environment (factory reset and disable "Background data" and "Auto-sync" functions). We also use monitoring tools (Wireshark [31]) to capture all-level packets to/from the phone. Whenever unwanted services observed, we rerun the experiment. For those tests that last for an extended period of time (e.g., three or six hours), we use Wireshark to filter out those unrelated packets in the trace analysis.

We use two tools to log data usage on mobile phones. The first is to use TrafficMonitor [7], a software tool available from Google Play. It records data volume for each application with 0.01KB accuracy. The second is to use our own tool, which is written via the TrafficStats class interfaces [8] in Android SDK to collect network traffic statistics. We record the number of packets and bytes transmitted and received on all interfaces and on a per-application basis. We use both tools to record the UE data usage and verify whether the usage is consistent or not. We further use Wireshark to log the traffic statistics at our server.

In each experiment, we record the data usage from operator V_{OP}, the one observed from mobile phone V_{UE}, and the one at server V_{SR} if used. We conduct each experiment for 5 –15 runs. The experimental results are quite stable in different runs. Due to space limit, we do not show results for individual runs, but only the average values unless explicitly stated.

4. WE PAY FOR WHAT WE DO NOT GET

The first finding is that, we might be charged for data that never reach us or the data we never deliver to the destination. The root cause is that, the data volume is recorded inside the cellular network core without taking feedback from the end device when making accounting decisions; it can be different from the volume received at end device. In the rest of this section, we first describe the results in the extreme cases, which represent some worst-case behaviors and may rarely occur in reality. We then describe the average cases, which show how applications and users behave in common usage scenarios. We elaborate on how large the difference between the user's usage and the operator's charge can reach in these scenarios, explain their root causes, and suggest quick fixes.

4.1 Extreme Cases

We first examine how bad overcharging can become in certain extreme conditions. The goal is to expose the potential downside of the largely successful 3G charging system. Note that these conditions do not represent the typical usage patterns in practice. They could occur in reality but only infrequently.

In these tests, data traffic is delivered using UDP between an Internet server and a mobile phone. Though most network applications run on TCP, recent traffic study [18] shows that, UDP is still used for data delivery in 10-20% applications, including video streaming, VoIP, and Virtual Private Network (VPN), etc. We consider downlink cases in both no-signal zone and weak-signal zone for the mobile device; these zones vary with locations. Our experiments show that, they are mainly caused by poor coverage by carriers. For example, we have experienced three to four dead zones or zones with very weak signals measured by RSSI on our office floor of the campus building. Mobile devices are unable to receive data when suddenly entering into a dead zone without signals. How-

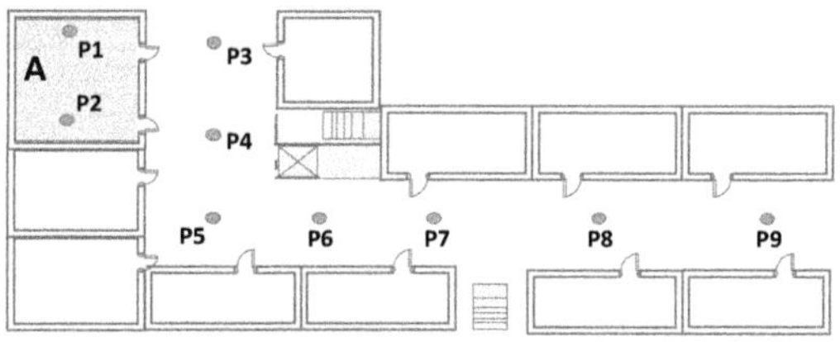

(a) Experimental floorplan

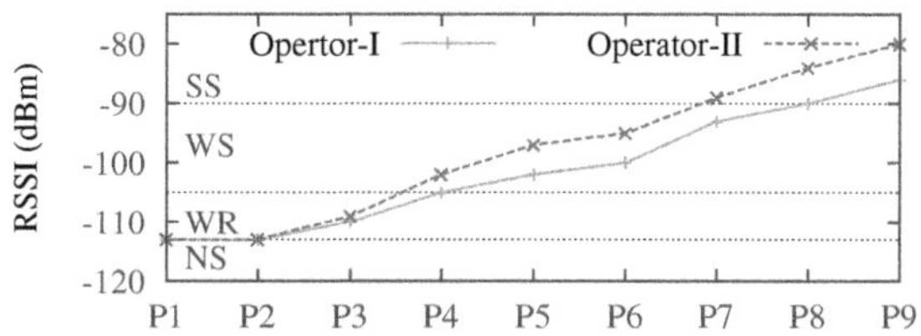

(b) RSSI measurement

Figure 5: Our indoor testbed, where room A is a dead zone (NS-zone) for both operators.

ever, they are still charged though such data never reach them. In the worst-case scenario, we have observed that charging proceed for more than three hours and result in more than 450MB data if the application has no control loop!

4.1.1 UDP in No-Signal Scenario

We conduct DL-NS experiments to put our phone into a dead zone without signals, and see what happens to the ongoing downlink UDP transmission from an Internet server to the UE. The goal is to examine whether the data usage charged by operators differs from that received by mobile phones.

Our experiments are conducted in an indoor environment shown in Figure 5(a). The coverage varies at locations for both operators. Figure 5(b) plots the medium of the measured received signal strength indicators (RSSIs). RSSI values vary from -113 dBm to -80dBm[4] at various spots and fluctuate within 3 dBm at each spot. Based on RSSI values, we divide the whole area into four zones: (1) *SS-zone* with strong signals (RSSI > -90 dBm); (2) *W-zone* with weak signals (-90 dBm ≥ RSSI >-105 dBm); (3) *WR-zone* with weaker signals (-105 dBm ≥ RSSI > -113 dBm); and (4) *NS-zone* (i.e., dead zone) with no signals (RSSI ≤ -113 dBm). Note that, different operators yield different coverage strength; Operator-II has stronger signal strength than Operator-I in this setting. However, Room A remains a NS-zone for both operators. We also conducted prior experiments (e.g., making a phone call) to ensure that the phone is indeed out of service in Room A.

DL-NS Experiment Setting: Figure 6 illustrates how to set up the DL-NS experiment step by step. First, at P9 (i.e., in the SS-zone), we send a UDP request from the mobile phone to our own server to start this experiment; Once the communication is ongoing, the server responds with an acknowledge message to the phone and sets a timer, which triggers UDP data transmission upon timeout. Upon receiving the ACK, we move the phone from the SS-zone to the NS-zone (i.e., Room A) (Step 2), hopefully before timeout. Since it takes about 30 seconds to walk into Room A, the timer

[4]-113 dBm is the lowest signal strength that a typical mobile phone can receive; it implies that the phone is out of service. In indoor environment, strong signal strength is much smaller (here, (-90, -80) dBm) than the outdoor one that usually reaches -65 dBm. dBm stands for the measured power ratio in decibels (dB) referenced to one mW and 0 dBm equals 1 mW.

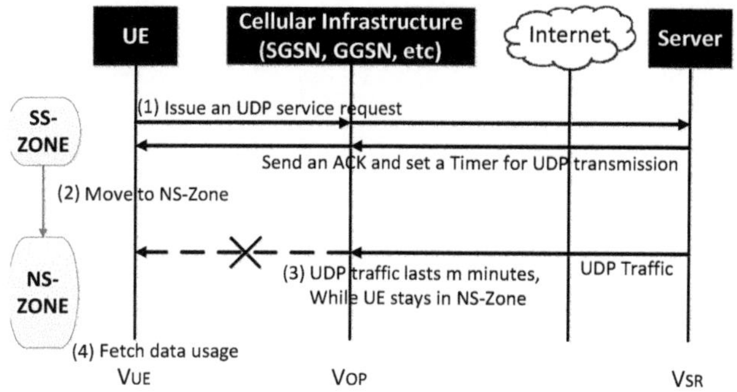

Figure 6: Procedure of DL-NS experiment.

is set as one minute to keep the server stay idle (no data delivery) during Step 2. Upon timeout, the server transmits UDP packets to the phone at a constant data speed s for another t minutes (Step 3); s and t are configurable parameters in the experiment. During Step 3, the phone remains in the NS-zone. We record data usage V_{SR}, V_{UE}, and V_{OP}, observed at the server, the UE, and the Operator, respectively.

Results: We first set the UDP source rate as $s = 50$ Kbps and the data transmission lasts for $t = 10$ minutes. Our server sends about 3.75 MB data ($50K \times 10 \times 60/8 = 3.75M$), similar to the volume charged by the operator (3.73 MB). The minor difference between these two volumes (i.e., $V_{SR} - V_{OP}$) is mainly caused by occasional packet loss. However, the mobile phone does not receive any such data, except the 80 B for one UDP request and one ACK message at the start. *This result shows that the charging infrastructure could charge mobile users of data that never reach them in case of a UDP-based application without control loop.* Moreover, we believe that, many mobile users might not be even aware of such a charge. It is quite common that mobile phone users unconsciously enter into a NS-zone in reality. They have no clue that roaming into the no-signal region may incur data volume charge by the operator, if the UDP sender is still transmitting.

4.1.2 Worst-Case Observations

We test with various source rates and different durations. The gap between the operator charge and the volume received by users (i.e., $V_{OP} - V_{UE}$) can be approximated by $s \times t$, which is exactly the volume of data sent by the server but never reached the phone:

$$\boxed{\text{DL-Volume-Gap} = V_{OP} - V_{UE} \approx s \times t.}$$

Our experiments show that the approximation still holds even when the speed s goes up at least 8 Mbps or the duration t lasts three hours! The above finding shows that, the operators charge mobile users based on the data volume sent by the server and arriving at the cellular core network, but not the volume that cellular networks have actually delivered to the users. This rule still applies no matter how large the gap could turn into. For example, the operators have charged us for 450 MB in one run when the server keeps sending downlink data at 1 Mbps for one hour ($1 \times 60 \times 60/8 = 450M$), even though no single data bit arrives at the mobile phone!

We change the source rate s from 50 Kbps to 8 Mbps to examine how the gap varies with high data speed. Figure 7 plots the DL-Volume-Gap for Operator-I using two sending servers with different link capacities. The transmission lasts one minute. The results are similar for Operator-II[5]. Note that in DL-NS experiments,

[5]The results for Operator-II will be omitted hereafter to save space if they are similar.

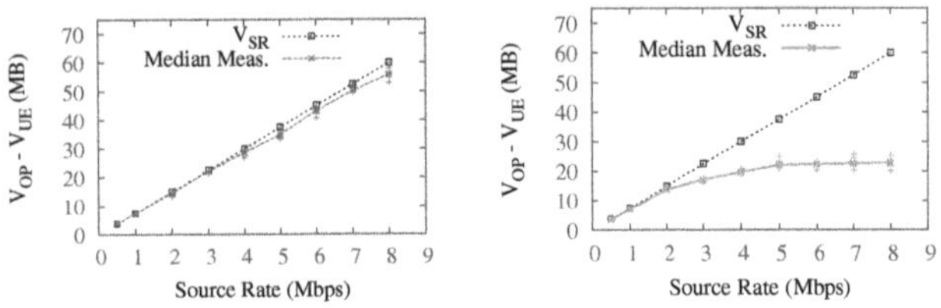

(a) Server 1 (b) Server 2

Figure 7: DL-Volume-Gap under various UDP source rates when $t = 1$ minute using two servers, where V_{SR} grows as $s \times t$, the medium values of DL-Volume-Gap, as well as each run observation (five runs), are given.

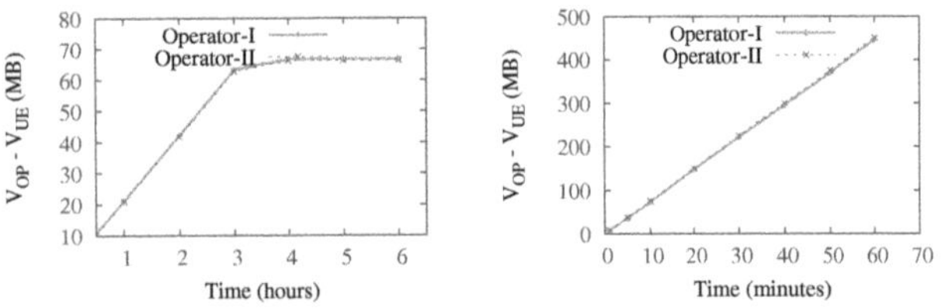

(a) $s = 50$ Kbps (b) $s = 1$ Mbps

Figure 8: DL-Volume-Gap with various in-NS-zone durations for 50 Kbps and 1 Mbps UDP flows.

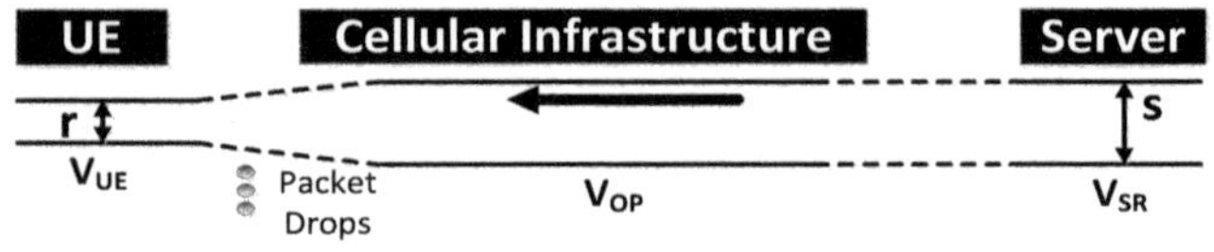

Figure 9: Illustration of DL-Volume-Gap creation in various wireless environments in DL-All experiments.

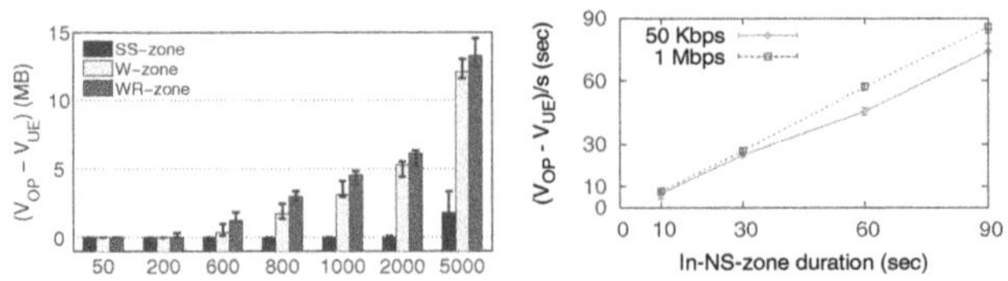

(a) Various signal strengths (b) Intermittent connectivity

Figure 10: DL-Volume-Gap in cases (a) under various signal strengths and (b) with intermittent connectivity.

the UE receives almost zero bits and the DL-Volume-Gap is approximately equal to the volume charged by operators V_{OP}. It is seen that the DL-Volume-Gap is in proportion to the UDP source rate s in Server-1 case (in Figure 7(a)). For Server-2, we find out that the gap is almost the same as V_{SR} when the data rate s is low ($\leq$ 2 Mbps); when the source rate increases ($>$ 2 Mbps), the one charged by operator is smaller. This is because Server-2 uses home Internet service and has bounded uplink speed. In contrast, using Server-1 with higher uplink bandwidth, the operator charges us for about 58.7 MB (close to V_{SR} = 60MB) in one minute at the 8 Mbps rate. *This test infers that, the operator charging practice is only based on how much data would arrive at the core network, no matter how fast it is.* Without much packet loss or congestion, DL-Volume-Gap grows in proportion to the UDP source rate.

Even worse, the operator may charge us for a long time. We put the phone in the NS-zone for different durations to see how long the gap may last. If the application layer tears down the session once the phone is out of service for a small duration of time, the gap would be small and not incur a large bill. However, if the application does not terminate, we find out that gap could last at least **three hours**! We run experiments for the slow session (50 Kbps) up to six hours and the fast session (1 Mbps) up to one hour. Figure 8 plots the DL-Volume-Gap when the in-NS-zone duration varies from one minute to six hours. Within the initial three hours, the gap for the slow session grows linearly with the duration t. Both operators stop recording when the usage reaches about 66.3 MB, which approximates about three-hour data transmissions. We do not run experiments for high-speed UDP sessions (e.g., 1 Mbps or even 8 Mbps) up to three hours, because the data usage probably goes up to 1.35 GB or 10.8 GB, which incurs a huge bill. In fact, the gap as large as 450 MB and the charging duration of about three hours are already significant enough.

Root Causes: We now explain the root cause. In the downlink case, traffic is delivered from the external server to the mobile device via cellular networks (e.g., GGSN, SGSN and RNC in turn). It is easy to see that, the observed data volume monotonically decreases along the downstream delivery path, i.e.,

$$V_{UE} \leq V_{SGSN} \leq V_{GGSN} \leq V_{SR}. \quad (1)$$

Due to unreliable packet delivery, packets might be dropped at any intermediate node, thus incurring the volume gap. In our DL-NS experiments, the last hop is broken, so no data would be delivered to the phone ($V_{UE} \approx 0$). As described in Section 2, the 3G/4G charging system obtains data usage based on the volume recorded by SGSN and GGSN. Therefore, those UDP packets, which arrive at GGSN or SGSN but never reach the UE, are still counted as the data usage by this UE. This results in a large gap between the actual data usage and the billing volume.

4.1.3 Still-Bad Case: Even With Signals

We next show that, the charging gap still exists even when the wireless link is not broken. The gap concerns the wireless environment in terms of available radio link rate. We conduct another DL-ALL experiment, where the mobile phone is statically placed in different zones with various signal strengths. Different from the DL-NS experiments, UDP packets are immediately transmitted once the handshake between the phone and the server is established.

Figure 10(a) plots the DL-Volume-Gap when the phone is placed in zones with different signal strengths under various source rates in Operator-I. Each data transmission lasts one minute. The figure shows that, *the DL-Volume-Gap becomes larger as the signal strength becomes weaker or as the source rate becomes larger.* Figure 9 illustrates why it happens, and Table 1 shows the detailed results for three examples of experimental traces.

We make three observations. First, the core network receives almost all data packets (i.e., $V_{OP} \approx V_{SR}$). The charging gap is still caused by the unsuccessful packet delivery from RAN to UE. Second, packets are dropped in RAN because the incoming source rate is much higher than the effective rate r of the wireless link to the phone (see Figure 9). The effective rate depends on wireless signal strength. For example, the effective rate is 168.1 Kbps in the WR-zone, much smaller than in the SS-zone (about 644.4 Kbps). Third, not all the mismatches between the source rate s and the effective rate r lead to packet drops. Take the example of SS-zone with s = 800 Kbps. It spends more time (about 74.7 seconds) and incurs large delay. We infer that this attributes to the buffer mechanism, which temporarily stores incoming packets (if too fast) and retransmits them if needed. However, as the source rate further increases, the speed mismatch becomes too large (especially, in the

Setting	V_{UE} (MB)	V_{OP} (MB)	V_{SR} (MB)	Link rate r (Kbps)	Finish time (sec)	DL-Volume-Gap (MB)
SS-zone (-84 dBm)	6.0	6.0	6.0	644.4	74.7	0
W-zone (-98 dBm)	2.90	6.0	6.0	326.7	71.0	3.10
WR-zone(-109 dBm)	1.46	6.0	6.0	168.1	69.5	4.54

Table 1: Example results for three DL-All experiments when source rate is s =800 Kbps and $t = 1$ minute.

WR-zone/W-zone) to be handled by buffers, leading to eventual packet drops. *Consequently, we still pay for bits that never reach us even though wireless links exist. The charging gap depends on the operating environment.*

We also consider the case with intermittent signals where mobile users may lose signals for a while but recover them shortly. This scenario is common with cases of mobility and special landscape (mountains or high buildings). Our findings show that, those packets that the phone miss in NS-zones still contribute to the charging gap, though the communication recovers soon and buffering and retransmission mechanisms reduce the charging gap to some extent. Figure plots the DL-Volume-Gap when the phone loses signals for 10, 30, 60, 90 seconds. In the meantime, the UDP server sends packets at a speed s. We can approximate the data volume that arrived in t time but finally received by the phone is $V_{back} = s \times t - (V_{OP} - V_{UE})$. Figure 10(b) plots $(V_{OP} - V_{UE})/s$ (i.e., $t - V_{back}/s$) under various in-NS-zone durations. The larger the duration, the fewer the received packets. The results imply that, buffering and retransmission do contribute to packet delivery (retrieving 15 out of 90 seconds data in 50 Kbps-UDP session). However, those packets not recovered are still counted into the mobile bills. *It also shows that the gap exists even when mobile users only lose signals for several seconds.*

4.2 Common Cases

We now study the common cases, which reflect the usage patterns by applications and users in their daily activities. Our study has three categories. The first is to see how TCP, the dominant transport protocol for applications, reacts in the no-signal and weak-signal zones. In the second category, we study five popular applications, including Web browsing, Skype for VoIP, YouTube, PPS streaming, and streaming over VPN tunnels. In the third category, we report the user-based, weekly accounting gap.

4.2.1 TCP Cases

We next study the charging behaviors of TCP flows, which turn out to be not too bad in terms of the overcharged volume. Intuitively, TCP behaves differently due to its built-in mechanisms of congestion control and reliable data transfer. Its feedback loop offers implicit coordination between the network and end devices. We conduct DL-NS experiments via TCP, but let the server deliver packets without timeout. We keep the mobile phone longer in the SS-zone before roaming into the NS-zone. As expected, the DL-Volume-Gap greatly reduces; it is seen to vary between 2.9 KB and 50 KB in our experiments. As we know, the charging gap is determined by how many bytes are delivered before automatic TCP session teardown. This is determined by the congestion window size before the UE enters into the NS-zone and the timer for automatic teardown. When the phone is out of service, packets in the congestion control window are still sent out; Since no more packets can be acknowledged and the window decreases; the unacknowledged packets are retransmitted until automatic connection teardown. Note that, some ACKs at the UE fail to be sent out due to the broken connection. Figure 11 plots an example of TCP segment records observed by the two ends (the gap is 45.7 KB here).

In reality, most current Internet applications are built on the top

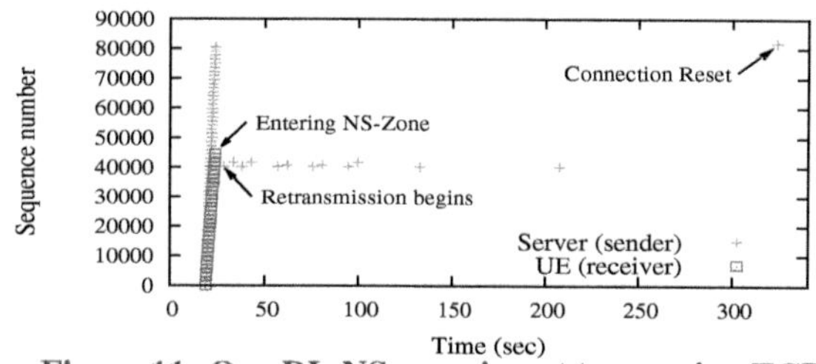

Figure 11: One DL-NS experiment trace using TCP.

of TCP. The inherent congestion control and automatic connection reset may release this connection once it fails for an extended period of time. This is why the charging gap is not large in reality. However, this is not the best practice from the 3G accounting perspective. It has to rely on higher-layer protocols such as TCP to handle abnormal behaviors of both connection teardown and overcharging. Moreover, UDP-based multimedia or other applications may still suffer if they do not implement control logic for automatic session teardown.

4.2.2 Application Behaviors

We carry out DL-NS experiments using five applications, including Web browsing, Skype, YouTube, PPS streaming [5], and VLC streaming [9] over VPN tunnels. In the web browsing test, we visit `www.cnn.com` at different times of the day; In Skype test, we make Skype video call, which uses UDP-based data delivery with built-in rate control [32]. Other three applications provide video streaming on phones. YouTube is TCP based, while PPS and VLC streaming are UDP based. PPS streaming is a very popular peer-to-peer application in China, whereas VLC + VPN offers one method to watch home HD TV. These applications have built-in control mechanisms, which can reduce rate or even tear down data delivery if the network connection degrades or breaks. We run each application for 5-15 runs with two US operators, except that VLC+VPN is blocked by the firewall of Operator-II. We start these applications in a SS-zone and enter into a NS-zone in several minutes.

Table 2 records the observed volume gap $(V_{OP} - V_{UE})$. The gap is negligible for Web browsing, and no more than 1MB for each Skype or Youtube run. However, it may reach up to 4.3 MB and 29.9 MB for PPS and VLC+VPN streaming, respectively. Interestingly, we observe that the gap be negative (but close to zero) for Web browsing; we figure out that it is because some packets (i.e., DNS packets) are free of charge, to be elaborated in Section 5. The difference between Web volume-gap of two operators is caused by different versions of Web pages. Mobile CNN page (about 0.2 MB) is fetched for Operator-II while the official CNN page (about 1.2MB) is fetched for Operator-I; the percentage of the volume gap is about 1.5–2.5%.

We also find that the volume gap varies significantly even for the same application (except Web browsing). For example, the gap varies from 0.1MB to 0.99MB for different Skype runs in Operator-II. It turns out that the average transmission rate during the ten seconds before entering into a NS-zone varies from 221 Kbps to 1.0 Mbps. We observe that the gap be bigger if the transmission rate before going into the NS-zone is larger. It is easy to understand, since the volume gap is contributed by the source rate s and durations t, where s is the average source rate during the period in the NS-signal zone and determined by rate control for each application. We also observe large gap for VLC+VPN. This is because its automatic teardown timer is larger. The teardown timer in VLC+VPN lasts from 30 seconds to several minutes (about 6-minute value was observed in our experiments, leading to 29.9 MB charging gap),

	Operator-I			Operator-II		
APPS	Med (MB)	Max (MB)	Min (MB)	Med (MB)	Max (MB)	Min (MB)
Web	-0.03	-0.04	0.00	-3KB	-4.6KB	0.6KB
Skype	0.88	0.99	0.40	0.68	0.99	0.10
YouTube	0.23	0.34	0.20	0.44	0.63	0.36
PPS	3.30	4.3	0.72	1.4	1.6	0.92
VLC + VPN	2.97	29.9	1.45	-	-	-

Table 2: Volume gaps for applications in DL-NS experiments.

	Operator-I				Operator-II		
User	1	2	3	4	5	6	7
Apps (excl. Web, Gmail)	Gmap	Stock Games	Skype, PPS FaceBook	YouTube PPS	ebook	-	YouTube PPS
V_{UE} (MB)	194.2	270.3	124.6	900.2	121.7	47.1	72.4
V_{OP} (MB)	192.6	270.0	129.4	948.4	120.9	47.3	77.6
Gap (MB)	-1.8	-0.3	4.8	48.2	-0.8	0.2	5.2
	-0.9%	-0.1%	3.9%	5.3%	-0.6%	0.4%	7.2%

Table 3: Volume gap for user studies during June 10-23, 2012 (User 7 had only one-day usage record on June 22, 2012).

whereas it is merely several seconds in Skype. These application tests again demonstrate that, though mobile data charging is largely successful in practice, nonnegligible overcharging is still observed due to problems in the current architecture.

We conduct another experiment to assess the performance over intermittent wireless channels. We watch videos via VLC streaming when we roam around the office area with several NS-zones. The wireless signals are intermittent, but usually recover within minutes. The video halts when we lose signals, but resumes once the wireless link is reestablished. In our experiment, we see that this video streaming never tear down and the observed volume gap reach up to 27.7 MB. Moreover, the gap depends on the number of SS-NS zone switches during our movement. We observe 11.8 MB and 27.7 MB charging gap for 10-minute and 30-minute movements, respectively. In DL-NS experiments, the observed gap for VLC streaming is at most 2.97 MB since the server tears down video streaming upon losing responses from the mobile client.

4.2.3 User-Based Usage Scenarios

We also study the accounting discrepancy for seven users (university students), who have data plans with two US carriers. We record the data usage observed by the mobile phone and the one charged by the operator for two weeks (June 10 - 23, 2012), except that User 7 had only one-day record on June 22. Note that, this small-sample user study may not well represent the common cases of daily usage for average users in our society. Instead, we intend to demonstrate how much the charging gap could be observed in reality, which depends on executed applications, usage patterns and locations. Among these users, the most popular applications are Web browsing and Gmail. Table 3 also shows other popular applications for each user, such as Gmap, Skype, YouTube, PPS, FaceBook, ebook reading, and games. Though data usage varies with users (from 47.1 MB to 900.2 MB) due to user behavior diversity, the volume gap is indeed small ($<$ 1 MB usually) for most users in reality. Big volume gap is not commonly observed in practice due to the built-in control mechanisms in many applications and infrequent encounters of NS-zones. However, we still observe that Users 4 and 7 have experienced volume gaps as large as 5.3% and 7.2%, respectively. User 4 once watched VLC streaming three times during a day while staying and roaming around his office area with several NS-zones; User 7 watched video using YouTube or PPS on the train to/from New York City, where there is a long tunnel without signals. During the round trip, User 7 transmitted and received 72.4 MB, but was charged by the operator for 77.6 MB, with the gap being 7.2%.

4.3 Recommended Quick Fix

We now recommend quick fix to the overcharging issues. The fundamental problem is that, the 3G network takes an SGSN/GGSN-based charging approach, which only records the data volume traversing these intermediate steps on the end-to-end delivery path. They do not coordinate with end devices when making accounting decisions. Specifically, they never take explicit feedback from the end systems. Therefore, when failures occur over the downstream path after SGSN/GGSN, the charging system is not aware of the status of the mobile device, thus incurring overcharge. We now suggest three feedback mechanisms that help to remedy the problems. Note that our proposals are also applicable to packet drops due to weak signals, not only in NS-zones.

In the first proposal, the charging system takes explicit feedback regarding the status of the end device. For the downlink case, our solution can be implemented within the 3G infrastructure without interacting with the UE. We use the feedback from RNC to obtain more accurate data usage delivered to the UE device. We use the field "*RNC Unsent Data Volume*," which records the data volume not delivered to UE, defined by the 3G standard [15]. RNC reports this record to SGSN, which computes the data volume successfully delivered to UE, i.e., $V_{succ} = V_{SGSN} - V_{RNC\_unsent}$. It thus enables the operator to charge the user based on the data volume delivered to UE. This way, the huge DL-Volume-Gap (e.g., the 450 MB) can be eliminated.

We next fix problems with the session teardown in the absence of signals, where the charging can stop early to avoid overcharging. Our suggested solution is to deactivate the PDP context soon after the UE device cannot be reached. This can be implemented by the soft-state mechanism on the PDP context. We set a timer with the PDP context for UE, and the timer, as well as the PDP context, will be refreshed via the data delivery to/from UE. Note in three-hour DL-NS experiments, PDP context is not released in time because there is incoming traffic associated with it. This implies that the operator probably makes wrong decision that the PDP context should be kept alive. We thus suggest refreshing the timer based on actual data delivery, or the paging of UE when the actual data usage is zero. This offers the 3G charging system an alternative feedback mechanism on the UE status, but may incur excessive control overhead.

The third feedback mechanism also helps to reduce overcharging. Whenever big data usage is generated, it should trigger an exception verification to check whether the charging makes sense or not. For example, when a data session lasts for an hour or produces about 100 MB data, the 3G core network should verify whether it is indeed normal charging practice. This can be done by sending a signaling message to RNC to query whether the UE status is normal. The RNC subsequently reports the UE status and facilitates SGSN/GGSN in its charging decision.

We note that Cisco has proposed overcharging protection for GGSN to be aware of lost radio coverage using the feedback of SGSN [1]. It is to assess the device status due to lost coverage.

4.4 For Other Carriers

We also run similar experiments with three other major carriers, one each in the US, China and Taiwan. All the observed results still hold in general. The minor difference is that, (1) three-hours charge for a 50 Kbps UDP flow in DL-NS experiments is observed for two major US operators, while at least 5.7 hour charge is observed for the third US operator, one hour charge is observed in China, and about 42 minutes occur in Taiwan; and (2) the maximal UDP source rate is smaller in China, e.g., the transition point in Figure 7(b) happens at 1 Mbps for the Chinese carrier. This is because the data rate supported over the wired Internet is smaller in China. We also conduct two-week usage studies for one user in China and two users in Taiwan; their observed gaps are negligible (<1 MB, within 0.5% error) because they mainly use them for Web, Gmail and SMS exchanges and the overall volume is small.

5. WE GET WHAT WE WANT FOR FREE

The second finding is just the opposite. We can take free rides to obtain "toll-free" data services without incurring any charge. The root cause is that, the current charging policy practiced by operators has loopholes, and can be exploited to build "free" data services. Our study shows that, both operators offer free Domain Name System (DNS) service via transport-layer port number 53. There is almost no enforcement mechanism to ensure that the packets going through this port are indeed DNS messages. Even worse, no effective mechanism exists to limit the traffic volume going through this port. Consequently, this free service can be readily abused to create "toll-free" data services. We have built a simple prototype, and demonstrated that it is feasible to offer various data services, e.g., file downloading or video streaming, through a special proxy server relaying data over the free transport-layer port. The process is similar to calling 800-like voice hotlines, but for free data access.

5.1 Loopholes in Charging Policy Practice

The 3G standards offer carriers enough freedom to define their own charging policies. Our experiments show that DNS packets are not charged by operators. We sent out 100 DNS queries (about 18KB in volume), and operators did not charge the incurred data usage. The operators' practice is also easy to understand, since DNS is a fundamental service to jump start Internet applications. Operators thus have every reason to offer it for free, to facilitate followup data usage by the services. Thus, free DNS service is well justified as a good policy practice.

However, the operator practice to offer free DNS service does have loopholes. Internet RFC 5966 stipulates that DNS service is offered using transport-layer port 53 via UDP or TCP [27]. To identify a data flow, the 3GPP standards define five-tuple flow ID composed of source and destination IP addresses, source and destination port numbers, and protocol ID (see Section 2). However, both operators do not strictly enforce this service via the standard five-tuple flow ID, but via only the destination port (plus protocol ID for Operator-II), thus exposing a loophole.

DNS-TEST Experiment Setting: We test DNS-related charging in five cases: (1) *DNS-Default*: send 100 DNS queries to the default DNS server provided by the operators; (2) *DNS-Google*: send 100 DNS queries to a Google DNS server (IP address: 8.8.8.8); (3) *TCP53-Google*: repeat (2) but via TCP at port 53; (4) *TCP53-Server*: send 50 random packets to our own server using TCP via port 53, and request the server to return the received packets; each packet (including IP/TCP headers) is 1KB; Source port number is randomly chosen; and (5) *UDP53-Server*: repeat (4) but using UDP. The goal of these experiments is to verify what factors the free DNS service depends on: (1) Does it depend on the server address? For example, is DNS free only via the operator DNS servers? (2) Does it depend on the protocol ID? For example, is it free for both UDP and TCP? (3) Does it depend on the source port number or check the DNS message semantics?

Results: Figure 12 plots the data volume observed by UE and the amount charged by Operator-I and Operator-II in all five cases. The results shows that,

Operator-I: Packets via **port 53** are FREE Operator-II: Packets via **UDP + port 53** are FREE

Specifically, UE sends and receives about 18.1 KB for 100 DNS queries and responses in both *DNS-Default* and *DNS-Google* tests. In the *TCP53-Google* test, the traffic volume increases to 48.1 KB due to TCP signaling overhead. In both *TCP53-Server* and *UDP53-Server* tests, the UE works as expected when sending/receiving 100 KB. Operator-I charges for free (i.e., $V_{OP} = 0$) in all cases while Operator-II charges those TCP cases. From these results, we learn that the free DNS service is implemented by Operator-I using only one field in the flow ID (i.e., the destination port 53). In contrast, Operator-II uses two tuples in the flow ID, i.e., UDP over destination port 53.

5.2 Building "Toll-Free" Data Services

We now exploit the loopholes in the free DNS service to enable "toll-free" data services. We deploy a proxy server (placed outside the cellular network), which exchanges data services with mobile phones through the 3G carrier. We use "DNS tunneling" between the phone and the proxy. Data communication between the proxy and mobile phones is carried in UDP at port 53 by encapsulating data packets in DNS messages, which traverse the 3G network free of charge. We can also build the DNS tunnel using TCP at port 53 for Operator-I. The design can be readily extended to communication between UE and an Internet server, where the proxy server will act as a hub to relay packets on behalf of both UE and the server. The idea of DNS tunneling is also used in the iodine tool [4], which is designed for data access in different scenarios where DNS queries are allowed but the Internet access is blocked.

We run experiments using the prototype to demonstrate that free data service is feasible. Figure 13 plots V_{UE} and V_{OP} for three scenarios: (I) UE sends one request to download a 5MB video clip from a public website, (II) UE uploads a 3MB file to our server, and requests to return the delivered packets, and (III) UE sends many small requests (100 B) to our server for an hour, and each demands a 1KB response. These cases are to validate whether our service supports unbounded traffic upon a single request, whether it supports large-volume uplink and downlink traffic, and whether it allows for long-lived sessions.

Our results show that, both operators can be exploited for free data services in these scenarios, except that Operator-I does not allow unbounded traffic for a fake "DNS" request. In the first test, Operator-I only allows to deliver 29 KB downlink data to the UE, while Operator-II delivers much larger file (up to 4 MB). We gauge that Operator-I might have enforced checking to verify the size of the response message, in which a DNS message size is typically bounded. However, this checking can be easily bypassed. The UE simply sends out many small, dumb packets over this session to increase the quota for downlink traffic. Then large downlink data can pass this checking. This has been validated in scenarios (II) and (III). In these tests, the gap between V_{UE} and the expected file size is mainly caused by unreliable transmissions via UDP. We can implement reliable transfer mechanisms over UDP to eliminate this loss. Since each scenario has ten or more experimental runs at the data rate from 100 Kbps to 1 Mbps, the total free data we have obtained from this DNS hack exceeds 200MB[6].

In addition to DNS hacking, other tricks exist for free data services by exploiting the loopholes in the charging policy. For example, in case some operators offer free Internet access to a given website, Web redirection from one free Web server to the target Web page is used to enable free data services [3]; using certain, free Access Point Name (APN), which is a configurable network

[6]We do not run into legal issues since we had unlimited data plan from Operator-II in the past and bought unlimited daily plan from Operator-I in our experiments.

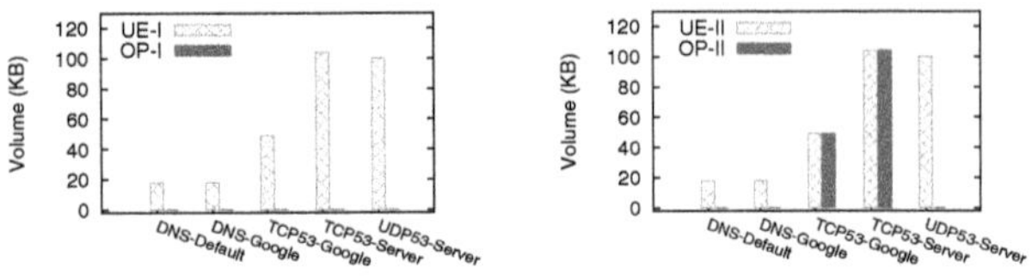

(a) Operator-I (b) Operator-II

Figure 12: V_{UE} and V_{OP} in DNS-TEST experiments.

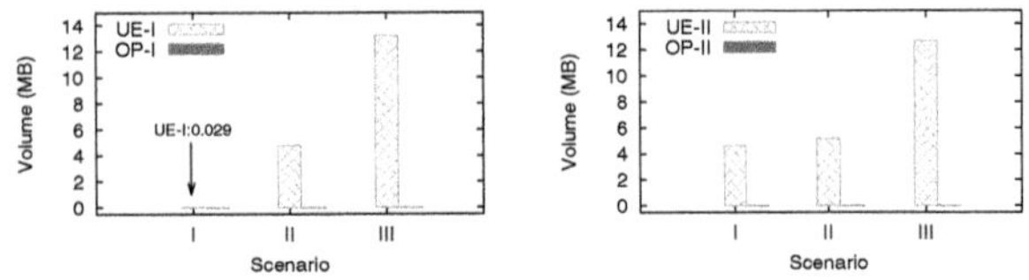

(a) Operator-I (b) Operator-II

Figure 13: Feasibility test of free data services.

ID used by a mobile device when connecting to a carrier, offers another way for free data service in some carriers, e.g., AirTel India [2] and UK Three [6]. These examples, including our DNS hacking, show that policy offers flexibility but may also be abused if not enforced properly.

5.3 Recommended Quick Fix

The simplest fix is to stop offering free DNS service or other forms of free data services inside cellular networks. Beyond this option, we suggest three possible remedies to this loophole while retaining free DNS service. The first solution is to enforce checking on the IP address of the DNS server. Therefore, free DNS services are only allowed if these messages go to designated DNS servers provided by operators or other DNS servers authorized by the operators. This enforcement eliminates the possibility to go through those fake DNS servers. The possible downsides include: extra effort is needed to authenticate DNS servers, not all DNS servers across the Internet can be directly accessed by UE, and workload at the designated DNS servers increases. Unfortunately, it is still possible for malicious users to deceive those resolvers/servers to forward fake DNS requests to a fake DNS server, but the cost is higher. The second fix is to provide quota for free DNS service. The quota for free DNS data can be set on a per-UE, per-week (or daily) basis in advance. The data usage beyond DNS quota will still be charged. The challenge for this approach is how to set an appropriate quota. Ideally, the quota should be estimated based on both normal usage and sudden surges. The third approach is to enforce message integrity check to verify the authenticity of each DNS message. However, it incurs excessive processing overhead on a per message basis.

5.4 Carriers in Other Regions

We run similar tests with other carriers. We indeed observe that the free DNS policy be operator dependent. The third US operator also offers UDP-based DNS for free, and behaves similar to Operator-II. However, for both carriers in China and Taiwan, the DNS service is not free. Both operators charge DNS messages identical to data traffic.

6. GRAY AREAS IN DATA CHARGING

We now describe charging cases in gray areas, where the users may be charged differently by the operators, compared with the actual data usage perceived at end hosts. However, there is no simple, accepted charging rule in these cases. We will show that the users may be charged for wrong or careless uplink operations; we also examine the impact of the middlebox deployment and Internet traffic congestion on mobile data charging. We finally assess charging on application overhead.

6.1 UDP Uplink to a Nonexistent Host

The worst uplink case is to use UDP packets to a nonexistent host (i.e., no packets can be successfully delivered). Our tests show that both operators still charge every bit sent by UE. The root cause still lies in the SGSN/GGSN based charging architecture, similar to the downlink case of Section 4. The good news is that this scenario is not very common unless the device is hijacked.

We also test UDP uplink traffic to our server under various wireless environments. It turns out that, there is no (obvious) gap between the data volume arriving at the receiver (i.e., our server) and the volume charged by the operator. However, our UE traces show that the UE does retransmit data over the wireless link (particularly in the WR-zone/WS-zone), i.e., $V_{UE} > V_{OP}$. Note that, these retransmitted data over the wireless link will not be observed by SGSN/GGSN, thus incurring no extra charge beyond those volume perceived by the receiver. The same conclusion also holds for uplink TCP sessions.

6.2 Effect of Middle-boxes

Middleboxes (e.g., proxy servers, CDN servers, NAT boxes, and firewalls) can be deployed inside 3G/4G networks for performance enhancement or extra service [30]. Indeed, our study confirms that proxy servers are placed in 3G networks. We find out that, Operator-I deploys proxy servers to handle popular applications, including HTTP and FTP. Consequently, the end-to-end data session between the UE and the server is split into two segments, one between the UE and the proxy, the other between the proxy and the HTTP/FTP server. We now assess the impact of such session splits due to middleboxes on charging. In the worst case, the UE interacts with the proxy rather than its intended server, and incurs overcharging. The user is charged though (s)he never receives any service! This can be illustrated by the following experiment.

We let the mobile phone connect to a non-existent host (i.e., an unallocated IP address) and upload a 1 MB file using TCP. Since the host does not exist, it should stop early (e.g., after several TCP SYN requests). To our surprise, we discover that, the data sessions for HTTP (80, 8080) and FTP (21) last much longer than expected. The delivered data volume for HTTP and FTP reaches 300 KB and 130 KB, respectively, in Operator-I. We also test with other popular applications, e.g., HTTPS(443) and SMTP(25), and unknown ports. Figure 14(a) plots V_{OP} (the same as V_{UE}) using different port numbers for both carriers. In Operator-I, we examined the TCP traces collected at the UE, and observed that those sent packets be acknowledged by TCP, though the IP address on the server side is nonexistent. This indicates that at least a middlebox has been deployed on the delivery path, which responds to UE requests on behalf of the server. The bad news is that, Operator-I also imposes charges though such HTTP/FTP data never reach the server side! We note that, the proxy is operator dependent and application specific. Other popular applications do not observe such proxy servers. Operator-II does not seem to have deployed such middleboxes in its core network even for HTTP/FTP.

6.3 Packet Drop over the Internet

Charging discrepancy exists when packets are dropped over the wired Internet, which is outside the cellular core network. We illustrate the scenario in Figure 14(b). The UE sends 200 KB data to

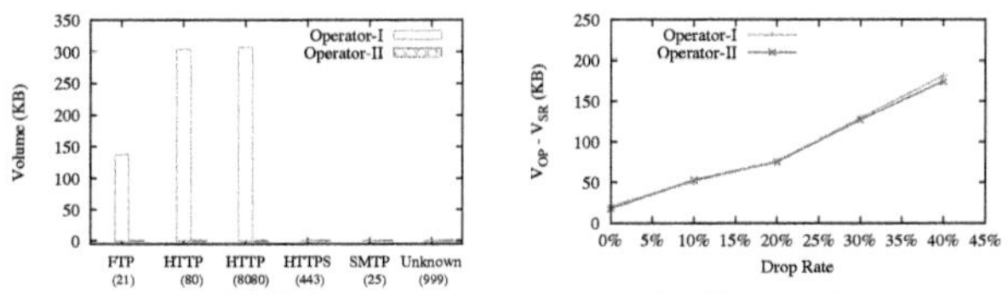

(a) Middlebox (b) Packet drop

Figure 14: Results when (a) connecting to nonexistent hosts using different TCP ports and (b) experiencing unreliable Internet packet delivery.

the Internet server via the cellular network, but packets are dropped right before reaching the server. Note that the Internet only offers best-effort service, so IP packet drops can be common, e.g., upon network congestion or malfunctioning routers. In this case, packets have been through the cellular network and recorded by SGSN/GGSN for charging. In our experiment, we upload an image file to our server from the mobile phone. We vary the packet-drop rate from 0% to 40% before reaching the TCP receiving end; we use it to emulate packet loss on the Internet. Since TCP will retransmit these lost packets from the UE, they incur additional charging volume at the carrier. We plot the charging gap (between the server and the operator) versus the drop percentage in Figure 14(b). The figure shows that, the charging volume increases almost in proportion to the drop rate. For operators, this makes perfect sense since the 3G network does deliver those packets. However, end users never receive those dropped data. The same phenomenon occurs for uplink UDP transmissions.

6.4 Charging for Application Signaling

Motivated by the previous DNS study, we also examine whether application signaling messages are also charged by 3G accounting and how much percentage they contribute when charged. Our findings show that, both 3G operators charge the signaling data (e.g., ICMP, SIP and RTSP) and protocol overhead. However, the actual signaling cost varies a lot across different applications. We consider three interesting cases.

FTP Signaling Channel: FTP uses two separate TCP sessions, with port 21 for command signaling and port 20 for data transfer. We conduct experiments to send messages mainly over the signaling channel (e.g., list a remote folder with 1 or 50 files), as well as data transfer (downloading one or ten 1MB files).

HTTP Redirect and Invalid Links: We run HTTP redirect cases, where the web page is redirected once or 15 times to reach the final content. We also access a web page with one invalid HTTP link. In the invalid link case, we access a web page that has one or 50 invalid image links.

Email and IM: We tested Yahoo Mail and Skype for Email and IM applications, respectively. We send a small or large email, login/logout skype, or remain idle for 10 minutes in Skype.

Results: The results are shown in Table 4. We make three observations. First, the application signaling messages, including FTP control commands, are indeed charged. Second, the signaling overhead percentage is particularly large when the content size is small, e.g., FTP listing, HTTP redirection. Command messages may only be a small percentage in the operator's charging volume, compared with protocol overheads (see FTP signaling); Third, signaling and protocol overhead do incur hidden costs (not perceived by average users). Note that, in the HTTP case, those invalid links or those redirects are never the content requested by the user. It explains why Alice is charged by an invalid click without accessing real content in the itemized bill of Figure 1.

Application	Test	Content (KB)	OP (KB)	Gap	Gap/OP (%)
FTP	Listing (1)	0.06	2.97	2.91	97.9
	Listing (50)	3.32	7.28	3.96	54.4
	Downloading (S)	1024	1190.5	166.5	14
	Downloading (L)	10240	10858.0	618.0	5.7
HTTP	Redirect (1)	0.05	1.9	1.85	97.3
	Redirect (15)	0.05	15.1	15.05	99.7
	Invalid (S)	0.13	2.05	1.92	93.6
	Invalid (L)	2.56	12	9.44	78.7
Email	Send (S)	0.02	13.0	12.98	99.8
	Send (L)	223.6	250.98	27.38	10.9
Skype	login/out	0	50.07	50.07	100
	idle (10mins)	0	5.05	5.05	100

Table 4: Signaling overhead of popular applications.

In a broader view, the above study makes us contemplate on who should pay for what. We have seen that free applications may raise more overhead for advertisements, while the paid ones may not. As the network moves toward content-based operations, lots of interesting issues and debates may arise.

7. DISCUSSIONS

We have described the accounting discrepancy in both extreme and common settings. The most optimistic view will claim that, the problem is not too bad, so we do not need to fix it; the built-in control mechanisms in TCP and applications at the end devices help to mitigate the damage. However, we believe that there are fundamental technical problems beneath these engineering missteps. We now discuss two issues: the architecture options and policy practice.

Rethinking Accounting Architecture: In general, there are three classes of accounting architecture: the network-based one such as the current 3G system, the end-system-based approach, and the collaborative one between the network and end devices.

Both the network-based and the end-system-based approaches have severe limitations. For the network-based, 3G charging system, we already observed that it should result in large accounting gap in the extreme cases. The fundamental problem is that, the network lacks coordination with end systems and makes the charging decision alone. This functions fine when everything goes well, but suffers when things go wrong. The built-in feedback loop in TCP and applications may help to certain extent. However, the concrete feedback mechanism and its operation accuracy are largely out of control to the accounting system. The charging system is not self-healing under failures and extreme conditions. On the other hand, the other extreme of end-system-based accounting will not work either. There is no easy mechanism to regulate users so that they will not cheat. The verification process of user-reported results is also challenging. Therefore, this approach is unrealistic in practice.

We believe that the coordinated charging system between the network and end systems is promising. The data delivery process is end to end, so both the network and end devices are players in data delivery. They need to make concerted decision in charging, too. Of course, there are various forms of coordination between the network and the device. We are not advocating the scheme that both parties play symmetric roles. Instead, we believe that the network has to take more central role in the charging process, while the end devices offer useful hints and feedback to the network. The network naturally has "centralized" views on users and flows, and more resources to control and regulate the charging decision. To build a more resilient charging system, several challenges arise along this direction: (1) What failures and losses does the accounting system have to handle? (2) What mechanisms are indispensable to coping with given failures? (3) When and how does the end device/server report delivery losses? (4) How does the system ensure that the feedback information is secure and trustworthy? (5)

How many mechanisms need to be placed into the future cellular network standards?

Policy as Double-Edged Sword: Policy practice is an inherent component of the accounting systems for mobile users. Policies can be good for both operators and end users! On one hand, policy practice offers carriers flexibility, while injecting dynamics into the market. It can serve as a viable mechanism to compete with other carriers when offering users better services at lower cost during certain times. On the other hand, users can also benefit from policy practice. As we have seen in the DNS case, users will pay less due to the free DNS service! However, we have to be prudent with policy practice. The policy choice needs to be conflict free. Moreover, its enforcement has to be strict. Otherwise, policy may open holes that operators may never anticipate.

8. RELATED WORK

Despite the popularity of 3G/4G data services, cellular network accounting remains a largely unaddressed area in the research community. [20] offers a nice survey on pricing, charging, billing methods for 3G systems in 2005. Among current industry efforts, Cisco proposed overcharging protection for GGSN to be aware of lost coverage at end devices based on the feedback of SGSN [1]. Using certain, free Access Point Name (APN) provides another way to obtain free data service in certain carriers, e.g., AirTel India [2] and UK Three [6]. DNS tunneling is also used in the iodine tool [4], designed for data access in scenarios different from ours, where DNS queries are allowed but the Internet access is blocked. In contrast, our work examines the current accounting practice in operational 3G networks. We use experiments to study various charging behaviors within the 3G accounting standards and policy practice by operators, identify root causes and propose fixes.

In the more general context, Internet accounting and pricing have been explored in the literature [19,22,28,29] (see [24] for a survey for work up to 2001). These prior efforts focus on the wired Internet. The proposed accounting solutions are quite different from the one used by current cellular networks.

9. CONCLUSION

The Internet is going wireless and mobile. Two underlying driving forces have been the explosive growth of smartphones/tablets and the rapid deployment of 3G/4G infrastructure. Unlike the wired Internet, cellular networks have implemented usage-based charging, rather than the simpler flat-rate charging. Going down this path, the 3G/4G standards finalize the accounting architecture, yet leave enough freedom for operators to define their own charging policy. In this work, we conduct experiments on operational 3G networks to study the implication of such an architecture and practice, and quantify the charging discrepancy between the operator's record and the user's observed volume.

Our study offers some insights. On the architecture side, the SGSN/GGSN element-based charging is easy to implement, yet poses limitations. When things go wrong outside the charging elements, the resulting data volume deviates from what is observed at end devices. The fundamental problem is that, the charging system mainly uses open-loop, but not closed-loop operations; it makes accounting decisions alone without taking feedback from end devices. On the other hand, policy offers flexibility, but is also mistake prone. Policy operators have to take extra care to make policy enforcement complete. On both fronts, really bad things can happen under extreme conditions, somewhat unexpectedly. Consequently, as shown by our extensive experiments, we may pay for what we never get and get what we want for free in worst-case scenarios. We hope our initial efforts will stimulate further research on this important topic from both academia and industry.

Acknowledgment

We greatly appreciate the insightful and timely comments from our shepherd, Dr. Dina Papagiannaki, and the anonymous reviewers for their constructive feedback. We also thank Mr. Xingyu Ma and other participants in the charging experiments.

10. REFERENCES

[1] Cisco ASR 5000 Series Serving GPRS Support Node Administration Guide. http://www.cisco.com/en/US/docs/wireless/asr_5000/12_0/OL-24828_SGSN_Admin.pdf.

[2] Free GPRS Hack For Reliance Mobile. http://www.megaleecher.net/Reliance_Internet_GPRS_Hack.

[3] Free Gprs Mobile Tricks. http://darkwap.mobi/gprs-tricks/Free-Gprs-Mobile-Tricks.

[4] Iodine. http://code.kryo.se/iodine/.

[5] PPS. http://www.pps.tv/.

[6] Three PAYG Mobile Internet for FREE. http://www.digitalworldz.co.uk/226311-three-payg-mobile-internet.html.

[7] Traffic Monitor - RadioOpt GmbH. https://play.google.com/store/apps/details?id=com.radioopt.widget.

[8] TrafficStats. http://developer.android.com.

[9] VLC Stream & Convert. http://traveldevel.com/.

[10] Wing FTP. http://www.wftpserver.com/.

[11] 3GPP. TS23.060: GPRS; Service description; Stage 2, Dec. 2006.

[12] 3GPP. TS23.125: Overall High Level Functionality and Architecture Impacts of Flow Based Charging, Mar 2006.

[13] 3GPP. TS32.240:Telecommunication management; Charging management; Charging architecture and principles, Sep. 2006.

[14] 3GPP. TS25.301: Radio Interface Protocol Architecture, 2008.

[15] 3GPP. TS25.413: UTRAN Iu interface RANAP Signaling, 2008.

[16] 3GPP. TS23.401: GPRS Enhancements for E-UTRAN Access, 2011.

[17] G. America. Global 3G Deployments UMTS HSPA HSPA+, 2010.

[18] Caida. http://www.caida.org/research/traffic-analysis/tcpudpratio/.

[19] R. Cocchi, S. Shenker, D. Estrin, and L. Zhang. Pricing in Computer Networks: Motivation, Formulation, and Example. *IEEE/ACM Transactions on Networking*, 1:614–627, 1993.

[20] Z. Ezziane. Charging and Pricing Challenges for 3G systems. *IEEE Communications Surveys and Tutorials*, 7(1-4):58–68, 2005.

[21] H. Holma and A. Toskala. *LTE for UMTS: Evolution to LTE-Advanced*. Wiley, 2011.

[22] M. Kouadio and U. Pooch. A Aaxonomy and Design Considerations for Internet Accounting. *SIGCOMM Comput. Commun. Rev.*, 32(5):39–48, Nov. 2002.

[23] mobiThinking. Global Mobile Statistics 2012. http://mobithinking.com/mobile-marketing-tools/latest-mobile-stats.

[24] A. Odlyzko. Internet Pricing and the History of Communications. *Computer Networks*, 36:493–517, 2001.

[25] OECD. Nearly Two-Thirds of US Broadband Subscribers are Wireless. http://www.websiteoptimization.com/bw/1012/.

[26] D. M. Reporter. Up to 20 million Americans 'Overcharged' by AT&T for Data Usage, 2011.

[27] RFC5966: DNS Transport over TCP - Implementation Requirements, 2010.

[28] S. Shakkottai, R. Srikant, A. E. Ozdaglar, and D. Acemoglu. The Price of Simplicity. *IEEE Journal on Selected Areas in Communications*, 26(7):1269–1276, 2008.

[29] D. Trossen and G. Biczók. Not Paying the Truck Driver: Differentiated Pricing for the Future Internet. In *Proceedings of the Re-Architecting the Internet Workshop*, ReARCH '10, 2010.

[30] Z. Wang, Z. Qian, Q. Xu, Z. Mao, and M. Zhang. An Untold Story of Middleboxes in Cellular Networks. In *SIGCOMM '11*, 2011.

[31] WireShark. http://www.wireshark.org/.

[32] X. Zhang, Y. Xu, H. Hu, Y. Liu, Z. Guo, and Y. Wang. Profiling Skype Video Calls: Rate Control and Video Quality. In *INFOCOM'12*, March 2012.

CloudIQ: A Framework for Processing Base Stations in a Data Center

Sourjya Bhaumik[1], Shoban Preeth Chandrabose[1], Manjunath Kashyap Jataprolu[1], Gautam Kumar[2], Anand Muralidhar[1], Paul Polakos[3], Vikram Srinivasan[1], and Thomas Woo[4]

[1]Alcatel-Lucent Bell Labs, India

[1]{sourjya.bhaumik, shoban.preeth_chandrabose, manjunath_kashyap.jataprolu, anand.muralidhar, vikram.srinivasan}@alcatel-lucent.com

[2]U. C. Berkeley

[2]gautamkumar.iit@gmail.com

[3]Cisco, USA

[3]ppolakos@cisco.com

[4]Alcatel-Lucent Bell Labs, USA

[4]thomas.woo@alcatel-lucent.com

ABSTRACT

The cellular industry is evaluating architectures to distribute the signal processing in radio access networks. One of the options is to process the signals of all base stations on a shared pool of compute resources in a central location. In this centralized architecture, the existing base stations will be replaced with just the antennas and a few other active RF components, and the remainder of the digital processing including the physical layer will be carried out in a central location. This model has potential benefits that include a reduction in the cost of operating the network due to fewer site visits, easy upgrades, and lower site lease costs, and an improvement in the network performance with joint signal processing techniques that span multiple base stations. Further there is a potential to exploit variations in the processing load across base stations, to pool the base stations into fewer compute resources, thereby allowing the operator to either reduce energy consumption by turning the remaining processors off or reducing costs by provisioning fewer compute resources. We focus on this aspect in this paper.

Specifically, we make the following contributions in the paper. Based on real-world data, we characterise the potential savings if shared homogeneous compute resources are used to process the signals from multiple base stations in the centralized architecture. We show that the centralized architecture can potentially result in savings of at least 22% in compute resources by exploiting the variations in the processing load across base stations. These savings are achievable with statistical guarantees on successfully processing the base station's signals. We also design a framework that has two objectives: *(i)* partitioning the set of base stations into groups that are simultaneously processed on a shared homogeneous compute platform for a given statistical guarantee, and *(ii)* scheduling the set of base stations allocated to a platform in order to meet their real-time processing requirements. This partitioning and scheduling framework saves up to 19% of the compute resources for a probability of failure of one in 100 million. We refer to this solution as CloudIQ. Finally we implement and extensively evaluate the CloudIQ framework with a 3GPP compliant implementation of 5 MHz LTE.

Categories and Subject Descriptors

C.3 [**Computer Systems Organization**]: Special-Purpose and Application-Based Systems—*Real-time and embedded systems,Signal processing systems*

General Terms

Algorithms, Design

Keywords

Cellular, Virtualization, Cloud RAN

1. INTRODUCTION

The proliferation of devices like tablets and smart phones, coupled with new types of applications, has resulted in a demand for high data rates in wireless communication. There are several innovative proposals to meet this demand for the explosion in wireless capacity [1]. These proposals require new architectures, protocols, and advanced signal processing techniques in a cellular base station. However, implementing these proposals increases the cost of delivering data to end users. Our approach is motivated by a proposal called Cloud Radio Access Network (Cloud RAN) that proposes a transition from distributed to centralized infrastructure for baseband processing of cellular signals. This is under consideration by NGMN [3]. Cloud RAN can potentially reduce the cost of delivering data and also enable signal processing techniques that improve network performance. Preliminary studies by China Mobile show that Cloud RAN has the potential to significantly reduce the total cost of operations [2].

In existing cellular networks, compute resources for performing baseband processing are located at each cell site. However, in the proposed centralized architecture, the compute resources are located in a central location. The architecture is realized by transporting the cellular signals (also known as IQ or In phase-Quadrature phase signals) over dedicated high speed fiber from the antennas at the cell site to the central location. The cell sites are situated in a radius of up to 20 km [3] from the central location. There are potential benefits to this approach which include (i) a savings in the operating expenses due to a reduction in the site visits for upgrade and maintenance, and possibly lower site lease costs, (ii) use of advanced coordinated signal processing techniques by jointly processing signals from neighboring base stations to significantly improve network performance [16], and (iii) a potential reduction in energy expenditure by exploiting the load variations and using only as many compute resources as required to process the base stations. This would exploit the variations in the processing load of base stations to multiplex multiple base stations on to a single compute resource. Energy reduction is a key consideration for operators in Asia. The operational expenditure from energy costs can be as high as 50% of the total operational expenditure in these regions. However the costs of transporting the IQ signals in the centralized architecture need to be factored in while evaluating the advantages of centralized processing, but this economic analysis is outside the scope of this paper. To address this, the cellular industry is actively considering some split processing alternatives, where some of the Layer 1 processing is done at the cell site and only user related Layer 1 processing is sent to the central location. This has the potential to significantly reduce the backhaul requirements and make it proportional to user load. Our goal in this paper is to evaluate the third benefit of the centralized architecture, i.e., we study the benefits of resource pooling across base stations and develop a framework to multiplex multiple base stations on a platform with a set of homogeneous compute resources. We refer to our approach as CloudIQ. CloudIQ is further motivated by emerging trends towards a more programmable wireless infrastructure. For example, many equipment manufacturers are migrating towards making the hardware more programmable and flexible by combining multiple hardware, DSPs, and programmable processor cores into single die (System on a Chip), while still maintaining the required cost/performance ratios. Further there is evidence that implementing wireless infrastructure on a general purpose processor (GPP) based platform is feasible for WiFi access points and LTE base stations [21]. Although GPP-based systems do not meet the current requirements in terms of performance-per-Watt, we evaluate the CloudIQ architecture on a homogeneous GPP-based platform since it provides a simple model to analyze the resource management problem and has wider applicability to resource sharing among a set of homogeneous compute resources (e.g. FPGAs, DSPs, or ASICs).

1.1 Objectives and Contributions

There have been several reports of trials of the Cloud radio access network concept by various vendors, to the best of our knolwedge, these demonstrations only backhaul the IQ samples to a central location and the rest of the processing is done as in existing cellular systems, i.e., each base station is assigned a single base band processing unit and base stations are not pooled on fewer baseband units to exploit energy savings or reduce the number of processing resources. Our paper is the first to do a systematic study of the gains possible due to resource pooling gains and provide a framework to precsiely achieve a trade-off between the statistical guarantees against failure and the savings (energy/cost) possible due to the Cloud RAN architecture. We elaborate on the contributions below.

1. **Analyze the benefits of CloudIQ on general compute platforms:** In traditional cellular systems, the compute resources are provisioned to handle the peak load at a base station, while meeting the real-time deadlines. However, in a CloudIQ architecture, the variations in the processing loads of base stations can be used to multiplex multiple base stations on a single compute resource. The variations in the load happen at small time scales (order of ms), but it is not possible to change the assignment of base stations to compute resources at such time scales. Hence, we allow changes in assignment of base stations to compute resources at larger time scales (order of minutes) and provide statistical guarantee on the base stations successfully meeting their real-time deadlines. We analyze real-world WCDMA data to study the feasibility and advantages of this approach. This data contains PHY layer logs from 175 base stations in a dense urban setting. We observe that, for realistic statistical guarantees, resource savings of at least 22% are possible in the CloudIQ architecture. See Section 3 for the details.

2. **Resource management framework:** In order to realize the potential savings of compute resources, we need to process multiple base stations in a single compute resource and meet their individual real-time constraints. We refer to this as *virtualization* of the base stations. There is a wealth of literature on scheduling of real-time systems, which includes scheduling on multi-processor platforms. But most of the existing work considers systems with *implicit deadlines* where the deadline coincides with the periodicity of the tasks. As we see later, this is unlike cellular systems where the deadline to process a task can exceed the periodicity of the task. Furthermore, we adopt a *separation principle* in the design of the system that adds to its simplicity and robustness. The separation principle decouples the problem of partitioning the base stations into subsets to be scheduled on each compute resource from the problem of designing algorithms to schedule a subset of base stations on a compute resource and meet their deadline constraints.

We design a resource management framework that adheres to the separation principle and can schedule multiple base stations on a homogeneous compute platform. Each homogeneous compute platform consists of multiple identical cores, where each core is equally capable of executing a task. This framework is simple to implement and the savings in the compute resources under the framework are close to those obtained by analyzing the real-world data. See Section 4 for the details of the framework.

3. **System Design and Evaluation:** We implement the CloudIQ framework on an Intel Xeon six core processor, where each core has a clock speed of 3.47GHz. We choose Linux with PREEMPT_RT patch as the operating system. We choose an open source implementation of the LTE standard called OpenAir to implement and demonstrate the resource management framework. We evaluate our implementation and show that two 5 MHz LTE base stations at peak load can be run on a single processor, while meeting their individual real time requirements. Peak load refers to the maximum load in processing the physical layer in 5 MHz LTE. We can schedule a maximum of four base stations in a single platform and meet their real-time requirements if their individual loads are below the peak processing load. See Section 5 for the details of the system implementation.

Section 2 contains a short primer on cellular systems and describes the OpenAir project. We overview the related work in Section 7 and outline some of the remaining challenges in Section 8.

2. PRELIMINARIES

2.1 Primer on cellular systems

In a cellular system, spectrum is treated as a resource that is shared among all the users. The spectrum is partitioned into *channel resources*, which are spreading codes in WCDMA or time-frequency resources referred to as physical resource blocks (PRB) in LTE. In each scheduling interval, the scheduler allocates certain number of channel resources and specifies a modulation and coding scheme (MCS) for a user to either receive or transmit data. As per LTE the smallest granularity for scheduling decisions is 1 ms, while this is much higher in WCDMA. In each scheduling decision epoch, the scheduled users transmit (receive) data in fixed time intervals called subframes. In LTE, a subframe has a duration of 1 ms, while in WCDMA, a subframe lasts for 2 ms. Each transmitted subframe is acknowledged (ACKed) in a subsequent subframe. In FDD-LTE, the ACK for subframe k is sent in subframe $k+4$. Since the subframes need to be decoded before an ACK is sent, the receiver in LTE has 3 ms to decode the subframe, while the receiver in WCDMA has more time to decode the subframes.

Observation 1. *In cellular systems, physical layer processing jobs arrive periodically every T units of time. The deadline d for processing a job is typically larger than T.*

2.2 OpenAir Project

We use an open source implementation of LTE called OpenAir [5] to demonstrate the CloudIQ framework that virtualizes base stations. This was developed by the OpenAir project and is a 3GPP compliant 5 MHz LTE implementation. We choose an implementation of LTE over readily available implementations of WiFi (for example, [21]) because our primary goal is to virtualize cellular systems. Also, cellular systems like LTE have a synchronous deterministic structure, while WiFi is asynchronous and data packets can be transmitted or received at arbitrary times.

2.2.1 *Profiling of OpenAir*

In this section, we want to compute the processing load offered by every function in the LTE physical layer and characterise the parameters that influence the load. A block diagram of the physical layer in a LTE base station is shown in Figure 1. We refer the interested reader to [1] for a detailed description of the various blocks in the figure. The key observation from this figure is that the Fast Fourier transform (FFT) and inverse FFT (IFFT) functions on the downlink and uplink chains respectively are performed in each subframe and are independent of the number of allocated PRBs or MCS. These functions impose a constant base processing load on the system. All the functions subsequent to the FFT or the IFFT depend on PRBs and the MCS allocated to users and the channel conditions.

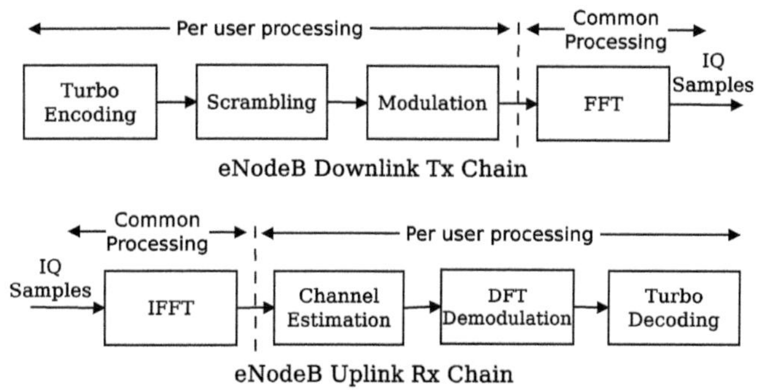

Figure 1: LTE base station PHY processing blocks

We wish to profile the physical layer of LTE to understand the relationship between the processing load and number of allocated PRBs and MCS. We study this relationship by scheduling a single user for transmission or reception. As we see later, the profiling results allow us to generalize the observations to multiple users. We profile the physical layer of the OpenAir implementation of LTE by varying the following parameters: *(i)* the number of PRBs used, *(ii)* the MCS, and *(iii)* the SINR for a given MCS value[1]. We also calculate the time taken for execution by varying these parameters for the downlink and the uplink scenarios.

Based on the profiling results in Figure 2, we make the following observations.

1. *The uplink dominates the downlink processing load and is about 2.5 times the downlink processing load for a given MCS, see Figure 2(a).* As shown in Figure 2(c), the uplink processing of a subframe is largely dominated by the turbo decoder, while the remaining functions offer a constant load. The turbo decoder is an iterative algorithm where every iteration improves the decoder's estimate of the transmitted codeword (see [8] for details). We can see from Figure 2(b) that there is no function in the downlink that matches the turbo decoder in processing complexity.

2. *The processing load is well approximated as a linear function of the MCS and channel resources (PRBs), see Figures 2(b), 2(c), and 2(d). The load is divided into two parts; a constant part (base load) that is independent of MCS, PRB,*

[1]Processing time of the turbo decoder depends on the number of iterations, which is in turn a function of the channel conditions. Hence, we choose the operating SINR for a MCS so that turbo decoder converges within a bounded number (7) of iterations.

and SINR[2]*, and a dynamic part that varies linearly with the MCS and PRBs.*

3. *Since the dynamic load is linear (in fact, convex) in the MCS and the PRBs, if multiple users are simultaneously scheduled for transmission or reception, then the total processing load is given by the sum of the individual loads and cannot exceed the processing load of a single user who is allocated all the PRBs at the highest MCS.*

4. *We can generalize the observations of the LTE profiling results to other cellular standards which use different channel resources but share other components of the physical layer. This applies to WCDMA where the channel resources are spreading codes, but the MCS and the turbo encoding/decoding blocks are similar to those in LTE.*

3. ANALYSIS OF REAL-WORLD DATA

In current cellular systems, the coverage area of an operator consists of many *cell sites* and each cell site is further subdivided into three equally sized *sectors*. Each sector is assigned a set of *carriers* for transmission, where a carrier corresponds to a band of frequencies in the spectrum assigned to the carrier. There is a limit on the number of users that can be supported by a carrier and sectors that experience high load are assigned multiple carriers. Therefore, a cell site can consist of many *sector-carriers*. In the rest of the section, we refer to a sector-carrier with the more commonly used term; *base station*.

In the traditional setup, each base station is provided with computing resources for performing the signal processing operations in cellular communications. This critical infrastructure is called the baseband processing unit (BBU). A cell site with multiple base stations has a separate BBU dedicated to each base station. The BBU is provisioned to handle the peak load in a base station. Our goal in this paper is to break away from this model and study the advantages of *pooling computing resources across base stations*. We allow sharing of a common pool of homogeneous compute resources across base stations and facilitate multiple base stations to perform their signal processing operations in a single computing resource. Since CloudIQ differs from the traditional approach, we do not refer to the compute resource as a BBU to avoid confusion. In this section, we use logs of real-world traffic to characterise the processing load distribution at a base station. Then, we study the potential savings in the number of compute resources in the CloudIQ architecture as a function of the processing load distributions.

3.1 Real-world cellular traffic

We obtained detailed logs of traffic in a WCDMA network from a cellular operator across 21 cell sites in a dense urban setting. We choose WCDMA over LTE, because WCDMA is widely deployed while LTE adoption is still in its infancy. Hence WCDMA logs are more likely to reflect true traffic patterns. We use these logs to estimate processing loads at base stations and the effect of resource pooling across different base stations. We analyze only processing time distributions for downlink traffic, i.e., transmissions from the base station to the mobile user. The logs had insufficient information on the traffic in the uplink direction.

[2]Independence of the base load from SNR is not shown in the plots in Figure 2 due to lack of space.

Each of the 21 cell sites has 4 to 12 sector-carriers (base stations). In total, we have 175 base stations spread over the cell sites. A carrier in WCDMA corresponds to a bandwidth of 5 MHz. The data is aggregated across 15 minute intervals. This is the minimum granularity at which we obtained logs. In WCDMA, there are 450000 sub-frames in a 15 minute interval and there are a total of 15 spreading codes which are shared between voice and data traffic. A spreading code cannot be simultaneously used for voice and data. The logs are solely based on mobile data usage and not voice.

The logs give us the following three pieces of information for each 15 minute interval:

- Total downlink data sent from the base station to the mobile users, in bits.
- Number of sub-frames in which each QAM modulation (QPSK, 16-QAM, or 64-QAM) is used in downlink.
- The number of spreading codes *available* for data traffic in each scheduling interval. Note that all the available codes *need not* be used for data transmission.

We face a challenge in using this aggregate logs at each base station to estimate the processing load distribution in a subframe. This distribution can be potentially different across time intervals (e.g., peak hours in the morning vs. late in the night). Our goal is to obtain a conservative estimate of the processing load distribution and avoid overstating benefits of the CloudIQ framework. In order to derive these estimates, we use the following steps:

1. We use the logs available on QAMs to estimate the probability that a particular QAM is used in a subframe.

2. Similarly, we use the logs on spreading codes *available* to estimate the probability that a certain number of codes are available in a subframe.

3. Next, we use the observation in Section 2 that the processing load is linear in the MCS and channel resources to estimate the distribution on the processing load. From the previous step, we know that this estimate is in fact an overestimate of the processing load since all the available codes need not be used for transmission.

4. In order to fix the issue of overestimating the processing load, we use the logs on the total traffic in the downlink direction in a 15 minute interval. Now, with the computed distributions of QAMs and available codes, we estimate the total traffic in the same 15 minute interval. We compare the estimated traffic with the actual traffic and use the difference between them to scale down the distribution on the codes available per subframe in the 15 minute interval. This further leads to a scaling down of the processing load distribution.

The details of this are outlined in Section 3.2.

In the steps outlined above, we make a crucial assumption that the distributions of QAMs and channel resources used in 5 MHz WCDMA is same as that in 5 MHz LTE. Although there are a few salient differences between WCDMA and LTE in terms of the sub-frame durations and the channel resources, our assumption is justified for the following reasons. The distribution of QAM modulations is a function of the relative location of the users with respect to the base station. Users close to the base station experience a good

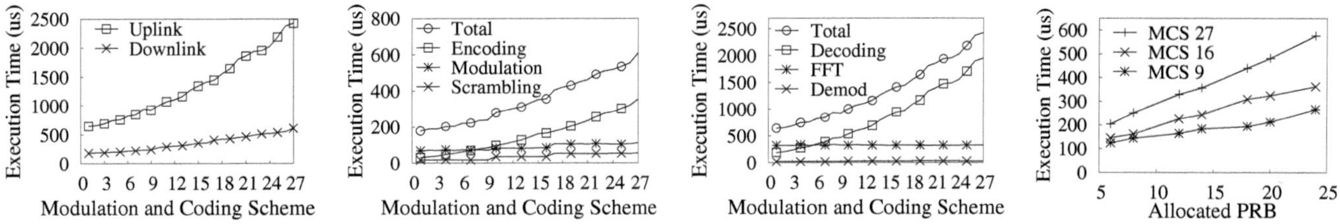

(a) Comparison of uplink and downlink processing times

(b) Downlink processing times

(c) Uplink processing times

(d) Downlink processing times as a function of PRBs

Figure 2: Results of profiling OpenAir code.

channel and observe high signal-to-noise-ratio (SNR), and are assigned higher order QAM modulations. Users experience signal degradation as they move away from the base station and will experience a reduction in the SNR. Consequently, they will be assigned lower order QAM modulations by the scheduler. Furthermore, the distribution of the user locations in LTE will be similar to the distribution for WCDMA as long as the cell sites and coverage areas remain the same. Hence, the same probability distribution of QAM modulations applies to both WCDMA and LTE. Similarly, the distribution of the codes across a 15 minute interval is a function of the data requirements of the users and their geographical locations. The data requirements of users may increase in LTE, but the distribution of the PRBs assigned to users in LTE will resemble the distribution of codes in WCDMA (although the absolute numbers might be higher in the next generation technology). This is because the objective of the scheduler in the base station in both the technologies is to ensure proportional fairness in the allocation of channel resources to users.

3.2 Distribution on the QAMs and codes

Since we are given the number of subframes in which each QAM modulation is used in a 15 minute interval, and there are 450000 subframes in a 15 minute interval, it is easy to compute the probability that a particular QAM is used in a subframe. Note that there are empty sub-frames where no QAM modulation scheme is used (and hence, no data was transmitted). We denote the distribution of QAMs as $(p_{\text{no-QAM}}, p_{\text{QPSK}}, p_{\text{16-QAM}}, p_{\text{64-QAM}})$. The distribution of QAMs is skewed in favor of QPSK (roughly 90%), with 16-QAM (roughly 9.9%) used less often, and 64-QAM used rarely (0.1%). Therefore the distribution of QAMs is already indicative of the fact that the typical processing load is far less than the peak processing load.

Similarly we can determine the distribution of the spreading codes available for data traffic and denote it by $(p_{\text{codes0}}, p_{\text{codes1}}, \ldots, p_{\text{codes15}})$. Recall that the log contains the codes available in each scheduling interval and can be a significant overestimate of the actual distribution of spreading codes used for transmission.

3.3 Computing distribution of processing load

In Section 2, we observed that the processing load can be split into two components: base processing load and dynamic processing load. The dynamic processing load is a function of the actual number of channel resources used in a sub-frame and MCS assigned to each channel resource. We assume that for each modulation, the highest coding rate is used in order to maximize the processing load for that modulation scheme. This also allows us to make a conservative estimate on processing loads. The corresponding MCS values for QPSK, 16-QAM and 64 QAM are 9, 16, and 27.

The number of bits transmitted per channel use under these three modulations is 2, 4, and 6 bits respectively. We translate these to workloads by assuming that 16-QAM (MCS = 16) takes twice as much dynamic processing load as QPSK (MCS = 9), and 64-QAM (MCS = 27) takes thrice as much effort as QPSK. This observation is also validated in Figure 2(b), where we can compare the processing loads for the dynamic part corresponding to turbo encoding.

We assume that QPSK modulation transmitted over one channel resource consumes 1 unit of dynamic processing load. Now, we can easily derive the processing load in any scenario. For example, 16-QAM with 4 codes will consume $2 \times 4 = 8$ units of dynamic load, while 64-QAM with 7 codes will consume $3 \times 7 = 21$ units of dynamic load. In order to compute the base processing load, we refer back to Figure 2(b) and observe that the intercept of the total load on the y axis corresponds to dynamic load at MCS = 16. By the earlier arguments, the base processing load will then equal the load of 16-QAM with 15 codes, i.e., $2 \times 15 = 30$ units. The total load for any allocation of QAMs and codes is determined by adding the base and the dynamic components and can be described by a simple formula:

$$\text{Total load} = \underbrace{\frac{(\log_2 \text{QAM})}{2} \times \text{Number of codes}}_{\text{dynamic}} + \underbrace{30}_{\text{base}}, \tag{1}$$

where QAM can take the value 4, 16, or 64 depending on whether QPSK, 16-QAM or 64-QAM was chosen. With the above formula, we can compute a distribution on the *estimated* processing load as a function of the distribution of QAMs and available codes by assuming that the choice of QAM modulations is independent of the number of codes. Since there are 3 choices for QAM modulations and 15 choices of codes, we have 45 combinations of MCS and channel resources used, with an extra option when no data is transmitted. We denote the tuple of 46 values as $(p_{\text{est}}(0), \ldots, p_{\text{est}}(45))$.

3.4 Scaling down the processing load

We first estimate the total data traffic in a 15 minute interval with the distributions on the QAMs and codes available and show that this exceeds the actual downlink traffic. Pick a 15-minute interval for a base station. With the distribution on the QAMs, we can compute the average number of bits/sub-frame/channel use, denoted by $\bar{Q}$, as

$$\bar{Q} := 2 \times p_{\text{QPSK}} + 4 \times p_{\text{16-QAM}} + 6 \times p_{\text{64-QAM}}. \tag{2}$$

Here, we use the fact that QPSK, 16-QAM, and 64-QAM can send 2, 4, and 6 bits of data per channel use respectively. Next we need to determine the number of channel uses from the distribution on the codes. First calculate the average number of codes in a sub-frame as

$$\bar{C} := 0 \times p_{\text{codes0}} + 1 \times p_{\text{codes1}} + \ldots + 15 \times p_{\text{codes15}}. \quad (3)$$

We assume that each of the 15 spreading codes occupies an equal portion of the available bandwidth, implying that each code corresponds to 5/15 MHz. A sub-frame lasts for a duration of 2 ms and the number of channel uses (s-Hz) corresponding to one spreading code in a sub-frame is

$$U := (2 \times 10^{-3}) \times (5/15)\, 10^6 \text{ (s-Hz)}. \quad (4)$$

Hence, from (3) and (4), the average number of channel uses in a sub-frame in the given 15 minute interval is

$$\bar{C} \times U. \quad (5)$$

From the above equation and (2), the average number of bits per sub-frame in the given interval is

$$\bar{Q} \times \bar{C} \times U \text{ (bits / sub-frame)}. \quad (6)$$

In a 15 minute interval, there are 450000 sub-frames, and the total number of bits transmitted in the interval is given by

$$D := \bar{Q} \times \bar{C} \times U \times 450000 \text{ bits}. \quad (7)$$

Hence D is the estimated downlink traffic in the chosen interval. The actual downlink traffic in the same interval is available as part of the aggregate logs and we compare it with D in figure 3. As we can see, the estimated downlink

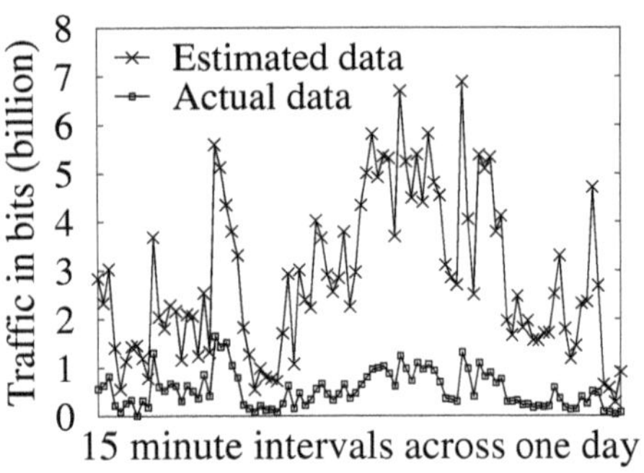

Figure 3: Comparing estimated with actual data.

traffic exceeds the actual downlink traffic. We compute the ratio $\kappa := \frac{D}{\text{Downlink traffic}}$ to capture the extent to which the estimate exceeds the actual data. The ratio κ is unique to a particular interval for a base station. We repeat this procedure across all the base stations, across all the 15 minute intervals, to compute the corresponding ratios. Clearly, we can use this information to get a better estimate of the processing load distribution. However the choice of how scaling should be done is not so straightforward.

For each 15 minute interval, κ is the scaling factor for the *average* processing load or the *first moment* of the processing load. However to compute the true processing load distribution, we will require to know the scaling factor for all the *moments* of the distribution. The scaling factor for the remaining moments is however impossible to compute with the available data. Hence there are an infinite number of processing load distributions whose average coincides with the scaled average computed earlier. Next we make a conservative estimate on the distribution. Note that the expected *estimated* processing load is given by $E_{\text{est}} = \sum_{j=0}^{45} j p_{\text{est}}(j)$. We can scale $\{p_{\text{est}}(j)\}$ in two natural ways to ensure that the new average processing load is $\frac{E_{\text{est}}}{\kappa}$. We could either scale each value of j by κ or we could scale each value of $p_{\text{est}}(j)$ by κ. Clearly scaling each processing load value j by κ will imply that the processing load never exceed $\frac{75}{\kappa}$, where 75 is the maximum processing load. This results in an optimistic estimate of the resource pooling gains. Instead, to obtain a conservative estimate, we scale the probabilities as $p_{\text{new}}(j) = \frac{p_{\text{est}}(j)}{\kappa}, j = 1, \ldots, 45$ and set $p_{\text{new}}(0) = 1 - \sum_{j=1}^{45} p_{\text{new}}(j)$ to ensure that we get a valid probability distribution.

Table 1: Reduction in aggregate computing load.

Prob. of failure (P_F)	utilization (%)	savings (%)
10^{-2}	59.99	40.01
10^{-4}	73.76	26.24
10^{-6}	76.74	23.26
10^{-8}	77.56	22.44

3.5 Gains from resource pooling

Next, we use the distribution on the processing load to calculate the aggregate computing load for base stations and check if there is a reduction in the aggregate load due to variability of the traffic across base stations. We focus on the busy hours in a day, which are from 0800 hours to 2200 hours. As mentioned before, in the traditional setup, the computing resource assigned to a base station is provisioned to handle a peak load when 64-QAM is the chosen modulation scheme and 15 codes are used for downlink transmission. This translates to a total load of 75 units. Hence, the resource can handle a peak load of 75 units in every sub-frame. As a baseline for comparison, we consider the peak load for 175 base stations, which is a total of $175 \times 75 = 13125$ units. Next, we fix a target probability of failure P_F and consider the distribution of processing loads for the base stations computed in the previous section. For each base station b, for each 15 minute interval t, we pick the maximum processing load $L_b(t)$ such that the processing load is guaranteed to be less than $L_b(t)$ for the chosen interval with probability at least $1 - P_F$. We calculate the aggregate processing load for the interval t as the sum of $L_b(t)$ for all base stations. We compute the ratio of this aggregate load with the peak load of 13125 units to compute the percentage utilization of resources when we permit statistical guarantees. We compute this ratio for different values of P_F and present the results in Table 1. We also compute percentage savings from resource pooling by subtracting the percentage consumption from 100.

We also plot the percentage utilization as a function of 15 minute intervals for the busy hours in a particular day in figure 4. As we can see from Table 1, the percentage savings

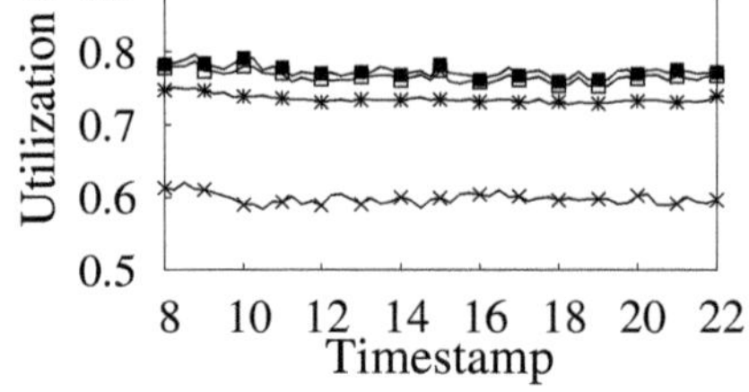

Figure 4: Utilization in busy hours.

monotonically increase with the probability of failure. Even with a very low probability of failure of 10^{-8}, we are obtain potential savings of 22.44 %. Our observations motivate us to investigate the design of virtualization mechanisms that allow multiple base stations to be pooled onto a single compute platform. Since we have made conservative estimates in calculating the processing load, we could potentially observe larger gains in practical systems.

REMARK 1. *We are modeling the cellular traffic with a probability distribution that we computed earlier, but we cannot validate this distribution based on the available real-world data. We have observed periodic patterns while computing the averages of various parameters in the available data, but extending this observation to higher order moments and to the distribution needs more fine grained data.*

4. CLOUDIQ FRAMEWORK

In this section we design the CloudIQ framework that allows pooling of compute resources across multiple base stations. Each compute resource consists of a set of identical cores. We do not assume the availability of specialized hardware as part of the platform like accelerators for computing FFT or for performing turbo decoding. In the previous section, we obtained bounds on the potential savings in the compute resources that can be obtained if we allow for statistical guarantees for the processing at base stations. In this section, our goal is two-fold: (*i*) partitioning the set of base stations into groups that are simultaneously processed on a single platform for a given statistical guarantee P_F, and (*ii*) scheduling the set of base stations allocated to a platform in order to meet their real-time constraints.

There is a wealth of literature on scheduling of real-time systems, which includes scheduling on multi-core platforms, see Section 7 for related work. But most of the work considers systems with *implicit deadline* where the deadline coincides with the periodicity of the process. This is unlike cellular systems where the deadline to process a subframe (3 ms for LTE) exceeds the periodicity (1 ms for LTE) of the subframe. Furthermore, we adopt a *separation principle* in the design of the system. We initially provision the system to handle the extreme scenario where every base station has the maximum processing load ($L_{\max}$) [3]. As we see later, this design leads to a simple solution of the above-mentioned partitioning problem with statistical guarantees. We use a standard bin-packing algorithm to pack base stations into bins of size $L_{\max}$. As a consequence of this design, the real-time processing requirements of the base stations are automatically guaranteed in the framework. This separation of the partitioning and scheduling problems is a robust and simple framework for critical real-time systems. To the best of our knowledge, such a separation is not possible under any of the existing solutions for real-time systems.

As a result of adopting the separation principle, we make the following design choices:

1. **One subframe one compute resource:** We assume that a subframe of a base station is completely processed on a single compute resource and meets its processing deadline requirement even at peak load. Different subframes of the same base station can be potentially scheduled to different compute resources. A compute resource is a logical entity and can consist of multiple physical compute resources. For example, since functions such as turbo decoding are extremely compute intensive, they may be parallelized to run across multiple physical cores.

2. **Offline schedule:** In order to adhere to the separation principle, under statistical guarantees, we allocate a certain set of base stations to a platform based on their individual processing loads. Once allocated to a platform, an offline algorithm determines the schedule of processing the subframes of the base stations and does not further optimize based on variations in the processing loads.

REMARK 2. *In this paper we do not explicitly consider signal processing techniques such as coordinated multi point scheduling (CoMP) which requires joint processing across multiple base stations. Typically however CoMP is performed across adjacent base stations and the techniques we describe here can be easily generalized to these scenarios also by constructing suitable "virtual base stations" comprising multiple physical base stations.*

4.1 Resource Management Algorithm

We develop a resource management framework that solves the partitioning and the scheduling problems mentioned above. First, we develop an algorithm to schedule base stations at peak load on a platform. This provides us with a natural framework to apply the separation principle to solve the partitioning problem. Next, under this framework, we develop an algorithm to schedule base stations with statistical guarantees. This is a standard bin packing algorithm that ensures that the base stations meet their processing deadlines.

4.1.1 A cyclic schedule

We develop a strategy that cyclically schedules base stations on a multi-core compute platform. Consider a compute platform with N cores, where the cores are ordered as $1, \dots, N$. Consider a set of M base stations ordered as $\{B_1, B_2, \dots, B_M\}$. The subframes at a base station arrive with a known periodicity and need to be processed within a deadline. Assume that the base stations are operating at peak load and the subframes take the maximum load $L_{\max}$ to complete processing. For the sake of exposition, let us assume that the periodicity is 1 unit while $L_{\max}$ can exceed 1. Let us view the subframes arriving for processing at a base stations as jobs and denote the j-th base station's job that arrives at time instant t as $B_j(t)$. The cyclic schedule is simple: *at time t, assign the job from the j-th base station, i.e., $B_j(t)$, to core $(tM + j) \mod (N) + 1$.*

There are two questions that immediately arise: how do we guarantee that the deadline requirements of the base stations are met and what is the maximum number of base stations that can be scheduled on a single processor. We will answer the questions next, but first we illustrate some of the properties of the cyclic schedule with an example. In the example in Figure 5, we consider a system with $N = 4$ cores which has to process $M = 3$ base stations. The subframes for each base station arrive at a periodicity of 1 unit and have a peak processing load of $L_{\max} = 4/3$ units. The resulting cyclic schedule for these parameters is shown in Figure 5. In

[3]Note that the time budget for processing is a function of the transmission delay in the network. In a fiber network, the typical transmission delay is 5μs per Km. Since the total time budget is 3 ms, the actual processing time is 3 - transmission delay ms

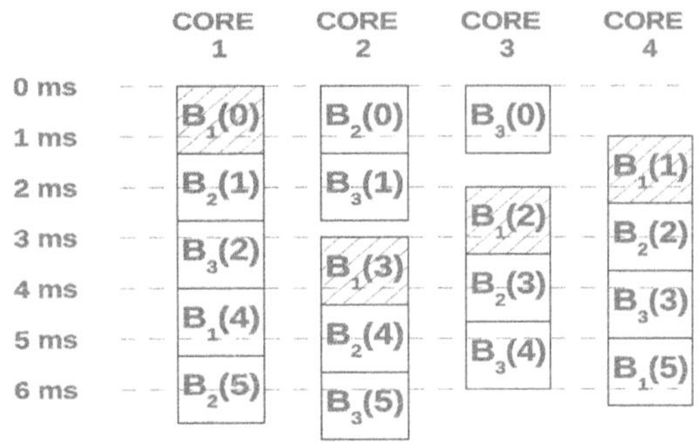

Figure 5: Example with 3 base stations and 4 cores

the figure, the horizontal axis indexes the four cores and the vertical axis indexes the time. We denote subframes by rectangles where the height of a rectangle denotes the processing load of the corresponding subframe (4/3 units).

In this example, observe the following:

- The order of the jobs processed by core 0 repeats every 4 units, it is always subframes of base station B_1 followed by B_2 followed by B_3.
- After an initial transient phase, the same schedule repeats at each core, i.e., each core sees jobs from base stations in the order B_1 followed by B_2 followed by B_3. Moreover, the jobs of a base station experience the same queuing delay for every subframe irrespective of the core they are scheduled on. In this example, the delays are 0, 1/3, and 2/3 ms for base stations B_1, B_2, and B_3 respectively.

The first property holds because $N = 3$ and $M = 4$ are coprime in the above example. If the number of cores M and the number of processors N are not coprime, it is easy to see that the sets of cores and base stations get partitioned so that each subset of cores processes a distinct subset of base stations. Let us assume that M and N are coprime for the sake of exposition. In this case, the multi-core scheduling problem essentially becomes a single core scheduling problem since the same schedule repeats across all the cores.

So far we have not dealt with the issue of synchronization between base stations at subframe boundaries. But we saw with the cyclic schedule that every base station experiences the same queuing delay for all the subframes, and this property can be used to our advantage when all the base stations are not synchronized among themselves. The subframes of different base stations need not arrive at the same time and might be delayed in relation to each other. We can handle this by closely matching the delay in receiving the subframe of a base station with the queuing delay it sees in the cyclic schedule. Revisiting the example in Figure 5, let us suppose that we have three base stations A_1, A_2, and A_3 which are not synchronized among themselves. Let us take the subframe boundaries of A_1 be the reference and assign A_1 to B_1 in the example. Then, either of A_2 or A_3 can be matched to B_2 or B_3 as long as their subframes see a relative delay of less than 1/3 and 2/3 units respectively. In the general case, the challenge is to order the base stations so that the offset in the arrival times of subframes is matched with the queuing delays and the deadline requirement of every base station is met, in other words we need to find the right permutation of base stations[4]. Also, as we later see, we can always ensure that M and N are coprime without much loss in optimality.

4.1.2 Maximizing the number of base stations

Our goal is to find the maximum number of base stations that can be scheduled on N cores, subject to the constraints that *every* subframe has the maximum load $L_{\max}$ and must be processed within a deadline $d > L_{\max}$. If a platform with N cores can schedule M base stations with load $L_{\max}$, then $ML_{\max} \leq N$ since in each unit of time the total load due to all the base stations is $ML_{\max}$. We obtain the upper bound $M \leq \lfloor \frac{N}{L_{\max}} \rfloor$ on the number of base stations. We can obtain a lower bound on M with a greedy provisioning of cores for each base station. We can assign $\lceil L_{\max} \rceil$ cores to each base station and one-by-one send subframes from the base station to each of the $\lceil L_{\max} \rceil$ cores. Clearly the deadline constraint of every base station is met with this assignment. Hence, we obtain the lower bound $M \geq \lceil \frac{N}{\lceil L_{\max} \rceil} \rceil$.

As mentioned before, if M and N are coprime, it is sufficient to analyze the schedule on a single core. We solve the scheduling problem by focusing on core 0. Let t_k denote the arrival time of the subframe for the k-th base station at core 0 and let s_k denote the delay it experiences before getting processed. As seen in the above recursion, the delay experienced by the k-th base station's subframe is 0 if the previous subframe corresponding to the $k-1$-th base station was processed by t_k. Otherwise, the delay induced due to the previous job is given by the difference between the time when the $k-1$-th job finishes and t_k. With a recursive computation, the $k-1$-th job finishes by $t_{k-1} + s_{k-1} + L_{max}$. Hence the delay experienced by the k-th base station is:

$$s_k = \max(0, s_{k-1} + L_{\max} - (t_k - t_{k-1})), \quad k = 1, \ldots, M-1 \tag{8}$$

$$s_0 = 0. \tag{9}$$

The constraints that have to be satisfied at each core under our framework can be written as follows:

$$\begin{aligned} \text{Maximize} \quad & M \\ \text{subject to} \quad & s_k + L_{\max} \leq d, \\ & s_{M-1} + L_{\max} \leq t_M - t_{M-1}, \\ & \lceil \frac{N}{\lceil L_{\max} \rceil} \rceil \leq M \leq \lfloor \frac{N}{L_{\max}} \rfloor, \\ \text{where} \quad & t_k = \lfloor \frac{kN}{M} \rfloor, M \in \mathcal{Z} \end{aligned}$$

where the fist constraint ensures that the deadline constraint is met for every base station. The second constraint ensures that the last base station B_M completes processing before the first base station B_1 in the the next cycle arrives. The last constraints are the previously obtained bounds on M.

It is easy to solve this optimization. We choose M to equal the upper bound on the number of base stations. We check if the constraints are satisfied for this choice, else we reduce M by 1 and repeat the check. We stop at the first case when all constraints are met and declare it to be the maximum number of base stations that can be accommodated. If the choice of M is not coprime with N, then we further reduce M till the result is coprime with N. The probability that both M and $M-1$ are not coprime with N is less than 2% for $M, N \leq 1000$, hence the loss in optimality is very low. Hence we can always ensure that M and N are coprime.

[4]Since the duration of each subframe is 1 ms, the relative delay between the arrival of subframes of two base stations is bounded by 0.5 ms.

4.1.3 Resource Pooling

Recall that in order to enable a centralized RAN, we have a two-fold goal of partitioning and scheduling base stations with statistical guarantees. In Section 4.1.1, we developed a framework to schedule base stations on a multi-core platform and computed the maximum number M of base stations that a platform with N cores can support. In this section, we solve the problem of partitioning the base stations into subsets, each of which can be scheduled on an N core platform. Each subset can contain more than M base stations. We will see that the cyclic schedule allows us to apply the separation principle and deadlines for all the base stations can be met with no additional effort.

We assume that the CDF of the processing load for base station B_i at time t is known and denoted by $F_{i,t}(x)$. Let P_F be the chosen statistical guarantee or the probability of failure. We compute the processing load $L_{i,t}(P_F)$ such that the probability that the base station's actual load exceeds $L_{i,t}(P_F)$ is less than P_F:

$$L_{i,t}(P_F) := \sup_{x:x\leq P_F} \{F_{i,t}^{-1}(x)\}, \forall\, i, t. \tag{10}$$

Next, we need to partition this set of $\{L_{i,t}\}_i$ for a given t, into smaller sets that can be scheduled on a GPP platform. Since each of the subsets is scheduled on a separate platform, we want to minimize the number of subsets that we create. We adopt the following approach for it: recall that in Section 4.1.2, we determined the maximum number of base stations with a processing load $L_{\max}$ that can be scheduled on a GPP platform. Conversely, we can determine the maximum feasible processing load $L_{\max}(M)$ for a given choice of M. Now we have a set of tuples for an N core platform

$$\{(M_1, L_{\max}(M_1)), (M_2, L_{\max}(M_2)), \ldots, (M_K, L_{\max}(M_K))\}, \tag{11}$$

where M_k is the maximum number of base stations that can be scheduled on a single GPP platform and $L_{\max}(M_k)$ is the corresponding maximum processing load.

Now consider a maximum processing load of $L_{\max}(M_k)$. We say a subset $\mathcal{J}$ of base stations $\{B_j\}_{j\in\mathcal{J}}$ is feasible for this load if the sum of the individual base station loads corresponding to a statistical guarantee P_F is less than $L_{\max}(M_k)$, i.e., $\sum_{j\in\mathcal{J}} L_{j,t}(P_F) \leq L_{\max}(M_k)$. These base stations can be combined together to create a *super* base station whose processing load is bounded by $L_{\max}(M_k)$. Now, we can schedule M_k super base stations on the GPP platform.

Our goal is to minimize the total number of platforms for scheduling the base stations. Instead, we partition the base stations into the minimum number of subsets, where each subset can be viewed as a super base station corresponding to a processing load $L_{\max}(M_k)$ for some k. Then, we group M_k of the super base stations corresponding to processing load $L_{\max}(M_k)$ into a single platform. We perform cyclic scheduling over the super base stations in order to process their subframes. If a super base station is scheduled for processing, then the subframes of the individual base stations constituting the super base station are processed in order. If the processing time of the subframe of base station B_j exceeds the allocated time $L_{j,t}(P_F)$, then the subframe is dropped from the queue. This approach represents the separation principle and the delay guarantees for each base station is met by virtue of the cyclic scheduling framework.

To compute the minimum number of super base stations, we note that each choice of processing load $L_{\max}(M_k)$ for some k can be viewed as a bin of size $L_{\max}(M_k)$. Hence, we have a variable size bin packing problem where the set of base stations with loads $\{L_{i,t}(P_F)\}_i$ have to be filled into minimum number of bins, where the size of the bins is from the set $\{L_{\max}(M_k)\}_{k=1}^{K}$. For certain mild assumptions satisfied here, the iterative first fit decreasing algorithm guarantees a $\frac{3}{2}$ approximation to the bin packing problem [17]. Since we group M_k of the bins (or super base stations) corresponding to $L_{\max}(M_k)$ into a platform, we might further step away from optimality if the number of bins of size $L_{\max}(M_k)$ is not a multiple of M_k. Firstly, this loss is bounded by K, which is the number of choices for bin size. Secondly, this loss can be further reduced by carrying out a second round of optimization on the platforms that are not fully occupied.

4.2 Simulation results

In Section 3, we evaluated the possible gains in resource pooling by analysing real-world data. We use the same data to evaluate the performance of our architecture. We consider two scenarios. First, we study the gains from resource pooling at each cell site by calculating the total number of compute resources required to provide a certain statistical guarantee. We compare this against the base case where each cell site is provisioned to ensure that all base stations with peak load are processed successfully. The second case is when the resource pooling for all base stations is done at a central location for a certain statistical guarantee. We compare this against the case where the central site is provisioned to process all the base stations at peak load. We study both these scenarios for the busy hours of the day which range from 0800 hours to 2200 hours. Clearly the gains of resource pooling will be higher in the second case than in the first case. But this comparison will allow cellular operators to analyse the cost-benefit trade-off of resource pooling at a central cloud versus performing it at the cell sites.

We report results based on provisioning for a five core platform. The choice of a five core platform is motivated by the Intel processor on which we implement our design and this will explained in detail in Section 5. The five cores can be provisioned to handle $2, 3, 4, 5$ or 6 base stations[5]. The corresponding $L_{\max}$ for each of these choices is given by $5/M_i$, i.e., if we provision 4 base stations, then $L_{\max}(4) = 5/4 = 1.25$. We assume that each base station has a peak load of $5/3 = 1.66$ ms, which means that for providing hard guarantees, we can provision 3 base stations in a single platform. We made this choice for two reasons: firstly, three base stations correspond to three sectors in a cell site. Secondly, we ensure that the utilization of all the cores in the platform is 100% for the base case with peak load. This gives us conservative estimates of the pooling gains from bin packing for the case with statistical guarantees.

In this setup, for the base case of scheduling 175 base stations at peak load, we require 63 platforms at the cell site and 59 platforms at the central location. Note that more compute resources are required at the cell site since one of the platforms at a cell site can be over provisioned, i.e., it processes fewer than the number of base stations it is provisioned for. The results for resource pooling gains under statistical guar-

[5]We limit the maximum number of base stations that can be scheduled on a compute resource since we do not factor all the overheads associated with scheduling multiple base stations like limits of the network interfaces which might constrain our options.

Table 2: **Reduction in load at the cell site vs. central location. The baseline for comparison is 59 platforms for central location and 63 for the cell site.**

Prob. of failure (P_F)	Number of Proc. at Cell Site (%Savings)	Number of Proc. at Central Location (% Savings)
10^{-4}	45.39 (27.95%)	44.51 (24.55%)
10^{-6}	49.03 (22.17%)	46.86 (20.57%)
10^{-8}	52.41 (16.8%)	47.49 (19.5%)

antees are reported in table 2. We obtain them by computing the number of processors required in each 15 minute interval and average the results. We observe from table 2 that most of the gains under resource pooling are obtained at the cell site itself. However, there might be other benefits to perform resource pooling at a central location like substantial savings in operations and management and capacity gains via advanced signal processing techniques involving cooperative transmissions/receptions from neighboring cell sites. Further we plot the number of total processors required as a function of time for a statistical probability of failure of 10^{-8} in figure 6.

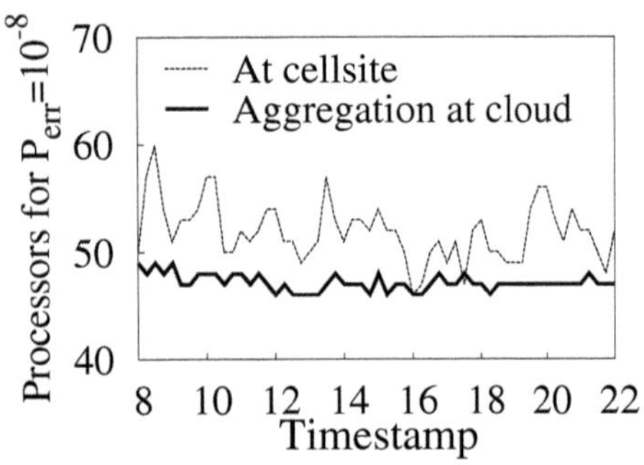

Figure 6: Number of processors required for aggregation at cell site or cloud at 15 minute intervals.

5. SYSTEM DESIGN

In this section we summarize our design exercise for implementing a multi-threaded LTE base-station. As mentioned earlier in Section 1, the basic building blocks of an LTE base station in our framework are ported from OpenAirInterface project [5]. The OpenAir project has developed a 3GPP compliant 5 MHz LTE implementation.The implementation executes each subframe in a single threaded fashion, i.e., it does not separate the user processing onto different threads as mentioned earlier. Furthermore it also runs the entire base station as a single thread and does not execute different subframes on different cores.

We implement our solution on an Intel Xeon W3690, with 6 physical cores and each core running at 3.47 GHz, 12 MB shared L3 cache and 256 KB L2 cache. Our platform uses Linux kernel 2.6.31 [9] with RT preemption patch (PREEMPT_RT) [4] whose default scheduling mechanism is Round-Robin [12]. But for better time predictability the threads can also be launched with FIFO scheduling option [12] and same *real-time priority* to prevent any preemption. We now describe some of the implementation challenges we faced and some design choices we made to overcome these issues.

Resident threads on cores: As explained in Section 4, our resource management framework dispatches different subframes of a single base station to different cores. Furthermore since LTE subframes arrive every 1 ms and the processing time for a subframe can exceed 1 ms, a base station can be concurrently processing multiple subframes. This implies that we require a multi threaded implementation for each base station that ensures that subframes can be processed concurrently without any conflicts. The central idea is to leverage the thread level parallelism (i.e. task parallelism) on a multicore general purpose processing platform to manage parallel processing modules of an LTE base station. There can be two approaches to this problem. First approach is to create a new thread for each subframe that has to be processed and assign this thread to a core according to the resource management algorithm (see Section 4. Although simple in design, this framework of transient threads has stability issues due to constraints of the underlying Linux kernel. We verified through experiments that creation and deletion of multiple threads at timescales of milliseconds overloads the system within a short span of 100 milliseconds. This is because the kernel is not able to free up resources at such timescales. Hence, we adopt the second approach of making each thread resident to a processor core and notifying the thread dynamically when a subframe needs to be processed.

Figure 7 explains our framework by showing how subframe processing jobs are launched in our system. For simplicity, we show only three cores out of which core 0 is always running the resource management algorithm (denoted by RM) and core 1 and 2 are dedicated for running LTE subframe processing. We assume that we are managing only two base-stations i.e. the RM process will launch two subframe processing jobs every millisecond. We also assume that with current processing loads, subframe processing for the first base-station (denoted by $BTS1$) will finish within a millisecond but the subframe processing for second base-station (denoted by $BTS2$) will not. Figure 7(a) shows that before time instance ($t = 0$) the threads running on core 1 and 2 are idle. In Figure 7(b) at time instance ($t = 0$), RM launches 2 subframe processing jobs to each of these core (denoted by solid arrow) and signals the idle threads to resume processing. In Figure 7(c) at time instance ($t = 1$), core 1 has finished the subframe processing of $BTS1$ but core 2 is still running the subframe processing of $BTS2$ and RM launches the next set of subframe processing jobs to these cores. At core 1, the new job is started immediately but at core 2 the job is enqueued (denoted by dashed arrow) after the previous job. Every resident thread maintains a *job-queue* where new jobs are enqueued by the RM. If the queue gets empty, the corresponding thread goes to sleep and the RM needs to wake-up the thread when a new job is dispatched to this thread.

FIFO order execution: Recall that the default scheduling mechanism of our Linux kernel is Round-Robin on each core. If two or more tasks are dispatched to a core and they are scheduled in Round-Robin fashion, it is difficult to predict their turn around times even if the individual processing loads are known. Therefore, we decided to incorporate FIFO order execution of multiple tasks on the same core. As mentioned earlier, this FIFO scheduling is implemented using an *job-queue* for each thread where new tasks are enqueued by RM. The thread goes to idle state if the queue is empty and needs to be signaled by RM after a new job is enqueued.

Circular shared buffers: To avoid conflicts between concurrent running processes, several shared buffers in OpenAir need to be duplicated (e.g. the IQ sample buffer for base station Rx processing modules). The degree of duplication however can be limited by the maximum number of concurrent

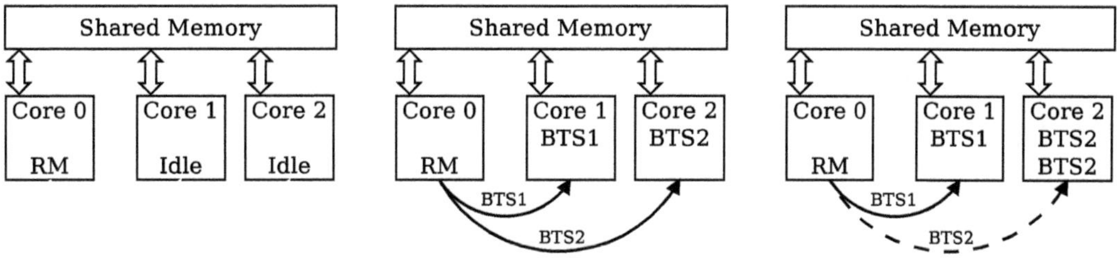

(a) Before Time $t = 0$ (b) At Time $t = 0$ (c) At Time $t = 1$

Figure 7: Resident Thread Framework and Signaling

Table 3: Average uplink subframe processing times. Single BTS and Dual BTS modes refer to one and two base stations scheduled on the processor respectively. Super base station mode refers to two base stations scheduled as a super base station. In Super BTS mode, processing time is the total time to process both the subframes.

Configuration		TBS	Uplink proc. time (ms)		
PRB	MCS	(Bytes)	Single BTS	Dual BTS	Super BTS
12	4	680	0.701	0.700	1.383
12	8	1480	0.794	0.788	1.557
12	12	2408	0.989	0.999	1.966
16	4	904	0.793	0.793	1.568
16	8	1928	0.899	0.911	1.786
16	12	3240	1.209	1.209	2.409
20	4	1160	0.883	0.885	1.756
20	8	2472	1.022	1.038	2.051
20	12	4008	1.526	1.525	3.035

threads in our system. We maintain a circular list of shared buffers and while launching a new task assign the next free block to it and advance in the queue in circular fashion. This helps us to keep the allocated memory to our system as low as possible.

6. EXPERIMENTS

We now present a subset of the experiments we conducted to evaluate our system. We study the effects of sharing the compute resources between multiple base stations and verify that there is negligible deterioration in average processing time of each base station. We also observe that the worst case processing time for a base station can increase if multiple base stations share the compute resource. This effect can be accounted for in the resource management algorithm.

The details of LTE are presented in Section 2 and Figure 2 shows the time taken to process uplink and downlink subframes under different configurations (allocated PRB and MCS). Since the uplink processing is more intensive than downlink (see Figure 2), we focus on the performance of the uplink (base station Rx) processing. All our experiments are done with a single UE per base station and a 2×1 MIMO configuration. Each UE gets connected to a base station and thereafter remains static and does not cause any hand-over or disconnection. After the connection is established, the UE always has a full buffer and all available PRBs in every subframe are used for data transfer. For simplicity and predictability, we override the CQI based MCS calculation in the scheduler and instead choose a pre-defined MCS.

Table 3 compares the uplink processing times for a single base station when either one or two base stations are scheduled on the processor. The total number of data bytes to be transmitted in a subframe is called the transport block size (TBS) and is specified by LTE as a function of the allocated PRB and MCS (as shown in Table 3). The average subframe processing times are calculated over 18000 LTE subframes. Note that the average processing time for a subframe remains almost unchanged irrespective of whether we schedule one or two base stations on the processor. Recall that for providing statistical guarantees, we can combine multiple base stations to form a super base station. Table 3 also includes the processing time for a super base station comprising of two base stations. This is roughly double than that of a single base station. Although, average processing times remain same

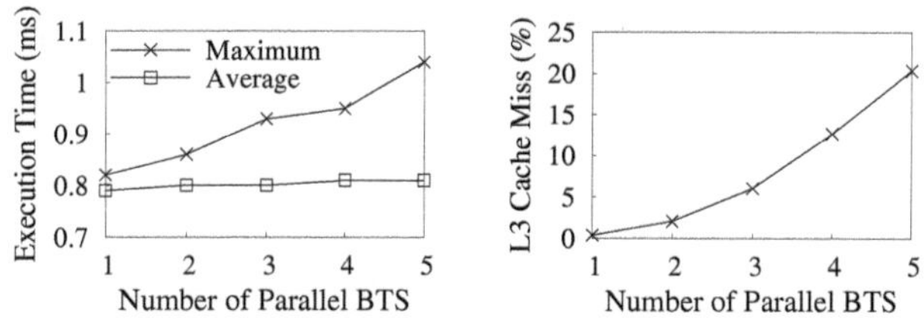

(a) Maximum and Average uplink processing time (b) Percentage of L3 Cache (shared) miss

Figure 8: Effect of shared cache misses

for multiple base stations running in parallel, we observed a steady increase in the maximum subframe processing time as we increase the number of base stations in the system. Figure 8(a) shows the average and maximum subframe (uplink) processing times for multiple base stations. All uplink subframes correspond to 16 PRBs and MCS 4. The average processing time stays almost constant around 0.8 ms but the maximum (worst-case) processing time observed among all active base stations increases from 0.81 ms to 1.04 ms as we increase the number of base stations running in the system simultaneously. We found by further investigations that increased number of shared cache misses are the primary reason behind this increase in maximum processing time. Figure 8(b) shows the percentage of L3 cache misses with increasing number of base stations running in parallel. Note that, with original OpenAir code this effect of shared cache misses on worst-case execution time of a base station was more severe and we were unable to schedule more than two base stations on 5 cores (same configuration). But with careful analysis of cache misses and restructuring of large data structures, we managed to avoid obvious cache-misses in few of the computation-heavy functions (e.g. turbo decoding) which enabled us to schedule as many as 5 base stations. Making all the data structures in OpenAir code cache-conscious [10] or tuning each of the functions to avoid cache misses [13, 14] is beyond the scope of this work but we strongly believe that such exercises will further decrease the gap between average and maximum processing time of multiple base stations.

7. RELATED WORK

We briefly review the literature in the area of real-time systems. Most of the literature in traditional real-time systems considers tasks with an implicit deadline equal to their period [19]. Furthermore, tasks are assigned priorities for completion. For the case of scheduling such tasks on a single processing core (uniprocessor), the Earliest Deadline First (EDF) scheduler is known to be optimal [11]. On multi-core (or multiprocessor) systems, however, more complex schedulers such as the Proportionally Fair scheduler are needed for optimal scheduling [15]. In the tasks of interest to us, the deadline for completion can exceed the time-period. Such systems are called arbitrary-deadline cases and have been studied in the literature [7] [6] [18]. We are interested in scheduling such tasks with arbitrary deadlines on a multi-core system, while ensuring the separation principle. To the best of our knowledge, there are no algorithms in existing real-time literature that satisfy our requirements.

The cellular industry is currently evaluating proposals to transition towards a centralized architecture for baseband processing of the signals, see [2, 3]. Similar ideas for a centralized compute resource were evaluated for distributed antenna systems (DAS) [20], though DAS considered the design of large cells with distributed antennas to provide coverage.

8. REFLECTIONS

This paper is the first attempt at providing a rigorous resource management framework that allows a cellular operator to trade-off between the quaility of the network and the cost of operating the network (in terms of either the energy or the cost of baseband compute resources). However, there are several research issues that must be addressed to engineer such a system. Some of them are outlined below.

Reducing Backhaul Costs: The cost of fiber to backhaul the IQ samples to a central location can be prohibitive. For example, a 20 MHz LTE system requires around 1 Gbps of bandwidth per antenna to backhaul the samples. Hence, 3 sectors with each having a 4×4 MIMO system might require up to 12 Gbps of bandwidth. The cost of laying fiber with such high capacity can be prohibitive. As an alternative, the industry is considering architectures where some of the PHY processing is done at the base station, while the user specific samples are transported to the central site for processing. This has the potential to significantly reduce the backhaul requirements, while still permitting advanced signal processing techniques like CoMP and availing other benefits of the CloudIQ architecture.

Network and Switching Backplane at Data Center: We have primarily addressed the question of how to pool compute resources to minimize the number of required compute resources. We have not addressed the issues of network backhaul and the topology of the data center.

Resource Pooling Time Scales: In this paper, we have naively assumed that the resource pooling time scale is of the order of several minutes. Further we have assumed that the processing load distributions are predictable. An alternate approach would be to see if these decisions can be made at finer time scales and understand the predictability of the processing load distributions at those time scales.

Heterogeneous Systems: We have designed a resource management framework for homogeneous systems. However, currently deployed systems use heterogeneous components like accelerators for compute intensive functions like turbo decoding and GPP's for less intensive signal processing functions. As part of the ongoing work, we are extending the CloudIQ framework to heterogeneous systems.

Acknowledgements

We thank Laurent Thomas and Laurent Roullet for help in setting up OpenAir, and Martin Havemann, Kazukata Murakami, Francis Dominique and Jonathan Segel for useful comments. We would also like to thank the Wireless Business Division, Alcatel-Lucent, for help in obtaining the data.

9. REFERENCES

[1] 3GPP TS 36.211: LTE physical channels and modulation (Release 10).

[2] C-RAN: The road towards green RAN, China Mobile Research Institute. `http://labs.chinamobile.com/article_download.php?id=104035`.

[3] *http://http://www.ngmn.org*.

[4] *https://rt.wiki.kernel.org/*.

[5] *http://www.openairinterface.org/overview/acquiring.en.htm*.

[6] Andersson, B., Bletsas, K., and Baruah, S. Scheduling arbitrary-deadline sporadic task systems on multiprocessors. In *Real-Time Systems Symposium* (2008), IEEE, pp. 385–394.

[7] Baruah, S., and Fisher, N. Global fixed-priority scheduling of arbitrary-deadline sporadic task systems. *Distributed Computing and Networking* (2008), 215–226.

[8] Berrou, C., and Glavieux, A. Near optimum error correcting coding and decoding: Turbo-codes. *Communications, IEEE Transactions on 44*, 10 (1996), 1261–1271.

[9] Bovet, D., and Cesati, M. *Understanding the Linux Kernel, Second Edition*, 2 ed. O'Reilly & Associates, Inc., Sebastopol, CA, USA, 2002.

[10] Chilimbi, T. M., Davidson, B., and Larus, J. R. Cache-conscious structure definition. In *Proceedings of the ACM SIGPLAN 1999 conference on Programming language design and implementation* (New York, NY, USA, 1999), PLDI '99, ACM, pp. 13–24.

[11] Dechter, R. *Constraint processing*. Morgan Kaufmann, 2003.

[12] Drozdowski, M. *Scheduling for Parallel Processing*. Computer communications and networks. Springer, 2009.

[13] Frigo, M., and Johnson, S. Fftw: An adaptive software architecture for the fft. In *Acoustics, Speech and Signal Processing, 1998. Proceedings of the 1998 IEEE International Conference on* (1998), vol. 3, IEEE, pp. 1381–1384.

[14] Frigo, M., Leiserson, C., Prokop, H., and Ramachandran, S. Cache-oblivious algorithms. In *Foundations of Computer Science, 40th Annual Symposium on* (1999), pp. 285–297.

[15] Garey, M., and Johnson, D. *Computers and intractability*, vol. 174. Freeman San Francisco, CA, 1979.

[16] Gesbert, D., Hanly, S., Huang, H., Shitz, S. S., Simeone, O., and Yu, W. Multi-cell mimo cooperative networks: A new look at interference. In *IEEE Journal on Selected Areas in Communications* (December 2010).

[17] Kang, J., and Park, S. Algorithms for the variable sized bin packing problem. In *European Journal of Operations Research* (2003).

[18] Lehoczky, J. Fixed priority scheduling of periodic task sets with arbitrary deadlines. In *Real-Time Systems Symposium, 1990. Proceedings., 11th* (1990), IEEE, pp. 201–209.

[19] Leung, J. *Handbook of scheduling: algorithms, models, and performance analysis*, vol. 1. CRC Press, 2004.

[20] Saleh, A., Rustako, A., and Roman, R. Distributed antennas for indoor radio communications. *Communications, IEEE Transactions on 35*, 12 (1987), 1245–1251.

[21] Tan, K., Zhang, J., Fang, J., Liu, H., Ye, Y., Wang, S., Zhang, Y., Wu, H., Wang, W., and Voelker, G. M. Sora: high performance software radio using general purpose multi-core processors. In *Proceedings of the 6th USENIX symposium on Networked systems design and implementation* (Berkeley, CA, USA, 2009), NSDI'09, USENIX Association, pp. 75–90.

ProfileDroid: Multi-layer Profiling of Android Applications

Xuetao Wei Lorenzo Gomez Iulian Neamtiu Michalis Faloutsos
Department of Computer Science and Engineering
University of California, Riverside
{xwei, gomezl, neamtiu, michalis}@cs.ucr.edu

ABSTRACT

The Android platform lacks tools for assessing and monitoring apps in a systematic way. This lack of tools is particularly problematic when combined with the open nature of Google Play, the main app distribution channel. As our key contribution, we design and implement ProfileDroid, a comprehensive, multi-layer system for monitoring and profiling apps. Our approach is arguably the first to profile apps at four layers: (a) static, or app specification, (b) user interaction, (c) operating system, and (d) network. We evaluate 27 free and paid Android apps and make several observations: (a) we identify discrepancies between the app specification and app execution, (b) free versions of apps could end up costing more than their paid counterparts, due to an order of magnitude increase in traffic, (c) most network traffic is not encrypted, (d) apps communicate with many more sources than users might expect—as many as 13, and (e) we find that 22 out of 27 apps communicate with Google during execution. ProfileDroid is the first step towards a systematic approach for (a) generating cost-effective but comprehensive app profiles, and (b) identifying inconsistencies and surprising behaviors.

Categories and Subject Descriptors

C.2.1 [**Computer-communication Networks**]: Network Architecture and Design—*Wireless communication*; D.2.8 [**Software Engineering**]: Metrics—*Performance measures*; D.4.8 [**Operating Systems**]: Performance—*Measurements*; H.5.2 [**Information Interfaces and Presentation**]: User Interfaces—*Evaluation/methodology*

General Terms

Design, Experimentation, Measurement, Performance

Keywords

Android apps, Google Android, Profiling, Monitoring, System

1. INTRODUCTION

Given an Android app, how can we get an informative thumbnail of its behavior? This is the problem we set to address in this paper, in light of the 550,000 apps currently on Google Play (ex Android Market) [1, 18]. Given this substantial number of apps, we consider scalability as a key requirement. In particular, we devise a profiling scheme that works even with limited resources in terms of time, manual effort, and cost. We define limited resources to mean: a few users with a few minutes of experimentation per application. At the same time, we want the resulting app profiles to be comprehensive, useful, and intuitive. Therefore, given an app and one or more short executions, we want a profile that captures succinctly what the app did, and contrast it with: (a) what it was expected or allowed to do, and (b) other executions of the same app. For example, an effective profile should provide: (a) how apps use resources, expressed in terms of network data and system calls, (b) the types of device resources (e.g., camera, telephony) an app accesses, and whether it is allowed to, and (c) what entities an app communicates with (e.g., cloud or third-party servers).

Who would be interested in such a capability? We argue that an inexpensive solution would appeal to everyone who "comes in contact" with the app, including: (a) the app developer, (b) the owner of an Android app market, (c) a system administrator, and (d) the end user. Effective profiling can help us: (a) enhance user control, (b) improve user experience, (c) assess performance and security implications, and (d) facilitate troubleshooting. We envision our quick and cost-effective thumbnails (profiles) to be the first step of app profiling, which can then have more involved and resource-intense steps, potentially based on what the thumbnail has revealed.

Despite the flurry of research activity in this area, there is no approach yet that focuses on profiling the behavior of an Android app *itself* in all its complexity. Several efforts have focused on analyzing the mobile phone traffic and show the protocol related properties, but they do not study the apps *themselves* [14,16]. Others have studied security issues that reveal the abuse of personal device information [22,31]. However, all these works: (a) do not focus on *individual* apps, but report general trends, or (b) focus on a single layer, studying, e.g., the network behavior or the app specification in isolation. For example, some apps have negligible user inputs, such as `Pandora`, or negligible network traffic, such as `Advanced Task Killer`, and thus, by focusing only on one layer, the most significant aspect of an application could be missed. In Section 5, we discuss and compare with previous work in detail.

We design and implement PROFILEDROID, a systematic and comprehensive system for profiling Android apps. A key novelty is that our profiling spans four layers: (a) static, i.e., app specification, (b) user interaction, (c) operating system, and (d) network. To the best of our knowledge, this is the first work[1] that considers all these layers in profiling individual Android apps. Our contributions are twofold. First, designing the system requires the careful selection of informative and intuitive metrics, which capture the essence of each layer. Second, implementing the system is a non-trivial task, and we have to overcome numerous practical challenges.[2]

We demonstrate the capabilities of our system through experiments. We deployed our system on Motorola Droid Bionic phones, with Android version 2.3.4 and Linux kernel version 2.6.35. We profile 19 free apps; for 8 of these, we also profile their paid counterparts, for a total of 27 apps. For each app, we gather profiling data from 30 runs for several users at different times of day. Though we use limited testing resources, our results show that our approach can effectively profile apps, and detect surprising behaviors and inconsistencies. Finally, we show that *cross-layer* app analysis can provide insights and detect issues that are not visible when examining single layers in isolation.

We group our most interesting observations in three groups, although it is clear that some observations span several categories.

A. Privacy and Security Issues.

1. Lack of transparency. We identify discrepancies between the app specification and app execution. For example, `Instant Heart Rate` and `Dictionary` use resources without declaring them up-front (Section 3.1).

2. Most network traffic is unencrypted. We find that most of the network traffic is not encrypted. For example, most of the web-based traffic is over HTTP and not HTTPS: only 8 out of the 27 apps use HTTPS and for `Facebook`, 22.74% of the traffic is not encrypted (Section 4.5).

B. Operational Issues.

3. Free apps have a cost. Free versions of apps could end up costing more than their paid versions especially on limited data plans, due to increased advertising/analytics traffic. For example, the free version of `Angry Birds` has 13 times more traffic than the paid version (Section 4.3).

4. Apps talk to "strangers". Apps interact with many more traffic sources than one would expect. For example, the free version of `Shazam` talks to 13 different traffic sources in a 5-minute interval, while its paid counterpart talks with 4 (Section 4.6).

5. Google "touches" almost everything. Out of 27 apps, 22 apps exchange data traffic with Google, including apps that one would not have expected, e.g., Google accounts for 85.97% of the traffic for the free version of the health app `Instant Heart Rate`, and 90% for the paid version (Section 4.7).

C. Performance Issues.

6. Security comes at a price. The Android OS uses virtual machine (VM)-based isolation for security and reliability, but as a consequence, the VM overhead is high: more than 63% of the system calls are introduced by the VM for context-switching between threads, supporting IPC, and idling (Section 4.4).

2. OVERVIEW OF APPROACH

We present an overview of the design and implementation of PROFILEDROID. We measure and profile apps at four different layers: (a) static, or app specification (b) user interaction, (c) operating system, and (d) network. For each layer, our system consists of two parts: a monitoring and a profiling component. For each layer, the monitoring component runs on the Android device where the app is running. The captured information is subsequently fed into the profiling part, which runs on the connected computer. In Figure 1, on the right, we show a high level overview of our system and its design. On the left, we have an actual picture of the actual system the Android device that runs the app and the profiling computer (such as a desktop or a laptop).

In the future, we foresee a light-weight version of the whole profiling system to run exclusively on the Android device. The challenge is that the computation, the data storage, and the battery consumption must be minimized. How to implement the profiling in an incremental and online fashion is beyond the scope of the current work. Note that our system is focused on profiling of an *individual* app, and not intended to monitor user behavior on mobile devices.

From an architectural point of view, we design PROFILEDROID to be flexible and modular with level-defined interfaces between the monitoring and profiling components. Thus, it is easy to modify or improve functionality within each layer. Furthermore, we could easily extend the current functionality to add more metrics, and even potentially more layers, such as a physical layer (temperature, battery level, etc.).

2.1 Implementation and Challenges

We describe the implementation of monitoring at each layer, and briefly touch on challenges we had to surmount when constructing PROFILEDROID.

To profile an application, we start the monitoring infrastructure (described at length below) and then the target app is launched. The monitoring system logs all the relevant activities, e.g., user touchscreen input events, system calls, and all network traffic in both directions.

2.1.1 Static Layer

At the static layer, we analyze the APK (Android application package) file, which is how Android apps are distributed. We use `apktool` to unpack the APK file to extract relevant data. From there, we mainly focus on the `Manifest.xml` file and the bytecode files contained in the `/smali` folder. The manifest is specified by the developer and identifies hardware usage and permissions requested by each app. The `smali` files contain the app bytecode which we parse and analyze statically, as explained later in Section 3.1.

[1]An earlier work [12] uses the term "cross-layer," but the layers it refers to are quite different from the layers we use.
[2]Examples include fine-tuning data collection tools to work on Android, distinguishing between presses and swipes, and disambiguating app traffic from third-party traffic.

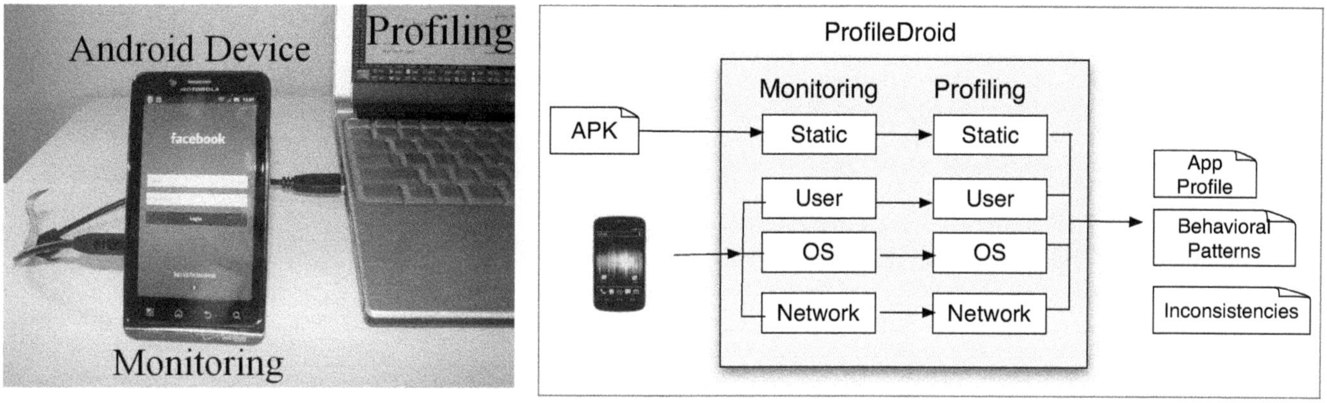

Figure 1: Overview and actual usage (left) and architecture (right) of PROFILEDROID.

2.1.2 User Layer

At the user layer, we focus on user-generated events, i.e., events that result from interaction between the user and the Android device while running the app. To gather the data of the user layer, we use a combination of the `logcat` and `getevent` tools of `adb`. From the `logcat` we capture the system debug output and log messages from the app. In particular, we focus on events-related messages. To collect the user input events, we use the `getevent` tool, which reads the `/dev /input/event*` to capture user events from input devices, e.g., touchscreen, accelerometer, proximity sensor. Due to the raw nature of the events logged, it was challenging to disambiguate between swipes and presses on the touchscreen. We provide details in Section 3.2.

2.1.3 Operating System Layer

At the operating system-layer, we measure the operating system activity by monitoring system calls. We collect system calls invoked by the app using an Android-specific version of `strace`. Next, we classify system calls into four categories: filesystem, network, VM/IPC, and miscellaneous. As described in Section 3.3, this classification is challenging, due to the virtual file system and the additional VM layer that decouples apps from the OS.

2.1.4 Network Layer

At the network layer, we analyze network traffic by logging the data packets. We use an Android-specific version of `tcpdump` that collects all network traffic on the device. We parse, domain-resolve, and classify traffic. As described in Section 3.4, classifying network traffic is a significant challenge in itself; we used information from domain resolvers, and improve its precision with manually-gathered data on specific websites that act as traffic sources.

Having collected the measured data as described above, we analyze it using the methods and the metrics of Section 3.

2.2 Experimental Setup

2.2.1 Android Devices

The Android devices monitored and profiled in this paper were a pair of identical Motorola Droid Bionic phones, which have dual-core ARM Cortex-A9 processors running at 1GHz. The phones were released on released September 8, 2011 and run Android version 2.3.4 with Linux kernel version 2.6.35.

2.2.2 App Selection

As of March 2012, Google Play lists more than 550,000 apps [2], so to ensure representative results, we strictly follow the following criteria in selecting our test apps. First, we selected a variety of apps that cover most app categories as defined in Google Play, such as Entertainment, Productivity tools, etc. Second, all selected apps had to be popular, so that we could examine real-world, production-quality software with a broad user base. In particular, the selected apps must have at least 1,000,000 installs, as reported by Google Play, and be within the Top-130 free apps, as ranked by the Google Play website. In the end, we selected 27 apps as the basis for our study: 19 free apps and 8 paid apps; the 8 paid apps have free counterparts, which are included in the list of 19 free apps. The list of the selected apps, as well as their categories, is shown in Table 1.

2.2.3 Conducting the experiment

In order to isolate app behavior and improve precision when profiling an app, we do not allow other manufacturer-installed apps to run concurrently on the Android device, as they could interfere with our measurements. Also, to minimize the impact of poor wireless link quality on apps, we used WiFi in strong signal conditions. Further, to ensure statistics were collected of only the app in question, we installed one app on the phone at a time and uninstalled it before the next app was tested. Note however, that system daemons and required device apps were still able to run as they normally would, e.g., the service and battery managers.

Finally, in order to add stability to the experiment, the multi-layer traces for each individual app were collected from tests conducted by multiple users to obtain a comprehensive exploration of different usage scenarios of the target application. To cover a larger variety of running conditions without burdening the user, we use *capture-and-replay*, as explained below. Each user ran each app one time for 5 minutes; we capture the user interaction using event logging. Then, using a replay tool we created, each recorded run was replayed back 5 times in the morning and 5 times at night, for a total of 10 runs each per user per app. The runs of each app were conducted at different times of the day to avoid time-of-day bias, which could lead to uncharacteristic interaction with the app; by using the capture-and-replay tool, we are able to achieve this while avoiding repetitive manual runs from the same user. For those apps that had both free and paid

App name	Category
Dictionary.com, Dictionary.com-$$	Reference
Tiny Flashlight	Tools
Zedge	Personalization
Weather Bug, Weather Bug-$$	Weather
Advanced Task Killer, Advanced Task Killer-$$	Productivity
Flixster	Entertainment
Picsay, Picsay-$$	Photography
ESPN	Sports
Gasbuddy	Travel
Pandora	Music & Audio
Shazam, Shazam-$$	Music & Audio
Youtube	Media & Video
Amazon	Shopping
Facebook	Social
Dolphin, Dolphin-$$	Communication (Browsers)
Angry Birds, Angry Birds-$$	Games
Craigslist	Business
CNN	News & Magazines
Instant Heart Rate, Instant Heart Rate-$$	Health & Fitness

Table 1: The test apps; app-$$ represents the paid version of an app.

App	Internet	GPS	Camera	Microphne	Bluetooth	Telephony
Dictionary.com	✓			I		I
Dictionary.com-$$	✓			I		I
Tiny Flashlight	✓		✓			
Zedge	✓					
Weather Bug	✓	✓				
Weather Bug-$$	✓	✓				
Advanced Task Killer	✓					
Advanced Task Killer-$$	✓					
Flixster	✓	✓				
Picsay	✓					
Picsay-$$	✓					
ESPN	✓					
Gasbuddy	✓	✓				
Pandora	✓				✓	
Shazam	✓	✓		✓		
Shazam-$$	✓	✓		✓		
YouTube	✓					
Amazon	✓		✓			
Facebook	✓	✓	I			✓
Dolphin	✓	✓				
Dolphin-$$	✓	✓				
Angry Birds	✓					
Angry Birds-$$	✓					
Craigslist	✓					
CNN	✓		✓			
Instant Heart Rate	✓		✓		I	I
Instant Heart Rate-$$	✓		✓		I	I

Table 2: Profiling results of *static* layer; '✓' represents use via permissions, while 'I' via intents.

versions, users carried out the same task, so we can pinpoint differences between paid and free versions. To summarize, our profiling is based on 30 runs (3 users × 10 replay runs) for each app.

3. ANALYZING EACH LAYER

In this section, we first provide detailed descriptions of our profiling methodology, and we highlight challenges and interesting observations.

3.1 Static Layer

The first layer in our framework aims at understanding the app's functionality and permissions. In particular, we analyze the APK file on two dimensions to identify app functionality and usage of device resources: first, we extract the permissions that the app asks for, and then we parse the app bytecode to identify intents, i.e., indirect resource access via deputy apps. Note that, in this layer only, we analyze the app without running it—hence the name *static layer*.

Functionality usage. Android devices offer several major functionalities, labeled as follows: Internet, GPS, Camera, Microphone, Bluetooth and Telephony. We present the results in Table 2. A '✓' means the app requires permission to use the device, while 'I' means the device is used indirectly via intents and deputy apps. We observe that Internet is the most-used functionality, as the Internet is the gateway to interact with remote servers via 3G or WiFi—all of our examined apps use the Internet for various tasks. For instance, `Pandora` and `YouTube` use the Internet to fetch multimedia files, while `Craigslist` and `Facebook` use it to get content updates when necessary.

GPS, the second most popular resource (9 apps) is used for navigation and location-aware services. For example, `Gasbuddy` returns gas stations near the user's location, while `Facebook` uses the GPS service to allow users to *check-in*, i.e., publish their presence at entertainment spots or places of interests. Camera, the third-most popular functionality (5 apps) is used for example, to record and post real-time news information (`CNN`), or for for barcode scanning `Amazon`. Microphone, Bluetooth and Telephony are three additional communication channels besides the Internet, which could be used for voice communication, file sharing, and text messages. This increased usage of various communication channels is a double-edged sword. On the one hand, various communication channels improve user experience. On the other hand, it increases the risk of privacy leaks and security attacks on the device.

Intent usage. Android intents allow apps to access resources indirectly by using deputy apps that have access to the requested resource. For example, `Facebook` does not have the camera permission, but can send an intent to a deputy camera app to take and retrieve a picture.[3] We decompiled each app using `apktool` and identified instances of the `android.content.Intent` class in the Dalvik bytecode. Next, we analyzed the parameters of each intent call to find

[3]This was the case for the version of the Facebook app we analyzed in March 2012, the time we performed the study. However, we found that, as of June 2012, the Facebook app requests the Camera permission explicitly.

the intent's type, i.e., the device's resource to be accessed via deputy apps.

We believe that presenting users with the list of resources used via intents (e.g., that the **Facebook** app does not have direct access to the camera, but nevertheless it can use the camera app to take pictures) helps them make better-informed decisions about installing and using an app. Though legitimate within the Android security model, this lack of user forewarning can be considered deceiving; with the more comprehensive picture provided by ProfileDroid, users have a better understanding of resource usage, direct or indirect [3].

3.2 User Layer

At the user layer, we analyze the input events that result from user interaction. In particular, we focus on *touches*—generated when the user touches the screen—as touchscreens are the main Android input devices. Touch events include *presses*, e.g., pressing the app buttons of the apps, and *swipes*—finger motion without losing contact with the screen. The intensity of events (events per unit of time), as well as the ratio between swipes and presses are powerful metrics for GUI behavioral fingerprinting (Section 4.2); we present the results in Figure 2 and now proceed to discussing these metrics.

Technical challenge. Disambiguating between swipes and presses was a challenge, because of the nature of reported events by the *getevent* tool. Swipes and presses are reported by the touchscreen input device, but the reported events are not labeled as swipes or presses. A single press usually accounts for 30 touchscreen events, while a swipe usually accounts for around 100 touchscreen events. In order to distinguish between swipes and presses, we developed a method to cluster and label events. For example, two events separated by less than 80 milliseconds are likely to be part of a sequence of events, and if that sequence of events grows above 30, then it is likely that the action is a swipe instead of a press. Evaluating and fine-tuning our method was an intricate process.

Touch events intensity. We measured touch intensity as the number of touch events per second—this reveals how interactive an app is. For example, the music app **Pandora** requires only minimal input (music control) once a station is selected. In contrast, in the game **Angry Birds**, the user has to interact with the interface of the game using swipes and screen taps, which results in a high intensity for touch events.

Swipe/Press ratio. We use the ratio of swipes to presses to better capture the nature of the interaction, and distinguish between apps that have similar touch intensity. Note that swipes are used for navigation and zooming, while touches are used for selection. Figure 2 shows that apps that involve browsing, news-page flipping, gaming, e.g., **CNN**, **Angry Birds**, have a high ratio of swipes to presses; even for apps with the same touch intensity, the swipe/press ratio can help profile and distinguish apps, as seen in the following table:

App	Touch intensity	Swipe/Press ratio
Picsay	*medium*	*low*
CNN	*medium*	*high*

Phone event intensity. The bottom chart in Figure 2 shows the intensity of events generated by the phone itself during the test. These events contain a wealth of contextual data that, if leaked, could pose serious privacy risks. The most frequent events we observed were generated by the accelerometer, the light proximity sensor, and for some location-aware apps, the compass. For brevity, we omit details, but we note that phone-event intensity, and changes in intensity, can reveal the user's proximity to the phone, the user's motion patterns, and user orientation and changes thereof.

App	Syscall intensity (calls/sec.)	FS (%)	NET (%)	VM& IPC (%)	MISC (%)
Dictionary.com	1025.64	3.54	1.88	67.52	27.06
Dictionary.com-$$	492.90	7.81	4.91	69.48	17.80
Tiny Flashlight	435.61	1.23	0.32	77.30	21.15
Zedge	668.46	4.17	2.25	75.54	18.04
Weather Bug	1728.13	2.19	0.98	67.94	28.89
Weather Bug-$$	492.17	1.07	1.78	75.58	21.57
AdvTaskKiller	75.06	3.30	0.01	65.95	30.74
AdvTaskKiller-$$	30.46	7.19	0.00	63.77	29.04
Flixster	325.34	2.66	3.20	71.37	22.77
Picsay	319.45	2.06	0.01	75.12	22.81
Picsay-$$	346.93	2.43	0.16	74.37	23.04
ESPN	1030.16	2.49	2.07	87.09	8.35
Gasbuddy	1216.74	1.12	0.32	74.48	24.08
Pandora	286.67	2.92	2.25	70.31	24.52
Shazam	769.54	6.44	2.64	72.16	18.76
Shazam-$$	525.47	6.28	1.40	74.31	18.01
YouTube	246.78	0.80	0.58	77.90	20.72
Amazon	692.83	0.42	6.33	76.80	16.45
Facebook	1030.74	3.99	2.98	72.02	21.01
Dolphin	850.94	5.20	1.70	71.91	21.19
Dolphin-$$	605.63	9.05	3.44	68.45	19.07
Angry Birds	1047.19	0.74	0.36	82.21	16.69
Angry Birds-$$	741.28	0.14	0.04	85.60	14.22
Craigslist	827.86	5.00	2.47	73.81	18.72
CNN	418.26	7.68	5.55	71.47	15.30
InstHeartRate	944.27	7.70	1.73	75.48	15.09
InstHeartRate-$$	919.18	12.25	0.14	72.52	15.09

Table 3: Profiling results: *operating system* layer.

3.3 Operating System Layer

We first present a brief overview of the Android OS, and then discuss metrics and results at the operating system layer.

Android OS is a Linux-based operating system, customized for mobile devices. Android apps are written in Java and compiled to Dalvik executable (Dex) bytecode. The bytecode is bundled with the app manifest (specification, permissions) to create an APK file. When an app is installed, the user must grant the app the permissions specified in the manifest. The Dex bytecode runs on top of the Dalvik Virtual Machine (VM)—an Android-specific Java virtual machine. Each app runs as a separate Linux process with a unique user ID in a separate copy of the VM. The separation among apps offers a certain level of protection and running on top of a VM avoids granting apps direct access to hardware resources. While increasing reliability and reducing the potential for security breaches, this vertical (app–hardware) and horizontal (app–app) separation means that apps do not

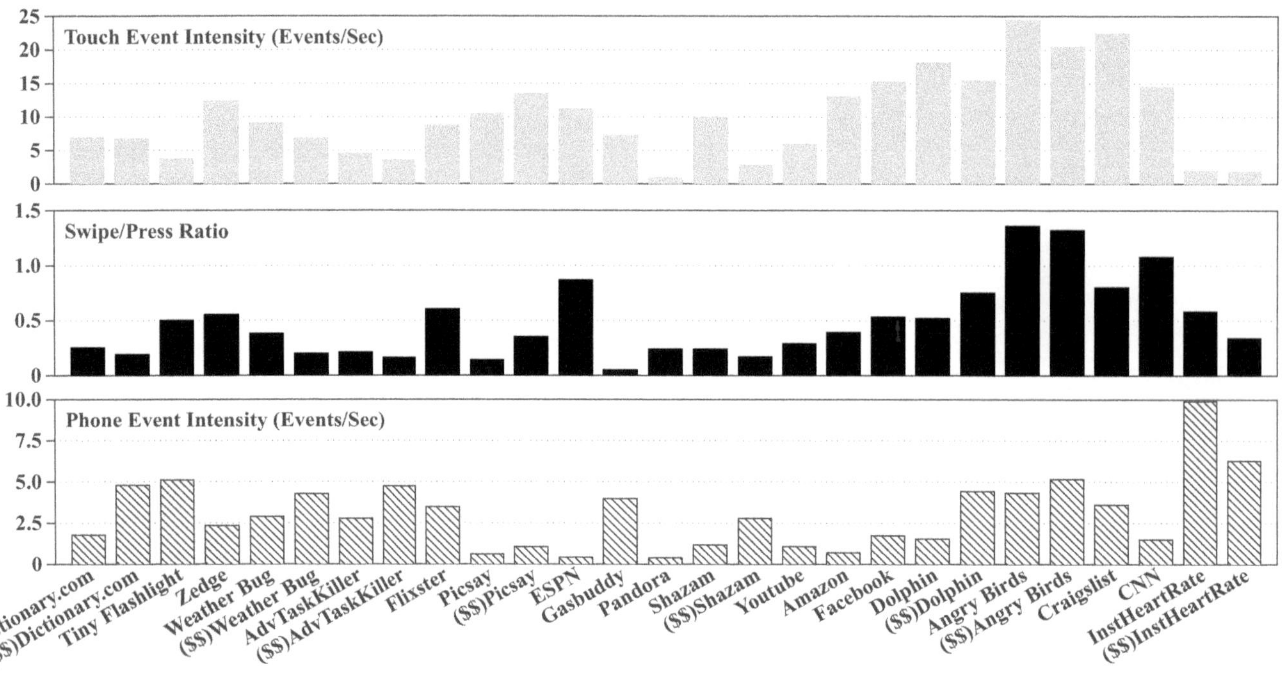

Figure 2: Profiling results of *user* layer; note that scales are different.

run natively and inter-app communications must take place primarily via IPC. We profile apps at the operating system layer with several goals in mind: to understand how apps use system resources, how the operating-system intensity compares to the intensity observed at other layers, and to characterize the potential performance implications of running apps in separate VM copies. To this end, we analyzed the system call traces for each app to understand the nature and frequency of system calls. We present the results in Table 3.

System call intensity. The second column of Table 3 shows the system call intensity in system calls per second. While the intensity differs across apps, note that in all cases the intensity is relatively high (between 30 and 1,183 system calls per second) for a mobile platform.

System call characterization. To characterize the nature of system calls, we group them into four bins: file system (FS), network (NET), virtual machine (VM&IPC), and miscellaneous (MISC). Categorizing system calls is not trivial.

Technical challenge. The Linux version running on our phone (2.6.35.7 for Arm) supports about 370 system calls; we observed 49 different system calls in our traces. While some system calls are straightforward to categorize, the operation of virtual filesystem calls such as `read` and `write`, which act on a file descriptor, depends on the file descriptor and can represent file reading and writing, network send/receive, or reading/altering system configuration via `/proc`. Therefore, for all the virtual filesystem calls, we categorize them based on the file descriptor associated with them, as explained below. FS system calls are used to access data stored on the flash drive and SD card of the mobile device and consist mostly of `read` and `write` calls on a file descriptor associated with a space-occupying file in the file system, i.e., opened via `open`. NET system calls consist mostly of `read` and `write` calls on a file descriptor associated with a network socket, i.e., opened via `socket`; note that for NET system calls, reads and writes mean receiving from and sending to the network. VM&IPC system calls are calls inserted by the virtual machine for operations such as scheduling, timing, idling, and IPC. For each such operation, the VM inserts a specific sequence of system calls. We extracted these sequences, and compared the number of system calls that appear as part of the sequence to the total number, to quantify the VM and IPC-introduced overhead. The most common VM/IPC system calls we observed (in decreasing order of frequency) were: `clock_gettime`, `epoll_wait`, `getpid`, `getuid32`, `futex`, `ioctl`, and `ARM_cacheflush`. The remaining system calls are predominantly `read` and `write` calls to the `/proc` special filesystem are categorized as MISC.

The results are presented in Table 3: for each category, we show both intensity, as well as the percentage relative to all categories. Note that FS and NET percentages are quite similar, but I/O system calls (FS and NET) constitute a relatively small percentage of total system calls, with the VM&IPC dominating. We will come back to this aspect in Section 4.4.

3.4 Network Layer

The network-layer analysis summarizes the data communication of the app via WiFi or 3G. Android apps increasingly rely on Internet access for a diverse array of services, e.g., for traffic, map or weather data and even offloading computation to the cloud. An increasing number of network traffic sources are becoming visible in app traffic, e.g., Content Distribution Networks, Cloud, Analytics and Advertisement. To this end, we characterize the app's network

App	Traffic intensity (bytes/sec.)	Traffic In/Out (ratio)	Origin (%)	CDN+Cloud (%)	Google (%)	Third party (%)	Traffic sources	HTTP/HTTPS split (%)
Dictionary.com	1450.07	1.94	–	35.36	64.64	–	8	100/–
Dictionary.com-$$	488.73	1.97	0.02	1.78	98.20	–	3	100/–
Tiny Flashlight	134.26	2.49	–	–	99.79	0.21	4	100/–
Zedge	15424.08	10.68	–	96.84	3.16	–	4	100/–
Weather Bug	3808.08	5.05	–	75.82	16.12	8.06	13	100/–
Weather Bug-$$	2420.46	8.28	–	82.77	6.13	11.10	5	100/–
AdvTaskKiller	25.74	0.94	–	–	100.00	–	1	91.96/8.04
AdvTaskKiller-$$	–	–	–	–	–	–	0	–/–
Flixster	23507.39	20.60	2.34	96.90	0.54	0.22	10	100/–
Picsay	4.80	0.34	–	48.93	51.07	–	2	100/–
Picsay-$$	320.48	11.80	–	99.85	0.15	–	2	100/–
ESPN	4120.74	4.65	–	47.96	10.09	41.95	5	100/–
Gasbuddy	5504.78	10.44	6.17	11.23	81.37	1.23	6	100/–
Pandora	24393.31	28.07	97.56	0.91	1.51	0.02	11	99.85/0.15
Shazam	4091.29	3.71	32.77	38.12	15.77	13.34	13	100/–
Shazam-$$	1506.19	3.09	44.60	55.36	0.04	–	4	100/–
YouTube	109655.23	34.44	96.47	–	3.53	–	2	100/–
Amazon	7757.60	8.17	95.02	4.98	–	–	4	99.34/0.66
Facebook	4606.34	1.45	67.55	32.45	–	–	3	22.74/77.26
Dolphin	7486.28	5.92	44.55	0.05	8.60	46.80	22	99.86/0.14
Dolphin-$$	3692.73	6.05	80.30	1.10	5.80	12.80	9	99.89/0.11
Angry Birds	501.57	0.78	–	73.31	10.61	16.08	8	100/–
Angry Birds-$$	36.07	1.10	–	88.72	5.79	5.49	4	100/–
Craigslist	7657.10	9.64	99.97	–	–	0.03	10	100/–
CNN	2992.76	5.66	65.25	34.75	–	–	2	100/–
InstHeartRate	573.51	2.29	–	4.18	85.97	9.85	3	86.27/13.73
InstHeartRate-$$	6.09	0.31	–	8.82	90.00	1.18	2	20.11/79.89

Table 4: Profiling results of *network* layer; '–' represents no traffic.

behavior using the following metrics and present the results in Table 4.

Traffic intensity. This metric captures the intensity of the network traffic of the app. Depending on the app, the network traffic intensity can vary greatly, as shown in Table 4. For the user, this great variance in traffic intensity could be an important property to be aware of, especially if the user has a limited data plan. Not surprisingly, we observe that the highest traffic intensity is associated with a video app, `YouTube`. Similarly, the entertainment app `Flixster`, music app `Pandora`, and personalization app `Zedge` also have large traffic intensities as they download audio and video files. We also observe apps with zero, or negligible, traffic intensity, such as the productivity app `Advanced Task Killer` and free photography app `Picsay`.

Origin of traffic. The origin of traffic means the percentage of the network traffic that comes from the servers owned by the app provider. This metric is particularly interesting for privacy-sensitive users, since it is an indication of the control that the app provider has over the app's data. Interestingly, there is large variance for this metric, as shown in Table 4. For example, the apps `Amazon`, `Pandora`, `YouTube`, and `Craigslist` deliver most of their network traffic (e.g., more than 95%) through their own servers and network. However, there is no origin traffic in the apps `Angry Birds` and `ESPN`. Interestingly, we observe that ony 67% of the `Facebook` traffic comes from Facebook servers, with the remaining coming from content providers or the cloud.

Technical challenge. It is a challenge to classify the network traffic into different categories (e.g., cloud vs. ad network), let alone identify the originating entity. To resolve this, we combine an array of methods, including reverse IP address lookup, DNS and `whois`, and additional information and knowledge from public databases and the web. In many cases, we use information from CrunchBase (crunchbase.com) to identify the type of traffic sources after we resolve the top-level domains of the network traffic [6]. Then, we classify the remaining traffic sources based on information gleaned from their website and search results.

In some cases, detecting the origin is even more complicated. For example, consider the `Dolphin` web browser— here the origin is not the Dolphin web site, but rather the website that the user visits with the browser, e.g., if the user visits CNN, then cnn.com is the origin. Also, YouTube is owned by Google and YouTube media content is delivered from domain 1e100.net, which is owned by Google; we report the media content (96.47%) as Origin, and the remaining traffic (3.53%) as Google which can include Google ads and analytics.

CDN+Cloud traffic. This metric shows the percentage of the traffic that comes from servers of CDN (e.g., Akamai) or cloud providers (e.g., Amazon AWS). Content Distribution Network (CDN) has become a common method to distribute the app's data to its users across the world faster, with scalability and cost-effectively. Cloud platforms have extended this idea by providing services (e.g., computation) and not just data storage. Given that it is not obvious if someone using a cloud service is using it as storage, e.g., as a CDN, or for computation, we group CDN and cloud services into one category. Interestingly, there is a very strong presence of this kind of traffic for some apps, as seen in Table 4. For example, the personalization app **Zedge**, and the video-heavy app **Flixster** need intensive network services, and they use CDN and Cloud data sources. The high percentages that we observe for CDN+Cloud traffic point to how important CDN and Cloud sources are, and how much apps rely on them for data distribution.

Google traffic. Given that Android is a product of Google, it is natural to wonder how involved Google is in Android traffic. The metric is the percentage of traffic exchanged with Google servers (e.g., 1e100.net), shown as the second-to-last column in Table 4. It has been reported that the percentage of Google traffic has increased significantly over the past several years [7]. This is due in part to the increasing penetration of Google services (e.g., maps, ads, analytics, and Google App Engine). Note that 22 of out of the 27 apps exchange traffic with Google, and we discuss this in more detail in Section 4.7.

Third-party traffic. This metric is of particular interest to privacy-sensitive users. We define third party traffic as network traffic from various advertising services (e.g., Atdmt) and analytical services (e.g., Omniture) besides Google, since advertising and analytical services from Google are included in the Google traffic metric. From Table 4, we see that different apps have different percentages of third-party traffic. Most apps only get a small or negligible amount of traffic from third parties (e.g., **YouTube**, **Amazon** and **Facebook**). At the same time, nearly half of the total traffic of **ESPN** and **Dolphin** comes from third parties.

The ratio of incoming traffic and outgoing traffic. This metric captures the role of an app as a consumer or producer of data. In Table 4, we see that most of the apps are more likely to receive data than to send data. As expected, we see that the network traffic from **Flixster**, **Pandora**, and **YouTube**, which includes audio and video content, is mostly incoming traffic as the large values of the ratios show. In contrast, apps such as **Picsay** and **Angry Birds** tend to send out more data than they receive.

Note that this metric could have important implications for performance optimization of wireless data network providers. An increase in the outgoing traffic could challenge network provisioning, in the same way that the emergence of p2p file sharing stretched cable network operators, who were not expecting large household upload needs. Another use of this metric is to detect suspicious variations in the ratio, e.g., unusually large uploads, which could indicate a massive theft of data. Note that the goal of this paper is to provide the framework and tools for such an investigation, which we plan to conduct as our future work.

Number of distinct traffic sources. An additional way of quantifying the interactions of an app is with the number of distinct traffic sources, i.e., distinct top-level domains. This metric can be seen as a complementary way to quantify network interactions, a sudden increase in this metric could indicate malicious behavior. In Table 4 we present the results. First, we observe that all the examined apps interact with at least two distinct traffic sources, except **Advanced Task Killer**. Second, some of the apps interact with a surprisingly high number of distinct traffic sources, e.g., **Weather bug**, **Flixster**, and **Pandora**. Note that we count all the distinct traffic sources that appear in the traces of multiple executions.

App	Static (# of func.)	User (events/ sec.)	OS (syscall/ sec.)	Network (bytes/ sec.)
Dictionary.com	L	M	H	M
Dictionary.com-$$	L	M	M	M
Tiny Flashlight	M	L	M	L
Zedge	L	M	M	H
Weather Bug	M	M	H	M
Weather Bug-$$	M	M	M	M
AdvTaskKiller	L	M	L	L
AdvTaskKiller-$$	L	M	L	L
Flixster	M	M	L	H
Picsay	L	M	L	L
Picsay-$$	L	M	M	M
ESPN	L	M	H	M
Gasbuddy	M	M	H	M
Pandora	M	L	L	H
Shazam	H	L	M	M
Shazam-$$	H	L	H	M
YouTube	L	M	M	H
Amazon	M	M	M	H
Facebook	H	H	H	M
Dolphin	M	H	M	H
Dolphin-$$	M	H	M	M
Angry Birds	L	H	M	M
Angry Birds-$$	L	H	H	L
Craigslist	L	H	H	H
CNN	M	M	M	M
InstHeartRate	M	L	H	M
InstHeartRate-$$	M	L	H	L

Table 5: Thumbnails of multi-layer intensity in the H-M-L model (H:high, M:medium, L:low).

The percentage of HTTP and HTTPS traffic. To get a sense of the percentage of secure Android app traffic, we compute the split between HTTP and HTTPS traffic, e.g., non-encrypted and encrypted traffic. We present the results in the last column of Table 4 ('–' represents no traffic). The absence of HTTPS traffic is staggering in the apps we tested, and even **Facebook** has roughly 22 % of unencrypted traffic, as we further elaborate in section 4.4.

4. PROFILEDROID: PROFILING APPS

In this section, we ask the question: *How can* PROFILEDROID *help us better understand app behavior?* In response, we show what kind of information PROFILEDROID can extract from each layer in isolation or in combination with other layers.

4.1 Capturing Multi-layer Intensity

The intensity of activities at each layer is a fundamental metric that we want to capture, as it can provide a thumbnail of the app behavior. The multi-layer intensity is a tuple consisting of intensity metrics from each layer: static (number of functionalities), user (touch event intensity), operating system (system call intensity), and network (traffic intensity).

Presenting raw intensity numbers is easy, but it has limited intuitive value. For example, reporting 100 system calls per second provides minimal information to a user or an application developer. A more informative approach is to present the relative intensity of this app compared to other apps.

We opt to represent the activity intensity of each layer using labels: H (high), M (medium), and L (low). The three levels (H, M, L) are defined relative to the intensities observed at each layer using the five-number summary from statistical analysis [10]: minimum (Min), lower quartile (Q_1), median (Med), upper quartile (Q_3), and maximum (Max). Specifically, we compute the five-number summary across all 27 apps at each layer, and then define the ranges for H, M, and L as follows:

$$Min < L \leq Q_1 \qquad Q_1 < M \leq Q_3 \qquad Q_3 < H \leq Max$$

The results are in the following table:

Layer	Min	Q_1	Med	Q_3	Max
Static	1	1	2	2	3
User	0.57	3.27	7.57	13.62	24.42
OS	30.46	336.14	605.63	885.06	1728.13
Network	0	227.37	2992.76	6495.53	109655.23

Note that there are many different ways to define these thresholds, depending on the goal of the study, whether it is conserving resources, (e.g., determining static thresholds to limit intensity), or studying different app categories (e.g., general-purpose apps have different thresholds compared to games). In addition, having more than three levels of intensity provides more accurate profiling, at the expense of simplicity. To sum up, we chose to use relative intensities and characterize a wide range of popular apps to mimic testing of typical Google Play apps.

Table 5 shows the results of applying this H-M-L model to our test apps. We now proceed to showing how users and developers can benefit from an H-M-L-based app thumbnail for characterizing app behavior. Users can make more informed decisions when choosing apps by matching the H-M-L thumbnail with individual preference and constraints. For example, if a user has a small-allotment data plan on the phone, perhaps he would like to only use apps that are rated L for the intensity of network traffic; if the battery is low, perhaps she should refrain from running apps rated H at the OS or network layers.

Developers can also benefit from the H-M-L model by being able to profile their apps with ProfileDroid and optimize based on the H-M-L outcome. For example, if ProfileDroid indicates an unusually high intensity of filesystem calls in the operating system layer, the developer can examine their code to ensure those calls are legitimate. Similarly, if the developer is contemplating using an advertising library in their app, she can construct two H-M-L app models, with and without the ad library and understand the trade-offs.

In addition, an H-M-L thumbnail can help capture the nature of an app. Intuitively, we would expect interactive apps (social apps, news apps, games, Web browsers) to have intensity H at the user layer; similarly, we would expect media player apps to have intensity H at the network layer, but L at the user layer. Table 5 supports these expectations, and suggests that the the H-M-L thumbnail could be an initial way to classify apps into coarse behavioral categories.

4.2 Cross-layer Analysis

We introduce a notion of *cross-layer analysis* to compare the inferred (or observed) behavior across different layers. Performing this analysis serves two purposes: to identify potential discrepancies (e.g., resource usage via intents, as explained in Section 3.1), and to help characterize app behavior in cases where examining just one layer is insufficient. We now provide some examples.

Network traffic disambiguation. By cross-checking the user and network layers we were able to distinguish advertising traffic from expected traffic. For example, when profiling the `Dolphin` browser, by looking at both layers, we were able to separate advertisers traffic from web content traffic (the website that the user browses to), as follows. From the user layer trace, we see that the user surfed to, for example, cnn.com, which, when combined with the network traffic, can be used to distinguish legitimate traffic coming from CNN and advertising traffic originating at CNN; note that the two traffic categories are distinct and labeled Origin and Third-party, respectively, in Section 3.4. If we were to only examine the network layer, when observing traffic with the source cnn.com, we would not be able to tell Origin traffic apart from ads placed by cnn.com.

Application disambiguation. In addition to traffic disambiguation, we envision cross-layer checking to be useful for behavioral fingerprinting for apps (outside the scope of this paper). Suppose that we need to distinguish a file manager app from a database-intensive app. If we only examine the operating system layer, we would find that both apps show high FS (filesystem) activity. However, the database app does this without any user intervention, whereas the file manager initiates file activity (e.g., move file, copy file) in response to user input. By cross-checking the operating system layer and user layer we can distinguish between the two apps because the file manager will show much higher user-layer activity. We leave behavioral app fingerprinting to future work.

4.3 Free Versions of Apps Could End Up Costing More Than Their Paid Versions

The Android platform provides an open market for app developers. Free apps (69% of all apps on Google Play [2]) significantly contributed to the adoption of Android platform. However, the free apps are not as free as we would expect. As we will explain shortly, considerable amounts of network traffic are dedicated to for-profit services, e.g., advertising and analytics.

In fact, we performed a cross-layer study between free apps and their paid counterparts. As mentioned in Section 2.1, users carried out the same task when running the free and paid versions of an app. We now proceed to describe findings at each layer. We found no difference at the static layer (Table 2). At the user-layer, Figure 2 shows that most of behaviors are similar between free and paid version of the apps, which indicates that free and paid versions have similar GUI layouts, and performing the same task takes similar effort in both the free and the paid versions of an app. The exception was the photography app `Picsay`. At first we found this finding counterintuitive; however, the paid version of `Picsay` provides more picture-manipulating functions than the free version, which require more navigation (user input) when manipulating a photo.

Differences are visible at the OS layer as well: as shown in Table 3, system call intensity is significantly higher (around 50%–100%) in free apps compared the their paid counterparts, which implies lower performance and higher energy consumption. The only exception is Picsay, whose paid version has higher system call intensity; this is due to increased GUI navigation burden as we explained above.

We now move on to the network layer. Intuitively, the paid apps should not bother users with the profit-making extra traffic, e.g., ads and analytics, which consumes away the data plan. However, the results only partially match our expectations. As shown in Table 4, we find that the majority of the paid apps indeed exhibit dramatically reduced network traffic intensity, which help conserve the data plan. Also, as explained in Section 4.6, paid apps talk to fewer data sources than their free counterparts. However, we could still observe traffic from Google and third party in the paid apps. We further investigate whether the paid apps secure their network traffic by using HTTPS instead of HTTP. As shown in Table 4, that is usually not the case, with the exception of Instant Heart Rate.

To sum up, the "free" in "free apps" comes with a hard-to-quantify, but noticeable, user cost. Users are unaware of this because multi-layer behavior is generally opaque to all but most advanced users; however, this shortcoming is addressed well by ProfileDroid.

4.4 Heavy VM&IPC Usage Reveals a Security-Performance Trade-off

As mentioned in Section 3.3, Android apps are isolated from the hardware via the VM, and isolated from each other by running on separate VM copies in separate processes with different UIDs. This isolation has certain reliability and security advantages, i.e., a corrupted or malicious app can only inflict limited damage. The flip side, though, is the high overhead associated with running bytecode on top of a VM (instead of natively), as well as the high overhead due to IPC communication that has to cross address spaces. The VM&IPC column in Table 3 quantifies this overhead: we were able to attribute around two-thirds of system calls (63.77% to 87.09%, depending on the app) to VM and IPC. The precise impact of VM&IPC system calls on performance and energy usage is beyond the scope of this paper, as it would require significantly more instrumentation. Nevertheless, the two-thirds figure provides a good intuition of the additional system call burden due to isolation.

4.5 Most Network Traffic is not Encrypted

As Android devices and apps manipulate and communicate sensitive data (e.g., GPS location, list of contacts, account information), we have investigated whether the Android apps use HTTPS to secure their data transfer. Last column of Table 4 shows the split between HTTP and HTTPS traffic for each app. We see that most apps use HTTP to transfer the data. Although some apps secure their traffic by using HTTPS, the efforts are quite limited. This is a potential concern: for example, for Facebook 77.26% of network traffic is HTTPS, hence the remaining 22.74% can be intercepted or modified in transit in a malicious way. A similar concern is notable with Instant Heart Rate, a health app, whose free version secures only 13.73% of the traffic with HTTPS; personal health information might leak in the remaining 86.27% HTTP traffic. We further investigate which traffic sources are using HTTPS and report the results in Table 6. Note how HTTPS data sources (Origin, CDN, Google) also deliver services over HTTP. These results reveal that deployment of HTTPS is lagging in Android apps—an undesirable situation as Android apps are often used for privacy-sensitive tasks.

App	HTTPS traffic sources	HTTP
Pandora	Pandora, Google	yes
Amazon	Amazon	yes
Facebook	Facebook, Akamai	yes
Instant Heart Rate	Google	yes
Instant Heart Rate-$$	Google	yes

Table 6: Traffic sources for HTTPS.

4.6 Apps Talk to Many More Traffic Sources Than One Would Think

When running apps that have Internet permission, the underlying network activity is a complete mystery: without access to network monitoring and analysis capabilities, users and developers do not know where the network traffic comes from and goes to. To help address this issue, we investigate the traffic sources; Table 4 shows the number of distinct traffic sources in each app, while Table 7 shows the number of distinct traffic sources per traffic category. We make two observations here. First, Table 4 reveals that most of the apps interact with at least two traffic sources, and some apps have traffic with more than 10 sources, e.g., Pandora and Shazam, because as we explained in Section 3.4, traffic sources span a wide range of network traffic categories: Origin, CDN, Cloud, Google and third party. Second, paid apps have fewer traffic sources than their free counterparts (3 vs. 8 for Dictionary.com, 4 vs. 13 for Shazam, 9 vs. 22 for Dolphin), and the number of third-party sources is 0 or 1 for most paid apps. This information is particularly relevant to app developers, because not all traffic sources are under the developer's control. Knowing this information makes both users and developers aware of the possible implications (e.g., data leaking to third parties) of running an app.

4.7 How Predominant is Google Traffic in the Overall Network Traffic?

Android apps are relying on many Google services such as Google maps, YouTube video, AdMob advertising, Google Analytics, and Google App Engine. Since Google leads the Android development effort, we set out to investigate whether Google "rules" the Android app traffic. In Table 4, we have presented the percentage of Google traffic relative to all traffic. While this percentage varies across apps, most apps have at least some Google traffic. Furthermore, Google traffic dominates the network traffic in the apps Tiny Flashlight (99.79%), Gasbuddy (81.37%) and Instant Heart Rate (85.97%), which shows that these apps crucially rely on Google services. However, some apps, such as Amazon and Facebook, do not have Google traffic; we believe this information is relevant to certain categories of users.

In addition, we further break down the Google traffic and analyze the ratio of incoming traffic from Google to outgoing traffic to Google. The ratios are presented in Table 7.

App	CDN+ Cloud	Google	Third party	Google In/Out
Dictionary.com	3	1	4	2.42
Dictionary.com-$$	2	1	0	1.92
Tiny Flashlight	0	1	3	2.13
Zedge	2	1	1	2.06
Weather Bug	5	1	7	4.93
Weather Bug-$$	3	1	1	13.20
AdvTaskKiller	0	1	0	0.94
AdvTaskKiller-$$	0	0	0	–
Flixster	4	1	4	0.90
Picsay	1	1	0	0.93
Picsay-$$	1	1	0	0.94
ESPN	1	1	3	3.84
Gasbuddy	2	1	2	17.25
Pandora	3	1	6	3.63
Shazam	3	1	8	2.61
Shazam-$$	1	1	1	0.84
YouTube	0	1	0	11.10
Amazon	3	0	0	–
Facebook	2	0	0	–
Dolphin	0	1	17	5.10
Dolphin-$$	0	1	4	2.99
Angry Birds	1	1	6	2.26
Angry Birds-$$	2	1	0	1.04
Craigslist	6	0	3	–
CNN	1	0	0	–
InstHeartRate	1	1	1	2.41
InstHeartRate-$$	1	1	0	1.21

Table 7: Number of distinct traffic sources per traffic category, and the ratio of incoming to outgoing Google traffic; '–' means no Google traffic.

We find that most apps are Google data receivers (in/out ratio > 1). However, **`Advanced Task Killer`**, **`Picsay`** and **`Flixster`**, are sending more data to Google than they are receiving (in/out ratio < 1); this is expected.

5. RELATED WORK

To our knowledge, none of the prior efforts have focused on multi-layer monitoring and profiling of *individual* Android app.

Smartphone Measurements and Profiling. Falaki et al. [14] analyzed network logs from 43 smartphones and found commonly used app ports, properties of TCP transfer and the impact factors of smartphone performance. Furthermore, they also analyzed the diversity of smartphone usage, e.g., how the user uses the smartphone and apps [16]. Maier et al. [13] analyzed protocol usage, the size of HTTP content and the types of hand-held traffic. These efforts aid network operators, but they do not analyze the Android apps themselves. Recent work by Xu et al. [26] did a large scale network traffic measurement study on usage behaviors of smartphone apps, e.g., locality, diurnal behaviors and mobility patterns. Qian et al. [12] developed a tool named ARO to locate the performance and energy bottlenecks of smartphones by considering the cross-layer information ranging from radio resource control to application layer. Huang et al. [19] performed the measurement study using smartphones on 3G networks, and presented the app and device usage of smartphones. Falaki et al. [15] developed a monitoring tool SystemSens to capture the usage context, e.g., CPU and memory, of smartphone. Livelab [9] is a measurement tool implemented on iPhones to measure iPhone usage and different aspects of wireless network performance. Powertutor [21] focused on power modeling and measured the energy usage of smartphone. All these efforts focus on studying other layers, or device resource usage, which is different from our focus.

Android Security Related Work. Enck et al. [30] presented a framework that reads the declared permissions of an application at install time and compared it against a set of security rules to detect potentially malicious applications. Ongtang et al. [24] described a fine-grained Android permission model for protecting applications by expressing permission statements in more detail. Felt et al. [5] examined the mapping between Android APIs and permissions and proposed Stowaway, a static analysis tool to detect overprivilege in Android apps. Our method profiles the phone functionalities in static layer not only by explicit permission request, but also by implicit Intent usage. Comdroid found a number of exploitable vulnerabilities in inter-app communication in Android apps [11]. Permission re-delegation attack were shown to perform privileged tasks with the help of an app with permissions [4]. Taintdroid performed dynamic information-flow tracking to identify the privacy leaks to advertisers or single content provider on Android apps [31]. Furthermore, Enck et al. [29] found pervasive misuse of personal identifiers and penetration of advertising and analytics services by conducting static analysis on a large scale of Android apps. Our work profiles the app from multiple layers, and furthermore, we profile the network layer with a more fine-grained granularity, e.g., Origin, CDN, Cloud, Google and so on. AppFence extended the Taintdroid framework by allowing users to enable privacy control mechanisms [25]. A behavior-based malware detection system Crowdroid was proposed to apply clustering algorithms on system calls statistics to differentiate between benign and malicious apps [17]. pBMDS was proposed to correlate user input features with system call features to detect anomalous behaviors on devices [20]. Grace et al. [23] used Woodpecker to examined how the Android permission-based security model is enforced in pre-installed apps of stock smartphones. Capability leaks were found that could be exploited by malicious activities. DroidRanger was proposed to detect malicious apps in official and alternative markets [33]. Zhou et al. characterized a large set of Android malwares and a large subset of malwares' main attack was accumulating fees on the devices by subscribing to premium services by abusing SMS-related Android permissions [32]. Colluding applications could combine their permissions and perform activities beyond their individual privileges. They can communicate directly [8], or exploit covert or overt channels in Android core system components [27]. An effective framework was developed to defense against privilege-escalation attacks on devices, e.g., confused deputy attacks and colluding attacks [28].

6. CONCLUSIONS

In this paper, we have presented ProfileDroid, a monitoring and profiling system for characterizing Android app behaviors at multiple layers: static, user, OS and network. We proposed an ensemble of metrics at each layer to capture the essential characteristics of app specification, user activities, OS and network statistics. Through our analysis of top free and paid apps, we show that characteristics and behavior of Android apps are well-captured by the metrics selected in our profiling methodology, thereby justifying the selection of these metrics. Finally, we illustrate how, by using PROFILEDROID for multi-layer analysis, we were able to uncover surprising behavioral characteristics.

Acknowledgements

We would like to thank the anonymous reviewers and our shepherd Srdjan Capkun for their feedback. This work was supported in part by National Science Foundation awards CNS-1064646, CNS-1143627, by a Google Research Award, by ARL CTA W911NF-09-2-0053, and by DARPA SMISC Program.

7. REFERENCES

[1] Google Play. https://play.google.com/store, May 2012.

[2] Androlib. Number of New Applications in Android Market by month, March 2012. http://www.androlib.com/appstats.aspx.

[3] A.P. Felt, E. Ha, S. Egelman, A. Haney, E. Chin, and D. Wagner. Android Permissions: User Attention, Comprehension, and Behavior. In *SOUPS*, 2012.

[4] A.P. Felt, H. Wang, A. Moshchuk, S. Hanna and E. Chin. Permission Re-Delegation: Attacks and Defenses. In *USENIX Security Symposium*, 2011.

[5] A.P.Felt, E. Chin, S. Hanna, D. Song, and D. Wagner. Android Permissions Demystified. In *ACM CCS*, 2011.

[6] B. Krishnamurthy and C. E. Willis. Privacy diffusion on the web: A longitudinal perspective. In *WWW*, 2009.

[7] C. Labovitz, S. Iekel-Johnson, D. McPherson, J. Oberheide, and F. Jahanian. Internet inter-domain traffic. In *ACM SIGCOMM*, 2010.

[8] C. Marforio, F. Aurelien, and S. Capkun. Application collusion attack on the permission-based security model and its implications for modern smartphone systems. In *Technical Report 724, ETH Zurich*, 2011.

[9] C. Shepard, A. Rahmati, C. Tossell, L. Zhong, and P. Kortum. LiveLab: Measuring Wireless Networks and Smartphone Users in the Field. In *HotMetrics*, 2010.

[10] D.C. Hoaglin, F. Mosteller and J. W. Tukey. Understanding robust and exploratory data analysis, 1983. Wiley.

[11] E. Chin, A. P. Felt, K. Greenwood, and D. Wagner. Analyzing Inter-Application Communication in Android. In *ACM MobiSys*, 2011.

[12] F. Qian, Z. Wang, A. Gerber, Z. Morley Mao, S. Sen, and O. Spatscheck. Profiling Resource Usage for Mobile apps: a Cross-layer Approach. In *ACM MobiSys*, 2011.

[13] G. Maier, F. Schneider, and A. Feldmann. A First Look at Mobile Hand-held Device Traffic. In *PAM*, 2010.

[14] H. Falaki, D. Lymberopoulos, R. Mahajan, S. Kandula, and D. Estrin. A First Look at Traffic on Smartphones. In *ACM IMC*, 2010.

[15] H.Falaki, R.Mahajan, and D. Estrin. SystemSens: A Tool for Monitoring Usage in Smartphone Research Deployments. In *ACM MobiArch*, 2011.

[16] H.Falaki, R.Mahajan, S. Kandula, D.Lymberopoulos, R.Govindan, and D.Estrin . Diversity in Smartphone Usage. In *ACM MobiSys*, 2010.

[17] I. Burguera,U. Zurutuza, and S. Nadjm-Tehrani. Crowdroid: behavior-based malware detection system for Android. In *SPSM*, 2011.

[18] IDC. Android- and iOS-Powered Smartphones Expand Their Share of the Market. http://www.idc.com/getdoc.jsp?containerId=prUS23503312, May 2012.

[19] J. Huang, Q. Xu, B. Tiwana, Z. M. Mao, M. Zhang, and P. Bahl. Anatomizing app Performance Differences on Smartphones. In *ACM MobiSys*, 2010.

[20] L. Xie, X. Zhang, J.-P. Seifert, and S. Zhu. pBMDS: A Behavior-based Malware Detection System for Cellphone Devices. In *ACM WiSec*, 2010.

[21] L. Zhang, B. Tiwana, Z. Qian, Z. Wang, R. Dick, Z. M. Mao, and L. Yang. Accurate Online Power Estimation and Automatic Battery Behavior Based Power Model Generation for Smartphones. In *CODES+ISSS*, 2010.

[22] M. Egele, C. Kruegel, E. Kirda, and G. Vigna. Detecting Privacy Leaks in iOS apps. In *NDSS*, 2011.

[23] M. Grace, Y. Zhou, Z. Wang, and X. Jiang. Systematic Detection of Capability Leaks in Stock Android Smartphones . In *NDSS*, 2012.

[24] M. Ongtang, S. McLaughlin, W. Enck and P. McDaniel. Semantically Rich Application-Centric Security in Android. In *ACSAC*, 2009.

[25] P. Hornyack, S. Han, J. Jung, S. Schechter, and D. Wetherall. These arenâĂŹt the Droids youâĂŹre looking for: RetroïňĄtting Android to protect data from imperious applications. In *ACM CCS*, 2011.

[26] Q. Xu, J. Erman, A. Gerber, Z. Morley Mao, J. Pang, and S. Venkataraman. Identify Diverse Usage Behaviors of Smartphone Apps. In *IMC*, 2011.

[27] R. Schlegel, K. Zhang, X. Zhou, M. Intwala, A. Kapadia, and X. Wang. Soundcomber: A Stealthy and Context-Aware Sound Trojan for Smartphones. In *NDSS*, 2011.

[28] S. Bugiel, L. Davi, A. Dmitrienko, T. Fischer, A. Sadeghi, and B.Shastry. Towards Taming Privilege-Escalation Attacks on Android . In *NDSS*, 2012.

[29] W. Enck, D. Octeau, P. McDaniel, and S. Chaudhuri. A Study of Android Application Security. In *USENIX Security Symposium*, 2011.

[30] W. Enck, M. Ongtang and P. McDaniel. On Lightweight Mobile Phone Application Certification. In *ACM CCS*, 2009.

[31] W. Enck, P. Gilbert, B. G. Chun, L. P. Cox, J. Jung, P. McDaniel, and A. N. Sheth. Taintdroid: An information-flow tracking system for realtime privacy monitoring on smartphones. In *OSDI*, 2010.

[32] Y. Zhou and X. Jiang. Dissecting Android Malware: Characterization and Evolution. In *IEEE S&P*, 2012.

[33] Y. Zhou, Z. Wang, Wu Zhou and X. Jiang. Hey, You, Get off of My Market: Detecting Malicious Apps in Official and Alternative Android Markets . In *NDSS*, 2012.

Building Efficient Spectrum-Agile Devices for Dummies

Eugene Chai, Kang G. Shin
The University of Michigan
{zontar, kgshin}@eecs.umich.edu

Jeongkeun Lee, Sung-Ju Lee, Raul Etkin
Hewlett-Packard Labs
{jklee, sjlee, raul.etkin}@hp.com

ABSTRACT

Spectrum management and device coordination for Dynamic Spectrum Access (DSA) networks have received significant research attention. However, current wireless devices have yet to fully embrace DSA networks due to the difficulties in realizing spectrum-agile communications. We address the practical hurdles and present solutions towards implementing DSA devices, answering an important question "what is a simple practical extension to current wireless devices that makes them spectrum-agile?" To this end, we propose RODIN, a general per-frame spectrum-shaping protocol that has the following features to support DSA in commercial off-the-shelf (COTS) wireless devices: (a) direct manipulation of passband signals from COTS devices, (b) fast FPGA-based spectrum shaping, and (c) a novel preamble design for spectrum agreement. RODIN uses an FPGA-based spectrum shaper together with a preamble I-FOP to achieve per-frame spectrum shaping with a delay of under $10\mu s$.

Categories and Subject Descriptors

C.2.1 [**Computer-Communication Networks**]: Network Architecture and Design—*Wireless Communications*; C.2.2 [**Computer-Communication Networks**]: Network Protocols

General Terms

Algorithms, Experimentation, Design, Measurement, Performance

Keywords

Software-defined radio, hybrid radio, spectrum agility, per-frame spectrum shaping, spectrum-agile preamble detection

1. INTRODUCTION

Dynamic spectrum access (DSA), or spectrum agility, has become a popular solution to the problem of spectrum scarcity in wireless networks [6]. New devices that are designed to use only a monolithic block of spectrum can no longer expect to increase throughput by simply increasing their bandwidth. In fact, the throughput of an 802.11n device operating at 40MHz can even be *lower* than its throughput at 20MHz when encountering a 20MHz interference from another 802.11g or 802.11n device [8, 22]. Numerous other studies [12, 18] have reported performance anomalies when rate or bandwidth is blindly increased in an attempt to wrest more throughput from an overcrowded spectrum. We can only expect such problems to compound with the introduction of 802.11ac that supports up to 160MHz bandwidth. While this example deals with WiFi networks for clarity in exposition, the infeasibility of enhancing throughput by merely increasing bandwidth is also prevalent in non-WiFi networks. For example, a study of GSM usage patterns [15] shows that a wideband device cannot operate within the GSM band without some form of spectrum agility.

However, despite this obvious problem and the list of well-studied solutions, building efficient spectrum-agile devices is still a challenge for two main reasons. First, the current crop of commercial wireless devices are ill suited for DSA networks as they are primarily designed to use static, monolithic spectra. For example, spectrum- and bandwidth-agile platforms, such as SampleWidth [5] and FLUID [20], all have channel-switch times on the order of milliseconds. Second, the protocol stack does not fully support spectrum-agile communications. As an example, consider 802.11n OFDM frames that are detected by exploiting the self-correlation property of the preamble. This approach fails if the preamble is spread out over a non-contiguous spectrum, or in the face of interference from narrower band devices. Non-contiguous OFDM (NC-OFDM) techniques can be applied, but synchronization can be performed if and only if the set of non-contiguous subcarriers is known at the receiver beforehand.

We argue that the key capability that is missing from current state-of-the-art radio hardware is per-frame spectrum shaping. This is an important functional primitive that allows a radio to adapt to challenging channel conditions at the smallest practical unit of transmission.

1.1 Why Per-Frame Spectrum Shaping?

WiFi Channels. 802.11 devices are known to suffer significant performance degradation due to narrowband interference [11]. The effects of narrowband interference include timing recovery failure, the automatic gain control (AGC) failure due to an unexpected introduction of interference energy, and Physical Layer Convergence Protocol (PLCP) header processing failure.

Rapid frequency hopping (FH) by an 802.11 device [11] has been shown to improve its performance in the presence of narrowband interference. However, FH cannot avoid interference from a FH interferer, such as Bluetooth, if the hopping sequences of the WiFi and the interferer are not properly synchronized. Furthermore, col-

The work reported in this paper was supported in part by the NSF under Grant CNS-1160775 lisions between multiple FH devices using different hopping sequences is a well-known challenge when scaling FH to a larger network [17].

This disadvantage of FH comes from the fact that it switches channels blindly, even when there is no interference on the channel it is currently using. This increases the possibility of the FH itself interfering with devices on other channels. We posit that a reactive approach to interference avoidance using per-frame spectrum shaping will enable 802.11 devices to avoid narrowband interference while maintaining high throughput and manageability. The use of per-frame spectrum shaping effectively re-allocates the spectrum of a transmission dynamically only when interference is detected on the channel. This minimizes the amount of spectrum touched by an 802.11 device and avoids the unnecessary channel-switch overhead when no interference is detected.

Non-WiFi Channels. Devices operating in non-WiFi channels have to contend with severe spectrum fragmentation due to multiple narrowband interferers. We illustrate this using spectrum traces [29] that took measurements from a 1.5GHz band and is centered at 770MHz frequency. This trace set thus covers multiple GSM and TV channels.

Fig. 1 shows the availability and outage durations of 1, 5 and 20MHz monolithic channels operating within this band. Consider, in particular, the 20MHz transmission that is typical of WiFi devices. At a first glance, the long median channel-availability duration of 3s can easily accommodate the channel-switch time of typical WiFi devices. However, we observe from Fig. 2 that monolithic 20MHz channels can transmit only about 6% of the time. This low availability is due to the presence of multiple uncoordinated narrow bandwidth interferers. Hence, in order to sustain a 20MHz transmission, multiple discontiguous 1MHz (or narrower) channels have to be bonded together. Given that the correlation between the different channels is low [15], such a device can expect to continuously reconfigure its set of bonded channels to avoid primary user interference. The otherwise long outage duration that it faces, as shown in Fig. 1, will severely degrade the quality of service. The ability to perform per-frame spectrum shaping is thus key for operating in non-WiFi channels as well.

1.2 The Limitation of SDRs

Software-defined radios (SDRs) have been used to develop the flexible RF interfaces required for DSA devices. However, SDR platforms face problems arising from poor efficiency and high complexity. SDR platforms, such as USRP [27] and SORA [26], are limited by the efficiency of a general-purpose platform in multitasking real-time DSP with other system tasks, while FPGA-based SDR platforms, such as WARP [28], are complex to work with. This complexity and inefficiency poses a significant challenge because it is necessary to re-implement the entire MAC/PHY protocol on the SDR platform in order to reap the advantage of PHY-layer flexibility.

1.3 The Limitation of COTS Devices

A commercial off-the-shelf (COTS) device that has its RF frontend separated from the MAC baseband chipset can facilitate easy integration between the SDR and COTS. However, COTS devices are increasingly implemented as single-chip solutions to improve power and space efficiency. This limits the flexibility of the RF frontends of COTS devices in supporting the various spectrum management policies required for per-frame spectrum shaping.

1.4 The Challenge

We take a very different approach to DSA and address an important question: *"What is a simple practical extension to current wireless devices that makes them spectrum agile?"* We stress that any solution must be *general* enough to apply to the majority of COTS wireless devices currently available, yet *simple* enough to minimize the additional overhead that are added to COTS devices.

The intuition behind this comes from the fact that neither COTS devices nor SDRs are individually capable of supporting the per-frame spectrum shaping necessary for DSA. Hence, a hybrid platform built using both SDRs and COTS devices is necessary. The SDR handles only the necessary PHY-layer manipulations, while the COTS device handles the main MAC/PHY processing. A practical DSA extension must have the following three important properties.

Property 1: Protocol independence. It must support as many current wireless protocols as possible. Hence, a COTS device should only have to be "plugged into" a DSA extension platform to gain spectrum agility. In reality, some modifications to the COTS platform may be necessary, but such changes must be minimal. Easy deployability of a DSA extension platform will naturally maximize the chance of its widespread acceptance. With this property, RODIN can be easily integrated into both OFDM and non-OFDM COTS devices.

Property 2: Per-frame spectrum shaping. Per-frame spectrum shaping is a general spectrum-shaping primitive that can be used to construct other spectrum-management protocols. In the absence of detailed knowledge about the behavior of other devices in the ISM or whitespace bands, a DSA platform must be able to adjust its spectral use on a frame-by-frame basis to react to unexpected transmissions by primary users.

Property 3: Fast spectrum agreement. Besides having the capability of per-frame spectrum shaping, the transmitter and receiver(s) must also agree on a common set of (possibly non-contiguous) spectrum bands before commencing transmission. Prior work on spectrum agreement made use of control channels [31], pre-defined backup channel lists [23], or centralized channel assignment [20]. Unfortunately, these approaches are too slow to meet the required delay bounds for per-frame spectrum shaping.

1.5 Rodin: Our Solution

We propose RODIN[2]—a hardware DSA extension to COTS devices. RODIN consists of three key components that enable it to serve as a drop-in DSA extension to arbitrary wireless devices.

Direct connection to COTS device. RODIN connects to a COTS device directly through the antenna port(s) on the COTS radio, thus upgrading unmodified COTS devices with spectrum agility.

Fast FPGA-based spectrum shaping. RODIN can split the spectrum of an unmodified signal from the COTS device into multiple non-contiguous spectrum subbands; the individual subbands are transmitted on unoccupied portions of the spectrum to avoid interference from other narrowband transmitters. RODIN does not decode the signals to and from the COTS device. Our hardware implementation achieves this spectrum subdivision of each frame within $2\mu s$ of detecting a passband signal from the COTS device.

Novel preamble design for spectrum agreement. A RODIN transmitter uses a novel preamble design to notify a RODIN receiver of the spectrum occupied by the accompanying spectrally-reshaped frame. With this preamble, RODIN eliminates the need for a separate control channel, backup channel lists or a centralized spectrum coordinator. This preamble, when combined with fast spectrum shaping, enables RODIN to rapidly adapt to any primary transmission pattern seen on channels.

[2]Named after Auguste Rodin, the French sculptor.

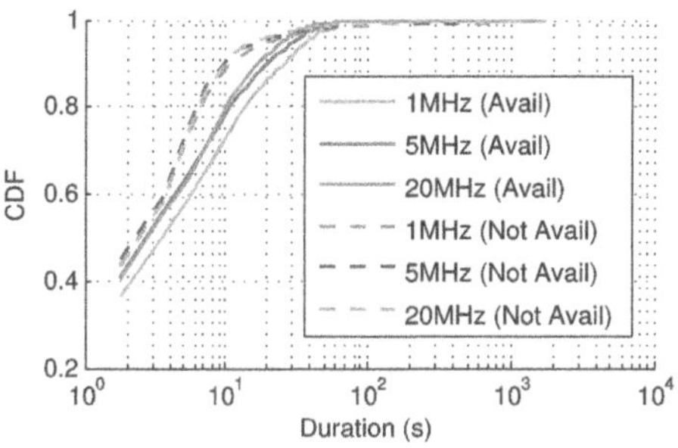

Figure 1: CDF of the channel busy and available durations.

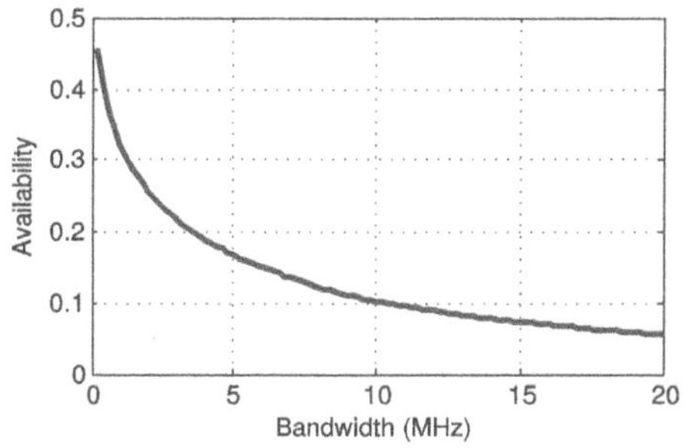

Figure 2: Channel availability of different transmission bandwidths.

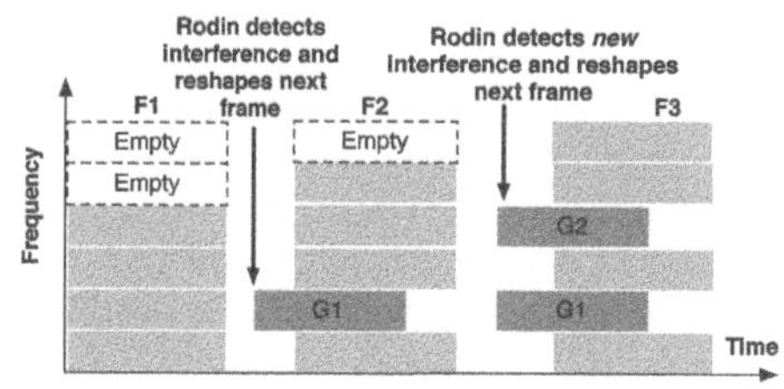

Figure 3: Transmission of 3 frames F_1, F_2 and F_3 using RODIN. RODIN reshapes the spectrum of F_2 and F_3 to avoid interference from G_1 and G_2, respectively.

To see how efficiently this can be done, consider shaping a 20MHz 802.11n frame over multiple 5MHz subbands. Spectrum agreement and shaping can be achieved in under $10\mu s$. This adds only 3.8% of additional overhead to the transmission time of an 802.11n frame without aggregation. The overhead will be even lower if frame aggregation is used. The negligible overhead enables RODIN to react to rapidly changing channel conditions on all types of channels.

RODIN is a novel RF frontend for COTS devices for cognitive spectrum management. In the short term, it extends the experimental capabilities of COTS devices but it can also be built into COTS devices to achieve integrated SDR-COTS hybrids in the future.

Our contributions in this paper are: (a) a detailed design of spectrum shaping and agreement in RODIN, (b) an evaluation of the real-world performance of RODIN via controlled experiments with FPGA-based implementations, and (c) an analysis of the performance of RODIN using detailed channel measurements.

2. OVERVIEW OF RODIN

RODIN is a general-purpose per-frame spectrum-sculpting platform designed for wideband frame-based COTS devices. In particular,

- RODIN is designed for wideband COTS devices that share the spectrum with other devices of narrower bandwidth. Examples of such scenarios include 160MHz 802.11ac or 40MHz 802.11n devices that share the same 5GHz band with 802.11a devices operating at 20MHz; UWB devices that share the spectrum with narrowband cellular networks.
- RODIN assumes that the maximum bandwidth of its SDR RF frontend is greater than the bandwidth of the transmitted COTS signal. RODIN shapes the spectrum of each frame while keeping the overall transmission bandwidth constant. Note that RODIN does not change the operating bandwidth of the COTS device.
- RODIN is designed for CSMA networks with multiple concurrent asynchronous transmitters that occupy non-overlapping spectra. This maximizes the frequency reuse of wireless channels. However, these channels are not perfectly orthogonal to each other due to non-ideal pulse shaping filters [16].

RODIN has three key features to function as a general per-frame spectrum-shaping platform for COTS devices: (a) capability for direct connection to the COTS device, (b) FPGA-based spectrum shaping, and (c) a novel preamble design for fast spectrum agreement.

RODIN divides its total RF bandwidth B into N subbands and shapes the spectrum of a frame that occupies N_F ($< N$) of these subbands. Fig. 3 shows an example of RODIN reshaping a wideband transmission, with $N = 6$ and $N_F = 4$, in the face of narrowband interference. Frame F_1 can be transmitted without any additional shaping since no interfering transmission is present. However, almost immediately after transmitting F_1, RODIN detects a narrowband interference G_1 that occupies one subband. It maps the spectrum of F_2 into the remaining subbands and transmits it without interfering with G_1. This frame-by-frame spectrum reshaping is repeated for F_3 to avoid interference from G_2.

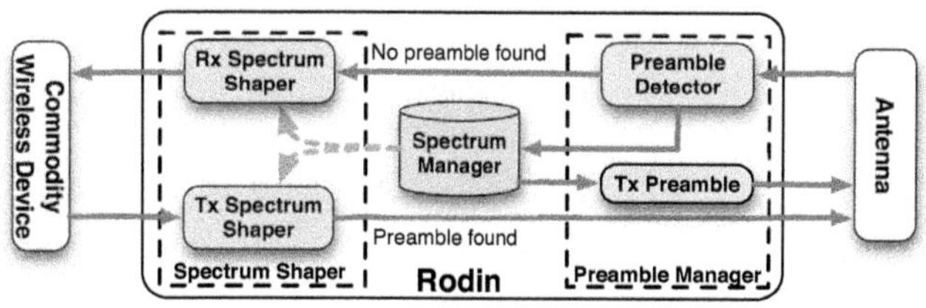

Figure 4: High-level architecture of RODIN.

If per-frame spectrum shaping is not used, a wideband transmission would be blocked by a narrowband transmission, or a wideband transmission collides with a narrowband transmission if the narrowband transmitter does not correctly detect the wideband transmission.

These features are realized with the system architecture shown in Fig. 4. The **Spectrum Shaper** reshapes the signal to and from the COTS wireless device in real time, while the **Preamble Manager**, consisting of a preamble detector and a preamble constructor, uses specially-constructed preambles to exchange spectrum information between RODIN devices. The **Spectrum Manager** executes a protocol that selects the best set of spectrum bands for a particular transmitter–receiver pair.

These components are detailed in the rest of this paper. For simplicity, our current design of RODIN is limited to SISO devices only, although an extension to MIMO devices is straightforward.

3. SPECTRUM SHAPING IN RODIN

Spectrum shaping divides the spectrum occupied by a COTS device into multiple discontiguous frequency bands. In order to realize real-time spectrum shaping, (a) the spectrum-shaping procedure must have low latency and (b) the spectrum shapers on the transmitter and the receiver must cooperate with minimal synchronization. Property (a) relates to the efficiency of the spectrum shaper — upon specification of the desired subbands, the shaper must quickly reshape the spectrum with minimal delay. In contrast, property (b) relates to the tolerance of the spectrum shaper to errors caused by channel distortion, timing, frequency shifts, etc. This is particularly important since different PHY protocols engage different measures to combat distortions. For example, DSSS-based protocols use Rake receivers and equalizers while OFDM-based protocols use the Schmidl-Cox algorithm. Obviously, it is not feasible for RODIN to support the wide variety of synchronization primitives to achieve protocol independence. Hence, RODIN focuses on spectrum shaping while leaving protocol-specific DSP functions (such as pilot handling) to the COTS device.

In the rest of this section, we only describe a two-band shaping process ($N > N_F = 2$) for the sake of clarity. This process can be easily extended to multi-band shaping.

3.1 Overview of Spectrum Shaping

Let $X(f)$ denote the original spectrum of the frame received by RODIN from the attached wireless device. The spectrum-shaping procedure for the frame *transmission* consists of the following components.

(a) Pre-filter modulation. RODIN only uses low-pass filters for spectrum shaping. Hence, the input signal $X(f)$ must be modulated to align the relevant portion of $X(f)$ with the passband of the filter $H(f)$. Let $m_1^{(a)}(t) = \exp\{j2\pi k_1 Bt/N\}$ and $m_2^{(a)}(t) = \exp\{j2\pi k_2 Bt/N\}$ be the time-domain complex-valued carrier used to modulate $X(f)$, with $k_i = 0, \ldots, N-1, \forall i = 1, 2$. The modulated spectrum is:

$$X_i^{(a)}(f) = X(f) * \delta(f - k_i B/N) = X(f - k_i B/N),\ \forall i = 1, 2 \quad (1)$$

where $\delta(\cdot)$ is the Dirac delta function.

(b) Filtering. Once the spectrum of the input signal has been appropriately modulated, a low-pass filter is applied to split the input spectrum into two separate subbands. Let $H_1(f)$ and $H_2(f)$ be the two low-pass filters used in this example. The two spectral subbands $X_1^{(b)}(f)$ and $X_2^{(b)}(f)$ are:

$$X_i^{(b)}(f) = H_i(f) X_i^{(a)}(f) = H_i(f) X(f - k_i B/N),\ \forall i = 1, 2 \quad (2)$$

(c) Post-filter modulation. Each filtered subband must be transmitted at a frequency that encounters minimum interference. This modulation step uses $m_1^{(c)}(t) = \exp\{j2\pi l_1 Bt/N\}$ and $m_2^{(c)}(t) = \exp\{j2\pi l_2 Bt/N\}$ as the modulating carrier, where $l_1, l_2 = 1, \ldots, N$. The second modulation step achieves, $\forall i = 1, 2$:

$$X_i^{(c)}(f) = X_i^{(b)}(f) * \delta(f - l_i B/N) = X_i^{(b)}(f - l_i B/N) = H_i(f - l_i B/N) X(f - (l_i + k_i)B/N) \quad (3)$$

(d) Combining spectra. Finally, the two subbands are added to produce a single spectrally non-contiguous frame. This results in a single time-domain data stream that is sent to the radio frontend of RODIN to be transmitted:

$$Y(f) = X_1^{(c)}(f) + X_2^{(c)}(f). \quad (4)$$

The RODIN *receiver* executes the same process as shown in Fig. 5 using the same low-pass filters but with the modulation sinusoids rearranged as:

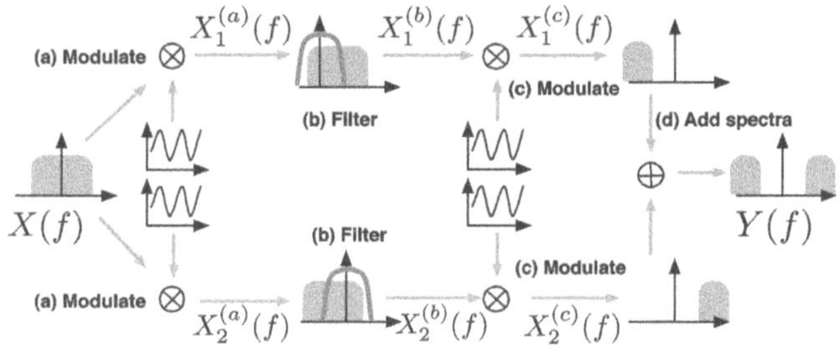

Figure 5: Shaping a frame occupying a contiguous spectrum $X(f)$ into two separate spectrum bands $Y(f)$. The shaping procedure is a 4-step process, labeled (a)-(d).

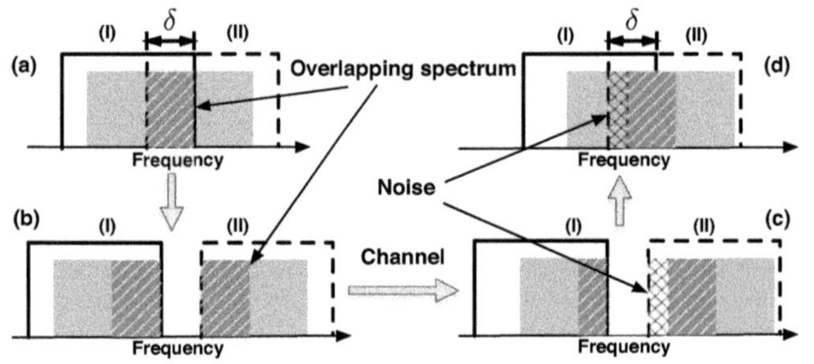

Figure 6: Spectrum shaping using two partially-overlapping filters. (a) Two subbands share an overlapping band δ. (b) After post-filter modulation, each subband contains a copy of the overlapping spectrum δ. (c) As a result of frequency drift at the receiver, only a portion of one subband is recovered while the other subband is recovered along with a noise band. (d) The overlapping spectrum δ ensures that the original spectrum can be reconstructed even if one subband is not recovered completely.

$$X(f) = \hat{Y}(f), Y(f) = \hat{X}(f),$$
$$m_1^{(a)}(t) = 1/m_1^{(c)}(t), m_2^{(a)}(t) = 1/m_1^{(c)}(t),$$
$$m_1^{(c)}(t) = 1/m_1^{(a)}(t), m_2^{(c)}(t) = 1/m_1^{(a)}(t)$$

where $\hat{Y}(f)$ is the spectrum of the received frame and $\hat{X}(f)$ is the spectrum of the reconstructed frame.

3.2 Filter Design for Spectrum Shaping

Prior work in spectrum shaping has largely adopted an OFDM-based approach [25,30,32]. While this approach draws upon many readily understood concepts similar to typical OFDM(A) modulation schemes, it has two significant disadvantages when applied to real-time spectrum shaping: (a) high overhead and complexity involved in maintaining strict time and frequency synchronization with pilot subcarriers, and (b) reduction in throughput due to the necessary use of a cyclic prefix to guard against inter-symbol interference.

RODIN mitigates these disadvantages with partially-overlapping finite-impulse response (FIR) spectrum shaping filters. Note that these FIR filters are only used for spectrum shaping. RODIN can support both OFDM and non-OFDM protocols using these FIR filters. RODIN itself is tolerant of timing drifts as time synchronization is handled by the attached COTS device as part of its PHY protocol; as long as the filtered spectrum encompasses the received frame, the COTS device can determine the appropriate frame boundary. RODIN is also resilient to frequency drifts by transmitting redundant spectral information through the use of partially-overlapping filters.

To understand this, consider the use of partially-overlapping filters to shape an input frame, as illustrated in Fig. 6. The two filters divide the spectrum into two portions, (I) and (II), that share a common overlapping subband of bandwidth δ, as shown in Figs. 6(a) and (b). A frequency shift at the receiver, as shown in Fig. 6(c), causes some spectrum to be lost from (I) and noise to be introduced into (II). Observe that when the two subbands are recombined, the spectral information missing from (I) can be recovered from its redundant copy in (II). The degree of resilience to frequency drift is governed by the overlapping bandwidth δ, which is a configuration parameter. We must ensure that the value chosen for δ is greater than the expected frequency drift. The lower bound on the overlapping bandwidth thus depends on the quality of the COTS device that RODIN is connected to. The effect of this noise is minimal since it is located at the very edge of the shaping filter and thus will be more heavily attenuated. Furthermore, this noise subband is typically very narrow as real-world measurements of actual frequency drift are shown to be small [4].

The overlapping bandwidth is also lower bounded by the amount of resources available on the FPGA: longer filters, which allow smaller overlapping bandwidths, require larger numbers of FPGA slices. The WARP platform used for our RODIN prototype can support a 64-tap filter.

The ideal requirements for a spectrum shaping filter are: (a) constant unit amplitude response and linear phase response in the passband, (b) narrow transition bandwidth, and (c) very high attenuation in the stopband. Unfortunately, neither the typical windowed-approach nor the Parks-McClellan algorithm can produce a filter that satisfactorily meets these three constraints. Thus, we adopt a *constrained least squares* algorithm [21] for filter design. We design our filters, using this algorithm, to have 64 taps, a passband ripple of 0.1dB and an overlapping spectrum bandwidth that is approximately 10% of the total filter bandwidth.

3.3 Spectrum-Shaping Latency

We have implemented the spectrum shaper using a 64-tap FIR filter on the FPGA of the WARP platform to both validate its functionality and study the latency incurred in real-time spectrum shaping. The FPGA on the WARP runs at 40MHz.

The modulation and spectral combination steps consists of time-domain multiplication and addition, respectively. Each step thus incurs a latency of 1 clock cycle. The filtering step consists of a 64-tap time-domain convolution, and incurs a latency of 64 cycles. Note that the filtering latency is *independent* of the number of subbands used since all filters run in parallel on the FPGA.

The total latency of real-time spectrum shaping is therefore $64 + 1 + 1 = 66$ cycles, or $1.65\mu s$ when running on the 40MHz FPGA. This spectrum-shaping latency is a mere 0.7% of the transmission time of a 1.5KB 802.11n frame sent at 54Mbps (Rodin currently only supports SISO). Hence, a real-time spectrum shaping extension to commodity wireless hardware is feasible.

4. PREAMBLE FOR SPECTRUM AGREEMENT

RODIN uses a unique preamble that is designed to indicate both the start of a frame as well as the spectrum bands it occupies.

4.1 Challenges to Spectrum Agreement

A frame sent by the transmitter can be decoded if and only if the spectrum occupied by the frame is known by the receiver. If the spectrum occupancy of a frame is unknown, the receiver can attempt to search for the frame over all the subbands. Assuming that a frame is known to occupy M out of N subbands, the receiver has to attempt to search for the frame over $N!/(M!(N-M)!)$ possible subband combinations; if the bandwidth of the frame is unknown, this search space increases to $\sum_{m=1}^{M} N!/(m!(N-m)!)$ subband combinations.

One might think of applying energy sensing to the subbands and decoding a frame using only the subbands with signal energy above a given threshold. This method, though simple, suffers from two serious limitations: (a) frequency-selective fading on the subband may result in a missed detection, and (b) in the case of multiple concurrent transmissions, each using a different set of subbands, it is impossible for a receiver to correctly map each occupied subband to its transmitter based on energy detection alone.

4.2 I-FOP Design

RODIN addresses this predicament by prepending a multi-subband preamble, I-FOP (In-Front Of Preamble), to the transmitted COTS frame. A unique preamble is assigned to each *flow* within the network, where a flow is simply a group of consecutive frames sent by the COTS device. This preamble must therefore be designed to (a) assign an address to each unique flow within the network, (b) specify the subband occupancy of each transmitted frame, and (c) enable the receiver to recover both the address and subband occupancy information of each frame without prior coordination with the transmitter. We stress that the spectrum occupancy can change from frame to frame even within the same flow.

A key feature that the preamble must possess is a strong correlation property — a receiver searching for a preamble P via correlations must encounter a large correlation peak if and only if P is present on the channel. Furthermore, this auto-correlation property must hold for a large set of sequences of the same length. This allows a different preamble to be assigned to each flow within a collision domain.

Zadoff-Chu (ZC) sequences [7] meet our requirements and are thus used in I-FOP. The length-L discrete ZC sequence is:

$$x_u[n] = \exp\left(-j\frac{\pi u n(n+1)}{L}\right) \tag{5}$$

where u is the sequence ID and $0 \leq n, u \leq L-1$. ZC sequences have strong correlation properties that make them ideal for I-FOP: (a) the auto-correlation of a length-L ZC sequence with a cyclically-shifted version of itself is zero if L is prime; (b) the cross correlation between two prime length ZC sequences is $1/\sqrt{L}$.

RODIN selects a set $\{p_0, \dots, p_{N_F-1}\}$ of ZC sequences to address a flow. The bandwidth of each frame within the flow occupies N_F subbands. RODIN applies a random cyclic shift to each sequence before constructing the preamble for the flow. The cross-correlation property reduces the chance of collision in the event that the same ZC sequence is selected by multiple transmitters. With this approach, there is a large set of L^2 ZC sequences of length-L that can be used to construct preambles.

Let $\mathbf{f} = \{f_0, \dots, f_{N_F-1}\}$ be the set of N_F subbands that RODIN uses to transmit a frame. The preamble constructed for this particular frame is specified by the set $\mathbf{S} = \{S_{f_k}^{p_k} : 0 \leq k \leq N_F - 1\}$, where $S_{f_k}^{p_k}$ indicates that sequence p_k is transmitted on the subband f_k and $f_0 \leq \ldots \leq f_{N_F-1}$. The time-domain representation of the preamble is:

$$y[n] = \sum_{k=0}^{N_F-1} x_{p_k}[n] \cdot e^{-j2\pi f_k n/N} \tag{6}$$

for $0 \leq n \leq L-1$.

4.3 I-FOP Detection

We assume, for now, that the transmitter and the receiver know the set of ZC sequences, $\{p_0, \ldots, p_{N_F-1}\}$, used to address the flow between them. The receiver faces the challenge of determining the set of subbands $\{f_0, \ldots, f_{N_F-1}\}$ occupied by the transmitted frame.

Let $\hat{\mathbf{S}} = \{\hat{S}_{f_k}^{p_k} : 0 \leq k \leq N_F - 1\}$ be the preamble that is detected by the receiver. This preamble detection procedure uses the following two properties of the transmitted preamble.

(a) The known order of the sequences. Given the set of ZC sequences, $\{p_0, \ldots, p_{N_F-1}\}$, used in the preamble, $\hat{\mathbf{S}}$ must be found such that $f_0 < f_1 < \ldots < f_{N_F-1}$. This increases the number of possible preambles by allowing for different preambles to be constructed using the same set of ZC sequences, but with different subband orders.

(b) Location of the correlation peaks. Multiple ZC sequences sent by the same transmitter as part of a single preamble will arrive at the receiver at approximately the same time. However, due to frequency-selective fading, the peaks may not be precisely aligned in time. To account for this, we use a threshold, ξ, to limit the range of acceptable separation between peaks—only sets of correlation peaks that are within ξ samples apart are considered as candidates for the preamble.

Algorithm 1: I-FOP detection.

Input : Set of ZC sequences $\mathbf{P} = \{p_0, \ldots, p_{N_F-1}\}$ RF sampling data stream, $\hat{y}[n]$, Correlation threshold, γ
Output: Occupied subbands $\mathbf{f} = \{f_0, \ldots, f_{N_F-1}\}$

1 **parallel-for** $k \in 0, \ldots, N-1$ **do**
2 /* Shift subband f_k to baseband */
3 $w_k[n] \leftarrow \hat{y}[n] \cdot e^{j2\pi f_k n/N}$;
4 **parallel-for** $l \in 0, \ldots, N_F - 1$ **do**
5 /* Correlate with p_l */
6 $\rho_{k,l}[n] \leftarrow (w_k \star p_l)[n]$;
7 $\lambda_{k,l} = \max_{0 \leq m \leq \xi} \rho_{k,l}[n-m]$;
8 **end-parallel-for**
9 /* Determine the ZC sequence on subband k */
10 $\sigma_k \leftarrow \arg\max_{0 \leq l \leq (N_F-1)} \lambda_{k,l}$;
11 $\eta_k \leftarrow \max_{0 \leq l \leq (N_F-1)} \lambda_{k,l}$;
12 **end-parallel-for**
13 $l \leftarrow 0$;
14 **for** $k \in 0, \ldots, N-1$ **do**
15 $f_l \leftarrow \infty$;
16 **if** $\sigma_k = p_l$ *and* $\eta_k > \gamma$ **then**
17 $f_l \leftarrow k$;
18 $l \leftarrow l + 1$;
19 **end**
20 **if** $l = N_F$ **then**
21 **return** $\mathbf{f} = \{f_0, \ldots, f_{N_F-1}\}$;
22 **end**
23 **end**
24 **return** $\mathbf{f} \leftarrow NULL$;

Algorithm 1 shows the pseudocode of the multi-preamble detection. In lines 1–12, RODIN searches for the ZC sequence that is transmitted in each subband. Observe that we use **parallel-for** loops for this search step since in an FPGA implementation, all iterations of these **parallel-for** loops can be executed concurrently to reduce the search time. In lines 13–24, RODIN searches for a set of subbands $\{f_0, \ldots, f_{N_F-1}\}$ that contain the sequences $\{p_0, \ldots, p_{N_F-1}\}$ such that $f_0 < \ldots < f_{N_F-1}$ must hold. Note that this **for** loop cannot be parallelized since the result of each iteration depends on the result of the previous iteration.

Table 1: Time required for preambles constructed with ZC of length 37, 73 and 113 to be transmitted at 5, 10 and 20MHz bandwidths.

		Preamble Length		
		37	73	113
BW (MHz)	5	$7.4\mu s$	$14.6\mu s$	$22.6\mu s$
	10	$3.7\mu s$	$7.3\mu s$	$11.3\mu s$
	20	$1.8\ us$	$3.65\mu s$	$5.56\mu s$

4.4 Inter-Subband Interference

Observe that RODIN does not apply any filter to isolate each subband before conducting a search for a ZC sequence. This choice is made to avoid the additional delay that comes with a filtering step. However, there is now a possibility that sequences on different subbands interfere with each other during the correlation-based search. This possibility is present regardless of the type of sequence used, e.g., Gold, ZC, Walsh-Hadamard, etc. However, we argue that the possibility of inter-subband collisions in our preamble design is very low.

A collision between two subbands can occur only if two or more different transmitters (a) select the same ZC sequence, (b) apply the same cyclic shift to the sequence, and (c) transmit at almost the same time. We posit that the probability of all three events occurring at even two non-colluding transmitters is very low. To gain some insight into this, first recall that in CSMA networks, the random backoff process undertaken by each transmitter minimizes the possibility of simultaneous transmissions. Even if simultaneous transmissions do occur, the set of ZC sequences can be made large enough to minimize the probability of collisions. For example, if we use ZC sequences of length 73, there are a total of $73 \times 73 = 5329$ possible sequences that can be used by RODIN. The probability of two devices picking the same sequence is a mere $(1/5329)^2 = 3.5 \times 10^{-8}$. Hence, inter-subband interference does not affect the performance of I-FOP.

4.5 I-FOP Delay

The spectrum-shaping delay incurred by I-FOP depends on two parameters: the length of the chosen ZC sequence, and the bandwidth at which each sequence is transmitted. Table 1 shows the transmission time required for each sequence built from ZC codes of 37, 73 and 113 samples long at 5, 10 and 20MHz. These subband bandwidths are suitable for use by 802.11 devices. The bandwidth of each transmitted sequence $S_{f_k}^{p_k}$ must be no larger than the bandwidth of each subband.

The delay at the receiver is due mainly to the processing time needed to find I-FOP. For every new sample, $\hat{y}[n]$, received by the detector in Algorithm 1, the **parallel-for** loops operate in constant $O(1)$ time while the search in lines 13-24 takes $O(N)$ time. With sufficient FPGA resources for full parallelism, the search can be completed in N clock cycles, or $(0.0225N)\mu s$ with a 40MHz FPGA.

As an example, if we spectrally shape a 20MHz 802.11n over a $B = 40$MHz RF bandwidth using the 64-tap filter from §3.3 and a preamble based on a length-37 ZC sequence, the overall delay is $1.65 + 7.4 = 9.05\mu s$. This is merely 3.8% of the transmission time of a 54Mbps 802.11n frame. The delay incurred by I-FOP may exceed the SIFS delay of WiFi COTS devices and trigger an

ACK timeout at the transmitter. However, these ACK timeouts can be easily changed in software [1] and do not pose a hurdle to SDR-COTS integration. This local SIFS modification allows the attached COTS device to account for the extra delay from I-FOP ; other non-RODIN WiFi devices can operate normally without modifications.

4.6 Preamble Address Assignment

RODIN devices must assign an address to each flow in a distributed manner before spectrum agreement between devices is completed. Addresses to new flows are assigned using an *association frame*.

An association frame is a control frame sent between RODIN devices, and is not passed to the COTS device. Each association frame is spectrally shaped to occupy only the available subbands and is prepended with a preamble constructed using a fixed set of ZC sequences. This set of ZC sequences is the *association set* and is known to all RODIN devices. The association frame contains only the IDs of the ZC sequences and the order in which they will be used.

A RODIN receiver searches all subbands for the association set. Once this association set is found, RODIN recovers the association frame using the spectrum shaper from §3. It then decodes the frame to obtain the ZC sequence information that will be used for subsequent frames from the same flow. Once an address has been assigned, all transmissions belonging to the same flow, even if they originate from different RODIN devices (e.g., DATA and ACK frames), use the same preamble address.

Since the information carried in the association frame is small, the size of the frame is small, especially when compared with the total size of the flow. Hence, the overhead of address assignment is negligible.

4.7 Subband Selection

The transmitter selects the subbands by choosing the N_F subbands that have the lowest energy levels at the point of frame transmission. We make use of an FFT (Fast Fourier Transform)-based energy detector — we take the FFT of incoming samples and measure the magnitude of the energy in each subcarrier. On the 40MHz FPGA, for example, a 128-bin FFT takes approximately $5\mu s$. Hence, energy values at any point in time are delayed by about $5\mu s$. This is acceptable since the channel state does not vary significantly over that short duration. Note that energy sensing delay decreases as the FFT length gets shorter.

On a faster and larger FPGA, we can also implement more advanced spectrum-scanning techniques, such as those based on the Spectrum Correlation Function [13]. This will enable RODIN to not only detect the currently occupied subbands, but also determine the protocol occupying them and predict future usage patterns of the interferer.

5. SPECTRUM MANAGEMENT

Algorithm 2 shows the pseudocode that defines the operation of the Spectrum Manager. RODIN is in the receive state until frames are detected from the COTS device. In this state, the RX spectrum-shaping filters are configured to span the occupied spectrum indicated by each received I-FOP.

When a frame is transmitted by the COTS device, RODIN first configures the TX spectrum-shaping filters and TX I-FOP to span the transmit spectrum subbands. The preamble is then transmitted while the samples from the COTS device are filtered and modulated. The spectrally shaped samples are transmitted after I-FOP transmission is complete.

Algorithm 2: Spectrum Manager.

while *True* **do**
  while *No frame from COTS device detected* **do**
    $\hat{y}[n] \leftarrow$ next sample from RF frontend;
    if *Preamble detected at* $\hat{y}[n]$ **then**
      Configure **Rx Spectrum Shaper** to span subbands of next frame;
    end
    Send $\hat{y}[n]$ to **Rx Spectrum Shaper**;
    Send output of **Rx Spectrum Shaper** to COTS wireless device;
  end
  while *Frame from COTS device detected* **do**
    Configure filters in **Tx Spectrum Shaper** to appropriate subbands, if necessary;
    Configure **Tx Preamble** to tag occupied subbands;
    Transmit preamble from **Tx Preamble**;
    $x[n] \leftarrow$ next sample from COTS device;
    Send $x[n]$ to **Tx Spectrum Shaper**;
    Send output of **Tx Spectrum Shaper** to RF frontend;
  end
end

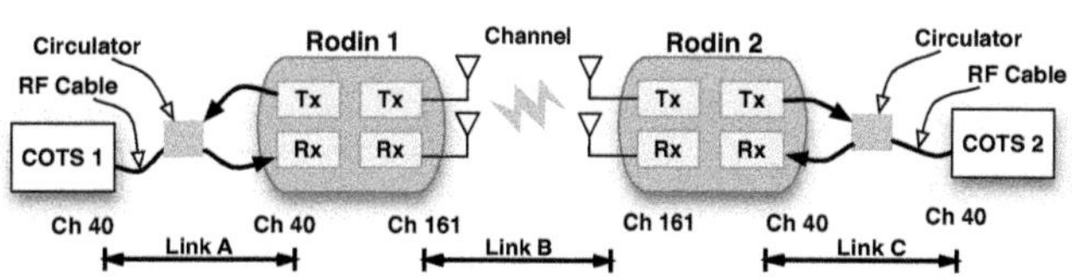

Figure 7: Experimental setup. Each RODIN device is connected to a COTS device via a coaxial cable.

6. EVALUATION: SPECTRUM SHAPING

6.1 Experiment Setup

Fig. 7 illustrates the setup used for evaluating the performance of individual RODIN devices. Each RODIN spectrum shaper is implemented in Verilog/VHDL and runs on the FPGA of a WARP platform with four radios. Each radio is permanently set to either the Tx or Rx mode. One pair of Tx/Rx radios from each WARP device is connected to a *circulator* that is then connected to a COTS device. These connections are made using coaxial cables. A circulator routes passband signals between the COTS device and the two radios on the WARP—analog signals coming from the COTS device is sent only to the Rx radio on the WARP, while signals from the Tx radio on the WARP is routed only to the COTS device. Signals between the Rx and Tx radios are blocked by the circulator.

The circulator is used here so that RODIN can receive frames from the COTS device without the Tx-Rx switching delay that will otherwise be incurred by the radio hardware if only one radio is connected to the COTS device. The other two Tx/Rx radios on each WARP device are connected directly to antennae. The two RODIN devices are placed approximately 2m apart. We have successfully used Ralink 802.11a WiFi card for COTS 1 and 2. However, to achieve finer-grained control of the transmitted signal for experimental purposes, we use WARP for COTS 1 and 2 for the rest of the experiments.

We send uncoded OFDM frames with a bandwidth of 10MHz between the two COTS devices. The spectrum of the OFDM frames

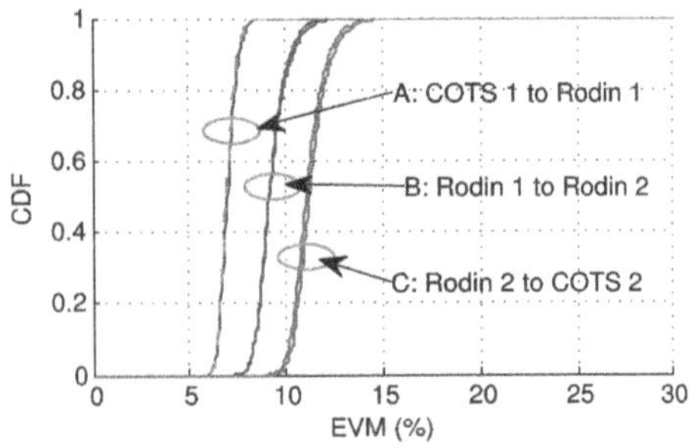

Figure 8: EVM of symbols in an OFDM frame with and without spectrum shaping. No interference.

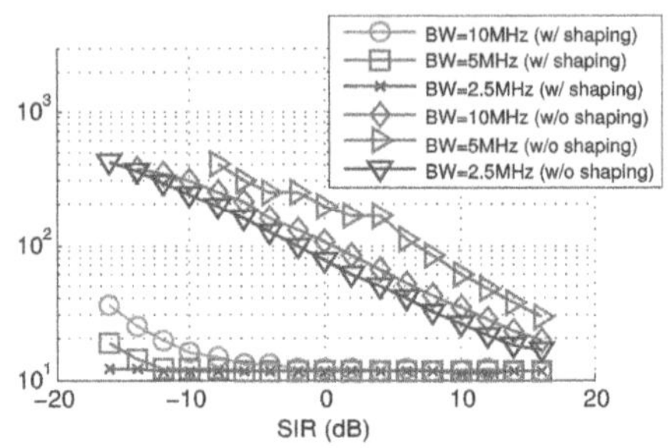

Figure 9: Mean EVM of OFDM frames measured at COTS 2 under different SIR levels.

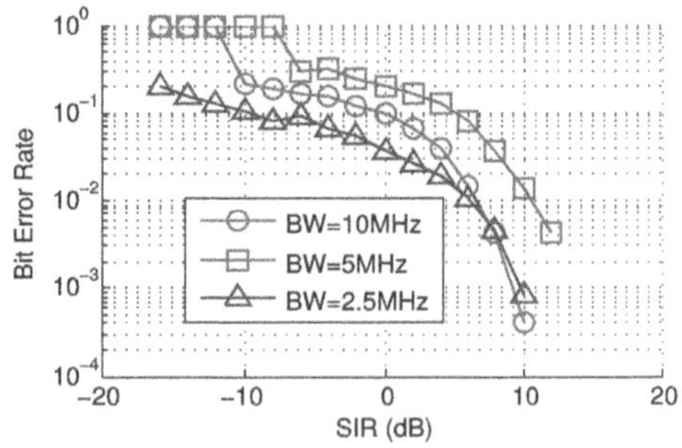

Figure 10: BER of OFDM frames measured at COTS 2 without shaping. No errors are encountered when spectrum shaping is used.

can be shaped to span any 10MHz of spectrum within the 20MHz maximum bandwidth supported by each radio. For all experiments in this section, we split the 10MHz OFDM frame into two subbands of 5MHz each. These subbands are transmitted with a 10MHz separation between them.

Each RODIN device detects transmissions from its attached COTS device by checking the RSSI of the Rx radio that is directly connected to the circulator. If the RSSI exceeds a predefined threshold, the COTS device is assumed to be transmitting. This can be done easily as the SNR of transmissions over the coaxial cable is high. At all other times, the Tx radio continuously transmits received signals to the COTS device for receiver processing. This maintains the capability of the COTS device to overhear transmissions from other devices that share the same discontiguous spectrum.

We use two metrics to measure the performance of the spectrum shaper: *Error Vector Magnitude* (EVM), which is shown as a percentage, and *Bit Error Rate* (BER), which is the fraction of bits received in error.

6.2 Spectrum Shaping Results

Without Interference. We transmit 2,000 OFDM frames using QPSK symbols from COTS 1 to COTS 2 using the setup in Fig. 7, and measure the mean EVM of the frames between each pair of directly connected devices. This experiment is conducted twice, once with and once without spectrum shaping. Fig. 8 shows the CDF of measured EVM. One important conclusion from this result is: *Spectrum shaping does not distort the signal.* The CDF of the EVM over each OFDM frame is identical with and without spectrum shaping of the transmitted OFDM frame. Hence, real-time spectrum shaping can be implemented in the FPGA without any loss of signal quality.

Direct manipulation of a signal from a COTS device with an attached RODIN platform does introduce some distortion into the signal. The median EVM of frames sent over Link A of Fig. 7 is 7% while median EVM of the frame that is spectrally shaped and sent over Link B is 9%. Finally, the transmission over Link C to COTS 2 increases the median EVM to 11%. (An EVM of 11% is small enough not to increase BER; BER of all frames transmitted in Fig. 8 is zero.) These additional distortions are introduced during (a) up and down signal modulation by the AD/DA converters at both COTS devices and the radios on the WARP, and (b) time and frequency offsets between the COTS device and its attached WARP. Both of these sources of distortion can be eliminated by tighter integration between RODIN and the COTS device: distortion due to up/down converters can be reduced by passing the baseband signal directly between RODIN and the COTS device; distortion due to time and frequency offsets can be mitigated by synchronizing RODIN with the clock used by the COTS device.

With Interference. We transmit an interfering signal using another WARP device. The transmission power of this signal is varied to achieve a range of Signal-to-Interference Ratios (SIR). At each interference power level, we transmit the interference at three different bandwidths—2.5, 5 and 10MHz. Fig. 9 shows the EVM of a 10MHz OFDM frame sent from COTS 1 to COTS 2 that experiences interference with bandwidth 2.5, 5 and 10MHz. This experiment is conducted over a range of SIR levels, with and without RODIN spectrum shaping.

We first consider the performance of spectrum shaping. The mean EVM of the OFDM transmission when SIR is greater than -2dB is 11%. This is equivalent to a spectrum-shaped OFDM transmission in the absence of interference, as shown in Fig. 8. At SIR levels lower than -2dB, the impact of interference on the OFDM transmission depends heavily on the interference bandwidth — interference with a 10MHz bandwidth increases the EVM to almost 40% while it remains at 11% when the bandwidth is 2.5MHz. This variation is due to the fact that filters used to generate the interference signal are not ideal. Hence, some energy leakage occurs at the edges of the filter. Although the two subbands of the spectrum-shaped OFDM frame are separated by 10MHz, they are still affected by the leaked interference energy. With a 10MHz interference bandwidth, the leakage energy is sufficient to distort the spectrum-shaped transmission. At 2.5MHz, the bandwidth of the interference is small enough that power leakage due to imperfect filters does not have a noticeable impact on the main OFDM transmission.

Without spectrum shaping, the narrowband interference has a significant impact on the OFDM transmission. For a given interference power, the smaller the interference bandwidth, the greater the interference power per subcarrier. The effect of this is seen from the fact that the distortion of the OFDM frames from the 5MHz interference is greater than that from the 10MHz frames—the increased interference power on fewer subcarriers is high enough to make up for the reduction in the number of subcarriers that encounter interference. When the interference bandwidth is at 2.5MHz, the small number of subcarriers affected allows the EVM to fall below that when a 10MHz interference is used.

This behavior is also evident when we consider the BER of the OFDM frames, as shown in Fig. 10. With spectrum shaping, the primary OFDM frames are sent on frequency bands that are not occupied by the interfering signal. The BER is thus zero for spectrum-shaped OFDM frames. Without spectrum shaping, the OFDM frame has a BER of 1.0 when it encounters a 10 or 5MHz interference at SIR below -12dB. The BER of the OFDM frame with a 2.5MHz in-

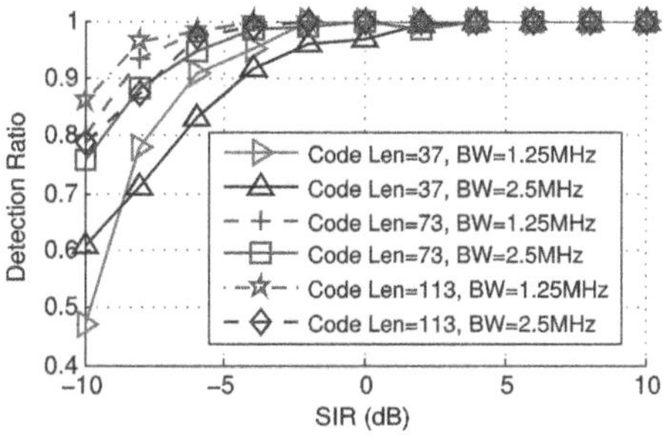

Figure 11: Preamble detection rate of three codeword lengths over $N = 8$ subbands on a 20MHz channel in the presence of interfering preambles. Each preamble is transmitted at 2.5MHz and 1.25MHz.

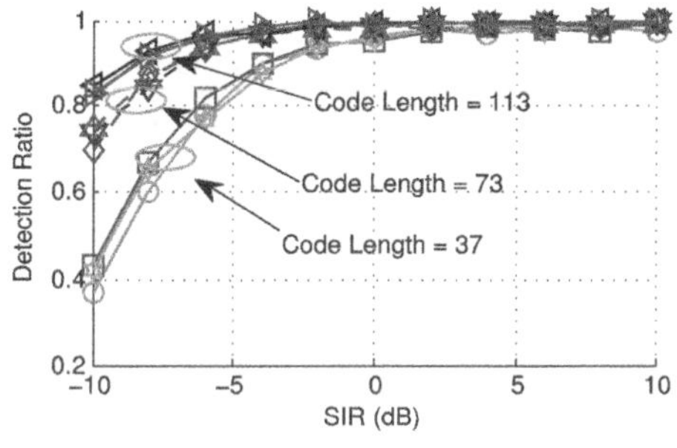

Figure 12: Preamble detection rate of three different codeword lengths over $N = 8$ subbands on a 20MHz channel. Each preamble is transmitted under 0, 12 and 20dB SNR.

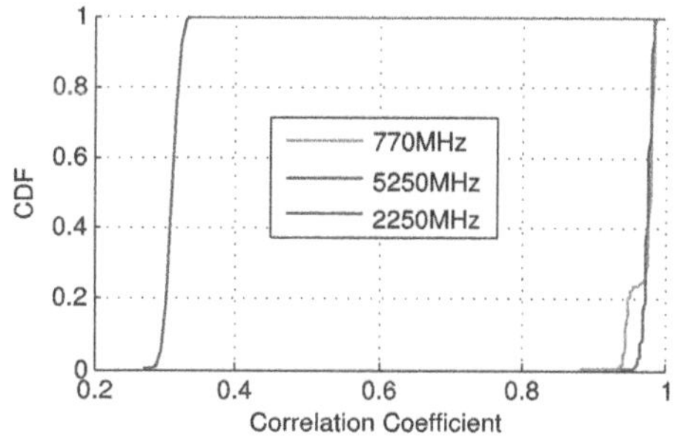

Figure 13: CDF of the correlation of the RSSI seen across all measurement slots over time.

terference is expectedly lower than that at interference bandwidths of 5 and 10MHz, but still stands at a high 1% at 8dB SIR.

7. EVALUATION: I-FOP

7.1 Experiment Setup

We evaluate I-FOP using five WARP devices placed at various locations around an office. Since the objective of this experiment is to evaluate the feasibility and performance of our preamble design, we run experiments using only WARPLab+MATLAB instead of an FPGA-based WARP implementation. The results obtained using WARPLab and an FPGA implementation will be identical.

The performance of I-FOP is evaluated under SIRs ranging from -10 to 10dB. This interference consists of different I-FOP that overlaps with the transmission of the primary I-FOP. The result for each SIR is the mean of 2,000 preamble transmissions. In each transmission, we select a random receiver, transmitter and interferer from five WARP devices. We use a 20MHz channel with $N = 8$ subbands (each subband is thus 2.5MHz wide). Three different preamble lengths are evaluated: 37, 73 and 113 samples. For every preamble, we randomly select $N_F = 4$ subbands and transmit a different ZC sequence on each one. All ZC sequences are transmitted at the same bandwidth.

The receiver searches for the known ZC sequences that belong to the primary preamble transmission using the procedure shown in Algorithm 1. If the set of ZC sequences is found in the specified order, the preamble is considered to be detected. Otherwise, a missed-detection is recorded.

We also evaluate the performance of the preamble under varying SNR levels. However, due to the difficulty of accurately controlling the noise level in the channel, SNR evaluations are conducted using a simulated 802.11 channel.

7.2 Experimental Results

Fig. 11 shows the detection probability of preambles with 3 different lengths, in the presence of overlapping interfering preambles. We run two experiments, with each one conducted over a range of SIR values. In the first experiment, each ZC sequence of every preamble (both the intended and interfering preambles) is sent at 2.5MHz (equal to the bandwidth of the subband); in the second experiment, each ZC sequence is sent at 1.25MHz, half the subband bandwidth. Interfering preambles are transmitted with a random time offset with respect to the non-interfering ones.

Under varying SIR levels. Observe that for preambles with the same length, the detection accuracy is greater as the bandwidth of each ZC sequence is reduced for two reasons. First, as the sampling rate of WARP is constant, the longer correlation period that results from a lower bandwidth ZC sequence gives a higher correlation peak magnitude when a match is found. Second, when ZC sequences are transmitted at 1.25MHz, there is a guard band between sequences on adjacent subbands. This reduces the inter-subband interference that arises due to energy leakage from adjacent subbands. No guard bands are present when the ZC sequences are sent at 2.5MHz.

Also, observe that the detection ratio increases with increasing ZC sequence length. This is because the peak auto-correlation magnitude is proportional to the sequence length L, while the cross-correlation magnitude of $1/\sqrt{L}$ actually *decreases* with increasing sequence length. These two effects cause the SNR of the correlation peak to increase with increasing ZC sequence length.

Under varying SNR levels. The accuracy of the preamble detector is similar over a wide range of SNR values, as shown in Fig. 12. For each ZC sequence length, we transmit the preamble at 0, 12 and 20dB SNR. Observe that accuracy is largely unaffected by the SNR level on the channel and is primarily dependent on the interference power.

In our experiments, the probability of detecting an I-FOP preamble when no I-FOP is present (false positive) is zero. False positives may occur due to ZC sequence collisions or more complicated channel fading scenarios. We can mitigate the effects of fading by using Rake correlators to search for the ZC sequences. However, false positives have limited impact on the operation of RODIN as the falsely received frame/signal are simply discarded by the COTS device.

8. EVALUATION: RODIN

We evaluate the performance of RODIN using simulations over detailed channel measurements from [29]. These channel measurements show the usage behavior of devices that operate on three separate bands. During periods when the channel RSSI is low, primary user activity is absent and spectrum agile devices can transmit opportunistically. Our objective is to show the efficacy of per-frame spectrum shaping in using these short-term transmission opportunities.

8.1 Simulation Setup

Trace data. Each channel measurement of [29] spans a 1.6 GHz bandwidth that is centered at three different frequencies 770, 2250 and 5250 MHz, so they cover the 2.4 GHz and 5 GHz ISM bands used by WiFi devices. Measurements were taken over several days at three different locations: for brevity, we only show results using the data set measured at rooftop of a school. Each sweep over the entire 1.6GHz bandwidth takes about 1.8s and captures 8,192 samples, with each sample spanning 200kHz. Although the measurement data does not capture channel usage patterns shorter than 1.8s, channel statistics have been shown to remain unchanged at shorter time scales [15]. This strongly suggests that we can expect such statistics to be present at sufficiently small time scales to make RODIN useful. Hence, our analysis using this data is still applicable even when considering finer-grained channel usage patterns.

Device models. We model three different types of wireless devices in our simulations; two that support spectrum shaping and one that does not. The maximum RF bandwidth of each device is 20MHz. The bandwidth of transmitted signal is 10 MHz, with the remaining 10MHz bandwidth used for spectrum reallocation. There are three models as follows.

(1) RODIN. This model uses per-frame spectrum shaping and the multi-subband preamble. At the beginning of each measurement slot (1.8s), RODIN measures the RSSI of all subbands and selects the N_F subbands with the lowest RSSI. This is equivalent to selecting the set of N_F subbands with the lowest interference powers. If all subbands have RSSIs lower than a predefined threshold, RODIN transmits a frame over those time slots. RODIN can carry out this measure-shape-transmit process within a single time slot due to its per-frame spectrum agreement and shaping capability. The performance of RODIN is modeled based on the I-FOP detection probability measured in the previous section.

(2) COTS-Spec. This model can bond multiple subbands for a single transmission, but cannot change the bonding on a per-frame basis. At the beginning of a time slot (1.8s), it selects the N_F subbands with the lowest RSSI as before. However, these selected subbands are used only in the *next* time slot. The set of subbands used for the current transmission is selected in the previous time slot. This represents the delay required by a COTS device to switch to a different set of subbands. Note that this is an optimistic model because (a) we do not consider the additional overhead required for spectrum agreement and (b) we assume that COTS-Spec can continue to transmit in the current time slot even as it is changing its set of bonded subbands.

(3) COTS-Mono. In this model, the COTS device makes use of the middle 10 MHz bandwidth for transmitting a frame, but no spectrum shaping is used. This represents a typical 802.11-type device that uses monolithic spectrum blocks for transmission.

Channel model. We are interested in finding the number of time slots during which each of these models can find a transmission opportunity. To evaluate the performance of each model, we partition each of the three traces into 20MHz channels (for a total of 81 channels) and simulate the operation of each model on all the channels. Each channel is divided into N subbands, of which N_F is used by the frame transmission. We repeat the experiments for RODIN and COTS-Spec using different subband bandwidths, b, and vary N and N_F to maintain $N \cdot b = 20\text{MHz}$ and $N_F \cdot b = 10\text{MHz}$. The threshold levels that we use for 770, 2250 and 5250MHz trace sets are -100, -90 and -90dBm, respectively. These are chosen to be similar to the 802.22 standard for 770MHz data set and the 802.11 standard for the others. We assume that there is only a single transmitter-receiver pair in each channel as it is sufficient to capture the behavior of the device models under a wide range of channel conditions. We leave the study of RODIN-to-RODIN interference to future work.

8.2 Simulation Results

Channel characteristics. The gain from per-frame spectrum shaping depends on the temporal variability—the more frequently the interference level on the channel changes, the greater the need for fast spectrum shaping. Fig. 13 shows the correlation coefficient of the RSSI on each measurement slot over time, for each trace set. Channels within the 5250MHz data set experience high temporal variability and have a median correlation coefficient of about 0.3. On the other hand, channels within the 770 and 2250MHz data sets experience minimal temporal variability, as seen by the high correlation coefficients. We expect the gain from per-frame spectrum shaping to thus be greater in the 5250MHz channels than in channels at other frequencies.

Transmission time slots. Fig. 14 shows the proportion of time slots in each channel in which the different devices can find transmission opportunities. Note that the channels are labeled in increasing order of their center frequencies. In the 5250MHz trace set, as shown in Fig. 14a, the high temporal variability of the channel means that subbands found to be available for transmission in one time slot are unlikely to still be available in the next time slot. Hence, COTS-Spec with 1MHz subbands can only transmit in up to 15% time slots. COTS-Spec with 2MHz subbands fails to find any transmission slots. A surprising result is that the performance of COTS-Mono is almost identical to that of COTS-Spec with 1MHz subbands. This shows that under highly varying channels, slow channel adaptation with narrow subbands performs almost identically to no spectrum adaptation; while slow channel adaptation with wider subbands fails to find any transmission opportunities.

The per-frame spectrum shaping of RODIN enables it to transmit on a significantly larger proportion of the time slots—up until 95% of the time slots in channel 81. Furthermore, we note that time slot utilization is increased when we use smaller subband bandwidths—RODIN using 1MHz subbands ($N = 20$, $N_F = 10$) can outperform the same device using 2MHz subbands ($N = 10$, $N_F = 5$) by more than 50% in some channels. Note that channels 1-50 in the 5250MHz data set fall into spectrum that is completely occupied by interferers. Hence, no slots can be found by any devices.

The performance of COTS-Spec improves under the low temporal variability of the 770 and 2250MHz trace sets. Fig. 14b shows that the fraction of time slots used by COTS-Spec is almost equal to that used by RODIN for transmissions. However, in Fig. 14c, we see that even in channels with high correlation coefficients, RODIN still finds more transmission opportunities than COTS-Spec at the same subband bandwidth. This is seen between channels 20 and 30. COTS-Mono performs poorly even on channels with low temporal variation, as shown in both Figs. 14b and 14c. Spectrum shaping is still necessary here as the low temporal channel variability does not imply the widespread availability of high bandwidth channels.

9. DISCUSSION

SDR-COTS hybrid. Our experimental setup in Fig. 7 is an example of an SDR-COTS hybrid, which enables interesting experiments impossible by either of COTS or SDR only. The SDR handles only the necessary PHY-layer manipulations, while the COTS device handles the main MAC/PHY processing. Since we need not support a full MAC/PHY protocol on RODIN, we use all the resources on the FPGA for a spectrum shaper and 64-tap spectrum correlation function. This hybrid architecture is applicable to both OFDM and non-OFDM-based COTS devices.

Interaction with COTS devices. Increasing the SDR-COTS integration can improve the performance of per-frame spectrum shaping. Using rate adaptation as an example, the SDR can provide the COTS hints on the SNR of other channels, so that the COTS device can immediately select the appropriate rate to match the per-frame spectrum when a spectrum reallocation is performed.

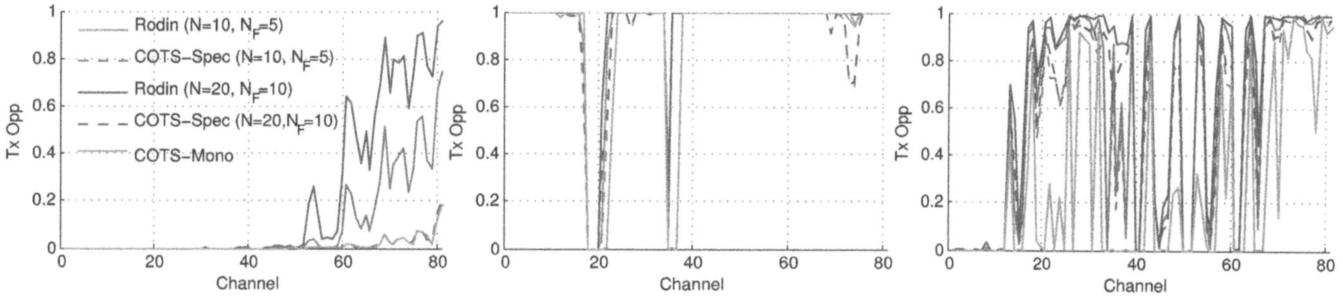

(a) 5250MHz. (b) 2250MHz. (c) 770MHz.

Figure 14: Proportion of time slots that each of the devices, RODIN, COTS-Spec and COTS-Mono, can transmit in.

Per-frame spectrum shaping in the network. RODIN transparently combines multiple spectrum fragments into a contiguous virtual channel that is seen by the COTS device. Since it obtains these spectrum fragments with a CSMA policy, we expect multiple RODIN nodes to interact without the need for more complex channel access protocols. Our current RODIN prototype is limited to single-link operation and we leave more detailed network-scale studies to future work.

COTS devices using non-contiguous spectrum. RODIN is designed for the case where the RF bandwidth of the SDR frontend is larger than that of the COTS device. At present, RODIN does not support COTS devices using non-contiguous bandwidths. As the SDR/ASIC platform evolves and supports larger bandwidths, RODIN can be extended to support non-adjacent frequency blocks.

RODIN **with more than two spectrum shaping filters.** Our experimental evaluation of spectrum shaping uses only two shaping filters due to FPGA resource constraints. However, given a larger FPGA, we can increase the number of shaping filters in RODIN. Furthermore, this can be accomplished while keeping the total overlapping bandwidth unchanged.

RODIN **with wideband COTS devices.** The variability in the channel response is known to increase with channel bandwidth. Hence if RODIN spreads a wideband spectrum (such as a 80MHz signal from an 802.11ac device) to an even wider band, additional processing steps such as RODIN-specific pilots might be necessary to compensate for the greater distortion seen on the channel. Other parameters, such as the overlapping bandwidth of the filters, might also need to be adjusted. However, since wideband COTS devices will already have built-in capability to accommodate the greater channel distortions, the modifications needed for RODIN might be minimal.

10. RELATED WORK

Spectrum Agility. WhiteFi [3] is a variable-bandwidth 802.11-based prototype that provides protocols that govern channel-switch triggers, channel probing and selection in whitespaces. This idea of variable-bandwidth communications is also used by FLUID [20] in enterprise networks. Jello [30] extends this variable bandwidth idea to support non-contiguous channel bonding in challenging networks. TIMO [10] adopts a different approach to handling interference on MIMO channels, treating interference as a single MIMO streams while simultaneously transmitting frames on the remaining MIMO streams. SVL [25] and Picasso [14] are both spectrum-shaping layers for general wireless devices. However, these solutions require tight integration with the COTS device's PHY and are not fast enough to support per-frame shaping. The new IEEE 802.11ac standard draft also specifies *non-contiguous 80+80 MHz* channel bonding as an optional feature [2], but does not support per-frame shaping. SWIFT [19] supports transmissions over non-contiguous bands while avoiding interference from narrowband devices. However, it differs from RODIN as it does not support per-frame spectrum shaping and agreement. Furthermore, it is not compatible with any available COTS devices and networks.

Spectrum Agreement. SIFTs [3], part of WhiteFi, is a single-channel bandwidth-independent signal detection algorithm used for determining the transmit bandwidth of an AP. FICA [24] uses binary amplitude modulation on multiple OFDM subcarriers, together with tight time synchronization, to enable each device to contend for different spectrum bands. Preamble detection on NC-OFDM networks [9] is useful for communications over disjoint spectral bands, but a separate mechanism must first be used to agree on the spectrum bands. Other typical uses for spectrum agreement include control channels [31] and backup channel lists [23].

11. CONCLUSION

We presented RODIN, a DSA extension platform that supports (a) direct connection to a COTS device, (b) fast FPGA-based spectrum shaping, and (c) I-FOP for fast and accurate spectrum agreement. A complete spectrum agreement and shaping operation can be carried out in about $10\mu s$, which adds only approximately 3.8% overhead to an 802.11n frame. We evaluated spectrum shaping and I-FOP using both simulations and real-world experiments, and demonstrated their efficacy even under low SIR levels.

12. ACKNOWLEDGEMENTS

We thank Suman Banerjee for shepherding our paper. We also thank Xinyu Zhang for the many insightful discussions on per-frame spectrum shaping.

13. REFERENCES

[1] Linux ath9k driver. `http://wireless.kernel.org/en/users/Drivers/ath9k`.

[2] Proposed TGac Draft Amendment (Draft 1.0) for IEEE 802.11 Wireless LANs, Jan. 2011.

[3] P. Bahl, R. Chandra, T. Moscibroda, R. Murty, and M. Welsh. White space networking with Wi-Fi like connectivity. *SIGCOMM*, 2009.

[4] V. Brik, S. Banerjee, M. Gruteser, and S. Oh. Wireless device identification with radiometric signatures. *MOBICOM*, 2008.

[5] R. Chandra, R. Mahajan, T. Moscibroda, R. Raghavendra, and P. Bahl. A case for adapting channel width in wireless networks. *SIGCOMM*, 2008.

[6] C.-T. Chou, N. Sai Shankar, H. Kim, and K. G. Shin. What and how much to gain by spectrum agility? *JSAC*, 25(3), Apr. 2007.

[7] D. Chu. Polyphase codes with good periodic correlation properties. *IEEE Trans. on Information Theory*, 1972.

[8] L. Deek, E. Garcia-Villegas, E. Belding, S.-J. Lee, and K. Almeroth. The impact of channel bonding on 802.11n network management. In *CoNEXT*, 2011.

[9] A. Dutta, D. Saha, D. Grunwald, and D. Sicker. Practical implementation of blind synchronization in NC-OFDM based cognitive radio networks. In *CoRoNet*, 2010.

[10] S. Gollakota, F. Adib, and D. Katabi. Clearing the RF smog: Making 802.11 robust to cross-technology interference. *SIGCOMM*, 2011.

[11] R. Gummadi, D. Wetherall, and B. Greenstein. Understanding and mitigating the impact of RF interference on 802.11 networks. *SIGCOMM*, 2007.

[12] M. Heusse and F. Rousseau. Performance anomaly of 802.11b. In *INFOCOM*, 2003.

[13] S. Hong and S. Katti. DOF : A Local Wireless Information Plane. In *SIGCOMM*, 2011.

[14] S. Hong, J. Mehlman, and S. Katti. Picasso: Full Duplex Signal Shaping to Exploit Fragmented Spectrum. *HotNets*, 2011.

[15] V. Kone, L. Yang, X. Yang, B. Y. Zhao, and H. Zheng. On the feasibility of effective opportunistic spectrum access. *IMC*, 2010.

[16] S. Lakshmanan, J. Lee, R. Etkin, S.-J. Lee, and R. Sivakumar. Realizing high performance multi-radio 802.11n wireless networks. In *SECON*, 2011.

[17] T. Lin and Y. Tseng. Collision analysis for a multi-Bluetooth picocells environment. *IEEE Communications Letters*, 7(10), 2003.

[18] M. Loiacono, J. Rosca, and W. Trappe. The snowball effect: Detailing performance anomalies of 802.11 rate adaptation. In *GLOBECOM*, 2007.

[19] H. Rahul, N. Kushman, D. Katabi, C. Sodini, and F. Edalat. Learning to share: narrowband-friendly wideband networks. *SIGCOMM*, 38(4):147–158, 2008.

[20] S. Rayanchu, V. Shrivastava, S. Banerjee, and R. Chandra. FLUID: Improving throughputs in enterprise wireless LANs through flexible channelization. In *MOBICOM*, 2011.

[21] I. Selesnick, M. Lang, and C. Burrus. Constrained least square design of FIR filters without specified transition bands. *IEEE Trans. on Signal Processing*, 1996.

[22] V. Shrivastava, S. Rayanchu, J. Yoon, and S. Banerjee. 802.11n Under the Microscope. *IMC*, 2008.

[23] J. So and N. Vaidya. Multi-channel MAC for ad hoc networks: handling multi-channel hidden terminals using a single transceiver. In *MOBICOM*, 2004.

[24] K. Tan, J. Fang, Y. Zhang, S. Chen, L. Shi, and J. Zhang. Fine-grained channel access in wireless LAN. In *SIGCOMM*, 2010.

[25] K. Tan, Y. He, H. Shen, J. Zhang, and Y. Zhang. Spectrum virtualization layer. Technical Report MSR-TR-2011-108, Microsoft Research, 2011.

[26] K. Tan, J. Zhang, J. Fang, H. Liu, Y. Ye, S. Wang, Y. Zhang, H. Wu, W. Wang, and G. M. Voelker. Sora: High performance software radio using general purpose multi-core. In *NSDI*, 2009.

[27] USRP. `http://www.ettus.com`.

[28] WARP. `http://mangocomm.com`.

[29] M. Wellens and P. Mähönen. Lessons learned from an extensive spectrum occupancy measurement campaign and a stochastic duty cycle model. *Mobile networks and applications*, 2010.

[30] L. Yang, W. Hou, L. Cao, B. Zhao, and H. Zheng. Supporting demanding wireless applications with frequency-agile radios. *NSDI*, 2010.

[31] Y. Yuan, P. Bahl, R. Chandra, and P. Chou. Knows: Kognitiv networking over white spaces. *DySPAN*, 2007.

[32] X. Zhang and K. G. Shin. Adaptive subcarrier nulling: Enabling partial spectrum sharing in wireless LANs. *ICNP*, 2011.

Asymmetric Caching: Improved Network Deduplication for Mobile Devices*

Shruti Sanadhya,[1] Raghupathy Sivakumar,[1] Kyu-Han Kim,[2] Paul Congdon,[2] Sriram Lakshmanan,[1] Jatinder Pal Singh[3]

[1] Georgia Institute of Technology, Atlanta, GA, U.S.A.
[2] Hewlett-Packard Laboratories, Palo Alto, CA, U.S.A.
[3] Xerox PARC, Palo Alto, CA, U.S.A.

shruti.sanadhya@cc.gatech.edu, siva@ece.gatech.edu, {kyu-han.kim, paul.congdon}@hp.com, sriram@ece.gatech.edu, jatinder@stanford.edu

ABSTRACT

Network deduplication (dedup) is an attractive approach to improve network performance for mobile devices. With traditional deduplication, the *dedup source* uses only the portion of the cache at the *dedup destination* that it is aware of. We argue in this work that in a mobile environment, the *dedup destination* (say the mobile) could have accumulated a much larger cache than what the current *dedup source* is aware of. This can occur because of several reasons ranging from the mobile consuming content through heterogeneous wireless technologies, to the mobile moving across different wireless networks.

In this context, we propose *asymmetric caching*, a solution that is overlaid on baseline network deduplication, but which allows the *dedup destination* to selectively feedback appropriate portions of its cache to the *dedup source* with the intent of improving the redundancy elimination efficiency. We show using traffic traces collected from 30 mobile users, that with asymmetric caching, over 89% of the achievable redundancy can be identified and eliminated *even when the dedup source* has less than *one hundredth of the cache size as the dedup destination*. Further, we show that the ratio of bytes saved from transmission at the *dedup source* because of asymmetric caching is over 6× that of the number of bytes sent as feedback. Finally, with a prototype implementation of asymmetric caching on both a Linux laptop and an Android smartphone, we demonstrate that the solution is deployable with reasonable CPU and memory overheads.

Categories and Subject Descriptors

C.2.m [**Computer-communication networks**]: Miscellaneous—*mobile networks, deduplication*

*This work was funded in part by the National Science Foundation under grant CNS-1017234 and the Georgia Tech Broadband Institute.

MobiCom'12, August 22–26, 2012, Istanbul, Turkey.

Keywords

Network deduplication, asymmetric caching, mobile traffic, mobile networks, mobile devices, bandwidth conservation

1. INTRODUCTION

Several recent efforts have established that there are considerable redundancies in network traffic [1, 2, 3] that are not fully leveraged by application layer approaches [4, 5, 6]. Network deduplication (dedup[1]) is a class of solutions that exploits such redundancies to improve network performance [1, 3, 7, 8]. Briefly, a dedup source (`dd-src`) intercepts traffic coming from the sender; segments the traffic into chunks of byte-sequences; and sends across only the hash of a byte-sequence if it is a repeating sequence. The dedup destination (`dd-dst`) then inflates any hashes back to the original byte-sequences and forwards the traffic to the eventual receiver. The reduction in the number of bytes sent between the `dd-src` and `dd-dst` results in improved network performance by allowing the network to sustain a greater traffic load and by reducing congestion levels for a given volume of traffic.

The application of dedup to wireless environments is an attractive proposition due to the ever-prevalent pressures on wireless capacities. Wireless service providers continually attempt to support more users and higher traffic loads without having to use additional spectrum, and dedup is a viable low-cost solution to do so. A straightforward deployment of dedup in a wireless environment would involve the `dd-src` residing in the wireless service provider's network (e.g. the SGSN in 3G) and the `dd-dst` residing on the mobile.

We argue in this paper that dedup faces a unique challenge when used in mobile environments. In static wireline environments, the `dd-src` is typically aware of the complete cache at the `dd-dst` by virtue of being on the data path to the destination. However, there exist several mobile scenarios where the `dd-src` is likely to have knowledge of only a small subset of the cache at the mobile. We elaborate on such scenarios in Section 2. However, if such an asymmetry exists between the caches at the `dd-src` and at the `dd-dst`, the efficiency of traditional dedup techniques is a restricted function of the smaller cache.

In this context, we consider the following question: *How can all of the past cached information at the mobile be successfully leveraged for dedup by any given* `dd-src`*?* In answering the

[1]For brevity we refer to network deduplication as dedup in rest of the paper.

question we categorize traditional dedup techniques as *symmetric caching* techniques where the dd-src and dd-dst maintain identical caches. We then introduce a new approach for dedup in mobile environments called *asymmetric caching*.

Fundamentally, asymmetric caching allows for the cache at the dd-dst to be larger than that at the dd-src. However, it enables the dd-dst to send feedback about portions of its cache to the dd-src. The feedback, sent in real-time, is selected to be pertinent to the ongoing traffic flow. The dd-src thus performs its operations not just based on its regular cache, but also based on the feedback received. The feedback selection problem is non-trivial because not only is the goal to increase redundancy elimination, but also to achieve a high feedback efficiency (ratio of bytes saved from transmission to bytes sent as feedback). Thus, the core of the asymmetric caching solution consists of an *application agnostic mechanism that can partition past and current traffic into contiguous byte sequences called* **flowlets** based on stationarity when the underlying byte stream is considered as a time series. Subsequently, feedback selection occurs by matching an arriving flowlet with a past flowlet, and then choosing content appropriately from the past flowlet to send as feedback. We elaborate on these mechanisms in Section 3.

Asymmetric caching achieves considerably better redundancy elimination by virtue of exploiting a much larger cache at the dd-dst. We show later with trace driven evaluations that asymmetric caching can increase redundancy elimination by over 100%, and provides such improvement even when the cache size at the dd-src is a fraction of that on the dd-dst. Furthermore, asymmetric caching achieves a feedback efficiency (ratio of bytes saved from transmission using feedback to bytes sent as feedback) of over $6X$. In other words, for every byte of feedback sent upstream, 6 *bytes* of downstream data are saved. In terms of adoption, asymmetric caching can be incrementally and independently deployed. A wireless service provider can thus deploy asymmetric caching to gain from all the past cached content accumulated on the mobile *without requiring any cooperation from other service providers the mobile might utilize*. Also, using prototype implementations of asymmetric caching on a laptop (Linux) and a smartphone (Android), we demonstrate that the CPU and memory overhead are quite reasonable.

In the rest of the paper we introduce the concept of asymmetric caching for dedup in mobile environments. We also answer several questions that arise including the following: How is the feedback chosen to make dedup perform better? Are the dedup benefits with asymmetric caching significant enough to justify the cost of the feedback? How much does asymmetric caching improve performance over a symmetric caching solution in a mobile environment? What are the overheads of implementing asymmetric caching on standard mobile platforms?

2. SCOPE AND MOTIVATION

2.1 Scope

The focus of this work is to enable better wireless network performance through the use of improved network deduplication for mobile devices. The technical contributions of the work broadly apply to a variety of mobile devices and networks. Nevertheless, we restrict the scope of the proposed work as follows:

- We specifically focus on **laptops** and **smartphones** as the mobile devices of interest, and **3G** and **WiFi** as the wireless technologies used at such devices.

- With regard to the wireless environment, since spectrum is inarguably more expensive in 3G environments, we primarily focus on 3G networks as the target environment for the proposed solution, but consider devices that consume content through both 3G and WiFi. However, our proposed solutions can be deployed in WiFi environments as well if required.

- While dedup can be deployed in an end-to-end fashion, we focus on a **last-hop layer 2.5 deployment** model for this work. In such a model, the dedup functionality is realized at entities on either side of the wireless link. While the mobile device is the only candidate deployment location for the dedup-dst, the dedup-src deployment location is likely to be a node such as the Serving GPRS Support Node (SGSN).[2] *In the rest of the paper, for brevity, we generically refer to the upstream dedup node as the* dd-src, *and the downstream dedup node as* dd-dst. Where the deployment location is relevant, we assume the SGSN as the deployment location for the dedup-src. However, the solution may be deployed in possibly other nodes (e.g. a dedicated dedup server) inside the wireless service provider's network.

- Finally, we restrict our focus in this work to dedup on the downstream and on only unencrypted traffic. Wireless traffic remains dominantly downstream and a significant portion of the traffic is not end-to-end encrypted. Hence, we believe that the contributions in the paper will have significant impact in spite of these restrictions. We leave for future work the extensions of the proposed strategies for upstream and end-to-end encrypted traffic.

2.2 Motivational Scenarios

Traditional network deduplication solutions require the dd-src to rely only upon portions of the dd-dst's cache that it is aware of. Such knowledge at the dd-src is implicitly accumulated when the corresponding data traffic flows through the dd-src en-route to the dd-dst. For static wireline hosts, such an arrangement is quite sufficient as the dd-src is always likely to be along the data-path to the dd-dst.

The basic premise of this work, however, is that for mobile devices using wireless connections, the dd-src is likely to be aware of only a fraction of the cache at the dd-dst. Thus, the dd-src is unable to perform deduplication to the fullest extent possible.

We now provide three scenarios in which the above *disconnect between the contents of the* dd-dst *cache and the* dd-src *cache* manifests itself.

- Multi-homed Devices: Most mobile devices today consume content through heterogeneous interfaces. WiFi is the preferred access technology when available due to its low cost and high data rate properties. However, 3G is the access technology used when users are not at locations with WiFi access. Recent studies of wireless data usage have profiled how both technologies are heavily used by mobile devices [9]. Moreover, cellular data offloading to WiFi is observed uniformly across both laptops and smartphones, and across different smartphone platforms [10].

  With traditional dedup, such data access offloaded to WiFi cannot be leveraged for redundancy elimination when the 3G interface is used, because the dd-src is different.

[2]We consider the SGSN as the point of upstream deployment as opposed to the Gateway GPRS Support Node (GGSN) since it already performs per-user functions such as encryption. The dedup-src can be Packet Data Service Node (PDSN) in CDMA networks, or the Access Point for WiFi networks.

- Resource Pooling: Cellular providers have increasingly started to perform IP core resource pooling that is part of the 3GPP standard. SGSN pooling is an example of this trend. In traditional GPRS networks, each SGSN, for example, is wholly responsible for its own service area. However, with SGSN pooling in 3G networks, all the SGSNs in the network work together, and the capacity load between them is distributed by the base station controllers (BSCs) and radio network controllers (RNCs). All BSCs and RNCs are connected to all SGSNs. Any mobile attached to the network is dynamically routed to the SGSN as per the current load distribution [11].

  Thus, the specific SGSN that serves a mobile at a certain point in time does not need to be the same SGSN that serves the mobile at a different time. If traditional dedup were to be used, the `dd-src` at a subsequent SGSN will be unable to use the entire data cache at the mobile because it has no knowledge of the cache entries accumulated through a different SGSN.[3]

- Memory Scalability: With traditional dedup, the `dd-src` dynamically creates a complete date cache for each associated `dd-dst`. A single SGSN typically serves 100,000-1,000,000 simultaneously attached users [12]. Even if SGSN pooling were not to be performed, requiring the SGSN to maintain persistent state across the different attachment sessions for a mobile is thus quite prohibitive. Thus, if the cache state per user is maintained only during the lifetime of that attachment session, then all data accumulated through past sessions will go unused.

All of the above scenarios point to the need for an approach that allows the `dd-src` to leverage the full extent of the cache at the mobile device, even if it might not have prior knowledge of the entire cache. In the rest of this section, we outline any additional goals we want such a solution to satisfy.

2.3 Goals

One approach to address the above-discussed problem is to enable the `dd-src` to fully leverage the cache at the mobile device, and therein increase the dedup efficiency for the downstream communication. However, the following additional goals are critical for the design of such a solution:

- Overall efficiency: While increasing dedup efficiency has the implicit result of helping in spectrum conservation by decreasing network load downstream, any solution has to be explicitly successful in using the overall spectrum (including both upstream and downstream) more efficiently.

- Application agnostic: Network deduplication is a generic technology that is application agnostic. Therefore, any solution to improve dedup has to remain application unaware, and hence applicable to any application used on the mobile device.

- Limited overheads: Both ends of the dedup solution (the SGSN or WiFi access point upstream, and the mobile device downstream) are resource constrained environments. Hence, any solution to improve dedup performance has to have deployable computational and memory complexities.

[3]SGSN pooling does not involve state transfer between SGSNs.

2.4 Background: Baseline Dedup

Network deduplication has been widely studied in related work, and in this paper we use the well known byte-sequence caching as the baseline dedup technique. [1] and [7] first introduced the concept of network deduplication, while [2] and [13] studied characteristics of real network traffic to establish that there indeed exists considerable amounts of redundancy that can be eliminated. More recently, [3] and [14] have proposed techniques to leverage redundancies for traffic reduction in an end-to-end fashion and by overhearing content in wireless networks, respectively. The byte-sequence based dedup is commonly used in both commercial WAN and storage optimization products (e.g., [8, 15]) and related research (e.g., [1, 2, 3, 7, 13, 14]).

Figure 1(a) depicts the operations of the byte-sequence caching. As shown in the figure, the byte-sequence caching algorithm essentially optimizes downlink traffic by replacing previously transmitted byte-sequences or segments of a packet with shorter hashes. Once the base-station receives a downstream packet destined to a mobile, it decomposes the packet into segments using Rabin Fingerprinting [16]. For each of the k segments of a packet, the hashes $[H_1, H_2, \ldots H_k]$ are computed using a known hashing algorithm such as Jenkins. If any of the hashes H_i is found in the cache, the corresponding segment in the packet is replaced by its hash. The resulting packet that includes both hashes and previously unsent data segments is then transmitted to the mobile. By virtue of the hashes being shorter, the load on the wireless link is reduced. At the mobile, the hashes for the data segments are computed and added to the hash table. Hashed segments are replaced with the corresponding original data before the packet is passed on to higher layers at the mobile.

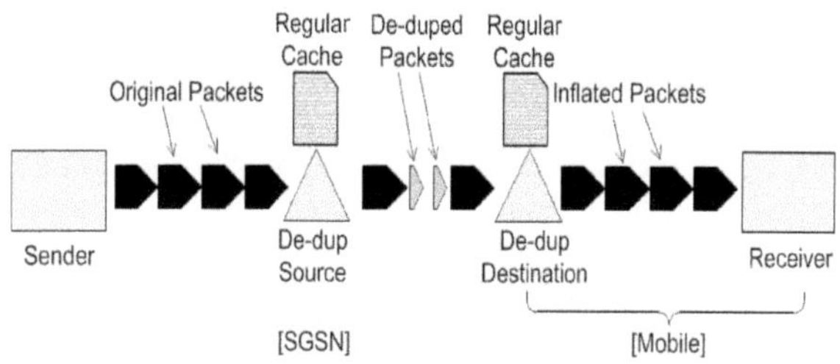

(a) Baseline dedup (byte-sequence caching)

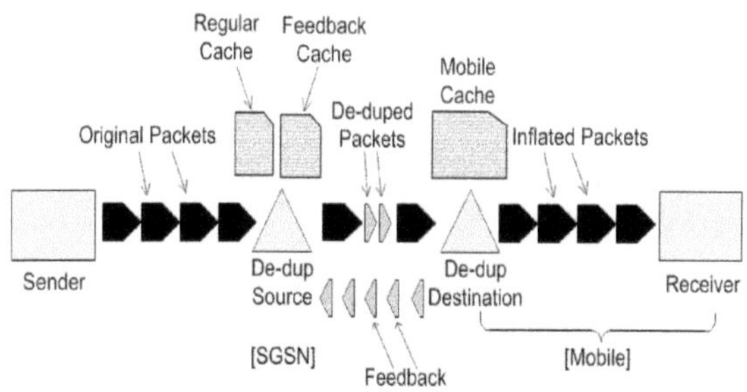

(b) Asymmetric caching

Figure 1: Baseline dedup vs. Asymmetric caching: Asymmetric caching is an overlay on baseline dedup.

3. ASYMMETRIC CACHING

In the rest of the paper, we present *asymmetric caching*, a solution that satisfies the goals identified earlier. Asymmetric caching performs dedup on the unencrypted downlink network traffic from the `dd-src` (at the SGSN) to the `dd-dst` (at the mobile) as shown in Figure 1(b). It is built atop a baseline dedup algorithm such as the one described in Section 2.4. At a high level, asymmetric caching enables the `dd-dst` to send timely feedback to the `dd-src` about selected portions of its cache. The feedback should be such that the redundancy elimination efficiency, when the `dd-src` uses both its regular cache and the feedback, approaches that of a scenario where the `dd-src` has complete knowledge of the `dd-dst`'s cache. While we present the details of the approach in Section 4, we describe the key design elements in the rest of this section.

3.1 When is the feedback sent?

Asymmetric caching uses a *reactive strategy* for sending feedback. Feedback is sent upstream only when data traffic is flowing to the destination. The matching of arriving content with past data in the cache is explicitly used for the selection of feedback that is likely to be most useful. Such a reactive strategy for sending feedback improves the redundancy elimination efficiency in the downstream *while maximizing the feedback efficiency*. An alternative proactive strategy that sends feedback even during idle downstream periods might have a better redundancy elimination performance by virtue of being able to send a larger volume of feedback. However, such an approach will not fare well in terms of feedback efficiency.

3.2 Where from is the feedback chosen?

The `dd-dst` cache, in asymmetric caching, is partitioned into *flowlets*. Each flowlet is a contiguous subset of a byte stream. The currently arriving traffic is partitioned into flowlets, and any arriving flowlet, say $flowlet_{arr}$ is matched with one of the past flowlets in cache. The feedback is then selected from that past flowlet.

The concept of flowlets is motivated by the fact that most content arriving at the destination is a composition of a collection of objects. For example, an HTTP connection carries different objects such as JPEG images, HTML files, CSS scripts, etc.; an SMB connection carries independent blocks of data as per the scope and sequence of requests; and a peer-to-peer application connection carries different chunks of data, not necessarily contiguous, as requested by the receiver. Thus, a desirable approach for matching arriving traffic to past content would be to match the currently arriving object to an object in the past, and then select feedback from that past object. This would enable the selection of relevant feedback and hence will be favorable to feedback efficiency. While objects may be identified easily if application knowledge is used, such an approach would violate the goal to remain application agnostic. Instead, in asymmetric caching, flowlets are considered as approximations of underlying objects in the data traffic, but are extracted using purely application unaware techniques. We elaborate on the approach next.

3.3 How are flowlets extracted?

In keeping with the application agnostic goal of the design, asymmetric caching employs statistical segmentation to break each downstream flow[4] into *flowlets* without any knowledge of the application. Asymmetric caching relies on the stationarity of the content of individual objects when considered as a time series to perform the segmentation. Thus, changes in the statistical distribution of the underlying byte stream is used to identify flowlet boundaries. Each flowlet is then stored as a sequence of hashes at the `dd-dst` cache. For any $flowlet_{arr}$, the `dd-dst` selects hashes from the past flowlet that matches the most.

[4]A flow consists of contiguous bytes/packets with same (source IP, destination IP, source port, destination port) tuple. A connection consists of an upstream and downstream flow.

The flowlet segmentation used in asymmetric caching is a variant of the strategy originally introduced in [17]. The segmentation strategy in [17] segments a piecewise stationary time series $\{X_1, X_2, ..., X_N\}$ into several separate time series $\{X_i, ..., X_j\}$, which are individually stationary. In asymmetric caching, the sequence of bytes in a flow is considered to be a time series, but the algorithm is simplified to grossly approximate position of the boundaries between the time series in return for a lowered computational complexity that is better suited for a resource-constrained mobile environment.

The approach takes a parameter l that is the minimum number of observations required to estimate reliable statistics of a series. At any given location $s(> l)$ in the series, three segments of the series are considered: segment from X_0 to X_s, segment from X_s to X_{s+l} and the aggregated segment from X_0 to X_{s+l}. An autoregressive (AR) model of order p is attempted to be fit on each segment, i.e.

$$X_i = \sum_{j=1}^{j=p} a_i X_{i-j} + \sigma\epsilon \qquad (1)$$

on each segment, where ϵ is a white noise (error term).

The gain ($\sigma^2_{a:b}$) of the white noise for the best fitting model on segment $\{X_a, ...X_b\}$ is computed from the sample covariance matrix of that segment. Next, a distance value $d_{0:s:s+l}$ is computed as:

$$d_{0:s:s+l} = (s+l)\log\sigma^2_{0:s+l} - s\log\sigma^2_{0:s} - l\log\sigma^2_{s:s+l} \qquad (2)$$

This is intuitively the extra power of the white noise (error) if the two segments are considered in one model as opposed to being in separate AR models. If this distance is more than a given threshold d_{thresh}, a boundary is said to exist between s and $s+l$. If a boundary is not detected, the next considered boundary is after l bytes. An empirical evaluation of the above solution shows that when presented with a mixed-source traffic consisting of different object types such as JPEG, TXT, XML, etc., the object byte boundaries are indeed approximately identified.

Finally, the flowlet in cache that has maximum number of matching hashes with the arriving traffic is identified to be $flowlet_{match}$, the flowlet from where feedback is selected.

3.4 How is the feedback selected?

Once $flowlet_{match}$ is identified, the specific feedback to be sent is selected based on two parameters: the location of the last matching hash between the $flowlet_{arr}$ and $flowlet_{match}$ and the latency for feedback on the upstream. Specifically, the location of the last matching hash in $flowlet_{match}$ offset by δ hashes, where δ is the number of segments that the `dd-src` is likely to have transmitted before the feedback reaches, is used as the start point for the feedback. δ depends on the average segment size, upstream data rate, and downstream data rate; all parameters computable at the `dd-dst`. γ hashes are selected from the start point, aggregated into a single packet and transmitted. Note that to ensure that the feedback is always new, the mobile keeps track of all the hashes that have occurred in past downstream packets from this `dd-src` or have been sent upstream in the past. Such hashes are explicitly removed from any feedback.

3.5 How is the feedback used?

Finally, the `dd-src` maintains a *feedback cache* in addition to its regular cache. Structurally, the feedback cache is identical to the regular cache and consists of a list of hashes available at the `dd-dst`. However, the feedback cache is populated only with hashes received through explicit feedback from the `dd-dst`. When

Algorithm 1 Operations at the `dd-src`

Input: *in_packet* = Packet received
Variables:
regular_cache = Regular cache at `dd-src`
feedback_cache = Feedback cache at `dd-src`
out_packet = Packet to be sent
pkt_chunks = List of chunks
pkt_hashes = List of hashes
seg_hash = Hash of a single chunk
on_flow = Flow to which *in_packet* belongs
shim_hdr = 16-bit header: first bit tells if the following sequence is a hash or original text, next 15 bits are used to specify the length of the sequence
Functions:
hash(chunk) = Compute hash of *chunk*
rabinFingerprints(string) = Return value based chunks of *string*
dd-srcDedup(in_packet)

```
1: if in_packet is going to dd-dst then
2:     pkt_chunks ← rabinFingerprints(packet)
3:     for each chunk in pkt_chunks do
4:         if hash(chunk) in regular_cache or hash(chunk) in feedback_cache then
5:             out_packet ← out_packet + shim_hdr + hash(chunk)
6:         else
7:             out_packet ← out_packet + shim_hdr + chunk
8:             add hash(chunk) to regular_cache
9:         end if
10:    end for
11:    send out_packet to dd-dst
12: else[packet is coming from dd-dst]
13:    pkt_hashes ← hashes in in_packet
14:    for each seg_hash in pkt_hashes do
15:        add seg_hash to feedback_cache
16:    end for
17:    out_packet ← IP and TCP headers of in_packet
18:    send out_packet to upstream node
19: end if
```

data arrives at the `dd-src`, each of its hashes is first looked up in the regular cache, and if there is no hit, the hash is added to the regular cache, but the same hash is then looked up in the feedback cache. If either of the cache lookups results in a hit, the hash is sent to the `dd-dst`. Otherwise, the original data segment is sent as-is. When a hash encounters a hit in the feedback cache, the hash is deleted after its first use since a corresponding entry would have been made into the regular cache.

4. SOLUTION DETAILS

This section presents the details of the asymmetric caching solution. We first describe its operations at the `dd-src` and the `dd-dst` respectively, and then discuss system details for the solution including packet formats and software architecture.

4.1 Operations at the `dd-src` (SGSN)

The operations of the asymmetric caching at the `dd-src` can be explained in two parts: downstream and upstream (see Algorithm 1). For every downstream packet, asymmetric caching first divides the packet into value based chunks using Rabin Fingerprinting (line 2). These chunks are then deduplicated using the *regular_cache* and *feedback_cache* (lines 3 to 10). Specifically, if a matching hash is found in either of the caches, the original chunk is replaced with a shim header and hash of the chunk. Otherwise, it is replaced with a shim header and the original chunk. For every new chunk, its hash is added to the *regular_cache* for future use. The new packet is then sent to the `dd-dst`.

Next, for an upstream packet, if it carries feedback from the `dd-dst`, asymmetric caching extracts all the hashes and inserts them into the *feedback_cache* (line 14 to 16). If the upstream packet was a piggybacked packet, the packet stripped of the feedback is forwarded upstream.

4.2 Operations at the `dd-dst` (mobile)

The asymmetric caching algorithm at the `dd-dst` can be explained in terms of three functions: cache maintenance, cache organization, and feedback selection.

First, for cache maintenance, asymmetric caching at the `dd-dst` (`dd-dstDedup`) maintains a local cache, called `dd-dst`*_cache*, that keeps chunks indexed by their hashes and all the flowlets seen thus far. When a packet is received from the `dd-src`, the `dd-dst` first reconstructs the original packet (line 3 to 10). The incoming packet is parsed into hashes and clear content by looking at the shim headers. The hashes are replaced by their corresponding chunks found in the `dd-dst`*_cache*, while clear content is copied as it is, without any shim headers. The reconstructed packet is sent up the network stack. The `dd-dst` then hashes all the Rabin chunks of the packet to create a list *seg_hashes* (lines 12 and 13).

Next, for cache organization, asymmetric caching uses the list of hashes, created and maintained by the above mechanism, to select relevant feedback for the ongoing flow. Specifically, the `dd-dst` first checks if a new flowlet has started in the current flow. This is done by the *updateFlowlets* module, which uses the segmentation approach presented in Section 3.3. It checks if a statistical boundary has occurred right before the current packet (line 6 and 7). If yes, a new flowlet is added in the current flow and the last flowlet of the flow is linked to it (line 8 to 11). The *gain* method is used to compute the power of white noise error term of any given series (i.e., sequence of bytes). The flow byte series is modified to remember the content of this packet (line 12). In case this is the first packet, a new flowlet is created for the ongoing flow and the flow byte series is set to be the packet's content (line 1 to 5)

Once the `dd-dst` has updated the current flowlet in the flow, the `dd-dst` can select and advertise feedback (*feedback selection*) using *seg_hashes*. To reduce the complexity of searching in the cache, the `dd-dst`*_cache* maintains a mapping from each hash to a list of past flowlets in which it had appeared. Only the past flowlets that have seen any of the hashes in the current flowlet are considered for feedback. After extracting these, *seg_hashes* are inserted in the `dd-dst` cache along with the current flowlet ID (line 17). Next, the *bestMatchedFlowlet* module takes the hashes of the packet and determines the number of hits seen in any past flowlet in the cache. The hit count for each past flowlet with the current flowlet is updated using the *subsequenceMatch* method. This method searches for the *seg_hashes* among ($win\_factor \times num\_hashes$) hashes in the past flowlet after the last matched hash in that past flowlet (line 5). The algorithm remembers where the current flowlet has last matched with the old flowlet. After these updates, the *bestMatchedFlowlet* selects the old flowlet with overall maximum hits for feedback. In case the best matching flowlet is the current flowlet itself, the module selects the second best matching flowlet that has $K\%$ of the maximum hits (line 5 to 9). Feedback is then selected from this old flowlet by the method *selectAdvertisement*.

The *selectAdvertisement* method keeps track of the last hash that matched between current flowlet and any old flowlet and also the last advertised hash for the pair. For the best matching flowlet, it first determines the later of these two pointers and then jumps δ hashes after that. This δ is the temporal offset to account for the feedback delay incurred by the underlying network. The larger

Algorithm 2 Operations at the `dd-dst`

Input: *in_packet* = Packet received
Variables:
d_thresh = Distance threshold to determine start of new flowlet
p = Order of AR() model to fit on the series
ar = Estimated coefficients of the *p* order AR model
win_factor = Region to be searched in the *past_flowlet*
`dd-dst`*_cache* = Extensive cache at `dd-dst`
out_packet = Packet to be sent
on_flow = Flow to which *in_packet* belongs
id_count = Number of flowlets seen so far, used as flowlet id
current_flowlet[on_flow] = Latest flowlet being created from *on_flow*
hit_count[flowlet][past_flowlet] = Redundant bytes between current flowlet and an old flowlet in `dd-dst`*_cache*
parsed_pkt = Mixed list of chunks and hashes in dedup packet
seg_hashes = List of hashes of chunks
adv_hashes = List of hashes to advertise
last_match[flowlet][past_flowlet] = Pointer to last matching hash in *past_flowlet* for *flowlet*
last_adv[flowlet][past_flowlet] = Pointer to last hash advertised from *past_flowlet* for *flowlet*
δ = Temporal offset to account for network delays
Functions:
hash(chunk): Compute hash of *chunk*
unhash(element): Fetch *chunk* (from cache) whose hash is *element*
rabinFingerprints(string) = Return value based chunks of *string*

gain(bseries)
1: $N \leftarrow len(bseries)$
2: covariance matrix $C \leftarrow [C_{i,j}]$, for $0 \le i,j \le p$, where $C_{i,j} \leftarrow \frac{1}{N-p}\sum_{k=p}^{N} bseries[k-i] * bseries[k-j]$
3: matrix $D \leftarrow [D_{i,j}]$, for $0 \le i,j \le p-1$, where $D_{i,j} \leftarrow C_{i+1,j+1}$
4: column vector $b \leftarrow [b_{i,0}]$, for $0 \le i \le p-1$, where $b_{i,0} \leftarrow C_{i+1,0}$
5: $\alpha \leftarrow D^{-1} * b$ ▷ estimate AR coefficients
6: vector $ar \leftarrow [1, \alpha_0, \ldots, \alpha_{p-1}]$
7: **return** $N * log(ar * C * ar^T)$

updateFlowlets(on_flow,packet)
1: **if** *series[on_flow]* is *null* **then**
2: *series[on_flow]* ← *packet* ▷ First packet in the flow
3: *current_flowlet[on_flow]* ← (++*id_count*)
4: **return**
5: **end if**
6: $d_{packet} \leftarrow gain(series[on\_flow] + packet) - gain(packet) - gain(series[on\_flow])$
7: **if** $d_{packet} > d_{thresh}$ **then**
8: *last_flowlet* ← *current_flowlet[on_flow]*
9: *current_flowlet[on_flow]* ← (++*id_count*)
10: link *last_flowlet* to *current_flowlet[on_flow]*
11: **end if**
12: *series[on_flow]* ← *packet*
13: **return**

subsequenceMatch(seg_hashes, flowlet, past_flowlet)
1: *num_hashes* ← *len(seg_hashes)*
2: **if** *last_match[flowlet][past_flowlet]* is *null* **then**
3: *last_match[flowlet][past_flowlet]* ← first hash in *past_flowlet*
4: **end if**
5: *max_seq* ← hashes in *seg_hashes* found among *win_factor* ∗ *num_hashes* hashes after *last_match[flowlet][past_flowlet]*
6: *last_match[flowlet][past_flowlet]* ← last hash in *max_seq*
7: **return** length of *max_seq*

bestMatchedFlowlet(flowlet,seg_hashes,old_flowlet_list)
1: **for** each *past_flowlet* in *old_flowlet_list* **do**
2: Add *subsequenceMatch(seg_hashes, flowlet, past_flowlet)* to *hit_count[flowlet][past_flowlet]*
3: **end for**
4: *best_past* ← *past_flowlet* with maximum *hit_count[flowlet][past_flowlet]*
5: **if** *best_past* is *flowlet* **then** ▷ Ongoing flowlet matches most
6: *top2* ← *past_flowlet* with 2nd most hits

Algorithm 2 Operations at the `dd-dst`(continued)

7: **if** *hit_count[flowlet][top2]* ≥ *K*% of *hit_count[flowlet][best_past]* **then**
8: *best_past* ← *top2*
9: **end if**
10: **end if**
11: **return** *best_past*

selectAdvertisement(flowlet,matched_flowlet)
1: *anchor* ← later of *last_match[flowlet][matched_flowlet]* and *last_adv[flowlet][matched_flowlet]*
2: *anchor* ← $\delta + 1$ hash after *anchor*
3: *adv_hashes* ← $MTU/(2 * hash\_length)$ hashes after *anchor* in *matched_flowlet*
4: Remove all hashes from *adv_hashes* which have been seen downstream or sent upstream before **return** *adv_hashes*

dd-dstDedup(in_packet)
1: **if** *in_packet* is coming from `dd-src` **then**
2: *on_flow* ← (src ip, dest ip, src port, dest port) of *packet*
3: *parsed_pkt* ← Parse *in_packet* into chunks and hashes
4: **for** each *element* in *parsed_pkt* **do**
5: **if** *element* is a hash **then**
6: *out_packet* ← *out_packet* + *unhash(element)*
7: **else**
8: *out_packet* ← *out_packet* + *element*
9: **end if**
10: **end for**
11: send *out_packet* to the application above
12: *pkt_chunks* ← *rabinFingerprints(out_packet)*
13: *seg_hashes* ← list of hashes of *pkt_chunks*
14: *updateFlowlets(on_flow, out_packet)*
15: *flowlet* ← *current_flowlet[on_flow]*
16: *old_flowlet_list* ← all the *past_flowlet* in `dd-dst`*_cache* in which any of *seg_hashes* have appeared
17: Insert *seg_hashes* with *flowlet* in `dd-dst`*_cache*
18: **if** *old_flowlet_list* is *null* **then**
19: *adv_hashes* ← *null*
20: **else**
21: *best_flowlet* ← *bestMatchedFlowlet(flowlet, seg_hashes, old_flowlet_list)*
22: *adv_hashes* ← *selectAdvertisement(flowlet, best_flowlet)*
23: **end if**
24: **else** ▷ packet is going to `dd-src`
25: **if** *adv_hashes* is not *null* **then**
26: *out_packet* ← *in_packet* + *adv_hashes*
27: **end if**
28: send *out_packet* to the `dd-src`
29: **end if**

the feedback delay, the more the offset must be for the feedback to be relevant when it reaches the `dd-src`. In our implementation we derive δ from the uplink and downlink data rates seen at the `dd-dst`. After including the temporal offset, the `dd-dst` selects $MTU/(2{*}hash\_length)$ hashes to advertise to the `dd-src`. We choose these many hashes as the minimum chunk size created by our *rabinFingerprints* method is $2 * hash\_length$, so $MTU/(2 * hash\_length)$ is the maximum number of chunks expected in a downstream packet.

This feedback is further optimized by removing from it all hashes that have occurred in the downstream or have been advertised upstream by the `dd-dst`. Note that if there are no hits in this packet, no feedback is generated (line 18 of `dd-dstDedup`). The feedback to the `dd-src` is opportunistically piggybacked on upstream data packets, e.g. TCP ACKs. If the `dd-dst` has some feedback to send in the form of hashes, it inserts these hashes into the payloads of upstream packets (line 26 in `dd-dstDedup`) and sends the packet upstream. If upstream data packets are not pending to be transmitted, a custom packet is constructed and transmitted.

Table 1: **Performance of hash algorithms in collision handling**

Hash algorithm	Digest size	Data set size	Collisions	TCP checksum detection
SHA1	20B	5GB	0	N/A
MD5	16B	5GB	0	N/A
Jenkins-8B	8B	5GB	0	N/A
Jenkins-4B	4B	5GB	0.02%	100%

4.3 Related Issues

Although the detailed algorithms presented above have established key insights for asymmetric caching, there are several issues associated with the design choices of the algorithms.

- *Hash function selection*: We use Bob Jenkins hash algorithm to create 8B hashes of the packet chunks on the `dd-src` and `dd-dst` [18]. The choice of hash is based on two conflicting goals: desire for more bandwidth savings by reducing the number of bytes sent on the network and minimum collision rate so that packets are not corrupted during dedup. Popular hash functions such as SHA1 and MD5 are typically computationally heavy and the digest size is large. We compare the aforementioned hash algorithms and the two versions of Jenkins hash (a 4B digest size and an 8B digest size) to hash the Rabin fingerprints generated over $5GB$ of traces. As shown in table 1, we observe no collisions with SHA-1, MD5 and Jenkins 8B hash, but 0.02% collision rate with Jenkins 4B hashing. Jenkins 8B provides a good trade-off between bandwidth savings and collisions, making it our choice for the implementation. Jenkins hash has also been used in prior work on network dedup [3, 14]
- *Handling hash collisions*: We use TCP checksum to detect hash collisions. A TCP checksum is computed on the header and the payload of a packet. As we do not mangle TCP headers, the checksum is sent unchanged. After reconstructing the original packet from a deduped packet the `dd-dst` device checks to see if the checksum matches. If not, it sends a control message upstream to the `dd-src`, requesting it to delete all the hashes that it had sent in the corresponding packet. The `dd-dst` also includes the hashes in the upstream packet. In our hash algorithm evaluation, we also apply this detection test to all cases. While the test was not invoked with SHA1, MD5 or Jenkins 8B, it was invoked by the Jenkins 4B hash. Table 1 shows that TCP checksum was able to detect 100% of the collision events.
- *Cache management*: Both the `dd-src` and `dd-dst` have finite cache space. Thus, asymmetric caching uses an LRU cache eviction policy. At the `dd-src`, LRU runs independently on the regular cache and the feedback cache, and evicts the least recently used hashes in each cache. As the `dd-dst` cache is organized in flowlets, the LRU on the `dd-dst` evicts least recently used flowlets at every run. All the state maintained for that flowlet is removed. The hashes (and chunks) seen in that flowlet are also removed unless they have also appeared in some other flowlet, which is still in the cache. A hash (and the original chunk) is evicted once the last flowlet referencing it is removed from the cache.

4.4 System Architecture

In the rest of this section we present a system architecture for asymmetric caching that is detailed in Figure 2. As shown in the figure, asymmetric caching (AC) works at layer 2.5 on the `dd-src` and `dd-dst`.

- *Dedup source:* This module (shown in the dotted-line box in Figure 2 (a)) captures the downstream packets at the base-station. The captured packet is then broken into chunks using Rabin Fingerprinting and hashes of each chunk are searched in the regular and feedback cache. If a matching hash is found, that chunk is replaced with its hash in the packet. If any packet is modified from its original, the IP options value is changed to reflect the same. We consider transport layer payload for dedup. Inside the transport layer payload, 2B[5] shim headers are inserted before every individual chunk and hash to demarcate original content from hashed content. Shim headers are not added if the entire packet is to be sent as original. This modified packet is then inserted back in the stack to be routed out.

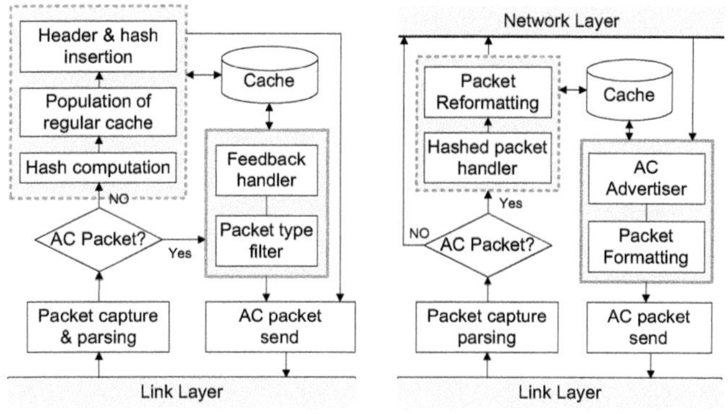

(a) Base-station (b) Mobile

Figure 2: Software prototype of asymmetric caching: It consists of `dd-src` at 2.5 layer of the base station and `dd-dst` at 2.5 layer of the mobile.

- *Dedup receiver:* At the mobile, the dedup packet is received by the receiver module (the dotted-line box in Figure 2 (b)). This module takes care of inflating the packet into its original form, updates flowlets in the caches, and selects the hashes to be advertised. It then passes the reconstructed packet to the higher layer (i.e., network layer).
- *Feedback source:* This module on the mobile device (the thick solid line box in Figure 2 (b)) is responsible for getting the hashes chosen by the dedup receiver and piggybacking the hashes on the next upstream packet. This module captures an outgoing packet, using Netfilter[19], modifies the IP options field and adds feedback to the packet. This packet is then inserted into the stack to be routed out.
- *Feedback receiver:* At the base-station, the feedback receiver (the thick solid line box in Figure 2 (a)) captures upstream IP packets with header options set. It then strips off the feedback from the payload, restores the original header and forwards the new packet to upstream nodes. The extracted hashes are inserted into the feedback cache at the base-station and used for further network deduplication.

5. PERFORMANCE EVALUATION

We evaluate asymmetric caching via trace-based analysis of real network traces. We first explain the trace collection environment, and then describe the trace analysis methodology. Finally, we present trace-based evaluation results as well as prototype-based experimental results.

[5]The first bit is set if it is a hash value and the rest of the bits indicate the length of the following chunk/hash

5.1 Collecting Network Traffic

We use real network traffic collected from 30 different mobile users (volunteers), 5 of whom are smartphone users and the rest are laptop users. We use these traces for performing the trace-based evaluation. Below are the details of the trace collection process.

- *Connectivity:* The laptop users relied only on WiFi connectivity for their network access. The smartphone users relied on both WiFi and 3G connectivity. The data collection spanned a period of 3 months and yielded over 26 *Gigabytes* of unsecured downlink data. Since we do not require any change in the user access pattern for the trace collection, users accessed the internet as per their normal behavior.
- *Devices and tools:* The laptop users ran Windows 7 and Linux operating systems and used Wireshark to collect their traces. The smartphone users used the Samsung Vibrant Galaxy phone and the HTC G2 phone, both running the Android 2.1 operating system on the T-Mobile network and relied on Tcpdump for trace collection. Users were able to parse trace files and remove any sensitive information before submitting them for analysis. Only unencrypted traffic was used in the analysis.
- *User demographics:* The volunteers included full-time employees at an industry research lab and an enterprise, as well as graduate students at a large university campus. The users span the age group of 21 to 50 and were spread over two different geographic regions.

5.2 Analysis Methodology

We use a custom trace analyzer to operate on the above traces. The analyzer models components of asymmetric caching presented in Section 4 and is configured and used for analysis, as follows:

Caches: The analyzer maintains three caches in the form of hash tables (i) `dd-src` regular cache (ii) `dd-src` feedback cache and (iii) `dd-dst` cache. It also maintains additional data structures required by the asymmetric caching algorithm at the `dd-dst`. We set the default `dd-src` cache size (i.e., regular + feedback) to 1MB and the default `dd-dst` cache size to 250MB. We explicitly study the sensitivity of the solution to cache sizes later in the section.

Past and Present Trace: To emulate the temporal history of the traces, the packet trace for each user is split equally into a past trace and a present trace (e.g., a 40MB trace was split into a 20MB *past* and a 20MB *present*. Then, the past trace is used as an input to the analyzer to populate the initial cache at the `dd-dst`. This is the memory collected at the `dd-dst` without the knowledge of the current `dd-src`. Next, in the present trace, a set of 30 connections is randomly selected and used for the second set of inputs to the analyzer.[6] We use a minimum threshold of 5KB for the size of connections under consideration to avoid very small (redundant) connections and to filter out insignificant connections from the analysis.

Metrics: Given the above set-up, we monitor three values for each user: (1) redundant bytes found in the `dd-src`'s regular cache, (2) redundant bytes found in the `dd-src`'s feedback cache and (3) total number of unique hashes sent as feedback from `dd-dst` to the `dd-src`.

Comparison: We also implement and run the byte-sequence caching algorithm (with average segment size of 128B) on a merged trace of the *past* and the 30 connections from the present for each user. This represents the scenario where the `dd-src` has all the hashes that the `dd-dst` has ever seen and thus gives a measure of 'ideally' achievable dedup.

[6]We have observed that the trend of redundancy is similar even when considering the entire present trace for each user.

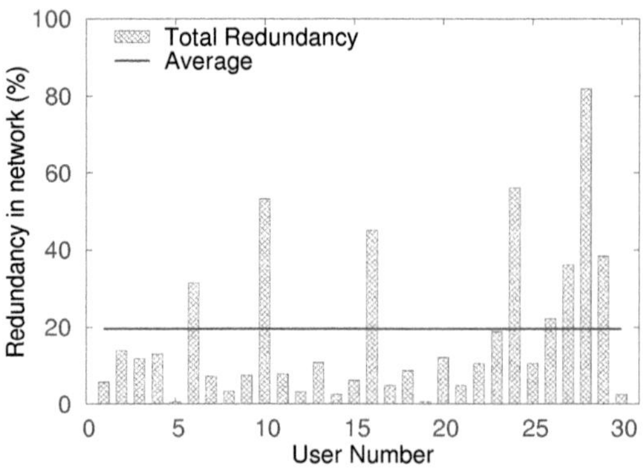

(a) Total network redundancy

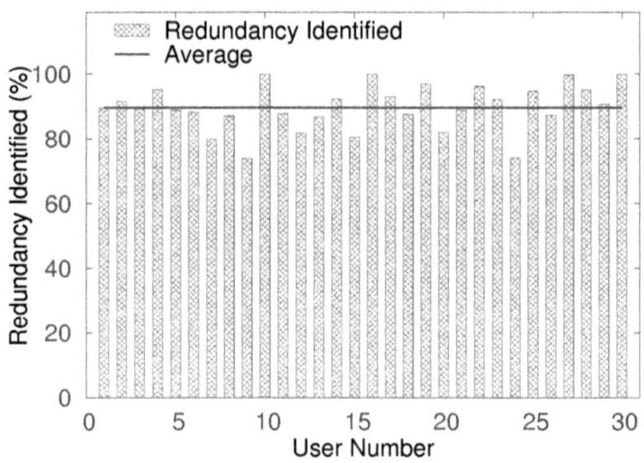

(b) Redundancy identified

Figure 3: Identifying network redundancy: (a) shows there exist network redundancy (avg., 19.6%) and (b) shows asymmetric caching finds most of network redundancy (avg., 89.7%).

5.3 Evaluation Results

5.3.1 *Identifying Network Redundancy*

We first show how much network redundancy exists in the collected traces and how much of that redundancy can be identified by asymmetric caching. Figure 3(a) plots total network redundancy that exists in the 30 user traces. Here, the total network redundancy is defined as the number of total cache hits at `dd-dst` for each chunk of a received packet, if the chunk exists in local cache (cache hit), we count the chunk as redundant bytes. As shown in the figure, there is indeed network redundancy of 19.6% on average.

Next, Figure 3(b) shows the percentage of the redundant bytes identified by asymmetric caching (found in Figure 3(a)). Specifically, we measure the number of cache hits at a `dd-src` based on both regular and feedback caches and use it for the redundant bytes identified. As shown in the figure, the asymmetric caching is able to identify on average 89.7% of the total redundancy, a considerable fraction of which is attributable to the feedback (see next subsection). Also note that its variance is small across 30 different users, owing to our fine-grained and adaptive advertisement scheme.

5.3.2 *Feedback Efficiency*

In this section, we present the feedback efficiency of asymmetric caching. Figure 4(a) plots the ratio (λ) of the redundant bytes

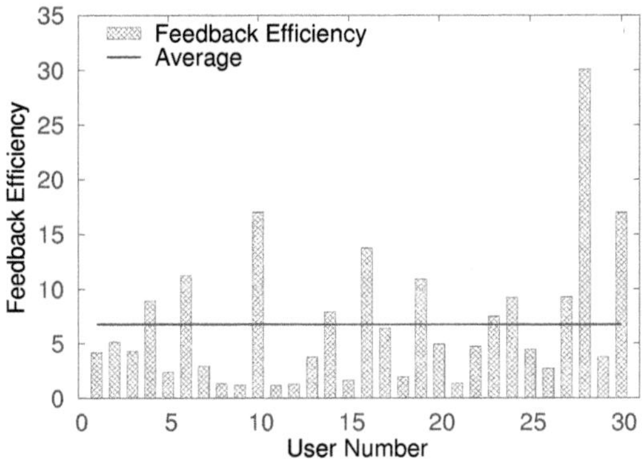

(a) Redundancy to feedback

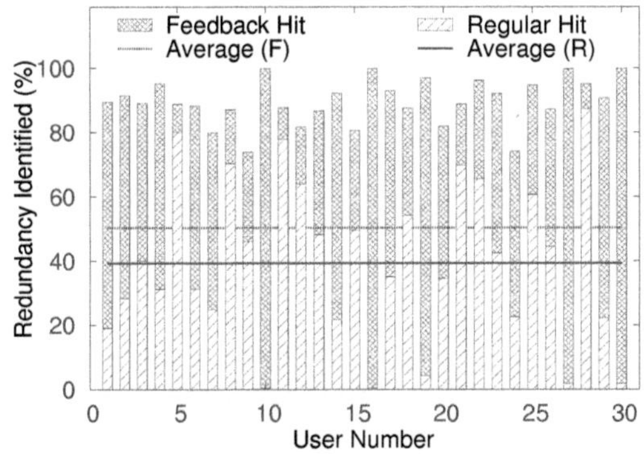

(b) Feedback hits vs Regular hits

Figure 4: Feedback Efficiency: (a) shows the ratio of the total redundancy identified to the feedback bytes (λ), and (b) shows how much each cache (feedback and regular) contributes to the redundancy detection.

identified to every feedback byte. If λ is greater than 1, asymmetric caching's feedback is effective in finding redundancy, and vice versa. As shown in the figure, the average λ value over 30 users is 6.74. One interesting observation is that the higher a user shows mobility (i.e., smartphone users including user 10, 28, 30), the higher λ is.

Next, we further study how much of the redundancy elimination is attributable to the feedback cache versus the regular cache. Recall that the redundant bytes are identified by searching its regular cache and then, if not found there, the feedback cache. For this, we analyze the redundancy found using only the feedback cache (F) and the redundancy found using only the regular cache (R). Figure 4(b) shows relative contribution from both R and F for each user. As shown in the figure, feedback cache largely contributes to 50.35% of the redundancy elimination, whereas the regular cache contributes to 39.3% of the redundancy elimination. Note that the hits in the regular cache can be considered as an indication of the performance of conventional dedup that relies only on symmetric caching. Hence, this result shows that asymmetric caching can improve the performance of dedup significantly.

5.3.3 *Sensitivity to Cache (Memory) Size*

In this section we measure the sensitivity of asymmetric caching to cache size. In this experiment, we first fix the cache size of the `dd-src` to 2MB. Then, while we increase the cache size of the `dd-dst` (from 5MB to >250MB), we analyze the percentage of

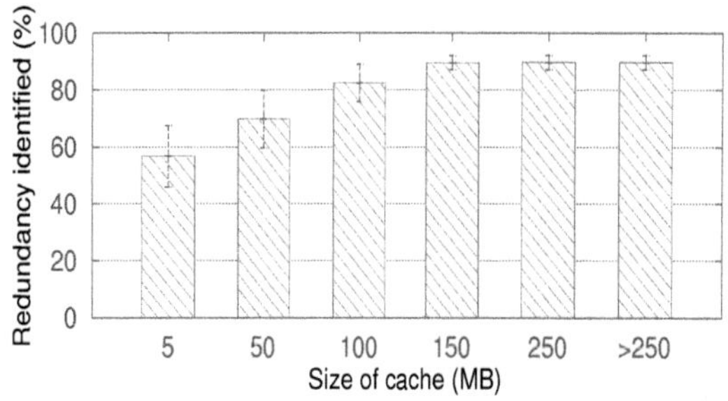

(a) Sensitivity to cache size at `dd-dst`

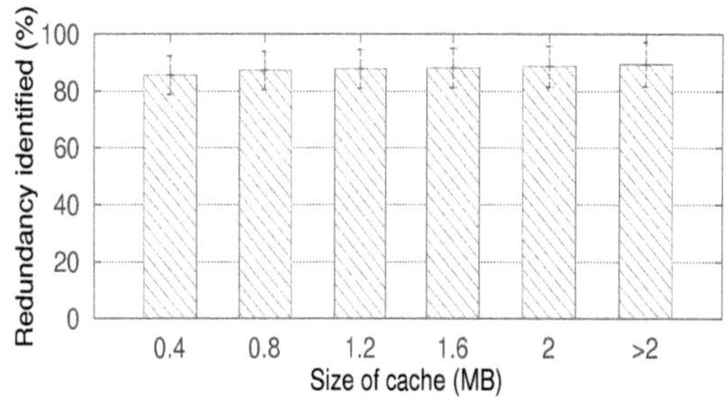

(b) Sensitivity to cache size at `dd-src`

Figure 5: Sensitivity to cache size: (a) shows that asymmetric caching can identify 89% of redundancy by using 150MB on the mobile device. (b) shows that the asymmetric caching requires only a small cache size at the `dd-src` (e.g., ~1MB) to achieve 85% of the redundancy detection.

redundancy identified by the asymmetric caching. We also analyze the opposite setting to study the sensitivity of `dd-src`'s cache size on its performance (250MB of `dd-dst` cache and from 0.4MB to >2MB of `dd-src` cache).

Figure 5 shows the results of the both experiments. First, as shown in Figure 5(a), given the constant size of cache at the `dd-src`, the redundancy identified by asymmetric caching increases with the increase in `dd-dst` cache size. This trend is a result of fewer cache evictions when using larger caches. In addition, with only 150MB of cache size at `dd-dst`, asymmetric caching is able to identify 89% of the redundancy.

Next, as shown in Figure 5(b), even with a small cache (e.g., ~1MB) at `dd-src`, asymmetric caching is able to identify more than 85% of redundancy. Finally, looking at the both figures in Figure 5, we can observe that for a 1:100 ratio in the cache sizes (e.g., ~1MB at the `dd-src` and 100MB at the `dd-dst`), asymmetric caching effectively detects and leverages redundancy. Furthermore, this ratio supports the design goal of asymmetric caching—the use of large cache at the `dd-dst` with a small cache requirement at the `dd-src`.

5.3.4 *Performance under varying data-rates*

So far we have assumed that the uplink and downlink data-rates are same, i.e. for every byte downstream, the mobile sends a byte of feedback upstream. In this section, we analyze the performance of asymmetric caching when the downlink and uplink data-rates are asymmetric. We vary the ratio of downlink to uplink data-rates from 1 through 5 and monitor the percentage of redundancy leveraged and feedback efficiency in each case. The results are shown in Figure 6.

We observe in Figure 6(a) that as downlink rate grows to 5$\times$ the uplink rate, the redundancy identified by asymmetric caching

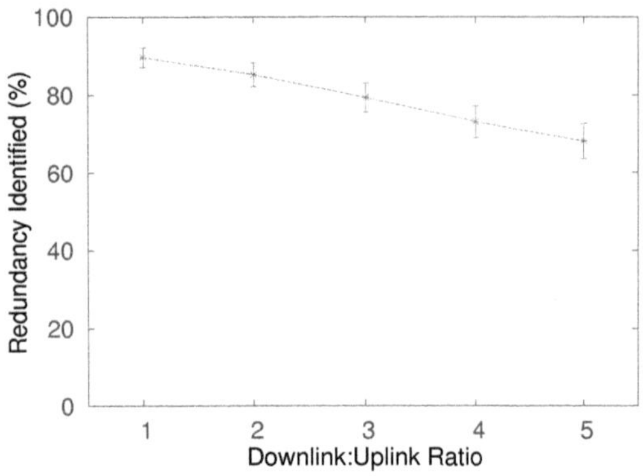

(a) Redundancy leveraged

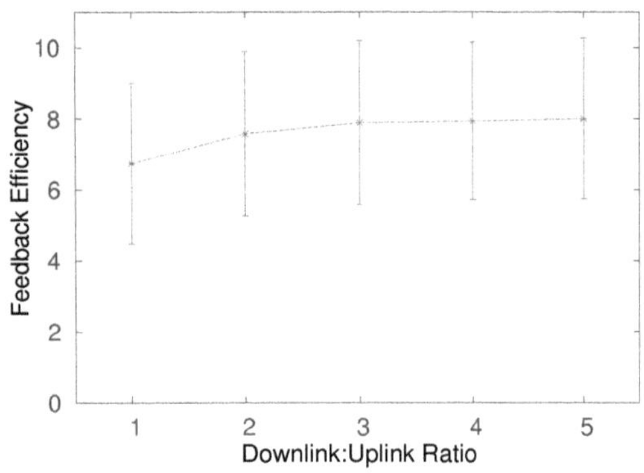

(b) Feedback efficiency

Figure 6: Performance with varying data-rates: (a) shows the % redundancy identified for increasing downlink to uplink data-rate ratio, and (b) shows the feedback efficiency in each case.

goes down to 68%. Interestingly, for the same ratio, the feedback efficiency grows to 8×, as shown in Figure 6(b). This shows that when asymmetric caching is restricted to send less upstream feedback, the redundancy elimination performance drops modestly and the feedback is more efficient.

5.3.5 Application-agnostic dedup

Finally, we measure how effectively asymmetric caching works for different types of application traffic. Recall that one of the design principles of asymmetric caching is to remain application agnostic— not require application information— but provide dedup performance regardless of application types. To this end, we analyze the trace of 30 users to calculate the percentage of redundancy identified by asymmetric caching over total network redundancy for each application. We classify the applications based on the port numbers used for the connections.

Table 2 shows the performance results of asymmetric caching

Table 2: Application-agnostic redundancy identification

Applications	Port Numbers	Redundancy Identified (%)
HTTP	80	98.76
ICSLAP	2869	80.48
Android Market	5228	53.10
McAfee, HP, SAP	5555	82.13
P2P and others	Ephemeral	97.70

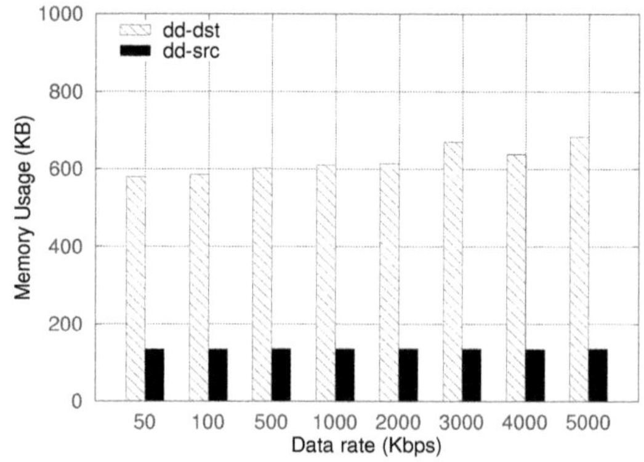

Figure 7: Memory consumed by Asymmetric Caching

under different types of applications. As shown in the table, the asymmetric caching is able to identify 53% to 97.76% of the network redundancy and it does not have radical performance penalty for specific types of application. This clearly supports our design goal of application-agnostic feature.

5.4 Prototype Results

To further validate the feasibility of asymmetric caching, we implement asymmetric caching in our testbed and have evaluated its CPU and memory overhead. We implement the `dd-src` on a desktop running Ubuntu OS with a 2GHz dual-core processor and 2GB memory. The desktop is equipped with an Atheros chipset based network card (NIC) with Madwifi [20] driver and talks to the mobile over wireless link. Using Madwifi, the `dd-src` is set in master mode. The desktop is connected to another desktop, which serves as server, over a wired link.

Next, the `dd-dst` is implemented on a Samsung Nexus S smartphone with 512MB RAM and 1GHz processor, running Android OS. The `dd-dst` has been implemented as a user space Linux module on the Android system. We use Netfilter framework on Linux to capture and modify downstream and upstream packets.

We use Iperf [21] to create packets with random content and send them over the network. We run Iperf at different data rates (from 50Kbps to 5Mbps) for 60s, and record the CPU and memory footprint of asymmetric caching on both `dd-src` and `dd-dst`, using Top utility. Our experiments show that the memory footprint of asymmetric caching is small enough to be deployed on a mobile phone and base-station. As shown in Figure 7, the physical memory taken by the `dd-dst` process stays around 600KB in our experiments while at the `dd-src` it is even lower at 140KB. This supports our goal of 'asymmetric' cache sizes, reducing the overhead on a `dd-src` serving millions of mobile devices.

In terms of CPU consumption, Figure 8 shows that the CPU usage on the mobile phone grows with growth in data-rate, but stays 15% even at 5Mbps data-rate. The CPU usage on the `dd-src` stays very minimal, less than 1% throughout. The CPU usage on the mobile increases with the data rate as much larger caches have to be searched to generate relevant feedback.

Overall, our user space implementation shows that asymmetric caching is feasible on both the ends of dedup identified in our motivational scenarios.

6. RELATED WORK

Network dedup approaches: The notion of network dedup was first presented in [7], where packets are decomposed into segments

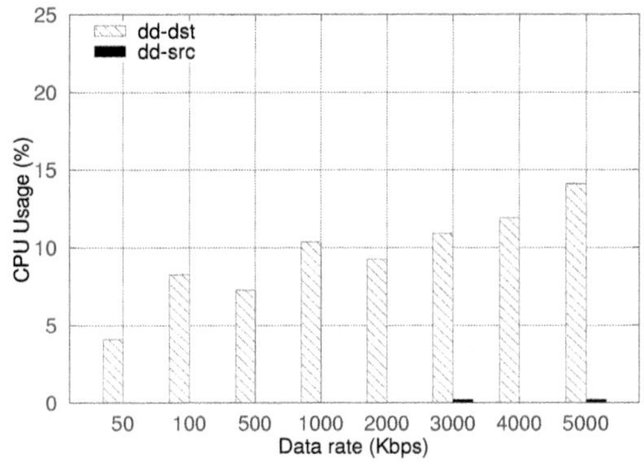

Figure 8: CPU occupied by Asymmetric Caching

using the Rabin fingerprinting algorithm so that partial-packet redundancy can be exploited. This approach was developed further in value-based web caching [1], where the data is cached on its value rather than its name. The idea of using packet caches on routers was introduced in [13]. In [2], the authors perform an experimental study of redundancy across 12 different enterprise networks. Both [13] and [2] identify that significant bandwidth savings can be achieved by using packet level dedup approaches, thereby motivating the current work. Similarly, EndRE [3] is an end-to-end solution for network dedup, which presents a new fingerprinting scheme called SampleByte. Recently, [14] proposed overhearing content in wireless networks to dedup across wireless users. Celleration [22] is another sender-driven dedup solution that leverages inter-user redundancy for a single point of attachment. All the above works are designed for static scenarios, do not work across points of attachment and do not support IP address changes due to mobility. Asymmetric caching is complementary to the above works and specifically optimizes wireless traffic, without requiring modification to internet servers.

Application layer dedup: Application layer works include caching http objects on browsers [23] and on proxy servers [24], delta encoding, file differencing (e.g. VCDIFF) [25], techniques for detecting duplicate transfers of the same file [4] and techniques such as base-instant caching [5], template caching [6] for enhanced cacheability of dynamic objects. More recent developments include content-delivery networks [26] and peer-to-peer caching solutions [27]. All these solutions operate at the granularity of files or application-objects and hence do not provide fine-grained redundancy elimination. Further, they are application layer solutions and have to be realized independently for every single application. Most importantly, using proxy or other intermediate caches while reducing the load on servers does not reduce the traffic on the wireless link. Asymmetric caching operates agnostic to different applications.

Transport layer dedup: Recently, Zohar *et* al [28] propose an end-to-end receiver driven dedup solution that extends TCP options. The receiver matches TCP stream chunks with its cache and sends predictions for future chunks in the ongoing flow to the sender. The dedup solution in [28] is similar to asymmetric caching in that both use receiver driven feedback to improve dedup performance. However, there are fundamental differences. The solution in [28] is closely tied to the TCP protocol and operates at a coarse data-granularity. Asymmetric caching on the other hand is transport protocol agnostic and operates at sub-packet level granularity. The solution in [28] is an end-to-end solution, whereas asymmetric caching is a last hop solution. This is important as wireless service providers who have the motivation to utilize their spectrum better can deploy asymmetric caching without any dependencies on the content provider. Also, the solution in [28] does not partition old connections into flowlets and hence maintains connections in their entirety. However, asymmetric caching partitions content into flowlets depending on their stationarity and this helps when the composition of connections changes in terms of a few objects or in terms of the ordering of the objects. Finally, the solution in [28] does not address how feedback might be chosen when chunks experience hits with multiple old connections. The feedback selection algorithm in asymmetric caching explicitly tackles this problem by choosing from multiple flowlets. This capability is especially important when operating at fine data granularities.

7. CONCLUSION

In this paper, we propose *asymmetric caching*, an improvement to baseline network deduplication that allows the *dedup destination* to selectively feedback appropriate portions of its cache to the *dedup source* with the intent of improving the redundancy elimination efficiency. We show using traffic traces collected from 30 mobile users, that with asymmetric caching, over 89% of the achievable redundancy can be identified and eliminated *even when the dedup source* has less than *one hundredth of the cache size as the dedup destination*. Further, we show that the number of bytes saved from transmission at the *dedup source* because of asymmetric caching is over 6× that of the number of bytes sent as feedback. Finally, with a prototype implementation of asymmetric caching on both a Linux laptop and an Android smartphone, we demonstrate that the solution is deployable with reasonable CPU and memory overheads.

8. ACKNOWLEDGEMENTS

The authors would like to thank Mr. Nitin Agarwal at the University of Illinois for early discussions and pointers on techniques to identify changes in stationarity in a time-series. The authors would also like to thank Dr. Ulas Kozat for acting as a shepherd for the camera-ready version of the paper.

9. REFERENCES

[1] Sean C. Rhea and Kevin Liang. Value-based web caching. In *The 12th Int. World Wide Web Conference*, 2003.

[2] Ashok Anand, Chitra Muthukrishnan, Aditya Akella, and Ramachandran Ramjee. Redundancy in network traffic: findings and implications. In *ACM SIGMETRICS*, 2009.

[3] Bhavish Aggarwal, Aditya Akella, Ashok Anand, Athula Balachandran, Pushkar Chitnis, Chitra Muthukrishnan, Ramachandran Ramjee, and George Varghese. Endre: an end-system redundancy elimination service for enterprises. In *NSDI*, 2010.

[4] Yee Man Chan, Terence Kelly, and Jeffrey C. Mogul. Design, implementation, and evaluation of duplicate transfer detection in http. In *NSDI*, 2004.

[5] Barron C. Housel and David B. Lindquist. Webexpress: a system for optimizing web browsing in a wireless environment. In *ACM MobiCom*, 1996.

[6] Fred Douglis, Michael Rabinovich, and Antonio Haro. Hpp: Html macro-preprocessing to support dynamic document caching. In *USENIX Symposium on Internet Technologies and Systems*, 1997.

[7] Neil T. Spring and David Wetherall. A protocol-independent technique for eliminating redundant network traffic. In *ACM SIGCOMM*, 2000.

[8] Riverbed, www.riverbed.com/us/solutions/wan_optimization/.

[9] Comscore Report: Digital Omnivore, October 2011. URL: www.comscore.com/Press_Events/Presentations_Whitepapers/2011/Digital_Omnivores.

[10] WeFi Analytics Report, August 2010. URL: mobilemarketingmagazine.com/sites/default/files/WeFi\%20Wi-fi\%20Data\%20Report\%20Q1\%202010.pdf.

[11] Ip core pooling tutorial, august 2006. URL: archive.ericsson.net/service/internet/picov/get?DocNo=1/28701-FGB101256.

[12] Ericsson SGSN-MME. URL: www.ericsson.com/ourportfolio/products/sgsn-mme?nav=fgb_101_256.

[13] Ashok Anand, Archit Gupta, Aditya Akella, Srinivasan Seshan, and Scott Shenker. Packet caches on routers: the implications of universal redundant traffic elimination. In *ACM SIGCOMM*, 2008.

[14] Ashok Anand Shan-Hsiang Shen, Aaron Gember and Aditya Akella. Refactor-ing content overhearing to improve wireless performance. In *ACM MobiCom*, 2011.

[15] EMC Data Domain, www.emc.com/backup-and-recovery/data-domain/data-domain.htm.

[16] M. O. Rabin. Fingerprinting by random polynomials. In *Technical Report TR 15-81, Department of Computer Science, Harvard University*, 1981.

[17] Ulrich Appel and Achim V. Brandt. Adaptive sequential segmentation of piecewise stationary time series. *Information Sciences*.

[18] Bob jenkins hash functions. URL: burtleburtle.net/bob/.

[19] Netfilter, www.netfilter.org/projects/libnetfilter_queue/.

[20] MadWifi Driver, www.madwifi.org.

[21] Iperf, iperf.sourceforge.net/.

[22] Eyal Zohar, Israel Cidon, and Osnat (Ossi) Mokryn. Celleration: loss-resilient traffic redundancy elimination for cellular data. In *ACM HotMobile*, 2012.

[23] RFC 2616: Hypertext Transfer Protocol – HTTP/1.1, www.w3.org/Protocols/rfc2616/rfc2616.html.

[24] The apache HTTP Proxy, httpd.apache.org.

[25] RFC 3284: The VCDIFF Generic Differencing and Compression Data Format, www.faqs.org/rfcs/rfc3284.html.

[26] The Akamai Content Delivery Network, www.akamai.com.

[27] On Saleh and Ma Hefeeda. Modeling and caching of peer-to-peer traffic. In *IEEE ICNP*, 2006.

[28] Eyal Zohar, Israel Cidon, and Osnat (Ossi) Mokryn. The power of prediction: cloud bandwidth and cost reduction. In *Proc. of the ACM SIGCOMM*, 2011.

Crowdsourcing to Smartphones: Incentive Mechanism Design for Mobile Phone Sensing *

Dejun Yang
Arizona State University
Tempe, AZ 85287
dejun.yang@asu.edu

Guoliang Xue
Arizona State University
Tempe, AZ 85287
xue@asu.edu

Xi Fang
Arizona State University
Tempe, AZ 85287
xi.fang@asu.edu

Jian Tang
Syracuse University,
Syracuse, NY 13244
jtang02@syr.edu

ABSTRACT

Mo ile one sensing is a ne ara ig ic ta es a - antage o t e er asi e s art ones to collect an ana- ly e ata eyon t e scale o at as re io sly ossi- le n a o ile one sensing syste t e *platform* re- cr its s art one *users* to ro i e sensing ser ice isting o ile one sensing a lications an syste s lac goo incenti e ec anis s t at can attract ore ser artici a- tion o a ress t is iss e e esign incenti e ec anis s or o ile one sensing e consi er t o syste o els: t e lat or -centric o el ere t e lat or ro i es a re ar s are y artici ating sers an t e ser-centric o el ere sers a e ore control o er t e ay ent t ey ill recei e or t e lat or -centric o el e esign an incenti e ec anis sing a Stac el erg ga e ere t e lat or is t e lea er ile t e sers are t e ollo ers e s o o to co te t e *unique Stackelberg Equilibrium* at ic t e tility o t e lat or is a i i e an none o t e sers can i ro e its tility y nilaterally e iating ro its c rrent strategy or t e ser-centric o el e esign an *auction-based incentive mechanism* ic is *com- putationally efficient individually rational profitable* an *truthful* ro g e tensi e si lations e e al ate t e er or ance an ali ate t e t eoretical ro erties o o r incenti e ec anis s

Categories and Subject Descriptors

Computer-Communication Networks : et or Arc itect re an esign

General Terms

Algorit esign cono ics

Keywords

ro so rcing Mo ile P one Sensing ncenti e Mec anis esign

* is researc as s orte in art y S grants an an A grant - - - e in or ation re orte ere oes not re ect t e osition or t e olicy o t e e eral go ern ent

1. INTRODUCTION

e ast e years a e itnesse t e roli eration o s art ones in eo le s aily li es it t e a ent o net or s an ore o er l rocessors t e nee s or la to s in artic lar a e eg n to a e S art one sales asse P s or t e rst ti e in t e nal arter o is in ection oint occ rre c ic er t an re- icte ic as s ose to e Accor ing to t e nternational ata or oration orl i e ar- terly Mo ile P one rac er it is esti ate t at illion s art ones ill e s i e orl i e in

o a ays s art ones are rogra a le an e i e it a set o c ea t o er l e e sensors s c as accelero eter igital co ass gyrosco e PS icro- one an ca era ese sensors can collecti ely onitor a i erse range o an acti ities an s rro n ing en iron- ent S art ones are n o te ly re ol tioni ing any sectors o o r li e incl ing social net or s en iron ental onitoring siness ealt care an trans ortation

all t e s art ones on t e lanet toget er constit te a o ile one sensing net or it o l e t e largest sensing net or in t e istory ne can le erage illions o ersonal s art ones an a near- er asi e ireless net or in rastr ct re to collect an analy e sense ata ar eyon t e scale o at as ossi le e ore it o t t e nee to e loy t o san s o static sensors

eali ing t e great otential o t e o ile one sensing any researc ers a e e elo e n ero s a lications an syste s s c as Sensorly or a ing cell lar i i net- or co erage a s ericell an rac or ro- i ing tra c in or ation P or calc lating erson- ali e en iron ental i act an e os re an ar-P one or creating noise a s or ore etails on o ile one sensing e re er intereste rea ers to t e s r ey a- er

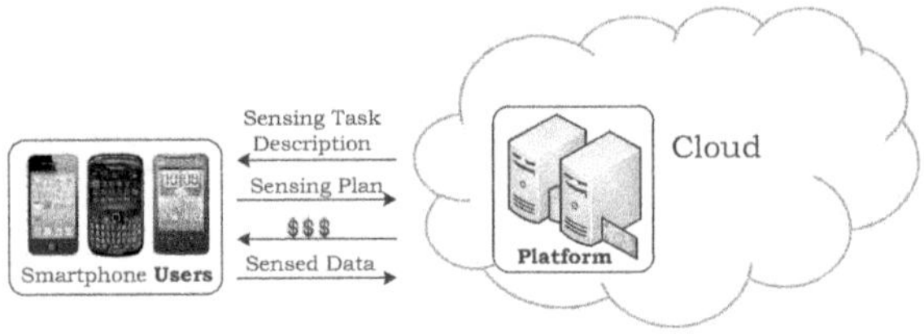

Figure 1: A mobile phone sensing system

As s o n in ig re a o ile one sensing syste con-sists o a o ile one sensing *platform* ic resi es in t e clo an consists o lti le sensing ser ers an any s art one *users* ic are connecte it t e lat or ia t e clo ese s art one sers can act as sensing ser ice ro i ers e lat or recr its s art one sers to ro i e sensing ser ices

Alt o g t ere are any a lications an syste s on o-ile one sensing ost o t e are ase on ol ntary artici ation ile artici ating in a o ile one sensing tas s art one sers cons e t eir o n re-so rces s c as attery an co ting o er n a ition sers also e ose t e sel es to otential ri acy t reats y s aring t eir sense ata it location tags ere ore *a user would not be interested in participating in mobile phone sensing, unless it receives a satisfying reward to compen-sate its resource consumption and potential privacy breach* it o t a e ate ser artici ation it is i ossi le or t e o ile one sensing a lications to ac ie e goo ser ice ality since sensing ser ices are tr ly e en ent on sers sense ata ile any researc ers a e e elo e i -erent o ile one sensing a lications t ey eit er o not consi er t e esign o incenti e ec anis s or a e neglecte so e critical ro erties o incenti e ec anis s o ll t is oi e ill esign se eral incenti e ec anis s to oti ate sers to artici ate in o ile one sensing a -lications

e consi er t o ty es o incenti e ec anis s or a o-ile one sensing syste : *platform-centric incentive mech-anisms* an *user-centric incentive mechanisms* n a lat or -centric incenti e ec anis t e lat or as t e a sol te control o er t e total ay ent to sers an sers can only tailor t eir actions to cater or t e lat or ereas in a ser-centric incenti e ec anis t e roles o t e lat or an sers are re erse o ass re itsel o t e otto -line ene t eac ser anno nces a reser e rice t e lo est rice at ic it is illing to sell a ser ice e lat or t en se-lects a s set o sers an ay eac o t e an a o nt t at is no lo er t an t e ser s reser e rice

1.1 Summary of Key Contributions

e ollo ing is a list o o r ain contri tions

- e esign incenti e ec anis s or o ile one sens-ing a ne sensing ara ig t at ta es a antage o t e er asi e s art ones to scale t e sense ata collection an analysis to a le el o at as re i-o sly i ossi le

  e consi er t o syste o els ro t o i erent er-s ecti es: t e lat or -centric o el ere t e lat-or ro i es a e re ar to artici ating sers an t e ser-centric o el ere sers can a e t eir reser e rices or t e sensing ser ice

- or t e lat or -centric o el e esign an incen-ti e ec anis sing a Stac el erg ga e e resent an e cient algorit to co te t e *unique Stackel-berg Equilibrium* at ic t e tility o t e lat or is a i i e an none o t e sers can i ro e its tility y nilaterally e iating ro its c rrent strat-egy

- or t e ser-centric o el e esign an a ction- ase incenti e ec anis ic is co tationally e - cient in i i ally-rational ro ta le an ore i - ortantly tr t l

1.2 Paper Organization

e re ain er o t is a er is organi e as ollo s n Section e escri e t e o ile one sensing syste o -els incl ing ot t e lat or -centric o el an t e ser-centric o el e t en resent o r incenti e ec anis s or t ese t o o els in Sections an res ecti ely e resent er or ance e al ations in Section an isc ss re-late or in Section e concl e t is a er in Section

2. SYSTEM MODEL AND PROBLEM FOR-MULATION

e se ig re to ai o r escri tion o t e o ile one sensing syste e syste consists o a o ile one sens-ing *platform* ic resi es in t e clo an consists o l-ti le sensing ser ers an any s art one *users* ic are connecte to t e lat or ia t e clo e lat or rst lici es t e sensing tas s Ass e t at t ere is a set $\mathcal{U}$ { , ,..., n} o s art one sers intereste in ar-tici ating in o ile one sensing a ter rea ing t e sensing tas escri tion ere $n \geq$ A ser artici ating in o-ile one sensing ill inc r a *cost* to e ela orate later ere ore it e ects a *payment* in ret rn or its ser ice a -ing cost an ret rn into consi eration eac ser a es its o n sensing lan ic co l e t e sensing ti e or t e reser e rice or selling its sense ata an s its it to t e lat or A ter collecting t e sensing lans ro sers t e lat or co tes t e ay ent or eac ser an sen s t e ay ents to t e sers e c osen sers ill con ct t e sensing tas s an sen t e sense ata to t e lat or is co letes t e ole o ile one sensing rocess

e lat or is only intereste in a i i ing its o n til-ity Since s art ones are o ne y i erent in i i als it is reasona le to ass e t at sers are *selfish* t *rational* ence eac ser only ants to a i i e its o n tility an ill not artici ate in o ile one sensing nless t ere is s cient incenti e e oc s o t is a er is on t e esign o incenti e ec anis s t at are si le scala le an a e ro a ly goo ro erties t er iss es in t e esign an i le entation o t e ole o ile one sensing syste is o t o t e sco e o t is a er Please re er to MA or energy sa ing iss es P SM or a lication e elo ing iss es an P PS an P or ri acy iss es

e st y t o o els: *platform-centric* an *user-centric* n t e lat or -centric o el t e sensing lan o an inter-este ser is in t e or o its sensing ti e A ser artici-ating in o ile one sensing ill earn a ay ent t at is no lo er t an its cost o e er it nee s to co ete it ot er sers or a e total ay ent n t e ser-centric o el eac ser as s or a rice or its ser ice selecte

Table 1: Frequently used notations

otation	escri tion
$\mathcal{U}, i, j$	set o sers an ser
n	n er o sers
R	re ar o t e lat or
t_i, t, t_{-i}	sensing ti e strategy o ser i
	strategy ro le o all sers
	strategy ro le e cl ing ser i s strategy
κ_i	cost nit o ser i
$\beta_i\ t_{-i}$	est res onse o ser i gi en t_{-i}
$\bar{u}_i, \bar{u}_0$	tility nction o ser i an t e lat or in t e lat or -centric o el
λ	syste ara eter in $\bar{u}_0$
Γ, Γ_i, τ_j	set o tas s set o ser i s tas s an tas
m	n er o tas s
ν_j	al e o tas τ_j
c_i, b_i	cost an i o ser i
$\mathcal{S}$	set o selecte sers
p_i	ay ent to ser i
$v\ \mathcal{S}$	total al e o t e tas s y $\mathcal{S}$
$\tilde{u}_i, \tilde{u}_0$	tility nction o ser i an t e lat or in t e ser-centric o el

t e ser ill recei e a ay ent t at is no lo er t an its as e rice nli e t e lat or -centric o el t e total ay ent is not e or t e ser-centric o el ence t e sers a e ore control o er t e ay ent in t e ser-centric o el

a le lists re ently se notations

2.1 Platform-Centric Model

n t is o el t erc is only one sensing tas e lat- or anno nces a total re ar $R >$ oti ating sers to artici ate in o ile one sensing ile eac ser eci es its le el o artici ation ase on t e re ar

e *sensing plan of user i* is re resente y t_i t e n er o ti e nits it is illing to ro i e t e sensing ser ice ence $t_i \geq$ By setting t_i ser i in icates t at it ill not artici ate in o ile one sensing e *sensing cost of user i* is $\kappa_i \times t_i$ ere $\kappa_i >$ is its *unit cost* Ass e t at t e re ar recei e y ser i is ro ortional to t_i en t e *utility of user i* is

$$\bar{u}_i \quad \frac{t_i}{\sum_{j \in \mathcal{U}} t_j} R - t_i \kappa_i,$$

i e re ar in s cost e *utility of the platform* is

$$\bar{u}_0 \quad \lambda \log \left(\quad \sum_{i \in \mathcal{U}} \log \quad t_i \right) - R,$$

ere $\lambda >$ is a syste ara eter t e log t_i ter re ects t e lat or s i inis ing ret rn on t e or o ser i an t e o ter log ter re ects t e lat or s i inis ing ret rn on artici ating sers

n er t is o el t e o ecti e o t e lat or is to e- ci e t e o ti al al e o R so as to a i i e ile eac ser $i \in \mathcal{U}$ sel s ly eci es its sensing ti e t_i to a i- i e or t e gi en al e o R Since no rational ser is illing to ro i e ser ice or a negati e tility ser i s all set t_i en $R \leq \kappa_i \sum_{j \neq i \in \mathcal{U}} t_j$

2.2 User-Centric Model

n t is o el t e lat or anno nces a set $\Gamma \quad \{\tau_1, \tau_2, \ldots, \tau_m\}$ o *tasks* or t e sers to select ac $\tau_j \in \Gamma$ as a *value* $\nu_j >$ *to the platform* ac ser i selects a s set o tas s $\Gamma_i \subseteq \Gamma$ accor ing to its re erence Base on t e selecte tas set ser i also as an *associated cost* c_i ic is ri ate an only no n to itsel ser i t en s its t e tas - i air Γ_i, b_i to t e lat or ere b_i calle ser i s *bid* is t e *reserve price* ser i ants to sell t e ser ice or on recei ing t e tas - i airs ro all t e sers t e lat or selects a s set $\mathcal{S}$ o sers as *winners* an eter ines t e *payment* p_i or eac inning ser i e *utility of user i* is

$$\tilde{u}_i \quad \begin{cases} p_i - c_i, & \text{i } i \in \mathcal{S}, \\ , & \text{ot er ise.} \end{cases}$$

e *utility of the platform* is

$$\tilde{u}_0 \quad v\ \mathcal{S} \ - \sum_{i \in \mathcal{S}} p_i,$$

ere $v\ \mathcal{S} \quad \sum_{\tau_j \in \cup_{i \in \mathcal{S}} \Gamma_i} \nu_j$

r o ecti e or t e ser-centric o el is to esign an incenti e ec anis satis ying t e ollo ing o r esira le ro erties:

- **Computational Efficiency**: A ec anis is co - tationally e cient i t e o tco e can e co te in olyno ial ti e
- **Individual Rationality**: ac artici ating ser ill a e a non-negati e tility
- **Profitability**: e lat or s o l not inc r a e cit n ot er or s t e al e ro g t y t e inners s o l e at least as large as t e total ay ent ai to t e inners
- **Truthfulness**: A ec anis is tr t l i no i er can i ro e its tility y s itting a i i erent ro its tr e al ation ic is cost in t is a er no atter at ot ers s it

e i ortance o t e rst t ree ro erties is o io s eca se t ey toget er ass re t e easi ility o t e incenti e ec anis Being tr t l t e incenti e ec anis can eli inate t e ear o ar et ani lation an t e o er ea o strategi ing o er ot ers or t e artici ating sers

3. INCENTIVE MECHANISM FOR THE PLATFORM-CENTRIC MODEL

e o el t e lat or -centric incenti e ec anis as a *Stackelberg game* ic e call t e MSensing ga e ere are t o stages in t is ec anis : n t e rst stage t e lat or anno nces its re ar R in t e secon stage eac ser strategi es its sensing ti e to a i i e its o n tility ere ore t e lat or is t e *leader* an t e sers are t e *followers* in t is Stac el erg ga e Mean ile ot t e lat or an t e sers are *players* e *strategy of the platform* is its re ar R e *strategy of user i* is its or - ing ti e t_i et $t \quad t_1, t_2, \ldots, t_n$ enote t e *strategy profile* consisting o all sers strategies et t_{-i} enote t e strat- egy ro le e cl ing t_i As a notational con ention e rite $t \quad t_i, t_{-i}$

ote t at t e secon stage o t e MSensing ga e itsel can e consi ere a non-coo erati e ga e ic e call t e Sensing i e eter ination S ga e i en t e MSensing ga e or lation e are intereste in ans ering t e ollo ing estions:

: or a gi en re ar R is t ere a set o sta le strategies in t e S ga e s c t at no ser as anyt ing to gain y nilaterally c anging its c rrent strategy
: t e ans er to is yes is t e sta le strategy set ni e en it is ni e sers ill e g arantee to select t e strategies in t e sa e sta le strategy set
: o can t e lat or select t e al e o R to a i i e its tility in

e sta le strategy set in corres on s to t e conce t o *Nash Equilibrium* in ga e t eory

Definition 1 **(Nash Equilibrium).** *A set of stra- te- gies* $t_1^{ne}, t_2^{ne}, \ldots, t_n^{ne}$ *is a Nash Equilibrium of the STD game if for any user* i,

$$\bar{u}_i \; t_i^{ne}, t_{-i}^{ne} \;\geq \bar{u}_i \; t_i, t_{-i}^{ne} \;,$$

for any $t_i \geq$ *, where* $\bar{u}_i$ *is defined (2.1).*

e *existence* o an is i ortant since an strat- egy ro le is sta le no layer as an incenti e to a e a nilateral c ange ereas a non- strategy ro le is n- sta le e *uniqueness* o allo s t e lat or to re ict t e e a iors o t e sers an t s ena les t e lat or to select t e o ti al al e o R ere ore t e ans er to e en s ea ily on t ose to an e o ti al so- l tion co te in toget er it t e o t e S ga e constit tes a sol tion to t e MSensing ga e calle *Stackelberg Equilibrium*

n Section e ro e t at **for any given $R >$, the STD game has a unique NE** an resent an e cient al- gorit or co ting t e n Section e ro e t at **the MSensing game has a unique Stackelberg Equi- librium** an resent an e cient algorit or co ting it

3.1 User Sensing Time Determination

e rst intro ce t e conce t o *best response strategy*

Definition 2 **(Best Response Strategy).** *Given* t_{-i}, *a strategy is user* i*'s best response strategy, denoted by* $\beta_i \; t_{-i}$, *if it maximizes* $\bar{u}_i \; t_i, t_{-i}$ *over all* $t_i \geq$.

Base on t e e nition o e ery ser is laying its est res onse strategy in an ro e no t at $t_i \leq \frac{R}{\kappa_i}$ eca se $\bar{u}_i$ ill e negati e ot er ise o st y t e est res onse strategy o ser i e co te t e eri ati es o $\bar{u}_i$ it res ect to t_i:

$$\frac{\partial \bar{u}_i}{\partial t_i} \qquad \frac{-Rt_i}{\sum_{j\in\mathcal{U}} t_j \;^2} \qquad \frac{R}{\sum_{j\in\mathcal{U}} t_j} - \kappa_i,$$

$$\frac{\partial^2 \bar{u}_i}{\partial t_i^2} \qquad -\frac{R\sum_{j\in\mathcal{U}\setminus\{i\}} t_j}{\sum_{j\in\mathcal{U}} t_j \;^3} < \;.$$

Since t e secon -or er eri ati e o $\bar{u}_i$ is negati e t e tility $\bar{u}_i$ is a *strictly concave function* in t_i ere ore gi en any $R >$ an any strategy ro le t_{-i} o t e ot er sers t e est res onse strategy $\beta_i \; t_{-i}$ o ser i is ni e i it e ists t e strategy o all ot er ser j / i is t_j t en ser i oes not a e a est res onse strategy as it can a e a tility ar itrarily close to R y setting t_i to a s ciently s all ositi e n er ere ore e are only intereste in t e est res onse or ser i en $\sum_{j\in\mathcal{U}\setminus\{i\}} t_j >$ Setting t e rst eri ati e o $\bar{u}_i$ to e a e

$$\frac{-Rt_i}{\sum_{j\in\mathcal{U}} t_j \;^2} \qquad \frac{R}{\sum_{j\in\mathcal{U}} t_j} - \kappa_i \qquad .$$

Sol ing or t_i in e o tain

$$t_i \qquad \sqrt{\frac{R\sum_{j\in\mathcal{U}\setminus\{i\}} t_j}{\kappa_i}} - \sum_{j\in\mathcal{U}\setminus\{i\}} t_j.$$

t e S rig t an si e o is ositi e is also t e est res onse strategy o ser i e to t e conca ity o $\bar{u}_i$ t e S o is less t an or e al to t en ser i oes not artici ate in t e o ile sensing y setting t_i to a oi a e cit ence e a e

$$\beta_i \; t_{-i} \qquad \begin{cases} \;, & \text{i } R \leq \kappa_i \sum_{j\in\mathcal{U}\setminus\{i\}} t_j \\ \sqrt{\frac{R\sum_{j\in\mathcal{U}\setminus\{i\}} t_j}{\kappa_i}} - \sum_{j\in\mathcal{U}\setminus\{i\}} t_j, & \text{ot er ise.} \end{cases}$$

ese analyses lea to t e ollo ing algorit or co - ting an o t e S ga e

Algorithm 1: o tation o t e

Sort sers accor ing to t eir nit costs $\kappa_1 \leq \kappa_2 \leq \cdots \leq \kappa_n$
$\mathcal{S} \leftarrow \{ \; , \; \}$ $i \leftarrow$
while $i \leq n$ *and* $\kappa_i < \frac{\kappa_i + \sum_{j\in\mathcal{S}} \kappa_j}{|\mathcal{S}|}$ **do**
  $\mathcal{S} \leftarrow \mathcal{S} \cup \{i\}$ $i \leftarrow i$
end
foreach $i \in \mathcal{U}$ **do**
  if $i \in \mathcal{S}$ **then** t_i^{ne} $\frac{(|\mathcal{S}|-1)R}{\sum_{j\in\mathcal{S}} \kappa_j} \left(\; - \frac{(|\mathcal{S}|-1)\kappa_i}{\sum_{j\in\mathcal{S}} \kappa_j} \right)$
  else t_i^{ne}
end
return t^{ne} $t_1^{ne}, t_2^{ne}, \ldots, t_n^{ne}$

Theorem *The strategy profile* t^{ne} $t_1^{ne}, \ldots, t_n^{ne}$ *com- puted by Algorithm 1 is an NE of the STD game. The time complexity of Algorithm 1 is* $O \; n \log n$.

PROOF. e rst ro e t at t e strategy ro le t^{ne} is an et n_0 $|\mathcal{S}|$ e a e t e ollo ing o ser ations ase on t e algorit : $\kappa_i \geq \frac{\sum_{j\in\mathcal{S}} \kappa_j}{n_0 - 1}$ or any $i \notin \mathcal{S}$ $\sum_{j\in\mathcal{S}} t_j^{ne}$ $\frac{(n_0-1)R}{\sum_{j\in\mathcal{S}} \kappa_j}$ an $\sum_{j\in\mathcal{S}\setminus\{i\}} t_j^{ne}$ $\frac{(n_0-1)^2 R\kappa_i}{\sum_{j\in\mathcal{S}} \kappa_j \;^2}$ or any $i \in \mathcal{S}$ e ne t ro e t at or any $i \notin \mathcal{S}$ t_i^{ne} is its est res onse strategy gi en t_{-i}^{ne} Since $i \notin \mathcal{S}$ e a e $\kappa_i \sum_{j\in\mathcal{U}\setminus\{i\}} t_j^{ne}$ $\kappa_i \sum_{j\in\mathcal{S}} t_j^{ne}$ sing an e a e $\kappa_i \sum_{j\in\mathcal{S}} t_j^{ne} \geq R$ Accor ing to e no t at $\beta_i \; t_{-i}^{ne}$

e t en ro e t at or any $i \in \mathcal{S}$ t_i^{ne} is its est res onse strategy gi en t_{-i}^{ne} ote t at $\kappa_i < \frac{\sum_{j=1}^{i} \kappa_j}{i-1}$ accor ing to Algorit e t en a e

$$n_0 - \quad \kappa_i \qquad i - \quad \kappa_i \qquad n_0 - i \; \kappa_i < \sum_{j=1}^{i} \kappa_j \qquad \sum_{j=i+1}^{n_0} \kappa_j,$$

ere $\kappa_i \leq \kappa_j$ or $i \quad \leq j \leq n_0$ ence e a e $\kappa_i < \frac{\sum_{i\in\mathcal{S}}\kappa_i}{n_0-1}$ rt er ore e a e

$$\kappa_i \sum_{j\in\mathcal{U}\setminus\{i\}} t_j^{ne} \quad \kappa_i \sum_{j\in\mathcal{S}\setminus\{i\}} t_j^{ne} \quad \kappa_i \frac{n_0 - \quad ^2 R\kappa_i}{\left(\sum_{j\in\mathcal{S}}\kappa_j\right)^2} < R.$$

Accor ing to

$$\beta_i \; t_{-i}^{ne} \quad \sqrt{\frac{R\sum_{j\in\mathcal{U}\setminus\{i\}} t_j^{ne}}{\kappa_i}} - \sum_{j\in\mathcal{U}\setminus\{i\}} t_j^{ne}$$

$$\frac{n_0 - \quad R}{\sum_{j\in\mathcal{S}}\kappa_j} - \frac{n_0 - \quad ^2 R\kappa_i}{\left(\sum_{j\in\mathcal{S}}\kappa_j\right)^2} \quad t_i^{ne}.$$

ere ore t^{ne} is an o t e S ga e

e ne t analy e t e r nning ti e o t e algorit Sorting can e one in $O \; n \log n$ ti e e ile-loo ines - re ires a total ti e o $O \; n$ e or-loo ines - re ires a total ti e o $O \; n$ ence t e ti e co le ity o Algorit is $O \; n \log n$ □

e ne t t eore s o s t e ni eness o t e or t e S ga e

Theorem *Let $R >$ be given. Let $\bar{t} \quad \bar{t}_1, \bar{t}_2, \ldots, \bar{t}_n$ be the strategy profile of an NE for the STD game, and let $\bar{\mathcal{S}} \quad \{i \in \mathcal{U} | \bar{t}_i > \; \}$. We have*

1) $|\bar{\mathcal{S}}| \geq \;$.

2) $\bar{t}_i \quad \begin{cases} \;, & \text{if } i \notin \bar{\mathcal{S}} \\ \frac{(|\bar{\mathcal{S}}|-1)R}{\sum_{j\in\bar{\mathcal{S}}}\kappa_j}\left(\; - \frac{(|\bar{\mathcal{S}}|-1)\kappa_i}{\sum_{j\in\bar{\mathcal{S}}}\kappa_j}\right), & \text{otherwise.}\end{cases}$

3) If $\kappa_q \leq \; a \;_{j\in\bar{\mathcal{S}}}\{\kappa_j\}$, then $q \in \bar{\mathcal{S}}$.

4) Assume that the users are ordered such that $\kappa_1 \leq \kappa_2 \leq \cdots \leq \kappa_n$. Let h be the largest integer in $\; ,n \;$ such that $\kappa_h < \frac{\sum_{j=1}^{h}\kappa_j}{h-1}$. Then $\bar{\mathcal{S}} \quad \{ \; , \; , \ldots, h\}$.

These statements imply that the STD game has a unique NE, which is the one computed by Algorithm 1.

PROOF. e rst ro e Ass e t at $|\bar{\mathcal{S}}|$ ser can increase its tility ro to $\frac{R}{2}$ y nilaterally c anging its sensing ti e ro to $\frac{R}{2\kappa_1}$ contra icting t e ass tion is ro es t at $|\bar{\mathcal{S}}| \geq$ o ass e t at $|\bar{\mathcal{S}}|$ is eans $\bar{t}_k >$ or so e $k \in \mathcal{U}$ an $\bar{t}_j$ or all $j \in \mathcal{U}\setminus\{k\}$ Accor ing to t e c rrent tility o ser k is $R - \bar{t}_k\kappa_k$ ser k can increase its tility y nilaterally c anging its sensing ti e ro $\bar{t}_k$ to $\frac{\bar{t}_k}{2}$ again contra icting t e ass tion ere ore $|\bar{\mathcal{S}}| \geq$

e ne t ro e et $n_0 \quad |\bar{\mathcal{S}}|$ Since e alrea y ro e t at $n_0 \geq$ e can se t e analysis at t e eginning o t is section it t re lace y $\bar{t}$ an $\mathcal{S}$ re lace y $\bar{\mathcal{S}}$ onsi ering t at $\sum_{j\in\mathcal{U}}\bar{t}_j \quad \sum_{j\in\bar{\mathcal{S}}}\bar{t}_j$ e a e

$$\frac{-R\bar{t}_i}{\sum_{j\in\bar{\mathcal{S}}}\bar{t}_j \;\; ^2} \quad \frac{R}{\sum_{j\in\bar{\mathcal{S}}}\bar{t}_j} - \kappa_i \quad , \quad i \in \bar{\mathcal{S}}.$$

S ing o er t e sers in $\bar{\mathcal{S}}$ lea s to $n_0 R - R \quad \sum_{j\in\bar{\mathcal{S}}}\bar{t}_j \cdot \sum_{j\in\bar{\mathcal{S}}}\kappa_j$. ere ore e a e

$$\sum_{j\in\bar{\mathcal{S}}}\bar{t}_j \quad \frac{n_0 - \quad R}{\sum_{j\in\bar{\mathcal{S}}}\kappa_j}.$$

S stit ting into an consi ering $\bar{t}_j$ or any $j \in \mathcal{U}\setminus\bar{\mathcal{S}}$ e o tain t e ollo ing:

$$\bar{t}_i \quad \frac{n_0 - \quad R}{\sum_{j\in\bar{\mathcal{S}}}\kappa_j}\left(\; - \frac{n_0 - \quad \kappa_i}{\sum_{j\in\bar{\mathcal{S}}}\kappa_j}\right)$$

or e ery $i \in \bar{\mathcal{S}}$ is ro es

e t en ro e By e nition o $\bar{\mathcal{S}}$ e no t at $\bar{t}_i >$ or e ery $i \in \bar{\mathcal{S}}$ ro $\bar{t}_i >$ i lies $\frac{(n_0-1)\kappa_i}{\sum_{j\in\bar{\mathcal{S}}}\kappa_j} <$ ere ore e a e

$$\kappa_i < \frac{\sum_{j\in\bar{\mathcal{S}}}\kappa_j}{|\bar{\mathcal{S}}| - }, \forall i \in \bar{\mathcal{S}}.$$

i lies t at

$$\underset{i\in\bar{\mathcal{S}}}{a} \; \kappa_i < \frac{\sum_{j\in\bar{\mathcal{S}}}\kappa_j}{|\bar{\mathcal{S}}| - }.$$

Ass e t at $\kappa_q \leq \; a \;_{j\in\bar{\mathcal{S}}}\{\kappa_j\}$ t $q \notin \bar{\mathcal{S}}$ Since $q \notin \bar{\mathcal{S}}$ e no t at $\bar{t}_q$ e rst-or er eri ati e o $\bar{u}_q$ it res ect to t_q en t $\bar{t}$ is

$$\frac{R}{\sum_{j\in\bar{\mathcal{S}}}\bar{t}_j} - \kappa_q \quad \frac{\sum_{j\in\bar{\mathcal{S}}}\kappa_j}{n_0 - } - \kappa_q > \underset{i\in\bar{\mathcal{S}}}{a} \; \{\kappa_i\} - \kappa_q \geq \; .$$

is eans t at ser q can increase its tility y nilaterally increasing its sensing ti e ro $\bar{t}_q$ contra icting t e ass tion o $\bar{t}$ is ro es

inally e ro e State ents an i ly t at $\bar{\mathcal{S}} \quad \{ \; , \; , \ldots, q\}$ or so e integer q in $\; ,n$ ro e concl e t at $q \leq h$ Ass e t at $q < h$ en e a e $\kappa_{q+1} < \frac{\sum_{j=1}^{q+1}\kappa_j}{q}$ ic i lies $\frac{\sum_{j=1}^{q}\kappa_j}{q-1} - \kappa_{q+1} >$ ence t e rst or er eri ati e o $\bar{u}_{q+1}$ it res ect to t_{q+1} en t $\bar{t}$ is $\frac{\sum_{j=1}^{q}\kappa_j}{q-1} - \kappa_{q+1} >$ is contra iction ro es $q \quad h$ ence e a e ro e as ell as t e t eore □

3.2 Platform Utility Maximization

Accor ing to t e a o e analysis t e lat or ic is t e lea er in t e Stac el erg ga e no s t at t ere e ists a ni e or t e sers or any gi en al e o R ence t e lat or can a i i e its tility y c oosing t e o ti al R S stit ting into an consi ering $t_i \quad i \notin \mathcal{S}$ e a e

$$\bar{u}_0 \quad \lambda \log\left(\quad \sum_{i\in\mathcal{S}} \log \quad X_i R \right) - R,$$

ere $X_i \quad \frac{(n_0-1)}{\sum_{j\in\mathcal{S}}\kappa_j}\left(\; - \frac{(n_0-1)\kappa_i}{\sum_{j\in\mathcal{S}}\kappa_j}\right)$.

Theorem *There exists a unique Stackelberg Equilibrium R^*, t^{ne} in the MSensing game, where R^* is the unique maximizer of the platform utility in (3.12) over $R \in \; ,\infty \;$, $\mathcal{S}$ and t^{ne} are given by Algorithm 1 with the total reward set to R^*.*

PROOF. e secon or er eri ati e o $\bar{u}_0$ is

$$\frac{\partial^2 \bar{u}_0}{\partial R^2} \quad -\lambda \frac{\sum_{i\in\mathcal{S}}\frac{X_i^2}{(1+X_iR)^2}Y \quad \left(\sum_{i\in\mathcal{S}}\frac{X_i}{(1+X_iR)}\right)^2}{Y^2} < \; ,$$

ere $Y \quad \sum_{i\in\mathcal{S}} \log \quad X_i R$ ere ore t e tility $\bar{u}_0$ e ne in is a strictly conca e nction o R or $R \in \; ,\infty$ Since t e al e o $\bar{u}_0$ in is or R an goes to $-\infty$ en R goes to ∞ it as a ni e a i i er R^* t at can e e ciently co te sing eit er isection or e ton s et o □

4. INCENTIVE MECHANISM FOR THE USER-CENTRIC MODEL

A ction t eory is t e er ect t eoretical tool to e-sign incenti e ec anis s or t e ser-centric o el e ro ose a re erse a ction ase incenti e ec anis or t e ser-centric o el An *auction* ta es as in t t e *bids* s - itte y t e sers selects a s set o sers as *winners* an eter ines t e *payment* to eac inning ser

4.1 Auctions Maximizing Platform Utility

r rst atte t is to esign an incenti e ec anis a i i ing t e tility o t e lat or o esigning an in-centi e ec anis eco es an o ti i ation ro le calle *User Selection* ro le : i en a set $\mathcal{U}$ o sers select a s - set $\mathcal{S}$ s c t at $\tilde{u}_0$ $\mathcal{S}$ is a i i e o er all ossi le s sets n a ition it is clear t at p_i b_i to a i i e $\tilde{u}_0$ $\mathcal{S}$ e tility $\tilde{u}_0$ t en eco es

$$\tilde{u}_0 \ \mathcal{S} \quad v \ \mathcal{S} \ - \sum_{i \in \mathcal{S}} b_i.$$

o a e t e ro le eaning 1 e ass e t at t ere e - ists at least one ser i s c t at $\tilde{u}_0$ $\{i\}$ $>$ n ort nately as t e ollo ing t eore s o s it is P-ar to n t e o ti al sol tion to t e ser Selection ro - le

Theorem *The User Selection problem is NP-hard.*

PROOF. e ill ro e t is t eore in t e a en i or a etter o o t e a er □

Since it is nli ely to n t e o ti al s set o sers e ciently e t rn o r attention to t e e elo ent o a - ro i ation algorit s o t is en e ta e a antage o t e *submodularity* o t e tility nction

Definition 3 (Submodular Function). *Let $\mathcal{X}$ be a finite set. A function f : $\mathcal{X} \mapsto \mathbb{R}$ is submodular if*

$$f \ \mathcal{A} \cup \{x\} \ - f \ \mathcal{A} \ \geq f \ \mathcal{B} \cup \{x\} \ - f \ \mathcal{B} \ ,$$

for any $\mathcal{A} \subseteq \mathcal{B} \subseteq \mathcal{X}$ and $x \in \mathcal{X} \setminus \mathcal{B}$, where $\mathbb{R}$ is the set of reals.

e no ro e t e s o larity o t e tility $\tilde{u}_0$

Lemma *The utility $\tilde{u}_0$ is submodular.*

PROOF. By e nition e nee to s o t at

$$\tilde{u}_0 \ \mathcal{S} \cup \{i\} \ - \tilde{u}_0 \ \mathcal{S} \ \geq \tilde{u}_0 \ \mathcal{T} \cup \{i\} \ - \tilde{u}_0 \ \mathcal{T} \ ,$$

or any $\mathcal{S} \subseteq \mathcal{T} \subseteq \mathcal{U}$ an $i \in \mathcal{U} \setminus \mathcal{T}$ t s ces to s o t at $v \ \mathcal{S} \cup \{i\} \ - v \ \mathcal{S} \ \geq v \ \mathcal{T} \cup \{i\} \ - v \ \mathcal{T}$ since t e secon ter in $\tilde{u}_0$ can e s tracte ro ot si es onsi ering $v \ \mathcal{S}$ $\sum_{\tau_j \in \cup_{i \in \mathcal{S}} \Gamma_i} \nu_j$ e a e

$$v \ \mathcal{S} \cup \{i\} \ - v \ \mathcal{S} \qquad \sum_{\tau_j \in \Gamma_i \setminus \cup_{j \in \mathcal{S}} \Gamma_j} \nu_j$$
$$\geq \sum_{\tau_j \in \Gamma_i \setminus \cup_{j \in \mathcal{T}} \Gamma_j} \nu_j$$
$$v \ \mathcal{T} \cup \{i\} \ - v \ \mathcal{T} \ .$$

ere ore $\tilde{u}_0$ is s o lar *As a byproduct, we proved that v is submodular as well* □

en t e o ecti e nction is s o lar onotone an non-negati e it is no n t at a gree y algorit ro- i es a $-$ /e -a ro i ation it o t onotonicity eige *et al.* a e also e elo e constant- actor a ro i-ation algorit s n ort nately $\tilde{u}_0$ can e negati e o circ ent t is iss e let f $\mathcal{S}$ $\tilde{u}_0$ $\mathcal{S}$ $\sum_{i \in \mathcal{U}} b_i$ t is clear t at f $\mathcal{S}$ $\geq$ or any $\mathcal{S} \subseteq \mathcal{U}$ Since $\sum_{i \in \mathcal{U}} b_i$ is a con-stant f $\mathcal{S}$ is also s o lar n a ition a i i ing $\tilde{u}_0$ is e i alent to a i i ing f ere ore e esign an a c-tion ec anis ase on t e algorit o calle ocal Searc -Base SB a ction as ill strate in Algorit e ec anis relies on t e local-searc tec ni e ic gree ily searc es or a etter sol tion y a ing a ne ser or eleting an e isting ser ene er ossi le t as ro e t at or any gi en constant $\epsilon >$ t e algorit can n a set o sers $\mathcal{S}$ s c t at f $\mathcal{S}$ $\geq$ $\frac{1}{3} - \frac{\epsilon}{n}$ f $\mathcal{S}^*$ ere $\mathcal{S}^*$ is t e o ti al sol tion

---

Algorithm 2: SB A ction

---

1 $\mathcal{S} \leftarrow \{i\}$ ere $i \leftarrow$ arg a $_{i \in \mathcal{U}}$ f $\{i\}$
2 **while** *there exists a user $i \in \mathcal{U} \setminus \mathcal{S}$ such that* f $\mathcal{S} \cup \{i\}$ $>$ $\frac{\epsilon}{n^2}$ f $\mathcal{S}$ **do**
3 | $\mathcal{S} \leftarrow \mathcal{S} \cup \{i\}$
4 **end**
5 **if** *there exists a user $i \in \mathcal{S}$ such that* f $\mathcal{S} \setminus \{i\}$ $>$ $\frac{\epsilon}{n^2}$ f $\mathcal{S}$ **then**
6 | $\mathcal{S} \leftarrow \mathcal{S} \setminus \{i\}$ **go to Line 2**
7 **end**
8 **if** f $\mathcal{U} \setminus \mathcal{S}$ $> f$ $\mathcal{S}$ **then** $\mathcal{S} \leftarrow \mathcal{U} \setminus \mathcal{S}$
9 **foreach** $i \in \mathcal{U}$ **do**
10 | **if** $i \in \mathcal{S}$ **then** $p_i \leftarrow b_i$
11 | **else** $p_i \leftarrow$
12 **end**
13 **return** $\mathcal{S}, p$

---

o goo is t e SB a ction n t e ollo ing e analy e t is ec anis sing t e o r esira le ro erties escri e in Section as er or ance etrics

- **Computational Efficiency**: e r nning ti e o t e ocal Searc Algorit is O $\frac{1}{\epsilon} n^3 m \log m$ ere e al-ating t e al e o f ta es O m ti e an $|\mathcal{S}| \leq m$ ence o r ec anis is co tationally e cient
- **Individual Rationality**: e lat or ays at t e inners i ence o r ec anis is in i i ally ratio-nal
- **Profitability**: e to t e ass tion t at t ere e ists at least one ser i s c t at $\tilde{u}_0$ $\{i\}$ $>$ an t e act t at f $\mathcal{S}$ strictly increases in eac iteration e g arantee t at $\tilde{u}_0$ $\mathcal{S}$ $>$ at t e en o t e a ction ence o r ec anis is ro ta le
- **Truthfulness**: e se an e a le in ig re to s o t at t e SB a ction is *not* tr t 1 n t is e a le $\mathcal{U}$ { , , } Γ $\{\tau_1, \tau_2, \tau_3, \tau_4, \tau_5\}$ Γ_1 $\{\tau_1, \tau_3, \tau_5\}$ Γ_2 $\{\tau_1, \tau_2, \tau_4\}$ Γ_3 $\{\tau_2, \tau_5\}$ c_1 c_2 c_3 S ares re resent sers an is s re resent tas s e n er a o e ser i enotes its i b_i e n er elo tas τ_j enotes its al e ν_j or e a le b_1 an ν_3 e also ass e t at ϵ .

e rst consi er t e case ere sers i tr t lly Since $f\{\}$ $v\Gamma_1 - b_1$ $\sum_{i=1}^{3} b_i$ $-$ $f\{\}$ an $f\{\}$ ser is rst selecte Since $f\{,\}$ $v\Gamma_2 \cup \Gamma_1 -$ b_2 b_1 $\sum_{i=1}^{3} b_i$ $>($ $\frac{0.1}{3^2})f\{\}$. ser is t en selecte e a ction ter inates ere eca se t e c rrent al e o f cannot e increase y a actor o $\frac{0.1}{9}$ ia eit er a ing a ser t at as not een selecte or re o ing a ser t at as een selecte n a ition e a e p_1 b_1 an p_2 b_2

e no consi er t e case ere ser lies y i ing δ ere $\leq \delta <$. Since $f\{\}$ δ $f\{\}$ an $f\{\}$ δ ser is rst selecte Since $f\{,\}$ $>($ $\frac{0.1}{9})f\{\}$ ser is t en selecte e a ction ter inates ere eca se t e c rrent al e o f cannot e increase y a actor o $\frac{0.1}{9}$ ia eit er a ing a ser or re o ing a ser ote t at ser increases its ay ent ro to δ y lying a o t its cost

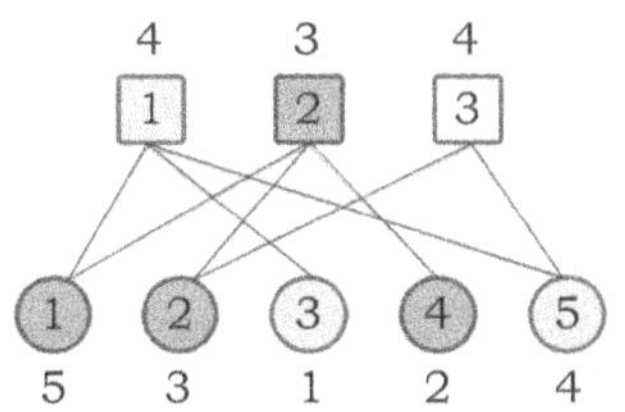

a sers i tr t lly

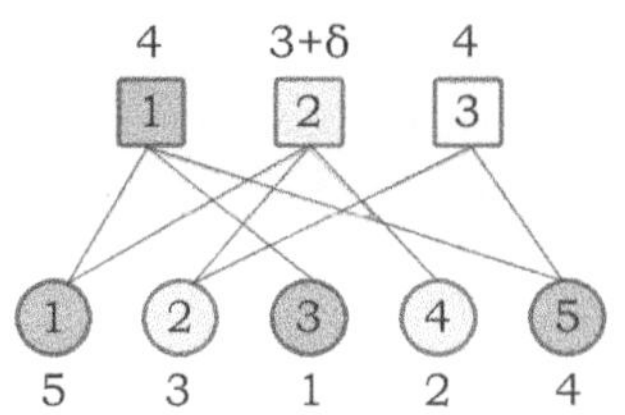

ser lies y i ing δ ere $\leq \delta <$.

Figure 2: An example showing the untruthfulness of the Local Search-Based Auction mechanism, where $\mathcal{U}$ $\{,,\}$, Γ $\{\tau_1,\tau_2,\tau_3,\tau_4,\tau_5\}$, Γ_1 $\{\tau_1,\tau_3,\tau_5\}$, Γ_2 $\{\tau_1,\tau_2,\tau_4\}$, Γ_3 $\{\tau_2,\tau_5\}$. Squares represent users. Disks represent tasks. The number above user i denotes its bid b_i. The number below task τ_j denotes its value ν_j. We also assume that ϵ . .

4.2 MSensing Auction

Alt o g t e SB a ction ec anis is esigne to a - ro i ately a i i e t e lat or tility t e ail re o g aranteeing tr t lness a es it less attracti e Since o r lti ate goal is to esign an incenti e ec anis t at oti ates s art one sers to artici ate in o ile one sensing ile re enting any ser ro rigging its i to ani late t e ar et e nee to settle or a tra e o e-t een tility a i i ation an tr t lness r ig est riority is to esign an incenti e ec anis t at satis es all o t e o r esira le ro erties e en at t e cost o sacri-cing t e lat or tility ne ossi le irection is to a e se o t e o -t e-s el res lts on t e *budgeted mechanism* esign e gete ec anis esign ro le is ery si ilar it o rs it t e i erence t at t e ay ent ai to t e inners is a constraint instea o a actor in t e o ecti e nction o a ress t is iss e it is int iti e t at e can l g i erent al es o t e get into t e -gete ec anis an select t e one gi ing t e largest tility o e er t is can otentially estroy t e tr t lness o t e incenti e ec anis

n t is section e resent a no el a ction ec anis t at satis es all o r esira le ro erties e esign rationale relies on Myerson s ell- no n c aracteri ation

Theorem *([24, Theorem 2.1]) An auction mechanism is truthful if and only if:*

- *The selection rule is monotone: If user i wins the auction by bidding b_i, it also wins by bidding $b_i' \leq b_i$;*
- *Each winner is paid the critical value: User i would not win the auction if it bids higher than this value.*

4.2.1 Auction Design

Base on eore e esign o r a ction ec anis in t is section ic is calle MSensing a ction ll strate in Algorit t e MSensing a ction ec anis consists o t o ases: t e *winner selection* ase an t e *payment determination* ase

Algorithm 3: MSensing A ction

```
1  // Phase 1: Winner selection
2  S ← ∅ i ← arg a j∈U vj S − bj
3  while bi < vi do
4  |  S ← S ∪ {i}
5  |  i ← arg a j∈U\S vj S − bj
6  end
7  // Phase 2: Payment determination
8  foreach i ∈ U do pi ←
9  foreach i ∈ S do
10 |  U' ← U \ {i} T ← ∅
11 |  repeat
12 |  |  ij ← arg a j∈U'\T vj S − bj T ← T ∪ {ij}
13 |  |  pi ← a {pi, in{vi T − vij T − bij , vi T }}
14 |  until bij ≥ vij
15 end
16 return S, p
```

e inner selection ase ollo s a gree y a roac : sers are essentially sorte accor ing to t e i erence o t eir arginal al es an i s i en t e selecte sers $\mathcal{S}$ t e arginal al e o ser i is $v_i\ \mathcal{S}$ $v\ \mathcal{S} \cup \{i\}$ $- v\ \mathcal{S}$ n t is sorting t e i t ser is t e ser j s c t at $v_j\ \mathcal{S}_i - b_j$ is a i i e o er $\mathcal{U}\setminus\mathcal{S}_i$ ere $\mathcal{S}_i$ $\{,\ ,\ldots,i\}$ an $\mathcal{S}_0$ $\emptyset$ e se v_i instea o $v_i\ \mathcal{S}_{i-1}$ to si li y t e notation onsi ering t e s o larity o v t is sorting i lies t at

$$v_1 - b_1 \geq v_2 - b_2 \geq \cdots \geq v_n - b_n.$$

e set o inners are $\mathcal{S}_L$ $\{,\ ,\ldots,L\}$ ere $L \leq n$ is t e largest in e s c t at $v_L - b_L >$

n t e ay ent eter ination ase e co te t e ay-ent p_i or eac inner $i \in \mathcal{S}$ o co te t e ay ent or ser i e sort t e sers in $\mathcal{U} \setminus \{i\}$ si ilarly

$$v'_{i_1} - b_{i_1} \geq v'_{i_2} - b_{i_2} \geq \cdots \geq v'_{i_{n-1}} - b_{i_{n-1}},$$

ere v'_{i_j} $v(\mathcal{T}_{j-1} \cup \{i_j\}) - v(\mathcal{T}_{j-1})$ enotes t e arginal al e o t e jt ser an $\mathcal{T}_j$ enotes t e rst j sers accor ing to t is sorting o er $\mathcal{U} \setminus \{i\}$ an $\mathcal{T}_0$ $\emptyset$ e arginal al e o ser i at osition j is $v_{i(j)}$ $v(\mathcal{T}_{j-1} \cup \{i\}) - v(\mathcal{T}_{j-1})$ et K enote t e osition o t e last ser $i_j \in \mathcal{U} \setminus \{i\}$ s c t at $b_{i_j} < v'_{i_j}$ or eac osition j in t e sorting e co te t e a i rice t at ser i can i s c t at i can e selecte instea o ser at jt lace e re eat t is ntil t e osition o t e rst loser n t e en e set t e al e o p_i to t e a i o t ese K rices

4.2.2 A Walk-Through Example

e se t e e a le in ig re to ill strate o t e MSensing a ction or s

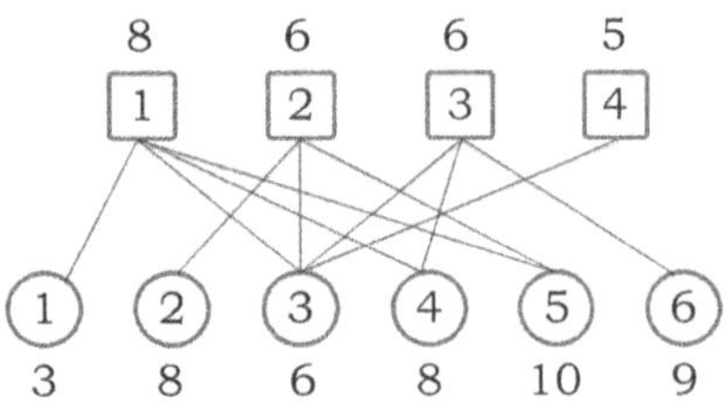

Figure 3: Illustration for MSensing

inner Selection:

◇ $\mathcal{S}$ $\emptyset$: $v_1(\emptyset) - b_1$ $v(\emptyset \cup \{\ \}) - v(\emptyset) - b_1$ ν_1 ν_3 ν_4 ν_5 – – – – $v_2(\emptyset) - b_2$ $v(\emptyset \cup \{\ \}) - v(\emptyset) - b_2$ $v_3(\emptyset) - b_3$ an $v_4(\emptyset) - b_4$

◇ $\mathcal{S}$ $\{\ \}$: $v_2(\{\ \}) - b_2$ $v(\{\ \} \cup \{\ \}) - v(\{\ \}) - b_2$ – – $v_3(\{\ \}) - b_3$ $v(\{\ \} \cup \{\ \}) - v(\{\ \}) - b_3$ an $v_4(\{\ \}) - b_4$ –

◇ $\mathcal{S}$ $\{\ ,\ \}$: $v_2(\{\ ,\ \}) - b_2$ $v(\{\ ,\ \} \cup \{\ \}) - v(\{\ ,\ \}) - b_2$ an $v_4(\{\ ,\ \}) - b_4$ –

◇ $\mathcal{S}$ $\{\ ,\ ,\ \}$: $v_4(\{\ ,\ ,\ \}) - b_4$ –

ring t e ay ent eter ination ase e irectly gi e inners en ser i is e cl e ro t e consi eration e to t e s ace li itations Also recall t at $v'_{i_j} > b_{i_j}$ or $j \le K$ an $v'_{i_j} \le b_{i_j}$ or $j \ge K$

Pay ent eter ination:

◇ p_1: inners are $\{\ \}$ $v_1(\emptyset) - v_2(\emptyset) - b_2$ $v_1(\{\ \}) - v_3(\{\ \}) - b_3$ $v_1(\{\ ,\ \})$ s p_1

◇ p_2: inners are $\{\ \}$ $v_2(\emptyset) - v_1(\emptyset) - b_1$ $v_2(\{\ \}) - v_3(\{\ \}) - b_3$ $v_2(\{\ ,\ \})$ s p_2

◇ p_3: inners are $\{\ \}$ $v_3(\emptyset) - v_1(\emptyset) - b_1$ $v_3(\{\ \}) - v_2(\{\ \}) - b_2$ $v_3(\{\ ,\ \})$ s p_3

4.2.3 Properties of MSensing

e ill ro e t e co tational e ciency e a t e in i i al rationality e a t e ro ta ility e a an t e tr t lness e a o t e MSensing a ction in t e ollo ing

Lemma *MSensing is computationally efficient.*

PROOF. in ing t e ser it a i arginal al e ta es $O(nm)$ ti e ere co ting t e al e o v_i ta es $O(m)$ ti e Since t ere are m tas s an eac inner s o l contri te at least one ne tas to e selecte t e n er o inners is at ost m ence t e ile-loo ines – t s ta es $O(nm^2)$ ti e n eac iteration o t e or-loo ines – a rocess si ilar to ines – is e ec te ence t e r nning ti e o t e ole a ction is o inate y t is or-loo ic is o n e y $O(nm^3)$ □

ote t at t e r nning ti e o t e MSensing A ction $O(nm^3)$ is ery conser ati e n a ition m is c less t an n in ractice ic a es t e r nning ti e o t e MSensing A ction o inate y n

Be ore t rning o r attention to t e roo s o t e ot er t ree ro erties e o l li e to a e so e critical o ser-ations: $v_{i(j)} \ge v_{i(j+1)}$ or any j e to t e s o larity o v $\mathcal{T}_j$ $\mathcal{S}_j$ or any $j < i$ $v_{i(i)}$ v_i an $v'_{i_j} > b_{i_j}$ or $j \le K$ an $v'_{i_j} \le b_{i_j}$ or $j \ge K$

Lemma *MSensing is individually rational.*

PROOF. et i_i e ser i s re lace ent ic a ears in t e it lace in t e sorting o er $\mathcal{U} \setminus \{i\}$ Since ser i_i o l not e at it lace i i is consi ere e a e $v_{i(i)} - b_i \ge v'_{i_i} - b_{i_i}$ ence e a e $b_i \le v_{i(i)} - v'_{i_i} - b_{i_i}$ Since ser i is a inner e a e $b_i \le v_i$ $v_{i(i)}$ t ollo s t at $b_i \le$ in$\left\{v_{i(i)} - v'_{i_i} - b_{i_i}, v_{i(i)}\right\} \le p_i$ □

Lemma *MSensing is profitable.*

PROOF. et L e t e last ser $j \in \mathcal{U}$ in t e sorting s c t at $b_j < v_j$ e t en a e $\tilde{u}_0$ $\sum_{1\le i\le L} v_i - \sum_{1\le i\le L} p_i$ ence it s ces to ro e t at $p_i \le v_i$ or eac $\le i \le L$ ecall t at K is t e osition o t e last ser $i_j \in \mathcal{U} \setminus \{i\}$ in t e sorting s c t at $b_{i_j} < v'_{i_j}$ et r e t e in e o t e ser s c t at

$$r \quad \arg \underset{1\le j\le K+1}{\text{a}} \quad \text{in}\left\{v_{i(j)} - v'_{i_j} - b_{i_j}, v_{i(j)}\right\}.$$

$r \le K$ e a e

$$p_i \quad \text{in}\left\{v_{i(r)} - v'_{i_r} - b_{i_r}, v_{i(r)}\right\}$$

$$v_{i(r)} - v'_{i_r} - b_{i_r} < v_{i(r)} \le v_i,$$

ere t e en lti ate ine ality is e to t e act t at $b_{i_r} < v'_{i_r}$ or $r \le K$ an t e last ine ality relies on t e act t at $\mathcal{T}_{j-1}$ $\mathcal{S}_{j-1}$ or $j \le i$ an t e ecreasing arginal al e ro erty o v r K e a e

$$p_i \quad \text{in}\left\{v_{i(r)} - v'_{i_r} - b_{i_r}, v_{i(r)}\right\} \quad v_{i(r)} \le v_i.$$

s e ro e t at $p_i \le v_i$ or eac $\le i \le K$ □

Lemma *MSensing is truthful.*

PROOF. Base on eore it s ces to ro e t at t e selection r le o MSensing is onotone an t e ay ent p_i or eac i is t e critical al e e onotonicity o t e selection r le is o io s as i ing a s aller al e can not s ser i ac ar s in t e sorting

e ne t s o t at p_i is t e critical al e or i in t e sense t at i ing ig er p_i co l re ent i ro inning t e a ction ote t at

$$p_i \quad \underset{1\le j\le K+1}{\text{a}} \quad \text{in}\left\{v_{i(j)} - v'_{i_j} - b_{i_j}, v_{i(j)}\right\}$$

$$\text{a}\left\{\underset{1\le j\le K}{\text{a}} \, v_{i(j)} - v'_{i_j} - b_{i_j}, v_{i(K+1)}\right\}.$$

ser i i s $b_i > p_i$ it ill e lace a ter K since $b_i > v_{i(j)} - v'_{i_j} - b_{i_j}$ i lies $v'_{i_j} - b_{i_j} > v_{i(j)} - b_i$ At t e K t iteration ser i ill not e selecte eca se $b_i > v_{i(K+1)}$ As K is t e osition o t e rst loser o er $\mathcal{U} \setminus \{i\}$ t e selection roce re ter inates □

e a o e o r le as toget er ro e t e ollo ing t e- ore

Theorem *MSensing is computationally efficient, individually rational, profitable and truthful.*

e ar : r MSensing A ction ec anis still or s en t e al ation nction is c ange to any ot er e - ciently co ta le s o lar nction e o r esira le ro erties still ol

5. PERFORMANCE EVALUATION

o e al ate t e er or ance o o r incenti e ec anis s e i le ente t e incenti e ec anis or t e lat or - centric o el t e ocal Searc -Base a ction enote y *LSB* an t e MSensing a ction enote y *MSensing*

Per or ance Metrics: e er or ance etrics incl e r nning ti e lat or tility an ser tility in general or t e lat or -centric incenti e ec anis e also st y t e n er o artici ating sers

5.1 Simulation Setup

e arie t e n er o sers n ro to it t e incre ent o or t e lat or -centric o el e ass e t at t e cost o eac ser as ni or ly istri te o er , κ_{max} ere κ_{max} as arie ro to it t e incre ent o e set λ to or t e ser-centric o el tas s an sers are ran o ly istri te in a × region as s o n in ig re ac ser s tas set incl es all t e tas s it in a istance o ro t e ser e arie t e n er o tas s m ro to it t e incre ent o e set ϵ to . or *LSB* e also a e t e ollo ing ass tions e al e o eac tas is ni or ly istri te o er , e cost c_i is $\rho|\Gamma_i|$ ere ρ is ni or ly istri te o er ,

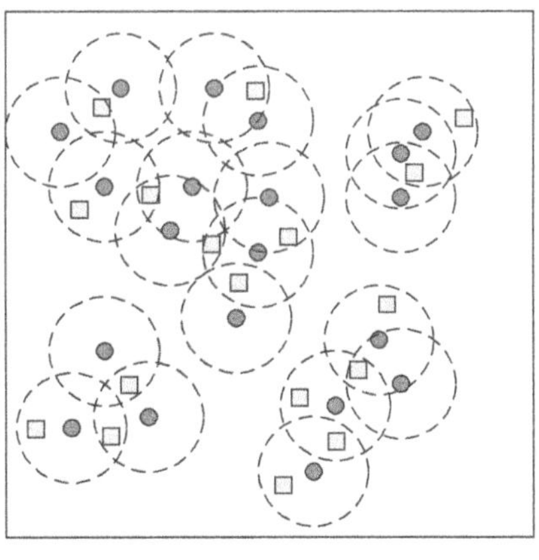

Figure 4: Simulation setup for the user-centric model, where squares represent tasks and circles represent users.

All t e si lations ere r n on a in ac ine it P an B e ory ac eas re ent is a erage o er instances

5.2 Evaluation of the Platform-Centric Incentive Mechanism

nning i e: e rst e al ate t e r nning ti e o t e incenti e ec anis an s o t e res lts in ig re e o ser e t at t e r nning ti e is al ost linear in t e n - er o sers an less t an × $^{-4}$ secon s or t e largest instance o sers As soon as t e sers are sorte an $\mathcal{S}$ is co te all t e al es can e co te sing close - or e ressions ic a es t e incenti e ec anis cry e cient

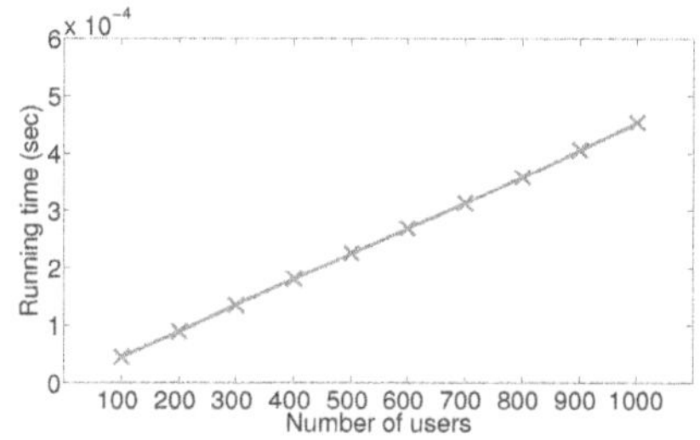

Figure 5: Running time

er o Partici ating sers: ig re s o s t e i - act o κ_{max} on t e n er o artici ating sers i e $|\mathcal{S}|$ en n is e at e can see t at $|\mathcal{S}|$ ecreases as t e costs o sers eco e i erse e reason is t at accor ing to t e ile-loo con ition i all sers a e t e sa e cost t en all o t e o l satis y t is con ition an t s ar- tici ate en t e costs eco e i erse sers it larger costs o l a e ig er c ances to iolate t e con ition

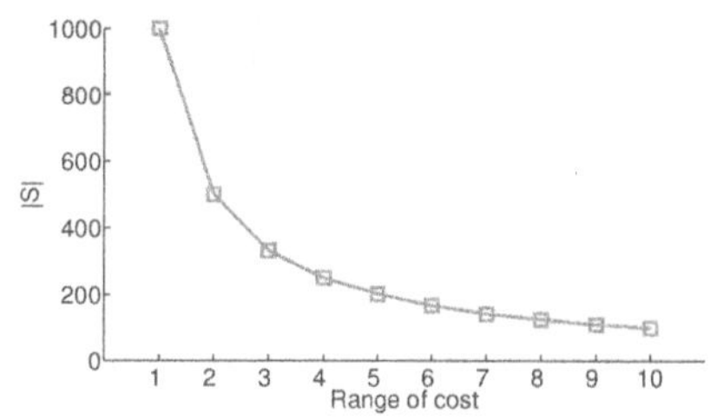

Figure 6: Impact of κ_{max} on $|\mathcal{S}|$

Plat or tility: ig re s o s t e i act o n an κ_{max} on t e lat or tility n ig re a e e κ_{max} e o ser e t at t e lat or tility in ee e onstrates i inis ing ret rns en n increases n ig re e e n it t e res lts in ig re it is e ecte t at t e lat or tility ecreases as t e costs o sers e- co e ore i erse

ser tility: e ran o ly ic e a ser an lot its tility in ig re e o ser e t at as ore an ore sers are intereste in o ile one sensing t e tility o t e ser ecreases since ore co etitions are in ol e

5.3 Evaluation of the User-Centric Incentive Mechanism

nning i e: ig re s o s t e r nning ti e o i er- ent a ction ec anis s ro ose in Section More s eci - ically ig re a lots t e r nning ti e as a nction o n ile m e can see t at *LSB* as etter e - ciency t an *MSensing* ote t at *MSensing* is linear in

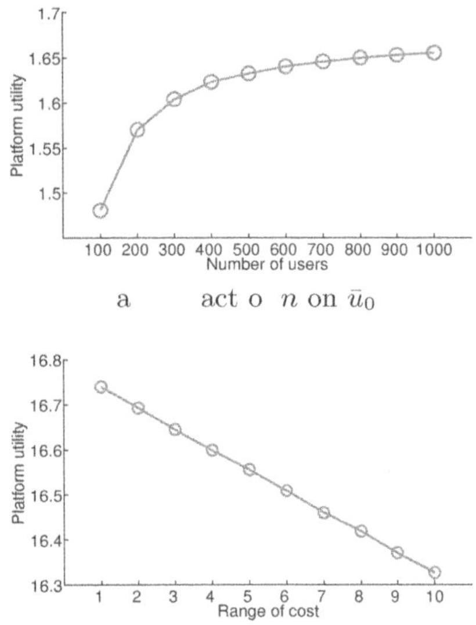

a act o n on $\bar{u}_0$

act o κ_{max} on $\bar{u}_0$

Figure 7: Platform utility

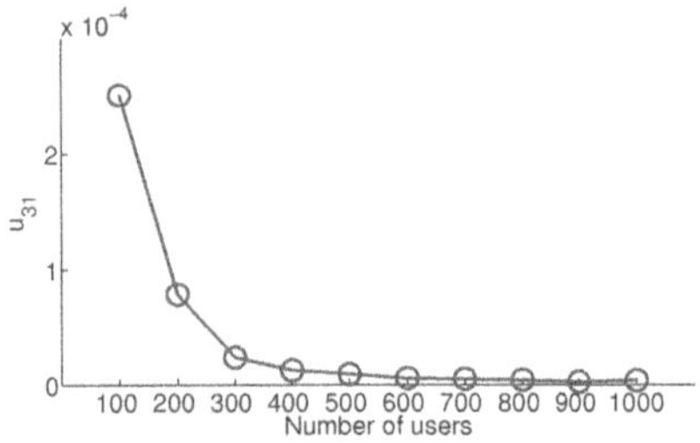

Figure 8: Impact of n on $\bar{u}_i$

n as e ro e in e a ig re lots t e r n-ning ti e as a nction o m ile n Bot *LSB* an *MSensing* a e si ilar er or ance ile *MSensing* o t er or s *LSB* slig tly

Plat or tility: o e s o o c lat or tility e nee to sacri ce to ac ie e t e tr t lness co are to *LSB* As s o n in ig re e can o ser e t e lat or tility ac ie e y *MSensing* is larger t an t at y *LSB* en t e n er o tas s is s all m is relation is re erse en m is large an t e sacri ce eco es ore se ere en m increases o e er note t at in ractice m is s ally relati ely s all co are to n e also o ser e t at si ilar to t e lat or -centric o el t e lat or tility e onstrates t e i inis ing ret rns as ell en t e n er o sers eco es larger

r t lness: e also eri e t e tr t lness o *MSensing* y ran o ly ic ing t o sers an an allo ing t e to i rices t at are i erent ro t eir tr e costs e ill strate t e res lts in ig re As e can see ser ac ie es its o ti al tility i it i s tr t lly b_{333} c_{333} in ig re a an ser ac ie es its o ti al tility i it i s tr t lly b_{851} c_{851} in ig re

6. RELATED WORK

n e y *et al.* e elo e recr it ent ra e or s to ena le t e lat or to i enti y ell-s ite artici ants

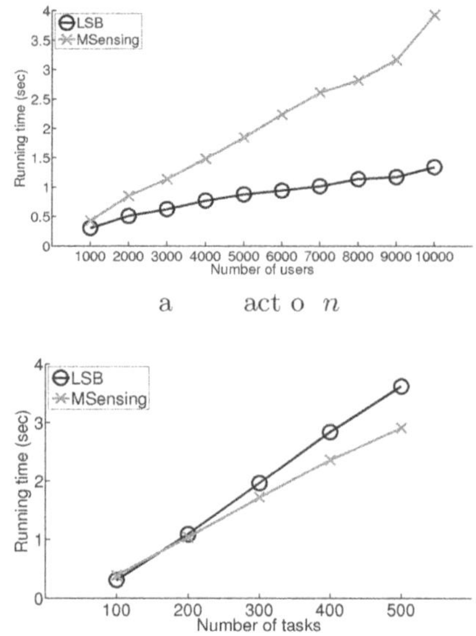

a act o n

act o m

Figure 9: Running time

or sensing ser ices o e er t ey oc se only on t e ser selection not t e incenti e ec anis esign o t e est o o r no le ge t ere are e researc st ies on t e in-centi e ec anis esign or o ile one sensing n ane is *et al.* e elo e a seale - i secon - rice a ction to oti ate ser artici ation o e er t e tility o t e lat or as neglecte in t e esign o t e a ction n ee an o esigne an e al ate a re erse a c-tion ase yna ic rice incenti e ec anis ere sers can sell t eir sense ata to t e ser ice ro i er it sers clai e i rices o e er t e a t ors aile to consi er t e tr t lness in t e esign o t e ec anis

e esign o t e incenti e ec anis as also st ie or ot er net or ing ro le s s c as s ectr tra ing an ro ting o e er none o t e can e irectly a lie to o ile one sensing a lications as t ey all consi ere ro erties s eci cally ertain to t e st ie ro le s

7. CONCLUSION

n t is a er e a e esigne incenti e ec anis s t at can e se to oti ate s art one sers to artici ate in o ile one sensing ic is a ne sensing ara ig allo ing s to collect an analy e sense ata ar eyon t e scale o at as re io sly ossi le e a e consi -ere t o i erent o els ro i erent ers ecti es: t e lat or -centric o el ere t e lat or ro i es a re-ar s are y artici ating sers an t e ser-centric o el ere eac ser can as or a reser e rice or its sensing ser ice

or t e lat or -centric o el e a e o ele t e in-centi e ec anis as a Stac el erg ga e in ic t e lat-or is t e lea er an t e sers are t e ollo ers e a e ro e t at t is Stac el erg ga e as a ni e e ili ri an esigne an e cient ec anis or co ting it is ena les t e lat or to a i i e its tility ile no ser can i ro e its tility y e iating ro t e c rrent strat-egy nilaterally

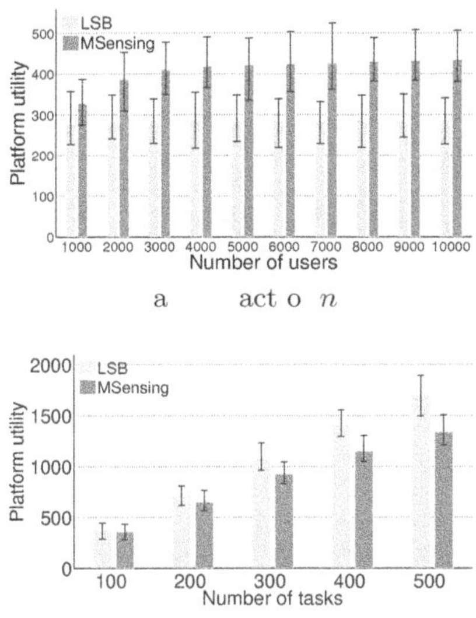

a act o n

act o m

Figure 10: Platform utility

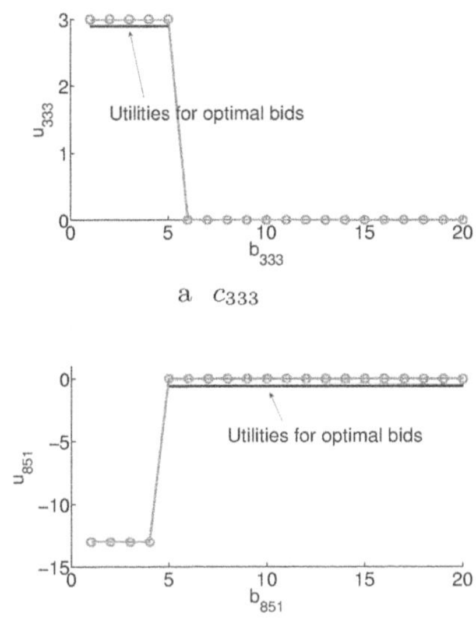

a c_{333}

c_{851}

Figure 11: Truthfulness of MSensing

or t e ser-centric o el e a e esigne an a ction ec anis calle MSensing e a e ro e t at MSensing is co tationally e cient eaning t at t e inners an t e ay ents can e co te in olyno ial ti e in i i ally rational eaning t at eac ser ill a e a non-negati e tility ro ta le eaning t at t e lat-or ill not inc r a e cit an ore i ortantly tr t- l eaning t at no ser can i ro e its tility y as ing or a rice i erent ro its tr e cost r ec anis is scala le eca se its r nning ti e is linear in t e n er o sers

ACKNOWLEDGMENT

e t an t e anony o s re ie ers an t es e er ose co ents an g i ance a e el e to signi cantly i ro e t e a er

8. REFERENCES

S Boy an an en erg e *Convex Optimization* a ri ge ni ersity Press

en ra in an P n t e a ro i a ility o get easi le ec anis s n *Proceedings of ACM-SIAM SODA* ages –

ort ne n stry rst: S art ones ass P s in sales
http://tech.fortune.cnn.com/2011/02/07/idc-smartphone-shipment-numbers-passed-pc-in-q4-2010/

er o A Balas ra anian - o A ol an S Saroi an ra an P Ba l MA : a ing s art ones last longer it co e o oa n *Proceedings of MobiSys* ages –

ane is S e is an An erson o c is location ri acy ort n *Proceedings of WEIS*

as P Mo an Pa ana an a ee an A S ar a P SM: lat or or re ote sensing sing s art ones n *Proceedings of ACM MobiSys* ages –

e risto aro an Soriente S ort a er: Pe si ri acy-en ance artici atory sensing in rastr ct re n *Proceedings of WiSec* ages –

eige S Mirro ni an on ra Ma i i ing non- onotone s o lar nctions *SIAM J. on Computing* : –

en erg an irole *Game theory* M Press

ao an ang Ma : M ltia ctioneer rogressi e a ction or yna ic s ectr access *IEEE Transactions on Mobile Computing* : – A g st

orl i es art one ar et e ecte to gro in an a roac s i ents o one illion in accor ing to http://www.idc.com/getdoc.jsp?containerId=prUS22871611

ris na *Auction Theory* Aca e ic Press

ane Mil o Pee les o ry an A a ell A s r ey o o ile one sensing *IEEE Communications Magazine* : –

ee an B o Sell yo r e eriences: A ar et ec anis ase incenti e or artici atory sensing n *Proceedings of IEEE PerCom* ages –

M Mee er Mary Mee er: S art ones ill s r ass P s i ents in t o years
http://techcrunch.com/2010/11/16/meeker-smartphones-pcs/

P Mo an Pa ana an an a ee ericell: ric onitoring o roa an tra c con itions sing o ile s art ones n *Proceedings of SenSys* ages –

M M n S e y S ilton a B r e strin M ansen o ar est an P Bo a P t e ersonal en iron ental i act re ort as a lat or or artici atory sensing syste s researc n *Proceedings of ACM MobiSys* ages –

Myerson ti al a ction esign *Math. of Operations Research* : –

e a ser olsey an M is er An analysis o t e a ro i ations or a i i ing s o lar set nctions *Mathematical Programming* : –

ana o S an ere B l s an ar one: An en -to-en artici atory r an noise a ing n *Proceedings of ACM/IEEE IPSN* ages –

S e y strin an M B Sri asta a ecr it ent ra e or or artici atory sensing ata collections n *Proceedings of Pervasive* ages –

ie erer rra illi A aintrea B ris na rt y an P o rig e or sale: yo r ata y: yo n *Proceedings of ACM HotNets* ages : – :

Sensorly Sensorly tt : sensorly co

Singer B get easi le ec anis s n *Proceedings of IEEE FOCS* ages –

A iagara an a in ranat a rts S Ma en Bala ris nan S ole o an ri sson trac : acc rate energy-a are roa tra c elay esti ation sing o ile ones n *Proceedings of ACM SenSys* ages –

ang B i an B iang istrict: racing local ar ets in tr t l s ectr o le a ctions n *Proceedings of IEEE SECON* ages –

S ong i i an ang n esigning incenti e-co ati le ro ting an or ar ing rotocols in ireless a - oc net or s: an integrate a roac sing ga e t eoretical an cry togra ic tec ni es n *Proceedings of ACM Mobicom* ages –

o S an i S S ri an eng eBay in t e s y: strategy- roo ireless s ectr a ctions n *Proceedings of ACM Mobicom* ages –

APPENDIX

A. PROOF OF THEOREM 4

e ro e t e P- ar ness o t e o ti i ation ro le y gi ing a olyno ial ti e re ction ro t e P- ar **Set Cover** ro le :

INSTANCE: A ni erse Z $\{z_1, z_2, \ldots, z_m\}$ a a ily $\mathcal{C}$ $\{C_1, C_2, \ldots, C_n\}$ o s sets o Z an a ositi e integer s

QUESTION: oes t ere e ist a s set $\mathcal{C}' \subseteq \mathcal{C}$ o si e s s c t at e ery ele ent in Z elongs to at least one e er in $\mathcal{C}'$

e constr ct a corres on ing instance o t e **User Selection** ro le as ollo s: et Γ e t e tas set corres on ing to Z ere t ere is a tas $\tau_j \in \Gamma$ or eac $z_j \in Z$ orres on ing to eac s set $C_i \in \mathcal{C}$ t ere is a ser $i \in \mathcal{U}$ it tas set Γ_i ic consists o t e tas s corres on ing to t e ele ents in C_i e set ν_j to n or eac tas τ_j an c_i to or eac ser $i \in \mathcal{U}$ e ro e t at *there exists a solution to the instance of the* Set o er *problem if and only if there exists a subset* $\mathcal{S}$ *of users such that* $\tilde{u}_0$ $\mathcal{S}$ $\geq nm - s$

e rst ro e t e or ar irection et $\mathcal{C}'$ e a sol tion to t e *Set Cover* instance e can select t e corres on ing set $\mathcal{S}$ o sers as t e sol tion to t e ec anis esign instance learly $\tilde{u}_0$ $\mathcal{S}$ $nm - |\mathcal{S}| \geq nm - s$ e t e ro e t e ac ar irection et $\mathcal{S}$ e a sol tion to ec anis esign instance e t en a e $\tilde{u}_0$ $\mathcal{S}$ $\geq nm - s$ e only ossi ility t at e a e s c a al e is en t e selecte ser set co ers all t e tas s eca se $s \leq m$ ere ore t e corres on ing set $\mathcal{C}'$ is a sol tion to t e *Set Cover* instance

Searchlight: Won't You Be My Neighbor?

Mehedi Bakht
Dept. of Computer Science
University of Illinois
Urbana-Champaign
mbakht2@illinois.edu

Matt Trower
Dept. of Computer Science
University of Illinois
Urbana-Champaign
mtrower2@illinois.edu

Robin Kravets
Dept. of Computer Science
University of Illinois
Urbana-Champaign
rhk@illinois.edu

ABSTRACT

The rapid deployment of millions of mobile sensors and smartphones has resulted in a demand for opportunistic encounter-based networking to support mobile social networking applications and proximity-based gaming. However, the success of these emerging networks is limited by the lack of effective and energy efficient neighbor discovery protocols. While probabilistic approaches perform well for the average case, they exhibit long tails resulting in high upper bounds on neighbor discovery time. Recent deterministic protocols, which allow nodes to wake up at specific timeslots according to a particular pattern, improve on the worst case bound, but do so by sacrificing average case performance. In response to these limitations, we have designed Searchlight, a highly effective asynchronous discovery protocol that is built on three basic ideas. First, it leverages the constant offset between periodic awake slots to design a simple probing-based approach to ensure discovery. Second, it allows awake slots to cover larger sections of time, which ultimately reduces total awake time drastically. Finally, Searchlight has the option to employ probabilistic techniques with its deterministic approach that can considerably improve its performance in the average case when all nodes have the same duty cycle. We validate Searchlight through analysis and real-world experiments on smartphones that show considerable improvement (up to 50%) in worst-case discovery latency over existing approaches in almost all cases, irrespective of duty cycle symmetry.

Categories and Subject Descriptors

C.2.1 [**Computer Systems Organization**]: COMPUTER-COMMUNICATION NETWORKS— *Network Architecture and Design*

Keywords

Neighbor Discovery, Ad-hoc Networks, Smartphone, Mobile Social Networks

1. INTRODUCTION

The world of social networking is being transformed by mobile devices. Where early applications required users to check into centralized servers to find and coordinate with friends (e.g., Google Latitude/DodgeBall [1], Gowalla [2], Foursquare [3], FaceBook Places [4], etc.), new applications are being built around the local communication capacity of all smartphones (e.g., Who's Near Me [5] MobiClique [6], , Aka-aki [7], BlueHoo [8]). One of the fastest growing areas of interest is the mobile gaming community, where users look for and play games with other players in their wireless network vicinity (e.g., Nintendo's StreetPass [9] and Sony's Vita [10]). Although the popularity of these proximity-based applications has increased, the network services needed to enable effective and energy-efficient performance have not caught up - especially when users want to operate in a purely local fashion. Essentially, the success of such applications depends on the availability of services that provide information about nearby users, or, more specifically, nearby users' devices.

While centralized services can provide much of the support needed for proximity-based applications, centralized services have significant problems. First, if the service is unavailable, the users are out of luck, even if they can see other users in their vicinity. Second, either a centralized location service must push location information to all users, or users must pull location information as they need it, resulting in excessive updates or control overhead. Third, most applications run their own proprietary location service, requiring users to register with many applications. Fourth, many users are reluctant to expose their location information to a centralized service, but may be willing to let people that are in their general vicinity or that can already see them discover their device and actual location. Finally, communicating through a central server instead of directly to near-by neighbors can introduce significant communication latency, limiting its usefulness for real-time and gaming applications. Therefore, local solutions are needed that do not rely connectivity to centralized servers and services.

To enable the full potential of these proximity-based applications, the applications need local services that find the direct neighbors in a device's communication range. Although neighbor discovery is also challenging in environments such as ad hoc and sensor networks, there is some expectation of network stability, so applications expect connectivity and deal with failure as it happens. There is also some expectation of sufficient density so that when a device has data to send, some number of neighbors will be available at the time of discovery. However, social networks do not have the same expectation of stability or device density. Similar to devices in a delay-tolerant network, devices in a social network search for neighbors and then decide to communicate or send data. Since most smartphones come with both Bluetooth and Wi-Fi and given the extreme differential in the energy needed for discovery between Bluetooth and Wi-Fi, most (if not all) current proximity-based applications limit proximity detection to the cheaper Bluetooth radio. (e.g., Aka-aki and Bluehoo). However, limiting neighbors to be within Bluetooth range obviously limits the reach and effectiveness of social networking applications. While there are many energy-efficient MAC protocols, there is still the need for more energy-efficient neighbor discovery protocols for higher-power communication technologies like Wi-Fi.

The key goal for energy-efficient neighbor discovery is to reduce energy consumption, while still enabling effective neighbor discovery, where effectiveness can be determined by both the ability to discover a neighbor as well as the latency to discovery that neighbor. While the impact of fewer neighbor discoveries is intuitive, discovery latency has some less obvious impacts. In stationary networks, high discovery latency can result in longer set up times, but does not adversely affect the normal operation of the network. In mobile environments, high discovery latencies can result in missing short, yet meaningful contact opportunities, resulting in reduced effectiveness of the application. Unfortunately, current discovery protocols, mostly aimed at sensor networks, have not been able to ensure both low average discovery latency while still guaranteeing a low maximum bound on that latency.

In response to these limitations, we have designed Searchlight, a highly effective and energy-efficient duty-cycle-based neighbor discovery protocol that provides improvements in average and worst-case discovery latency while still providing a significant decrease in energy consumption. Searchlight is designed around the observation that two energy conserving devices follow well-defined patterns to enable discovery, resulting in a stable relationship between the patterns of any two devices, which allows Searchlight to significantly reduce the worst-case bound on discovery latency. By further employing a random/probabilistic component to the protocols, Searchlight considerably reduces average discovery latency as well. By slightly increasing each awake time, Searchlight can eliminate a significant number of redundant discoveries to reduce total awake time, and so total energy consumption for discovery, by almost 50%. Finally, Searchlight can be applied to devices with different duty cycles if all of the chosen duty cycles are a power of two multiple of the smallest duty cycle (i.e., x, $2x$, $4x$, $8x$, ...).

To validate the design and feasibility of Searchlight, we evaluated it through simulation and tested a real implementation on commercial off-the-shelf hardware. Our results show that Searchlight significantly reduces the worst-case discovery latency bound by 50% for a given energy budget. Searchlight also improves average discovery latency by up to 25% for low duty cycles. To gauge the potential of Searchlight in a real network setting, we implemented Searchlight on a testbed consisting of Nokia N900 smartphones. Since most recent duty-cycle-based neighbor discovery algorithms were targeted at and implemented on sensor motes (i.e., U-Connect [11]), our implementation provides valuable new insight into issues related to the execution of such protocols on smartphones with Wi-Fi interfaces to support mobile social networking applications. However, it is important to note that Searchlight is a general discovery protocol that provides significantly improved performance in both mobile and sensor environments.

The rest of this paper is organized as follows. Section 2 briefly describes existing approaches to neighbor discovery and discusses why they fail to meet the performance goals. Section 3 describes Searchlight in detail, including analysis that shows how it improves the worst case bounds for the symmetric case and how this improves the average latency for both cases. Section 4 presents our simulation-based performance evaluation of Searchlight in comparison to other neighbor discovery protocols. A prototype implementation of Searchlight on a smartphone testbed is described in Section 5. We conclude and outline the direction of our future research in Section 6.

2. ENERGY-EFFICIENT NEIGHBOR DISCOVERY FOR PROXIMITY-BASED APPLICATIONS

The key to all proximity-based applications is continuously determining who is near each other, even if the phone is not being used and is in a user's pocket. Although neighbor discovery services are being integrated into smartphone platforms - as seen by the emergence of Qualcomm's AllJoyn proximity-enhanced mobile software environment [12] and the integration of proximity-based services into the latest version of the Android OS [13] using Wi-Fi-Direct [14] - the challenge still remains of balancing battery life and connectivity. In general, the more devices these applications can discover, the better they can perform. However, for a device running on a battery-and thus on a limited energy budget, it is not practical to continuously search for neighbors. Existing approaches such as Bluetooth's inquiry method were designed with user initiated requests in mind and do not fare well under continuous discovery. A more feasible approach is to adopt the techniques from the sensor network community and keep the wireless interface in a sleep state most of the time and periodically wake it up for discovery.

The success of such duty-cycling schemes depends on ensuring that the wakeup times of two neighboring devices overlap. This is not hard to achieve when device clocks can be synchronized, for example through GPS [15, 16]. However, synchronization through GPS is usually too energy-expensive for mobile sensors [17] and smartphones [18]. This has necessitated the design of periodic schemes that ensure such overlap within a reasonable time bound while operating at low duty-cycles and requiring *no clock synchronization.*

While there are many asynchronous communication protocols (e.g., BMAC [19], SMAC [20]), the majority of these assume that all devices (which we call nodes from here on) have *symmetric* sleep patterns (i.e., run at the same duty cycle). However, realistic scenarios, for both mobile social networking and sensor networks, are more likely to result in nodes with varying energy requirements and so *asymmetric* duty cycles. While nodes may start with the same duty cycle, as the network is used, the available energy to each node will change, resulting in asymmetric duty cycles. In response, a new class of asynchronous and asymmetric slotted discovery protocols has evolved.

Asynchronous neighbor discovery algorithms mostly work on a time-slot basis, where time is assumed to be divided into slots of equal size and all nodes agree on the size of a slot. Based on the protocol used, nodes decide to remain awake during specific slots, called *active* slots, and sleep during the remaining slots. During an active slot, the node may send/receive or do both, depending on application requirements. Successful discovery takes place between two neighboring nodes whenever their active slots overlap. To be energy efficient, a discovery scheme needs to use as few active slots as possible to discover neighbors within a reasonable time limit. Current approaches to energy-efficient asynchronous neighbor discovery fall broadly into two categories - probabilistic and deterministic. The relative strengths and weaknesses of these existing schemes can be judged by comparing their average and worst-case latency, as well as considering how well they handle duty-cycle asymmetry.

Both average and worst-case latency are important to opportunistic applications. While it is clear that these applications can benefit from a reduced average latency, most such applications are designed to handle missed contacts and so may seem like they would not benefit from a lower worst-case bound. However, successful discoveries still determine the effectiveness of these applications. Since many encounters may be short, lower worst-case bounds can enable much higher discovery success ratios, and so more contact opportunities.

Most well-known among probabilistic approaches is a family of "birthday protocols" [21] where nodes transmit/receive or sleep with different probabilities. Given the probabilistic nature of this type of approach, such scheme work well in the average case and allow asymmetric operation, since there is no restriction on duty cycle length. However, the main drawback of the birthday protocol is its failure to provide a bound on the worst case discovery latency, leading to long tails on discovery probabilities.

Deterministic protocols, by either ensuring a "quorum" or using prime values for duty cycle lengths, provide strict guarantees on worst-case discovery latency. In the Quorum-based protocols [22, 23], time is divided into sets of m^2 contiguous intervals. These m^2 intervals are arranged as a 2-dimensional $m \times m$ array and each host can pick one row and one column of entries as awake intervals. This pattern ensures that no matter what row and column are chosen, any two nodes have at least two overlapping awake intervals. While the Quorum protocol provides a reasonable bound on worst-case latency, it performs much worse than the probabilistic approach in the average case. Since m is a global parameter, the initial Quorum protocol [23] only supports symmetric operation. Lai et al. [22] provided an improvement to handle asymmetric cases when there are only two different schedules in the entire network. However, allowing only two duty cycles, and so only two energy-saving levels, is too restrictive. Another approach that only works for the symmetric case is the application of block design using difference sets [24]. For the asymmetric case, designing the appropriate schedule following the proposed scheme becomes similar to the vertex-cover problem, which is an NP-complete problem. Hence, these kinds of approaches are limited to symmetric networks.

Prime-based deterministic protocols overcome this limitation and can handle both symmetric and asymmetric operation while still providing a strict bound on worst-case latency. In Disco [25], each node chooses a pair of prime numbers such that the sum of their reciprocals is as close as

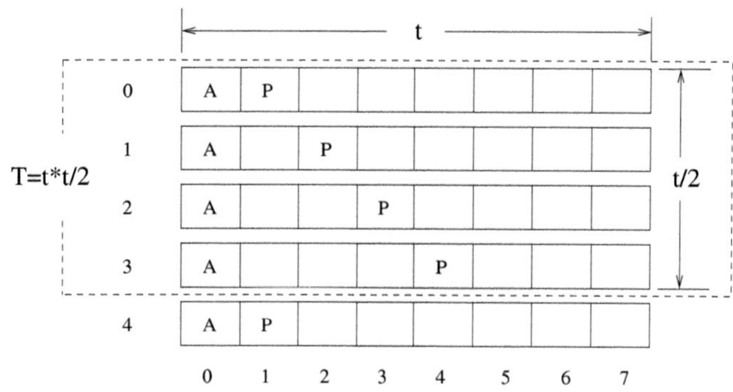

Figure 1: Searchlight with sequential probing (t=8)

possible to the desired duty cycle. The nodes then wake up at multiples of the individual prime numbers. If one node chooses primes p_1, p_2 and another node chooses p_3, p_4, the worst-case discovery latency between these two nodes will be $min\{(p_1 \cdot p_3), (p_1 \cdot p_4), (p_2 \cdot p_3), (p_2 \cdot p_4)\}$, provided the two primes in the pair are not equal. A more recent deterministic approach, U-Connect [11], uses a single prime per node. Instead of just waking up only 1 slot every p slots, the nodes also wake up $\frac{p+1}{2}$ slots every p^2 slots. The worst-case latency for U-Connect is p^2 which is similar to Disco. However, for the energy-latency product, a metric proposed by the authors to evaluate the energy-efficiency of asynchronous neighbor discovery protocols, U-Connect fares better than Disco in the symmetric case.

Although these deterministic protocols have good worst-case performance, in the average case, they are considerably worse than the birthday protocol. To successfully meet all of the goals of energy-efficiency, asymmetry, and good average-case and worst-case latency, we present a new protocol named Searchlight. Searchlight follows a deterministic approach and provides a strict bound on worst-case latency, which is *lower* than existing protocols for both symmetric and asymmetric cases. While deterministic protocols guarantee discovery within a certain bound by using a particular pattern to determine the awake slots, these protocols overlook hidden sub-patterns within these patterns. When the awake slots occur periodically, Searchlight leverages the constant offset between the awake slots of any two nodes using the same period. Additionally, the best performing protocols frequently result in multiple discoveries, where one is sufficient. Searchlight reduces these extraneous discoveries, resulting in a 50% reduction in active slots in symmetric scenarios and only slightly less in asymmetric scenarios. Finally, Searchlight incorporates randomization techniques that significantly improve average discovery latency in the all symmetric cases or even when only some nodes as symmetric.

3. SEARCHLIGHT

Searchlight is an asynchronous periodic slot-based discovery protocol, where a period consists of t contiguous slots as determined by the target duty cycle for a given node. Nodes sleep during the majority of the slots and can only discover each other if they are active during the same slot. Searchlight can operate in two modes: symmetric, where all nodes have the same t, and asymmetric, where nodes may have different ts. The insight into the design of Searchlight comes from the very simple observation that when two nodes wake up periodically with the same interval, the temporal distance between these awake slots always remains the same (assuming negligible clock drift, which we discuss at the end of this section). Searchlight leverages this constant temporal relationship to provide better worst-case bounds on discovery latency than any existing protocol, and then improves the energy cost further by reducing the number of active slots required to ensure discovery by almost half by allowing active slots to "overflow". In addition, Searchlight also has the option to employ probabilistic techniques that considerably improves discovery latency in the average case when all nodes operate on same duty cycle.

In this section, we describe Searchlight in detail, explaining the design of systematic probing, "overflowing" awake slots, and the introduction of randomized probing. Through analysis, we derive the worst-case latency bounds for both the symmetric and asymmetric cases and show that Searchlight reduces worst-case discovery latency by 50% in the symmetric case.

3.1 Systematic Probing

In Searchlight, each node has two active slots in t. The first active slot, called the *anchor* slot, is the first slot in the period. In the symmetric case, since the position of this anchor slot is fixed in t but the start times for the ts for different nodes vary, the anchor slots for two nodes only overlap if the difference between the start times of the two periods is less than one timeslot. For all other offsets, assuming negligible clock drift, the two anchor slots would never meet since the offset remains constant. Essentially, the relative position of the anchor slot of one node always remains the same with respect to that of the other node and is in the range $[1, t-1]$. Since most anchor slots miss each other, Searchlight introduces a second active slot in t called the *probe* slot, which systematically searches for the anchor slot of the other node. It is important to note that anchor slots and probe slots perform exactly the same operation - the different names are used to merely convey the fact that one remains fixed in each period, while the other one moves around in search of the anchor slot of the other node. Hence, probe-anchor, probe-probe and anchor-anchor overlaps all result in discovery.

Given the stability of the relative offset, no two nodes can be offset by more than $\frac{t}{2}$ slots. Therefore, Searchlight only needs to probe the first half of the t slots to guarantee an overlap between the probe slot of one node and the anchor slot of the other node. To achieve this goal, the position of the probe slot is determined by a counter that starts at 1, increments by 1 every period, ends at $\lfloor\frac{t}{2}\rfloor$ and then starts at 1 again (see Figure 1). In other words, if P_i denotes the position of the probe slot in the i-th period, then

$$P_{i+1} = ((P_i) \mod \lfloor\frac{t}{2}\rfloor) + 1. \tag{1}$$

The position of the probe slot actually follows the periodic pattern $\{1, 2, ..., \lfloor\frac{t}{2}\rfloor\}$ and this pattern gets repeated every $\lfloor\frac{t}{2}\rfloor$ periods, which we call the *hyper-period* T. For example, for $t = 8$, Searchlight uses the periodic pattern {1,2,3,4} to determine the position of the probe slot in each period.

3.2 Discovery Latency

In this section, we present the analysis for worst-case discovery latency for Searchlight with symmetric duty cycles.

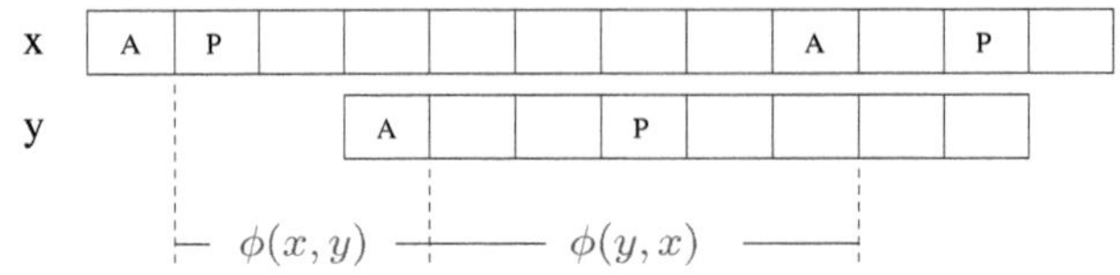

Figure 2: Phase offsets (t=8)

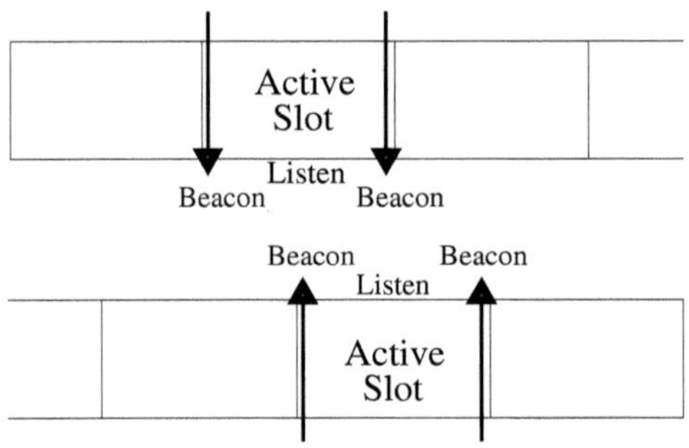

Figure 3: Beaconing in an active slot

LEMMA 3.1. *The worst-case discovery latency for Searchlight with parameter t is equal to $t \cdot \lfloor\frac{t}{2}\rfloor$ slots.*

Proof: For two nodes x and y, let $\phi(x, y)$ be the phase offset (in slots) from the anchor slot of x, which we call A_x, to the anchor slot of y, which we call A_y. Similarly, let $\phi(y, x)$ be the phase offset from A_y to A_x (see Figure 2). Clearly,

$$\phi(x, y) + \phi(y, x) = t. \tag{2}$$

In the symmetric case, where both nodes use the same t, $\phi(x, y)$ and $\phi(y, x)$ remain constant during the contact. It follows from Equation (2) that

$$min(\phi(x, y), \phi(y, x)) \leq \lfloor\frac{t}{2}\rfloor.$$

The probe slot goes through all positions from 1 to $\lfloor\frac{t}{2}\rfloor$ every $\lfloor\frac{t}{2}\rfloor$ periods. Let us denote the probe slots of x and y as P_x and P_y respectively. P_x will meet A_y within $\lfloor\frac{t}{2}\rfloor$ periods as long as

$$1 \leq \phi(x, y) \leq \lfloor\frac{t}{2}\rfloor.$$

Similarly, P_y will meet A_x within $\lfloor\frac{t}{2}\rfloor$ periods as long as

$$1 \leq \phi(y, x) \leq \lfloor\frac{t}{2}\rfloor.$$

So, as long as

$$min(\phi_(x, y), \phi_(y, x)) \leq \lfloor\frac{t}{2}\rfloor,$$

discovery will take place within $\lfloor\frac{t}{2}\rfloor$ periods or $t \cdot \lfloor\frac{t}{2}\rfloor$ slots.

3.3 Slot non-alignment

For illustration purposes, we have shown the slots to be aligned. However, Searchlight is designed as an asynchronous protocol and hence does not assume or rely on the alignment of slot boundaries at different nodes. To ensure that an overlap between two active slots always leads to discovery, Searchlight employs the same beaconing strategy as

Disco [25]. A beacon gets sent both at the beginning and end of an active slot and the node remains in listening mode in the intermediate period (see Figure 3). Because of this approach, non-alignment of slot boundaries actually results in lower discovery latency in general, since each slot overlaps with two slots of the other node and discovery can happen from either of the two overlaps.

However, two overlaps are redundant when discovery can be ensured with just one overlap. Based on this observation, we next describe striped probing, a strategy that improves the performance of Searchlight significantly by eliminating redundant probes.

3.4 Striped probing

Probing of every slot from 1 to $\lfloor \frac{t}{2} \rfloor$ guarantees two overlaps every $\lfloor \frac{t}{2} \rfloor$ periods whenever the slot boundaries of two nodes are not aligned. To reduce this redundancy, the probe slot can instead probe every even slot, by using a counter that starts at 2, increments by 2 every period up to $(2 \cdot \lceil \frac{\lfloor \frac{t}{2} \rfloor}{2} \rceil)$ and then starts at 2 again. In other words, if P_i denotes the position of the probe slot in the i-th period, then

$$P_{i+1} = ((P_i) \mod (2 \cdot \lceil \frac{\lfloor \frac{t}{2} \rfloor}{2} \rceil)) + 2. \tag{3}$$

It is important to note that striping cannot be applied to the birthday protocol or Disco since neither of these protocols use consecutive slots. Since U-Connect has an additional set of $\frac{p+1}{2}$ sequential active slots every p^2 slots, striping can be used to reduce this to $\frac{p+1}{4}$, resulting in 10% fewer active slots. We evaluate this striped version of U-Connect in our implementation.

However, striped probing would not work for either protocol in the rare case when slot boundaries are completely aligned. To handle perfect alignment, each active slot "overflows" by δ, a small amount that is sufficient to receive a leading beacon for another node. This basically means that if a regular slot size is x, then every active slot (both anchor and probe) become $x(1+\delta)$ long, and length of the regular slot that follows is reduced to $x(1-\delta)$. The value of δ should be the smallest possible overlap that guarantees discovery. Because of factors like message transmission time, clock drift, etc., the specific value of δ will be largely platform-dependent. As a guideline, δ should be a small fraction of the slot size (i.e., 1%) or a fixed minimum value based on the device specifications.

The main advantage of striped probing is that it significantly reduces the worst-case bound. Using the approach detailed in Section 3.2, it is easy to show that the worst-case bound for the symmetric case changes to $t \cdot \lceil \frac{\lfloor \frac{t}{2} \rfloor}{2} \rceil$ slots.

3.5 Asymmetry

The main design of Searchlight is based on the constant relative offset between the anchor slots of any two neighboring nodes. Unfortunately, this constant offset does not hold in the asymmetric case, when nodes can choose different values of t based on their own energy requirements.

One possible approach to guarantee overlap withing a fixed bound is to require that t should always be prime. Essentially, restricting t to be a prime number, similar to Disco and U-Connect, would ensure that for any two nodes operating at different duty cycles, their period lengths t_1 and t_2 will be relatively prime, i.e., they will have no common factors other than 1. Since period lengths are relatively prime, it follows from the Chinese Remainder Theorem [26] that the two anchor slots would overlap at least once every $t_1 \cdot t_2$ slots, where t_1 and t_2 are the two different period lengths. However, restricting t to prime number does not maintain the necessary constant offset.

Instead of using primes, Searchlight takes a novel approach to duty cycle asymmetry. It is possible to maintain the constant offset for the anchor slots of any two nodes if the larger valued period is an integer multiple of the smaller one. To see how this works, consider two nodes, node A with period $3t$ and node B with period t (see Figure 4). B's anchor slots always have the same relative distance from all of A's anchor slots. Since A only probes half of its slots, this ensures that the probe slot of A will eventually meet with an any anchor slot of B that falls in the first half of A's cycle. In this example, B always probes the first $3t/2$ slots. Since A's anchor slots always align with slots 2 and 10 for B, discovery is guaranteed. This example can be generalized to A having any value nt.

While any multiple works for a given pair of nodes, to maintain the constant offset for all nodes, Searchlight restricts the duty cycle to a power-multiple of the smallest duty cycle (i.e. 2, 4, 8, 16 or 3, 9, 27, 81, etc.), guaranteeing that any two nodes' duty cycles are multiples of each other. This means that if t is the smallest duty cycle, the other nodes can select $2t$, $4t$, $8t$, etc, for any t. Although this seems to restrict the choice of energy conservation levels, it is important to note that the use of primes is also very restrictive. While prime-based approaches are restricted to only primes, Searchlight can be set up to use any base value of t to fit the requirements of the system. Additionally, the use of the power of 2 multiples provides the most flexibility at the shorter duty cycles, giving applications more choices where it is needed. Finally, Searchlight can use any power of x multiple, providing a little more flexibility if necessary.

Along with flexibility, the power-multiple-based approach also provides the best worst-case latency bound, which is based on the time needed for A's probe slot to meet an anchor slot of B. A's probe slot probes every other slot from 2 to $\lceil \frac{\lfloor \frac{t_A}{2} \rfloor}{2} \rceil$. Let us assume that within that range, there are m anchor slots of B. In the worst-case, when A meets B, A's probe slot would be just past the $m-th$ anchor slot. To overlap, the probe slot will still have to go up to $\lceil \frac{\lfloor \frac{t_A}{2} \rfloor}{2} \rceil$, and then start from 2 again and reach the 1st anchor slot of B. That means the probe slot would have to travel $t_B + (\lceil \frac{\lfloor \frac{t_A}{2} \rfloor}{2} \rceil \mod t_B)$ slots. With striped probing, it would require

$$\frac{t_B + (\lceil \frac{\lfloor \frac{t_A}{2} \rfloor}{2} \rceil \mod t_B)}{2}$$

periods, or

$$\frac{(t_B + (\lceil \frac{\lfloor \frac{t_A}{2} \rfloor}{2} \rceil \mod t_B))}{2} \cdot t_A$$

slots.

It is possible to combine the prime-based and power-multiple-based approaches if necessary, as long as the chosen primes are relatively prime with the base t. For example, when four different duty cycles are needed, three allowable t val-

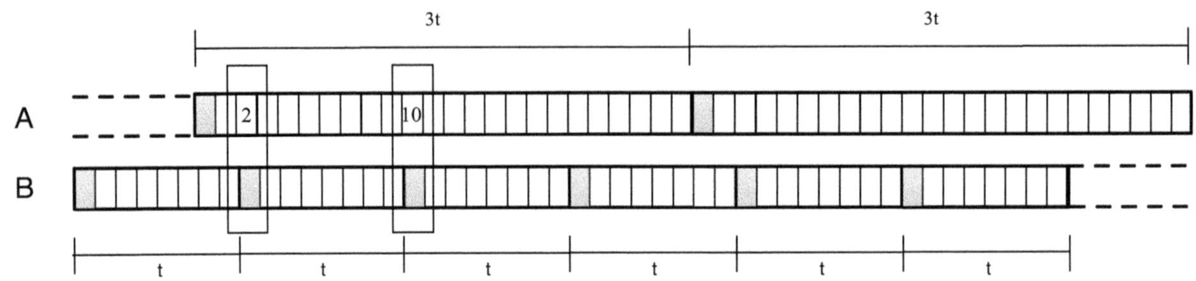

Figure 4: Searchlight with Asymmetric Duty Cycles (t and $3t$)

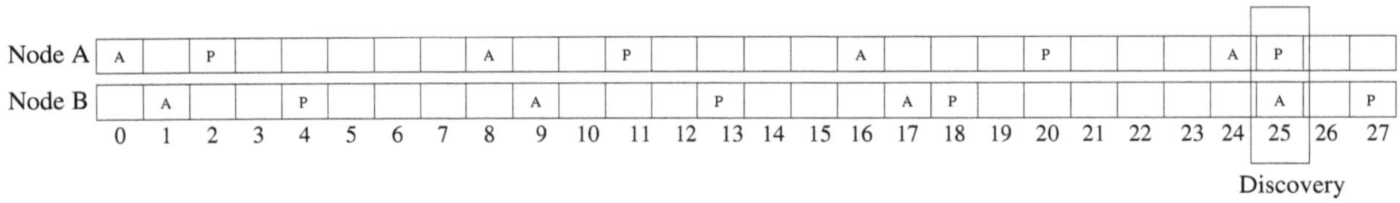

Figure 5: Overlap with sequential probing (t=8)

Protocol	Parameters	Duty Cycle	Worst-case bound Symmetric case
Disco	p_1, p_2	$\frac{p_1+p_2}{p_1 \cdot p_2}$	$p_1 \cdot p_2$
U-Connect	p	$\frac{3p+1}{2p^2}$	p^2
Searchlight	t	$\frac{2\cdot(1+\delta)}{t}$	$t \cdot \lceil\frac{\lfloor\frac{t}{2}\rfloor}{2}\rceil$

Table 1: Deterministic Protocols

Protocol	Worst-case bound(slots)
Disco	$4x^2$
U-Connect	$\frac{9x^2}{4}$
Searchlight	$2x(1+\delta))\lceil\frac{\lfloor\frac{2x(1+\delta)}{2}\rfloor}{2}\rceil \approx x^2(1+\delta)^2$

Table 2: Worst-case bounds for different protocols operating at the same duty cycle

ues can be $t_{base}, 2 \cdot t_{base}, 4 \cdot t_{base}$ and the fourth value can be any *prime* other than 2 that is relatively prime with t_{base}.

3.6 Duty Cycle and Energy Consumption

So far we have looked at the worst-case bounds on the discovery latency provided by Searchlight for different scenarios. However, it is important to consider the energy cost of achieving a particular bound. Since the primary cost we are concerned about is energy, we can look at the duty cycle. For a given t, there are two active slots every t slots. Taking into consideration the cost of extending the slots by δ, the duty cycle for any given t is $\frac{2\cdot(1+\delta)}{t}$. Table 1 summarizes the cost (duty cycle) and performance (worst-case bound) of the three main deterministic protocols. However, based on that information, it is hard to compare the different protocols since they use different parameters.

To provide a common baseline, it is interesting to keep the duty cycle fixed so that all protocols consume energy at the same rate and then look at the worst-case bound given by the different protocols. By doing so, we can compare the performance of different protocols when they incur the same cost. Disco does the best in the symmetric case when the two primes p_1 and p_2 are as close as possible. For ease of analysis, let us assume that $p_1 = p_2 = p$. Next, we express the parameters of each protocol as functions of x, where $\frac{1}{x}$ is the common duty cycle.

For Disco,

$$\frac{p_1 + p_2}{p_1 \cdot p_2} \approx= \frac{p+p}{p \cdot p} = \frac{1}{x}$$

and hence, $p = 2x$. Similarly, for Searchlight,

$$t = 2x(1+\delta)$$

. Finally, for U-Connect,

$$\frac{3p+1}{2p^2} = \frac{1}{x}$$

and so $p \approx \frac{3x}{2}$. Now, we can express the worst-case bounds for all protocols as $f(x)$ (see Table 2). It is clear from the table that even with a conservative value of $\delta = 0.1$ ($\frac{1}{10}$-th of the actual slot size), Searchlight reduces the worst case bound by around 50% in comparison to U-Connect, the current best performing deterministic protocol.

3.7 Randomized Probing

Sequential probing does a good job of ensuring discovery by finding the anchor slot of the other node (i.e., probe-anchor overlap). However, the overlap of two probe slots should also result in a successful discovery. While probe-probe overlap is possible in sequential probing, the probe slots of two nodes follow the same pattern, and hence they are often in sync with each other, greatly reducing the probability of a probe-probe overlap (see Figure 5). Consider two nodes A and B with a relative phase offset of one slot

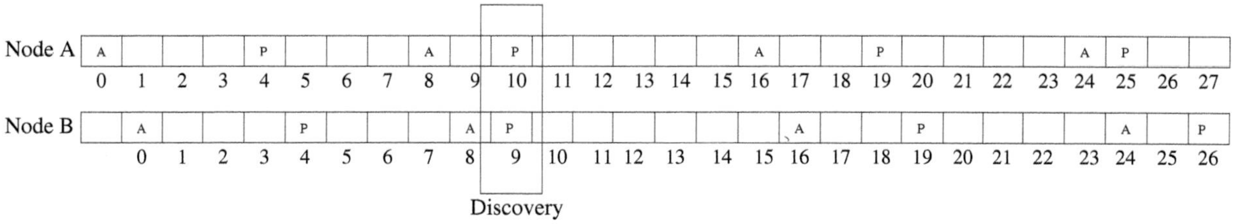

Figure 6: Overlap with randomized probing (t=8)

and $t = 8$. For sake of simplicity, let us assume that they are not using striped probing. When they first meet, A's probe slot is at position 2, while B's is at position 3 (with respect to its own anchor slot). Since both A and B follow the pattern {1,2,3,4}, the next positions of the probe slots will be 3 and 4 respectively. Thus, A and B's probe slots "chase" each other without ever overlapping at any point in time.

To increase the probability of a probe-probe overlap, Searchlight introduces a probabilistic component, similar to the Birthday protocols. Instead of being restricted to the pattern $1, 2, 3 ... \lfloor \frac{t}{2} \rfloor$, nodes can *randomly* pick any probe slot pattern that is a permutation of values from 1 to $\lfloor \frac{t}{2} \rfloor$. With striped probing, the range will change to $2, 4 ... (2 \cdot \lceil \frac{\lfloor \frac{t}{2} \rfloor}{2} \rceil)$). To differentiate, we call this version of Searchlight with randomized probing Searchlight-R, while the one with sequential probing is called Searchlight-S in the rest of the paper.

In the context of the last example, instead of being restricted to just {1,2,3,4}, Searchlight-R allows nodes to randomly choose any pattern that is a permutation of the integers 1,2,3 and 4 (e.g., {1,4,3,2}, {1,2,4,3}). In this case, A randomly chooses the periodic pattern {1,4,2,3} while its neighbor B chooses the periodic pattern {1,3,2,4} (see Figure 6). When they first meet, the probe slots of A and B are at positions 4 and 4 respectively and their relative offset is 1 slot. Because of the phase offset, the probe slots miss each other initially but meet in the next period when A's probe slot moves to position 2 and B's probe slot moves to position 1. Thus, the use of different patterns for probe slot selection results in quicker discovery through probe-probe overlaps.

This probabilistic approach essentially increases the possibility of discovery through a probe-probe overlap without diminishing a probe slot's ability to find an anchor slot in the symmetric case. However, in the asymmetric case, randomized probing increases the worst-case bound since in the worst case, the probe slot of the node with the bigger t can go through all regular slots first before meeting an anchor slot. In that case, the worst-case bound would be $t_A \cdot \lceil \frac{\lfloor \frac{t_A}{2} \rfloor}{2} \rceil$ slots, where t_A is the period of the node with bigger period length (smaller duty-cycle). However, the bound can be kept the same if we restrict the permutations that can be chosen. A simple approach would be to limit randomness within blocks of t_{base} slots but this would decrease average-case performance for symmetric duty cycles. We next present more intelligent permutations that would preserve the worst-case bound but allow more randomness.

3.8 Restricted Randomized Probing in the Asymmetric Case

The worst-case bound for sequential probing is based on the observation that A's probe slot is guaranteed to meet an anchor slot of B every t_B periods. However, no such guarantee can be provided for randomized probing. To better understand this point, let us divide the slots of A into t_B buckets. Assuming that the anchor slot is at position 0, the buckets would be as follows:

Bucket 1: $\{1,\ 1 + t_B,\ 1 + 2 \times t_B,\ ...\ \}$
Bucket 2: $\{2,\ 2 + t_B,\ 2 + 2 \times t_B,\ ...\ \}$
....
Bucket t_B: $\{t_B,\ 2 \times t_B,\ 3 \times t_B,\ ...\ \}$

Based on the offset between A and B, B's anchor slots would fall into any of the above t_B buckets. For example, if the offset between A and B is 3 slots, then B's anchor slots with respect to A's anchor slot would be at slots 3, $3 + t_B$, $3 + 2 \times t_B$ and so on (Bucket 3). With sequential probing, it is guaranteed that the movement pattern followed by the probe slot will contain one slot from each of the above buckets every t_B slots. Unfortunately, with randomized probing, no such guarantee can be provided. We next present a slightly restricted form of randomized probing that addresses this issue.

The basic goal of restricted randomized probing is to ensure that only those movement patterns would be chosen for probing where there is exactly one slot from each bucket every t_B slots. We achieve this goal with two-step randomization. First, a random permutation of the buckets gets chosen. This determines the order in which slots from different buckets would appear. The next step is to randomly choose a permutation of slots within each bucket. We illustrate the strategy with the following example.

Let us assume that $t_A = 60$ and $t_B = 10$. With sequential probing without striping, A's probe slot will follow the pattern (left to right, top to bottom):

1	2	3	4	5	6	7	8	9	10
11	12	13	14	15	16	17	18	19	20
21	22	23	24	25	26	27	28	29	30

Each column represents one bucket. The first step of restricted randomized probing is to randomize the order of these buckets. One such valid ordering can be:

6	9	1	10	2	5	7	8	4	3
16	19	11	20	12	15	17	18	14	13
26	29	21	30	22	25	27	28	24	23

The second and final step is to randomize the order of slots within each bucket:

16	9	1	10	22	15	7	18	4	23
6	29	11	20	12	25	17	8	24	3
26	19	21	30	2	5	27	28	14	13

The above pattern ensures that there is exactly one slot

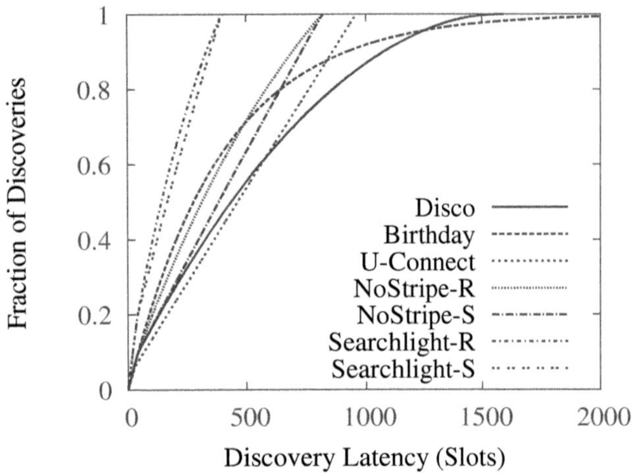

Figure 7: CDF for Discovery Latency - 5% duty cycle

from each bucket every t_B slots. Thus, by adopting this approach, restricted randomized probing achieves the same worst-case bound as sequential probing in the symmetric case.

3.9 Effects of clock drift

Searchlight's systematic probing relies on the assumption of a constant offset between the anchor slots of any two nodes. In reality, clocks drift and the offset between anchor slots will change over time. However, if the drift is sufficiently small, the worst case bound will not change.

To understand the effect of clock drift on the worst case bound, let us consider the worst case for two node's schedules. We will assume that no probe- probe discoveries occur and concentrate on probe-anchor discovery. In Section 3.2, we showed that this kind of discovery must exist given any two schedules. The least amount of overlap will occur between an anchor slot and a probe slot when the anchor slot starts immediately following the probe slot. Due to the striped nature of Searchlight, the anchor slot of size $x(1+\delta)$ will start at the same time as the inactive slot of size $x(1-\delta)$ producing a total overlap of $x \cdot (2\delta)$ with the next active slot.

Since this anchor-probe overlap is the only overlap that our worst case analysis relies upon, if the anchor slot were to drift to the point where it was no longer overlapping with this probe slot within one hyper-period, discovery could not be guaranteed. Thus the clock drift must be less than $\frac{x \cdot (2\delta)}{T}$ for the worst case bound to hold. Given this constraint, δ can be chosen based on the clock drift for a given device such that the bound holds.

4. EVALUATION

The primary goal of our evaluation is to show that Searchlight achieves significant performance improvements over other asynchronous neighbor discovery protocols by virtue of its systematic probing. Specifically, we evaluate how long it takes for different protocols to discover neighbors when *they spend the same amount of energy*, i.e., operate at the same duty cycle. The time from when two nodes first enter each other's transmission range to the time when they actually discover each other is known as the *discovery latency*. We look at the CDF of discovery latencies to understand the overall trend of a protocol. Since absolute latency in terms of time units is dependent on slot-width and slot-width is platform dependent and the same for all protocols, we use the number of slots as the metric for latency. We compare the performance of the variants of Searchlight with the deterministic protocols, Disco and U-Connect, and with the probabilistic Birthday protocol.

Since duty cycle directly translates to energy, we have not included separate results for energy consumption. Instead, we normalized all protocols to consume the same amount of energy and compared the performance of all of the protocols when they operated at the same duty cycle. Therefore, our results show how different protocols perform in terms of latency when they use the same amount of energy. The other possibility would be to hold the latency fixed and show how much energy is consumed to achieve that latency. However, this results are simply the inverse of the latency results.

For any random pair of nodes, the discovery latency can vary widely based on the relative offset. To fully characterize the performance of the different protocols, including the worst-case bound, we evaluated the protocols across all possible offsets through a state-based simulation.

Except for the Birthday protocol, all other protocols follow a discovery schedule to determine when to sleep and when to wake up. This schedule repeats every T slots, which we call the hyper-period. When a node is in a particular slot in its schedule, that slot index can be considered the *state* of the node at that point and is always in the range $[0, T-1]$. When two nodes with hyper periods T_1 and T_2 come into each other's transmission range, their states have one of $T_1 \cdot T_2$ possible combinations. For a given combination, the discovery latency is always the same. For Disco, U-Connect and Searchlight-S, we use this observation to loop through all possible combinations and determine the latency for each case. The same approach is not feasible for Searchlight-R since for a given t, a node can choose any of the $(\lfloor \frac{t}{2} \rfloor - 1)!$ permutations to determine the schedule of its probe slot. For scalability purposes, we run the protocol 1000 times with different seeds. At each run, we generate a new schedule for both nodes. Then, for that particular schedule pair, we loop through all possible state combinations like we do for Searchlight-S. Since for Disco and U-Connect, all slots are of equal size, we assume the slot boundaries to be completely aligned for the worst case (non-alignment leads to more overlap). On the other hand, since Searchlight uses "overflowing" active slots, complete slot boundary alignment creates the best case scenario. Therefore, we consider the slot boundaries to be offset by half a slot-width for simulating Searchlight. For the Birthday protocol, we use a closed form expression for determining the CDF and the expected value of discovery latency [21].

4.1 Symmetric case

First, we compare the performance of different protocols when nodes operate at the same duty cycle. We look at the cumulative distribution of discovery latencies for all protocols operating at a 5% duty cycle (see Figure 7). Based on the protocol design as described in respective papers to achieve a 5% duty cycle, Disco uses the primes (37,43) and U-Connect uses the prime 31. Birthday uses probability = 0.05. Both versions of Searchlight use the integer 40, which results in a 5% duty cycle as described earlier. To gauge the effectiveness of striped probing, we also evalu-

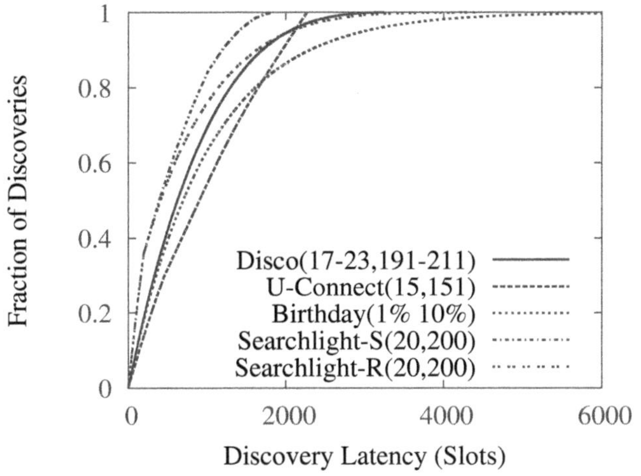

Figure 8: CDF of Discovery Latency - Asymmetric Duty Cycles - 1%-10%

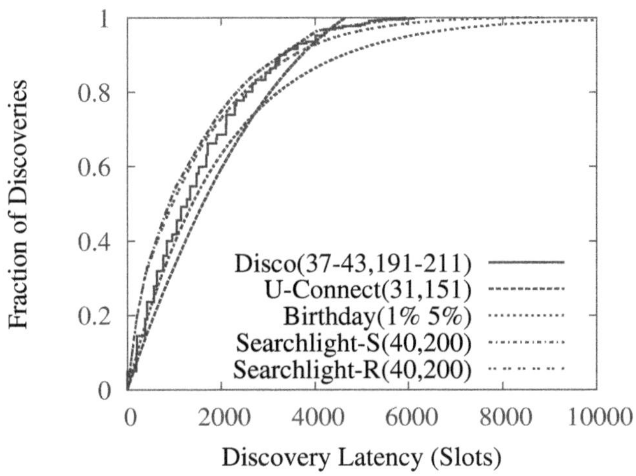

Figure 9: CDF of Discovery Latency - Asymmetric Duty Cycles - 1%-5%

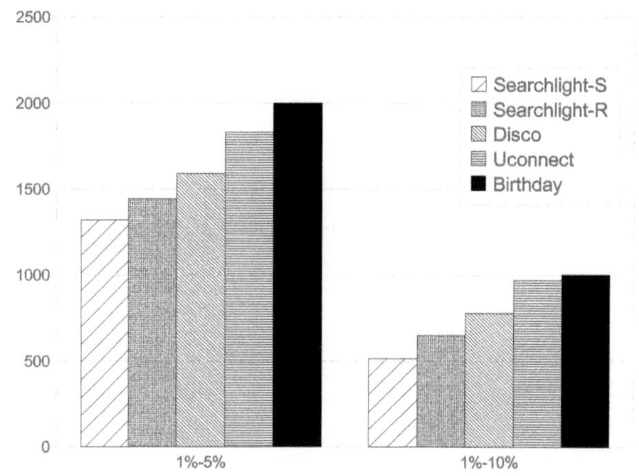

Figure 10: Average Latency - Asymmetric Duty Cycles - 1%-5% and 1%-10%

ate both sequential and randomized versions of Searchlight without striped probing. They are denoted by NoStripe-S and NoStripe-R respectively. Both versions of Searchlight *always* achieve the lowest latency in the worst-case, with Searchlight-R doing better in the average case by virtue of its probabilistic component. Without striping, the number of slots needed for discovery doubles. Still, for 65% of the time, NoStripe-R performs on par with the Birthday protocol or slightly lags behind. Beyond that, the probabilistic nature of the Birthday protocol leads to a long tail and NoStripe-R achieves the lowest latency. In comparison to U-Connect, NoStripe-R achieves better latency all along. For both versions of Searchlight, maximum discovery latency is 400 slots, followed by 800 slots for the NoStripe versions. U-Connect comes closest with a maximum discovery latency of 961 slots, which is more than a 100% increase over Searchlight.

4.2 Asymmetric case

The evaluation of the asymmetric case is particularly important because the worst-case latency is different for Searchlight in this scenario. In this section, we evaluate two scenarios. In the first, one node operates at 1% duty cycle and the other at 10%. In the second, the second node operates at 5% duty cycle. As part of our future work, we plan to explore the space of asymmetry and how the degree of asymmetry affects all of the protocols.

When two nodes operate at 1% and 10% duty cycles respectively (see Figure 8), Searchlight-S performs best all along and its worst-case discovery latency is almost half of U-Connect. Searchlight-R initially performs similar to Searchlight-S, but because of the higher worst-case bound starts to deviate after around the 65-th percentile.

For the asymmetric scenario when the two duty cycles are 1% and 5% (see Figure 9) U-Connect outperforms Searchlight-S in terms of worst-case discovery latency. In the asymmetric case, we know the worst-case bound for Searchlight-S is

$$\frac{(t_B + (\lceil \frac{\lfloor \frac{t_A}{2} \rfloor}{2} \rceil \mod t_B))}{2} \cdot t_A,$$

where t_A and t_B are the two periods and $t_A > t_B$. For the 1%-10% case, $\lfloor \frac{t_A}{2} \rfloor = 100$ was completely divisible by the smaller period $t_B = 20$, resulting in lowest worst-case latency. But for 1%-5%, 100 is not completely divisible by $t_B = 40$, which has led to the higher worst-case bound.

To further investigate this issue, we look at the average latency for both scenarios with 1%-5% and 1%-10% asymmetric duty cycles (see Figure 10). In both cases, Searchlight-S performs best, followed by Searchlight-R, while U-Connect performs even worse than Disco. This shows that in terms of average latency, Searchlight-S still performs best in all asymmetric scenarios.

4.2.1 *Effect of restricting Permutation*

We next evaluate the effect of restricting the set of allowable permutations for randomized probing. When two nodes operate at 1% and 10% duty cycle respectively, the worst-case latency for Searchlight-S is less than 50% of the worst-case latency for Searchlight-R (see Figure 8). But when we restrict the set of movement patterns that can be chosen, the worst-case performance of such restricted randomized probing (Searchlight-R+) becomes almost the same as Searchlight-S (see Figure 11).

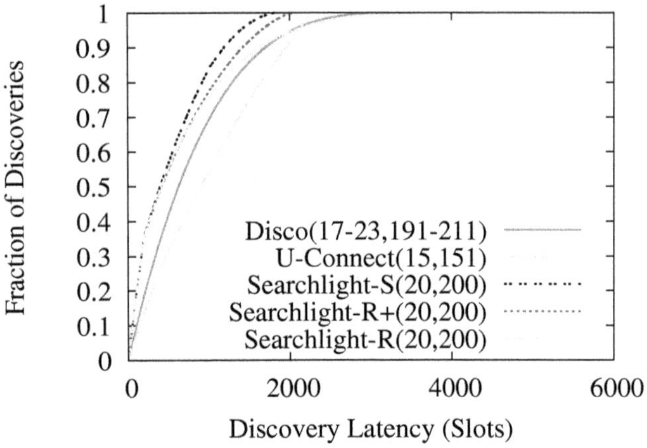

Figure 11: Effect of Restricted Randomized Probing: 1%-10% duty cycle

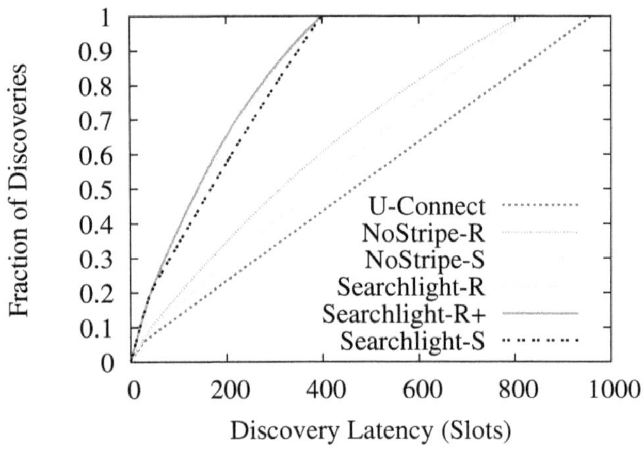

Figure 12: Effect of Restricted Randomized Probing: 5% duty cycle

While such restricted randomization helps in the asymmetric case, it is important to evaluate whether it decreases the advantage of Searchlight-R in the symmetric case. We next look at the comparative performance of Searchlight-R+ when all protocols operate at 5% duty cycle. Since 5% duty cycle corresponds to a period of 40 slots for Searchlight, we use 10 slots as the base period for restricting randomization. As we can see from Figure 12, Searchlight-R+ performs the same as Searchlight-R. This shows that the reduction in randomization does not affect the average-case behavior of randomized probing in the symmetric case.

5. IMPLEMENTATION

One of the main objectives for designing asynchronous neighbor discovery protocols is to facilitate proximity-based communication between handheld devices like smartphones. To gauge how Searchlight achieves this goal in practice, we implemented the protocol in Python and deployed it on the Nokia N900, a smartphone by Nokia that supports the open source Maemo 5 mobile device platform [27], and a Dell Latitude E6400 laptop with an Intel 5100AGN Wi-Fi card.

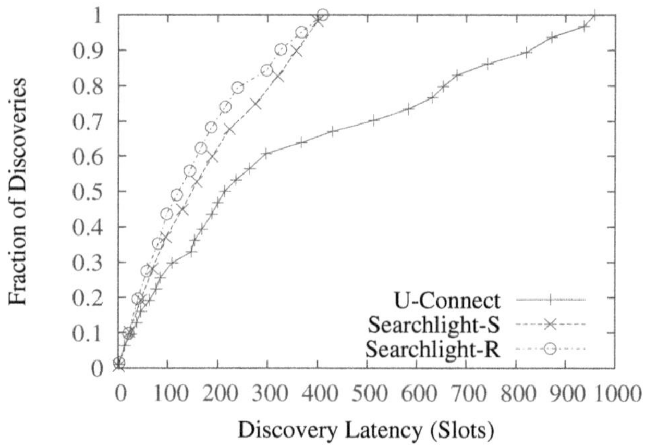

Figure 13: Implementation: CDF of Latency for Symmetric Duty Cycles

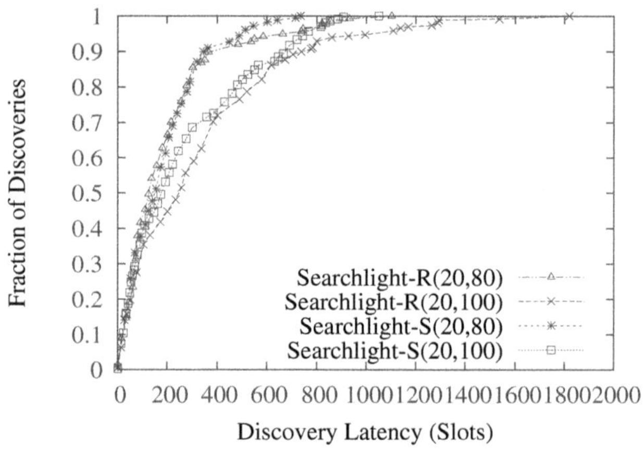

Figure 14: Implementation: CDF of Latency for Asymmetric

Results shown are collected strictly from the Nokia devices. We first describe different implementation issues and then present some preliminary results.

5.1 Implementation Issues

5.1.1 Slot Duration

We implemented Searchlight to use the Wi-Fi radio of the N900 phone for neighbor discovery. Earlier protocols for asynchronously discovering neighbors were all implemented on sensor nodes, allowing them to use small slots on the order of milliseconds [25] or even microseconds [11]. However, unlike sensor radios, Wi-Fi radios have a non-negligible transition latency from sleep to transmit/receive. On the N900 phone, we found out that from the application level, the time to bring the wireless interface up is around 1 second. Because of this latency, we chose a slot size of 2 seconds. Other phones we have looked at, including the Android G1, also have startup latencies on the order of seconds. These 2 second slots are quite large and so result in longer discovery delays. However, this is a limitation of the current hardware and not of our protocol. Although we expect this delay to be improved in future generations of smartphones, our implementation evaluates the current state-of-the art.

5.1.2 Pre Slots

Because of the non-negligible start-up time, we introduced the notion of *pre* slots. A *pre* slot basically precedes any active slot and switches on the interface. For example, assume that a node needs to be active during slot 10. Now, if the command to wake up the radio is issued at the beginning of slot 10, it will take almost the whole slot duration for that command to return and the effective awake time during that slot will be reduced to 1 second. To get around this problem, the wake up command now gets issued at the start of the preceding slot which is slot 9 in this case. Such slots are called *pre* slots and their positions are determined based on the active slot schedule. *Pre* slots are kept shorter than normal slots to allow for the larger active slots. In U-Connect, two active slots can be neighbors. In this case, the radio is left on rather than using a *pre* slot.

5.1.3 Active slot

When the protocol starts up, it creates an active slot schedule based on the desired duty cycle. In an active slot, a hello message containing the node id gets sent at the very beginning and at the very end of the slot. In between, the node continuously listens for hello messages from other nodes. When it gets a hello message, it adds the name of the sender to a discovered list. The size of the slot extension for active slots, which guarantees slot overlap, was set to be 5 milliseconds. This more than accommodates a hello message transmission and jitter from timer scheduling. As newer smartphones improve device wakeup times, we will evaluate the use of shorter slots and also less overflow, with the goal of maintaining enough overlap to compensate for completely aligned slots with requiring large slot sizes.

5.2 Evaluation

We implemented Searchlight-S and Searchlight-R on five N900 phones and logged the discovery latency for around 200 runs with 5% duty cycle. We also implemented U-Connect since it performs best among existing protocols in the symmetric case. Since U-Connect has consecutive active slots, we implemented a striped version of U-Connect, which reduces the number of active slots by about 10%.

For the implementations, Searchlight-R and Searchlight-S fare much better than U-Connect (see Figure 13), similar to our simulation results (see Figure 7). Even with striped probing, U-Connect's still falls short of Searchlight. We also looked at the asymmetric case of Searchlight with t values of (20,80) and (20,100) which represent 10%, 2.5%, 10%, and 2% duty cycles respectively. Searchlight-S outperforms its randomized version as was predicted in the asymmetric case (see Figure 14).

The implementation results actually turned out to be better than the state-based simulation results, with more discoveries taking place at lower latencies. Since the devices were not synchronized, this is probably an artifact of the slots being non-aligned (see section 3.3). However, the overall trends agree with the simulation results.

6. CONCLUSIONS AND FUTURE WORK

Solving the problem of energy efficient asynchronous neighbor discovery is an important pre-condition for more widespread use of proximity-based communication between mobile devices, including sensors and smartphones. In this paper, we present Searchlight, a new asynchronous neighbor discovery protocol. By adopting a systematic approach that has both deterministic and probabilistic components, Searchlight achieves average-case performance comparable to the probabilistic protocols and significantly improves worst-case bounds for symmetric operation in comparison to the current best deterministic protocols. In the asymmetric case, Searchlight still has the best performance. However, if multiple devices share the same duty cycle, even when other devices use different duty cycles, Searchlight exploits this partial symmetry and provides even better performance. Searchlight was also successfully implemented on a smartphone testbed, and showed performance trends similar to the simulation results.

As future work, we would like to integrate neighbor discovery with data transmission. It is not clear when a node should send a packet to an already discovered neighbor. Should it wait for the next overlap to take place and send a packet after receiving a beacon from the node it wishes to communicate with? This will obviously increase delay and we will have to consider the energy-latency trade-off.

Finally, we would like to investigate how Searchlight can dynamically adapt to energy requirements, contact patterns and other factors to adjust its duty cycle. For example, when neighbor count is low, a higher duty cycle might be required to find enough neighbors that meet application requirements. On the other hand, in a crowded place where the number of co-located nodes is high, a lower duty cycle might suffice to find a reasonable number of neighbors.

7. REFERENCES

[1] "Google latitude," http://www.google.com/latitude.
[2] "Gowalla," http://gowalla.com.
[3] "Foursquare," http://foursquare.com.
[4] "Facebook places," http://www.facebook.com/about/location.
[5] "Who's near me," http://www.synergetechsolutions.com/whos-near-me.aspx.
[6] A. K. Pietiläinen, E. Oliver, J. Lebrun, G. Varghese, and C. Diot, "MobiClique: middleware for mobile social networking," in *WOSN '09: Proceedings of the 2nd ACM workshop on Online social networks*, 2009, pp. 49–54.
[7] "Aka-aki," http://akakicom.
[8] "Bluehoo," http://bluehoo.com.
[9] "Nintendo 3ds - streetpass." http://www.nintendo.com/3ds/hardware.
[10] "Sony ps vita - near." http://us.playstation.com/psvita.
[11] A. Kandhalu, K. Lakshmanan, and R. R. Rajkumar, "U-connect: a low-latency energy-efficient asynchronous neighbor discovery protocol," in *Proceedings of IPSN '10*, 2010, pp. 350–361.
[12] "Qualcomm alljoyn," https://www.alljoyn.org/.
[13] "Android - an open handset alliance project." http://developer.android.com.
[14] "Wi-fi direct," http://www.wi-fi.org/discover-and-learn/wi-fi-direct

[15] T. Liu, C. M. Sadler, P. Zhang, and M. Martonosi, "Implementing software on resource-constrained mobile sensors: experiences with impala and zebranet," in *MobiSys 2004*. ACM, 2004, pp. 256–269.

[16] H. Jun, M. H. Ammar, M. D. Corner, and E. W. Zegura, "Hierarchical power management in disruption tolerant networks using traffic-aware optimization," *Comput. Commun.*, vol. 32, pp. 1710–1723, October 2009.

[17] J. Elson and K. Römer, "Wireless sensor networks: a new regime for time synchronization," *SIGCOMM Comput. Commun. Rev.*, vol. 33, pp. 149–154, January 2003.

[18] J. Paek, J. Kim, and R. Govindan, "Energy-efficient rate-adaptive gps-based positioning for smartphones," in *Proceedings of MobiSys '10*. ACM, 2010, pp. 299–314.

[19] J. Polastre, J. Hill, and D. Culler, "Versatile low power media access for wireless sensor networks," in *Proceedings of SenSys '04*. ACM, 2004, pp. 95–107.

[20] W. Ye, J. Heidemann, and D. Estrin, "An energy-efficient mac protocol for wireless sensor networks," in *INFOCOM 2002*, vol. 3, 2002, pp. 1567–1576 vol.3.

[21] M. J. McGlynn and S. A. Borbash, "Birthday protocols for low energy deployment and flexible neighbor discovery in ad hoc wireless networks," in *Proceedings of MobiHoc '01*. ACM, 2001, pp. 137–145.

[22] S. Lai, B. Ravindran, and H. Cho, "Heterogenous quorum-based wakeup scheduling in wireless sensor networks," *IEEE Transactions on Computers*, vol. 99, no. PrePrints, 2010.

[23] Y.-C. Tseng, C.-S. Hsu, and T.-Y. Hsieh, "Power-saving protocols for ieee 802.11-based multi-hop ad hoc networks," in *INFOCOM*, 2002.

[24] R. Zheng, J. C. Hou, and L. Sha, "Asynchronous wakeup for ad hoc networks," in *MobiHoc 2003*. ACM, 2003, pp. 35–45.

[25] P. Dutta and D. Culler, "Practical asynchronous neighbor discovery and rendezvous for mobile sensing applications," in *SenSys '08: Proceedings of the 6th ACM conference on Embedded network sensor systems*, 2008, pp. 71–84.

[26] I. Niven and H. S. Zuckerman, *An Introduction to the Theory of Numbers*. John Wiley and Sons (WIE), 1991.

[27] "Maemo - open source operating system." http://maemo.org.

Distinguishing Users with Capacitive Touch Communication

Tam Vu, Akash Baid, Simon Gao, Marco Gruteser, Richard Howard, Janne Lindqvist,
Predrag Spasojevic, Jeffrey Walling
WINLAB, Rutgers University
{tamvu, baid, gruteser, reh, janne, spasojev} @winlab.rutgers.edu
{simongao, jeffrey.s.walling}@rutgers.edu

ABSTRACT

As we are surrounded by an ever-larger variety of post-PC devices, the traditional methods for identifying and authenticating users have become cumbersome and time-consuming. In this paper, we present a capacitive communication method through which a device can recognize who is interacting with it. This method exploits the capacitive touchscreens, which are now used in laptops, phones, and tablets, as a signal receiver. The signal that identifies the user can be generated by a small transmitter embedded into a ring, watch, or other artifact carried on the human body. We explore two example system designs with a low-power continuous transmitter that communicates through the skin and a signet ring that needs to be touched to the screen. Experiments with our prototype transmitter and tablet receiver show that capacitive communication through a touchscreen is possible, even without hardware or firmware modifications on a receiver. This latter approach imposes severe limits on the data rate, but the rate is sufficient for differentiating users in multiplayer tablet games or parental control applications. Controlled experiments with a signal generator also indicate that future designs may be able to achieve datarates that are useful for providing less obtrusive authentication with similar assurance as PIN codes or swipe patterns commonly used on smartphones today.

Categories and Subject Descriptors

C.2.1 [**Computer-Communication Networks**]: Network Architecture and Design—*Wireless Communication*

General Terms

Design, Human Factors, Measurement, Experimentation, Performance, Algorithms

Keywords

SignetRing, Touchscreen Communication, User Identification, Capacitive Touch Communication, Distinguishing Users

MobiCom'12, August 22–26, 2012, Istanbul, Turkey.

1. INTRODUCTION

Mobile devices now provide us ubiquitous access to a vast array of media content and digital services. They can access our emails and personal photos, open our cars [41] or our garage doors [13], pay bills and transfer funds between our bank accounts, order merchandise, as well as control our homes [10]. Arguably, they now provide the de-facto single-sign on access to all our content and services, which has proven so elusive on the web.

As we increasingly rely on a variety of such devices, we tend to quickly switch between them and temporarily share them with others [26]. We may let our children play games on our smartphones or share a tablet with colleagues or family members. Sometimes a device may be used by several persons simultaneously, as when playing a multi-player game on a tablet, and occasionally, a device might fall into the hands of strangers.

In all these situations, it would be of great benefit for the device to know who is interacting with it and occasionally to authenticate the user. We may want to limit access to age-appropriate games and media for our children or prevent them from charging our credit card.[1] We desire to hide sensitive personal information from strangers, colleagues, or perhaps even an curious spouse [23, 26]. Or, we may simply want to enjoy an enhanced user experience from the multi-player game that can tell who touched the screen.

Unfortunately, user identification and authentication mechanisms available on today's mobile devices have been largely adopted from PC software and have not followed the versatility of the usage and sharing possibilities. For example, several mobile devices (e.g. iPad or iOS devices) do allow to restrict access to device functions, but the devices do not provide any easy way to quickly change, let alone authenticate, users. They provide PIN codes, passwords, for authentication, and a number of other techniques have been proposed by researchers [11]. Yet they remain cumbersome and very few people enable these security features on their phones.

In this paper, we will explore a form of "wireless" communication, that we term *capacitive touch communication* to address this issue. The key idea is to exploit the pervasive capacitive touch screen and touchpad input devices as receivers for an identification code transmitted by a hardware identification token. While the token can take many forms, we consider here an example realization as a ring, inspired by the signet rings used since ancient times. The token transmits electrical signals on contact with the screen, either direct contact or indirect contact through the human skin.

The major contributions of this paper are as follows:

- painting a vision to use the near-ubiquitous capacitative touch sensors to distinguish and possibly authenticate users.

[1] Apple is facing a law suit over children's in-app credit card purchases [18].

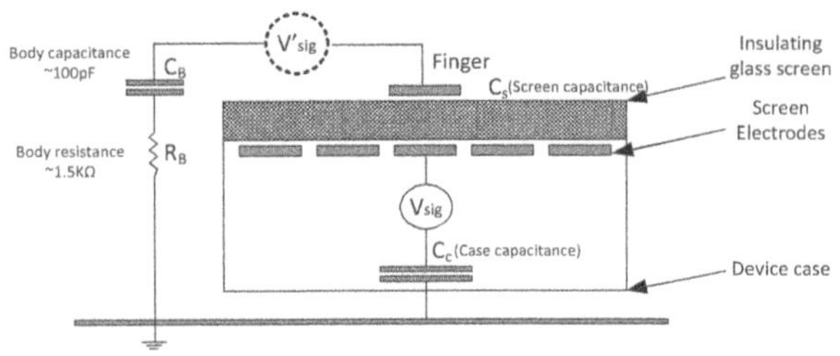

Figure 1: Schematic of a basic capacitive touchscreen

- introducing and exploring the concept of capacitive touch communication as one mechanism to distinguish users.
- showing how the output of an off-the-shelf touchscreen system can be affected by electrical signals generated in a token that is in contact with the screen. We also show how such signals can be transmitted through the human skin.
- designing and implementing a prototype transmitter in the form of a signet ring and receiver software for communicating short codes through an off-the-shelf capacitative touch screen

2. BACKGROUND

Touchscreen technology was first developed in the 1960's for air traffic control systems [25] and is now a popular user interface technology on devices ranging from ATMs and self-service terminals in grocery stores or airports, to cars, smartphones, and tablets. Even the touchpads used in laptops are based on similar technology. These products employ different touchscreen implementations, including analog resistive, surface capacitive, projected capacitive, surface acoustic wave, infrared and optical technology to mention a few. On mobile devices, however, capacitive touchscreens have emerged as the main technology and we focus our work on those.

2.1 Capacitive Touchscreen Technology

A capacitive screen in most commercial tablets and smart phones consists of an array of conducting electrodes behind a transparent, insulating glass layer which detects a touch by measuring the additional capacitance of a human body in the circuit. Figure 1 shows a schematic of one possible realization of such a system [47]. When a user touches the screen, her finger acts as the second electrode in a capacitor with the screen as the dielectric. The touchscreen electrodes are driven by an AC signal (V_{sig}) which sends a current through the screen capacitance C_s passing through the body capacitance C_B, and then back into the tablet through the case capacitance C_c. This change in voltage measured at one or more screen electrodes is then passed to the screen controller for processing. Because all of the relevant capacitance values are small (hundreds of picofarads [19]) environmental noise makes direct measurement of this current impractical. Instead, the charge integration circuitry in Figure 2 is used to measure the excess capacitance associated with a finger touch. In this case, a digital signal, V_{sig}, is synchronized with a pair of switches and a charge integrator. Switch S_3 is first closed to discharge capacitor C_i and then opened. Next, switch S_1 is closed and S_2 opened while V_{sig} is high. This charges the series combination of the C_B, C_c, and C_s. Then S_1 is opened and S_2 closed, transferring this charge to C_i. After a fixed number of cycles, the voltage on C_i is directly proportional to the ratio between C_i and the series combination of C_B, C_c, and C_s. This voltage is then used to detect touch and, through the matrix addressing of the electrodes, position of the touch. Hence, even when a finger moved across the screen surface without lifting it, the finger triggers this detection at different positions on the electrode array.

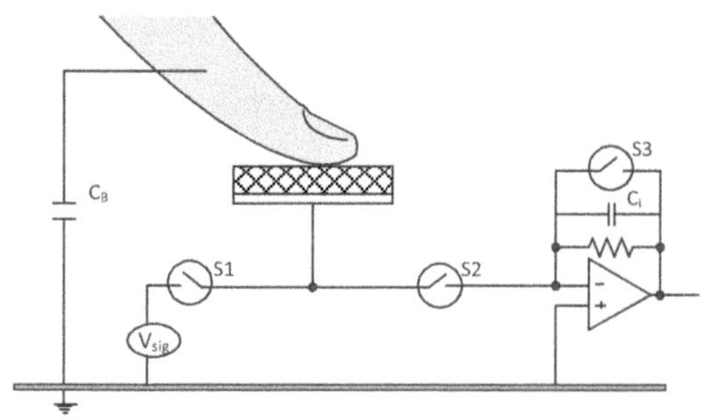

Figure 2: Internal touch detection circuit

2.2 Related Work

The most closely related projects to our work are Touché [39], DiamondTouch [14], Signet [44], IR Ring [38], Magkey/Mickey [12]. Proposed in 2001 as one of the first efforts toward differentiating touches of different users interacting with the same surface, DiamondTouch uses a table to transmit capacitively coupled signals through users, chairs, and finally to the receiver. This approach requires extensive hardware infrastructure which make it impossible to apply to mobile scenarios. Touché proposes a technique, called Swept Frequency Capacitive Sensing, that can recognize human hand and body configurations. While the technique could enable a new way of human computer interaction, it would require additional special hardware component to be manufactured onto the devices. Signet uses physical patterns of conductive material as unique inputs for authentication through a capacitive touch screen. In contrast, our work focuses on using arbitrary programmable sequences of bits through direct use of the user's fingers. As such, it makes the solution non-intrusive and applicable to wider classes of applications.

There are several ways to authenticate a user, which in general can be divided into 1) what you know, 2) what you have, and 3) who you are. PINs, passwords and swipe patterns are the most widely spread authentication mechanism for mobile phones [11, 16]. These methods are easy to implement and require no special hardware, but are easily observable by an adversary and usually have very low information entropy. For example the usual 4 bit numeric PINs used in most phones have a theoretical potential entropy of $\log_2(10^4) = 13.3$ bits. Practical entropy for 4-digit PINs is likely to be much lower, as is the case with passwords [46]. The second type of authentication mechanisms ("what you have") are often also referred to as authentication tokens, examples include Magkey/Mickey [12], RFID or other wireless tokens such as transient authentication [36], and IR Ring [38]. Magkey and Mickey are tokens that use magnetic fields and acoustic signals that are received by the phone's compass and microphone respectively. RFID, NFC and other wireless-based techniques are prone to eavesdropping and suffer from interference among multiple radio signal sources. One example technique belonging to that category is RingBow [37], a wearable hardware token in the form of a ring, which communicates with the mobile device using Bluetooth. This type of communication is insecure during the pairing period and does not allow touchscreen-enabled devices to associate touch events to their users. And finally, IR Ring demonstrated the possibility to use infra-red and IR video cameras to authenticate users on a multitouch display, which is not directly applicable to today's mobile devices due to its additional hardware requirement.

Examples of "who you are" include iris recognition, face recognition and voice recognition all of which are being actively prototyped and tested on mobile devices. Motorola Atrix claims to be the first phone in the western market to have a fingerprint sensor [34] while Sony is developing a novel finger-vein pattern matching technique [42]. Both these techniques require specialized hardware which adds to the cost and form-factor of handheld devices and are prone to known vulnerabilities [30, 20]. On the other hand, face, iris and voice recognition utilizes the in-built sensors and most of the feature set required are already implemented in mobile devices for other applications [27, 33]. While these techniques can leverage the abundance of past research in the respective fields, they also suffer from the well known spoofing mechanisms [17, 3]. For example both high-quality photograph of the eye and printed contact lenses have been used to achieve close to 100% spoof acceptance rates for iris recognition systems [45].[2] Similar results hold for face detection and voice detection although large strides are also being made for spoof detection in biometric authentication systems (see Jain et al. [24] and the references therein). More recently, innovative uses of the various sensors available in most smart phones have led to a number of *unconventional techniques*. For example, there are proposals [8, 28] for in-air gesture based authentication mechanism which uses the accelerometer sensors of the mobile device. Being easily visible to an adversary, such a scheme suffers from an unpleasant tradeoff between coming up with complex gestures and being susceptible to copy attacks, and can also be socially awkward. Implicit authentication is a similar approach which aims to authenticate mobile users based on everyday actions such as number/duration of calls, location, connectivity pattern, etc. and keeps a multi-variable continuous authentication score of the user. As is obvious, this requires a continuous modelling and logging of data from a variety of sensors and has a high energy cost.

Today's consumer electronic devices often include some form of parental control mechanisms, which are usually limited to locking out some functionalities of the device or service, e.g. adult content. Parental control mechanisms are an overlooked area of research, however, recent studies indicate that there would be demand for flexible access control mechanisms at home [32]. Our presented work can be seen as an easy to use enabler for parental access control mechanisms.

The problem of device pairing is also closely related to secure authentication and solution approaches often overlap. The general objective in this case is to enable two devices with no prior context to securely associate with each other in the presence of man-in-the-middle adversary. The short-range and frequency hopping nature of Bluetooth makes it a robust authentication mechanism, however several recent works expose a key vulnerability - passive sniffing of the PIN during the pairing process [40]. Similarly, for near-field communications (NFC) [2] based pairing, eavesdropping using directional antennas has been shown to be a critical security threat [22]. Novel use of the accelerometer sensor in mobile devices have recently been shown to provide a secure method of device pairing [31]. While robust for two equipped mobile devices, the requirement of shaking prevents its use from cases which require pairing of a mobile device with a fixed device. Further, replication of the movement by an adversary is possible through careful observation of the pairing process. Finally, a recent approach uses

[2]The face recognition system available in the new Google Android based Galaxy Nexus platform can be compromised just by showing a picture taken with another smartphone [1].

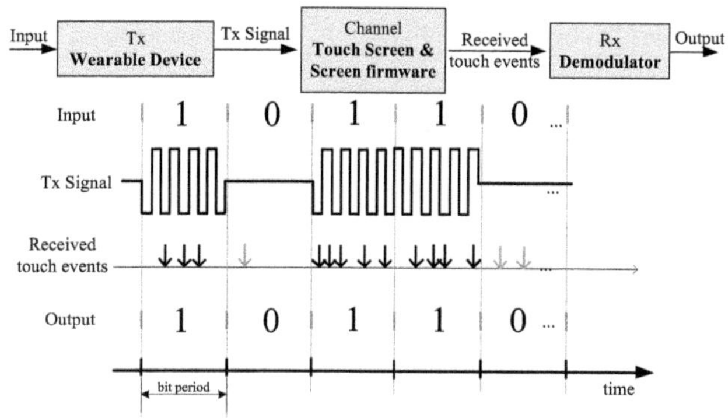

Figure 3: Capacitive touch screen communication showing OOK modulation and variations in number and timing of generated events

public RF signals such as TV and FM broadcasts to derive cryptographic keys for secure pairing between close-by devices [29].

Auxiliary channels to established shared secrets have been studied extensively in the domain of secure pairing since the resurrecting duckling model [43]. Examples include using infrared [9] or humans [21]. More recently secure pairing efforts have focused on using the same channel for authentication and data, and deriving the keying material based on the local environment, e.g [29]. In contrast, our approach provides a seamless way to both securely pair the device and authenticate later.

3. CAPACITIVE TOUCH COMMUNICATION

To allow mobile devices to identify their users in a less obtrusive manner, we explore a novel form of "wireless" communication in which a touch panel acts as a receiver and a small ring-like device worn by the user serves as the transmitter. This type of communication, which we term *capacitive touch communication*, could have wide applicability since touch panels are now ubiquitous.

While it would be interesting to also consider modifications to the touch sensor hardware and firmware to facilitate such communication, we focus this first study on exploring to what extend the communication can be achieved with off-the-shelf touch sensor systems. This means we will only have access to the touch events exported by the screen's driver, not the raw voltages measurements. It imposes very stringent requirements on the communication protocols, as we will see in the next sections. We believe, however, that this is a useful point solution within the design space of capacitive touch communication, since this approach would allow more rapid deployment on existing devices.

3.1 Creating Artificial Touch Events

Motivated by this goal, we discovered a technique for "spoofing" the screen detection algorithm by causing the system to alternately register touch/no touch conditions even when the finger is not moving. This allows us to send a digital signal into the touchscreen.

Referring again to Figure 1, one possible method for artificially creating touch events is by injecting a synchronized signal (V'_{sig}) into the circuit with the proper amplitude and phase to increase or decrease the charge integrated on C_i. Unfortunately, the signal in the device, V_{sig}, is not available to the external user, so such synchronization would be extremely difficult. As such, we use an unsynchronized lower frequency signal of high amplitude which

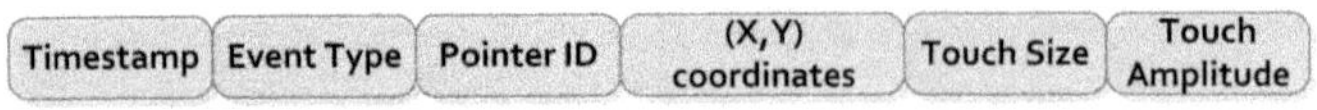

Figure 4: Touch event structure retrieved by touch screen controller

charges and discharges C_i asynchronously, leading to repetitive, but irregular, touch/no touch events captured by the touchscreen controller. This process essentially "spoofs" the touch detection mechanism by injecting high level repetitive signals and introduces a technique to send a low bit rate signal into the tablet. With precise knowledge of the proprietary touch sensor systems it should be possible to create much more fine-grained signaling methods. For the purpose of this feasibility study, however, we will now consider how this coarse technique can be leveraged for designing a user identification system.

3.2 Communication System Overview

The communication scheme we are proposing can be modeled as a classical communication system with a transmitter, a receiver and a complex channel connecting the two, as shown in Figure 3.

Transmitter: The transmitter in our system is a wearable battery-powered hardware token. One possible form that such a token could take is that of a ring, essentially a digital version of the signet rings carried by nobility in earlier times[3]. While many other forms of tokens are possible, we will use the *ring* concept as a running example throughout the paper.

The ring would contain a small flash memory that stores a bit sequence or a message, which could be a user identifier or a secret key that authenticates a user. It also has a simple processor that reads the bit sequence and generates a On-Off keying (OOK) [15] modulated signal. That is, bit *one* is represented by turning *on* a carrier signal; and bit *zero* by switching *off* that carrier signal, as the *Tx Signal* shown in Figure 3. When the ring is pressed against the screen, it acts as a voltage source(V'_{sig} in figure 1) which creates a set of touch events with timestamps following the bit sequence being transmitted.

Channel: Since the events generated follow the bit sequence being transmitted, these events can be used to reconstruct the original bit sequence, which is unknown to the screen otherwise. Thus, in this setting, the channel can be thought of as the combination of all hardware and software components that affect the relationship between the transmitted bit sequence and the events registered: (i) the series of capacitances, (ii) the firmware that comes with the screen, and (iii) the proprietary driver that is a part of the device's operating system.

Unfortunately, due to the internal switching frequency inside the touch panel, non-deterministic amount of charge accumulation and the firmware/driver artifacts, the number and the timing of the events do not directly follow the input sequence. For example, in figure 3, when the first bit *one* is transmitted, three touch events are triggered, while in the succeeding *ones* five and four events are produced. Furthermore, even though transmission of a *zero* should not trigger touch events, one and two events are registered in the two *zeros* presented in this example respectively. In addition, the channel adds a variable and unknown delay between the transmitted sequence and the touch event registered.

Receiver: The *Tx Signal* transmitted by the ring generates touch events represented by the 6-tuple structure depicted in Figure 4 (a detailed description of this structure is presented in Section 5). Because the only information we can use is the timestamps of the events registered by the screen driver, the system requires an unconventional receiver design. Instead of the usual practice of looking at the amplitude (*Touch Amplitude* field) of the received signal, which in this case is not related to the transmitted data, we use the *number of event registered* for demodulating. That is, the software component receives a bit 1 if the number of events which appeared in that bit period is greater than a certain threshold and receives a bit 0 otherwise.

[3]A finger ring bearing a hard-to-fake engraved pattern, which serves as a seal of authority, a signet.

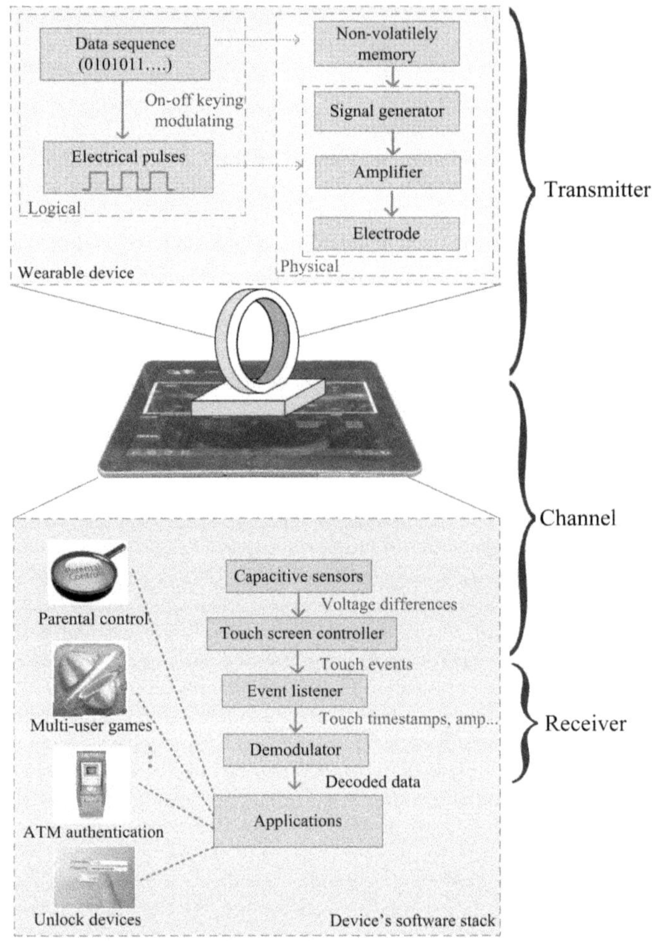

Figure 5: Overall architecture of the capacitive touch communication system

We note that there is a variable delay from the moment that touch events were registered to the kernel until it is handed to the application-level software, which in our case is the application-level demodulator. This delay makes demodulating less accurate. The time variance, we suspect, is due to the queueing and processing delays incurred when the event information travels up the software stack, from the touch-event handler in the Android kernel to the application level. To mitigate this inaccuracy, our demodulator looks at the touch event timestamps at the kernel level (using a few *printk* commands in the touchscreen driver of our prototype).

The key challenge is to handle the variance in the number and timing of the events that is introduced by the channel. To address this issue, we characterize the expected behaviour of the channel, reflected in terms of event counts, for decoding of the received se-

Bit Rate (bps)	4	5	8	10	12	15
Expected no. of events in $ones$ (1e)	11.3	9.2	5.8	4.5	3.6	3.3
Expected no. of events in $zeros$ (0e)	2.0	1.6	1.2	1.0	0.7	0.7
One-Zero threshold	7	6	4	4	2	2

Table 1: One-Zero threshold and expected number of events in bit one and $zero$ for different bit rates

quence, as described in Section 4.1. Specifically, we apply a joint decoding-synchronization technique that uses a threshold-based and distance-based method to simultaneously synchronize and decode the received sequence.

Figure 5 shows the high level architecture of the system and how components interact. Note that since we do not have access to the touchscreen controller, the touch events, not the underlying physical voltage differences, are the input to the receiver.

Indirect Communication: Even without direct contact with the screen, the ring can communicate with the touchscreen device as long as the ring bearing finger is in contact with the screen. In particular, the electrical pulses that are transmitted through a human finger's skin from the ring create the same effect of changing the screen capacitance to register artificial touch events. However, we found that due to the skin resistance, the number of events generated through this type of indirect contact is only enough for detecting the presence of the ring, but not stable and regular enough for reliably decoding the data being transmitted. We can leverage this capability of the communication system to enable a novel technique to differentiate two users simultaneously interacting with the same touchscreen, for example in a shared-screen two player game. The detection algorithm used for this mode of communication is described in Section 4.3.

4. DECODER DESIGN

The proposed capacitive touch communication system allows users to send messages to the application layer of the device. This unconventional use of the touchscreen, especially under the constraint of using commercial off-the-shelf devices without lower layer access, poses a number of challenges:

1. We observed that the receiver responds differently to the same input following a different bit pattern; this could be due either to the physical layer or the software that is optimized for detecting touch events from a human finger. For example, the number of events registered to the screen when bit 1 is sent after a long sequence of 0s is different from that of a bit 1 that comes after a sequence of 1s. The normal solution is to code the data to avoid this pattern dependent effect. Rather than adopting a typical bit-by-bit decoding solution, our data rate is already so limited that we developed our own code optimized specifically for the observed pattern dependence.

2. There is a variable delay between the transmission of a symbol and its reception at the receiver after processing through all layers of firmware and software. This jitter significantly increases the difficulty of detection. Since the communication channel has low bandwidth and high jitter, no traditional symbol synchronization schemes can be directly applied. We overcome the bit synchronization challenge by simultaneously synchronizing and demodulating the signal.

3. The channel adds an unknown delay between receiver and transmitter; this problem is classically solved using a frame synchronization which requires using a preamble. Since we have a low bandwidth channel and would like to transmit the message in only a few seconds, the message can only include limited number of bits. Thus, we can not afford to add the preamble. Instead, we use constrained bit patterns that are unique under cyclic shifts caused by unsynchronized frames.

The conversion from touch events to a sequence of binary digits is based on the principle of On-Off keying; the touchscreen driver produces several events when a binary one is transmitted and only a few events when a $zero$ is transmitted. The key challenge is to handle the variance in the number of events associated with $ones$ and $zeros$. In the coming sections, we describe an off-line calibration procedure to characterize the expected behaviour of the channel, which is then used in the online phase to classify touch responses as $zero$ or one transmissions. Once a sequence of bits is decoded, we use a "closeness" metric to determine the distance of the received message from the set of all possible messages of the same length. This process corrects for uncertainty in timing and event number. Details about the design of the closeness metric and the decoding process are presented in the next sections.

Algorithm 1: Threshold selection algorithm

input : $E_{discrete}$ - Event sequence in time domain
$TxBitSeq$ - Original transmitted bit sequence
$BitRate$ - Transmission bit rate (bps)
output: $1e$ and $0e$ - Expected number of events in $ones$ and $zeros$

bitPeriod $\leftarrow \frac{1000}{BitRate}$
oneC $\leftarrow$ 0 // Event counter for all ones
zeroC $\leftarrow$ 0 // Event counter for all zeros
//Convert discrete events to event vector in time series
for $i = 1 \rightarrow max(E_{discrete})$ **do**
    if *exist* $E_{discrete}[j] == i$ **then** $E_t[i]$= 1
    else $E_t[i]$ = 0
//Find the starting position that gives the max $1e0e$ Ratio
for startPos =1 →bitPeriod **do**
    for j =1 → *length*($TxBitSeq$) **do**
        eCount = sum(E_t[startPos + (j − 1) ∗ bitPeriod, startPos + j ∗ bitPeriod])
        if $TxBitSeq(j) == 1$ **then**
            oneC = oneC + eCount
        else zeroC = zeroC + eCount
    // Update $1e0e$ Ratio
    current$1e$ = oneC/*no. bit* 1 *in* $TxBitSeq$
    current$0e$ = zeroC/*no. bit* 0 *in* $TxBitSeq$
    if *current1e/current0e* > maxRatio **then**
        maxRatio = current1e/current0e
        $1e$ =current$1e$; $0e$= current$0e$
return $1e$ and $0e$;

4.1 Determination of Expected Number of Events for $ones$ and $zeros$

To determine the number of touch events associated with a one or $zero$, it is necessary to calibrate the device at each data rate before use. This calibration to determine thresholds only needs to be performed once per device, at initialization; thereafter it can be stored in a lookup table and adjusted during self calibration depending on an estimate of the data rate of the incoming data sequence or fetched as an input from applications. To determine the counting threshold for each data rate, a sequence of $ones$ and $zeros$ is repeatedly transmitted in a prescribed pattern. On the receiver side, event sequence is detected and recorded to a log file. Threshold selection algorithm, algorithm 1, takes the log file and the prescribed pattern as input to compute the two expected counter thresholds $1e$

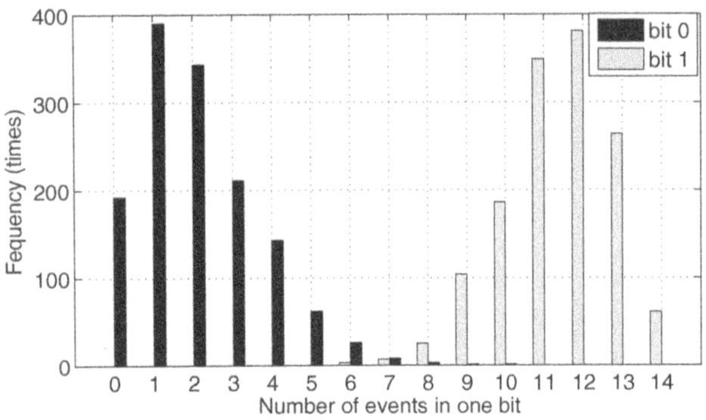

Figure 6: Number of events in bit one **and bit** $zero$ **for transmissions at 4 bits/s**

and $0e$. We devise the algorithm 1 to simultaneously demodulate the received event sequence and find the bit starting point. The intuition behind the algorithm is that the correct bit synchronization maximizes total number of events in all $ones$ and minimize number of event in all $zeros$ We define *1e0e* ratio as being the normalized ratio between the *total number of events in all* $ones$ and *total number of events in all* $zeros$:

$$1e0e\ Ratio = \frac{\frac{\Sigma(Event\ Counters\ in\ Ones)}{Number\ of\ Ones}}{\frac{\Sigma(Event\ Counters\ in\ Zeros)}{Number\ of\ Zeros}}$$

This ratio is maximized when bit synchronization is correct. The ideal synchronization, for example, should have total number of events in all $zeros$ close to 0, and number of events in all $ones$ close to the total number of events in the whole event sequence, in which case *1e0e* ratio reaches its maximum.

Algorithm 2: Min Distance Demodulation Algorithm

```
input  : Et - Event sequence in time domain
         BitRate - Transmission bit rate (bps)
         MessageLength - Original message length
         PosMessageVec - Possible message vector
         1e and 0e - Expected number of events in ones and zeros
output: RxBitSeq - Received bit sequence
bitPeriod ← 1000 / BitRate
minDistance ← MAX-INT
for startPos = 1 → bitPeriod do
    foreach message in PosMessageVec do
        rotatedMesgVec = getAllCyclicVersions(message)
        for j = 1 → MessageLength do
            eCount[j] = sum(Et[startPos + (j − 1) ∗ bitPeriod, startPos + j ∗ bitPeriod])
        foreach rotatedInstance in rotatedMesgVec do
            currentDist = 0;
            foreach i^th bit in rotatedInstance do
                if i^th bit == 1 then
                    currentDist = currentDist + max(0, 1e - eCount[i])
                else
                    currentDist = currentDist + max(0,eCount[i] - 0e)
            // Update Min distance
            if currentDistance < minDistance then
                minDistance = currentDist;
                RxCandidate = rotatedInstance;
return RxCandidate
```

Algorithm 1 first converts the discrete timing event information to a event/no-event time series data. That is, if the received sequence of event is $E_{discrete} = \{E_1, E_2, ..., E_m\}$ in which E_i is i^{th} event, it will be represented by a vector in the form: $Et = [Et_1, Et_2, ...Et_{t_{max}}]$ where $Et_i = 1$ if there exists an event E_k such that $E_k = i$, and 0 otherwise. In the second step, the algorithm tries all possible bit starting points within the first bit period, with each trial involving a counting of the number of events in all bit periods of the sequence. The starting point that leads to the highest ratio is considered the correct bit sync position, while the bit sequence corresponding to that starting point is the demodulated result of the event sequence. At the end of this process, since the total number of events in all $ones$ and total number of events in all $zeros$ is found, the expected number of events in $ones$ and $zeros$, $1e$ and $0e$ can be derived and stored in memory for future demodulation. Figure 6 shows the distribution of the number of touch events registered corresponding to the transmissions of $zero$ and one evaluated by using algorithm 1 for a 3000 bit sequence of alternating $zeros$ and $ones$. The variations due to the transmission bit rate is recorded in table 1, which shows that the event count threshold required for decoding varies from 7 events for 4 bits/s to 2 events for 15 bits/s.

4.2 Minimum Distance Demodulation

Using the counter thresholds from the previous section, algorithm 2 demodulates the timing event sequence to get the data sequence sent by the transmitter. Sharing the same synchronization challenge with the threshold detection algorithm, this algorithm has to detect the point in time at which the data is transmitted. At the same time, it demodulates the event sequence to get the information that has been transmitted. Note that simply relying on the first event to determine the starting point is not enough since there is a fair amount of timing uncertainty in the communication channel. Intuitively, the algorithm traverses the sequence to try all possible starting points. At each point, it gauges the "distance" between the event sequence and all messages. It then ranks the positions with "similarity" value and selects the one that has highest "similarity" index. The message corresponding to that index will be the decoded value of the event sequence.

So the question remains as to how to measure the similarity between two sequences. We define a distance metric as following: let $D(i, j)$ be the distance between a event sequence that has a starting point at point i and the message, K_j, with $j = 1..number\ of\ messages$. Using the same notations as defined in the previous algorithm, in which $Et = [Et_1, Et_2, ...Et_{t_{max}}]$ is the event vector re-sampled along the time domain, an event counter, eC_p, for bit at p_{th} position from the starting point can be computed by:

$$eC_p = \sum_{q=(p-1)*bit\ period}^{p*bit\ period} Et_q$$

Then distance $D(i, j)$ can be derived as:

$$D(i, j) = \sum_{k=1}^{message\ length} d_k$$

with

$$d_k = \begin{cases} max(0, eC_p - 0e) & \text{if the } k^{th} \text{ bit on message } K_j \text{ is 0} \\ max(0, 1e - eC_p) & \text{if the } k^{th} \text{ bit on message } K_j \text{ is 1} \end{cases}$$

We note that since messages are cyclically transmitted, the algorithm does not only compute the distance of a sequence to a message but it does so for all unique *rotated* version of that message.

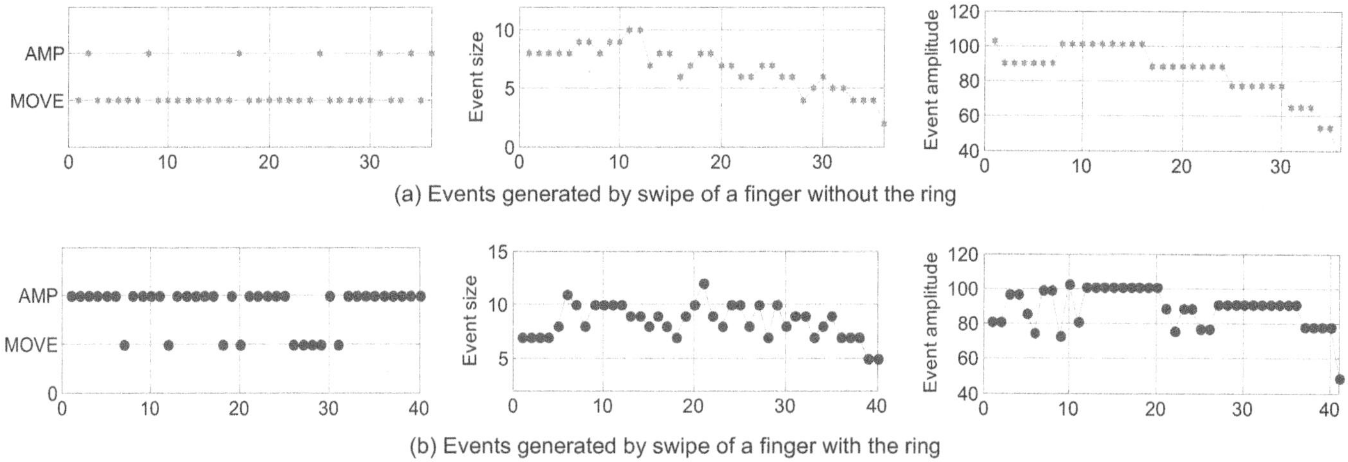

Figure 7: *Type, Size* and *Apmlitude* values generated from finger swipes with and without the ring. A ring bearing finger produces many *AMP* events while swipe without ring induces correlation between *Size* and *Amplitude*

The intuition behind this metric is that it rewards starting points that make the decoded sequence look similar to one of the messages in the message vector. The smaller the distance, the closer the decoded sequence to the message. Hence, smallest $D(i,j)$ will tell which position on the sequence is correct synchronization position and which message is the event sequence representing.

We note that when the number of possible messages is small (order of hundreds), it is feasible to apply Algorithm 2 to exhaustively search through the whole message space to demodulate. However, when the number of possible messages is large, the above exhaustive algorithm can become prohibitively expensive or impossible. In such cases, more efficient algorithm assuming no knowledge of the message becomes handy. That algorithm shares the same intuition with Algorithm 1, in that it tries all possible starting points. However, at each possible position, it directly converts the sequence to data bit sequence by counting number of events in each bit period and select the one that yields the highest 1e0e ratio.

Other demodulation schemes. In the process of finding the most suitable demodulation scheme, we experimented with several other demodulations schemes such as *Non-thresholding modulation*, 1e0e *ratio demodulation* and *maximum key correlation*. Non-thresholding modulation scheme does not require any training to learn expected number of events in zeros (*0e*) and ones (*1e*), instead it looks at all possible starting positions and compares them with all possible keys to find the best match. It simultaneously picks the synchronization point and decodes the sequence of events by selecting the starting point that gives the highest correlation with one of the possible keys. The maximum key correlation method takes an approach that is similar to minimum distance modulation but has a different evaluation function. For that we defined our own correlation coefficient function to take the noisy channel into account. The function gives one point to a bit that is equal to the bit at the same position on the correct key. It gives half a point to the bit that is not correctly decoded but has a number of events close to the $One - Zero$ threshold. Lastly, 1e0e ratio demodulation becomes useful when the possible message space is unknown or so large that it is prohibitively expensive to conduct an exhaustive search to find minimum distance or maximum correlation. The Min Distance Algorithm presented earlier outperformed these techniques in our experiments.

4.3 Ring Detection for Indirect Communication

As mentioned in Section 3, an indirect mode of communication is enabled when instead of the ring, a ring bearing finger is in direct contact with the touchscreen. In such cases, only the presence of a ring needs to be detected. However detecting the ring in the presence of finger movements (or finger swipes) is challenging since the events generated due to the movement of the finger and those by the ring cannot be easily distinguished.

Figure 7 shows three fields of the touch event outputs: *Type*, *Size* and *Amplitude*, generated when a user swipes a finger across the screen with and without the ring. We leverage two key observations from the patterns observed for designing a robust detection algorithm: (i) events generated by the finger movement without the ring are mostly of type *MOVE* while those generated by the ring are mostly of type *AMP*, however due to the excess pressure exerted from the drag force of the finger on the touchscreen, a few *AMP* events can also be generated during finger swipe movements without the ring; (ii) in the absence of the ring, the sequence of *Size* and *Amplitude* values are correlated since increasing the pressure brings more surface area of the finger in contact with the screen. We confirm these observations by collecting data from a large number of swipe movements, both with and without the ring from 5 different users.

Since both the presence of a large number of *AMP* events and the absence of correlation between *Size* and *Amplitude* indicate the presence of a ring, we define a metric p_{ring}, which relates to the normalized number of *AMP* events registered (n_{amp}) and the correlation coefficient between the *Size* and *Amplitude* values (c_{SA}) as:

$$p_{ring} = \alpha \times n_{amp} + (1 - \alpha) \times (1 - c_{SA})$$

where $\alpha \in [0, 1]$ is parameter which signifies the relative contributions of n_{amp} and c_{SA} in determining the p_{ring} value. Given a set of generated events, a detection threshold λ_{th} is then used on the p_{ring} value to classify the presence or absence of the ring. We determined the values of the two parameters α and λ_{th} through a training set consisting of 1000 swipes from 3 different users, using traditional least square minimization. After the training, α and λ_{th} were determined to be 0.83 and 0.5 respectively.

5. EXPLORING THE PARAMETER SPACE

Data transmission using capacitive touchscreen communication is an unexplored mode of communication. In this section, we explore the dynamic ranges of frequency, voltage and signal types that can be used for triggering usable events through the screen driver. Having picked the most suitable set of parameters, we then study the performance of the communication system for different use cases.

In order to conveniently vary the input signal parameters, as required in this analysis, we placed a flat, rounded copper piece on the screen surface of a Samsung Galaxy Tab 10.1, which uses a Atmel maxTouch touch panel [7], and attached it to the output of a AFG 3000 Series function generator [4]. This setup simulates the touch of the ring on the screen surface while offering two main benefits over the battery-powered prototype described in Section 6: (i) it alleviates the microcontroller's limitation in generating arbitrary waveforms and (ii) it greatly expands the scale of repeated experiments (order of tens of thousands of logging runs) which would be otherwise limited by time and human effort.

5.1 Triggering Touch Events

The inner-workings of the touch screen are proprietary and not available for use in designing either our hardware or software. A main task is to determine what type of electrical signal will be interpreted as a touch event when it is injected into the touchscreen. To answer this, we inject different signals from a function generator through an attached electrode approximately the size of a finger to the surface of the touchscreen.

The touch events retrieved by the tablet's operating system, Android 3.2, are represented in a 6-tuple structure depicted in Figure 4. Indicated through *Event Type* field, touches are classified into one of the following types: *MOVE*, *AMP*, *MOVEAMP*, *PRESS*, *RELEASE* and *SUPPRESS*. For example, a *MOVEAMP* event is registered when both touch pressure and X,Y-coordinates change at the same time; and a *SUPPRESS* event happens when the touch pressure exceeds a predefined threshold. Note that such touch events are triggered when a finger first touches the panel, when the position of the finger on the screen changes, when the pressure changes, and when the finger leaves the screen. *Touch Size* and *Touch Amplitude* specify the size and amplitude of the touch respectively. *Pointer ID* is used to differentiate the presence of two or more points of contact at the same time, or multi-touch. A physical touch causes voltage changes at many different electrodes, but the firmware and driver aggregate them to output a single touch event to the operating system. Since the aggregation algorithm is proprietary, the conversion from electrical signals of our interest to such touch events can only be empirically learnt.

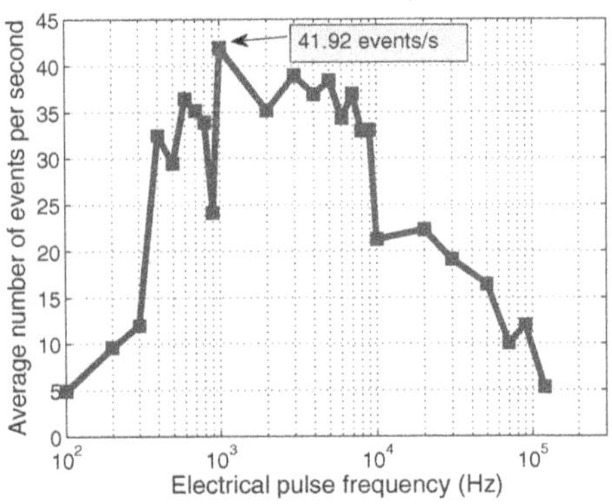

Figure 8: Touchsreen responses to 10 Volt peak-to-peak square wave signals with different frequencies ranging from 100Hz to 120KHz (log scale)

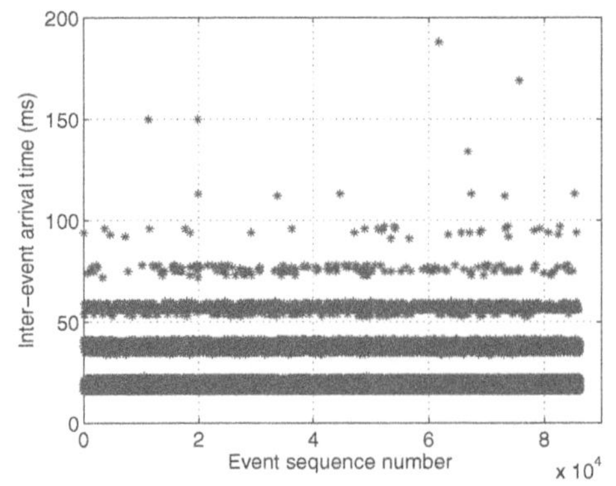

Figure 9: Distribution of inter-event arrival times of events generated by a 10 Volt peak-to-peak 1KHz square wave signal captured in kernel level log files

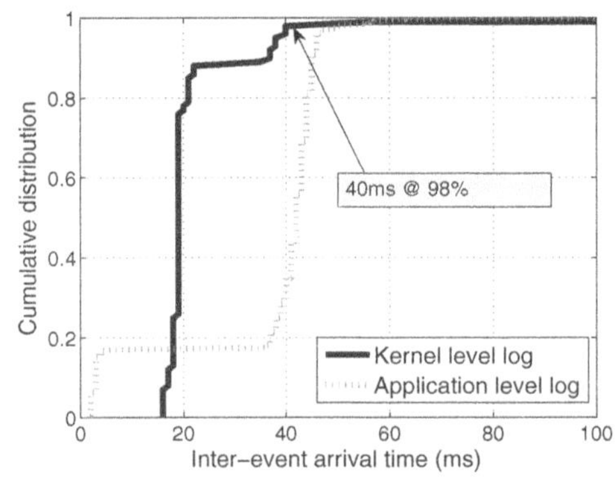

Figure 10: CDF of inter-event arrival times of events captured at application level and kernel level log files with 10 Volt peak-to-peak 1KHz square wave input signal

An important aspect of the system is the maximum possible data rate through the screen, which depends on two key characteristics of the screen: (i) the highest rate at which the driver and firmware allows touch events to be registered and (ii) the internal switching frequency of the sensing hardware. Atmel mXT1386's datasheet specifies a maximum of 150 *raw touch events* per second [7]. However, due to the driver in Android's software stack, the maximum rate is significantly reduced. We conducted many experiments to gauge the actual maximum event detection rate. We transmit signals with different waveforms, at different frequencies and voltage levels to a screen. With frequencies ranging from 100 Hz to 120 KHz, we observed that a 10 Volt peak-to-peak square wave signal at a frequency of 1 KHz can register the maximum rate of 41 events/second (i.e. average inter event arrival time of $\frac{1}{41} = 24$ ms). In particular, we began with finding the frequency that the touch-screen was most responsive. To do so, we set the Tektronix digital function generator to generate square wave of different frequencies at 10 Volt peak-to-peak amplitude. The frequency varies from 100 Hz to 1 KHz with 100 Hz difference, from 1 KHz to 10 KHz with 1KHz difference, and from 10 KHz to 120 KHz with 10s

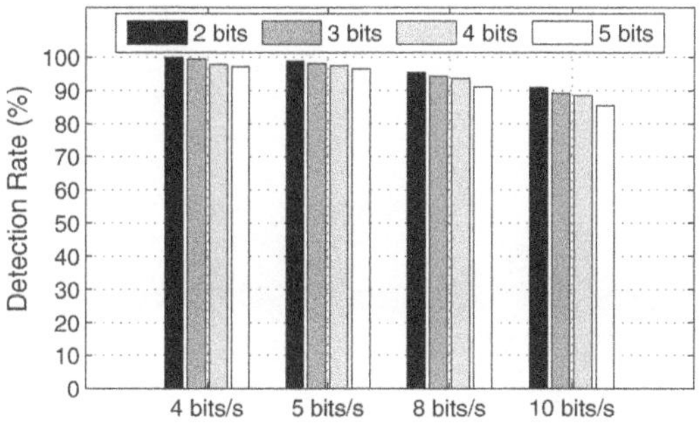

(a) Detection rate for different message lengths and bit rates

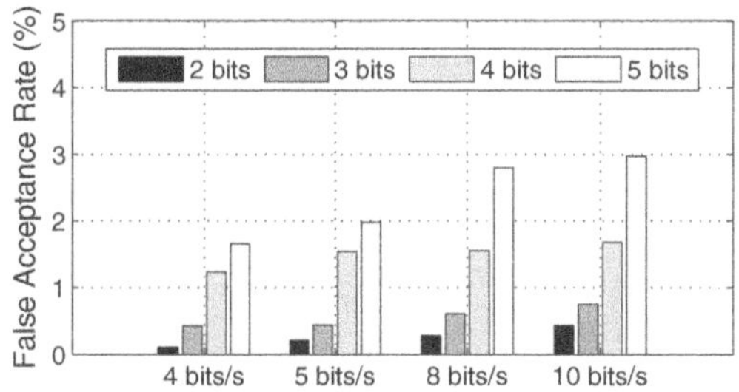

(b) False acceptance rate for different key lengths and bit rates

Figure 11: Detection rate of 99.4% for 3-bit message transmitted at 4 bits/s

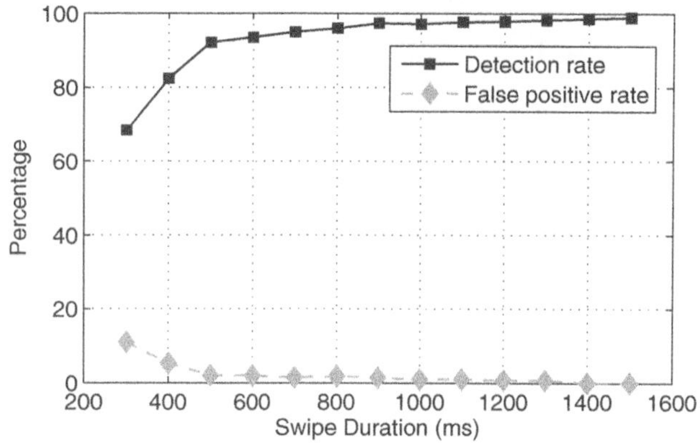

Figure 12: Multi-user games: Swipe detection rate

KHz difference. To collect the signals, we wired the output from the function generator to a flat solder electrode, then put the electrode on the surface of the Samsung Galaxy Tab 10.1 touchscreen. To make the electrode stable on the surface, we taped the it tightly to the touchscreen to avoid unintended movement. For each frequency, we collected the data for 200 seconds. Then record the number of event collected from the kernel. The average number of events is shown in Figure 8, which suggests that the screen best responses to signal at 1 KHz.

Figure 10 shows CDF of inter-event arrival times at kernel and application level log files. While almost 90% of the times, 2 consecutive events captured by the kernel log happen within 20 ms with very little variation, that number widely varies from 3 ms to 48 ms in the case of application level log. That observation indicates that using the timing information from kernel level log would could improve the demodulation results which mainly relies on event timestamps. In additional, we also observed that sinusoidal or triangular signals do not register any events. We further test with signals with different amplitudes and noticed that if the voltage amplitude of the signal is too high, the screen blocks all subsequent touch events for a short period of time and sends an error event to the operating system, which is the SUPPRESS event mentioned above. If the voltage is too low, the signal is either not detected or detected at a very low rate by the touch screen. We do not present the results of those variations in this paper due to the space limit.

A scatter plot of 86,200 events collected over 1850 seconds, Figure 9, illustrates the distribution of inter-arrival times, i.e. the time difference between two consecutive events, captured in kernel level log. An interesting pattern can be observed in Figure 11 is: most of the inter arrival times fall into specific narrow bands which we believe to be due to firmware throttling. Its cumulative distribution shows that 98% of the time, the inter-event arrival time is less than 40 ms. Note that this event detection rate is more than 7 times lower than the rate of 150 raw touch events per second specified by the manufacturer [7]. Without access to the physical layer and the proprietary driver, we cannot determine the origin of this discrepancy. However, data rate could be at least 7 times faster than what we currently achieve if we have access to the driver; and an even higher data rate might be possible with direct access to the lower physical layer.

5.2 Bitrate vs. Detection Rate Trade-off

The main performance metrics here are the detection rate and the false acceptance rate. The detection rate signifies the probability of correct decoding of a message while the false acceptance rate characterizes the probability of a wrong message being incorrectly decoded as the original message. As explained in section 4 and shown in table 1, there exists a trade-off between the detection rate and the bit rate at which messages can be decoded from the touchscreen event logs. Correspondingly, since there are higher chances of incorrect decoding at higher bit rates, the number of false positives increase as the bit rate increases. In order to quantify this phenomenon, we use the setup described above to repeatedly transmit messages of different length at different bit rates. Figure 11 shows the detection and false acceptance rates observed. To derive the detection rate for each message length and bit rate, we transmit each message of that length 5000 times and present the average percentage of messages that are correctly decoded. Similarly, the false acceptance rate is derived by sequentially fixing each message as the correct message and transmitting all other messages of the same length 5000 times.

The trends in Figure 11 indicate a gradual decline in the detection rate with the increase in either the transmission bit rate or message length. We note that for simple parental control applications, a 99% detection rate can be achieved by using 2 or 3 bit messages at 4 bits/s. For applications that have a less stringent detection rate requirement, a much higher bit rate can be used to speed up the required data transmission time.

5.3 Indirect Communication Results

The next set of results are targeted towards detection of individual users in an indirect communication scenario. In this scenario, while the bit rate required is not very high, touching the ring to the touchscreen would hinder in the game-playing process. As such we leverage the fact that even if the finger-tip of the ring bearing finger touches the screen, the patterns in the registered event logs can be

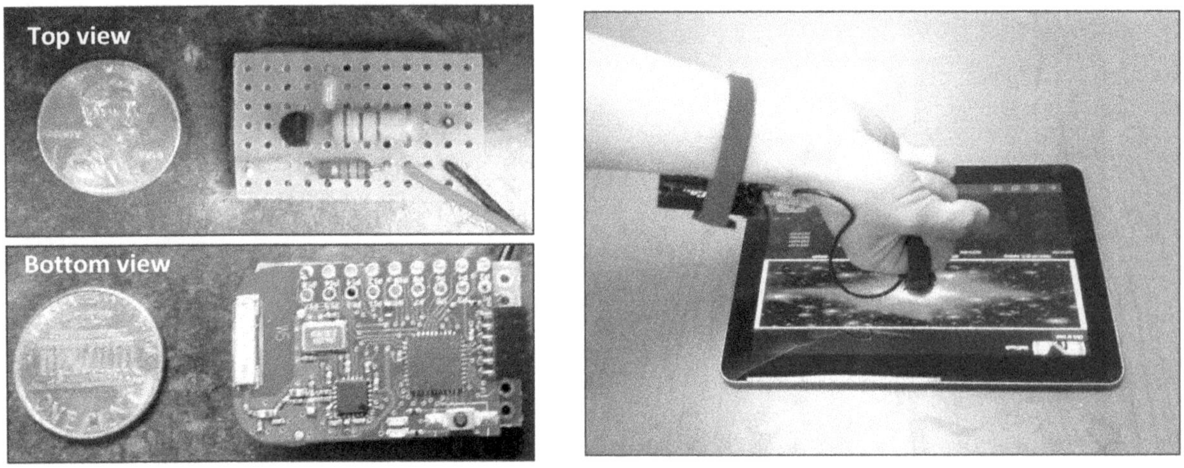

(a) Prototype circuit board

(b) Usage of the prototype ring

Figure 13: The prototype ring and its usage for transmitting short messages from the ring to a touchpad

used to differentiate between a user with a SigRing and one without it.

In order to quantify the performance of this algorithm, we collected a total of 6,000 swipes from 3 different users with half the swipes with a ring on. We asked the users to vary the swipe duration between 300ms to 1.5 seconds but since making a swipe last for precisely a given time is hard, we bucket the collected swipes into 100ms durations starting from 250ms to 1550ms and discard swipes outside of this range. The swipe duration of all swipes within a bucket are approximated by the mean value of the bucket. Using the move events registered in this dataset, we calculated the detection rate of ring bearing users and the percentage of swipes without rings which were wrongly classified as one with rings, i.e., the false acceptance rate.

The resulting values shown in Figure 12 shows that the detection rate increases with the duration of the swipe, at first sharply from ~68% for 300ms swipes to ~92% for 500ms swipes and then gradually after that. Thus if the swipes used in a multi-player game is more than 700ms, the users can be classified correctly with a 95% confidence level.

6. RING PROTOTYPE AND EVALUATION

The use of this communication channel with touchscreen device requires a hardware token to generate the appropriate electrical signal and inject it into the touch sensing circuitry. In this section, we describe our prototype of the ring which uses off-the-shelf components.

6.1 Hardware Prototype

The core of the token is the low cost, low power microprocessor, TI-MSP430F2722 [35] that was programmed to generate modulated 3 Volt square waves at a frequency of 1KHz. Figure 14 shows the schematic of the custom-built ring. This square wave is modulated with On-Off keying to trigger artificial touch events in the screen's firmware. The microprocessor is mounted on a 18 mm x 30 mm off-the-shelf board, part of TI-MSP430 eZ430 development kit, as shown on Figure 13(a)-bottom view. We specify the transmission data rate and data sequence by programming the microprocessor through the USB interface that comes with the kit. The square wave and its parameters were selected through experiments with a function generator, as described in section 5. Since

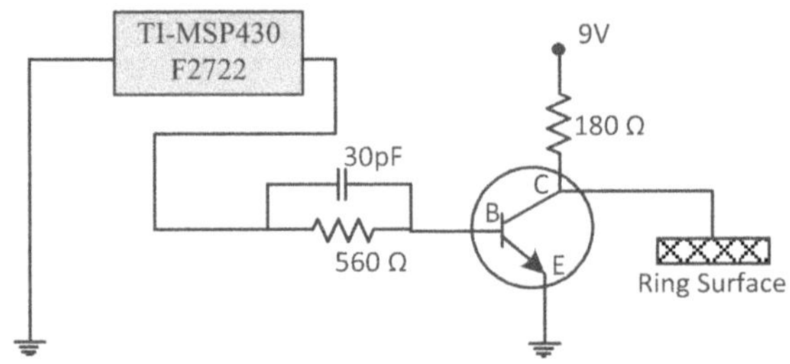

Figure 14: Schematic of the custom-built ring

we found that 3 Volt was not adequate for generating touch events, we amplify the output of the microprocessor using a single bipolar transistor, BC548B [5], with the supply voltage of 9 Volt (Figure 13(a)-top view). One of the most challenging parts of the prototype was to design the electrode configuration that would allow the signal to be injected in series with the touchscreen and the body capacitance of the user. The best point in the circuit to inject the signal, V'_{sig} in figure 1, would be in series with the finger and the rest of the body at a point close to the screen. This has obvious anatomical difficulties and the low internal resistance of the body makes injection between two closely spaced electrodes, as on the inside surface of a ring, impractical. We opted for a system where the user would wear an insulating ring where V'_{sig} was injected between electrodes on the inside and outside of the dielectric band. The inner electrode was connected with the finger and, through the body capacitance C_B and case capacitance C_c (as described in section 2), to the internal circuitry in the tablet. The outer electrode on the ring was directly pressed on the screen, forming C_s to complete the circuit.

Because a uniform and reproducible contact between the touchscreen and the ring is essential to minimize the error rate, we choose to use a flexible conductive material to make the electrode and design the face of the ring to control the compression of that material. If the pressure is too high, the screen bends and its capacitance, C_s, increases which in turns can introduce errors. We control this pressure by surrounding the electrode with an insulating spacer of the

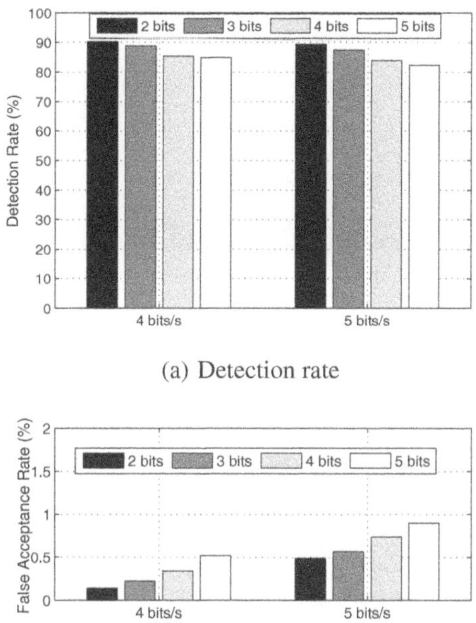

(a) Detection rate

(b) False acceptance rate

Figure 15: Detection rate and False acceptance rate using ring prototype for different message lengths and bit rates

correct thickness to properly control the compression of the flexible electrode.

6.2 Preliminary Prototype Performance

Using the prototype ring, we experimented with injecting messages through the Samsung Galaxy Tablet 10.1 touchscreen. While conducting the experiment, we noticed that at times no events were triggered during the transmission of a 1 bit or vice versa. This lead to unreliable decoding of messages but we were still able to distinguish two codes with a larger hamming distance. We used one ring with the code '1110' and one ring with the code '1000' and touched each of the 50 times on the tablet display. A simple threshold algorithm that uses as input the number of touchscreen events generated was able to identify the first ring correctly 44 times and the second ring 43 times, leading to an overall detection rate of 87%. We suspect that the quality of the contact between the ring and the touch panel plays a critical role in these experiments.

To eliminate this variance due to contact differences from touch to touch, we experimented with transmitting multiple messages while the ring was held steady on the display. Here, we used message lengths between 2 and 5 bits transmitted at the rates of 4 bits/s and 5 bits/s from which the detection rate (DR) and false acceptance rate (FAR) are evaluated. For each message at each data rate, we put the ring down onto the screen 3 times and keep it there long enough so that 200 repetitions of the message are transmitted from the ring to the screen. We show in Figure 15 the DR and FAR results over the 200 repetitions from best case (presumably best contact) of these three trials. Each bar represents the average rates over different data rates and message lengths. We observed that the detection rate decreases with the increase of both the message length and bit rate. Note, however, that the overall detection rate could be improved through retransmissions of the message. Therefore, even the lower detection rate of 82% may still be adequate for some of our targeted applications. For the user identification application, for example, up to 3 seconds of continuous repeated message transmission would results in less than 6 errors per 1000 uses. These results illustrate what can be achieved with this transmitter if the reliability issues are worked out.

We believe that another source of error in this prototype stems from the relatively long rise time of the square wave since the touchscreen events appear to be triggered by the edges in the input signal. It is also important to note that both the electronics and the firmware of the screen, which we do not have access to, are optimized for the relatively slow movement of a human finger. Thus, the screen driver deliberately throttles the maximum rate of touch events, which reduces touch error in normal use but limits our system to very low bit rate transmission. We believe that the transmission rate could be improved substantially with access to the touchscreen controller firmware, which should allow processing internal touchscreen measurements, e.g. physical voltage differences.

7. DISCUSSION

Let us briefly consider several remaining issues related to the energy consumption and security applications of this technique.

Energy Consumption. The current prototype implementation is based on a 3 Volt microprocessor driving a 9 Volt high speed bipolar transistor amplifier to generate a continuous signal. Energy consumption and some of the synchronization issues in our prototype could be significantly reduced by incorporating a switch under the contact surface that powers up the ring when pressed against the touch screen. To estimate the cost and battery life of such a ring version, we use the smallest readily available lithium primary battery, the CR2025 which is 10 mm diameter and supplies 3.0 volts with a 30 mA-h capacity. The typical current drain in standby with RAM-retention of a modern microprocessor (e.g. the TI MSP430 family) is about 0.1 microamps. Even with this small battery, this would provide over 3 decades of standby lifetime for the ring electronics. Once awake, the processor will use significantly more current, but the minimal computing requirements result in this being low, also. The smaller MSP-430 processors typically use about 220 microamp at 1 MHz, so even if shifting out the short code takes 100 cycles of the CPU, this battery will still provide enough energy for over 5000 uses.

Since the capacitances are very small, the current also will be low and a simple buck-boost dc-dc converter with one miniature inductor will be quite adequate to supply the 9 Volt[6]. Assuming only a 10 % charge conversion efficiency for the converter, this circuit still uses only about 2 nanocouloumbs/charge-discharge cycle. Modulating at 1 KHz and sending 10 bits/second, this allows the battery to supply over 50 million bits, far in excess of any of the other limits in the system. The cost of such a system will be dominated by the processor, several tens of cents, but in high volume that can be replaces by a simple sequence generator, either read-only or flash, for only a few cents.

Security considerations. The current limits on datarate only allow transmission of very short codes and thus allow only weak authentication at best. Improvements in datarate through modifications in the touchscreen firmware could alleviate these limits, however. The low carrier frequency of our system, between 5-10 kHz, would then also offers additional protection against eavesdropping. Since antenna size should be proportional to the wavelength of the signal, transmission of this signal into the RF domain would require an antenna much larger than the size of the human body. While we cannot rule out that some signal can be received with customized resonant antennas, however, the level of effort required would be much higher than for picking up a e.g. 2.4 GHz signal used in WiFi and Bluetooth. If such eavesdropping ever were an issue, it could also be addressed by transmitting a noise signal from the receiving device.

Finally, the design could be enhance with a feedback channel using a photodetector. The ring could receive information from the mobile device through this visual channel, where the device encodes the information in the pixel intensities. This would enable a challenge response protocol, which could greatly enhance the security of an authentication system.

8. CONCLUSION

We have presented the design and implementation of a technique to transmit information through a capacitive touchscreen. Our method triggers touch events in the touchscreen device by injecting an electric signal that affects the capacitance measurements of the screen. Our experiments show that this is feasible even with an off-the-shelf touchscreen system, albeit at very low bitrates. Controlled experiments with a signal generator demonstrates data rates of 5-10 bps. While some reliability challenges remains, we also achieved up to 4-5 bps with a wearable transmitter token in the form of a small signet ring and demonstrated that some signals can be transmitted through the human skin. Transmission of information via small physical tokens can be used to distinguish who is interacting with a mobile device, and can be useful for parental control, multiuser games (particularly when played on a single device), and possibly play a role in authentication solutions. It differs from other short-range communication systems in that it requires physical touch for communication, which can be an advantage if multiple potential users are so close that they cannot be differentiated with the other short-range systems. The technique could also be used to distinguish different devices touching the screen such as styluses or boardgame tokens. We believe that significantly higher data rates could be achieved by designing receiver capabilities into touch screens and few this work as a first step towards exploring how this touch sensor can participate in the exchange of information between mobile devices.

Acknowledgement

We are thankful to Nilanjan Paul, Joshua Devasagayaraj, and Ivan Seskar for their help with developing prototypes of the signet ring. This material is based in part upon work supported by the National Science Foundation under Grant Nos. 0845896 and 1040735. Any opinions, findings, and conclusions or recommendations expressed in this material are those of the authors and do not necessarily reflect the views of the National Science Foundation.

9. REFERENCES

[1] Face recognition on galaxy nexus is very neat, but easily compromised with the picture? http://tiny.cc/gj50gw, 2011.

[2] Nearfield communications forum. http://www.nfc-forum.org/, 2011.

[3] A. Adler. Vulnerabilities in biometric encryption systems. In *Proc. of AVBPA*, 2005.

[4] Afg3000 series arbitrary function generators. http://www.tek.com/AFG3000.

[5] Amplifier transistors bc548b. http://onsemi.com.

[6] An mcu-based buck-boost converter for battery chargers. http://www.st.com/.

[7] Atmel maxtouch mxt1386 specifications. http://www.atmel.com/devices/MXT1386.aspx.

[8] G. Bailador, C. Sanchez-Avila, J. Guerra-Casanova, and A. de Santos Sierra. Analysis of pattern recognition techniques for in-air signature biometrics. *Pattern Recognition*, 44(10-11):2468 – 2478, 2011.

[9] D. Balfanz, D. K. Smetters, P. Stewart, and H. C. Wong. Talking to strangers: Authentication in ad-hoc wireless networks. In *Proc. of NDSS*, Feb. 2002.

[10] Belkin. Wemo home automation system.

[11] N. Ben-Asher, N. Kirschnick, H. Sieger, J. Meyer, A. Ben-Oved, and S. Möller. On the need for different security methods on mobile phones. In *Proc. of MobileHCI*, Aug. 2011.

[12] H. Bojinov and D. Boneh. Mobile token-based authentication on a budget. In *Proc. of HotMobile*, Feb. 2011.

[13] Craftsman. Assurelink garage doors openers.

[14] P. Dietz and D. Leigh. Diamondtouch: a multi-user touch technology. In *Proc. of UIST*, Nov. 2001.

[15] Digital Modulation Schemes. http://robotics.eecs.berkeley.edu/ sastry/ee20/modulation/node5.html.

[16] P. Dunphy, A. P. Heiner, and N. Asokan. A closer look at recognition-based graphical passwords on mobile devices. In *Proc. of SOUPS*, July 2010.

[17] M. Faundez-Zanuy. On the vulnerability of biometric security systems. *Aerospace and Electronic Systems Magazine, IEEE*, 19(6):3 – 8, June 2004.

[18] C. Foresman. Apple facing class-action lawsuit over kids' in-app purchases. *Ars Technica*, Apr. 2011.

[19] Fundamentals of electrostatic discharge. http://www.esda.org/fundamentalsp5.html, 2010.

[20] J. Galbally-Herrero, J. Fierrez-Aguilar, J. D. Rodriguez-Gonzalez, F. Alonso-Fernandez, J. Ortega-Garcia, and M. Tapiador. On the vulnerability of fingerprint verification systems to fake fingerprints attacks. In *Proc. of Carnahan Conferences on Security Technology*, Oct. 2006.

[21] C. Gehrmann, C. J. Mitchell, and K. Nyberg. Manual authentication for wireless devices. *CryptoBytes*, Spring 2004.

[22] E. Haselsteiner and K. Breitfuss. Security in Near Field Communication (NFC). In *Proc. of the Workshop on RFID Security*, 2006.

[23] K. Hawkey and K. M. Inkpen. Examining the content and privacy of web browsing incidental information. In *Proc. of WWW*, May 2006.

[24] A. K. Jain, K. Nandakumar, and A. Nagar. Biometric template security. *EURASIP J. Adv. Signal Process*, 2008:113:1–113:17, Jan. 2008.

[25] E. A. Johnson. Touch Displays: A Programmed Man-Machine Interface. *Ergonomics*, 10(2):271–277, 1967.

[26] A. K. Karlson, A. B. Brush, and S. Schechter. Can i borrow your phone?: understanding concerns when sharing mobile phones. In *Proc. of CHI*, Apr. 2009.

[27] S. Kurkovsky, T. Carpenter, and C. MacDonald. Experiments with simple iris recognition for mobile phones. In *Proc. of Information Technology: New Generations*, pages 1293 –1294, april 2010.

[28] J. Liu, L. Zhong, J. Wickramasuriya, and V. Vasudevan. User evaluation of lightweight user authentication with a single tri-axis accelerometer. In *Proc. of MobileHCI*, Sept. 2009.

[29] S. Mathur, R. Miller, A. Varshavsky, W. Trappe, and N. Mandayam. Proximate: proximity-based secure pairing using ambient wireless signals. In *Proc. of MobiSys*, June 2011.

[30] T. Matsumoto, H. Matsumoto, K. Yamada, and S. Hoshino. Impact of Artificial "Gummy" Fingers on Fingerprint Systems. In *Proc. of SPIE Vol. 4677*, 2002.

[31] R. Mayrhofer and H. Gellersen. Shake well before use: Intuitive and secure pairing of mobile devices. *IEEE TMC*, 8(6):792 –806, Jun. 2009.

[32] M. L. Mazurek, J. P. Arsenault, J. Bresee, N. Gupta, I. Ion, C. Johns, D. Lee, Y. Liang, J. Olsen, B. Salmon, R. Shay, K. Vaniea, L. Bauer, L. F. Cranor, G. R. Ganger, and M. K. Reiter. Access control for home data sharing: Attitudes, needs and practices. In *Proc. of CHI*, May 2010.

[33] Mobile biometry: European funded project (fp7-2007-ict-1). http://www.mobioproject.org/, 2011.

[34] Motorola atrix: Technical specifications. http://bit.ly/f7nfEX, 2011.

[35] Msp430 ultra-low power 16-bit microcontrollers. http://msp430.com.

[36] A. Nicholson, M. D. Corner, and B. D. Noble. Mobile Device Security using Transient Authentication. *IEEE TMC*, 5(11):1489–1502, November 2006.

[37] Ringbow. http://ringbow.com/ringbow/.

[38] V. Roth, P. Schmidt, and B. Güldenring. The IR ring: authenticating users' touches on a multi-touch display. In *Proc. of UIST*, Oct. 2010.

[39] M. Sato, I. Poupyrev, and C. Harrison. Touche: enhancing touch interaction on humans, screens, liquids, and everyday objects. In *Proc. of CHI*, pages 483–492, New York, NY, USA, 2012.

[40] Y. Shaked and A. Wool. Cracking the bluetooth pin. In *Proc. of MobiSys*, pages 39–50, 2005.

[41] E. Sofge. Can a smartphone make your car smarter? *MSN Autos*, 2010.

[42] Sony - mofiria finger-vien pattern recognition. http://bit.ly/cIHGuj, 2009.

[43] F. Stajano and R. J. Anderson. The resurrecting duckling: Security issues for ad-hoc wireless networks. In *Proc. of Security Protocols*, Apr. 2000.

[44] Technology review article: Pushing the limits of the touch screen, 2011.

[45] L. Thalheim and J. Krissler. Body check: Biometric access protection devices and their programs put to the test. *ct Magazine*, Nov. 2002.

[46] M. Weir, S. Aggarwal, M. Collins, and H. Stern. Testing metrics for password creation policies by attacking large sets of revealed passwords. In *Proc. of CCS*, Oct. 2010.

[47] W. Westerman and J. G. Elias. US 0232567 Patent - Capacitive Sensing Arrangment, 2006.

MuVi: A Multicast Video Delivery Scheme for 4G Cellular Networks

Jongwon Yoon*
University of Wisconsin
Madison, WI
yoonj@cs.wisc.edu

Honghai Zhang
NEC Labs America Inc.
Princeton, NJ
honghai@nec-labs.com

Suman Banerjee
University of Wisconsin
Madison, WI
suman@cs.wisc.edu

Sampath Rangarajan
NEC Labs America Inc.
Princeton, NJ
sampath@nec-labs.com

ABSTRACT

Although wireless broadband technologies have evolved significantly over the past decade, they are still insufficient to support the fast-growing mobile traffic, especially due to the increasing popularity of mobile video applications. Wireless multicast, aiming to exploit the wireless broadcast advantage, is a viable approach to bridge the gap between the limited wireless networking capacity and the ever-increasing mobile video traffic demand.

In this work, we propose MuVi, a Multicast Video delivery scheme in OFDMA-based 4G wireless networks, to optimize multicast video traffic. MuVi differentiates video frames based on their importance in reconstructing the video and incorporates an efficient radio resource allocation algorithm to optimize the overall video quality across all users in the multicast group. MuVi is a lightweight solution with most of the implementation in the gateway, slight modification in the base-station, and no modification at the clients. We implement MuVi on a WiMAX testbed and compare its performance to a Naive wireless multicast scheme that employs the most robust MCS (Modulation and Coding Scheme), and an Adaptive scheme that employs the highest MCS supportable by all clients. Experimental results show that MuVi improves the average video PSNR (Peak Signal-to-Noise Ratio) by up to 13 and 7 dB compared to the Naive and the Adaptive schemes, respectively. MuVi does not require modification to the video encoding scheme or the air interface. Thus it allows speedy deployment in existing systems.

Categories and Subject Descriptors

C.2.1 [**Network Architecture and Design**]: Wireless Communication

*Jongwon Yoon interned in the Mobile Communications and Networking department at NEC Laboratories America Inc, Princeton during this work.

MobiCom'12, August 22–26, 2012, Istanbul, Turkey.

General Terms

Algorithms, Design, Experimentation, Performance

Keywords

Video Multicast, Cellular Networks, OFDMA

1. INTRODUCTION

Mobile video streaming (e.g., Youtube [1], Hulu [2]) is one of the most popular applications in recent years and the amount of video traffic destined for mobile devices is increasing rapidly as the number of hand-held devices (e.g., smart phones, iPad) grows. According to the Cisco Visual Index, video accounted for about 40 percent of consumer Internet traffic in 2010 and will reach 62 percent by the end of 2015 [3]. Global mobile data traffic is expected to increase 26 times from 2010 to 2015. Moreover, high-definition (HD) video will surpass standard definition video by the end of 2012 and will become the dominant form of video traffic.

Current WiFi systems can not provide satisfactory quality of video streaming services due to the small coverage and relatively limited bandwidth as the number of mobile users increases. Even worse, WiFi networks are not robust enough to sustain user mobility. 3G mobile networks such as CDMA and UMTS can provide more robust wireless connections to mobile users. However, their bandwidth is not adequate to support applications with high bandwidth requirements (e.g., HD video). Limitations of current WiFi and 3G systems naturally turn our attentions to emerging 4G cellular networks.

4G cellular networks, such as WiMAX and LTE, have emerged as alternatives that can provide much higher bandwidth, spectrum efficiency, and extended coverage. 4G networks are more robust to user mobility compared to WiFi systems, so that they can provide seamless real-time video streaming services. More specifically, 4G technologies can provide peak data rate of 100 Mbps for high mobility applications and 1 Gbps for nomadic applications [12]. Despite the much higher bandwidth provided by 4G technologies, efficient resource utilization is still needed for meeting the high bandwidth and stringent delay requirements of video applications, because the wireless spectrum is shared by multiple users.

Multicast is an efficient way of leveraging the shared nature of wireless spectrum to deliver traffic to multiple clients simultaneously while minimizing the wireless resource usage. The opportunity for wireless multicast arises in many scenarios: **(i) Live sport programs.** In US, numerous sport programs are broadcast live on the Internet. In fact, ESPN provides applications to allow smart phones and tablets to watch its live programs. These programs can be provided to mobile clients using *multicast services*. **(ii) Breaking news.** When breaking news happens such as the Indonesia Tsunami, Japan earthquake, or 911 terrorist attack, many people click on the same video clips released by major news agencies. These requests that are made around the same time can be batched, stream-merged, and fulfilled using multicast transmissions. **(iii) EMBS and MBMS Services.** LTE and WiMAX define EMBS (Evolved Multicast Broadcast Service) and MBMS (Multimedia Broadcast Multicast Service) as separate services independent of regular unicast data services while sharing the same devices and spectrum with the unicast traffic. Both the services can be deployed readily to provide mobile TV programs or other commercial video programs, which will require multicast transmission of video content over the wireless systems.

In this work, we develop a Multicast Video delivery scheme (MuVi) in 4G OFDMA-based mobile broadband wireless networks. MuVi is motivated by several recent theoretical works [13, 16, 18] on multicasting scalable videos in OFDMA-based wireless networks. These works make an ideal assumption that the video has *multiple layers* with fixed sizes and do not consider the practical issue of *packetization of video frames with variable sizes*. Although scalable videos are interesting to study, most Internet videos today are not scalable and are encoded using traditional technologies such as MPEG4 or H.264. Therefore, the schemes in [13, 16, 18] cannot apply directly to most popular Internet videos (such as Youtube). Fortunately, we identify similar dependency relationship between P frames in a non-scalable video to that between different layers in a scalable video, and leverage a dynamic-programming technique similar to the one used for scalable video multicasting in [18], for *non-scalable* video wireless multicasting. By doing so, we also address the practical issue of variable video frame sizes naturally.

MuVi Overview: MuVi is designed as a proxy in the Access Service Network (ASN) gateway as depicted in Figure 1. There are four key design elements in MuVi. **1)** MuVi collects the feedback of wireless channel quality information of all clients in a multicast group through the base-station and obtains the supportable MCS for each client. **2)** MuVi prioritizes video frames by setting a utility value for each frame based on the frame type. The utility may also depend on the user profile. Thus, MuVi supports user differentiation. **3)** MuVi performs efficient resource allocation to maximize the system utility considering both the available radio resources and the video packets to be transmitted. **4)** MuVi assigns Modulation and Coding Schemes (MCS) for the video packets and hands them over to the base-station, which sends the video packets over the air using the assigned MCSs.

Different from several recent works on video streaming [8, 14] which provide graceful video delivery, MuVi does NOT modify the video encoding nor the wireless transmission schemes. It performs resource allocation in operational wireless systems and video frame prioritization with existing video encoding technologies to optimize the resource utilization and video delivery. MuVi is also different from several WiFi-specific video optimizations which utilize the notion of differential value of data packets, e.g., Medusa [21], in that 1) MuVi is designed for OFDMA-based systems, and 2) MuVi does not require client side modifications. Since it does not modify the air interface or the clients, MuVi is standards compatible and allows immediate deployment in commercial wireless systems (such as WiMAX).

We implement MuVi on a PicoChip [11] WiMAX testbed including a video server, an ASN gateway, a PicoChip basestation, and several mobile clients. We evaluate the performance of MuVi and compare it with a Naive multicast scheme that uses the most robust MCS and the Adaptive scheme proposed in DirCast [9], which picks the highest MCS that is supportable by all clients. We have the following findings.

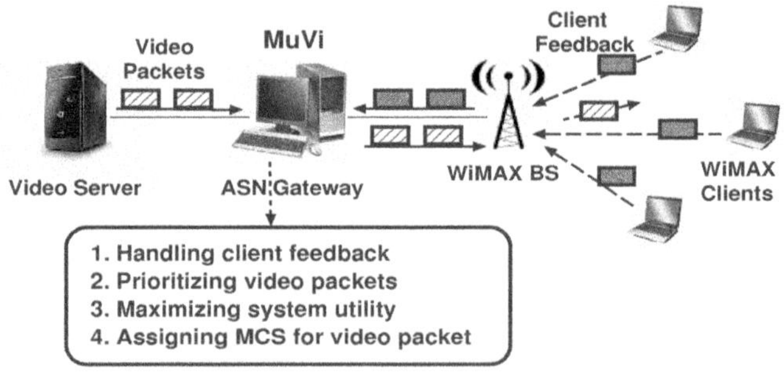

Figure 1: MuVi performs four operations while receiving video packets from the media server, and then it delivers video packets to the WiMAX base-station.

- MuVi improves the average video quality by up to 13 dB and 7 dB in terms of PSNR compared to the Naive scheme and the Adaptive scheme, respectively.
- MuVi reduces the inter-packet arrival delay by up to 80% and 55%, compared to the Naive and the Adaptive schemes, respectively. This indicates significantly fewer glitches and video stalls for MuVi.
- MuVi is a gateway solution and does not require any modification on the client side. It can be easily incorporated in cellular networks with only minimum modifications on the base-stations.
- Although it is implemented in WiMAX, MuVi is also applicable to other OFDMA-based systems, e.g., LTE and LTE-Advanced systems.

The rest of the paper is organized as follows. Section 2 provides background on WiMAX and OFDMA systems. We describe basic operations of MuVi and functional building blocks in Section 3. Section 4 describes the prototype implementation and details of our WiMAX testbed. In Section 5, we evaluate the performance of MuVi using our testbed and present detailed experimental results. We discuss limitations and related work in Section 6 and 7, respectively. Section 8 concludes the paper.

2. BACKGROUND

WiMAX Preliminaries: While our work applies to Orthogonal Frequency Division Multiple Access (OFDMA)-based networks in general, we implement MuVi and conduct experiments on a WiMAX (802.16e [5]) testbed due to its availability[1]. Here, we give a brief overview of OFDMA and WiMAX networks (The detailed 802.16e specifications can be found in [5]). A WiMAX frame is a two-dimensional template across the time and the frequency domains. A slot, which is the minimum allocation unit in the frame, is made up of one sub-channel in the frequency domain and one, two, or three symbols in the time domain. Data to multiple clients are scheduled as rectangular bursts of slots in a frame. In TDD (time division duplexing) mode, a WiMAX frame is divided into two sub-frames: a downlink sub-frame and an uplink sub-frame, separated by Transmit-To-Receive and Receive-To-Transmit Gaps (TTG and TRG, respectively). An example of a WiMAX TDD frame is shown in Figure 2. Under the most commonly deployed

[1] LTE platform is not available yet.

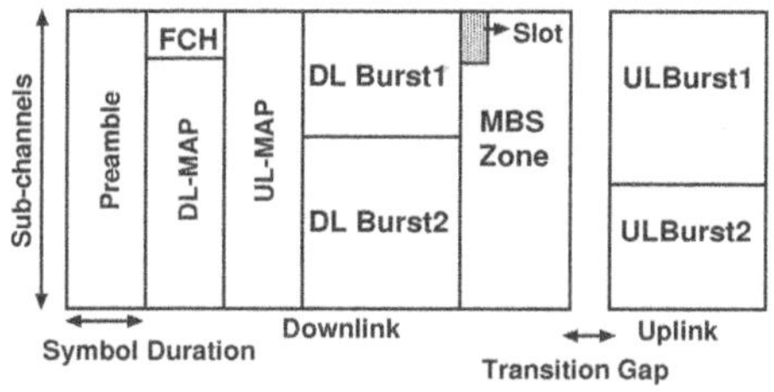

Figure 2: WiMAX frame structure.

Index	Modulation and Coding Rate
0	BPSK
1	QPSK(3/4)
2, 3	16 QAM(1/2, 3/4)
4, 5, 6, 7	64 QAM(1/2, 2/3, 3/4, 5/6)

Table 1: MCS indices available for data modulation and coding rates.

profile with 5 MHz and 10 MHz bandwidth, a WiMAX frame has a frame duration of 5 milliseconds.

A frame consists of a preamble, control data and payload data. The preamble is transmitted with power boost and allows mobile clients to lock on to the base-station. The control signaling consists of a Frame Control Header (FCH) and a MAP. The FCH contains the system control information and the information about the MAP. The DL-MAP indicates where each burst is placed in the frame, which client it is intended for, and what modulation and coding scheme (MCS) decodes it. Table 1 shows the different modulation levels employed in a WiMAX base-station. Similarly the UL-MAP indicates where the client should place its data on the uplink frame. The uplink sub-frame has dedicated sub-channels for HARQ, which are used by clients to explicitly acknowledge (ACK/NACK) the reception of each data burst. Clients also use the uplink sub-frame to report the instantaneous channel state information (CSI) to the base-station. Base-stations use a dedicated data burst zone, called MBS zone (in Figure 2), for multicast and broadcast services. While data bursts can be modulated using any MCS, the control parts (FCH and DL/UL MAPs) are always transmitted using QPSK and typically with multiple repetitions. The preamble is transmitted with a higher power compared to the other parts of the frame.

MPEG4: We describe the characteristics of MPEG4/H.264 video frames, since we use MPEG4 as example video encoding schemes. In MPEG4 or H.264, a video sequence is partitioned into Group of Pictures (GOP). Figure 3 shows an example of frame sequence in a GOP. Each GOP consists of a certain number of I, P, and B video frames, which are then further divided into packets typically with fixed length except for the last packet of each video frame. I frames are intra-coded pictures and it can be decoded independently. P frames are forward predicted pictures and require their preceding I or P frames to be decoded. B frames are bidirectionally predicted pictures and require both preceding and succeeding I or P frames for decoding. The arrows in Figure 3 represent the dependencies between frames. A typical GOP consists of 30 frames of repeated sequences (e.g., IBBPBBPBB...PBB).

3. DESIGN AND OPERATION OF MuVi

Our proposed system, MuVi, comprises a media server, media-

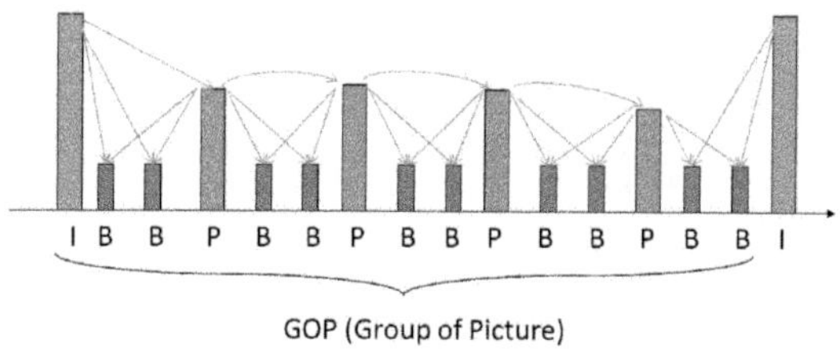

Figure 3: An example video frame sequence in a GOP. An arrow indicates that the frame at the head of the arrow depends on that at the tail of the arrow.

aware gateway, base-station, and multiple clients in the multicast session as depicted in Figure 1. The basic operations of MuVi are as follows:

1) The WiMAX clients send the channel state information (CSI) to the base-station periodically. Providing CSI is a standard mechanism in 802.16e (and all mobile broadband systems). Upon receiving the channel feedback, the base-station computes appropriate moving average for each client, aggregates the averaged feedback, and forwards them to the ASN gateway. The ASN gateway then determines the supportable MCS for each client based on their average channel conditions, along with an MCS table (details in subsection 3.1).
2) The media server sends the video packets to the ASN gateway, which performs packet scheduling and possibly drops some video packets based on the available radio resources and the frame types. The packet scheduling includes packet re-ordering and determining the Modulation and Coding Scheme (MCS), i.e., the PHY rate, employed at the base-station for each video packet (details in subsection 3.3).
3) After receiving the video packets, the base-station applies the MCS selected by the ASN gateway and transmits them over the air using multicast (details in subsection 3.4).
4) When the clients receive the packets, they perform packet re-ordering and then pass the received packets to the video player (e.g., VLC [7]). Packet re-ordering is typically implemented in upper-layer protocols such as RTP (Realtime Transport Protocol).

To execute the above operations, we modify the ASN gateway and the base-station. Our major modifications lie at the ASN gateway, which consist of 1) collecting the supportable MCS for each client, 2) classifying packet types (i.e., video frame types), and 3) determining the PHY transmission rate for each video packet.

The ASN gateway can perform deep-packet-inspection to find the type of video frames, but this may cause extra overhead at the ASN gateway. Hence, we slightly modify the video player application (VLC player) running on the media server to ease the job at the ASN gateway. The customized VLC player inserts the video frame type information in the IP header (e.g., DiffServ field) before sending video packets to the ASN gateway. The ASN gateway simply looks up the IP header to categorize received packets.

The base-station needs to make small modifications to 1) forward the client channel feedback to the ASN gateway, and 2) transmit the packets using the MCS instructed by the ASN gateway. We believe these are within the scope of what BS vendors are willing to do. The clients are not required to change given that re-ordering can be performed by upper-layer protocols such as RTP, which is typically employed in real-time multicast streaming.

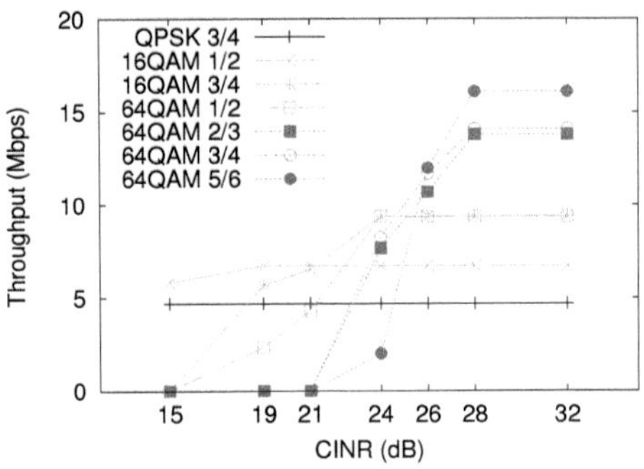

Figure 4: MuVi can find the supportable MCS for a client based on its CINR feedback.

In the following subsections, we describe the details of each design element.

3.1 Handling Client Feedback

The wireless channel suffers from packet losses and errors caused by interference and noise. It is hard to determine the target MCS (i.e., PHY transmission rate) in a wireless multicast system because of the different channel conditions amongst multiple users in the multicast group. Client channel condition is an important factor to determine the target PHY rate to multicast while avoiding resource under-utilization. To guarantee data delivery over wireless channels, traditional wireless multicast systems use the most robust MCS (i.e., the lowest PHY transmission rate). However, this approach leads to under-utilization of available radio resources when most multicast users can support higher MCS than the lowest one.

A carefully selected MCS can improve the throughput and the video quality of a client. The most appropriate MCS for each client depends on its channel conditions. Using a lower MCS leads to resource under utilization and using a higher MCS renders high packet loss rate. MuVi leverages the information of the supportable MCS for each client to make efficient use of radio resources when multicasting the video traffic.

To find the relationship between the client channel condition and the supportable MCS, we measure the client throughput from various locations where the channel conditions vary. We deploy the clients on several locations where clients experience different levels of channel conditions in terms of Carrier to Interference plus Noise Ratio (CINR). For each location, base-station sends UDP traffic using *iperf* to the client while trying out all possible MCSs and the client records the obtained UDP throughput with respect to each MCS used. We plot the measured throughput for all locations where each location is represented as a point in Figure 4.

Using this figure, we can obtain a MCS-CINR table, as shown in Table 2. By looking up the table, we can easily figure out the highest supportable MCS for a client based on its reported channel feedback. For example, a client can successfully decode packets transmitted with MCS 4 (64-QAM with 1/2 coding) or lower when its observed CINR is 24 dB. We comment that constructing the MCS-CINR table only needs to be done once.

For clients, providing channel feedback (CINR report) to base-station is part of the IEEE 802.16e (WiMAX) standard [5]. We modify the MAC code of the base-stations to compute the moving average of the CINR values for each client, aggregate the averaged CINR values of all clients, and forward them to the ASN gateway periodically. Once the ASN gateway receives the client feedback from the base-station, it can identify the highest MCS that can be decoded by each client from Table 2.

MCS index	CINR range (dB)
7	$(28, \infty)$
6	(26, 28]
5	(24, 26]
3, 4	(20, 24]
2	(15, 20]
1	$(-\infty, 15]$

Table 2: CINR range and the corresponding MCS.

3.2 Packet Values

The video sequence consists of different frame types (i.e., I, P, and B frames) for MPEG4 or H.264 encoded video. Each video frame contains different video/audio information. For example, P and B frames carry only difference information so they depend on some other frames (i.e., I and P frames) to be decoded successfully.

MuVi prioritizes the packets based on their dependency and assign the priorities as a function of number of packets depending on them. The more packets depend on it, the higher priority is assigned to a packet. Typically, I frames have the highest priority and B frames have the lowest priority. P frames may have different priorities depending on their positions in a GOP. For example, a P frame in the earlier part of a GOP has more number of dependent frames compared to that in the later part of a GOP, hence the former has higher priority than the latter.

In addition, users may have different priorities. For example, a high-profile user may pay higher subscription fee and expect high video quality. So we allow a video frame (i.e., the packets) to have different priority values for different clients.

3.3 Utility Maximization

The main media optimization engine lies in the ASN gateway. It collects the channel feedback for each client in the multicast group from the base-station and determines the resource allocation for the video frames in order to maximize the total system utility.

The problem setup is as follows. There are Q available slots in a WiMAX frame for the video multicast session, where a slot is a two-dimensional minimum allocation unit in the time and frequency domains in an OFDMA frame. The OFDMA frame duration is τ ms ($\tau = 5$ in most WiMAX systems). There are M Modulation and Coding Schemes (MCS) representing different transmission rates. For MCS m, one OFDMA slot can deliver R_m bytes and the minimum required CINR for decoding MCS m is $\bar{\gamma}_m$, where $1 \leq m \leq M$. There are K clients in the multicast session and their CINR values are $\gamma_1, \gamma_2, \cdots, \gamma_K$.

We consider the transmission of a Group of Picture (GOP) with J video frames. A GOP includes an I frame, a sequence of P frames, and a group of B frames as depicted in Figure 3. The jth video frame has length L_j. Receiving frame j at client k obtains a utility u_j^k given that all frames it depends on are also received by client k. By including a superscript k in the utility function, we allow different priorities among different users. We assume that the inter-frame time is Δ second and thus the total number of slots available for transmitting the GOP is $T = \lfloor \Delta JQ/\tau \rfloor$. The objective is to maximize the total utility received by all clients subject to the total slot constraint.

The dependency relationship between I/P frames and B frames is different. All frames in the GOP need to refer I frame when decoding. P frames depend on their preceding P frames and are needed by both intermediate B frames and succeeding P frame. B frames are not referred by other frames. Considering different frame dependencies, we first schedule groups of I/P frames and B frames separately and then optimize the overall utility across the two groups.

3.3.1 I/P Frame Scheduling

We first consider the problem of scheduling only I and P frames in a GOP. For notational convenience, we assign a special sequence number P_0 to the only I frame in the GOP and hereafter, we treat I frame as a special P frame P_0. So we have a sequence of P frames P_j, where $j = 0, 1, \cdots, G$, such that frame P_{j+1} depends on P_j $(0 \leq j \leq G-1)$, where G is the number of P frames. Let the total number of available slots for P frames be $t \leq T$ and the objective is to maximize the total utility subject to the total slot constraint t for all P frames.

To ensure minimum video quality at every client, we require that the first $j_0 \geq 0$ P frames be received by all clients, and set aside t_1 slots for transmitting these P frames using the highest MCS that can be decoded by all clients. Thus the available number of slots for the remaining P frames is $t' = t - t_1$. Without loss of generality, we assume $j_0 = 0$ (i.e., no P frame is required by all clients) in the following.

Let binary variable x_{jm} indicate that frame P_j is transmitted with MCS m and z_j^k indicate that frame j is valid for client k, which means that all frames on which frame j depends, including j, can be decoded by client k, so

$$z_j^k = \begin{cases} 1, & \text{if for all } 0 \leq l \leq j, \text{ there exists } m \\ & \text{such that } x_{l,m} = 1 \text{ and } \bar{\gamma}_m \leq \gamma_k, \\ 0 & \text{otherwise.} \end{cases}$$

Now the problem of utility maximization can be written as

$$\begin{aligned} \max \quad & \sum_{j=0}^{G} \sum_{k=1}^{K} z_j^k u_j^k \\ s.t. \quad & \sum_{m=1}^{M} x_{jm} \leq 1, \\ & \sum_{m} \left\lceil \sum_{j} \frac{x_{jm} L_j}{R_m} \right\rceil \leq t, \end{aligned} \tag{1}$$

where in the last equation, we assume that frames transmitted using the same MCS can be bundled together for the allocation of radio resources (i.e., slots). We slightly abuse the notations by using u_j^k to denote the utility of frame P_j for client k.

Interestingly, problem (1) is similar to the problem of maximizing sum of video quality when multicasting SVC-encoded videos by viewing the jth P frame as the jth layer in an SVC video. Indeed, the dependency relationship between P frames in non-scalable videos are equivalent to that between different layers in SVC videos. Therefore, we can leverage a dynamic programming algorithm similar to the one used in SVC video multicasting [18] to solve problem (1).

Define $U_P(j, m, t)$ as the optimal utility with the P frames $P_l, l = 0, \cdots, j$ with MCS up to m and at most t slots. Let $\tau_{j_1, j_2, m} = \lceil \sum_{l=j_1}^{j_2} L_l / R_m \rceil$ be the number of slots required to deliver frames $P_l, l = j_1, \cdots, j_2$ using MCS m. To compute the optimal utility $U_P(j, m, t)$, only the last few frames ending at P_j may choose MCS m. Assuming that frames $P_l, l = i+1, \cdots, j$, are transmitted with MCS m (note that if $i = j$, no frame is transmitted with MCS m), the utility $U_P(j, m, t)$ is then the summation of the optimal utility of the first i frames using MCS up to $m-1$ with $t - \tau_{i+1,j,m}$ slots and the utility obtained by transmitting frames $i+1$ to j using MCS m. Maximizing over all possible i, we obtain the recursive equation for $U_P(j, m, t)$ as follows.

$$U_P(j, m, t) = \max_{0 \leq i \leq j} \left[U_P(i, m-1, t - \tau_{i+1,j,m}) + \sum_{l=i+1}^{j} \sum_{k \in S_m} u_l^k \right] \tag{2}$$

$$q(j, m, t) = \text{argmax}_i \left[U_P(i, m-1, t - \tau_{i+1,j,m}) + \sum_{l=i+1}^{j} \sum_{k \in S_m} u_l^k \right] \tag{3}$$

where S_m is the set of users who can decode MCS m. In Eq. (3), $q(j, m, t)$ keeps track of the best parameter i that maximizes Eq. (2) and it is used for finding the optimal resource allocation later.

The initial conditions for $U_P(j, m, t)$ are

$$\begin{aligned} U_P(j, m, t) &= -\infty, \text{ if } t < 0 \\ U_P(j, 0, t) &= -\infty, \text{ if } j \geq 0, t \geq 0 \\ U_P(-1, m, t) &= 0, \text{ if } m \geq 0, t \geq 0 \end{aligned} \tag{4}$$

The first two equations state that $t < 0$, or $m = 0$ and $j \geq 0$ is not a valid choice for the utility function $U_P(j, m, t)$. We note that the valid MCS choices are 1 to M and $m = 0$ is a dummy MCS used for initialization. In the last equation above, $j = -1$ with MCS $m \geq 0$ indicates that all frames ($j \geq 0$) have been considered and the dummy frame $j = -1$ has zero utility.

From Eq. (2), we can see that $U_P(j, m+1, t) \geq U_P(j, m, t)$. Therefore, the optimal utility is always achieved at $m = M$. For each t, we can then compute the optimal utility U_P^* for each available time slot t:

$$\begin{aligned} U_P^*(t) &= \max_{j \geq 0} U_P(j, M, t) \\ j^*(t) &= \text{argmax}_{j \geq 0} U_P(j, M, t) \\ m^*(t) &= \min\{m : U_P(j^*(t), m, t) == U_P^*(t)\} \end{aligned} \tag{5}$$

where $j^*(t)$ achieves the optimal utility, indicating that P frames $j > j^*(t)$ are dropped. The dynamic-programming procedure for I/P frame scheduling is illustrated in Algorithm 1.

Algorithm 1 I/P frame scheduling

1: Use Eq. (4) to compute the utility $U_P(j, m, t)$ at the boundary.
2: **for** all j, m, t **do**
3: Compute $U_P(j, m, t)$ iteratively using Eq. (2).
4: **end for**
5: Find the optimal utility $U_P^*(t)$ for all $0 \leq t \leq T$ using Eq. (5).

3.3.2 B Frame Scheduling

Given that t out of total T slots are allocated for transmitting I and P frames, $T - t$ slots are left for B frames. Since some P frames may be dropped after the P frame scheduling, transmitting the B frames that depend on those dropped P frames does not produce any benefit. Hence, we only consider the set of B frames that are still useful. Let

$$\mathcal{B}(t) = \{b : \text{ B-Frame } b \text{ does not depend on P frame } j > j^*(t)\}.$$

As B frames are less important than P frames, we use higher (or equal) MCS for B frames than the last transmitted P frame $j^*(t)$ for efficient use of given resources (lower MCS requires more number of slots for transmission). Assume that B frames in $\mathcal{B}(t)$ are naturally ordered by their decoding time (or display time).

The problem of B frame scheduling can be formulated as, selecting a subset of B frames and an appropriate MCS $m \geq m^*(t)$ for each B frame, to maximize the total utility, such that the total number of slots does not exceed $T - t$. The problem can easily be seen as a multiple-choice knapsack problem and is NP-hard. Nevertheless, most B frames have relatively small sizes and small utility compared to I and P frames.

We use the following algorithm to obtain a sub-optimal solution by imposing the requirement that all B frames use the same MCS. To be specific, for each MCS $m \geq m^*(t)$, we find maximum number b_m of B frames that can be transmitted using MCS m under the slot constraint. Finally, we pick the MCS that maximizes the total utility. B frame scheduling algorithm is described in Algorithm 2.

Algorithm 2 B frame scheduling

```
for all m*(t) ≤ m ≤ M do
    Find the maximum b_m such that the first b_m B frames in
    B(t) can be transmitted with MCS m in T − t slots.
    The resulting utility is U_B(m, T − t) = Σ_{b=1}^{b_m} Σ_{k∈S_m} u_b^k.
end for
Find the optimal m_0 = argmax_m U_B(m, T − t) and obtain the
utility U*_B(T − t) = U_B(m_0, T − t).
```

3.3.3 *Joint I/P/B Frame Scheduling*

To perform the joint scheduling, we find the optimal resource allocation between I/P frames and B frames. Let

$$\begin{aligned} U^* &= \max_t U_P^*(t) + U_B^*(T-t) \\ t^* &= \text{argmax}_t U_P^*(t) + U_B^*(T-t) \end{aligned} \tag{6}$$

be the optimal total utility and the optimal number of slots allocated to I/P frames, respectively. $T - t^*$ is the optimal number of slots allocated to B frames.

3.4 MCS assignment

Finally, MuVi determines the number of I/P frames and B frames transmitted and the MCS for each transmitted frame based on the result of utility optimization obtained in the previous subsection. The following steps are executed in order. Note that I frame is viewed as the special P frame with index 0.

1. t^* and $T - t^*$ (obtained from Eq. (6)) are the number of slots allocated to P frames and B frames, respectively.
2. The first $j^*(t^*)$ P frames (from Eq. (5)) are transmitted and the rest P frames are discarded.
3. $t = t^*, j = j^*(t^*), m = m^*(t^*), i = q(j, m, t)$.
4. P frames $P_l, l = i + 1, \cdots j$ are transmitted with MCS m (if $i == j$, no frames are transmitted with MCS m).
5. If $i < 0$, go to Step 6. Otherwise, $t = t - \tau_{i+1,j,m}$, $j = i, m = m - 1, i = q(j, m, t)$, go to step 4.
6. $m_0 = \text{argmax}_m U_B(m, T - t^*)$. The first b_{m_0} B frames are transmitted with MCS m_0 and the rest B frames are dropped.

Once the MCS for each video frame is determined, the packets belonging to each frame are marked with the assigned MCS index in the DiffServ field of the IP header.

4. WIMAX TESTBED AND PROTOTYPE IMPLEMENTATION

In this section, we describe our WiMAX network testbed and the prototype implementation on it.

Testbed: Our WiMAX testbed consists of four components: a WiMAX femto base-station, an Access Service Network (ASN) gateway, a video server, and several WiMAX clients as depicted

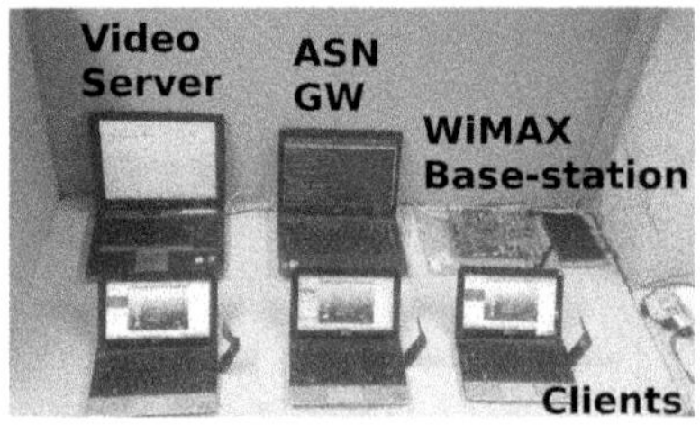

Figure 5: WiMAX testbed.

in Figure 5. The base-station is a PicoChip [11] WiMAX platform based on IEEE 802.16e standard [5]. The PicoChip base-station is tuned to operate in a 10 MHz bandwidth with the center carrier frequency of 2.59 GHz, for which we have obtained an experimental license to transmit WiMAX signals over the air. Both the ASN gateway and the video server run on typical Linux machines with a 2 GHz processor and 1 GB memory. The WiMAX clients are Windows laptops with commercial USB dongle WiMAX cards [10] or Beceem PCMCIA [4] interfaces.

The clients can be associated with the WiMAX base-station through the ASN gateway. The ASN gateway controls and maintains both uplink and downlink connections between the WiMAX base-station and the clients through configuring service flows which are unidirectional flows of data traffic. All uplink and downlink traffic between the base-station and the video server are tunneled through the ASN gateway.

The base-station manages the scheduling for both downlink and uplink traffic. Since multiple users share the OFDMA wireless resources (i.e., the WiMAX frames), the base-station incorporates a scheduler for efficient resource management. The downlink and uplink scheduler assigns a certain number of slots to each flow and informs the clients about the resource allocations through DL/UL MAPs.

Prototype Implementation: Our prototype implementation includes modifications at the video server, the ASN gateway, and the WiMAX base-station. The main components of MuVi (handling client feedback, optimizing resource allocation, and frame re-ordering) are implemented on the ASN gateway using the Click Modular Router [19] as a user-level module to handle all video frames to the base-station. We significantly extend and modify the click module (about 2000 lines of C++ code) to realize MuVi's building blocks.

The modified version of VLC media player [7] runs on the video server to send media traffic to the group of multicast users. The VLC module first inserts the frame information in each video packet's IP header (i.e., DiffServ field) and then sends them to the ASN gateway. This marking process helps the ASN gateway classify the video packets easily without deep packet inspection. After classifying the packet types (i.e., video frame types), the ASN gateway puts them in different queues based on their types. Then, ASN gateway schedules video frames to the base-station after applying the packet re-ordering and rate selection algorithms as described in Section 3. In broadband cellular networks, gateways are typically sophisticated servers managing hundreds of base-stations. Since all downlink and uplink traffic go through the gateway, assigning a MCS for all packets at the gateway causes very little overhead, therefore, the overhead of MuVi at the gateway is minimal (evaluation in subsection 5.4). As a result of the scheduling algorithm, each packet is marked with the assigned MCS index in the DiffServ field in the IP header and is delivered to the base-station.

Now that the scheduling process is offloaded from the base-station to the ASN gateway, the base-station simply packs incoming video

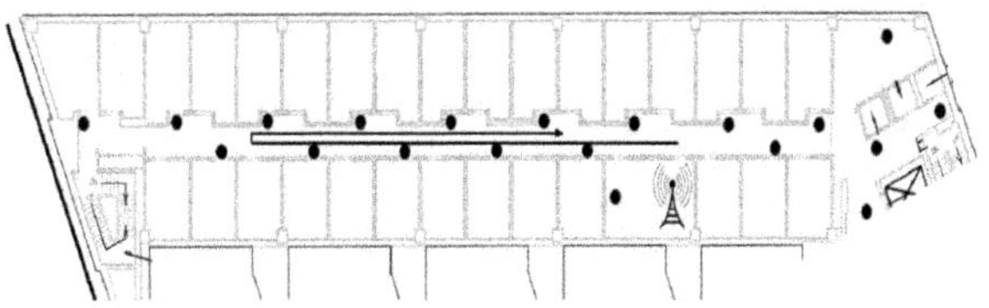

Figure 6: Dots represent the client locations for all experiments. The line represents the path of a mobile client under the mobility experiment.

frames into the WiMAX data bursts and transmits them to the clients in a multicast group. The reference design of PicoChip platform does not involve sophisticated scheduling routines and provides just a working link between the base-station and the clients. We modify the base-station in two aspects to incorporate MuVi. First, it reads the DiffServ Field of the IP header to extract the MCS information determined by the scheduling process in the ASN gateway and transmits the packets using the specified MCS. Second, we modify the base-station MAC code to provide the client feedback to the ASN gateway once every 100 milliseconds. Client channel conditions are periodically reported from the clients to the base-station as part of the standard operation, so the base-station just aggregates the feedback and forwards to the ASN gateway. The size of each channel feedback is relatively small (less than 40 Bytes), so this routine causes very little overhead at the base-station.

5. PROTOTYPE EVALUATION

We evaluate the efficacy of MuVi on our WiMAX testbed by comparing it with existing wireless multicast schemes. We first describe our experiment setup, two reference schemes, evaluation metrics, and test video we used for experiments.

Experiment Setup: Our WiMAX testbed (Figure 5) is deployed in a typical indoor environment as depicted in Figure 6. To generate various topologies, we deploy clients in multiple locations, where the channel quality varies in terms of Carrier to Interference plus Noise Ratio (CINR). Each client is exposed to a different level of channel condition. All clients have the same priority and the utility of each frame is set to the number of frames depending on it (including itself).

For the confidence results, we repeat each set of experiments more than 5 times and present the averaged results with a 95% confidence interval, except for the microscopic and mobility results, where we present the results of a single run with finer granularity. We maintain the same client topologies while running different video multicast schemes to ensure fair comparisons.

Reference Schemes: We compare the performance of MuVi with that of Adaptive and Naive approaches. The Naive scheme is a traditional WiFi/WiMAX multicast approach that uses the most robust (lowest) MCS for delivering data to the group of clients. This approach does not consider the client channel conditions and only guarantees successful deliveries without considering the real-time requirement.

For the Adaptive scheme, we only adapt transmission bit-rates for video frames like DirCast [9]. Based on the client channel conditions, Adaptive scheme uses the highest MCS which can be supported by all clients in the multicast group. The MCS used in Adaptive scheme to deliver the multicast data reflects instantaneous client channel conditions.

While Medusa [21] also determines the MCS based on video frame types, it requires the clients send the reception reports (i.e., ACK/NACKs) to the Medusa proxy and requires modifications at the clients. Thus it is not amenable to implementation in cellular networks, so we do not compare MuVi against it.

Video quality (MOS)	PSNR range
Excellent	> 37
Good	31-37
Fair	25-31
Poor	20-25
Bad	< 20

Table 3: The relationship between video quality (MOS) and the PSNR range.

Evaluation Metrics:

- *PSNR:* PSNR (Peak Signal-to-Noise Ratio) is a standard metric of video quality and is a function of the mean square error between the original and the received video frames. If a video frame is dropped or past the deadline, it is considered lost and is concealed by copying from the last received frame before it. The relationship between the user perception expressed in Mean Opinion Score (MOS) and the PSNR values [20] is summarized in Table 3.
- *Inter-packet delay:* We measure inter-packet delay of received packets to quantify the jitter of delivered video stream. To successfully display a streaming video on the client, all video packets belonging to the same video frame need to be received within a given deadline. High jitter values between packets cause bad visual quality (e.g., glitches and stopped video frames when displaying).
- *Ratio of packets past the deadline:* PSNR measures video quality based on the received video frames regardless of the deadline of video frames. However, the video frames past the deadline can not be used to display real-time streaming media. We measure the ratio of packets which miss the deadline to reflect the real-time streaming video quality.
- *MCS:* MCS selection for each video frame is important to satisfy the guaranteed delivery to all clients that are exposed under different channel conditions. MCS selection is even more important to prevent the under-utilization of given resources (i.e., number of slots in WiMAX frame). We measure the average MCS used for delivering video frames which reflects the efficacy of WiMAX frame resources usage.

Test Video: We use MPEG4-encoded [6] video streaming for system evaluations. In the experiments, the video is encoded at 2.5 Mbps and is multicasted to the clients. Although we consider non-scalable video sequences in the experiments, our scheme can also be applied to scalable videos by assigning packets with different utility based on the layers they belong to.

5.1 Performance under Various Resource Constraints

We evaluate the performance of MuVi and compare it to Adaptive and Naive schemes in the presence of cross-traffic (i.e., background traffic) while varying the number of available resources (i.e., slots) in a WiMAX frame for multicast transmissions. We use 2.5 Mbps video stream for all experiments.

The maximum data throughput depends on both the number of slots allocated in each WiMAX frame and the MCS level chosen in the WiMAX system. For example, allocating 60 slots per frame with the highest MCS (64-QAM with 5/6 coding) can yield 2.6 Mbps throughput, theoretically. In practical experiments, MCS

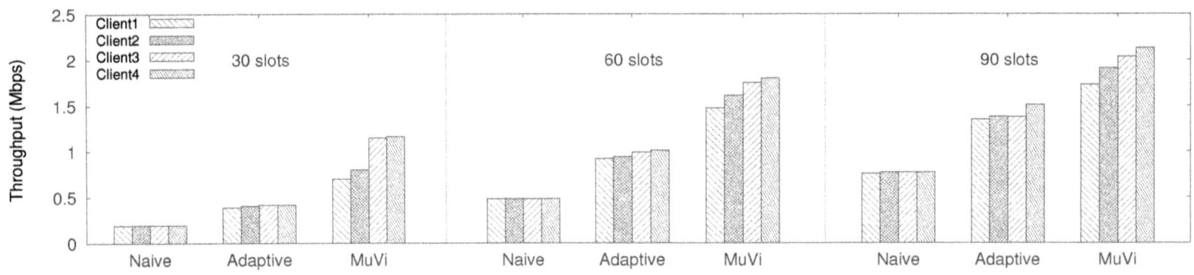

Figure 7: Measured throughput for all clients under different resource constraints. MuVi provides at most x4.9 and x2.35 throughput improvement comparing to Naive and Adaptive schemes, respectively.

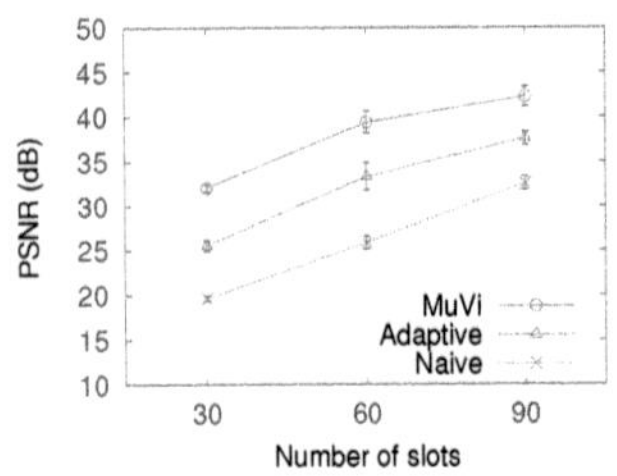

(a) The average PSNR value and confidence interval

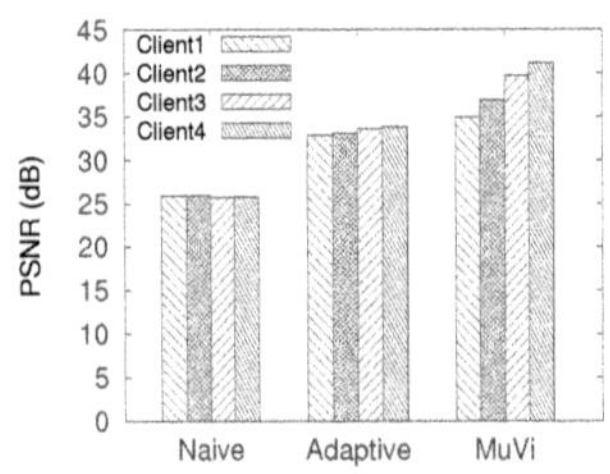

(b) PSNR value for each client under multicast to four clients.

Figure 8: (a) MuVi provides best video quality regardless of available resources. (b) MuVi provides differentiated service to the clients depends on their channel conditions.

used for multicast may be less than the highest one considering various channel conditions of the users in the multicast group. Hence, 60 slots may not be sufficient to successfully deliver the video encoded with 2.5 Mbps. MCS selection for transmitting multicast data depends on the clients' channel conditions, the available radio resources, and the optimization algorithm as described in Section 3. We vary the number of slots in a WiMAX frame to generate different resource constraints introduced by the background traffic. In the experiments, 30 and 60 slots per frame represent very tight resource constraints and 90 slots per frame represent just enough resources for delivering the video at 2.5 Mbps.

5.1.1 PSNR and Throughput

The average PSNR values and 95% confidence intervals for each multicast scheme obtained from 10 experiments are presented in Figure 8(a). For each multicast scheme, we aggregate the PSNR values from all clients in the multicast group and present the average of them with confidence interval from multiple runs. Figure 8(a) shows MuVi significantly outperforms the other two schemes under all resource constraints we evaluated. Specifically, the average PSNR value is excellent for MuVi, good for Adaptive, and only fair for Naive when 90 slots per frame for multicasting is used. Even when the radio resources are severely insufficient (i.e., 30 slots per frame), MuVi still provides good video quality while the other two schemes suffer. In average, MuVi improves the video quality by up to 13 dB and 7 dB, compared to the Naive scheme and the Adaptive scheme, respectively.

Similar subjective results are observed when we watch streaming video on the client side. We notice more frequent glitches and stalls with Adaptive and Naive schemes, while we see much smoother streaming with MuVi. The reason that MuVi outperforms the other two schemes is mainly due to the packet scheduling (e.g., resource optimization) and MCS assignment discussed in Section 3.3. Using the most robust MCS can guarantee successful deliveries, but it leads to very low efficacy of resource utilization and eventually lots of frames cannot be delivered before their display or decoding deadlines. Instead, MuVi selectively drops some less important video frames, and judiciously assigns the MCS of packets based on their priorities in order to achieve the maximum overall system utility while guaranteeing that the deadlines of all packets are met.

In the experiments, we deploy the clients in different locations to create distinct channel conditions. Specifically, clients 1, 2, 3, and 4 are placed in the CINR range of 19-21 dB, 22-24 dB, 25-27 dB, and 28-32 dB, respectively. Figure 8(b) shows the PSNR values of each client. The Naive scheme uses the most robust MCS (i.e., MCS index 0) for delivering all video frames, hence, the PSNR values for different clients are almost equal. Similarly, the Adaptive scheme uses a single MCS for all video frames limited by the client who has the poorest channel conditions. Therefore, all clients also experience similar video quality. On the other hand, MuVi uses different MCS levels for different video frames, and hence the client's PSNR values highly depend on their channel conditions. Higher video quality can be expected with better channel condition due to different MCS employed on different video frames. The client 4, which has the best channel condition, experiences the highest video quality amongst all clients.

During the same experiments, we measure the throughput for each client and present them in Figure 7 with respect to the different resource constraints. The throughput pattern is very similar to that of PSNR (Figure 8(b)), since throughput is highly related to the video packet receptions. The more packets a client receives, the higher throughput and PSNR value are observed. The throughput results for both Naive and Adaptive schemes are linearly correlated to the resource constraints. By contrast, MuVi provides differentiated service to the clients and guaranties relatively high throughput even under very limited resource conditions (i.e., 30 slots).

5.1.2 MCS Selection

To understand the MCS usage, we present the CDF of selected MCSs for all video frames obtained from multiple runs of various topologies in Figure 9(a), where the curve MuVi30 (60, 90) represents the experiment using MuVi with 30 (60, 90) slots per WiMAX frame. MuVi uses higher MCSs than the other schemes and provides higher PSNR values as we see in Figure 8(a). The Adaptive scheme transmits 50% of frames using MCS 3 or higher while MuVi transmits 50% of frames using MCS 6 or 7. Recall that, both

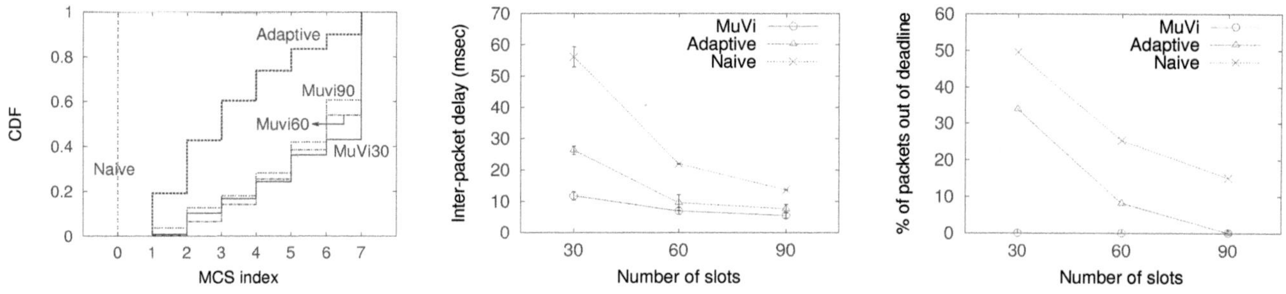

(a) CDF of selected MCS for all frames (b) Average inter-packet delay (c) Ratio of packets missing a deadline

Figure 9: (a) MuVi uses higher MCS compared to the Adaptive and Naive schemes. (b) and (c) MuVi outperforms the Adaptive and the Naive schemes in terms of inter-packet delay and percentage of packets missing the deadline under various resource constraints. Especially, MuVi guarantees on-time delivery of all transmitted video packets.

MuVi and Adaptive schemes tune the MCS for frames instantaneously based on the clients' feedback of their channel conditions. The Adaptive scheme keeps the same MCS for all video frames, however MuVi assigns different MCS to each frame reflecting its system utility discussed in Section 3.3. The frames with higher utilities (e.g., I and P frames) are transmitted with lower, more reliable MCSs which can be supported by most of clients. Higher MCSs are used for the frames with lower utilities (e.g., B frames) which only target the clients in relatively good channel conditions. Some clients cannot receive all video frames due to their bad channel conditions and the higher MCSs employed with some frames, but it helps to improve the video quality for other clients with good channel conditions. This is especially beneficial when the resource constraint is very tight (e.g., 30 slots per frame). This explains why MuVi selects higher MCSs than the Adaptive and the Naive schemes on average and provides differentiated service to the clients.

The other interesting fact we notice is that MuVi aggressively uses higher MCS when there are insufficient resources for delivering all video frames (e.g., MuVi30 and MuVi60 curves in Figure 9(a)). Specifically, MuVi selects 40% of frames with MCS 7 when 90 slots are available per frame. However, it selects 60% of frames with MCS 7 when 30 slots are available per frame. The reason behind this is that when the resources are severely insufficient, MuVi tends to sacrifice some users with weak channel conditions by selecting higher MCSs and reducing the transmission time of selected video packets.

5.1.3 Inter-packet Delay

The average inter-packet delay for the three schemes with respect to various resource constraints is presented in Figure 9(b). We see that MuVi keeps the inter-packet delay low compared to the other two schemes regardless of the number of available slots. The inter-packet delay depends on the MCSs used for transmitting packets. A lower MCS leads to longer transmission time, which in turn results in larger inter-packet delay. The gap between MuVi and other schemes is larger under tight resource constraints (e.g., 30 slots), and it becomes smaller as the number of slots increases. Increasing radio resources significantly increases the number of packets packed in a frame and reduces the delay for the Naive and the Adaptive schemes. However, the improvement (reduced inter-packet delay) due to the increased resources is marginal for MuVi because MCS selection for each video frame and packet scheduling are already optimized.

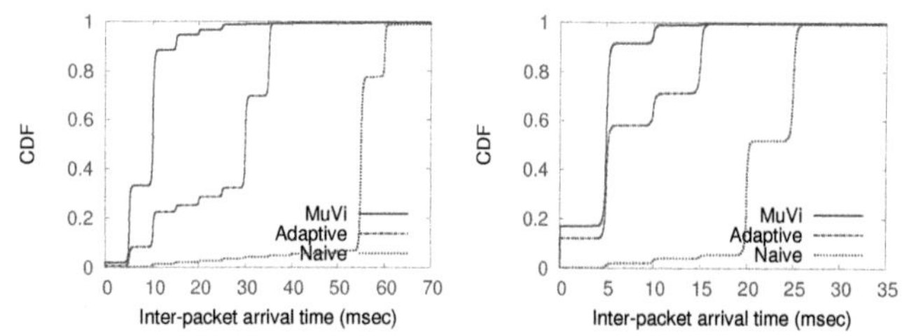

(a) 30 slots (per frame) (b) 60 slots (per frame)

Figure 10: CDF of inter-packet arrival time under different resource constraints. MuVi keeps inter-packet time close to 5 milliseconds (same as WiMAX frame interval) regardless of the available resources.

Figure 10 shows the CDF of inter-packet arrival time from the same set of experiments as shown in Figure 9(b) with respect to the different resource constraints. As we can see from Figure 10(b), for MuVi with 60 slots per frame, more than 94% of packets have inter-packet arrival time less than or equal to 5 milliseconds, which is the WiMAX frame duration. This shows that MuVi delivers most of the packets in one WiMAX frame. Even when the radio resources are insufficient (i.e., 30 slots), MuVi delivers 90% of the frames within 10 milliseconds (Figure 10(a)). On the other hand, the inter-packet delay for the Adaptive and the Naive schemes highly depends on the number of slots available. Hence, the performance of the Adaptive and the Naive schemes is limited by the amount of available resources.

Figure 9(c) shows the percentage of received packets that miss their deadlines. This pattern is very similar to the average inter-packet delay presented in Figure 9(b) because higher inter-packet delay leads to later arrival time. MuVi guarantees delivery of video frames within their deadlines, and hence all packets meet their deadlines (no packets out of deadline regardless of resource constraints).

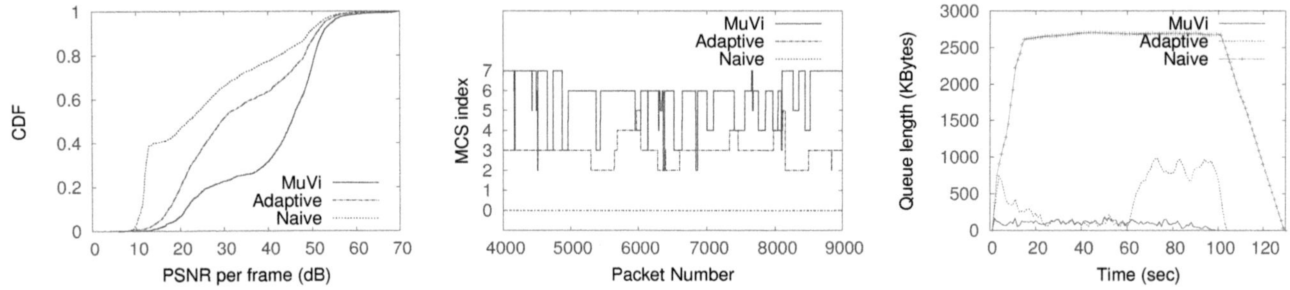

(a) CDF of per frame PSNR (b) MCS for each packet (c) Queue length at the WiMAX BS

Figure 11: (a) MuVi's per frame PSNR is higher than the other two schemes. (b) MuVi uses higher MCS for transmitting frames. (c) MuVi keeps the queue size significantly smaller than the other two schemes.

5.2 Micro-benchmark

We have confirmed that MuVi provides the best video quality amongst the three schemes. For further understanding, we present the microscopic view of the performance of the three schemes. All results are obtained from the experiments with 60 slots for multicast traffic (we observe similar behavior for 30 and 90 slots and omit them for the sake of brevity).

5.2.1 Per frame PSNR

In Figure 11(a), we plot the CDF of per frame PSNR from a single experiment. The video frame rate is 30 frames per second and each video frame consists of multiple packets. Per frame PSNR value is determined by the reception of video packets. We pick a client in the multicast group whose CINR values are in the range of 25-27 dB and present the CDF of per frame PSNR of the selected client. And we observe similar patterns from all other runs and for clients with different channel conditions.

We can see that the per frame PSNR of MuVi is significantly higher than that of the other two schemes. The median of per frame PSNR for MuVi, Adaptive and Naive schemes is 45, 29 and 21 dB, respectively. Moreover, the PSNR value of 75% of video frames in MuVi is greater than 33 dB (considered good in terms of MOS). This leads to the higher average PSNR values as we have seen in the previous subsection.

5.2.2 Per packet MCS

In Figure 11(b), we present the MCS used for each packet from the same experiment as above. Here again, we observe similar patterns from all other experiments. MuVi aggressively selects higher MCS than the Adaptive and the Naive schemes. We also notice that the variation of chosen MCS for MuVi is higher than that of the other two schemes because MuVi adaptively changes the MCS for each video frame based on the system utility and the client's feedback. Typically, higher MCSs (64-QAM with 2/3 coding or higher) are used for transmitting B frames and lower MCSs are used for transmitting I and P frames in MuVi. The Adaptive scheme uses MCS 3 for most of the packets while adapting it for some packets based on the clients' channel conditions.

MCS used for MuVi is higher than the other schemes, but MuVi provides better video quality as presented in Figure 11(a). Although the Adaptive and the Naive schemes use lower MCSs for robust delivery regardless of client's channel conditions, this requires longer time to deliver the video packets. As a result, many packets miss their deadlines and are discarded at the client, which results in lower PSNR values. In other words, Adaptive and Naive schemes waste the given resources while focusing on robust transmission rather than resource optimization.

5.2.3 Queue Length

Figure 11(c) shows the instantaneous queue size (in KBytes) at the base-station for the three schemes. The queue length of MuVi is constantly below 100 KBytes and is significantly smaller than the other two schemes. The queue length highly depends on the MCS (i.e., the transmission rate) used for transmission. When the base-station uses higher transmission data rate, it can quickly send data and empty the queue[2]. As we see in Figure 11(b), MCS used for MuVi is higher, and hence it leads to smaller queue length than Adaptive and Naive schemes. The duration of video we used for the experiments is 100 seconds. With given resources (60 slots), MuVi can deliver all incoming video traffic in time, while Adaptive scheme completely empties its queue at around 104 seconds. Even worse, Naive scheme finishes at 125 seconds. The frames which are received late is shown as the number of frames missing their deadlines in Figure 9(c).

5.3 Client Mobility

We have seen the performance improvement of MuVi under static conditions where all clients remain stationary. In this subsection, we evaluate the performance of MuVi under client mobility. In this set of experiments, the multicast group contains two clients, with one stationary and the other moving at walking speed. The mobile client starts from a location close to the base-station, moves away from it, and then moves back, according to the route depicted in Figure 6.

Figure 12(a) shows the observed CINR of the mobile client. We see that the CINR decreases till 15 seconds and then increases. The CINR increase between 8 and 13 seconds and multiple downside spikes are probably caused by random channel fading. Figure 12(b) plots the PSNR value of each frame for both MuVi and the Adaptive scheme under the same mobile trajectory. We see from Figure 12 that the PSNR values under both MuVi and the Adaptive scheme are highly correlated to the CINR of the mobile client, but the Adaptive scheme obtains much lower average PSNR values and

[2]In this work, we do not consider re-transmissions, which are typically not employed for multicast traffic to avoid feedback explosion.

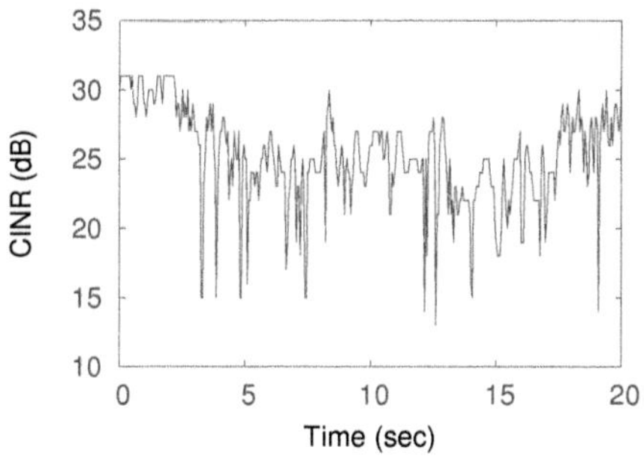

(a) Instantaneous CINR of the mobile client

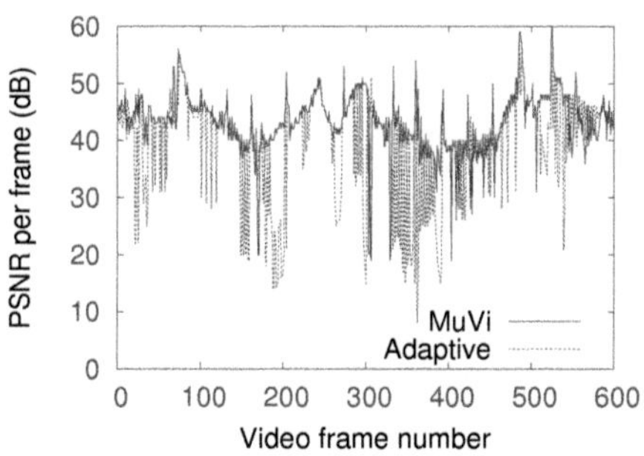

(b) PSNR values of first 600 frames

Figure 12: Mobility experiments. The video frame rate is 30 frames per second, so the total number of video frames generated in 20 seconds is about 600.

much higher variation than MuVi. Although MuVi obtains low PSNR values for some frames, most of them are due to dropped B frames, which do not affect other frames. On the contrary, under the Adaptive scheme, some P frames may get lost, which will affect several other dependent frames. This indicates that instantly adapting MCS based on the weakest client's feedback is not robust enough to sustain user mobility. Simultaneously optimizing resource allocations and differentiating packet types by using different MCSs can provide more robust performance under client mobility.

WiMAX base-station receives the CINR report (which is part of the WiMAX standard) from all clients periodically (e.g., every 5 msec) through the uplink channel. MuVi leverages this information to determine the highest MCS for each client. Since MuVi incorporates the client's channel conditions in real-time, it provides the best performance even for mobile clients.

5.4 Overhead

MuVi's core engine lies at the ASN gateway like Opal [15]. Typically, a single ASN gateway manages multiple base-stations in a cellular network. The overhead at the ASN gateway could be problematic considering the number of base-stations associated with it. Hence, any implementations or additional work load on the ASN gateway will require appropriate provisioning.

To investigate the computational overhead of MuVi's operations (optimizing resource allocation, determining MCS for each packets and processing packet re-ordering), we measure the execution time of MuVi while delivering video frames to the multicast group. The total execution time for MuVi's algorithm is 83.63 milliseconds while the total experimental duration (the same as the video length) is 100 seconds. It amounts to less than 0.1% of the total CPU time. Therefore, the overhead of MuVi is almost negligible for handling a single base-station and would be small even if several hundred base-stations are managed by a single gateway.

The process of handling client's CINR report for getting supportable MCS for each client will be done by referring the MCS-CINR table we summarized in Table 2. The base-station has a MCS-CINR table and hence whenever base-station receives the CINR report from the clients it can easily read the supportable MCS value accordingly. The complexity of this process is O(1). The uplink channels are used to send the client's CINR report to the base-station, it would not affect the downlink resources for video traffic to the multicast clients.

6. DISCUSSION

In this work, we consider non-scalable (e.g., MPEG4) video and differentiate video frames based on their types. MuVi's optimization algorithm is also applicable to SVC-encoded videos in a similar manner. Scalable video consists of one base layer which provides minimum video quality and multiple enhancement layers for refined video quality. To implement MuVi with scalable videos, we can prioritize frames based on their layers and apply similar optimization algorithm for efficient resource allocation. Video packets that belong to lower layers will be assigned relatively higher priority comparing to the other frames in enhancement layers. Implementing a multicast system for SVC-encoded videos in 4G networks (including both the media server and the player for clients) is an interesting topic for our future work.

MuVi employs scheduling, resource allocation and PHY rate selection for optimizing multicast video delivery. Other MAC and PHY techniques, such as packet re-transmission, power control, and beam-forming, are also possible to improve video delivery performance, which are part of our future work.

7. RELATED WORK

Video multicast in wireless (WiFi): There are many prior works on wireless multicast that aim to improve the performance through PHY and MAC layer design. DirCast [9] applies association control to minimize the total multicast delay, and within each AP (Access Point), DirCast chooses the transmission rate based on the channel condition of the "worst" client in the multicast group.

A recent work Medusa [21] employs a proxy-based solution to improve media streaming performance in wireless LAN. Medusa focuses on a content-dependent PHY rate selection and packet value awareness for WiFi multicast. However, Medusa is designed for asynchronous, WiFi systems where the wireless media is shared via contention-based random channel access. As a result, Medusa does not perform radio resource allocations and only employs a heuristic-based rate adaptation algorithm. Moreover, Medusa requires that clients periodically send reception reports about the packets transmitted previously and performs retransmissions using network coding, and thus it requires modifications at the client side, which are challenging in cellular networks (because it involves modifications of the wireless standards).

By contrast, MuVi is designed for OFDMA-based, synchronous broadband mobile wireless systems (e.g., WiMAX, LTE) where the radio resources are allocated by the base-stations. As a result, MuVi employs a near-optimal resource allocation algorithm to maximize the total system utility by intelligent resource allocation (e.g., frame dropping) and PHY rate adaptation of each video packet. MuVi does not require packet reception reports or perform re-transmissions. Instead, MuVi incorporates CINR reports that are available in current cellular networks to adjust the OFDMA resource allocation and PHY rate adaptation. Thus it does not require modification at the client-side or the air interface, and has minimum control overhead.

Scalable video multicast in wireless systems: In [13], Deb *et al.* studied the problem of multicasting scalable video (SVC) in WiMAX cellular networks with the goal of maximizing the system utility. They developed a greedy algorithm to allocate radio resources and adaptively assign the MCS for each transmitted video layer. In [16, 17], the authors use dynamic programming approach to find the optimal system utility and to assign modulation and coding schemes for scalable video traffic in mobile cellular networks. In [18], Li *et al.* considered joint-layer resource allocation to further improve multicast performance and developed approximation algorithms to trade off algorithm complexity with performance. All these works were only evaluated through theoretical analysis and simulations based on fixed layering sizes. The dynamics of video traffic across different video frames were not considered and no system implementation has been conducted.

Channel-unaware wireless video transmission: A couple of recent works proposed wireless video encoding and transmission schemes that need not be aware of the wireless channel conditions. SoftCast [14] proposed a joint channel encoding and video source coding scheme for mobile video transmission. This is essentially an analog approach for delivering video over the wireless. FlexCast [8] modified the MPEG4 video codec and incorporated rateless coding for efficient video streaming in wireless systems. Neither SoftCast nor FlexCast requires any feedback about the wireless channel conditions. The received video quality automatically adjusts depending on the channel quality at the clients. As a result, these two schemes provide natural support for wireless video multicast although they are not specifically designed for it. Nevertheless, both SoftCast and FlexCast require heavy modifications to video source coding, the air interface, and the mobile clients. By contrast, MuVi does not require any changes in these elements. MuVi requires long-term channel feedback and optimizes the video multicast transmission via efficient radio resource allocation and frame prioritization under the existing video and wireless standards. Thus it allows speedy deployment.

8. CONCLUSION

This paper presents MuVi, a wireless video multicast scheme in OFDMA-based cellular networks. We design and implement MuVi using a WiMAX testbed to show its efficacy in real systems. MuVi incorporates the clients' channel feedback and different video frame priorities to adapt the MCS for each video packet. MuVi's efficient resource allocation scheme helps to optimize the overall video quality in the multicast group. MuVi also allows providing differentiated services among users in the multicast group by assigning different utility to different users for the same packet while optimizing the overall video quality in the multicast group. MuVi does not modify the video encoding nor the wireless transmission schemes. Instead, it only involves resource allocations along with video frame prioritizations. MuVi is also applicable to other OFDMA-based wireless technologies (such as LTE and LTE-Advanced) and it does not require any modifications at the mobile clients, so it is readily deployable with commercial off-the-shelf devices.

Acknowledgments

We thank our shepherd, Dr. Thyaga Nandagopal, for his guidance through the camera-ready submission process. We thank Rajesh Mahindra from NEC Labs for invaluable support and suggestions during the preparation of this work. Jongwon Yoon and Suman Banerjee have been supported in part by the following grants of the US National Science Foundation: CNS-1040648, CNS-0916955, CNS-0855201, CNS-0747177, CNS-1064944, and CNS-1059306.

9. REFERENCES

[1] YouTube, *http://www.youtube.com/.*

[2] Hulu, *http://www.hulu.com/.*

[3] Cisco Visual Networking Index: Forecast and Methodology, 2010-2015, *http://www.cisco.com.*

[4] Beceem wimax chips for mobile applications, *http://www.beceem.com.*

[5] IEEE 802.16e-2005 Part 16: Air Interface for Fixed and Mobile Broadband Wireless Access Systems, *IEEE 802.16e standard.*

[6] MPEG4 standard group, *http://mpeg.chiariglione.org/standards/mpeg-4/mpeg-4.htm.*

[7] VLC media player, *http://www.videolan.org/vlc.*

[8] S. T. Aditya and S. Katti, "FlexCast: Graceful Wireless Video Streaming", *ACM MobiCom*, 2011.

[9] R. Chandra, S. Kandula, T. Moscibroda, V. Navda, J. Padhye, R. Ramjee and L. Ravindrananth, "DirCast: A Practical and Efficient Wi-Fi Multicast System", *IEEE ICNP*, 2009.

[10] Accton, http://www.accton.com/.

[11] PicoChip femtocell solutions, http://www.picochip.com/.

[12] J. M. Costa, "More Frequencies Needed for Mobiles - Terrestrial Spectrum Sought for IMT", *ITU News, No3*, 2007.

[13] S. Deb, S. Jaiswal and K. Nagaraj, "Real-time Video Multicast in WiMAX Networks", *IEEE Infocom*, 2008.

[14] S. Jakubczak and D. Katabi, "A Cross-Layer Design for Scalable Mobile Video", *ACM MoiCom*, 2011.

[15] R. Kokku, R. Mahindra, S. Rangarajan and H. Zhang, "Opportunistic Alignment of Advertisement Delivery with Cellular Basestation Overloads", *ACM MobiSys*, 2011.

[16] P. Li, H. Zhang, B. Zhao and S. Rangarajan, "Scalable Video Multicast in Multi-carrier Wireless Data Systems", *IEEE ICNP*, Oct. 2009.

[17] P. Li, H. Zhang, B. Zhao and S. Rangarajan, "Fast Optimal Resource Allocation for Scalable Video Multicast in Broadband Wireless Networks", *IEEE ICC*, 2010.

[18] P. Li, H. Zhang, B. Zhao and S. Rangarajan, "Scalable Video Multicast with Joint Layer Resource Allocation in Broadband Wireless Networks", *IEEE ICNP*, 2010.

[19] R. Morris and E. Kohler and J. Jannotti and M. F. Kaashoek, "The click module router", *SIGOPS Operating System Rev.*, volume 33(5), pages 217-231, 1999.

[20] Z. Orlov, "Network-driven Adaptive Video Streaming in Wireless Environments", *PIMRC*, 2008.

[21] S. Sen, N. K. Madabhushi and S. Banerjee, "Scalable WiFi Media Delivery Through Adaptive Broadcasts", *USENIX NSDI*, 2010.

Temporal Quality Assessment for Mobile Videos

An (Jack) Chan, Amit Pande, Eilwoo Baik and Prasant Mohapatra
University of California, Davis, CA 95616, USA
{anch, pande, ebaik, pmohapatra}@ucdavis.edu

ABSTRACT

Video quality assessment in mobile devices, for instances smart phones and tablets, raises unique challenges such as unavailability of original videos, the limited computation power of mobile devices and inherent characteristics of wireless networks (packet loss and delay). In this paper, we present a metric, Temporal Variation Metric (TVM), to measure the temporal information of videos. Despite its simplicity, it shows a high correlation coefficient of 0.875 to optical flow which captures all motion information in a video. We use the TVM values to derive a reduced-reference temporal quality assessment metric, Temporal Variation Index (TVI), which quantifies the quality degradation incurred in network transmission. Subjective assessments demonstrate that TVI is a very good predictor of users' Quality of Experience (QoE). Its prediction shows a 92.5% of correlation to subjective Mean Opinion Score (MOS) ratings. Through video streaming experiments, we show that TVI can also estimate the network conditions such as packet loss and delay. It depicts an accuracy of almost 95% in extensive tests on 183 video traces.

Categories and Subject Descriptors

H.5.1 [**Multimedia Information Systems**]: Video

General Terms

Performance

Keywords

Video quality, Quality of Experience, mobile devices

1. INTRODUCTION

In 2011, video traffic has accounted for more than 50% of the total traffic in mobile networks [4]. Quality of Experience (QoE) of watching videos over mobile devices, such as smart phones and tablets, has been attracting interest from content providers and network service providers.

Effective and efficient video quality metrics are highly desirable for service providers. The metric can help them to extract quick feedback from end-users and can enable them to "turn knob" at their end (if possible) to enhance the quality of service. Many tools have evolved for evaluating video quality delivered to end users. Most of them re-use the techniques in image quality measurements which evaluate the spatial quality, such as Peak Signal to Noise Ratio (PSNR [10]) and Structural SIMilarity index (SSIM [25]) of each video frame. However, video quality comprises of both spatial and temporal quality.

Unfortunately, unlike its counterpart, the temporal quality assessment cannot re-use the existing static image quality measurement techniques. Videos over mobile devices (e.g. smart phones & tablets) are usually delivered through wireless such as Wi-Fi or cellular networks. This includes a host of applications including interactive multimedia applications, video-on-demand (VOD), video chat & HDTV streaming. We call these videos as ***"mobile videos"***. Temporal quality degradation due to packet loss and delay is more pervasive in wireless networks than in wired scenarios. Mobile devices also pose unique constraints such as limited computation power, small memory, short battery life and low video resolution. Existing temporal quality assessment schemes tend to depend on the availability of the original video (i.e. full-reference metrics) [16] or incur heavy computational cost in computing motion information [23], neither of which is possible in mobile context.

In this paper, we present a novel scheme for temporal quality assessment. We first measure the motion information of a video by comparing the consecutive frames of a video. Considering both the accuracy and the low computational cost, we design a temporal information metric, we call it Temporal Variation Metric (TVM). TVM evaluates the difference of the corresponding pixel values in the two neighboring frames to estimate the motion of objects in the video. Compared to other frame comparison techniques proposed in video quality research community, TVM has the shortest running time and a relatively low memory usage. Such properties make it suitable for using in mobile videos. Despite its simplicity, TVM shows an average Pearson correlation coefficient of 0.875 with the optical flow, while the optical flow is widely used to measure the actual temporal motion in a video.

From TVM, we also design a new reduced-reference temporal quality metric, Temporal Variation Index (TVI). TVI measures the temporal quality degradation between the source (original) and the received videos. From subjective video quality tests, we find that TVI strongly correlates with Mean Opinion Score (MOS) of a pool of human observers watching the video. It is worth noting that TVI evaluates the temporal quality of the received video without the original copy of the video. TVI can quantify the perceived video quality degradation caused by coding or communication impairments.

In addition to the correlation between TVI and MOS, through extensive experiments, we also find that the correlation between TVI and network impairments, in particular packet loss rate and end-to-end delay. With such correlation, by measuring TVI in the video application layer at the user end, we can estimate the packet loss rate and delay in the wireless networks. This information will be extremely useful to network service providers to improve the video streaming quality (and thus Quality of Experience (QoE)) by adjusting video coding and allocating network resources.

The main contributions of this paper are as follows:

1. We propose Temporal Variation Metric (TVM) to measure the temporal information of mobile videos. TVM has low computation and memory requirements to suit the mobile devices and it closely relates to motion information in videos.

2. From TVM, we derive a new temporal quality metric, Temporal Video Index (TVI). TVI quantifies the perceived quality degradation between the source and received videos.

3. Subjective tests demonstrate the strong correlation between TVI and end-user QoE.

4. We also show that TVI can estimate the network impairments, in particular packet loss rate and end-to-end delay.

Although TVM and TVI are video quality metrics, they can help facilitate many other technologies and applications. For example, TVM and TVI are very useful for content distribution in multicast groups to maximize users' QoE. They are also useful in scalable video delivery for quality-oriented bandwidth allocation in wireless networks. Mobile videos in surveillance applications can also be monitored to guard against network impairments. TVI and TVM can be used to monitor video quality, to benchmark video processing systems and to be embedded in communication systems to optimize algorithms and parameter settings for content delivery.

The rest of paper is organized as follows: Section 2 gives the motivation for developing a new temporal metric. We propose TVM in Section 3. To show the advantages of TVM, we also discusses other potential candidate metrics for estimating the motion information of a video. In Section 4, we derive TVI from TVM. We show that TVI demonstrates a strong correlation to the MOS values obtained from subjective tests. In Section 5, we further explore the relationship between TVI and network metrics in particular packet loss rate and end-to-end delay. Section 6 concludes the paper.

2. MOTIVATION

The most accurate way to obtain the end-user Quality of Experience (QoE), such as Mean Opinion Score (MOS) ratings [12] or Crowdsourcing [3], is to conduct subjective tests involving human subjects. Such experiments have to be conducted offline and incurs huge cost in terms of labor and time. More importantly, we cannot use the subjective measurement for real-time monitoring to guarantee the quality of video delivery. Therefore, an objective measurement of QoE is necessary to feedback and maintain the quality of delivered video over mobile devices.

2.1 Getting Video Temporal Information

A video is composed of both spatial and temporal information. The spatial information comprises of appearance of objects, resolution, smoothness, etc. The degradation of spatial quality in videos causes blocking and blurring. A number of metrics derived from static image quality assessment can be re-used to measure spatial quality across a video. Such examples include Peak Signal-to-Noise Ratio (PSNR), Structural SIMilarity index (SSIM [25]), VQM [29], and Blocking[14].

Temporal information is the measure of the motion of objects in a video or movement of background including scene changes. Temporal quality measurement needs motion information in a video, so existing static image quality evaluation metrics cannot be re-used for this purpose. Motion information is formally obtained by calculating optical flow. It is based on the movement of an object in a video frame caused by a relative motion between an observer and the scene [27]. Optical flow field is represented by vectors of the object points. The length of the vectors represents the magnitude of the motion while the direction of the vectors represents the direction of the motion [9]. Motion vectors, which are widely used for motion estimation in video coding, are a special case of optical flow when vectors are computed in a macro block basis [13]. Calculating the optical flow or motion vectors is a computationally intensive task which is unsuitable for embedded devices. Pauwels and Hulle [20] reported that a GPU (GeForce FX 5800) and CPU (Pentium-4 2.8 GHz) implementation of optical flow calculation for low resolution 256x256 images take around 13 ms and 35 ms of dedicated computation time respectively.

Unfortunately, most existing temporal quality assessments [23, 30] still depend on measuring temporal information by optical flows and motion vectors. To reduce the computation cost of extracting optical flow or motion vectors, the authors in [16] assume that the motion vectors during video encoding can be re-used in temporal quality assessment. However, such assumption is not always valid. In many cases, the codecs may be closed source (consider for example Skype video). Furthermore, it may be very difficult to account for optical flow information from the codec. Take Scalable Video Coding (SVC) codec as an example. There are tiers of encoding/decoding done to achieve scalability, so it is very difficult to assess how motion vectors in different layers are to be assimilated and compared. Thus, it is desirable to build a temporal quality metric for videos which can provide measurements without using motion vectors nor being dependent on the underlying video codecs.

2.2 Desired Properties of a Video Quality Metric

The degradation of temporal quality is primarily observed as delay, freezing(stop-motion), blockiness and blackout. A desired temporal quality metric should be able to capture these effects alongside with motion information.

There is a gap between the temporal quality evaluation

and end-user QoE which is how people actually perceive the video quality. There has been little effort to quantify the correlation between end-user QoE and temporal quality metric measurements. Therefore, a desired temporal quality metric, in addition of capturing the video temporal quality degradation, should also have a great correlation to the end-user QoE.

Although the optical flow gives the accurate temporal information of a video, mobile devices are computational power limited, making them impossible to re-compute the optical flow from reconstructed frames. Therefore, a desired video temporal quality metric should have low computational cost.

To make the video temporal quality metric be readily applied to the end-user mobile devices, the metric should be video codec independent. Of course, video content providers can always customize the metric to their own codecs for the optimized use.

In some wireless networks, for example Wi-Fi, the network impairments such as the packet loss rate and the delay are easily measured. (These network impairments are only the measurement of Quality of Service (QoS) of the network not the QoE of the end users.) In service providers' point of view, it is also interesting to know the network impairments so that they can "tune" the parameters, for example network bandwidth allocation, to increase QoS and QoE. The bad news is in mobile networks these network impairments are not always easily available. For instance, to measure the packet loss rate in video streaming over cellular networks, we require packet transmission information from the base station. Comparing how many packets sent from the base station, and how many packets received in the mobile device, we can calculate the packet loss rate. More importantly, to access the network layer metrics in mobile devices, e.g. smart phones, we need the root access privilege which is usually not open to general public, such as application developers.

Therefore, it is desirable if an application-layer video quality metric can also be an indication of network impairments which lead to quality degradation. The network metrics such as packet loss and delay can be estimated using the video quality metric obtained at the end user, allowing the service providers to modify the channel resources allocation or video coding to guarantee the quality of videos delivered.

In addition, existing temporal quality metric tend to depend on the availability of the original video (i.e. full-reference metrics) [16]. This is not feasible for on-line computations where it is not possible to have original video for referencing.

In summary, a video temporal quality metric should have the following desired features:

- Captures the degradation of the temporal quality;
- Correlates with user QoE;
- Low computational cost;
- Codec independent;
- Estimates the network impairments such as packet loss and delay; and
- Does not require the original copy of the video.

2.3 Related Work

There is a large amount of existing efforts on video spatial quality assessments [5, 26]. However, there are few works about video temporal quality which becomes important when video applications move to wireless context. MOVIE index [23] integrates both spatial and temporal aspects of distortion assessment using a full-reference technique which requires the original copy of the video. The metric is designed based on modeling of human visual system. It shows a good performance but also incurs a high computational cost.

Moorthy and Bovik [16] present a video quality assessment algorithm and enable the motion vector re-use in the decoding stage to reduce the computational complexity. But the drawback is that not all video codecs have the motion vectors reusable during the decoding process. The metric is also full-reference.

Yang et al. [30] consider various factors such as frame dropping, scene boundary, motion activity and motion mapping in the video to estimate the temporal quality. It reuses motion activity and motion mapping information from video codec and packet loss information from network. Again, it depends on the reusability of motion information during the decoding process. It also requires a cross-layer design to get the packet loss information from the network layer.

Vidal and Gicquel [19] presents a no-reference quality metric for detecting fluidity impairments due to frame losses in video. This mainly accounts for frame dropping in the video. The subjective experiments by [11] suggest that content, motion magnitude and frame rate are some factors affecting perception of temporal artefacts. This motivates us to develop a new metric, which quantifies the end user perception of the video and accounts for these factors.

2.4 Our Approach

Considering the large number of video codecs and service providers, we look for a codec independent solution for temporal quality assessment. We compare the similarity of two neighboring (consecutive) frames of video to estimate the motion in video. A large similarity between two frames indicates a slow motion of objects during the time period of these two frames. Instead of recording the difference of each pixel, we propose only record the average similarity score of a whole frame. This decreases the complexity compared to those in optical flow or motion vector calculations. Furthermore, consecutive frames comparison can be done in a no-reference manner without a copy of source video at the receiver.

We call our metric measuring the motion of objects in the video as Temporal Variation Metric (TVM). TVM is suitable for mobile devices because of low computational power requirement and small memory overhead. Based on TVM values between the source and the received videos, we design a reduced-reference temporal quality metric, Temporal Variation Index (TVI). TVI measures the temporal quality degradation in mobile video streaming. It is worth noting TVI only needs the TVM of the original video but not the whole copy of the original video. This makes it easy to be deployed in mobile devices and services.

We conduct extensive subjective tests to demonstrate the correlation between TVI and the subjective user experience. Based on the subjective tests, we related the TVI and the Mean Opinion Score (MOS) which is the end-user QoE.

With the linear regression technique, we derive the equations to estimate MOS from TVI. In addition, we find a strong correlation between TVI and network packet loss. We also derive equations to estimate the packet loss rate from TVI. By feeding back TVI to the source, the content providers and network services providers can adjust the video coding or network resources to improve the end-user experience of video streaming.

3. MEASURING TEMPORAL INFORMATION

To develop a temporal quality metric, accurately measuring the temporal information of a video is important. We design Temporal Variation Metric (TVM) to measure such information.

3.1 Temporal Variation Metric (TVM)

Temporal Variation Metric (TVM) measures temporal information of a video by comparing the content amongst consecutive frames in a video. Two successive frames are compared to obtain a single score by calculating the Peak Signal to Error Ratio in the consecutive frames. This is analogous to Peak Signal-to-Noise Ratio (PSNR) measurement used in comparing reconstruction quality of images and videos. Different from PSNR, we do not use any original video frame, but instead use previous frame of the video as a reference and calculate temporal motion in the video. Unlike PSNR which is an indicator of noise introduced in the image due to communication, TVM is an indicator of motion or temporal variation in the current frame as compared with previous frame.

TVM only uses the two neighboring frames for measuring temporal variation, so if any temporal distortion happens in the current frame will only be recorded by the current and the next TVM calculations. This effect in the current and the next TVM values will not propagate to the subsequent TVM calculations.

Numerically, TVM is calculated as log of mean square value of difference between two consecutive frames (F_{p-1} and F_p) of the video (measured in dB).

$$TVM_p = 10\log_{10}\left(\frac{k^2}{d}\right) \quad (1)$$

where k is a constant for a video with a particular color depth. It is the maximum pixel value of the video frame. For example, if the color depth of the video is 8-bit, $k = 255$. d is the mean square difference of the corresponding pixels in two frames, F_{p-1} and F_p. Logarithm is used to compensate the non-linearity of human visual system.

$$d = \frac{1}{MN}\sum_{i=0}^{M-1}\sum_{j=0}^{N-1}(F_p(i,j) - F_{p-1}(i,j))^2$$

where $F_p(i,j)$ is the value of the pixel in i^{th} row and in j^{th} column in the pth frame. Each frame is of size $M \times N$ pixels.

3.2 Consecutive Frames Comparison

Besides the peak signal-to-error ratio which we used in TVM, there are other spatial quality metrics which can be modified to measure the difference between two frames. Thus, they can also be potential candidates for measuring temporal information. In this section, we compare other candidates with TVM to evaluate their appropriateness to be a measurement of the temporal information in mobile videos. We select three representative candidates from the research literature: Structural Similarity Index (SSIM), Edge Similarity Score (ESS) and Visual Signal-to-Noise Ratio (VSNR). These three metrics represent a wide choice amongst existing research. Instead of using them directly, we apply these metrics on *consecutive* frames of a same video to measure the motion in the video. Thus, the derived three potential temporal metrics are respectively called TSSIM, TESS and TVSNR.

TSSIM (Temporal Structure Similarity Index) - [24] proposes SSIM to measure the structural similarity of two frames. SSIM first measure the luminance which is the intensity of the pixel values of the frames. Then, the luminance is removed from the frames and the contrast which is the standard deviation between two frames is obtained. Then the contrast measure is also removed to measure the structural similarity between two frames. We derive TSSIM to measure the SSIM of the two consecutive frames. It is robust to common spatial artifacts such as mean-shifts, contrast-stretch, compression (codec) losses, blurring and salt-pepper noise.

TESS (Temporal Edge Similarity Score) - [28] proposes ESS to reflect the way that human perceives visual information. The important information extracted by human visual system includes spatial-luminance information, edge and contour information [7]. TESS measures the edge and contour change information of the consecutive frames to evaluate the motion. The frames are first partitioned into macro blocks and then Sobel's mask is applied to get the edge information along each pixel along both horizontal and vertical directions. Edge direction information is obtained from these values and quantizing it into one of the eight representative directions that are equally spaced by $\pi/8$ radians (from $-\pi/2$ to $\pi/2$). The dominant edge direction for each macroblock is computed and the average cosine of the difference in edge directions measures the consecutive edge similarity score of the consecutive frames.

TVSNR (Temporal Visual Signal to Noise Ratio) - VSNR metric is proposed in [2] to capture the human visual system properties. It operates in two stages. (1) It computes the contrast thresholds for the detection of distortions via wavelet-based models of visual masking and visual summation. Then it determines the visibility of distortions in an image. If the distortions are below the threshold, the distorted image is deemed to be of perfect visual fidelity (VSNR = infinite) and no further analysis is required. (2) To measure supra-threshold distortions, the low-level visual property of the perceived contrast and the mid-level visual property of global precedence are used. These two properties are modeled as Euclidean distances in the distortion-contrast space using a multi-scale wavelet decomposition. Linear sum of these distances is computed for consecutive frames in our case to get TVSNR value.

3.3 Testing with Videos

To evaluate whether the proposed TVM is a good metric to measure the temporal information of videos, we test it with six different videos and compared the result with TSSIM, TESS and TVSNR. The videos we used are listed in Table 1. They have different resolutions, lengths and degrees of motion. A long video will constitute a combination of scenes of all degrees of motion. Therefore, by separately

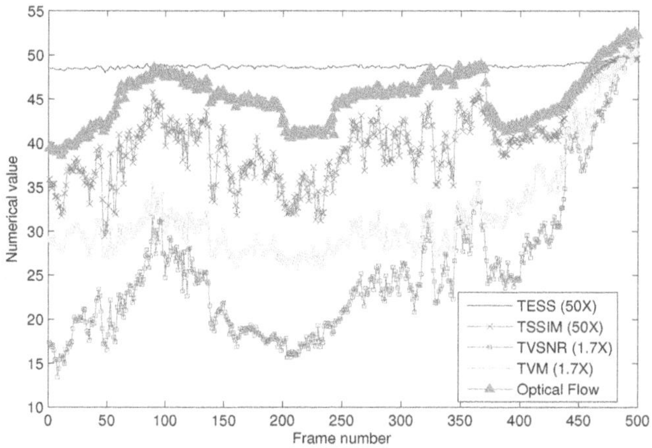

Figure 1: Plot showing the correlation of TVM, TESS, TSSIM and TVSNR with Optical Flow (dB)

Table 1: Details of sample videos

Name	Length	Resolution	Motion
Old Woman	30 sec	1920 × 1080	1(relatively slow)
Cows	60 sec	1920 × 1080	2
Foreman	10 sec	352 × 288	3
Duck	10 sec	1920 × 1080	4
Park	10 sec	1920 × 1080	5
Intersection	60 sec	352 × 288	6(relatively fast)

considering these scenes of different motion degrees, we have actually considered the variety in existence. These few video clips are the representative video sequences chosen. In the experiments presented later, we have used hundreds of traces with varying network conditions.

For each video, we calculate TVM, TSSIM, TESS TVSNR and the optical flow. Figure 1 shows the variation of the optical flow and different temporal information measurements for the video sequence *Duck*. In this video, TESS has low correlation while the other metrics show high correlation. TSSIM curve shows many uncorrelated high frequency components. Table 2 shows the Pearson Correlation Coefficients of the different temporal information measurements with the optical flow. We find that both TVSNR and TVM have the highest correlation coefficients with the optical flow.

Then we evaluate the computation time and memory overhead required by different temporal information measurements. Table 3 gives the details of computation time required by those metrics on a 12GB 12-core Intel Xeon (dual-core) processor running Matlab release 2011. TVM gives the shortest running time which shows its low computational cost. As expected, due to its huge computational cost, the optical flow calculation requires the longest time.

In memory overhead measurements, we exclude the optical flow and only focus on the four different temporal information measurements. We find that different video content or degree of motion has no obvious effect on the memory requirements. Instead, the resolution of the video greatly influences the memory overhead of temporal information measurements. Table 4 shows the memory overhead of temporal information measurements. We scale the video sequence *Park* into three different resolutions for comparison. We find that TVM and TESS have the smallest memory overhead in operation.

From the above tests and comparisons, we have shown that TVM is the most suitable candidate for measuring temporal information considering its accuracy and suitability for mobile devices.

Table 2: Pearson Correlation Coefficients of Temporal Information Measurements with Optical Flow

Video	TVM	TSSIM	TESS	TVSNR
Old Woman	0.7719	**0.8033**	0.7500	**0.8054**
Cows	**0.8679**	0.7476	0.7408	**0.9153**
Foreman	**0.8598**	0.7720	0.7390	**0.9156**
Duck	**0.9106**	**0.9183**	-0.2461	0.7477
Park	0.6072	**0.7828**	0.7200	**0.9178**
Intersection	**0.9241**	0.5706	0.6107	**0.9230**

Table 3: Running Time (in seconds in 3 significant figures) of Different Temporal Information Measurements in Matlab

Video	OP	TVM	TSSIM	TESS	TVSNR
Old Woman	46600	**9.65**	99.9	6230	45.8
Cows	83100	**11.6**	129	11700	84.5
Foreman	16700	**0.983**	8.38	1980	13.6
Duck	29700	**3.70**	21.5	3280	23.9
Park	29600	**2.97**	20.0	3290	24.6
Intersection	84000	**12.7**	117	11500	84.5

OP - optical flow.

Table 4: Memory Overhead (in KB in 3 significant figures) of Different Temporal Information Measurements

Resolution	TVM	TSSIM	TESS	TVSNR
Low (352x288)	**428**	23300	**315**	3240
Medium (960x640)	**1930**	129000	**1120**	19000
High (1920x1080)	**6200**	428000	**3440**	67800

4. ESTIMATING VIDEO TEMPORAL QUALITY

With the help of TVM, we derive a new reduced-reference temporal quality metric, Temporal Variation Index (TVI). We then show that TVI has a strong linear relationship with end-users' quality of experience (QoE). We can use TVI to accurately predict QoE.

4.1 Temporal Variation Index (TVI)

TVM is a measure of temporal information between two neighboring frames in a video. TVM of the original video measures the motion of the objects in the video. In addition to the motion, TVM of the received video in the end user also indicates the temporal quality degradation. For example, a large TVM value can imply either the objects in the video scene moves very fast or there are some losses in the video frame sequence. But by comparing the temporal information of the received video with that of the original video, we can measure the temporal information change. Such temporal information change is due to a degradation of temporal quality in video processing, such as wireless video streaming. Therefore, we design a reduced-reference temporal quality metric, Temporal Variation Index (TVI), to estimate the temporal quality degradation. TVI removes the "motion" part which is in the TVM values.

TVI of a video at time t, $TVI(t)$, is defined as follows.

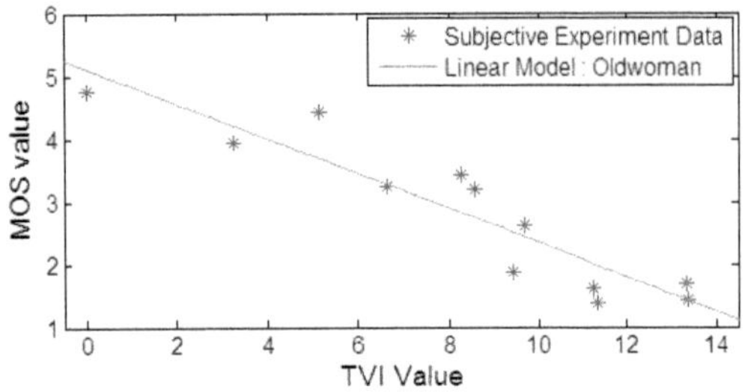

(a) *Slow Motion* (Old Woman)

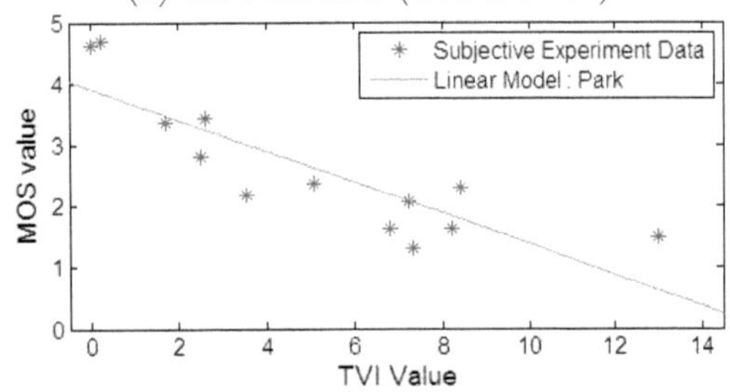

(b) *Moderate Motion* (Park).

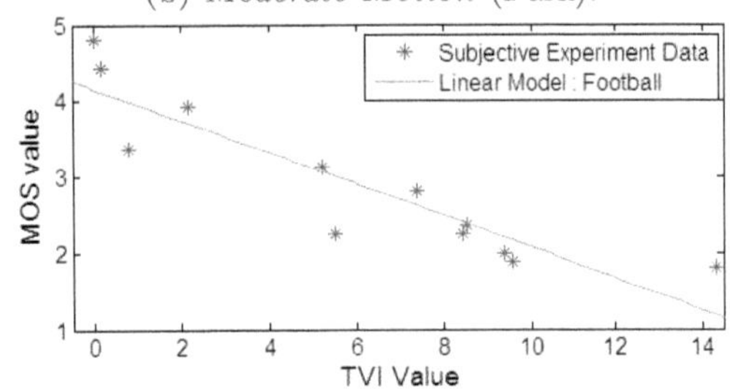

(c) *Fast Motion* (Football).

Figure 2: Linear fits of all three video categories for MOS prediction.

$$TVI(t) = \frac{|TVM_s(t) - TVM_r(t)|}{TVM_s(t)} \quad (2)$$

where $TVM_r(t)$ is the TVM value for the received frame at current time, t. For every received video frame, the mobile device records its time stamp, t. It then calculates TVM, $TVM_r(t)$, of the received video at time t by comparing frame at that time, f_t, with the preceding frame $f_{t-\delta}$. $\delta = \frac{1}{fps}$, where fps is the frame per second of the video. The receiver will search the corresponding TVM value $TVM_s(t)$ of the source video from the control data, and calculate $TVI(t)$.

In case the original video itself has some temporal noise or distortion, TVI would not indicate such temporal distortion. Therefore, TVI merely measures the temporal distortion or noise incurred in the transmission, such as streaming, process. Generally speaking, TVI can measure the temporal distortion incurred in any video processing procedure, for example, video transcoding. In this paper, we focus on wireless video transmission.

It is worth noting that TVI does not require the copy of the original video, but TVM of the original video. We assumed that the sender computes the TVM values of the original video and sends them to the receiver through the control channel during the streaming. This assumption is reasonable as TVM calculation has low computation and storage requirement. For each pair of video frames, a TVM value is represented by a 32-bit float number. So, the size of the temporal information of a 2-hour movie with 25 frame-per-second (fps) is only around 0.7MB. If we compress it (zip), the size is merely a few KB. Therefore, the overhead for adding the TVM values into the video during streaming is very small. In a typical video streaming, this temporal information can be sent through out-of-band control channel, such as Real-Time Control Protocol (RTCP) sender report (SR) [22], so that the receiver can get this information for every pair of neighboring frames.

Although TVI only captures the temporal quality degradation of the received video, it is a very good predictor for the end-users' QoE. In the next subsections, we carry out subjective tests and show that TVI accurately predict the Mean Opinion Score (MOS) of different videos.

4.2 Subjective Experiments

To test the TVI with videos having different degrees of motion, we add another video sequence, *Football*, into our video samples. *Football* has scenes with very fast motion. We select three videos with different degrees of motion for the experiments. They are *Old Woman*, *Park* and *Football*, where *Old Woman* has the slowest motion while *Football* has the fastest motion.

We first stream these three videos over our wireless testbeds with different channel conditions. We collect a total of 50 samples of these videos with different quality. There are 17 samples of *Old Woman*, 17 samples of *Park* and 16 samples of *Football*. We then write an Android application on Nexus S smartphone and load the collected sample videos on the phone. We engaged 17 volunteers as subjects to watch the videos on the smartphone [1].

The Android app asked each subject to score the watched video on a standard five-grade scale [12]. Score 1 is for a video with the worst quality and it means the impairment in the video is very obvious and very annoying. Score 5 is for a video with the best quality and it means the video is perfect.

When a subject rates video quality while viewing a video, his/her rating changes as the video plays. In a long video, the subject would give several different quality scores to the different portions of the video. Basically, it is same as watching several short clips of video, and the subject gives one score for each clip. Therefore, in these subjective tests and the subsequent experiments in Section 5, we use short video clips, each plays around 10 seconds, which respectively contain slow, moderate and fast motion scenes. It simplifies our experimental settings and procedures.

Our test was performed according to the ITU single-stimulus (SS) method [12]. The standard videos with the five different scores were shown to the viewer at the beginning of the test. During the test, only the videos to be scored were shown without any display of the standard videos. For each video, the android app records the quality scores given by the subjects and obtain a mean score that is the Mean Opinion Score (MOS).

4.3 Estimating QoE

After the tests, we group the video samples according to their content. i.e. there are three groups, *Old Woman*, *Park* and *Football*. They respectively correspond to three motion

[1] According to ITU-R BT.500-11 subjective assessment standard [12], 15 subjects would be enough for subjective quality evaluation

Table 5: Estimation of the linear model for each video category.

Predicted MOS	Linear Model $(\hat{\beta}_0, \hat{\beta}_1)$	95% Confidence Interval of $\hat{\beta}_1$	ρ_{MOS}
$\widehat{TMOS}_{slow}$	(5.1, -0.28)	[-0.346 -0.205]	0.8708
$\widehat{TMOS}_{moderate}$	(3.9, -0.25)	[-0.345 -0.161]	0.9396
$\widehat{TMOS}_{fast}$	(4.2, -0.21)	[-0.268 -0.146]	0.9655

categories, *slow*, *moderate* and *fast*. We also measure TVI for each video sample. For each category, we will predict the MOS with TVI measurement.

For each category, we randomly select four samples and separate them from the others. These four samples will serve as a validation set. The remaining samples are used to derive an equation to predict the MOS with TVI. We use the linear model in the derivation. We propose a two-parameter linear model to predict MOS of each video category.

$$TMOS_{md} = \beta_0 + \beta_1 \overline{TVI} \quad (3)$$

for some constants β_0 and β_1. In this linear model, we use the average TVI, $\overline{TVI}$ of the video as the predictor variable. $TMOS$ is the predicted MOS, not the actual MOS that is evaluated from the human subjects. Hence, $TMOS$ is an objective video quality metric based on $\overline{TVI}$. We use a subscript md to indicate the motion degree of the video. In our categorization, md can be *slow*, *moderate* or *fast*.

It is interesting to note that during the TVI measurement, TVI values for some frames in the video are infinite. It is because that the TVM values of the corresponding frames in the received video are infinite. When there are frames lost and delay, the receiver playback mechanism will duplicate the last received frame. Therefore, the TVM value evaluated goes to infinite. We will explain this phenomenon in more detail in Section 5. For MOS prediction, we take infinite TVI value as 1. In fact, most TVI values are smaller than 1.

If we predict a quality score, Y, given by a particular user, we have

$$Y = \beta_0 + \beta_1 X + \epsilon \qquad E[\epsilon] = 0 \quad (4)$$

where X can be any predictor variable. There is also an error term, ϵ, added in the regression analysis. But, we are only interested in predicting the mean value of Y that is $TMOS$, hence we ignore the error term [15]. The same principle is also applied to the prediction of network impairment that we will discuss in Section 5.

Figure 2 shows the linear fits of the estimated $TMOS$, $\widehat{TMOS}$, for each video category. We use the *linear model* package of the statistics tool, ***R*** [21], to derive $\hat{\beta}_0$ and $\hat{\beta}_1$, that are respectively the estimates of β_0 and β_1 in Equation (3). We have the following general equation for the linear model estimation.

$$\widehat{TMOS}_{md} = \hat{\beta}_0 + \hat{\beta}_1 \overline{TVI} \quad (5)$$

The resultant $\widehat{TMOS}_{md}$, $\hat{\beta}_0$ and $\hat{\beta}_1$ for each video category are shown in Table 5.

The results shows that TVI has a very strong linear relation with MOS which is the users' QoE. Take *slow*-motion video as an example, its 95% confidence interval for $\hat{\beta}_1$ is (-0.346, -0.205). The small interval indicates that the sample size (number of videos) in the training set is large enough for a good estimation. Mean of $\hat{\beta}_1$ (-0.28) is significant to $\overline{TMOS}$ prediction. This justifies our decision of including $\overline{TVI}$ in our linear model for predicting MOS.

With $\hat{\beta}_0$ and $\hat{\beta}_1$ in Table 5, we then use the TVI of the video samples in the validation set to estimate TMOS of each sample. Comparing the estimated TMOS and the actual MOS we measured from the subjective experiments, we calculate the Pearson Correlation Coefficient between TMOS and MOS. Table 5 also shows such correlation coefficients (ρ_{MOS}). We find that across the different video categories, TMOS is highly correlated with MOS with an average correlation coefficient of 0.925. It shows TVI is a very good predictor of QoE.

5. NETWORK EVALUATION

A good video quality metric for mobile videos should not only estimate QoE accurately but also predict the network impairments, such as packet loss and delay. With extensive video streaming experiments, we demonstrate how TVI computations can be used to accurately estimate delay and packet loss.

5.1 Experimental Setup

We set up a single-hop wireless testbed. The WLAN configuration for the testbed is IEEE802.11n and the average minimum end-to-end delay is 0.604 milliseconds. Like in Section 4, we use three representative videos, one for each motion degree (slow, moderate and fast), for this set of experiments. We use FFmpeg [6] for video coding and VideoLan [18] for video streaming.

During the streaming, we use IPFireWall tool [8] to deliberately introduce packet loss and delay into the wireless link. Packet loss is injected as uniform distributed or as bursty pattern or a combination of both. Delays are injected randomly. Details of packet loss rate and delay are specified in each subsection below. Packet loss and delay in the network lead to the loss of temporal quality in the video. We see different extent of blocking, blurring and freezing in the receiver end when the packet loss rate and delay change.

For each streaming, both the sender and receiver calculate the TVM values of the video it sends and receives. The receiver also calculates the TVI values by matching the corresponding TVM values in received video with source video. TVI, TVM, packet loss rate and delay values are collected for each video streaming session. We then analyse the collected data sets with Matlab and R software.

5.2 Detecting Packet Loss

We first conduct experiments to observe the variation in TVM and TVI values with packet losses in the network. We vary the packet loss rate (PLR) from 0.1% to 3% for high-resolution videos and from 5% to 50% for low-resolution videos.

A low uniform PLR causes more significant distortions in high-resolution videos than in low-resolution ones. We can see an example in Figure 3. In Figure 3, *Old Woman* is an high-resolution video while *Foreman* is a low-resolution video. *Old Woman* with 3% PLR seems to have the same distortion as *Foreman* with 50% PLR. This is because high-resolution videos have larger spatial coding interdependencies than low-resolution videos. The frame size is large in a high-resolution video. If the frame size is larger than Maxi-

(a) 0.5% Loss; Old Woman (b) 3% Loss; Old Woman

(c) 5% Loss; Foreman (d) 5% Loss; Foreman

Figure 3: Comparing low and high resolution videos with different network packet loss effects

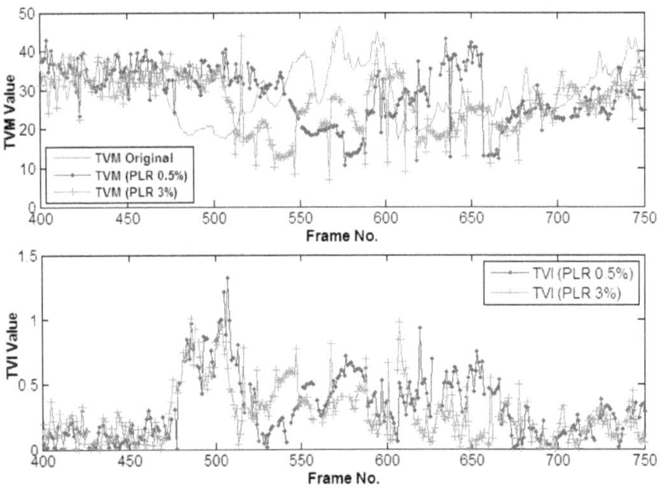

Figure 4: TVM and TVI computations for Video *Old Woman*

mum Transmission Unit (MTU), the frame is split into several packets during streaming. A packet lost in the network will cause incomplete reconstruction of the frame in the receiver, leading to display distortion. However, if the video resolution is low, a frame can be represented by a single packet during streaming. A packet loss results in a whole frame loss. In the perspective of human visual system, an occasional single frame loss is not as obvious to visual artifacts. Therefore, low-resolution videos are more "immune" to packet loss.

From TVI measurements, we find that increasing PLR leads to higher TVI values, indicating larger temporal quality degradation. Figure 4 shows the variation of TVM and TVI values for a high resolution video (*Old Woman*) when PLR is 0.1% and 3%. Figure 5 shows those values for a low resolution video (*Foreman*) when PLR is 5% and 50%. The variation of TVI of *Old Woman* with 3% PLR looks even greater than the variation of TVI of *Foreman* with 50% PLR due to visual distortions introduced by packet loss.

If we look at Figure 4 and Figure 5 more closely, we find there are many unconnected dots in Figure 5. This is due the infinite (INF) values of TVM and TVI. When we compute TVM, the current frame f_t becomes exactly the same as the preceding frame, $f_{t-\delta}$, where $\delta = \frac{1}{fps}$, so the value of $TVM_r(t)$ goes to INF according to Equation (1). When $TVM_r(t)$ is INF, according to Equation (2), TVI also becomes INF. These "INF" values are represented by breaks or discontinuity in Figure 5.

If the original video has some stationary scenes, $TVM_s(t)$ during those scenes will be recorded as INF. If there is no impairment in the streaming process, the received video will also have the stationary scenes, so $TVM_r(t)$ will also have corresponding INF values. It creates subtlety in arithmetic in Equation (2). To avoid that, we treat INF as a symbolic arithmetic variable in Equation (2). If there are two INFs in the numerator, they will cancel each other. In this way, any INF values in TVI indicate freezing due to packet losses.

Uniform and burst losses

Uniform and burst losses have different effect on temporal video quality. We demonstrate such effect by introducing uniform packet loss and burst packet loss in the same video streaming session but in different times. The uniform packet loss was applied from 1 second to 3.5 second of the video at PLR of 30%. In this uniform loss period, the packets are dropped randomly with a probability of 30%. The dropped

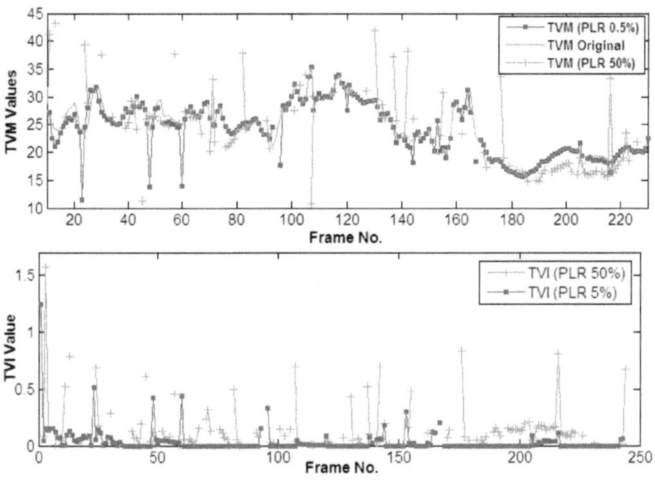

Figure 5: TVM and TVI computations for Video *Foreman*

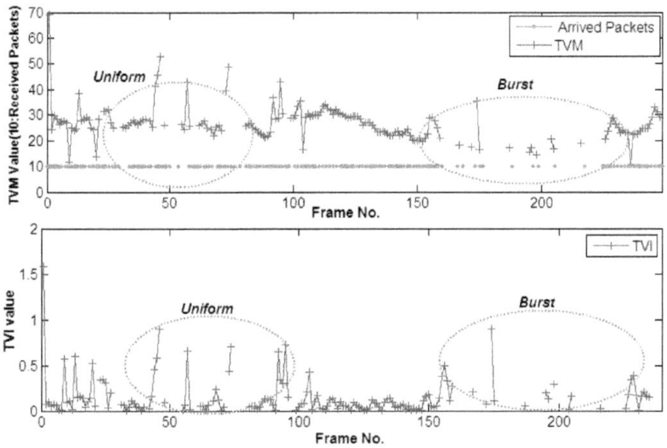

Figure 6: TVM and TVI values at PLR(30%) Types:Uniform and Burst(dense);

packets are not necessary adjacent to each other. Burst packet loss happens from 5 second to 7 second. Within the burst loss period, the system drops the packets with an average probability of 30%. But once a packet is selected to be dropped, its neighboring packets will also be dropped with a probability of 90% to 95%.

Figure 6 shows the TVM and TVI values for a sample video. It can be clearly observed that there are more INF values (discontinuities) for burst loss period than that in uniform loss period. This is attributed to the fact that burst losses cause more frame losses while a uniform loss may only impair the quality of frame without frame loss. That is for burst loss period, there will be more freezing of the video. An end-user gets a very annoying experience with subsequent freezing of video frames (number of INF values in our case).

5.3 Predicting Packet Loss Rate

In Section 5.2, we discuss how to use TVI to detect packet loss in a video streaming. Generally, when TVI value increases (including the number of INF increases), the packet loss rate also increases. Therefore, we expect there is a lin-

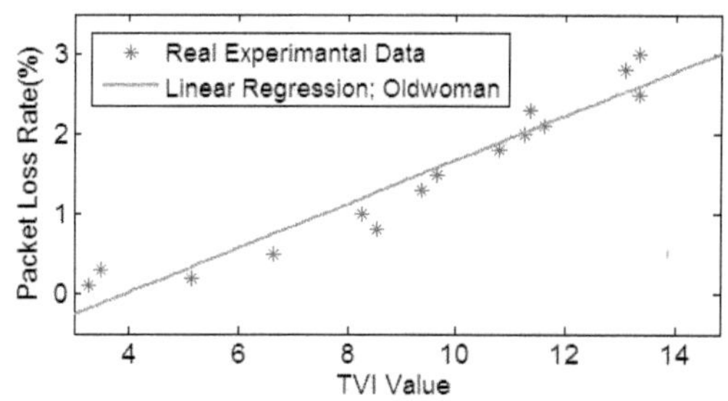

(a) *Slow-Motion* videos.

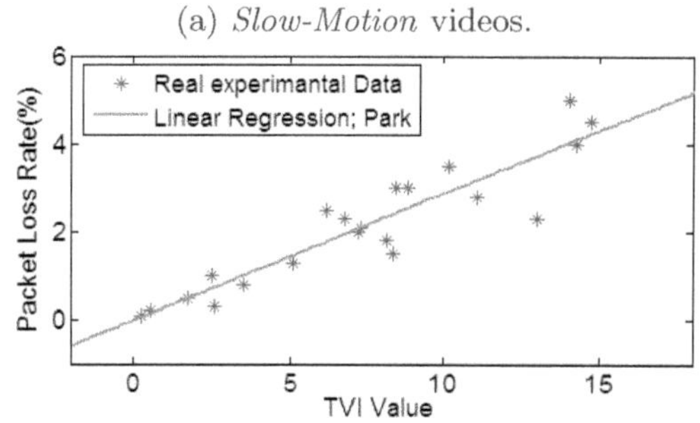

(b) *Moderate-Motion* videos.

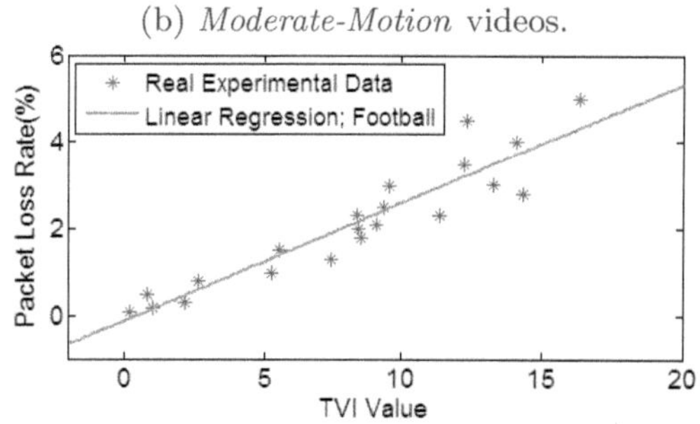

(c) *Fast-Motion* videos.

Figure 7: Linear fits of all three video categories for PLR prediction.

ear relationship between TVI and packet loss rate (PLR). To confirm this relationship, we carried out an extensive video streaming experiments. In the experiment, different videos are streamed over our wireless testbed (refer to Section 5.1) and different PLR are introduced during streaming. The received videos are collected and their TVI are measured.

We choose three videos for the experiment: the videos *old woman* has slow-motion, *park* has moderate-motion, and *football* is in the fast-motion category. For slow-motion video, we collect 60 samples from the experiment, among them 40 are for linear regression model while 20 are for validation. For moderate-motioned video, we collect 78 samples: 42 are for regression model and 36 are for validation. We collect 45 samples for fast motion video: 30 are for regression model while 15 are for validation.

We propose to have have following linear model for TVI to predict PLR.

$$TPLR_{md} = \beta_0 + \beta_1 \overline{TVI} \quad (6)$$

for some constants β_0 and β_1. In this linear model, we use the average TVI, $\overline{TVI}$ of the video as the predictor variable. *TPLR* is the predicted PLR, not the actual PRL that is measured from the network metrics [2]. Similarly to the derivation of *TMOS*, we use a subscript *md* to indicate the

[2]In our case, PLR is predefined as a system input

Table 6: Reliablilty to expect actual packet loss; Validation with linear regression models.

Predicted PLR	Linear Model $(\hat{\beta}_0, \hat{\beta}_1)$	95% Confidence Interval of $\hat{\beta}_1$	ρ_{PLR}
$\widehat{TPLR}_{slow}$	(-1.091, 0.277)	[0.234 0.319]	0.9655
$\widehat{TPLR}_{moderate}$	(-0.002, 0.287)	[0.236 0.343]	0.9498
$\widehat{TPLR}_{fast}$	(-0.104, 0.271)	[0.224, 0.318]	0.9046

motion degree of the video. Again, md can be *slow*, *moderate* or *fast*. Like the $TMOS$ derivation, we take TVI INF value as 1 if there is any.

Figure 7 shows the linear fits of the estimated $TPLR$, $\widehat{TPLR}$, for each video category. We use the *linear model* package of the statistics tool, ***R*** [21], to derive $\hat{\beta}_0$ and $\hat{\beta}_1$, that are respectively the estimates of β_0 and β_1 in Equation (6). We have the following general equation for the linear model estimation.

$$\widehat{TPLR}_{md} = \hat{\beta}_0 + \hat{\beta}_1 \overline{TVI} \quad (7)$$

The resultant $\widehat{TPLR}_{md}$, $\hat{\beta}_0$ and $\hat{\beta}_1$ for each video category are shown in Table 6.

The results shows that TVI has a very strong linear relation with PLR. Take *slow*-motion video as an example, its 95% confidence interval for $\hat{\beta}_1$ is (0.234, 0.319). The small interval indicates that the sample size (number of videos) in the training set is large enough for a good estimation. Mean of $\hat{\beta}_1$ (0.28) is significant to $\overline{TPLR}$ prediction. This justifies our decision of including $\overline{TVI}$ in our linear model for predicting PLR.

With $\hat{\beta}_0$ and $\hat{\beta}_1$ in Table 6, we then use the TVI of the video samples in the validation set to estimate TPLR of each sample. Comparing the estimated TPLR and the actual PLR we measured from the streaming experiments, we calculate the Pearson Correlation Coefficient between TPLR and PLR. In Table 6, ρ_{PLR} indicates the Pearson's Correlation Coefficient from the validation data. We find that across the different video categories, TPLR is highly correlated with PLR with an average correlation coefficient of 0.940. It shows TVI is a very good predictor of PLR.

5.4 Detecting and Predicting Delay

Delay causes freezing in streaming video services. In this section, we discuss the relationship between end-to-end delay and TVI. We first describe the impairment of the video quality caused by delay.

The frames transmitted by the source are split into packets and transmitted into the network. The packets received in the client are reconstructed into frames and played by decoder. Any delay introduced in the network will lead to delay in packets being received by the receiver and subsequently in freezing of frames. Figure 8 shows this effect. The delay in network causes modifications in TVI computations and INF values are generated in the TVI measurements at receiver. Unlike the INF values generated by packet losses, these INF values are followed by a lagged version of TVM values. The receiver estimates the best match for TVM values at receiver to the values at source (using a sequence alignment like algorithm, see [1, 17] for details).

In order to predict the delay effect, TVM values are calculated first in the receiver end. In the bottom of figure 8, after the first two frames, there is delay between frame $f2$ and frame $f3$. After $f2$ arrived well, decoder cannot play the very next frame on time so it has to wait until it receives next frame. This brings freezing or stop-motion in the display. Therefore, INF are produced during the delay because there is no frame to be calculated so it processes with the previous one. This does not necessary mean that another $f2$ is made to calculate the corresponding TVM.

Based on the TVM values of the receiver side, The TVM values are compared with TVM of source video to get TVI values. There are two delays in the figure 8. The first delay causes 3 INFs and the second one brings 2 INFs in this example.

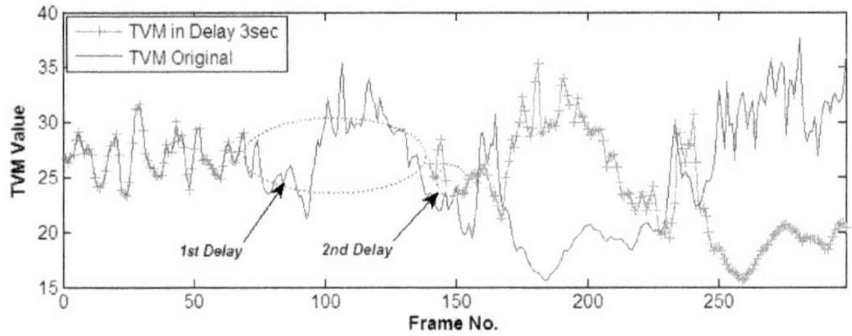

Figure 9: TVM values of Video *Foreman* with a delay of 3 seconds

Table 7: Estimation of Network Delay using TVI computations

Video	fps	D(actual)	# INF	$\hat{D}$	Error(%)
Duck	50	2.00	99	1.98	1
Foreman	25	3.00	75	3.00	0
Park	50	2.00	99	1.98	1
Oldwoman	30	4.50	155	5.17	15

Figure 9 shows TVM values of original and delayed video *foreman*. The dotted circle indicates vacant part of the graph, which are INFs in TVM computations of the delayed video. The TVM values of received video are shifted to right and INFs appear in between. An prediction of end-to-end delay (D) can be obtained by counting these INFs.

$$\hat{D} = \frac{I}{fps} \quad (8)$$

where fps denotes the frame rate of the video. I is the number of INFs in between two non-INF TVMs at the receiver end.

Table 7 shows average results for the four videos we selected to test the accuracy of end-to-end delay prediction. As only number of INFs in TVM is counted, unlike MOS and PLR prediction, the delay prediction is video motion independent. Therefore, we just randomly choose the four different video contents from our video pool. The inaccuracy of the delay prediction mainly come from the inaccuracy of TVM matching (the application of sequence alignment algorithm).

In Table 7, # INF means the total number of INFs (pertaining to delay) in the whole playing time. The estimation has the accuracy of over 95%.

In practical scenarios, we calculate TVI as the video streaming is ongoing. We will first detect the delay by applying existing sequence alignment methodologies [1, 17]. After

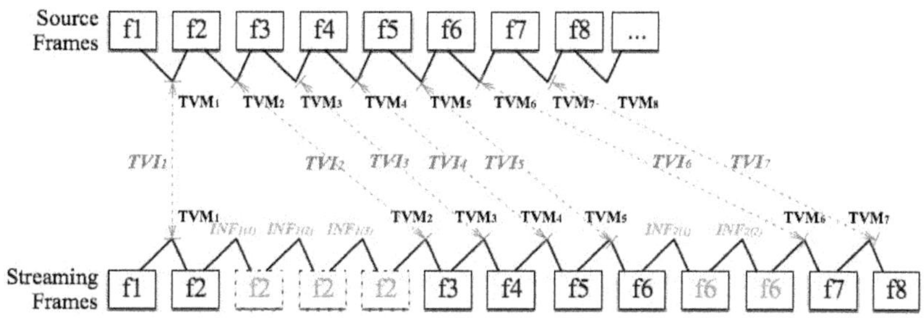

Figure 8: TVM/ TVI computations in presence of network delay

excluding those delay-caused INFs, we detect and predict the packet loss rate based on TVI values.

6. CONCLUSIONS

Accurately evaluating video streaming quality over networks always attracts great attention from content providers and network service providers. However, video quality evaluation itself is a challenging topic as users' QoE cannot be measured with low cost. When the video applications move to mobile devices, evaluating video quality becomes even more challenging. Wireless networks cause packet loss and delay to the video streaming while mobile devices have limited computation power and memory. In this paper, we have proposed a novel reduced-reference temporal quality metric, Temporal Variation Index (TVI) for mobile videos. Despite its simplicity, it shows strong linear relationship with users' Mean Opinion Score (MOS). Therefore, it is a very good predictor for users' QoE on mobile videos. Its prediction of MOS has a correlation coefficient of 0.925 with the actual MOS measured from subjective tests. TVI also predicts network impairments, such as packet loss and end-to-end delay. From our extensive experiments of video streaming over a wireless network testbed, we find that the packet loss rate predicted by TVI has a correlation coefficient of 0.94 with the measured loss rate. During the computation of TVI, we can also predict the end-to-end delay with an accuracy of around 95%. These appealing features make TVI a very useful tool for content and network service providers to offer the future high quality video streaming services.

Although temporal quality is a unique and important aspect for video, it is also very interesting to integrate TVI with other spatial quality metrics. We believe that such integration will provide a comprehensive video quality evaluation metric for the future use.

7. ACKNOWLEDGEMENTS

This research was supported in part by the National Science Foundation through the Grant No. CNS-0831914.

8. REFERENCES

[1] A. Chan, K. Zeng, P. Mohapatra, S. Lee, and S. Banerjee. Metrics for evaluating video streaming quality in lossy ieee 802.11 wireless networks. In *INFOCOM, 2010 Proceedings IEEE*, pages 1–9. IEEE, 2010.

[2] D. Chandler and S. Hemami. Vsnr: A wavelet-based visual signal-to-noise ratio for natural images. *Image Processing, IEEE Transactions on*, 16(9):2284–2298, 2007.

[3] K. Chen, C. Chang, C. Wu, Y. Chang, and C. Lei. Quadrant of euphoria: a crowdsourcing platform for qoe assessment. *Network, IEEE*, 24(2):28–35, 2010.

[4] CiscoSystem. Cisco visual networking index: Global mobile data traffic forecast update, 2010 - 2015, 2011.

[5] U. Engelke and H.-J. Zepernick. Perceptual-based quality metrics for image and video services: A survey. In *Next Generation Internet Networks, 3rd EuroNGI Conference on*, pages 190 –197, may 2007.

[6] FFMPEG. Ffmpeg. `http://www.ffmpeg.org`, 2012.

[7] D. Field, A. Hayes, and R. Hess. Contour integration by the human visual system: Evidence for a local. *Vision Research*, 33(2):173–193, 1993.

[8] F. Group. Freebsd ipfirewall. `http://www.freebsd.org/doc/en_US.ISO8859-1/books/handbook/`, 2012.

[9] B. Horn and B. Schunck. Determining optical flow. *Artificial Intelligence*, 1981.

[10] Q. Huynh-Thu and M. Ghanbari. Scope of validity of psnr in image/video quality assessment. *Electronics Letters*, 44(13):800 –801, 19 2008.

[11] Q. Huynh-Thu and M. Ghanbari. Temporal aspect of perceived quality in mobile video broadcasting. *Broadcasting, IEEE Transactions on*, 54(3):641 –651, sept. 2008.

[12] ITU. *Methodology for the subjective assessment of the quality of television pictures.* International Telecommunication Union: Recommendation ITU-R BT.500-11, 2002.

[13] D. Kim, S. Cho, and J. Jeong. A motion vector recovery algorithm for temporal error concealment using optical flow in h.264 video coding. *Multimedia and Expo, IEEE International Conference on*, 0:1713–1716, 2006.

[14] A. Leontaris and A. Reibman. Comparison of blocking and blurring metrics for video compression. In *Proceedings ICASSP*, 2005.

[15] N. Matloff. *From Algorithm to Z-Scores: Probabilistic and Statistical Modeling in Computer Science.* Creative Commons Lincense, `http://heather.cs.ucdavis.edu/matloff/probstatbook.html`, 2009.

[16] A. K. Moorthy and A. C. Bovik. Efficient video quality assessment along temporal trajectories. *IEEE Trans. Circuits Syst. Video Techn.*, 20(11):1653–1658, 2010.

[17] D. W. Mount. *Bioinformatics: sequence and genome analysis.* CSHL press, 2004.

[18] V. Organization. Videolan. `http://www.videolan.org`, 2012.

[19] R. Pastrana-Vidal and J. Gicquel. Automatic quality

assessment of video fluidity impairments using a no-reference metric. In *Proc. of Int. Workshop on Video Processing and Quality Metrics for Consumer Electronics*, 2006.

[20] K. Pauwels and M. V. Hulle. Realtime phase-based optical flow on the gpu. In *Computer Vision and Pattern Recognition Workshops*. IEEE Computer Society, 2008.

[21] R. The r project for statistical computer. `http://www.r-project.org`, 2009.

[22] H. Schulzrinne, S. Casner, R. Frederick, and V. Jacobson. Real time control protocol (rtcp) attribute in session description protocol (sdp). RFC 3550, The Internet Society, 2003.

[23] K. Seshadrinathan and A. C. Bovik. Motion tuned spatio-temporal quality assessment of natural videos. *IEEE Trans. on Image Processing.*, 2010.

[24] Z. Wang, A. Bovik, H. Sheikh, and E. Simoncelli. Image quality assessment: From error visibility to structural similarity. *Image Processing, IEEE Transactions on*, 13(4):600–612, 2004.

[25] Z. Wang, L. Lu, and A. Bovik. Video quality assessemnt based on structural distortion measurement. *Signal Processing: Image Communication*, 19(2):121–132, February 2004.

[26] Z. Wang, H. Sheikh, and A. Bovik. Objective video quality assessment, 2003.

[27] D. H. Warren and E. R. Strelow. *Electronic Spatial Sensing for the blind: contributions from perception.* Springer, 1985.

[28] C. Wu and C. Kuo. Design of integrated multimedia compression and encryption systems. *Multimedia, IEEE Transactions on*, 7(5):828–839, 2005.

[29] F. Xiao et al. Dct-based video quality evaluation. *Final Project for EE392J*, page 769, 2000.

[30] K.-C. Yang, C. Guest, K. El-Maleh, and P. Das. Perceptual temporal quality metric for compressed video. *Multimedia, IEEE Transactions on*, 9(7):1528 –1535, nov. 2007.

ParCast: Soft Video Delivery in MIMO-OFDM WLANs

Xiao Lin Liu*, Wenjun Hu†, Qifan Pu†*, Feng Wu† and Yongguang Zhang†

Microsoft Research Asia† University of Science and Technology of China*

{wenjun, fengwu, ygz}@microsoft.com {lin717, puqf}@mail.ustc.edu.cn

ABSTRACT

We observe two trends, growing wireless capability at the physical layer powered by MIMO-OFDM, and growing video traffic as the dominant application traffic. Both the source and MIMO-OFDM channel components exhibit non-uniform energy distribution. This motivates us to leverage the source data redundancy at the channel to achieve high video recovery performance. We propose ParCast that first separates the source and channel into independent components, matches the more important source components with higher-gain channel components, allocates power weights with joint consideration to the source and the channel, and uses analog modulation for transmission. Such a scheme achieves fine-grained unequal error protection across source components. We implemented ParCast in Matlab and on Sora. Extensive evaluation shows that our scheme outperforms competitive schemes by notable margins, sometimes up to 5 dB in PSNR for challenging scenarios.

Categories and Subject Descriptors

C.2 [**Computer-Communication Networks**]: Miscellaneous

General Terms

Algorithms, Design, Performance

Keywords

MIMO-OFDM, Video, Joint source-channel design

1. INTRODUCTION

MIMO-OFDM technologies have become the default building blocks for next generation wireless networks. As wireless capability continues to grow, so does the application demand. According to the Cisco Visual Networking Index [ ], traffic from wireless devices will exceed traffic from wired devices by 2015. Internet video is a major source of such traffic growth, which accounted for 40% of consumer Internet traffic in 2010, and will reach 62% by the end of 2015. Video-on-demand traffic is expected to triple by 2015, and video unicast constitutes the bulk of the volume. As a result, considerable efforts have been devoted to improving video delivery quality over wireless links, especially unicast. Supporting in-home high-definition video streaming is precisely one of the motivations for 802.11ac.

Traditionally, video sources are first compressed, represented digitally as a bit stream, and then transmitted in the same way as other binary data, as shown by MPEG [ ]. Conventional video coding schemes normally include a Discrete Cosine Transform (DCT) to compact energy across all pixels and remove correlation between them. Non-uniformly distributed DCT coefficient energy levels imply statistical features, or *redundancy*, in the source data, and suitable entropy coding could remove such redundancy. Less redundancy means more power per unit compressed data, and hence better error resilience and overall decoding performance. Given the expected channel condition, we can then separately determine the compression rate at the source and the transmission rate at the channel. However, if the actual channel condition turns out worse than expected, the predetermined source and channel rates would have been too aggressive and cause bit errors in decoding. Since digital rates do not have a graceful fallback behavior in the face of bit errors, glitches will occur. This can happen frequently for wireless links, due to unpredictable channel conditions.

While wireless links are loss prone, the original video need not be received in its entirety to ensure a good visual quality. A synergy between the two sides could therefore be more effective than separately optimizing the source and the channel coding. This has motivated an industry standard effort, Wireless High Definition Interface (WHDI) [ ], which sends uncompressed high definition videos over wireless links. Nevertheless, this is still inefficient due to the sheer volume of uncompressed video data for transmission, since the redundancy within the data is not utilized at either the source encoder or the decoder.

A few recent cross-layer approaches aim to better exploit the redundancy in the source to improve video delivery quality over wireless. They are motivated by the observation that such source redundancy could be used at the channel to protect against errors. *SoftCast* [ ] starts with a similar DCT transform to that in MPEG, but then allocates power weights based on the redundancy level of the source components before transmission. The entire codec uses a series of linear transforms to ensure the mean squared error of the reconstructed pixel luminance is proportional to the channel noise throughout. While conventional systems perform *lossy compression over lossless digital communication*, SoftCast performs *lossless compression over lossy analog communication*. *FlexCast* [ ] retains much of the MPEG encoding process, but replaces the entropy coding stage with a rateless code. The source blocks that contribute more to the overall distortion, i.e., the high-entropy components, are represented with more bits. Note that allocating different amounts of power or bits to different source components is a form of unequal error protection (UEP). Both SoftCast

*The work was done while Xiao Lin Liu and Qifan Pu were research interns with Microsoft Research Asia.

MobiCom'12, August 22–26, 2012, Istanbul, Turkey.

and FlexCast use a single code to simultaneously compress and protect the source. Their performance shows that it is unnecessary to optimally compress the source, provided the amount of residual source redundancy matches the required error protection over the channel.

However, these approaches were designed with single-antenna links in mind or channel oblivious for broadcast. OFDM (Orthogonal Frequency Division Multiplexing) decomposes a wideband channel into a set of mutually orthogonal *subcarriers*, and the channel gains across these subcarriers are usually different [ ], sometimes by as much as 20 dB. With MIMO (Multiple-Input Multiple-Output), each subcarrier is further divided into a set of *spatial subchannels*, again with different channel gains. Furthermore, a channel dependent *precoding* operation is often necessary to make the spatial subchannels on the subcarrier mutually orthogonal, or *independent*, so that signals do not interfere with one another along different subchannels. As a result, even for unicast, there are several issues with running SoftCast or FlexCast directly over a MIMO-OFDM link. In particular, error behavior differs across subchannels. If a one-size-fits-all code rate is used for a few or all subchannels, it generally needs to be conservative, and hence suboptimal, to accommodate the worst subchannels. In this sense, unicast over a MIMO-OFDM link resembles a broadcast channel. The need for precoding makes it difficult to ensure that a single channel oblivious error protection scheme can perform well across MIMO-OFDM links.

Given that both the source DCT component energy and subchannel gains are non-uniformly distributed, if we encode the video source such that the more important, high-energy DCT components will be transmitted in high-gain subchannels, and less important parts in lower-gain subchannels, we may better utilize the overall channel. Note, however, that the DCT and the precoding steps are essential at the source and channel respectively to avoid interference between source components and between subchannels. This suggests that we should match DCT components to subchannels based on the respective sorted order of energy level, and then perform joint source-channel power allocation to optimize per DCT component and subchannel error performance. This is unequal error protection operating at a finer granularity than conventional layered video coding.

With these in mind, we present ParCast (**Par**allel video uni**Cast**), which tailors the unicast video delivery quality to the MIMO-OFDM channel. The key features are: (1) obtaining independent source components and subchannels, (2) matching important source components to high-gain subchannels, (3) scaling the source data with power weights computed with joint source-channel considerations, and (4) transmission using analog modulation.

We implement the video codec in Matlab, and the channel dependent modules of ParCast both as a Matlab simulation and on Sora [ ]. We run experiments on Sora to validate the Matlab simulation, and then run channel trace driven simulations to compare ParCast against alternatives. Results show that ParCast outperforms the best conventional digital scheme, sometimes by 5 dB, for videos with fast motion and a large energy spread over links with a large subchannel gain spread. This is a challenging case for conventional schemes, but favors ParCast.

To summarize, our main contributions are: First, we highlight the analogy between the energy distributions for video

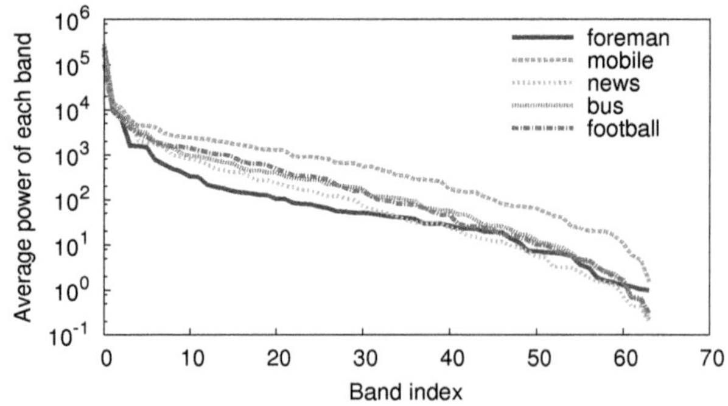

Figure 1: 2D 8×8 block DCT coefficient energy.

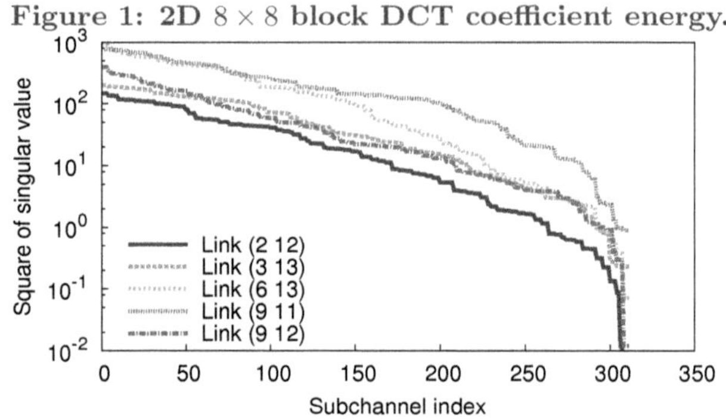

Figure 2: 3×3 MIMO-OFDM subchannel gains.

sources and MIMO-OFDM channels, and show how imperfect distributions for both can work in synergy. Second, we show the importance of decorrelating both the source and the channel and matching them accordingly at a fine granularity. Third, we show that a simple scheme could achieve significant benefit by implementing the above steps. Extensive evaluation shows that this is indeed an effective video delivery mechanism over MIMO-OFDM links.

2. BACKGROUND AND MOTIVATION

2.1 Source and channel characteristics

Source. Video sources exhibit spatial correlation within a frame and temporal correlation across frames. A DCT transform is often a key step of video coding to remove redundancy. If we divide a frame into 8×8 blocks, and plot the energy distribution of the block DCT coefficients, we typically see very non-uniform distributions (Figure 1). The energy often spreads across 5 or 6 orders of magnitude, and the high energy end drops very quickly.

The non-uniform distribution implies statistical features, and hence remaining redundancy, in the DCT coefficients. If left as is, such residual source redundancy is considered harmful, because there would be more information to transmit, with less power per unit of information. Therefore, traditional video compression schemes follow DCT with entropy coding to further compress the DCT coefficients. The high-entropy coefficients naturally require more bits to be fully represented. The end result is that the energy distribution over each bit is the same and the encoded source appears Bernoulli distributed across bits.

Unequal error protection for the source is often a key element of video coding. Each DCT coefficient has a different contribution to the total distortion and should be protected accordingly. Entropy coding could be viewed as implicit unequal error protection, precisely because more bits are allocated to more complex, high-entropy signals. Since all bits are equal, more bits imply more power needed for transmitting the complex signal, which means more protection.

Channel. A MIMO-OFDM link can be viewed as a set of narrowband MIMO channels for each subcarrier of OFDM. Each subcarrier channel can be represented with a channel matrix H. We can obtain its singular value decomposition (SVD), $H = U\ S\ V^H$, where U and V are two unitary matrices, S is a diagonal matrix with the singular values of H on the diagonal, and $(\cdot)^H$ means conjugate transpose, or Hermitian. Each non-zero singular value s_i corresponds to a subchannel i with channel gain s_i^2. The diagonal form of S means the subchannels are mutually orthogonal. In general V is not diagonal, and neither is H, which means we cannot directly leverage the mutually orthogonal subchannels when sending a signal vector $\mathbf{x}$ unless all s_i are the same. If, instead, we always transmit $V\mathbf{x}$, or *precode* the signal vector $\mathbf{x}$ with a precoding matrix V, then the received signals would be mutually orthogonal and can be decoded after a rotation, i.e., multiplying by U^H. Therefore, with SVD based precoding for each subcarrier, a MIMO-OFDM link can simply be viewed as a set of independent subchannels.

Figure 2 shows sorted subchannel gains for a few 3×3 MIMO-OFDM links, with 3 antennas on either end of the link. Each link is labeled (<source ID> <destination ID>). We can also see a large spread between the strongest and the weakest subchannels, though faster dropoff at the low gain end. Having more antennas increases the spread but flattens the high end slightly.

The diverse subchannels naturally merit different quantities of information each. This is the overarching goal of channel coding, so that the energy per bit sent on any subchannel would appear the same. Consequently, rate increase across a MIMO-OFDM channel is no longer monotonic with the overall channel SNR, unlike for single-antenna channels, which are scalar.

Source-channel separation and challenges. With fully random source bits and fully random channel bits, Shannon's source-channel separation theorem then shows we get the optimal result.

However, a capacity-achieving channel code must have an infinite block length. Given wireless channels are error prone, optimally loading source bits to each subchannel, along with sufficient protection bits, is impractical. Furthermore, the transmitter must know the exact subchannel gains and also notify the receiver how to decode on each subchannel. The former is susceptible to channel variation and the latter step would incur a high overhead.

2.2 Issues with existing approaches

Naturally, jointly performing source and channel coding would resolve some of the issues above, by retaining some source redundancy as channel protection. Such schemes typically still design separate, but matching codes for the source and the channel.

Instead, two recent systems, SoftCast and FlexCast, each designs a single code that compresses and protects simultaneously. SoftCast builds on an analog code that adjusts the compression-protection tradeoff with power allocation, whereas FlexCast achieves a similar goal with bit allocation.

For single-antenna channels, both schemes were shown to achieve almost optimal unicast performance, although SoftCast was designed for broadcast. However, directly running these over MIMO, without precoding, cannot avoid interference between received signals on different subchannels.

Even with precoding, neither distinguishes between different subchannel gains. Note also that FlexCast is currently built without knowledge of the channel, whereas precoding requires channel information. Once there is channel information, the source could narrow down the transmission rate to within a few choices, and the rateless code in FlexCast would be largely unnecessary.

Although there have been other digital unequal error protection schemes over MIMO or MIMO-OFDM (e.g., [[illegible]]), they are typically based on the scalable video coding (SVC) extension of H.264 [[illegible]]. These schemes divide the source into layers of different importance levels and allocate the most important layer to the best spatial subchannel. There is strong dependency between the layers. If the most important layer is not correctively decoded, any further layers are useless. Furthermore, there are usually more subchannels than layers. Limited digital modulation choices could constrain the performance when a layer needs to be transmitted over several heterogeneous subchannels.

2.3 Source-channel similarities

We observe a few similarities between the operations of source and channel coding, and therefore a joint source-channel scheme that exploits the synergy between the two is likely to work well.

First, SVD based precoding for a MIMO channel can be viewed as compacting energy to diagonalize the channel matrix, whereas DCT at the source is also expected to diagonalize the correlation matrix between source pixels. Without precoding at the channel, the received spatial signals would interfere with one another and hurt decoding performance, as some signal strength is used to cancel out the inter-signal interference before decoding. With correlation between the source components, power or bit allocation to them would be suboptimal. Therefore, it is important to avoid correlation at both the source and the channel.

Second, Figures 1 and 2 show a similar spread between the highest- and lowest-energy components, although the dropoff rate behavior differs. Therefore, it seems natural to match both sides, so that the high-energy DCT components are transmitted on the high-gain subchannels to avoid them acting against each other. Furthermore, the large number of subchannels allows fine-grained error protection levels.

Third, power weights can be allocated to transmitted source components to minimize the overall distortion. On the other hand, power allocation between different subchannels would also affect the achieved channel rates, which would eventually translate to the reconstructed source distortion based on a suitable rate-distortion function. Therefore, a joint source-channel power allocation strategy could optimize the overall recovery performance.

3. ParCast DESIGN

3.1 Overview

ParCast encodes and transmits a video sequence via a series of linear transforms.

The source encoder first computes 3D-DCT [[illegible]] coefficients for each 4-frame group of pictures (GOP). These DCT coefficients are grouped into chunks, the total number of which matches that of the subchannels in the MIMO-OFDM link, and a variance is computed for the energy of each chunk. The chunks are assigned to subchannels such that high-energy DCT coefficients would be sent on high-gain subchannels in successive packets, which is a permutation transform. Power whitening is then performed per chunk to even out power distribution.

The power allocation stage formulates an optimization problem to minimize the total distortion, and scales the magnitude of the DCT coefficients in each chunk by a weight. These power weights are computed by taking into account both the variance of the DCT coefficients and the corresponding subchannel gain.

Pairs of coded values are combined into complex symbols. Those for different spatial subchannels on each subcarrier are then precoded, so that they will be transmitted on orthogonal spatial channels and arrive at the receiver uninterfered by other spatial streams. The precoded values are then transmitted without further channel coding over the MIMO-OFDM channel.

The encoding and transmission process can be written as:

$$\mathbf{Y} = H\,Q\,G\,M\,R\,\mathbf{X} = C\,\mathbf{X} \tag{1}$$

where H is the overall channel matrix of all the subcarriers stacked into one, Q is the precoding matrix, G is a diagonal matrix with the power weights on the diagonal, M represents per-stream whitening, R implements the mapping between DCT coefficient streams to subchannels, and $\mathbf{X}$ are the DCT coefficients.

The decoder does not perform regular MIMO processing to first decode the received signals on each antenna. Instead, it forms an aggregate matrix with the received signals on all the spatial and frequency domain subchannels, before invoking a Linear Least Square Estimator (LLSE) decoder to recover the DCT coefficients. Then, it reconstructs the original frames by taking the inverse of the 3D-DCT.

3.2 Effects of a MIMO-OFDM channel

Per-subcarrier view. Let us consider a 2×2 MIMO-OFDM link as an example. By design, the signals on different OFDM subcarriers are independent from one another. On a particular subcarrier, each transmit antenna can send a complex symbol composed of two real values, and therefore 4 real values can be sent at a time. Let $H_{complex}$ be the 2×2 complex channel matrix for the subcarrier, with each matrix element h_{ij} to denote the path gain from transmit antenna j to receive antenna i. Let $x_i = m_{i1} + j\,m_{i2}$ be the complex symbol on the ith transmit antenna, and y_i be the complex symbol received on the ith receive antenna. Disregarding channel noise for now, we have

$$\begin{pmatrix} y_1 \\ y_2 \end{pmatrix} = \begin{pmatrix} h_{11} & h_{12} \\ h_{21} & h_{22} \end{pmatrix} \begin{pmatrix} m_{11} + j\,m_{12} \\ m_{21} + j\,m_{22} \end{pmatrix}$$

Separating the real and imaginary parts on both sides:

$$\begin{pmatrix} \Re(y_1) \\ \Im(y_1) \\ \Re(y_2) \\ \Im(y_2) \end{pmatrix} = \begin{pmatrix} \Re(h_{11}) & -\Im(h_{11}) & \Re(h_{12}) & -\Im(h_{12}) \\ \Im(h_{11}) & \Re(h_{11}) & \Im(h_{12}) & \Re(h_{12}) \\ \Re(h_{21}) & -\Im(h_{21}) & \Re(h_{22}) & -\Im(h_{22}) \\ \Im(h_{21}) & \Re(h_{21}) & \Im(h_{22}) & \Re(h_{22}) \end{pmatrix} \begin{pmatrix} m_{11} \\ m_{12} \\ m_{21} \\ m_{22} \end{pmatrix}$$

Essentially, the vector $(m_{11}, m_{12}, m_{21}, m_{22})^T$ has traversed an equivalent channel represented by the real matrix, H_{real}, that can be derived from the complex channel, $H_{complex}$.

Precoding at the channel. Recall that precoding based on the Singular Value Decomposition (SVD) can decorrelate the spatial subchannels to be mutually orthogonal. Since we can now view the channel as either a real or a complex matrix, there are two options for calculating the precoding matrix. The SVD of both, $H_{real} = U_{real}S_{real}V_{real}^H$ and $H_{complex} = U_{complex}S_{complex}V_{complex}^H$, are closely related. There are twice as many singular values in S_{real}, but just the same numbers from $S_{complex}$ and each with a duplicate copy. Since a singular value represents a subchannel, each *complex subchannel* of $H_{complex}$ corresponds to two equal gain *real subchannels* of H_{real}. V_{real} and $V_{complex}$ are related similar to H_{real} and $H_{complex}$.

Therefore, we simply compute $S_{complex}$ and $V_{complex}$, and convert them to S_{real} and V_{real} for later steps.

The overall link. For the overall link, we can stack together all the individual subcarrier matrices to form a large block diagonal matrix. Since each chunk is mapped to a subchannel on a subcarrier, we can still decode on each subcarrier separately before combining the chunks to reconstruct the video frame. However, the per-chunk whitening creates dependency between the symbols on the same subchannel in different packets (time slots), and therefore we should first stack the matrices across time on the same subcarrier.

3.3 Managing bandwidth requirement

We determine the number of source chunks per video frame based on the number of real subchannels and the available channel time, and then compute the required number of time slots to send all data in the source chunk. For each chunk, we calculate the variance of the energy of all data, λ. The λ is the metadata per chunk to be used at the decoder. More chunks would mean more metadata, which could help improve the reconstruction performance though at a higher overhead. That said, we need only enough chunks so that the distribution of λ has the desirable fast dropoff behavior as shown in Figure 1.

The product of the number of OFDM symbols, the number of packets (i.e., time slots), and the number of chunks per GOP reflects the bandwidth requirement of sending a GOP.

Insufficient bandwidth. When there are insufficient time slots, we should discard the least important chunks, since they contribute least to the overall reconstruction quality.

Bad subchannels. If some subchannels are very bad, it would be better to skip those. This way, the power budget can be expended on the remaining, more important chunks in higher gain subchannels. If this causes bandwidth shortage, we simply discard the least important source chunks.

Redundant bandwidth. If there is more bandwidth than we need to transmit the video sequence, we could also skip the bad subchannels and allocate the power to the remaining good channels. With each skipped bad subchannel, ParCast can benefit from a diversity gain.

3.4 Error protection and power allocation

The distribution of per source chunk energy variance, λ, is analogous to that of the 8×8 block DCT coefficients in Figure 1. If the chunk data are sent verbatim, their energy helps guard against the channel noise and serves as implicit channel coding. However, it is likely that the high-energy source components would consume too much transmission power budget and leave the remaining components more sensitive to the channel noise. Therefore, we need to re-distribute the power among source components before transmission.

Power weights. With the precoding operation at the encoder using V_{real} and equalization based on the matrix U^H

at the receiver, the MIMO-OFDM channel is separated into a set of orthogonal subchannels. The optimization problem in ParCast is to minimize the expected mean square error (MSE) under the transmission power constraint with budget P. Assuming inverse decoding, i.e., using C^{-1} from Equation (1), we can formulate the problem as:

$$\min \quad MSE = \sum_i \frac{\sigma^2}{s_i^2 g_i^2} \tag{2}$$

$$\text{subject to:} \quad trace\ E[(QG\mathbf{x})(QG\mathbf{x})^T] = trace\ E[(G\mathbf{x})(G\mathbf{x})^T] = \sum_i g_i^2 \lambda_i \leq P$$

where σ^2 is the average noise power, s_i is the singular value on subchannel i, Q is the channel precoding matrix, and G is the diagonal, power weight matrix, with g_i as its ith diagonal entry. By using Lagrange multipliers, the solution is:

$$g_i = \sqrt{\frac{P}{\sqrt{\lambda_i s_i^2} \sum_j \sqrt{\lambda_j / s_j^2}}} \tag{3}$$

and the derived MSE is

$$\frac{\sigma^2}{P}\Big(\sum_i \sqrt{\lambda_i / s_i^2}\Big)^2 \tag{4}$$

The MSE (4) is determined by the mapping of λ to s, i.e., the matrix R in Equation (1). By the Rearrangement Inequality [ ], the best R lets the λ_i follow the same sorted order as s_i. Hence, the chunk with the largest variance is assigned to the highest-gain subchannel.

Strictly speaking, the above is an approximation for when the optimal LLSE decoder is used. This is because inverse decoding could magnify the channel noise. When the channel condition is good, however, the two decoders have comparable performance, and the optimization of MSE of inverse decoding is much simpler to solve.

If we re-formulate the optimization using LLSE instead, the derived g_i depend on the channel noise σ^2 [ ]. If λ_i/s_i^2 is too low, the corresponding g_i would be zero in the new optimization result. This means the subchannel and the corresponding chunk should be discarded, and their power returned to the overall power budget for others. This is analogous to the water-filling power allocation strategy in MIMO systems to achieve the channel capacity.

Per-chunk whitening. The data within a chunk normally carry different amounts of energy. If these are assigned to OFDM symbols directly, there could be a large power difference between symbols and cause issues at the receiver. Therefore, we first mix data within a chunk with a random orthogonal matrix to achieve the same effect as a Hadamard transform[1], such that the average power per OFDM symbol is the same. This also ensures the same average power across transmit antennas and eases the requirement on the dynamic range of the power amplifier at the transmitter.

3.5 Decoder

Given the linear encoder, we use the LLSE decoder at the receiver, which is the optimal decoder [ ].

$$D_{LLSE} = \Lambda C^T (C \Lambda C^T + \Sigma)^{-1} \tag{5}$$

where C is the overall encoder from Equation (1), and Σ and Λ are both diagonal matrices. The ith diagonal element of Σ is the channel noise experienced by the packet carrying the ith row of the received $\mathbf{Y}$. The ith diagonal element of Λ is λ_i/s_i^2, where λ_i is the variance of the ith chunk and s_i is the gain on the corresponding subchannel.

After decoding the DCT coefficients, ParCast reconstructs the original frames by taking the inverse 3D-DCT.

3.6 Managing metadata

Source metadata. The λ per chunk is the metadata from the video source, which should be sent to the video decoder. The video sender transmits this metadata per GOP in a regular packet at the lowest rate before sending encoded video data. The overhead is less than 0.005 bits per pixel.

Channel state information (CSI). The video sender needs to know the channel detail to perform the chunk-subchannel mapping, power allocation, and precoding. Therefore, the channel state information can be viewed as metadata from the channel. Since the upcoming WiFi standard 802.11ac and LTE both expect to use channel dependent precoding, it is reasonable to assume the availability of CSI.

The CSI updates are *cumulative*. Since later packets would be precoded, the video receiver would then measure a precoded channel, $H_{measured} = H_{actual}\ V_{previous}$. $H_{measured}$ has the same singular values as the actual channel H_{actual}, so that we can assign chunks to subchannels and calculate power weights directly. However, typically $V_{actual} \neq V_{measured}$, so $V_{measured}$ alone cannot be used as the precoding matrix. We have two options: (1) the sender computes the actual channel as $H_{actual} = H_{measured} \cdot V_{previous}^H$, get V_{actual} as the new precoding matrix; or (2) the sender gets V_{new} from $H_{measured}$, and use $V_{previous} \times V_{new}$ as the precoding matrix. Although $V_{actual} \neq V_{previous} \times V_{new}$ in general, the two sides differ by a multiplication by a unitary block diagonal matrix, and achieve the same precoding effect. We have verified both options and currently use the first one.

Currently, the video receiver would measure the channel using the regular channel estimation technique, e.g., that for 802.11n MIMO-OFDM packets. It then feeds the CSI back to the sender by sending a regular packet. A CSI update every 100 ms is normally considered acceptable overhead. If there is also traffic from the receiver to the sender, we could leverage channel reciprocity by letting the sender estimate the channel. Therefore, the CSI update overhead can be easily managed.

3.7 ParCast vs SoftCast

ParCast is inspired by SoftCast and follows a similar linear transform process, but with several modifications and additions. Although the same techniques are adopted in several steps, they serve different goals.

The main stages in SoftCast include 3D-DCT, power weighting, and per-frame whitening via a Hadamard matrix, transmission over an uncoded OFDM channel, and decoding using LLSE. The matrix view of the overall encoding process is:

$$\mathbf{Y} = H\ H_a\ G\mathbf{X} = C_{SoftCast}\mathbf{X} \tag{6}$$

where H is the overall channel matrix, H_a is the Hadamard matrix, G is the diagonal power weight matrix, and $\mathbf{X}$ is the DCT coefficients.

ParCast also starts with 3D-DCT, not only because it achieves compression, but it allows us to generate equal-sized, independent chunks. Otherwise, different numbers of OFDM symbols would be required to transmit the chunks,

[1]Hadamard matrices can only have certain dimensions due to their orthogonal property, whereas our data can be of arbitrary dimensions.

necessitating a further power re-distribution. The independence between chunks allows us to gain performance from any successfully delivered and decoded chunk. We will return to this point in Section 5.

SoftCast treats each channel as a whole, whereas ParCast specifically optimizes for each subchannel individually. The power allocation formulation and the LLSE decoder in ParCast incorporate channel details, whereas those in SoftCast consider the source property only. At low SNR, ParCast naturally discards the worst subchannels and the corresponding source chunks. The diversity gain thus achieved mitigates the lack of error correction capability of analog modulation. This is unlike SoftCast, which can only send duplicate data to improve reliability if there are extra time slots.

SoftCast performs Hadamard *across DCT chunks* for each GOP to guard against packet loss, whereas the corresponding transform in ParCast operates *within a chunk* and mainly balances power across OFDM symbols for the same chunk and power across transmit antennas. Notice that dividing the source into chunks turns $\mathbf{X}$ in Equations (6) and (1) into a diagonal matrix. The overall encoding matrix, C, in Equation (1) is diagonal after a rotation, but $C_{SoftCast}$ is not even if H is. H_a specifically mixes $\mathbf{X}$. Therefore, any source correlation will manifest as channel correlation.

SoftCast uses analog modulation to achieve graceful degradation across different broadcast receivers, whereas ParCast uses this per subchannel of a unicast link to harness the channel capacity with a single modulation. ParCast effectively turns a unicast session into many parallel sessions, as if over a broadcast channel.

4. IMPLEMENTATION

4.1 ParCast implementation

We divide the whole system into the video codec at the application layer and physical layer operations.

At the video sender side, the video encoder performs the 3D-DCT transform, generates a whitened video data stream and the variance per chunk, and passes these to the PHY module. The number of chunks is determined based on the bandwidth availability. The PHY module performs the chunk-subchannel mapping, power allocation, and precoding, before grouping enough OFDM symbols into individual packets for transmission. In addition, the PHY module on the video sender sends the source metadata and processes the channel state feedback from the video receiver to calculate the SVD of each subcarrier channel matrix.

Note that although we currently implement the entire source encoder at the transmitter, this is not strictly necessary. The 3D-DCT transform is independent of the channel and could easily take place on a remote video server connected with the wireless transmitter over the wire. If the server is notified of the number of chunks, the division into chunks, chunk variance calculation, and the per-chunk power whitening could also be performed remotely, leaving only the PHY operations to be performed at the wireless transmitter.

At the video receiver, the PHY module performs channel estimation, compensation for carrier frequency offsets (CFO), and OFDM pilot phase tracking without compensating for the channel, before passing the received complex samples to the video decoder. It also decodes the control packet with the source metadata and feeds back the estimated channel matrices to the video sender. The video decoder then finds the LLSE solution to the received signal vector, decodes video data in each chunk, and finally performs the inverse DCT to reconstruct the video source.

We use Sora [ ] as our experimental platform. Since the current Sora platform can support real-time 20 MHz wideband packet transmission and reception, but not quite single-device MIMO, we need to emulate MIMO with multiple single-antenna boxes. Due to our analog modulation, this means that we would need to send raw signal samples to the sending nodes separately, and copy the received signals from the receiving nodes to one box, both via Ethernet. The latency from moving the samples across the network makes it currently infeasible to run the whole system in real time. However, since CSI updates can be sent using a single antenna, we can still run the channel dependent components in real time to assess the latency of PHY processing.

Therefore, we currently implement the PHY modules using Sora SDK 1.5 to be run in real time. The actual video codec is implemented in Matlab and interfaces with the PHY module via the generated video data as real values and the received complex samples. The video data packets follow 802.11n like packet structure, so that we can leverage techniques for CFO compensation, channel estimation and precoding in 802.11n. Unlike in 802.11n, however, the pilot subcarriers for each antenna are used alternately across different OFDM symbols, similar to the approach in [ ], so that we can update all the estimated channel coefficients for one transmit antenna per symbol. Noise is estimated from the error vector magnitude of the signal field in the packet preamble, and passed to the video decoder to be plugged into the LLSE decoder.

We also implement a channel simulation that could act in place of the Sora based PHY modules but replay measured channel traces. This allows us to study a larger variety of channel instances and compare different schemes more easily and fairly.

4.2 Schemes for comparison

The implementation of ParCast can be viewed as a superset of SoftCast. In addition, we consider two schemes for comparison: (1) a version of omniscient H.264/MPEG-4 [ ] over 802.11n like channel rates (Omni-MPEG) as an optimal case, and (2) Layered Scalable Video Coding (Layered SVC) [ ] with unequal error protection as a practical case. Both are based on MPEG-4.

MPEG-4 takes a sequence of raw frames and encodes it into three types of frames. The *I-frame* is an independently encoded still image, used as a reference frame. A *P-frame* encodes changes in the image from the previous frame, whereas a *B-frame* includes changes from both the previous and the future frames.

The raw I-frame is first divided into blocks. The blocks in consecutive raw frames are then compared to each other to identify the closest matches and the corresponding motion vector. The differences between the matching blocks are then stored in the P- and B-frames. The blocks for all frames then go through DCT transform, quantization, and entropy coding for further compression.

Omni-MPEG. Typically, the source encoder needs to know the expected channel throughput in order to select an appropriate quantization levels, and the physical layer transmitter should determine the most suitable bit rate accordingly. We alter the standard MPEG-4 encoding and transmission process to optimally choose the source and channel rates based on the given CSI.

We implement MPEG based on the reference implementation JM16.1 [ ]. The source first generates encoded data under different Quantization Parameter (QP) values. We derive a table of PSNRs under various QPs, the corresponding data stream size and the channel throughput required.

Data is transmitted using 802.11n like modulation and coding, and we select the channel rate based on the effective SNR of the given channel [ ]. Depending on the number of antennas on both sides, a sender could choose between sending 1 to 4 concurrent spatial streams (i.e., packet fragments) at a time. To use a good rate, we precode the transmitted streams and allow a different modulation and coding combination per transmitted spatial stream. This allows a higher overall rate than those provided by unequal modulation across streams in 802.11n [ ]. If the worst spatial stream is too weak to support even BPSK, we allocate the power to a stronger spatial stream instead, using a high rate where possible. If any stream is decoded incorrectly, the entire packet is dropped at the receiver and retransmitted.

To compare Omni-MPEG with ParCast under the same bandwidth constraint, we select an encoded data size which requires a transmission time closest to ParCast, look up the appropriate QP, and select the corresponding PSNR as the performance of Omni-MPEG.

Layered SVC. In order to adapt to varying channel conditions, the H.264 scalable video coding extension provides spatial, temporal, and quality scalability. For SVC, the entropy coded bits are separated into layers with varying importance levels. The base layer data is always expected to be correctly received to achieve an acceptable video quality, while a higher quality is derived from the enhancement layers. There have been several proposals for unequal error protection between layers.

The SVD-based video transmission scheme [ ] is an example design for a narrowband MIMO channel. The base layer would normally be assigned to the best spatial subchannel and sent using QPSK. The enhancement layers can use denser modulations. Power is carefully allocated to the subchannels to ensure a target error rate for the base layer and the highest possible rates for other layers. We extend it to run over 802.11n like OFDM for our comparison.

The source codec is implemented based on JSVM [ ]. We use quality scalability, which adopts a large QP for the base layer, and smaller QPs for enhancement layers. Each enhancement layer is generated by using a smaller QP to encode the quantized residue from the previous layer. Therefore, there are strong dependencies between layers. When one transmitted packet can not be correctly decoded for the base layer data, the whole frame and any other frame that is dependent on the current frame has to be discarded.

The original proposal in [ ] aimed to achieve an *uncoded* BER of 10^{-5} for the base layer and 10^{-3} for the enhancement layers. Our initial trials suggest these settings are too optimistic and would produce bad video quality. Therefore, we set the target *coded* BER for all layers to 10^{-5}. A Reed-Solomon code (239, 255) is used for all layers at the application layer, along with a $\frac{3}{4}$ channel code to provide better channel protection. We also skip the antenna selection suggested, since it often over-aggressively improves the enhancement layer quality at the expense of the base layer.

FlexCast. A comparison against FlexCast would be interesting but unfair. First, FlexCast is built to achieve a different goal. For our setting, the equivalent would be H.264 along with a capacity achieving channel code over MIMO-OFDM. Second, FlexCast was designed for single-antenna systems. Extending it to MIMO is beyond the scope of this paper. In particular, we are not aware of a rateless code for non-ideal MIMO channels, since MIMO rate increase is not monotonic with the overall channel SNR, as shown by the 802.11n rates [ ].

5. EVALUATION

5.1 Experimental setup

Performance metric. We evaluate video delivery quality with the standard Peak Signal-to-Noise Ratio (PSNR) metric. $PSNR = 20\log_{10}\frac{2^L-1}{\sqrt{MSE}}$, where L is the number of bits used to encode pixel luminance, typically 8, and MSE is the mean squared error between all pixels of the decoded video and the original. We average the PSNR across frames to produce a single value for each video sequence.

Test videos. We select a few representative reference video sequences [,] with different characteristics: *foreman* (little motion, large energy spread), *news* (even more stationary, large energy spread), *bus* (more motion, smaller energy spread), *mobile* (faster motion, larger motion area, even smaller energy spread), and *football* (significant motion, large energy spread). Figure 1 shows the energy distribution of the first frame of each sequence. The distribution of the variance of the chunk energy after the 3D-DCT is similar.

We use a video frame size of 352 × 288 pixels for all experiments. For a 2 × 2 link, we could send up to 2 values per complex symbol, 2 symbols per subcarrier, and 52 data subcarriers, or 208 values total in a single 802.11n OFDM symbol. We currently send 192 values (i.e., 48 chunks per video frame) in each OFDM symbol and skip the worst few subchannels, so that an entire video frame takes 528 OFDM symbols. For a fair comparison with other schemes, we define packet size in terms of number of OFDM symbols. Similarly, we use 72 chunks per video frame for a 3 × 3 link, and send 288 real values in one OFDM symbols. Calculations show that, if all subchannels within and between the corresponding 2 × 2 and 3 × 3 links have the same gain, the PSNR will increase by 0.5 dB from 192 chunks to 288 chunks, and 0.8 dB from 192 to 396 chunks.

We take 32 frames per sequence for microbenchmarks and 200 frames per sequence for the mobile experiments.

Sora setup. We use four single-antenna Sora boxes to emulate a 2 × 2 MIMO link, and move the antennas to different locations to get different channel instances. The carrier frequencies of all four radios are configured to be within a few hundred Hz of one another. This is the precision we can achieve given the configuration tool, and we have verified that pilot tracking can correct residual carrier frequency offsets between each transmit-receive pair sufficiently well.

We calibrate the CPU clocks of all boxes offline and use a SourceSync [ ] like approach to achieve synchronized transmissions from the two transmit antennas.

Channel traces. We use two sets of channel traces of 3 × 3 802.11n MIMO-OFDM links from previous work [ ] for trace-driven simulations, one including around 120 stationary links and the other a mobile trace. These were collected on commodity Intel `iwl5300` NICs. Figure 2 shows

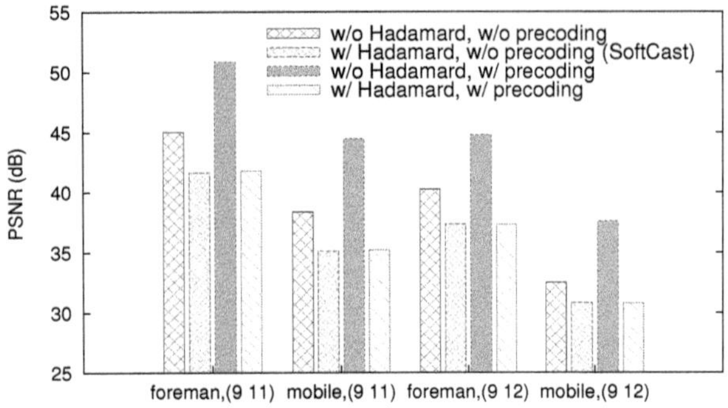

Figure 3: Effects of Hadamard and precoding.

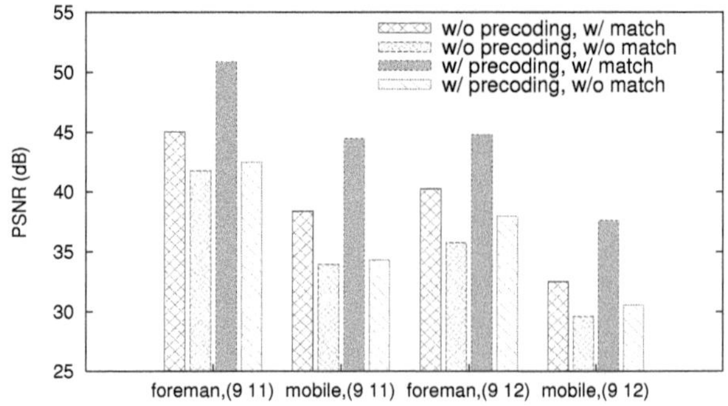

Figure 4: Effects of matching important chunks with high gain subchannels.

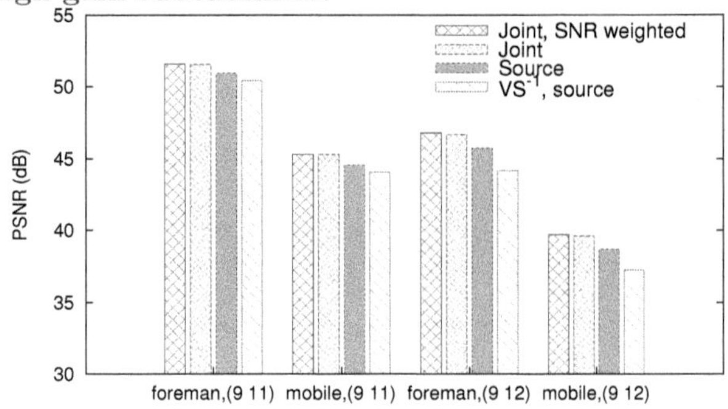

Figure 5: Effects of joint source-channel power allocation.

the squared singular value distributions for example links. Each link is identified by (<source ID> <destination ID>).

For each stationary trace, there is one packet approximately every 6 ms. Experiments show that results differ mainly for whether there is CSI feedback delay, and is not sensitive to the delay period. Therefore, we simply present the results for one-packet delay, or around 6 ms. For the mobile trace, we select a CSI sample approximately every 10 ms to ensure channel variation.

5.2 Microbenchmarks from Sora experiments

The main purposes of the Sora experiments are to validate the simulation implementation and assess the complexity of several operations. Subsequent experiments are all done using trace-driven simulation, since that makes comparison fairer and the mobility experiment more manageable.

Minimum CSI delay. In our current implementation, the time cost for an explicit CSI update is less than 1 ms, which is much smaller than the typical update period.

SVD calculation latency. Given all subcarriers are independent, we only need to calculate SVD for 52 4×4 real matrices. This takes around 300μs for a box with a 3.40 GHz CPU frequency.

Residual CFO. Since the two numbers in a complex symbol are correlated, any residual CFO after pilot tracking could induce an error in the decoding process. However, our experiments show that the error is insignificant at low channel SNR, and within 0.5 dB even for a 200 Hz residual CFO at high SNR.

Limited sample precision. Although the power allocation step results in a large energy spread in the frequency domain, the IFFT of OFDM has a mixing effect so the time domain signal energy is better evened out. Therefore, limited precision of transmitted time-domain samples causes an error of around 0.2 dB to the final PSNR.

Computational complexity. The computational overhead of 3D-DCT in ParCast is the same as for SoftCast, and was evaluated previously [ ]. The main overhead in the remaining steps of ParCast is in computing the precoding matrix (SVD), evaluated above. Current wireless transmitters need to do this anyway in order to apply channel-specific precoding. Precoding in a broad sense (i.e., channel independent spatial expansion or mapping) is in the 802.11n standard and already performed on commercial cards (e.g., by default on Intel `iwl5300`), which shows that the overhead is acceptable.

5.3 ParCast microbenchmarks

We run a few benchmarks to assess the individual contributions of the operations of ParCast, one of which also serves as a comparison against SoftCast. We use two representative video sequences, *foreman* and *mobile*, which feature a large and small spread of energy distribution respectively. We also choose two representative 3×3 channel traces, links (9 11) and (9 12), which feature a very small and large spread of squared singular values respectively. The CSI is delayed unless otherwise stated.

Spatial correlation over MIMO. SoftCast includes a Hadamard transform to even out energy between packets and provide resilience to packet loss. However, applying it naively over a MIMO channel would cause interference between the signals on different spatial subchannels of the same subcarrier and degrade the decoding performance. Instead, a subcarrier-specific precoding matrix should be used.

Figure 3 shows that Hadamard does not help when we *do not* precode the channel, but hurts when we *do* precode[2]. Without precoding, the PSNR without Hadamard is on average 2.8 dB higher than with. With precoding, the PSNR without Hadamard is a further 5.4 dB higher. The reason is that Hadamard acts as *random* precoding, which is statistically equivalent to not precoding at all [ ]. Note that the source-channel matching is implicit for 'without Hadamard', but there is no matching for 'with Hadamard'. All are without the joint source-channel power allocation but use the source aware power allocation as in SoftCast.

The importance of matching. We next show the importance of sending the more important source components on high gain subchannels. Since this can be affected by whether there is precoding, we compare between ParCast with different combinations of with and without precoding and matching. Without matching means that chunks are simply allocated to subchannels on each subcarrier in a ran-

[2] We let SoftCast divide each video frame into 72 chunks.

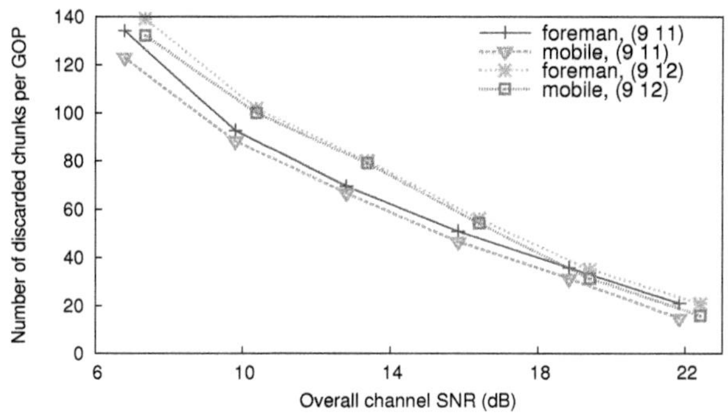

Figure 6: ParCast discards source chunks if needed.

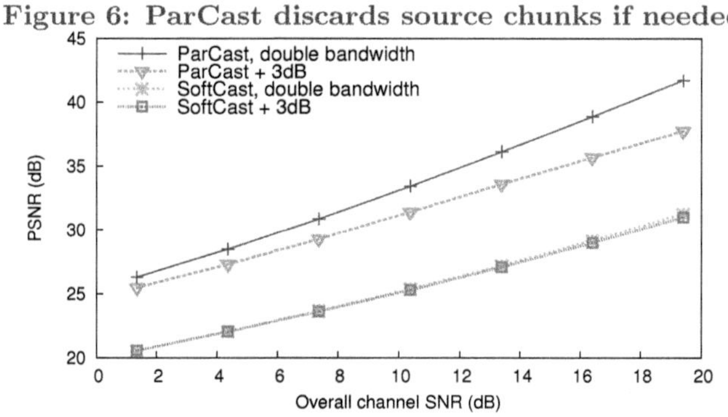

Figure 7: Effects of extra bandwidth.

dom order. Again, all variants use the source aware power allocation only.

Figure 4 shows that matching alone could improve the performance by around 3.5 dB even without precoding, and by 8.1 dB with precoding. Note that the bars for 'without matching' should not be compared directly across videos or channels due to their random nature.

There are some interesting similarities between Figure 3 and Figure 4. The *without precoding, with Hadamard* bar in Figure 3 is similar to the *without precoding, without match* bar in Figure 4. The former is slightly better due to Hadamard mixing the data for different subcarriers. The *with precoding, with Hadamard* bar in Figure 3 and the *with precoding, without match* bar in Figure 4 are also similar, but the latter could be better if the data chunks happen to match subchannels *on the same subcarrier.*

Joint source-channel power allocation. With precoding and matching, we next compare ParCast with source aware power allocation only and with joint source-channel allocation. Note that, in effect, the source-only allocation in SoftCast assumes all subchannels are equivalent, which would correspond to precoding with VS^{-1}, where V and S are from the SVD of $H = U\ S\ V^H$. Therefore, we also consider this case, i.e., precoding with VS^{-1} combined with source-only allocation. We use the ground-truth CSI for these experiments, and defer the discussion on the effects of CSI inaccuracy to the next subsection.

Figure 5 shows that, depending on the channel characteristics, the non-weighted joint power allocation outperforms the source-only approach by 0.6 to 0.9 dB. However, precoding with VS^{-1} is similar to using H^{-1}, and is especially bad for a link with a large singular value spread. As expected, the trend is the same for both videos for the same channel, since this comparison highlights the benefit from leveraging channel details.

The non-weighted joint power allocation is most effective under three conditions: (1) the source chunks exhibit a large energy variation, (2) the MIMO-OFDM channel is very spatially correlated and frequency selective, and (3) CSI is fairly accurate. The first condition is often met. The latter two result in a large variation between subchannel gains, which is a challenging case for MIMO-OFDM to work well. The joint power allocation strategy is thus a way to take advantage of otherwise challenging conditions, although the second condition is a double-edged sword when CSI is inaccurate.

The SNR weighted approach is most helpful at low SNR by skipping the worst subchannels and the corresponding source chunks. Figure 6 shows the number of chunks discarded per 4-frame GOP as the channel SNR lowers, assuming there is no spare bandwidth. Recall that a chunk-subchannel pair is discarded if its λ/s^2 is too low (Section 3.4). This is more affected by the chunk energy at high channel SNR, and by the subchannel gain at low channel SNR.

Extra bandwidth. Finally, we study the PSNR of ParCast and SoftCast when given extra bandwidth. Figure 7 shows the results for 16-frame *mobile* and Link (9 12). The trends for other videos and links are similar and hence omitted.

The '+3dB' lines show the performance of both schemes with exact bandwidth but twice the power. SoftCast can only repeat the same data when given double the bandwidth, which is equivalent to using twice the power but no extra bandwidth. In contrast, ParCast benefits from a subchannel diversity gain when given extra bandwidth, although the gain is less pronounced at low SNR when some bad subchannels (and source chunks) would be skipped anyway.

5.4 Comparison against alternatives

Overall. We first compare ParCast, Omni-MPEG, and Layered SVC over all stationary 3×3 links, with precoding and delayed CSI feedback. This represents the most common scenario. Figure 8 shows that ParCast often achieves a much higher PSNR than the other schemes, especially for *football,* which is a very challenging case for traditional video compression techniques. We next zoom into individual traces and video sequences.

Channel dimension. For ideal channels, with equal gain on all subchannels, an $N \times N$ MIMO system would scale the SISO capacity by approximately N times, or reduce the bandwidth requirement by a factor of N. However, this N factor is hard to achieve for real channels. In fact, more antenna may not help if not used optimally. We therefore compare the performance of all three schemes for 2×2 and 3×3 configurations for the same source-destination pair, both with delayed CSI. Note that the bandwidth requirement for 3×3 is only $\frac{2}{3}$ of that for 2×2 links.

We see in Figure 9 that ParCast normally achieves comparable performance for both 2×2 and 3×3, whether with a large subchannel gain spread or not. This shows that ParCast can harness the capacity scaling potential of the additional channel dimensions. The slight drop from 3×3 to 2×2 is due to different numbers of chunks and how good the 2×2 part compares with its 3×3 counterpart. Layered SVC often benefits from 3×3 by being able to use a denser modulation for the enhancement layer than in 2×2. Omni-MPEG is more sensitive to the channel detail. The 2×2 channel for (9 11) is *nicer* than its 3×3 counterpart by allowing high rates to be selected at a higher accuracy, and hence the PSNR is much higher for 2×2 than for 3×3.

It is interesting to note that ParCast often outperforms

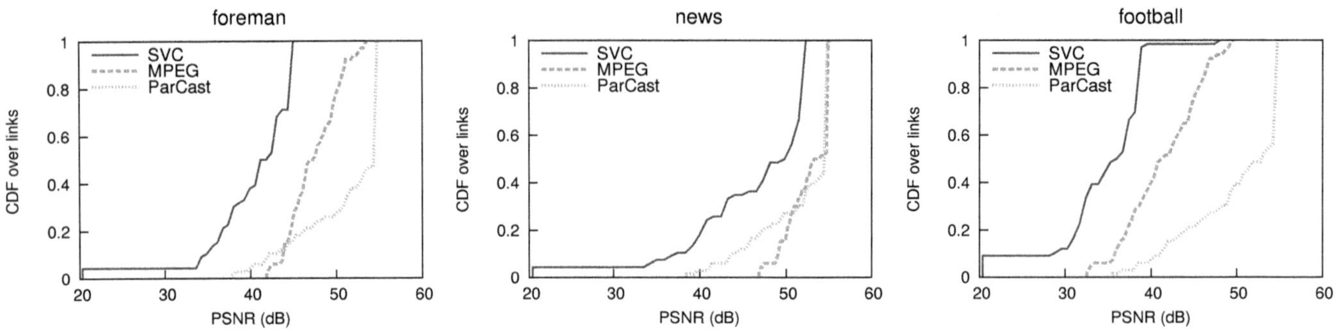

Figure 8: PSNR for H.264, SVC, and ParCast for all stationary traces.

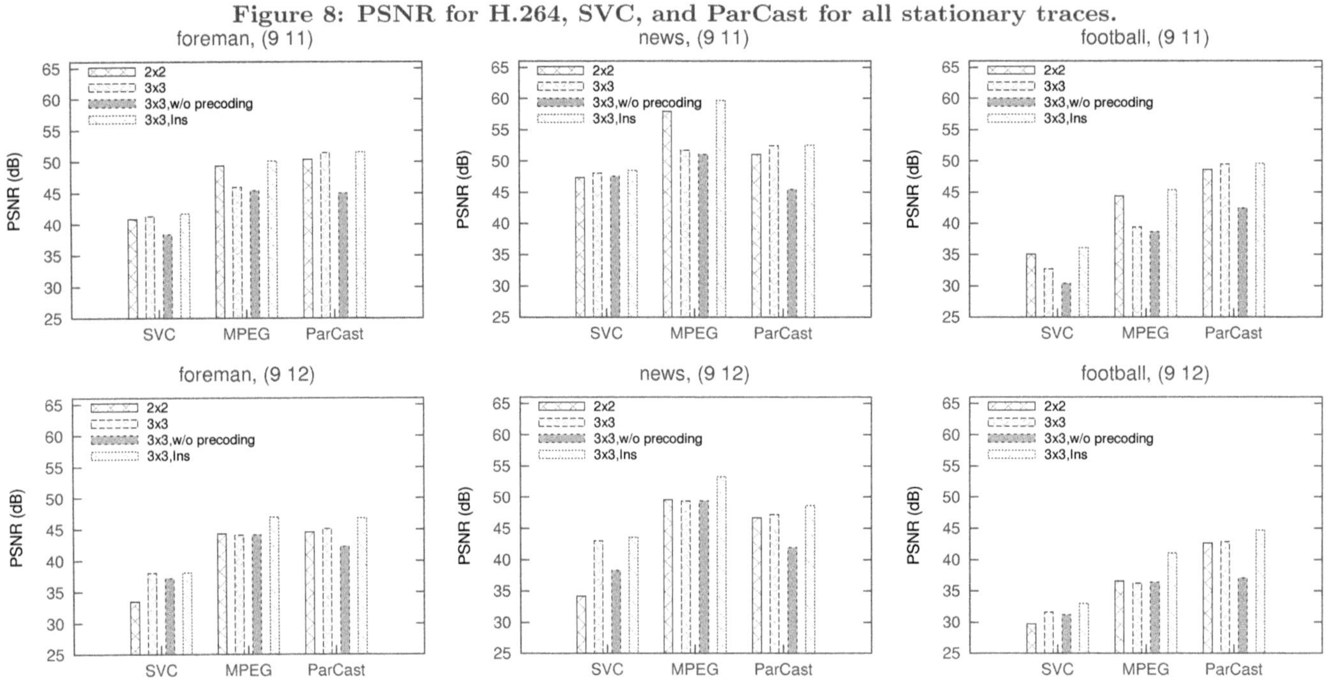

Figure 9: Effects of channel dimension and CSI accuracy.

Omni-MPEG by a few dBs, except for the video *news*. This is because the digital modulations used for Omni-MPEG cannot easily achieve the channel capacity. In contrast, the analog modulation approach in ParCast is very helpful in MIMO-OFDM situations to fully utilize the diverse channel gains without much overhead. Otherwise, the transmitter would also need to signal to the receiver how to demodulate each subchannel, and even a FARA [ ] like scheme may not work well. *news* is fairly compressible and favors traditional compression approaches. Even the non-capacity-achieving 802.11n like rates suffice for such compressible videos.

CSI accuracy. The main steps of ParCast, precoding, matching, and power allocation, all depend on the SVD of the channel matrix, which is sensitive to the CSI accuracy. Therefore, we next assess the effects of CSI accuracy on the overall performance. We first study the 3×3 stationary traces and then consider the mobile trace in the next part. Figure 9 compares the performance with delayed (the 3×3 bar) and instantaneous CSI (labeled 3×3, *Ins*).

For the stationary traces, CSI inaccuracy is mainly caused by estimation noise and precision errors. The inaccuracy in our trace indicates the amount of error margin manageable on commodity NICs.

The issue with delayed CSI is that precoding would not be able to entirely decorrelate spatial streams. Our data show that such spatial correlation generally weakens the strongest spatial subchannel, but strengthens the weakest slightly. The larger the singular value spread, the larger the effects of inaccurate CSI. This could cause modulation selection to be too optimistic, which affects both Omni-MPEG and Layered SVC. Inaccurate modulation selection could cause decoding errors. Therefore, the Omni-MPEG performance is about the same with or without precoding (3×3 vs 3×3, *w/o precoding* in Figure 9).

In contrast, ParCast does not adapt the encoding of the chunk after selecting a subchannel, and the effective subchannel gain variation would just translate to slight variation in the mean squared error for the pixel. Hence, ParCast can degrade gracefully at a subchannel level. In particular, since Link (9 11) has a smaller singular value spread, the PSNR difference between ParCast with instantaneous CSI (the ground truth) and delayed CSI is only about 0.1 dB regardless of the video. For Link (9 12), the PSNR drop for ParCast due to inaccurate CSI is between 1.4 and 1.8 dB for the three videos. The bulk of that is due to precoding. If we use the ground-truth CSI for matching and power allocation, then precoding with delayed CSI causes ParCast to lose 1.1 to 1.4 dB in PSNR. The matching step alone loses about 0.2 dB in a similar comparison. SNR weighted power allocation alone is hardly affected, but can interact with the

Video	No delay		10ms		100ms		1s		5s	
	Ours	MPEG	Ours	MPEG	Ours	MPEG	Ours	MPEG	Ours	MPEG
foreman	48.9	46.8	43.6	43.5	42.2	43.3	41.8	43.2	41.4	42.7
mobile	42.8	40.5	37.4	35.6	36.0	35.3	35.6	35.2	35.2	34.6
news	50.5	52.2	44.8	48.4	43.2	48.2	42.7	48.1	42.3	47.5
bus	45.9	42.7	40.1	37.7	38.7	37.3	38.3	37.1	37.7	37.1
football	50.6	43.7	44.7	38.4	43.2	38.1	42.8	37.9	42.3	37.1

Table 1: PSNR (dB) for videos given different CSI delay, ParCast vs MPEG.

precoding step adversely to cause more PSNR degradation. In particular, the non-weighted joint power allocation evaluated previously can be worse than the source-only allocation.

As we saw in the previous subsection, matching chunk with subchannel is more important than additional channel-aware power allocation. Therefore, even without very accurate CSI, matching alone could still help as long as the relative subchannel gain order is not affected. Similarly, matching the base layer with a good subchannel also helps Layered SVC, but different layers carry different number of bits and require different modulations. Omni-MPEG over 802.11n means that the implicit unequal protection achieved in the entropy coding stage (Section 2) is not aligned with the channel, even with MIMO precoding. What does help is that precoding reduces the difference between the SNRs on the corresponding spatial subchannel across subcarriers.

Even though delayed CSI makes precoding less effective, it is still better than not precoding at all for both Layered SVC and ParCast. Without precoding, each received signal on an antenna has interference from other spatial streams. Reducing such interference could significantly improve the decoding performance.

Mobility. We next use ParCast and Omni-MPEG to deliver 200-frame video sequences over the mobile trace[3]. Inaccurate CSI is a main feature of the trace, but likely due to changing channel as opposed to estimation noise. If the channel itself changed too much, the matching would be wrong in ParCast, and power allocation based on the wrong matching would hurt further. We consider CSI delay periods of 0 (No delay), 1 packet (10 ms), 10 packets (100 ms), 100 packets (around 1 s), and 500 packets (around 5 s). Trials suggest that the joint power allocation in ParCast and precoding for Omni-MPEG are sometimes too aggressive for the mobility case. Therefore, we only use the source-aware power allocation for ParCast[4] and no precoding for MPEG.

Table 1 summarizes the results. The error margin is about ± 0.6 dB for MPEG and ± 0.9 dB for ParCast for any delay period. We see that the main difference is again between whether there is delay. Omni-MPEG degrades less with CSI delay, though it cannot benefit from more accurate CSI information. ParCast performance degrades slightly more.

If we compare the ParCast PSNRs for 1s delay with no delay, matching alone incurs about 1 dB loss, and the remaining loss can be attributed to inaccurate precoding. The bulk of the drop is likely caused by the estimation error on the Intel NICs. Sora based experiments show PSNR degradation of 1 to 2 dB with delayed CSI.

6. DISCUSSION AND FUTURE WORK

Limitation of analog modulation. Analog modulation lacks error-detection or correction capabilities. However, several design decisions in ParCast can mitigate these.

First, ParCast is designed for heterogeneous MIMO-OFDM channels. Our results suggest analog modulation can work with MIMO-OFDM in complementary ways. At low SNR, MIMO-OFDM provides better reliability and error resilience via spatial and frequency diversity. At high SNR, analog modulation is an easy mechanism to harness the channel capacity. The latter is more useful a setting for MIMO for getting higher rates, which motivates HD video streaming.

Second, ParCast encodes video frames independently, so that the distortion for any GOP does not affect other GOPs. Distributed video coding has a similar property and could be an alternative. If instead, the video frames are encoded based on one another as the B- and P-frames in MPEG-4, errors might accumulate over frames due to the distortion in reconstructing the reference frame.

Third, the DCT and precoding steps in ParCast aim to remove correlation between any source chunks and between subchannels. Without either, there would interference between received signals. The LLSE decoder would need to first cancel out the inter-signal interference during decoding and incur performance loss.

Source division. There are several other ways of dividing the source into layers to assign to subchannels, such as using the I-, B-, and P-frames in MPEG. The number of frames of each type is very source dependent and the size of different frame types can differ based on channel conditions. In contrast, the source coding stage of ParCast and SoftCast divides source independent of the source detail. Furthermore, ParCast allocates the same number of symbols to all subchannels, while the amount of decodable information naturally scales with the subchannel SNR. Therefore, the symbol modulation is also independent of the channel.

Source-channel dissimilarities. Since there is still disparity between the distribution of DCT coefficient energy and the subchannel gains as shown in Figures 1 and 2, there might still be more residual redundancy at the source than is effectively used at the channel. Moreover, 3D-DCT does not fully capture motion in the video. Further refinement to the 3D-DCT or LLSE could improve the overall performance. However, the former should be designed with the subchannel characteristics in mind similar to the joint source-channel power allocation step, whereas the latter could be more flexible based on actual channel conditions.

7. RELATED WORK

ParCast is related to work on unequal error protection, joint source-channel coding, and graceful degradation.

Unequal error protection (UEP) over MIMO-OFDM. As mentioned earlier, unequal error protection is often implicit in entropy coding. There have also been many explicit UEP schemes specifically for MIMO or MIMO-OFDM chan-

[3]Only 148 frames for *bus*, because the sequence is shorter.
[4]PSNRs with the SNR weighted joint power allocation are about 1 dB lower.

nels, for example, involving more forward error correction for the worst subchannels [ ].

The fine-grained scalable (FGS) bit-stream extension of MPEG-4 can be used to generate many layers for different MIMO-OFDM subchannels [ ]. However, FGS is rarely used in practice, and it assumes unequal digital modulation across subchannels, which is hard to implement in practice. Another proposal [ ] is similar in principle, though with far fewer, coarse-grained layers and only the subchannels of a narrowband MIMO channel. Together with the scheme we studied earlier [ ], these schemes mainly differ in how they divide the source into layers and the granularity of layers and subchannels used. Nevertheless, common to all, the layers are not independent, and an incorrectly decoded base layer would render all other layers useless.

ParCast is similar in matching importance source data with high-gain subchannels, but does so at a fine granularity and for independent source components. Unlike in any SVC based schemes, the importance and implicit redundancy levels of the source components in ParCast are *aligned*, i.e., the most important component also embeds the highest level of redundancy. Therefore, the power allocation process actually tries to better protect the less important components.

Graceful degradation. Hierarchical modulation is a common digital approach to mitigate glitches and reduce the cliff effects [,]. It can be combined with multiple antennas, as in [,]. These work typically focus on how to correctly decode part of the complex symbol.

Apex [ ] recognizes that the bits in a regular QAM symbol have different error probability, and allocates important data to the bits with lower error probability.

Although ParCast is not explicitly optimized for graceful degradation, its linear codec performance degrades with noise in a manner similar to SoftCast.

Joint source-channel coding. Besides SoftCast and FlexCast, there have been a number of designs for lossy single-antenna channels (e.g., [, ,]), though few for MIMO. ParCast follows the same motivation to jointly consider source compression and error resilience at the channel, but specifically designs a scheme to leverage the properties of both the source and the MIMO-OFDM channel with a single code.

8. CONCLUSION

ParCast is motivated by non-uniform energy distribution at both the source and the channel. It sends independent source components on independent subchannels, with the more important components on higher-gain subchannels, joint source-channel power allocation, and analog modulation to optimally leverage source redundancy for error protection at the channel. We show that such a design can significantly outperform conventional digital schemes by turning challenging MIMO-OFDM links into opportunities.

9. ACKNOWLEDGMENTS

We sincerely thank Chong Luo, Ji-Zheng Xu, and the anonymous reviewers for invaluable feedback.

10. REFERENCES

[1] http://trace.eas.asu.edu/yuv/.

[2] ftp://ftp.tnt.uni-hannover.de/pub/svc/testsequences/.

[3] H.264/AVC JM Reference Software. http://iphome.hhi.de/suehring/tml/.

[4] High-Definition Wireless. AMIMON-WHDI Technology Overview. http://www.amimon.com/technology.shtml.

[5] IEEE Std. 802.11n-2009: Enhancements for Higher Throughput.

[6] SVC Reference Software. http://ip.hhi.de/imagecom_G1/savce/downloads/.

[7] Cisco visual networking index: Forecast and methodology 2010-2015. http://www.cisco.com/en/US/solutions/collateral/ns341/ns525/ns537/ns705/ns827/white_paper_c11-481360.pdf, June 2011.

[8] S. Aditya and S. Katti. FlexCast: Graceful wireless video streaming. In *ACM MobiCom*, 2011.

[9] B. Barmada, M. Ghandi, E. Jones, and M. Ghanbari. Prioritized transmission of data partitioned H. 264 video with hierarchical QAM. *IEEE Signal Processing Letters*, 12(8):577–580, 2005.

[10] R. K. W. Chan and M. C. Lee. 3D-DCT quantization as a compression technique for video sequences. In *International Conference on Virtual Systems and MultiMedia*, 1997.

[11] D. Halperin, W. Hu, A. Sheth, and D. Wetherall. Predictable 802.11 packet delivery from wireless channel measurements. In *ACM SIGCOMM*, 2010.

[12] G. Hardy, J. Littlewood, and G. Pólya. Inequalities, (2nd ed.), Cambridge Mathematical Library, 1989.

[13] S. Jakubczak and D. Katabi. A cross-layer design for scalable mobile video. In *ACM MobiCom*, 2011.

[14] Z. Ji, Q. Zhang, W. Zhu, Z. Guo, and J. Lu. Power-Efficient MPEG-4 FGS Video Transmission over MIMO-OFDM Systems. In *IEEE ICC*, 2003.

[15] K. Lee and D. Petersen. Optimal linear coding for vector channels. *IEEE Trans. on Communications*, 24(12):1283–1290, 1976.

[16] Y. Noh, H. Lee, and I. Lee. Design of Unequal error protection for MIMO-OFDM systems. In *VTC*, 2005.

[17] H. Rahul, F. Edalat, D. Katabi, and C. G. Sodini. Frequency-aware rate adaptation and MAC protocols. In *ACM MobiCom*, 2009.

[18] H. Rahul, H. Hassanieh, and D. Katabi. SourceSync: A Distributed Wireless Architecture for Exploiting Sender Diversity. In *ACM SIGCOMM*, 2010.

[19] I. Richardson. *H. 264 and MPEG-4 video compression*, volume 20. Wiley Online Library, 2003.

[20] M. F. Sabir, R. Heath, and A. Bovik. An unequal error protection scheme for Multiple Input Multiple Output systems . In *Asilomar*, 2002.

[21] A. Schertz and C. Weck. Hierarchical modulation-the transmission of two independent DVB-T multiplexes on a single frequency. *EBU Technical review*, 2003.

[22] H. Schwarz, D. Marpe, and T. Wiegand. Overview of the scalable video coding extension of the H. 264/AVC standard. *IEEE Trans. on Circuits and Systems for Video Technology*, 17(9):1103–1120, 2007.

[23] S. Sen, S. Gilani, S. Srinath, S. Schmitt, and S. Banerjee. Design and Implementation of an "Approximate" Communication System for Wireless Media Applications. In *ACM SIGCOMM*, 2010.

[24] M. Skoglund, N. Phamdo, and F. Alajaji. Hybrid Digital-Analog Source-Channel Coding for Bandwidth Compression/Expansion. *IEEE Trans. on Information Theory*, 52(8):3757–3763, 2006.

[25] D. Song and C. Chen. Maximum-throughput delivery of SVC-based video over MIMO systems with time-varying channel capacity. *Journal of Visual Communication and Image Representation*, 19(8):520–528, 2008.

[26] D. Song and C. W. Chen. Scalable H.264/AVC Video Transmission Over MIMO Wireless Systems With Adaptive Channel Selection Based on Partial Channel Information. *IEEE Trans. on Circuit Systems and Video Techonology*, 17(9), 2007.

[27] M. Stoufs, A. Munteanu, J. Barbarien, J. Cornelis, and P. Schelkens. Optimized scalable multiple-description coding and FEC-based joint source-channel coding: A performance comparison. In *IEEE 10th Workshop on Image Analysis for Multimedia Interactive Services*, 2009.

[28] K. Tan, J. Zhang, J. Fang, H. Liu, Y. Ye, S. Wang, Y. Zhang, H. Wu, W. Wang, and G. M. Voelker. Sora: High Performance Software Radio Using General Purpose Multi-core Processors. In *NSDI*, 2009.

[29] D. Tse and P. Viswanath. *Fundamentals of Wireless Communication*. Cambridge University Press, 2005.

[30] Q. Xu, V. Stankovic, and Z. Xiong. Distributed joint source-channel coding of video using Raptor codes. *IEEE JSAC*, 25(4):851–861, 2007.

[31] G.-H. Yang, D. Shen, and V. Li. Unequal Error Protection for MIMO Systems with a Hybrid Structure. In *ISCAS*, 2006.

MARVEL: Multiple Antenna based Relative Vehicle Localizer

Dong Li†, Tarun Bansal†, Zhixue Lu† and Prasun Sinha
Department of Computer Science and Engineering
The Ohio State University
Columbus, OH 43210
{lido, bansal, luz, prasun}@cse.ohio-state.edu
†Co-primary authors

ABSTRACT

Access to relati e location o near y e icles on t e local roa s or on t e ree ays is se l or ro i ing critical alerts to t e ri ers t ere y en ancing t eir ri ing e erience as ell as re cing t e c ances o acci ents e ro le o eter ining t e relati e location o t o e icles can e ro en into t o s aller s ro le s: i elati e lane locali ation ere a e icle eter ines i t e ot er e icle is in le t lane sa e lane or rig t lane it res ect to it an ii elati e ront-rear locali ation ere it nee s to e e-ter ine ic o t e t o e icles is a ea o t e ot er on t e roa n t is a er e ro ose a no el antenna i er-sity ase sol tion MA t at sol es t e t o ro le s o eter ining t e relati e location o t o e icles MA - as t o co onents: i a s art one an ii o r ireless ra ios nli e e isiting tec nologies MA can also eter ine relati e location o e icles t at are not in t e i e iate neig or oo t erc y ro i ing t e ri er it ore ti e to react rt er e to ini al ar - are re ire ents t e e loy ent cost o MA is lo an it can e easily installe on ne er as ell as e isting e icles sing res lts ro o r real ri ing tests e s o t at MA is a le to eter ine t e relati e lane location o t o e icles it acc racy ro g trace- ri en si lations e also s o t at y aggregating in or ation across i erent e icles MA is a le to increase t e locali ation acc racy to

Categories and Subject Descriptors

Computer-Communication Networks : istri te Syste s *Distributed Applications* **Computer Applications** : o ters in t er Syste s *Consumer Products* **Discrete Mathematics** : ra eory *Graph Algorithms*

General Terms

esign eri entation Meas re ent Algorit s Per-or ance

Keywords

S art one ireless anging ocation lassi cation

1. INTRODUCTION

e se o PS e i e s art ones as een increasing ra i ly B t t e PS e ices on t e ones o not a e s cient acc racy to locali e t e e icles to t e lane-le el A aila ility o tra c in or ation at t e icro-le el or lane-le el gran larity is se l or lti le a lications t at not only re ces t e c ances o acci ents t also en ances t e ri ing e erience So e a lications are as ollo s: i Alerting t e ri ers o co ing o stacles or ot oles t at are in t e sa e lane as t e e icle an rt er g i ing t e ri er to o e to t e a ro riate lane to a oi t e ii Alerting t e ri ers i t ere is a e icle in t e lin one or i t is e icle is tailgating anot er e icle t ere y re cing t e c ances o collision iii Alerting t e ri er t at t e e icle a ea is slo ing o n i t e t o e icles are in t e sa e lane i etecting t e lane-le el location o slo o ing lnera le e icles an eter ining t e i erences in s ee s o i erent lanes to assist in tra c lanning

e ro le o *Vehicular Localization* as t o ariants: i *Relative Localization* ere or t o e icles t e o ec-ti e is to eter ine t e location o a e icle it res ect to t e ot er e icle an ii *Absolute Localization* ere or a single e icle t e o ecti e is to eter ine its a sol te location n t is a er e e elo a no el sol tion calle MA M lti le Antenna ase elati e e icle ocal-i er t at ro i es *relative localization* o on-roa e icles S eci cally or t o gi en e icles V_1 an V_2 MA ses antenna i ersity to ena le ot o t e to eter ine t e relati e osition o t e ot er e icle a ong t e si i erent ossi le regions

Alt o g PS tec nology is i ely se or e ic lar locali ation ario s actors s c as signal lti at n-no n elays e to ionos ere an tro os ere error in t e cloc s o PS e ices an inacc racies in t e locations o satellites re ce its acc racy e ice an act rers s c as ar in re ort t e a erage PS acc racy to e e-ters e en or e ices e i e it ne er AAS i e

MobiCom'12, August 22–26, 2012, Istanbul, Turkey.
Copyright 2012 ACM 978-1-4503-1159-5/12/08 ...$15.00.

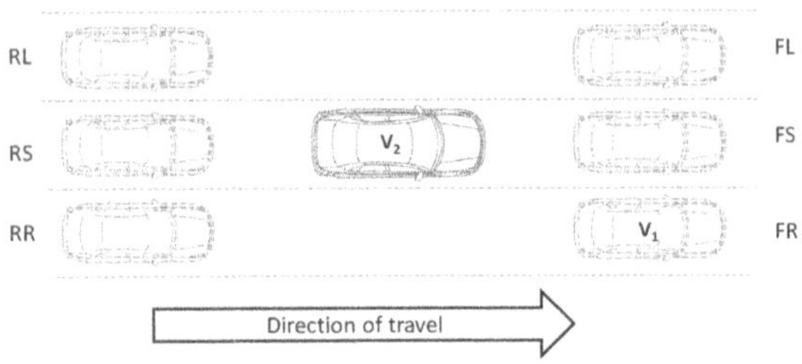

Figure 1: Vehicle V_1 can be in six different relative positions with respect to V_2: Front-Left, Front-Same, Front-Right, Rear-Left, Rear-Same and Rear-Right. Vehicle V_1 is said to be in *front* of V_2 if *front* of V_1 is ahead of that of V_2 along the direction of travel.

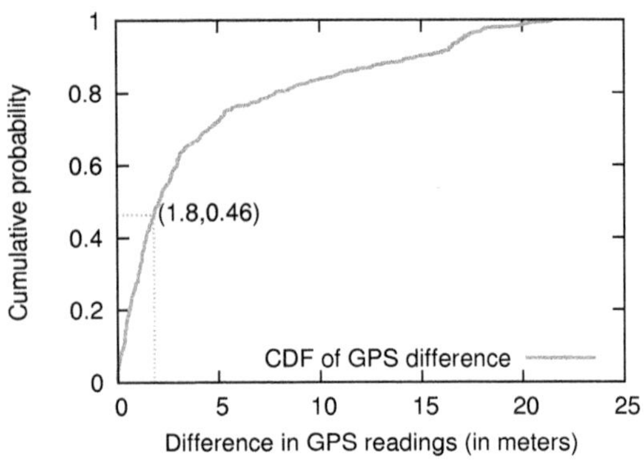

Figure 2: Comparison of GPS trace of two smartphones located in the same car: Cumulative distribution function of GPS error (in meters).

Area A g entation Syste an PS i erential PS tec nology e PS trace collecte y s or t o ones Motorola Atri an iP one o nte ne t to eac ot er ori ontal istance ≤ c on t e in s iel o t e sa e car s o e si ilar inacc racies ig re s o s t e c - lati e ro a ility istri tion o istance et een t e loca-tions o t e e icle along t e i t o t e roa as re orte y t e t o s art ones e i t o t e lanes as reco - en e y AAS on local an arterial roa s is eters So i t e error is ore t an eters across t e irection o tra el PS ill isco te t e *relative lane location* le t rig t or sa e lane o t e ot er e icle ic a ens in o t e cases Si ilarly PS ay also in-correctly co te t e *relative front vs. rear location* o t e ot er e icle i t e t o e icles are close to eac ot er n r an canyons e to lti at e ects t e acc racy o PS is e ecte to ecrease e en rt er *In this paper, we propose an antenna diversity based novel solution for both relative lane localization and relative front-rear localization*

ario s e icle an act rers are eginning to roll o t ne car o els it e ic lar tec nologies s c as o ler ra ar an ca eras or relati e locali ation ese tec nolo-gies or e en en t e ar are is e loye on only one e icle an not t e ot er e icles in t e icinity t t ey a e se eral li itations: i Bot t e o ler ra ar an ca era can acc rately etect only t ose e icles t at are in t e i e iate neig or oo ii ear y o stacles s c as ar e e icles store ronts an cras arriers re ce t e acc racy o t e ra ars as ell as ca eras iii ese tec nologies cannot e easily retro tte to e isting e icles res lting in lo er ar et enetration an i ere is generally a tra eo et een acc racy an n er o ra ars se More ra ars can co er ore regions ile ro-i ing ig locali ation acc racy o e er t is increases t e e loy ent cost

MA is ini al in esign an consists o t o co -onents: i A s art one t at co l e t e sa e as one carrie y t e ri er an ii o r o -t e-s el ireless ra-ios o nte on t e si es o t e e icle e s art one r ns o r locali ation algorit an is lays rele ant alerts to t e ri er e ireless ra ios on i erent e icles sen *lane discovery* eacons to eac ot er an re ort t e SS ecei e Signal Strengt n icator o t e recei e eacons to t e s art one MA eter ines t e relati e lane location o e icles y le eraging t e i erences in SS eas re ents e to s atial antenna i ersity MA is ase on t e o ser ation t at at loss o t e lin s e-t een ra ios is asy etrical en t e e icles are in i er-ent lanes an it is sy etrical en t e e icles are in t e sa e lane MA satisi es t e ollo ing esire re ire-ents:

- se l in or ation s c as resence o ot oles a -lying ar ra es or resence o slo o ing lner-a le tra c ro e icles t at are in t e sa e lane as t is e icle can ro i e t e ri er it ore ti e to react r syste is a le to etect relati e loca-tion o e en t ose e icles t at are not in t e i e i-ate neig or oo t s in isi le to ca era an ra ar t ere y ro i ing t e ri er it ore in or ation to re ce t e inci ence o collisions is in or ation can also e se or i ro ing t e esign o a tono o s cr ise control ec anis s
- t or s in a i e ariety o en iron ental con itions s c as a eat er rain sno etc ario s lig t con itions ar s n glare etc as ell as in resence o r an noise e g ar e cars an store ronts on roa si e cras arriers so n alls etc
- t is lo cost e to i ity o s art ones an t e lo cost o ireless ra ios
- t can e easily e loye on ol er as ell as ne er e icles
- r ri ing res lts er or e sing i erent e icles an n er arying s ee an tra c con itions s o t at MA co tes t e relati e location it an a erage acc racy o By aggregating in or ation across lti le e icles MA is a le to increase t e locali ation acc racy to

is a er is organi e as ollo s: n Section e isc ss relate or in t is area Section escri es MA in etail n t e ne t t o sections e escri e t e res lts ro o r real ri ing e eri ents n Section e escri e t o algorit s t at rt er i ro e t e locali ation acc racy y aggregating in or ation across i erent e icles n Section e escri e t e res lts ro o r ns- an S M ase si lations n Section e isc ss i erent et o s t at can a e MA ore realistic an se l inally in t e last section e concl e t e a er

2. RELATED WORK

e to ics o e icle etection an lane recognition a e een st ie in t e literat re o a tono o s e icles or eca es a ar laser an aco stic ase sensors an ca eras are co on e ices in s c st ies a ar laser an aco stic sensors etect t e istance o o - ects y eas ring t e ro n tri ti e o signals ose sensors are co only se or e icle etection an lin s ot etection or instance e e s - ra ar sensors into t e ac si e er to onitor t e lin s ots o t e ost e icle an triggers a ision arning signal i so e e icles are etecte o e er t ose sensors are s ally li ite in range to line o sig t i c lt to install es ecially on e isting e icles an e i it a tra eo et een acc racy an cost a eras a e roa er a lication in e icle etection an lane recognition e to t eir lo costs e e isting sol tions a e se ca era or recogni ing e icles o stacles an lane ar s t ro g i age recognition tec ni es o e er sing ca era or e icle etection an lane recognition is ig ly s sce ti le to errors e to ario s actors s c as: i Ba lig t con itions e g nig t ti e s n glare ea lig t glare s a o s ro near y il ings ii ro er eat er con itions e g sno rain an iii S rro n ing noise e g a e lane ar s e icles ar e on roa si e roa si e cras arriers trees store ronts etc

n t e ot er an t e co tational o er o s art ones is eing tili e in ario s a lications s c as assiste ri ing an roa in rastr ct re onitoring or e a le as recently ro ose ays to eter ine i t e ser o t e s art one is t e ri er or a assenger in t e e icle ic co l e se to a st t e e a ior o t e one ase on t e o ner s ty e e i ea o sing ireless sensors to etect an trac e icles as een st ie in ese or s ainly oc s on etecting an trac ing t e o e ents o e icles in a ireless sensor net or e loye in a gi en area n contrast t e o ecti e o t is a er is to *determine relative vehicle locations* sing *multiple wireless radios installed on vehicles*

3. SYSTEM DESIGN

n t is section e ill rst escri e t e o ecti es o o r sol tion en e gi e an o er ie o o r syste esign ol lo e y a list o c allenges inally e escri e MA in etail

3.1 Problem Statement

e o ecti e o o r syste is to eter ine t e relati e osition o t o e icles S eci cally gi en t o e icles see ig re V_1 an V_2 e see to eter ine: i *Relative lane localization:* V_1 is in le t lane sa e lane or in rig t lane it res ect to V_2 s lane an ii *Relative front-rear localization:* V_1 is in ront or rear it res ect to V_2 e call t e co ine ro le o elati e ane ocali ation an elati e ront-rear locali ation as *Relative Vehicle Localization* e icle V_1 is sai to e in *front* o V_2 i t e *front* o V_1 is a ea o t at o V_2 along t e irection o tra el is e nition also ta es into acco nt t e case en one e icle is in t e lin one o t e ot er

3.2 Overview and Challenges

MA co rises o t o co onents or e ery e icle e rst co onent is a s art one t at co l also e t e ersonal s art one o t e e icle s ri er e secon co onent co rises o o r ireless ra ios locate at ario s ositions on t e lateral si es o t e e icle e ass e t at eac e icle as a ni e *vehicle id* ic co l ossi ly e eri e ro its license late n er e o t e e icle V_i is no n to t e s art one insi e t e e icle as ell as t e o r ireless ra ios locate on t e e icle

e ass e t at t e s art one is e i e it an accelero eter an is i ca a le e s art one or t e MA a lication r nning on it is se to co nicate it ot er s art ones in ot er e icles as ell as to is lay alert essages A e icle ay a e lti le s art ones resent insi e o e er e ass e t at in eac e icle only one s art one is r nning t e relati e locali ation algorit So e ass e t at t e ni e *phone ID* o t e s art one in e icle V_i is P_i t e s art one as PS ca a ility t en t e elocity o t e e icle can e o taine ro t e PS is in or ation along it t e relati e lane in or ation can e se to e elo ario s ot er a lications see Section

r algorit r nning on t e s art one ee s trac o relati e locations o all near y e icles it o t e s art one can co nicate or eac e icle it trac s: i e ni e e icle o t e ot er e icle an ii elati e location o t e ot er e icle a ong t e ossi le regions see ig re t is also ossi le to e ten t is in or ation to - o neig ors y sing lti o co nication is o l ro i e e tra in or ation to t e ri er an t e s art one at t e cost o ig er essage o er ea o e er in t is a er e ass e t at t e s art one aintains in or ation o e icles t at are only it in one o

e secon co onent o o r syste co rises o o r ireless ra ios locate on t e lateral si es o t e e icle n Section e ill s o t at lacing t e ra ios close to eac eel o t e e icle a i i es t e acc racy o locali ation e o r ireless ra ios co nicate it ra ios locate on ot er e icles as ell as it t e s art one locate in t e sa e e icle e ra ios ay se i or ig ee to co nicate it eac ot er ile to co nicate it t e one t ey can se i Bl etoot or ig ee e en ing on t e ca a ility o t e ra ios an t e s art one e also ass e t at t e ireless ra ios are a are o t e osition ere t ey are o nte on t e e icle

n o r e icle locali ation algorit one o t e t o ones irects its o r ireless ra ios to sen eacons ile t e ra ios on t e ot er e icle listen or t e eacons t ere y esti ating t e at loss et een t e airs o ireless ra ios ig re s o s a si ler case ere only t o ireless ra ios are o nte on t e e icle one on eac si e r e icle locali ation algorit is ase on t e o ser ation t at en t e t o e icles are in sa e lane t e t o lin s *AD* an *BC* are sy etrical see ig re o e er en t e e icles are in i erent lanes t e lin s *AD* an *BC* are assy etrical isc sse in Section in ore etail o e er esigining a e icle locali ation syste ase on SS rea ings o t e lin s in ol es lti le c allenges: i e est location to e loy t e ireless ra ios t at a i i e t e locali ation acc racy nee s to e eter ine ii a e icle c anges its lane t e relati e lane location o ot er e icles s o l e ate it ini latency ile li iting t e n er o ac ets trans itte y ra ios so as to ini i e congestion an energy cons tion iii Pac ets trans itte ay e lost e to collision it ot er trans issions

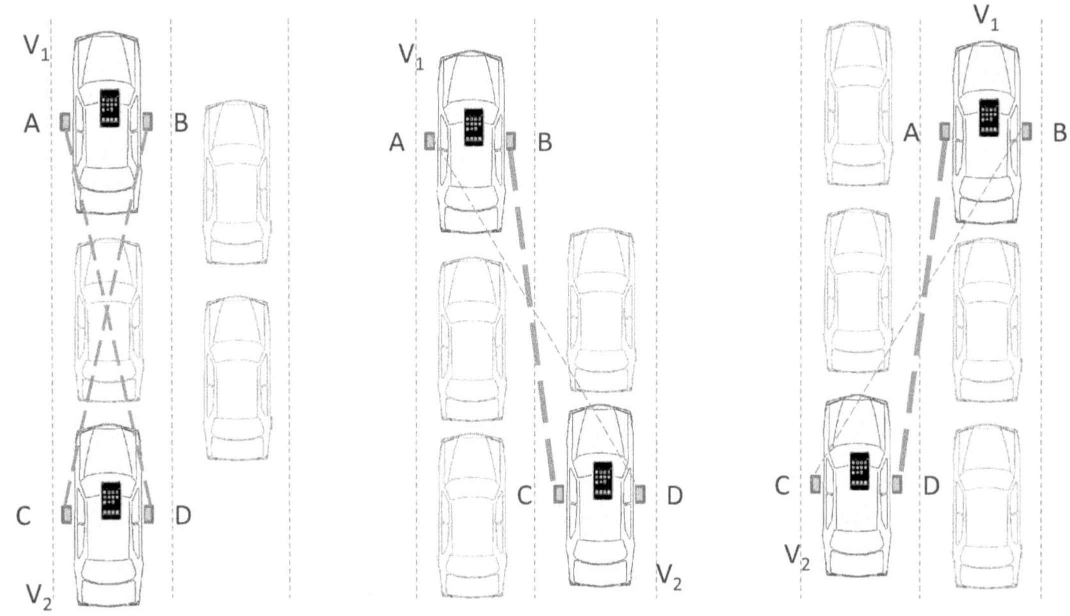

(a) Both vehicles in same lane (b) Front vehicle in left lane (c) Front vehicle in right lane

Figure 3: RSSI based Relative Lane Localization with two radios: A, B, C, and D are radios mounted on vehicles V_1 and V_2. The expected RSSI for two links AD and BC are shown for three different cases with thicker lines respresenting links with higher RSSI values. (a) When V_1 and V_2 are in the same lane, then the path losses for links AD and BC are almost symmetrical. (b) When V_1 is to the left of V_2, then BC is a direct line of sight link while AD passes through bodies of 2 vehicles (total of 4 walls and 1 vehicle's heavy machinery compartment). (c) When V_1 is right of V_2, link AD is stronger but BC is weaker. Nearby vehicles may affect the signal strengths slightly due to multipath.

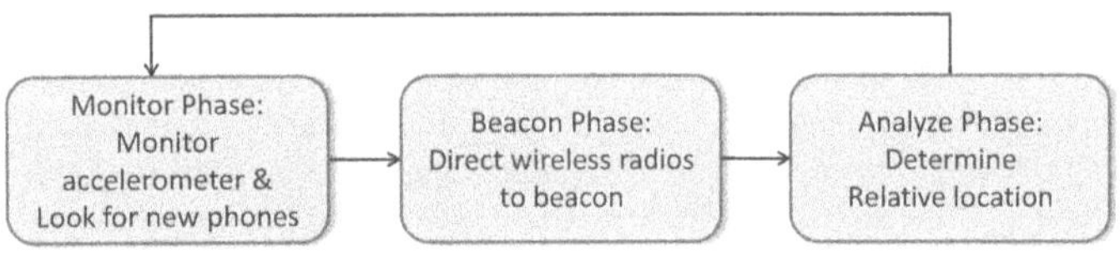

Figure 4: Transitions made by the smartphone among the three phases.

an i M lti at ro agation ay a ect t e acc racy o relati e locali ation es ecially i t e sen er an t e recei er are se arate y a large istance

3.3 Lane Localization Algorithm

n o r e icle locali ation algorit t e s art one tran- sitions et een t ree i erent ases see ig re e ill no escri e t e t ree ases o t e s art one an t e lo- cali ation algorit in ore etail

3.3.1 Monitor Phase

ne nai e ay to trigger t e lane locali ation algorit is to in o e it at erio ic inter als o e er t is o l in ol e consi era le ac et trans issions ro t e ireless ra ios t e sel es t ere y a ecting t eir attery li eti e rt er it re ces t e ro a ility t at a eacon trans it- te is recei e correctly y a recei ing ra io Since lane c anges or t rns are not ery re ent e ents t ere ore in t e rst ase t e s art one itsel *monitors conditions* t at in icate t e ollo ing: i A ossi le c ange in rela- ti e location o t e e icle it res ect to ot er e icles an ii Arri al o a ne e icle in t e icinity n er ot con itions t e s art one ill in or its neig ors o a ossi le lane c ange e ent an o e to t e eacon ase A c ange in relati e location o a e icle it res ect to anot er e icle ay a en e to t o actors: i t e e icle itsel c anges lanes or t rns to a ne street or ii t o erta es anot er e icle e ill no isc ss o t e s art one etects all s c ossi le e ents

Detecting Possible Lane Change: o ta e into ac- co nt t e rst con ition t e s art one i e o r a lica- tion r nning on t e s art one contin o sly onitors t e rea ings ro t e accelero eter o t e one to etect os- si le lane c anges or t rns t e s art one o ser es t at t e accelero eter rea ing is eyon a certain t res ol t e one o es to t e secon ase o e er as e ill s o in Section t e accelero eter rea ings are not co letely relia le e to actors s c as ot oles s an slo es in t e roa s r aces an c r at re o roa s ere ore as isc sse later in Section MA co tes t e i er- ence et een *maximum* an *minimum* accelero eter rea - ings o er a in o o ration secon s an ses t at to etect a lane c ange e ent

Detecting Possible Overtaking: e to i erences in s ee s o i erent e icles it is ossi le t at a e icle V_1 t at as initially in one o t e t ree rear regions o V_2 later o erta es V_2 t ere y c anging its relati e osition to one o t e t ree ront regions o V_2 n s c a case t e relati e locali ation algorit nee s to e triggere y eit er V_1 or V_2 to ate t e relati e locations ne ay to o t is o l e to trigger t e relati e locali ation algorit at erio ic inter als o e er t is o l s stantially re ce t e li e- ti e o t e ra ios Anot er et o t at co l e e lore is ere t e ones t e sel es sen eacons to eac ot er at reg lar inter als an also onitor t e SS o t e corre- s on ing recei e eacons A c ange in t e SS rea ing o t e eacon in icates c anges in t e istance et een t e t o e icles A attern in t e recei e SS ere it c anges ro lo to ig an t en ac to lo in icates t at t e ot er e icle is getting closer an t en again getting ar- t er a ay t e one o ser es s c a attern it o es to P ase an triggers t e locali ation algorit

Discovering New Vehicles in Vicinity: o esta - lis connections it ne s art ones t at enter its icin- ity an to eter ine t e relati e location o t ose e icles t e s art one roa casts erio ic *discovery beacons* sing P e one also listens or isco ery eacons roa - cast y ot er ones t e one ears a isco ery eacon ro a re io sly n no n one t e algorit o es to P ase e cing t is isco ery ti e ena les ones to eter ine t e relati e lane location o neig oring e icles aster n or er to re ce t is isco ery ti e ile anaging t e s art one s attery cons tion MA can or in con nction it any o t e neig or isco ery algorit s t at are alrea y a aila le in t e literat re

3.3.2 Beacon Phase

n t e eacon ase t e ireless ra ios on t e e icle trans it or recei e eacons et V_1 e t e e icle it one P_1 co ting its relati e osition an V_2 it one P_2 e any e icle a ong its set o neig ors en P_1 ill instr ct its o r ireless ra ios to sen *localization beacons* ile P instr cts its ra ios to listen to t ese eacons ac ireless ra io ears lti le locali ation eacons an re orts t e SS o t e recei e eacons to its associate one P_2

e lane locali ation eacon roa cast y ra ios on V_1 contains t ree el s: i V_1 ii ocation o t e ra io a ong t e o r ossi le locations on V_1 an iii rans ission o er le el o t e eacon e ac ets trans itte y ra- ios on V_2 to t e s art one P_2 contains e el s: i V_1 ii V_2 iii ocation o t e ra io sen ing t is ac et on V_2 i rans ission o er le els o t e o r lane local- i ation eacons recei e an i SS al es o t e o r eacons recei e By incl ing t e e icle s in t e es- sages MA is a le to ens re t at t e ra ios an t e s art one only rocess essages t at are eit er irecte at t e or t e roa cast essages rt er t e ra io on V_2 ay not e a le to recei e eacons ro all t e o r ra ios o V_1 n t at case t e corres on ing al es in t e ac et are le t e ty

n o r e eri ents e o ser e t at e to collisions ac et losses ay a en n or er to a e t e algorit ore ro st e re ire eac ra io to sen lti le lane locali ation eacons t at are ran o ly se arate e re- cei ing ra io co tes an a erage SS al e or eac lin ile ignoring o tliers e ore re orting it to t e one is ro i es a goo tra co et een acc racy an o er- ea

3.3.3 Analyze Phase

n t c t ir ase s art one P_2 rocee s to co te t e relati e location on t e asis o at loss o lin s P_2 can co te t e at loss or t e lin s o r al es ro eac ra io ase on t e in or ation recei e ro t e ireless ra ios ig re s o s o sing t o ra ios it is ossi le to isting is t e t ree ossi le cases o t o cars in t e sa e lane or in i erent lanes ser e t at en t e t o cars are in t e sa e lane t en lin s AD an BC are ro g ly sy etrical s t e at loss al es o t ese t o lin s o l e si ilar o e er en t e cars are in i erent lanes t e lin s are not sy etrical S eci cally en V_1 is in t e le t lane o V_2 lin BC as lo at loss co are to at loss on lin AD since BC is a irect line o sig t at it no o stacles ile AD asses t ro g o ies o e icles or alls an t e engine co art ent o one o t e e icles

So it is ossi le t at t e relati e signal strengt o t ese t o lin s can e se to isting is t e t ree scenarios t ere y sol ing t e relati e lane locali ation ro le n t e sa e set a ing t o ore ra ios to eac e icle si ilarly ro i es s it ore in or ation t at can e ti- li e to sol e relati e ront-rear locali ation ro le ater in Section e s o o t e a erage SS al es are se as in t to a S ort ector Mac ine S M ase iscri inator to i enti y t e si relati e locations see ig- re A ter P_2 co tes t e relati e location o V_1 it res ect to V_2 it in or s P_1 to ate accor ingly

e ne t t o sections escri e t e res lts ro o r real ri ing e eri ents ere e e al ate o MA er- or s in resence o lti le e icles n Section e e- scri e res lts ro o r ns- an S M ase si lations ere e e al ate t e er or ance o MA it l- ti le e icles n er lti o co nication scenarios

4. LANE CHANGE AND TURN DETECTION

o re ce o er cons tion e to erio ic eacon trans- issions see Section e ro ose tting t e ra ios into a slee o e an engaging t e or trans itting or recei ing o eacons y t e one only en eit er a lane- c ange or a t rn e ent is etecte or en a neig or re- ests or locali ation is also re ces t e n er o ac ets trans itte t ere y increasing t e c ances t at t e trans itte eacons are recei e correctly y t e recei ing ra ios ane-c ange e ents can e etecte y accelero e- ters it in t e one o ar t is e esign a t res ol ase algorit or accelero eters to etect t e lane-c ange e ents sing ata collecte ro o r e eri ents e s o t at it is ossi le to etect lane-c ange e ents it ig acc racy

n t is e eri ent e se a Motorola Atri s art one it An roi . to collect t e accelero eter ata e one as a ilt-in -A is accelero eter an as lace on t e as oar o t e car o si li y t e roce re o ata analysis t e one as oriente in a ay so t at only rea ings along t e -a is ill e a ecte it in t e ra- tion o lane-c ange t e one is lace in an ar itrary orientation t en irt al reorientation can e se to co te t e acceleration along t e i t o t e e icle n a erage t e accelero eter in t e one gi es rea ings e ery secon ic are s sce ti le to noise o cancel o t t e noise an inacc racy o t e accelero eter e co te t e rea ing at any oint y ta ing an a erage o rea ings recei e in t e re io s . secon ig re s o s an e - a le ra ata set an t e corres on ing s oot e ata

r algorit is ase on t e o ser ation t at en a lane-c ange e ent a ens t e orce co te y t e accelero eter along t e i t o t e e icle as a s eci c attern Ass ing t at t e e icle is c anging its lane to ar s t e irection o t e accelero eter s ositi e -a is t en t e accelero eter rea ing rst increases to a ig al e an t en ecreases ac to a lo er al e

o ca t re t is attern o r algorit or s as ollo s e one aintains t e *maximum* an *minimum* rea ings o t e accelero eter along t e i t o t e e icle it in a in o o last secon s is ration o t e in o is al o t e a i ration ta en y ri ers to er or a lane c ange[1] ene er a ne rea ing is a aila le it - ates ot t e *maximum* an *minimum* al es an c ec s t eir i erence t e i erence is larger t an a s eci e t res ol τ t e one re orts a lane-c ange e ent co - nicates it neig oring ones as ing t e to engage t eir ra ios an t en o es to t e Beacon P ase an triggers t e lane locali ation algorit

As s o n ne t e o n . m/s^2 to e an a ro riate t res ol to etect lane-c anges an t rns on ot local roa s an ig ays Base on a total o lane-c ange e ents er or e in o r e eri ents e o n see ig re t at a tra eo e ists et een t e *recall* $\frac{TruePositive}{GroundTruth}$ an *precision* $\frac{TruePositive}{TotalPredictions}$ o t e re iction o a e s re t at t e etection algorit oes not iss any lane-c ange e ents a ig er recall rate is esire ro ig re it is clear t at it a t res ol . m/s^2 recall is ossi le it ro g ly recision or local an recision or ree ay is ens res t at ene er a lane-c ange e ent a ens t e s art one is a le to etect it At t e sa e ti e it ay also generate so e alse ositi e e ents ro g ly — or e ery re ictions o e er since a ore acc rate relati e locali ation is er or e in P ase an P ase s c alse ositi es o l not a ect t e acc racy o o r sol tion n ractice e o not e ect too re ent lane c anges t s e elie e t at t e recision is acce ta le

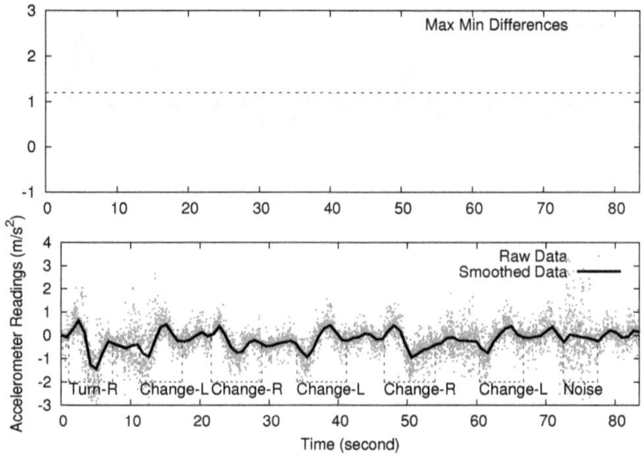

Figure 5: Bottom graph shows the raw and smoothed accelerometer readings along the X-axis. Left and right lane Change events are shown by "Change-L" and "Change-R", respectively while left and right turn events bu "Turn-L" and 'Turn-R", respectively. Top graph shows the variation in the difference of values of *maximum* and *minimum* smoothed accelerometer readings within the window of last 4 seconds.

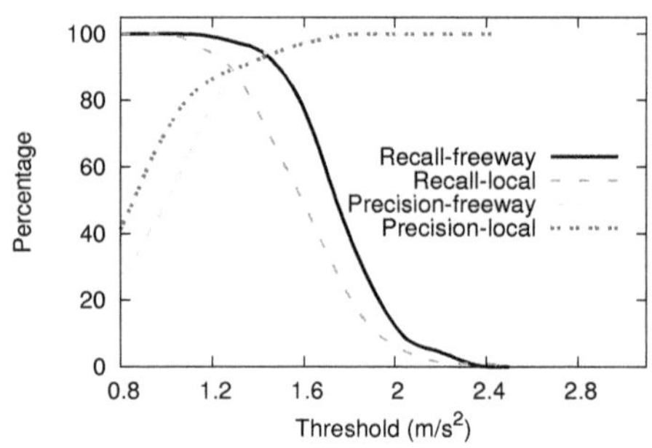

Figure 6: The tradeoff between recall and precision of predictions with variation in detection threshold τ on freeway and local drives.

5. REAL DRIVING EXPERIMENTS

n t is section e ill escri e t e res lts ro o r real ri ing e eri ents er or e it i erent e icles n er arying tra c an roa ty es e r ose o t e e eri ents as: i o eter ine t e est o nting osition o ireless ra ios ii o e al ate t e correctness an generali a ility o MA n er i erent roa ty es as ell as tra c con itions an iii eter ine i MA s acc - racy e en s on t e ty e o e icle se e concl sions ro t e e eri ents are escri e in Section

5.1 Experiment Setup and Data Processing

or o r ri ing e eri ents e se o r i erent e icles: i ar : oyota Solara o e ii ar : Ma a M Se an iii ar : issan Sentra Se an an i S : y n ai Santa e o e al ate t e acc racy o MA on roa s it i erent s ee li its e also c ose t o i erent in s o roa s: i A lane local s ee li it roa ay in an r an en iron ent it ar e e icles an store ronts on ot t e si es an ii ree ay s ee li it it so n alls an cras arriers on t e si es o t e roa at e laces ra c as o - ser e to e *moderate* in all t e e eri ents e ce t Section ere e e al ate t e generali a ility o o r sol tion it arying tra c con itions Since t e e eri ents ere er or e n er ncontrolle settings so eti es ot er e- icles ay inter ene et een o r t o test e icles

e se t e co ercially a aila le elosB otes as t e ireless ra ios o nte on t e si es o t e e icles e o y o t e ote along it atteries as o ser e to e aro n c t ic it a c antenna Since t ese i ensions are ery s all co are to t e i t o t e e icle e o not antici ate any signi cant re ction in e ciency o t e car e to t e a itional in rag e range o t e elosB to elosB co nication as o ser e to e eters en t ere are no o stacles in et een in ar -

[1] n o r e eri ents e o ser e t at it > ro a- ility all t e lane c anges are co lete it in secon s

ing lots o e er e co l not eas re t e range ring ri ing e to na aila ility o a ro riate e i ent or t e r ose o e eri ents instea o t e s art one e se anot er elosB ote connecte it a la to insi e t e e icles to collect SS rea ings ro t e elosB otes o nte o tsi e o an le ossi le ac et losses e to collisions elosB otes sen locali ation eacons at a ran- o s acing o - s e at loss al e or a lin is t en co te y ta ing an a erage o t e al es

e i enti y t e ro le o n ing anot er e icle s o- sition a ong t e *six* ossi le ositions see ig re as a s er ise classi cation ro le an a ress it sing a- c ine learning algorit s or t is e se t e a i Miner so t are ersion or training an testing n a i Miner e c ose t e S M s i le entation li S M it t e S M ty e set as -S an ernel ty e as lin- ear e al es o C cac e si e an e silon ere set to . an . res ecti ely e e ne t e *Classification accuracy* as t e ercentage o t e testing ata or ic t e relati e location is correctly re icte y t e classi cation o el or i erent e eri ents e se i erent training an testing tec ni es to e al ate t e acc racy o MA - n so e cases e i i e t e collecte ata set into t o arts train on t e rst an test on t e secon o - e er in so e ot er cases e se in e en ent test sets i e e train t e syste on one in o ata e g or local roa s an test it on anot er in o ata e g ree ays

rrors in classi cation acc racy ay arise e to a a- riety o actors: i Presence o near y ri ing e icles on t e roa ay an r an o stacles store ronts cras ar- riers ar e cars so n alls etc ay a ect t e al e o t e signal strengt ii rans issions ro t e roa si e i access oints ay a ect t e ac et ro rate o t e trans itte eacons iii So eti es a car ay stra le t o lanes an i i erent e icles ay a e i erent ysi- cal ro le i t organi ation o ac inery n er t e car oo etc ic ay a ect t e at loss al es

e res lts s o n in t is a er are ase on o r iles o ri ing ring t e ri ing e eri ents one o t e as- sengers recor e t e gro n tr t y an ally recor ing t e ti e an t e relati e e icle ositions ile t e ri ers c ange lanes at reg lar inter als e ran o ly s lit t e nally collecte ata set into *testing data set* an *training data set* t as ens re t at eac o t e classes corre- s on ing to si i erent relati e ositions ro ig re a e e al a o nt o ata in ot t e training sets an t e testing sets

5.2 Determining the best configuration

n t is e eri ent e arie t e n er an ositions o ra ios on t e e icle to eter ine t e con g ration t at a i i es t e locali ation acc racy n o r e eri ent e e lore can i ate con g rations on t e e terior o a car to o nt t e elosB otes ig re s o s o r s c os- si le con g rations along it t eir la els an acc racies or rest o t e res lts lease re er to o r tec nical re ort

n eac ri ing test e o nte t e s eci e n - er o ra ios on t e t o cars e total ration o ri ing or eac test as aro n in tes it a o t in tes o local ri ing an in tes o ree ay ri ing n t is s section e se o r ty es o ri ing ata or eac ra io osition co ination: i ata et een t o se ans on local roa s ii ata et een t o se ans on ree ay iii ata

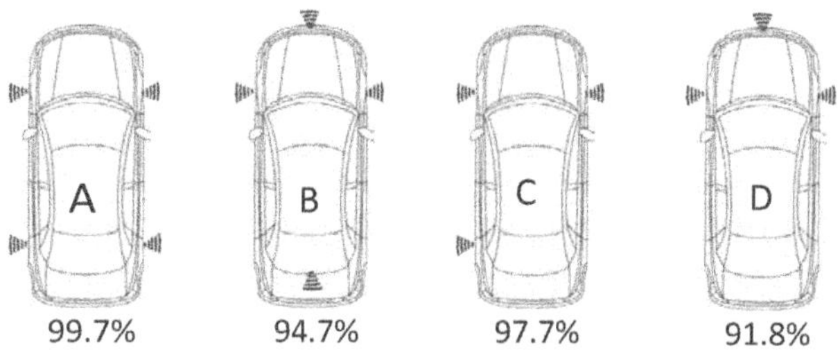

Figure 7: The different combinations of placement of the wireless radios and the corresponding relative lane localization accuracy. A and B have four radios attached while C and D have 3 radios. The classifier from configuration A has the highest prediction accuracy.

et een t e co e an S on local roa s an i ata et een t e co e an S on ree ay e ata as t en i i e into e al-si e training set an a testing set

e o ser e t at en o r ra ios are lace ertically a o e t e o r eels on t e e icle s o y e get t e a - i acc racy o . n t e later e al ations e only se con g ration A nless entione ot er ise Also ro ig re e can o ser e t at sing ore ra ios increases t e classi cation acc racy since ore ra ios ca t re ore in or ation a o t t e relati e locations o e icles e res lts o t e e eri ent also in icate t at t e osition o t e ra ios a ect t e acc racy o t e classi er or e a - le an ot a e t ree ra ios t t eir acc racy is signi cantly i erent

5.3 Evaluation with varying road types

e r ose o t is e eri ent as to e al ate t e gen- erali a ility o MA it arying roa ty es: i ocal roa s an ii ree ays e s ee li it on local roa s an ree ay ere MP an MP res ecti ely e tra c as o ser e to e *moderate* en collecting t e t o ata sets e se t e co e an t e S s ri ing ata as ell as t e t o se an s ri ing ata in t is e eri ent A ter t e e eri ent e create a local ri ing ata set *Local* an a ree ay ri ing ata set *Freeway*

e rst se ata ro *Local* to train an S M classi er en teste on ata ro *Freeway* t e classi er s acc racy as o n to e . n t e ne t test e traine anot er S M classi er sing *Freeway* ata an teste it on *Local* ata is ti e t e classi er ga e an acc racy o . is ig classi cation acc racy in icates t at t e at loss atterns on local roa s an ree ays are si ilar an t at training is in e en ent o roa ty e an s ee

5.4 Evaluation under varying traffic conditions

n t is e eri ent e ant to eter ine et er MA - is ro st to ariations in tra c con itions an t en eci e et er e nee to incl e t e ri ing ata o i- erent tra c con itions to train t e classi er or t is e create a ea y tra c ata set calle *Heavy* an a lig t tra c ata set calle *Light* *Heavy* contains t e ata e- t een t e co e an S on a i o sy local roa s an sy ree ays e sa e t o e icles ere also ri en n er lig t tra c con itions to create t e ataset *Light*

e se *Heavy* to train an S M classi er an t en on

testing it it *Light* t e classi cation acc racy as o n to e . n t e ne t test e se *Light* as t e training set en teste y *Heavy* t e ne classi er a acc racy o . n ot cases t e acc racy is ite lo is in icates t at t e ra io s at loss attern in lig t an ea y tra c con itions are i erent is is eca se in ea y tra c re ently t ere are lti le ot er e icles et een t e t o testing e icles is res lt is consistent it o r o ser ation t at t e o y o a car can a ect t e at loss o t e ireless ra io signal

o e er e n t at s litting *Heavy* into t o e al si e sets *Heavy* an *Heavy* training t e classi er on *Heavy* an testing it on *Heavy* increases t e classi cation acc - racy to . Si ilarly i *Light* is s lit into t o sets *Light* an *Light* it training er or e on one an testing on t e ot er it also increases t e classi cation acc - racy to . is res lt in icates t at ot ea y tra c ata an lig t tra c ata a e clear an consistent atterns ere ore to esti ate t e relati e osition in i erent tra - c con itions one ay is to incl e t e tra c con ition as an in t to t e classi er

e create a i e tra c ata set *Mix* y co ining e al a o nts o ata ro ot *Heavy* an *Light* en e s lit *Mix* into t o e al si e ata sets *Mix* an *Mix* traine t e S M classi er on *Mix* an teste it se arately on *Mix* *Light* an *Heavy* is classi er s acc racy as o ser e to e . . an . res ecti ely e i e tra c con ition classi er as ig acc racy in i- cating t at or ig acc racy classi cation in ario s tra c con itions it is se l to train it n er ot lig t an ea y tra c con itions is re o es t e re ire ent to ro i e t e tra c con ition in or ation as an in t

5.5 Evaluation with variation in vehicle types

n t is e eri ent e teste o t e o y o a car a ects t e er or ance o t e classi er e call t e ata collecte y sing t e co e an t e S on ot local roa s an ree ays as *SUVData* an t e ata collecte sing t o se ans on sa e roa s as *SedanData* or ot ty es o roa s t e tra c as o ser e to e o erate

e rst se *SedanData* to train an S M classi er an on testing it it *SUVData* e o ser e an acc racy o . Si ilarly in t e ne t test e traine t e S M classi er it *SUVData* an teste it sing *SedanData* an o ser e t e acc racy to e . o are it t e res lts in Section t e res lts ere are aro n lo er is in icates t at i erent ty es o e icle o ies create i erent signal strengt atterns o e er t e ari- ation in t e atterns is c s aller t an t e ariation e to i erent tra c con itions e can see ro t e res lt in Section t at en e incl e i erent ty es o e icles into t e training set t e acc racy as ery ig .

is in icates t at t e classi cation o el oes not a e a ias ro le it res ect to e icle ty e t ere ore it is not re ire to ro i e e icle o y as an in t to t e o el

5.6 Evaluation with variation in wireless radio position

As isc sse in Section t e est osition or o nting t e ireless ra ios is a o e t e eels o e er constraints on t e o y o t e e icle an an errors ay lea to installations in ositions t at are not necessarily o ti al n t is e eri ent e st ie t e sensiti ity o t e res lts

Table 1: Test results of the classifier trained by mixing all the driving data.

Testing data set	Accuracy
Mixed data	96.8%
Two sedans, Local, Moderate traffic	98.5%
Two sedans, Freeway, Moderate traffic	97.9%
Coupe and SUV, Local, Moderate traffic	97.7%
Coupe and SUV, Freeway, Moderate Traffic	99.4%
Coupe and SUV, Local, Heavy Traffic	92.8%
Coupe and SUV, Freeway, Heavy Traffic	95.7%
Two sedans on a freeway with curves	94.0%

it res ect to t e e ice osition is ti e or collect- ing ata e ro e t o se ans on local as ell as ree ay an collecte t ree ata sets: i *Correct*: a ios o nte at correct osition it correct orientation ii *Incorrect10*: a ios o nte any ere it in c o t e correct o- sition it ran o orientation an iii *Incorrect20*: a- ios o nte any ere it in c o t e correct osition it ran o orientation e t en i i e *Correct* in t o e al arts *Correct1* an *Correct2* By training t e S M classi er on *Correct1* an testing on *Correct2* *Incorrect10* an *Incorrect20* e o taine acc racies o an res ecti ely e t e i e all t e t ree ata sets to get *Mix* an se it or training en teste on *Correct* *Incorrect10* an *Incorrect20* t e S M classi er ga e acc racies o an res ecti ely

5.7 Mix all types of driving data together

n t is section e traine t e classi er on a ata set co - ine ro all ri ing tests an e al ate its acc racy on i erent ty es o ri ing ata n t e co ine training set e incl e: i t e t o se an s local an ree ay ri - ing ata in o erate tra c an ii t e co e an S s local an ree ay ri ing ata n er ot o erate an ea y tra c con itions e testing as one on i erent sets o ri ing ata se arately as ell as on anot er ata set ro a ri e on a *curved freeway* e to na aila ility o a local long eno g roa it lti le lanes an c r es e s i e testing t e classi er on *curved local roads* ro a le e can see t at or all t e cases t e i e o e s acc racy is at least . en testing on c r e ree- ays t e acc racy is slig tly lo er as t e classi er re icts so e o t e sa e lane cases as le t lane or rig t lane e o erall acc racy as o ser e to e is res lt con- r s t at MA is ro st to ost o t e ariations in ri ing cases

5.8 Conclusion from the experiments

ro t e e eri ent res lts Section e can a e t e ollo ing concl sions: i e est con g ration a ong all t e con g rations e teste is A ig re ere t e o r ra ios are lace a o e t e o r eels o t e e icle ii ri ing s ee oes not a ect t e acc racy signi cantly iii ea y tra c con itions an lig t tra c con itions a e i erent at loss atterns t t eir o n atterns are con- sistent ic allo s s to train t e o el it i e tra c ata an still get ig classi cation acc racy i e clas- si er s re iction acc racy aries it ariation in e icle s o y t a ter co ining t e training ata o i erent car o ies it is ossi le to ac ie e ig re iction acc racy an r esign is ro st to lace ent errors o t e ra- ios inally a ter i ing all t e training ata an testing it in i i ally on eac o t e testing ata sets e o ser e t at t e classi er as . e iction acc racy en teste n er o erate tra c con itions n ea y tra c con ition an c r e ri e t e acc racy ill ecrease t t e o erall acc racy re ains a o e .

6. INCREASING ROBUSTNESS THROUGH AGGREGATION

n Section e isc sse o t e ra ios on t e e icle can e se to eter ine t e relati e location o t o e i- cles o e er so eti es it ay lea to incorrect res lts n t is section e e lain o t e acc racy o locali ation can e rt er increase t ro g aggregation e g i V_1 is le t o V_2 an V_2 is le t o V_3 t en it is ossi le to in er t at V_1 is also le t o V_3 Si ilar in o aggregation can e se or ront an rear relations i s Acc racy o rela- ti e locali ation can e i ro e y a lying aggregation r les se arately to *left-same-right* an *front-rear* relation- s i s et een neig oring e icles onse ently in t is section e isc ss t o algorit s e rst one i ro es t e acc racy o lane locali ation ile t e secon i ro es ront an rear locali ation acc racy Bot t e algorit s are istri te in nat re an are e ec te y e ery e icle in t e net or ene er t e one or t e corres on - ing e icle eter ines its relati e osition it res ect to so e neig oring e icle it e ec tes ot t e algorit s to rt er i ro e t e ro stness o e er esinging an aggregation algorit t at i ro es locali ation acc racy in ol es t e ollo ing c allenges: i o a oi e en ence on a centrali e ser er it is re era le to se a istri te algorit ii e set o neig ors o a e icle ay c ange it ti e a ing it ar to n a e set o aggregating e icles an iii e relati e location eter ine y t e S M classi er ay e incorrect in so e cases

6.1 Aggregating relative lane localization information

o i ro e t e acc racy o le t-sa e-rig t relations i e assign e ery e icle V_i a *coordinate system* C_i a *virtual lane number* L_i corres on ing to t at coor inate sys- te an a ro a ility α_i t at V_i is in irt al lane L_i irt al lane n ers are *comparable* a ong e icles t at elong to t e sa e coor inate syste i e any t o e icles t at elong to t e sa e coor inate syste a e t e sa e irt al lane n ers *iff* t ey are in t e sa e ysical lane o t e roa ay Si ilarly a e icle on t e le t si e as ig er irt al lane n er t an a e icle on t e rig t si e i ot o t e elong to sa e coor inate syste e irt al lane lo- cation is co te sing t e relati e locations it res ect to lti le neig ors t at elong to t e sa e coor inate syste s t e sage o t e coor inate syste can also errors t at ay arise in relati e locali ation rt er it also allo s t o neig oring e icles t at a e ne er irectly co nicate it eac ot er e ore to easily eter ine i t eir relati e lane location can e co te y si ly co aring t eir irt al lane n ers

A coor inate syste C_a is enote y t e t le T_a, I_a ere T_a is t e ti esta en t e coor inate syste as create an I_a is t e o t e e icle t at create t e coor- inate syste A coor inate syste C_a is sai to e s aller t an coor inate syste C_b i one o t e ollo ing con itions is tr e: i $T_a < T_b$ or ii T_a T_b an $I_a < I_b$ o ens re t at t e irt al lane n ers o e icles is co ara le o r algorit tries to o e all e icles to t e sa e coor inate syste or t is in o r algorit e ery e icle oins t e *smallest* coor inate syste a ong all its neig ors ote t at *smallest* is re erre o er *largest* since a ne e icle oining an e isting coor inate syste ill re ire only t at e icle to c ange its coor inate syste

A e icle ay also create a ne coor inate syste en- e er its irt al lane n er eco es n no n or e a - le t is ay a en: i en t e a on t e one is starte ii en t e a etects t at t e e icle is in o- tion a ter eing ar e or ore t an T_{stop} ti e or iii t e ata ro t e accelero eter in icates a ossi le t rn e ent or a lane c ange e ent[2] e creation o a ne coor- inate syste a ter a t rn e ent ens res t at t e coor inate syste o t e t rning e icle is larger t an t ose o t e e- icles t at are resent on t e ne roa is orces t e t rning e icle to oin t e coor inate syste o t e ot er e icles s only one e icle ay c ange its coor inate syste instea o all t e e icles t at ere resent on t e ne roa is not only re ces t e co nication o er- ea t also a es t e syste ore sta le

on creation o a ne coor inate syste it t e c rrent ti esta an its o n t e e icle say V_i initiali es its lane n er L_i in t e ne coor inate syste to e an t e corres on ing ro a ility α_i as A ter initiali ing its coor inate syste V_i ay see to oin t e coor inate syste o anot er e icle or t is it co nicates it all its neig ors to o tain t e in or ation o t eir c rrent coor inate syste s t eir lane n ers an t e corres on - ing ro a ility al es e e icle t en oins t e s allest coor inate syste a ong all its neig ors ines - o Al- gorit is ens res t at e ent ally all neig oring e icles ill elong to t e sa e coor inate syste t ere y increasing t e c ances t at t eir irt al lane n ers are co ara le en oining t e ne coor inate syste V_i also eter ines its lane n er it t e el o its neig - ors enote y S_{low} t at are in t e s allest coor inate syste ines - en V_i eter ines its relati e loca- tion it res ect to all e icles in S_{low} sing t e algorit isc sse in Section ine en it rocee s to co - te its lo est ossi le lane *min* an ig est ossi le lane *max* n ers ine ic are one lo er an one ig er t an t e lo est an ig est lane n ers o all e icles in S_{low} res ecti ely e e icle ay e assigne a irt al lane n er o *min* only i it is to t e le t o all its neig - oring e icles

en or e ery lane a ong t e ossi le lanes ro *min* to *max* it co tes t e con ence t at t e e icle is lo- cate in t at lane ines - o co te t e con ence o eing in a artic lar lane say l it co tes t e relati e location o l it t e lane n er o e icle V_j ines - is con ence is co te o er all e icles an t eir s *confi*$_l$ re ects t e con ence t at V_i is in l ase on in or ation o all e icles in S_{low} ine ere c is t e o t t o t e S M ase classi er see Section an its al e lies et een an inally t e lane t at a i- i es t e *confi*$_l$ is esignate as t e lane n er o V_i an

[2]Since it ay not e ossi le to isting is a t rn e ent ro a lane c ange e ent as ell as to eter ine o any lanes t e e icle act ally c ange

its ro a ility al e α_i is also ate y nor ali ing it across all ossi le lanes ines -

t is also ossi le t at t o e icles t at are ysically s earate y lti le lanes are assigne consec ti e irt al lane n ers is ay later ca se con icts en a ne e icle o es to t e lane in et een t ese t o e icles o ta e into acco nt s c cases an or increase ro stness eac e icle in o r algorit also in o es Algorit eri- o ically to ate its irt al lane n er

Since eac e icle no as a irt al lane n er t o e- icles can no arri e at a ore acc rate res lt or t eir rel- ati e locali ation y st co aring t eir irt al lane n - ers Also since eac irt al lane n er corres on s to a ysical lane t ere ore t e n er o irt al lanes are li ite y t e n er o ysical lanes on t e roa

Algorithm 1: or a gi en e icle V_i istri te algo- rit or ating coor inate syste C_i lane n er L_i an corres on ing ro a ility α_i or V_i

```
For every neighbor V_j, obtain V_j's coordinate system (C_j),
  V_j's most probable virtual lane number (L_j) and probability
  that V_j is in L_j (α_j)
C_low ← arg min_{j:V_j is a neighbor of V_i} C_j
S_low ← Set of neighboring vehicles that are in C_low
if C_low ≤ C_i then
    C_i ← C_low, L_i ← φ
    Determine relative lane location with respect to each
    vehicle in S_low.
    min ← min_{j:V_j∈S_low} L_j − 1, max ← max_{j:V_j∈S_low} L_j + 1
    for l = min to max do
        confi_l ← 0
        for all Vehicle V_j ∈ S_low do
            if L_j − l > 0 then
                c ←Confidence that relative location of V_j
                with respect to V_i is left
            if L_j − l = 0 then
                c ←Confidence that relative location of V_j
                with respect to V_i is same lane
            if L_j − l < 0 then
                c ←Confidence that relative location of V_j
                with respect to V_i is right
            confi_l ← confi_l + α_j × c
    maxLane ← arg max_i confi_i
    L_i ← maxLane, α_i ← confi_maxLane / Σ_{l=min}^{max} confi_l
```

6.2 Aggregating front-back localization information

Si ilarly in or er to i ro e t e acc racy o t e ront- ac locali ation eac e icle o tains relations i in or a- tion it its neig ors sing t at in or ation an its o n relati e osition it res ect to its neig ors it constr cts a relations i gra G W, E ere W is t e set o e icles See Algorit rt er t ere is an e ge ro e icle V_a to V_b *iff* V_a elie es t at V_b is in its ront[3] Presence o a irecte cycle in G i lies t at relati e location or at least one air o e icles is incorrect See or roo ere- ore y re o ing cycles ro G Algorit increases t e acc racy o relati e locali ation

[3] i alent to saying t at V_b elie es t at V_a is in rear since t e e icle t at co tes relati e location asses on t e res lt to t e ot er e icle Section

Since co nting t e n er o cycles in a gra is an P- ar ro le See or roo t ere ore e e ne a e risitcal etric β G t at ca t res t e n er o cycles in G o eter ine i a irecte e ge V_i, V_j is a art o so e cycle e only nee to c ec i V_i is reac a le ro V_j

$$\beta \; G \quad \{|e : e \in E \text{ an } e \text{ is art o so e irecte cycle}|\}$$

o re o e cycles ro G it o t losing any relati e lo- cation in or ation Algorit re erses e ges in G ntil G eco es acyclic or t is at eac ste it iterati ely re- erses t e e ge t at re ces t e n er o cycles in G y t e largest a o nt e ersing an e ge V_a, V_b in icates t at ase on in or ation o neig oring e icles it is ore li ely t at V_a is act ally in ront o V_b o ini i e t e n er o e ges re erse Algorit rst tries to n a single e ge t at can e re erse to re ce t e n er o cy- cles ines - o e er it is ossi le t at it is not a le to n any s c e ge e en t o g t e gra is cyclic n t at case it ic s a erte an re erses all t e inco ing e ges ines - or all t e o tgoing e ges ines - Algo- rit er or s t is searc o er all ertices an er or s t e o eration on t e erte t at a i ies t e n er o cycles re ce ines - t can e s o n t at Algo- rit e ent ally ter inates an as a olyno ial ti e co le ity or roo see

Algorithm 2: or a gi en e icle V_i istri te al- gorit or ating ront- ac relations i o V_i it res ect to its neig ors

```
G = (W, E) ← Front-back graph based on neighborhood info
  where
W ← {V_i}∪ Neighbors of V_i
E ← {(V_a, V_b) : V_a believes that V_b is in its front }
while β(G) > 0 do
    maxdiff ← 0, edgeToReverse ← φ
    for all e ∈ E do
        E' ← E\{e} ∪ {ē}, G' ← (W, E')
        if β(G) − β(G') > maxdiff and ē is not a part of
        some directed cycle in G' then
            maxdiff ← β(G) − β(G'), edgeToReverse ← e
    if maxdiff > 0 then
        E ← E\{edgeToReverse} ∪ {edgeToReverse‾}
    else
        edgesToReverse ← φ
        for all V_i ∈ W do
            F ← {(V_j, V_i) for some V_j ∈ W}
            F̄ ←Obtained after reversing all edges in F
            E' ← E\F ∪ F̄, G' ← (W, E')
            if β(G) − β(G') > maxdiff then
                maxdiff ← β(G) − β(G'),
                edgesToReverse ← F
            F ← {(V_i, V_j) for some V_j ∈ W}
            F̄ ←Obtained after reversing all edges in F
            E' ← E\F ∪ F̄, G' ← (W, E')
            if β(G) − β(G') > maxdiff then
                maxdiff ← β(G) − β(G'),
                edgesToReverse ← F
        E ← E\{edgesToReverse} ∪ {edgesToReverse‾}
```

7. SIMULATIONS

n Section e escri e t e e eri ent res lts o taine on a ariety o e icles o e er e al ating t e e a ior an er or ance o o r algorit or a larger set o e i- cles as la or intensi e as e ery e icle re ires one ri er

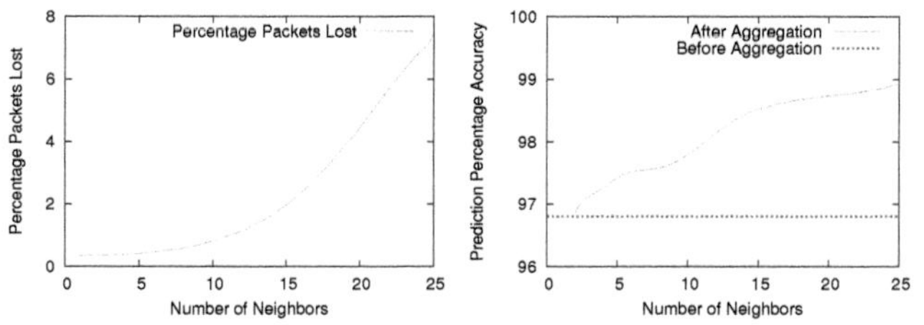

a Percentage Pac ets ost   ocali ation acc racy e- ore an a ter er or ing aggregations

Figure 8: Simulation results with varying number of neighbors.

an one assenger to recor t e gro n tr t So e er- or e *trace-driven simulations* sing ns- an S M S M is a si lator or A s ic gi en a roa net- or generates a re- eter ine n er o tri s or e- icles or eac tri it c ooses a ran o starting oint a ran o en ing oint an generates a ro te or t e e i- cle co leting t at tri S M is a icrosco ic-le el roa tra c si lator ic i lies t at S M o els lti- le lanes in t e roa ays an e icles in S M er or a to atic lane-c anges an o erta ing o e icles as t ey o e ro t eir starting oint to t eir en ing oint in t e a e o ecti es o t e si lations ere as ollo s: i o eter ine t e increase in acc racy ac ie e y sing aggregation algorit s an ii o eter ine t e ac et loss rate an its a ect on acc racy en lti le e icles are r nning MA

or o r si lations e c ose a iles × iles area aro n o nto n o A stin[4] rt er e se S M to generate tri s or e icles an t eir corres on ing ro tes e si lation as e ec te or secon s ic as eno g or e ery e icle to reac its estination e o t t o S M osition o e ery e icle at eac instant o ti e as se or generating t e locations o corres on ing no es in ns-

n ns- e ery e icle consiste o no es: a s art one an t e ireless ra ios e o ility o t e no es is generate ro t e S M s o t t e s art ones sen P *discovery beacons* e ery secon s t at are ear y neig oring s art ones e range o ra io to ra io co - nication as set to eters ile t e range o s art- one to s art one co nication as set to eters e ireless ra ios trans itte *localization beacons* en irecte y t e s art one o si late t e e ect o e- icle s o y an t e neig oring r an str ct res ar e cars ot er e icles an store ronts on t e signal strengt e too t e signal strengt al es at t e recei ing ireless ra ios ro t e ata o taine t ro g e eri ents corre- s on ing to t e gro n tr t location o t e e icles o e er t e ac ets ro one ireless ra io or s art- one to anot er ere still ro e y ns- i so e ot er neig oring no e as also o ser e to e trans itting at t e sa e ti e Si ilar to e eri ents ere also e ery ireless no e trans itte *localization beacons* at ran o s acings o - s e s art one co te t e relati e location o its e icle it res ect to neig oring e icles y ass- ing t e a erage SS al es to t e S M classi er or t e si lations e ass e t at t e s art one triggers lane c ange e ents it recall an recision See Section

Simulation Results: ig re a s o s t at ac et loss rate increases it increase in n er o neig ors e a erage ac et loss rate across all no es as o ser e to e Since in MA ra io no es roa cast *lane discovery* eacons lti le ti es it is e ecte t at t is lo loss rate o l not ecrease t e acc racy o e er it increase in n er o neig ors it is ossi le or MA to increase acc racy y a ing se o aggregation ig re s o s t e ariation in locali ation acc racy e ore an a ter e er or e aggregation sing t e t o algorit s it aggregation MA is a le to i ro e t e locali ation acc racy to en t e n er o neig ors is ery ig en n er o neig ors is lo aggregating in or ation across e icles ro i es li ite ene t er all e icles in t e si lation e o ser e t at aggregation i ro e acc racy to

8. DISCUSSION AND FUTURE WORK

n t is section e isc ss so e ec anis s t at can a e MA ore acc rate an ractical

8.1 Increasing lifetime of wireless radios

By sing t e accelero eter o t e one to trigger t e relati e locali ation algorit e a e een a le to a oi e- rio ic triggering o ireless ra ios t ere y increasing t eir li eti e es lts ro o r e eri ents an si lations s o t at t e stan ar elosB otes e i it a attery li eti e o at least ont s or an a erage o o rs aily ri ing an a a i reaction ti e o secon a ter t e lane c ange e ent is triggere See or etails o re o e t e e en ence on atteries t e ossi ility o connecting t e ireless ra ios irectly to t e e icle s o er s ly can also e e lore

8.2 Incremental Deployment

MA can eter ine relati e location only i ot e- icles are e i e it ireless ra ios t cannot eter ine t e relati e location i only one o t e e icles is e i e it ra ios o e er MA can ro i e incre ental ene t to e icles t at are e i e it ra ios n SA as alrea y reser e M o s ectr or e i- cate S ort ange inter- e icle o nication S t is e ecte t at in t e t re all e icles ill e e i e it at least one antenna or S r e eri ents s o t at i one o t e e icles is e i e it only one antenna locate on t e e icle s o y near t e ront assenger si e eel an t e ot er it antenna MA correctly re icts t e relati e location it acc racy ere ore as S eco es ore i es rea e icles it antennas o l e a le to se MA it an acc racy en t ey enco nter ot er e icles it a single antenna an it o r antennas res ecti ely[5] sing si - lations e also o ser e t at i o t e e icles a e antenna ile t e ot er a e antennas t en a erage locali ation acc racy as or one antenna- o r antenna

[4] an o ly c osen city

[5] is is n er t e ass tion t at t e location o S antenna is on t e si e o t e e icle A centrally locate antenna ay gi e i erent res lts

e icle airs an or o r antenna- o r antenna e i-cle airs s it aggregation it is ossi le to ac ie e ig er locali ation acc racy e en en all e icles are not e i e it o r ra ios

8.3 Training Cost

o ac ie e ig er locali ation acc racy it is ene cial to train MA or t e s eci c e icle As e sa in Section it still ro i es an acc racy o en traine on a e-icle it i erent ysical ro le Also in o r e eri ents e sa t at testing a classi er on se an-se an air t at is traine it se an-co e air still gi es acc racy is i lies t at it is not necessary to train MA se arately or e icles t at a e si ilar ysical ro les alt o g to get ig er acc racy it is ene cial to train MA on only t ose e icles t at a e signi cantly i erent ro les

9. CONCLUSION

n t is a er e ro ose a no el antenna i ersity ase sol tion calle MA or eter ining t e relati e loca-tion o t o e icles on roa ays elati e location in or-ation as t e otential to not only en ance t e ri ing e erience y ro i ing rele ant alerts t also re ce t e c ance o collisions MA as lo cost an is easy to install on ne er as ell as e isiting e icles es lts ro o r ri ing e eri ents er or e n er arying con itions s o t at MA re icts relati e location o t o e i-cles it an a erage acc racy o MA is a le to eter ine t e relati e osition o e icles t at are not in t e i e iate neig or oo t ere y gi ing t e e icle ri er ore ti e to react o re ce energy cons tion o ireless ra ios an to re ce t e n er o ac ets trans-itte e also ro ose sing t e one s accelero eter to trigger t e locali ation algorit e resente t o algo-rit s t at increase t e locali ation acc racy y aggregating in or ation across lti le e icles r trace- ri en si -lations s o t at y aggregating in or ation MA is a le to increase t e locali ation acc racy to

10. ACKNOWLEDGEMENTS

is aterial is ase on or artially s orte y t e ational Science o n ation n er rants S-an S- e o l also li e to t an Pro Marco r teser or s e er ing t e a er

11. REFERENCES

e ns- net or si lator tt : nsna org
aleo ayt eon Syste s aleo co
AAS *A Policy on Geometric Design of Highways and Streets* –
M Be risc Bie er r ann an ra e ic S M - Si lation o r an M ility: An er ie n *Proc. of Third International Conference on Advances in System Simulation (SIMUL)*
M Bis o *Pattern Recognition and Machine Learning* S ringer
ella a ian an eng e icle etection an rac ing sing Aco stic an i eo Sensors n *Proc. of IEEE Int'l Conf. Acoustic, Speech, and Signal Processing*
eng B S eng P seng an an ane etection it Mo ing e icles in t e ra c Scenes *IEEE Transactions on Intelligent Transportation Systems* : –
P tta an ller Practical Async rono s eig or isco ery an en e o s or Mo ile Sensing A lications n *Proc. of ACM SenSys* ages –
ar in at is PS tt : gar in co a o t PS in e t l
A ern ran e an P e i A ance ane ecognition- sing ision an a ar n *Proc. of IEEE Intelligent Vehicles Symposium*
ei es an agel o ar s Acti e Mac ine- ision-Base ri er Assistance or r an Areas *Int'l J. Computer Vision* : –
A an al a s anan an a ar -connect: A o - atency nergy- cient Async rono s eig or isco ery Protocol n *Proc. of ACM IPSN*
i Bansal an P Sin a MA : M lti le Antenna ase elati e e icle ocali er ec nical re ort io State ni ersity S - S - -
Miers a M rst lin en erg M Sc ol an ler A : a i Prototy ing or o le ata Mining as s n *Proc. of ACM SIGKDD*
P Mo an Pa ana an an a ee ericell: sing Mo ile S art ones or ic Monitoring o oa an ra c on itions n *Proc. of ACM EMNETS*
SA al ation o ane ange ollision A oi ance Syste s sing t e ational A ance ri ing Si lator S S e art ent o rans ortation May
Pa a at i S an ga riya an M alai ani A St y on e icle etection an rac ing sing ireless Sensor et or s en Access *Wireless Sensor Network* : –
S Par i S ang an eon A o el Signal Processing ec ni e or e icle etection a ar *IEEE MTT-S Int'l Microwave Symp. Digest*
P Piran M rt y an P Ba e ic lar A oc an Sensor et or s: Princi les an allenges *International Journal of Ad hoc, Sensor and Ubiquitous Computing* : –
ang or e an A S e a ar-Base etection an rac ing o Mo ing ects ro a ro n e icle at ig S ee s n *Proc. of IEEE Intelligent Vehicle Symp.*
ang S S ang an ri er Assistance Syste or ane etection an e icle ecognition it ig t ision n *Proc. of IEEE Intelligent Robots and Systems*
ang S Si o an rase aran i ecan en M r teser an P Martin etecting ri er P one se e eraging ar S ea er n *Proc. of ACM MobiCom*

MIDU: Enabling MIMO Full Duplex

Ehsan Aryafar[1], Mohammad (Amir) Khojastepour[2], Karthikeyan Sundaresan[2], Sampath Rangarajan[2], and Mung Chiang[1]
[1]Princeton University, Princeton, NJ, USA [2]NEC Laboratories America, Princeton, NJ, USA
[1]{earyafar, chiangm}@princeton.edu, [2]{amir, karthiks, sampath}@nec-labs.com

ABSTRACT

Given that full duplex (FD) and MIMO both employ multiple antenna resources, an important question that arises is *how to make the choice between MIMO and FD?* We show that optimal performance requires a combination of both to be used. Hence, we present the design and implementation of MIDU, the first MIMO full duplex system for wireless networks. MIDU employs antenna cancellation with symmetric placement of transmit and receive antennas as its primary RF cancellation technique. We show that MIDU's design provides large amounts of self-interference cancellation with several key advantages: (i) It allows for two stages of additive antenna cancellation in tandem, to yield as high as 45 dB self-interference suppression; (ii) It can potentially eliminate the need for other forms of analog cancellation, thereby avoiding the need for variable attenuator and delays; (iii) It easily scales to MIMO systems, therefore enabling the coexistence of MIMO and full duplex.

We implemented MIDU on the WARP FPGA platform, and evaluated its performance against half duplex (HD)-MIMO. Our results reveal that, with the same number of RF chains, MIDU can potentially double the throughput achieved by half duplex MIMO in a single link; and provide median gains of at least 20% even in single cell scenarios, where full duplex encounters inter-client interference. Based on key insights from our results, we also highlight how to efficiently enable scheduling for a MIDU node.

Categories and Subject Descriptors

C.2.1 [**Computer-Communication Networks**]: Network Architecture and Design—*Wireless Communication*

General Terms

Design, Experimentation, Measurement, Performance

Keywords

Full Duplex, Multi-User MIMO

MobiCom'12, August 22–26, 2012, Istanbul, Turkey.

1. INTRODUCTION

A full duplex wireless device is one that can transmit and receive at the same time in the same frequency band, and typically it requires at least one Tx and one Rx antenna. The key challenge in realizing such a device lies in the self-interference (SI) generated by the Tx antenna at the Rx antenna. As an example, consider a Wi-Fi signal with a transmit power of 20 dBm. A Tx–Rx antenna separation of about 6–8 inches results in a path loss of about 40 dBm (depending on channel characteristics), resulting in a self-interference of at least -20 dBm. With a noise floor around -93 dBm, one would further require a self-interference cancellation of at least 73 dB to be able to decode the desired received signal. While one can solely employ digital interference cancellation techniques, current ADC's do not have a resolution to pass a received signal that is 73 dB less than the noise floor. Hence, several practical full duplex (FD) systems [7, 8, 13] have been proposed that couple RF cancellation along with digital cancellation to achieve the desired level of SI suppression.

RF cancellation may include a combination of antenna cancellation and analog cancellation. In [7], antenna cancellation was achieved by placing two Tx antennas asymmetrically at ℓ and $\ell + \frac{\lambda}{2}$ distance from the Rx antenna. This allows the transmit signals to add π out of phase and hence cancel each other. On the other hand, analog cancellation involves the generation of a π phase shift internally, coupled with the estimation and compensation of the SI channel [8, 13]. This allows for π phase shifters with a better frequency response over a wide-band channel (e.g., BALUN in [13]) to be employed, in contrast to the strong dependence on frequency (λ) posed by the antenna cancellation in [7]. While the existing schemes employ at least two antennas, one can also envision FD with a single antenna [14], where a circulator is used to isolate the Tx and Rx signals. However, owing to the lack of both path loss attenuation and RF cancellation, the required level of SI cancellation is significantly higher and hence hard to realize.

Given that at least 2 antennas are needed for a practical implementation of FD at Wi-Fi Tx power, an alternate approach would be to employ the multiple antennas as in half duplex (HD)-MIMO to increase the link capacity. With next generation wireless devices (access points, base stations, etc.) equipped with multiple antennas (often more than two), it is both timely and important to understand *how to best employ the available spatial degrees of freedom (antennas).* When comparing FD with MIMO, we consider two models that are of practical interest – *antenna conserved*

(AC) and *RF chain conserved* (RC). While the AC model allows for FD to be realized in legacy MIMO nodes (# Tx/Rx chains = # antennas), the RC model allows for nodes that are designed with full duplex in mind (# Tx/Rx chains $\leq$ # antennas). We show that the relative merits of FD and MIMO differ significantly depending on the model considered.

However, irrespective of the model considered, it turns out that there are several scenarios where the best strategy is not MIMO or FD in isolation but a combination of both. This in turn brings us to the next question as to *how to effectively realize a joint MIMO+FD system?*

We observe that the existing antenna cancellation [7] and analog cancellation [13] approaches cannot be readily extended to MIMO systems. Although one might envision an extension of [7] using two Tx and one Rx antenna for every transmitted/received MIMO stream, this would require antennas to be placed such that each of the Tx pairs (for each stream) lead to SI signals that are 180° out of phase at *every* Rx antenna, which is hard to realize. On the other hand, analog cancellation in [13], when extended to N stream MIMO, requires one to estimate the self-interference channel between every pair of N^2 Tx–Rx antennas. This in turn results in the use of N^2 variable delays and attenuators, each of which has to be auto-tuned to track the N^2 SI channels, which is hardly practical, specifically in frequency selective wide-band channels.

In addressing the above challenges, we design and prototype *MIDU*[1] – a node with joint MIMO and (full) DUplexing capabilities. MIDU employs antenna cancellation with *symmetric* placement of antennas as its primary RF cancellation technique. Specifically, for a single stream transmission, it employs antenna cancellation with a symmetric placement of either two Rx antennas and one Tx antenna (which we refer to as Rx antenna cancellation), or in a dual manner, two Tx antennas and one Rx antenna (which we refer to as Tx antenna cancellation). In addition to avoiding the shortcomings of asymmetric antenna cancellation [7], we show that such a design provides large amounts of SI cancellation with several key advantages:

- It allows for a two-level design, whereby Tx antenna cancellation is followed by Rx antenna cancellation. Using antenna cancellation theory, we show that such a design has the potential to double the antenna cancellation gains because of its additive nature.
- It potentially eliminates the need for any other form of analog cancellation, which seems limited in practice due to the need for variable attenuators and delay elements and its subsequent lack of scalability to MIMO systems.
- More importantly, it scales very easily to MIMO systems, thereby enabling the co-existence of MIMO with FD.

We have prototyped MIDU on WARP radios [2] for a 3×3 narrow-band MIMO + FD system. Briefly, our experiments reveal that each level of antenna cancellation can contribute to 25–30 dB of SI suppression, while their combination can yield up to 45 dB of suppression, thereby eliminating the dependence on analog cancellation. Coupled with 25–30 dB of SI suppression from digital cancellation techniques as in [11, 13, 16], MIDU can enable FD for Wi-Fi transmit powers. We also evaluate the performance of MIDU as a relay between two hidden nodes and as an access point (AP) serving multiple single antenna clients. For each of these schemes, we compare the performance of MIDU against half duplex MIMO (or MU-MIMO). Our results reveal that MIDU increases relay capacity up to 80%, and even in the presence of uplink-downlink interference in single cells, increases the median capacity by at least 20%.

[1]pronounced as "MyDu"

Finally, we highlight the implications of a MIDU node on the design of the MAC itself. While the design of a scheduler (MAC) for single cell MU-MIMO (including client selection and precoding) is challenging in its own right, extending it to a MIDU system takes it to another level – spatial degrees of freedom now have to be carefully split between downlink and uplink for FD, with client selection and precoding jointly addressed for MU-MIMO in each direction. Since the search space for the problem is very large, we use key insights from analysis and experimental data to identify regions of pronounced FD gains, and hence provide guidelines for an efficient scheduling strategy with a tractable search space.

In summary, we are driven in this paper by three questions concerning full duplex MIMO: *Is it feasible? Is it scalable? What are the tradeoffs with half duplex MIMO?* In addressing these questions, we make the following contributions:

- Designed MIDU, a system that can enable both MIMO and FD in tandem.
- Enabled two stages of antenna cancellation with additive gains to yield as much as 45 dB suppression, thereby alleviating the dependence on analog cancellation and SI channel estimation.
- Built a prototype of MIDU and showcased the additive benefits of its two level cancellation as well as joint operation of 3×3 MIMO + FD in practice.
- Provided guidelines for the design of an efficient MAC for single cells employing MIDU nodes.

The next section describes the related work. The joint merits of FD and MIMO are discussed in the following section. Thereafter, we present the design of MIDU, performance evaluation, guidelines on scheduling, discussions, and finally, the concluding remarks.

2. RELATED WORK

Full Duplex. Recent work has proposed several techniques to enable FD at Wi-Fi transmit powers [7, 8, 13]. Fig. 1(b) shows the various components of a FD system that contribute to SI suppression, namely: antenna cancellation, analog cancellation, and digital cancellation.

Antenna cancellation arranges Tx and Rx antennas in such a way that SI is reduced at the Rx antenna. While [7] employed asymmetric placement of Tx antennas to generate a π phase shift between the transmitted signals at the RX, [9] employed directional Tx antennas that place a null at the Rx antenna. For a MIMO capable FD node, antenna placement must cancel SI at all of the receiving antennas, which is hard to realize in the proposed schemes.

Analog cancellation requires knowledge of the SI channel to create a copy of the SI signal in the RF domain and cancel it before the signal is digitized. While [16] uses a noise canceler chip for the purpose to achieve about 20-25 dB of SI suppression, [13] employs a combination of BALUN (balanced to unbalanced) transformer (to generate a negative copy of transmit signal) and variable delay and attenuators

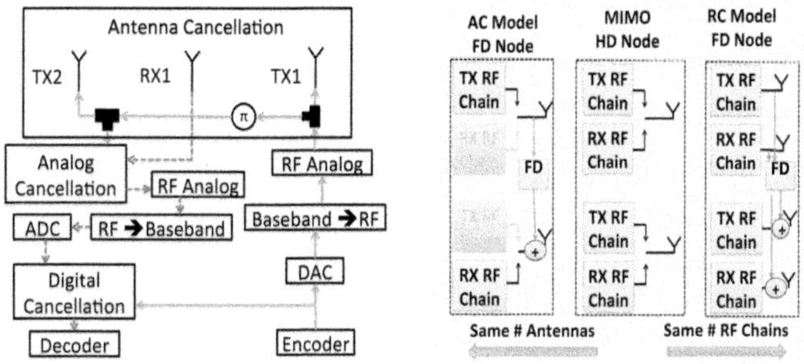

Figure 1: (a) FD Architecture, (b) Comparison Model.

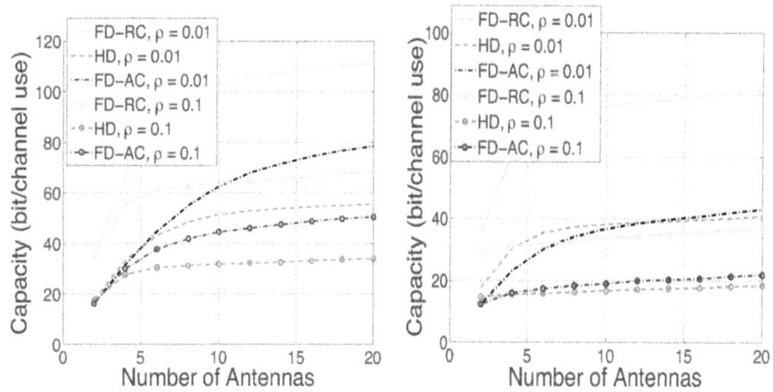

Figure 2: FD vs. MIMO Performance: (a) Perfect SI suppression, (b) SI suppression loss: 6 dB.

(to track the SI channel), to yield a SI suppression of 25–30 dB in practice. In [8], a pseudo analog cancellation technique is introduced, in which an additional RF chain creates a canceling signal in RF from a digital estimate of SI in the base band, removing 35 dB of SI.

All these schemes require an estimation of SI channel between a Tx and Rx antenna, which becomes a scalability bottleneck for MIMO systems with FD.

Digital cancellation utilizes the digital samples of the transmitted signal in the digital domain and subtracts them from the received samples, removing up to 25 dB of SI [11, 12, 13]. However, the pseudo analog cancellation approach adopted in [8] limits the additional suppression from digital cancellation to only 4–5 dB. This restricts its total cancellation to less than 39 dB, limiting its applicability to small–medium range communications only.

While MIDU can also benefit from digital cancellation (as in [11, 12, 13]), in contrast to the above works, the large potential for SI suppression from MIDU's antenna cancellation (albeit complementary to existing schemes), allows it to break away from the dependence on analog cancellation. This allows MIDU to easily scale to MIMO systems.

MIMO. In single user MIMO (SU-MIMO) systems (e.g., 802.11n [4] and BLAST [10]), the capacity of a point-to-point communication link is (theoretically) expected to scale with the number of antennas at the Tx and Rx (say N). However, in practice, the number of antennas at a base station or access point (N) is much more than those at the clients (n), limiting the performance of SU-MIMO to scale only with n. Multi-User MIMO (MU-MIMO) can be employed to overcome the limitation in such scenarios. Recent work [6, 17] has implemented MU-MIMO schemes, in which an AP can communicate with a number of clients simultaneously by utilizing the antennas that belong to a group of clients. As a baseline for comparison, we compare the performance of MIDU against such half duplex (HD)-MIMO schemes.

3. MIMO OR FD, OR BOTH?

To understand why we need a combination of MIMO and FD, we need to study the relative merits of FD and MIMO. While our primary focus is on performance, hardware complexity must also be taken into account for a fair comparison. We define two models that are of practical interest: *antenna conserved* (AC) and *RF chain conserved* (RC) models. The models are defined with respect to a legacy MIMO node as shown in Fig. 1 (b), where each antenna is associated with a pair of Tx and Rx RF chains, and are both important from different perspectives.

AC Model: Here, the node employing FD has the same number of antennas (and RF chain pairs) as the legacy MIMO node and hence represents the case where FD is enabled directly on legacy MIMO nodes. Hence, when FD is enabled, depending on the split of antennas between downlink and uplink, a mix of Tx and Rx RF chains will be utilized in the FD mode. The total number of streams in FD (including both uplink and downlink) is the same as that in MIMO (either downlink or uplink). However, the full potential of FD is not leveraged in this case (half of the RF chains are not used as shown in Fig. 1(b)).

RC Model: Here, the node employing FD has more antennas than the number of RF chain pairs. Note that in HD-MIMO nodes, due to the half duplex nature, only half the number of RF chains (Tx or Rx) can be used in any transmission. However, this is not the case with FD, where all the RF chains can be effectively used as long as there are sufficient number of antennas available. For example, compared to a two antenna MIMO node supported by two pairs of Tx–Rx RF chains and capable of sending (or receiving) two streams, a corresponding FD node can use all the four RF chains through four antennas to send and receive two streams simultaneously. On the other hand, having the additional antennas on the legacy MIMO node would only contribute to diversity.

Note that the processing complexity of a transceiver lies predominantly in its RF chains and not its passive antennas. Hence, adding more antennas, while an issue for *form factor* and hence mobile devices, is not an obstacle for base stations and access points, which is where we expect FD to be predominantly employed. Therefore, when a node is designed taking FD into account, one can provision it with more passive antennas than the number of RF chains to leverage the potential of FD, which in turn is captured by the RC model.

3.1 FD vs MIMO Performance

Note that when a node has multiple antennas and employs more than two streams, FD by definition refers to a combination of FD and MIMO. This is because, while the available antennas can be split between downlink and uplink in FD, one would still need to employ MIMO within the available number of antennas in each direction to maximize the number of streams. Hence, the actual question that we are interested in understanding here is that, *given a set of antennas and RF chains, should one consider splitting the antennas between downlink and uplink through FD (with MIMO within each direction), or should one directly employ all the antennas solely towards MIMO in just one direction?*

To address this question, we contrast the performance of

FD-MIMO vs. HD-MIMO in a single link by comparing their respective capacities under both the AC and RC models. We also consider both perfect SI cancellation at the FD node, as well as when there is a remaining SI of 6 dB. We also incorporate the correlation between antenna elements on a node (ρ) that arises in practice and influences MIMO performance ($\rho = 1$ indicates perfect correlation). The results are presented in Fig. 2 as a function of the number of antennas in the HD-MIMO mode (corresponding to the same number of antennas in the AC model and half of the antennas in the RC model) at either ends of the link, as well as the antenna correlation factor and FD SI suppression. Two key observations can be made.

- In the AC model, even small antenna correlations ($\rho = 0.01$) saturates MIMO performance with increasing antennas. Hence, while FD does not have an advantage over MIMO theoretically in terms of number of stream transmissions, in practice with good SI suppression capability, it can deliver better performance even with a moderate number of antennas (about 4-6 in Fig. 2(a)). However, with imperfect SI suppression (loss of 6dB), such a transition happens at a higher number of antennas (12 in Fig. 2(b)) unless the antenna correlation is high.
- In the RC model, there is an advantage for FD as it has the potential to transmit twice as many streams as MIMO. This, coupled with MIMO's saturating performance, allows FD to deliver significant gains with a strong SI suppression capability (Fig. 2(a)). However, even for a moderate SI suppression capability (with a loss of 6dB), gains can be observed, even for a moderate number of antennas (two in Fig. 2(b)) at low antenna correlations.

In summary, we find that *irrespective of the model considered, the use of available antenna resources towards a combination of FD and MIMO is critical for optimal performance.* While the above comparison is with respect to a single point-to-point link, applicable to scenarios involving cellular backhaul links, mesh network links, relay links, etc., we will also discuss the relative impact in single cell point-to-multipoint scenarios in Section 5.4. We next detail how to realize such a joint MIMO and FD node.

4. DESIGN OF MIDU

MIDU employs antenna cancellation achieved through symmetric antenna placement (SAP) as its primary RF cancellation technique. We first outline the rationale behind our choice by identifying the advantages of SAP. Then, with the help of antenna cancellation theory, we show how to enable two levels of antenna cancellation (Tx and Rx antenna cancellation) with additive benefits. Finally, we show how it scales easily to incorporate MIMO, thereby completing the design of MIDU.

4.1 Symmetric Antenna Placement

Our antenna cancellation approach is based on a symmetric placement of antennas. Figure 3(a) illustrates our Rx antenna cancellation, where two Rx antennas are placed symmetrically at a distance ℓ from the Tx antenna. The signal received from one of the receive antennas is phase shifted internally using a fixed π phase shifter before being

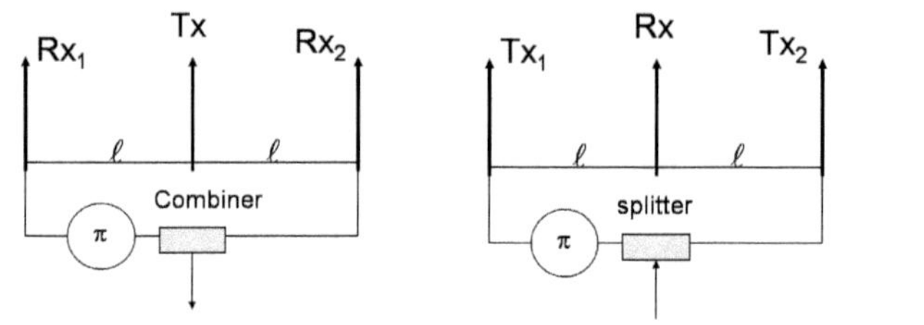

Figure 3: (a) Receive Cancellation, (b) Transmit Cancellation.

combined with the other receive signal to help nullify the self-interference signal. Similar to Rx antenna cancellation, we can also have an analogous Tx antenna cancellation as shown in Fig. 3(b).

While the basic antenna configuration for cancellation is simple, we now highlight its significant potential to not only address the limitations of existing FD schemes, but also to allow for two levels of antenna cancellation and to leverage MIMO in tandem. Compared to the transmit antenna cancellation in [7], where the π phase shift was realized with asymmetric placement of Tx antennas (ℓ and $\ell + \frac{\lambda}{2}$), SAP has the following advantages:

- Bandwidth Dependence: Moving the π phase shift internally alleviates the bandwidth dependence (due to λ) of asymmetric antenna placement. Further, fixed π phase shifters have significantly better frequency responses over wide bandwidths compared to variable ones.
- Tuning: Since the received powers are similar, this avoids the need for tuning of the attenuation and phase of the self-interference signal, which are otherwise required to counteract the power difference due to asymmetric antenna placement.
- Scalability: Symmetric antenna placement allows for easy realization of several null points, therefore scaling to MIMO systems. Further, when compared with schemes such as [13] that do not consider antenna cancellation, SAP does not require an estimation of the SI channel between every pair of Tx and Rx antennas, which becomes a scalability issue for MIMO systems.

One limitation that was raised in [7] with respect to symmetric antenna placement, is its potential destructive impact on the far-field. However, the simulations used to highlight this observation used a free space path loss model. Note that while the SI channel can be modeled as free space, it is well known [18] that the far-field channels (indoors or outdoors) from the transmit antennas experience independent fading at any far-field receive point with sufficient separation (greater than λ which is 12.5 *cm* at 2.4 GHz) between the transmitting antennas (also validated in our experiments in Section 5.2). Hence, asymmetric antenna spacing does not provide an advantage over a symmetric placement with respect to impact on far-field. An analogous argument holds for signals received from far-field.

4.2 Understanding Antenna cancellation

To leverage antenna cancellation effectively, it is important to understand the notion of *signal nulling.* A signal is said to be "nulled"' when two copies of the signal add π out of phase to cancel each other, thereby pushing the received signal strength to or below the noise floor. Let us consider transmit antenna cancellation for explaining the concepts.

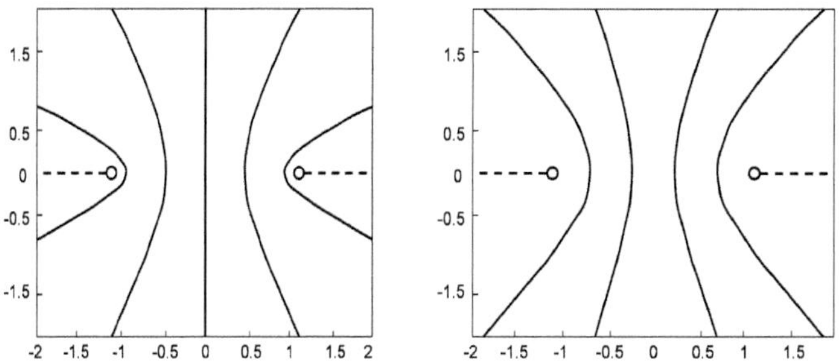

Figure 4: Loci of Null Points: (a) Phase Offset + Antenna Spacing, (b) Only Antenna Spacing.

There are two parameters affecting the nulling process: relative phase and amplitude of the transmitted signals at the receiver. The relative phase between the two signals could be further controlled either by directly introducing a phase offset (ϕ) to one of the signals and/or by varying the relative distance between the transmit antennas with respect to the receive antenna.

Let d_t be the distance between the two Tx antennas, with d_1 and d_2 denoting the distance of the two transmit antennas with respect to a receive point, respectively. First, we consider the set of *potential* receive null points, when there is a phase offset $\phi = \pi$. Whether these null points can be realized in turn depends on the relative amplitude of the signals as well, which is discussed subsequently. Now the set of potential null points in a two-dimensional plane can be defined as the locus of the points satisfying $|d_1 - d_2| = k\lambda$, for some integer k, and includes the following (see Fig. 4(a)).

- The Perpendicular Bisector (PB) of the line joining the transmit antennas (i.e. $d_1 = d_2 = \frac{d_t}{2}$).
- A set of hyperbolas with the transmit antennas as the focal points. Each hyperbola intersects the line connecting the two transmit antennas at points that are $\frac{k\lambda}{2}$, $k \in \mathcal{Z}_+, \geq 1$ from the mid-point towards either one of the transmit antennas.
- If $d_t = m\lambda$, in addition to the above points, all points on the line passing through the two transmit antennas, besides those lying in between them, also contribute to the set of potential null points.

To understand scenarios where the relative phase is controlled only with just the help of antenna spacing (i.e., phase offset, $\phi = 0$), we note that the locus of the potential null points is now defined as those satisfying $|d_1 - d_2| = \frac{(2k+1)\lambda}{2}$ and consist of (see Fig. 4(b)),

- A set of hyperbolas with the transmit antennas as the focal points. Each hyperbola intersects the line connecting the two transmit antennas at points that are $\frac{(2k+1)\lambda}{4}$, $k \in \mathcal{Z}_+, \geq 1$, from the mid-point towards either one of the transmit antennas.
- If $d_t = \frac{(2m+1)\lambda}{2}$, in addition to the above points, all points on the line passing through the two transmit antennas, besides those lying in between them, also contribute to the set of potential null points.

Now for a potential null point to be realized, the two transmit signals must arrive at the receive point with *equal* amplitude but π out of phase. Due to symmetry, this can be easily achieved on the PB with an equal transmit power from the two transmit antennas. Hence, all null points on the PB are realizable. However, for a null point on a hyperbola, it is easy to see that different transmit powers will be required from the two transmit antennas. Further, this will vary from one point to another on the same hyperbola as well as across hyperbolas. Hence, for fixed (and potentially different) transmit powers from the two transmit antennas, at most two null points on each hyperbola may be realizable. Note that we do not have null points on the PB when $\phi \neq \pi$. Given that the null points on the hyperbolas are hard to realize, this limits the applicability of asymmetric antenna spacing based approaches (e.g., [7]) to two level antenna cancellation. This limitation is compounded in the case of MIMO. This important property of realizing null points on the PB, when transmit signals are phase shifted by π, is leveraged for two purposes: (1) extend the transmit antenna cancellation to a two-level transmit and receive antenna cancellation scheme, and (2) to realize FD communication together with MIMO.

4.3 Two Level Antenna cancellation

Given that the above properties of transmit antenna cancellation analogously apply to receive cancellation as well, we can easily extend our proposed scheme to employ two stages (transmit and receive) of antenna cancellation in tandem. In the first stage two transmit antennas transmit at equal power and π out of phase signals that destructively interfere at any point on the PB of the transmit antennas. Now, place two Rx antennas symmetrically on the PB of the transmit antennas as shown in Fig. 5(a), such that the Tx and Rx sets of antennas are on each other's PB. While the transmit signals add destructively at each Rx antenna, the signals received from the two Rx antennas are further combined 180 degrees out of phase to provide the second level of antenna cancellation. Although four antennas are employed to achieve two levels of antenna cancellation, the number of RF chains used is still only two (one for forward and another for reverse streams).

The isolation (in dB) achieved by these two stages of cancellation are additive in theory, although in practice the cancellations might not be perfectly additive. In fact under ideal conditions even a one stage cancellation should provide a perfect null. However, gain imbalance or a slight phase offset between the signals may prevent one from achieving a perfect null, wherein a residue of the self-interference signal remains. We can now establish the following property.

PROPERTY 1. *Under small gain imbalance and/or phase offset (from imprecise antenna placement or imperfect RF devices) between the transmit and receive cancellation paths, the self-interference cancellation provided by two levels of antenna cancellation are additive (in dB scale).*

For details refer to [3].

4.4 Scaling to MIMO systems

Realizing null points on a straight line is critical because it facilitates the design of MIMO transmit and receive antenna arrays. Hence, our proposed two-level antenna cancellation solution based on phase offset can be readily extended to MIMO systems by using standard linear antenna array configurations. In particular, to generate an $N \times M$ MIMO + FD system, we start by placing two sets of antennas (N transmit and M receive) on two perpendicular axis to allow for $N \times M$ MIMO (in each direction of FD) as shown in Fig.

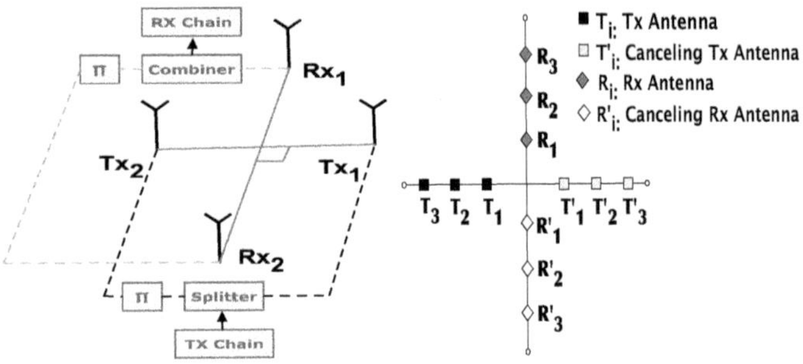

Figure 5: (a) 2-Level cancellation, (b) MIMO.

5(b) where $(N, M) = (3, 3)$. Then, to enable this $N \times M$ system with full duplex, we use an equal number of transmit (N) and receive (M) canceling antennas and place them in a symmetric position on the opposite side of their respective axis. The MIMO transmit streams from the N transmit and their respective canceling antennas will add out of phase at each of the receive antennas in the first stage of cancellation. The composite received signals at each of the M receive antennas are then further combined out of phase with their respective canceling antennas to provide the second level of cancellation. It is worth pointing out that only such symmetric antenna configurations can be extended to generic MIMO systems without the need for variable attenuators and delay elements. Again note that, while $2(N + M)$ antennas are employed for achieving two levels of antenna cancellation with FD, the total number of RF chains required is only $N + M$, which is the minimum required to enable $N \times M$ HD-MIMO communication in either direction. Fig. 5(b) shows the antenna structure for a 3×3 MIMO + FD node.

To summarize, the key aspect in MIDU's design is its symmetric antenna placement for antenna cancellation that not only provides two levels of additive RF cancellation, thereby alleviating the dependence on analog cancellation, but also realizes it through a structure that scales seamlessly to MIMO systems.

5. IMPLEMENTATION AND EXPERIMENTAL EVALUATION

In this section we provide experimental results on the performance of MIDU. We first perform extensive channel measurements to verify some of the assumptions in the design of MIDU. Next, we evaluate the cancellation design by measuring the amount of self-interference cancellation for each of the two levels of antenna cancellation and for the case when the two levels are combined. Finally, we compare the performance of MIDU to an equivalent half duplex MIMO system.

5.1 Measurement Setup

We performed channel measurement experiments using the WARP Vertex-4 FPGA boards. Our implementation is based on the WARPLab framework [2]. In this framework, all WARP boards are connected to a host PC through an Ethernet switch. The host PC is responsible for baseband PHY signal processing, while WARP boards act as RF front-ends to send/receive packets over the air. Since the baseband processing is performed in MATLAB, there is a 50 ms delay between consecutive transmissions.

We construct an access point (AP) in an indoor open space (low multi-path) environment. There is a distance of at least 20 m between our setup and the nearest wall. We mount our antennas on a wooden board, which itself is at the height of approximately 1.8 m from the earth surface. The carrier frequency was centered at 2.484 GHz (2.4 GHz, channel 14), and we implemented a single subcarrier narrowband system with a bandwidth of 625 KHz. We used 2.4 GHz 3dBi dipole antennas. We discuss the implications of rich scattering environment, wider bandwidths and antenna type in Section 7. In our experiments we construct 100-bit packets, and use the BPSK modulation scheme. The beginning of each packet is augmented with a preamble header. Each preamble is composed of training samples and an appropriate pilot tone for channel estimation between a transmitter and the receiver.

5.2 MIDU Feasibility

Three assumptions are critical in order for our full duplex system to work effectively: (i) Channel between a Tx and Rx antenna is only dependent on the distance between the two in the near field; (ii) Channel symmetry is stable over time; (iii) Symmetric antenna placement does not create nulls in the far-field. We now proceed to verify each of these assumptions.

The channel between a Tx and Rx antenna is only dependent on the distance between the two in the near field. The first assumption implies that an Rx antenna with similar distances from two Tx antennas would observe channels with equal amplitude and phase. This allows us to realize several null points in the BP of the two Tx antennas through the use of a π phase shifter, and thus enable MIMO.

We mount two antennas on our AP. The two antennas are attached to a rod with adjustable distance. We fix the two Tx and Rx antennas at a distance of 40 cm and send 40 back-to-back packets and measure the average channel (amplitude and phase) across all the packets.

Once the channel measurements are taken, we rotate the rod (with Tx and Rx antennas still attached to it) on a circle around the Rx antenna, and perform the same experiment at 7 other locations. These locations are shown as 8 dark squares on the circle around the Rx as shown in the picture of Fig. 6(a). Once these measurements are taken, we perform the same experiment on 3 more circles around the Rx by increasing the distance between the Tx and Rx antennas.

For each circle, we calculate the average and standard deviation of the *average channels* measured over the 8 selected points. Fig. 6(b) depicts the corresponding results. For all the circles, standard deviation of gain is less than 1%. Similarly, the standard deviation of phase is around 2° out of 360°. The results in Table 6(b) confirm that in an open-space environment, the near field channel is dependent only on the distance between the transmitter and receiver.

Symmetry stability over time. We now measure the amount of variation in the self-interference channel between two Tx and two Rx antennas over time. While in the design of MIDU two TX(RX) antenna pairs are connected to only one TX(RF) chain, such a setup would not allow us to measure all of the four self-interference channels simultaneously. Therefore, for benchmarking, we perform an experiment in which we connect each antenna to a separate RF chain. We next construct a preamble header in which the training samples are followed by two non-interfering pilot tones for cor-

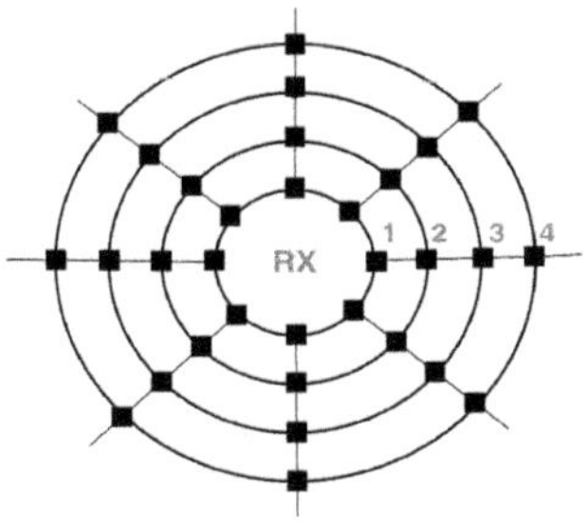

Circle	$\|\bar{h}\|$	$\sigma_{\|\bar{h}\|}$	$\angle\bar{h}$	$\sigma_{\angle\bar{h}}$
1	1.10	0.005	55.1°	1.51°
2	0.82	0.006	9.84°	2.08°
3	0.64	0.01	-45.8°	1.76°
4	0.39	0.01	120.9°	2.03°

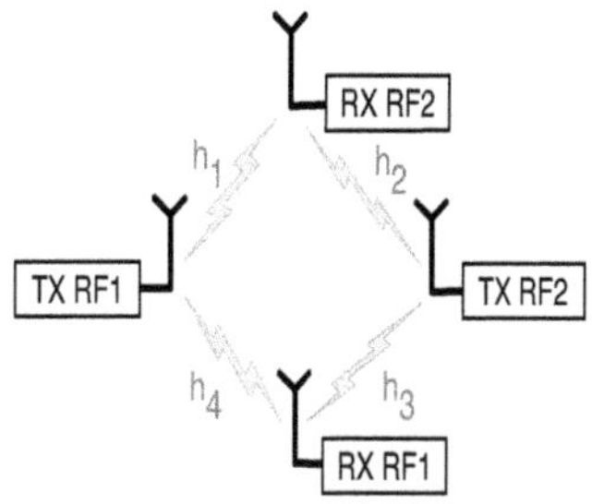

	$\sigma_{\|h_i\|}$	$\sigma_{\angle h_i}$	$\sigma_{\frac{\|h_i\|}{\|h_j\|}}$	$\sigma_{\angle h_i - \angle h_j}$
i=1 ; j=2	0.003	3.23°	0.004	0.56°
i=4 ; j=3	0.002	3.69°	0.003	1.05°
i=1 ; j=4	0.003	3.23°	0.002	0.99°
i=2 ; j=3	0.004	3.11°	0.003	0.87°

Figure 6: (a) h-dist measurement setup, (b) h-dist variation results, (c) stability setup, (d) stability results.

rect channel estimation between all of the Tx and Rx pairs. Fig. 6(c) shows our experiment's setup. We next store the packet with the modified preamble in the two Tx RF chains, and send the same packet continuously for a duration of ten minutes. For each packet transmission, we measure the four self-interference channels (h_i, $i = 1...4$).

Fig. 6(d) presents the stability results among the SI channels by measuring the variation of each SI channel (h_i) and the ratio between them. Fig. 6(d) shows that each SI channel observes small variations (less than 1% for both phase and gain). This is intuitive as the two antennas are fixed at a close distance. Fig. 6(d) further reveals that the variation in phase difference between two SI channels is less than 0.3%, implying that when the two transmissions are synchronized, the channel observed is even more stable.

Impact of symmetric antenna placement on the far-field clients. We have performed experiments to verify that symmetric antenna placement at the Rx or Tx would not cause severe reduction in SNR at far-field locations. We observed similar results with Rx cancellation, thus here we only present the results with Tx cancellation. We perform an experiment in which an Rx client moves in two different circles around the AP with a distance of $10m$ and $3m$ from the AP respectively. In each circle, the client performs SNR measurements at 10 different equally spaced locations. In each location, the AP first transmits ten packets using a single antenna. Next, we change the antenna configuration by pairing the original antenna with a canceling transmit antenna (consisting of a transmit antenna and a 180 ° phase shifter). In this setup, the canceling antenna is at a distance of 40 cm from the first antenna, and the two Tx pairs are able to cancel 25 dB of SI for a carefully placed Rx antenna on the AP (discussed more in the next section). We repeat the experiment by sending ten back-to-back packets and measuring the SNR at the client's location. The experiment is repeated for all the 20 client locations.

Fig. 7(d) compares the received SNR for the two antenna configuration schemes. It shows that the achieved SNR can be up to 4 dB lower than the single antenna scheme. However, Fig. 7(d) also shows that Tx cancellation's achieved SNR can be up to 4 dB higher than the single antenna scheme. Note that Tx cancellation uses two antennas for transmission. If the second antenna used by Tx cancellation has a higher gain than the first antenna, the combined effect can even increase the SNR. Fig. 7(d) confirms that there is no particular correspondence to increase/decrease in SNR, as one would expect from far-field fading [18].

5.3 Self-Interference Cancellation Evaluation

We now investigate MIDU's SI cancellation performance for Tx cancellation and Rx cancellation separately, as well as when the two levels of cancellation are combined. Finally, we measure the amount of cancellation when more than one Tx pair is active.

Cancellation over wire. Prior to our over-the-air measurements, we first verify if a commercial phase shifter can cancel two equal signals over wires. For the transmit signal, we used the Agilent MXG vector signal generator that creates a sinusoid signal at the 2.4 GHz center frequency. The signal was passed through a splitter; one of the resulting copies was next passed through our phase shifter; and then the two paths were combined using a combiner. The output of the combiner is next connected to an Agilent CSA spectrum analyzer to observe the received signal power. In this setup, we observed that for a 0 dBm transmitted signal power, one level of cancellation was able to achieve 35 dB of cancellation. We next employed two phase shifters in our system (one on each path), with one set to 0° and the other set to 180° of phase shift. With this setup we were able to cancel 90 dB of self-interference. *This implies that using a phase shifter on each path is necessary to maintain the correct symmetry and account for phase shifter induced insertion loss and delay.*

Wireless experiment setup. We use the antenna configuration setup depicted in Fig. 5(b). There are 6 transmit antennas symmetrically placed on a straight line, and 6 Rx antennas on the bisector perpendicular. The distance between T_1T_1' is set to 40 cm, and the same distance is set between R_1R_1'. The rest of the antennas are placed at 20 cm distance from the adjacent antennas. Each antenna pair is further connected to two phase shifters, with phase shifter values set to 0° and 180°, respectively. The 6 Tx antennas and the 6 Rx antennas are connected to 3 Tx RF chains and 3 Rx RF chains, respectively. We use the WARPLab experimental setup of Section 5. However, now we set the transmitting RF chains in a continuous transmission mode for a duration of 10 sec, in order to correctly measure the received signal power on the spectrum analyzer. In continuous transmission mode, a transmitting WARP board continuously transmits the same packet without any delay between back to back packets.

Transmit Cancellation. In this experiment, we set the transmit antenna pair fixed as T_1T_1', and measure the amount of transmit self-interference cancellation at each of the 6 receive antennas as a function of transmit power. Specifically, for each Rx antenna we first measure the Rx signal strength when only T_1 is active. Next, we measure the received signal strength on the spectrum analyzer when both T_1 and T_1' are active simultaneously. We repeat this set of measurements by varying the transmit power from -5 to 15 dBm, and for each of the 6 Rx antennas.

Fig. 7(a) shows the amount of Tx cancellation for each of

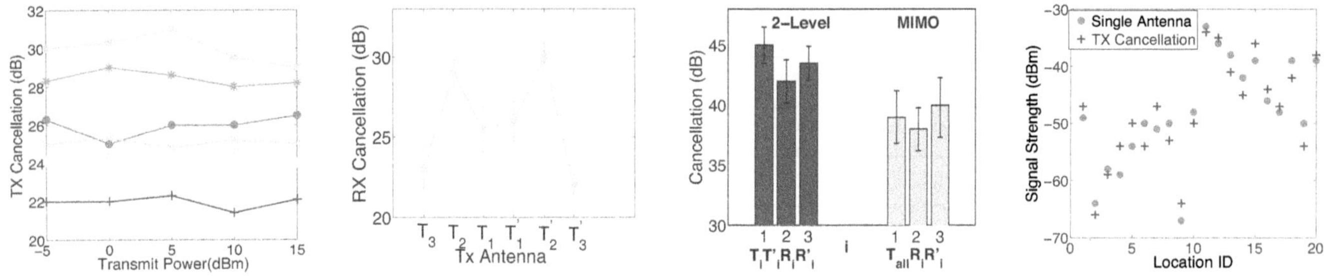

Figure 7: (a) Tx cancellation, (b) Rx cancellation, (c) 2-Level and MIMO cancellation; (d) Impact in far-field.

the Rx antennas as a function of transmit power. We observe that a one level Tx cancellation results in 22–30 dB of self-interference cancellation for each of the Rx antennas. Similar Tx cancellation results were also observed in [7]. We further observe that the cancellation results remain relatively flat across different transmit powers, suggesting that the amount of cancellation does not depend on the Tx power but on the precise placement of the Tx and Rx antennas.

Receive Cancellation. In this experiment, we fix the receiving antenna pair as R_1R_1', and measure the amount of receive SI cancellation for each of the transmitting antennas. Since we observed in Fig. 7(a) that the amount of cancellation does not depend on the transmit power, we use a fixed transmit power of 15 dBm. We first measure the received signal power when only R_1 is connected to the spectrum analyzer. Next we measure the received signal power when R_1 and R_1' are combined. We calculate the amount of Rx cancellation by subtracting the two values. The experiment is repeated for each of the Tx antennas.

Fig. 7(b) shows the measured Rx cancellation values. Similar to Tx cancellation results of Fig. 7(a), we observe that one level Rx cancellation can provide cancellation values between 20 to 30 dB. We also observe that although cancellation measurements are performed at different time instants, symmetric transmitting antennas observe almost similar cancellation values.

2-Level and MIMO cancellation. We now perform experiments to validate if the two levels of cancellations are additive, and further measure the cancellation results when more than one Tx antenna pair is transmitting (i.e., MIMO). We measure the 2-Level cancellation results, by measuring the resulting amount of cancellation for each $T_iT_i'R_iR_i'$ combination. We measure MIMO cancellation by activating all transmit antennas while using R_iR_i' for $i = 1 \ldots 3$ (i.e., 3x1 MIMO). MIMO cancellation results in Fig. 7(c) are shown as $T_{all}R_iR_i'$.

Fig. 7(c) shows the 2-Level and MIMO cancellation results. The x-axis in Fig. 7(c) is the i variable, denoting the receiving antenna pair for both 2-Level and MIMO cancellation. Fig. 7(c) reveals that the 2-Level combination can provide up to 45 dB of cancellation. This in turn verifies that *Rx cancellation after Tx cancellation provides additive cancellation gains.* Fig. 7(c) also shows that when all three antenna pairs are active, the 2-Level cancellation gains reduce by up to 5 dB. Note that if the second active antenna pair makes interference levels equal to the noise power at R_1R_1', the resulting observed cancellation at R_1R_1' would decrease by 3 dB. In order for T_2T_2''s addition to have no impact on R_1R_1', the resulting interference should be far lower than the noise power. Further, note that while one additional transmit pair can cause 3–4 dB of additional interference, with 2 additional transmit pairs the observed reduction is 5 dB. This implies that each additional transmit pair would only slightly reduce the amount of self-interference cancellation.

5.4 Comparing MIDU with MIMO

In this section we compare the performance of HD-MIMO and MIDU systems in a relay architecture, and a point-to-multipoint (PtMP) single-cell architecture. In the relay setup, all nodes are MIMO capable. In the point-to-multipoint architecture, the AP has multiple antennas, whereas the clients each have a single antenna. We first consider the RC model in which both MIDU and HD-MIMO have the same number of TX/Rx RF chains, before turning to the AC model in Section 5.4.2.

5.4.1 Relay Architecture

We first consider a setup in which a HD-MIMO node (N_2) acts as a relay between two hidden HD-MIMO nodes (N_1 and N_3). Fig. 8(a) depicts our experiment setup. We next consider the case in which the relay node (AP) is a MIDU node. Note that when a MIDU node is considered as a relay, N_1 and N_3 can be active simultaneously (with one transmitting and the other node receiving). Thus, the architecture would be similar to a Point-to-Point (PtP) MIDU link with zero self-interference at the second node.

Performance Metric. We use the signal to noise plus interference ratio (SINR) observed by each receiving antenna, or the corresponding Shannon capacity (C) as our performance metric. For a PtP MIMO link with N Tx and N Rx antennas, we measure the SINR for each receive antenna and calculate the corresponding capacity. We calculate the link capacity as the summation of per-antenna capacities.

For the HD-MIMO relay architecture, we assume a TDMA scheme with an equal amount of time for each of the two links of Fig. 8(a). Thus, the aggregate capacity of the HD-MIMO system is equal to $\frac{C_{N_1N_2}^{HD-MIMO}}{2} + \frac{C_{N_2N_3}^{HD-MIMO}}{2}$. When a MIDU node is deployed as a relay, the two links can be active at the same time. Thus, the aggregate capacity is equal to $C_{N_1N_2}^{MIDU} + C_{N_2N_3}^{MIDU}$.

Note that the overall end to end throughput of a system is dependent on the specific MAC protocol implementation (including rate adaptation, etc.) and is an active research area. Shannon capacity is a measure of physical layer capacity and is used here to provide an upper bound on the throughput that would be achieved by any MAC protocol.

HD-MIMO Implementation In our implementation, *all the physical layer processing is done at the relay node (N_2).* Nodes N_1 and N_3 are located such that they have a strong SNR at the relay station (AP), however, they are

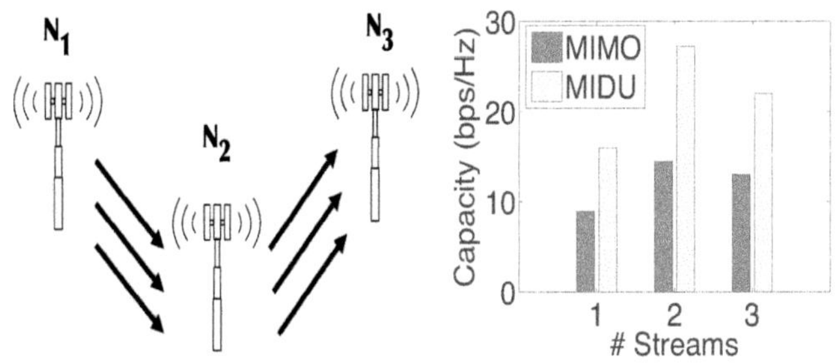

Figure 8: (a) Relay architecture, (b) Capacity.

hidden from each other. Each of the nodes in the relay architecture of Fig. 8(a) has 3 Tx RF chains and 3 Rx RF chains.

During $N_1 \rightarrow N_2$ transmission, each independent data stream is transmitted from a separate antenna at N_1. The relay node first obtains the channel information between its three Rx antennas and the transmitting antennas at N_1. It then employs a standard technique termed *zero force (ZF)* [18] filtering to separate the received streams. During $N_2 \rightarrow N_3$ transmission, N_3 receives interference free streams on separate receive antennas. Similarly, the relay node first obtains the channel information between its three antennas and the receive antennas at N_3. It then uses ZF beamforming to transmit independent streams to each of N_3's receiving antennas. In our ZF implementation, equal power is assigned to each ZF weight vector (for details of ZF filter/beamformer refer to [18]).

$N_2 \rightarrow N_3$ SINR Measurement. We use the received signal strength (RSS, in dBm) value reported by the radio boards for our SINR measurements. We take the following approach to measure the SINR for a receiving antenna k at N_3. The MIMO transmitter (N_2) first performs ZF beamforming by maintaining the power associated to k, and setting the rest of the powers to zero. The measured RSS value would then correspond to the signal power. We next re-run the experiment by setting the power associated to k as zero and maintaining the powers associated to the other receive antennas. The measured RSS value would then correspond to the noise and inter-stream interference. By subtracting the two RSS values, we measure the SINR. We take 10 such SINR measurements and report the average value for each data point.

$N_1 \rightarrow N_2$ SINR Measurement. During N_1 to N_2 transmission, the AP is responsible for separating the data streams in the base band. In order to measure the SINR for a stream k, the MIMO transmitter first only transmits stream k, and the receiver applies the ZF weights to obtain the received signal samples ($\mathbf{y_k}$). The base-band signal power (P_s) is then calculated as $E\{\mathbf{y_k y*_k}\}$ (* denotes the conjugate transpose). Next, the MIMO transmitter (N_1) sets the power allocated to stream k as zero while maintaining the power associated to the other streams, and the receiver uses the previous weights to obtain the interference samples (z_k) caused by other streams. The base-band interference power (P_i) is then calculated as $E\{\mathbf{z_k z*_k}\}$. We report the $10 \times log_{10} \frac{P_s}{P_i}$ as the SINR value.

MIDU Implementation. We change the antenna configuration at the AP to a 2-Level antenna cancellation scheme as described in Section 4. We keep nodes N_1 and N_2 at the same locations. Similar to HD-MIMO implementation, all of the physical layer processing is performed at the relay node. However, unlike the HD-MIMO mode, the transmission and reception is performed simultaneously.

SINR Measurement. $N_2 \rightarrow N_3$ SINR Measurement in MIDU is similar to HD-MIMO. However, the imperfect self-interference cancellation can reduce the performance of $N_1 \rightarrow N_2$ transmission. In our implementation of MIDU on the WARP boards, we observed only 4-5 dB of SI remaining above the noise floor as reported by the WARP boards.

This remaining SI is due to the presence of multi-path components in our environment, which can be suppressed by employing conventional digital cancellation techniques [11, 12, 13]. However, given the remaining small margin for SI suppression with WARP, we do not consider it in our implementation. Thus our results would be a lower bound on MIDU's performance.

We measure the $N_1 \rightarrow N_2$ SINR similar to HD-MIMO SINR measurement. However, during noise measurement for stream k, we also activate the transmitting N_2 streams to measure the combined impact of SI and received inter-stream interference.

Evaluation. Fig. 8(b) depicts the aggregate capacity of HD-MIMO and MIDU as a function of the number of independent data streams. We first observe that MIMO capacity does not scale linearly as the number of data streams is increased. In fact, while 2 streams increase the aggregate capacity by 80% compared to having only 1 stream, 3 streams decrease the aggregate capacity by 7% compared to having only 2 streams. This is due to MIMO capacity saturation when all degrees of freedom are used as discussed in Section 3. When all degrees of freedom are used to enable spatial multiplexing, per-link SINR can potentially decrease, reducing the benefits of spatial multiplexing.

Fig. 8(b) also compares MIDU's performance to the HD-MIMO scheme. We observe similar capacity saturation trend due to the incorporation of MIMO. However, we observe that MIDU on average increases HD-MIMO's capacity by over 80%.

5.4.2 Single Cell Architecture

We now compare the relative gains of HD-MIMO and MIDU in a single cell scenario.

Setup. We consider a setup in which we deployed 6 client nodes scattered around the AP. Each of the clients has only a single antenna for transmission or reception. The clients are placed in a manner that they all have a strong link to the AP, while some of them can be hidden from each other.

The AP is equipped with multiple antennas and has 3 Tx and 3 Rx RF chains. In HD-MIMO mode, the AP uses three fixed antennas. In addition, the AP uses all of its antennas to transmit (receive) to (from) up to 3 clients. We denote an $M \times N$ transmission strategy by the AP, as a scheme in which the HD-MIMO AP transmits to M clients during downlink, and receives from N clients during uplink. For a given M and N, we select M out of the six clients as our downlink clients, and N among the remaining clients as our uplink clients, and measure the resulting capacity. We denote the resulting capacity by $C_{M \times N}^{HD-MIMO}$. For a given M and N, there are $\binom{6}{M} \times \binom{6-M}{N}$ different selections. We repeat the same experiment for all possible client selections, and denote the average capacity as $\bar{C}_{M \times N}^{HD-MIMO}$. Finally, we repeat the experiment for all selections of $M = 1, 2, 3$ and $N = 1, 2, 3$ leading to a total of 500 sub-topologies. Our capacity measurements are based on the measurement setup in

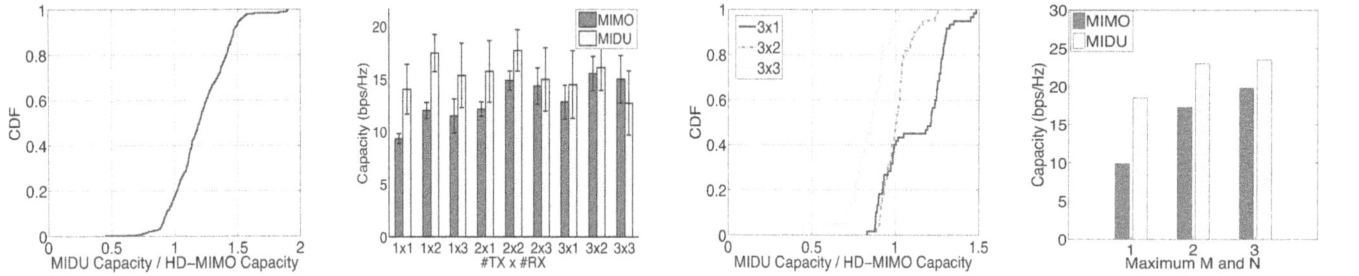

Figure 9: (a) CDF Overall, (b) $C_{M\times N}$ Comparison, (c) CDF with 3 downlink clients; (d) Capacity with maximum allowed M and N.

the previous subsection. However, instead of obtaining the channel matrix from multiple antennas at a single node, we obtain the channel information from multiple single antenna clients and use the ZF technique accordingly.

We next change the antenna configuration to the setup described in Fig. 5(b), and perform the same set of experiments with a MIDU AP. Note that in an $M \times N$ transmission scheme by MIDU, the M downlink clients and N uplink clients are active at the same time. We measure the resulting capacities based on our measurement setup in the previous section. However, since not all of the clients are hidden from each other, uplink transmission in MIDU can cause interference on the downlink clients. Thus, when measuring inter-stream interference for each of the downlink clients, we also activate all of the selected uplink transmissions to accurately measure the resulting interference at the downlink nodes. We denote MIDU's average $M \times N$ capacity as $\bar{C}^{MIDU}_{M\times N}$.

Evaluation. Fig. 9(a) plots the cdf of all the $\frac{C^{MIDU}}{C^{HD-MIMO}}$ values across all possible 500 sub-topologies. According to the results shown in Fig. 9(a), MIDU achieves a better performance compared to HD-MIMO in over 80% of all the sub-topologies. In addition, for almost 50% of all sub-topologies, MIDU gains over HD-MIMO by at least 20%.

Fig. 9(b) shows the average $\bar{C}_{M\times N}$ results for different M and N values. We first observe that MIDU provides larger gains compared to HD-MIMO when total number of uplink (UL) and downlink (DL) streams is small (less than 4). Note that with a small number of UL and DL clients, the total interference present in the system is lower and thus full duplex (FD) can provide significant gains.

In order to better understand the impact of UL→DL interference problem in FD we also plot: (1) cdf of the $\frac{C^{MIDU}}{C^{HD-MIMO}}$ for $M = 3$ and varying number of uplink clients in Fig. 9(c), and (2) the maximum HD-MIMO and MIDU capacities over all $M \times N$ sub-topologies as a function of the maximum allowed M and N. Two key inferences can be made:

From Fig. 9(c) we observe that *in order to leverage the maximum gains form FD, the number of streams on DL and UP must be kept dis-proportionate (asymmetric)* The reason is as follows. If the number of UL (DL) streams from (to) HD clients is increased for a fixed number of DL (UL) streams, then it increases the interference faced by the clients receiving DL streams, thereby limiting FD performance significantly. While this limitation is alleviated with FD clients, it cannot be avoided.

From Fig. 9(d) we observe that *FD yields diminishing gains as the number of streams is scaled in either direction.* Note that, when the number of streams is small, it is possible to pick DL and UL clients that do not interfere with each other, allowing the increased number of streams from FD to yield maximal gains. However, with more streams, the impact of interference dominates and the few additional streams in FD (due to asymmetry) tend to contribute less gain compared to the baseline (larger) number of streams in MIMO.

FD gains in perspective. While these inferences indicate the restricted nature of FD gains in sigle cell MU-MIMO (due to UL → DL interference), in practice the total number of multi-user streams, in either DL or UL direction, is small compared to the number of antennas at an AP. This is due to many practical reasons such as feedback overhead, computational complexity, minimum SNR supported, etc. For example, Arraycomm [1] APs (base stations) only support *up to* 4 streams simultaneously, with 12 antennas at the AP. This implies that FD can provide significant gains under the RC (RF Chain Conserved) model in realistic PtMP architectures. The impact of interference between UL and DL on FD gains is more pronounced in AC (Antenna Conserved) model, making it less favorable for adoption in such single cell MU-MIMO scenarios.

To contrast these results with the single-link results in Section 5.4.1, FD has significant potential under both AC and RC models in single-link scenarios. These scenarios include back-haul, relay, mesh network links (owing to the lack of DL-UL interference), and single-cell SU-MIMO (which is still predominantly considered in cellular networks due to the complexity of scheduling and realization of MU-MIMO gains).

6. IMPLICATIONS FOR MAC DESIGN

Challenges: Our results in Fig. 9 clearly reveal the need for scheduling on a MIDU node to identify opportunities for FD gain. While the design of a scheduler (MAC) for single cell MU-MIMO (client selection and precoding) is challenging in its own right, extending it to a MIDU system takes it to another level - spatial degrees of freedom now have to be carefully split between downlink and uplink for FD, with client selection and precoding being jointly addressed for MU-MIMO in each direction. While the search space for the problem is very large, we can use the two insights from our evaluation in Section 5.4.2 to make the search space tractable.

Insights: To ensure that our inferences are not specific to the FD implementation in MIDU, we have also evaluated the relative performance of ideal FD-MIMO and HD-MIMO through extensive simulations in larger topologies and hundreds of channel realizations.

We perform simulations with 8 uplink and 8 downlink

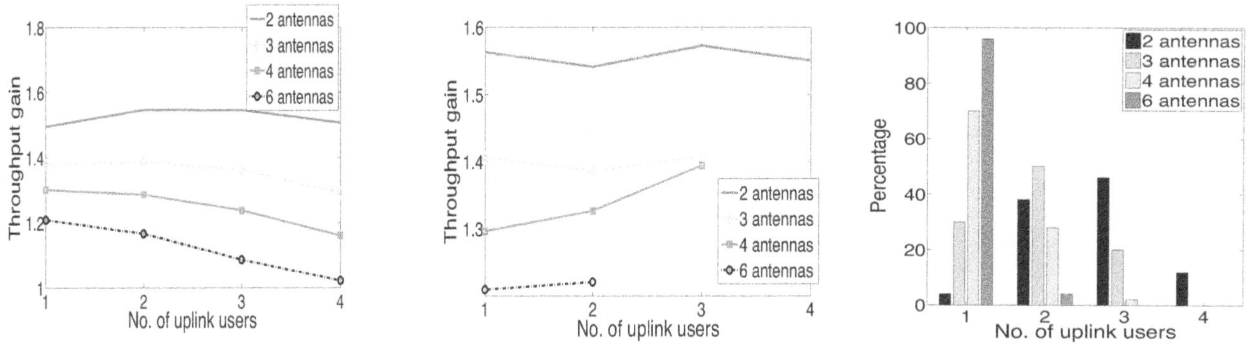

Figure 10: Large-scale simulations of FD gains in a single-cell architecture: (a) FD gains with forced uplink users, (b) Maximum FD gains, (c) Number of uplink users that result in maximum FD gains.

clients, and with varying number of antennas at the AP. Since we assume an equal number of users in the uplink and downlink, the average throughput in the uplink and downlink are similar due to channel reciprocity. Therefore, we define the *ergodic DL capacity* [15] for a given number of antennas as the HD-MIMO capacity. Next, for a given selection of UL users, we calculate the FD capacity that is the sum of the ergodic downlink and uplink capacities while considering the impact of UL clients interference on the downlink clients. FD capacity is then averaged over all UL user combinations for a given total number of UL users and across all different channel realizations.

Fig. 10(a) depicts the corresponding throughput gain of FD to HD-MIMO. From Fig. 10(a), we observe that as the number of antennas and therefore the number of DL streams increases, the FD gains decrease. We further observe that, as the number of the uplink clients increases, the FD gains start to decrease.

We next characterize the number of uplink users that result in maximum FD gains and the corresponding gains across all different channel realizations. For a given number of antennas and for each channel realization, we calculate the maximum FD gain, as well as the number of uplink clients that result in the maximum FD gains. Fig. 10(b) depicts the average maximum FD gain as a function of number of uplink clients, while Fig. 10(c) depicts the number of uplink users that result in maximum FD gains. From these results, we observe that as the number of streams in DL direction is increased, only a small number of clients in the reverse direction can be scheduled simultaneously. In fact, the results in Fig. 10(c) show that with six downlink streams, in more than 90% of all channel realizations only one uplink client provides additional FD gains, and none of the realizations achieved FD gains with three or more uplink clients.

These simulation results re-emphasize our earlier inferences in Section 5.4.2. Hence, for maximal benefits in single-cell networks, where MU-MIMO operation is feasible, it is better to *operate FD in small, asymmetric configurations – where the maximum number of streams is kept small in either direction, while the active number of streams between DL and UL is dis-proportionate.*

Scheduling Strategy: Guided by these conclusions, one can devise the scheduler to look at asymmetric configurations for FD only when the number of streams is small. This significantly reduces the search space for client selection on DL and UL for FD, making it comparable to MU-MIMO scheduling. Thereafter, conventional MU-MIMO scheduling and precoding algorithms [5] can be directly employed in either direction to evaluate the relative benefits of FD-MIMO over HD-MIMO and select the better strategy. On the other hand, when the number of streams is large, the scheduler can directly employ MU-MIMO.

Freq (GHz)	2.484	2.474	2.464	2.454
Cancellation (dB)	45	38	17	9

Table 1: Phase shifter's frequency response impact on cancellation.

7. DISCUSSION

Wide Bandwidth. Our implementation of MIDU cancels 45 dB of SI over a 625 KHz bandwidth signal. However, most of the current and future technology standards use bandwidths much wider than that (e.g., up to 20 MHz LTE and 40 MHz 802.11n). Therefore, it is important for SI cancellation techniques to cancel SI over wide-band.

We perform an experiment in which we change the center frequency of the WARP boards and measure the resulting SI cancellation at different frequencies. The phase shifters are tuned such that they provide 45 dB of SI cancellation at 2.484 GHz. Table. 1 shows MIDU's SI cancellation results. We observe that MIDU provides 37 dB of SI cancellation at 2.474 GHz, implying that MIDU can *still provide over 35 dB of SI cancellation for a 20 MHz bandwidth signal.* However, at 2.464 GHz, SI cancellation is only 17 dB, limiting the SI cancellation to less than 20 dB for a 40 MHz bandwidth signal. While channel symmetry holds across different frequency bands, a phase shifter's ability to create a signal inverse is precise only for a small bandwidth signal. Hence, the performance of MIDU is limited by the manner in which a signal inverse is generated. Recent research [13] suggested that a transformer (termed BALUN) can provide π phase shift over large bandwidths. Therefore, one can potentially employ BALUNs in lieu of phase shifters in MIDU's implementation to provide SI cancellation over wider bandwidths.

Impact of multi-path interference. MIDU is able to provide 45 dB of SI cancellation in an open-space indoor (less multi-path) environment and can potentially yield higher cancellation outdoors. However when measured in an indoor multi-path rich environment, we observed only 15 dB of SI cancellation. While MIDU can help with SI cancellation of the LoS component, channel symmetry may not hold for strong NLoS components. In such scenarios, if the additional level of suppression required is less than 25 dB, digital cancellation would suffice. However, if more cancella-

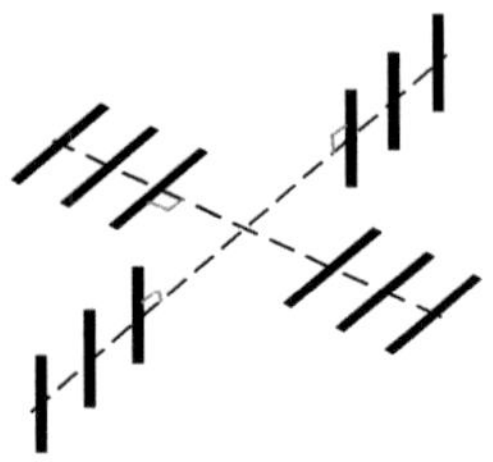

Figure 11: Example Tx and Rx antenna placement with orthogonal polarization.

tion is required, then estimation and compensation of the SI channel becomes necessary. One can employ a pseudo analog cancellation approach [8] for this purpose. While this approach by itself is not sufficient to provide the desired level of cancellation (as discussed in Section 2), it can be combined with MIDU to address multi-path without comprising on its scalability. However, other analog cancellation approaches (e.g.,[13]) cannot be employed as they incur the limitation of scalability, and sensitivity of SI compensation for wide-band frequency-selective channels.

Antenna Type and Polarization. In our implementation of MIDU, all Tx and Rx antennas were placed such that they have similar polarization. Due to the symmetric structure of MIDU, one can envision a placement of antennas in which Tx and Rx antennas have orthogonal polarization, potentially increasing the amount of SI cancellation. Fig. 11(b) depicts an example of such antenna placement.

Further, as MIDU requires only similar radiation patterns by Tx or Rx antennas, it can easily work with an array of directional antennas (e.g., directional/sectorized antenna placement in 3G). As part of our future work, we are investigating these schemes in order to provide additional 20–30 dB SI cancellation, thereby potentially enabling FD-MIMO in pico and macro cellular towers.

8. CONCLUSION

We presented the design and implementation of MIDU, the first MIMO full duplex wireless system. We showed that MIDU's design allows for 2 levels of antenna cancellation, eliminates the need for variable attenuators and delays, and, most importantly, easily scales to MIMO. We implemented a prototype of MIDU, and compared its performance to HD-MIMO. Our results demonstrated that MIDU has significant potential in both point-to-point and point-to-multipoint schemes. We also provided guidelines for scheduling in practical MIMO full duplex implementations.

9. ACKNOWLEDGEMENTS

We would like to thank our shepherd, Yunhao Liu, and the anonymous reviewers for their valuable feedback, which helped in improving the presentation of the paper. This work was in part supported by NSF grant CNS-1011962 and AFOSR grant FA9550-09-1-0643.

10. REFERENCES

[1] ArrayComm. Available at: `http://www.arraycomm.com`.

[2] Rice University WARP project. Available at: `http://warp.rice.edu`.

[3] Technical Report. Available at: `http://www.nec-labs.com/~karthiks/midu.pdf`.

[4] Antenna selection and RF processing for MIMO systems. *IEEE 802.11-04/0713r0*, 2004.

[5] W. Ajib and D. Haccoun. An overview of scheduling algorithms in mimo-based fourth-generation wireless systems. *IEEE Network*, 19(5):43 – 48, Oct 2005.

[6] E. Aryafar, N. Anand, T. Salonidis, and E. Knightly. Design and experimental evaluation of multi-user beamforming in wireless LANs. In *Proceedings of ACM MobiCom*, Sep 2010.

[7] J. Choi, M. Jain, K. Srinivasan, P. Levis, and S. Katti. Achieving single channel, full duplex wireless communication. In *Proceedings of ACM MobiCom*, Sep 2010.

[8] M. Duarte, C. Dick, and A. Sabharwal. Experiment-driven characterization of full-duplex wireless systems. 2011. Avaiable at: `http://warp.rice.edu/trac/wiki/TransWireless2011_FullDuplex`.

[9] E. Everett, M. Duarte, C. Dick, and A. Sabharwal. Empowering full-duplex wireless communication by exploiting directional diversity. In *Forty-Fifth Asilomar Conference on Signals, Systems, and Computers*, Nov 2011.

[10] G.D. Golden, R.A Valenzuela, P.W. Wolniansky, and G.J. Foschini. V-BLAST: an architecture for realizing very high data rates over the rich-scattering wireless channel. In *Proccedings of URSI International Symposium on Signals, Systems, and Electronics*, Pisa, Italy, Sep 1998.

[11] S. Gollakota and D. Katabi. ZigZag decoding: combatting hidden terminals in wireless networks. In *Proceedings of ACM SIGCOMM*, Aug 2008.

[12] D. Halperin, T. Anderson, and D. Wetherall. Taking the sting out of carrier sense: interference cancellation for wireless LANs. In *Proceedings of ACM MobiCom*, Sep 2008.

[13] M. Jain, T.M. Kim, D. Bharadia, S. Seth, K. Srinivasan, P. Levis, S. Katti, and P. Sinha. Practical, real-time, full duplex wireless. In *Proceedings of ACM MobiCom*, Sep 2011.

[14] M. Knox. *Self-jamming Cancellation Networks for Full Duplex Communication*. PHD Thesis, Polytechnic University, 2008.

[15] L. Li and A. J. Goldsmith. Capacity and optimal resource allocation for fading broadcast channels - part I: ergodic capacity. *IEEE Transactions on Information Theory*, 47(3):1083–1102, Mar 2001.

[16] B. Radunovic, D. Gunawardena, P. Key, A. Proutiere, N. Singh, V. Balan, and G. Dejean. Rethinking indoor wireless mesh design: Low power low frequency, full-duplex. Technical Report, MSR-TR-2009-27, Microsoft Research.

[17] K. Tan, H. Liu, J. Fang, W. Wang, J. Zhang, M. Chen, and G. Voelker. SAM: enabling practical spatial multiple access in wireless LAN. In *Proceedings of ACM MobiCom*, Beijing, China, Sep 2009.

[18] D. Tse and P. Viswanath. *Fundamentals of Wireless Communication*. Cambridge University Press, 2005.

Locating in Fingerprint Space: Wireless Indoor Localization with Little Human Intervention

Zheng Yang, Chenshu Wu, and Yunhao Liu

School of Software and TNList, Tsinghua University
{yang, wu, yunhao}@greenorbs.com

ABSTRACT

Indoor localization is of great importance for a range of pervasive applications, attracting many research efforts in the past decades. Most radio-based solutions require a process of site survey, in which radio signatures of an interested area are annotated with their real recorded locations. Site survey involves intensive costs on manpower and time, limiting the applicable buildings of wireless localization worldwide. In this study, we investigate novel sensors integrated in modern mobile phones and leverage user motions to construct the radio map of a floor plan, which is previously obtained only by site survey. On this basis, we design LiFS, an indoor localization system based on off-the-shelf WiFi infrastructure and mobile phones. LiFS is deployed in an office building covering over 1600m^2, and its deployment is easy and rapid since little human intervention is needed. In LiFS, the calibration of fingerprints is crowdsourced and automatic. Experiment results show that LiFS achieves comparable location accuracy to previous approaches even without site survey.

Categories and Subject Descriptors

C.2.4 [**Computer-Communication Networks**]: Miscellaneous

General Terms

Design, Experimentation, Performance

Keywords

Indoor Localization, Floor Plan, RSS Fingerprint, Smartphones, Site Survey

1. INTRODUCTION

The popularity of mobile and pervasive computing stimulates extensive research on wireless indoor localization. Many solutions are introduced to provide room-level location-based services, for example, locating a person or a printer in an office building.

The majority of previous localization approaches utilize Received Signal Strength (RSS) as a metric for location determinations. RSS fingerprints can be easily obtained from most off-the-shelf wireless network equipments, such as WiFi- or ZigBee-compatible devices. In these methods, localization is divided into two phases: training and operating. In the first stage, traditional methods involve a site survey process (a.k.a. calibration), in which engineers record the RSS fingerprints (e.g., WiFi signal strengths from multiple Access Points, APs) at every location of an interested area and accordingly build a fingerprint database (a.k.a. radio map) in which fingerprints are related with the locations where they are recorded. Next in the operating stage, when a user sends a location query with his current RSS fingerprint, localization algorithms retrieve the fingerprint database and return the matched fingerprints as well as the corresponding locations.

Although site survey is time-consuming, labor-intensive, and vulnerable to environmental dynamics, it is inevitable for fingerprinting-based approaches, since the fingerprint database is constructed by locationally labeled fingerprints from on-site records. In the end of 2011, Google released Google Map 6.0 that provides indoor localization and navigation available only at some selected airports and shopping malls in the US and Japan. The enlargement of applicable areas is strangled by pretty limited fingerprint data of building interiors.

The development of wireless and embedded technology has fostered the flourish of smartphone market. Nowadays mobile phones possess powerful computation and communication capability, and are equipped with various functional built-in sensors. Along with users round-the-clock, mobile phones can be seen as an increasingly important information interface between users and environments. These advances lay solid foundations of breakthrough technology for indoor localization.

On this basis, we reassess existing localization schemes and explore the possibility of using previously unavailable information. Considering user movements in a building, originally separated RSS fingerprints are geographically connected by user moving paths of locations where they are recorded, and they consequently form a high dimension *fingerprint space*, in which the distances among fingerprints, measured by footsteps, are preserved. In addition, we reform the floor plan of a building to the *stress-free floor plan*, a high dimension space in which the distance between two locations reflects their walking distance according to the real floor plan. The spatial similarity of stress-free floor plan and fingerprint space enables fingerprints labeled with real locations, which would be done only by site survey previously. These observations motivate us to design practical, flexible, and rapidly deployed localization approaches with little human costs and intervention.

In this study, we propose *LiFS* (Locating in Fingerprint Space), a wireless indoor localization approach. By exploiting user motions from mobile phones, we successfully remove the site survey process of traditional approaches, while at the same time, achieve competitive localization accuracy. The key idea behind LiFS is that human motions can be applied to connect previously independent radio fingerprints under certain semantics. LiFS requires no prior knowledge of AP locations, which is often unavailable in commercial or office buildings where APs are installed by different organizations. In addition, LiFS' users are in no need of explicit participation to label measured data with corresponding locations, even in the training stage. In all, LiFS transforms the localization problem from 2D floor plan to a high dimension fingerprint space and introduces new prospective techniques for automatic labeling.

To validate this design, we deploy a prototype system and conduct extensive experiments in a middle-size academic building covering over $1600m^2$. Experiment results show that LiFS achieves comparable location accuracy to previous approaches even without site survey. The average localization error is 5.8 meters, while the room-level localization error is about 11%.

The rest of the paper is organized as follows. We discuss the state-of-the-art of indoor localization technology and multi-dimensional scaling in Section 2. Section 3 presents the system overview of LiFS. The construction of stress-free floor plan is introduced in Section 4. Section 5 shows how to transform RSS fingerprints into high-dimension fingerprint space. In Section 6, we promote several techniques to establish the relationship between stress-free floor plan and fingerprint space. Design details and limitations are discussed in Section 7. The prototype implementation and experiments are discussed in Section 8. We conclude the work in Section 9.

2. RELATED WORK

Wireless Localization

In the literature of indoor localization, many techniques have been proposed in the past two decades. Generally, they fall into 2 categories: fingerprinting-based and model-based.

Fingerprinting-based techniques. A large body of indoor localization approaches adopt fingerprint matching as the basic scheme of location determination. The main idea is to fingerprint the surrounding signatures at every location in the areas of interests and then build a fingerprint database. The location is then estimated by mapping the measured fingerprints against the database. Researchers have striven to exploit different signatures of the existing devices or reduce the mapping effort. Most of these techniques utilize the RF signals such as RADAR [2], Horus [35], improved upon RADAR, LANDMARC [20], ActiveCampus [11], PlaceLab [15] and OIL [23]. SurroundSense [1] performs logical location estimation based on ambience features including sound, light, color, WiFi, etc. In two recent works, FM radio [5] and Channel Frequency Response [27] are explored to use as fingerprints. All these approaches require site survey over areas of interests to build a fingerprint database. The considerable manual cost and efforts, in addition to the inflexibility to environment dynamics are the main drawbacks of fingerprinting-based methods.

Model-based techniques. These schemes calculate locations based on geometrical models rather than search for best-fit signatures from pre-labeled reference database. The prevalent log-distance path loss (LDPL) model, for instance, builds up a semi-statistical function between RSS values and RF propagation distances [16, 6]. These approaches trade the measurement efforts at the cost of decreasing localization accuracy. [31] investigates several approaches based on AP locations and radio propagation models, and reports average error greater than 5 meters. Apart from power-distance mapping, Time of Arrival (ToA) [36], Time Difference of Arrival (TDoA) [24], and Angle of Arrival (AoA) [21, 37] have brought a host of alternative perspectives to characterize geometric relationship between signal transmitters and receivers.

Simultaneous Localization and Mapping (SLAM)

While the robotics and computer vision communities have developed techniques for jointly estimating the locations of a robot and a map of an environment, the nature of wireless signal strength prohibits the use of standard SLAM techniques [30, 19]. These techniques typically depend on two facts: 1) the ability to sense and match discrete entities such as landmarks or obstacles detected by sonar or laser range-finders; 2) precisely controlled movement of robots to depict discovered environments. Both of them are unreasonable for smartphone-based localization [32].

WiFi-SLAM [9] uses the Gaussian process latent variable models to relate RSS fingerprints and models human movements (displacement, direction, etc.) as hidden variables. When a small portion of RSS measurements are tagged with the real coordinates, semi-supervised localization [25] estimate the others' locations according to RSS dissimilarity. GraphSLAM [12] further improves WiFi-SLAM regarding computing efficiency and relying assumptions. Similar in leveraging human mobility, Zee [26] devises techniques for accurate dead-reckoning using smartphones and places recorded user paths into an indoor map according to the constrains imposed by the map (e.g., that a user cannot walk through a wall or other barrier marked on the map), such that wireless fingerprints are related to locations.

Different from previous SLAM solutions and [26], LiFS only measures walking steps and is free of using dead-reckoning based on noisy inertial sensors of smartphones. In the proposed solution, neither digital compass nor gyroscope is involved. Instead, we use accelerometer (as pedometer) to record only the number of footsteps, which can be accurately measured by nowadays smartphones, with respect to the displacement and directions of users' movements. Locations are computed through the deterministic MDS method. The mapping of discovered world and the ground-truth one has not been specifically discussed in SLAM and the solution relies on global references. In contrast, LiFS exploits the geometry of fingerprint space to construct fingerprints databases.

Multidimensional Scaling

Multidimensional scaling (MDS) [4] is a set of related statistical techniques often used in information visualization for exploring similarities or dissimilarities in data. An MDS algorithm starts with a matrix of item-item dissimilarities, then assigns a location to each item in d-dimensional space, where d is specified a priori. For sufficiently small d ($d = 2, 3$), the resulting locations may be displayed in a 2D graph or a 3D structure.

Seeing inter-device distances as a metric of dissimilarity, many approaches of network localization adopt MDS as a tool for calculating the locations of wireless devices [28, 8]. For example, in wireless sensor networks, sensor nodes are capable of measuring the distances to neighboring nodes by RSS, ToA, TDoA, etc. MDS is used to assign a coordinate to each node such that the measured inter-node distances are as much preserved as possible. Some researchers propose MDS to figure out WiFi AP locations [14]. In their approach, AP-AP distances are determined by a radio attenuation model. Although being similar to our solution in terms of the usage of MDS, it is neither for user localization nor fingerprinting-based.

3. OVERVIEW

3.1 Data Collection

User participation is essential in the initial period at the online stage. Untrained users walk in a building following daily activities. Mobile phones, carried by users, collect WiFi RSS characteristics (a.k.a. RSS fingerprints or signatures) at various locations along user movement paths, and the walking distances are also recorded. Walking distances are measured as footsteps from the readings of integrated accelerometers in mobile phones. Similarly, accelerometers also infer the starting and finishing moments of user paths. LiFS harnesses the walking distance between two endpoints (denoted by corresponding fingerprints) along a user path to establish the geographical relationship among fingerprints. During data collection, users can be even unaware of the collection task in which they are actually involved.

3.2 System Architecture

In this subsection, we present the system architecture of LiFS, as shown in Figure 1. The working process of LiFS consists of two phases: training and operating. The major output of training phase is a fingerprint database in which an RSS fingerprint and its corresponding location are associated. The fingerprint database is further used in operating phase to process location requests. We describe the training and operating phases in detail next.

Training Phase. The core task of training phase is to build the fingerprint database. We divide this task into 3 steps: (1) transforming floor plan to stress-free floor plan; (2) creating fingerprint space; (3) mapping fingerprints to real locations.

A floor plan shows a view of a building structure from above, including the relationships between rooms, spaces, and other physical features. The geographical distance between two locations in a floor plan is not necessary to be the walking distance between them due to the block of walls and other obstacles. Hence, we propose stress-free floor plan, which puts real locations in a floor plan into a high dimension space by MDS [4], such that the geometrical distances between the points in the high dimension space reflect their real walking distances. Through stress-free floor plan, the walking distances collected by users can be accurately and carefully utilized.

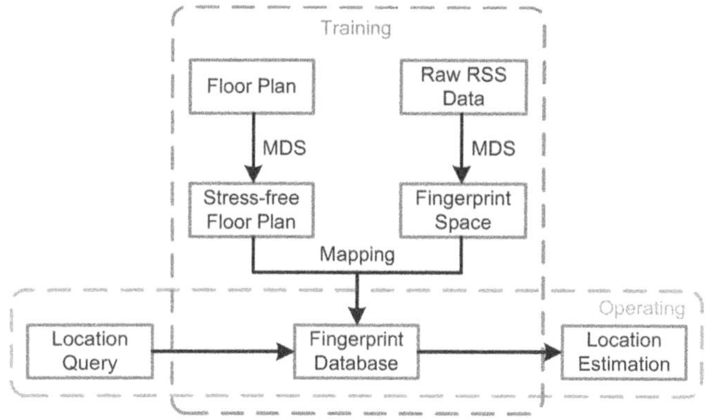

Figure 1: System architecture.

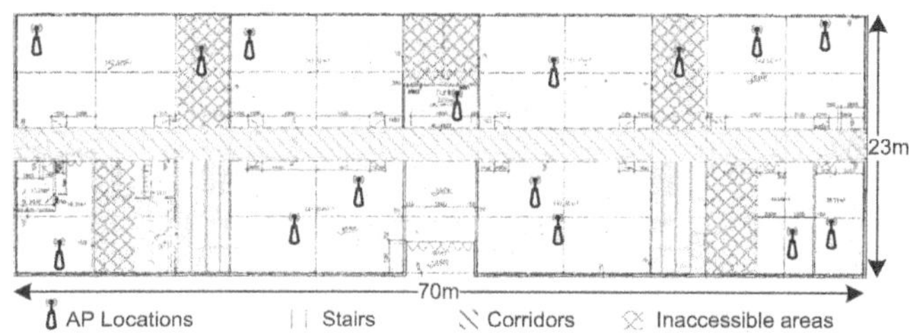

Figure 2: Floor plan of the experiment field.

Fingerprint space is a unique component in LiFS, different from traditional approaches. According to the inter-fingerprint distances, MDS is used to create a high dimension space, in which fingerprints are represented by points, and their mutual distances are preserved. In traditional approaches, fingerprints are geographically unrelated, losing the possibility of building fingerprint space.

In fingerprint database, fingerprints are associated with their collecting locations (i.e., fingerprints are labeled with locations). Such associations are achieved by mapping fingerprint space (fingerprints) to stress-free floor plan (locations) . As shown in Figure 1, fingerprint database, as the core component, connects training and operating phase.

Operating Phase. When a location query comes, usually an RSS fingerprint sent by a user, LiFS takes it as a keyword and searches the fingerprint database. The best matched item is viewed as the location estimation and sent back to users. To find the best matches, many searching algorithms can be used. In this design, we adopt a simple one, the nearest neighbor algorithm. More specifically, we assume that a fingerprint f is collected at the same location as f', if f' is the most similar to f in the fingerprint database.

4. STRESS-FREE FLOOR PLAN

In architecture and building engineering, a floor plan is a diagram, showing a view from above of the relationships between rooms, spaces, and other physical features at one level

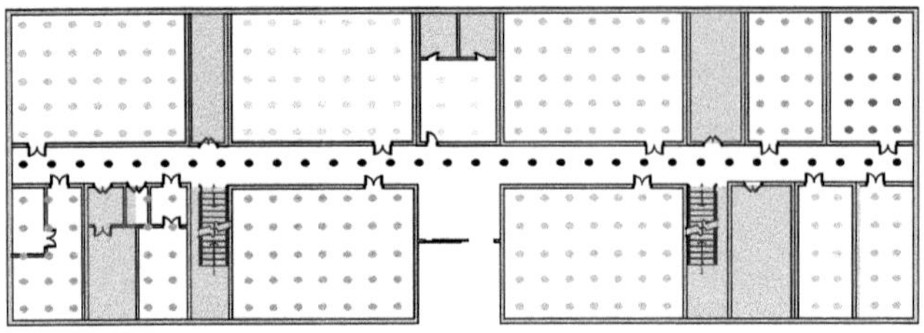

Figure 3: Floor plan with sample locations.

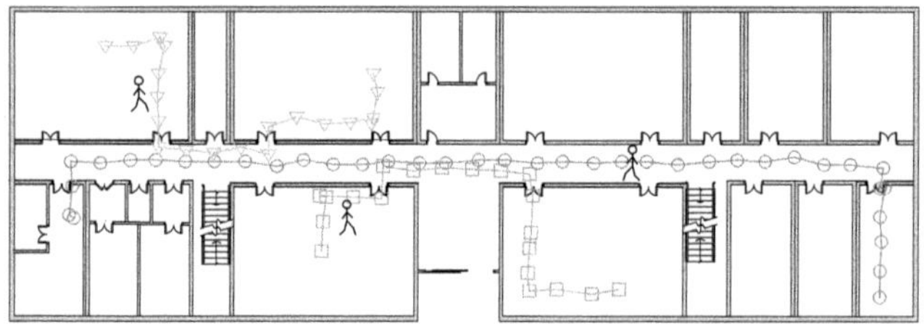

Figure 4: Moving paths.

of a structure. Dimensions are usually drawn between the walls to specify room size and wall length. The floor plan of our experiment field is shown in Figure 2. The geographical distance between two locations in a floor plan does not necessarily equal to the walking distance between them due to the block of walls and other obstacles. Hence, ground-truth floor plans come into conflict with the measured distances during data collection. Figure 2 also illustrates the distance mismatch phenomenon. The walking distance of two marked locations is greatly larger than their straight-line distance since walls are not easily passed through by users.

To address the distance mismatch problem, we propose the concept of *stress-free floor plan*. We sample an area of interests at the intersecting locations of a mesh of grids in a floor plan, as shown in Figure 3. The length l of a grid can be 1-3 meters according to the general performance of fingerprinting-based localization methods. Overmuch large or small values of l will decrease location accuracy or gain marginally or even scarcely. In our experiment, we set $l = 2$m. By calculating the distances between all pairs of sample locations, we have the distance matrix $D = [d_{ij}]$, where d_{ij} is the walking distance between two sample locations p_i and p_j in the floor plan. Using D as an input, MDS maps all p_is into a d-dimension Euclidean space. In a stress-free floor plan, the Euclidean distance between a pair of points reflects the walking distance of their corresponding locations in a real floor plan. Stress-free floor plans are often hardly embeddable in a low dimension space due to excessive distance constraints. For the convenience of observation, we set $d = 2, 3$ and the resulting stress-free floor plans in 2D and 3D visualization are shown in Figure 5 and 6, respectively, where points with the same color represent the sample locations from the same area.

5. FINGERPRINT SPACE

This section discusses the techniques for constructing fingerprint space according to the data collected by users.

5.1 Fingerprint Collection

Suppose m APs in an area A. For each location in A, the RSS fingerprint at this location can be denoted as a vector $f = (s_1, s_2, \ldots, s_m)$, where s_i is the RSS of the ith AP and $s_i = 0$ if the signal of the ith AP cannot be detected. Let d'_{ij} denote the distance between the positions of f_i and f_j. We set $d'_{ij} = +\infty$ temporarily if the distance record between f_i and f_j is not available. We measure d'_{ij} as follows. Suppose at somewhere a mobile phone records f_i; Along with walking users, it moves to another position and records f_j. In this case, d'_{ij} is the number of footsteps during the movement.

RSS fingerprints are collected during users' routine indoor movements. Users walk in a building and their mobile phones record RSS fingerprints along their walking paths, as well as the footsteps between every pairs of two consecutive fingerprints. As illustrated in Figure 4, fingerprints (denoted as squares, circles, or triangles) are recorded along three walking paths and the line segments between fingerprints indicate their distances in terms of footsteps.

After fingerprint collection, we have a set of fingerprints $F = \{f_i, i = 1 \ldots n\}$ (n is the number of records) and a distance matrix $D' = [d'_{ij}]$, both of which are essential for constructing the fingerprint space.

5.2 Pre-processing

As user movements are usually arbitrary and ruleless, walking paths might be intersectant and accordingly the fingerprints might be overlapped. Hence data pre-processing is necessary to merge similar fingerprints, which means they are likely from the same (or very close) locations in the floor plan.

Generally, for two fingerprints $f_i = (s_1, s_2, \ldots, s_m)$ and $f_j = (t_1, t_2, \ldots, t_m)$, define RSS difference (dissimilarity) δ between f_i, f_j as follows:

$$\delta_{ij} = \| f_i - f_j \|_1 = \sum_{k=1}^{m} |s_k - t_k|$$

For f_i and f_j, if their dissimilarity δ_{ij} is smaller than a predefined threshold ϵ, then they are merged as a same point in the fingerprint space to be generated. Otherwise, if $\delta_{ij} > \epsilon$, f_i and f_j are treated as two different points. The determination of epsilon is based on the fingerprint samples collected at a given location (when phones are not moving). Several other works like [34, 32] adopt the similar solution as well.

Moreover, the raw data from accelerometer readings are pre-processed to obtain walking distance measurements. Theoretically the distance traveled can be calculated by integrating acceleration twice with respect to time. However due to the presence of noise in accelerometer readings, error accumulates rapidly and can reach up to 100 meters after one minute of operation [33].

To avoid accumulation of measurement errors, we adopt the individual step counts as the metric of walking distance instead, like a pedometer. Figure 7 shows the magnitude of acceleration during walking for ten steps. We employ a local variance threshold method [13] to detect the number of steps. The method is based on filtering the magnitude of acceleration followed by applying a threshold on the variance of acceleration over a sliding window. Step counting is accurate and in our experiments the measured steps are almost exactly what they actually are.

We understand that stride lengths vary from person to person. Previous solutions like [32] assume a fixed stride length of a person according to his weight and height, and

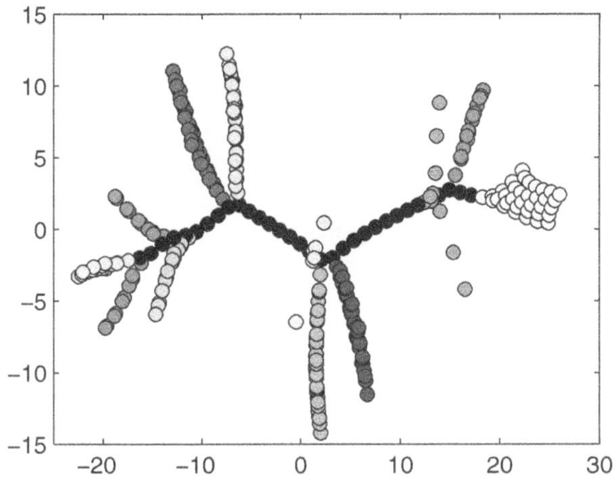

Figure 5: 2D stress-free floor plan.

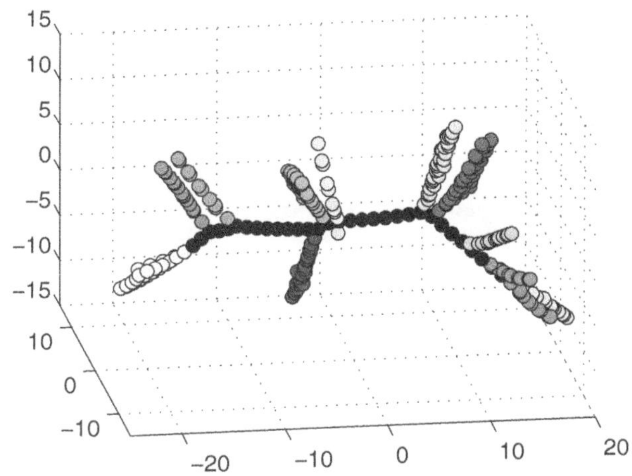

Figure 6: 3D stress-free floor plan.

achieve accurate results. In our solution, the variation of stride length can be efficiently alleviated through the fact that MDS tolerates measurement errors gracefully, due to its over-determined nature [28, 29].

5.3 Fingerprint Space Construction

To construct an accurate and informative fingerprint space, adequate fingerprints and their distance measurements are required. In our experiments, the operating phase of LiFS starts when the number of collected fingerprints reaches 10 times of the number of the sample locations in the construction of stress-free floor plans. Another possible way is to assign the first several days of LiFS' pilot run for training because routine activities exhibit certain repetitiveness day after day. Actually the running of operating phase does not mean the end of data collection. It is reasonable to select a less conservative starting point and refine the current fingerprint database uninterruptedly in operating phase according to newly coming data.

If no user path passes through a pair of fingerprints f_i and f_j, the direct measurement of the distance d'_{ij} is unavailable. However, all user paths constitute a network of fingerprints in which f_i and f_j are connected via more than one user paths. Hence, the value of d'_{ij} can be approximated as the length of the shortest path between f_i and f_j by passing several user path segments. Note that some measured values of d'_{ij} can also be updated under this intuition. For example, if $d'_{ij} > d'_{ik} + d'_{kj}$ for some k, then d'_{ij} is updated to $d'_{ik} + d'_{kj}$. Such updates can eliminate the negative distance estimates caused by the circuitous paths or the adverse (to LiFS) user habit of pacing back and forth.

We adopt the Floyd-Warshall algorithm [10] to compute all-pair shortest paths of fingerprints. It takes $O(n^3)$ running time and n is the number of fingerprints. For convenience, we still use D' to denote the distance matrix after the above-mentioned refinements on the original D'. So far D' is dense and meaningful.

Similar to constructing stress-free floor plan, using D' as an input, MDS maps all f_i into a d-dimension Euclidean space. Figure 8 and 9 demonstrate the 2D and 3D visualization of fingerprints, respectively.

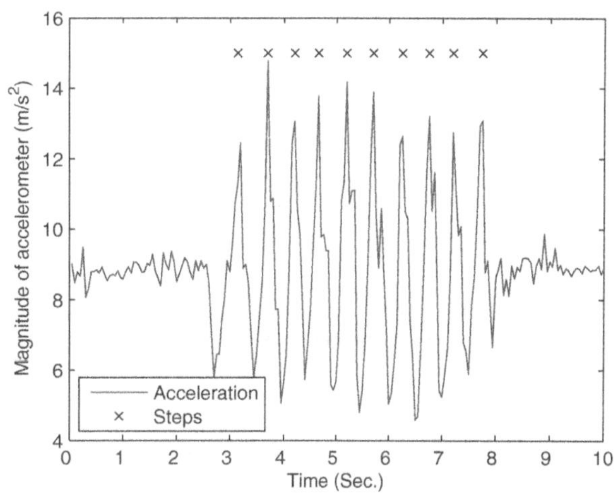

Figure 7: The acceleration pattern for 10 steps.

6. MAPPING

If all fingerprints correspond with the sample locations in the stress-free floor plan, we are able to label each fingerprint with a real location. Such correspondence comes from the spatial similarity between stress-free floor plan and fingerprint space.

6.1 Feature Extraction

6.1.1 Corridor Recognition

Generally speaking, corridors in a building connect all other office rooms like hubs in a network. When people walk from one room to another, they need to pass through corridors. Such characteristics in real life are reflected in both stress-free floor plan and fingerprint space, as shown in Figure 5, 6, 8, and 9. We observe that fingerprints collected at corridors reside in core positions in fingerprint space. In terms of graph centrality [22], these fingerprints have a relatively large centrality values.

In graph theory, vertex centrality can be valued by degree,

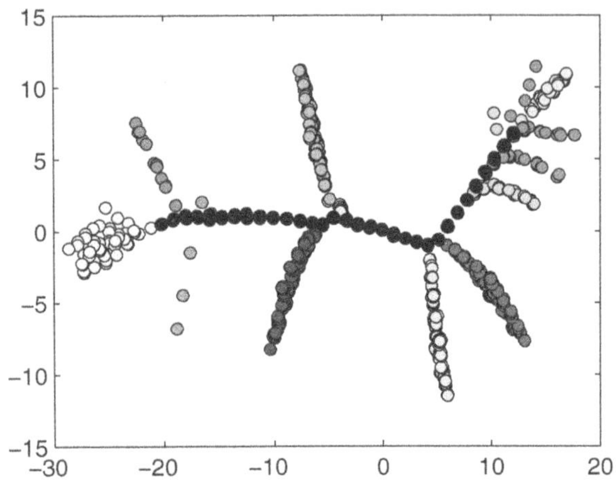

Figure 8: 2D fingerprint space.

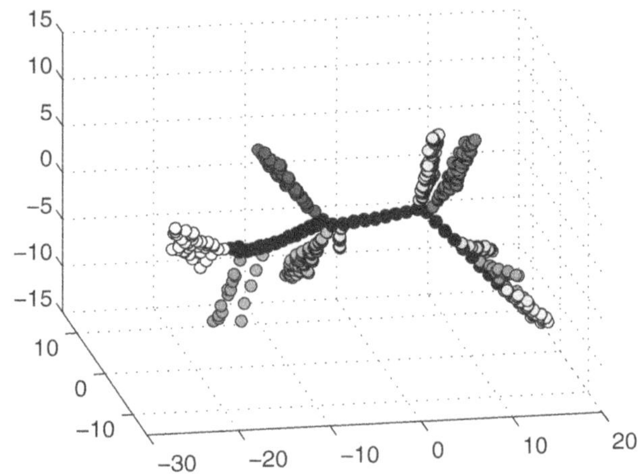

Figure 9: 3D fingerprint space.

betweenness, closeness, etc [22]. In our context, we adopt the betweenness centrality to identify corridor fingerprints. Conceptually, vertices that have a high probability to occur on a randomly chosen shortest path between two randomly chosen nodes have a high betweenness. Formally, in a graph $G = (V, E)$ of vertices V and edges E, the betweenness centrality of a vertex $v \in V$ is defined as

$$B(v) = \sum_{s \neq v \neq t \in V} \frac{\sigma_{st}(v)}{\sigma_{st}},$$

where σ_{st} is the number of shortest paths from s to t, and $\sigma_{st}(v)$ is the number of shortest paths from s to t that pass through a vertex v.

Our solution first recognizes the fingerprints collected in corridors in the fingerprint space. According to the distances among fingerprints, we build the *Minimum Spanning Tree* (MST) [7] T that connects all fingerprints in F, as illustrated in Figure 10. In addition, we compute the vertex betweenness for all vertices (fingerprints) in T and then distinguish fingerprints from corridors and other areas based on a betweenness watershed. The betweenness watershed value is determined by two parameters: 1) the area ratio c of corridors to entire floor plan (i.e., $c = \text{size(corridor)}/\text{size(all)}$), which is available when generating the stress-free floor plan; and 2) the large gap of betweenness values of fingerprints. In our experiment, the resulting cumulative distribution of betweenness is shown in Figure 11. Roughly, nearly 8.6% have their betweenness larger than 8,200; while others all less than 7,000 (shown in Figure 11). Obviously two groups of fingerprints are formed and we regard the one of larger betweenness as coming from corridors considering the structure shown in Figure 2. Let F_c denote the set of fingerprints that are estimated collected from corridors.

6.1.2 Room Recognition

Removing F_c from the fingerprint space, we observe from both Figure 8 and 9 that the remaining fingerprints form several clusters that are apparently spatially separated. To gather the fingerprints that are sufficiently close to each other, the k-means algorithm [17] (a classic clustering method) is chosen due to its computational efficiency. Thus all fingerprints in $F - F_c$ are classified into k clusters (denoted by F_{R_i}, $i = 1, 2, \ldots, k$) and in the k-means algorithm k is set to be the number of rooms in real floor plan. After clustering, all fingerprints of a same F_{R_i} are considered from the same real rooms, though we cannot tell which specific room they are from. The next subsection focuses on this mapping problem.

6.1.3 Reference Point Mapping

After characterizing corridors and rooms, we are able to establish relationships between stress-free floor plan and fingerprint space, and we think doors are the keys. Particularly, we are intended to identify the fingerprints that are collected near doors. We define $\hat{f}_i$ and $\hat{f}'_i$ as follows:

$$(\hat{f}_i, \hat{f}'_i) = \operatorname*{arg\,min}_{f \in F_{R_i}, f' \in F_c} \| f - f' \|,$$

where $\| \cdot \|$ denotes the 2-Norm in the fingerprint space.

Specifically, $\hat{f}_i$ and $\hat{f}'_i$ locate as close as possible to a door in the floor plan but in opposite sides ($\hat{f}_i$ inside the room and $\hat{f}'_i$ outside the room). Let $F_D = \{\hat{f}'_i, i = 1, 2, \ldots, k\}$ denote the set of key corresponding points. Actually, the fingerprints in F_D can be organized in a chain in the MST T, as shown in Figure 10. So we present F_D in a vector form as $F_D = (f_1, f_2, \ldots, f_k)$.

While in the stress-free floor plan, let $P_D = (p_1, p_2, \ldots, p_k)$ denote the set of sample locations in the corridor that are the closest to every door. The order of sample locations in P_D are in accord with their appearance from one side to the other side along the corridor. There are two possible ways ($\sigma_1, \sigma_2 : F_D \rightarrow P_D$) mapping F_D to P_D:

$$\sigma_1 : f_i \mapsto p_i;$$

$$\sigma_2 : f_i \mapsto p_{k-i+1}.$$

In fact only one of σ_1 and σ_2 is the ground-truth. We use the distance constraints in both stress-free floor plan and fingerprint space to eliminate the ambiguity. We define $l = (l_1, l_2, \ldots, l_{k-1})$ and $l_i = \| p_{i+1} - p_i \|$. Similarly, l' is defined as $l' = (l'_1, l'_2, \ldots, l'_{k-1})$ and $l'_i = \| f_{i+1} - f_i \|$. The values of l_i and l'_i can be determined according to the distance matrix

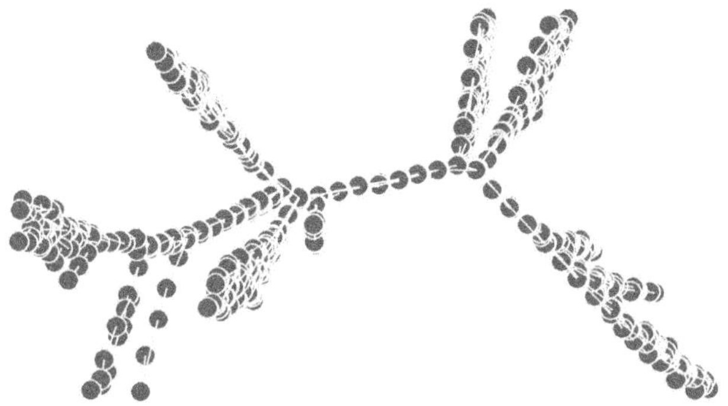

Figure 10: MST in 3D fingerprint space.

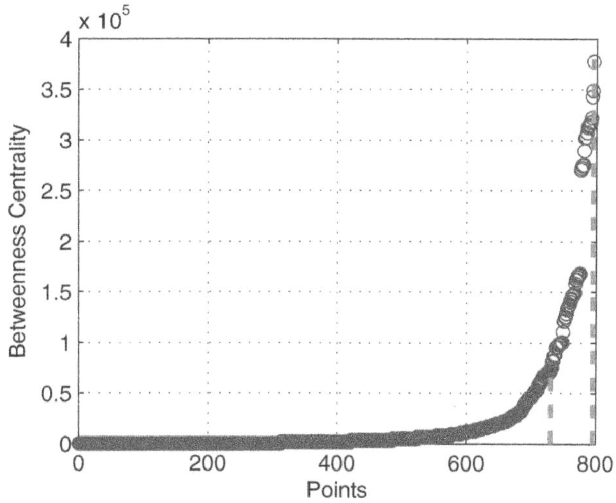

Figure 11: Betweenness distribution.

D and D', respectively. The cosine similarity of l and l', denoted by s_1, is calculated by

$$\frac{l \cdot l'}{\| l \| \| l' \|}.$$

While the similarity of l and the reverse of l', denoted by s_2, is also calculated. If $s_1 \geq s_2$, we adopt σ_1, otherwise σ_2. Without loss of generality, σ_1 is chosen in the following discussion.

Up to now, a group of fingerprints (F_D) are labeled with real locations. The relationship between F_D and P_D can be further used to map other fingerprints to real locations.

6.2 Space Transformation

In this section, we discuss how to map fingerprints (fingerprint space) to locations (stress-free floor plan). We initially try floor-level transformation and then turn to room-level transformation for better accuracy.

6.2.1 Floor-level Transformation

From the visualization of the stress-free floor plan and the fingerprint space, we observe that they are structurally similar but under trivial variations, including translation, rotation, or reflection. We use a transform matrix to solve such trivial variations.

Suppose a fingerprint $f_i \in F_D$ has its coordinate in the form of $x_i = [x_i^1\ x_i^2\ \dots\ x_i^d]^{\mathrm{T}}$, where d is the dimension of the fingerprint space. And its corresponding location $p_i \in P_D$ has a coordinate $y_i = [y_i^1\ y_i^2\ \dots\ y_i^d]^{\mathrm{T}}$ in the stress-free floor plan. Let A denote the $d \times d$ transformation matrix and $B = [b_1\ b_2\ \dots\ b_d]^{\mathrm{T}}$. We have $k = |F_D|$ following equations

$$y_i = Ax_i + B.$$

We re-write the k equations as

$$H_i z = G_i,$$

where $H_i = [x_i^{\mathrm{T}}\ 1]$, $z = [A\ B]^{\mathrm{T}}$, and $G_i = y_i^{\mathrm{T}}$. Combining k equations as a matrix equation, we have

$$Hz = G,$$

where H_i and G_i are the ith row of H and G, respectively.

The least square estimation [3] of above k equations gives

$$\bar{z} = (H^{\mathrm{T}}H)^{-1}H^{\mathrm{T}}G,$$

which minimizes $\| G - Hz \|$.

So far, the transformation matrices A and B can be determined by $\bar{z}$; thus we are able to map any fingerprint to the stress-free floor plan with a fixed location. For a fingerprint f with the coordinate $x = [x^1\ x^2\ \dots\ x^d]^{\mathrm{T}}$, the sample location that is closest to $Ax + B$ is estimated as the real location of f.

6.2.2 Room-level Transformation

From the experiment results, the unsatisfactory performance of floor-level transformation motivated us to design a fine-grained mapping solution. As previously mentioned, doors and fingerprints near doors are related, which further indicates that the rooms and the fingerprints from corresponding rooms are also related since a door belongs to only one room (i.e., the mapping from doors to rooms is injective). This fact enables room-level mapping instead of floor-level mapping.

Using MDS, the fingerprints from one room are transformed to d-dimension space. In the same way, the sample locations from the corresponding room are also mapped to d-dimension stress-free floor plan. Using doors and room corners as reference points, the fingerprints and sample locations are linked determinately by the transformation matrix above discussed. We perform the above step one room by one room and finally achieve a full mapping for all fingerprints after multiple steps of room-level transformation.

7. DISCUSSION

7.1 Global Reference Point

Global reference points include the last reported GPS location [6], AP's location [18], similar surrounding sound signature [1], feature-distinct public area, etc. Though we do not use global reference point in this design, they can be integrated into LiFS, resulting in a more robust mapping

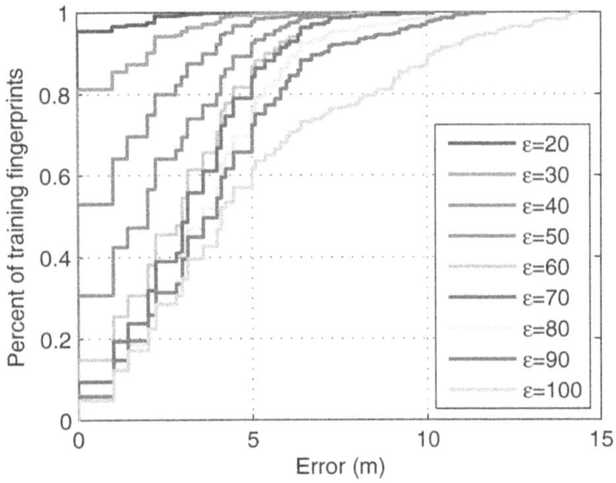

Figure 12: CDF of location error on different ϵ.

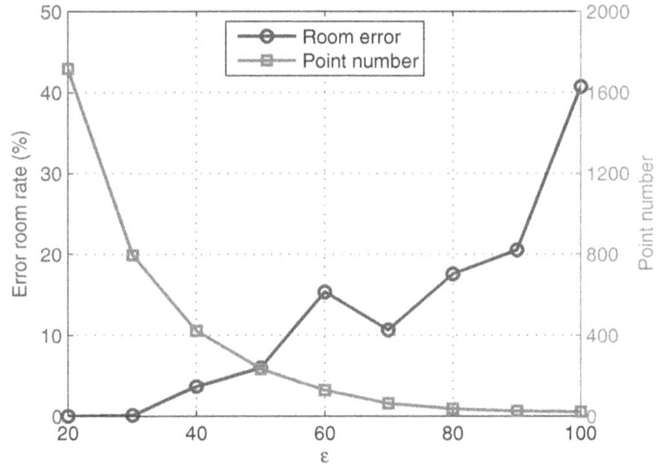

Figure 13: Room error rate vs. ϵ.

solution, while LiFS still works in case of deficient global information. Global reference points are also the key in case of symmetric floor plans or multi-floor buildings.

7.2 Building Type

Our experiment field is one floor of an academic building. The corridor in the middle connects all other office rooms that lie on both sides of the corridor. According to such layout, we try to distinguish corridors and rooms based on user traces. This solution fits a majority of office buildings but may fail in large open environments, such as hall, atrium, gymnasium, or museum, in which users' movements are difficult to characterize. We envision that the recognition of different functions of areas helps to regionalize spaces and model users' movements, as our future work.

8. EXPERIMENTS

8.1 Experiment Design

We develop the prototype of LiFS on the increasingly popular Android OS and on two Google Nexus S phones which support WiFi and accelerometer sensors. We conduct the experiment on one floor of a typical office building covering 1600m^2, with the length of 70m and width of 23m. As shown in Figure 2, the building contains 16 office rooms, of which 5 are large rooms of 142m^2, 7 are small ones with different sizes and the other 4 are inaccessible. Totally $m = 26$ APs are installed, of which 15 are with known locations and are denoted in Figure 2.

We sample the experiment floor plan approximately every 4m^2 (2m×2m grid) and obtain 292 sample locations over all accessible areas. Afterwards, we conduct MDS and the results in 2D and 3D are depicted in Figure 5 and 6, respectively.

The experiment lasts five hours by 4 volunteers. Each volunteer holds a mobile phone in hand and walk through areas of interests. LiFS records the accelerometer readings to count walking distances and picks up RSS values along the paths. Fingerprints are recorded every 4~5 steps during moving, which corresponds 2~3m under normal walking styles. Accelerometers work in two different frequencies: when detecting movements, they record sensory data with short intervals; otherwise a relatively long interval is adopted. WiFi is only scanned when the users are detected to be moving.

Totally 600 user traces along with 16,498 fingerprint records are collected. These traces cover most of the areas of the experimental field. The small and large rooms are covered by at least 5 and 10 paths, respectively. In addition, the corridor is covered by more than 500 paths. Different paths vary not only in the areas they covered but also in lengths. We select a half of these data for training and use the rest in operating phase.

8.2 Performance Evaluation

8.2.1 Fingerprint Space Generation

Before generating the fingerprint space, we obtain fingerprint points (i.e., points in the fingerprint space) and their pairwise distances from raw sensory data. Each point has a set of fingerprints. Fingerprints are distinguished by their RSS dissimilarities. Fingerprints with similar RSS features are attached to the same fingerprint points while fingerprints with large dissimilarities are sticked to different points. As user traces may be overlapped in the floor plan, fingerprints collected from different traces may be attached to the same point in the fingerprint space. In addition, fingerprints from a same sample location may be bounded to different points due to the RSS fluctuation. Hence, the threshold value of ϵ can affect the fingerprint space generation a lot.

To obtain an appropriate ϵ for generating fingerprint space, we try different values of ϵ from 10 to 100 stage-by-stage with a step length of 5. *Location error* and *room error* defined as follows are used to examine the effects of ϵ.

$$Location\_Error = ||L(f) - L'(f)||,$$

$$Room\_Error = \frac{1}{N}\sum_{f \in F} I(R(f) \neq R'(f)),$$

where f is a fingerprint, $L(f)$ ($R(f)$) and $L'(f)$ ($R'(f)$) represent the ground truth location (room) in floor plan and

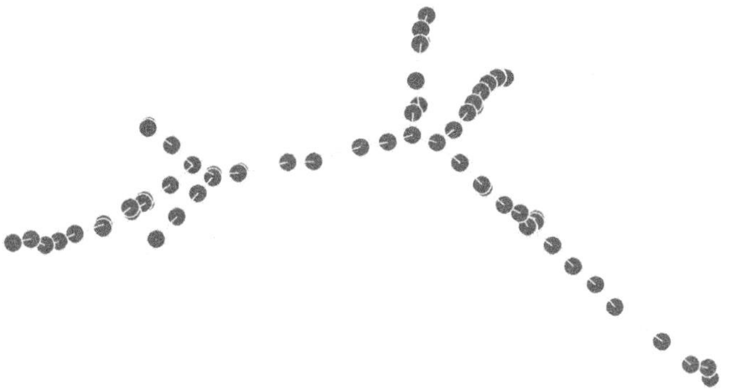

Figure 14: MST of corridor points extracted using betweenness centrality.

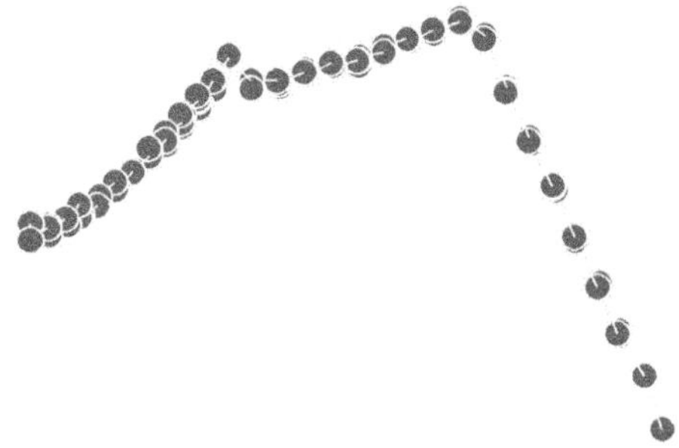

Figure 15: MST of corridor points.

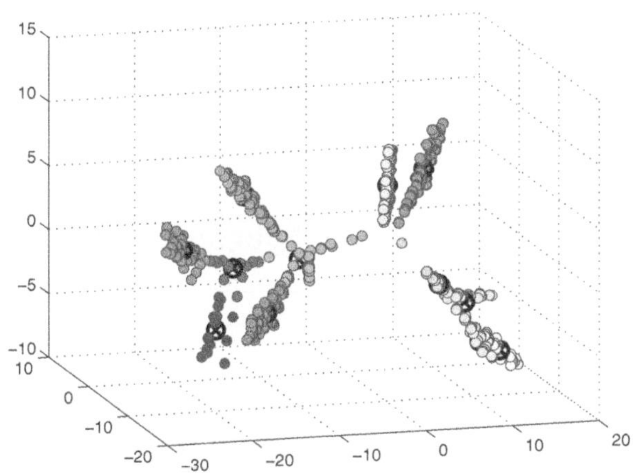

Figure 16: Clustering results by K-Means.

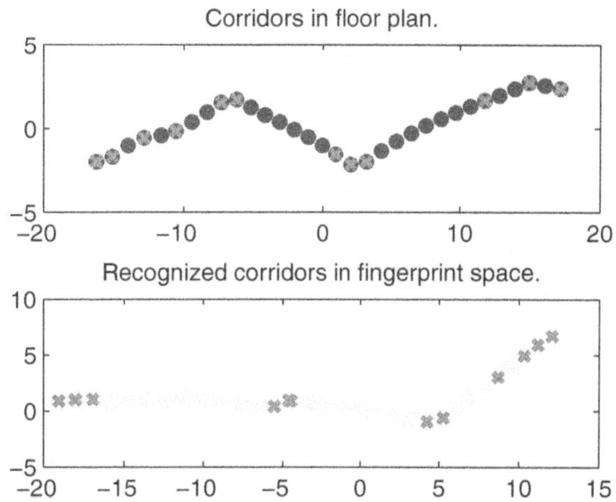

Figure 17: Floor plan corridor vs. Recognized corridor. Points marked with 'X' are reference points.

in fingerprint space respectively, N is the number of fingerprints, F is the set of fingerprints, and I is an indicative function. For each fingerprint, its ground truth location (room) in fingerprint space is determined as the labeled-location of those predominant fingerprints with the same location (room) label.

We plot the cumulative distribution (CDF) of location error in Figure 12. The impact of ϵ on room error and the number of points are illustrated in Figure 13. As from the results, location error and room error both increase when ϵ changes from 10 to 100, while the number of points decreases from about 1,600 to 1. Too small or large values of ϵ deteriorate the performance as fingerprints will be wrongly clustered. We choose $\epsilon = 30$ for further experiments, since 80% of fingerprints are accurate when $\epsilon = 30$.

To obtain walking distances of fingerprints, we first evaluate the step counts estimation using the local variance threshold method by accelerometers. Paths with different lengths (from 5 to 200 footsteps) are designed for testing. Experiment results show an error rate of 2% in the number of detected steps. Although different users have various step sizes which results in different distances of the same number of steps, it will be shown later that MDS has outstanding performance in tolerance to measurement errors. The accumulative error of long paths brings about unobvious performance drop as only path segments (inter-fingerprint distances) are used by MDS and the distance of far-away points are calculated by aggregating many paths.

Totally, 795 points are generated for fingerprint space when $\epsilon = 30$. First we assign the pairwise distances of these points with their measured walking distances and thus we get a connected network. By performing the Floyd-Warshall algorithm on the network, we obtain all the pairwise distances of 795 points. Finally, we conduct 2D and 3D MDS on these points and the results are shown in Figure 8 and 9, where each color denotes one room (or the corridor) in the floor plan. As seen from the figures, real floor plan structure is well reflected by MDS under constraints of walking distances.

8.2.2 Mapping Performance

We build the MST of the fingerprint points (Figure 10) to calculate the betweenness centrality of each point. We sort all points by betweenness centrality in Figure 11 and select those points with higher betweenness than the watershed value (8,000 in our experiments). All selected points are

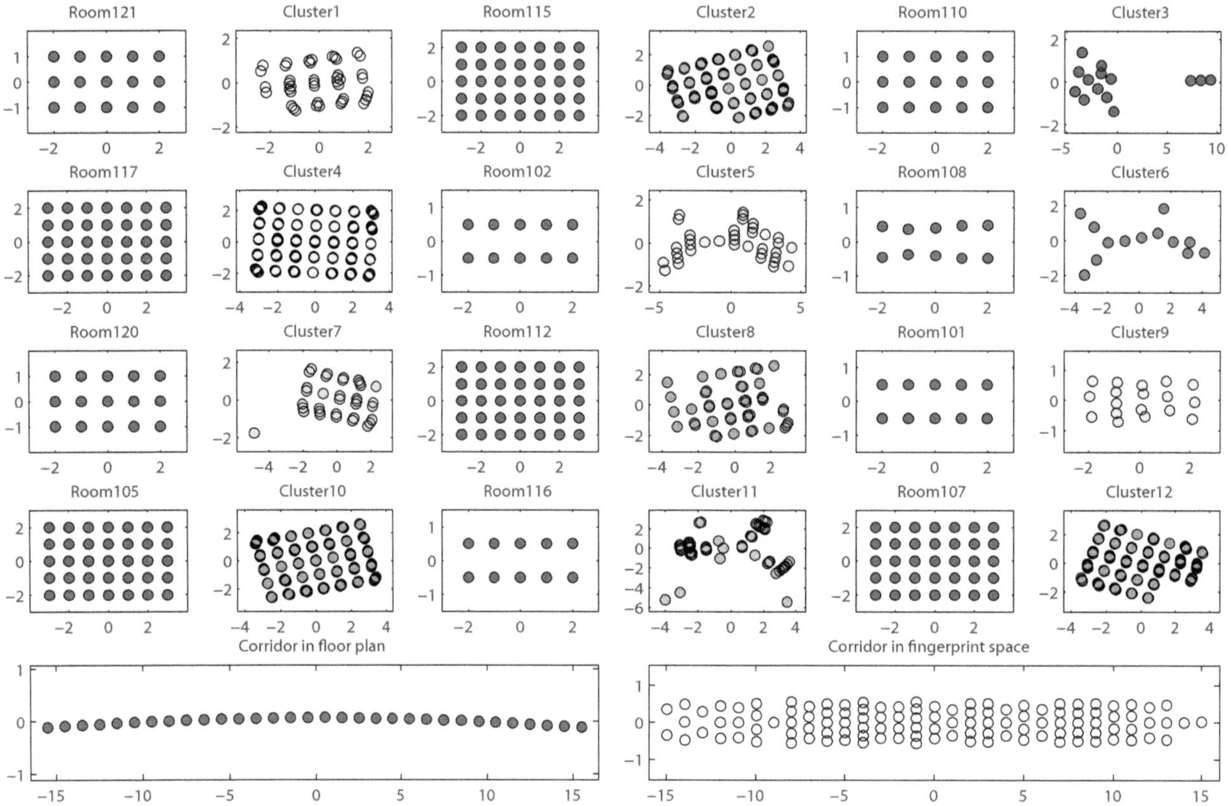

Figure 18: Fingerprints clusters vs. floor plan rooms.

estimated from the corridor recognition. As illustrated in Figure 14, most of the candidate corridor points are correctly extracted. However, some room points are also mixed among them and on the other hand, some true corridor fingerprints are not included. Hence, we refine the corridor recognition by iteratively performing MST and sifting low betweenness points until the MST of the remaining points form a single line, i.e., each point has at most one parent and one child in the MST. The final corridor points are depicted in Figure 15.

The rest of fingerprint points, most of which are actually collected from rooms, are then clustered into 12 clusters (equal to the room number) using k-Means. The clustering results are shown in Figure 16, where each different color indicates a cluster. The figure shows that most rooms can be recognized correctly while only a small portion of corridor points are mixed.

For each cluster, we identify a point in the corridors that has the shortest distance to all the points in a cluster as the reference point. The 12 reference points for 12 clusters are shown in Figure 17. Some clusters may take the same point as its reference point, which is caused by the clustering errors. The reference points in the floor plan which link rooms to corridors are also presented in Figure 17. The reference point sequences of floor plan and fingerprint space are $l = \{2.37, 3.36, 9.40, 1.18, 1.16, 8.07, 1.17, 3.82, 2.51, 2.55, 1.25\}$ and l'={0.33, 2.12, 12.98, 1,31, 1,31, 10.17, 1.24, 10.17, 1.24, 5.99, 3.69, 1.18, 0} respectively. Let l'' be the reverse of l'. The cosine similarities of l and l' and l and l'' are $s_1 = 0.97$ and $s_2 = 0.67$, respectively. Since $s_1 > s_2$, l' is adopted.

Up to now, the corresponding relationship of clusters to rooms is achieved. We then conduct the room-level transformation below. To understand the cluster-room mapping, we plot the 2D MDS results of each cluster and each room in Figure 18. The mapping relations of 12 clusters and their corresponding rooms are also illustrated in Figure 18. As seen from Figure 18, the stress-free rooms are the same as in the floor plan while the 2D fingerprint points especially those from small rooms are a bit rambling. This is because the points are from multiple rooms and the measured distances are of errors.

We then map the points in each cluster to sample locations in its corresponding room by choosing the nearest neighbor for each point. As shown in Figure 19, the mapping results are satisfactory as the location error of up to 96% points is lower than 4 meters. In addition, the average mapping error of is only 1.33 meters.

8.2.3 *Localization Error*

Two metrics are designed for localization performance: location error and room error. Location error is defined as the Euclidean distance from the estimated location to the ground truth one. Room error means the error rate of fingerprints that are estimated to be in incorrect rooms. As the final outputs of LiFS, the RSS noises and mapping errors are simultaneously taken into account. We emulate 8,249 queries using real data on LiFS, and integrate all the local-

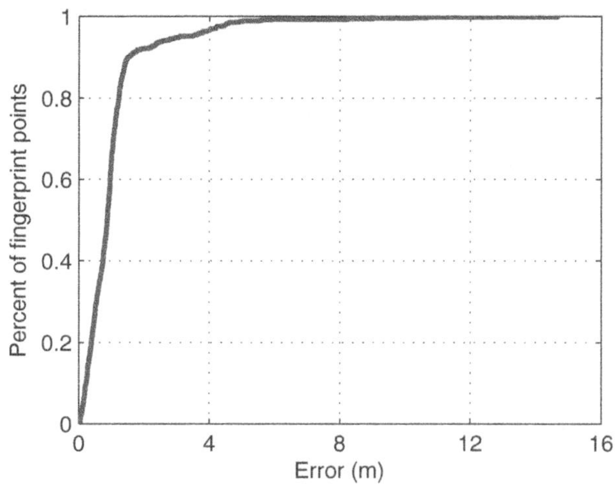

Figure 19: CDF of mapping error.

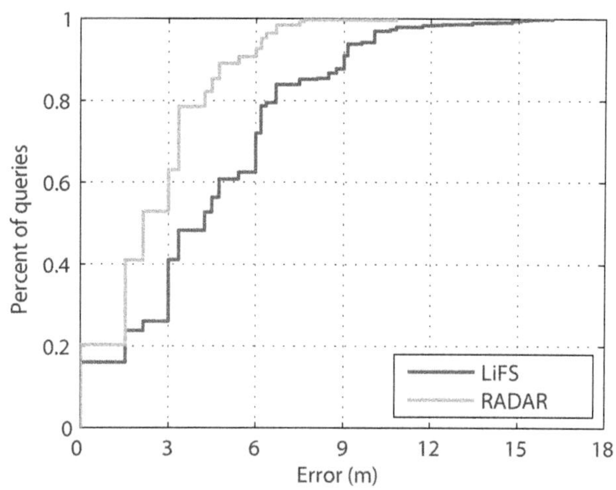

Figure 20: CDF of localization error.

ization results, as shown in Figure 20. Each query contains a fingerprint and LiFS returns an estimated location. We also implement RADAR [2] and compare its performance with LiFS on the same experiment data. The average localization error of LiFS is 5.88 meters, which is larger than RADAR (3.42 meters). The performance of LiFS is comparable to the state-of-the-art model-based approaches (larger than 5 meters) reported in [31] and outperforms EZ (larger than 7 meters) [6]. As shown in Figure 20, localization error of 80% of fingerprints is under 9 meters while about 60% is under 6 meters. Some location errors are caused by the symmetric structure of rooms, but they are relatively small and will not contribute to room error. This accuracy is fairly reasonable, though not much impressive, as LiFS needs no site survey and no specific infrastructure.

We analyze the room error of all queries on LiFS and find that the room error rate is only 10.91%. In summary, LiFS achieves acceptable localization accuracy and quite good logical localization performance, which is competitive with traditional approaches.

9. CONCLUSION

By utilizing the spatial relation of RSS fingerprints, we are able to create fingerprint space in which fingerprints are distributed according to their mutual distances in real world. On this basis, we design and implement LiFS, an indoor localization system based on off-the-shelf WiFi infrastructure and mobile phones. The preliminary experiment results show that LiFS achieves low human cost, rapid system deployment, and competitive location accuracy. This work sets up a novel perspective to cut off human intervention of indoor localization approaches. Our ongoing research focuses on making LiFS feasible and pervasive to various applied environments and buildings.

10. ACKNOWLEDGMENTS

We would like to thank the anonymous reviewers and our shepherd, Prof. Romit Roy Choudhury, for providing valuable comments and introducing the latest advances [26] to us. This work is supported in part by the NSFC Major Program 61190110, NSFC under grant 61171067 and 61133016, National High-Tech R&D Program of China (863) under grant No. 2011AA010100, National Basic Research Program of China (973) under grant No. 2012CB316200, and the NSFC Distinguished Young Scholars Program under Grant 61125202.

11. REFERENCES

[1] M. Azizyan, I. Constandache, and R. Roy Choudhury. Surroundsense: mobile phone localization via ambience fingerprinting. In *Proceedings of ACM MobiCom*, pages 261–272, 2009.

[2] P. Bahl and V. N. Padmanabhan. RADAR: an in-building RF-based user location and tracking system. In *Proceedings of IEEE INFOCOM*, volume 2, pages 775–784, 2000.

[3] Å. Björck. *Numerical methods for least squares problems*. Number 51. Society for Industrial Mathematics, 1996.

[4] I. Borg and P. Groenen. *Modern multidimensional scaling: Theory and applications*. Springer Verlag, 2005.

[5] Y. Chen, D. Lymberopoulos, J. Liu, and B. Priyantha. Fm-based indoor localization. In *Proceedings of the ACM MobiSys 2012*, MobiSys '12, pages 169–182, New York, NY, USA, 2012. ACM.

[6] K. Chintalapudi, A. Padmanabha Iyer, and V. N. Padmanabhan. Indoor localization without the pain. In *Proceedings of ACM MobiCom*, pages 173–184, 2010.

[7] T. Cormen. *Introduction to algorithms*. The MIT press, 2001.

[8] J. Costa, N. Patwari, and A. Hero III. Distributed weighted-multidimensional scaling for node localization in sensor networks. *ACM Transactions on Sensor Networks (TOSN)*, 2(1):39–64, 2006.

[9] B. Ferris, D. Fox, and N. Lawrence. Wifi-slam using gaussian process latent variable models. In *Proceedings of IJCAI*, pages 2480–2485, 2007.

[10] R. Floyd. Algorithm 97: shortest path. *Communications of the ACM*, 5(6):345, 1962.

[11] W. G. Griswold, P. Shanahan, S. W. Brown, R. Boyer, M. Ratto, R. B. Shapiro, and T. M. Truong. ActiveCampus: experiments in community-oriented ubiquitous computing. *Computer*, 37(10):73–81, 2004.

[12] J. Huang, D. Millman, M. Quigley, D. Stavens, S. Thrun, and A. Aggarwal. Efficient, generalized indoor wifi graphslam. In *IEEE International Conference on Robotics and Automation, ICRA 2011, Shanghai, China, 9-13 May 2011*, pages 1038–1043. IEEE, 2011.

[13] A. Jiménez, F. Seco, C. Prieto, and J. Guevara. A comparison of pedestrian dead-reckoning algorithms using a low-cost MEMS IMU. In *Intelligent Signal Processing, IEEE International Symposium on*, pages 37–42, 2009.

[14] J. Koo and H. Cha. Autonomous construction of a WiFi access point map using multidimensional scaling. *Pervasive Computing*, pages 115–132, 2011.

[15] A. LaMarca, Y. Chawathe, S. Consolvo, J. Hightower, I. Smith, J. Scott, T. Sohn, J. Howard, J. Hughes, F. Potter, et al. Place lab: Device positioning using radio beacons in the wild. *Pervasive Computing*, pages 301–306, 2005.

[16] H. Lim, L. C. Kung, J. C. Hou, and H. Luo. Zero-configuration indoor localization over IEEE 802.11 wireless infrastructure. *Wireless Networks*, 16(2):405–420, 2010.

[17] J. MacQueen et al. Some methods for classification and analysis of multivariate observations. In *Proceedings of the fifth Berkeley symposium on mathematical statistics and probability*, volume 1, page 14, 1967.

[18] D. Madigan, E. Einahrawy, R. P. Martin, W. H. Ju, P. Krishnan, and A. S. Krishnakumar. Bayesian indoor positioning systems. In *Proceedings of IEEE INFOCOM*, volume 2, pages 1217–1227, 2005.

[19] M. Montemerlo, S. Thrun, D. Koller, and B. Wegbreit. Fastslam: A factored solution to the simultaneous localization and mapping problem. In *Proceedings of AAAI*, pages 593–598, 2002.

[20] L. M. Ni, Y. Liu, Y. C. Lau, and A. P. Patil. LANDMARC: indoor location sensing using active RFID. *Wireless Networks*, 10(6):701–710, 2004.

[21] D. Niculescu and B. Nath. Ad hoc positioning system (APS) using AOA. In *Proceedings of IEEE INFOCOM*, volume 3, pages 1734–1743, 2003.

[22] T. Opsahl, F. Agneessens, and J. Skvoretz. Node centrality in weighted networks: Generalizing degree and shortest paths. *Social Networks*, 32(3):245–251, 2010.

[23] J. Park, B. Charrow, D. Curtis, J. Battat, E. Minkov, et al. Growing an organic indoor location system. In *Proceedings of ACM MobiSys*, pages 271–284, 2010.

[24] N. B. Priyantha, A. Chakraborty, and H. Balakrishnan. The cricket location-support system. In *Proceedings of ACM MobiCom*, pages 32–43, 2000.

[25] T. Pulkkinen, T. Roos, and P. Myllymäki. Semi-supervised learning for wlan positioning. *Artificial Neural Networks and Machine Learning*, pages 355–362, 2011.

[26] A. Rai, R. Sen, K. K. Chintalapudi, and V. Padmanabhan. Zee: Zero-effort crowdsourcing for indoor localization. In *Proceedings of ACM MobiCom*, 2012.

[27] S. Sen, B. Radunovic, R. R. Choudhury, and T. Minka. You are facing the mona lisa: spot localization using phy layer information. In *Proceedings of the ACM MobiSys 2012*, MobiSys '12, pages 183–196, New York, NY, USA, 2012. ACM.

[28] Y. Shang and W. Ruml. Improved MDS-based localization. In *Proceedings of IEEE INFOCOM*, volume 4, pages 2640–2651, 2004.

[29] Y. Shang, W. Ruml, Y. Zhang, and M. Fromherz. Localization from mere connectivity. In *Proceedings of ACM MobiHoc*, pages 201–212, 2003.

[30] S. Thrun, W. Burgard, and D. Fox. *Probabilistic Robotics*. MIT Press, 2005.

[31] D. Turner, S. Savage, and A. Snoeren. On the empirical performance of self-calibrating wifi location systems. In *Local Computer Networks (LCN), 2011 IEEE 36th Conference on*, pages 76 –84, oct. 2011.

[32] H. Wang, S. Sen, A. Elgohary, M. Farid, M. Youssef, and R. Choudhury. No need to war-drive: Unsupervised indoor localization. In *Proceedings of ACM MobiSys*, 2012.

[33] O. Woodman and R. Harle. Pedestrian localisation for indoor environments. In *Proceedings of ACM UbiComp*, pages 114–123, 2008.

[34] C. Wu, Z. Yang, Y. Liu, and W. Xi. WILL: wireless indoor localization without site survey. In *INFOCOM, 2012 Proceedings IEEE*, pages 64–72, Mar. 2012.

[35] M. Youssef and A. Agrawala. The horus WLAN location determination system. In *Proceedings of ACM MobiSys*, pages 205–218, 2005.

[36] M. Youssef, A. Youssef, C. Rieger, U. Shankar, and A. Agrawala. Pinpoint: An asynchronous time-based location determination system. In *Proceedings of ACM MobiSys*, pages 165–176, 2006.

[37] Z. Zhang, X. Zhou, W. Zhang, Y. Zhang, G. Wang, B. Y. Zhao, and H. Zheng. I am the antenna: accurate outdoor ap location using smartphones. In *Proceedings of the 17th annual international conference on Mobile computing and networking*, MobiCom '11, pages 109–120, New York, NY, USA, 2011. ACM.

Centaur: Locating Devices in an Office Environment

Rajalakshmi Nandakumar, Krishna Kant Chintalapudi, Venkata N. Padmanabhan
Microsoft Research India

ABSTRACT

We consider the problem of locating devices such as laptops, desktops, smartphones etc. within an office environment, without requiring any special hardware or infrastructure. We consider two widely-studied approaches to indoor localization: (a) those based on Radio Frequency (RF) measurements made by devices with WiFi or cellular interfaces, and (b) those based on Acoustic Ranging (AR) measurements made by devices equipped with a speaker and a microphone. A typical office environment today comprises devices that are amenable to either one or both these approaches to localization. In this paper we ask the question, *"How can we combine RF and AR based approaches in synergy to locate a wide range of devices, leveraging the benefits of both approaches?"* The key contribution of this paper is Centaur, a system that fuses RF and AR based localization techniques into a single systematic framework that is based on Bayesian inference. Centaur is agnostic to the specific RF or AR technique used, giving users the flexibility of choosing their preferred RF or AR schemes. We also make two additional contributions: making AR more robust in non-line-of-sight settings (EchoBeep) and adapting AR to localize speaker-only devices (DeafBeep). We evaluate the performance of our AR enhancements and that of the Centaur framework through microbenchmarks and deployment in an office environment.

Categories and Subject Descriptors

C.2.m [**Computer Systems Organization**]: COMPUTER - COMMUNICATION NETWORKS—*Miscellaneous*

Keywords

Indoor localization, WiFi, acoustic ranging, Bayesian inference

1. INTRODUCTION

Locating mobile devices such as laptops and smartphones in office environments, can enable several pervasive computing applications. Furthermore, even locating less portable devices such as desktop computers and printers can be useful for applications such as asset tracking.

A large body of research over the past two decades has been dedicated to enabling indoor localization. Much of this research has fallen into one of two categories based on the nature of measurements that participant devices are expected to perform: (a) *Radio Frequency* (RF)-based techniques that rely on measuring the strength of signals from proximate RF sources (such as WiFi access points or cell towers) [2, 8, 16, 5], and (b) *Acoustic Ranging* (AR)-based techniques that rely on range measurements between proximate devices [15, 12]. Devices with WiFi or cellular interfaces are amenable to RF-based localization while those with speakers and microphones can be localized using acoustic ranging.

A modern day office environment has devices with one or both of WiFi interfaces and speakers/microphones. Devices such as laptops, equipped with both WiFi interfaces, and speakers/microphones, are amenable to both approaches to localization. Devices such as desktop computers are typically equipped with speakers and can participate only in AR-based localization techniques, although the absence of a microphone in a desktop computer is a key challenge that needs to be overcome. Further, each of these approaches to localization is associated with its own set of constraints and error characteristics. RF-based localization techniques have a wide coverage, given the ubiquity of WiFi in indoor environments. However, the inherent variability in RF propagation characteristics limits the accuracy of such techniques to the range of a few meters to a few tens of meters. On the other hand, AR-based trilateration is typically very accurate and can enable localization to the sub-meter level. However, these techniques have limited coverage since these depend on a very high device deployment density, requiring three or more devices with known locations to be within audible range (typically 10m) to enable trilateration. *In this paper we ask the question: "How can we combine both these approaches in synergy and achieve the benefits of both?" The key contribution of this paper is* Centaur, *a systematic framework that fuses RF and AR based localization techniques, using Bayesian inference.*

The Centaur framework comprises three different categories of devices. First, there are *anchors*, which are fixed devices whose locations are known a priori (e.g., desktop PCs assigned to specific employees) and which at least have a speaker and so can be made to emit sound to aid in the localization of other devices. Second, there are *dual-mode devices*, which have WiFi interfaces, as well as speakers/microphones, and thus are amenable to both methods of localization (e.g., laptops, smartphones). Finally, there are *speaker-only devices*, which can only emit a sound in order to enable their localization (e.g., desktop computers whose locations are not known a priori).

WiFi measurements performed by dual mode devices help obtain an initial probability distribution of the possible locations of these devices. Acoustic measurements amongst dual-mode devices and

MobiCom'12, August 22–26, 2012, Istanbul, Turkey.

between these devices and anchors, provide geometric constraints (e.g. a distance constraint) that relate the locations of these devices. Centaur, then uses all available initial probability distributions of dual-mode devices and inter-device geometric constraints to simultaneously refine the initial probability distributions of dual-mode devices and compute the probability distribution of speaker-only devices.

Acoustic measurements provide two kinds of geometric constraints in Centaur namely, *Inter-device Distance Constraints* (IDC) and *Inter-device Distance Difference Constraints* (IDDC), and each of these presents interesting challenges. When two devices within audible proximity each has a microphone and a speaker (e.g., laptops and smartphones), techniques such as BeepBeep [11] can be used to estimate the distance between them, providing an IDC for each such pair. However, one significant challenge we faced in our deployment is that devices are seldom in direct line of sight because of walls, furniture, and clutter in an office environment. State-of-the-art acoustic ranging techniques do not perform well in such a setting because the direct line-of-sight acoustic signal could be much weaker than a delayed but stronger reflected signal. To address this problem, in Section 4 we propose an improved ranging scheme, *EchoBeep*, that performs significantly better than existing ranging schemes when the line-of-sight signal is much weaker than the reflected "echoes".

In contrast to laptops and smartphones, devices such as desktops are typically equipped with only speakers but not microphones, and so cannot measure distances on their own. To accommodate such devices, we propose a novel scheme, *DeafBeep*, that relies on estimating the *difference* in the distance between a speaker-only "deaf" device and two other devices equipped with microphones, providing an IDDC for each such triplet. Centaur then uses all the available IDC and IDDC constraints, in combination with WiFi measurements, to estimate the most likely locations of all devices in the office.

Centaur thus combines the ubiquity of RF-based localization with the lower errors of AR techniques to provide superior localization overall. Another key attribute of Centaur is that it is agnostic to the specific RF-localization or acoustic ranging technique used. Thus, it provides the flexibility to pick and combine any RF-based localization scheme with any AR scheme. In summary, our contributions are:

- Centaur: a systematic framework based on Bayesian inference that allows unification of two distinct approaches to localization, RF and AR based schemes.
- EchoBeep: an improved acoustic ranging scheme for non-line-of-sight (NLoS) settings involving a pair of devices, each equipped with both speakers and microphones.
- DeafBeep: a novel scheme for locating a device that is equipped with a speaker but no microphone, based on estimating distance differences.

We present evaluation made through microbenchmarks and deployment in a large office building. We show the improved robustness of EchoBeep in NLoS settings and the ability of DeafBeep to estimate distances differences with a speaker-only device. We show that, by combining RF (WiFi) and AR measurements (including distance difference estimates), Centaur not only improves the localization accuracy over that attainable with any one of the schemes, it also enables localization of devices, such as speaker-only PCs, that were hitherto not amenable to either WiFi-based localization or to acoustic ranging.

2. RELATED WORK

We briefly survey related work, focusing on the WiFi and acoustic techniques most closely related to ours.

2.1 Wireless LAN-based Localization

Wireless LAN-based techniques can be broadly classified as fingerprint based and model-based. Fingerprint based techniques employ a training phase to associate an RF fingerprint (typically RSSI-based) with each of a set of known locations, where the fingerprint could either be expressed in a deterministic form (e.g., RADAR [2]) or in a probabilistic form (e.g., as in Horus [16]). To perform location lookup, the measured RSSI values are compared with the fingerprints recorded previously, to find the closest match.

On the other hand, model-based techniques look to avoid the need for extensive measurements during the training phase by employing a model for RF propagation and estimating the model parameters through a limited set of measurements, either made at the clients (e.g., EZ [6]) or at the APs (e.g., WiGEM [7]).

While wireless LAN-based techniques have been used to localize WiFi-enabled devices such as smartphones and laptops, these sometimes suffer from localization errors of several meters and are also inapplicable to devices such as PCs that are not WiFi-enabled.

2.2 Acoustic Ranging and Localization

Acoustic ranging is attractive because the relatively slow speed of sound (about 332 m/s at sea level) means that the time-of-flight of an acoustic signal, and hence the distance between the transmitter and the receiver, can be measured precisely. In some systems (e.g., Active Bat [9], Cricket [12]), localization is based on the "thunder-and-lightening" principle, wherein a much faster RF signal is used to first synchronize the transmitter and the receiver, and thereafter the one-way time-of-flight of the much slower acoustic signal is measured, to estimate distance. These systems have typically used special-purpose ultrasound hardware on the clients as well as in ceiling-mounted units to enable precise distance measurement with a strong, line-of-sight acoustic signal. In contrast, Centaur focuses on existing devices such as laptops and PCs, typically in locations (e.g., on desks) where the line-of-sight acoustic signal could be blocked by obstructions.

Beep-Beep [11] introduced a novel way to do acoustic ranging, without requiring the devices to be synchronized, which we discuss further in Section 3. However, Beep-Beep suffers from a couple of limitations: it is targeted at proximate devices, where the direct acoustic signal is strong, and it assumes devices that include both a speaker and a microphone, both of which assumptions might not hold in an office environment. We address these limitations in Sections 4 and 5.

Acoustic information has also been used for purposes other than ranging, e.g., for acoustic fingerprinting of locations [13].

2.3 Graphical Model for Localization

Madigan et al. [10] introduced the idea of using a Bayesian graphical model for indoor localization based on WiFi measurements. They introduced various models to incorporate prior knowledge of WiFi signal propagation, thereby reducing the dependence on empirical measurement. Centaur draws inspiration from such prior work, but the specific Bayesian graphical models it employs incorporate the geometric constraints arising from acoustic range measurements, in addition to WiFi constraints. These geometric constraints also render the Gibbs sampling technique employed in [10] impractical, as discussed in Section 9.

2.4 Multimodal Localization

There has also been work on combining wireless and acoustic localization. WALRUS [4] provides room-level localization by having an in-room PC (anchor, in our terminology) emit an ultrasound pulse to signal its presence while concurrently transmitting the identity of the room over wireless. While attractive because of its simplicity, the room-level focus could break down, either because of the non-availability of a PC to serve as the beacon (e.g., in a lecture room) or because the "room" is large (e.g., an open cubicle setting). Centaur avoids these difficulties by using ranging (not just proximity detection) and doing so with respect to both anchor and non-anchor devices.

The Berkeley Multimodal Localization project [1] includes a scheme to combine WiFi-based localization with acoustic localization [14]. This scheme depends on the availability of microphones in devices to enable distance measurement and also requires each of the two approaches — WiFi-based and audio-based — to produce its own location estimate. In contrast, Centaur can accommodate devices without microphones and also inter-device distance and distance difference measurements even if these (partial) measurements by themselves do not permit localization.

3. PRIMER ON LOCALIZATION

3.1 WiFi-based Localization

WiFi-based localization relies on the variation of RF signal strength with location. There are two phases in a typical WiFi-based localization scheme. First, there is a *training phase*, wherein data is collected from various known locations to capture the RSS characteristics of the various APs in the target (indoor) environment. Second, there is the *localization phase* in which a device desiring to locate itself records RSS from various WiFi APs and then uses this measurement coupled with the RSS characteristics determined in the training phase to locate itself. We shall now describe two WiFi-based localization schemes that we make use of in this paper.

HORUS [16]: The basic HORUS system employs an RF fingerprinting based approach to localization. During the training phase, HORUS captures the RSS characteristics of the indoor space through a probability distribution, $P(rss_{AP_k} = r|\mathbf{x} = \mathbf{x}_i)$, which is the probability of seeing an RSS value of r from access point AP_k at location $\mathbf{x}_i$. During the localization phase of HORUS, a device obtains a vector of RSS measurements, $\mathbf{R} = < r_1, r_2, \cdots, r_m >$, where r_i is the RSS from AP_i. The joint probability of observing $\mathbf{R}$ at a location $\mathbf{x}_i$ is computed as,

$$P(\mathbf{R}|\mathbf{x} = \mathbf{x}_i) = \prod_k P(rss_{AP_k} = r_k|\mathbf{x} = \mathbf{x}_i) \qquad (1)$$

Bayesian inference is then used to compute the posterior, $P(\mathbf{x} = \mathbf{x}_i|\mathbf{R})$. The final location estimate can then be either the location with the highest probability (*maximum likelihood location*) or computed as an expectation over all locations (*expected location*).

EZ [6]: In contrast, EZ uses an RF modeling based approach to localization. It avoids the need for extensive manual measurement using crowdsourcing without requiring active user participation. Most of these measurements are from unknown locations while a few are from known locations, e.g., near windows where a GPS lock can be obtained. EZ processes the RSS measurements from both known and unknown locations to estimate the locations of the observed APs and the parameters of the log-distance path loss (LPDL) model for each such AP. Localization is then performed using multi-lateration, by converting the RSS measurements to distances from the corresponding APs.

3.2 Acoustic Ranging based Localization

The basic idea here is to perform localization using acoustic ranging with respect to multiple landmarks in known locations. A straightforward way of implementing ranging between two devices A and B is as follows: device A transmits a sound pulse at time t_A and say device B receives the sound pulse at t_B. The time difference $\Delta t_{AB} = t_B - t_A$ then represents the time required by the sound pulse to traverse the distance between device A and B. The distance d_{AB} between the devices is then $c\Delta t_{AB}$, c being the speed of sound.

The above scheme has three key implementation challenges. First, *acoustic signal detection i.e.,* device B must accurately identify the precise point where device A's sound pulse reached B even in the presence of ambient noise. Second, *precise time synchronization i.e.,* the clocks of devices A and B must be in precise synchronization; each 3 ms error in time-synchronization between A and B's clocks results in about 1 m error. Third, *software delays*, e.g. delays due to OS multi-tasking could be large and highly variable, so kernel support is typically required to record the exact time of arrival of the sound pulse.

The first challenge is typically addressed by using sound sequences with very sharp auto-correlation properties (*e.g.,*chirp, Pseudo Random (PN) sequences) agreed upon a priori by both devices A and B. Device B, then determines the exact point of reception of device A's transmission by correlating it with the known sound sequence in its received sound samples. While prior work have addressed the issues of time-synchronization and software delays through the use of special hardware/software support, BeepBeep [11] circumvents both these problems altogether by having both devices A and B transmit sound pulses to each other and then using the sampling rate of the sound-card to keep the measure of time. BeepBeep thus allows acoustic ranging to be implemented as an application without requiring any special kernel support or additional hardware.

BeepBeep [11] : Figure 1 depicts the typical operation of BeepBeep. Initially, both devices A and B turn on their microphones and begin to record ambient sounds. First, device A emits a known chirp signal. Then, after an arbitrary wait, device B emits another chirp signal. Both devices correlate the recorded sequence with the known chirp signal and determine the sample number at which the chirp was recorded. In Figure 1, N_X^Y represents the sample number at which device Y's microphone detected the chirp emitted by device X's speaker. Note that since devices A and B could have started recording on their respective devices at arbitrary times, sample counts across the two devices will not be synchronized. The key idea in BeepBeep is to estimate propagation delay by only using differences in sample counts at each device instead of absolute sample counts. Given the sampling frequency F of the sound card, BeepBeep computes the propagation delay Δt_{AB} in seconds as,

$$\Delta t_{AB} = \frac{\Delta N}{F} = \frac{1}{2F}\left[(N_B^A - N_A^A) - (N_B^B - N_A^B)\right] \qquad (2)$$

Challenges with BeepBeep and our solution: There are two key, practical challenges that BeepBeep faces. First, we find that the acoustic signal detection used in BeepBeep performs poorly in non-line-of-sight (NLoS) scenarios, which is quite common in office environments. To address this challenge, we devise an enhanced version called *EchoBeep*, which uses an improved signal detection scheme better suited for NLoS scenarios (Section 4). Second, some devices are only equipped with a speaker but not a microphone (e.g., desktop PCs). To accommodate the participation of such devices in acoustic localization, we device a novel scheme, *DeafBeep* (Section 5), which adapts the ideas in BeepBeep to es-

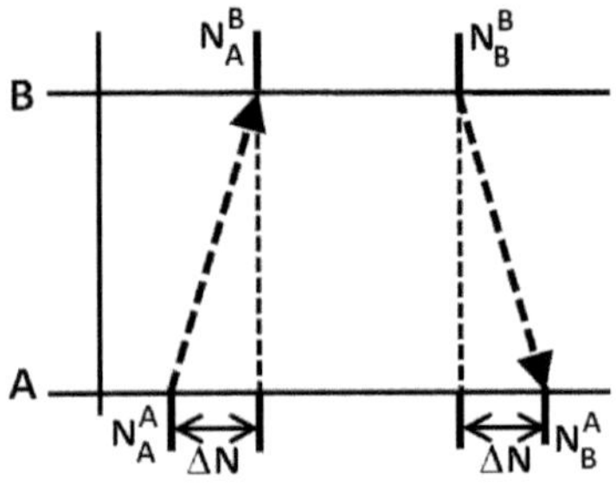

Figure 1: Beep Beep Ranging

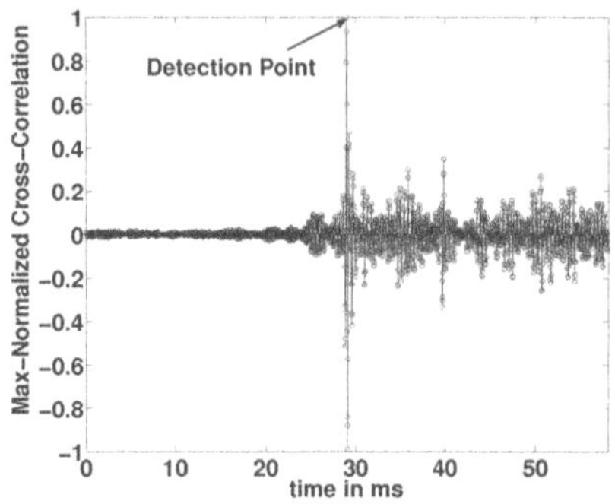

Figure 2: Correlation in BeepBeep for LoS

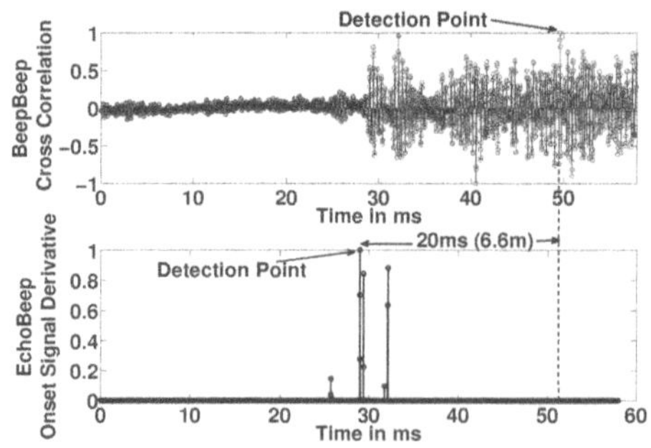

Figure 3: Comparison of EchoBeep and BeepBeep for NLoS

timate distance differences rather than absolute distances. These distance difference measurements are then used for localization.

4. ECHOBEEP: DISTANCE CONSTRAINTS

While BeepBeep provides extremely accurate ranging, to within a few cm, in LoS environments, we find that it performs poorly in NLoS environments, often resulting in more than 200% errors. In a typical office environment, where devices are often placed on desks within cubicles separated by wooden partitions and obstructed by other clutter, NLoS is a very common scenario. Our analysis reveals that the key reason for the inaccuracy is that the acoustic signal detection component in BeepBeep performs poorly in multipath environments, in the absence of LoS. To overcome this problem, we have developed *EchoBeep*, an enhanced version of BeepBeep that uses an improved acoustic signal detection scheme.

Acoustic signal detection in BeepBeep: In BeepBeep, devices transmit an a priori agreed upon chirp signal (which is a signal with a linearly increasing frequency) to each other. The receiver performs a normalized cross-correlation on the received signal with the known chirp signal in order to detect the arrival of a chirp signal. Figure 2 shows the result of this cross-correlation in a typical LoS scenario, where devices were 6m apart. As seen in Figure 2, the cross-correlation (normalized by maximum) exhibits a very sharp correlation and is easy to detect.

In contrast, Figure 3 depicts the cross-correlation for two devices 6m apart in an NLoS scenario. As seen from Figure 3, unlike the LoS scenario, there are several large peaks in the cross-correlation due to the multiplicity of paths that sound takes to reach the receiver. Furthermore, the sound signal arriving via the shortest path is typically weaker than the reflections because the former has to traverse through obstructions. For example, in Figure 3, the maximum correlation occurs almost 20ms after the arrival of the shortest path signal, which leads to a large error (i.e., overestimation) in the estimation of the time of flight of the acoustic signal. To overcome this problem, BeepBeep looks for shorter peaks that are within a 100-samples window of the tallest peak and at least 85% as tall as the tallest peak. However, there are large errors even with this heuristic.

Acoustic signal detection in EchoBeep : While EchoBeep also uses a chirp signal and cross-correlation at the receiver just as BeepBeep does, it performs additional processing on the normalized cross-correlation to determine weak correlation peaks reliably. The intuition is as follows: the receiver goes from (a) receiving noise to (b) receiving the first, albeit weak, copy of the chirp signal via the direct path, and finally (c) receiving the strongest copy of the chirp signal via a reflected path. When cross-correlation is performed, the difference between the weak signal and noise (i.e., (b) and (a)) would likely be much larger than that between the strongest signal and the weak signal (i.e., (c) and (b)).

Accordingly, let $C(n)$ be the normalized cross-correlation of the received signal with the chirp. EchoBeep performs two additional processing steps on $C(n)$. First, it computes the *onset signal* $O(n)$ given by,

$$O(n) = \max\left(C(k)\right) \forall n \geq k > n - W. \tag{3}$$

$O(n)$ has the property that it captures sudden increases in the cross-correlation (e.g., when going from (a) to (b)) while ignoring the decreases. W is a window that must be larger than twice the duration of the chirp signal (50ms in our implementation). Then, in the second step, we compute the first derivative of $O(n)$ given by,

$$\Delta O(n) = O(n) - O(n-1) \tag{4}$$

This first derivative, makes the transition from noise to the first, albeit weak, peak (i.e., (a) to (b)) stand out compared to subsequent transitions to higher peaks (e.g., (b) to (c)). Thus, this procedure enables the reliable detection of the first peak and hence the shortest acoustic path.

Figure 3 shows an example of $\Delta O(n)$ for the same NLoS setting as in case of BeepBeep. We observe that the first jump in cross-correlation stands out clearly as a spike in $\Delta O(n)$. In order to detect the correct spike, EchoBeep maintains an estimate of maximum height of spike s seen in the absence of a chirp signal (i.e., when cross-correlating with noise) and uses $2s$ as the threshold to eliminate spurious spikes. The earliest non-spurious spike is then deemed as the point of arrival of the chirp. As described in Section 6, the ranging errors in EchoBeep are in the order of 50cm-1m in a typical NLoS office setting compared to an error of 3-5m with BeepBeep. The more accurate ranging provides tighter distance constraints for the Centaur framework.

5. DEAFBEEP: DISTANCE DIFFERENCE CONSTRAINTS

As described in Section 3.2, to estimate the distance between two devices A and B, BeepBeep and EchoBeep require both A and B to be equipped with speaker and microphone each. In a typical office environment, however, devices such as desktop PCs are equipped with only a speaker but no microphone. To allow such speaker-only devices to participate in localization, we have developed a novel scheme called *DeafBeep*. Instead of estimating the distance between two devices, DeafBeep estimates the *difference* in distances of a speaker-only device from two devices that have

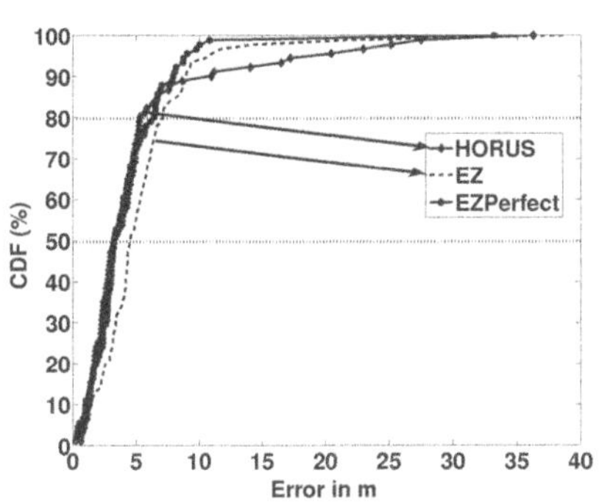

Figure 4: Benchmarking WiFi-based localization

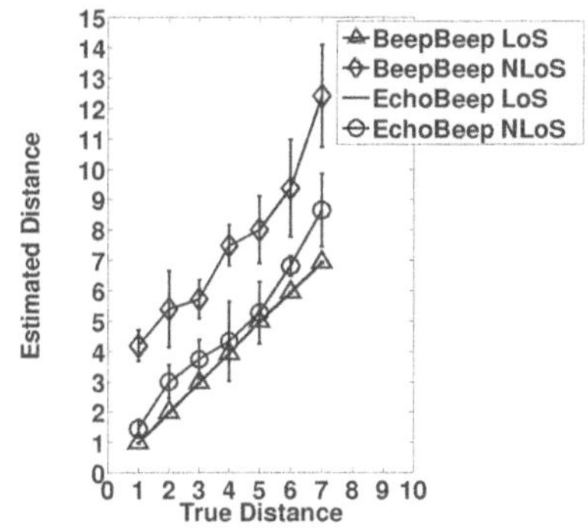

Figure 5: Benchmarking for Acoustic Ranging

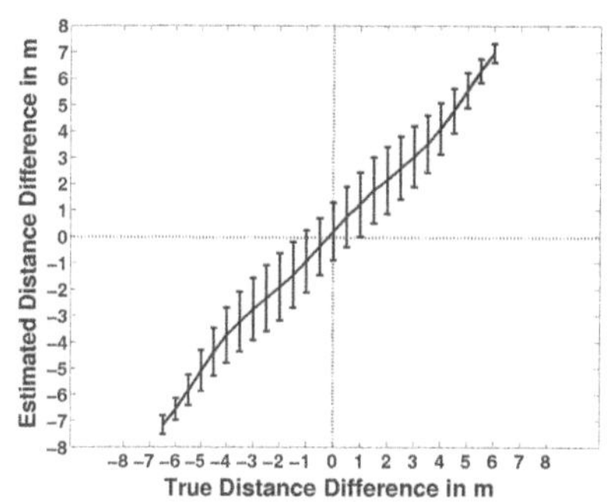

Figure 6: Distribution of Distance Difference Errors

both a speaker and a microphone. Centaur then uses these distance difference constraints to estimate device locations.

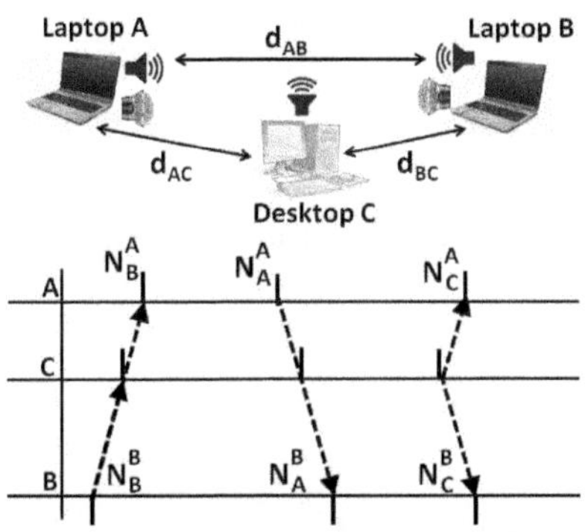

Figure 7: Functioning of DeafBeep

The most basic configuration in DeafBeep comprises three devices located within audio range of each other: one speaker-only device (Desktop C) and two other dual-mode devices equipped with speaker and microphone each (Laptops A and B) as depicted in Figure 7. As in BeepBeep, laptops A and B turn on their microphones and start recording ambient sounds. Then, each of these three devices transmits a chirp sequence at a different point in time. Laptops A and B then analyze the collected sound samples at their respective microphones and detect the reception of each of the three chirp signals using the acoustic signal detection technique used by EchoBeep (Section 4).

In Figure 7, Laptop X detects a chirp from device Y at sample number N_Y^X. Note that, since all three devices could have started recording on their respective microphones at arbitrary times, their sample counts will not be synchronized in time. Let Δt_{XY} be the propagation time for sound to travel from device X to device Y. Given the sampling rate F of the sound cards, DeafBeep then estimates the difference in propagation delays of Desktop C from the Laptops A and B as,

$$\Delta_{ABC}^2 = (\Delta t_{AC} - \Delta t_{BC}) = \frac{1}{F}\left[\begin{array}{c}(N_C^A - N_C^B) - \\ \frac{1}{2}\left[(N_B^A - N_B^B) + (N_A^A - N_A^B)\right]\end{array}\right] \quad (5)$$

This equation follows from applying a BeepBeep-like computation to two pairs — AC and BC — and then factoring out the unknowns (i.e., the reception times at C, which are not known because C does not have a microphone). We present a derivation in Appendix A.

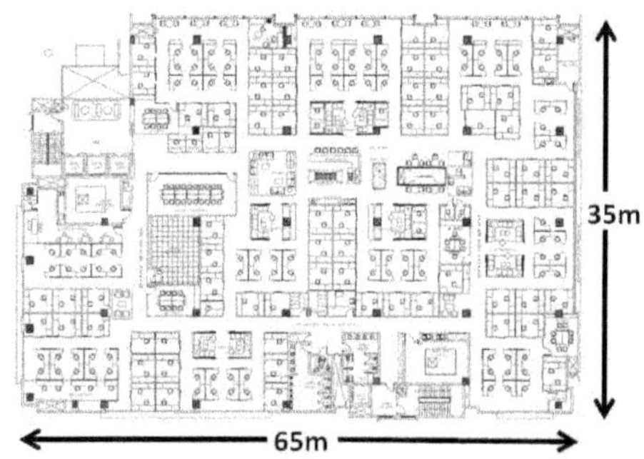

Figure 8: The large office floor used for evaluations

Each distance difference relationship Δ_{ABC}^2 results in a geometric constraint between the locations of devices A, B and C, which Centaur then uses to estimate and refine the location estimate of various devices as described in Section 8.

6. MICROBENCHMARKS

In this section, we evaluate WiFi-based localization, acoustic ranging, and acoustic distance differencing individually, to set the stage for Centaur (Section 8), where we apply these techniques in unison. Our evaluation is performed on a large office floor measuring 65m × 35m (Figure 8).

6.1 WiFi Localization

Our training data set was generated by collecting RSS measurements at 117 grid locations spanning the entire floor, with an approximate spacing of 3m between locations and about 12000 beacons gathered per AP at each location. For testing, we measured WiFi RSS at a separate set of 91 test locations within the floor, with just a few tens of beacons gathered per AP.

We evaluated three different WiFi-based localization schemes: HORUS, EZ, and EZPerfect. For HORUS, we used the training data set to create the $P(rss_{AP_i} = r|\mathbf{x})$, i.e., the probability of seeing a certain RSS value r dBm at a location $\mathbf{x}$ within the floor. For EZ, out of the 117 training locations in all, 24 locations near the windows were chosen as known locations since we could obtain a GPS lock there, and the rest were deemed as unknown. Finally, to determine a best-case bound on the accuracy of EZ, we treat all 117 training locations as known and called this EZPerfect.

Figure 4 depicts the Cumulative Distribution Function (CDF) of the localization error of these three schemes over the 91 test points. HORUS and EZPerfect had a median error of 3.3m while EZ had a median error of about 4.5m. The 80%ile errors for HORUS, EZPerfect, and EZ are 5.3m, 6.4m and 6.9m, respectively. Around the 90%ile mark, the error shoots up to 15-40m for all three schemes. As we discuss later, the incorporation of acoustic ranging constraints in Centaur helps eliminate this long tail in the distribution of localization error.

6.2 Acoustic Ranging using EchoBeep

We tested BeepBeep and EchoBeep in both LoS and NLoS scenarios when devices were placed at 1m to 10m from each other. For LoS scenarios for each distance, we tested BeepBeep and EchoBeep at 90 different locations within the office space by keeping two laptops in LoS of each other. As seen from Figure 5, *both BeepBeep and EchoBeep range extremely accurately with error under 10cm even at a distance of 10m for LoS scenarios.*

To evaluate BeepBeep and EchoBeep in NLoS scenarios, we placed pairs of laptops on tables inside cubicles, separated by various distances through wooden partitions. For each distance we repeated the experiment at 90 different location pairs within the office. As seen from Figure 5, BeepBeep's accuracy degenerates significantly in NLoS environments and can result in an error of over 200% at short distances of 1-3m and errors of up to 100% at greater distances. On the other hand, EchoBeep performs significantly better, yielding an error of 0.5-1m over distances of 1-8m. Note that, in Figure 5, although the *mean* distances estimated by BeepBeep appear to be merely shifted with respect to those estimated by EchoBeep, the former estimates have a much higher *variance*, reflecting the diversity of NLoS paths measured.

6.3 Acoustic Distance Difference

Figure 6 depicts the distribution of errors in the distance difference estimated using DeafBeep over 900 different location triplets corresponding to 3 participating devices in NLoS conditions. The variance of errors is maximum at about ±1.5m when the true distance differences are close to zero, but it drops to about ±50cm as magnitudes of true distance difference increases to 8m. This trend stems from the fact that a distance difference $\Delta^2_{ABC} = 0$, *i.e.*, $d_{AC} - d_{BC} = 0$ can occur in many more ways (e.g., the permissible values are $d_{AC} = d_{BC} \in (0, 10)$) than $\Delta^2_{ABC} = 8$ (e.g., the permissible values are $d_{AC} \in (8, 10)$ and $d_{BC} \in (0, 2)$, or vice versa). Note that the above ranges are bounded by 10m, which is the reliable detection range with 50ms chirps.

7. CENTAUR – EXAMPLE SCENARIOS

Centaur accommodates both *anchors*, whose locations are known a priori (e.g., desktop PCs assigned to specific employees), and *non-anchors*, whose locations need to be estimated (e.g., laptops, temporarily allocated PCs). Centaur uses probabilistic reasoning to combine WiFi measurements with geometric constraints (i.e., inter-device distances and inter-device distance differences) obtained from applying acoustic techniques with respect to known anchors as well as amongst non-anchors, to estimate or refine the non-anchor devices' locations. We present some simple examples to provide an intuition for how Centaur works.

7.1 Simple Example

Consider two laptops A and B that are located within audible distance of each other at unknown locations $\mathbf{x}_A$ and $\mathbf{x}_B$. For ease of exposition in this example, we shall consider locations to be 1D rather than 2D or 3D. Each of these laptops is equipped with a WiFi card, a speaker, and a microphone. Consequently, these can each perform WiFi localization independently and also perform acoustic ranging to estimate the distance d_{AB}, between them.

Using its WiFi RSS measurements ($WiFi_A$), laptop A computes $P(\mathbf{x}_A = \mathbf{x}|WiFi_A)$ – the probability that location of laptop A is $\mathbf{x}$ based on the RSS measurements $WiFi_A$. Similarly, laptop B computes $P(\mathbf{x}_B = \mathbf{x}|WiFi_B)$. These distributions are depicted in Figure 9. As depicted in Figure 9, based on $WiFi_A$ alone, laptop A can lie anywhere between 0m to 5m; its location estimate (maximum likelihood/expected) being around 2.5m. Similarly, based on $WiFi_B$ alone, laptop B can lie anywhere between 9m to 14m; its location estimate being around 11.5m. By performing acoustic ranging, the laptops learn that the distance between them $d_{AB} = 5m$. This distance constraint further limits the possible locations of A and B to $\mathbf{x}_A \geq 4m$ and $\mathbf{x}_B \leq 10m$. Figure 9 also depicts $P(\mathbf{x}_A = \mathbf{x}|WiFi_A, WiFi_B, d_{AB})$ (the probability distribution of laptop A's location based on $WiFi_A$ and the acoustic ranging measurement d_{AB}) and $P(\mathbf{x}_B = \mathbf{x}|\ WiFi_B, WiFi_A, d_{AB})$. The new location estimates of the laptops A and B are thus, 4.5m and 9.5m respectively. *Thus, acoustic ranging information can help decrease the uncertainty inherent to WiFi localization, thereby increasing localization accuracy.* While in this example we assumed that d_{AB} was measured exactly, in general d_{AB} would also have an estimation error, as described in Section 6. In practice, therefore, Centaur also models d_{AB} as a random variable while performing probabilistic inference.

This basic idea can be easily extended to several laptops on a large floor as depicted in Figure 10. Pairs of laptops that are within audible range of each other perform acoustic ranging to obtain a set of distance constraints shown as arrows between pairs of laptops. These distance constraints are then used by Centaur to refine the location estimates of laptop locations obtained from WiFi measurements.

7.2 Composite Example

In this example we try to show the various scenarios that arise in Centaur in a single example.

As depicted in Figure 11 there are two laptops (A and B), and three desktop PCs (C, D and E). The laptops are each equipped with WiFi, speaker and a microphone. Desktops C and D belong to permanent employees and their locations are know a priori and thus are to be treated as anchors, while Desktop E's location is not known a priori. Desktop C has both a speaker and a microphone, whereas desktops D and E have only a speaker on them. All devices are within audible range of each other.

As in the simple example above, the laptops can estimate their locations using WiFi measurements along with the inter-device distance d_{AB}. In addition, since desktop C is an anchor and has both a speaker and a microphone, laptops A and B can further measure distances d_{AC} and d_{BC}. In other words, laptop A must lie on a circular band (with the width of the band dictated by the ranging error) of radius d_{AC} around desktop C. Thus, knowing the location of C, A can further narrow down its location and similarly so can B. Desktop D only has a speaker and so Centaur uses DeafBeep to estimate the distance differences Δ^2_{ABD}, Δ^2_{ACD} and Δ^2_{BCD}. Note that each of these distance difference constraints is associated with a probability distribution of error. Since the locations of C and D are known, Centaur uses these constraints to further narrow down the locations of A and B. Finally, the location of Desktop E is unknown. Since E is only equipped with a speaker, it can estimate the distance difference constraints Δ^2_{ABE}, Δ^2_{ACE} and Δ^2_{BCE}. Using the probability distributions of the locations of A and B and the exact location of C, E uses these geometric constraints to locate itself.

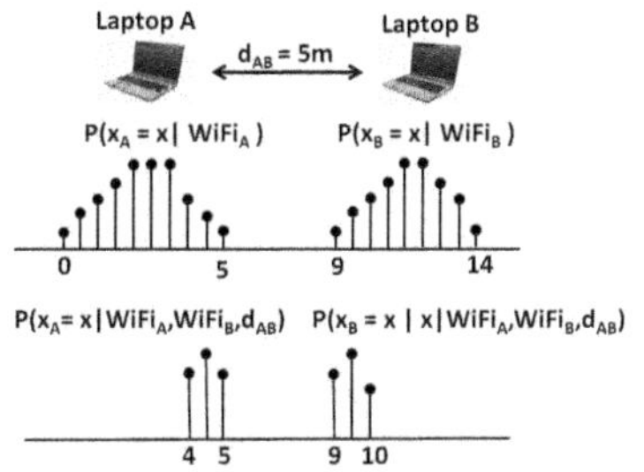

Figure 9: Simple example: probabilistic inferencing for two-laptop case

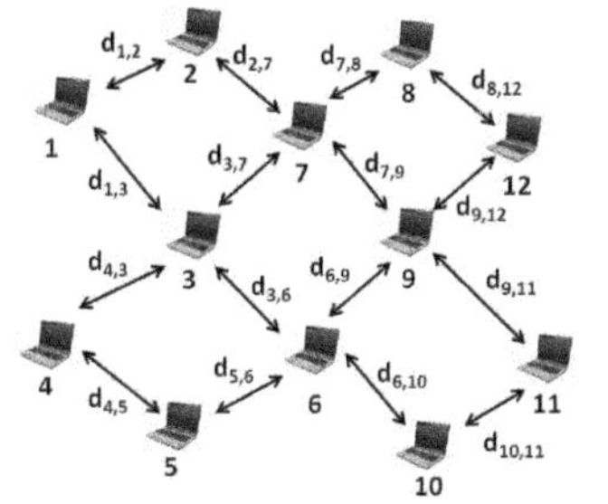

Figure 10: Example with several laptops on a floor

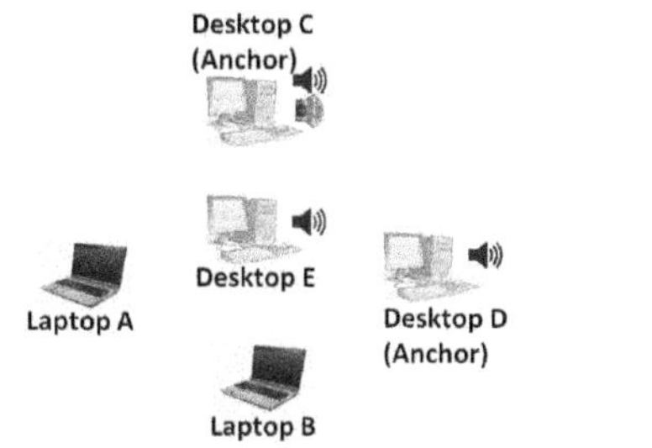

Figure 11: Composite example with anchors, dual-mode, and speaker-only devices

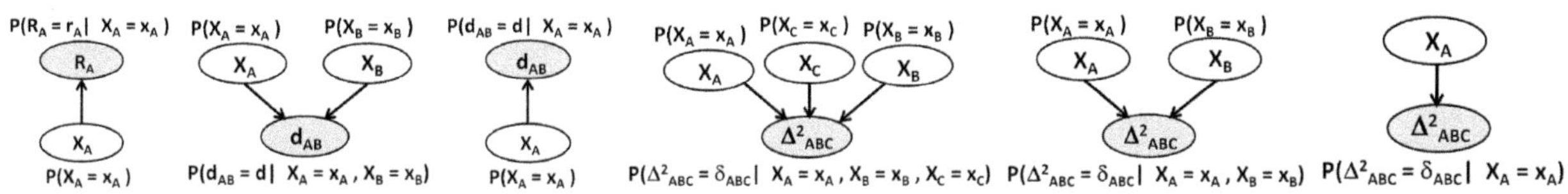

(a) *WiFi* (b) *Range measurements between non-anchors* (c) *Range measurements to an anchor* (d) *DD measurement with non-anchors* (e) *DD measurement with a single anchor* (f) *DD measurement from two anchors*

Figure 12: Graphical models for various measurements in Centaur

In summary, as conveyed in these examples, Centaur uses all manner of WiFi measurements, inter-device distance estimates and inter-device distance difference estimates in a single probabilistic inference framework, to estimate the location of devices.

8. BAYESIAN NETWORKS IN CENTAUR

Centaur uses probabilistic inference to locate devices using the available WiFi measurements, and the inter-device distance and distance difference estimates obtained from acoustic ranging. To accomplish this, Centaur first constructs a Bayesian graphical model to capture all the WiFi measurements and the geometric constraints arising from acoustic measurements. After constructing the Bayesian graph, Centaur infers the most likely set of locations for the devices that satisfies both the geometric constraints and the WiFi measurements. Here we describe Centaur's construction of the Bayesian graph, and then discuss its inferencing technique in Section 9.

8.1 Bayesian Graphs

In a Bayesian graph, random variables are modeled as nodes. In Centaur, all unknown device locations, WiFi measurements, and distance as well as distance difference measurements represent nodes in the graph. Random variables that can be measured are called *evidence* variables (or nodes). So WiFi, distance, and distance difference measurements are the evidence variables in Centaur. Each non-evidence node is also associated with a *prior* — the probability distribution of the variable in the absence of any evidence (i.e., measurements). Finally, directed edges between nodes in the graph represent the relationships between them in the form of conditional probabilities. In the rest of this section, we describe how each measurement is modeled in Centaur as a Bayesian sub-graph and present an example that combines all the sub-graphs together to form the complete Bayesian graph.

8.2 WiFi RSS Measurements

The graph for a WiFi measurement by device A is depicted in Figure 12(a). The prior for location of device A, $P(\mathbf{X}_A = \mathbf{x}_A)$, is the probability, in the absence of any evidence, that the location $\mathbf{X}_A$ is equal to $\mathbf{x}_A$. Centaur uses as prior, a uniform distribution over the entire indoor space. In other words, in the absence of any measurements, a device is equally likely to be located anywhere in the floor. The edge connecting the WiFi measurement node to the location of device A is characterized by the probability distribution, $P(\mathbf{R_A} = \mathbf{r}_A | \mathbf{X_A} = \mathbf{x}_A)$, which is the probability of a certain RSS vector $\mathbf{r}_A = < rss_{AP_1}, \cdots, rss_{AP_n} >$ (rss_{AP_k} being the RSS from AP_k) being measured by device A when it is located at $\mathbf{x}_A$. As described in Section 3.1, this distribution is constructed using training data.

8.3 Distance Between Non-Anchors

The graph in Figure 12(b) depicts an EchoBeep measurement d_{AB} between devices A and B. If the locations $\mathbf{X}_A$ and $\mathbf{X}_B$ of devices A and B are $\mathbf{x}_A$ and $\mathbf{x}_B$ respectively, then distance d_{AB}^{true} would be the Euclidean distance between them. EchoBeep however is not free from errors (Figure 5), so it will measure a distance d characterized by the probability distribution $P(d_{AB} = d | \mathbf{X}_A = \mathbf{x}_A, \mathbf{X}_B = \mathbf{x}_B)$.

8.4 Range to an Anchor

If device B is an anchor then $\mathbf{x}_B$ is a known constant. Consequently, the graphical model for this measurement is the same as that for range between non-anchors except with the node corresponding to B removed since it is not a random variable. Figure 12(c) depicts this graph characterized by the probability distribution $P(d_{AB} = d | \mathbf{X}_A)$.

8.5 Distance Difference between Non-anchors

If devices A, B and C are located at $\mathbf{x}_A$,$\mathbf{x}_B$ and $\mathbf{x}_C$, then the distance difference $\Delta^{2true}_{ABC} = d^{true}_{AB} - d^{true}_{AC}$. DeafBeep can be used to estimate this distance difference. The graphical model for this constraint is depicted in Figure 12(d) and is characterized by $P(\Delta^2_{ABC} = \delta_{ABC} | \mathbf{X}_A = \mathbf{x}_A, \mathbf{X}_B = \mathbf{x}_B, \mathbf{X}_C = \mathbf{x}_C)$.

8.6 Distance Difference with Anchors

In a distance difference measurement Δ^2_{ABC}, when one or more devices are anchors, their locations are treated as constants. The graphical model in such cases, is exactly the same as the graphical model for the distance difference constraint for non-anchor devices, but with the anchor devices removed. Figures 12(f) and 12(e) are examples corresponding to distance difference measurements with one and two anchors, respectively.

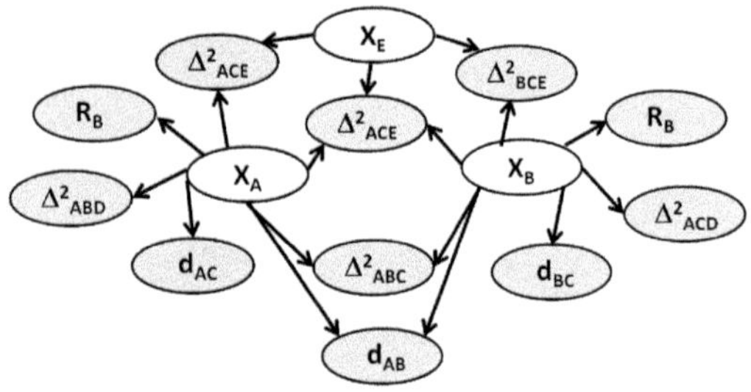

Figure 13: The Bayesian graph for composite example in Section 7

8.7 Putting it all together

Having constructed all the subgraphs for the measurements, Centaur simply combines these together to obtain the overall graph. As an example, Figure 13 depicts the Bayesian graphical model for the composite example described in Section 7.

9. INFERENCE IN CENTAUR

Once the Bayesian graph has been constructed as described in Section 8, Centaur infers the most likely locations for all the non-anchor devices given all the measurements. In this section we describe how Centaur performs this inference on the Bayesian graph.

9.1 Inference in Centaur is NP-Hard

Let $\mathbf{G} = (\mathbf{X}, \mathbf{\Phi}, \mathbf{E})$ be a Bayesian graph, where $\mathbf{X} = \{\mathbf{X}_i\}_{i=1}^{i=N}$ are all the non-evidence variables (nodes), $\mathbf{\Phi} = \{\mathbf{\Phi}_i\}_{i=1}^{i=M}$ is the set of all evidence nodes and $\mathbf{E}$ is the set of all edges in the graph. *The inference problem in G is to determine the posterior probability* $P(\mathbf{X}_i = \mathbf{x}_i|\mathbf{\Phi})$, $\forall i = 1, \cdots, N$, – the probability that a non-evidence variable $\mathbf{X}_i$ takes a value $\mathbf{x}_i$ given all the measured values for non-evidence variables. This probability can be expressed as,

$$P(\mathbf{X}_k = \mathbf{x}_k|\mathbf{\Phi}) \propto \sum_{\forall \mathbf{x}_i, i \neq k} P\left(\begin{array}{c} \mathbf{X_1} = \mathbf{x}_1, \cdots, \mathbf{X_k} = \mathbf{x}_k, \\ \cdots, \mathbf{X_N} = \mathbf{x}_N \end{array} |\mathbf{\Phi}\right). \tag{6}$$

Here, $P(\mathbf{X}|\mathbf{\Phi}) = P(\mathbf{X_1} = \mathbf{x}_1, \cdots, \mathbf{X_N} = \mathbf{x}_N)$ is the joint distribution of the all non-evidence variables given all the measurements. This joint distribution can then be expressed as,

$$P(\mathbf{X}|\mathbf{\Phi}) = \prod_{\forall \mathbf{X}_i \in Ch(\mathbf{G})} P(\mathbf{X}_i = \mathbf{x}_i|\Psi(\mathbf{X_i})) \prod_{\forall \mathbf{X}_i \in \mathbf{X}} P(\mathbf{X}_i = \mathbf{x}_i) \tag{7}$$

In Eqn 7, $Ch(\mathbf{G})$ is the set of all variables that have at least one parent (i.e., are child nodes in the Bayesian graph), and $\Psi(\mathbf{X}_i)$ is the set of all of node $\mathbf{X}_i$'s parents. Further, all evidence variables in Eqn 7 are set to their measured values.

Inference in Bayesian graphical networks is a well studied area. When a Bayesian graph can be represented as a Poly-tree (i.e., the undirected version of the graph has no loops), the Pearl's message passing algorithm [3] solves the inference problem with $O(N)$ complexity, N being the number of nodes in the graph. For a general graph with loops, however, inference has been proven to be NP-Hard [3]. The graphs generated in Centaur almost always have several loops as seen in Figure 13, consequently exact inference in Centaur is an NP-Hard problem.

There are three popular approaches to approximate inference in Bayesian networks – loopy belief propagation [3], sampling based techniques [3] and variational methods [3]. We found that using loopy belief propagation on graphs constructed by Centaur often neither converged nor yielded correct results. We further discovered that the large number of geometric constraints in a typical Bayesian graph in Centaur makes it particularly ill-suited to using sampling techniques such as Gibbs sampling [3] and importance sampling, and required impractically large running time for convergence. This is because, the large number of geometric constraints limits the feasible set of device locations to a very narrow region in the $2N$ dimensional space of possible device locations that is hard to draw samples from. Finally, we found that the distribution of RSS measurements at several locations is non-Gaussian and not amenable to an analytical form tractable to variational methods. Consequently, we devised an approximate inference technique specifically tailored to the graphs generated by Centaur.

9.2 The Maximum log-likelihood approach

An alternative to solving the inference problem is be content with finding device locations that maximize the log-likelihood of the joint distribution, $\log P(\mathbf{X}|\mathbf{\Phi})$. This is especially suitable in Centaur since Centaur only attempts to estimate the device locations. The maximization problem is also NP-Hard and requires searching over all possible combinations of values of $\mathbf{X}$. Further, the log-likelihood function in Centaur is a multi-modal function and is not amenable to iterative schemes such as Newton-Raphson or gradient decent that rely on finding a local maxima. One approach is to use a Genetic Algorithm (GA) to search efficiently. However, we found that the GA took a long time to converge to a solution. This is for the same reason that sampling-based inference techniques perform poorly in the context of Centaur, as discussed previously.

9.3 The Centaur Approach

In Centaur, we found that a hybrid, two-step approach of search and inference performs the best. In the first step, *partial inference step*, Centaur performs an exact inference using Pearl's algorithm, accommodating as much evidence as possible without having loops in the graph. The goal of this step is to narrow down the space of possibilities and make maximum log-likelihood search tractable. The second step, *maximum log-likelihood step*, then takes a maximum log-likelihood approach by searching over space of possibilities given by the first step. In the rest of this section, we describe this scheme.

9.3.1 Partial Inference Step

The key idea in this step is to partition the entire Bayesian graph $\mathbf{G}$ into a sequence of loop-free subgraphs $< \mathbf{G}_1, \mathbf{G}_2, \cdots, \mathbf{G}_n >$. The key property in the partition is that no two sub-graphs have any common evidence nodes and the union of all evidence nodes across all sub-graphs covers the entire set of evidence nodes in $\mathbf{G}$. Let $\mathbf{\Phi}_i$ be set of all evidence nodes in sub-graph $\mathbf{G}_i$. The scheme starts by performing exact inference on $\mathbf{G}_1$. Then for each successive graph $\mathbf{G_k}$, the results of the inference performed on the preceding sub-graphs, $\mathbf{G}_1, \cdots, \mathbf{G_{k-1}}$, is used as the prior to perform exact inference on $\mathbf{G}_k$. The scheme is algorithmically described below.

1: $PartialInference()$
2: $< \mathbf{G}_1, \cdots, \mathbf{G}_k >= Partition(\mathbf{G})$.
3: Set all prior $P(\mathbf{X}_i)$ in $\mathbf{G}_1$ to the same as $\mathbf{G}$.

Compute $P_{\mathbf{G}_1}(\mathbf{X}_i|\mathbf{\Phi}_{\mathbf{G}_1})\ \forall i$ using Pearl's algorithm.
for $k = 2$ to n **do**
    Set all priors $P(\mathbf{X}_k)$ in $\mathbf{G}_k$ to the same as $P_{G_{k-1}}(\mathbf{X}_i|\mathbf{\Phi}_{\mathbf{G}_{k-1}})$.
    Compute $P_{\mathbf{G}_k}(\mathbf{X}_i|\mathbf{\Phi}_{\mathbf{G}_k})$ using Pearl's algorithm.
end for

$Partition()$
$\mathbf{G}_{rem} = \mathbf{G}$.
$k = 1$
repeat
    Find the set of edges $\mathbf{E}_k$ that need to be removed to eliminate cycles from $\mathbf{G}_{rem}$.
    Remove all evidence nodes associated with these edges in $\mathbf{E}_k$ to obtain $\mathbf{G}_k$.
    Remove all evidence nodes considered in $\mathbf{G}_k$ from $\mathbf{G}_{rem}$.
until $\mathbf{G}_{rem}$ has not evidence nodes

The complexity of $Partition()$ is $O\left((|E| + N)N\right)$. The complexity of the loop in $PartialInference()$ is $O(|E|)$. This is because, at each step in the loop, the complexity of Pearl's algorithm is $O(|E_i|)$, $|E_i|$ being the number of edges in G_i. Since sum of all edges across all the subgraphs is equal to $|E|$, i.e., the number of edges in the graph G, the complexity of the loop is $O(|E|)$. In a practical setting, since the running time of $Partition()$ is insignificant compared to that of the Pearl's algorithm, the practical complexity of $PartialInference()$ is $O(|E|)$.

9.3.2 *Maximum log-likelihood Step*

In this step we run a GA to determine the set of device locations that maximize the log-likelihood. In all our settings we found that the GA typically converges in under a minute (as evaluated in Section 10). This is because the partial inference step reduces the space of possibilities dramatically.

10. RESULTS

Given its goal of combining acoustic and WiFi localization in synergy to improve localization error, we evaluate Centaur with a view to answering the following key questions under various deployment scenarios:

- How much can WiFi localization benefit from acoustic distance and distance difference measurements?
- How much can acoustic distance difference and distance measurement based localization in turn benefit from WiFi measurement?
- How well can we use Centaur to locate speaker-only devices with no microphone or WiFi?
- What are the typical running times for Centaur?

We implemented and deployed Centaur in a large office floor depicted in Figure 8 (Section 6). Based on our deployment, we try and quantitatively answer the above questions.

10.1 Implementation and Deployment

The Centaur software comprises a Centaur server and a client. The clients were deployed on devices such as laptops and desktop PCs. The clients connect to the server whenever the devices are turned on and report to the server as to whether they have a WiFi-card, speakers and/or a microphone. In case of anchor nodes, the anchors additionally report their known locations to the server. The Centaur server then, depending on the capabilities of the device, schedules the devices to perform WiFi measurements, transmit Chirp sequences, and record ambient sounds for EchoBeep and DeafBeep measurements. The clients transmit all recorded data (WiFi measurements and recorded sound samples) to the server. Centaur then uses EchoBeep and DeafBeep to convert the recorded samples to inter-device distance and distance difference estimates. It then constructs the Bayesian graph (Section 8) and uses the inferencing algorithm described in Section 9.3 to estimate the most likely locations of all devices.

10.2 Exp I: WiFi localization with acoustic ranging among non-anchors

In this experiment we ask the following question, *"In the absence of any anchors (devices with known locations), can the performance of WiFi localization be improved by using inter-device distance measurements?"* This scenario corresponds to the simple example (Figures 9 and 10) in Section 7. First, laptops were placed at various locations within the floor and their locations were estimated only using WiFi localization (HORUS). Then, pairs of laptops estimated their inter-device distances using EchoBeep. A laptop can, however, only estimate distance to another laptop if it is within audible range. So in successive experiments, we increased the number of laptops that any given laptop could estimate distance with using acoustic ranging. When there are N laptops in audible proximity of each other, there are $\frac{N(N-1)}{2}$ distance measurements among the N laptops to be combined with N WiFi measurements made at the individual laptops. Figure 14 depicts the CDF of the localization error, with N increasing from 1 to 5 laptops and compares it to the case where laptops used on WiFi localization.

As seen from Figure 14, even with just one other laptop in audible range, there is an improvement in localization performance compared to using only WiFi measurements. The 50%ile error decreases from 4.2m to 3.4m while the 80%ile error decreases from 7m to 5m. The most significant improvement occurs in the tail of the localization errors. Errors as large as 20m that occurred in WiFi are completely eliminated with the help of just one acoustic distance measurement. This is because, the likelihood that both devices will have a large localization error at the same time is low and the node with a high WiFi localization error benefits from the other. As we increase the number of laptops within audible proximity, the performance further improves. The 50%ile error reduces to 2.4m and the 80%ile error reduces to 4m. However, increasing the number of proximate laptops beyond 3 does not bring significant improvements. *This experiment demonstrates how WiFi localization error can benefit from inter-device distance estimation even when there are no known locations.*

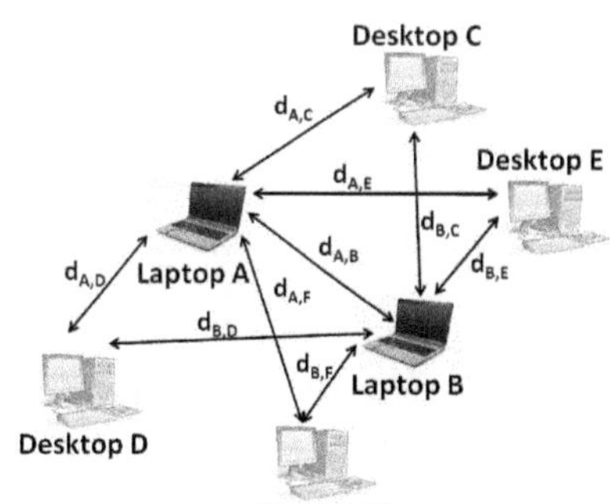

Figure 17: Setup for evaluating fusion of WiFi with speaker-only anchor devices

10.3 Exp II: WiFi localization with acoustic ranging to speaker-only anchor devices

In a typical office environment the location of several desktop PCs can be known a priori and thus these devices can be treated as

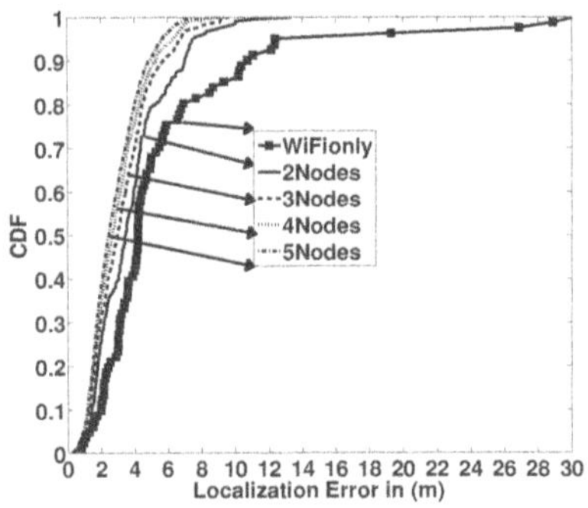

Figure 14: Performance of WiFi localization with ranging among non-anchors

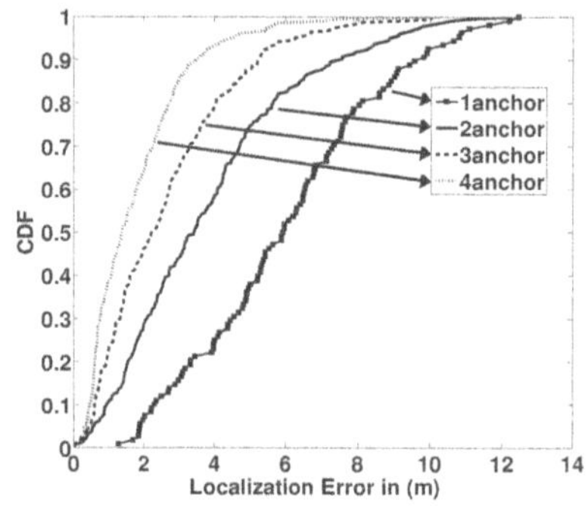

Figure 15: Performance of localization with speaker-only anchors

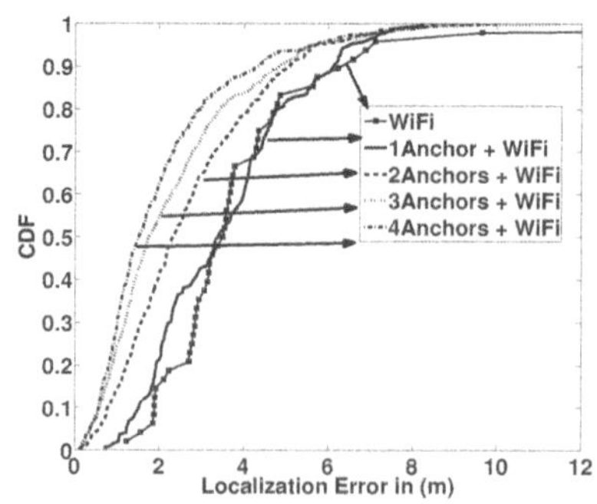

Figure 16: Performance of localization with WiFi and speaker-only anchors

anchors. However, most desktop PCs are typically equipped only with a speaker but no microphone. These desktop PCs however can aid in localization by measuring acoustic distance difference measurements using DeafBeep. In this section we first ask the question, *"How much can WiFi localization benefit from distance difference measurements?"*

In order to answer this question, we use the setup depicted in Figure 17. Since at least two laptops are required to measure distance difference, our set up comprises of two laptops and N ($N = 1, \cdots, 5$) desktop devices equipped with only speakers. We evaluated this set up in 41 different configurations spread across the entire office floor.

First, we quantify the performance of distance difference localization by itself, by locating the laptops using distance difference measurements only. The distribution of localization errors for different number of anchors is depicted in Figure 15. As the number of anchors increases from 1 to 5, the 50%ile and 80%ile errors localization error decreases from 6m to 1m and 8m to 2m respectively. Next, we use Centaur to combine WiFi measurements at the laptops with the distance difference measurements. As seen from Figure 16, using distance difference information from 2 or more anchors, WiFi localization benefits significantly. While the 50%ile and 80%ile errors for WiFi itself are 3.6m and 4.4m respectively, they decrease to 1m and 2.4m with the help of 4 speaker-only anchors. Another interesting observation comparing Figures 15 and 16 is the fact that even acoustic localization benefits from WiFi measurements. The errors of using 1,2 and 3, anchors without aid of WiFi measurements are higher than those that result from using WiFi measurements. For example, the 50%ile error for localization using 3 anchors by themselves is 2.4m and decreases to 1.8m when combined with WiFi. *These experiments indicate that both WiFi localization and acoustic localization benefit in terms of increased localization accuracy by combining each other's information.*

10.4 Exp III: WiFi localization with acoustic ranging to anchors with speaker and mic

In the previous set of experiments anchors were equipped with only speakers and could be used only in measuring distance differences. In these experiments we ask the question, *"How can WiFi and acoustic localization gain from each other if acoustic ranging to anchors was possible?"* In order to answer this question we used the configuration depicted in Figure 18 and tested it at several locations across the office floor.

First, we evaluate the performance of only acoustic localization without WiFi. Figure 19 depicts the distribution of localization errors with increasing number of anchors. As seen from Figure 19,

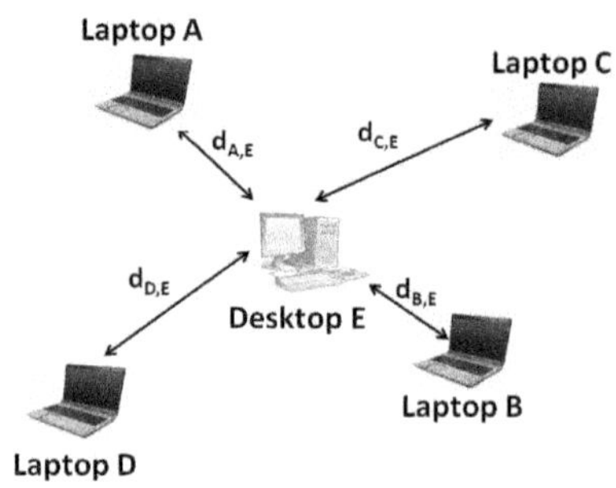

Figure 18: Setup for evaluating fusion of WiFi with ranging to anchor devices

ranging based localization significantly outperforms distance difference based localization or WiFi based localization, with 50%ile and 80%ile errors of 60cm and 80cm respectively. This is expected since acoustic ranging is extremely accurate. Next, we evaluate the performance of Centaur when WiFi and acoustic ranging information are combined. Figure 20 depicts the distribution of localization errors with increasing number of anchors. As seen from Figure 20, there is a significant improvement in WiFi performance with the aid of acoustic ranging even to a single anchor and improves further with increasing number of anchors. However, the performance of the combination of WiFi and acoustic ranging is almost the same as using only acoustic ranging. In fact we found that there was only a 1% decrease in average error by adding WiFi measurements to acoustic range measurements. *Hence, we conclude that given acoustic range measurements to anchors, adding WiFi information does not bring any significant improvement in localization performance.*

10.5 Exp IV: Locating speaker-only devices without any anchors

In some cases it may be required to locate desktop PCs which are equipped with only a speaker but no microphone. If there are laptops equipped with both WiFi cards they can locate themselves and then use distance difference localization to locate the desktop PC. Note that these speaker only devices cannot be localized in any other way without combining WiFi and acoustic measurements. In these experiments we ask the question, *"How accurately can we locate speaker-only devices using devices that are equipped with WiFi-cards, speaker and a microphone?"* The experimental setup for these experiments is the same as depicted in Figure 18 except that, the desktop PC is equipped with a speaker but no microphone and its location is unknown. Thus, the laptops measure distance dif-

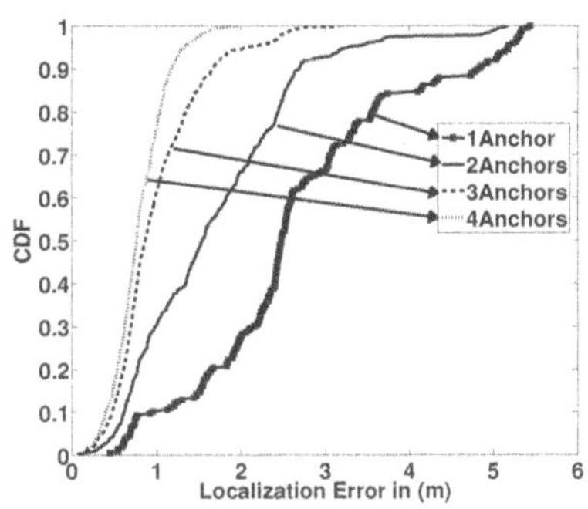

Figure 19: Performance of acoustic localization by ranging to anchors.

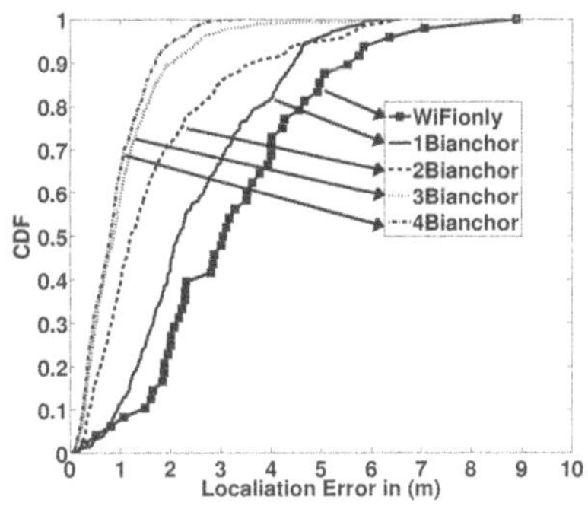

Figure 20: Performance of WiFi + acoustic localization by ranging to anchors

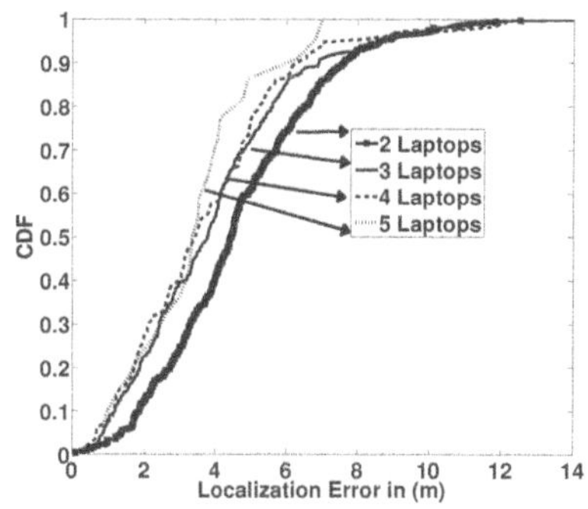

Figure 21: Performance of localizing speaker-only devices without any anchors

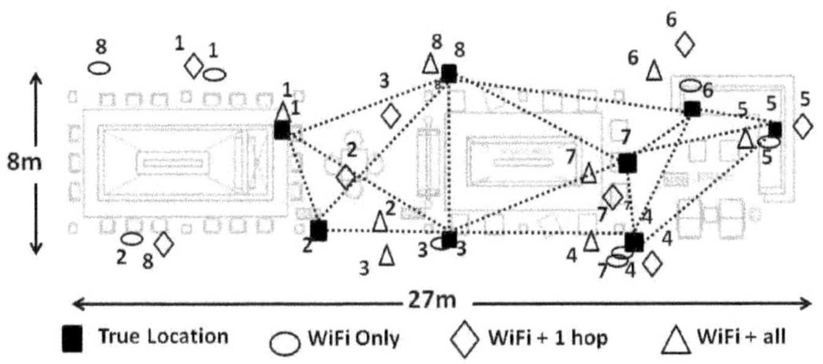

Figure 22: Locating 8 dual-mode devices (laptops), using WiFi only, WiFi coupled with AR with respect to 1-hop neighbors, and WiFi coupled with all available AR information.

ferences to the desktop, while locating themselves using WiFi and acoustic ranging to each other. Figure 21 depicts the distribution of localization errors of the desktop PC with increasing number of laptops in audible proximity. As seen from Figure 21, the 50%ile and 80%ile errors are 2m and 4m respective when the speaker only device has 4 laptops in its audible proximity.

Device	Error in m		
	WiFi Only	WiFi + 1 Hop	WiFi + All
1	4.5	4.0	0.7
2	6.3	2.1	2.9
3	0.7	4.9	3.5
4	0.7	1.1	2.4
5	0.7	1.2	1.6
6	1.3	3.5	2.4
7	4.1	1.5	2.8
8	13.2	10.7	1.4

Table 1: Location estimation errors seen for Experiment V

10.6 Exp V: Locating several dual-mode devices simultaneously

"How will Centaur perform in a realistic setup, where devices are typically distributed in various cubicles of an office floor, not all within acoustic range of each other?" In order to answer this question we conducted an experiment in a realistic setting comprising 8 laptops located across the floor. Figure 22 depicts the true locations of all the laptops (indicated as solid squares). Laptops that are within acoustic range of each other are connected by dotted lines in Figure 22 and are able to range to each other.

In order to evaluate Centaur's performance in this setting, we consider three different scenarios: (i) WiFi-only, where each node is localized only using its own WiFi measurements, (ii) WiFi + 1-hop, where WiFi localization is combined with AR with respect to 1-hop acoustic neighbors, and (iii) WiFi + all, where WiFi localization is combined with all available AR information across the 8 nodes.

Figure 22 depicts the final locations of the devices obtained for each of the three scenarios, while Table 1 provides the exact values of the localization error (in meters). We observe that WiFi-only localization can result in a significant error: 4.5m, 6.3m, and 13.2m, respectively, for nodes 1, 2, and 8. Adding AR constraints, whether from the 1-hop neighborhood or from the entire network, helps reign in these outliers.

In the WiFi+1-hop case, we compute a location estimate for each node separately, by fusing its WiFi location estimate with AR constraints relative to neighboring nodes that it is within acoustic range of. For example, as shown in Figure 22, the 1-hop location estimate for node 8 is computed by fusing the WiFi-based location estimate of 8 with the WiFi-based location estimates of, and the acoustic range relative to, each of its acoustic neighbours: 1, 2, 3, and 7. As the figure shows, these additional constraints, even from the local neighborhood alone, cause the location estimate of node 8 to be pulled in close to the node's true location, with its localization error decreasing from 13.2m in the WiFi-only case to 10.7m in the WiFi + 1-hop case. However, the grossly incorrect location estimate for node 8 coupled with the AR constraint between it and node 3 also causes the latter to be pulled away from its true location (and towards the incorrect location estimate for 8), with the localization error for node 3 increasing from 0.7m in the WiFi-only case to 4.9m in the WiFi + 1-hop case.

When we consider all the WiFi and AR constraints and solve for the locations of all nodes simultaneously (the WiFi + all case), we generally obtain the best results. The large error due to outliers is eliminated (e.g., the localization error for node 8 drops to 1.4m) and no error is worse than about 3.5m (node 3 at 3.5m has the worst error across all nodes).

Thus, Centaur's approach of holistically fusing WiFi and AR information across a set of nodes is beneficial, even when not all nodes are within "earshot" (i.e., acoustic range) of each other.

10.7 Running Time of Centaur

Finally, we try and answer the question, *"How long does Centaur require to perform localization?"*. Table 2 shows the running times for Centaur for various configurations. As described in Section 9, the running time of centaur is $o(e)$ complexity, e being the number of measurements (acoustic as well as WiFi). Table 2 provides the times taken by a Intel Xeon CPU E5405 at 2.0 GHz with 2 processors, 8 GB RAM and a 64 bit operating system in running each experiments I through IV. As seen from Table 2, typical running times for experiments I through III are within a few seconds. However, the running time of experiment IV is a few minutes. Analysis indicated that generating the probability distributions for distance difference constraints in the absence of anchors was the key computation intensive operation in experiment IV. Each distance difference constraint without anchors generates a family of hyperbolae across the floor. Having no analytical means of computing these distributions required us to generate them using sampling techniques that lead to the larger computation time. Experiment V, which corresponds to a realistic scenario, took 15.9s. These are reasonable times to locate devices in an office environment for asset tracking.

Configuration	No Edges	No of Nodes			
		2	3	4	5
Exp I	$N + \frac{N(N-1)}{2}$	2.8s	5.0s	9.0s	15.1s
Exp II	$3 + 2N$	1.2s	1.9s	2.6s	3.7s
Exp III	$1 + N$	2.8s	2.9s	3.4s	5.4s
Exp IV	$2N + \frac{N(N-1)}{2}$	45.8s	80.7s	179s	285s

Table 2: Running times for various experiments

11. CONCLUSION

We have presented Centaur, which provides a novel, Bayesian framework for combining WiFi-based localization with the geometric constraints arising from acoustic ranging. We have also presented two new acoustic techniques — EchoBeep for ranging in non-line-of-sight settings, and DeafBeep for locating speaker-only devices. Through extensive experiments in an office environment, we have shown the effectiveness of EchoBeep and DeafBeep. Further, Centaur was able to eliminate instances of large WiFi based localization error arising from RSS variation.

12. ACKNOWLEDGEMENTS

We thank our shepherd, Moustafa Youssef, and the anonymous reviewers for their constructive comments.

13. REFERENCES

[1] The Berkeley Multimodal Loalization Project. http://mmle.icsi.berkeley.edu/mmle/project.html.

[2] P. Bahl and V. N. Padmanabhan. RADAR: An Inbuilding RF-based User Location and Tracking System. In *INFOCOM*, 2000.

[3] C. M. Bishop. *Pattern Recognition and Machine Learning (Information Science and Statistics)*. Springer, 1 edition, 2007.

[4] G. Borriello, A. Liu, T. Offer, C. Palistrant, and R. Sharp. WALRUS: Wireless Acoustic Location with Room-Level Resolution using Ultrasound. In *MobiSys*, 2005.

[5] Y.-C. Cheng, Y. Chawathe, A. LaMarca, and J. Krumm. Accuracy Characterization for Metropolitan-scaleWi-Fi Localization. In *MobiSys*, 2005.

[6] K. Chintalapudi, A. P. Iyer, and V. N. Padmanabhan. Indoor Localization Without the Pain. In *Mobicom*, 2010.

[7] A. Goswami, L. E. Ortiz, and S. R. Das. WiGEM : A Learning-Based Approach for Indoor Localization. In *CoNEXT*, 2011.

[8] A. Haeberlen, E. Flannery, A. M. Ladd, A. Rudys, D. S. Wallach, and L. E. Kavraki. Practical Robust Localization over Large-Scale 802.11 Wireless Networks. In *MobiCom*, 2004.

[9] A. Harter, A. Hopper, P. Steggles, A. Ward, and P. Webster. The Anatomy of a Context-Aware Application. In *Mobicom*, 1999.

[10] D. Madigan, E. E., R. P. Martin, W. Ju, P. Krishnan, and A. Krishnakumar. Bayesian Indoor Positioning Systems. In *Infocom*, 2005.

[11] C. Peng, G. Shen, Z. Han, Y. Zhang, Y. Li, and K. Tan. A beepbeep ranging system on mobile phones. In *SenSys*, 2007.

[12] N. B. Priyantha, A. Chakraborty, and H. Balakrishnan. The Cricket Location-Support System. In *MobiCom*, 2000.

[13] S. Tarzia, P. Dinda, R. Dick, and G. Memik. Indoor Localization without Infrastructure using the Acoustic Background Spectrum. In *MobiSys*, 2011.

[14] O. Vinyals, E. Martin, and G. Friedland. Multimodal Indoor Localization: An Audio-Wireless-based Approach. In *IEEE Semantic Ccmputing*, 2010.

[15] A. Ward, A. Jones, and A. Hopper. A New Location Technique for the Active Office. *IEEE Per. Comm.*, 4(5):42–47, 1997.

[16] M. Youssef and A. Agrawala. The Horus WLAN Location Determination System. In *MobiSys*, 2005.

APPENDIX

A. DERIVATION OF DEAFBEEP

Suppose that devices A and B are equipped with a microphone and a speaker while device C is equipped with only a speaker but no microphone. As depicted in Figure 23, the microphones of devices A and B recording samples at times t_0^A and t_0^B respectively. Each of the three devices now takes turns to transmit an acoustic chirp signal at times t_1^A, t_1^B and t_1^C respectively. As depicted in Figure 23 device X receives this signal from device Y at time t_Y^X (the time of arrival is determined using EchoBeep described in Section 4). Let the distance between between devices X and Y be d_{XY} (d_{XX} is the distance between the speaker and microphone of the same device X). Further, let the speed of sound be V. Then,

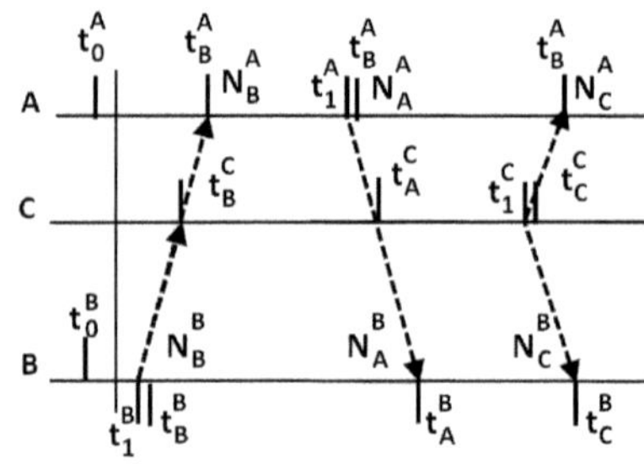

Figure 23: Derivation of DeafBeep

$$N_Y^X = S\left(t_1^X - t_0^X + \frac{d_{XY}}{V}\right) \tag{8}$$

In Equation 8, $S = \frac{1}{F}$ is the sampling rate of the sound card in samples/sec and F its sampling frequency in Hertz. From Equation 8, we obtain,

$$N_Y^A - N_Y^B = S\left[\frac{d_{AY} - d_{BY}}{V} + \left(t_1^A - t_0^A\right) - \left(t_1^B - t_0^B\right)\right] \tag{9}$$

From Equation 9 we can deduce that,

$$\begin{aligned}\left[N_C^A - N_C^B\right] - \tfrac{1}{2}\left[\left(N_A^A - N_A^B\right) + \left(N_A^A - N_A^B\right)\right] \\ = \tfrac{S}{V}\left[\left(d_{AC} - d_{BC}\right) + \left(d_{AA} - d_{BB}\right)\right]\end{aligned} \tag{10}$$

Typically, d_{AA} and d_{BB} will each be very small (in the order of a few centimeters). Further, if device A and B are identical, $d_{AA} = d_{BB}$. Consequently, for all practical scenarios, $d_{AA} - d_{BB}$ will be very small if not almost zero. Equation 5 in Section 5 thus, directly follows from Equation 10.

Zee: Zero-Effort Crowdsourcing for Indoor Localization

Anshul Rai†, Krishna Kant Chintalapudi†, Venkata N. Padmanabhan†, Rijurekha Sen‡*
†Microsoft Research India ‡Indian Institute of Technology, Bombay

ABSTRACT

Radio Frequency (RF) fingerprinting, based on WiFi or cellular signals, has been a popular approach to indoor localization. However, its adoption in the real world has been stymied by the need for site-specific calibration, i.e., the creation of a training data set comprising WiFi measurements at known locations in the space of interest. While efforts have been made to reduce this calibration effort using modeling, the need for measurements from known locations still remains a bottleneck. In this paper, we present Zee – a system that makes the calibration zero-effort, by enabling training data to be crowdsourced without any explicit effort on the part of users.

Zee leverages the inertial sensors (e.g., accelerometer, compass, gyroscope) present in the mobile devices such as smartphones carried by users, to track them as they traverse an indoor environment, while simultaneously performing WiFi scans. Zee is designed to run in the background on a device without requiring any explicit user participation. The only site-specific input that Zee depends on is a map showing the pathways (e.g., hallways) and barriers (e.g., walls). A significant challenge that Zee surmounts is to track users without any a priori, user-specific knowledge such as the user's initial location, stride-length, or phone placement. Zee employs a suite of novel techniques to infer location over time: (a) *placement-independent step counting and orientation estimation*, (b) *augmented particle filtering* to simultaneously estimate location and user-specific walk characteristics such as the stride length, (c) *back propagation* to go back and improve the accuracy of localization in the past, and (d) *WiFi-based particle initialization* to enable faster convergence. We present an evaluation of Zee in a large office building.

Categories and Subject Descriptors

C.2.m [**Computer Systems Organization**]: COMPUTER - COMMUNICATION NETWORKS—*Miscellaneous*

Keywords

Indoor localization, WiFi, crowdsourcing, inertial tracking

*The author was an intern at Microsoft Research India during the course of this work.

1. INTRODUCTION

RF fingerprinting of WiFi signals is a popular approach to indoor localization. Typically, there is an initial training or calibration phase during which received signal strength (RSS) measurements from multiple WiFi access points are recorded at known locations. Then, when a device is to be located, RSS measurements from proximate APs are matched against the training data, either deterministically [3] or probabilistically [32], to estimate location.

The need for calibration is a key bottleneck since it is labour-intensive. Further, it needs to be repeated for each new space and also every time there is a significant change in the space (e.g., when new APs are added or existing ones repositioned). While efforts have been made to reduce the calibration effort using RF modeling, these suffer from various limitations, including the need for at least some data from known locations [9], the need for control over the APs and knowledge of their locations [11], and loss in accuracy because measurements are made at fewer points than ideal to save effort. These limitations also come in the way of a crowdsourcing-based approach to training because, for instance, on a mall floor, the locations of APs installed by multiple providers and stores would not be known, and obtaining a GPS lock might not be feasible at any location.

In this paper we enable *zero-effort crowdsourcing* of WiFi measurements in indoor spaces by developing a system called *Zee* (name derived from the first syllable of "zero"). Our vision is of users carrying smartphones who walk around in the indoor space of interest in normal course (e.g., stroll through a mall), with each user traversing a subset of the paths in the space. We do not assume knowledge of where within the space a user walks or even the starting point of the user's walk. As well, we do not assume knowledge of the placement of a user's smartphone, i.e., whether it is in their hand, shirt pocket, bag, or elsewhere, which also means that we do not know the orientation of the phone relative to the user's direction of motion. All of these elements accord well with the needs of crowdsourcing, where little can be assumed about users, and explicit input or other action from users is best avoided.

The only external input that Zee depends on is a map of the indoor space of interest, which we do not view as onerous since a map would be needed anyway for the purposes of location-based applications such as navigation. Armed with just the map, Zee uses WiFi and inertial sensor measurements crowdsourced from the users' smartphones to automatically infer location over time and thereby construct a WiFi training set (i.e., WiFi RSS measurements annotated with location information).

The key idea behind the automatic inferencing of location in Zee is to combine the sensor information with the constraints imposed by the map (e.g., that a user cannot walk through a wall or other barrier marked on the map), thereby filtering out infeasible locations over time and converging on the true location. As an example, the inertial sensors such as accelerometer and compass might indicate that the user walked in a zigzag path, taking a certain number of steps in a certain (unknown) direction, then turning 90 ° to the right and continuing the walk, and finally turning 90 ° to the left to take a few more paces and then stopping. While the above information does not, by itself, reveal location, it could when viewed together with a floor map. For instance, the map might indicate that there is only one pathway on the floor that could accommodate the kind of zigzag trajectory observed, say the path from the entrance of a mall to a particular store. Thus, at the conclusion of the walk, we can infer that the user's ending location must be the store, and then we can trace back and infer that their starting location must have been the entrance.

To codify the above intuition in Zee, we incorporate the uncertainty arising from sensing and the constraints imposed by the map, into a novel *augmented particle filtering* framework. Whereas particle filtering in the context of localization has typically used particles only to represent the uncertainty in location, we create multi-dimensional particles that also incorporate the uncertainty in other aspects such as the stride length of a user and their direction of walk. Augmented particle filtering then enables the estimation of these latter variables concurrently with the estimation of location. To speed up the convergence of particle filtering, we use two techniques to estimate better priors for the variables being estimated: *placement-independent motion estimation* to estimate the step count and the approximate orientation (or heading offset) of a device relative to the direction of walk, and *WiFi-based initialization* to leverage partial WiFi information to make an initial guess of the location(s) where a device might be. Finally, since the uncertainty in location would tend to reduce with time as a user takes a longer walk with more turns, we use *backward belief propagation* to take advantage of the greater certainty in location at a later point in time to trace back and reduce uncertainty in location at earlier times, post facto.

Concurrently with estimating location, Zee performs WiFi scans and records the results indexed by time. As and when the location estimate for a particular time becomes available, the corresponding WiFi measurement is annotated with the estimated location, thereby adding a record to the WiFi training set. Thus, Zee provides a way to crowdsource WiFi measurements without requiring any explicit effort on the part of users. To evaluate the quality of the crowdsourced training data set, we feed it into Horus [32], a well-known WiFi fingerprinting-based localization technique, and EZ [9], a newer modeling-based technique. We find that with the crowdsourced training data set, Horus and EZ achieve a median localization error of about 3m, which is comparable to the median localization error of 3.5m achieved with a training data set that is explicitly measured.

Thus, Zee offers a truly zero-effort solution for crowdsourcing WiFi data for the purpose of indoor localization, by leveraging the walks that users take through the space of interest in normal course. We view this as a significant contribution of our work, one that could be a key enabler of WiFi-based localization in real-world settings. As well as providing a way of constructing a training data set for later use, another key contribution of Zee is a way to perform accurate tracking of a walking user for the purposes of real-time applications such as indoor navigation.

2. BACKGROUND AND RELATED WORK

Zee draws on prior work in multiple areas, chiefly WiFi-based localization, robotic navigation, and inertial sensing.

2.1 Infrastructure-Based Localization Systems

Early systems required the deployment of special-purpose infrastructure in the indoor space to enable localization. The inability to use GPS indoors has led to myriad approaches based on alternative signals, ranging from infrared [26] to acoustic [27, 19] and visual [28]. There have also been localization systems based on a deployment of RF transmitters and sniffers [15] or RFID [18]. While each of these approaches offers certain advantages (e.g., high accuracy in the case of acoustic ranging), the need for special-purpose hardware and infrastructure is a significant challenge.

2.2 RF Fingerprinting based Localization

Localization based on measuring the RF signal of a wireless LAN has the significant cost advantage of leveraging an existing infrastructure. A popular approach, pioneered by Radar [3], is to employ received signal strength (RSS) based fingerprinting of locations in the space of interest, where typically multiple access points (APs) are heard at each location. While Radar used a simple, deterministic fingerprinting and matching scheme, Horus [32] developed a more sophisticated and accurate approach wherein the RSS measurements corresponding to each location and AP are represented as a probability distribution and matching is performed using the maximum likelihood criterion. SurroundSense [2] extends this idea and builds a map using several features found in typical indoor spaces such as ambient sound, light, color, etc., in addition to WiFi RSS. Several other improvements over and extensions of the basic RF fingerprinting based localization have been proposed, such as the incorporation of mobility constraints [12] and an extension to outdoor settings [8].

The above approaches depend on calibration of the space of interest to construct a training data set comprising RSS measurements at *known* locations. Such calibration tends to be onerous, more so because it has to be repeated for every new space and each time there is a significant change in a given space (e.g., a change in AP placement). Zee is aimed squarely at eliminating the need for such explicit calibration effort.

2.3 Modeling instead of Calibration

An alternative to empirical calibration is to use an RF propagation model to estimate the RSS, P_x, at a given location x based on the the transmit power, P_0, and the distance, d_{0x} between the transmitter and the location x. A popular model is the log-distance path-loss model, which models the RSS (in dBm) as $P_x = P_o - \gamma log(d_{0x}) + N$, where N represents a noise term [20]. Extensions of this model have incorporated the presence of and the attenuation due to obstructions such as walls.

Radar [3] included a model-based variant, which estimated RSS at various locations using knowledge of the AP locations and transmit powers, and a floor map. Radar also proposed the idea of RF environment profiling [4], where measurements between the (stationary) access points is used to characterize the changing RF environment. This idea was leveraged in [17], to develop a zero-configuration localization system, where APs make RSS measurements, with respect to clients and also each other. The measurements made between the APs are used to construct a model that maps RSS to distance. This model is then used locate clients through trilateration. This scheme, however, requires the AP software to be modified and so cannot make use of off-the-shelf APs.

More recently, efforts have been made to use modeling with minimal assumptions. EZ [9] only requires measurements at a few known client locations and WiGem [11] only requires knowledge of AP locations and the ability to measure the client signals as received at the APs. The reduced measurement effort with a modeling based approach typically comes at the cost of reduced accuracy.

Zee avoids the need for measurement at any known location or any knowledge or control over AP locations and measurements. This, we believe, makes it much more amenable to a crowdsourcing approach because, for instance, a public space such as a mall might have APs deployed by multiple providers and moreover the absence of GPS coverage might make it challenging to make any measurements at all at known locations. Equally importantly, Zee is able to avoid the loss of accuracy inherent in modeling based on limited data. This is particularly relevant in the context of accurate but measurement-intensive techniques such as Horus [32].

2.4 Alternatives to RSS-based Localization

Several alternatives to RSS have also been considered. In particular, detailed physical layer information has been used to fingerprint both devices [7, 5] and locations [23]. We view the contribution of Zee as being orthogonal to these; the above non-RSS-based approaches also require a training set to be constructed and could benefit from the zero-effort calibration made possible by Zee.

On the other hand, there has also been research on leveraging non-RSS information from RF beacons in a way that does not require any calibration. For instance, [13] simply estimates a client's location as the centroid of the known locations of the APs heard, without regard to the RSS, which leads to loss in accuracy. There has also been work on leveraging time of flight [1] or angle of arrival [31, 22] relative to APs, whose locations are assumed to be known. However, these approaches either require specialized and potentially expensive hardware [1, 31] or require special human effort, e.g., taking a spin while walking [22]. In Zee, we avoid these disadvantages, making our approach more amenable to crowdsourcing albeit with reduced accuracy compared to the approaches based on precise measurement.

2.5 Robotic Navigation

In the robotics community, there has been much work on the Simultaneous Localization and Mapping (SLAM) problem, which dates from the mid-1980s [24, 16]. A robot equipped with sensors, such as laser-based ranging and cameras, is assumed to be exploring the space of interest, e.g., an unmapped building. The space is assumed to have landmarks, which are typically artificially inserted (e.g., barcode pasted on walls or a particular pattern painted on the ceiling). The "mapping" problem is to determine the locations of the landmarks relative to each other whereas the "localization" problem is to determine the location of the robot relative to the landmarks. Estimates are updated based on an action model, which helps relate the new location to the previous location (e.g., based on the number of revolutions of the robot's wheels), as well as sensor data (e.g., visual landmarks captured by a camera sensor).

While the early work on SLAM used Kalman Filters for estimation, subsequent work has been based on Markov localization, which is a better match in practice since it allows the robot's position to be modeled as multi-modal and non-Gaussian probability density functions. Of particular interest to us is the Monte Carlo Localization (MCL), or *particle filtering* based, approach [10], wherein the idea is to represent the belief in the robot's location as weighted random samples, or particles. The particles are then evolved based on the action model and the sensor readings.

In Zee, we build on the key idea of particle filtering by augmenting particles to incorporate not only location but also other unknowns such as the stride length of a particular user. However, in other respects, Zee differs from the SLAM approach. In Zee, it is humans, not robots, that are moving around in the space of interest. This means that we have to contend with all of the complexities associated with human locomotion (e.g., measuring walking is much more complex than odometry based on wheel revolutions). Also, Zee does not depend on additional sensors since these are either not present in typical consumer devices (e.g., laser ranging) or even if present are not amenable to use in a crowdsourcing setting (e.g., users walking through a mall would not normally be taking pictures with their smartphone camera). On the other hand, unlike SLAM, Zee assumes the availability of a map and leverages the constraints imposed by the map for the purposes of localization.

2.6 Inertial Sensing

Finally, we discuss the problem of inertial sensing of human motion for the purposes of localization. This approach is interesting because consumer mobile devices such as smartphones are increasingly being equipped with sensors such as magnetometer (or compass), accelerometer, gyroscope, and barometer. These sensors, respectively, enable measurement of direction, acceleration, rotational velocity, and altitude. Knowing the starting location, a device can, in principle, be tracked using dead-reckoning, wherein the inertial sensor measurements are integrated over time [6]. However, a significant challenge is that even small errors in inertial sensing could be magnified by integration. Zee addresses this problem by leveraging the constraints imposed by the map to filter out erroneous measurements (e.g., a measurement that has a particle passing through an obstruction such as a wall would be filtered out). A complementary approach to prevent the accumulation of errors is proposed in UnLoc [25], where virtual landmarks are created using existing sensing modalities such as WiFi.

Another significant challenge is the complexity of human locomotion. People tend to have different stride lengths. Although there has been work on estimating stride length based on careful modeling of a step using accelerometer data [14], such estimation tends to be sensitive to the placement of the sensor, as does other estimation such as step counting. Indeed, the accelerometer signal tends to be strongest and most distinctive when the sensor is foot-mounted [21], but this does not accord with the typical placement of a device such as a smartphone. Also, in general, the device could be in an arbitrary orientation relative to the user's body and their direction of motion, which also makes estimation challenging.

Some of the above challenges have been addressed in [29, 30]. The use of map constraints for particle filtering, WiFi-based initialization of particles, and augmenting the particles with orientation information are all aspects inherited by Zee. In addition, this prior work also considers transition between floors, which Zee does not in its present incarnation. However, the prior work assumes a foot-mounted IMU that reports step events, the stride length, and the change in heading. These assumptions are problematic in a smartphone-based sensing context, for the reasons noted above.

To avoid assumptions regarding the placement of the IMU or processed step information being returned by it, we employ two techniques in Zee. First, rather than working with sensing thresholds or careful modeling that tend to be placement-dependent, we leverage the fundamental, placement-independent property of walking — its repetitiveness and hence periodicity — to estimate unknowns such as the step count. Second, we augment particle filtering to also estimate unknowns such as the stride length based, in part, on the constraints imposed by the map. In addition, Zee employs a novel backward belief propagation technique to trace back in time and infer location information post facto, thereby enabling more effective WiFi crowdsourcing.

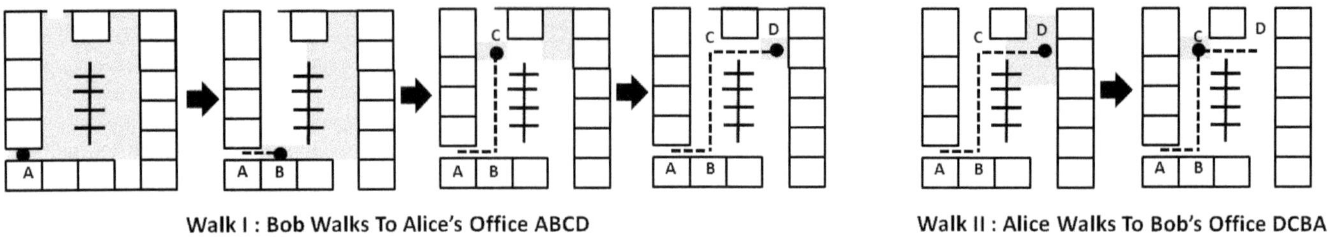

Figure 1: Example Scenario

3. ZEE EXAMPLE SCENARIO

All WiFi-based indoor localization schemes require a training data set — a set of tuples (WiFi measurement,location) of WiFi RSS measurements annotated with the indoor locations where the measurements were made. *The goal of Zee is to enable smartphone-based crowdsourcing of this training data set without requiring any active user participation.* In fact, Zee can run as a background process on smart phones without affecting the user in any way. In this section, we walk through an example scenario to provide an overview of and intuition for how Zee achieves zero-effort crowdsourcing.

Inferring a user's location. Bob is sitting in his office at work at location A (Figure 1). He downloads the Zee client and continues with his work as usual. At some point he decides to walk to Alice's office at location D. Initially, Zee does not know where Bob is located and hence Zee initializes Bob's locations as a probability distribution uniformly across the entire floor, as depicted by the gray region corresponding to Walk I in Figure 1). To walk to Alice's office, Bob takes the path ABCD indicated in Figure 1.

As Bob traverses this path, Zee uses the accelerometer, compass and the gyroscope on Bob's smartphone to continuously infer the direction and distance walked by him. Then using the floorplan of the indoor space, Zee updates the probability distribution of Bob's location by eliminating possibilities that would require him to violate the physical constraints imposed by the floorplan, such as walking through walls. Thus, as Bob reaches point B and then goes on to point C (as depicted in Figure 1), the spread of possible locations that Bob can be at, shrinks. Eventually, when Bob reaches D, Zee is able to narrow down the possible locations of Bob to his correct location. The key reason for this convergence is that there is only one possible path in the shape of ABCD that can be accommodate within the indoor space. *Thus, even without knowing Bob's initial location, but by simply tracking his movements, Zee is able to eliminate all alternative possibilities and eventually determine his location.*

Backward belief propagation. At this point, having narrowed down Bob's location using the sequence of his movements, Zee traces back the entire path taken by Bob and infers post facto that he must have taken the path ABCD. *Thus, Zee can also trace the entire history of locations at which Bob was present.*

Recording WiFi measurements. As Bob was traversing the path from A to D, Zee was also periodically scanning for proximate WiFi Access Points (APs) and recording the Received Signal Strength (RSS) from these APs. Knowing the entire path ABCD taken by Bob, Zee can associate with each RSS measurement the corresponding location on the path ABCD where the measurement must have been made. Thus, Zee obtains its first set of tuples $< RSS\ Measurements, Locations >$. Also, from this point on, having located Bob, Zee can track Bob's future movements and hence his locations. As more WiFi measurements are made, Zee on Bob's phone obtains more location-annotated WiFi measurements over the rest of the day. *Zee is thus is able to obtain location-annotated WiFi measurements from Bob's walks, without having any a priori knowledge of his initial location.*

Using past WiFi measurements to locate subsequent users. At this point Alice learns about Zee from Bob and installs it in her phone. She then decides to walk to Bob's office from her office along the path DCBA. This time, however, the Zee server has obtained some WiFi measurements from Bob's walk. Thus, Zee first performs a WiFi scan and obtains RSS measurements from proximate APs. Rather than initializing Alice's locations uniformly across the entire floor (as in Walk I in Figure 1), Zee uses these RSS measurements and the database to obtain a confined probability distribution, as depicted by the gray region corresponding to Walk II in Figure 1. In other words, the WiFi database obtained from Bob's walk helps Zee to narrow down the possibilities for Alice's initial location. Now as Alice walks, her location estimate converges much more quickly to her true location (by the time she reaches C in Figure 1) than it did in Bob's case. *Thus, each new walk in Zee benefits from the WiFi measurements accumulated from prior walks and in turn benefits the localization of future walks. In fact, after enough walks, Zee will be able to accurately locate a new user simply from a WiFi scan, using WiFi-based localization.*

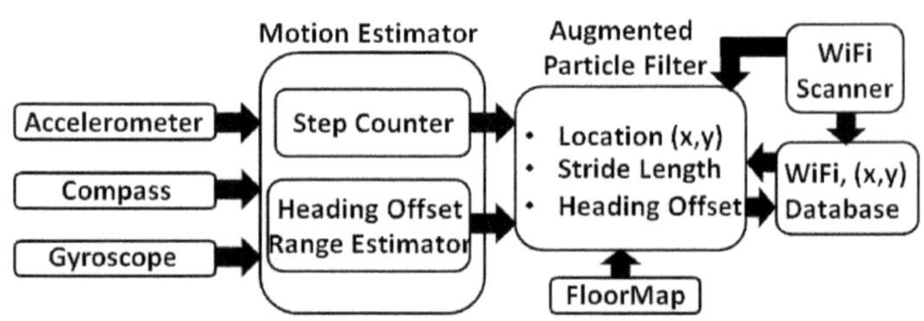

Figure 2: Zee Architecture

4. ZEE ARCHITECTURE

Figure 2 depicts a pictorial overview of Zee's architecture. There are two key components in Zee: *Placement Independent Motion Estimator* (PIME) and *Augmented Particle Filter* (APF). PIME uses mobile sensors such as the accelerometer, compass, and gyroscope to estimate the user's motion. The APF uses the motion estimates from PIME and the floormap as input to track the user's location on the floor.

Placement Independent Motion Estimator (PIME). PIME uses the accelerometer, compass and gyroscope data to perform three key functions. First, it reliably determines whether or not the person is walking. Second, when the user is walking, it generates an event each time a step occurs and reports this event to the APF. Third, it provides APF with a rough estimate of the *Heading Offset (HO)*, i.e., the angle between the orientation of the phone and the user's direction of motion. The heading offset arises due to the combination of two reasons: (i) in general, the phone might not be oriented along the direction of the user's walk. For example, the user might be walking along the north-south direction while holding the phone laterally, so that it points east-west, which is a common scenario when phone users watch video while walking, and (ii) the presence of magnetic materials often affects the phone's compass. The APF then starts with the rough HO information and refines it to arrive at an accurate HO estimate.

A key feature of PIME is that it is independent of device placement, i.e., *whether the user is carrying the phone in his shirt pocket, trouser pocket, hand,* etc. This is a design requirement for Zee since it must be able to crowdsource in the background without any active user participation. To achieve this, Zee includes novel techniques for placement independent step detection (described in Section 5) and heading offset range estimation (described in Section 6).

Augmented Particle Filter. The key function of the APF is to track the probability distribution of a user's location as he/she walks on the floor. In order to convert steps into distance, the APF needs to estimate the stride length of the user, i.e., the distance traversed per step. Further, to track the user's location, estimating the direction of walking is crucial. While compass measurements provide the phone's orientation, the APF also estimates the HO to correctly compute the user's direction of motion. To simultaneously estimate location, stride length, and heading offset, the APF takes a novel approach – it maintains a four-dimensional joint probability distribution function in the form of a particle filter, comprising 2D location, stride length and, HO, and learns all these values as the user walks on the floor. The APF also implements backward-belief propagation, as described in Section 3.

Creating the WiFi Database. The APF runs belief back-propagation to correct the user's path history. This yields a time-indexed sequence of the user's estimated location during their walk. These location estimates are used to annotate the time-indexed WiFi information that Zee obtains through periodic scans. Thus, Zee's WiFi database comprises WiFi measurement annotated with location information.

WiFi-based initialization in APF. After the WiFi database has been initialized using the data from the first user, for subsequent users, APF uses information from WiFi scans and the current WiFi database, to obtain a confined initial location distribution, as described in Section 8. Confining the initial distribution instead of spreading it uniformly across the floor, enables a quicker convergence to the actual location, as noted in Section 3.

Refinement of the WiFi database. The training data obtained from each subsequent walk is in turn used to refine the existing WiFi database, thus making it more accurate for the next walk. In this manner, both APF-based user tracking and WiFi-based localization work in tandem, benefitting each other and progressively helping refine the WiFi database through crowdsourcing.

5. COUNTING STEPS

As described in Section 4, Zee uses step counting to estimate the distance traversed by the user. There are two tasks for any step counting algorithm: first, to reliably ascertain that the user is indeed walking, and second, to count the number of steps. Zee's step counting algorithm is designed to function irrespective of device placement *i.e.,* how the device is carried by the user.

Typical mobile phone placement scenarios. In order to find out how people typically carry their phones we interviewed 30 employees in an office. Based on our interviews we found that while most men carry their phones in front pant pockets, some carry in their shirt pockets or rear pant pockets and pouches attached to their belt. Women most often carry them in their hands or in handbags and sometimes in pant pockets. Further, both men and women may hold the phone in their hand while using it or to their ear while talking. In designing and evaluating our schemes we used these inputs as a guide.

Data collection. In order to test our schemes, six different people (four men and two women) were given smartphones to collect accelerometer data. Men collected data for five different placements: shirt pocket, pant front pocket, pant rear pocket, in hand while not using the phone, and in hand while using the phone. Women collected data for three different placements: handbag, in hand while using the phone and in hand while not using the phone. For each of these scenarios, data was collected when the user was not walking as well as when the user was walking.

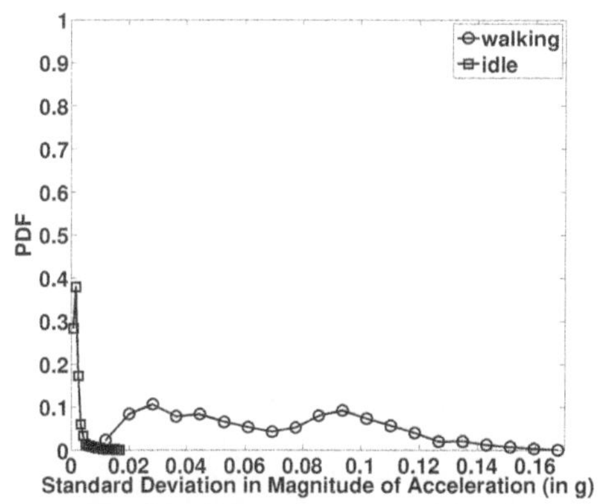

Figure 3: Distribution of standard deviation of acceleration magnitudes during idle and walking states.

Idle versus Motion. When the user is idle, it is expected that their phone will register little acceleration. Thus, the standard deviation in the magnitude of acceleration would be a good indicator of whether there is any movement. Figure 3 depicts the probability density function (PDF) of the standard deviation in the magnitude of acceleration experienced by the phone over a 1s period, in both idle and walking scenarios. The PDF was constructed from a total of 12500 data points for idle and 17000 points for walking. As seen from Figure 3, while the standard deviation is under 0.01g with 99% probability when the user is idle, it is over 0.01g with almost 100% probability when the user is walking. Using the standard deviation by itself, however, is not sufficient to ascertain that the user is walking. For example, sudden movements made by users while they are idle (*e.g.,*hand gestures, turning or shifting in the chair *etc.*) could also result in a large acceleration. To distinguish walking from such sudden movements, we use a novel scheme that exploits a very fundamental property of walking, namely its repetitive nature.

Repetitive nature of walks. Figure 4 depicts the acceleration values seen along the three axes by a mobile phone carried by two different users, one carrying it in his shirt pocket and the other in his front pant pocket. As seen from Figure 4, the acceleration along each axis and in each case exhibits a very repetitive pattern. The repetitive pattern arises because of the rhythmic nature of walking, with a sequence of two steps (one left and one right) constituting the period. While these patterns may be very different across users and placements, *the acceleration pattern for a given user with a*

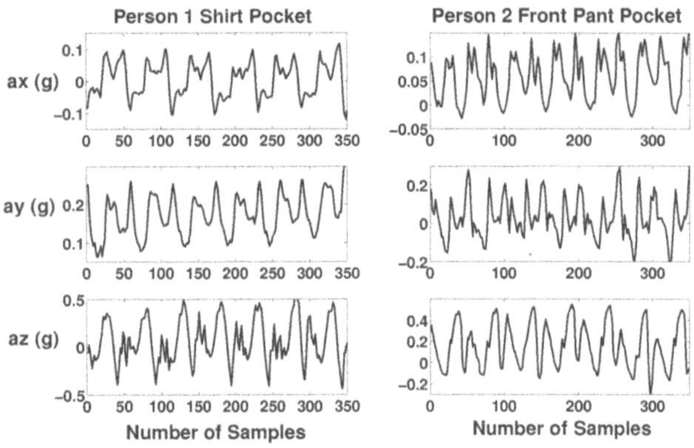

Figure 4: Walking is repetitive

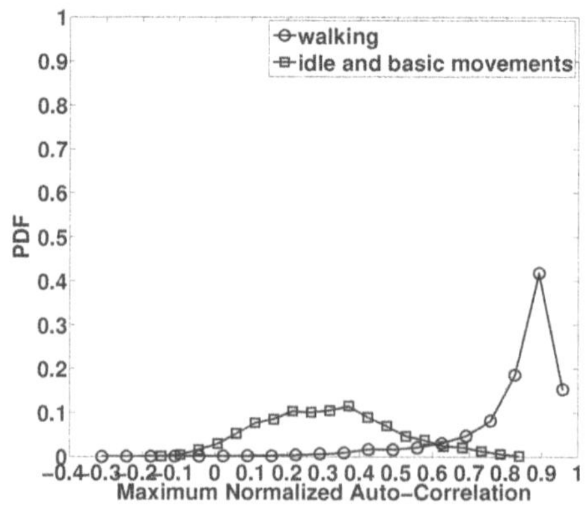

Figure 5: Distribution of Maximum Normalized Auto-Correlation during idle and walking states.

particular device placement repeats. Zee uses this observation to not only count steps but also to ascertain that the user is indeed walking.

Normalized Auto-correlation based Step Counting (NASC). The intuition behind NASC is that if the user is walking, then the auto-correlation will spike at the correct periodicity of the walker. Thus, given an acceleration signal $a(n)$, Zee computes the normalized auto-correlation for lag τ at the mth sample as,

$$\chi(m,\tau) = \frac{\sum_{k=0}^{k=\tau-1}\left[\begin{array}{c}(a(m+k)-\mu(m,\tau))\\(a(m+k+\tau)-\mu(m+\tau,\tau))\end{array}\right]}{\tau\sigma(m,\tau)\sigma(m+\tau,\tau)} \quad (1)$$

In Eqn 1, $\mu(k,\tau)$ and $\sigma(k,\tau)$ are the mean and standard deviation of the sequence of samples $< a(k), a(k+1), \cdots a(k+\tau-1) >$.

When the person is walking and τ is exactly equal to the period of the acceleration pattern, the normalized auto-correlation will be close to one. Since, the value of τ is not known a priori, NASC tries values of τ between τ_{min} and τ_{max} to find the value of τ for which $\chi(m,\tau)$ becomes maximum. Thus,

$$\psi(m) = max_{\tau=\tau_{min}}^{\tau=\tau_{max}}\left(\chi(m,\tau)\right). \quad (2)$$

$\psi(m)$, the maximum normalized auto-correlation, simultaneously provides two pieces of information. A high value (close to 1) suggests that the person is walking and the corresponding value of $\tau = \tau_{opt}$ gives the periodicity of the person's walk.

Since the sampling frequency of our accelerometer was 50Hz, two step duration of most people lies between 40 and 100 samples. Consequently, in our implementation, the initial search window (τ_{min}, τ_{max}) is set to $(40, 100)$. However, once the periodicity of the person's walk is found to be τ_{opt}, the search window is reduced to a few samples around τ_{opt}. In our implementation, after finding the user's periodicity, we used $\tau_{min} = \tau_{opt} - 10$ and $\tau_{max} = \tau_{opt} + 10$. NASC continuously updates the value of τ_{opt} to account for small changes in the user's walking pace.

Figure 5 depicts the distribution of $\psi(m)$ for idle and walking states. The idle state included movements such as hand gestures, transition from sitting to standing and vice versa, and spinning in a chair. As seen from Figure 5, when ψ is 0.7 or higher, the probability that the person is idle is extremely low (less than 1%). Note that NASC will also detect other repetitive activity such as running, but in this paper we did not evaluate such activities.

Walk versus Idle decision in Zee To decide whether or not the user is walking, we use a combination of both standard deviation in the magnitude of acceleration $\sigma_{||a||}$ and the maximum normalized auto-correlation ψ. Zee transitions between the IDLE and WALKING states as follows:

- If $\sigma_{||a||} < 0.01$ then $state = IDLE$.
- Else If $\psi > 0.7$ then $state = WALKING$.
- Else no change in current value of $state$.

Counting steps. Zee uses the periodicity estimated by NASC for step counting. Having estimated τ_{opt}, NASC generates a step occurred event every $\frac{\tau_{opt}}{2}$ samples while the person in the WALKING state.

	Hand While Using	Pant Front Pocket	Pant Back Pocket	Hand Not Using	Shirt Pocket	Hand bag	Over (all)
False +ive	0%	0%	0%	0%	0%	0%	0%
False -ive	2%	0%	0%	0%	0%	0%	0.6%
True +ive	100%	100%	100%	100%	100%	100%	100%
True -ive	98%	100%	100%	100%	100%	100%	99.4%

Table 1: Performance of step counting in Zee

Evaluation of step counting. Table 1 presents the findings from our evaluation of step counting in Zee. A false positive means that an extra step was counted while the user was in idle state, while a false negative means that a step was missed while the user was walking. Table 1 shows that these error rates are very low, often zero, for various placements of the phone across users.

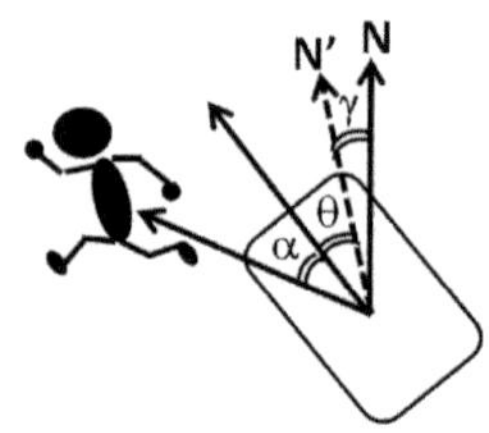

Figure 6: Heading Offset

6. ESTIMATING HEADING OFFSET RANGE

To track the user's path, Zee must estimate the user's direction of walking. The mobile phone's compass provides orientation of the phone relative to the perceived magnetic north (i.e., the angle θ in Figure 6). In general, however, the compass reading might not be aligned with the direction of motion of the user. We refer to this difference between the compass reading and the direction of motion of the user as the *heading offset (HO)*. HO arises due to a combination of two factors: *magnetic offset* and *placement offset*.

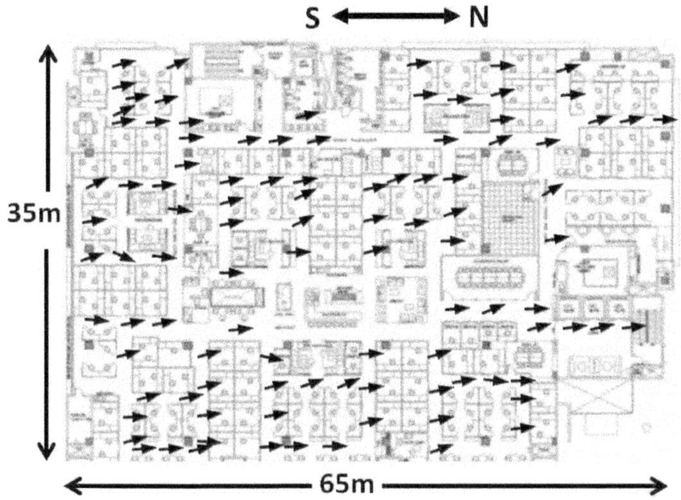

Figure 7: Direction of North as shown by the compass across the floor of a large office

Magnetic Offset: The presence of magnetic materials (e.g., metal) in close proximity of the mobile phone can disturb its perception of North, leading to an offset error in the compass measurement. We refer to this difference between the true north and the north perceived by the compass as the *magnetic offset*. As illustrated in Figure 6, the phone perceives north to be towards N' while the true north is along N. The magnetic offset is depicted as the angle γ in Figure 6. We have found that the magnetic offset is usually a characteristic of a given location, depending on the construction and other materials in the vicinity, and typically remains stable with time. Figure 7 depicts the North direction as measured by a phone's compass at various locations on the floor of a large office building, and Figure 8 shows the distribution of magnetic offset measured at 100 different locations in the floor. The magnetic offset is within $\pm 15°$ in 90% of the locations, and occasionally as high as $30°$.

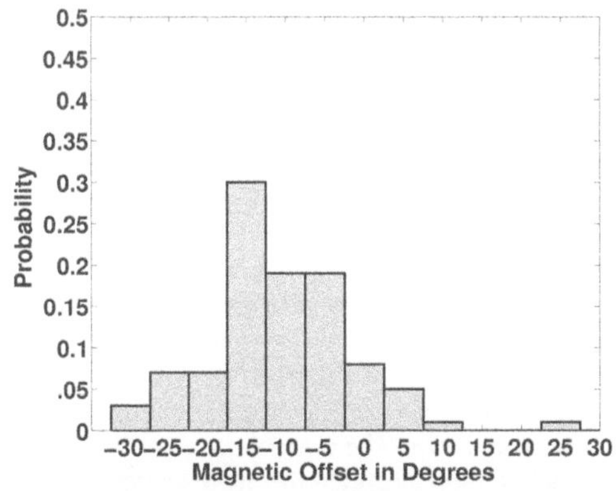

Figure 8: Distribution of magnetic offsets

Placement offset: The phone's compass measures the angle of orientation of the phone with respect to the perceived north. The phone, however, might not be oriented along the direction of motion of the user. For example, when the user walks while watching a video or a photo with the phone held laterally, the orientation of the phone would be at an angle of $90°$ relative to the direction of motion of the user. We refer to this difference between the phone's orientation and the direction of motion of the user as the *placement offset*. In the example of Figure 6, the placement offset is depicted by the angle α. We found that for most placements such as in the shirt or pant pockets, the placement offset is typically $\pm$ $45°$. However, when the phone is placed in a pouch or a handbag, the placement offset can be arbitrary. Nevertheless, the placement offset typically remains unchanged even when the user takes a turn and changes the direction of walking.

Heading Offset: As depicted in the Figure 6, the direction of motion of the user is $\alpha + \theta + \gamma$ relative to the true magnetic north. In other words, the heading offset in this scenario is $\alpha + \gamma$ — the sum of the magnetic offset and the placement offset. While the placement offset is typically constant during a single walk (unless the user changes how he/she is carrying the phone in the middle of the walk), the magnetic offset can change as the user moves through different locations. Consequently, Zee must accommodate both these components of the HO. To estimate HO, Zee takes a two-step approach. It first estimates the HO in a broad range (sector) based on the acceleration experienced by the phone. Then, the APF uses this range of values as the prior for the HO distribution and proceeds to refine its estimate of the HO as the user walks. In the remainder of this section, we describe our technique for estimating the HO range; we defer discussion of the APF to Section 7.

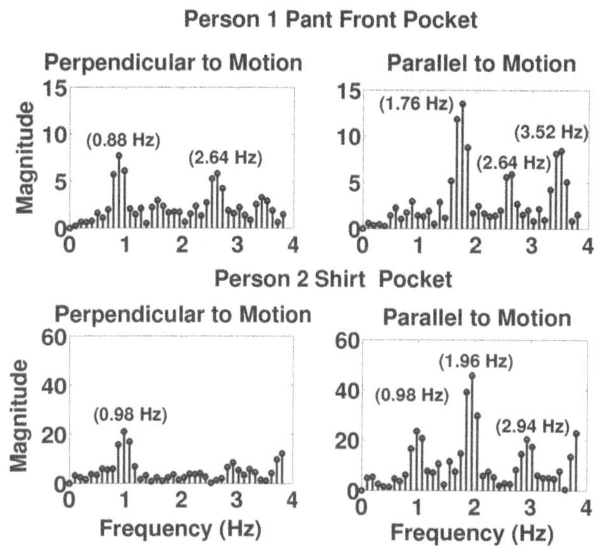

Figure 9: Spectrum of walking

The spectrum of a typical walk. Figure 9 depicts the magnitude of the Fourier transform of the accelerometer signal registered by the phone along directions parallel and perpendicular to the direction of walking, for two different people. Person 1 carries the phone in his shirt pocket while person 2 carries the phone in his front pant pocket. As seen in Figure 9, the dominant frequencies are multiples (harmonics) of a fundamental frequency. For example, in the case of person 1, the fundamental frequency is 0.88 Hz while for the second person it is 0.98 Hz. This fundamental frequency corresponds to the periodicity of two steps, i.e., the combination of a left step and a right step.

After observing the Fourier transform across several users and placements, we discovered an interesting fact: *the second harmonic (two times the fundamental frequency, corresponding to the periodicity of single steps) is either completely absent or is extremely weak in the accelerations experienced by the phone in the direction perpendicular to the user's walk. It is however always present and dominant in the direction parallel to the user's walk.* This rather curious phenomenon arises for the following reason. In the direction parallel to the walk, each step (left or right) registers as a strong, repetitive acceleration signal. However, this is not so in the perpendicular direction, because the user's sideways or lateral movement (e.g., hip sway) only repeats every two steps. Indeed, as can be seen in Figure 9, the second harmonic frequencies corresponding to 1.76 Hz (person 1) and 1.96 Hz (person 2), while being present in the direction parallel to the user's walk, are absent in the perpendicular direction.

Heading Offset (HO) range estimation. Suppose the magnitude of the second harmonic in the Fourier transform along north is F_y and that along west is F_x. Almost the entire contribution to this harmonic is from the direction of walking, so F_x and F_y must be its components. Thus, $\alpha + \theta + \gamma = \arctan \frac{F_x}{F_y}$ (Figure 6). However, since we work with the (unsigned) magnitude of the Fourier transform, we cannot tell whether the person is walking forward or backward along this angle. Consequently, it is equally likely that $\alpha + \theta + \gamma = \arctan \frac{F_x}{F_y} + 180^o$. Knowing θ from the compass then, the two possible values of $\alpha + \gamma$ (HO) can be estimated.

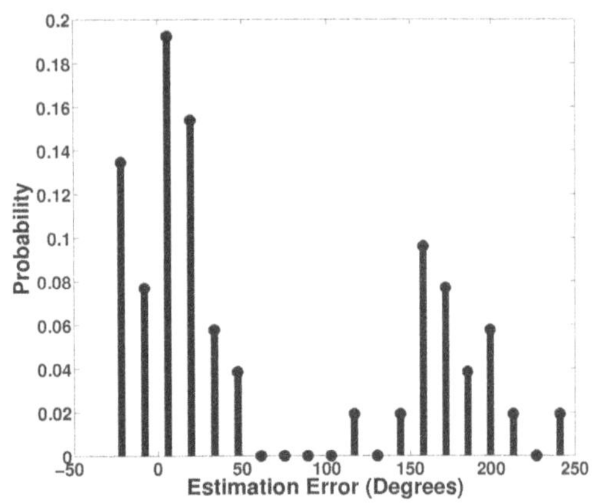

Figure 10: Error distribution of HO estimation

We characterize the error in our HO estimation scheme by testing it on the same walk data collected across six different people, as in Section 5. Figure 10 depicts the probability distribution of the error in the HO estimation. The distribution is bimodal around $0°$ and $180°$, because, as noted above, the HO estimation scheme cannot distinguish between the forward and backward directions. Further, the error spans about $60°$ around each of $0°$ and $180°$. Thus, the error in our HO estimation is large and by itself cannot be used to track the user. Nevertheless, it helps narrow down the possibilities to two diametrically opposite $90°$ sectors (expanded from $60°$ to accommodated the possibility of larger errors) centered about $\arctan \frac{F_x}{F_y}$. The APF then uses this estimate as the prior to efficiently converge on the correct HO, as discussed next.

7. TRACKING USING AUGMENTED PARTICLE FILTER (APF)

In this section we describe how Zee uses APF to track the users' paths as they walk around in an indoor environment.

The key idea. *The key idea of Zee is that as a user continues to walk in an indoor environment, navigating through hallways and turning around corners, the possibilities for the user's path and location shrink progressively.* For example, if a person walks 10m north followed by 12m east and then 20m south in an indoor space with hallways and walls, there will likely be only a few paths in the indoor environment that could accommodate such a walk without having the user run into a wall or other barrier. The longer the walk, the more concentrated the possibilities. As a user walks in the indoor environment while going about their routine, Zee continually eliminates possibilities that violate wall constraints, until eventually only one possibility remains. To drive towards such convergence, Zee maintains a probability distribution of possible locations of the user and updates it for every step taken by the user.

Zee uses the stride length to convert each step into the corresponding distance traversed. However, people generally have different stride lengths, because of differences in height and walking style, so Zee must estimate the stride length of each user, while simultaneously attempting to locate them. Further, as described in Section 6, Zee must also estimate the HO accurately. To accomplish these tasks, Zee augments the standard particle filter to include not only of the location of the user but also the stride length and the HO as two additional unknowns. Thus, Zee continuously maintains and updates a four-dimensional joint probability distribution over the 2D location, stride length, and HO of the user.

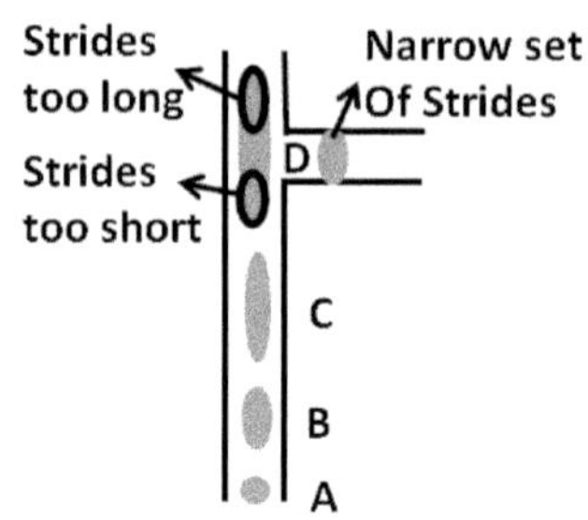

Figure 13: Stride estimation in Zee

Example of stride length estimation. For intuition on how stride length estimation works, consider a simple example, where the user's initial location is assumed to be known but not their stride length. Initially, Zee considers the full range of humanly feasible stride lengths (0.5m to 1.2m, in our current implementation). The user's initial location is shown in Figure 13 as the particle concentration (gray region) labeled A. As the user takes steps and moves to B and then onto C, the gray region becomes elongated. The reason is that since Zee consider a range of possible stride lengths, particles corresponding to a longer stride length travel farther than those with a shorter stride length. At point D, when the user takes a right turn into a passage way, particles with stride lengths that are either too large or too short attempt to move to the right but are hindered by the walls on either side of the passage way. Consequently, as the user walks into the passage way, the possibilities for the user's stride length narrow down. The more the user walks and navigates around corners, the more accurate the estimate becomes. While in this simple example we assumed that the initial location is known, even if the initial location were unknown, Zee can estimate location and stride length simultaneously as the user continues to walk.

In practice, we found that within a single walk, users exhibit up

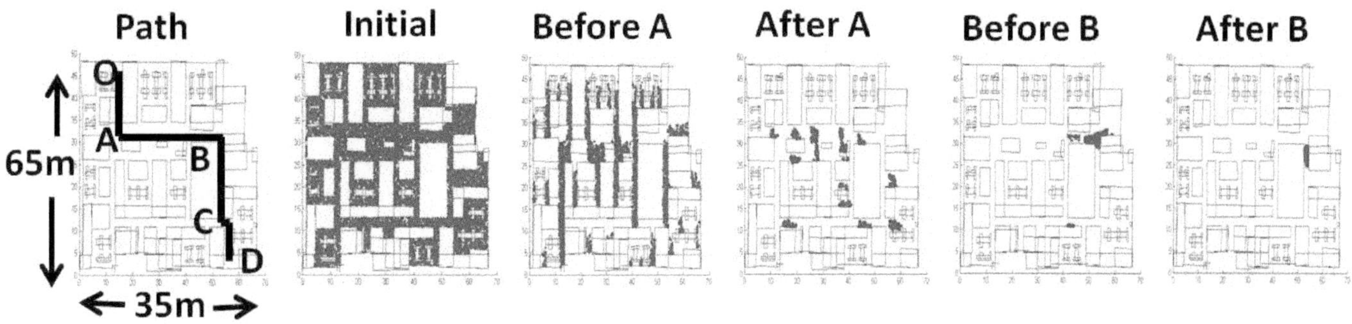

Figure 11: An example run of Zee

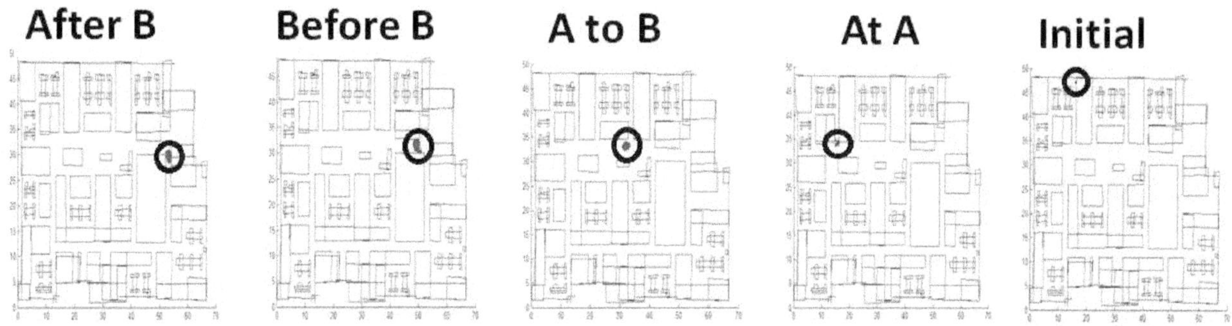

Figure 12: Backward belief propagation in Zee

to ±10% variation in their stride length. To account for this variation, we add a random error δ uniformly distributed in the range of ± 10% of the stride length (used in Equation 3 below).

Working with heading offset. Zee estimates the HO in a similar manner as it estimates the stride length. Initially, the heading offset is uniformly distributed within the two 90° sectors suggested by the HO range estimation scheme in Section 6. As the user walks, incorrect HO values are eliminated due to the wall constraints. While this approach accounts for the errors in the constant component of HO (due to phone placement), the APF must also account for the changes in HO due to different magnetic offsets at different locations. In order to accommodate this variation, the APF models this component as a compass measurement error — a Gaussian random variable β that is added to the compass measurement (used in Equation 3 below). We found that using a zero mean Gaussian with a standard deviation of 5° typically was sufficient in our floor.

The particle filter. The APF maintains a four-dimensional joint probability distribution as a particle filter, with a set of particles (samples), $\mathbf{X} = (\mathbf{X}_1, \mathbf{X}_2 \cdots, \mathbf{X}_N)$ representing the probability distribution. Here $\mathbf{X}_i = (x_i, y_i, s_i, \alpha_i)$, with (x_i, y_i) being the 2D location, s_i the stride length, and α_i the placement offset. After the user takes the kth step, the ith particle is updated as,

$$x_i^k = x_i^{k-1} + (s_i + \delta_i)\cos(\alpha_i + \theta + \beta_i) \quad (3)$$

$$y_i^k = y_i^{k-1} + (s_i + \delta_i)\sin(\alpha_i + \theta + \beta_i) \quad (4)$$

As noted earlier, s_i is perturbed by δ_i to account for variation in the stride length, and θ (the compass reading) is perturbed by β_i to account for compass measurement error and variation in the magnetic offset.

After each update, particles are tested to see if they violate any wall constraints. If the line joining (x_i^{k-1}, y_i^{k-1}) and (x_i^k, y_i^k) intersects a wall then the particle is eliminated. In order to replace each eliminated particle, a new particle is randomly chosen from the particle set at the $k-1^{th}$ step and updated. Note that s_i and α_i are not updated, rather incorrect values only get eliminated.

In our implementation we found that within one step, often it is possible to sample the compass several times. Accordingly, we use an average value of all the samples to arrive at the θ for the step. Sometimes, however, when the user turns, θ might change by a large amount within one step. Consequently, if the compass reading changes by more than 20° within a single step, we perform incremental updates by subdividing the step into fractional sub-steps.

An example walk in an office building. Figure 11 shows how Zee works in a large office floor, measuring 65m × 35m. The user was asked to walk from O to D, along the path shown in Figure 11. There are three turns in the path, at points A, B and C. Initially, particles were instantiated at all possible locations with all possible stride lengths and heading offsets spread uniformly over the two 90° sectors suggested by the HO range estimator. As the user walks to point A and then turns, the spread of particles dramatically decreases, as several possibilities are eliminated. Further, as the user walks towards point B, even more possibilities are eliminated. Finally, as the user navigates around the turn at B, Zee has narrowed down to the user's correct location.

Backward belief propagation. After the turn taken at B, the position of the user is localized to a single concentration of particles. Tracing these particles back in time would, therefore, allow us to infer the past locations of the user accurately. To enable such a trace-back, each particle $\mathbf{X}_i^k$ after the kth step maintains a link to its parent particle, *i.e.*, the particle $\mathbf{X}_j^{k-1}$ that it was generated from. In the backward belief propagation step, out of the set of particles $\mathbf{X}^{k-1} = \{\mathbf{X}_1^{k-1}, \mathbf{X}_2^{k-1}, \cdots, \mathbf{X}_N^{k-1}\}$, the particles with no children surviving in the succeeding step (i.e., in the set $\mathbf{X}^k$), are eliminated. This helps ensure that possibilities that are filtered out in future steps are not retained in the past steps. Figure 12 depicts the results of applying backward belief propagation on the user's walk along the reverse direction, starting from after the turn at B and going back to the initial location O. As the figure shows, using backward belief propagation, Zee is able to accurately trace the user's path back to the starting location O.

8. PUTTING IT ALL TOGETHER: ZERO-EFFORT CROWDSOURCING

Zee periodically scans for beacons from proximate WiFi Access Points (AP) and records the Received Signal Strength (RSS) measured, along with a timestamp. Also, as Zee tracks the path taken by users, whether in the forward direction or through backward belief propagation, it annotates the path with timestamps, indicating the times at which the user was located at particular points in the path. Thus, Zee can determine where in the floor a certain WiFi measurement was taken and thereby generate location-annotated WiFi measurements of the form (*location, WiFi RSS*). This database of measurements can then be used to locate new users using existing WiFi localization techniques. We now describe the two WiFi localization schemes used in our evaluation.

HORUS. HORUS uses a set of location-annotated RSS measurements to construct a probability distribution, $P(rss_{AP_k} = r|\mathbf{x} = \mathbf{x}_i)$, i.e., the probability of measuring an RSS value of r from access point AP_k at location $\mathbf{x}_i$. $\mathbf{x}$ here is a 2-dimensional location. In order to locate a device using HORUS, a device measures a vector of RSS measurements, $\mathbf{R} = < r_1, r_2, \cdots, r_m >$, where r_i is the RSS from AP_i. The probability of observing $\mathbf{R}$ at a location $\mathbf{x}_i$ is then computed as,

$$P(\mathbf{R}|\mathbf{x} = \mathbf{x}_i) = \prod_k P(rss_{AP_k} = r_k|\mathbf{x} = \mathbf{x}_i) \qquad (5)$$

Using Bayesian inference, HORUS computes $P(\mathbf{x} = \mathbf{x}_i|\mathbf{R})$. The location of the device is then estimated as either the *maximum likelihood location* (i.e., the location with the highest probability) or the expectation over all locations (*expected location*).

EZ. EZ relies on a widely-used RF propagation model for WiFi received signal strength (RSS) in indoor environments — the Log Distance Path Loss (LDPL) model. The LDPL model estimates RSS $rss^k_{\mathbf{x}}$ (in dBm) measured at a location $\mathbf{x}$ of the signal from AP_k placed at a location $\mathbf{c}_k$ as,

$$rss^k_{\mathbf{x}} = rss^k_0 - 10\gamma_k \log\left(d\left(\mathbf{x}, \mathbf{c}_k\right)\right) + N(\mathbf{x}). \qquad (6)$$

In Eqn 6, rss^k_0 is the RSS from AP_k at the reference distance of 1m, γ_k is the path loss exponent, and $d(\mathbf{x}, \mathbf{c}_k)$ is the distance between locations $\mathbf{x}$ and $\mathbf{c}_k$. $N(\mathbf{x})$ captures the random fluctuations in RSS due to multi-path effects. EZ uses WiFi measurements (a few annotated with location information but most not) from within the indoor space data to construct the LDPL model for each WiFi AP. Then, to locate a device, it converts the RSS measurement obtained from an AP to the estimated distance from that AP using,

$$d(\mathbf{x}, \mathbf{c}_k) = 10^{\left(\frac{rss^k_0 - r_k}{10\gamma}\right)}. \qquad (7)$$

Standard trilateration is then used to locate the device, once its distance from three or more APs has been estimated.

Using existing measurement database for subsequent crowdsourcing. In the absence of prior information, Zee starts by spreading the probability distribution of a user's location across the floor. However, once "enough" location-annotated WiFi measurements have been gathered, Zee can use existing localization schemes such as HORUS or EZ to initialize the probability distribution in a more localized area rather than over the entire floor. When using HORUS, Zee draws random samples from $P(\mathbf{x} = \mathbf{x}_i|\mathbf{R})$ and initializes the locations of the particles accordingly. When using EZ, Zee perturbs the measured RSS values to simulate the effect of multipath (using a Gaussian distribution with mean 0dB and standard deviation 5dB) and then generates sample locations for the particle filter. Using such a localized distribution rather than one spread across the entire floor helps speed up the convergence of the particle filter, thereby aiding subsequent crowdsourcing.

Dealing With Paths With Unconverged Particles. Not all paths may lead to a unique location in Zee, e.g., short paths that do not include enough turns to uniquely establish the user's trajectory. In such cases, the spread of particles will typically be large instead of being confined to a small region. Zee identifies such unconverged paths using measures such as the variance in the locations of particles and the number well-separated clusters, and discards the WiFi measurements corresponding to such walks.

9. EVALUATION

We evaluate the performance of Zee in a large office building, with a 65m × 35m floor plan, as depicted in Figure 14. We seek to answer several questions, including: (a) how well is Zee able to track users, (b) if Zee had been provided the user's initial location or had accurate knowledge of the heading offset, would its performance have improved significantly, (c) how well does Zee estimate quantities such as the stride length and the heading offset of the user, and (d) how well do existing localization schemes perform when using the location-annotated data generated by Zee versus using manually collected data.

Experimental methodology. In a real-world setting, we expect several users to be running the Zee client on their phone as they walk through various sections of an indoor space. Further, users are unlikely to be walking continuously; they would typically walk between locations of interest and dwell at certain locations (e.g., a store) for a significant length of time. However, for our evaluation, we did not have the luxury of several users. Instead, we handed a phone running the Zee client to a user, who kept it with himself for about 15 hours continuously. Our experiment included elements aimed at emulating the real-world characteristics noted above.

To recreate walks with breaks at various locations, the user was asked to walk to various parts of the office floor whenever he had the time and leave the phone there for periods ranging from 20 minutes to a half-hour. Zee was left running on the phone continuously. Upon detecting no walking activity for more than 10 seconds, the accelerometer data was not processed until walking was detected again. WiFi data, however, is more valuable when collected from a single location over an extended period, since it helps capture the variability in the WiFi signal at that location. Consequently, WiFi RSS measurements were collected even during the time when the device was stationary. To record the ground truth for location, the user made note of the locations on a map where he had stopped and the corresponding time periods.

In order to evaluate the benefit of using data from previous users for subsequent data collection, the user in our experiment stopped Zee briefly, walked over to a completely different location on the floor, and then restarted Zee. In this manner four different "walks" were created by restarting Zee four times. Each time, Zee would initialize the user's location based on the WiFi data gathered from the previous walks.

Figure 14 shows the paths covered by the user during the 15-hour period (shown as the dark lines in the figure). Several of the locations were visited multiple times over the course of the experiment.

Zee's tracking performance. We first show the performance of

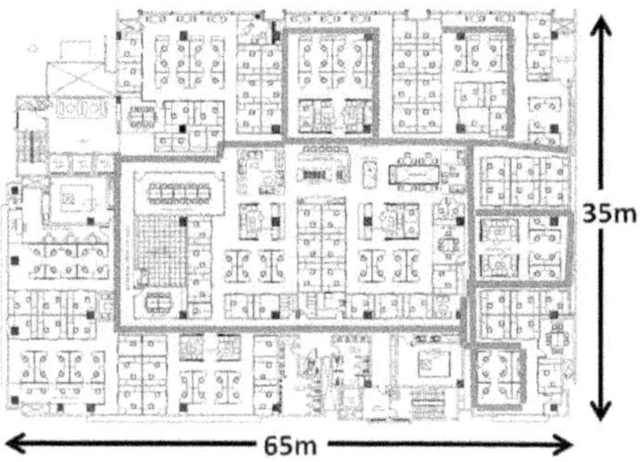

Figure 14: Total area covered during the 15 hrs by the user

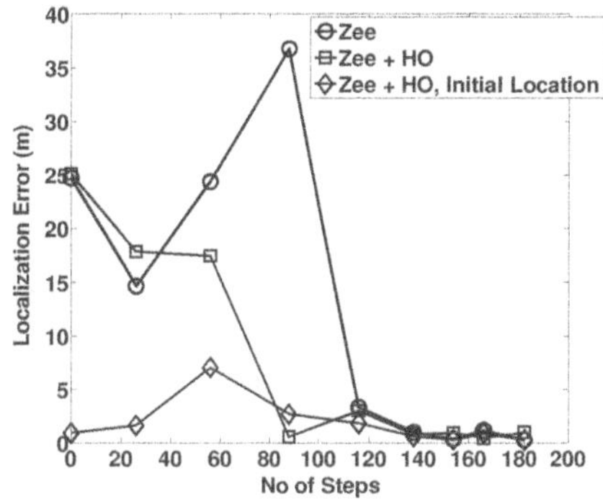

Figure 15: Location errors seen by Zee during walk 1.

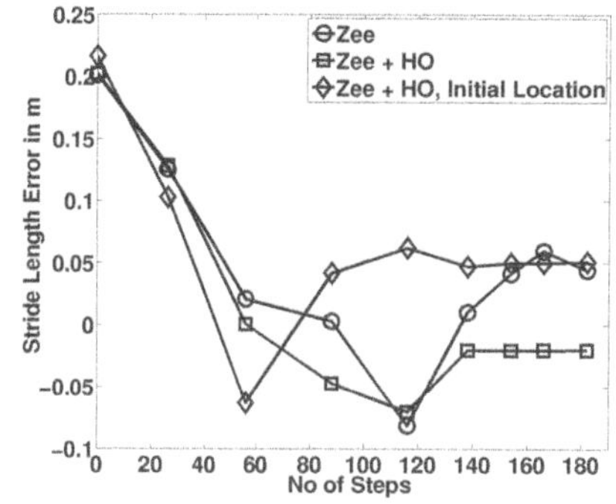

Figure 16: Stride length errors seen by Zee during walk 1.

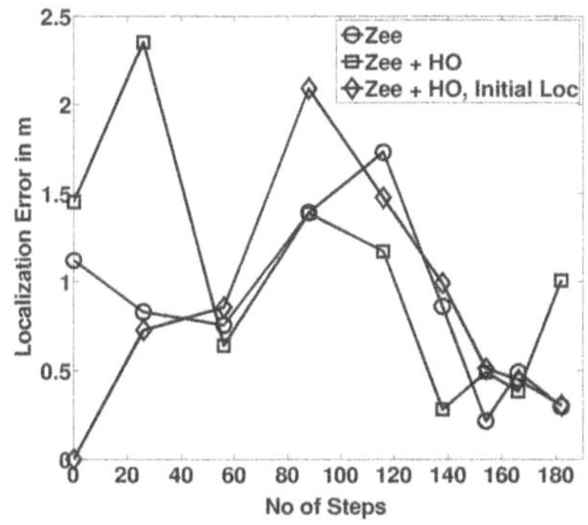

Figure 17: Location errors seen after backward belief propagation by Zee during walk 1

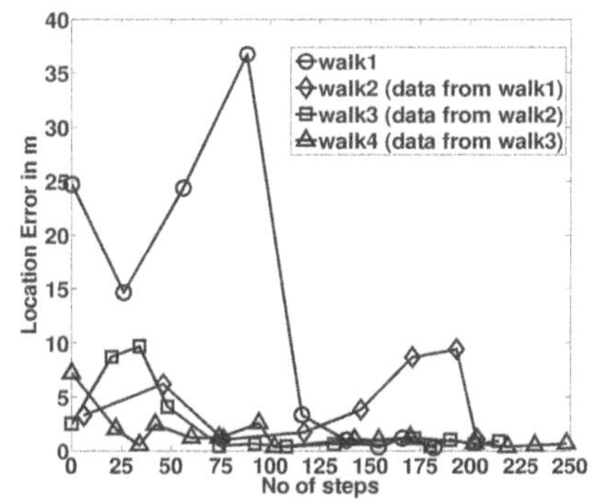

Figure 18: Using WiFi measurements from previous walks to start subsequent walks

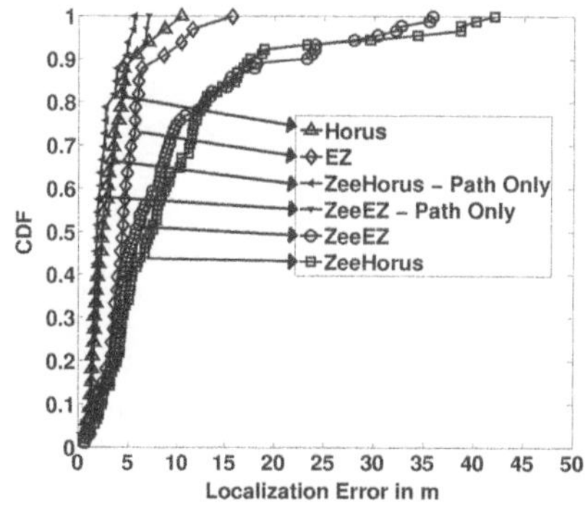

Figure 19: Performance of WiFi localization on Zee data

Zee on the first walk when there was no prior data from the floor. Note that this was not a continuous walk but instead was punctuated by several long stops, each lasting 20-30 minutes, as the user went to different locations on the floor. The walk lasted a total of 3 hours. Figure 15 depicts the location error seen by the particle filter at 9 different checkpoints (i.e., predetermined locations, where the ground truth was recorded) as a function of the step count during the walk. The location error was computed by estimating the user's location as the average location over all surviving particles, and then finding its distance from the ground truth location.

Figure 15 shows the location error with the particle filter run in the forward direction. As can be seen from the curve labeled Zee, initially the average error is extremely high. This is because the initial location is unknown and particles are spread all across the floor. Somewhere between steps 80 and 100, the user took a turn that eliminated all spurious possibilities and the error dropped sharply. At the end of the walk, the location error was under a meter; in other words, Zee's location estimate was correct to the last step!

Figure 17 depicts the location error after backward belief propagation was applied to the walk. As can be seen from the figure, the location error along the entire path is under 2m after this step.

Zee's performance with initial location and HO being known. In this experiment, we tested two different scenarios. In the first scenario, the initial location of the user was kept as unknown while heading offset was treated as known. In the second scenario, both heading offset and initial location were treated as known. Stride length, however, was assumed to be unknown in both cases.

As seen from Figure 15, with the HO known but initial location unknown, Zee converged must faster to the correct location compared to not knowing the HO, which is as expected. When both HO and initial location were known, initially the error increased because the stride length was unknown. However, as soon as Zee learned the stride length, the localization error reduced.

Figure 17 depicts the localization error for both these scenarios after backward belief propagation is applied. As seen from the figure, the location error along the entire path remains under 1m in both the scenarios. Further, the key point to note here is that *Zee was able to perform similarly well despite not knowing HO and initial location as when this information was known a priori, the only difference being that it took longer to converge in the former case, which is as expected.*

Zee's estimation of stride length. To evaluate how well Zee estimates stride length, we first measured the user's stride length by making him walk 20 steps in a straight line and directly measuring the distance covered. Then, during the actual experiment, we estimated the stride length at any step as the average stride length over the surviving particles. Figure 16 shows that the error in stride length estimation drops to around 5cm by the end of the walk.

Using WiFi measurements from previous walks for subsequent walks. For each successive walk, Zee was able to use the WiFi data gathered from the prior walks to estimate the distribution of the user's starting location. To compute this distribution, we used

EZ, as described in Section 8. To emulate a new user for each walk, the stride length and HO were deemed as being unknown at the beginning of each successive walk. Figure 18 shows that by using WiFi information from prior walks, Zee is able to converge much faster and have smaller location errors than otherwise.

Performance of WiFi localization using Zee-based crowdsourcing. To evaluate the performance of existing WiFi-based localization schemes when fed with Zee-based crowdsourced data, we first set up the baseline by collecting WiFi measurements manually at 117 locations, spaced about 3m apart and spread across the floor. At each location approximately 1000 WiFi beacons were collected for each proximate AP. This data was used to train HORUS and EZ. Additionally, we collected data from 91 separate test locations spread across the entire floor. Figure 19 depicts the distribution of localization error seen by EZ and HORUS at these test locations (labeled as EZ and HORUS in the figure).

Next, instead of the data gathered manually, we used the data obtained from Zee over the four walks to train EZ and HORUS, and tested these across all the 91 test locations. The corresponding distributions of localization error are also shown in Figure 19 (labeled as ZeeEZ and ZeeHORUS). We see that while ZeeEZ and ZeeHORUS have almost the same 50%ile errors as EZ and HORUS, respectively, the 80%ile errors are significantly larger. This was because, at the test locations that were far from all of the walk paths (which defined the spatial extent of crowdsourcing in this experiment), ZeeEZ and ZeeHORUS performed much worse than EZ and HORUS (which had the benefit of manually-gathered training data from across the floor).

To evaluate the performance of Zee more fairly, we tested Zee only on the test locations that were within 5m of any of the four walk paths. Figure 19 depicts the corresponding error distributions (labeled as ZeeEZ Path Only and ZeeHORUS Path Only). We see that the 50%ile and 80%ile errors are 1.2m and 2.3m, respectively, which are comparable to those seen when manually-gathered data is used for training (the curves labeled EZ and HORUS). This encouraging result suggests that Zee-based crowdsourcing could be effective and could enable localization with high accuracy, provided the space is well-covered by users during the course of their crowdsourcing walks.

10. CONCLUSION

In this paper, we have presented Zee, a system that enables crowdsourcing of location-annotated WiFi measurements in indoor spaces, using the mobile phones carried by users in normal course. A key attribute of crowdsourcing with Zee is that it is "zero-effort", not requiring any active user intervention in terms of location input, placement of the phone, or other aspect. Zee employs a set of novel techniques to resolve ambiguity in location during crowdsourcing, using inertial and WiFi measurements, and a map of the indoor space as the inputs. The data thus gathered can help train existing WiFi-based localization algorithms. Our evaluation on a large office floor shows that existing WiFi-based localization schemes, both fingerprinting-based and modeling-based, are able to perform accurate localization when trained with data that is crowdsourced automatically using Zee.

11. ACKNOWLEDGEMENTS

We thank Gursharan Sidhu for suggesting that WiFi data could be crowdsourced with the help of inertial tracking. We also thank our shepherd, Romit Roy Choudhury, and the anonymous reviewers for their constructive comments.

12. REFERENCES

[1] Time Domain. http://www.timedomain.com/.

[2] M. Azizyan, I. Constandache, and R. Roy Choudhury. SurroundSense: Mobile Phone Localization via Ambience Fingerprinting. In *MobiCom*, 2009.

[3] P. Bahl and V. N. Padmanabhan. RADAR: An Inbuilding RF-based User Location and Tracking System. In *INFOCOM*, 2000.

[4] P. Bahl, V. N. Padmanabhan, and A. Balachandran. Enhancements to the RADAR User Location and Tracking System, Feb 2000. Microsoft Research Technical Report MSR-TR-2000-12.

[5] K. Bauer, D. Mccoy, B. Greenstein, D. Grunwald, and D. Sicker. Using Wireless Physical Layer Information to Construct Implicit Identifiers. In *HotPETs*, 2008.

[6] S. Beauregard and H. Haas. Pedestrian Dead Reckoning: A Basis for Personal Positioning. In *WPNC*, 2006.

[7] V. Brik, S. Banerjee, M. Gruteser, and S. Oh. Wireless Device Identification with Radiometric Signatures. In *Mobicom*, 2008.

[8] Y.-C. Cheng, Y. Chawathe, A. LaMarca, and J. Krumm. Accuracy Characterization for Metropolitan-scaleWi-Fi Localization. In *MobiSys*, 2005.

[9] K. Chintalapudi, A. P. Iyer, and V. N. Padmanabhan. Indoor Localization Without the Pain. In *Mobicom*, 2010.

[10] D. Fox, W. Burgard, F. Dellaert, and S. Thrun. Monte Carlo Localization: Efficient Position Estimation for Mobile Robots. In *AAAI*, 1999.

[11] A. Goswami, L. E. Ortiz, and S. R. Das. WiGEM : A Learning-Based Approach for Indoor Localization. In *CoNEXT*, 2011.

[12] A. Haeberlen, E. Flannery, A. M. Ladd, A. Rudys, D. S. Wallach, and L. E. Kavraki. Practical Robust Localization over Large-Scale 802.11 Wireless Networks. In *MobiCom*, 2004.

[13] N. B. John, J. Heidemann, and D. Estrin. GPS-Less Low Cost Outdoor Localization For Very Small Devices. *IEEE Personal Communications Magazine*, 7:28–34, 2000.

[14] J. W. Kim, H. J. Jang, D.-H. Hwang, and C. Park. A step, stride and heading determination for the pedestrian navigation system. *Journal of Global Positioning Systems*, 3:273–276, 2004.

[15] P. Krishnan, A. S. Krishnakumar, W. H. Ju, C. Mallows, and S. Ganu. A System for LEASE: Location Estimation Assisted by Stationery Emitters for Indoor RF Wireless Networks. In *Infocom*, 2004.

[16] J. Leonard and H. F. Durrant-whyte. Simultanoues Map Building and Localization for an Autonomous Mobile Robot. In *IROS*, pages 1442–1447, 1991.

[17] H. Lim, L. Kung, J. Hou, and H. Luo. Zero-Configuration, Robust Indoor Localization: Theory and Experimentation. In *Infocom*, 2006.

[18] L. Ni, Y. Liu, C. Yiu, and A. Patil. LANDMARC: Indoor Location Sensing Using Active RFID. In *WINET*, 2004.

[19] N. B. Priyantha, A. Chakraborty, and H. Balakrishnan. The Cricket Location-Support System. In *MobiCom*, 2000.

[20] T. Rappaport. *Wireless Communications Principles and Practice*. Prentice Hall, 2001.

[21] P. Robertson, M. Angermann, and B. Krach. Simultaneous Localization and Mapping for Pedestrians using only Foot-Mounted Inertial Sensors. In *Ubicomp*, 2009.

[22] S. Sen, R. R. Choudhury, and S. Nelakuditi. SpinLoc: Spin once to know your location. In *ACM HotMobile*, 2012.

[23] S. Sen, B. Radunovic, R. R. Choudhury, and T. Minka. Precise Indoor Localization using PHY Layer Information. In *ACM HotNets*, 2011.

[24] R. Smith and P. Cheeseman. On the Respresentation and Estimation of Spatial Uncertainty. *The International Journal of Robotics Research*, 5, Winter 1986.

[25] H. Wang, S. Sen, A. Elgohary, M. Farid, M. Youssef, and R. R. Choudhury. No Need to War-Drive: Unsupervised Indoor Localization. In *Mobisys*, 2012.

[26] R. Want and et al. The Active Badge Location System. *ACM Transactions on Information Systems*, Jan 1992.

[27] A. Ward, A. Jones, and A. Hopper. A New Location Technique for the Active Office. *IEEE Per. Comm.*, 4(5):42–47, 1997.

[28] A. Williams, D. Ganesan, and A. Hanson. Aging in Place: Fall Detection and Localization in a Distributed Smart Camera Network. In *ACM Multimedia*, 2007.

[29] O. Woodman and R. Harle. Pedestrian Localisation for Indoor Environments. In *Ubicomp*, 2008.

[30] O. Woodman and R. Harle. RF-Based Initialisation for Inertial Pedestrian Tracking. *Lecture Notes in Computer Science*, 5538:238–255, May 2009.

[31] J. Xiong and K. Jamieson. ArrayTrack: A Fine-Grained Indoor Location System. In *ACM HotMobile*, 2012.

[32] M. Youssef and A. Agrawala. The Horus WLAN Location Determination System. In *MobiSys*, 2005.

Push the Limit of WiFi based Localization for Smartphones

Hongbo Liu[†], Yu Gan[†], Jie Yang[†], Simon Sidhom[†], Yan Wang[†],
Yingying Chen[†], Fan Ye[*]

[†]Stevens Institute of Technology, Hoboken, NJ 07030, USA
[†]{hliu3, ygan, jyang, ssidhom, ywang10, yingying.chen}@stevens.edu

[*] IBM Thomas J. Watson Research Center, Hawthorne, NY 10532, USA
[*]fanye@us.ibm.com

ABSTRACT

Highly accurate indoor localization of smartphones is critical to enable novel location based features for users and businesses. In this paper, we first conduct an empirical investigation of the suitability of WiFi localization for this purpose. We find that although reasonable accuracy can be achieved, significant errors (e.g., $6 \sim 8m$) always exist. The root cause is the existence of distinct locations with similar signatures, which is a fundamental limit of pure WiFi-based methods. Inspired by high densities of smartphones in public spaces, we propose a peer assisted localization approach to eliminate such large errors. It obtains accurate acoustic ranging estimates among peer phones, then maps their locations jointly against WiFi signature map subjecting to ranging constraints. We devise techniques for fast acoustic ranging among multiple phones and build a prototype. Experiments show that it can reduce the maximum and 80-percentile errors to as small as $2m$ and $1m$, in time no longer than the original WiFi scanning, with negligible impact on battery lifetime.

Categories and Subject Descriptors

C.2.4 [**Computer-Communication Networks**]: Distributed Systems—*Distributed Applications*; C.3 [**Special-Purpose and Application based Systems**]: *Real-time and embedded systems*; H.5 [**Information Interfaces and Presentation(e.g., HCI)**]: Sound and Music Computing

General Terms

Design, Experimentation, Measurement, Algorithms, Performance

Keywords

Smartphone, Peer Assisted Localization, WiFi fingerprint localization

1. INTRODUCTION

Indoor localization is a critical enabler for location based smartphone applications. In many environments (e.g., airport terminals, railway stations and shopping malls), the location helps users access navigation, merchandise and promotion information; businesses need it to understand the patterns of customer visit and stay, such as the popularity of different sections in the store, or the spatial-temporal distribution of passenger flows.

Accurate indoor localization on smartphones, however, remains elusive. Although there have been some recent commercial offerings such as Google Maps 6.0 and Shopkick [1], they either have errors up to 10 meters [9], or only locate at the granularity of stores. There has been a plethora of academic work on indoor localization. Those achieving high accuracy usually require special hardware not readily available on smartphones [20, 21], or infrastructure expensive to deploy [6, 17]. WiFi-based localization leverages prevalent wireless access points, thus avoiding such drawbacks. But most studies [3, 27] have been largely based on laptops with quite different antenna forms and possibly radio characteristics, whereas recent work on smartphone indoor localization [2, 15, 24] achieves room or floor level accuracies. The feasibility of leveraging the most prevalent WiFi infrastructure for high accuracy localization on smartphones is still an open question.

In this paper, we first conduct a set of experiments to empirically study the impact of various factors on the accuracy of WiFi localization on smartphones. We find that although reasonable accuracy (e.g., $3 \sim 4$m) can be achieved, there always exist large errors (e.g., $6 \sim 8$m) unacceptable for many scenarios. Similar or much larger errors (e.g., >15m) have been reported in previous studies [3, 27]. One work [7] found that high accuracy (e.g., sub-meter median and 2m maximum) is possible but only under hundreds of APs, infeasible in practical settings. Such errors may cause a passenger make a wrong turn leading to a different train platform, or a store erroneously stock up for a section with much less real customer interests. Our investigation on these large errors reveals the insight that they are caused by possibly faraway locations with similar WiFi signatures, an intrinsic phenomenon of the radio signal propagation and fundamental limit of WiFi methods.

On the other hand, we observe that smartphones are gradually woven into our social life and usually a high density of them exist in public spaces. The relative positions of nearby peer devices could be used as physical constraints on the possible location of a smartphone. Inspired by this observation, we set out to study how to exploit the unique physical constraints among smartphones to reduce large errors and push the limit of WiFi based approaches. We propose a peer-phone assisted localization approach that leverages the acoustic ranging between peers, without requiring special hardware yet producing highly accurate location estimates.

In particular, the peer-phone assisted localization can be carried out concurrently with WiFi localization or when a smartphone has obtained a rough location from WiFi but needs further improvements. The targeting smartphone exchanges sound signals with nearby peer devices. A server collects such ranging estimates and constructs a graph of the relative positions among peers. It applies a localization algorithm that maps the vertices of the graph against the WiFi signature database to locate all peers jointly. Experiments using data from various environments, including the airport, train station, and shopping mall, have shown that our approach can reduce 80 percentile error to about 1m, and limit the maximum error to about 2m, demonstrating the feasibility of WiFi for high accuracy localization.

Specifically, we make the following contributions:

- We discover the root cause of large errors as the existence of faraway locations sharing similar radio signatures, which is due to the intrinsic dynamic propagation of the radio signal, thus presenting a fundamental limit of WiFi methods.
- We propose a peer-phone assisted localization approach utilizing minimum auxiliary COTS sound hardware for reducing large errors and push the limit of WiFi approaches. We devise a peer-assisted localization algorithm that leverages acoustic ranging and locates peer phones jointly for greatly improved accuracy. We identify the frequencies, sound signal design, detection and emission scheduling methods appropriate for fast ranging among multiple peers that are unobtrusive, robust to noise and have minimum impact on users' regular activities.
- We prototype our system and carry out real world experiments. The results demonstrate that our approach greatly reduces the maximum error from $6 - 8$m to 2m, and limit 80 percentile error to 1m, which were shown empirically possible but only under hundreds of APs [7]. The assistance finishes in time no longer than a few seconds of WiFi scanning, and poses negligible impact on battery lifetime.

The rest of this paper is organized as follows. In Section 2, we perform a systematic evaluation on WiFi localization on smartphones and report our findings. Section 3 introduces our system design and present the detailed peer assisted localization algorithm. We study the frequencies, sound detection methods for fast concurrent ranging among multiple peer phones in Section 4. We describe the system implementation and report evaluation results in Section 5. We discuss related issues and survey related work in Section 6 and 7. Finally, Section 8 concludes the paper.

2. PERFORMANCE OF USING WIFI ALONE FOR SMARTPHONE LOCALIZATION

To understand the practical performance of smartphone localization using WiFi, we first conduct a systematic study on the impact of various factors (e.g., orientation, holding position, time of the day and number of samples). We find that reasonable accuracy can be achieved in many cases (e.g., $\sim 4m$). However, large errors (e.g., beyond $6 \sim 8m$) always exist. Further investigation reveals two root causes: static environmental effects, and dynamic obstacles or interferences, both of which pose fundamental limits on WiFi localization accuracy.

2.1 Methodology

Fingerprint Based Localization. Fingerprint based method was pioneered by Bahl et. al [4] and is the most popular WiFi localization approach. It first measures the "fingerprint", the WiFi signal strengths from various access points (APs) at a number of known locations and stores them as *training data*. A device samples the signal strengths from various APs to obtain *testing data*. Then an algorithm finds the "closest fingerprints" in the training data to the sample, using Euclidean distance in the signal space where each dimension is for a different AP. A location estimation is given based on the locations of the closest fingerprints (e.g., the centroid of a few "closest" fingerprint locations). Recent work [8, 24] showed that the training data could be constructed without extensive site survey, making this approach even more attractive. In the test, we build a fine-grained signature map as training data, using interpolation between locations with actual measurements.

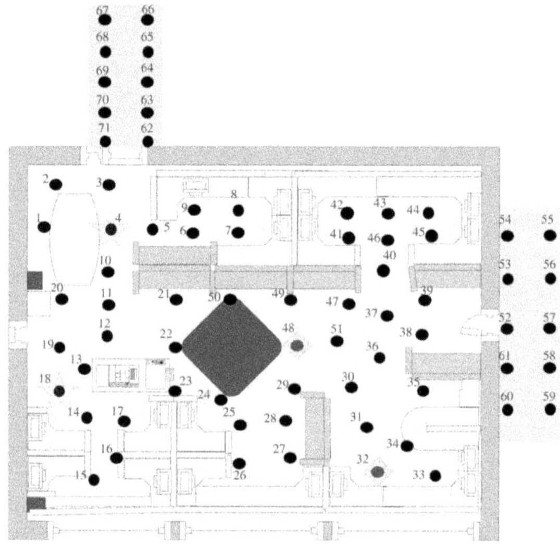

Figure 1: Each dot in the floor map represents a location where the RSS fingerprint is measured.

Experimental Setup We conduct the study in an office environment as shown in Figure 1 with a set of Android phones. The experimental area is 12 m × 11 m with hallways, office wall dividers and furnitures, such as desks, shelves and chairs. The phone takes 60 Received Signal Strength (RSS) samples at each of the 71 known locations from 14 APs. Each location can observe signals from 8-9 APs on average. We repeat the above process for each of the 4 factors, orientation, holding position, time of the day and number of samples (shown in Table 1, to understand how they impact the localization performance. For each test, the default parameters are south for orientation, normal style for holding position, 60 samples and morning time.

2.2 Impact of Various Factors

Figure 2 presents the cumulative distribution function (CDF) of the localization error under various factors. We first examine how the orientation affects the RSS readings due to the blocking and reflection of radio signals by the human body. In outdoor cases,

Factors	Description	Value
Orientation	The direction to which the user is facing	West, North, East, South
Holding Position	The way the user holds the phone	Holding bottom, Holding middle
Number of Samples	The number of samples taken at each location	3,5,10,60
Time of the Day	Different times that we conduct the measurements	Morning, Noon, Night

Table 1: Factors under study

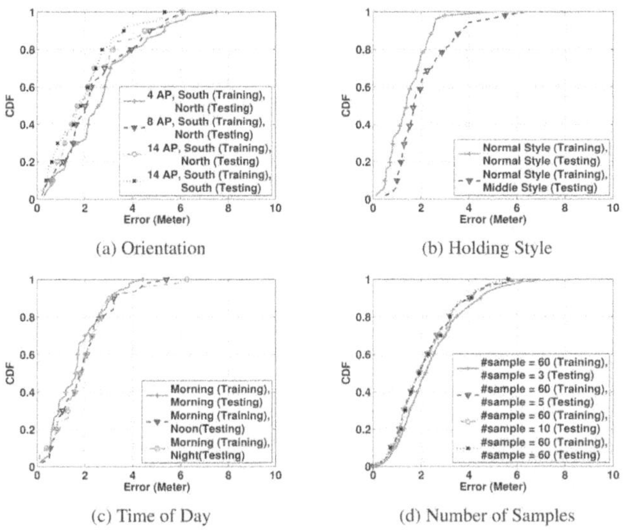

Figure 2: WiFi localization error under various factors.

10dB RSS difference for certain APs was observed on smartphones [29]. In our indoor environment, we recorded an average of 4dB difference. We believe this reduced variation is due to stronger multipath effects of indoor environments, thus the lack of direct line of sight does not attenuate the signal as much. In particular, as shown in Figure 2 (a), when a mismatched training data set is used (e.g., south-facing training set for north-facing user), long tails of CDF curves exhibit large errors ranging from 6m to 8m. Even when the matching training set is applied, errors beyond 5m still exist.

Due to the small size, how the user's hand holds the phone can affect the received radio signal as well. We tried two holding positions: bottom and middle. Figure 2 (b) shows that using mismatched training and testing data (e.g., bottom-holding as training to localize middle-holding phones) can lead to large error beyond $6m$. In addition, the localization results are evaluated under three different times of the day representing morning, noon and evening in Figure 2 (c). We again observe long tails beyond 5 meters when mismatched training and testing data are used.

Finally, more samples lead to more reliable measurements as the input for localization, but at the cost of higher energy and latency overhead. To understand the accuracy-overhead tradeoff, we depict the localization error when varying the number of samples from 3 to 60 in Figure 2 (d) under a training set of 60 samples per location. We find that using more than 5 samples does not significantly improve 90-percentile accuracy. We thus choose 5 samples in our WiFi localization study throughout the paper.

2.3 Root Cause of Large Errors

We make one critical observation from the above investigation: although reasonable accuracy can be achieved in many cases, large errors $6m$ or more always exist. In many applications these large errors can cause problems, such as giving the user incorrect navigation instructions, or wrong statistics about the visits of customers to different sections inside a store.

After a close examination of those large errors, we find the root cause. In essence, two physically distant locations happen to share similar WiFi signal strength measurements, thus a testing sample is erroneously localized to a physically faraway location with short Euclidean distance in the signal space. This can be classified into two cases: (1) permanent environmental settings such as walls, furniture placement, which affect radio signal propagation and create persistent similar signal reception; and (2) transient factors or measurement mismatch between training and testing data. Such transient variation in RSS reception is due to dynamic changes in the environment, such as a nearby moving object or wireless interference from other electronic devices, while the mismatch can be in orientation, holding style, time of the day or number of samples.

We illustrate case 1 by three locations in Figure 1: 18 and 13 are close to each other whereas 4 is farther at the other side of the room (marked as red stars). However, during testing we find that locations 18 and 4 share similar WiFi fingerprints. The distance between their fingerprints in the signal space is only 1.98 dB/AP, whereas the fingerprint at closeby location 13 has a distance of 2.44 dB/AP to that of location 18. In this case, the office wall dividers cause the large localization error beyond 6m at location 18.

The example for case 2 are location 32, 34 and 48 (marked as blue squares in Figure 1). Locations 32 and 34 are close to each other. However, we find that the WiFi fingerprint at location 32 becomes similar to that at location 48 at night when less people are around. Thus when testing at location 32 using the training data collected at night, location 32 will be matched to 48, instead of 34, resulting in large errors of over 6m. Through our study, we find that the percentage of large errors resulted from Case 1 is 60% to 70% while that from Case 2 accounts for the rest. Both cases are caused by irregular multipath reflections, an intrinsic character of radio signals. They present fundamental limits for WiFi localization to achieve high accuracy.

3. PEER ASSISTED LOCALIZATION

From the previous investigation, WiFi as-is is not a suitable candidate for high accuracy localization due to large errors. However, is it possible to address this fundamental limit without the need for additional hardware or infrastructure? Our answer is yes: by exploiting acoustic ranging, a phone can use nearby peer phones as reference points and obtain its relative positions to them. This imposes unique physical constraints on the possible location of the phone, thus reducing the uncertainty and improving the accuracy.

Such an idea is motivated by two observations. First, in many public indoor environments (e.g., airport terminals, railway stations, shopping malls and museums), there are usually a high density of users, thus smartphones. Each neighboring peer has a unique physical location for restraining the location uncertainty of a smartphone. Second, a number of research work [19, 26] has shown that highly accurate relative ranging can be achieved within a car (passenger vs. driver side) or between a pair of mobile devices (at centimeter accuracy) by using sound signals.

3.1 Design Goals and Challenges

The above concept may sound quite simple. However, building such a peer-assisted localization system involves a number of great challenges in both the design and implementation:

- **Peer assisted localization algorithm.** How to utilize the physical constrains imposed by the neighborhood peers to reduce the large errors incurred from WiFi localization? Given only the relative distances among peers and their location estimates are available in real scenarios, exactly what is the algorithm?

- **Concurrent acoustic ranging of multiple phones.** Previous work on acoustic position estimation was for one or two devices only. When there are multiple devices and they all do acoustic ranging, how can we tell which distance measure is

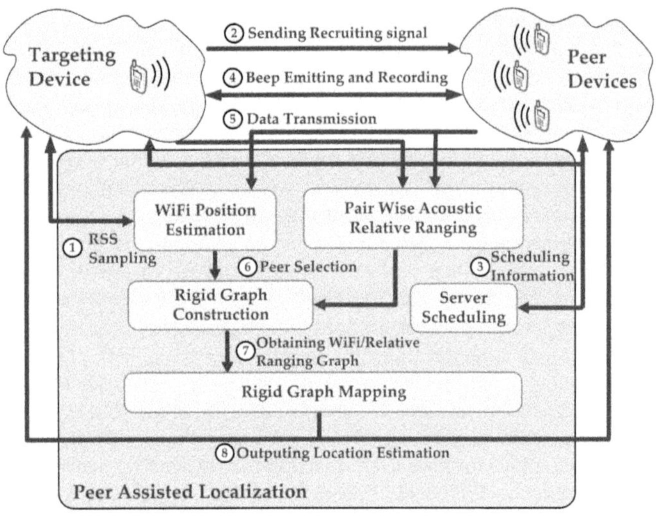

Figure 3: The workflow of our system.

for which pair? How to design and detect the sound signal, so that the system is robust to noises in different environments?

- **Ease of use.** The peer assistance process should complete in short time; otherwise users may have moved to different locations. The sound emitted should not annoy or distract users from their regular activities.

In the rest of this section, we will present an overview of the system, then present the details of the peer assisted localization algorithm. Section 4 addresses challenges in concurrent acoustic ranging and ease of use.

3.2 System and Algorithm Overview

Our system works as follows (shown in Figure 3): when a target phone needs to further improve accuracy or desires high accuracy upfront, it broadcasts a special audio signal to "recruit" nearby peers. Those receiving the recruiting signal send their identifiers to a server. The server comes up with a schedule about which phone should emit a beep signal at which time slot. The phones involved, both the target and peers, emit the beeps accordingly. They also record the beeps from others and send the files back to the server. All peers also conduct WiFi sampling and send the measurements to the server. The server determines the locations of peers from WiFi samples, and distances among them from acoustic ranging based on the recorded sound files. Finally the server computes the new location estimate of the target using our peer assisted localization algorithm, and sends back the result to the target.

The intuition underlying the peer assisted localization algorithm is to construct a graph based on the relative distances among devices, then "superimpose" the graph onto the signature map based on the initial WiFi location estimates. The algorithm "rotates" and translationally [1] "moves" the graph against the signature map, such that the vertices are placed "closest" to the true locations, as measured by certain metric. Where the vertices are placed become the new location estimates.

In the algorithm, we use as the metric the sum of RSS Euclidean distances between the WiFi samples of each device and the WiFi signature of where its graph vertex is placed. Such a metric quantifies the aggregate "closeness" between WiFi samples and new estimated locations of all devices. We use the example in Figure 4 to illustrate the intuition. Initially, each phone has a location estimate (vertices in the dashed-line graph), e.g., from WiFi localization. Because the acoustic ranging is highly accurate to identify the relative distances between peers (in terms of centimeters), the shape of the solid-line graph is quite close to that of the ground truth (the dotted-line graph). Such additional relative distance constraints "force" the new location estimate of the target to move closer to its real location, thus reducing large localization errors and achieving higher accuracy.

[1]"Translationally" means moving the graph without rotation.

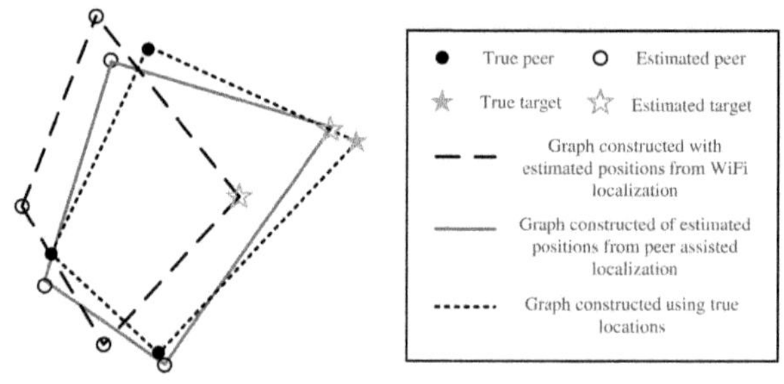

Figure 4: Illustration of using physical constraints to perform peer assisted localization.

3.3 Peer Assisted Localization Algorithm

Our peer assisted algorithm comprises of two main parts: (a) graph orientation estimation: estimating the range of graph orientation angle ϕ [2] by combining the acoustic ranging information and initial WiFi localization results. This is important because we find that its performance is sensitive to the orientation (shown in Section 5.5). (b) new location estimation: search for the optimal combination of graph orientation angle and translational position, such that the RSS distance summation is minimized.

The algorithm takes two inputs: 1) $\{p_i\}, i = 0, \cdots, M$: initial WiFi location estimations of the target phone (p_0) and its peers ($p_1, \cdots, p_M$); 2) $\{l_{i,j}\}, i, j = 0, \cdots, M$: pairwise acoustic ranging measurements between device i and j. The algorithm produces $\{q_i\}$, the new location estimate of each involved device i. In the following, the first 3 steps estimate the orientation range, and the last two find the new locations.

Step 1: Compute edge directions from acoustic ranging. The server constructs a graph G' with the pairwise ranging measurements, $l_{i,j}, i, j \in \{0, \cdots, M\}$. It rotates G' such that the direction of its longest edge is parallel to the X axis. Then it calculates the direction vector $v_l' = \frac{p_i'-p_j'}{|p_i'-p_j'|}, (i < j)$ for the L (e.g., 3)[3] longest edges, where p_i', p_j' are the locations of the two vertices of an edge.

Step 2: Compute edge directions from initial WiFi localization. The server also computes the direction vectors for the same L edges using initial WiFi locations: $v_l = \frac{p_i-p_j}{|p_i-p_j|}, (i < j)$.

Step 3: Graph Orientation Estimation. The server rotates G' to find the optimal graph orientation Φ that maximizes the inner

[2]We define ϕ as the angle between the X axis and the direction of the graph's longest edge, from the vertex of the smaller ID to the other.

[3]When choosing L longest edges, we experimented various values (e.g., 2, 3 and 4) and found different values of L have little impact on the final localization results. We used the value of 3 as an example in our evaluation.

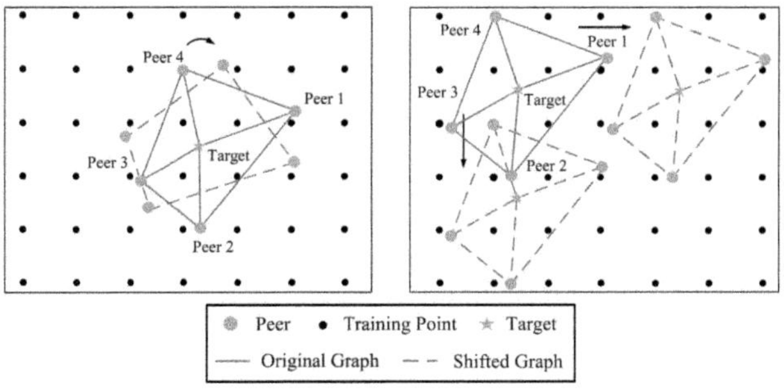

(a) Rotation (b) Translational movements

Figure 5: The translational movements or rotation of a graph preserve the distance constraints between vertices.

product summation between $\{v_l\}$ and $\{v'_l\}$:

$$\Phi = argmax \sum_{l=1}^{L} v'_l v_l^T \tag{1}$$

The inner product is a monotonically decreasing function of the angle difference between two vectors. Thus Equation 1 minimizes the angle differences of L edges obtained from acoustic ranging and initial WiFi localization.

Due to ranging errors, the real orientation may differ from Φ. To ensure that the true orientation is covered, the search of $\{q_i\}$ is conducted in an orientation range of $[\Phi - \Delta\Phi, \Phi + \Delta\Phi]$, where $\Delta\Phi = 20°$ from our empirical study in Section 5.5. We choose the L longest edges because the longer an edge, the less the impact on its direction caused by the same ranging error.

Step 4: Set the Search Scopes. Based on the initial WiFi position estimation p_i, the server tries to superimpose G' onto the WiFi signature map, such that each vertex i except the target is restricted inside a small circle A_i centered at p_i with radius r_i during the searching process in following step. We set r_i at $2m$ based on empirical study (see Section 5.1).

Step 5: Joint location estimation. Finally, with a small movement step of α meters and rotation step of β degrees (set to $0.1m$ and $2°$ based on empirical study), the server searches for the optimal location and orientation combination to superimpose the graph against the fingerprint map using the following objective function, where the moving and rotating operations are illustrated in Figure 5 (a) and (b) respectively:

$$\underset{\{q_i\},\ q_i \in S}{\arg\min} \sum_{i=0}^{M} [f(q_i) - f(p_i)]\,[f(q_i) - f(p_i)]^T \tag{2}$$

where $f(x) = [RSS_x^1\ RSS_x^2 \cdots RSS_x^k]$ is the WiFi fingerprint or measurement at location x (i.e., the RSS measurements of k detected access points).

Equation 2 produces the final results by finding a set of locations $\{q_i\}, i = 0, 1, ..., M$ from the WiFi fingerprint database S, so as to minimize the summation of RSS distances, each of which is between device i's WiFi measurement $f(p_i)$ and the signature $f(q_i)$ at its new location estimation q_i.

One comment we want to make is that the distances between some pairs of vertices may not determine the shape of a graph uniquely. For example, a square is *flexible* since its vertices can rotate against each other and form a family of rhombi while preserving the edge lengths, whereas the shape of a triangle is "rigid" (i.e., uniquely determined) given the lengths of the three edges. The rigid graph theory [14] describes under what conditions a graph is *rigid*: A complete graph with a distance between any two vertices is rigid. The concurrent ranging among multiple devices and the resulting pairwise distances give a complete, thus rigid graph. [4]

Extensive evaluation shows that our algorithm greatly improves the localization accuracy of the target phone (details in Section 5). The improvement is limited only under some rare cases (e.g., when peers are aligned almost on a straight line, or clustered together and located far away from the target), which we discuss in Section 5.4.1.

4. ACOUSTIC RELATIVE RANGING

Our peer-assisted approach requires acoustic ranging among multiple phones. This raises two new issues: First, how to ensure that the signals from different peer phones do not interfere. Sound signals do not carry a MAC address so there is no way to tell which phone emitted which sound. Second, the whole process has to complete in short time. Otherwise users may have moved, causing inaccurate ranging results. In this section, we describe our signal design, detection, and scheduling techniques that satisfy the requirements of concurrent multi-peer ranging.

4.1 Beep Signal Design and Detection

4.1.1 Acoustic Ranging Principle

In principle, ranging can be done by Time-of-Arrival (TOA) method that estimates the sound travel time from one device to another. The difficulty is in uncertainties: both the emitting and detecting have variable delays difficult to measure. The lack of clock synchronization between devices further adds to the problem. To address these issues, we have each involved peer device emit signals and the uncertainties will cancel out each other in calculation, similar to Beepbeep [19]. By doing so, only delay measurements from the same device are needed, thus circumventing the above issues.

4.1.2 Beep Design

We need to design the acoustic ranging signal carefully. The signal should be robust to various background noise (e.g., human conversations, PA announcements and music) that make it hard to detect. It should have minimal disturbance to people's normal activities. Based on several considerations, we choose the frequency band between 16kHz and 20kHz for the signal.

The majority of the background noise are located at the lower frequency band (e.g., conversation between 300Hz to 3400Hz, music from 50Hz to 15kHz, which covers almost all naturally occurring sounds). 16-20kHz is still audible to human ears [11], but much less noticeable and thus present less disturbances. We also found that the current cell phone microphones are more sensitive to high frequency sounds than human ears [26]. A high frequency beep at the edge of the microphone's frequency response curve make it both easier to filter out noise and renders the signal unnoticeable to most people.

Our sound signal consists of several evenly paced beeps of equal lengths. We study how to set the number, length and frequency of the beeps, and the length of the intervals in Section 4.2.1. They all impact ranging accuracy and time. For example, too short a beep may not be picked up by the microphone, while too long a beep will add more delay to the ranging.

[4]Due to outliers in acoustic ranging, we may miss some pairwise distances with a small probability. The small search scopes set in step 4 of the algorithm help ensure the rigidity.

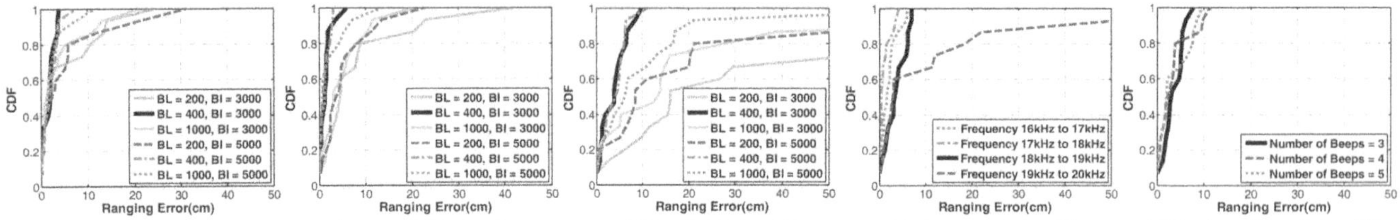

(a) BL and BI design, 100cm (b) BL and BI design, 200cm (c) BL and BI design, 300cm (d) Frequency design, 300cm (e) Number of beeps design, 300cm

Figure 6: Beep Design: results obtained from HTC EVO.

4.1.3 Beep Detection

Beep detection determines exactly when the beep signal arrives and it is critical to the accuracy of relative ranging. We investigate the behaviors of two most common signal detection methods, *change-point detection* and *correlation-based* under high frequency band.

Change-Point Detection Method: This method requires the beep have distinct energy uniformly distributed over a short frequency band (e.g., 16-18kHz). Given that band, this method first filters out the background noise using a Short Time Fourier Transform (SIFT). It then identifies the first strong signal that deviates from the noise in the targeted frequency band. The observed energy distributions of the recorded signal differ significantly before and after the arrival of the beep signal. Sequential change-point detection technique is adopted to identify the arrival of the signal, i.e., the exact time point of changed distribution [26].

Correlation-based Method: This method uses a chirp signal of a much shorter length. It correlates the emitted chirp sound with the recorded signal using L2-norm cross-correlation, and picks as the signal detection time when the correlation value reaches the maximum. To be robust to multipath, the earliest sharp peak in the correlation values is used as the signal detection time [19].

4.2 Detailed Design

Through extensive experiments we find that change point has much higher accuracy than correlation. We will present the signal design for change point first, then compare the two methods under different environments.

4.2.1 Sound Signal Parameters

Experimental Set Up: We used HTC EVO and Android Developer Phone 2 (ADP2) phones and 44.1 kHz sampling rate in recording. During each test, two smartphones of the same model are used. One phone emits the sound first. Upon receiving the signal, the other phone emits the sound and both phones record. The relative ranging is computed by obtaining the time difference based on the signal detection in the recorded sounds by the two phones as described in Section 4.1.1. We spaced the two phones apart at 100cm, 200cm, and 300cm respectively. Each curve in our results is obtained with 30 runs.

Beep Length (BL) and Beep Interval (BI) analysis: Figure 6 (a) - (c) presents the cumulative distribution function (CDF) of the ranging error when the beep length is set to 200, 400 and 1000 samples and the beep interval is set to 3000 and 5000 samples, respectively. The frequency band is 18kHz to 19kHz. It is clear that beep length of 400 samples achieves the best performance: the ranging error is around 10cm even when the two phones are placed 300cm away, and the error is similar with beep interval of 3000 and 5000 samples. We thus choose beep length of 400 samples and beep interval of 3000 samples in our system implementation.

Beep Frequency Band: With the above settings, we examine the performance of different frequency bands beyond 15kHz. Figure 6 (d) shows the results when varying the frequency band from 16kHz to 20kHz with two HTC EVO phones 300cm apart (better results are obtained in 100cm and 200cm cases). The ranging errors are less than 10cm when the beep frequency is below 19kHz. Similar experiments are conducted for ADP2 and we find that the highest frequency band producing comparable ranging results is 16-17kHz. We choose 16-17kHz range for ADP2 and 18-19kHz for HTC EVO in our system.

Number of Beeps (NB): Outliers can occur in ranging results due to dynamic factors in the environment. The sound signal should contain multiple beeps, so as to obtain multiple ranging estimates and filter out outliers. However, too many beep in the sound signal may make it more susceptible to multi-path distortions. Figure 6 (e) shows the ranging error when different numbers of beeps are employed under frequency band 18-19kHz with BL = 400 samples and BI = 3000 samples. Basically, comparable performance (less than 10cm) is observed across three to five beeps. In our implementation, we used three beeps, which is robust up to two outliers on opposite side of the true distance. We observe that the probability of getting reliable results is more than 95%.

4.2.2 Robustness to Various Noises

To evaluate the robustness of the design, we conduct a comprehensive study under various environments including lab, shopping mall, train station, and airport. There are all kinds of noises during our test such as human talking, radio broadcasting, dog barking, and trolley rolling. In each scenario, two phones are placed 100cm, 200cm, and 300cm apart.

Figure 7 shows the median and 90th percentile ranging error using change-point detection method and correlation-based method respectively under high frequency band. For change-point, we use 16-17kHz for ADP2 and 18-19kHz for HTC EVO. The sound signal follows the previous parameters in Section 4.2.1. For correlation method, we use 16kHz-20kHz for both phone types and the chirp signal length is 50ms [19]. We find that change-point detection significantly outperforms correlation in all cases: it has consistent low median errors around 10cm, while that of correlation ranges over 20cm to 40cm; its 90th percentile error is always below 20cm, while that of correlation can go up to 65cm. We thus choose change-point detection method for acoustic ranging in our system.

4.3 Server Based Emission Schedule

Identify Nearby Peers. A target phone needs to find which peers are nearby to get their help. Since large errors may exist in the WiFi location estimates for the target and peers, using WiFi

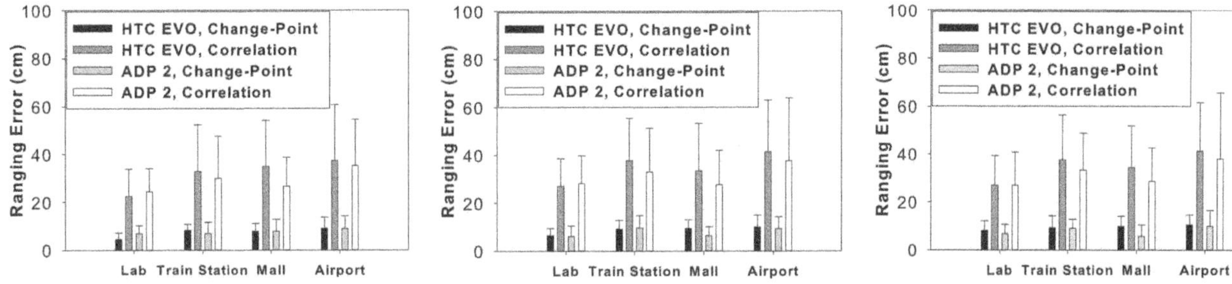

(a) 100cm (b) 200cm (c) 300cm

Figure 7: Ranging errors under different environments

localization alone is not reliable. We let the target phone emit a customized sound signal containing a single beep to identify truly nearby peers. Only phones really close enough can detect the signal. Those willing to help can send their IDs to the server. The server will then instruct these peers when they should emit the ranging signal.

Beep Emission Strategy: To speed up the ranging process, multiple phones need to emit sounds in a short time. Due to the lack of synchronized clocks, this may lead to interference among beeps from different phones, thus prohibiting accurate signal detection. There are two options to avoid such interference: time-division multiplexing or frequency-division multiplexing. We have tested concurrent emission of signals of different frequency bands, but find that interference still exists due to energy leakage among nearby frequency bands.

We employ time-division multiplexing based on server scheduling. [5] We divide the time into slots of length t_s long enough for the beep emission of one phone (e.g., 14000 samples or about $0.3s$, for the previous signal parameters and a beginning padding of 3000 samples). After receiving the IDs of all phones, the server comes up with a back-to-back schedule for them. It sends the schedule to them, with the ith phone starting beeping after a delay of $T_p + i \times t_s$. T_p is some extra buffer to accommodate small variations in the reception of the schedule at different phones. It ensures that all phones have received the schedule before its execution.

Each phone records sound from others when its time slot has not come and plays the sound during its time slot. After finishing the whole schedule, all phones send their IDs and recorded files to the server, which performs signal detection and estimates distances.

5. IMPLEMENTATION AND EVALUATION

In this section, we first describe how to detect the presence of large errors in the initial WiFi localization results in Section 5.1. We present the experimental methodology, testing scenarios and evaluation metrics in Section 5.2. We implement a prototype of our proposed system and report its performance in Section 5.3, including the accuracy, localization latency, and incurred energy consumption using Android phones. To capture the statistical performance of our scheme, we apply a trace-driven approach to conduct an extensive study of the impact of the quality and quantity of peers in Section 5.4. Finally, we study the sensitivity of our algorithm to orientation estimation in Section 5.5 and show the algorithm's performance under different real-world environments, such as train station, shopping mall and airport, in Section 5.6.

[5] We investigate an alternative where each phone randomly choose a timeslot to emit sound, but find that much more time slots are needed and it is difficult to distinguish the signals from different phones.

5.1 Detecting the Presence of Large Errors

The system needs to detect the presence of large errors in the initial WiFi position estimation for two purposes. First, target devices need this to help decide whether they should trigger peer assistance. Second, when selecting peers, we find that only those with small initial errors can serve as reliable reference points. However, the system does not know the true location, how can it tell whether large errors exist in a position estimation? Based on the insights on large errors described in Section 2, we use 3 simple rules for reliable detection.

Rule I: Examine the probability of large errors at the initial estimated location. [6] If a location has a large probability (e.g. > 0.8) for an error beyond a threshold (e.g., 4.5m), declare the presence. This rule deals with consistent large errors at specific locations caused by permanent environmental settings (i.e., Case 1 in Section 2.3). For each location in the training data, we randomly pick 5 from the 60 samples to compute a location estimate, and repeat 400 times to compute the probability of large error occurrence.

Rule II and III: Examine the two sets of APs detected in the samples and in the training data at the estimated location. If the ratio of the number of common APs to the number of all detected APs is below a certain threshold, declare the presence; Compare the standard deviations of each AP in the samples and the training data at the initial estimated position. Among all common APs, if the fraction of APs having larger standard deviations in the samples exceeds a threshold, declare the presence. These two rules deal with dynamic changes caused by transient factors (i.e., Case 2 in Section 2.3). They measure the statistical difference between the samples and the training data at the estimated location.

Intuitively, the system should be aggressive in declaring the presence for large errors, i.e., upon the detection from any of the 3 rules. Thus most targets having large errors are detected, and a peer with slight suspicion of having large errors is excluded. The cost is a few targets with smaller errors unnecessarily triggering peer assistance (but further improving accuracy), or peers with small errors excluded from assistance.

The threshold values for the three rules are set such that they minimize false classification on the training data. For example, for detecting bad targets, we can use 4.5m, 62%, 82%; for excluding bad peers, 1.5m, 90% and 42%. We find that collectively, 87% of bad targets are correctly identified, and among peers selected, 90% indeed have errors less than 1.5m.

[6] The nearest location in the interpolated training data set is actually used. Similar for the other two rules.

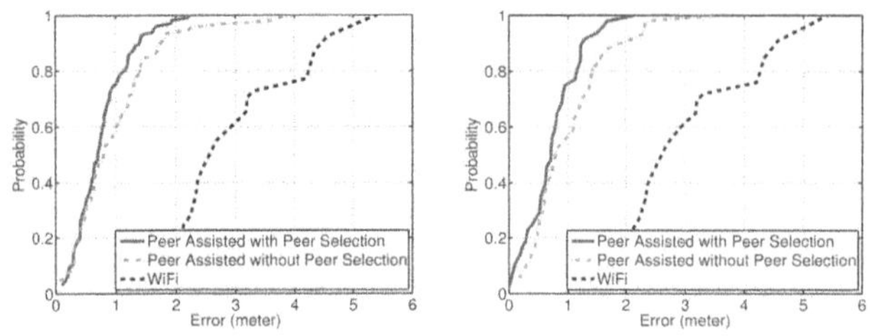

(a) 5 Peers HTC results (b) 5 peers ADP2 results

Figure 8: Localization performance for peer assisted method with and without peer selection rule

5.2 Evaluation Methodology

We conduct experiments using Android Developer Phone 2 and HTC EVO. Both phones support 44.1kHz audio sampling. HTC EVO uses 512 MB RAM and 1 GHz Qualcomm QSD8650 processor, while ADP2 192 MB RAM and slower 528MHz MSM7200A processor. [7]

We generate the sound file based on the design in Section 4.1.2, which consists of three beeps, each created by uniformly distributing white noise and then bandpass filtered to 16-17kHz for ADP2 and 18-19kHz for HTC EVO.

We implement a system prototype including an Android smartphone app and a backend server, both written in Java. The server runs on Lenovo Thinkpad X201 with Intel Core i5 2.53GHz processor and 4GB DDR3 RAM. The app does WiFi RSS sampling, acoustic emitting and recording. The measurements are sent to the server for processing. The app has been tested for both ADP2 and HTC EVO. Both the server and app are multi-threaded so that they can perform multiple tasks in parallel to speed up the whole process.

Experimental Scenarios. We use the prototype to validate our algorithm design and measure practical performance. In particular, we select 10 target phone locations with large errors, each with different combinations of peer locations from the office setup in Figure 1. Then we repeat the peer assisted localization for each target/peer location combination 10 times to measure the localization accuracy, total latency and energy consumption. In total we have 100 combinations of target/peer locations for the prototype experiments.

Trace-driven Statistical Performance. To capture the overall statistical performance under factors that we do not have enough manpower/device (e.g., more numbers of peers) or control (e.g., noise in different environments), we conduct trace-driven experiments. We use 200 combinations of target/peer locations in our office setup, feed the training data as WiFi samples, and distances perturbed with errors following the same distribution as results obtained in real environments (e.g., train station, shopping mall, and airport), as input to the peer-assisted algorithm. We study its the effectiveness of peer selection rules (Section 5.1), and the impact of various factors such as the number of peers, quality vs. quantity of peers, orientation estimations.

Metrics: We use the localization error to quantify the accuracy. We also measure the total time needed to complete the peer assistance process, and decompose the total time to understand the dominant factors. We measure the energy consumption using tools in [28] that is shown to achieve accuracy within 5%.

[7] We encountered some problem doing concurrent sound playing and recording on iPhone and are still investigating the issue.

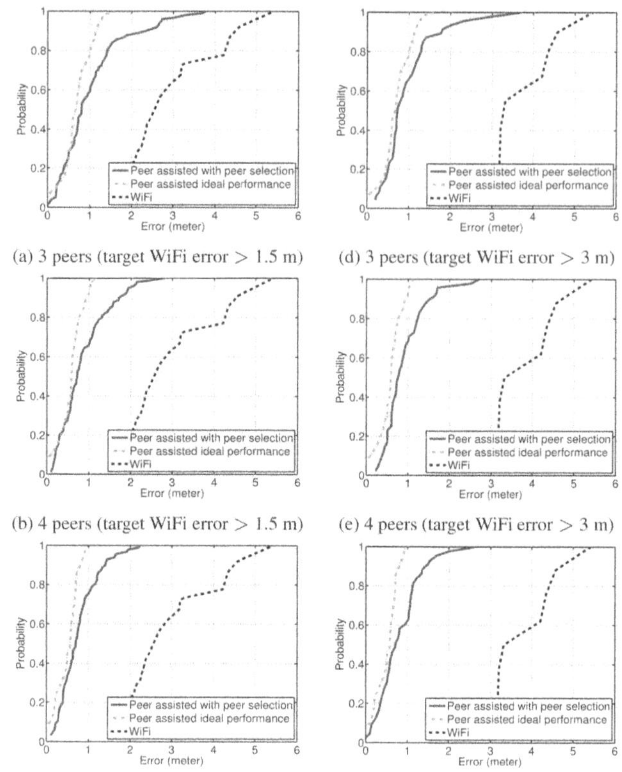

(a) 3 peers (target WiFi error > 1.5 m) (d) 3 peers (target WiFi error > 3 m)
(b) 4 peers (target WiFi error > 1.5 m) (e) 4 peers (target WiFi error > 3 m)
(c) 5 peers (target WiFi error > 1.5 m) (f) 5 peers (target WiFi error > 3 m)

Figure 9: Peer-Assisted localization performance for target phone under different number of peers for HTC EVO. The first colomn is results when the target WiFi error is greater than 1.5m and the second column with target WiFi error greater than 3m.

5.3 Prototype performance

Localization Accuracy We evaluate the accuracy in the lab for 3 and 4 phones (2 and 3 peers) due to limited manpower and device availability. Peers have WiFi localization errors under 1.5m, and the target over 3 meters. As shown in Table 2, we can observe that the peer assisted method can reduce the maximum error from 6.7m to 4.3m and 3 m with 2 peers and 3 peers respectively. For median error, there is at least 1.5m accuracy improvement. Furthermore, more peers benefit the localization performance: 3 peers further reduces 1.3m on the maximum error compared with 2 peers.

Overall Latency and Decomposition We report the latency measurements of major components in the peer assistance process. The recruiting takes about 1s, which is independent of the number of peers involved. Then the scanning of 5 WiFi samples takes about 4.8s on phones (HTC EVO), while in parallel the beep emitting takes 1.5s (4 phones, 0.3s time slot, 1 extra buffer slot), the file uploading 1s, the TOA on server 1.2s. After WiFi and TOA results, the algorithm takes another 2s on server. In total that is 7.8s (i.e., $1 + max(4.8, 3.7) + 2$). We want to point out that we have not optimized the system yet. E.g., the algorithm is mostly matrix operation, which can speed up at least one order of magnitude using proper libraries [8]. That would reduce the algorithm time to 0.2s, making the total time down to 6s. Since most other components

[8] http://stackoverflow.com/questions/529457/performance-of-java-matrix-math-libraries shows almost 40 times speedup.

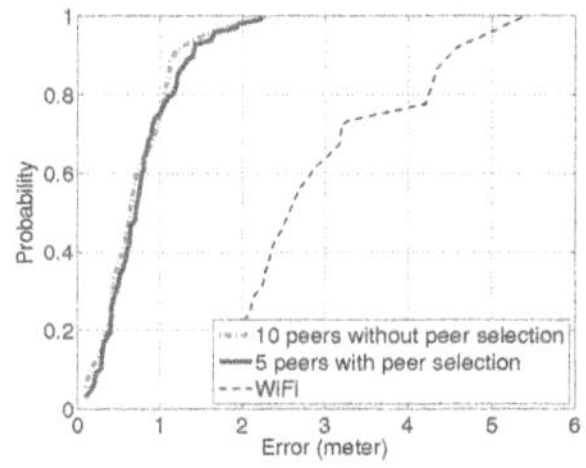

(a) target WiFi error > 1.5 m

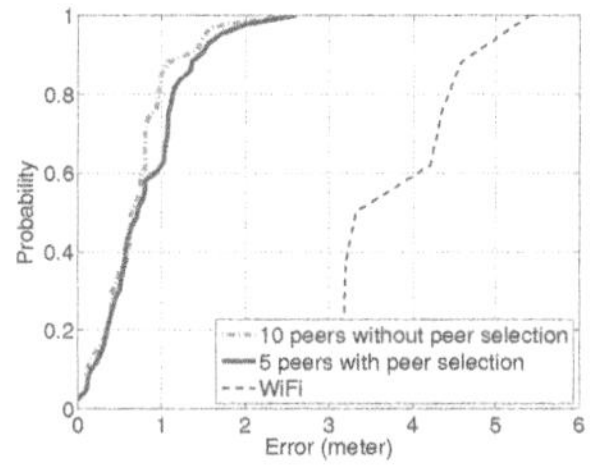

(b) target WiFi error > 3 m

Figure 10: Impact of peer quality vs. peer quantity.

can finish in parallel and faster than WiFi scan, our system does not pose much more latency than required in the original WiFi localization.

Energy consumption We also estimate the energy overhead using tools and methodology in [28]. The WiFi scan, sound emitting/recording and file uploading are the three major components. WiFi scan of 4.8s takes about 0.12 Joule. The beeps occupy about 0.18s in the 0.3s timeslot while the recording during the 1.5s schedule (4 phones) results in a file of 100KB. These audio parts take about 0.35 Joule. File uploading is about 2.1 Joules. In total it is 2.57 Joules, with file uploading dominating at 80%. All these happen in about 8s, translating into about 320mW additional power consumption. This is smaller than the average power of a phone (e.g., HTC Evo lasts 12.7 hours with average power of 450mW). We believe such overhead does not pose a burden for the battery life.

5.4 Overall Statistical Performance

We next use trace-driven experiments by feeding the results of initial WiFi estimates and error-perturbed acoustic ranging results as inputs to capture the statistical behavior of our peer-assisted method.

Effect of Peer Selection. We first study the effect of the 3 peer selection rules (Section 5.1) on localization accuracy. Figure 8 shows the localization error CDFs of those testing points with initial WiFi localization errors exceeding 1.5m for both HTC EVO and ADP2. We observe great performance improvements with and without peer selection with CDF curves shifting to the left significantly for both phones. In particular, the median error exhibits a 75% improvement under peer assistance with and without peer selection. We find that peer selection is essential for improving localization accuracy of points with large errors: the long CDF tail has been reduced 60% from 5.5m to 2.2m. This shows that peers filtered by the rules are more reliable reference points and constitute more accurate constraints on the target location. For the rest of the paper, we only present results with peer selection.

Impact of the number of peers. Figure 9 shows the localization accuracy for the target phone when varying the number of peers at 3, 4 and 5. In our experiments, the median error of initial WiFi localization is 1.5m, which indicates that about 50% of the testing

Localization Error (meter)	Mean	Median	90%	Maximum
Peer assisted (3 peers)	1.6	1.5	2.8	3
Peer assisted (2 peers)	1.9	1.8	3.1	4.3
WiFi	3.7	3.3	5.6	6.7

Table 2: Localization accuracy of the prototype: 2 and 3 peers for a target phone with large initial errors (> 3 m)

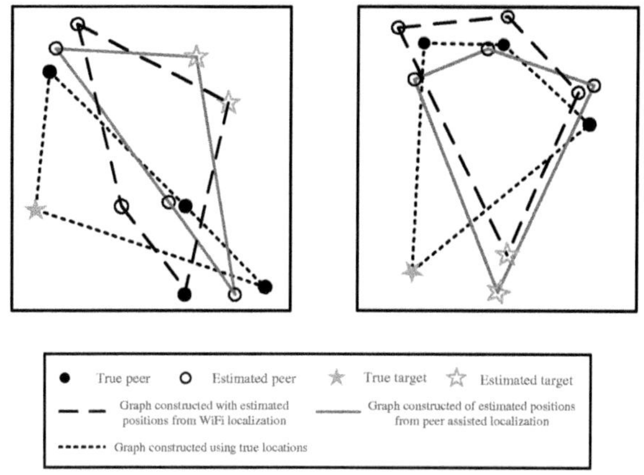

(a) Mirroring outliers (b) Rotation outliers

Figure 11: Illustration of larger errors for peer assisted localization method.

target points have small errors below 1.5m which are acceptable in most applications, while the other 50% of the target phones experiences errors beyond 1.5m. Based on this observation, we perform two sets of tests to evaluate the effectiveness of our peer-assisted approach: the first test applies to the 50% of the target phones with errors beyond 1.5m as shown in the first column of Figure 9, and the second test applies to those suffering from large errors exceeding 3m (about 30% of target phones) as shown in the second column of Figure 9.

The most encouraging result is that our approach show consistently great performance improvement for both test cases, suggesting that our algorithm is highly effective to reduce large errors. In general, more peers lead to more improvements in accuracy: When the number increases from 3 to 5 in Figure 9, the maximum error in the peer-assisted curve reduces from about 3.8m to 2.4m for both test cases, while the median error stays more or less the same around 0.7m. This is because more peers pose more constraints and less uncertainty on the location of the target. We also show the ideal performance where the true graph orientation and distances between phones are known for the algorithm. The maximum and median error are 1.5m and 0.5m. For about 80% of the case, the accuracy is very close (within 0.4m) to the ideal performance. This indicates that our orientation angle estimation is quite accurate.

Quality vs. Quantity of Peers One interesting question is how important is the quality versus the quantity of peers. Figure 10 presents the localization error CDF when using peer assisted localization with 10 peers without selection and 5 peers with selection (based on the rules designed in Section 5.1) respectively. We find that the performance of using 5 peers with selection is comparable to that of using 10 peers without selection when applying to scenarios of target phones with WiFi error greater than 1.5m, and only slightly worse (only for the bottom 40% cases and differ by about 0.2m) when the target WiFi error is greater than 3m. Thus, by utilizing peer selection only half of the peers is needed to achieve a similar performance as when adopting peers blindly. This observation strongly suggests that peer quality is more important than peer quantity and peer selection is critical.

5.4.1 *Insights on Remaining Errors*

During our initial experiments we analyzed what happened for the points towards the tail of the CDF curve under peer assistance. We found two reasons limiting the accuracy improvement. The first

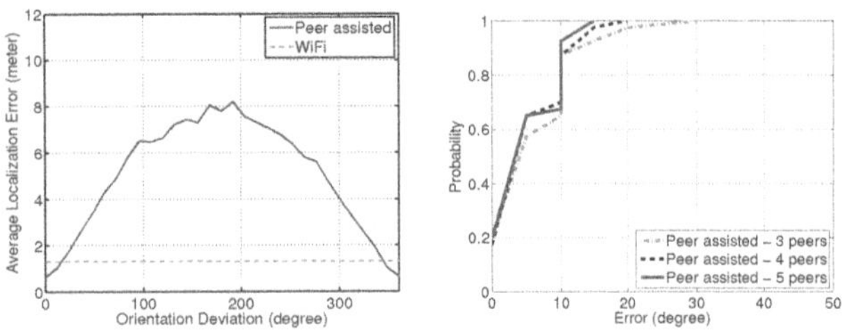

(a) (b)

Figure 12: (a) Average localization error under different orientation deviations; (b) Orientation estimation performance with different number of peers

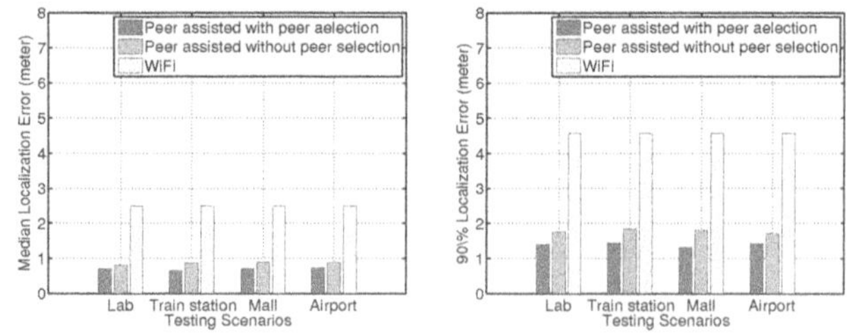

(a) Median Error (b) 90-Percentile Error

Figure 13: Localization performance under different environments with 5 peers.

is called "mirroring" (shown in Figure 11 (a)). When the peers are almost aligned on a straight line, the constructed rigid graph could be a flipped "mirror" of the ground truth. If the mirrored graph happens to have a smaller RSS distance summation, the target would be located to the opposite side of the line. By comparing the ranging measurements, we can tell whether such alignment happens and avoid using such peer combinations. We have addressed this problem in all previous results.

The second is due to errors in orientation when the target is relatively faraway from peers (shown in Figure 11 (b)). When the peers are close to each other, a small orientation deviation can move the target far away from its true location while the peers are still estimated close to their true locations. We tried some heuristics by comparing the relative distances among peers and to the target and found encouraging preliminary improvements (maximum error reduction from 3.6m to within 2m). We are still investigating for a complete solution to this problem.

5.5 Orientation Estimation and Sensitivity

The peer-assisted localization accuracy is affected by the orientation estimation. we take a rigid graph of 4 phones constructed using their true distances to showcase the sensitivity of localization results to orientation errors. We rotate the graph so it deviates from its true orientation at a fixed degree, then we move the graph translationally to find the optimal locations. The results are shown in Figure 12 (a). We find that when the orientation deviation is larger than 24°, the average localization error becomes larger than that of initial WiFi localizations ($1.3m$ in this case); it increases sharply to 8 meters when the orientation deviation approaches 180°. We examine cases with other numbers of peers, and find that generally the orientation deviation shall not exceed 20°, otherwise the improvement will be very limited. Figure 12 (b) shows the accuracy of orientation estimation under different number of peers. We find that our orientation estimation method is indeed very effective, which can restrict the error within $\pm 20^\circ$ and $\pm 10^\circ$ around the true angle for over 98% and 90% of the testing scenarios.

5.6 Impact of Various Environments

We further study the peer-assisted localization performance for the bottom 50% target points (i.e., those with $>$ 1.5m initial errors) with 5 peers, using ranging results perturbed with errors under various environments. In Figure 13, we find that our localization performance only varies slightly when comparing real-world environmens (e.g., train station, shopping mall and airport) to the lab environment. The median errors are all around 0.7m whereas the 90% errors are about 1.4m; even without peer selection, the median and 90% errors are around 0.8m and 1.7m respectively. Compared to the target's WiFi results of median error 2.5m and 90% error 4.5m, our proposed method is quite robust to noises in different environments.

6. DISCUSSION

Peer Involvement. Since peers need to spend energy and bandwidth for helping the target, we envision that whichever application that leverages our localization solution may have certain rewards (e.g., points, virtual currency) exchanged to compensate peers that help out. Such incentives are already present in many mobile applications. In our system, peers need to run a recruiting signal detection thread to know whether they are needed. To avoid running this thread continuously, the server may send push notifications to peers in a large area around a target upon its request for help. The notification triggers the thread; if no recruiting signal is detected after a timeout threshold, a peer will stop the thread. We note that the improvement in accuracy depends on the number and quality of peers. In general, more high quality peers lead to greater improvements. But even with just a few (e.g., 3) quality peers, there is still significant reduction in error.

Movements of users. Peers might be moving during the acoustic ranging. Movements affect the accuracy only when they occur during the sound-emitting period, which happens concurrently with and takes less time than WiFi scanning (1.5s vs 4.8s in our prototype). Thus we do not pose more constraints on movements than existing WiFi methods. If the peers are static or do not move too much in this 4.8s interval, both the ranging and WiFi measurements are still accurate. In many public spaces it is common for many users to remain still for a short while (e.g., resting in airports, window-shopping in stores). Movements before and after the WiFi scanning (and beeping) do not affect the accuracy. We recognize that less peers may be suitable due to movements and plan to examine additional input such as acceleration to filter out those that have moved too much. The main latency bottleneck is the WiFi scanning time, which depends on both hardware and OS that we do not have control. Our experiments show that the acoustic ranging design is robust to noises under different environments. One issue we plan to further investigate is the effects of clothes and human body. When phones are placed in pockets, the clothes or human body may attenuate the sound signal.

Triggering peer assistance. Our work provides the technology for peer assistance. However, it is eventually up to the users to decide when they desire such help. The rules provide some hints to users about the likelihood of large errors to help them make the decision. Users may also set up certain policies about under what kinds of conditions assistance is needed. One that always requires help gets high accuracy, at the cost of paying more points or virtual currency to others. In our system, more than one target phone may request assistance at the same time and the acoustic signals may collide if these targets are close to each other. Since users do not need assistance continuously, they request help only once in a while, e.g., when wondering which hallway to take to a train platform. So the likelihood of two nearby users requesting help simultaneously is small. When this happens, the server can distinguish them and provide assistance one at a time. This may delay the localization to some targets, but it still provides accurate results.

7. RELATED WORK

Smartphone indoor localization has attracted tremendous interests recently. Methods in prior work mainly focus on employing more sensing modalities, less infrastructure support, and reducing the efforts of building signal maps [2,8,15,24]. SurroundSense [2] utilizes multiple sensing modalities (e.g., cameras, microphones and accelerometers) available on smartphones to perform logical localization (e.g., different stores) via ambiance fingerprinting by combining optical, acoustic, and motion attributes. WiGEM [13] proposes a learning based approach that uses the Gaussian Mixture Model (GMM) and employs Expectation Maximization (EM) to estimate the model parameters without relying on labor-intensive "training." It is robust to multiple factors (including device and power level variability, mobility, and changes of indoor spaces) that many training-based systems are susceptible to. WILL [24] aims to perform indoor logical localization without the need of building radio signature maps ahead of time. It exploits abrupt signal changes through walls and accelerometers to infer user movements and achieve room level accuracy. EZ [8] is another work that targets configuration-free indoor localization by utilizing genetic-based algorithms. Large errors still exist with 50 and 80 percentile errors at 7m and 10m, respectively. WiGEM, WILL and EZ are complementary to our system for reducing the efforts of building the radio signal map during training.

Few studies have leveraged the unique peer constraints to assist smartphone localization. Virtual Compass [5] utilizes both WiFi and Bluetooth and results in a median error of 3-4m. Other studies require special hardware or infrastructure not readily available on smartphones [17,18]. Our work leverages abundant peers in public spaces to reduce large errors. It aims at the most prevalent WiFi infrastructure and do not require any special hardware, which is essential for easy deployment.

There have been quite some work for acoustic positioning techniques such as Cricket [20], Bat system [16], ENSBox [12], and WALRUS [6]. Recently, several proposals have studied ranging between cell phones using low frequency bands (e.g., 2-6kHz). Beepbeep [19] proposes a ToA based acoustic-ranging method between two phones. It can achieve 1 or 2cm accuracy in a range of 10m. Whistle [25] leverages multiple receivers with well-known locations to receive two sounds signals from the target. It avoids tight synchronization usually required in Time Difference of Arrival (TDoA) methods. Qiu *et al.* [22] utilizes the acoustic signal to estimate the relative position between two phones to support phone-to-phone games and apps.

Compared to the above work, we face a different challenge of fast and accurate ranging among multiple peer phones. We use server scheduling to coordinate multiple peer phones to avoid signal interference and collision for almost concurrent pairwise ranging. We propose signal design and detection methods that are lightweight in computation, robust to noisy environments, and much less perceptible to the human ear due to the adoption of high frequency band (e.g., 16-19kHz). Our system utilizes minimum auxiliary COTS sound hardware to reduce large errors incurred from general WiFi-based approaches.

For smartphone based localization using acoustic signals, Tarzia *et al.* [23] introduces a technique based on ambient sound fingerprint called Acoustic Background Spectrum. They exploit acoustic signals as fingerprints instead of measuring the ranging information between phones, and their localization granularity is at room level. Constandache *et al.* [10] deploys inaudible sound beacons randomly placed in the building, as a reference frame for correcting users' movement traces captured by the accelerometer and compass. However, this system requires an extra acoustic infrastructure and has localization errors around 8m on average. We exploit the constraints from nearby peer phones to achieve much higher localization accuracy, without the need for extra acoustic infrastructure.

8. CONCLUSION

Indoor localization on smartphones is critical to enable novel features for location based applications. However, existing approaches have yet to prove that they can satisfy what is desired in many business scenarios. Due to the prevalence of WiFi infrastructure, we set out to study the accuracy that WiFi localization can practically achieve on smartphones. We find that despite reasonable accuracies in many cases, the dynamic radio propagation poses fundamental limits and causes large errors. Inspired by the idea of relative positions of nearby peer devices as unique physical constraints on the possible location of a smartphone, we propose a peer assisted localization approach that leverages much more accurate distance estimate through acoustic ranging. Extensive experiments have demonstrated our approach successfully pushes further the limit of WiFi localization accuracy to what is empirically possible only under hundreds of APs, making WiFi a candidate for high accuracy localization. Our system does not pose much more latency than required in the original WiFi localization and has negligible impact on the battery lifetime.

9. ACKNOWLEDGMENTS

We sincerely thank our shepherd Samir R. Das and the anonymous reviewers for their invaluable feedback which helped improve this paper. This work is supported in part by the National Science Foundation Grants CNS0954020, CNS1016303, and CCF1018270.

10. REFERENCES

[1] Shopkick application. http://www.shopkick.com/.

[2] M. Azizyan, I. Constandache, and R. Roy Choudhury. Surroundsense: mobile phone localization via ambience fingerprinting. In *Proceedings of the 15th annual international conference on Mobile computing and networking (MobiCom'09)*, pages 261–272, 2009.

[3] P. Bahl, V. Padmanabhan, and A. Balachandran. Enhancements to the RADAR User Location and Tracking System. Technical Report Technical Report MSR-TR-2000-12, Microsoft Research, February 2000.

[4] P. Bahl and V. N. Padmanabhan. RADAR: An in-building RF-based user location and tracking system. In *Proceedings of the IEEE International Conference on Computer Communications (INFOCOM'00)*, pages 775–784, March 2000.

[5] N. Banerjee, S. Agarwal, P. Bahl, R. Chandra, A. Wolman, and M. D. Corner. Virtual compass: Relative positioning to sense mobile social interactions. In *Proceedings of the 8th International Conference on Pervasive Computing (Pervasive'10)*, pages 1–21, 2010.

[6] G. Borriello, A. Liu, T. Offer, C. Palistrant, and R. Sharp. WALRUS: wireless acoustic location with room-level resolution using ultrasound. In *Proceedings of the 3rd international conference on Mobile systems, applications, and services (MobiSys'05)*, pages 191–203, 2005.

[7] G. Chandrasekaran, M. A. Ergin, J. Yang, S. Liu, Y. Chen, M. Gruteser, and R. P. Martin. Empirical evaluation of the limits on localization using signal strength. In *Proceedings of the 6th Annual IEEE communications society conference on Sensor, Mesh and Ad Hoc Communications and Networks.*

[8] K. Chintalapudi, A. Padmanabha Iyer, and V. N. Padmanabhan. Indoor localization without the pain. In *Proceedings of the sixteenth annual international conference on Mobile computing and networking (MobiCom '10)*, pages 173–184, 2010.

[9] CNN. New version of google maps brings indoor floor plans to your phone. http://www.cnn.com/2011/11/30/tech/mobile/new-version-google-maps-indoors-wired/.

[10] I. Constandache, X. Bao, M. Azizyan, and R. R. Choudhury. Did you see bob?: human localization using mobile phones. In *Proceedings of the sixteenth annual international conference on Mobile computing and networking (MobiCom'10)*, pages 149–160, 2010.

[11] S. Gelfand and H. Levitt. *Hearing: An Introduction to Psychological and Physiological Acoustics*. Marcel Dekker New York, 2004.

[12] L. Girod, M. Lukac, V. Trifa, and D. Estrin. The design and implementation of a self-calibrating distributed acoustic sensing platform. In *Proceedings of the 4th international conference on Embedded networked sensor systems (Sensys '06)*, pages 71–84. ACM, 2006.

[13] A. Goswami, L. E. Ortiz, and S. R. Das. WiGEM: a learning-based approach for indoor localization. In *Proceedings of the 7th International Conference on emerging Networking EXperiments and Technologies*, 2011.

[14] J. Graver, B. Servatius, and H. Servatius. *Combinatorial rigidity*. Graduate studies in mathematics. American Mathematical Society, 1993.

[15] Y. Haibo, G. Tao, Z. Xiaorui, X. Jingwei, T. Xianping, L. Jian, and J. Ning. Ftrack: Infrastructure-free floor localization via mobile phone sensing. In *Proceedings of the IEEE International Conference on Pervasive Computing and Communications (PerCom'12)*, 2012.

[16] A. Harter, A. Hopper, P. Steggles, A. Ward, and P. Webster. The anatomy of a context-aware application. In *Proceedings of the 5th annual ACM/IEEE international conference on Mobile computing and networking (Mobicom '99)*, pages 59–68, 1999.

[17] M. Hazas, C. Kray, H. Gellersen, H. Agbota, G. Kortuem, and A. Krohn. A relative positioning system for co-located mobile devices. In *Proceedings of the 3rd international conference on Mobile systems, applications, and services (MobiSys'05)*, pages 177–190, 2005.

[18] M. Minami, Y. Fukuju, K. Hirasawa, and S. Yokoyama. DOLPHIN: a practical approach for implementing a fully distributed indoor ultrasonic positioning system. *Proceedings of 6th international conference on Ubiquitous Computing (UbiComp'04)*, 2004.

[19] C. Peng, G. Shen, Y. Zhang, Y. Li, and K. Tan. Beepbeep: A high accuracy acoustic ranging system using cots mobile devices. In *In Proceedings of the 5th international conference on Embedded networked sensor systems (Sensys'07)*, pages 1–14, 2007.

[20] N. Priyantha, A. Chakraborty, and H. Balakrishnan. The cricket location-support system. In *Proceedings of the ACM International Conference on Mobile Computing and Networking (MobiCom'00)*, pages 32–43, Aug 2000.

[21] A. Prorok, P. Tome, and M. Alcherio. Accommodation of NLOS for Ultra-Wideband TDOA Localization in Single- and Multi-Robot Systems. In *Proceedings of the International Conference on Indoor Positioning and Indoor Navigation*, 2011.

[22] J. Qiu, D. Chu, X. Meng, and T. Moscibroda. On the feasibility of real-time phone-to-phone 3d localization. In *Proceedings of the 9th ACM Conference on Embedded Networked Sensor Systems (SenSys'11)*, pages 190–203, 2011.

[23] S. P. Tarzia, P. A. Dinda, R. P. Dick, and G. Memik. Indoor localization without infrastructure using the acoustic background spectrum. In *Proceedings of the 9th international conference on Mobile systems, applications, and services (MobiSys'11)*, pages 155–168, 2011.

[24] C. Wu, Z. Yang, Y. Liu, and W. Xi. WILL: Wireless Indoor Localization Without Site Survey. In *Proceedings of the IEEE International Conference on Computer Communications (INFOCOM'12)*, pages 64–72, 2012.

[25] B. Xu, R. Yu, G. Sun, and Z. Yang. Whistle: Synchronization-free tdoa for localization. In *Proceedings of the 2011 11st International Conference on Distributed Computing Systems (ICDCS'11)*, pages 760–769, 2011.

[26] J. Yang, S. Sidhom, G. Chandrasekaran, V. Tam, H. Liu, N. Cecan, Y. Chen, M. Gruteser, and R. P. Martin. Detecting driver phone use leveraging car speakers. In *Proceedings of The 17th Annual International Conference on Mobile Computing and Networking (MobiCom'11)*, pages 97–108, 2011.

[27] M. Youssef and A. Agrawala. The horus wlan location determination system. In *Proceedings of the 3rd international conference on Mobile systems, applications, and services (MobiSys'05)*, pages 205–218, 2005.

[28] L. Zhang, B. Tiwana, Z. Qian, Z. Wang, R. P. Dick, Z. M. Mao, and L. Yang. Accurate online power estimation and automatic battery behavior based power model generation for smartphones. In *Proceedings of the eighth IEEE/ACM/IFIP international conference on Hardware/software codesign and system synthesis.*

[29] Z. Zhang, X. Zhou, W. Zhang, Y. Zhang, G. Wang, B. Y. Zhao, and H. Zheng. I am the antenna: Accurate outdoor ap location using smartphones. In *Proceedings of The 17th Annual International Conference on Mobile Computing and Networking (MobiCom'11)*, pages 109–120, 2011.

Empowering Developers to Estimate App Energy Consumption

Radhika Mittal[‡], Aman Kansal[†], Ranveer Chandra[†]
[‡]IIT Kharagpur, Kharagpur, India
radhikamittal.iitkgp@gmail.com
[†]Microsoft Research, Redmond, WA
{kansal,ranveer}@microsoft.com

ABSTRACT

Battery life is a critical performance and user experience metric on mobile devices. However, it is difficult for app developers to measure the energy used by their apps, and to explore how energy use might change with conditions that vary outside of the developer's control such as network congestion, choice of mobile operator, and user settings for screen brightness. We present an energy emulation tool that allows developers to estimate the energy use for their mobile apps on their development workstation itself. The proposed techniques scale the emulated resources including the processing speed and network characteristics to match the app behavior to that on a real mobile device. We also enable exploring multiple operating conditions that the developers cannot easily reproduce in their lab. The estimation of energy relies on power models for various components, and we also add new power models for components not modeled in prior works such as AMOLED displays. We also present a prototype implementation of this tool and evaluate it through comparisons with real device energy measurements.

Categories and Subject Descriptors

C.4 [**Computer Systems Organization**]: Performance of Systems; D.2.2 [**Software**]: Software Engineering—*Design Tools and Techniques*

General Terms

Algorithms, Experimentation, Measurement, Performance, Design

Keywords

energy efficiency, display power, developer tools

1. INTRODUCTION

Poorly written apps can sap 30 to 40% of a phone's battery [10]. Battery lifetime is a common cause of frustration in smartphone users. Several efforts in research and industry are investigating techniques to improve the battery life, such as through use of higher battery density, dedicated low power processors for offloading computations from the application processor, or even off-loading to the cloud. However, such platform layer improvements do not suffice by themselves since poorly written software can callously eat into the resultant extra battery juice.

A significant portion of the battery is used when the phone is actively used in interactive foreground apps. While OS designers spend significant effort in optimizing the battery impact of background OS services, the foreground apps are largely controlled by app developers. Very few tools are available for developers to improve the energy efficiency of their apps. In fact, most developers are not even aware of the amount of energy their app consumes under a typical usage pattern. While they could use a power meter to measure the energy draw for their app during test runs, such an approach is not used in practice because it is tedious and expensive. It is very difficult to perform these measurements for multiple devices, network conditions, user settings for screen brightness, and so on. Even if measurements are performed, they are biased to the network conditions at the developer's location. Measurements by themselves do not give any insight into how much each component (CPU, network, or display) contributed to battery drain, making it harder to focus the optimization effort.

One alternative for developers is to use the "battery use" tool on Android phones, the Nokia Energy Profiler (NEP) or power modeling tools such as eProf [15]. While these tools can circumvent the need for power metering equipment, they are still limited in exploring multiple configurations, user settings, or network conditions. The profiled behavior remains biased to conditions at the developer's location.

In this paper we present WattsOn, a system that allows a developer to *estimate the energy consumed by her app in the development environment itself.* WattsOn can (i) identify energy hungry segments during the app run, and (ii) determine which component (display, network or CPU) consumes the most energy. For instance, an developer may determine if the battery drain is dominated by the download of a high resolution image is consuming more energy, the app display using a white background, or the computation that the app is doing. The developer may compare the energy impact of design choices such as developing a portable browser based app that downloads most content on the fly or a native app that only downloads incremental updates.

Furthermore, WattsOn allows *what-if* analyses, to answer questions, such as: How much energy is consumed on a different phone model? How does energy consumption change if the user has a 2G or a 3G network? What if the brightness is set to high? or, What if the app is used under a low signal strength area?

Our current WattsOn prototype emulates the power consumption for the display, network, and CPU, since these are the dominant energy consumers on a smartphone, consuming between 800mW to 1500mW in their highest power states. Other components such

MobiCom'12, August 22–26, 2012, Istanbul, Turkey.

as the GPU (peaking to 250mW - 350mW) and A-GPS (160mW - 350mW, across various chip sets), while significant, account for a smaller fraction of the total power consumption for interactive foreground apps. These components can be added to our framework provided appropriate resource consumption counters, relevant power models, and emulation strategies are available.

Specifically, we make the following contributions.

First, we present the first system that can estimate an appŠs energy consumption in different operating conditions (carrier, signal strength, brightness) without requiring expensive lab equipment and explicit measurements with repeated runs for each operating condition.

Second, we enable energy to be profiled within the development environment without requiring a specific mobile device. To achieve this, we scale the development machine's resources including the CPU and the network to match the characteristics of a real phone. The app thus behaves as if on a real phone and its energy impact on all components, such as idle display power-on time while waiting for a download, is correctly emulated. The combination of power modeling and resource scaling required for WattsOn design implies that only those power models and scaling techniques can be applied that are mutually compatible. WattsOn allows the app to use external resources such as web services from the Internet. These requirements are in contrast to other energy emulators, such as Power TOSSIM [22], based on event driven simulation, where all communicating nodes are within the simulator. Hence techniques such as virtual clocks cannot be applied in WattsOn for resource scaling. Processor power models based on architecture specific performance counters are also not amenable to use with scaled resources.

Third, we expand the catalogue of power models available for mobile devices. Power models for many components have been established previously in [2,4,6,8,17,20] at varying levels of complexity and accuracy. We use previously proposed models for CPU and WiFi. For displays, while LCD and OLED displays have been modeled in detail [5,25], prior work did not provide a model for AMOLED displays. Hence we develop a new power model for AMOLED. For cellular networks, while the power model exists [18], measurements for model parameters with varying signal strengths and operators were not available in published literature. We perform new measurements to fill in some of the missing data.

Finally, we have validated WattsOn with multiple applications, devices, network conditions, and carrier networks. Average energy error varied from 4% to 9% across the apps tested. The accuracy of energy estimation when compared with the variability in hardware energy measurement for the same task over multiple runs indicates that WattsOn can offer a better energy estimate by eliminating several variable factors such as background activity on the mobile device. We also show how the component energy break-down produced by WattsOn can help application designers.

While the concepts of power modeling and resource scaling have been developed before, to the best of our knowledge, this is the first work that investigates the suitable selection of modeling and scaling techniques for development-time mobile app energy emulation, and validates their use through fine grained hardware measurements.

2. WattsOn SYSTEM DESIGN

Most mobile development toolchains provide an emulator to assist in app development such as the Android Emulator, the iOS Simulator, or the Windows Phone Emulator. While the emulators do not accurately reproduce all mobile device characteristics, the low overhead of their use makes them very beneficial for a variety of tests. WattsOn extends existing emulators to estimate app battery consumption.

The two major techniques used in WattsOn design are power modeling and resource scaling.

Power Modeling: One way to measure app energy is to use power metering equipment [7]. However, requiring every app developer to install such equipment is an arduous ask. Also, measurements have variability due to differences in network conditions, background activities running on the phone, not all of which can disabled by the developer. For instance, the measured energy after an energy reducing change made by the developer may turn out to be higher than before the change, due to degraded network quality at the time of the new measurement. Further, the measurement does not separate out the impact of display, network, and CPU that can be important for the developer to make their optimization decisions. To overcome these limitations, WattsOn computes energy from the resource utilization counters using power models [2,4,6,8,17,20].

Resource Scaling: Resources consumed by the app on the developer's workstation are very different from those on the phone. This leads to two challenges. First, the resource counters measured on the developer workstation cannot be fed directly into the phone power models. Second, timing of events might be different when running the app on the emulator than on the phone. Network packets may arrive much faster on the development machine, causing the user clicks for the subsequent tasks to be occur sooner, drastically changing the time spent on application tasks. Resource scaling addresses these challenges.

2.1 Design Overview

A block diagram of WattsOn is shown in Figure 1. The leftmost blocks represent the measurement of real device power characteristics required for *power model generation*[1]. These measurements may be performed by the smartphone manufacturers, mobile OS platform developers, or even volunteers using automated modeling methods [6]. The mobile app developer simply downloads the appropriate models.

On the developer machine, the app code for the mobile device runs in a mobile device emulator. We insert *resource scaling* techniques between the emulator and the actual hardware. As the app is executed on the emulator, we monitor its resource consumption using *resource profiling* methods available on the development workstation. The resource consumption monitored on the scaled resources is used in the *energy calculation* block to estimate the app energy using power models.

We describe resource scaling and power modeling methods for each of the modeled components below.

2.2 Cellular Network (3G)

The cellular network interface [17] consumes significant energy and we emulate it as follows.

2.2.1 Resource Scaling

The goal of resource scaling is to obtain the network resource consumption of the app as if executed on a real cellular data link. Multiple scaling options may be considered:

Virtual Clock: Event driven simulators such as ns-3 and Power TOSSIM [22] simply record time in ticks and the ticks can be mapped to the real time for the network nodes of interest while they may run much faster (or slower) on the simulation workstation.

[1]Power models may be developed for all mobile devices of interest; the number of models required may be reduced by considering representative devices in various device classes with different screen sizes and cellular network types.

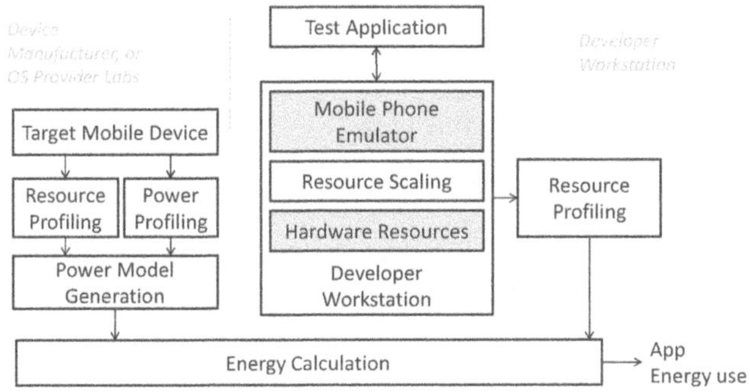

Figure 1: Block diagram of WattsOn components. Blocks shaded gray are used in existing systems; WattsOn adds the other blocks for energy emulation.

This technique is not suitable for WattsOn because the emulated app uses external resources such as web services on the Internet that are not operating on simulated tick time.

Trace Stretching: Another possibility is to capture the packet activity over the high speed network connection of the developer's machine and then stretch the timing characteristics to match those on a lower speed link. For example, suppose a 10 second trace would take 30 seconds over the cellular interface, one could multiply all time intervals in the Ethernet packet trace by 3. One could even replay the trace from the high speed interface on a lower speed interface simulated using a fine grained packet level simulator, such as QualNet or ns-3 and get more accurate stretching. However, this technique has several drawbacks. A slower interface will affect other parameters of the network flow, such as the TCP window size, which are not captured by stretching the high speed packet trace. Also, stretching the network trace alone will not produce the corresponding impact on other resources such as the timing of user clicks or other tasks in the app that depends on network activity.

Link Shaping: The approach we have taken is to shape the network link bandwidth and latency such that the emulated network activity in terms of packets sent and received is similar to the activity that would be observed on a cellular data link for the same task. We introduce our resource scaling code at layer 2.5 in the network stack, that is, between IP and the MAC layer. The delay, loss and bandwidth parameters are chosen to mimic different network conditions. The other components of the network remain the same as used by the app on a real device.

The latency, bandwidth and loss parameters for cellular networks have been studied in several prior works in wireless communications, and we rely on the existing literature to guide our choice of these parameters. Using measured distribution based models, rather than physical measurements, has the advantage that the estimate is much *more representative for a wide population of users.* It is not biased to the developer's specific phone location, and does not show uncontrolled fluctuations from measurement to measurement. Parameter values may be updates as technology evolves. In our prototype, we selected the parameter values as follows.

Latency: The measurement and modeling work in [9, 12] has characterized the cellular data link latency for 3G networks using a normal distribution (with mean = 200ms, and standard deviation = 100ms) and we use this characterization.

Bandwidth: We use the measurements from [23] that have experimentally characterized 3G HSDPA download and upload bandwidths. The bandwidth varies due to various reasons including changes in network congestion, wireless channel quality at the phone location, and other factors. To keep the number of varying conditions manageable, we bin the network quality into three levels, denoted *good, average*, and *poor*, and based on [23], set the parameter values as shown in Table 1.

Network quality	Download (kbps)	Upload (kbps)
Good	2500	1600
Average	1500	900
Poor	500	200

Table 1: Network scaling parameters for bandwidth.

Loss: We model losses using the well known Gilbert Elliot Channel Model and the corresponding parameters measured for 3G links from [26]. According to this model, the network is assumed to be in one of two states, denoted good and bad, each with a different loss rate. The model also describes the transition probabilities between the two states, that allows simulating the losses over time during an emulation run.

In certain instances, when the developer is using a slow Internet connection for their developer machine, such as a home Internet connection, then the underlying network may itself have non-negligible latency and losses. In this case, WattsOn should first probe and estimate the network latency using known methods [11] and then add on any additional resource scaling to the measured characteristics.

2.2.2 *Power Modeling*

The power model for the 3G network must model not only the active energy consumption when communicating data but also the "tail" time, or the time for which the radio interface remains in a higher power state after finishing the communication activity. Part of the tail time is spent in the active state (DCH) itself, and part in an intermediate state (FACH), where the radio power consumption is reduced but any further communication requires a small channel acquisition overhead. Some radios may have a second lower powered intermediate state (PCH). This model, called ARO model, was studied in depth in [17, 18], and allows back-calculating the radio power state from a network packet trace.

However, the number of mobile operators for which the model parameters have been measured is limited [18]. Second, the energy consumed at varying received signal strengths has not been reported for the ARO model, though dependence on signal strength is well known [21]. To fill in the above gaps, we set up two experiments (Figure 2).

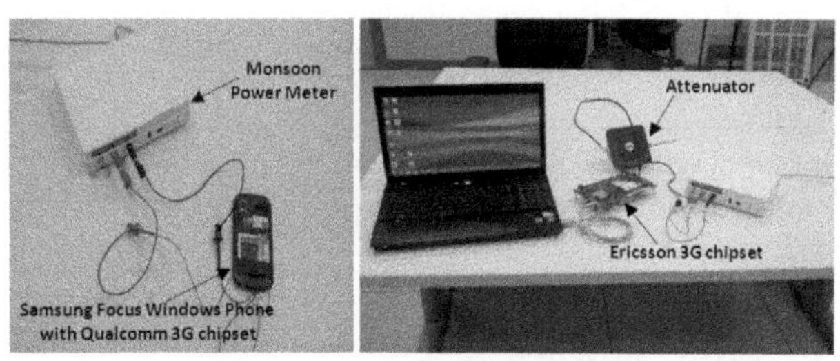

Figure 2: Experimental setups for the network power measurements.

Signal Strength. We used an Ericsson cellular data development board that exposes the antenna ports and allows controlling the received signal strength via an RF signal strength attenuator. The setup was located where we typically observed good signal strength. Measurements were performed on a weekend when the network was lightly loaded. We varied the amount of data downloaded and uploaded at different signal strengths and measured the DCH and FACH power and tail times. Since a developer will likely only emulate their application with a small number of signal strength variations, we discretize the signal strength to three levels (Table 2). These power measurements are taken using a radio interface board and not a phone, implying that the idle power of the board can be different from that of the phone. However, the differences between the power levels at different signal strengths comes primarily from the radio and can be used to adapt the power model for varying signal strengths. The tail times for DCH and FACH did not vary with signal strength.

Signal Strength	DCH (mW)	FACH (mW)
High	600	300
Medium	800	300
Low	1500	400

Table 2: Cellular interface power variation with signal strength for the AT&T network.

Mobile Operators. The second setup uses a power meter attached to the battery terminals of a smartphone to measure power. We connected multiple devices from different mobile operators available in our region: AT&T (Samsung Focus), T-Mobile (HTC HD7), Verizon (HTC Trophy), and Sprint (HTC Arrive). T-Mobile and AT&T use GSM based networks with the UMTS standard from 3GPP for data. Verizon and Sprint use the Evolution-Data Optimized (EVDO) standard from the CDMA2000 family provided by 3GPP2. A sample power trace for a download using Sprint's network is shown in Figure 3. As expected, the radio stays in a high power state long after the data communication has completed. Similar measurements were performed for other operators and all measured tail times are listed in Table 3.

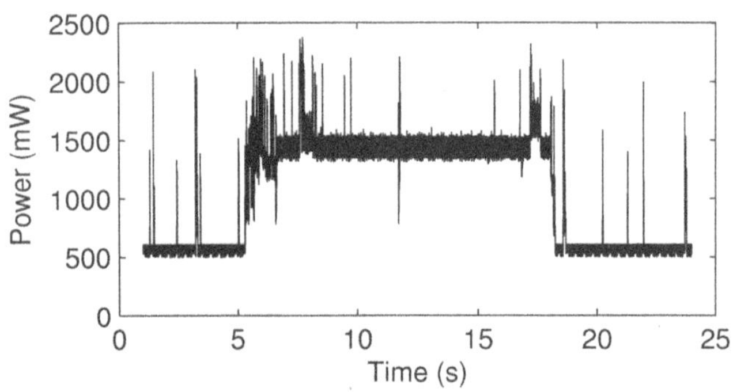

Figure 3: Network tail energy measurement for Sprint. The data communication ends near time = 7s along the x-axis but the radio stays in a higher power state for an additional 10 seconds after that.

2.3 WiFi Network

The resource scaling and power model methods for WiFi used are as follows.

Resource Scaling: For cases where the developer machine is connected to the Internet using WiFi (as is common for laptops), resource scaling is not needed. Otherwise, resource scaling is performed using the same layer 2.5 approach as for the cellular network. The scaling parameters are based on well-studied WiFi characteristics [1].

Operator	DCH	FACH	PCH
AT&T (3G)	5s	12s	0
T-Mobile (3G)	5s	1s	1s
T-Mobile (4G HSPA)	4s	2s	1s
Verizon (3G)	6s	0	0
Sprint (3G)	10s	0	0

Table 3: Tail state times for different operators. SOme operators do not use all intermediate states resulting in zero tail times in those states.

Power Model: The WiFi power model uses the PSM state model described in [13]. The model uses four states - Deep Sleep (10mW), Light Sleep (120mW), Idle (400mW), and High (600mW). The power consumption is slightly different for transmit vs. receive but since the difference is small and switching between these states is very frequent, we use a common power value, denoted as the high state.

When not communicating, the interface remains in Deep Sleep. Brief power spikes of 250mW at intervals of 100ms are observed in this state, corresponding to reception of beacons from an associated AP. When a packet is to be transmitted, the interface moves to the High state immediately. If a packet is to be received, the radio learns about it at the next beacon, and moves to the High power state. Once the transfer is completed, the radio moves to Idle. From Idle it can immediately move to High in case of a transmit or receive. If no network activity occurs in Idle state for 1s the radio moves to the Light Sleep. The Light Sleep tail time is 500ms, after which, if no network activity occurs, it falls back to Deep Sleep. If network activity does occur in either the Idle or Light Sleep state, the state changes to High. Both Idle and Light Sleep states also have regular spikes of additional 250mW of power spaced at 100ms to receive beacons.

The above power state transitions can be re-created, and energy can hence be computed, using the (scaled) network packet trace captured using a network sniffing library.

2.4 Display

As for other components we need to resource scale the display and model its power consumption. Fortunately, existing mobile device emulators already perform resource scaling for the display: the emulated app is provided only a small screen area representing the mobile device display. The only scaling needed is that the emulator window may be re-sized to a larger view by the developer, changing the number of pixels in the display and to overcome this, one may simply multiply the number of pixels by the appropriate scaling ratio.

While the peak power of the display is similar to that of the CPU and the network, the fraction of energy consumed by the display can be much larger than the other components since the display is constantly active throughout an application's use. An accurate estimation is thus important for this most dominant energy consumer. The power model for the display depends on the display technology used. Prior work has provided power models for LCD and OLED displays [5, 25]. However, several modern mobile devices use Active Matrix OLED (AMOLED) displays, that do not fit existing models.

The OLED power model shows linear and additive properties: the energy consumption of the display as a whole is the sum of the energy consumption of the R,G, and B components of the individual pixels. Further, the power model for the R,G, and B components is linear, provided the colors are converted from the standard RGB (sRGB) color space to linear RGB. This does not hold for

AMOLED. Figure 4 shows the power measured for the AMOLED display (on a Samsung Focus smartphone) set to different colors. The R,G and B components labeled are converted to linear RGB. While the linear model holds at low magnitudes, it breaks down at high magnitudes (the colors in the surface labeled B=255).

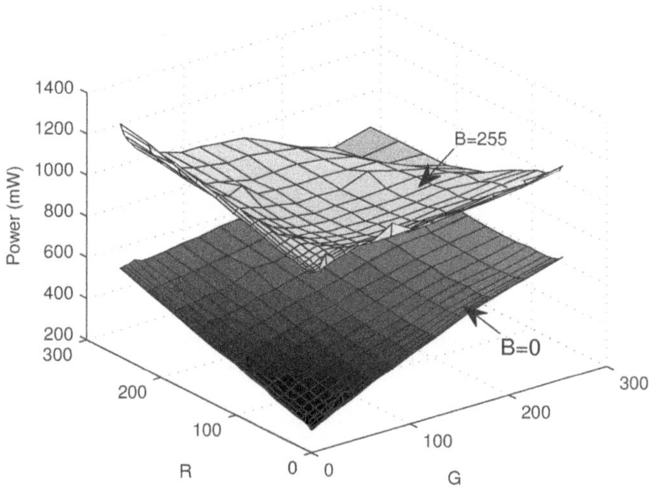

Figure 4: Power measurements for different colors on an AMOLED display.

Both the additive and linearity properties break down. As seen in the figure, power can increase or decrease with increasing color component values, implying that linearity does not hold. More measurements reveal that the power consumed is not just a function the color of a pixel but also depends on the properties of the other pixels in the image. Thus, the additivity does not hold either. Figure 5 shows the power consumption at varying fractions of the screen set to white. The graph shows the power predicted using the additive OLED model where the power would increase linearly with the area, as well as the observed ground truth. Power decreases when a greater portion of the screen is emitting brighter light levels.

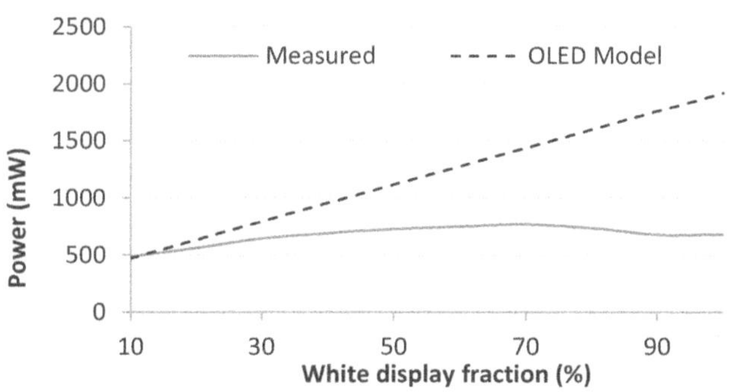

Figure 5: AMOLED power changes as the fraction of white colored pixels changes.

Further measurements reveal that reduction in power compared to the OLED model not only depends on the area but varies by color. Figure 6 shows an example measurement for two colors, compared with the OLED model based predicted values. While the predicted values are different for the two colors, what is important to note is that the difference between the measured and predicted values is also different for the two colors: $s2$ is significantly greater than $s1$.

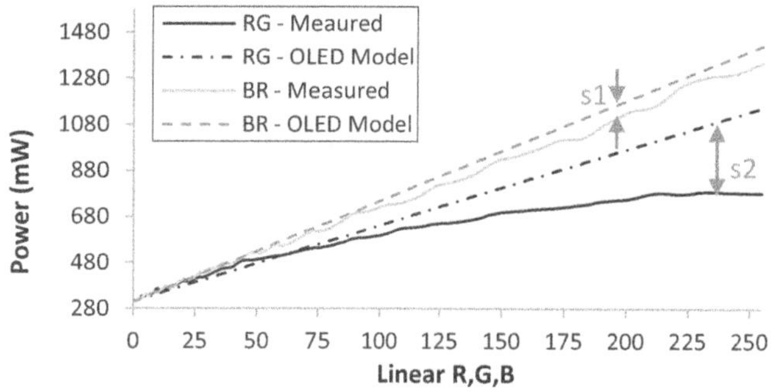

Figure 6: Change in power with color.

With display technologies being highly proprietary, the exact mechanisms used to control power are not widely available. Given the dependence on color and lack of explicit knowledge of the underlying power optimizations, the AMOLED power model becomes difficult to parameterize. Thus, we use a lookup table spanning the entire R,G,B gamut of values, discretized to 16 color magnitudes per component, yielding a table with $16 \times 16 \times 16 = 4096$ entries. The entries of this table contain the power value measured if the entire screen was set to that color, and in effect models the power scale down compared to the additive model at 100% area coverage for that color. The lookup table cannot be used directly however because the power value does not follow the lookup table when a color occupies less than the entire screen, and then its power consumption is closer to that predicted by the additive OLED model. Rather than fitting 4096 individual curves to model the area dependant power variation for the basis colors, we make a simplifying approximation that is based on the observation that the deviation from the OLED model is only significant for higher color magnitudes. The deviation does become significant at different threshold for different colors and we approximate that using four different thresholds: one for R+G+B magnitude, and one each for R+G, G+B, and B+R magnitudes, denoted $\tau_{RGB}, \tau_{RG}, \tau_{GB}$ and τ_{BR} respectively. The power scale down is assumed to be proportional to the fraction of such high intensity pixels.

The final model is summarized as follows. Suppose the lookup entry for a color indexed by $[r, g, b]$ by finding the nearest index in the lookup table is denoted $l(r, g, b)$. Let the power consumption of the entire screen, obtained by summing up the powers over all pixels of a screen-shot $\mathbf{s}$, be denoted $L(\mathbf{s})$. Suppose the fraction of pixels in $\mathbf{s}$ that exceed either of the thresholds $\tau_{RGB}, \tau_{RG}, \tau_{GB}$ and τ_{BR} is denoted $\beta(\mathbf{s})$. Also, suppose that the power consumption for the display using the additive OLED power model is computed as $O(\mathbf{s})$. Then the power consumption when the screen display matches $\mathbf{s}$ is computed as:

$$P_{display} = \beta(\mathbf{s}) * L(\mathbf{s}) + (1 - \beta(\mathbf{s})) * O(\mathbf{s}) \qquad (1)$$

One further optimization to the above model is that instead of considering all pixel values on the screen, we could randomly select a smaller fraction of the pixels and use those as representative of the entire screen. The accuracy of this model along with sub-sampling is evaluated in Section 4.

From an implementation perspective, the pixel information of the display is easy to obtain when an app is being emulated. We simply capture the developer machine's screen and extract the region corresponding to the emulator's display area.

For completeness, we also measured the power consumption of LCD displays with varying screen sizes at two different color levels (Table 4). The LCD can be modeled using simply the brightness level as the variation with color is not significant. As expected, devices with different screen sizes do consume different amounts of energy.

Brightness	**Color**	3.6in	**4.3in**
Low	Black	132 mW	294 mW
	White	127 mW	330 mW
Medium	Black	363 mW	557 mW
	White	359 mW	573 mW
High	Black	559 mW	778 mW
	White	554 mW	790 mW

Table 4: Power measurements for LCD displays for HTC Arrive (3.6in screen size) and HTC HD7 (4.3in screen size).

2.5 CPU

Mobile devices use much lower power processors due to their battery constraints compared to the processors used on the developer workstations. The processor frequency, cache hierarchy, and various computational units are different, resulting in very different execution performance across the two processors for the same computation. Since processor designs have been extensively studied, many techniques for emulating performance and power have been developed for CPUs. We select resource scaling and power modeling methods that are mutually compatible and work with the full system app energy emulation constraints.

2.5.1 Resource Scaling

One approach to resource scaling would be to use a cycle-accurate simulation of the processor. Detailed cycle-accurate simulation has excessive computational overheads, and is impractical, especially since it may not run in real time and result in inaccuracies and delays in the timing of other components. The existing Android Emulator uses instruction set simulation through binary translation, built upon QEMU [16]. The Windows Phone Emulator similarly uses processor virtualization with support for memory and GPU emulation. While these lighter weight approaches followed by Android and Windows Phone emulators allow capturing certain CPU characteristics and memory limitations of the mobile device, they *do not preserve the timing characteristics* and do not model the processor architecture in sufficient detail to provide accurate resource counters. The Android Emulator allows for developer driven scaling of speed through a delay parameter that can take values between 0 and 1000 to add a non-deterministic delay to the application execution. However, the choice of the delay input is left to the developer.

In WattsOn, we scale down the performance of the emulated app running on the development machine by restricting the number of processor cycles available to the mobile device emulator. The goal is primarily to preserve the timing characteristics on the CPU. This is only an approximation because the nature of computation can affect how the number of cycles on one processor map to the other processor, given that the processor architectures are vastly different and aside from the number of cycles, the size of processor caches, accelerators and bus speeds will matter as well. We compare the execution time on a mobile device processor (Samsung Focus with a 1GHz Scorpion CPU) and a development machine using a 2.7GHz Intel Core-2 Quad-core processor for a few simply computational tasks including floating point computations, fixed point computations, and memory intensive compute tasks. The slow-down in execution time did not vary greatly across these tasks and for this pair of processors, the slow down was a factor of 7.2, implying that a 100% CPU utilization on the phone processor can be approximated using a $100/7.2 = 13.8\%$ utilization of one core on the developer machine processor[2]. The emulator can now be restricted to a fraction of the overall processor that yields the same slow down in execution speed. The fraction of restricted cycles is not exactly equal to the desired slow down ratio since the emulator may have additional overheads other than the execution of the test app[3].

The actual mechanism to restrict the number of cycles allocated to the emulator depends on the operating system used on the developer's workstation. For instance, in Linux, one may use the `cpulimit` utility. In Windows the same can be achieved via restricting the emulator to a virtual machine and restricting the CPU fraction allocated to it.

2.5.2 Power Modeling

Power models for the CPU are available in the literature [20] and we used a simple utilization based power model where the CPU power is expressed as a linear function of the phone's CPU utilization:

$$P_{cpu} = \alpha * u_{cpu} \quad (2)$$

Here, u_{cpu} represents the phone CPU utilization and α is a power model parameter. The CPU utilization measured by WattsOn for the emulator process on the developer machine is scaled according to the scaling factor in Section 2.5.1 to obtain u_{cpu}.

We measured the value of α on the Samsung Focus device for different computations. The above linear model being one of the simplest CPU power models is not perfect, and actual power consumption at the same peak utilization varied between 665mW to 781mW (a range of 16% w.r.t. the average) depending on type of computation performed. However, this model has the advantage that the scaling of the resource counter as described above works directly.

Other processor power models that are more accurate than equation (2) are available in the literature. While they can be used for run time energy profiling, their use in emulation is non-trivial because they require the use of multiple processor performance counters. Scaling all performance counters is challenging since most of the additional counters are highly dependent on processor architectures that vary a great deal between the developer workstation and the mobile device. For instance, the last level cache (LLC) miss counter is often found useful to improve power model accuracy. However, given that the cache hierarchy varies drastically between the two processors, scaling the LLC miss counter becomes impractical. Using the linear model from (2) enables a reasonable approximation at emulation time.

GPU: Some of the other components are also significant from an energy standpoint. The most notable one is the Graphics Processor (GPU). The GPU is mostly hardware managed and no software observable metrics regarding its utilization level are easily available. Power models for the GPU are thus not widely studied in the literature. Resource scaling for the GPU will also be non-trivial. We

[2]The scaling factor can easily be adapted for a new developer machine by measuring the execution time for a known processing task for which the phone size execution time is already known

[3]This approach assumes that the multi-threading characteristics of the app are preserved across the phone and workstation processors, i.e., if the app uses n threads on the phone, it will not use more than n threads on the developer workstation and the slow down is thus governed by the same bottleneck thread. The mobile device emulator generally ensures that this is indeed the case, especially if the instruction set of the phone processor is being simulated.

largely omit the GPU from our model. We measured the power consumption on a mobile device for several video playback tests and found that while some portion of the power measurement can be explained using the CPU and display, a significant portion remains attributable to the GPU: on an average 231mW across three different videos. Thus, if the emulated app is detected to be playing video, we can add this power value for the duration that the video is played. Other uses of the GPU such as graphics intensive games, are not covered by this approximation.

3. IMPLEMENTATION

Our implementation of WattsOn integrates the resource scaling and power modeling techniques described above, with the Windows Phone Emulator. Mobile emulators for other platforms such as th Android and iOS also execute on the developer workstation and most of our power models and network resource scaling techniques apply to them as well. However, Android uses a different CPU emulation strategy and preservation of timing characteristics may require additional changes.

WattsOn also adds a graphical user interface that allows the developers to explore multiple operating conditions for the run, such as network conditions (good, medium, bad), display brightness (low, medium, high), phone brand and model, network carrier (AT&T, Verizon, T-Mobile, or Sprint), and received signal strength. Since display brightness and signal strength do not affect the timing of the app execution, these parameters can even be changed after the test run and the energy estimates are updated on the fly using a previously obtained resource consumption trace.

The inputs to WattsOn include the power model parameters and the resource scaling parameters, encapsulated for a specific device into an XML file. Energy use across multiple devices may be emulated by simply changing the parameter file, without requiring access to a large number of physical devices.

The output of WattsOn includes: (i) a time series of power consumed for every component, and (ii) the total energy consumed. Using the battery capacity specification, we can also estimate how long the battery would last if a user continuously ran the app on her phone. A screenshot of WattsOn output is shown in Figure 7. The UI elements on the left allow the developer to change operating conditions while the 4 graphs show the energy used on the network, display, CPU, and the overall device respectively.

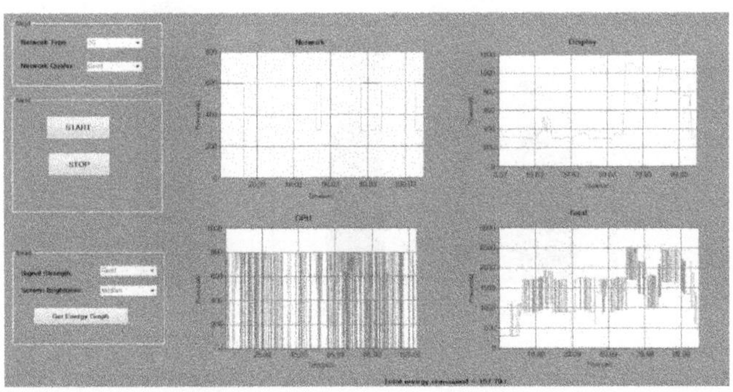

Figure 7: WattsOn output showing energy data. The UI elements on the left allow varying emulated operating conditions and technology characteristics.

While the prototype implementation includes power models for various environmental conditions such as multiple network signal strengths, user settings such as screen brightness, and also multiple hardware technologies such as LCD and AMOLED, additional configurations of test conditions are likely to emerge. The primary overhead of including any new configurations is that a one time measurement of that configuration must be performed with a hardware power meter, and provided for download to all developers using the tool.

4. PERFORMANCE EVALUATION

The accuracy of the energy estimate produced by WattsOn depends both on the correctness of resource scaling as well as the accuracy of the power models used. To evaluate the overall performance, we compare the estimated energy use as well as the measured energy use for multiple tasks performed on the mobile device. Since apps cannot be downloaded to the emulator from the app marketplace, we developed our own apps to perform these tests.

Application 1: Display only. We begin our evaluations with the display power model. While the LCD and OLED power models are available in prior work, the AMOLED model is new and we evaluate its accuracy using an application that consumed energy only in the display.

We first test the accuracy of our discretized lookup table that only stored 4096 colors out of the possible $255 \times 255 \times 255 = 16581375$, which is approximately only 0.024% of all possible color values. This test is performed with 100 random colors that are not in the lookup table, set to occupy 100% of the display area. This tests the accuracy of the modified AMOLED model for color dependent power scaling. Figure 8 shows both the estimated an the measured power and suffers from very little error.

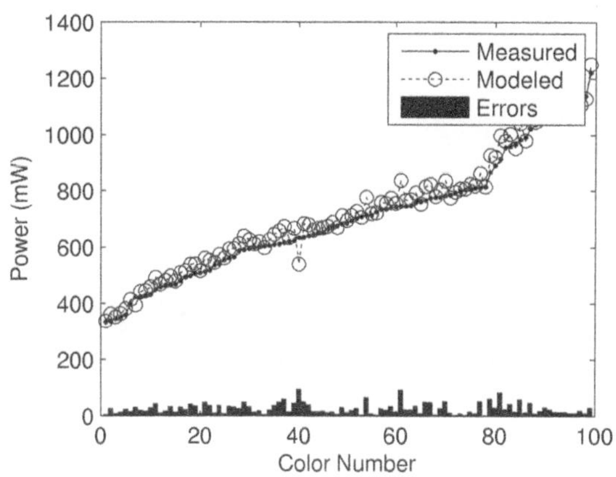

Figure 8: Testing the AMOLED display power model with 100 random colors.

For a more realistic test, we use an application that simply displays a static image and performs no CPU or network activity. Thirty different images encompassing various typical app displays including simple GUIs to rich textures and photographic imagery were included in this set. Figure 9 shows the measured and estimated energy (using equation (1)) for all 30 images. The energy estimated using 1% randomly selected pixels instead of considering all pixels is also shown. Again the model has very little error.

Another parameter worth exploring is the loss in accuracy suffered if we sub-sample only a small fraction of the pixels rather than recording all pixels. Sub-sampling can reduce the display tracing data overhead and may be important for long running tests. We report the incremental increase in error sub-sampling at 1% and 0.1% of the pixels in Figure 10. Reducing to 1% from 100% does

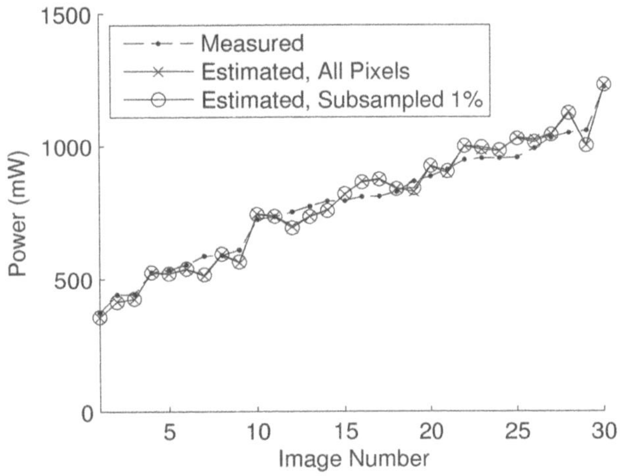

Figure 9: Measured and emulated energy for Application 1, with 30 different images. Images are sorted by the measured energy used.

not degrade the accuracy significantly (the gray bars indicating the loss in accuracy are very small).

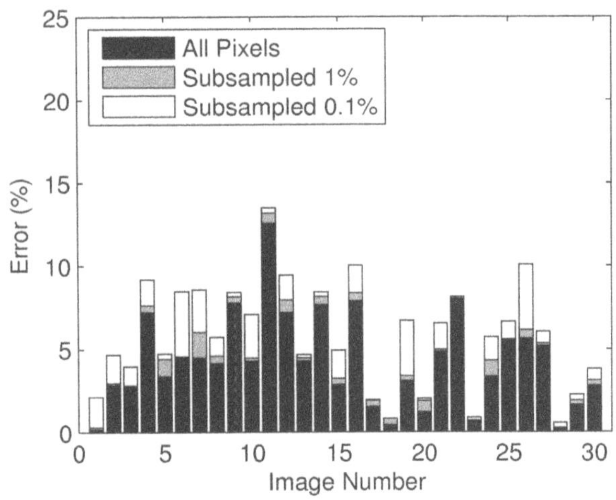

Figure 10: Display model error when all pixels are considered, or only 1% or 0.1% of the pixels are considered.

Application 2: Local Computation. This test is designed to model applications that use the processor and display, without network use or heavy graphics. The test app has a simple GUI, and user clicks perform computations executing over 5s to 20s. An example power trace captured using the power meter is showed in Figure 11(a). For the same task, the measured power trace averaged over 1s intervals as well as power emulated using WattsOn is shown in Figure 11(b). Power was measured with the mobile device in the airplane mode to avoid cellular network background activity. However, the device does have some background activity as seen in the raw trace near time 13s, and such activity, being independent of the app, should be excluded from consideration for the app developer. This is difficult in the measured power but is easy when using WattsOn.

Performing such an evaluation for the three different compute units, the total energy consumed over a 30s period, as measured and emulated is shown in Figure 12. Each measurement is averaged over 5 random runs and the standard deviations are shown as error bars. Since the resource scaling can behave differently from one run to another, the emulated energy can also vary slightly across runs. Overall, the emulated energy is lower since certain background activities are absent from the emulated trace. The mean error across all these runs is 9.3%. The primary factors leading to the error include any errors in the power model, the fact that background activities cause the measured energy to be artificially higher, and the error in resource scaling that may cause the emulated CPU usage to be different from measured.

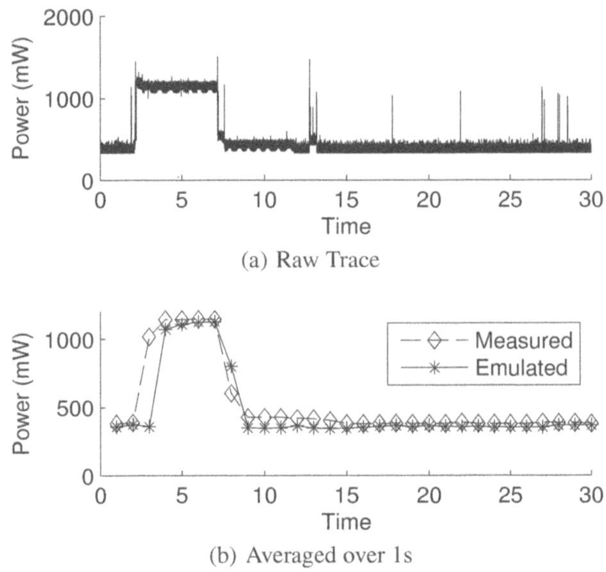

Figure 11: Comparing measured and emulated energy for Application 2.

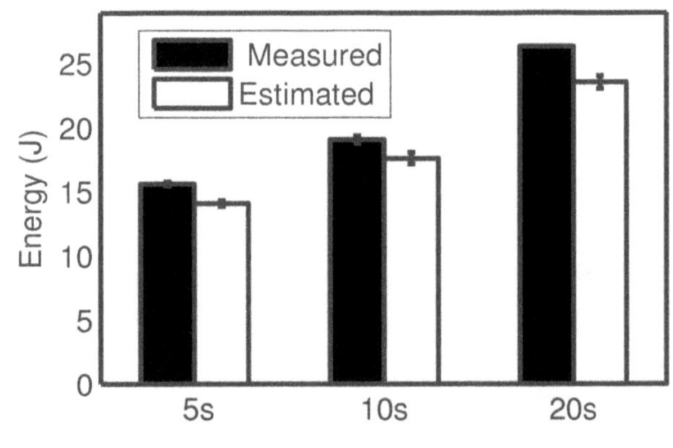

Figure 12: Experimental evaluation of emulation accuracy for Application 2.

Application 3: Networked Apps. Next we consider applications that use the network in addition to the CPU and display. A simple application is developed that can download files of different sizes from a predetermined web server. A sample power measurement is shown in Figure 13 and depicts the total power consumed by the display, network and CPU during one run. Performing the above test for different downloaded file sizes ranging from small to

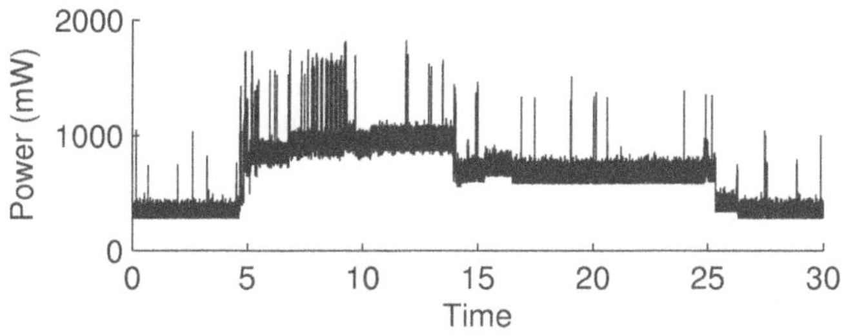

Figure 13: Example data capture for Application 3.

large, and repeating each test 5 times, the measured and estimated energy use is shown in Figure 14. The emulation energy is close enough for the developer to make the correct design choices based on the estimates provided by WattsOn. Average error is 4.73% across all tests.

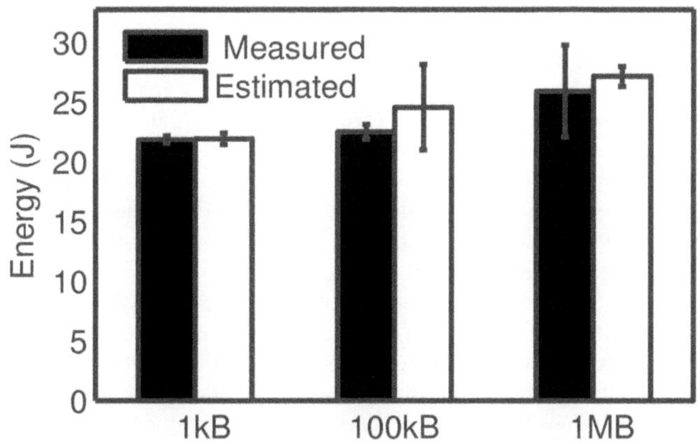

Figure 14: Estimated and measured energy for Application 3 with varying download sizes.

Application 4: Internet Browsing. This application downloads a webpage and renders it on the display. We test emulation accuracy over five different web pages (Table 5) differing in content size and complexity. The web pages are chosen to be static pages without variable advertisement content to make the experiment repeatable across a real device and WattsOn.

Web page	Size	Images
AMOLED Wiki Page	357kB	9
ACM	353kB	16
MSDN Mobile Apps	828kB	9
Google	117kB	4
NPS/Yose	532kB	14

Table 5: Webpages used for testing Application 4.

Energy is measured over a common session length of 50s for each URL. A variable portion of the session is spent on fetching the webpage for each case. There are variations across multiple runs due to network and web server variability. The measured and estimated energy is shown in Figure 15. The average error over all these experiments is 4.64% which is very similar to the variability in measurement one may observe even with real device measurements due to changing network conditions and device background activity.

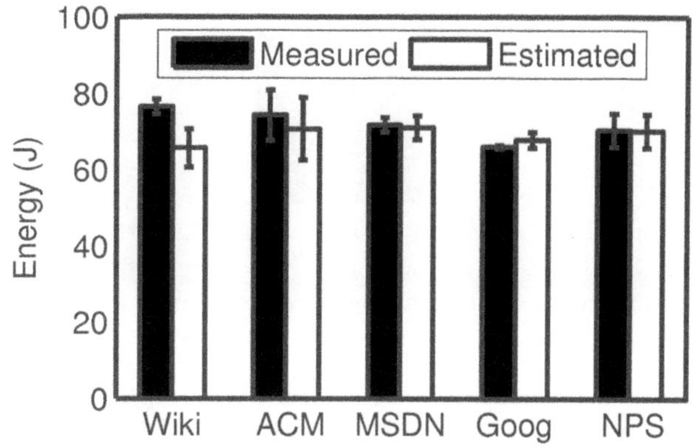

Figure 15: Accuracy evaluation for Application 4.

4.1 Case Study: Multi-Component Optimization

Let us consider an application that uses multiple components (CPU, network, display) and has design options to increase or reduce their use. A simple weather application suffices to demonstrate the trade-offs involved. While the application is simple, it allows illustrating how the break-down of energy use across multiple components provided by WattsOn helps understand the design choices more clearly.

The developer has multiple design decisions to make that may impact energy use:

1. **Portability:** The app may always download the entire content to be displayed from the web. Such application designs are sometimes referred to as hybrid apps and using the HTML5 standard, such apps can operate seamlessly on multiple platforms. Alternatively, the app may locally supply bulk of the content including weather images and only download succinct weather data from its web server. The latter approach reduces the amount of data downloaded from the Internet at run time but requires a platform specific app implementation on each mobile platform of interest.

2. **Rich Graphics:** The app may use different images for depicting the same weather condition: a simple cloud icon (such as Figure 16-(a)) only 18kB in file size, or a rich photographic cloud image which requires a larger 138kB file to be stored or downloaded (Figure 16-(b)). Richer graphics may allow for more sophisticated application UI designs but at the cost of higher data sizes.

3. **Animation:** The developer could even use an animated image of moving clouds to create a more engaging user experience, e.g., an animated image with two frames, one frame shown in Figure 16-(c), using a 90kB file. Animation would use up processor resources, and possibly require a larger image file than a simpler icon.

A quantitative energy cost estimate for each of the above three app features is necessary for the developer to determine whether to include them, and WattsOn can provide that with very low development overhead.

The decision depends on understanding energy consumption in multiple dimensions. Simply optimizing the app for lower CPU utilization will not necessarily reduce energy. Also, the display power model depends on color implying that even two images that

Figure 16: Sample images considered for an app.

appear similarly bright may consume different energy. Trying the multiple options in the app and recording the WattsOn output, while running the app for the same duration in each case, we obtain the data summarized in Figure 17.

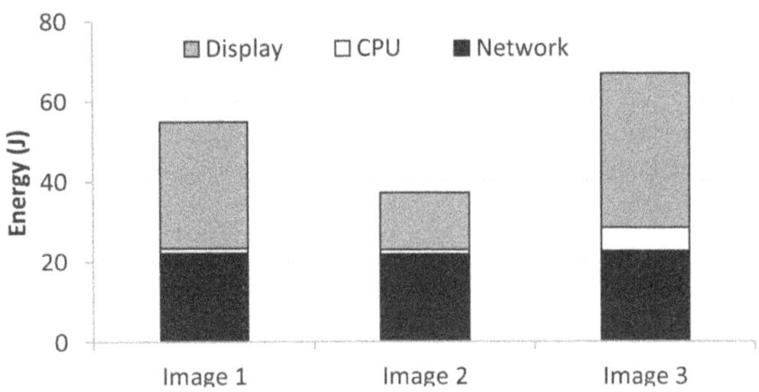

Figure 17: Energy breakdown into multiple components.

Several facts become obvious from this data. First, even though Image 1 is much simpler and has a smaller file size, it does not help save any energy on the network. This happens because for such small downloads the network energy is dominated by the tail states of the radio. A related implication is that saving the images locally and only downloading the text data related to weather will not help save much energy. Hence, including *rich graphics* and designing for *portability* has negligible energy overhead for this app.

Second, the display is consuming the largest fraction of the energy for this scenario. Clearly an image that saves energy on the display is likely to be beneficial, and other factors are less important to worry about during app energy optimization.

Third, looking in detail at the component breakdowns, one also notices that the CPU energy consumption of the third image is the highest. Looking at the emulation output from WattsOn (not shown for brevity) also shows that, while for the first two images the CPU energy is nearly zero, after the download completes, the CPU is consuming an average of 150mW continually while Image 3 (animation) is displayed. Since WattsOn shows the exact overhead of animation, the developer can determine if animation is important for their app scenario or not. Armed with such data, the developer can also decide if the app should highlight important information, such as a hazardous weather alert, using brighter colors or animation.

5. RELATED WORK

WattsOn builds upon a large body of work on energy modeling for phones [3, 6, 25]. The work in [3] focused specifically on a Palm device and measured several components including the CPU, LCD, device buttons, touch pen, and the serial link which was the primary communication channel on that device. On similar lines [25] is a more recent work that presents tools for automated execution of test benchmarks and measurement of power, to enable generation of power models for various components on the mobile device. The modeling approach was enhanced further in [6] to eliminate the need for external power measurement equipment, by using the battery drain measurements available on the mobile device itself. Aside from full system models, specific models for key components have also been studied in depth, including OLED displays [4], LCD displays [25], 3G cellular networks [17–19,21], and WiFi networks [19]. We leverage these established power modeling techniques in our design, and where needed, expand the set of models to cover additional technology variations.

Prior work has also looked at app energy accounting at run time. A key challenge for run time methods is the attribution of energy among multiple applications simultaneously using the device. In [24] resource usage of each component is tracked per application, and power models from the data sheet specs are used to estimate app energy. PowerScope [7] tracks the application with the active context on the processor (a single core CPU without hyper-threading was used) and measures the power at fine time resolution. The total power consumed when the application had active processor context is attributed to that application. Another powerful modeling approach, eProf [15], traces system calls made by applications and uses power state models for various components to infer energy used. It can incorporate tail-energy use, such as when a component remains in a higher energy state after the application is closed. We leverage similar power models but rather than profiling on a real mobile device we extend the work to enable energy emulation on the developer machine itself.

Energy emulation at development time has previously been developed in Power TOSSIM [22]. However, the event based simulation approach used in that work does not directly apply to mobile app emulation due to the interaction with external resources such as web services and interdependence among the timing characteristics of the network and other components. Different resource scaling techniques and compatible power models are hence required for WattsOn. Energy emulation methods using fine grained processor architecture models have also been developed [2]. However, such models are not available for all key mobile device components and have an extremely high resource overhead. Their use for full system emulation during app development with real time external interactions remains an open problem.

6. CONCLUSIONS AND FUTURE WORK

In this paper we presented a system to estimate the energy consumption of apps during development. This enables app developers to use this feedback to write more energy efficient code. WattsOn scales down the emulation environment (network, CPU, display) to mimic the phone and applies empirically-derived power models to estimate app energy consumption. This approach gives us the flexibility to test an app's energy consumption under various scenarios and operating conditions. While we leverage known power modeling and resource scaling concepts, their combination for app energy emulation introduces certain constraints that requires a careful selection of compatible modeling and scaling techniques. We discussed such suitable techniques and experimentally validated their use.

We have prototyped WattsOn for the Windows Phone platform and shown its effectiveness for a variety of apps. Moving forward, WattsOn is just the first step in improving the energy efficiency of apps. Other useful steps include obtaining energy measurements from the wild to provide developer feedback based on real world usage patterns. Techniques to augment power models with real measurement data to overcome modeling limitations [14] are also of interest. Together, we believe that these efforts will help developers produce apps that will significantly increase the battery lifetime of mobile phones.

7. REFERENCES

[1] A. Balasubramanian, R. Mahajan, and A. Venkataramani. Augmenting mobile 3g using wifi. In *Proceedings of the 8th international conference on Mobile systems, applications, and services*, MobiSys '10, pages 209–222, 2010.

[2] D. Brooks, V. Tiwari, and M. Martonosi. Wattch: a framework for architectural-level power analysis and optimizations. In *ISCA*, pages 83–94, 2000.

[3] T. L. Cignetti, K. Komarov, and C. S. Ellis. Energy estimation tools for the palm. In *Proceedings of the 3rd ACM international workshop on Modeling, analysis and simulation of wireless and mobile systems*, MSWIM '00, pages 96–103, 2000.

[4] M. Dong, Y.-S. K. Choi, and L. Zhong. Power modeling of graphical user interfaces on oled displays. In *Proceedings of the 46th Annual Design Automation Conference*, DAC '09, pages 652–657, 2009.

[5] M. Dong and L. Zhong. Chameleon: a color-adaptive web browser for mobile oled displays. In *Proceedings of the 9th international conference on Mobile systems, applications, and services*, MobiSys '11, pages 85–98, 2011.

[6] M. Dong and L. Zhong. Self-constructive high-rate system energy modeling for battery-powered mobile systems. In *Proceedings of the 9th international conference on Mobile systems, applications, and services*, MobiSys '11, pages 335–348, 2011.

[7] J. Flinn and M. Satyanarayanan. Powerscope: A tool for profiling the energy usage of mobile applications. In *Proceedings of the Second IEEE Workshop on Mobile Computer Systems and Applications*, WMCSA '99, pages 2–, 1999.

[8] R. Fonseca, P. Dutta, P. Levis, and I. Stoica. Quanto: tracking energy in networked embedded systems. In *Proceedings of the 8th USENIX conference on Operating systems design and implementation*, OSDI'08, pages 323–338, Berkeley, CA, USA, 2008. USENIX Association.

[9] J. Huang, Q. Xu, B. Tiwana, Z. M. Mao, M. Zhang, and P. Bahl. Anatomizing application performance differences on smartphones. In *Proceedings of the 8th international conference on Mobile systems, applications, and services*, MobiSys '10, pages 165–178, 2010.

[10] S. Jha. Poorly written apps can sap 30 to 40% of a phone's juice., June 2011. CEO, Motorola Mobility, Bank of America Merrill Lynch 2011 Technology Conference.

[11] H. V. Madhyastha, T. Anderson, A. Krishnamurthy, N. Spring, and A. Venkataramani. A structural approach to latency prediction. In *Proceedings of the 6th ACM SIGCOMM Conference on Internet Measurement*, IMC '06, 2006.

[12] J. Manweiler, S. Agarwal, M. Zhang, R. Roy Choudhury, and P. Bahl. Switchboard: a matchmaking system for multiplayer mobile games. In *Proceedings of the 9th international conference on Mobile systems, applications, and services*, MobiSys '11, pages 71–84, 2011.

[13] J. Manweiler and R. Roy Choudhury. Avoiding the rush hours: Wifi energy management via traffic isolation. In *Proceedings of the 9th international conference on Mobile systems, applications, and services*, MobiSys '11, pages 253–266, 2011.

[14] J. C. McCullough, Y. Agarwal, J. Chandrashekar, S. Kuppuswamy, A. C. Snoeren, and R. K. Gupta. Evaluating the effectiveness of model-based power characterization. In *Proceedings of the 2011 USENIX annual technical conference*, USENIXATC'11, pages 12–12, 2011.

[15] A. Pathak, Y. C. Hu, M. Zhang, P. Bahl, and Y.-M. Wang. Fine-grained power modeling for smartphones using system call tracing. In *Proceedings of the sixth conference on Computer systems*, EuroSys '11, pages 153–168, 2011.

[16] Qemu: open source processor emulator. `http://wiki.qemu.org/Main_Page`.

[17] F. Qian, Z. Wang, A. Gerber, Z. Mao, S. Sen, and O. Spatscheck. Profiling resource usage for mobile applications: a cross-layer approach. In *Proceedings of International Conference on Mobile Systems, Applications, and Services*, MobiSys '11, pages 321–334, 2011.

[18] F. Qian, Z. Wang, A. Gerber, Z. M. Mao, S. Sen, and O. Spatscheck. Characterizing radio resource allocation for 3g networks. In *Proceedings of the 10th annual conference on Internet measurement*, IMC '10, pages 137–150, 2010.

[19] M.-R. Ra, J. Paek, A. B. Sharma, R. Govindan, M. H. Krieger, and M. J. Neely. Energy-delay tradeoffs in smartphone applications. In *Proceedings of the 8th international conference on Mobile systems, applications, and services*, MobiSys '10, pages 255–270, 2010.

[20] S. Rivoire, P. Ranganathan, and C. Kozyrakis. A comparison of high-level full-system power models. In *Proceedings of the 2008 conference on Power aware computing and systems*, HotPower'08, pages 3–3, 2008.

[21] A. Schulman, V. Navda, R. Ramjee, N. Spring, P. Deshpande, C. Grunewald, K. Jain, and V. N. Padmanabhan. Bartendr: a practical approach to energy-aware cellular data scheduling. In *Proceedings of the sixteenth annual international conference on Mobile computing and networking*, MobiCom '10, pages 85–96, 2010.

[22] V. Shnayder, M. Hempstead, B.-r. Chen, G. W. Allen, and M. Welsh. Simulating the power consumption of large-scale sensor network applications. In *Proceedings of the 2nd international conference on Embedded networked sensor systems*, SenSys '04, pages 188–200, 2004.

[23] W. L. Tan, F. Lam, and W. C. Lau. An empirical study on the capacity and performance of 3g networks. *IEEE Transactions on Mobile Computing*, 7:737–750, June 2008.

[24] H. Zeng, C. S. Ellis, A. R. Lebeck, and A. Vahdat. Ecosystem: managing energy as a first class operating system resource. In *Proceedings of the 10th international conference on Architectural support for programming languages and operating systems*, ASPLOS-X, pages 123–132, 2002.

[25] L. Zhang, B. Tiwana, Z. Qian, Z. Wang, R. P. Dick, Z. M. Mao, and L. Yang. Accurate online power estimation and automatic battery behavior based power model generation for smartphones. In *Proceedings of the eighth IEEE/ACM/IFIP international conference on Hardware/software codesign and system synthesis*, CODES/ISSS '10, pages 105–114, 2010.

[26] X. Zhao, Y. Dong, H. tao Zhao, Z. Hui, J. Li, and C. Sheng. A real-time congestion control mechanism for multimedia transmission over 3g wireless networks. In *Communication Technology (ICCT), 2010 12th IEEE International Conference on*, pages 1236 -1239, nov. 2010.

FLIGHT: Clock Calibration Using Fluorescent Lighting

Zhenjiang Li[1,4], Wenwei Chen[1], Cheng Li[1], Mo Li[1], Xiang-yang Li[2], Yunhao Liu[3,4]
[1]Nanyang Technological University, Singapore
[2]Illinois Institute of Technology, USA
[3]Tsinghua University, China
[4]Hong Kong University of Science and Technology, Hong Kong
{lzjiang, chen0746, cli6, limo}@ntu.edu.sg, xli@cs.iit.edu, yunhao@greenorbs.com

ABSTRACT

In this paper, we propose a novel clock calibration approach called FLIGHT, which leverages the fact that the fluorescent light intensity changes with a stable period that equals half of the alternating current's. By tuning to the light emitted from indoor fluorescent lamps, FLIGHT can intelligently extract the light period information and achieve network wide time calibration by referring to such a common time reference. We address a series of practical challenges and implement FLIGHT in TelosB motes. We conduct comprehensive experiments using a 12-node test-bed in both static and mobile environments. Over one-week measurement suggests that compared with existing technologies, FLIGHT can achieve tightly synchronized time with low energy consumption.

Categories and Subject Descriptors

C.2.2 [**Computer Communication Networks**]: Network Protocols; C.2.4 [**Computer Communication Networks**]: Distributed Systems

General Terms

Design, Algorithms, Performance

Keywords

Clock calibration, Fluorescent lighting, Energy efficiency

1. INTRODUCTION

Maintaining a common notion of time is one fundamental service in many distributed networking systems. A variety of applications depend on the availability of tightly synchronized time cross network nodes. For instance, the emerging High-speed Locational, Phone-to-Phone (HLPP) gaming [1] with a real-time requirement would benefit from such a service. A series of network-wide events must follow a strict common time order on different phones. In body area networks and healthcare monitoring networks [2–4], sensory data from multiple devices are usually processed to cooperatively analyze certain body movements or events of interest. Consistent time clocks among different devices are necessary for correct time alignment and data processing. Other typical application examples include wireless sensor networks, mobile ad hoc networks, etc.

Retaining a common time in the network, however, faces substantial challenges. Due to the low-cost design, CMOS crystal oscillators serve as the most privileging signal source to generate on-board clocks. The frequency of an oscillator is not stable and it fluctuates with the surrounding environment, affected by various factors like temperature, humidity, voltage, pressure, etc. For example, the $32KHz$ oscillator adopted by popular sensor platforms has a clock drift rate of 30-50 ppm [5]. The clock drift in smart phones normally ranges from 5-100 ppm [6, 7]. Thus, clocks on different nodes need to be precisely calibrated for correcting above errors such that consistent time can be maintained. Due to the inherent uncertainty, no matter how accurately clocks are initially calibrated, they will ultimately tick towards divergency. Hence, frequent clock calibrations are necessary in most network systems. Existing networks may typically comprise hundreds or even more individual nodes, and those nodes are usually wireless interconnected in a multi-hop manner. As a direct consequence, frequently calibrating clocks prohibitively incurs high overhead and significant energy consumption for coordinating the entire network.

Great efforts have been made in the past decade to address above issues. Proposed solutions mainly rely on heavy intercommunications between nodes to exchange their local time references, including RBS [8], TPSN [9], FTSP [10], etc. The calibration error, however, gets accumulated hop by hop exponentially as the clock calibration spreads to the entire network [11]. In addition, those solutions lead to excessively high communication overhead and significant power drain in the system. Such facts largely prohibit the scalability of those solutions in practice. Against the problems, an emerging type of solutions have been recently proposed, which make use of certain external signal sources as common time references. Typical examples are using power lines [12], FM radio [13], radio stations [14], Wi-Fi [15]. Although such newly emerging solutions can dramatically reduce the communication cost, they also utilize radios for the clock calibration at the expense of higher power consumption. Even if clock calibration in [12–15] can be performed less frequently, the energy consumption of *each* calibration could remain high. Besides, most of those solutions need specific hardware components to generate or capture periodical signals, which introduce extra cost and design complexity.

In this paper, we propose a new clock synchronization approach that allows network nodes to receive a globally available clock reference by tuning to the fluorescent lighting. Our approach is based on the following observations and facts. Alternating Current (AC) is a periodical signal with a frequency of 50 or $60Hz$ and the frequency has adequate stability of $5 \cdot 10^{-5}$ measured by the Allan variance [12, 16]. In order for power to be delivered efficiently across the country, the phase difference between any two points remain highly constant [12, 16]. Powered by AC, the fluorescent light intensity thus changes with a stable period that equals half of the alternating current's. Commonly available in most indoor environments like universities, airports, hospitals, and supermarkets, the fluorescent light provides a universal period reference. On the other hand, the light sensor or camera is widely embedded in commodity wireless platforms, e.g., sensor nodes, and smart phones. This offers a great opportunity for a plenty of indoor applications to maintain common time by referring to the light emitted from fluorescent lamps.

The proposed solution has several key advantages. First, it does not require any extra hardware support. By sampling the light sensor or camera with a sufficiently high rate, accurate period information can be detected for clock calibration. Hence, our solution is widely available on most existing commodity platforms. Second, taking advantage of the stability of the AC frequency, the detected light intensity, even from different lamps, exhibits a consistent and stable period, which ensures high synchronization accuracy. Third, compared with prior radio operation based solutions, sampling the light sensor consumes substantially less energy, which provides us a lightweight solution. Finally, since our approach is independent to the network message exchange, time synchronization can be retained even when the network is temporarily disconnected. When individual devices are moving, they can still obtain desired periodical patterns for clock calibration. Such characteristics particularly suit various mobility-enabled scenarios.

Successfully implementing our idea is non-trivial. We first empirically study the feasibility of detecting periods from the fluorescent lighting. Starting from static deployment in the laboratory, our measurements validate that the fluorescent lighting is able to serve as a viable reference for time calibration. We then propose the Fluorescent LIGHTing (FLIGHT) approach for achieving a low-power and accurate clock calibration. FLIGHT can intelligently extract the period information from the fluorescent light in both static and dynamic environments. In addition, to achieve a lightweight FLIGHT implementation, we address a sequence of practical issues including responsive light sensor sampling, period detection with interference from other light sources, etc. To evaluate our design, we implement FLIGHT in TelosB motes and conduct extensive experiments using a 12-node test-bed in the laboratory and further with mobility across the university main academic building of over $20,000m^2$. The results demonstrate that FLIGHT can achieve tightly synchronized time with low energy consumption.

The rest of this paper is organized as follows: In Section 2, we introduce the preliminary information and perform a measurement study. The architecture and the design detail of FLIGHT are given in Sections 3 and 4, respectively. In Section 5, we evaluate FLIGHT's performance through extensive experiments. In Section 6, we introduce the limitations and future works of this paper. Related works are reviewed in Section 7 and we conclude this paper in Section 8.

2. PRELIMINARY

In this section, we introduce preliminary concepts of this paper. We first discuss the clock calibration problem that this paper studies and compare it with the traditional clock synchronization problem. We then detail the target applications for clock calibration. We illustrate the principle of fluorescent lighting and present our initial empirical measurement study that motivates this work. To facilitate the discussions, we tabulate key notations in Table 1.

Table 1: Notation

$c_n(t)$	Native clock at time t
$c_g(t)$	Global reference clock at time t
$c_l(t)$	Logic clock at time t
$f_n(t)$	Frequency of the native clock at time t
f_g	Frequency of the fluorescent light
$\alpha(t)$	Frequency ratio: $f_n(t)/f_g$

2.1 Clock calibration v.s. synchronization

A *native clock* is the clock driven by a node's built-in crystal oscillator directly. We denote the native clock as $c_n(t_0+t)$, referring to the measured time duration from an initial time t_0 until t by the native clock. Without loss of generality, we assume t_0 to be zero; hence the native clock can be written as $c_n(t)$ for short. The periodical pattern of the fluorescent light intensity can be treated as a *global reference clock*, which is denoted as $c_g(t)$. In our system, the AC frequency is $50Hz$; thus, the frequency of the global reference clock f_g is $100Hz$ and the time unit of the global reference clock is $1/(100Hz) = 10ms$. In addition, each node maintains a *logic clock*, written as $c_l(t)$. The logic clock is used by upper-layer applications and it advances as follows:

$$c_l(t) = c_l(0) + \int_0^t r(\tau)d\tau, \quad (1)$$

where $r(\tau)$ is the instant rate of the native clock at time τ. The goal of FLIGHT is to ensure **logic clocks** consistent among different nodes.

Clock synchronization in prior literatures [8–10] ensures the absolute clock values of different nodes to be consistent, while in this work we primarily focus on *clock calibration*, which mainly ensures that different clocks advance with a same speed. Referring to Eq. (1), clock synchronization essentially first compensates the difference of the starting time $c_l(0)$ of each node and further maintains consistent clock ticks for $\int_0^t r(\tau)d\tau$. On the contrary, clock calibration only guarantees a consistent clock rate for $\int_0^t r(\tau)d\tau$ despite of different node starting times. Therefore, with accurate clock calibration, clock synchronization can be easily achieved by compensating the initial time differences of nodes, while sole clock calibration already suffices to support many practical applications. In this paper, we first propose FLIGHT to achieve an accurate clock calibration, and then we discuss how FLIGHT can be integrated with existing mechanisms (e.g., FTSP [10]) to achieve precise clock synchronization with much reduced communication and energy overhead.

2.2 Target applications

Nowadays, the fluorescent lighting resource is commonly available in most indoor environments. Due to the stability of AC, fluorescent lighting can provide a universal period reference in such environments. Meanwhile, the light sensor or camera has been widely embedded in commodity wireless platforms. This offers a great opportunity for a plenty of indoor applications to maintain common time by referring to the light emitted from fluorescent lamps. The primary applications of FLIGHT are those, in which a precise timing service across indoor environments is desired with minimal communications. For instance, body area networks and healthcare monitoring networks need to detect and report a series of gestures, movements, and accidents (e.g., falling down of a patient). Clock calibration could be adopted in such applications since event detections rely on an accurate data timing sequence.

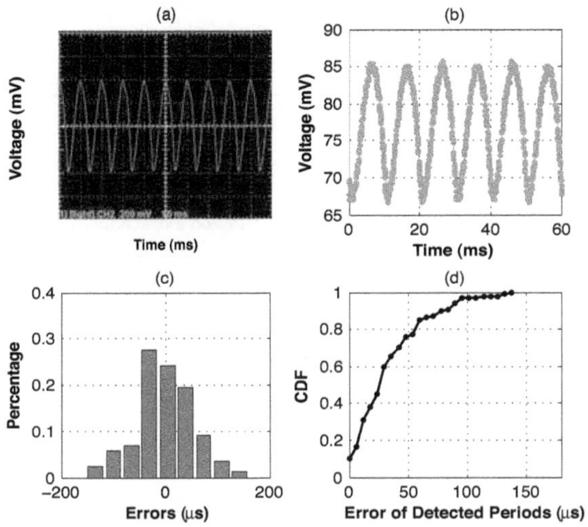

Figure 1: Single-lamp experiment in the laboratory

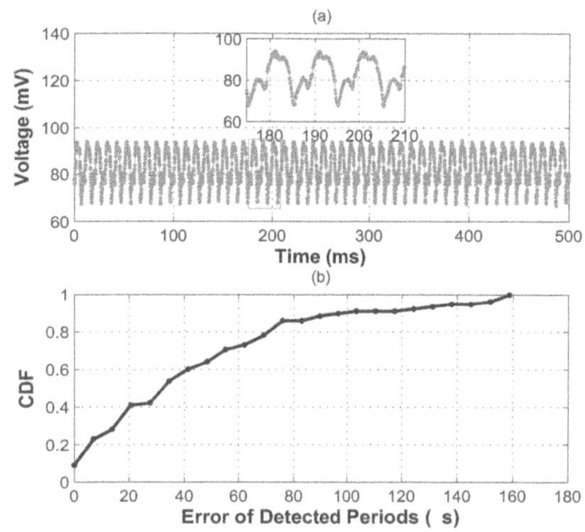

Figure 2: Multi-lamp experiment in the laboratory

Absolute time values of sensory data, however, are not necessarily needed. In such a scenario, nodes are attached to human bodies. The miniature requirement on the node size prevents prior external signal based approaches, e.g., [12–14], from being applied due to the extra hardware requirement. On the other hand, the node density in such applications could be high. Frequent message exchanges of radio-based approaches, like [8–10], can influence the regular event detection and report. Against such issues, body area and healthcare monitoring networks would greatly benefit from the design of FLIGHT.

In wireless sensor networks, clock calibration can support many services as well, like the target tracking, the topology control, and the event logging for diagnosis where the order of events rather than their precise time matters [17–19]. Sensor networks normally consist of hundreds, or even thousands of nodes [20]. Extra hardware components [12–14] may dramatically increase the deployment and maintenance costs. On the other hand, the useable bandwidth resources of sensor nodes are usually limited (e.g., due to low data rates [21] and duty-cycled operation mode [22]). FLIGHT is parallel to the wireless communication, which saves the bandwidth resource and simplifies the MAC protocol design.

Emerging HLPP gaming can also benefit from FLIGHT, such as multi-player games by using phones as props (e.g., playing table tenis [1]). With the help of clock calibration, the captured movements on different devices can be ordered to precisely infer gestures. Sensory data from players can also be analyzed based on such time sequences to make accurate judgement. Different players may be in different rooms or buildings, and radio-based approaches thus fail to work in such applications. Due to the inconvenience brought by the extra hardware in the game, solutions like [12–14] may not be suitable candidates either. FLIGHT, on the other hand, can overcome both barriers in this type of applications.

There is no strict requirement on the space and time coverage of fluorescent lamps in FLIGHT. As we will show later, the period information can be extracted even when the detected light intensity is very low. One fluorescent lamp, thus, can coordinate plenty of surrounding nodes. Fluorescent lighting from different lamps can exhibit a stable and consistent period. Hence, lamps can be sparingly used in the field. In our experiment, only a small number of lamps can coordinate all the nodes in our laboratory of $600m^2$. Moreover, lamps can also coordinate nodes in different rooms even different buildings. Therefore, in order for the proper working of FLIGHT, the only requirement is to turn on several fluorescent lamps, which is always available in many places, like universities, hospitals, airports, supermarkets, etc.

2.3 Empirical measurement study

The fluorescent lamp is a gas discharge lamp using electricity to excite mercury vapor. When the lamp is turned on, the electric power heats up the cathode to emit electrons. Electrons quickly jump from a higher energy level to a lower level and photons will be emitted. Emitted photons are absorbed by electrons in the atoms of the interior fluorescent coating of lamp, leading to physical reactions with emission of visible light. After being heated up, the gas conductivity inside the lamp rapidly rises, allowing the alternating current to flow through and continuously emit light. From the principle of fluorescent lighting, the emitted light is expected to exhibit a periodical pattern since its AC power source is a periodical signal. As the period of AC is adequately stable and phase-coherent, we expect that the light generated by fluorescent lamps can be used as a periodic reference for clock calibration.

We focus on evaluating following two properties of the fluorescent lighting strategy: *stability* and *accuracy* of detected light periods. In the rest of this paper, we mainly take standard TelosB motes as a vehicle to present the measurement results and the design principle. Due to the limited computation capacity and buffer size, design challenges of FLIGHT for TelosB motes are more general. Dealing with the performance optimization in other high-end devices, like smart phones and laptops, can further serve as a promising future work of this paper. In the experiments, 12 TelosB motes are deployed in the laboratory and the longest distance between any two pairs of sensor nodes is up to $15m$. We report the results based on one week measurement. Among those 12 deployed TelosB nodes, the minimum distance to a light lamp is around $20cm$ and the maximum distance is approximately $7m$. The detected light intensity is measured by the voltage value of two pins of the light sensor.

We start from examining a simple case, in which only one lamp is open. As a benchmark, we plot the instant voltage value from two frontend pins of the light sensor using an oscilloscope. Through one week measurement, we observe that the reading from oscilloscope is quite stable and exhibits a regular fluctuation with a period

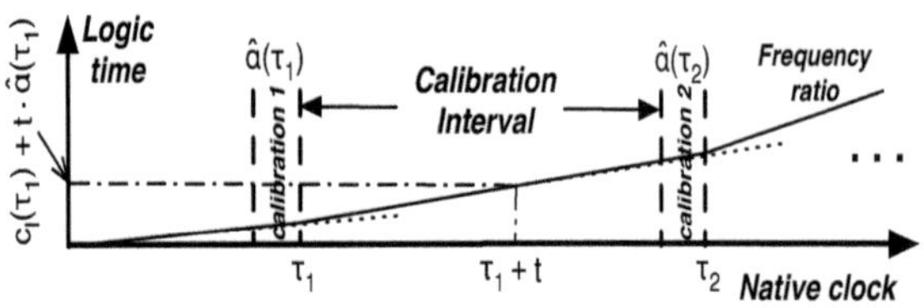

Figure 3: Illustration of periodical calibrations

of $10ms$ (depicted in Figure 1(a)). In practice, sensor nodes need to sample the light sensors. After sampling, we see that the sampling results exhibit the same periodical property as depicted in Figure 1(b), which suggests that the obtained light intensity pattern can be used for clock calibration. Figure 1(c) summarizes the statistics of the detected period lengths from the trace. We observe that the detected period lengths are highly concentrated within the range centered at $10ms$. The error of detection is less than $50\mu s$ for most cases and it is bounded by $145\mu s$ as depicted in Figure 1(d).

We further examine the case where there are more than one lamps. During the measurement, 200 fluorescent lamps in our laboratory are all turned on. Nodes may extract period patterns from different lamps. Similar to the single-lamp case, readings from different nodes all exhibit good periodical patterns of $10ms$ duration. Different distances to a lamp only incur different amplitudes of the detected patterns. The periodical property always exists. In Figure 2(a), we plot a portion of the trace from the sensor node with the largest distance to its closest lamp. Note that in one period, three local maximum points do not mean the light pattern in Figure 2(a) is from exact three different lamps. They only represent phase differences. Each local maximum point of the light intensity might be composed of the light from multiple lamps with the same phase. In Figure 2(b), we further depict the statistical result of detected period lengths of all the nodes with respect to different lamps. Based on the trace, the error is normally smaller than $65\mu s$.

In summary, our initial empirical studies validate the stability and accuracy of fluorescent lighting in practice. Later we will make use of such lighting period information for clock calibration in FLIGHT.

3. SYSTEM OVERVIEW

In this section, we first introduce the basic idea of the fluorescent lighting based clock calibration scheme. We then give an overview of the design challenges as well as the FLIGHT architecture and implementation.

3.1 Principle of clock calibration

The objective of FLIGHT is to ensure logic clocks consistent among different nodes. In Section 2.1, Eq. (1) has stated that the logic clock evolves as $c_l(t) = c_l(0) + \int_0^t r(\tau)d\tau$, where $r(\tau)$ is the instant rate of the native clock at time τ. Ideally, we have $\int_0^t r(\tau)d\tau = c_n(t)/f_n$, where f_n represents the claimed frequency of the built-in crystal oscillator. In practice, however, f_n is not stable and it fluctuates due to the variances of surrounding environmental parameters as mentioned in Section 1. In other words, f_n is not constant in real systems and keeps varying. Such uncertainty is also known as **skew** [6,23]. We utilize $f_n(t)$ to capture the time-varying nature of f_n. By defining the **frequency ratio** as $\alpha(t) = f_n(t)/f_g$,

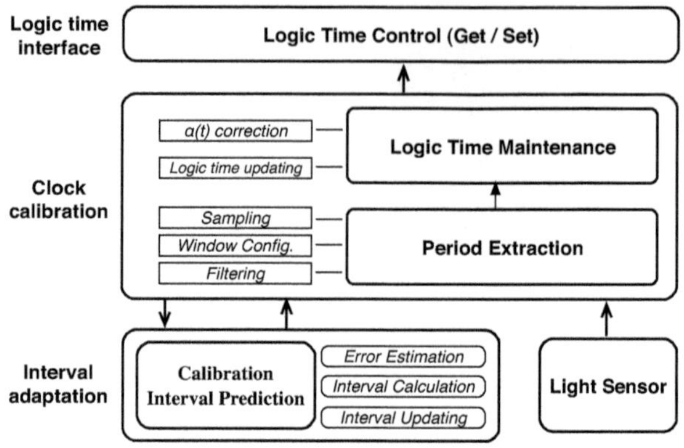

Figure 4: Illustration of the FLIGHT architecture

Eq. (1) can be rephrased as follows:

$$
\begin{aligned}
c_l(t) &= c_l(0) + c_n(t)/(\frac{1}{t}\int_0^t (\alpha(\tau)\cdot f_g)d\tau), \\
&= c_l(0) + c_n(t)/(\overline{\alpha}(t)\cdot f_g). \qquad (2)
\end{aligned}
$$

where $\overline{\alpha}(t)$ is the average value of $\alpha(t)$ from time 0 to t. As unveiled in existing works [6,11,23], clock skew can accumulate with time. Thus, FLIGHT should calibrate clocks to minimize its negative impact by controlling the differences of each $c_n(t)/(\overline{\alpha}(t)\cdot f_g)$ among different nodes; otherwise, logic times will rapidly exhibit heterogeneous advancing rates respect to the global reference and among themselves. As a convention, the time difference caused by the skew accumulation is also referred to as **drift**.

Eq. (2) essentially indicates that by continuously estimating the frequency ratio, the logic time can be precisely maintained. As a matter of fact, the variance of the logic time is limited within a short period of time and most applications can tolerate certain amount of such errors [24,25]. Hence, instead of continuously measuring the frequency ratio, we predict the logic time by using the native clock and the estimated frequency ratio. To further guarantee the accuracy of such a prediction, we configure the clock calibration in a periodical fashion, as depicted in Figure 3.

In Figure 3, we illustrate two calibration instances. During *calibration 1*, the node estimates the real frequency ratio $\overline{\alpha}(\tau_1)$ as $\hat{\alpha}(\tau_1)$. The slope of the straight line between *calibrations 1* and *2* can be viewed as $\hat{\alpha}(\tau_1)$. After the first calibration, the node relies on $\hat{\alpha}(\tau_1)$ to calculate its logic time until *calibration 2*. For example, t native clock ticks (units) after time τ_1 (before τ_2) will be mapped to $t\cdot\hat{\alpha}(\tau_1)$ time units in the logic time domain. If the estimated frequency ratio can well compensate the native clock uncertainty, the mapped time stamp in the logic time domain will be consistent in

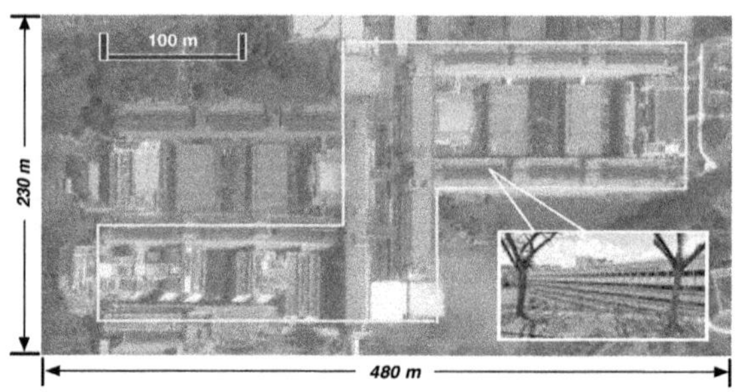

Figure 5: Main academic building

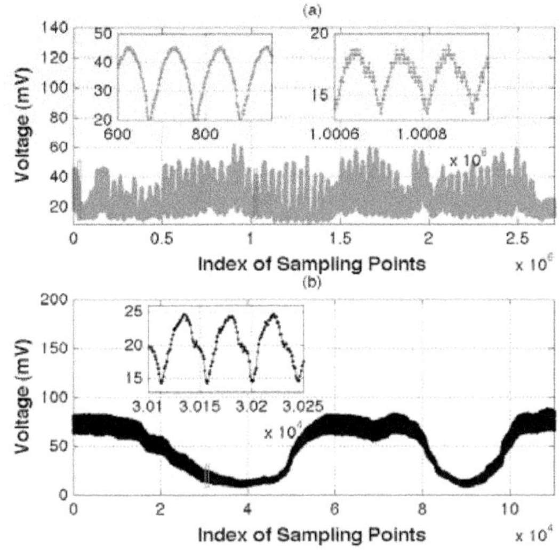

Figure 6: Experiments with mobility and sparsely deployed lamps

each node. However, in practice skew cannot be fully compensated and it is inevitable for drift to accumulate gradually. If such a drift can be (even partially) compensated in each calibration, the differences of $\int_0^t r(\tau)d\tau$ in the entire system will be largely reduced and the consistent logic time with a longer time span can be maintained by clock calibration. Hence, clock calibration in FLIGHT needs to precisely compensate the clock frequency and properly eliminate drift.

3.2 System architecture and implementation

Figure 4 depicts the system architecture. There exist three major components in FLIGHT: clock calibration, logic time interface, and interval adaptation. Based on the readings from the light sensor, the period generation module in the calibration component is launched to produce the global reference clock for later calibration. The output of the period generation module acts as the input of the logic time maintenance module, which is responsible for updating the frequency ratio $\alpha(t)$ and correcting the accumulated drift between two calibrations. The newly obtained logic time can be used by upper layer applications. Meanwhile, the output from the logic time maintenance module can be adopted to determine the interval length for the next calibration.

We implement FLIGHT in TinyOS 2.1x on the TelosB platform. The native clock is driven by the built-in crystal oscillator with a $32KHz$ frequency. The AC frequency in our experiments is $50Hz$. Thus, the frequency of the global reference clock is $100Hz$. The kernel of FLIGHT includes 600 lines of NesC code that was complied to 2242 bytes of RAM and 16114 byes of ROM. As a result, FLIGHT only occupies 23% of the total buffer and the remaining space is available for other applications. We use the built-in light sensor on the TelosB mote in our experiments. The light sensor utilizes an ADC module to measure the light intensity and records it as a corresponding voltage value in the register. The standard TinyOS packages the ADC module and relies on the MSP430 Timer to control the ADC sampling. The maximum sampling rate of the MSP430 Timer is around $1KHz$, which is not fast enough for FLIGHT. To overcome this issue, we write a driver to directly access the register of the ADC module. By doing so, the effective sampling rate can break the $1KHz$ barrier and achieves up to

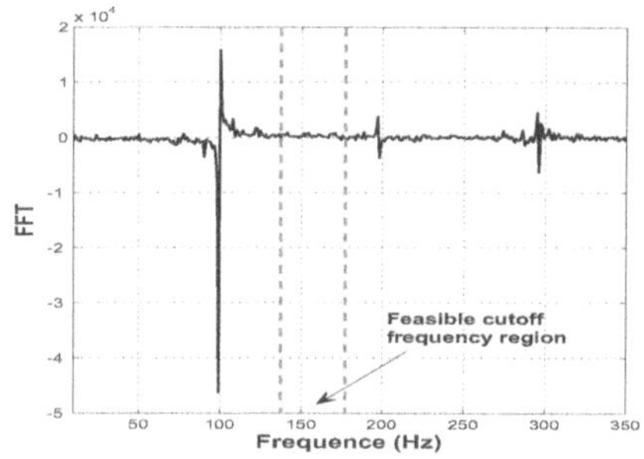

Figure 7: Frequency domain property of the harvested trace

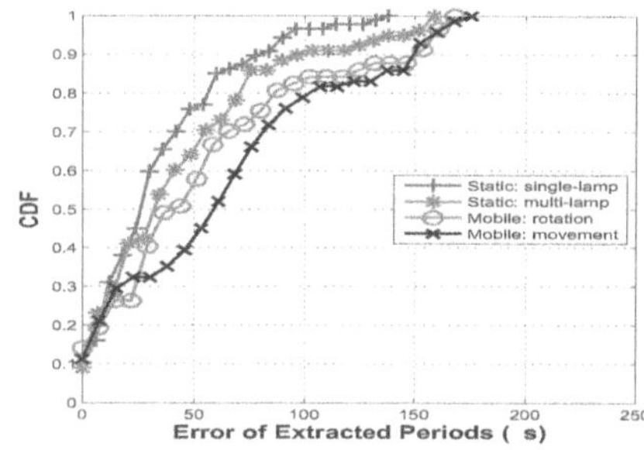

Figure 8: CDF of detected period errors in different scenarios

$83KHz$ in the implementation. There are some system parameters that we have not touched yet and we will later discuss each of them in corresponding sections.

4. SYSTEM DESIGN

In this section, we introduce the design details of FLIGHT.

4.1 Period generation

After sampling the light sensor, nodes need to extract the light period and further generate a series of signals with the same period to serve as the common time reference. In this subsection, we discuss how to efficiently and precisely generate such global reference signals.

4.1.1 Challenging issues

Assume the light sampling rate to be $10KHz$. Ideally, $100 = 10ms \times 10KHz$ samples are detected within one light period. A greedy searching solution for counting the maximum (or minimum) points cannot work here, since there exist multiple local maximum points in one light period for the general multi-lamp scenario. A more sophisticated method may record samples of one period and then apply the greedy searching. When one local maximum point is detected, if its value is close to the recorded maximal value, such a point can be viewed as a period delimiter. Nevertheless, as we will show next in the mobile case, although the light period is still stable, the detected light intensity can fluctuate significantly.

We perform experiments to verify such a point in the mobile environment. In the experiments, three persons roam in the main academic building of an University and each of them holds one TelosB mote and one laptop computer. Due to the limited buffer size, motes are connected to computers through a USB 2.0 port for storing sampled data. Particularly, they move cross different floors. During the movement, sensor nodes might be rotated, shaken, or even temporarily blocked (not sight-in-line to lamps). Figure 5 depicts the

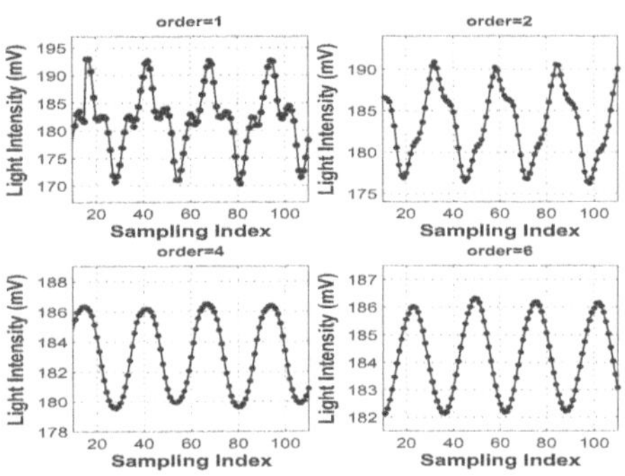

Figure 9: Filtering results with different orders

academic building where the whole area counting different floors is over $20,000m^2$. The moving duration of each person lasts around 120 minutes. During the movement, many places that we passed by are only sparsely covered by lamps, i.e., some places are fully covered by the light while some places are relatively dark. To clearly present our harvested trace, we depict one representative clip in Figure 6(a), which contains the portions collected in both **light** and **dark** places. From the zoom in sub-figures in Figure 6(a), however, we can observe that the detected pattern always demonstrates a good periodical property in both light and dark scenarios. The only difference is the observed light intensity.

To further exploit the sensitivity of the detected light intensity, we record the detected pattern of one node when it is rotated. We hold the sensor node in the multi-lamp laboratory and its distance to the closest lamp is around $4m$. Initially, the light sensor faces to the lamp. After a 180-degree rotation, the light sensor almost faces to the ground and the incident light is mainly the scattering light. Figure 6(b) summarizes the result during a 540-degree rotation. In the zoom in sub-figure, we can see that even when the light sensor is opposite to the lamp and the detected light intensity is low, the detected pattern can exhibit a periodical property as well. As we will show later, one lamp can cover up to around $100m^2$ floor size such that fluorescent lamps can be sparingly used in FLIGHT. Now the question is how we can capture such period information under such intense light fluctuation.

One possible way for the period generation is to use existing DSP techniques [26–28], e.g., self-correlation, folding, etc. In principle, they are effective for extracting periodic signals in both static and mobile environments. In practice, however, the extensive computation burdens involved in those techniques will occupy the MCU of nodes for a long time, which may disturb the processing of other important tasks, and inevitably consume more energy.

4.1.2 *Filtering solution*

To tackle such challenges, we propose to utilize a lightweight filtering method to extract periods. In Figure 7, we perform the Fast Fourier Transformation (FFT) for the trace depicted in Figure 6(a) and find that the trace demonstrates a dominant frequency response around $100Hz$. Such a result indicates that the light intensity indeed changes with a stable period of $10ms$. However, due to noises, the light samples are also mixed with other frequency band components. Figure 7 implies if the trace is processed by a low-pass filter with a cutoff frequency between $135Hz$ and $175Hz$, the trace can be used to generate stable periodical signals.

Denote $x[i]$ to be input function and $y[i]$ to be the output function, where i is the sample index. In general, the first-order filter can be expressed as:

$$y[i] = a \cdot x[i] + b \cdot y[i-1], \tag{3}$$

where a and b are filter parameters and $a+b=1$. Based on Eq. (3), a higher order filter can be further constructed. For instance, the second-order filter is $y[i] = a \cdot x[i] + b \cdot (a \cdot x[i-1] + b \cdot y[i-2])$. A general expression for any m-order filter, where $m \geq 1$, can be obtained iteratively. The settings of a and b have been discussed in [29]. We conduct an experiment to evaluate performances of the period extraction with filters of different orders. Figure 9 shows that if the filter order is small, the trace still exhibits multiple maximum points. When the filter order is sufficiently high, it becomes viable to apply a simple greedy searching scheme to accurately extract periods. In our experiments, we find that a 6-order filter is adequate in FLIGHT (as shown by Figure 9 and Section 5). In such a case, a node only needs to buffer 6 latest samples for the filtering operation. The involved computation burden contains merely 12 multiply operations, 6 plus operations and the greedy searching, which is particularly beneficial for resource constrained mobile devices.

In Figure 8, we summarize the period detection accuracy of the filtering method. We have shown the detection errors in static scenarios in Section 2. For the mobile scenario in Figure 6(a), Figure 8 shows that the error is bounded by $165\mu s$ and smaller than $100\mu s$ most of time. The detection errors in our rotation experiments are also within $165\mu s$ and less than $65\mu s$ for majority cases.

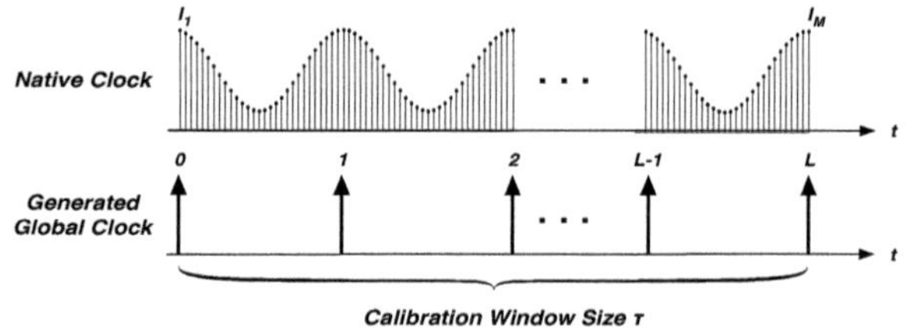

Figure 10: Illustration of the clock calibration

4.2 Logic time maintenance

To maintain the logic time, frequency ratio must be updated based on the extracted light period in each calibration.

4.2.1 *Frequency ratio calibration*

According to the output from the period generation component, the generated signal can be viewed as a common time reference. Suppose the start time of the calibration and the calibration window size are t_c and τ, respectively. For the presentation simplicity, we assume $t_c = 0$ without loss of generality. The basic principle of clock calibration is illustrated in Figure 10. In Figure 10, there are totally M samples between the first and the last generated periods. Each I_j represents the sample index where $1 \leq j \leq M$. We denote the native clock frequency in τ as $\overline{f}_n(\tau)$. Based on the generated global reference, $\overline{f}_n(\tau)$ can be measured by $\hat{f}_n(\tau)$ as follows:

$$\frac{L}{f_g} = \frac{I_M - I_1}{\hat{f}_n(\tau)}, \tag{4}$$

where L is the number of light periods detected in the calibration window τ. Based on the definition, the frequency ratio can be calibrated by the following equation:

$$\hat{\alpha}(\tau) = \frac{\hat{f}_n(\tau)}{f_g} = \frac{I_M - I_1}{L}. \tag{5}$$

The rationale behind the calibration is that native clocks of different nodes run with different speeds. By referring to the global reference, frequency ratios are calculated to compensate those differences. If the measured native clock $\hat{f}_n(\tau)$ in Eq. (4) is accurate, the frequency ratio in Eq. (5) can precisely capture the real situation when predicting the logic time after the calibration.

4.2.2 Calibration window configuration

In Figure 10, M native clock ticks have been elapsed during the calibration window τ. Due to the instability of the native clock, $M/\overline{f}_n(\tau)$ may not exactly equal to M/f_n, where f_n is the claimed clock frequency and $\overline{f}_n(\tau)$ is the average value of the native clock within τ. We denote this offset in the calibration window as $o(\tau) = M/\overline{f}_n(\tau) - M/f_n$ and it can be viewed as a relative drift of the native clock with respect to the global reference since both f_n and f_g are constant. The relative skew, thus, can be expressed as $s(\tau) = o(\tau)/\tau$.

On the other hand, due to jitters, each sampling point (except the first one I_1 since we artificially fix it and view it as the starting point) actually further exhibits a tiny shift, h, compared with the ideally stable sampling point. Prior studies have shown that such a shift follows $h \sim \mathcal{N}(0, \sigma_h^2)$. Therefore, the really observed offset in the calibration window τ can be expressed as:

$$\begin{aligned} \hat{o}(\tau) &= \sum_{j=2}^{M}[(\frac{I_j}{\overline{f}_n(\tau)} + h_j) - (\frac{I_{j-1}}{\overline{f}_n(\tau)} + h_{j-1})] - \frac{L}{f_g}, \\ &= o(\tau) + h_M, \end{aligned} \quad (6)$$

where each h_j indicates the shift for the sampling point I_j for $2 \leq j \leq M$. We can show that if $\hat{s}(\tau) = \hat{o}(\tau)/\tau$ is an accurate estimation of $s(\tau)$, $\hat{f}_n(\tau)$ will precisely estimate its true value $\overline{f}_n(\tau)$. In other words, the frequency ratio $\overline{\alpha}(\tau)$ can be well measured by the observed ratio $\hat{\alpha}(\tau)$ in Eq. (5), yielding a good calibration result. We utilize $Var[(\hat{s}(\tau) - s(\tau))]$ to quantify the estimation accuracy. According to the definition of skew, we have:

$$Var[(\hat{s}(\tau) - s(\tau))] = Var[\frac{(\hat{o}(\tau) - o(\tau))}{\tau}] = \sigma_h^2/\tau^2. \quad (7)$$

Eq. (7) suggests that it is beneficial to set the calibration window to be large. However, we argue that in practice, the solution is beyond such a simple strategy and there exist practical limitations. For example, the buffer size limitation may prohibit to obtain enough data during the calibration. Devices usually have buffers with limited capacity for multiple applications, which may not be able to hold enough samples during the calibration window. For instance, the buffer size of a TelosB mote is $10kB$, in which $8kB$ can be allocated to store sampled data at the most. If each sample occupies two bytes, the buffer can hold up to 4000 samples. If the sampling rate is $10KHz$, only 40 periods can be collected, which may not be enough for an accurate calibration.

To tackle such an issue in FLIGHT, we propose to parallel the sampling and period generation operations. After initial raw data have been filtered, we can drop those data and only store 1) the number of periods that have been detected, and 2) the total amount of samples within those periods. By doing so, the buffer size is virtually increased so as to accommodate adequate data for the calibration. On the other hand, the real buffer occupancy is small and it will not conflict with other applications. To implement such an idea in FLIGHT, we insert the computation process of filtering between the interval of two consecutive samplings. Whenever a new sample has been obtained, it will be processed immediately. We claim that the selection of the sampling rate and the order of the filter should guarantee that the filtering computation must be completed in between two consecutive samplings; otherwise, the data loss will lead to errors in the generated global reference signals. In Section 5, we will verify such a statement.

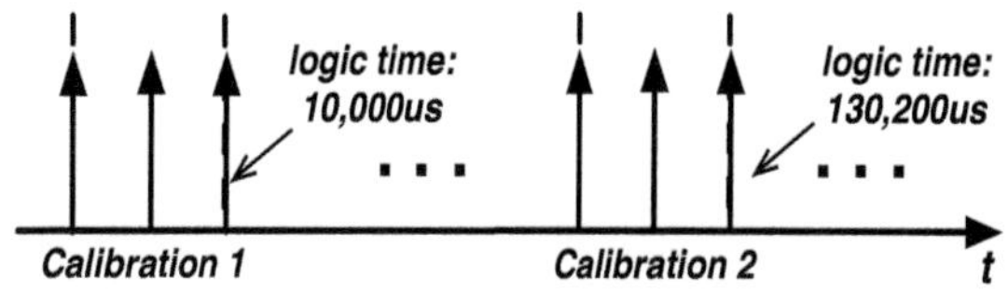

Figure 11: Example of the logic time error

4.2.3 Logic time updating

After the frequency ratio has been updated as $\hat{\alpha}(\tau)$, the logic time afterward can be predicted by using such a new frequency ratio and the native clock till the next calibration window. To maintain an accurate logic time, in addition to the calibration, we also need to correct the drift accumulated between two consecutive calibrations. We utilize Figure 11 to illustrate such a point. In the first calibration interval, we set the logic time to $10,000\mu s$ at the end of the last detected period. Since the period length of light is $10ms = 10,000\mu s$, the logic time should increase $10,000\mu s$ whenever one period has elapsed. After the first calibration, we assume that after 10 light periods, the node performs the next calibration. If the clock is stable, the logic time at the last detected period in the second calibration should be $130,000\mu s$. However, the real logic time might be $130,200\mu s$ due to the clock uncertainty. Such an error cannot be corrected by the frequency ratio updating, while it will naturally deteriorate the time consistency among different nodes.

The insight obtained from Figure 11 is that if the accumulated error can be controlled within a certain range, the last several digits contains the information about the drift. In Figure 11, if the accumulated drift is smaller than $10,000\mu s$, in principle, the logic time can be corrected by resetting the last four digits to zero, i.e., $130,200\mu s$ will be changed to $130,000\mu s$. If we denote t_{c_i}, L_i, and n_i to be the finish time of the ith calibration, the number of light periods detected in the ith calibration, and the number of light periods between the ith and $i+1$th calibration respectively, we can essentially adjust the logic time at the last detected period to $c_l(t_{i+1}) = c_l(t_i) + (n_i + L_i)/f_g$ to eliminate the drift. Now we formally derive the logic time in FLIGHT. If $t = t_{c_i}$, then

$$c_l(t) = \sum_{j=1}^{i}(L_j + n_j)/f_g. \quad (8)$$

As a matter of fact, each n_j is unknown. The operation in Eq. (8) can be achieved by resetting the last several digits of the logic time to zero (the exact digit number is related to the required error bound), or n_j can be estimated via $\hat{\alpha}(t_{c_{j-1}}) \times n_{j-1}$. On the other hand, if $t_{c_i} < t < t_{c_{i+1}}$, we have:

$$c_l(t) = c_l(t_{c_i}) + (c_n(t) - c_n(t_{c_i}))/(\hat{\alpha}(t_{c_i}) \cdot f_g). \quad (9)$$

Now, by directly tuning to the light period at the end of each calibration, the logic time can be precisely aligned to the global reference and the drift accumulated in the previous calibration interval is thus eliminated.

4.3 Calibration intervals

In this section, we describe how we can further utilize the calibration interval to reduce the number of samplings. We also discuss how to determine the interval length between two calibrations.

In Figure 11, if a proper calibration window size should at least contain 13 light periods, a naive method is to extract 13 light periods for each calibration window. However, Figure 11 implies that

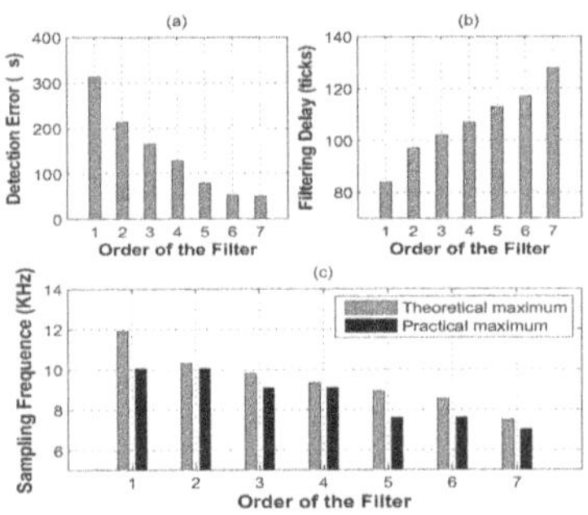

Figure 12: Filtering configurations

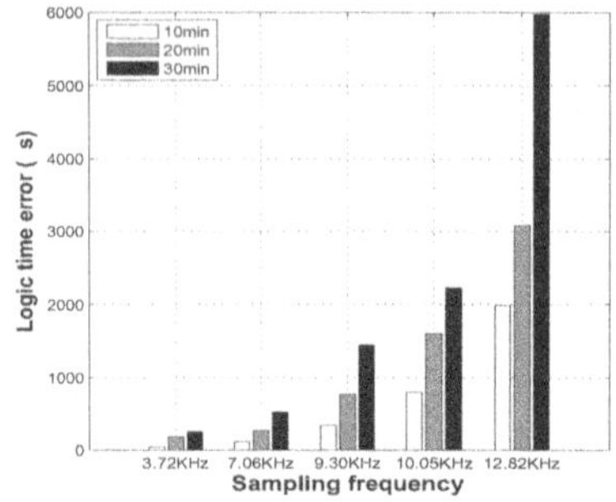

Figure 13: Logic time error v.s. sampling frequency

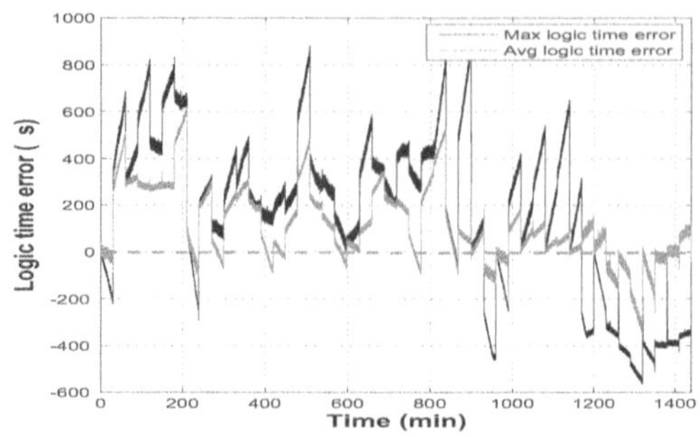

Figure 14: Maximum and average logic time error in the stable environment

if we know that 10 light periods will elapse between two consecutive calibration windows, we only need to sample three additional periods in the second calibration window (*Calibration 2*). By using the similar method in Eq. (8), the node can estimate how many light periods have been elapsed between *Calibrations 1* and *2*. In addition to the number of the light periods have been detected in *Calibration 2*, the total number of light periods between two calibration windows can be obtained. On the other hand, the number of the native clock ticks are always recorded (to calculate any instant logic time). As a result, the frequency ratio can still be calibrated using Eq. (5). Benefited from such a method, although only three light periods have been extracted in *Calibration 2*, the frequency ratio can be accurately calibrated based on 13 light periods. In practice, the calibration interval normally spans more than 20 minutes. Hence, nodes are able to sample a small number of periods while still achieve high accuracy in each calibration.

In Section 5, we examine the calibration interval selection to balance the calibration accuracy and cost. In this study, we also propose a dynamic interval adjustment scheme. We introduce a drift controlling factor (u) to restrict the total amount of logic time errors accumulated between two calibrations. Suppose the drift detected at the last period in the ith calibration is e_i and the calibration interval length between the ith and the $i+1$th calibrations is d_i. The next calibration interval d_{i+1} in FLIGHT can be predicted as:

$$d_{i+1} = d_i \times \frac{u}{|e_{i+1}|}. \tag{10}$$

Initially, d_0 is set to be $20min$. Eq. (10) states that if the drift rapidly accumulates, the calibration interval should be short such that the drift can be corrected in time; vise versa. In Section 5, we will investigate the impact of the factor u.

4.4 Upgraded to time synchronization

Clock calibration only guarantees a similar increasing rate of the maintained logic clock despite of different starting times. As mentioned in Section 2.2 and [12, 13, 15], clock calibration itself serves for a variety of useful applications. Sometimes, people may want to upgrade to time synchronization for a finer understanding of the system performance in the time domain. For instance, nurses in the healthcare networks may need to understand timely situations of monitored patients. System operators in wireless sensor networks may need to record the exact time of certain events. FLIGHT can be easily used to facilitate such an upgrading. To eliminate the clock offset globally, we may simply adopt the FTSP protocol [10]. With the FLIGHT calibration, we do not need to run the full synchronization protocol with heavy intercommunications between nodes. Instead, we only need to cancel the initial time difference with one communication round. We note that the communication overhead and energy consumption to maintain an absolute clock value can be dramatically reduced when clocks of nodes are already precisely calibrated with respect to a common time reference.

5. PERFORMANCE EVALUATION

In this section, we evaluate the system performance of FLIGHT through extensive experiments.

5.1 Filter configuration

In Figure 12, we evaluate the impact of the order of the filter by varying it from 1 to 7. We output a series of samples to computers (due to the limited buffer, we do not store such information in sensor motes) and analyze the error of extracted periods. Figure 12(a) shows that when $m = 1$, the period generation component leads to the largest detection error. From statistics, we observe that the average error reaches as high as $323\mu s$. As we increase m, errors of generated periods reduce dramatically. From Figure 12(a), we can also notice that if we increase m after 5, the error reduction is slight, implying that the order of the filter can be kept relatively low without compromising the achieved period accuracy. In Figure 12(b), we further examine the processing delay to filter one sample respect to different filter orders. As what we expect, the filtering delay becomes linearly increased when m increases as depicted in Figure 12(b). Such a processing delay essentially limits the maximum sampling rate that can be used for each specific m; otherwise, the filtering and sampling operations cannot be parallelized.

To facilitate the sampling rate selection for the practical development, we plot the maximum sampling rates allowed for each m in principle (i.e., theoretical maximum in Figure 12(c)). Since feasible frequencies in practice are not continuous, in Figure 12(c), we plot the maximum frequency that can be adopted for each m in the TelosB platform as an example. Based on the results in the figure, we find that the pair of $m = 6$ and $f_s = 7.06KHz$ can produce both a small period detection error and a sufficiently high sampling rate.

In Section 4.2.2, we claim that the filtering computation should be completed within two consecutive samplings; otherwise, the inaccurately detected periods may incur serious logic time errors. To validate such an argument, we conduct an experiment using two TelosB motes. In this experiment, $m = 6$ and the sampling rate changes from $3.72KHz$ to $12.42KHz$. After the initial calibration, we examine their logic time error in 10, 20, and 30 minutes. From Figure 13, we can easily observe that when f_s is greater than $7.06KHz$, the error increases dramatically. On the other hand, due to the extra computation burden from the period delimiter searching and the interruptions of the MCP to other tasks, the $7.06KHz$ sam-

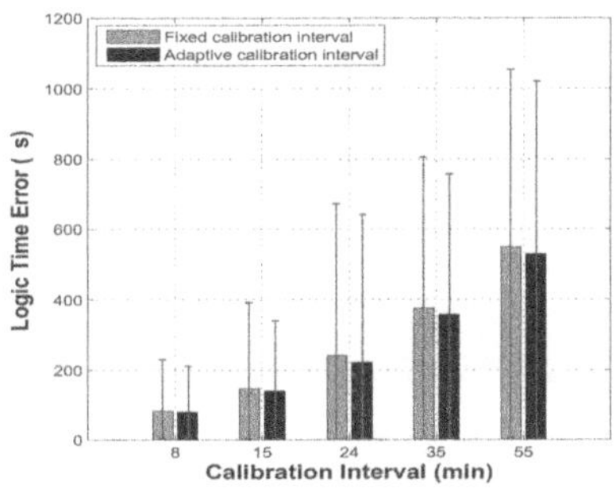

Figure 15: Logic time error v.s. calibration intervals

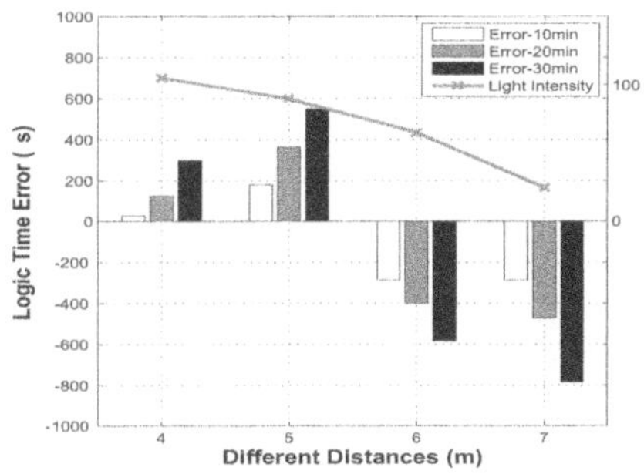

Figure 16: Logic time error v.s. light intensity

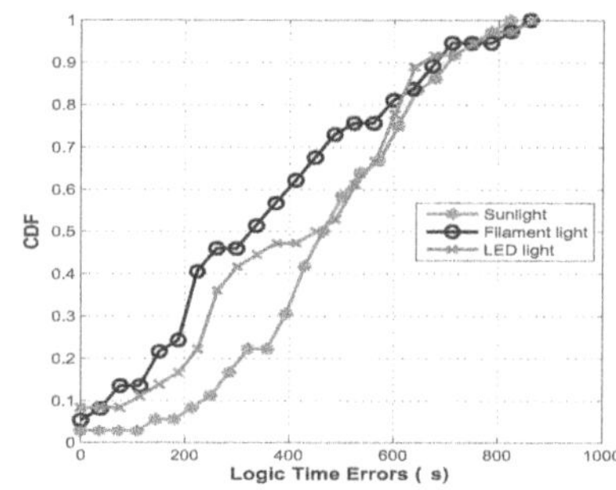

Figure 17: CDF of logic time errors mixed with other types of lights

pling rate may not always provide sufficient safe guarding region to parallel the sampling and computation. As a result, the synchronization accuracy might be impacted. In our experiment, we use a smaller frequency value $3.72KHz$ that offers adequate safe guarding region and reliably provides a small logic time error. We use the $3.72KHz$ sampling rate in the following experiments.

5.2 Logic time accuracy

In this section, we examine the logic time differences among different nodes in both static and dynamic environments.

5.2.1 Static environment

In the static environment, 12 sensor nodes are uniformly distributed in our laboratory. One beacon node is placed in the middle of the laboratory and broadcasts beacon messages to trigger each node logging its current logic time. The experiment has been conducted for one week and we summarize the performance from Figures 14 to 17.

Figure 14 depicts the max and average logic time errors with respect to a randomly selected reference node. We plot one 24-hour trace clip for a clear presentation. Overall, the trace evolves following a zigzag pattern. The sudden drops of two curves are due to the logic time updating in each calibration window. From Figure 14, we can observe that on average, different clocks have been tightly synchronized. From statistics, we find that the average error keeps less than $600\mu s$ all the time and it is smaller than $200\mu s$ for over 80% of time. The maximum logic time error is also well controlled in the experiment. Statistics show that the maximum error is always smaller than $950\mu s$ and it is less than $350\mu s$ most time. Figure 14 indicates that different clocks can be highly calibrated by FLIGHT.

In Figure 15, we evaluate the adaptive calibration interval scheme compared with the deterministic interval setting. We choose five different error controlling factors to be 5, 10, 15, 25, and 30 native clock ticks and each native clock tick corresponds to $30.515\mu s$. For each factor, the experiment lasts 5 hours. When the experiment terminates, we calculate the average length of the calibration interval to be 8, 15, 24, 35 and $52min$, respectively. Then, we repeat the experiment for the same duration by using the deterministic calibration interval policy with the calculated interval length. In Figure 15, we plot the average, maximum, and minimum logic time errors observed in the experiment. From the figure, we find that the deterministic policy generally incur a larger error than the adaptive policy. In order to control the average error below $500\mu s$ as [13,15], we choose 25 as the default value of u in FLIGHT since such a setting can achieve the accuracy requirement and reduce the number of calibrations at the same time.

In Figure 16, we examine the impact of the light intensity on

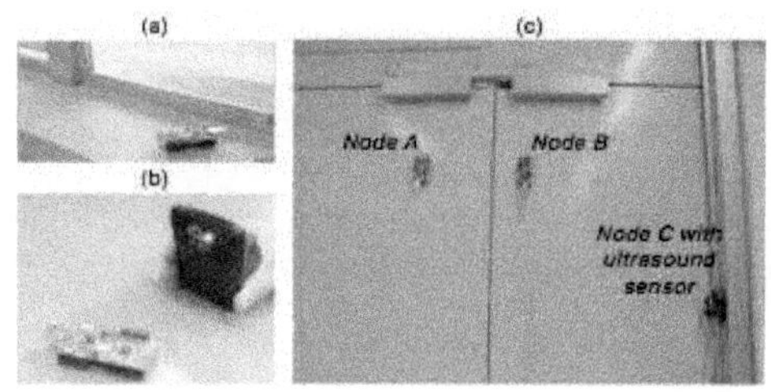

Figure 18: Illustration of several experimental settings

the logic time consistence. In this experiment, we open one lamp array in the rear of our laboratory and switch off all other lamps. Then, we place sensor nodes in a line covering both light region and dark region. In Figure 16, the x-axis represents the approximate distance of one node to the rear of the laboratory. We first plot the received light intensity in Figure 16. As the node becomes far away from the rear region, the detected light intensity drops. From Figure 16, we can observe that when the light intensity is higher than $25mV$, logic times are accurately maintained, e.g., the logic time errors are smaller than $600\mu s$ after 30 minutes. Only after the light intensity drops below $25mV$, the logic time error reaches $780\mu s$. However, such a performance is still acceptable in many systems with the $1ms$ error bound [15]. As a matter of fact, in most indoor environments, the node can normally observe $50mV$ above light intensity. Therefore, one lamp can cover a large-scale field such that fluorescent lamps can be sparingly used in FLIGHT.

Finally, we examine the impact of the interference from other light sources. We first mix the fluorescent light with the sunlight, which commonly occurs when the node is deployed close to the window (Figure 18(a)). The sunlight is a direct signal. We observe that the combined reading of the node consists of a $800mV$ direct component and a $90mV$ alternating component, and find that the periodical pattern from fluorescent lighting can still be accurately detected. The result shows that up to 80% of time, the logic time error is less than $600\mu s$. In addition, we also mix the fluorescent light with the LED signal from a smart phone (Figure 18(b)), which is the most widely existing noise for FLIGHT in the indoor environment. Although the generated light from LED is relative stable, due to the slight shaking of the smart phone, the amplitude of the combined signal varies, and the amplitude of the LED signal is greater than that of the fluorescent light. Similar to Figure 6, FLIGHT can still capture the frequency component close to $100Hz$ and accurately detect the lighting period delimiters. From statistics, we observe that the logic time is still accurate. Figure 17 shows that the

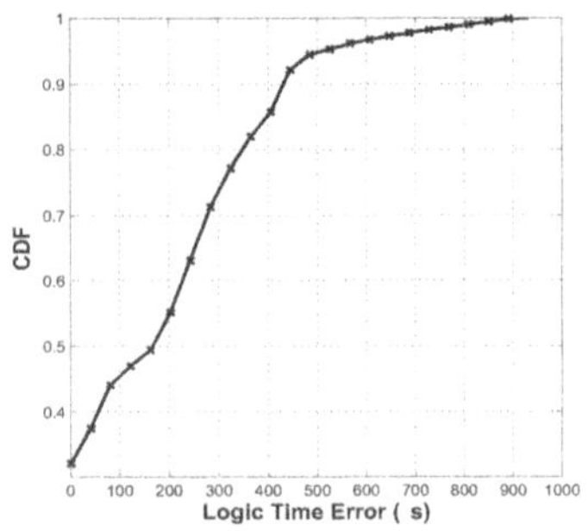

Figure 19: Time error with controlled mobility

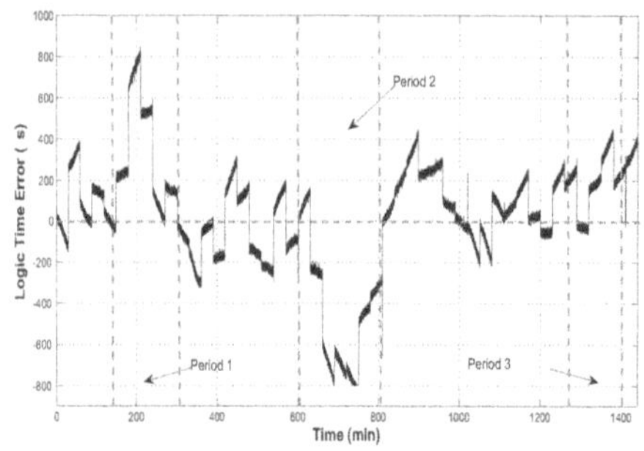

Figure 20: Nodes' logic time error during the movement

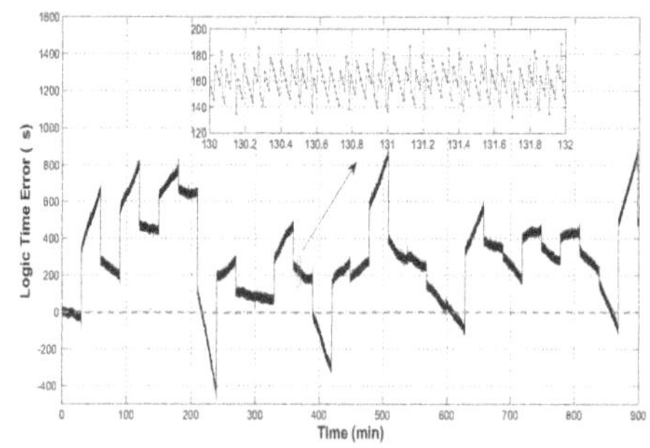

Figure 21: Nodes' logic time error in different buildings

error is controlled within 900μs and in most cases the error is less than 610μs. Finally, we mix the fluorescent light with the filament light that is another common noise for FLIGHT. In this scenario, the logic time is also accurate according to Figure 17. Therefore, we believe that FLIGHT is robust to external interferences from other light sources during clock calibration.

5.2.2 *Dynamic environment*

In this subsection, we conduct three trails of experiments to evaluate FLIGHT in dynamic environment. In the first experiment, two sensor nodes are attached to the gates of the laboratory as shown in Figure 18(c). Notice that only the gate with sensor node A can be open. Both of the two nodes can synchronize their clocks with the same lamp. To avoid disturbing the entrance of students to the laboratory, node A transmits its recorded logic times through wireless communications. This experiment lasts 6 hours and we also deploy a node (node C) with the ultrasonic wave sensor to count the number of people passing through the gate to approximate the times of gate opening. During the experiment, the final value of the counter is 158 and we estimate the gate has been opened more than 70 times during the experiment. We summarize the system performance in Figure 19. We find that the average clock difference between node A and B is smaller than $1,000\mu s$ all the time and it is less than 400μs most of time. Compared with the static environment in Figure 13, the average error increasing is slight and accounts for 12% performance degradation.

In the second trail of experiment, three persons take turns to roam in the academic building. In particular, during periods 1, 2, and 3 in Figure 20, they roam in the office area, the class room area, and the laboratory, respectively. In each period, each person holds three TelosB motes and roams for around one hour. Two sensor nodes are used to be calibrated and the third one periodically sends out beacon messages to trigger the logic time recording. To store recorded logic times, the to-be-synchronized motes are connected to a laptop inside a bag carried by the person. Except the three highlighted periods, sensor nodes are placed in the laboratory stilly. From Figure 20, we can observe that when the motes roam outside the laboratory (i.e., periods 1 and 2), the logic time error is mainly distributed within $[600, 800]\mu s$. The increasing of the logic time error in the two periods is mainly due to the noises introduced by the mobility and surrounding environments. In addition, some places during the roaming are not well covered by light, which also contributes to the performance deterioration. Differently, logic times in period 3 suffers from a much less error during the movement. Compared with the static performance, the logic error in period 3 only slightly increases.

In the third experiment, we place two sensor nodes in two different buildings that are around 150m away from each other. We place one beacon node in between. The beacon node periodically sends out a beacon message with the maximum power to trigger the two experimenting nodes to record their instant logic times. During $[350, 550]min$, two nodes roam within their local regions (constrained by the transmission range of the beacon node) separately. We plot their logic time differences in Figure 21. Figure 21 shows that the average error is around 400μs. When the mobility is enabled, the error is limited within $1,000\mu s$ all the time. The overall performance crossing different buildings is similar to the results obtained in previous two experiments.

5.3 Energy consumption

To understand the energy efficiency of FLIGHT, we first evaluate FLIGHT under different calibration intervals and then compare it with some recent approaches including ROCS, WizSync, and FTSP [10, 13, 15]. We directly measure the energy consumption of FLIGHT from the hardware and report the result in Figure 22. The working power of FLIGHT is around 5.394mW and each calibration in FLIGHT can be finished within 100ms. The energy consumption of FLIGHT with different intervals is plotted in Figure 22. We see that the energy consumption of FLIGHT is usually less than 5μW, which is highly desired for power-constrained mobile devices. For the comparison study, we directly measure the energy consumption of FTSP as well. For ROCS and WizSync, on the other hand, since bot two methods need specific-designed hardware and algorithms, we refer to the results reported in their corresponding literatures. We configure all four schemes with consistent settings and report the results in Figure 23. We evaluate the energy consumptions of FLIGHT, ROCS, and WizSync, by varying the calibration interval from 10min to 30min. From Figure 23, we can see that the communication-based protocol indeed incurs a high-energy consumption. Compared to FLIGHT, ROCS, and WizSync, the power consumption of FTSP is much higher even when we set a much larger calibration interval. As we can see from Figure 23, the energy consumption of ROCS and WizSync is similar with the same calibration interval setting. Due to the uncertainty of the Wi-Fi beacon period, the calibration interval of WizSync is usually configured to be 10min in practice and the corresponding power consumption is around 50μW. The calibration interval of FLIGHT and ROCS can be set around 30min, so their energy efficiencies are around 40 times and 6 times higher than WizSync, respectively. Since sampling the light sensor in FLIGHT consumes much less energy than operating the FM receiver module in ROCS, FLIGHT can achieve higher energy efficiency.

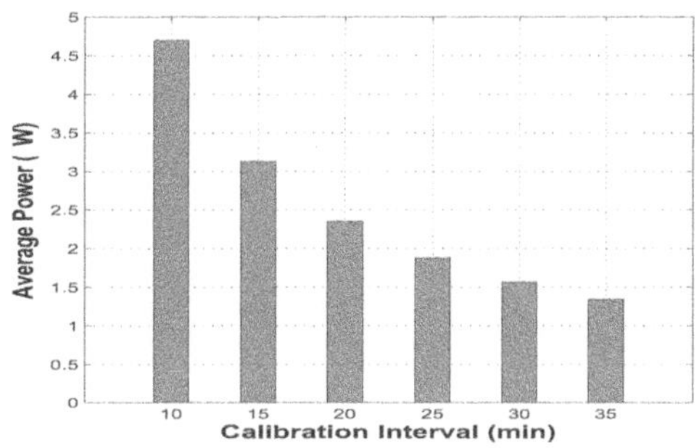

Figure 22: Energy consumption of FLIGHT with different calibration intervals

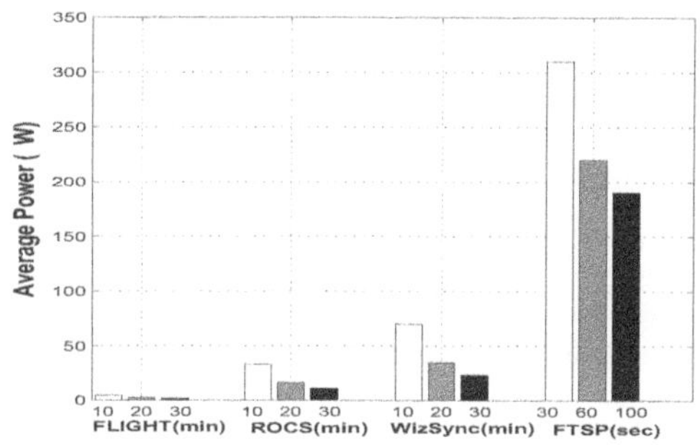

Figure 23: Energy consumption comparison of different schemes

6. LIMITATIONS AND FUTURE WORKS

- **Lighting availability**: FLIGHT requires the availability of fluorescent lighting. As we have shown in Section 5, the period information can be extracted even when the detected light intensity is very low. One fluorescent lamp, thus, can coordinate plenty of surrounding nodes. Fluorescent lighting from different lamps can exhibit a stable and consistent period. Hence, lamps can be sparingly utilized in the field, which is available across almost all indoor places, e.g., universities, hospitals, airports, supermarkets, etc. FLIGHT might generate extra cost while the cost is generally slight.

- **Sensor availability**: In FLIGHT, light sensors should be exposed to the environment during the calibration. It can be achieved for the majority of the applications that have been introduced in Section 2, such as the emerging HLPP gaming, the event detection and diagnosis in sensor networks and so on. In some other applications, like the body area networks, if certain nodes are permanently placed inside pockets, those nodes have to synchronize their clocks with their neighbors via communication based protocols, e.g., FTSP, while those exposed nodes can still use FLIGHT to correct clocks. Nevertheless, benefiting from the existing clock calibration with FLIGHT, further communication is minimized and synchronization overhead can be dramatically reduced.

- **Enhancing robustness**: In the current filter design for time period generation in FLIGHT, we focus on filtering the noises outside the $100Hz$ frequency band. It is mainly because the $100Hz$ component in the frequency domain serves as an indication for the fluorescent lighting in most indoor environments. Such a design, however, could be vulnerable to noise near $100Hz$. Some advanced digital techniques, e.g., PLL, can be applied to further improve the reliability of FLIGHT.

7. RELATED WORK

Great efforts have been made for achieving common time notion across the networking systems over the past decade. Most existing works rely on wireless communications and mainly focus on improving the synchronization accuracy. RBS [8] eliminated the sender-side delay for synchronization. TPSN [9] further canceled out the propagation delay. By using the MAC-layer stamping technique, FTSP [10] largely increases the synchronization accuracy. However, the authors in [11] find that errors among different clocks exponentially increase with the network diameter. [30] proposes a fast flooding scheme and [24] studies the synchronization accuracy in low-duty-cycle sensor networks. [22] estimates clock uncertainty for duty-cycled sensor networks. On the other hand, some other works also focus on addressing the clock uncertainty and reducing the time synchronization cost. ACES [23] suggested to track skew by using Kalman filter and ODS [6] introduces the on-demand accuracy synchronization. Both [31] and [32] explore the temperature compensated scheme to mitigate the clock skew. In [24] and [25], authors introduce to maintain a common time by using two clocks on each individual node with different accuracies.

Recently, several emerging studies exploit the external signal source with a stable period for clock calibration and synchronization. Communication overhead and energy consumption can be reduced compared to protocols mentioned above. RT-Link [33] and ROCS [13] both explore to synchronize clocks to a certain type of radios. Specific hardware are required in those proposals to generate periodical signals. The majority of those solutions utilize radios for the clock calibration at the expense of higher power consumption. Although clock calibration can be performed periodically, the energy consumption for each calibration could not be reduced and the accumulated power draining for time synchronization is still high. [14] designs passive radio receivers for synchronizing nodes to radio stations. In [15], authors introduce to synchronize clocks to Wi-Fi beacons without the extra hardware support. Such a global reference (i.e., Wi-Fi beacons), however, may not be stable enough due to channel contention and collisions. In [12], the authors propose to use power lines for clock calibration. However, due to dramatically decay, the periodical signals can only be detected in the vicinity of power lines. Extra hardware is required as well in [12]. On the contrary, FLIGHT is not constrained by those limitations and FLIGHT can precisely calibrate clocks in an energy-efficient manner without any specific hardware. In [34], the compact fluorescent light bulbs have been used for the proximity sensing. The technique in [34] utilizes the special property of the bulbs when they are illuminated, while we focus on the special property of the fluorescent light itself.

8. CONCLUSION

In this paper, we develop a new clock synchronization approach called FLIGHT, which leverages the fact that the light intensity from fluorescent lamps varies with a stable period. FLIGHT does not require any extra hardware and radio operations. By sampling light sensor or camera, FLIGHT can intelligently extract periods for the clock calibration and retain a common notion of time in the system. We give in-depth analysis on various practical issues to ensure the calibration accuracy. We implement FLIGHT in TelosB motes and conduct experiments using a 12-node test-bed under both static and mobile settings. During one week measurements, Over one-week measurement results show that FLIGHT can achieve a tightly synchronized time with low energy consumption.

Acknowledgement

We would like to thank our shepherd, Dr. Kyle Jamieson, as well as the anonymous reviewers for providing constructive feedbacks and valuable input for improving the quality of this paper. We acknowledge the support from NTU Nanyang Assistant Professorship (NAP) grant M4080738.020 and Microsoft research grant FY12-RES-THEME-001. The research of Xiang-Yang Li is partially supported by NSF CNS-0832120 and NSF CNS-1035894. The research of Yunhao Liu is supported by the NSFC Distinguished Young Scholars Program under grant No. 61125202 and NSFC Major Program 61190110.

9. REFERENCES

[1] J. Qiu, D. Chu, X. Meng, and T. Moscibroda, "On the feasibility of real-time phone-to-phone 3d localization," in *Proc. of ACM Sensys*, 2010.

[2] T. Gu, Z. Wu, X. Tao, H. Pung, and J. Lu, "epsicar: An emerging patterns based approach to sequential, interleaved and concurrent activity recognition," in *Proc. of IEEE PerCom*, 2009.

[3] G. Virone, A. Wood, L. Selavo, Q. Cao, L. Fang, T. Doan, Z. He, and J. Stankovic, "An advanced wireless sensor network for health monitoring," in *Proc. of D2H2*, 2006.

[4] A. Natarajan, B. de Silva, K. Yap, and M. Motani, "Link layer behavior of body area networks at 2.4 ghz," in *Proc. of ACM Mobicom*, 2009.

[5] J. Polastre, R. Szewczyk, and D. Culler, "Telos: enabling ultra-low power wireless research," in *Proc. of ACM/IEEE IPSN*, 2005.

[6] Z. Zhong, P. Chen, and T. He, "On-demand time synchronization with predictable accuracy," in *Proc. of IEEE Infocom*, 2011.

[7] "Product List of Seiko Instruments Inc., 2009," http://speed.sii.co.jp/pub/compo/quartz/productListEN.jsp.

[8] J. Elson, L. Girod, and D. Estrin, "Fine grained network time synchronization using reference broadcasts," in *Proc. of ACM OSDI*, 2002.

[9] S. Ganeriwal, R. Kumar, and M. Srivastava, "Timing sync protocol for sensor networks," in *Proc. of ACM Sensys*, 2003.

[10] M. Maroti, B. Kusy, G. Simon, and A. Ledeczi, "The flooding time synchronization protocol," in *Proc. of ACM Sensys*, 2004.

[11] C. Lenzen, P. Sommer, and R. Wattenhofer, "Optimal clock synchronization in networks," in *Proc. of ACM Sensys*, 2009.

[12] A. Rowe, V. Gupta, and R. Rajkumar, "Low-power clock synchronization using electromagnetic energy radiating from ac power lines," in *Proc. of ACM Sensys*, 2009.

[13] L. Li, G. Xing, L. Sun, W. Huangfu, R. Zhou, and H. Zhu, "Exploiting fm radio data system for adaptive clock calibration in sensor networks," in *Proc. of ACM Mobisys*, 2011.

[14] Y. Chen, Q. Wang, M. Chang, and A. Terzis, "Ultra-low power time synchronization using passive radio receivers," in *Proc. of ACM/IEEE IPSN*, 2011.

[15] T. Hao, R. Zhou, G. Xing, and M. Mutka, "Wizsync: Exploiting wi-fi infrastructure for clock synchronization in wireless sensor networks," in *Proc. of IEEE RTSS*, 2011.

[16] D. Allan and H. Machlan, "Time transfer using nearly simultaneous reception times of a common transmission," *26th Annual Symposium on Frequency Control*, 1972.

[17] M. Li, D. Ganesan, and P. Shenoy, "Presto: feedback-driven data management in sensor networks," *IEEE/ACM Transactions on Networking (TON)*, 2009.

[18] Y. Wu and X. Wang, "Achieving full view coverage with randomly-deployed heterogeneous camera sensors," in *Proc. of IEEE ICDCS*, 2012.

[19] X. Chang, R. Tan, G. Xing, Z. Yuan, C. Lu, Y. Chen, and Y. Yang, "Sensor placement algorithms for fusion-based surveillance networks," *IEEE Transactions on Parallel and Distributed Systems (TPDS)*, no. 99, pp. 1–1, 2011.

[20] S. Zhao, L. Fu, X. Wang, and Q. Zhang, "Fundamental relationship between node density and delay in wireless ad hoc networks with unreliable links," in *Proc. of ACM Mobicom*, 2011.

[21] A. Kashyap, S. Ganguly, and S. Das, "A measurement-based approach to modeling link capacity in 802.11-based wireless networks," in *Proc. of ACM Mobicom*, 2007.

[22] S. Ganeriwal, I. Tsigkogiannis, H. Shim, V. Tsiatsis, M. Srivastava, and D. Ganesan, "Estimating clock uncertainty for efficient duty-cycling in sensor networks," *IEEE/ACM Transactions on Networking (TON)*, 2009.

[23] B. Hamilton, X. Ma, Q. Zhao, and J. Xu, "Aces: adaptive clock estimation and synchronization using kalman filtering," in *Proc. of ACM Mobicom*, 2008.

[24] J. Koo, R. Panta, S. Bagchi, and L. Montestruque, "A tale of two synchronizing clocks," in *Proc. of ACM Sensys*, 2009.

[25] T. Schmid, P. Dutta, and M. Srivastava, "High resolution, low power time synchronization an oxymoron no more," in *Proc. of ACM/IEEE IPSN*, 2010.

[26] Z. Zhang, X. Zhou, W. Zhang, Y. Zhang, G. Wang, B. Zhao, and H. Zheng, "I am the antenna: accurate outdoor ap location using smartphones," in *Proc. of ACM Mobicom*, 2011.

[27] A. Bhartia, Y. Chen, S. Rallapalli, and L. Qiu, "Harnessing frequency diversity in wi-fi networks," in *Proc. of ACM Mobicom*, 2011.

[28] S. Rallapalli, L. Qiu, Y. Zhang, and Y. Chen, "Exploiting temporal stability and low-rank structure for localization in mobile networks," in *Proc. of ACM Mobicom*, 2010.

[29] J. R. Barry, *Digital Communications*, 2003.

[30] F. Ferrari, M. Zimmerling, L. Thiele, and O. Saukh, "Efficient network flooding and time synchronization with glossy," in *Proc. of ACM/IEEE IPSN*, 2011.

[31] T. Schmid, Z. Charbiwala, Z. Anagnostopoulou, M. Srivastava, and P. Dutta, "A case against routing-integrated time synchronization," in *Proc. of ACM Sensys*, 2010.

[32] Z. Yang, L. Cai, Y. Liu, and J. Pan, "Environment-aware clock skew estimation and synchronization for wireless sensor networks," in *Proc. of IEEE Infocom*, 2012.

[33] A. Rowe, R. Mangharam, and R. Rajkumar, "Rt-link: A time-synchronized link protocol for energy-constrained multi-hop wireless networks," in *Proc. of SECON*, 2006.

[34] S. Gupta, K. Chen, M. Reynolds, and S. Patel, "Lightwave: using compact fluorescent lights as sensors," in *Proc. of ACM UbiComp*, 2011.

Energy-based Rate Adaptation for 802.11n

Chi-Yu Li[1] Chunyi Peng[1] Songwu Lu[1] Xinbing Wang[2]

[1] University of California, Los Angeles, CA 90095, USA [2] Shanghai Jiao Tong University, China

{lichiyu, chunyip, slu}@cs.ucla.edu, xwang8@sjtu.edu.cn

ABSTRACT

Rate adaptation (RA) has been used to achieve high goodput. In this work, we explore to use RA for energy efficiency in 802.11n NICs. We show that current MIMO RA algorithms are not energy efficient for NICs despite ensuring high throughput. The fundamental problem is that, the high-throughput setting is not equivalent to the energy-efficient one. Marginal throughput gain may be realized at high energy cost. We propose EERA, an energy-based RA solution that trades off goodput for energy savings at NICs. Our experiments have confirmed its energy savings at NICs while keeping the cost at the device level and across clients acceptable.

Categories and Subject Descriptors

C.2.1 [**Computer-Communication Networks**]: Network Architecture and Design—*Wireless communication*

General Terms

Algorithms, Design, Experimentation, Performance

Keywords

Energy Efficiency, IEEE 802.11n, MIMO, Rate Adaptation

1. INTRODUCTION

Rate adaptation (RA) is a popular mechanism [6,8,11,14,19,21–23] to improve the performance of wireless network interface card (NIC). It dynamically selects the best physical-layer configuration (e.g., various modulation and coding schemes) depending on time-varying channel conditions. The traditional goal of RA is to achieve high goodput (i.e., effective throughput). In this work, we explore to use RA to ensure energy efficiency on recent 802.11n NICs.

Our research is motivated by two factors. First, 802.11n devices are increasingly popular. The shipment reached 5.9M in the second quarter of 2010. It is expected to accelerate at an annual rate of 15% in upcoming years [5]. Battery-powered smartphones and tablets have become the next target for 802.11n [3]. Second, an 802.11n device consumes much more power than its legacy 802.11a/b/g NIC. Our measurements show that, an 802.11n 3x3 MIMO receiver consumes about twice the power of 802.11a during active transmission, and 1.5 times power when idle. Therefore, energy efficiency becomes a critical issue for 802.11n NIC operations.

Existing RA solutions are effective to ensure high goodput but not energy savings. We observe that, two popular 802.11n RA algorithms ARA [22] and MiRA [19] incur per-bit energy waste at an NIC as large as 54.5% and 52.9%, respectively. The energy waste still exists with/without the power-saving mechanisms of SMPS and PSMP. The root cause is that, current RAs obtain high goodput at whatever energy cost. Marginal goodput gain is realized by powering on more antennas, more streams, and higher MCS rates. The fundamental problem is that, conflicts arise between high-goodput settings and low-energy settings in 802.11n NICs.

We propose EERA, a new RA algorithm that trades off goodput for energy savings at an 802.11n client NIC. EERA searches for the MIMO setting consuming less per-bit energy, rather than achieving higher goodput. It thus slows down communication to save energy. However, this slowdown is contained by two conditions: EERA must accommodate its data source rate and not affect other clients using traditional RAs. EERA supports both single-client and multi-client operations. In the former case, it abstracts the problem as multi-dimensional search, and exploits ternary search and MIMO characteristics to speed up its runtime convergence. The latter case builds on top of single-client design. Each client is periodically allocated a fair share of airtime, and can only use up this airtime share but no more. Fair sharing of extra airtime protects each client through isolation, thus enabling coexistence of EERA and other MIMO RA clients. Moreover, EERA is configurable. Each client specifies a tuning knob, which controls multi-client interference, and energy balance at NIC and other device components. EERA reverts to traditional goodput-optimizing RA if needed.

In all test scenarios, EERA consistently outperforms other RA algorithms in terms of NIC energy efficiency. It saves about 30% energy compared with ARA and MiRA in all scenarios. It saves 6-36% in static settings and 20-24% in mobility and field tests, compared with another energy-saving proposal MRES [20]. At the device level, EERA does not incur nonnegligible energy increase for Web, VoIP and video applications, but incurs 1.77% waste in FTP compared with ARA in our tested scenario. This is because FTP application program stops consuming more energy once a file transfer completes; faster transmission helps to reduce energy consumption on FTP application. EERA also compares well with ARA in multi-client scenarios. It increases packet delay by about 0.07-0.19ms in tested cases. For contention, it increases retries by at most 3.4% (0.009 retries/ms), but avoids up to 34% retries (0.08 retries/ms) when the traffic source rate is low.

The rest of the paper is structured as follows. Section 2 in-

MobiCom'12, August 22–26, 2012, Istanbul, Turkey.

troduces the background on 802.11n MIMO and its power-saving schemes. Section 3 uses a case study to examine the limitations of 802.11n RA in NIC energy efficiency, and Section 4 models the 802.11n NIC energy. Sections 5, 6, and 7 describe the design, implementation, and evaluation of EERA, respectively. Section 8 discusses the related work and Section 9 concludes the paper.

2. BACKGROUND

In a nutshell, rate adaptation (RA) offers an effective mechanism to exploit the multi-rate, adaptive modulation capability at the physical layer. The design of RA is more complex in 802.11n than in the legacy 802.11a/b/g systems. Given the wireless channel condition, it has to select the appropriate configuration in three dimensions, the modulation and coding scheme (MCS), the chain setting (i.e., the chosen numbers of transmit and receive antennas at the sender and the receiver), and the number of spatial streams in the 802.11n scenario.

The 802.11n specification defines a large parameter space, thus posing *scaling* issues for RA design. The MCS rates span from 6Mbps to 600Mbps. Each sender/receiver can *activate* one to four transmit/receive antennas. The standard also supports multiple-stream operations, i.e., single-stream (SS), double-stream (DS), triple-stream (TS), and quadruple-stream (QS). The streams are bounded by the smaller number of transmit and receive antennas. Each setting is denoted by $N_t \times N_r/rateStream$, with N_t and N_r being transmit and receive antennas, $rate$ being the MCS rate, and $Stream$ being the number of streams. For example, $3{\times}3/13.5$SS defines the setting using a single stream on three transmit/receive antennas each, with the MCS rate being 13.5 Mbps.

Power Saving Mechanisms in 802.11n: Since our interest is on energy savings, we briefly introduce two power-saving mechanisms defined by the 802.11n standard, i.e., Spatial Multiplexing Power Save (SMPS) and Power Save Multi-Poll (PSMP). Both reduce power consumption during the non-active (i.e., *idle* or *sleep*) period without data transmission. SMPS reduces the idle power consumption at the receiver by activating only one receive chain. The standard supports two modes. In the *Static* mode, the client statically retains a single receive chain. In the *Dynamic* mode, the receiver switches to multiple receive chains during data transmission, but shifts back to one chain afterwards. PSMP allows for a receiver to sleep during its non-active period (e.g., when AP transmits to other clients). The 802.11n standard also supports two modes: Scheduled PSMP (S-PSMP) and Unscheduled PSMP (U-PSMP). In S-PSMP, AP periodically initiates a PSMP sequence to schedule the transmission. In U-PSMP, AP starts an unscheduled sequence and delivers to those wakeup clients. Both SMPS and PSMP complement our RA scheme. Our design primarily handles the *active* period for data transfer and strikes balance between active and non-active energy consumption to ensure NIC energy efficiency.

3. 802.11N RA LIMITATION: AN ENERGY EFFICIENCY PERSPECTIVE

We now use a simple case study to examine the limitations of current 802.11n RA algorithms in terms of energy efficiency. We show that, current solutions are effective to achieve high goodput, but may not ensure energy efficiency. There exists fundamental conflicts between the best goodput setting and the most energy-efficient setting for 802.11n NICs.

3.1 Experimental Setting

Our goal is to quantify the energy consumption of 802.11n NICs under various RA algorithms. We select two existing 802.11n RA algorithms for comparison. Atheros RA (ARA) algorithm [22] is used by Atheros 802.11n NICs, while MiRA [19] is a new proposal for 802.11n radios. Both ARA and MiRA apply *sequential* search to probe different settings and locate the best setting eventually. ARA probes from the medium setting (i.e., the highest MCS rate in the DS mode). If the probe succeeds, it switches to the TS mode; otherwise, it goes down through MCS rates in DS and SS modes. Its implementation excludes half of rates during the probing process. In contrast, MiRA uses zigzag probing and starts from the highest MCS in the TS mode, and then switches to DS and SS until it succeeds. Both algorithms can be implemented in the current platforms using available 802.11n chipsets.

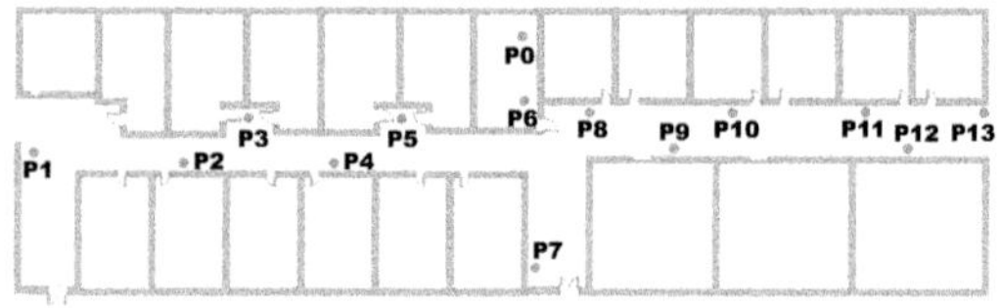

Figure 1: Experimental floorplan.

	EE	ARA	MiRA
E_b (nJ/bit)	19.2	29.7	29.4
Gap (%)	–	54.5%	52.9%
Goodput (Mbps)	35.4	52.4	52.5
#Bits (Mbits)	3598	3576	3598
Energy (J)	69.0	106.2	105.6

Table 1: Per-bit energy consumption, goodput, number of bits, and energy consumption for ARA, MiRA and the optimal EE setting at Location P1. The UDP source rate is 30 Mbps.

We conduct our experiments in a controlled laboratory environment over the 5GHz band; no external interference is observed on the used channel. Both AP and clients operate in an office building (see Figure 1 for the floor plan); Spots P1 to P13 represent different locations for the client, whereas AP is always at P0. We consider the infrastructure mode only, the dominant deployment in reality. Both the AP and clients are programmable 802.11n devices, which use Atheros AR9380 2.4/5 GHz MIMO chipset and support three transmit/receive antennas. The platform supports SS, DS, and TS modes, with transmission rates up to 450Mbps over 40MHz bands. We also use Intel 5300 wireless NIC on the client side. We plot results for downlink transmissions, with the client being the receiver. In each test, we send the UDP traffic generated by *iperf* at constant source rate (say, 30 Mbps for the tests in Table 1) for 120 seconds and collect measurement results over five runs.

We use power meter Agilent 34401A to record the consumed power. For PSMP, we collect traces and use the ideal doze power to simulate its energy value, since PSMP is not implemented in current drivers. Energy saving from PSMP is overestimated without counting its processing overhead. To quantify the energy efficiency of RA algorithms, we use *per-bit energy consumption* E_b as the evaluation metric, defined as the consumed energy when exchanging each bit given a setting. This metric represents the energy consumption when transferring each bit, including consumed energy during both active and non-active periods [20].

3.2 Case Findings

Our experiments show that, both ARA and MiRA incur large energy waste, compared with the most energy-efficient fixed setting. Table 1 gives the *per-bit energy consumption* by both ARA and MiRA, as well as the best fixed setting (called "EE" in the table) achieving highest energy efficiency among all settings at P1 (see Figure 1). The results show that, ARA and MiRA incur per-bit energy waste as large as 54.5% and 52.9%, respectively, when

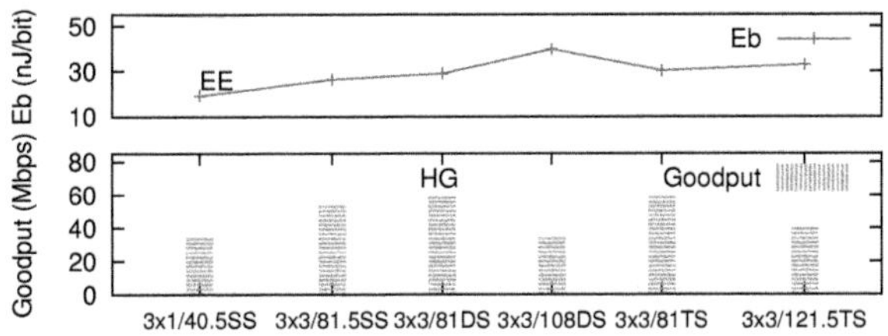

Figure 2: Goodput and per-bit energy of various settings at P1.

Setting	3x1/ 40.5SS	3x3/ 81SS	3x3/ 81DS	3x3/ 108DS	3x3/ 81TS	3x3/ 121.5TS
Active Power (mW)	580.6	812.3	975.0	982.5	1046.4	1063.4
Idle Power (mW)	541.2	765.6				

Table 2: Power consumption of the most energy-efficient setting and those used by MiRA and ARA at P1.

compared with the best setting. Interestingly, we also observe that MiRA and ARA ensure higher goodput during active data transmissions, about 52.5Mbps and 52.4Mbps, respectively, compared with the EE setting that yields only 35.4Mbps. Note that all can sustain the 30Mbps UDP data source.

We next find why current algorithms incur energy waste for NICs. It turns out that, both ARA and MiRA are able to achieve high goodput during the active period, but these settings are not among the most energy-efficient ones. To this end, we first compute the goodput, as well as the *per-bit energy consumption* for those selected settings, and plot them in Figure 2. The setting yielding highest goodput (marked with "HG" in the figure) is 3x3/81DS. However, this HG setting is not the most energy-efficient one (i.e., 3x1/40.5SS, marked with "EE"). The gap in *per-bit energy consumption* between these two settings reaches 11.1 nJ/bit, incurring energy waste as large as 57.8% when using the HG setting. Figure 3(a) further plots the rate distribution (in percentage) of ARA and MiRA. We see that both algorithms are effective in reaching high goodput. This observation is consistent with the primary goal of their design. ARA mainly selects two settings, 3x3/81DS and 3x3/108DS, whereas the selection of MiRA spreads over 5 settings. ARA chooses fewer settings because it considers only half of the rate settings. However, these high-goodput settings consume more energy per bit, as large as 39.7 nJ/bit (3x3/108DS), about twice the *per-bit energy consumption* of the EE setting 3x1/40.5SS. They hence make ARA and MiRA deviate from the EE setting by as large as 54.5% and 52.9%, respectively, in Table 1.

We further examine why the highest-goodput settings cannot ensure best energy efficiency. It turns out that, at these high-goodput settings, the per-bit energy cost to obtain the marginal goodput gain is pretty high. To achieve higher goodput, we have to activate more antennas and more streams no matter how much extra power could be consumed. Here, the "HG" setting (3x3/81DS) consumes additional 394.4 mW and 224.4 mW in active and non-active periods, compared with the "EE" setting (3x1/40.5SS) (shown in Table 2). Figure 3(b) plots goodput and *per-bit energy consumption* for various settings at P1. It shows that, E_b does not monotonically decrease as the goodput increases; Several dips appear in these settings. The marginal goodput gain at the cost of extra energy consumption becomes smaller or even negative for some high-rate settings. As a result, it becomes less energy efficient to chase for higher goodput.

Moreover, our study reveals that current RA algorithms may not have fast convergence when locating the best setting. Both ARA and MiRA apply sequential search to locate the best setting by sequentially probing feasible settings. These sequential search operations may result in slower convergence. This can be illustrated by

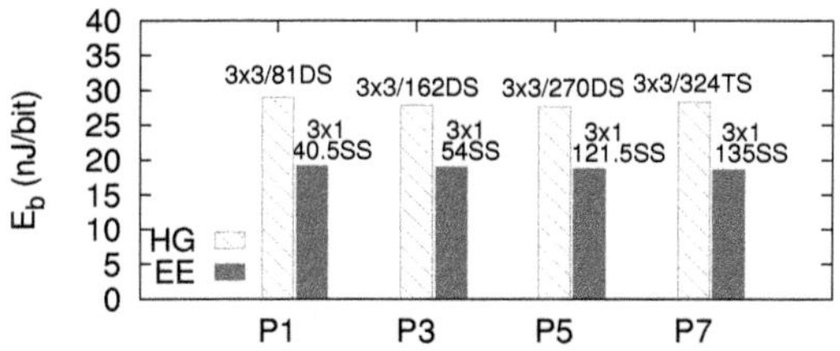

Figure 4: Per-bit energy by RA and EE settings at various locations with 3 transmit antennas and 30 Mbps source.

the detailed probing process of ARA and MiRA shown in Figure 3(c). The sequential search in ARA and MiRA further faces the scaling issue when the number of settings becomes larger. Assume three antennas at the AP. The number of receive antennas can be one (SS), two (SS/DS), and three (SS/DS/TS). The client thus supports $1+2+3=6$ (i.e., $N_r(N_r+1)/2$, where N_r is the maximum number of receive antennas) modes, and each mode allows for multiple MCSes (8 MCSes for 802.11n). The total number of settings can reach $6\times 8=48$. The scaling issue becomes more prominent when the number of antenna grows to eight and the number of MCSes per mode reaches ten in the upcoming 802.11ac [4]. This leads to the overall search space of 360 ($=10\times 8(8+1)/2$) choices.

3.3 Dynamics of Energy-Efficient Settings

We next show that, the most energy-efficient (EE) setting varies with several factors, including location, data source, power-saving schemes, and the number of activated AP antennas.

Location dependence: We now show that the energy waste by current RA schemes is location dependent. The fundamental reason is that, the EE setting varies with locations, but it is not the same as the HG one in general. Figure 4 shows *per-bit energy consumption* for HG and EE settings at various locations; the HG setting is where current RA schemes would stay at. The data source is set as 30 Mbps and AP uses three transmit antennas. The HG setting underperforms the EE setting in terms of energy efficiency. Specifically, HG consumes 51.8%, 46.3%, 47.6%, and 52.2% extra energy compared with EE at P1, P3, P5, and P7 respectively.

Effect of power-saving schemes: Current RA schemes still incur energy waste, whether or not we use the power-saving schemes. However, power-saving schemes may reduce the waste percentage because of reduced power during the non-active (idle/sleep) state. The root cause is that, the EE setting varies with the use of power-saving schemes, while the HG setting remains invariant as long as the channel condition remains unchanged. Assume three antennas at AP. When we use different power-saving schemes (i.e., SMPS and PSMP), the EE setting turns into 3x1/40.5SS and 3x1/54SS, respectively, still different from the HG setting. Note that, the *per-bit energy consumption* of the HG and EE settings indeed decreases due to smaller energy consumption at idle/sleep states. The difference thus becomes smaller. However, the gap is still as large as 31.8% at these two locations as shown in Figure 5(a).

Effect of data source rates: We also study the impact of source rates on energy efficiency. We note that the energy inefficiency of current RAs also varies with data source. The reason is that, the EE setting varies with source rates, while the HG setting remains unchanged. For instance, when the source rate increases from 10 Mbps to 50 Mbps, the EE setting at P1 changes from 3x1/54SS to 3x2/54SS. The energy waste by HG still reaches 44.7%, 51% and 21.7%, for three source rates, as shown in Figure 5(b).

Effect of activated antennas: The energy efficiency of current RA schemes also changes with the number of activated antennas.

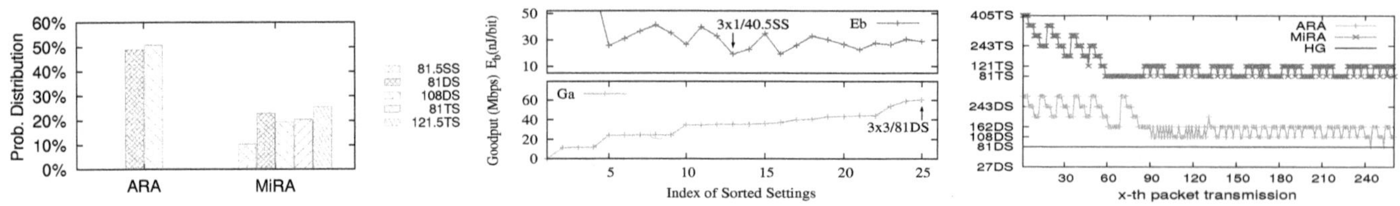

(a) Selected rate distribution (b) Goodput vs. per-bit energy (c) RA Probing process

Figure 3: Performance of ARA and MiRA at P1. All the settings in (a) use three receive antennas.

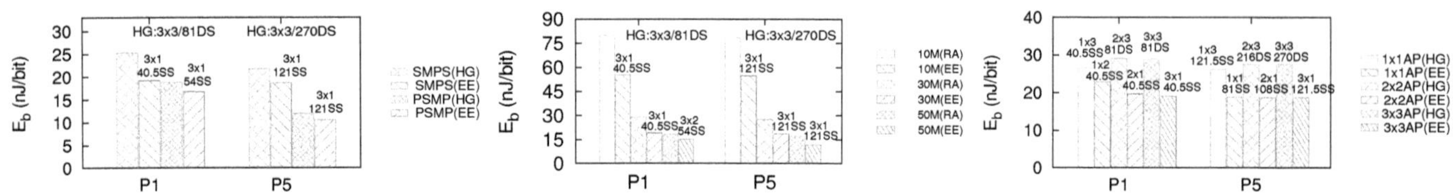

(a) Power saving schemes (b) Source rates (c) Number of AP antennas

Figure 5: Per-bit energy consumption of the HG and EE settings at P1 and P5.

When the number of AP antennas reduces from 3 to 1, the HG setting at P5 changes from 270DS to 216DS, and finally to 121.5SS using three receive antennas, with less per-bit energy consumption. However, Figure 5(c) shows that, the difference between HG and EE is still as large as 47.6%, 47.6% and 39.4%, respectively. In general, the fewer the number of AP antennas, the smaller the gap between HG and EE settings. It depends on how much more receive antennas contribute to goodput improvement and power increase. When the receiver is closer to AP, the marginal goodput gain is small when activating an extra receive antenna; it is ineffective to use more receive antennas in terms of energy efficiency. However, an extra receive antenna may bring higher marginal gain when the client is far away from AP.

3.4 Impact on Device-Level Energy Efficiency

While the most energy-efficient setting saves energy at the 802.11n NIC, it may have negative impact on the device. Since it deviates from the HG setting, it slows down data transmission and prolongs the delivery time. For instance, the portion of active period increases from 56% to 85%, when the EE setting 3x1/40.5SS, but not the HG setting 3x3/81DS, is used in Section 3.2. This slowdown might negatively increase energy consumption of other components (e.g., display, CPU, disk, memory), since they may have to stay active longer for data transmission over NICs.

To assess the impact on device-level energy efficiency, we study two issues: (1) Can energy consumption of other components be decoupled from the NIC status? (2) If not, how much is the incurred energy increase at these components? We focus on display and CPU, since they contribute to the dominant portion of energy consumption at laptops and smartphones [7, 18]. In our experiments, we measure the energy consumption over the same duration when EE or HG is used. The duration always lasts until the slower transmission by EE has completed. The device enters its default power-saving mode after HG completes its transmission first. We use ASUS F8S laptop with Intel Core2 Duo T8300 processor in our measurements. Its display power is from 2.7 W to 9.4 W, much bigger than that of smartphones and other mobile devices [7].

We find out that the display energy consumption is mostly independent of the NIC status. It is mainly decided by the degree of brightness and how long users want to interact. We measure the display energy with different brightness (from 0% to 100%) in four 802.11n NIC states: off, idle, high-rate reception (e.g., 3x3/81DS at P1), and low-rate reception (e.g., 3x1/40.5SS at P1). We have

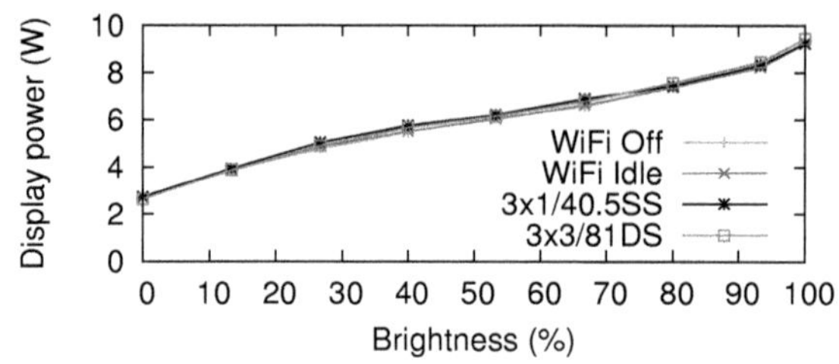

Figure 6: Display energy varies with degree of brightness.

CPU State	C0	C1	C2	C3
EE (3x1/40.5SS)	5.8%	0%	26.0%	66.4%
HG (3x3/81DS)	5.5%	0%	42%	52.0%
Power@800MHz (W) (from [2])	16.8 – 21.3	16.8 – 21.3	10.3 – 13.0	9.8 – 12.4

Table 3: CPU status statistics using HG and EE at P1.

20 runs for each NIC state. Figure 6 plots the average display power consumption over the same measurement period; the standard derivation of each point is within 0.15. The display energy remains invariant at different 802.11n NIC states (including the OFF state), as long as the brightness and duration are fixed. This is also consistent with other studies on display energy [7, 18]. Moreover, it concurs with user experience. The power-saving operations and active duration of the display are mainly determined by user interaction (e.g., screen off after 15-second idle period) but not NIC or other components; it can power off at any time even when NIC is still transmitting. We also tested more cases with various source rates, locations and wireless rates. The results are similar.

We also discover that, the CPU status can be slightly changed due to slower wireless transmissions. Given the same measurement duration, the incurred power increase is roughly in order of tens of milliwatts. We use Intel PowerTop [1] to record CPU status when HG and EE settings are used in our experiments. The CPU has five power states: C0 (normal), C1 (normal), C2 (stop-grant, i.e., sleep), C3 (deep sleep) and C4/C6 (deeper sleep) [2]. In our traces, no C4/C6 state is observed. Table 3 shows an example trace of the CPU status. We observe that, the EE setting slightly increases the CPU duration at C0 (by 0.3%) when the transmission time increases from 56% to 85%. In our experiments, we observe that the CPU mainly work at 800 MHz or in its idle mode. We also list the CPU power consumption for different states at 800 MHz using the data sheet of [2]. The incurred CPU power is no more than 60 mW (at most 21.3 W × 0.3%) over the measure interval.

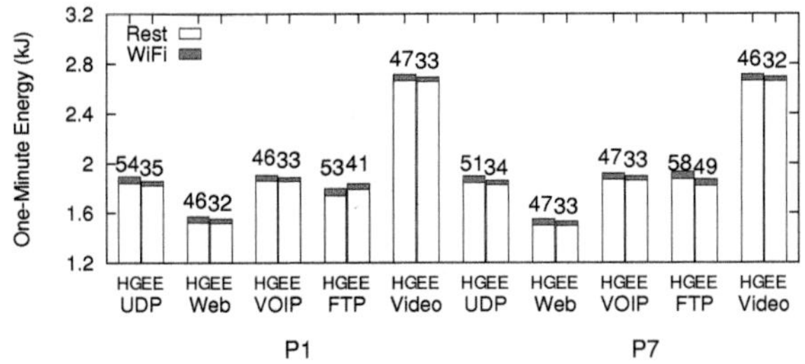

Figure 7: Device and NIC energy consumption in various application scenarios.

In fact, it could be smaller, since less energy is consumed at C2 and C3 states and CPU works at a lower frequency during idle. We also test more cases and obtain similar results.

We further quantify the device energy consumption, as well as NIC, with five applications. They include (1) UDP: start 30 Mbps downlink flow; (2) Web: fetch a 3.8 MB webpage five times within a minute; (3) VoIP: chat for two minutes; (4) FTP: download a 721.9 MB file; and (5) Video streaming: play a 10-minute 1080p HD video. In our experiments, VoIP includes both uplink and downlink traffic, its average rate is 163 Kbps, and the average packet size is 162B. The average source rate of Video streaming is 3.9 Mbps, and the average packet size is 1277 Bytes, while the client side buffering (5-second buffer) is used. Figure 7 plots the average device energy consumption over one minute, where the number above each bar being the NIC energy consumption in Joul. We make two observations. First, the 802.11n HG/EE settings have negligible impact on the energy consumption of other components at the device except in the FTP case. Second, NIC energy saving is the major source to device-level energy efficiency. Though we only observe a small percentage of device-level energy savings, the savings are about 42–54% for NIC alone except in the FTP case (29% and 18% at P1 and P7, respectively). Since NIC consumes only 520–900 mW but the laptop consumes about 30W, the saving percentage at the device level is not large. The gain percentage will be much larger for smartphones [7].

In the FTP case, the HG setting downloads this file faster (104 seconds), compared with 148.2 seconds using EE at P1. Therefore, the FTP application stops early and consumes 4.4 KJ, smaller than the consumed energy of 4.5 KJ in EE over 148.2 seconds. For other applications, there is no significant change even though wireless transmission finishes early, thus imposing little impact on the energy consumption of other components. Note that, device energy consumption is also location dependent in the FTP scenario; device-level energy saving is still achieved at P7. In all cases, EE ensures energy savings for 802.11n NICs.

4. MODELING 802.11N NIC ENERGY

We now model the power consumption of an 802.11n NIC.

Per-Bit Energy: The NIC energy efficiency is quantified by *per-bit energy consumption* E_b, i.e., the consumed energy when transmitting/receiving each bit. Assume that the data source can be accommodated by the setting. Given the time interval of interest T, the per-bit energy is calculated as

$$E_b = Energy/N_{Bits} = (P \times T)/(S \times T) = P/S, \quad (1)$$

where P represents the average power consumption and S represents the data source rate over the entire period T. When data source rate S is smaller than the achievable goodput G, i.e., $S \leq G$, the client experiences both active and non-active (i.e., idle or sleep) modes. Then we have, $P \cdot (T_a + T_{na}) = P_a \cdot T_a + P_{na} \cdot T_{na}$,

Platform	α_1	α_2	α_3	$f(N_{ss})$			P_f	i_1	i_2
				SS	DS	TS	(mW)		
Atheros 9380	2.31	19.8	0.3	0.6	4.6	7	429.0	2.31	19.8
Intel 5300	2.95	195	0.33	3.3	4.1	4.3	496.8	2.9	195

Table 4: Receive power model for Atheros 9380 and Intel 5300.

where P_a and P_{na} are the power consumption by the 802.11n receiver during the active period T_a and non-active period T_{na}, respectively. Since data delivery occurs during the active period, we have $G \cdot T_a = S \cdot T$. Hence, E_b can be computed as

$$E_b = \frac{P_a \times T_a + P_{na} \times T_{na}}{S \times T} = \frac{P_a - P_{na}}{G} + \frac{P_{na}}{S}. \quad (2)$$

Therefore, the data source S is determined by higher-layer applications, while the goodput G is decided by rate settings and wireless channel. We next model the active and non-active power consumptions, which depend on the NIC state including both rate settings and power-saving modes.

Power Model of an 802.11n NIC: The power consumption of an 802.11n receiver can be decoupled as $P_{rx} = P_{rc} + P_{rb}$, where P_{rc} is the power consumption of MIMO circuitry, and P_{rb} is that of baseband signal processing [9]. P_{rc} includes power consumption of all circuit paths, each of which contains all the circuits from RF to analog to digital converter (ADC), e.g., frequency synthesizer, low/band-pass filter, mixer, low-noise amplifier and variable gain amplifier. Based on the MIMO power model of [9, 10], the ADC power can be estimated as a linear function of bandwidth whereas the remaining circuits consume constant power. The baseband power consumption P_{rb} scales with bandwidth and also depends on the number of receive chains. The decoder power correlates with the number of streams and rate settings; P_{rb} can be approximated as a linear function of the number of receive antennas, channel width, and rate. All these models have been validated by our measurements [17]; the details are omitted due to space limit.

Now we model the receiver power at three states: *active*, *idle* and *sleep*. The receiver active power P_{ra} includes both P_{rc} and P_{rb} for circuit and baseband processing. It can be modeled as:

$$P_{ra} = (\alpha_1 \cdot N_r + f(N_{ss})) \times BW + \alpha_2 \cdot N_r + \alpha_3 \cdot r + P_f,$$

where N_r is the number of receive antennas, N_{ss} is the number of streams, BW is the channel width (MHz), r is the rate setting (Mbps). P_f is a constant in mW, and $\alpha_1, \alpha_2, \alpha_3$ are model coefficients. These parameters are platform dependent. Table 4 list the estimated coefficients for Atheros 9380 and Intel 5300 from our measurements [17]. The active power lies in 550−1200 mW. Note that, power consumption increases with N_{ss} and N_r. For example, the power for 3x1/81SS, 3x2/81SS and 3x3/81SS is 588.7 mW, 700.5 mW and 812.3 mW, respectively. However, given the same N_r and N_{ss}, the gap using different MCSes is negligible. For instance, 3x3/13.5SS and 3x3/135SS consume 798.8 mW and 823.1 mW, respectively, close to 812.3 mW for 3x3/81SS.

The idle power P_{ri} roughly equals to the circuitry power (P_{rc}) because of almost no processing during idle. It is thus estimated as

$$P_{ri} = i_1 \cdot N_r \times BW + i_2 \cdot N_r + P_f,$$

where i_1 and i_2 are idle power coefficients. The idle power depends on the number of antennas and channel width, but not the number of streams. For example, Atheros 9380 consumes about 541.2 mW, 653.4 mW and 765.6 mW when using 1, 2, and 3 receive antennas during idle (BW = 40MHz).

In sleep mode, we discover that few components remain active. Both Atheros 9380 and Intel 5300 NICs consume constant power (158.4 mW and 166.5 mW), saving at least 2/3 of idle energy.

The 802.11n transmit power mainly varies with the number of transmit antennas (N_t) and channel width. In fact, the power consumption for power amplifier dominates transmit power and it is in proportion to N_t. The measured transmit power for Atheros 9380 is 1.16 W, 1.88 W and 2.64 W in case of one, two, and three transmit antennas used. The transmit power for Intel 5300 is given by [13].

5. EERA DESIGN

EERA trades off goodput for energy savings at an 802.11n client NIC. It seeks to find the MIMO setting that consumes less per-bit energy at NIC, rather than that achieves higher goodput. EERA runs at AP in its default operation mode, which transmits downlink data and reduces per-bit energy at each client receiver. Note that EERA does not offer a holistic solution that minimizes entire device energy; It only saves NIC energy from the RA perspective.

EERA is a configurable RA algorithm. Each client i specifies a threshold parameter $R_{c,i}$, which defines the minimum goodput that EERA cannot go below when selecting settings. It specifies how much a client is willing to slow down to save its NIC energy. The parameter is set in the percentage (say, 90%) of the highest goodput to the client. When $R_{c,i}$ is chosen as 100%, EERA reverts to traditional goodput-optimizing RA.

The per-client parameter $R_{c,i}$ serves as a tuning knob for EERA. It offers flexible tradeoffs between goodput and energy savings along multiple dimensions. It may help to balance energy budgets between NIC and other components of the mobile device due to communication slowdown at each client. It also facilitates to mitigate cross-client interference. Since it limits on how much an EERA client may slow down, a client can configure this parameter to reduce effect on other traditional RA clients. Moreover, this parameter may take into account application requirements (e.g., minimum throughput needed by video streaming service).

The overall idea of EERA is to let each client select the most energy-efficient setting from its feasible candidates when slowing down. However, this slowdown is contained by two factors: it must accommodate its data source rate, and not affect other clients when they were to choose their highest-goodput settings. EERA supports both single-client and multi-client operations. In the single-client case, it abstracts the problem as multi-dimensional search, and exploits ternary search and MIMO characteristics to speed up its runtime convergence. The multi-client case builds on top of single-client design, but requires additional operations. Note that each client cannot slow down too much to affect others. This is done by setting a limit, expressed in airtime share, for each client. The client cannot select a setting such that its extra communication time (due to slowdown) exceeds its airtime share, no matter how energy efficient this setting can be. To this end, AP calculates the temporal fair share (defined in airtime) for each EERA client. Assume that each client uses its highest-goodput rate and extra air time is available thereafter. The extra air time is then fairly allocated among all active clients. Each client uses EERA to minimize energy based on its fair airtime share. This way, an EERA client cannot be arbitrarily slow to hurt others. We now present detailed designs on both single-client and multi-client cases.

5.1 Single-Client Case

We first consider the simple, single-client scenario. In this case, EERA formulates the energy-efficient RA as a multi-dimensional search problem that locates the low-energy MIMO setting. It organizes settings into a multi-level tree, and then applies the ternary search scheme over each branch. At each setting, EERA uses probing to obtain the per-bit energy. The probing is "in band" by using multiple data frames sent from the AP to the client. By further

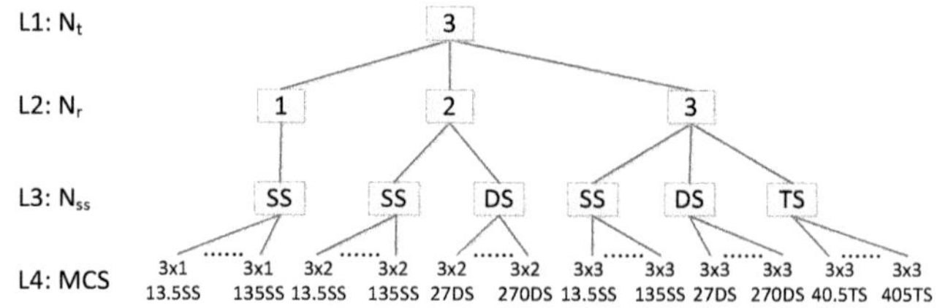

Figure 8: An example of the MIMO search tree

exploiting the MIMO communication features, EERA can simultaneously prune multiple branches at runtime, thus eliminating those probings deemed unnecessary. Its runtime efficiency is even better than ternary search. The solution also works with/without complementary power-saving schemes (e.g., SMPS and PSMP). There are three key issues: (a) How to organize the settings into a search graph for fast lookup? (b) How to prune branches and reduce probing at runtime? and (c) How to estimate the per-bit energy for each setting? We next elaborate on these details.

5.1.1 RA as Multidimensional Search

In addition to MCS rates, MIMO RA in 802.11n has to consider more dimensions: the number of transmit and receive antennas activated, and the number of data streams used. We thus abstract the problem of RA as multidimensional search. The goal is to find the setting given pre-specified optimality criteria. The traditional objective for a RA is high goodput, whereas the goal for EERA is reduced energy consumption. Since low-energy settings also depend on the data source (shown in Section 3.3), we pose RA as the following problem. Given the data source, EERA searches for lower-energy settings that can sustain the source along four dimensions: the number of transmit antennas N_t, the number of receive antennas N_r, the number of data streams N_{ss}, and the various MCS options N_{MCS}. The devised algorithm needs be efficient in its runtime complexity, as well as incurring low probing overhead. The search has to scale to large space. The search space for 802.11n, which supports four antennas, can have 80 settings. It doubles for the upcoming 802.11ac standard, which supports eight antennas.

We organize the search graph as a four-level tree, where each node denotes a setting with its estimated per-bit energy. As shown in Figure 8, the hierarchy of the tree is built following the order of N_t, N_r, N_{ss}, and N_{MCS}. Specifically, the first level is organized using the number of transmit antennas N_t, with the second level being the number of receive antennas N_r. Since the number of data streams N_{ss} is the minimum of N_t and N_r, we use it as the third level of the tree. The bottom level is the MCS options, which typically have the largest number of choices.

As the first heuristic, the AP uses the maximum number of antennas. This eliminates the top level and reduces to a three-level tree. The rationale is as follows. Our goal is to reduce energy consumption at the client with full collaboration from the AP. Therefore, it is easy to show that, given the maximum number of antennas activated at AP, the client has the largest number of choices, thus leading to better search results on energy savings.

The search tree-based abstraction also illustrates how traditional RAs work. They typically follow sequential search (e.g., MiRA and ARA) or randomized search (e.g., the MIMO version of SampleRate algorithm [19]). Consequently, these algorithms have the complexity of $O(N_t N_r N_{ss} N_{MCS})$. Take MiRA as an example at P1. The number of search steps is about 35. Assume 30Mbps data source and no PS mode with 3x3 AP at P1. The per-bit energy of all the settings is shown in Figure 3(b). MiRA will go all steps shown in Table 5 to reach the low-energy setting 3x1/40.5SS. Given each

Branch	Probing Sequence: MCS (E_b nJ/bit)	Steps
*3x1/SS	135SS(∞) $\cdots$ $\rightarrow$ **40.5SS(19.2)** $\rightarrow$ 27SS(25.7)	7
3x2/DS	270DS(∞) $\cdots$ $\rightarrow$ **54DS(27.6)** $\rightarrow$ 27DS(36.6)	8
3x2/SS	108SS(∞) $\cdots$ $\rightarrow$ **54SS(22.7)** $\rightarrow$ 40.5SS(22.9)	4
3x3/TS	405TS(∞) $\cdots$ $\rightarrow$ **81TS(30.3)** $\rightarrow$ 40.5TS(33.0)	8
3x3/DS	162DS(∞) $\cdots$ $\rightarrow$ **81DS(29)** $\rightarrow$ 54DS(30.2)	4
3x3/SS	121.5SS(∞) $\cdots$ $\rightarrow$ **81SS(26.4)** $\rightarrow$ 54SS(26.5)	4
Total		35

Table 5: Search steps of MiRA sequential search at P1.

branch with the same N_r and N_{ss}, sequential search needs to keep on probing until reaching the setting after the optimal one.

5.1.2 Ternary Search in Each Branch

Given the multidimensional search, EERA uses a novel solution technique, called ternary search with simultaneous pruning, to locate low-energy setting in EERA. The resulting algorithm is more efficient than sequential or randomized search. We start the search at the lowest-level, i.e., all the MCS rates given fixed receive antennas and the number of streams. The proposed ternary search uses the following property (its proof is in [17]):

Property I: The per-bit energy E_b is a unimodal function with respect to the MCS rate, given fixed number of chains and fixed number of streams.

The above property makes case for ternary search. Note that binary search cannot be applied since E_b is not a monotonic function with respect to MCS rates. We sort the MCS rates in the increasing order based on their indices, say, $[l, r]$, and find the MCS rate that yields lower per-bit energy. In ternary search, we select two intermediate points that partition the interval into three equal segments, i.e., $m_1 = l+(r-l)/3$; $m_2 = r-(r-l)/3$, as shown in Figure 9. There are three cases: (1) if $f(m_1) < f(m_2)$, then the minimum cannot be on the right side $[m_1+1, r]$. We then search only the left side $[l, m_1]$; (2) if $f(m_1) > f(m_2)$, then the situation is similar. The minimum cannot be on the left $[l, m_2 - 1]$, so we go to the right side - $[m_2; r]$; (3) If $f(m_1) = f(m_2)$, then the search should be conducted in $[m_1, m_2]$. It can be solved recursively by referring to the first two cases.

An illustrative example is given in Figure 10. We apply ternary search on each branch. Take the branch 3x1/SS as an example. Initially, the indices of the MCS rates are $[0, 7]$, and we choose two intermediate rate settings, 40.5SS and 108SS, to partition the branch into three segments, i.e., $m_1 = 2$; $m_2 = 5$. The former setting can achieve 19.2 nJ/bit, whereas the latter gets high per-bit energy due to high loss. Therefore, the minimum of per-bit energy is located in the interval $[0, 4]$. Then, the next intermediate points picked for probing are 54SS and 27SS. Finally, we can locate the optimal setting 40.5SS over this branch. After each branch is traversed, the best setting, 3x1/40.5SS, can be reached. The number of total search steps is 25.

5.1.3 Simultaneous Pruning of Branches

It turns out that EERA can be more efficient than the above tree-based scheme, which uses ternary search over each lowest level. The technique is to simultaneously prune the search space and reduce probing at each step. Consequently, it not only eliminates some branches for further lookup based on runtime search results, but also reduces the range for the ternary search (i.e., l or r). The technique exploits the MIMO communication characteristics. It has two concrete cases at each search step, depending on whether the MCS rate used by the setting has exceeded the channel capacity.

Low-loss probing The first case is when the probing of the rate at the current search step does not result in high packet loss,

Branch	$[l, r]$	Steps	# Pruned Settings
3x3/SS	$[0, 7]$	4	23
3x3/DS	$[2, 4]$	3	2
3x3/TS	$[2, 3]$	2	0
3x2/SS	$[2, 4]$	3	2
3x2/DS	$[2, 3]$	2	0
3x1/SS	$[2, 4]$	3	0
Total		17	27

Table 6: Ternary search steps with simultaneous pruning at P1.

i.e., it yields reasonable goodput. In such a case, we can apply the first rule to eliminate some lower MCS rates from further ternary search, given the fixed numbers of chains and streams. It uses the following property:

Property II: The lower bound of a setting's per-bit energy can be estimated from its loss-free goodput.

Specifically, whenever current probing finds a new setting with lower per-bit energy, we use the above rule to eliminate those lower-MCS-rate settings, which cannot have the same per-bit energy even in the loss-free case. The rationale is that, when a setting achieves its maximum goodput (in the loss-free case), it obtains the lowest per-bit energy. When such bounds cannot beat the current setting, these MCS rates can be removed from the search process.

High-loss probing The second case is when the probe of the rate incurs high packet loss (say, larger than a threshold such as 90%), thus giving very low goodput. This tells us that the current MCS rate exceeds the channel capacity. We then eliminate some settings from further search based on the following property:

Property III: Loss monotonically increases with (1) MCS rate, given the same N_r, N_{ss}; (2) decreasing N_r, given the same MCS rate and N_{ss}; (3) N_{ss}, given the same MCS rate and N_r.

Since the probe at the current MCS m fails, then we can eliminate two more scenarios, both of which would yield higher loss (i.e., further exceeding the channel capacity). The first is the settings with the MCS higher than or equal to m and the number of chains lower than N_r. These settings would also fail in probing, given the same N_{ss}. The second scenario concerns those settings with MCS higher than or equal to m, the number of streams higher than N_{ss}, and the number of chains lower than or equal to N_r. These settings would also fail.

With both pruning heuristics, we further reduce the search steps to 17 by excluding settings at runtime, as shown in Table 6. During the search of the 3x3/SS branch, we prune up to 23 settings at other branches. For example, high-loss probing at 3x3/108SS triggers the following 15 settings to be pruned based on *Property III*: those higher than or equal to 3x3/216DS, 3x3/324TS, 3x2/108SS, 3x2/216DS, and 3x1/108SS. Moreover, based on *Property II*, the per-bit energy of 3x3/81SS (i.e., 26.4 nJ/bit) helps to remove the following 8 settings, with their loss-free goodput lower than 3x3/81SS: the ones lower than or equal to 3x3/54DS, 3x3/81TS, 3x2/27SS, 3x2/DS, and 3x1/13.5SS. The pruning incurred by the 3x3/SS branch results in smaller search space, $[2, 4]$, at 3x3/DS. The continuous pruning reduces the remaining search space at each branch to only 2 to 3 MCS rates.

5.1.4 Estimation at Each Setting

We need to calculate two metrics: the per-bit energy for each setting, and the data source rate. For a given setting, we need to estimate its consumed per-bit energy E_b. In EERA, we compute instantaneous E_b upon receiving every aggregate frame at the receiver using Equation (2). The active and non-active power at a given setting can be measured *a priori*, since they do not change at runtime. The goodput is computed upon the arrival of an aggre-

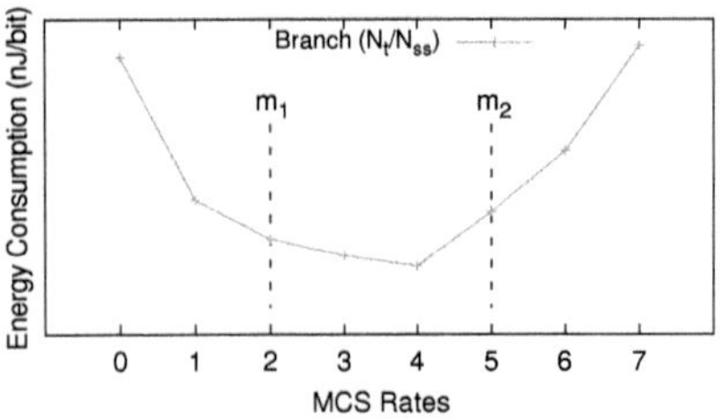

Figure 9: Ternary search.

Branch	Probing Sequence: MCS (E_b nJ/bit)	Steps
*3x1/SS	**40.5SS(19.2)** $\rightarrow$ 108SS(∞) $\rightarrow$ 54SS(19.3) $\rightarrow$ 27SS(25.7)	4
3x2/DS	**81DS(22.9)** $\rightarrow$ 216DS(∞) $\rightarrow$ 108DS(∞) $\rightarrow$ 54DS(27.6)	4
3x2/SS	40.5SS(22.9) $\rightarrow$ 108SS(∞) $\rightarrow$ **54SS(22.7)** $\rightarrow$ 81SS(26.0)	4
3x3/TS	121.5TS(32.9) $\rightarrow$ 324TS(∞) $\rightarrow$ 162TS(∞) $\rightarrow$ **81TS(30.3)** $\rightarrow$ 40.5TS(33.0)	5
3x3/DS	**81DS(29)** $\rightarrow$ 216DS(∞) $\rightarrow$ 108DS(39.7) $\rightarrow$ 54DS(30.2)	4
3x3/SS	40.5SS(26.6) $\rightarrow$ 108SS(∞) $\rightarrow$ 54SS(26.5) $\rightarrow$ **81SS(26.4)**	4
Total		25

Figure 10: Search steps of EERA ternary search at P1.

gate frame. The source rate is also estimated following the procedure described next. Once we obtain the instantaneous E_b for each probe frame, we estimate the moving average E_b at time t, denoted by $\overline{E_b}(t)$, using the instantaneous per-bit energy $E_b(t)$ and the standard procedure $\overline{E_b}(t) = (1-\alpha)\cdot\overline{E_b}(t-1) + \alpha \cdot E_b(t)$, where $\alpha = \frac{1}{8}$ is the weighting factor. This can smoothen out transient variations while tracking the evolving trend.

We also estimate the data source rate, which affects the energy-efficient setting. The estimation is implemented at the transmitter buffer. Upon each frame arrival or departure at the buffer, the instantaneous source rate can be estimated as $S(t) = G(t) + \Delta Q(t)$, where $G(t)$ is the outgoing goodput, and $\Delta Q(t)$ is the buffer change at t. We then compute the moving average of $S(t)$, using procedures similar to $\overline{E_b}(t)$. This way, we estimate the source rate by monitoring the change of data buffer and outgoing goodput.

We further make a somewhat counter-intuitive observation. In the 802.11n context, buffering packets at the source and transmitting them in a batch might not always help in energy savings in EERA. Note that, EERA selects the proper energy-efficient setting to afford the traffic source. Amortizing transmissions in batches tends to over-estimate the source rate, typically resulting in another higher-goodput setting that is less energy efficient. Consider that EERA uses a setting 3×1/13.5SS for a 5Mbps source. When we buffer this flow for 90% time, it becomes a flow of 50 Mbps for 10% time and idle for the rest 90% time. Consequently, EERA has to select a higher-goodput setting 3×3/81DS to accommodate this source. Though this setting transmits faster, the overall energy during both active and non-active periods may still be larger than the one at the setting of 3×1/13.5SS. The formal statement of this results in a theorem format is available in [17]. Note that this is different from frame aggregation in 802.11n, which EERA always uses to amortize transmission overhead.

It should be noted that EERA can work with or without the other power-saving schemes, such as SMPS and PSMP. These schemes only change the per-bit energy at a given setting by having smaller non-active power P_{na}. They work together with EERA since these two complement with each other by primarily managing the active and idle periods, respectively. EERA also has mobility and interference handling mechanisms, similar to the design in the literature [19]. We omit the details due to space limit. Finally, EERA has nice runtime complexity (its proof is in [17]):

THEOREM 1. *(Search Complexity) Assume that the increase of power consumption at the MIMO receiver with the MCS rate is negligible. EERA has a worst-case search complexity no worse than* $O(\log N_{MCS} \cdot N_r \cdot N_{ss})$.

5.2 Multi-Client Case

In the multi-client scenario, EERA uses additional mechanisms to prevent its clients from hurting others (running EERA or traditional RA schemes such as ARA/MiRA). An EERA client selects lower-goodput (but more energy-efficient) settings only if it does not affect other clients' transmissions when they were to use their highest-goodput rates. Specifically, a client is given certain amount of extra airtime it can use to slow down for energy savings. The extra airtime is allocated through a temporal fair share, which helps to isolate one client from another during transmission. Each client can only use up its fair share of airtime to slow down for energy savings, but cannot spend more time designated for other clients.

Specifically, EERA runs over regular time intervals (called epoch, and its duration is T_{ep}) periodically. During each epoch, it has three phases of operations. In the first phase, AP probes each client for its highest-goodput setting. This can be done via a traditional MIMO RA algorithm such as ARA or MiRA. We can refine ARA and MiRA by eliminating their sequential search. We abstract the problem as multi-dimensional search, and apply binary search over each tree branch, similar to Sections 5.1.1 and 5.1.2, but for goodput instead of energy. During the second phase, AP calculates the temporal fair share for each client. The fair airtime share stipulates how much extra time each client may spend when slowing down to save energy. During the third phase, EERA selects the most energy-efficient setting given the constraints set by the airtime share and pre-configured threshold $R_{c,i}$ for client i.

The fair share calculation is as follows. Assume every client uses its highest-goodput rate setting during epoch k. Given the highest goodput $G_{c,i}$ and the source rate $S_{k,i}$ for client i, its used airtime percentage is given by $\frac{S_{k,i}}{G_{c,i}}$. The unused airtime (in percentage) by all n clients during epoch k is thus obtained as $1 - \sum_{i=1}^{n} \frac{S_{k,i}}{G_{c,i}}$. EERA equally allocates this extra airtime among all n clients. Therefore, during epoch k with duration T_{ep}, each client i is allocated airtime share $F_{k,i}$ as $(1 - \sum_{i=1}^{n} \frac{S_{k,i}}{G_{c,i}}) \cdot \frac{T_{ep}}{n}$. If client i cannot use up its airtime share (say, limited by its parameter $R_{c,i}$), we then allocate fair share based on the celebrated max-min fairness [12]. Note that other fairness index (e.g., proportional fairness) may also be used to allocate the extra airtime in EERA.

Once each client is allocated its airtime share, it can effectively apply operations during each epoch, similar to the single-client case of Section 5.1. The minor difference is that, tree branches can be further pruned by both parameters of pre-configured threshold $R_{c,i}$ and fair share $F_{k,i}$. The rule is that the selected setting cannot exceed the airtime share, nor yields goodput lower than $R_{c,i}$ percent, compared with its highest-goodput setting during current epoch.

5.3 Other Issues

There are a number of additional issues to consider in EERA.

Coexistence of EERA and other MIMO RA clients EERA may coexist with clients running other traditional 802.11n RA algorithms. Its client ensures that other clients can always use their highest-goodput rates. Only when extra airtime is available (e.g., sources of other clients are not fast enough), does it slow down to save NIC energy. When clients running ARA/MiRA are greedy, EERA reverts to goodput-optimizing RA mode.

Greedy clients Greedy clients (e.g., those running TCP flows) always demand highest goodput. For these clients, we set the pre-configured parameter $R_{c,i}$ as a fixed percentage (say, 90%). The value $1 - R_{c,i}$ is interpreted as the percentage of goodput the client is willing to give up for NIC energy savings during each epoch.

Uplink case EERA can be extended to support the uplink case. Using client transmit power model, EERA minimizes $\frac{P_a(tx)-P_i}{G_{UL}}$, where $P_a(tx)$ is the active transmit power, G_{UL} is the uplink goodput, and P_i, is the idle power. Transmit power is dominated by the power amplifier, while baseband processing is much smaller. Hence, at each setting, power consumption of different MCS rates varies slightly, less than 5% based on our measurements. The mixed uplink and downlink case can also be supported similarly. AP acts as the coordinator by collecting the required information to calculate fair share for each uplink/downlink client. It then notifies each uplink client its airtime share.

Ad-hoc mode EERA currently does not support ad-hoc operations. There are two associated challenges: (1) How to allocate fair share of airtime in the multihop setting? (2) How to coordinate RA operations among multiple clients in a fully distributed manner? We plan to study these issues in the future.

6. IMPLEMENTATION

We implement EERA in the open-source driver, ath9k, for Atheros 802.11n WiFi chipsets. EERA resides at the transmitter. To save energy at the client, AP coordinates clients to configure their receive RF chains for downlink transmissions, whereas each client configures its transmit chains for uplink traffic. During the association phase, a client enables EERA at AP by issuing a request with mandatory parameters, including the maximal number of receive antennas, power parameters, as well as its non-active power parameters under different power-saving schemes. All such messages are exchanged by a new 802.11n management frame.

Two technical issues arise in EERA operations. First, how to work with other power-saving schemes? Once the client changes its power-saving scheme (e.g., enabling SMPS), it notifies its AP (e.g., via sending a SMPS frame). Upon receiving this notification, the EERA module at AP automatically updates the client's power-saving mode and estimates its per-bit energy. Second, transient loss may occur during the switching process of the client's receive chains due to the inconsistent views of the chains between both sides. For example, after the client switches its receive chain setting from three to two, EERA at AP may still use the TS rates before the acknowledgement from the client is received. Consequently, the client cannot decode these packets with only two receive chains. During chain switching, EERA uses rates accepted by both settings.

7. PERFORMANCE EVALUATION

We conduct extensive experiments to assess EERA performance. The main criterion is energy savings at the client NIC. In addition to ARA and MiRA algorithms, we also compare EERA with MRES [20], a recent proposal to improve 802.11n energy efficiency by adjusting the number of RF chains. The proposed MiRA and MRES work with maximum two receive chains and DS modes. We extend both to support three chains in TS mode. We conduct extensive tests in static office environment (see Figure 1), with various factors of client location, wireless configuration and traffic pattern. We also examine EERA in more scenarios of mobility, interference, uplink traffic, multiple-client settings, and field trials. In our experiment, both AP and the client support three antennas, working on 40MHz channel over 5GHz band. The default setting is to use UDP-based downlink transmission to a client without enabling any power-saving mode.

	ARA	MiRA	MRES
Static UDP	(13.4-35.6) %	(14.3-36.1) %	(5.8-26.8) %
Static TCP	(5.1-20.5) %	(10.4-32.3) %	(7.3-23.8) %
Application	(26.5-33.9) %	(26.6-35.2) %	(6.7-36.5) %
Mobility	27.8 %	30.1 %	20.3 %
Field Trials	31.7 %	33.1 %	24.1 %

Table 7: NIC energy savings of EERA over other designs.

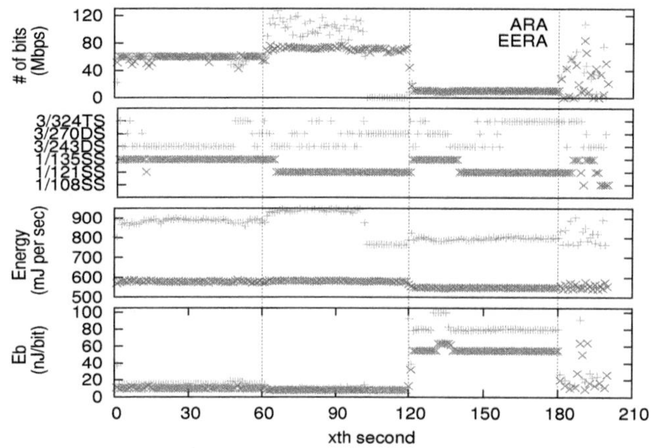

Figure 11: Trace of EERA and ARA in hybrid-traffic example.

A quick summary of EERA performance is as follows. In all test scenarios, EERA consistently outperforms other algorithms in terms of NIC energy efficiency. Table 7 summarizes its energy-saving percentage in major test settings, which also includes field trials in an office building. In general, EERA saves about 30% energy compared with ARA/MiRA in all scenarios (equivalent to energy waste of 43% by ARA/MiRA). Compared with MRES, EERA saves about 6-36% NIC energy in static settings and 20-24% in mobility and field tests.

7.1 An Example of EERA Performance

We first test EERA at a static location P5 with hybrid traffic patterns. It runs 210 seconds, including the first minute with 60 Mbp UDP traffic, the second minute with 500 MB TCP-based file downloading, the third minute with 10 Mbps UDP traffic, and the last 30 seconds for ten small (<10MB) file downloading. Figure 11 plots the traces of the delivered bits, the selected major rate settings, energy consumption, and per-bit energy over time for both ARA and EERA. The NIC energy saving of EERA over ARA and MiRA ranges between 29.2–35.1%, whereas the gain over MRES is 6.8–20.8% (6.8% for 10M UDP). Therefore, EERA outperforms all three other algorithms in terms of NIC energy consumption. This is because EERA selects a more energy-efficient setting (typically a lower-rate setting, 3x1/135SS, 3x1/121SS, or 3x1/108SS), which balances energy and goodput. EERA outperforms MRES because it quickly locates the energy-efficient setting, while MRES incurs more overhead since it runs on top of conventional RA algorithms.

7.2 Single Client Under Various Factors

We now evaluate energy efficiency of EERA in various single-client scenarios, including at different locations, with different wireless configurations (e.g., the number of AP antennas, frame aggregation, and power-saving modes), under a variety of traffic sources (e.g., source rates, packet size, applications), and in settings with mobility, interference and uplink transmissions. The detailed results are reported in [17]. We next summarize our main findings.

Client locations: We vary the client location (from P1 to P13) to examine how EERA performs under various wireless channels

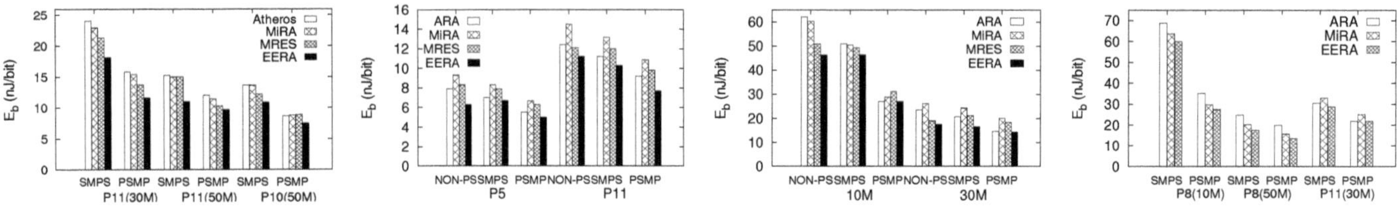

(a) Power saving modes (b) TCP (c) Interference (d) Uplink UDP

Figure 12: Per-bit energy consumption of a single client under various factors.

with 30 Mbps source. EERA consistently outperforms other algorithms, with more than 30% energy savings over both ARA and MiRA, and 8.6–22.2% savings over MRES. The trace analysis shows that, EERA always chooses and stays at the energy-efficient setting adaptive to wireless channels. Note that the gain becomes smaller at far locations. It is because the wireless link becomes too weak, leaving little room to trade off goodput for energy savings.

Number of AP antennas: We change the number of AP antennas from one to three, and observe its impact on EERA. Our results indicate that, EERA always outperforms other algorithms. For instance, for the client with a 30 Mbps source at P3, the gain is 18.5%, 33%, 33% over ARA and MiRA, and 16.5%, 25.2% and 9.6% over MRES, when one, two, and three AP antennas are activated, respectively. It confirms our finding in Section 3.3, which states that, fewer AP antennas may reduce the per-bit energy gap between the highest-goodput and most-energy-efficient settings.

Frame aggregation: We study the impact of frame aggregation on EERA by limiting the maximum aggregation level (from 0% to 100%). Frame aggregation has little effect on EERA energy efficiency unless the EE setting chosen by EERA cannot sustain its traffic source at lower aggregation. A rare example is observed when the maximum aggregation decreases from 40% to 20%, given a 30 Mbps source at far location P11. In this case, EERA has to pick up a higher-rate setting to sustain its traffic source since frame aggregation is low. Its energy savings drop from 32.8% to 12.1% subsequently. In general, frame aggregation is beneficial to EERA since it helps to reduce transmission overhead.

Power-saving modes: We also evaluate EERA when the client enables different power-saving modes of SMPS and PSMP. Figure 12(a) plots the per-bit energy at two locations P10 and P11 under two source rates. Compared with ARA, MiRA and MRES, EERA still yields energy savings from 13.37% to 28%, from 14.31% to 26.6%, from 5.96% to 26.8%, respectively. When compared with the case without power-saving modes, the reduction in energy savings is attributed to decreased power at the idle/sleep state. Smaller non-active power favors faster transmission so that the client enters into idle/sleep mode early. However, we note that, racing to sleep cannot ensure highest energy efficiency at NIC. If the active rate is not selected properly, energy waste during active period cannot be offset by the reduction in idle/sleep energy consumption.

Traffic sources: We now assess EERA under various traffic patterns. We vary UDP source rates (from 10 Mbps to 80 Mbps) and packet sizes (from 100 B to 1400 B) at different locations. EERA outperforms all other algorithms. The source rate change does affect the per-bit energy consumption (the slower source, the larger E_b), but not much on the saving gain; EERA achieves about 30% energy savings over both ARA and MiRA, and 5.1–12.4% over MRES at near and far clients (i.e., at P8 and P11, respectively). Regarding packet size, we observe that they impose negligible impact on per-bit energy of all RA algorithms, because frame aggregation minimizes the potential overhead due to small packets.

We also conduct experiments with TCP flows. Figure 12(b) demonstrates that, EERA consistently outperforms others no matter whether any power-saving mode is enabled or not. EERA yields energy savings at NIC from 5.1% to 20.5% over ARA, more than 19% over MiRA, and from 7.3% to 23.8% over MRES. This shows that, EERA estimates dynamic TCP source rate well and locates the EE setting quickly.

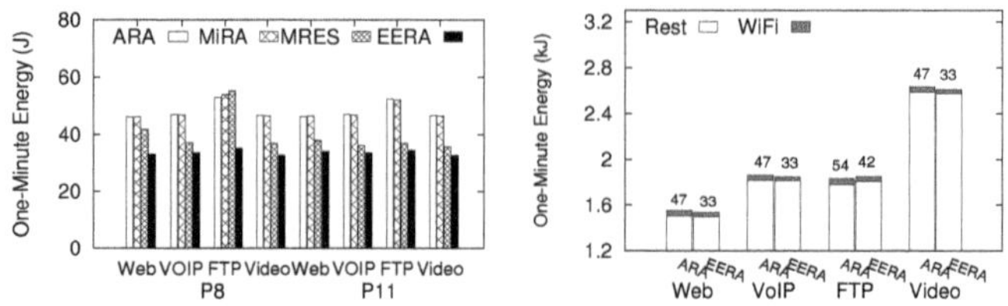

(a) NIC (b) Device at P10

Figure 13: Energy consumption under various applications.

Applications: We further gauge EERA for four popular applications: Web, VoIP, FTP and Video streaming. Their settings are described in Section 3.4. Figure 13(a) plots the per-bit energy of clients at P8 and P11 for these applications. EERA outperforms all three algorithms, since it handles a variety of traffic patterns in terms of source rate, traffic dynamics and different packet sizes. Its saving percentage at NIC ranges between 26.5–33.9%, 26.6–35.2%, and 6.7–36.5% over ARA, MiRA and MRES, respectively.

We also measure device-level energy consumption under these applications. The results are shown in Figure 13(b). At the device level, all components (except NIC) consume 4.1J, 1.9J, and 9.8J less energy when EERA runs for Web, VoIP and Video streaming applications, compared with ARA. These correspond to the energy gap of 0.27%, 0.11% and 0.39%, respectively. For FTP, EERA wastes 1.77% energy compared with ARA, i.e., about 31.5J during one minute. This is because FTP application program stops consuming additional energy once a file transfer completes. In contrast, Web, Video and VoIP applications need to remain active after data transfer for user interaction. For NIC only, EERA outperforms ARA for all four applications of Web, VoIP, Video, and FTP, with savings of 0.95%, 0.75%, 0.54%, 0.66%, respectively. It correspondingly saves 14.3J, 13.6J, 14.0J, 11.8J during a minute.

Mobility: In order to measure the efficiency of EERA and its probing effectiveness, we move a client from P6 to P1 through P4 and P2, and then go back to P6 at approximately constant, pedestrian speed of 1 m/s. AP sends 30Mbps UDP source to the client. In this mobility case, EERA outperforms ARA, MiRA, MRES with NIC energy savings of 27.8%, 30.1%, and 20.3%, respectively. Moreover, Table 8 lists the major rate settings selected by all RA algorithms, at each location. It shows that EERA is still able to locate the energy-efficient settings during mobility.

We further study its probing cost. EERA, as well as MiRA and MRES, uses a single aggregate frame to probe each setting. Per setting, there are up to four transmissions until they succeed. Our trace analysis shows that, EERA is able to exclude most less-energy-efficient settings with simultaneous pruning. For example,

	P6 → P4 → P2 → P1 → P2 → P4 →P6	E_b
EERA	3x1/108SS → 3x1/108SS → 3x1/81SS → 3x2/54SS → 3x1/81SS → 3x1/108SS → 3x1/108SS	19.7
MRES	3x1/135SS → 3x1/121.5SS → 3x2/108DS → 3x2/54SS → 3x1/81SS → 3x1/121.5SS → 3x1/135SS	24.7
ARA	3x3/324TS → 3x3/243DS → 3x3/162DS → 3x3/108DS → 3x3/108DS → 3x3/216DS → 3x3/324TS	27.3
MiRA	3x3/324TS → 3x3/243DS → 3x3/162DS → 3x3/108DS → 3x3/162DS → 3x3/243DS → 3x3/324TS	28.2

Table 8: Selected rate settings over locations during mobility.

EERA needs to probe four settings to locate the optimal 3x1/108SS at P4, whereas MiRA and MRES probe five and ten settings, respectively. Moreover, EERA starts from lower-rate settings with a higher chance to success. In fact, only 7 frame transmissions are needed in EERA, whereas 15 and 29 frame transmissions are for MiRA and MRES, respectively. The lower probing overhead also contributes to energy efficiency of EERA compared with MRES.

Interference: We gauge the performance of EERA in the interference scenario by placing the client into the crowded 2.4GHz band (Channel 11), on which we sniff more than 10 APs. The channel width is switched from 40 MHz to 20 MHz. We evaluate this case for different traffic sources and power-saving schemes at P12. Results are shown in Figure 12(c). Although external interference reduces the gain of EERA due to more packet losses and collisions, the NIC energy savings still reach up to 25.8%, 33.3% and 22.1%, compared with ARA, MiRA and MRES, respectively.

Uplink traffic: We also evaluate EERA for uplink traffic in the power-saving modes of SMPS and PSMP. In this experiment, the clients at P8 and P11 send UDP traffic to the AP. Figure 12(d) shows that, EERA outperforms others with energy savings up to 32%. Large gain is observed at close location P8. It is because EERA mainly uses one transmit chain for uplink transmission, whereas ARA and MiRA always activate three transmit chains.

7.3 Multi-Client Scenarios

We now evaluate EERA in multi-client scenarios. We consider three cases: (1) multiple EERA clients associated with the same AP, (2) EERA and non-EERA (ARA is used here) clients coexisting within an AP, and (3) EERA and ARA clients coexisting with two co-located APs. We test with two clients and three clients in these these cases. The detailed configuration is shown in Table 9. The benchmark scenario is when all clients use the same ARA algorithm. The goal is to see whether and how an EERA client affects others when slowing down its transmission. We are particularly interested in assessing how well the fair share mechanism in EERA works. To this end, we evaluate two versions of EERA, with and without the fair share mechanism (called F-EERA and Naive EERA (N-EERA), respectively).

Settings	C_1	C_2
2C-Bench	ARA	ARA
2C-EERA	EERA	EERA
2C-Coexist	ARA	EERA

Settings	C_1	C_2	C_3
3C-Bench	ARA	ARA	ARA
3C-EERA	EERA	EERA	EERA
3C-Coexist	ARA	EERA	EERA

Table 9: Experimental settings for multiple clients (Left: two clients; Right: three clients).

Multiple EERA clients: We examine the overhead incurred by EERA in terms of packet delay D_p, and then evaluate its energy efficiency. In the experiment, clients C_1, C_2 and C_3 are placed at P6, P8 and P11, respectively; the traffic source at C_2 varies its rate from 10 Mbps to 70 Mbps; the source of other clients is 10 Mbps. Figures 14(a) and 14(b) plot the average packet delay in 2C-EERA and 3C-EERA scenarios, compared with their ARA benchmark settings. The result shows that, F-EERA is able to achieve comparable packet delivery latency as ARA does. The latency increase in F-EERA is within 0.2 ms per packet in the 2C-EERA case, and 0.13 ms per packet in the 3C-EERA scenario. As the source rate of client C_2 increases, F-EERA has to accommodate multi-client traffic demand and allocate extra air time among all clients in a fair manner. Client C_1 is consequently allocated with smaller airtime share, and F-EERA forces it to select a higher-goodput setting compared with the case using N-EERA. It thus adapts to multi-client traffic load with reasonable overhead. In contrast, N-EERA further increases packet latency when client C_1 slows down for its NIC energy efficiency. It also hurts other clients in packet latency under high traffic demand. Figure 14(c) plots the per-bit energy consumption for both clients in the two-client scenario. As the source rate increases, energy-saving gain also decreases. F-EERA saves 30.5% and 29.7% for C_1 and C_2, respectively, given a 10 Mbps data source. The saving reduces to 5% for C_1 when data source increases to 70 Mbps, in which case the aggregate traffic demand is close to the link capacity. Similar results are also observed in the 3C-EERA case. In summary, F-EERA chases for energy savings when the aggregate traffic demand is low, and behaves similarly to conventional RA algorithms when the traffic demand is high.

Coexistence of EERA and ARA clients: Now EERA and ARA clients coexist with the same AP. The settings are identical to above experiments, except that client C_1 runs ARA. We seek to examine whether EERA clients affect other conventional RA clients. We also explore how the energy efficiency of EERA clients is affected. Figure 14(d) plots the average packet delay at C_1 when other clients run F-EERA. The delay gap between benchmark and coexistence cases is below 0.08 ms per packet, thus negligible. Figures 14(e) and 14(f) plot the per-bit energy for F-EERA clients (clients C_2 and C_3) in the 2C-Coexist and 3C-Coexist settings. Similar to Figure 14(c), energy saving decreases as the aggregate traffic demand grows. For instance, energy-saving gain reduces from 30.8% to 0.1% in the 2C-Coexist case, and decreases from 31.0% to 14.6% in the 3C-Coexist case. .

Coexistence of EERA/ARA clients with two APs: We further study how an EERA client coexists with other RA (ARA is used here) client in the two-AP scenarios. In the test settings, client C_1 always interacts with an AP running ARA. However, client C_2 receives data from another AP running ARA, N-EERA, and F-EERA, respectively, at P5. These two APs are co-located in spatial proximity and can hear each other. Both contend for the same channel. Figures 14(g) and 14(h) plot the average packet delay and the number of retries per millisecond at client C_1. The number of retries serves as an indicator for channel contention between two AP-client pairs. The results show that, F-EERA incurs comparable overhead as ARA, with the packet-delay gap being at most 0.12 ms. As for the number of retries, F-EERA performs even better than ARA. We gauge that it is because more active transmission time reduces the likelihood of contentions.

In summary, EERA compares well with ARA in multi-client scenarios. In case of multiple EERA clients, it increases per-packet delay by at most 14.2% (0.19ms) and 8.7% (0.13ms), but saves NIC energy by 30.8% and 31%, in 2-client and 3-client cases, respectively. For co-located EERA and ARA clients, it increases packet delay by at most 5.3% (0.07ms) and 5.2% (0.08ms) in 2-client and 3-client coexistence settings, respectively. For contention, EERA increases at most 3.4% retries (0.009 retries/ms), but avoids up to 34% retries (0.08 retries/ms) when the traffic source rate is low.

8. RELATED WORK

Numerous RA algorithms [6,8,11,14,19,21–23] have been proposed in the literature. All aim to achieve high goodput, rather than energy efficiency. Moreover, these proposals typically use sequential or randomized search; the search does not scale well to 802.11n/ac where search space is bigger. Other MIMO RA propos-

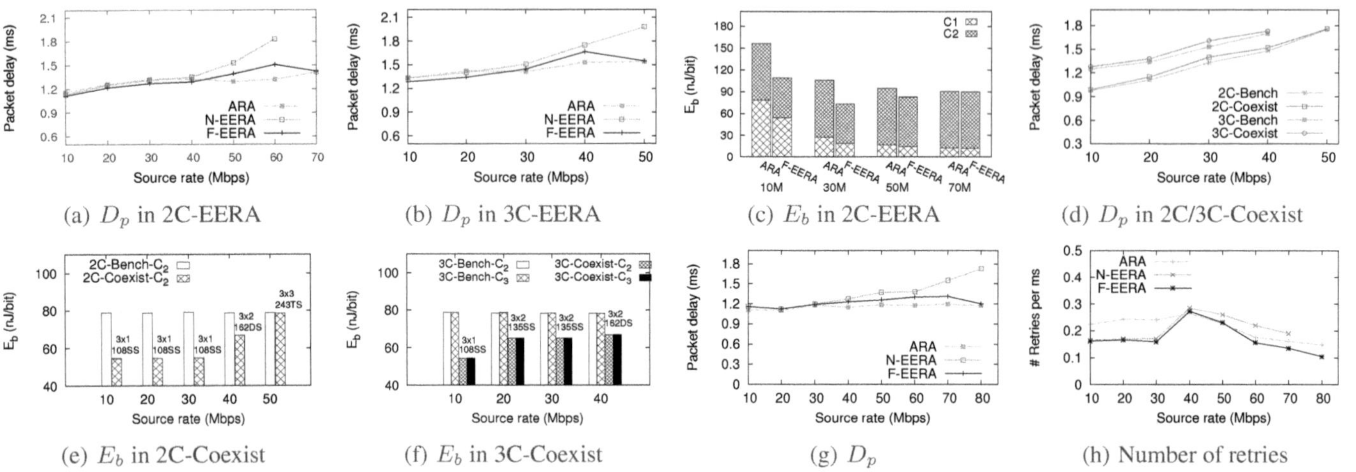

(a) D_p in 2C-EERA (b) D_p in 3C-EERA (c) E_b in 2C-EERA (d) D_p in 2C/3C-Coexist

(e) E_b in 2C-Coexist (f) E_b in 3C-Coexist (g) D_p (h) Number of retries

Figure 14: Energy efficiency and packet delay in multi-client scenarios.

als (e.g., ESNR [14] and Soft-Rate [21]) cannot be implemented in current commodity 802.11n platforms.

Recent theoretical studies on energy-efficient MIMO systems seek to find the crossover point, which trades off MIMO gains at the cost of increased power consumption [9, 16]. Both cannot be used on commodity platforms. Several research studies [10, 13, 15, 20] have focused on energy savings in MIMO systems. [10] seeks to find the most energy-efficient settings only for transmission period, using their MIMO-OFDM based software-defined radio; it is not 802.11n standard compliant. [13] identifies factors that affect energy consumption on 802.11n commodity hardware. MRES [20] examines the strength and limitation of SMPS, and proposes a energy-saving solution through dynamically adjusting chain settings. Snooze [15] schedules client sleep time, and configures chains for energy savings. All these efforts do not address the problem from RA perspective. Moreover, EERA also complements and works with power-saving mechanisms such as SMPS and PSMP.

9. CONCLUSION

Rate adaptation for 802.11n devices is more complex than that in legacy 802.11a/b/g systems, since it has to adjust over multi-dimension PHY parameter space. Various proposals [6, 8, 11, 14, 19, 21–23] have so far focused on improving goodput. However, we show that, this is possibly achieved at higher energy cost at NICs. In the race for higher speed in wireless technologies (e.g., 802.11n and 802.11ac WLAN, and 4G LTE WWAN to name a few), we believe that energy efficiency is equally important. The technology has to balance between energy and speed. EERA reports our effort on adapting RA to improve NIC energy efficiency.

10. ACKNOWLEDGMENT

We highly appreciate the constructive and insightful comments by our TPC shepherd and the anonymous reviewers, and the professional help from TPC co-chair Dr. Alex C. Snoeren during the shepherding process.

11. REFERENCES

[1] Intel PowerTop 2.0. https://01.org/powertop/.
[2] Intel Core 2 Duo Processor Datasheet, Jan. 2008.
[3] Atheros, Marvell to Push Fast 11n Into Phones, Feb. 2011. http://www.pcworld.com/article/219395/atheros_marvell_to_push_fast_11n_into_phones.html.
[4] IEEE P802.11 Wireless LANs Proposed TGac Draft Amendment, Jan 2011.
[5] ABI Research. Demand for 802.11n Spurs Wi-Fi Equipment Market to 18 Million Units in 2Q 2010, Nov. 2010.
[6] P. A. Acharya, A. Sharma, E. M. Belding, K. C. Almeroth, and K. Papagiannaki. Rate Adaptation in Congested Wireless Networks through Real-time Measurements. *IEEE Transactions on Mobile Computing*, 9(11), Nov. 2010.
[7] A. Carroll and G. Heiser. An Analysis of Power Consumption in a Smartphone. In *USENIX Annual Technical Conference*, 2010.
[8] X. Chen, D. Qiao, J. Yu, and S. Choi. Probabilistic-Based Rate Adaptation for IEEE 802.11 WLANs. In *IEEE Globecom*, 2007.
[9] S. Cui, A. J. Goldsmith, and A. Bahai. Energy-constarined Modulation Optimization. *IEEE Trans. Wireless Communications*, 4(5), 2005.
[10] W. Gabran and B. Daneshrad. Hardware and Physical Layer Adaptation for a Power Constrained MIMO OFDM System. In *IEEE ICC*, 2011.
[11] A. Gudipati and S. Katti. Strider: Automatic Rate Adaptation and Collision Handling. In *SIGCOMM*, 2011.
[12] E. L. Hahne. Round-Robin Scheduling for Max-Min Fairness in Data Networks. *IEEE Journal on Selected Areas in Communications*, 9:1024–1039, 1991.
[13] D. Halperin, B. Greenstein, A. Sheth, and D. Wetherall. Demystifying 802.11n Power Consumption. In *HotPower*, 2010.
[14] D. Halperin, W. Hu, A. Sheth, and D. Wetherall. Predictable 802.11 Packet Delivery from Wireless Channel Measurements. In *SIGCOMM*, 2010.
[15] K.-Y. Jang, S. Hao, A. Sheth, and R. Govindan. Snooze: Energy Management in 802.11n WLANs. In *ACM CoNEXT*, 2011.
[16] H. Kim, C.-B. Chae, G. De Veciana, and R. W. Heath. A Cross-layer Approach to Energy Efficiency for Adaptive MIMO Systems Exploiting Spare Capacity. *IEEE Trans. on Wireless Communication*, 8:4264–4275, Aug. 2009.
[17] C. Li, C. Peng, and S. Lu. Achieving 802.11n NIC Energy Efficiency through Rate Adaptation. Technical report, UCLA Computer Science, 2012.
[18] A. Mahesri and V. Vardhan. Power Consumption Breakdown on a Modern Laptop. In *PACS*, 2004.
[19] I. Pefkianakis, Y. Hu, S. H. Wong, H. Yang, and S. Lu. MIMO Rate Adaptation in 802.11n Wireless Networks. In *MOBICOM*, 2010.
[20] I. Pefkianakis, C.-Y. Li, and S. Lu. What is Wrong/Right with IEEE 802.11n Spatial Multiplexing Power Save Feature? In *ICNP*, 2011.
[21] M. Vutukuru, H. Balakrishnan, and K. Jamieson. Cross-layer Wireless Bit Rate Adaptation. In *SIGCOMM*, 2009.
[22] M. Wong, J. M. Gilbert, and C. H. Barratt. Wireless LAN using RSSI and BER Parameters for Transmission Rate Adaptation, 2008. US patent, 7,369,510.
[23] S. H. Y. Wong, H. Yang, S. Lu, and V. Bharghavan. Robust Rate Adaptation for 802.11 Wireless Networks. In *MOBICOM*, 2006.

Faster GPS via the Sparse Fourier Transform

Haitham Hassanieh Fadel Adid Dina Katabi Piotr Indyk
Computer Science and Artificial Intelligence Laboratory
Massachusetts Institute of Technology
{haithamh, fadel, dk, indyk}@mit.edu

ABSTRACT

GPS is one of the most widely used wireless systems. A GPS receiver has to lock on the satellite signals to calculate its position. The process of locking on the satellites is quite costly and requires hundreds of millions of hardware multiplications, leading to high power consumption. The fastest known algorithm for this problem is based on the Fourier transform and has a complexity of $O(n \log n)$, where n is the number of signal samples.

This paper presents the fastest GPS locking algorithm to date. The algorithm reduces the locking complexity to $O(n\sqrt{\log n})$. Further, if the SNR is above a threshold, the algorithm becomes linear, i.e., $O(n)$. Our algorithm builds on recent developments in the growing area of sparse recovery. It exploits the sparse nature of the synchronization problem, where only the correct alignment between the received GPS signal and the satellite code causes their cross-correlation to spike.

We further show that the theoretical gain translates into empirical gains for GPS receivers. Specifically, we built a prototype of the design using software radios and tested it on two GPS datasets collected in the US and Europe. The results show that the new algorithm reduces the median number of multiplications by 2.2× in comparison to the state of the art design, for real GPS signals.

Categories and Subject Descriptors C.2 [**Computer Systems Organization**]: Computer-Communications Networks

General Terms Algorithms, Design, Performance, Theory

Keywords GPS, Synchronization, Sparse Fourier Transform

1. INTRODUCTION

The global positioning system (GPS) is one of the most pervasive wireless technologies. It is incorporated in more than one billion smartphones world-wide [17], and embedded in a wide variety of devices, including personal navigation systems [37], sensors [5], digital cameras [26], and even under-the-skin bio-chips [13]. The key functionality of a GPS receiver is to calculate a position, called a fix. Computing a fix involves locking on the GPS satellite signals and decoding satellite orbit and time data. Most GPS receivers, however, are embedded with some other radio (e.g., WiFi, cellular, or ZigBee) and, hence, can download the content of the GPS signal from assisted GPS (A-GPS) servers instead of decoding it from the satellite signals [19].[1] With assisted GPS used widely in phones and other GPS-capable devices [9], the bulk of what a GPS receiver does is to lock on the satellite signal (i.e., synchronize with it). This allows the receiver to calculate the sub-millisecond synchronization delay necessary for computing its position [10]. The importance of locking is further emphasized by the fact that current GPS receivers are typically duty-cycled [34, 2]; hence, they need to re-synchronize with the satellite signals regularly. Synchronizing with the satellite signal, however, is a costly process that requires tens of millions to a few billion digital multiplications [36]. Many GPS-enabled devices (e.g., mobile phones, sensors, etc.) have strict power limitations and would significantly benefit from reducing the complexity of this process.

In this paper, we aim to reduce the cost of synchronizing with weak signals like GPS. At a high level, GPS synchronization works as follows: each satellite is assigned a CDMA code. For each satellite, the receiver needs to align the corresponding CDMA code with the received signal. The process is complicated because GPS signals are very weak (about 20 dB *below* the noise level [31]). To find the right alignment of each satellite, a GPS receiver conducts a search process. It computes the correlation of the CDMA code with the received signal for all possible shifts of the code with respect to the signal. The correct shift is the one that maximizes the correlation.

So, how does a GPS receiver compute all these shifted correlations? The traditional approach *convolves* the received signal with the CDMA code of each satellite in the time domain. The correct alignment corresponds to the one that maximizes this convolution. This approach has a computational complexity of $O(n^2)$, where n is the number of samples.[2] More recent GPS receivers lock on the satellite using frequency domain computation. This approach lever-

[1]The data includes almanac, ephemeris, reference time. AGPS may also provide other optional assistance data [19].

[2]The CDMA code consists of 1023 chips transmitted at 1.023 MHz. For a GPS receiver that samples at 5 MHz, the computational complexity of the shifted correlation is $(1023 \times 5/1.023)^2$, which is about 25 million multiplications of complex signal samples. The GPS receiver has to repeat this process for multiple satellites (between 4 to 12 satellites) and multiple Doppler shifts (between 21 to 41 shifts) for each satellite, which brings the number of multiplications to over a billion. Further, correlating with one block of the signal may not be sufficient. For weak signals, the receiver may need to repeat this process and sum up the output [20].

MobiCom'12, August 22–26, 2012, Istanbul, Turkey.

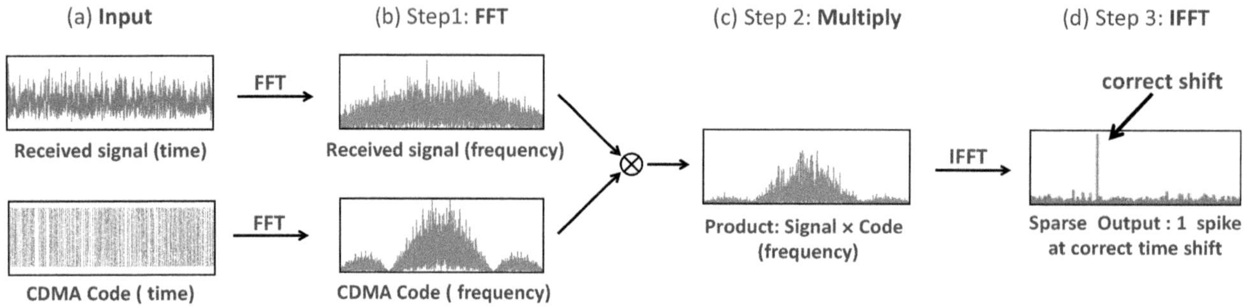

Figure 1—The steps performed by the FFT-based synchronization algorithm. The algorithm multiplies the FFTs of the received signal with the FFT of the code, and takes the IFFT of the resulting signal. The output of the IFFT spikes at the shift that correctly synchronizes the code with the satellite signal.

ages the fact that convolution in the time domain corresponds to multiplication in the frequency domain. It proceeds in the following three steps, shown in Fig. 1: 1) The receiver takes the FFT of the received signal; 2) It multiplies the output of this Fourier transform by the FFT of the CDMA code; and 3) It performs the inverse FFT on the resulting signal. This 3-step process is mathematically equivalent to convolving the signal with the code; thus, the output of the inverse FFT will spike at the correct shift that synchronizes the code with the received signal, as shown in Fig. 1d. The computational complexity of this approach is $O(n \log n)$. For the past two decades, this has been the algorithm with the lowest computational complexity for synchronizing a GPS receiver [36].

This paper introduces the lowest complexity GPS synchronization algorithm to date. Our synchronization algorithm is based on the following observations:

- First, we note that since the output of the synchronization process has a single major spike at the correct shift, as shown in Fig. 1d, the inverse FFT is very sparse. The problem of sparse FFT has recently received much attention in the computer science theory community, which resulted in new algorithms that can compute the FFT (or inverse FFT) of a sparse signal in sublinear time [14, 15, 11].[3] We build on these advances to significantly reduce the runtime of the GPS synchronization algorithm. However, existing sparse FFT algorithms use relatively complex filters (e.g., Dirichlet [11], Gaussian [14], or Dolph-Chebyshev filters [15]) to deal with the interaction of multiple potential spikes at the output of the transform. In contrast, we exploit the fact that the synchronization problem produces only one spike, and design a simple sublinear algorithm that uses only aliasing to filter the signal. This allows us to reduce the complexity of the IFFT step in Fig. 1d to sublinear time.
- Although the output of the inverse FFT is sparse and can be quickly computed, the GPS signal in the frequency domain is not sparse (Fig. 1b) and, hence, the runtime of the forward FFT cannot be reduced by applying a sparse transform. Thus, simply using sparse inverse FFT does not reduce the overall complexity of the problem (which is still $O(n \log n)$ due to the forward FFT). To address this issue, we note that the FFT in Fig. 1b is just an intermediate step that will be used as an input to the sparse IFFT. Since sparse IFFT algorithms (including ours) operate only on a subset of their input signal, we do not need to compute the values of all frequencies at the output of the forward FFT. We leverage this property to compute only a subset of the frequencies and reduce the complexity of the FFT step.

We provide an algorithm that, for any SNR, is as accurate as the original FFT-based (or convolution-based) algorithm, but reduces the computational complexity from $O(n \log n)$ to $O(n\sqrt{\log n})$. Further, when the noise in the received signal can be bounded by $O(n/\log^2 n)$, we prove that the same algorithm has a linear complexity, i.e., $O(n)$.[4]

We implement our design and test it on two datasets of GPS signals: We collected the first dataset in the US using software radios. The second dataset was collected in Europe.[5] The datasets cover both urban and suburbun areas. We compare our design against an FFT-based synchronization algorithm. Our design reduces the number of multiplications for detecting the correct shift by a median of $2.2\times$. Since a large fraction of GPS power is consumed by the synchronization process (30% [28] to 75% [29] depending on the required accuracy), we expect the new design to produce a significant reduction in GPS power consumption.

Contributions: This paper makes algorithmic and system contributions, which can be summarized as follows:

- The paper presents the fastest algorithm to date for synchronizing GPS receivers with satellite signals. The algorithm has multiple features: 1) It is adaptive, i.e., it can finish faster if the SNR is higher; 2) it continues to work at very low SNRs; and 3) it is general, i.e., it can be used to synchronize any signal with a random (or pseudo random) code.
- The paper provides an implementation and an empirical evaluation on real GPS signals, demonstrating that the algorithmic gains translate into a significant reduction in the number of operations performed by a GPS receiver.

2. GPS PRIMER

The key functionality of a GPS receiver is to calculate its position using the signal it receives from the GPS satellites. To do so, the receiver computes the time needed for the signal to travel from each satellite to itself. It then multiplies the computed time by the speed

[3]Sparse FFT algorithms are designed for the case where the output of the Fourier Transform contains only a small number of spikes. They are applicable to both sparse FFTs and sparse IFFTs. For a more detailed description see section 3.

[4]Note that n is not a constant and varies across GPS receivers. Specifically, different receivers sample the GPS signal at different rates, hence obtaining a different number of samples per codeword. For example, for a receiver whose sampling rate is 5MHz, n=5000, whereas for a 4MHz receiver, n=4000.

[5]The Europe dataset is courtesy of the GNSS-SDR team [8]. at the Centre Tecnologic de Telecomunicacions de Catalunya (CTTC).

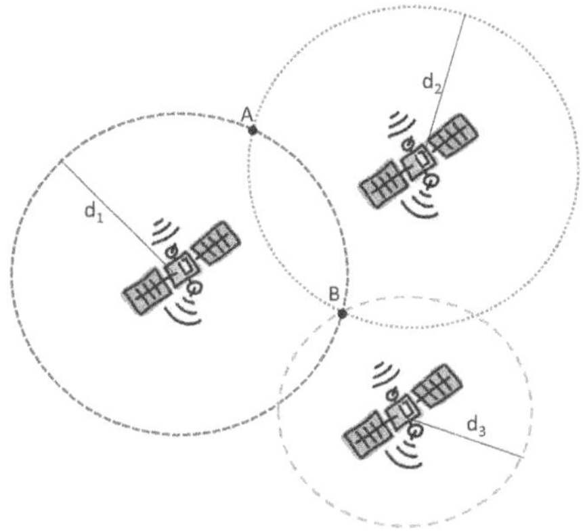

Figure 2—Trilateration: After determining the distance to different satellites, the receiver can draw spheres centered at each of the satellites and whose radii are the respective distances. These spheres should intersect at the receiver's position. A GPS receiver needs four satellites to uniquely determine its position [20]. Extra satellites can be used to correct for the lack of very tight synchronization between the receiver's clock and those of the satellites.

of light to obtain its distance from each satellite. As a result, the receiver knows that it lies on a sphere centered at that satellite and whose radius is the computed distance. It then determines its position as the intersection of several such spheres through a method called trilateration [20] shown in Figure 2.

But how does the receiver compute the propagation time from the satellites? The propagation time is obtained using a synchronization algorithm that allows the device to lock on the received signal. Specifically, each satellite has its own CDMA code, called the C/A code, which consists of 1023 chips [20]. Assuming the receiver's and satellites' clocks are perfectly synchronized, a GPS receiver generates the satellites' codes at the same time as the satellites. Due to propagation delay, however, the signal arrives in a shifted version at the receiver by exactly the amount of time it took the signal to travel from the satellite. By correlating with shifted versions of the satellite's code, the receiver calculates the propagation time as the shift at which the correlation spikes [36]. In practice, the receiver's clock is not fully synchronized with that of the satellites; this, however, can be compensated for by increasing the number of satellites used in the trilateration process.[6]

The motion of the satellites introduces a Doppler shift in the received signal. The signal does not correlate with the C/A code unless the Doppler shift is corrected. To deal with this issue, a GPS device typically performs a 2-dimensional search on the received signal [20]: one for time (code shifts), and one for Doppler shifts. Specifically, the receiver tries all possible code shifts, and 41 equally spaced Doppler shifts within +/-10 kHz of the center frequency [36], as shown in Fig. 3. Finally, the GPS satellites repeat the code 20 times for each data bit to enable the GPS receiver to decode very weak signals. The receiver tries to use one code repetition to synchronize. However, if the signal is too weak, the receiver repeats the 2D-search for multiple codes and sums the result [20].

[6]All GPS satellites use atomic clocks and are fully synchronized with each other [20]. Hence, a GPS receiver will have the same clock skew with respect to all satellites and all the estimated propagation delays will have the same error ϵ. However, trilateration needs only 4 satellites to estimate the position and thus extra satellites can be used to to estimate and correct ϵ.

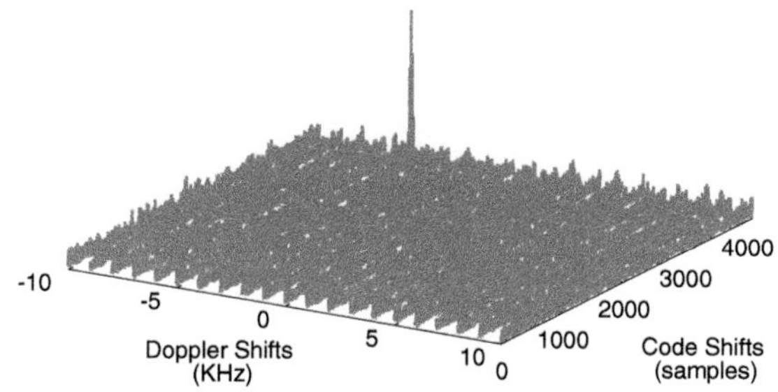

Figure 3—2D search for correlation. The plot shows the result of correlating with a C/A code for a satellite whose signal is present in the received signal. On the x-axis, we search 4000 different code shifts and on the y-axis 21 different Doppler shifts.

3. QuickSync

We describe QuickSync, a synchronization algorithm for GPS receivers. The algorithm works in the frequency domain similar to the FFT-based algorithm described in §1. QuickSync, however, exploits the sparse nature of the synchronization problem, where only the correct alignment between the received GPS signal and the satellite code causes their cross-correlation to spike. QuickSync harnesses this property to perform both the Fourier and inverse Fourier transforms in a time faster than $O(n \log n)$, therefore reducing the overall complexity of GPS synchronization.

The next subsections formalize the problem and detail the algorithm.

3.1 Problem Formulation

The synchronization problem can be formulated as follows: Given a spreading code $\mathbf{c} = c_0, \ldots, c_{n-1}$ of size n and a received signal $\mathbf{x} = x_0, \ldots, x_{n-1}$, find the time shift $\hat{t}$ that maximizes the correlation between $\mathbf{c}$ and $\mathbf{x}$, i.e., compute:

$$\hat{t} = \arg\max \mathbf{c}_{-n} \circledast \mathbf{x}, \qquad (1)$$

where $\circledast$ is a circular convolution and $\mathbf{c}_{-n}$ is the time reversed code; i.e. $\mathbf{c}_{-n} = c_{n-1}, \ldots, c_0$. Computing this convolution in the time domain requires performing n correlations each of size n and thus has complexity $O(n^2)$. However, convolution in the time domain corresponds to element-by-element multiplication in the frequency domain. Therefore computing the convolution in Eq. 1 can be done more efficiently by performing FFT on each of the code and the signal, multiplying those FFTs, then performing an inverse FFT (IFFT) as shown below:

$$\arg\max_t \mathbf{c}_{-n} \circledast \mathbf{x} = \arg\max_t \mathcal{F}^{-1}\{\mathcal{F}\{\mathbf{c}\}^* \cdot \mathcal{F}\{\mathbf{x}\}\}, \qquad (2)$$

where $\mathcal{F}(.)$ is the FFT, $\mathcal{F}^{-1}(.)$ is the IFFT, $*$ is the complex conjugate and t is any time sample in the output vector of the convolution. This reduces the complexity of the synchronization process to $O(n \log(n))$. Accordingly, in the remainder of this paper, we only consider the FFT-based synchronization algorithm as a baseline for evaluating QuickSync's performance.

3.2 Basics

Before introducing our synchronization algorithm, we remind the reader of a basic property of the Fourier transform, which we rely on in our design: *Aliasing a signal in the time domain is equivalent*

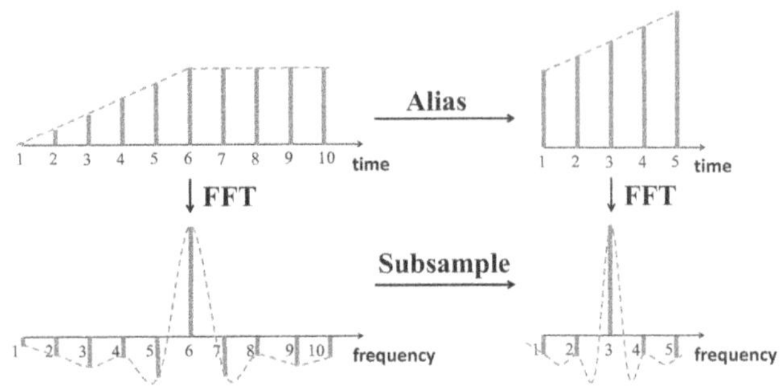

Figure 4—The duality of aliasing and subsampling. Aliasing in the time domain corresponds to subsampling in the frequency domain and vice versa. Folding (aliasing) the time domain signal in the top left results in the signal in the top right; specifically, time samples 1 and 6 add into sample 1 in the aliased signal, samples 2 and 7 into sample 2, etc. In the Fourier domain, the FFT of the aliased signal is a subsampled version of the FFT of the initial signal; namely, sample 1 in the bottom right signal corresponds to sample 2 in the bottom left, sample 2 corresponds to sample 4, etc.

to subsampling it in the frequency domain, and vice versa. Figure 4 illustrates this property.

Formally, let $\mathbf{x}$ be a discrete time signal of length n, and $\mathbf{X}$ its frequency representation. Let $\mathbf{x}'$ be a version of $\mathbf{x}$ in which adjacent windows of size B (where B divides n) are aliased on top of each other (i.e., samples that are $p = n/B$ apart are summed together). Then, for $t = 0 \ldots B-1$:

$$\mathbf{x}'_t = \sum_{j=0}^{n/B-1} \mathbf{x}_{t+jB}. \tag{3}$$

Thus, $\mathbf{X}'$, the FFT of $\mathbf{x}'$ is a subsampled version of $\mathbf{X}$, and for $f = 0 \ldots B-1$

$$\mathbf{X}'_f = \mathbf{X}_{pf}, \tag{4}$$

where $p = n/B$, and the subscript in $\mathbf{X}_{pf}$ refers to the sample whose index is $p \times f$.

3.3 The QuickSync Algorithm

We describe how QuickSync operates on a received GPS signal to synchronize it with an internally generated C/A code. For simplicity, we assume that the input signal neither exhibits a carrier frequency offset nor a Doppler shift; in later sections, we extend the algorithm to deal with these frequency offsets. Furthermore, in this section, we describe the algorithm in the context of synchronizing the GPS receiver with the signal of only one satellite; the algorithm can be easily adapted for synchronizing with multiple satellites.

The key insight to our algorithm is that the IFFT performed in step 3 of the FFT-based synchronization algorithm is sparse in the time domain, i.e., it has only one spike and, hence, can be performed in sub-linear time [14]. Further, a sub-linear time algorithm for computing the sparse IFFT would require a sub-linear number of samples as input; thus, there is no need to perform a full $n \log n$ FFT on the received GPS signal and obtain all of its n frequency samples. Rather, we only need to compute the frequency samples that will be used to perform the sparse inverse FFT.

Below, we explain how we exploit these ideas to reduce the complexity of both the IFFT and FFT performed to synchronize the signal with the code. We then put these components together in a complete algorithm.

(a) Sparse IFFT.

Inspired by recent work on sparse Fourier (inverse Fourier) transform [14, 11], we develop a simple algorithm to efficiently perform the IFFT step of GPS synchronization and quickly identify the spike of the correlation between the received signal and the CDMA code. To do so, our algorithm uses a sublinear number of samples of the signal.

The sparse IFFT algorithm proceeds as follows. It first subsamples the frequency domain signal of size n by a factor of p. It then computes the IFFT over these n/p frequency samples. Recall that subsampling in the frequency domain is equivalent to aliasing in the time domain. Thus, the output of our IFFT step is an aliased version of the output in the original IFFT step shown in Fig 1. Aliasing here can be viewed as a form of hashing, where the n original outputs samples, i.e. time shifts, are hashed into n/p buckets. Time shifts which are p apart will be summed and hashed together in the same bucket at the output of our IFFT. Since there is only one correlation spike in the output of the IFFT, the magnitude of the bucket it hashes into will be significantly larger than that of other buckets where only noise samples hash to. Hence, the algorithm chooses the bucket with the largest magnitude among the n/p buckets at the output of our IFFT.

Out of the p time shifts that aliased (or hashed) into this chosen bucket, only one is the actual correlation spike. To identify the spike among these p candidate shifts, the algorithm correlates the received signal with each of those p shifts of the CDMA code. The shift that produces the maximum correlation is the right spike.

Using aliasing as a form of hashing in our synchronization algorithm, as opposed to other forms of hashing used in sparse FFT algorithms (such as Gaussian and Dirichlet bucketization [15, 11]), has two major benefits:

- **Simplicity:** Subsampling can be implemented simply by taking a small IFFT of size n/p on a subset of the input and does not require any additional complex processing such as random hashing.
- **No Leakage:** In subsampling, each non-zero sample is hashed into exactly one bucket and does not leak power into any other bucket. In alternative forms of hashing [14, 15], each non-zero sample is hashed into a bucket where it is dominant but still leaks power into other buckets, which reduces the performance of the sparse IFFT algorithm.

(b) Subsampled FFT.

With the sparse IFFT step in place, the algorithm does not need the whole n-point FFT of the signal. Specifically, all the IFFT requires is a subsampled version of this signal. Thus, rather than taking a full n-point FFT, QuickSync aliases the received signal in the time domain before taking its FFT, as in Eq. 3 (Said differently, QuickSync sums up blocks of size n/p and then computes a smaller FFT of size n/p.) The output of this FFT, expressed in Eq. 4, is exactly the samples we need at the input of the sparse IFFT, described above.

A subsampled input to the IFFT (as described in §3.3(a)) results in an output spike of smaller magnitude relative to the noise bed. To compensate for this loss, we alias $p \times n$ samples instead of n into blocks of size n/p before performing the FFT.

(c) Full Algorithm.

The QuickSync algorithm proceeds in the following steps:

1. **Aliasing:** Alias $p \times n$ samples of the GPS signal into $B = n/p$ samples as described in Eq. 3, where $p = \sqrt{\log n}$.

2. **Subsampled FFT:** Perform an FFT of size n/p on the aliased time signal. This is effectively equivalent to performing an FFT of size pn and subsampling the output by p^2 according to Eq. 4.

3. **Multiplying with the code:** Subsample the FFT of the satellite CDMA code of length n by p, and multiply the resulting samples by the n/p samples at the output of step 2, above. Note that the algorithm can precompute the FFT of the CDMA code and store it in the frequency domain.

4. **Sparse IFFT:** Perform an IFFT on the n/p resulting samples. Since the input of this IFFT was subsampled, its output is aliased in the time domain. Specifically, each of the n/p buckets at the output of this stage is effectively the sum of p aliased time samples [7] as described in §3.3a.

5. **Find the unique solution:** Find the bucket with the maximum magnitude among the n/p buckets. Then, check the correlation of each of the p possible time shifts which are aliased into this bucket, and pick the shift that gives the maximum correlation. Checking the correlation can be done using only n/p samples as per Lemma A.7; therefore, it takes a total of $p \times n/p = n$ to perform the correlation of the p shifts and pick the one that maximizes the correlation.

(d) Runtime.

The running time of the QuickSync algorithm may be computed as follows. Step 1 performs np additions. Step 2 performs an FFT which takes $n/p \log(n/p)$. Step 3 performs n/p multiplications. Step 4 takes $n/p \log(n/p)$ to perform the IFFT, and finally Step 5 performs n operations to compute the correlations and find the solution. Thus, the complexity of QuickSync is $O(pn + (n/p)\log(n/p))$. To minimize this complexity, we set $p = \sqrt{\log n}$ which makes the overall running time of QuickSync $O(n\sqrt{\log n})$.

(e) Scaling with the SNR.

If the signal is too weak, GPS receivers repeat the synchronization algorithm on subsequent signal samples and sum up the output to average out the noise [36]. This approach allows the receiver to scale the computation with the SNR of the signal. The approach can be applied independent of the algorithm; hence, we also adopt it for QuickSync. However, QuickSync operates on blocks of size pn whereas the traditional FFT-based algorithm operates on blocks of size n. Both QuickSync and the traditional FFT-based algorithm compare the magnitude squared of the largest spike to the noise variance in the received signal. If the largest spike's squared magnitude exceeds the noise variance by a desired margin, the algorithm terminates the search and declares the time shift corresponding to the largest spike as the correct alignment. Otherwise, the algorithm repeats the same process on the subsequent signal samples, and sums the new output with the previous one. Since the spike corresponding to the correct synchronization is at the same time shift in each run, it becomes more prominent. In contrast, noise spikes are random and hence they tend to average out when combining the output of multiple runs of the synchronization algorithm.

[7] Note that we only get p candidate shifts (and not p^2) because the actual code is of size n; hence, all shifts mod n are the same. Thus, although the total number of samples is np and they are aliased into n/p buckets, we only have p distinct shifts per bucket.

(f) Linear Time Algorithm.

The algorithm described in §3(c) above can be made linear-time by modifying Step 1: instead of taking pn samples, we take only n samples and alias them into n/p buckets, where $p = \log n$. The rest of the steps are unmodified. This reduces the complexity of Step 1 to n, and the total complexity of the algorithm to $O(n + (n/p)\log(n/p)) = O(n)$.

This linear-time algorithm has weaker guarantees than the above super-linear algorithm and may not always work at very low SNRs, as detailed in §4. One can try this algorithm first. If a spike is detected with the required margin, the algorithm terminates. Otherwise, one can fall back to the super-linear algorithm in §3(c).

4. GUARANTEES

In this section, we analyze the performance of the baseline and QuickSync algorithms (both the linear and super-linear variants), under natural probabilistic assumptions about the input signal $\mathbf{x}$. In particular, we show that both the baseline and the super-linear QuickSync are correct under the same asymptotic assumptions about the variance of the noise in the signal $\mathbf{x}$. At the same time, the running time of our algorithm is equal to $O(pn + (n/p)\log(n/p))$, where p is the number of blocks used. This improves over the baseline algorithm which has $O(n \log n)$ runtime as long as the term pn is smaller than $(n/p)\log(n/p)$. In particular, by setting $p = \sqrt{\log n}$, we achieve the running time of $O(n\sqrt{\log n})$.

(a) Assumptions.

Recall that we use $\mathbf{c} = c_0 \ldots c_{n-1}$ to denote the spreading code. We use $\mathbf{c}^{(t)}$ to denote the code $\mathbf{c}$ shifted by $t = 0 \ldots n-1$, i.e., $c_i^{(t)} = c_{t+i \bmod n}$. We have that $\mathbf{x} = \mathbf{c}^{(t)} + \mathbf{g}$ for some shift t, where $\mathbf{g}$ denotes the noise vector. We make the following assumptions:

1. The coordinates $g_0 \ldots g_{n-1}$ of the noise vector $\mathbf{g}$ are independent and identically distributed random variables that follow a normal distribution with zero mean and variance σ. That is, we assume additive white Gaussian noise (AWGN).

2. The coordinates $c_0 \ldots c_{n-1}$ of the spreading code $\mathbf{c}$ are independent and identically distributed random variables with values in $\{-1, 1\}$, such that $\Pr[c_i = 1] = \Pr[c_i = -1] = 1/2$. This assumption models the fact that the CDMA code, $\mathbf{c}$, is *pseudorandom*.

(b) Combining Multiple Runs.

As described in §3(e), both the baseline and our algorithm can sum the output of multiple runs to average out the noise and increase the probability of identifying the correct spike. The analysis of such multi-run scenario can be derived directly from a single run. Specifically, say the algorithm runs L times and sum up the outputs of these L runs. This is equivalent to reducing the noise variance to $\sigma' = \sigma/L$. Therefore, the L-run scenario can be analyzed by reducing it to the case of a single run, with variance divided by L.

(c) Guarantees.

We start by stating the sufficient condition for the baseline algorithm to work with probability approaching 1.

THEOREM 4.1. *Assume that $\sigma \leq c(n)n/\ln n$ for $c(n) = o(1)$. Then the baseline algorithm is correct with probability $1 - o(1)$.*

The above condition is tight. Specifically,

THEOREM 4.2. *There exists a constant $c > 0$ such that for $\sigma \geq cn/\ln n$, the baseline algorithm is incorrect with probability $1 - o(1)$.*

We then proceed with the analysis of the two variants of the QuickSync algorithm. The first statement holds for the super-linear variant, and shows that the algorithm works with probability approaching 1 under the same condition as the baseline algorithm, while being faster.

THEOREM 4.3. *Assume that $\sigma \leq c(n)n/\ln n$ for $c(n) = o(1)$, and that $p = o(n^{1/6})$. Then, the QuickSync algorithm that aliases p blocks of size n into n/p buckets is correct with probability $1-o(1)$. The running time of the algorithm is $O(pn+(n/p)\log(n/p))$, which is $O(n\sqrt{\log n})$ for $p = \sqrt{\log n}$. Moreover, the algorithm performs only $O(n + (n/p)\log(n/p))$ multiplications, for any p.*

Finally, we analyze the linear-time variant of the QuickSync algorithm.

THEOREM 4.4. *Assume that $\sigma \leq c(n)\frac{n}{p \ln n}$ for $c(n) = o(1)$, and that $p = o(n^{1/6})$. Then, the QuickSync algorithm that aliases one block of n samples into n/p buckets is correct with probability $1 - o(1)$. The running time of the algorithm is $O(n + (n/p)\log(n/p))$, which is $O(n)$ for $p > \log n$.*

See Appendix for the proofs.

5. DOPPLER SHIFT & FREQUENCY OFFSET

GPS satellites orbit the Earth at very high speeds. Consequently, the GPS signal arrives at the receiver with a Doppler shift. This shift is modeled as a frequency offset f_d which is a function of the relative speed of the satellite (see Chapter 2 in [12] for exact calculations). Furthermore, the discrepancy between the RF oscillators of the GPS satellite and the GPS receiver induces a carrier frequency offset Δf_c. The total frequency offset $\Delta f = f_d + \Delta f_c$ typically ranges from -10 kHz to 10 kHz [36] and is modeled as a phase shift in the received samples. Formally, if $\mathbf{x}$ and $\tilde{\mathbf{x}}$ are respectively the signal without and with a frequency offset then:

$$\tilde{x}_t = x_t e^{j2\pi\Delta f t}, \quad (5)$$

where t is time in seconds.

Like past synchronization algorithms, QuickSync must search and correct for the frequency offset in the received GPS signal in order for the correlation to spike at the correct code shift. However, since QuickSync processes $p \times n$ samples as opposed to n samples in past algorithms(see §3), it needs to deal with larger phase shifts that accumulate over pn samples. In order to overcome this limitation, QuickSync performs a finer grained frequency offset search, which introduces an overhead to the 2D search. This overhead, however, is amortized across all satellites in the GPS signal since correcting for this frequency offset is done on the received signal before it is multiplied by each satellite's C/A code. In §7.2, we show that despite this overhead, QuickSync still provides a significant reduction in the computational complexity of GPS synchronization. Furthermore, the frequency offset changes slowly (see §7.2); hence, the receiver can cache its value from recent GPS readings, and does not need to search for it for every GPS synchronization event.

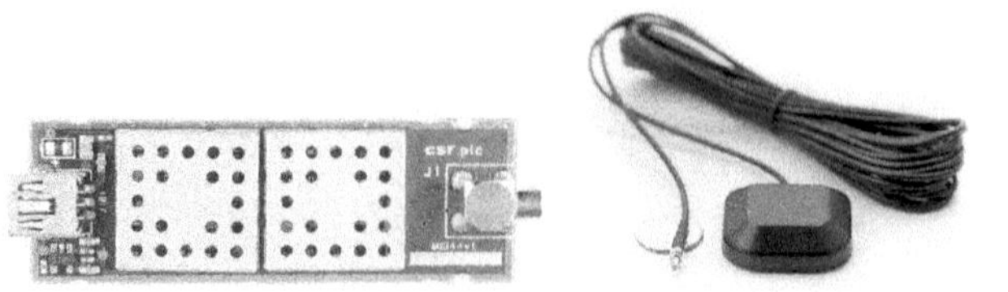

(a) Sampler (b) Antenna

Figure 5—The SciGe GN3S Sampler. The sampler is used to collect raw GPS data. It downcoverts the received signal and delivers the I and Q samples to the computer.

6. TESTING ENVIRONMENT

(a) Data Collection. We test our algorithm on a data set consisting of 40 GPS signal traces captured from urban and suburban areas in US and Europe. The traces in US are collected using the SciGe GN3S Sampler v3 [7] shown in Fig. 5(a). The GN3S is a form of software radio that collects raw complex GPS signal samples. We set the sampling rate of the GN3S to 4.092 MHz and its carrier frequency to 1575.42 MHz. The traces from Europe are collected using the USRP2 software radio [18] and the DBSRX2 daughterboard, which operates in the 1575.42 MHz range and is capable of powering up active GPS antennas [18]. The Europe traces are collected with a sampling frequency of 4 MHz. We also use a 3V magnetic mount active GPS antenna shown in Fig. 5(b). These datasets allow us to test the performance of QuickSync in different geographical areas and for different GPS receiver hardware.

(b) Baseline Algorithm. We compare our algorithm agains a baseline that uses the traditional FFT-based synchronization [39]. The baseline algorithm operates on blocks of size n. If a spike is not detected after processing the first block, the algorithm repeats the computation on the next block, i.e., the next set of n samples, and sums up the output of the IFFTs. The algorithm keeps processing more blocks until the magnitude of the peak crosses a certain threshold (as described in §7.1). Note that the algorithm must sum up the magnitudes of the output of the IFFTs rather than the actual complex values; otherwise, samples would combine incoherently due to the accumulated phase caused by the doppler shift (see §5).

(c) Implementation. We implement both QuickSync and the FFT-based algorithm in Matlab and run them on the collected GPS traces. Both algorithms use the FFTW [1] implementation internally to compute the Fourier transform (though the baseline computes an n-point transform while QuickSync computes an n/p-point transform).

(d) Metrics. We use two metrics for comparing the algorithms: number of multiplications, and number of floating point operations (FLOPs). We mainly focus on the number of real multiplications executed until an algorithm finds the synchronization offset. This metric is particularly important for hardware-based GPS synchronization, where multiplications are significantly more expensive than additions [32], and serve as standard metric to estimate the complexity of a potential hardware implementation [4, 35].

Some GPS-enabled devices do not have a full fledged GPS receiver hardware to reduce cost and form factor [23]. They use a GPS radio to collect signal samples, but offload the synchronization algorithm to the main CPU of the device, where it is done in software. To evaluate QuickSync's performance on software-based GPS receivers, we count the number of FLOPs executed by both QuickSync and the baseline. FLOPs is a standard metric

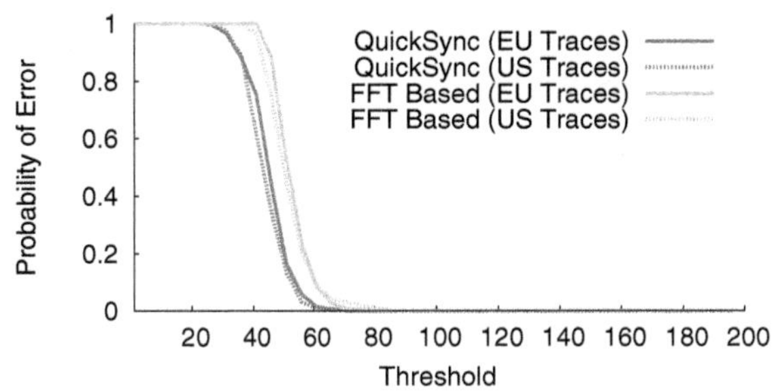

Figure 6—Probability of error versus threshold. The plot shows that the probability of error decreases sharply for both algorithms, and that a threshold of 90 for QuickSync and 100 for the baseline produce a zero error probability.

used to evaluate software implementations of algorithms, including FFTW [1]. It includes both multiplications and additions.

We count the FLOPs using OProfile, a standard profiler for Linux systems [27]. We run the code in Matlab R2011b under Ubuntu 11.10 on a 64-bit machine with Intel i7 processor. We run OProfile from within Matlab in order to profile the part of the code executed by each algorithm, and get a more accurate estimate of the number of FLOPs. We program OProfile to log the counter INST_RETIRED (the number of executed floating point operations on the Intel i7 processor [27]).

7. RESULTS

7.1 Setting the Synchronization Threshold

As explained in §3(e), both QuickSync and the FFT-based synchronization algorithm check that there is a sufficient margin between the detected maximum spike and the noise level, before accepting the spike as the one that identifies the correct alignment. Specifically, they check that the ratio of the spike's magnitude squared to the noise variance exceeds a particular threshold. This threshold defines how large the spike has to be in comparison to the bed of noise to ensure enough confidence that the spike is not due to noise and is indeed due to the code matching. Hence, the threshold is a measure of the SNR of the spike and is not dependent on the data. In particular, if the GPS data is noisy as in an urban area, the algorithm will continue processing more data until the threshold is crossed (as discussed in section §6(b)). In contrast, if the GPS data is less noisy as in an open suburban area, the algorithm will terminate early on since the spike will cross the threshold after processing one or two blocks.

In this section, we aim to verify that there is such a threshold that works for all datasets. Thus, we perform the following experiment. We vary the threshold between a value of 1 and 200, and for each of those values, we run both algorithms on a subset of the GPS traces from both datasets. We define the probability of error as the ratio of runs that output a false positive (i.e., in which the algorithm terminates by returning an invalid shift) to the total number of runs at a given threshold.

Fig. 6 plots the probability of errors versus the preset threshold. The plot shows that setting the threshold to 90 for QuickSync and 100 for the baseline produces a zero error probability. The baseline has a slightly higher error probability than QuickSync. This is because the baseline takes an n-point IFFT and, hence, has to ensure that none of the $n-1$ noise spikes exceeds the correct spike that corresponds to the proper alignment. In contrast, QuickSync takes an n/p-point IFFT and hence has fewer noise spikes that have to be kept below the threshold.

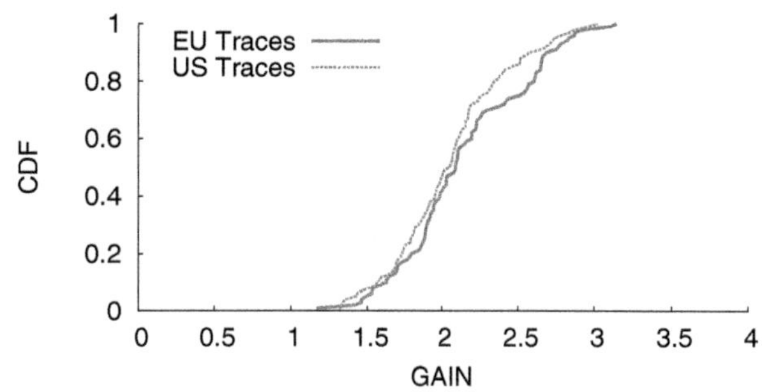

Figure 7—Gain of QuickSync over the FFT-based algorithm in number of multiplications. The two curves show the CDFs of the QuickSync's gains for the US and Europe datasets. QuickSync achieves a median gain of around 2.2× and a maximum gain of 3.3×.

The figure also shows that the used metric is stable, i.e.: (1) the metric is consistent across traces captured from two continents, and (2) the probability of error decreases monotonically as the threshold increases. This shows that the threshold is independent of the location of the GPS receiver.

In the experiments that follow, we set the thresholds to 90 and 100 for QuickSync and the baseline respectively. We also use a different set of traces from those used in testing for this threshold to ensure separation between testing and training.

7.2 Performance in Terms of Hardware Multiplications

We start by evaluating the performance gain of QuickSync over FFT-based synchronization in terms of the number of hardware multiplications. We run each of QuickSync and the FFT-based algorithm on both traces collected in US and Europe. We run the experiment 1000 times; each time taking a different subset of samples from these datasets. We compare the total number of multiplications required by each of the algorithms to synchronize with the signals of satellites present in the GPS traces. Fig. 7 shows a CDF of the gain. The gain is calculated as the number of multiplications needed by the FFT-based algorithm divided by the number of multiplications required by QuickSync. The figure shows that QuickSync always outperforms the FFT-based synchronization on both the US and EU traces with a median gain of 2.2×. This means that QuickSync can save on average twice the number of hardware multiplications.

To better understand the performance of QuickSync we zoom in on the number of multiplications required by each algorithm for each of the satellites. Specifically, each point in the CDFs in Fig. 7 corresponds to a full GPS reading with all satellites. However, because different satellites have different Doppler shifts and signal strengths, we expect the gains to vary from one satellite to another. Specifically, for each of the satellites detected in the Europe traces, and for each GPS reading, we measure the number of multiplications required by both algorithms to perform the synchronization. We repeat this experiment 1000 times on different subset of the samples and plot the average results in Fig. 8.

Fig. 8 shows that each of the satellites, on average, requires less multiplications using QuickSync. However, the gains vary considerably among those satellites. For example, satellites 5 and 18 have an average gain of 4× whereas satellites 14 and 15 have an average gain of only 1.5×. Examining the Doppler shifts of these satel-

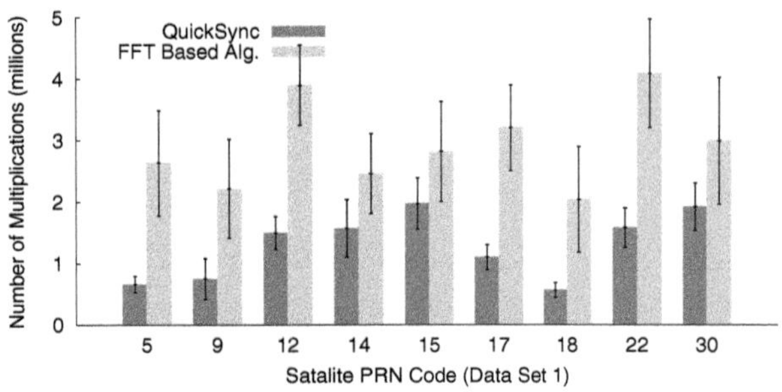

Figure 8—Number of multiplications on a per satellite basis for the Europe trace. The gain of QuickSync over the FFT-based algorithm varies among different satellites and ranges between and 1.5× and 4×.

Satellite Code:	9	12	14	15	18	21	22	25	27
Mean (Hz):	75	100	150	175	75	75	175	25	125
Max (Hz):	300	200	300	300	300	200	300	100	300

Table 1—Variation in the Doppler Shift in the US traces. For a given satellite, the Doppler shift of the received signal varies very little over a period of 2 hours and in an area of 2-mile diameter.

lites we find that satellites 5 and 18 have Doppler shifts of 6000 Hz and 1000 Hz respectively while satellites 14 and 15 have Doppler shifts of 600 Hz and 6800 Hz. This shows that the latter require a finer grain Doppler search as explained in §5. However, because QuickSync is opportunistic, it first attempts to search at courser grain shifts (the same as the baseline), but falls back to finer resolutions when it fails to detect a peak that passes the threshold. Even in such scenarios, however, it consistently outperforms the baseline as the figure shows.

In many scenarios, the receiver knows the Doppler shift a priori. The reason for this is that the Doppler shift varies only slightly between nearby locations and over a time period of about an hour. In order to test how much the Doppler shift varies, we measure the Doppler shift of satellite signals in the GPS traces captured at different locations within a 2-mile diameter geographical area over a period of 2 hours. For each of those satellites, we calculate the mean and the maximum variation in the Doppler shift of all those signals and record them in Table 7.2. The mean change is around 100 Hz and the maximum is 300 Hz. Accordingly, since GPS receivers are duty cycled, whenever the receiver wakes up, it may use the Doppler shift it calculated before going to sleep rather than performing an exhaustive search for it. Alternatively, assisted GPS receivers may download the measured Doppler shift from an adjacent base station [6]. In both of these situations, the GPS receiver can significantly reduce the overhead of searching for the right Doppler shift.

In order to measure the gains of QuickSync without the Doppler search, we repeat the first experiment but this time by providing each of the synchronization algorithms with the correct Doppler shift for each satellite. Fig. 9 shows a CDF of QuickSync's gain over the FFT-based algorithm in terms of number of multiplications over all the runs on both traces. For both traces, QuickSync achieves a median gain of 4.8×. This shows that QuickSync's gains increase when the receiver caches the correct Doppler shift across readings. We note that some of the traces benefit from QuickSync much more than others; the reason is that these runs have higher SNRs such that QuickSync can synchronize to their signals using the linear-time algorithm without falling back to the super-linear variant.

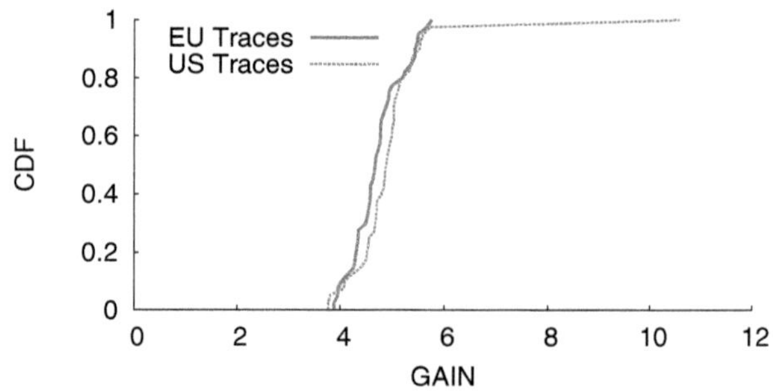

Figure 9—Gain of QuickSync over the FFT-based algorithm when the Doppler shift is known. The two curves show the CDFs of the gain in number of multiplications for both of our GPS traces. QuickSync achieves a median gain of 4.8×.

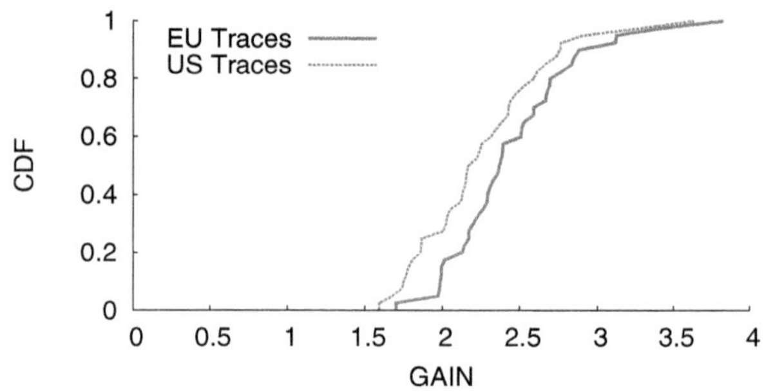

Figure 10—Gain of QuickSync over the FFT-based algorithm in FLOPs. This metric illustrates the gains of QuickSync for a software based implementation. The CDFs show a median gain about 2.2× and a maximum gain of around 3.7×.

7.3 Performance on software based GPS receivers

In this section, we test the performance of QuickSync on software based GPS receivers in terms of the number of floating point operations (FLOPs). We run QuickSync and the FFT-based algorithm on the US and Europe traces and use OProfile to count the number of FLOPs as described in §6. We run the experiment 1000 times with a different subset samples of the traces and calculate the gain as the ratio of the number of FLOPs required by the FFT-based algorithm to the number of FLOPs required by QuickSync. We do not assume in this experiment that the Doppler shift is known and we let both algorithms search for the right Doppler shift. Fig. 10 shows a CDF of the gains. QuickSync achieves a median gain of 2× and 2.3× over the FFT-based algorithm for the US and Europe traces respectively. This shows that QuickSync can reduce the number of CPU computation on average by half in software based GPS receivers.

8. RELATED WORK

FFT-based GPS synchronization was first proposed by Nee et al. [39] who showed that it reduces synchronization complexity from $O(n^2)$ to $O(nlog(n))$, where n is the number of samples per C/A code. QuickSync builds on this technique and recent advances in sparse FFT to further reduce the synchronization complexity.

Our approach is related to past work on GPS block-averaging [36, 24], which sums up consecutive signal blocks before performing the FFT. QuickSync however differs from that work along two axes: First, on the algorithmic front, past work performs a full size FFT of n points. One cannot simply replace this n-point

FFT with sparse FFT because, as explained in §1, the output of the FFT is not sparse. In contrast, QuickSync introduces a design that can harness sparse FFT. This enables QuickSync to operate with a smaller FFT of size n/p, which provides faster synchronization. Second, past work on block-averaging focuses on weak GPS signals and does not provide an adaptive design that works for the whole range of GPS SNRs. Applying their approach to the whole SNR range can incur unnecessary overhead. This is because they average and pre-process many blocks independent of the SNR. As a result, these schemes increase the synchronization delay for scenarios in which only one (or a few) blocks are sufficient for detecting the spike. In contrast, our algorithm adaptively processes more data when the spike is too low for detection, and hence gracefully scales with the SNR of the received GPS signal.

Signal synchronization is a general problem that applies to other wireless technologies, e.g., WiFi and cellular. However, synchronization in these technologies is simpler because the noise level in the received signals is much lower than in GPS. For example, WiFi receivers can lock on the signal simply by detecting an increase in the received power [16]. This is not possible in GPS since the signal is received at 20 dB *below* the noise floor [31]. Cellular systems also operate at much higher SNRs than GPS, which allows them to synchronize with relatively low overhead [21].

QuickSync is also related to the general class of work on reducing GPS power consumption. The most common approach uses assisted-GPS [19, 33], which involves connecting to an assisted GPS server through a WiFi or cellular network. The server provides the GPS receiver with the GPS data decoded from each satellite signal, which allows the receiver to avoid decoding the GPS signal. The device can also offload GPS computations to the server after acquiring the GPS signal [9]. The latter approach, however, reduces the complexity of the device but still consumes much power because it requires the device to transmit the undecoded noisy GPS signal to the cellular tower (thus consuming transmission power and even bandwidth [9]). Other approaches for reducing GPS power consumption leverage WiFi, sensors, or the cellular signal to localize the receiver [38, 3, 25]. These schemes typically are less accurate than GPS and are more constrained in terms of where they can be deployed. Our work contributes to this effort by tackling the complexity of the GPS receiver itself.

Finally, QuickSync builds on recent advances in sparse FFT algorithms [11, 14, 15]. These algorithms are designed for the case where the output of the Fourier Transform contains only a small number of spikes k, and use a variety of signal filters (Dirichlet, Gaussian, or Dolph-Chebyshev filters) to isolate and identify each of the spikes. The running times range from $O(\log n\sqrt{nk\log n})$ [14] to $O(k\log^4 n)$ [11] and $O(k\log n\log(n/k))$ [15]. The synchronization problem differs from the sparse Fourier problem in that only the IFFT is sparse. Hence, simply applying past sparse FFT algorithms does not reduce synchronization complexity.

9. CONCLUSION

This paper presents the fastest synchronization algorithm for GPS receivers to date. The gains of the algorithm are also empirically demonstrated for software GPS implementations as well as potential hardware architectures. Because synchronization consumes a significant amount of power, we expect the reduced complexity to decrease GPS power consumption. Further, we believe that the sparse synchronization algorithm we introduced has other applications in signal processing and pattern matching. We plan to explore those applications in future work.

Acknowledgments: We thank Jue Wang, Nate Kushman, Hariharan Rahul, Lixin Shi, Arthur Berger, Swarun Kumar, the reviewers and our shepherd Kyle Jamieson for their insightful comments. We thank Jonathan Eastep for explaining the proper profiling. We also thank Carles Fernandez and Javier Arribas from the GNSS-SNR project for providing us with the Europe dataset. This research is funded by NSF.

10. REFERENCES

[1] FFTW 3.2.3. http://www.fftw.org.

[2] B. Buchli, F. Sutton, and J. Beutel. GPS-equipped wireless sensor network node for high-accuracy positioning applications. In *EWSN 2012*, Trento, Italy.

[3] N. Bulusu, J. Heidemann, and D. Estrin. GPS-less low-cost outdoor localization for very small devices. *IEEE Personal Communications*, Oct. 2000.

[4] C. Cheng and K. Parhi. Low-cost fast VLSI algorithm for discrete fourier transform. *IEEE Transactions on Circuits and Systems*, April 2007.

[5] Dexter Industries. dGPS for LEGO MINDSTORMS NXT. http://dexterindustries.com.

[6] G. Djuknic and R. Richton. Geolocation and assisted GPS. *Computer*, 34(2):123 –125, feb 2001.

[7] S. Electronics. Sige gn3s sampler v3. http://www.sparkfun.com.

[8] C. Fernandez-Prades, J. Arribas, P. Closas, C. Aviles, and L. Esteve. GNSS-SDR: an open source tool for researchers and developers. In *ION GNSS Conference*, 2011.

[9] G. Fleishman. How the iphone knows where you are. Macworld, Aug. 2011.

[10] G. Fleishman. Inside assisted GPS: helping GPS help you. Arstechnica, Jan. 2009.

[11] A. C. Gilbert, S. Muthukrishnan, and M. Strauss. Improved time bounds for near-optimal sparse fourier representations. In *SPIE Wavelets XI*, 2003.

[12] A. Goldsmith. *Wireless Communications*. Cambridge University Press, 2005.

[13] S. Gossett. GPS implant makes debut. WND, May 2003.

[14] H. Hassanieh, P. Indyk, D. Katabi, and E. Price. Nearly optimal sparse fourier transform. In *STOC 2012*.

[15] H. Hassanieh, P. Indyk, D. Katabi, and E. Price. Simple and practical algorithm for sparse fourier transform. In *SODA'12*.

[16] J. Heiskala and J. Terry. *OFDM Wireless LANs: A Theoretical and Practical Guide*. Sams Publishing, 2001.

[17] A. Hepburn. Infographic: Mobile stats & facts 2011. Digital Buzz, April 2011.

[18] E. Inc. Universal software radio peripheral. http://ettus.com.

[19] J. L. Jani Jarvinen, Javier DeSalas. Assisted GPS: A low-infrastructure approach. GPSWorld, March 2002.

[20] E. D. Kaplan. *Understanding GPS Principles and Applications*. Artech House Publishers, Feb. 1996.

[21] M. Karim and M. Sarraf. *W-CDMA and CDMA2000 for 3G mobile networks*. McGraw-Hill, 2002.

[22] J. Matousek. On variants of the johnson-lindenstrauss lemma. *Random Structures & Algorithms*, 33(2), 2008.

[23] Maxim IC. MAX2745 single-chip global positioning system front-end downconverter. http://www.maxim-ic.com.

[24] R. J. L. Mohamed Sahmoudi, Moeness G. Amin. Acquisition of weak gnss signals using a new block averaging pre-processing. In *IEEE/ION Position, Location and Navigation Symposium 2008*.

[25] D. Niculescu and B. Nath. Ad hoc positioning system (APS). In *IEEE GLOBECOM*, 2001.
[26] Nikon USA. Coolpix p6000. http://www.nikonusa.com.
[27] OProfile. Linux profiler. http://oprofile.sourceforge.net.
[28] OriginGPS. ORG447X series datasheet. http://www.acaltechnology.com.
[29] Perthold Engineering LLC. SkyTraq Venus 6 GPS Module. http://www.perthold.de.
[30] G. Pisier. *The Volume of Convex Bodies and Banach Space Geometry*. Cambridge University Press, May 1999.
[31] D. Plausinaitis. GPS receiver technology mm8. Danish GPS Center, http://kom.aau.dk.
[32] J. Rabaey, A. Chandrakasan, and B. Nikolic. *Digital integrated circuits*. Prentice-Hall, 1996.
[33] H. S. Ramos, T. Zhang, J. Liu, N. B. Priyantha, and A. Kansal. Leap: a low energy assisted GPS for trajectory-based services. In *ACM Ubicomp 2011*, pages 335–344, Beijing, China.
[34] D. Raskovic and D. Giessel. Battery-Aware embedded GPS receiver node. In *IEEE MobiQuitous*, 2007.
[35] T. Tan, G. Bi, Y. Zeng, and H. Tan. Dct hardware structure for sequentially presented data. *Signal Processing*, 81(11), 2001.
[36] A. C. Team. *Fundamentals of Global Positioning System Receivers: A Software Approach*. Wiley-Interscience, 2000.
[37] Ted Schadler. GPS: Personal Navigation Device. Texas Instruments. http://www.ti.com.
[38] A. Thiagarajan, L. Ravindranath, H. Balakrishnan, S. Madden, and L. Girod. Accurate, low-energy trajectory mapping for mobile devices. In *NSDI 2011*.
[39] D. Van Nee and A. Coenen. New fast GPS code-acquisition technique using FFT. *Electronics Letters*, 27(2), Jan 1991.

APPENDIX

A. PROOFS

A.1 Analysis of the baseline algorithm

The baseline algorithm computes $a_k = \mathbf{c}^{(k)} \cdot \mathbf{x}$ for all shifts $k = 0 \ldots n-1$. The probability that the algorithm reports an incorrect output is equal to

$$P(\sigma) = \Pr[a_t \leq \max_{k \neq t} a_k]$$

To estimate the probability of success as a function of σ, we derive the distribution of the coordinates a_k. From our assumptions we have that $a_k = \mathbf{c}^{(k)} \cdot (\mathbf{c}^{(t)} + \mathbf{g}) = v_k + u_k$, where $v_k = \mathbf{c}^{(k)} \cdot \mathbf{c}^{(t)}$ and $u_k = \mathbf{c}^{(k)} \cdot \mathbf{g}$. Note that u_k has normal distribution with zero mean and variance

$$\text{Var}[u_k] = \text{Var}[\sum_{i=0}^{n-1} c_i^{(k)} g_i] = n\text{Var}[c_1^{(k)}]\text{Var}[g_1] = n\sigma$$

Regarding v_k, we have the following two cases:

- If $k = t$, i.e., for the correct value of the shift, we have $v_t = \mathbf{c}^{(t)} \cdot \mathbf{c}^{(t)} = n$.
- If $k \neq t$, i.e., for an incorrect value of the shift the expectation of v_k is 0.

We need to bound $P(\sigma) = \Pr[n + u_t \leq \max_{k \neq t} v_k + u_k]$. The following theorem establishes the sufficient (and as we show later, necessary) condition for the baseline algorithm to be correct with probability $1 - o(1)$, i.e., $P(\sigma) \to 0$ as $n \to \infty$.

LEMMA A.1. *Assume that $\sigma \leq c(n)n/\ln n$ for $c(n) = o(1)$. Then $P(\sigma) = o(1)$.*

PROOF: We will bound the probabilities of the following events: E_1: $\exists_{k \neq t} u_k \geq n/3$; E_2: $u_t \leq -n/3$; E_3: $\exists_{k \neq t} v_k \geq n/3$. If none of the events hold then the algorithm output is correct. We will show that $\Pr[E_1] + \Pr[E_2] + \Pr[E_3] = o(1)$.

To analyze E_1 and E_2 recall the following fact:

FACT A.2. *Let $\Phi(s)$ be the c.d.f. of the normal distribution with zero mean and unit variance. Then for $s > 0$*

$$1 - \Phi(s) \leq e^{-s^2/2}/s$$

We can now bound

$$\Pr[E_1] \leq n \Pr[u_k \geq n/3] = n(1 - \Phi(\frac{n/3}{\sqrt{\text{Var}[u_k]}}))$$

where k is any index distinct from t. Since $\text{Var}[u_k] = n\sigma \leq c(n)n^2/\ln n$, we have

$$\Pr[E_1] \leq n(1 - \Phi(\sqrt{\ln n/(9c(n))})) \leq e^{\ln n} \cdot e^{-\frac{\ln(n)}{18c(n)}} = o(1)$$

The probability $\Pr[E_2]$ can be bounded in the same way.

To bound $\Pr[E_3]$, assume without loss of generality that $t = 0$. In this case

$$\begin{aligned} v_k &= \mathbf{c}^{(k)} \cdot \mathbf{c} \\ &= (c_k c_0 + c_{k+1} c_1 + \ldots + c_{n-1} c_{n-k-1}) \\ &+ (c_0 c_{n-k} + \ldots + c_{k-1} c_{n-1}) \\ &= S_k + S'_k \end{aligned}$$

The terms in the sum $S_k + S'_k$ are in general *not* independent. In particular, if $k = n/2$, then $S_k = S'_k$. However, we observe the following.

CLAIM A.3. *Each of S_k and S'_k is a sum of independent random variables taking values in $\{-1, 1\}$ with probability $1/2$.*

The claim enables us to bound each sum separately. We will bound $\Pr[S'_k \geq n/6]$ first. If $k < n/6$ then the probability is zero. Otherwise, by applying the Chernoff bound we have

$$\Pr[S'_k \geq n/6] \leq e^{-(n/6)^2/(2k)} \leq e^{-n/72}$$

The probability $\Pr[S_k \geq n/6]$ can be bounded in the same way. Hence $\Pr[E_3] \leq ne^{-n/72} = o(1)$. □

Theorem 4.1 follows from Lemma A.1.

A.2 Tightness of the variance bound

In this section we show that the assumption on the noise variance used in Theorem 4.1 is asymptotically tight. Specifically, we show that if $\sigma \geq cn/\ln n$ for some large enough constant $c > 0$, then there with probability $1 - o(1)$ the output of the baseline algorithm is incorrect. This will prove Theorem 4.2.

Recalling the notation in Section A.1, we have $a_t = n + u_t$, $a_k = v_k + u_k$ for $k \neq t$. We need to show that $P(\sigma) = \Pr[a_t \leq \max_{k \neq t} a_k]$ approaches 1 for c large enough. To this end, we first observe that (i) $v_k \geq -n$ holds always and (ii) $\Pr[u_t \geq n] = o(1)$ since u_t is a normal variable with variance $n\sigma = O(n^2/\ln n)$, so the desired bound holds e.g., by Chebyshev inequality. Hence it suffices to show that

$$\Pr[\sup_{k \neq t} u_k \leq 3n] = o(1) \tag{6}$$

The main difficulty in proving Equation 6 is the fact that the random variables u_k are *not* independent. If they were, a simple calculation would show that the expected value of the maximum of n independent normal variables with variance $n\sigma$ is at least $\sqrt{n\sigma \ln n} = cn$, which is larger than $a_t = n$ for a large enough c. This would then ensure that the reported shift is distinct from t with constant probability. Unfortunately, the independence could be guaranteed only if the shifted codes $c^{(k)}$ were orthogonal, which is not the case.

Instead, our argument utilizes the fact that the shifted codes $\mathbf{c}^{(k)}$ are "almost" orthogonal. Specifically, let $C = \{\mathbf{c}^{(k)} : k \neq t\}$. Since (as shown in the earlier section in the context of the event E_3) the probability that for any pair $\mathbf{c} \neq \mathbf{c}' \in C$ we have $\mathbf{c}\cdot\mathbf{c}' \leq n/3$ is $o(1)$, it follows that $\|\mathbf{c} - \mathbf{c}'\|_2^2 \geq n$ for all such $\mathbf{c}, \mathbf{c}' \in C$ with probability $1 - o(1)$.

We can now use a powerful inequality due to Sudakov [30] to show that the random variables u_k are "almost" independent, and thus the expected value of the maximum is still $\sqrt{n\sigma \ln n}$. In our context, the inequality states the following.

FACT A.4. *There exists a constant $c_2 > 0$ such that if D is the Euclidean distance between the closest pair of vectors in C, then:*

$$E = E[\max_{\mathbf{c}\in C} \mathbf{c} \cdot \mathbf{g}] \geq c_2 D \sqrt{\sigma \ln n}$$

Since $D = \sqrt{n}$, we obtain that

$$E \geq c_2 \sqrt{n} \sqrt{cn/\ln n \cdot \ln n} = c_2 n$$

The lower bound on the expected value of the maximum can be then converted into an upper bound on probability that the maximum is much lower than its expectation (this follows from simple but somewhat tedious calculations). This leads to Equation 6 and completes the proof of Theorem 4.2.

A.3 Analysis of the QuickSync algorithm

In this section we show that the probability of correctness for the QuickSync algorithm that aliases into n/p buckets exhibits a similar behavior to the baseline algorithm, albeit with the bucket variance larger by a factor of $O(p)$. At the same time, the running time of our algorithm is equal to $O(pn + (n/p)\log(n/p))$. This improves over the baseline algorithm which has $O(n \log n)$ runtime as long as the term pn is smaller than $(n/p)\log(n/p)$.

Recall that the algorithm first computes the aliased spreading code $\mathbf{c}(p)$ and signal $\mathbf{x}(p)$, defined as

$$c(p)_i = \sum_{q=0}^{p-1} c_{i+qn/p} \text{ and } x(p)_i = \sum_{q=0}^{p-1} x_{i+qn/p}$$

for $i = 0 \ldots n/p - 1$. The aliased noise vector $\mathbf{g}(p)$ is defined in an analogous way.

The application of FFT, coordinate-wise multiplication and inverse FFT computes

$$a(p)_k = c(p)^{(k)} \cdot x(p)$$

for all shifts $k = 0 \ldots n/p - 1$. The algorithm then selects $a(p)_k$ with the largest value. The last step of the algorithm fails if for $t' = t \mod n/p$ we have $a(p)_{t'} \leq \max_{k\neq t'} a(p)_k$. Let $P'(\sigma)$ be the probability of this event. We will show that as long as $\sigma = o(pn/\ln n)$ we have $P'(\sigma) = o(1)$.

LEMMA A.5. *Assume that $\sigma \leq c(n)\frac{n}{p \ln n}$ for $c(n) = o(1)$, and that $p = o(n^{1/6})$. Then $P'(\sigma) = o(1)$.*

PROOF: We start by decomposing each term $a(p)_k$:

$$\begin{aligned} & a(p)_k \\ = & \sum_{i=0}^{n/p-1} \left(\sum_{q=0}^{p-1} c^{(k)}_{i+qn/p}\right) \left(\sum_{q=0}^{p-1} c^{(t')}_{i+qn/p} + \sum_{q=0}^{p-1} g_{i+qn/p}\right) \\ = & \sum_{i=0}^{n/p-1} \left(\sum_{q=0}^{p-1} c^{(k)}_{i+qn/p}\right) \left(\sum_{q=0}^{p-1} c^{(t')}_{i+qn/p}\right) \\ & + \sum_{i=0}^{n/p-1} \left(\sum_{q=0}^{p-1} c^{(k)}_{i+qn/p}\right) \left(\sum_{q=0}^{p-1} g_{i+qn/p}\right) \\ = & v_k + u_k \end{aligned}$$

Consider the following events: E_0: $v_{t'} \leq n - n/4$; E_1: $\exists_{k\neq t'} u_k \geq n/4$; E_2: $u_{t'} \leq -n/4$; E_3: $\exists_{k\neq t'} v_k \geq n/4$. If none of the events hold, then the algorithm output is correct. We need to show that $\Pr[E_0] + \Pr[E_1] + \Pr[E_2] + \Pr[E_3] = o(1)$.

Events E_1 and E_2.

Let $\hat{c}_i = \sum_{q=0}^{p-1} c_{i+qn/p}$, $\hat{g}_i = \sum_{q=0}^{p-1} g_{i+qn/p}$ and $m = n/p$. Observe that $|\hat{c}| \leq p$ and $\hat{g}_i$'s are i.i.d. random variables chosen from the normal distribution with mean zero and variance $p\sigma$. Conditioned on the choice of $\hat{c}_i$s, the random variable $u_k = \sum_{i=0}^{m-1} \hat{c}_{i+k}\hat{g}_i$ has normal distribution with variance $\mu p\sigma$, where $\mu = \sum_{i=0}^{m-1} (\hat{c}_{i+k})^2$. We first show that $\mu \leq 4pm$ with very high probability, and then bound the tail of a normal random variable with variance $4pm$.

The following fact is adapted from [22], Theorem 3.1.

FACT A.6. *Let R_{ij}, $i = 0 \ldots m-1$ and $j = 0 \ldots p-1$, be i.i.d. random variable taking values uniformly at random from $\{-1, 1\}$. Let $T_i = 1/\sqrt{m}\sum_{j=0}^{p-1} R_{ij}$. There is an absolute constant C such that*

$$\Pr[\sum_{i=0}^{m-1} T_i^2 \leq 4p] \geq 1 - 2e^{-m/C}$$

Applying the fact to $R_{ij} = c^{(k)}_{i+jn/p}$, we conclude that with probability $1 - o(1)$ we have $\mu \leq m\sum_{i=0}^{m-1} T_i^2 \leq 4pm = 4n$. In that case $\mathrm{Var}[u_k] \leq 4np\sigma \leq 4npc(n)\frac{n}{p \ln n} = 4n^2 c(n)/\ln(n)$. We then follow the proof of Lemma A.1.

Event E_0.

Observe that $v_{t'}$ is a sum of terms $q_i = (\sum_{q=0}^{p-1} c^{(t')}_{i+qn/p})^2$, where each q_i is a square of a sum of p independent random variables taking values in $\{-1, 1\}$ with probability $1/2$. It follows that $E[q_i] = p$, and therefore $E[v_{t'}] = n/p \cdot p = n$. To bound the deviation of $v_{t'}$

from its mean n, we first compute its variance.

$$\begin{aligned}\mathrm{Var}[v_k] &= \mathrm{Var}[\sum_{i=0}^{n/p-1} q_i] \\ &\leq n/p \cdot E[q_1^2] \\ &= n/p \cdot E[(\sum_{q=0}^{p-1} c^{(t')}_{1+qn/p})^4] \\ &= n/p \cdot (\binom{4}{2} \sum_{q \neq q'} E[(c^{(t')}_{1+qn/p})^2 (c^{(t')}_{1+q'n/p})^2] \\ &\quad + \sum_q E[(c^{(t')}_{1+qn/p})^4]) \\ &= n/p \cdot (6p(p-1)+p) \leq 7pn\end{aligned}$$

We can now use Chebyshev's inequality:

$$\Pr[|v_{t'} - n| \geq n/4] \leq \mathrm{Var}[v_{t'}]/(n/4)^2 \leq 7pn/(n/4)^2 = o(1)$$

Event E_3.

It suffices to bound $\Pr[E_3]$. To this end we bound $\Pr[v_k \geq n/4]$. Without loss of generality we can assume $t = 0$. Then $v_k = \sum_{i=0}^{n/p-1} \hat{c}_i \hat{c}_{i+k}$, where $i+k$ is taken modulo n/p.

We first observe that in each term $\hat{c}_i \hat{c}_{i+k}$, the random variables $\hat{c}_i$ and $\hat{c}_{i+k}$ are independent (since $k \neq 0$). This implies $E[\hat{c}_i \hat{c}_{i+k}] = E[\hat{c}_i]E[\hat{c}_{i+k}] = 0$, and therefore $E[v_k] = 0$.

To bound $\Pr[v_k \geq n/4]$, we compute the fourth moment of $v_{t'}$ (using the second moment does not give strong enough probability bound). We have

$$\begin{aligned}E[v_k^4] &= E[(\sum_{i=0}^{n/p-1} \hat{c}_i \hat{c}_{i+k})^4] \\ &= \sum_{i_1,i_2,i_3,i_4=0}^{n/p-1} E[\hat{c}_{i_1}\hat{c}_{i_1+k} \cdot \hat{c}_{i_2}\hat{c}_{i_2+k} \cdot \hat{c}_{i_3}\hat{c}_{i_3+k} \cdot \hat{c}_{i_4}\hat{c}_{i_1+k}]\end{aligned}$$

Observe that the expectation of any term in the above sum that contains an odd power of $\hat{c}_i$ is zero. Hence the only remaining terms have the form $E[\hat{c}_{j_1}^2 \hat{c}_{j_2}^2 \hat{c}_{j_3}^2 \hat{c}_{j_4}^2]$, where $j_1 \ldots j_4$ are not necessarily distinct. Let I be a set of such four-tuples (j_1, j_2, j_3, j_4). We observe that for (j_1, j_2, j_3, j_4) to belong in I, at least two disjoint pairs of indices in the sequence $i_1, i_1+k, i_2, i_2+k, i_3, i_3+k, i_4, i_4+k$ must be equal. This means that $|I| = C(n/p)^2$ for some constant C. Since $|\hat{c}_i| \leq p$, we have $E[v_k^4] \leq C(n/p)^2 p^8 \leq Cn^2p^6$. Thus

$$\Pr[v_k \geq n/4] \leq E[v_k^4]/(n/4)^4 \leq \frac{Cn^2p^6}{(n/4)^4} = C \cdot 4^4 \cdot p^6/n^2$$

This implies $\Pr[E_3] \leq n \Pr[v_k \geq n/4] \leq C \cdot 4^4 \cdot p^6/n$ which is $o(1)$ if $p = o(n^{1/6})$. □

We now show that if the noise variance σ is "small" then one can check each shift using few time domain samples.

LEMMA A.7. *Assume that* $\sigma \leq c(n)\frac{n}{p \ln n}$. *Consider an algorithm that, given a set* K *of* p *shifts, computes*

$$a'_k = \mathbf{c}^{(k)}[0 \ldots T-1] \cdot \mathbf{x}[0 \ldots T-1]$$

for $T = n/p$, *and selects the largest* a_k *over* $k \in K$. *Then*

$$P''(\sigma) = \Pr[a_t \leq \max_{k \neq t} a_k] = o(1)$$

PROOF: The argument is similar to the proof of Lemma A.1. We verify it for an analog of the event E_1; the proofs for E_2 and E_3 are straightforward syntactic modifications of the original arguments.

First, observe that $\mathbf{c}^{(t)}[0 \ldots T-1] \cdot \mathbf{c}^{(t)}[0 \ldots T-1] = T$. Let $u'_k = \mathbf{c}^{(k)}[0 \ldots T-1] \cdot \mathbf{g}[0 \ldots T-1]$. Consider the event $E'_1 : \exists_{k \in K} u'_k \geq T/3$. We can bound

$$\Pr[E'_1] \leq p \Pr[u'_k \geq T/3] = n(1 - \Phi(\frac{T/3}{\sqrt{\mathrm{Var}[u'_k]}}))$$

Since $\mathrm{Var}[u'_k] = T\sigma \leq c(n)\frac{n^2}{p^2 \ln n}$, we have

$$\Pr[E'_1] \leq p(1 - \Phi(\sqrt{\ln n/(9c(n))})) \leq e^{\ln n} e^{-\frac{\ln(n)}{18c(n)}} = o(1) \quad \square$$

Proofs of Theorem 4.3 and Theorem 4.4.

To prove Theorem 4.3, recall that given a signal $\mathbf{x}$ consisting of p blocks, each of length n and with noise variance σ, QuickSync starts by aliasing the p blocks into one. This creates one block of length n, with noise variance σ/p (after normalization). We then apply Lemmas A.5 and A.7.

Theorem 4.4 follows from the assumption that $\sigma = c(n)\frac{n}{p \ln n}$ and Lemma A.5.

Every Bit Counts - Fast and Scalable RFID Estimation

Muhammad Shahzad and Alex X. Liu
Department of Computer Science and Engineering
Michigan State University
East Lansing, Michigan, USA
{shahzadm, alexliu}@cse.msu.edu

ABSTRACT

Radio Frequency Identification (RFID) systems have been widely deployed for various applications such as object tracking, 3D positioning, supply chain management, inventory control, and access control. This paper concerns the fundamental problem of estimating RFID tag population size, which is needed in many applications such as tag identification, warehouse monitoring, and privacy sensitive RFID systems. In this paper, we propose a new scheme for estimating tag population size called *Average Run based Tag estimation* (ART). The technique is based on the average run-length of ones in the bit string received using the standardized framed slotted Aloha protocol. ART is significantly faster than prior schemes because its estimator has smaller variance compared to the variances of estimators of prior schemes. For example, given a required confidence interval of 0.1% and a required reliability of 99.9%, ART is consistently 7 times faster than the fastest existing schemes (UPE and EZB) for any tag population size. Furthermore, ART's estimation time is observably independent of the tag population sizes. ART is easy to deploy because it neither requires modification to tags nor to the communication protocol between tags and readers. ART only needs to be implemented on readers as a software module. ART works with multiple readers with overlapping regions.

Categories and Subject Descriptors

C.2.1 [**Computer-Communication Networks**]: Network Architecture and Design – *Wireless communication*; C.2.8 [**Mobile Computing**]: Algorithm Design and Analysis

General Terms

Algorithms, Design, Performance, Experimentation

Keywords

RFID, Tags, Estimation

1. INTRODUCTION

1.1 Motivation and Problem Statement

RFID systems are widely used in various applications such as object tracking [12], 3D positioning [19], indoor localization [13], supply chain management [9], inventory control, and access control [4, 11] because the cost of commercial RFID tags is negligible compared to the value of the products to which they are attached (*e.g.*, as low as 5 cents per tag [15]). An RFID system consists of tags and readers. A tag is a microchip with an antenna in a compact package that has limited computing power and communication range. There are two types of tags: (1) passive tags, which are powered up by harvesting the radio frequency energy from readers (as they do not have their own power sources) and have communication range often less than 20 feet; (2) active tags, which have their own power sources and have relatively longer communication range. A reader has a dedicated power source with significant computing power. It transmits a query to a set of tags and the tags respond over a shared wireless medium.

This paper concerns the fundamental problem of estimating the size of a given tag population. This is needed in many applications such as tag identification, privacy sensitive RFID systems, and warehouse monitoring. In tag identification protocols, which read the ID stored in each tag, population size is estimated at the start to guide the identification process. For example, for tag identification protocols that are based on the framed slotted Aloha protocol (standardized in EPCGlobal Class-1 Generation-2 (C1G2) RFID standard [3] and implemented in commercial RFID systems), tag estimation is often used to calculate the optimal frame size [7]. In privacy sensitive RFID systems, such as those used in parks for continuously monitoring the number of visitors in different areas of a park to plan the guided trips efficiently, readers may not have the permission to identify human individuals. In warehouses with RFID-based monitoring systems, managers often need to perform a quick estimation of the number of products left in stock for various purposes such as the detection of employee theft. Note that although tag population size can be accurately measured by tag identification, the speed will be too slow.

We formally define the tag estimation problem as: *given a tag population of unknown size t, a confidence interval $\beta \in (0, 1]$, and a required reliability $\alpha \in [0, 1)$, a set of readers needs to collaboratively compute the estimated number of tags $\tilde{t}$ so that $P\left\{|\tilde{t} - t| \leq \beta t\right\} \geq \alpha$.* When the number of readers is one, we call this problem *single-reader estimation*; otherwise, we call this problem *multi-reader estimation.*

A tag estimation scheme should satisfy the following three requirements:

1. *Reliability*: The actual reliability should always be greater than or equal to the required reliability. The reliability α given as input is called the *required reliability*. The reliability that an estimation scheme achieves is called its *actual reliability*.
2. *Scalability*: The estimation time needs to be scalable to large population sizes because in many applications, the number of passive tags can be very large due to their low cost, easy disposability, and powerless operation.
3. *Deployability*: The estimation scheme needs to be compliant with the C1G2 standard and should not require any changes to tags.

1.2 Proposed Approach

In this paper, we propose a new scheme called *Average Run based Tag estimation* (*ART*), which satisfies all of the above three requirements. The communication protocol used by ART is the standardized framed slotted Aloha protocol, in which a reader first broadcasts a value f to the tags in its vicinity where f represents the number of time slots present in a forthcoming frame. Then each tag randomly picks a time slot in the frame and replies during that slot. Thus, the reader gets a binary sequence of 0s and 1s by representing a slot with no tag replies as 0 and a slot with one or more tag replies as 1. The key idea of ART is to estimate tag population size based on the average run size of 1s in the binary sequence. We show that the average run size of 1s in a frame monotonously increases with the increase in the size of tag population. Thus, average run size of 1s is an indicator of tag population size.

1.3 Advantages of ART over Prior Art

ART is advantageous in terms of speed and deployability. For speed, ART is faster than all prior schemes. For example, given a confidence interval of 0.1% and the required reliability of 99.9%, ART is consistently 7 times faster than the fastest existing schemes (*i.e.*, UPE [7] and EZB [8]) for any tag population size. The reason behind ART being faster than prior schemes is that the new estimator that we propose in this paper, namely the average run size of 1s, has significantly smaller variance compared to the estimators used in prior schemes (such as the total number of 0s [7, 8] and the location of the first 1 in the binary sequence [6]), as we analytically show in Section 4.2. An estimator with small variance is faster because the estimation process needs to be repeated fewer times to achieve the required reliability. *The intuitive reason that our estimator has smaller variance than prior estimators is that our estimator uses more information available in the bit sequence.* Furthermore, ART estimation time is observably independent of tag population sizes. In contrast, as tag volume increases, the estimation time of some prior schemes (*e.g.*, FNEB [6]) increases.

For deployability, ART neither requires modification to the tags nor to the communication protocol between tags and readers. ART only needs to be implemented on the reader side as a software module without any hardware modifications. ART also does not demand any unpractical system parameters beyond the C1G2 standard. In contrast, some prior schemes require modification to tags and some demand unrealistic system parameters. For example, the scheme in [14] requires each tag to store thousands of hash functions, which is not practical to implement on passive tags and is not compliant with the C1G2 standard. As another example, the scheme in [6] uses increasingly large frame sizes as population size increases (*e.g.*, the frame size required by the scheme in [6] is greater than half of tag population size), which soon exceeds the maximum limit allowed by the C1G2 Standard.

2. RELATED WORK

The first tag estimation scheme, called Unified Probabilistic Estimator (UPE), was proposed by Kodialam and Nandagopal in 2006 [7]. UPE uses the framed slotted Aloha protocol and makes estimation based on either the number of empty slots or that of collision slots in a frame. Besides this estimator having larger variance than ART, UPE requires the differentiation among empty, single, and collision slots, which takes significantly more time than differentiating between empty and non-empty slots. According to C1G2, a reader requires $300\mu s$ to detect an empty slot, $1500\mu s$ to detect a collision, and $3000\mu s$ to complete a successful read. In [8], Kodialam *et al.* proposed an improved framed slotted Aloha protocol based estimation scheme called Enhanced Zero Based (EZB) estimator, which performs estimation based on the total number of 0s in a frame. While UPE estimates population size in each round and averages the estimated sizes when all rounds are finished, EZB only records the total number of 0s in each frame and at the end of all rounds, EZB first averages the recorded values and then uses it to do estimation.

In [14], Qian *et al.* proposed an estimation scheme called Lottery Frame (LoF). Compared to UPE and EZB, LoF is faster; however, it is impractical to implement as it requires each tag to store a large number (*i.e.*, the number of bits in a tag ID times the number of frames, which can be in the scale of thousands) of unique hash functions. LoF needs to modify both tags and the communication protocol between readers and tags, which makes it non-compliant with C1G2. Han *et al.* proposed a tag estimation scheme called First Non Empty Based (FNEB) estimator, which is based on the size of the first run of 0s in a frame [6]. FNEB is based on an unrealistic assumption that frame size can be arbitrarily large. Li *et al.* proposed an estimation scheme called Maximum Likelihood Estimator (MLE) for active tags with the goal of minimizing power consumption of active tags [10]. In [17], Shah and Wong proposed a multi-reader tag estimation scheme which is based on an unrealistic assumption that any tag covered by multiple readers only replies to one reader.

3. ART — SCHEME OVERVIEW

3.1 Communication Protocol Overview

ART uses the framed slotted Aloha protocol specified in C1G2 as its MAC layer communication protocol. In this protocol, the reader first tells tags the frame size f, which is typically no more than 512 slots for practical reasons [16], and a random seed number R. Later in the paper, we will see how a simple use of seed number R will make it straightforward to extend our estimation scheme to use multiple readers with overlapping regions. Each tag within the transmission range of the reader then uses f, R, and its ID to select a slot in the frame by evaluating a hash function $h(f, R, ID)$ whose result is in $[1, f]$ following a uniform distribution. Each tag has a counter initialized with the slot number it chose to reply. After each slot, the reader first transmits an end of slot signal and then each tag decrements its counter by one. In any given slot, all the tags whose counters are equal to 1 respond to the reader. In essence, each tag picks a random slot from 1 to f following a uniform distribution. If no tag replies in a slot, it is called an *empty slot*; if exactly one tag replies, it is called a *singleton slot*; and if two or more tags reply, it is called a *collision slot*.

3.2 Estimation Scheme Overview

At the end of a frame, the reader obtains a sequence of 0s and 1s by representing an empty slot with 0 and a singleton or collision slot with 1. In this binary sequence, a *run* is a subsequence where all bits in this subsequence are 0s (or 1s) but the bits before and after the subsequence are 1s (or 0s), if they exist. For example, frame 011100 has 3 runs: 0, 111, and 00.

ART uses the average run size of 1s to estimate tag population size. The intuition is that as tag population increases, the average run size of 1s increases (and that of 0s decreases). We illustrate this intuition using the simulation results in Figure 1, which shows that the average run size of 1s increases as tag population size increases from 0 to 160. The markers in this figure are the average of 100 runs. The lines above and below each marker show the standard deviation of the experiments. This figure shows that given a tag population size and a frame size, there is a distinct expected value of the average run size of 1s. The expected value of the average run size of 1s is a monotonic function of the number of tags, which means that a unique inverse of this function exists. Thus, given the observed average run size of 1s, using the inverse function, we can get the estimated value $\tilde{t}$ of tag population size t. Similar to other tag estimation schemes, ART also uses multiple frames obtained from multiple rounds of the framed slotted Aloha protocol to reduce its estimation variance and therefore increase its estimation reliability. Using different seed values for different frames, in each frame, the same tag will choose a different slot to respond.

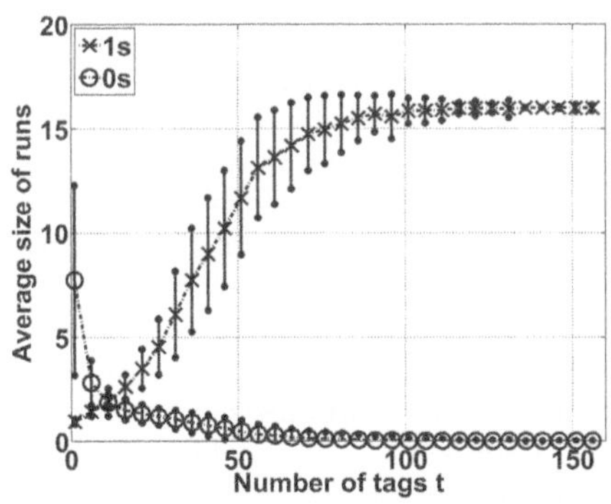

Figure 1: Average run size vs. t ($f = 16$)

To scale to large tag population sizes, ART uses a persistence probability p by which a tag decides whether it should reply to the reader in a given frame. The persistence probability was first introduced in [7]. To avoid making any modification to tags, this probability is implemented by "virtually" extending frame size $1/p$ times, *i.e.*, the reader announces a frame size of f/p but terminates the frame after the first f slots. According to C1G2, the reader can terminate a frame at any point. By adjusting p, ART is able to estimate tag population of arbitrarily large sizes.

3.3 Formal Development: Overview and Assumptions

To formally develop an estimator, we first need to derive the equation for the expected value of average run size of 1s as a function of frame size f, tag population size t, and persistence probability p. We then use the inverse of this function to get the estimated value $\tilde{t}$ from the observed value of the average run size of 1s. To achieve the required reliability and minimized estimation time, we optimize f, p, and the number of rounds n so that the total number of slots $(f+3) \times n$ is minimized while satisfying $P\{|\tilde{t}-t| \leq \beta t\} \geq \alpha$. We add 3 to f because at the start of each frame, the reader waits for about $1ms$ (≈ 3 empty slots) to let the tags get energized [3,16]. We use $\overline{f}$ to denote $f + 3$.

To make the formal development tractable, we assume that instead of picking a single slot to reply at the start of frame of size f, a tag independently decides to reply in each slot of the frame with probability $1/f$ regardless of its decision about previous or forthcoming slots. Vogt first used this assumption for the analysis of framed slotted Aloha protocol for RFID and justified its use by recognizing that this problem belongs to a class of problems known as "occupancy problems", which deals with the allocation of balls to urns [18]. Ever since, the use of this assumption has been a norm in the formal analysis of all Aloha based RFID protocols [2,6–8,10,14,17,18,20].

The implication of this assumption is that when a tag independently chooses a slot to reply, it can end up choosing more than one slots in the same frame or even not choosing any at all, which is not in accordance with C1G2 standard that requires a tag to pick exactly one slot in a frame. Note that even with the independence assumption, the expected number of slots that a tag chooses in a frame is still one. As we draw our estimate from a large number of frames to achieve required reliability, we can expect to observe this expected number. Therefore, the analysis with the assumption of independence is asymptotically the same as that without the independence assumption. Bordenave *et al.* further explained in detail why this independence assumption in analyzing Aloha based protocols provides results just as accurate as if all the analysis was done without this assumption [1]. Note that this independence assumption is made only to make the formal development tractable. The simulations in Section 6 are based on the C1G2 standard where a tag chooses exactly one slot at the start of frame.

4. ART — ESTIMATION ALGORITHM

Next, we first focus on the single-reader version of ART. In Section 5.6, we present a method to extend ART to handle multiple-readers with overlapping regions. Table 1 lists the symbols used in this paper.

4.1 Average Run Based Tag Estimation

For ART, in each round of the Aloha protocol, we calculate the average run size of b. For example, the average run size of 1 in frame 01110011 (which has two runs of 1, *i.e.*,

Table 1: Symbols used in the paper

Symbol	Description
t	actual tag population size
t_m	upper bound on # of tags
t_M	maximum # of tags that can be estimated
$\hat{t}$	estimated # of tags
α	required reliability
β	required confidence interval
f	frame size
f_{op}	optimal frame size
$\overline{f}$	$f+3$ to include delay between two consecutive frames
n	# of rounds (*i.e.*, frames)
n_{op}	optimal # of rounds (*i.e.*, frames)
p	persistence probability
p_{op}	optimal persistence probability
R	random seed from reader
$h(f, R, ID)$	unform hash function in $[1, f]$
b	value of a slot: $b = 0$ or $b = 1$
$\overline{b}$	1-b
q_b	probability that a slot is b
$S_{b,i}$	random variable for run size of b starting at i
$E[.]$	expected value
Var(.)	variance
Cov(.)	covariance
Y_b	random variable for # of b slots in frame
y	element of sample space of Y_b
R_b	random variable for # of runs of b in frame
r	element of sample space of R_b
X_b	random variable for average run size of b in a frame
$\mu\{.\}$	expected value of X_b
$\sigma\{.\}$	standard deviation of X_b
$\xi\{f, y, r\}$	number of ways in which y occurrences of b and $f-y$ occurrences of $\overline{b}$ can be arranged in f slots while ensuring that the number of runs of b are r

111 and 11) is $(3+2)/2 = 2.5$. After n rounds, we obtain n average run sizes of b and then calculate the average of these n values. This final value is then used as the expected value of the average run size of b in a frame to estimate the tag population size.

The probability that a slot in a frame is b, where $b = 0$ or 1, can be calculated using Lemma 1.

LEMMA 1. *Let t be the actual tag population size, f be the frame size, p be the persistence probability (*i.e., *the probability that a tag participates in a frame), and q_b be the probability that a slot in a frame is b. Thus:*

$$q_b = \begin{cases} (1-\frac{p}{f})^t & \text{if } b = 0 \\ 1-(1-\frac{p}{f})^t & \text{if } b = 1 \end{cases} \tag{1}$$

PROOF. The probability that a tag chooses a given slot in a frame is p/f. The probability that it does not choose that slot is $1-\frac{p}{f}$. The probability that none of the tags choose that slot is $(1-\frac{p}{f})^t$, which is the value of q_0. As the tags choose the slots independently, q_b is the same for each slot of the frame. The probability that a slot is chosen by at least one tag is $1-q_0$, which is the value of q_1. □

Let $S_{b,i}$ be the random variable representing the run size of b starting at location i in a frame. If the i-th slot is not the starting point of a run, then $S_{b,i} = 0$. The expectation and variance of $S_{b,i}$ can be calculated using Theorem 1.

THEOREM 1. *Let $S_{b,i}$ be the random variable representing the run size of b starting at location i in a frame of size f. Let $a = f - i + 1$. The expectation and variance of $S_{b,i}$ are:*

$$E[S_{b,i}] = \frac{q_b}{1-q_b}(1-q_b^a) \tag{2}$$

$$\text{Var}(S_{b,i}) = \frac{q_b + (2a+1)(q_b^{a+2} - q_b^{a+1}) - q_b^{2(a+1)}}{(1-q_b)^2} \tag{3}$$

PROOF. To calculate the expected value of $S_{b,i}$, we first need its probability density function $P\{S_{b,i} = s\}$. The probability that a run starting at location i will be of length s is the product of the probability that s consecutive slots are b and $s + i^{th}$ slot is $\overline{b}$, where $\overline{b} = 1 - b$. Thus,

$$P\{S_{b,i} = s\} = \begin{cases} q_b^s(1-q_b) & \text{if } s < a \\ q_b^s & \text{if } s = a \\ 0 & \text{if } s > a \end{cases}$$

where $1 \le i \le f$ and $1 \le s \le a$. To derive the expected value and variance of $S_{b,i}$, we use the moment generating function:

$$\phi(\tau) = E[e^{\tau S_{b,i}}] = \sum_{s=1}^{a} e^{\tau s} P\{S_{b,i} = s\}$$

$$= (q_b e^\tau)^a + (1-q_b)\sum_{s=1}^{a-1}(q_b e^\tau)^s$$

Differentiating both side w.r.t. τ, we get

$$\phi'(\tau) = a(q_b e^\tau)^a + (1-q_b)(q_b e^\tau)\sum_{s=1}^{a-1} s(q_b e^\tau)^{s-1}$$

$$= a(q_b e^\tau)^a + (1-q_b)(q_b e^\tau)\frac{d}{d(q_b e^\tau)}\Big(\frac{1-(q_b e^\tau)^a}{1-q_b e^\tau} - 1\Big)$$

$$= a(q_b e^\tau)^a + \frac{(1-q_b)\Big((a-1)(q_b e^\tau)^{a+1} - a(q_b e^\tau)^a + q_b e^\tau\Big)}{(1-q_b e^\tau)^2}$$

Evaluation of $\phi'(\tau)$ at $\tau = 0$ gives us Equation (2). To find $\text{Var}(S_{b,i})$, we calculate $E[S_{b,i}^2]$ by taking the second derivative of $\phi'(\tau)$ and setting $\tau = 0$. Thus,

$$E[S_{b,i}^2] = \frac{q_b + q_b^2 + (2a-1)q_b^{a+2} - (2a+1)q_b^{a+1}}{(1-q_b)^2}$$

Evaluating $E[S_{b,i}^2] - E^2[S_{b,i}]$ gives Equation (3). □

Let X_b be the random variable representing the average run size of b in a frame. Next, we calculate the expectation and variance of X_b using the expectation and variance of $S_{b,i}$ from Theorem 1. The expectation of X_b will be used to estimate the tag population size and the variance of X_b will be used to calculate optimal values for f, p, and n. Let Y_b be the random variable representing the number of times b occurs in a frame and R_b be the random variable representing the number of runs of b in a frame. By definition, $X_b = \frac{Y_b}{R_b}$ holds for any frame. Next, we first calculate $E[Y_b]$, $\text{Var}(Y_b)$, $E[R_b]$, $\text{Var}(R_b)$, and $\text{Cov}(Y_b, R_b)$ in Lemmas 2 and 3. Then, we use them to calculate $E[X_b]$ and $\text{Var}(X_b)$ in Theorem 2. Using Equation (14) in Theorem 2, replacing $E[X_b]$ by the average run size of b from n frames, we obtain an equation with only one variable t. Finally, we use Brent's method to obtain the numerical solution of this equation. The result is the estimated tag population size. Since ART uses X_b to estimate the tag population size, we call X_b the *estimator* of ART.

LEMMA 2. *Let Y_b be the random variable representing the number of times b occurs in a frame and R_b be the random variable representing the number of runs of b in a frame. Given tag population size t, frame size f, and persistence probability p, we have:*

$$E[Y_b] = fq_b \quad (4)$$
$$\text{Var}(Y_b) = fq_b(1-q_b) \quad (5)$$
$$E[R_b] = q_b\big(q_b + f(1-q_b)\big) \quad (6)$$
$$\text{Var}(R_b) = f(q_b - 4q_b^2 + 6q_b^3 - 3q_b^4) + (3q_b^2 - 8q_b^3 + 5q_b^4) \quad (7)$$

PROOF. The expected size of a run of b starting at location i is $E[S_{b,i}]$. A run of b starting at location i means that the slot $i-1$ is $\bar{b}$. Thus,

$$E[Y_b] = E[S_{b,1}] + \sum_{i=2}^{f} E[S_{b,i}] \times P\{\text{slot } i-1 \text{ is } \bar{b}\}$$
$$= \frac{q_b}{1-q_b}(1-q_b^f) + \sum_{i=2}^{f}\frac{q_b}{1-q_b}(1-q_b^{f-i+1}) \times (1-q_b)$$
$$= \frac{q_b}{1-q_b}(1-q_b^f) + q_b\Big[(f-1) - q_b^{f+1}\Big(\frac{q_b^{-2} - q_b^{-(f+1)}}{1-q_b^{-1}}\Big)\Big]$$
$$= fq_b$$

Each slot i of frame f has probability q_b of being b. So $Y_b \sim \text{Binom}(f, q_b)$. Thus, $\text{Var}(Y_b)$ is simply the variance of a binomial distribution with parameters f and q_b:

$$\text{Var}(Y_b) = fq_b(1-q_b)$$

Let $\gamma_1, \gamma_2, \ldots, \gamma_f$ represent the sequence of binary random variables representing the value of each slot in a frame of size f. Since each tag randomly and independently picks a slot in the frame, all γ_i are identically distributed. Furthermore, $P\{\gamma_i = b\} = q_b$. Let I_i be the indicator random variable whose value is 1 if a run of b begins at γ_i.

$$I_i = \begin{cases} 1 & \text{if } (\gamma_i = b, i = 1) \lor (\gamma_i = b \land \gamma_{i-1} = \bar{b}, i > 1) \\ 0 & \text{otherwise} \end{cases}$$

Thus,

$$R_b = \sum_{i=1}^{f} I_i$$

Because

$$E[I_i] = \begin{cases} P\{\gamma_i = b\} = q_b & \text{if } i = 1 \\ P\{\gamma_{i-1} = \bar{b}, \gamma_i = b\} = q_b(1-q_b) & \text{if } i > 1 \end{cases}$$

we get

$$E[R_b] = \sum_{i=1}^{f} E[I_i] = q_b + \sum_{i=2}^{f} q_b(1-q_b) = q_b\big(q_b + f(1-q_b)\big)$$

As R_b is the sum of f identically distributed random variables, we use the general expression for the variance of the sum of identically distributed random variables to obtain the variance of R_b.

$$\text{Var}(R_b) = \text{Var}(\sum_{i=1}^{f} I_i)$$
$$= \sum_{i=1}^{f}\text{Var}(I_i) + 2\sum_{j=2}^{f}\sum_{\forall i<j}\text{Cov}(I_i, I_j)$$

Here we used the fact that the frame size is always greater than 1 during the estimation process whenever the information about runs is used. As $I_i \sim \text{Bernoulli}(q_b)$, its variance is that of a bernoulli random variable given by

$$\text{Var}(I_i) = E[I_i](1 - E[I_i]) \quad (8)$$

Note that I_i and I_j are dependent on each other if and only if (iff) $i = j-1$ because I_{j-1} and I_j can not both be 1 in the same frame. Other than that, $\forall i < j-1$, I_i and I_j are independent. Thus,

$$\text{Cov}(I_i, I_j) = \begin{cases} 0 & \text{if } i < j-1 \\ -E[I_i]E[I_j] = -E[I_i]q_b(1-q_b) & \\ & \text{if } i = j-1 \end{cases}$$

Hence we have:

$$\text{Var}(R_b) = \text{Var}(I_1) + \sum_{j=2}^{f}\text{Var}(I_j) + 2\text{Cov}(I_1, I_2)$$
$$+2\sum_{j=3}^{f}\text{Cov}(I_{j-1}, I_j)$$
$$= q_b(1-q_b) + (f-1)q_b(1-q_b)\{1 - q_b(1-q_b)\}$$
$$-2q_b^2(1-q_b) - 2(f-2)q_b^2(1-q_b)^2$$
$$= f(q_b - 4q_b^2 + 6q_b^3 - 3q_b^4) + (3q_b^2 - 8q_b^3 + 5q_b^4) \quad \square$$

LEMMA 3. *Given tag population size t, frame size f, and persistence probability p, we have:*

$$\text{Cov}(Y_b, R_b) = \sum_{y=0}^{f}\sum_{r=0}^{\lceil \frac{f}{2} \rceil} yrq_b^y(1-q_b)^{f-y}.\xi\{f, y, r\}$$
$$-E[Y_b]E[R_b] \quad (9)$$

where

$$\xi\{f,y,r\} = \begin{cases} \binom{y-1}{r-1}\Big[\binom{f-y-1}{r-2} + 2\binom{f-y-1}{r-1} + \binom{f-y-1}{r}\Big] \\ \text{if } r > 1 \land 0 < y < f \land r \le y \land r \le f-y-1 \\ \\ \binom{y-1}{r-1}\Big[2\binom{f-y-1}{r-1} + \binom{f-y-1}{r}\Big] \\ \text{if } r = 1 \land 0 < y < f \land r \le y \land r \le f-y-1 \\ \\ 1 \quad \text{if } r = 1 \land y = f \\ \\ 1 \quad \text{if } r = 0 \land y = 0 \\ \\ 0 \quad \text{otherwise} \end{cases}$$

PROOF. By definition, we have

$$\text{Cov}(Y_b, R_b) = \sum_{y=0}^{f}\sum_{r=0}^{f} yrP\{Y_b = y, R_b = r\} - E[Y_b]E[R_b] \quad (10)$$

Here $P\{Y_b = y, R_b = r\}$ represents the probability that exactly y out of f slots in the frame are b and at the same time the number of runs of b is r. This probability is difficult to evaluate directly, but conditioning on Y_b simplifies the task.

$$P\{Y_b = y, R_b = r\} = P\{R_b = r | Y_b = y\} \times P\{Y_b = y\} \quad (11)$$

As $Y_b \sim \text{Binom}(f, q_b)$, we have:

$$P\{Y_b = y\} = \binom{f}{y} q_b^y(1-q_b)^{f-y} \quad (12)$$

Now we calculate $P\{R_b = r|Y_b = y\}$ *i.e.*, the probability of having r runs of b in a frame of size f given that y out of f slots are b. As tags choose the slots independently, each occurrence with r runs having y slots of b is equally likely. Therefore, we determine the total number of ways, denoted by $\xi\{f, y, r\}$, in which y occurrences of b and $f - y$ occurrences of $\bar{b}$ can be arranged such that the number of runs of b is r. We treat this as an ordered partition problem. First, we separate all the y occurrences of b from the frame and make r partitions of these y occurrences. Then, we create appropriate number of partitions of $f - y$ occurrences of $\bar{b}$ such that between consecutive partitions of b, the partitions of $\bar{b}$ can be *interleaved*. For r partitions of b, there are 4 possible partitions of $\bar{b}$.

1. The frame starts with b and ends with b, implying that there are $r-1$ partitions of $\bar{b}$, each interleaved between adjacent partitions of b.

2. The frame starts with b and ends with $\bar{b}$, implying that there are r partitions of $\bar{b}$.

3. The frame starts with $\bar{b}$ and ends with b, implying that there are r partitions of $\bar{b}$.

4. The frame starts with $\bar{b}$ and ends with $\bar{b}$, implying that there are $r + 1$ partitions of $\bar{b}$.

We can make r partitions of y occurrences of b in $\binom{y-1}{r-1}$ ways and r partitions of $f - y$ occurrences of $\bar{b}$ in $\binom{f-y-1}{r-1}$ ways. Similarly, we can make $r+1$ partitions of $f - y$ occurrences of $\bar{b}$ in $\binom{f-y-1}{r}$ ways and $r - 1$ partitions of of $f - y$ occurrences of $\bar{b}$ in $\binom{f-y-1}{r-2}$ ways. The equation of $\xi\{f, y, r\}$ in the lemma statement follows from this discussion. The total number of ways in which y zeros can be arranged among f slots is $\binom{f}{y}$. Thus, we get

$$P\{R_b = r|Y_b = y\} = \frac{\xi\{f, y, r\}}{\binom{f}{y}} \tag{13}$$

Substituting values from Equations (12) and (13) in (11) and (10) results in Equation (9). □

THEOREM 2. *Given tag population size t, frame size f, and persistence probability p, we have:*

$$E[X_b] = \frac{E[Y_b]}{E[R_b]} - \frac{\text{Cov}(Y_b, R_b)}{E^2[R_b]} + \frac{E[Y_b]}{E^3[R_b]}\text{Var}(R_b) \tag{14}$$

$$\text{Var}(X_b) = \frac{\text{Var}(Y_b)}{E^2[R_b]} - \frac{2E[Y_b]}{E^3[R_b]}\text{Cov}(Y_b, R_b) + \frac{E^2[Y_b]}{E^4[R_b]}\text{Var}(R_b) \tag{15}$$

PROOF. Let $g(Y_b, R_b) = X_b = \frac{Y_b}{R_b}$. The Taylor series expansion of g around (θ_1, θ_2) is given by:

$$g(Y_b, R_b) = \sum_{j=0}^{\infty}\Big\{\frac{1}{j!}\Big[(Y_b - \theta_1)\frac{\partial}{\partial Y_b'} + (R_b - \theta_2)\frac{\partial}{\partial R_b'}\Big]^j \times g(Y_b', R_b')\Big\}_{\substack{Y_b'=a_1 \\ R_b'=a_2}}$$

According to Bienaymé-Chebyshev inequality, we have $\theta_1 = E[Y_b]$ and $\theta_2 = E[R_b]$. Therefore, we get the following expansion of the Taylor series of $g(Y_b, R_b)$:

$$g(Y_b, R_b) = g(\theta_1, \theta_2) + \Big[(Y_b - \theta_1)\frac{\partial g}{\partial Y_b} + (R_b - \theta_2)\frac{\partial g}{\partial R_b}\Big] + \frac{1}{2}\Big[(Y_b - \theta_1)^2\frac{\partial^2 g}{\partial Y_b^2} + 2(Y_b - \theta_1)(R_b - \theta_2)\frac{\partial^2 g}{\partial Y_b \partial R_b} + (R_b - \theta_2)^2\frac{\partial^2 g}{\partial R_b^2}\Big] + O(j^{-1})$$

Taking the expectation of both sides, we get

$$E[g(Y_b, R_b)] \approx \frac{1}{2}\Big[\text{Var}(Y_b)\frac{\partial^2 g}{\partial Y_b^2} + 2\text{Cov}(Y_b, R_b)\frac{\partial^2 g}{\partial Y_b \partial R_b} + \text{Var}(R_b)\frac{\partial^2 g}{\partial R_b^2}\Big] + g(\theta_1, \theta_2) \tag{16}$$

Evaluating the partial derivatives of g as required in Equation (16), we get

$$\frac{\partial^2 g(Y_b, R_b)}{\partial Y_b^2}\Big|_{\substack{Y_b=\theta_1 \\ R_b=\theta_2}} = 0$$

$$\frac{\partial^2 g(Y_b, R_b)}{\partial Y_b \partial R_b}\Big|_{\substack{Y_b=\theta_1 \\ R_b=\theta_2}} = -\frac{1}{\theta_2^2}$$

$$\frac{\partial^2 g(Y_b, R_b)}{\partial R_b^2}\Big|_{\substack{Y_b=\theta_1 \\ R_b=\theta_2}} = 2\frac{\theta_1}{\theta_1^3}$$

Putting these values in Equation (16) and using $\theta_1 = E[Y_b]$ and $\theta_2 = E[R_b]$, we get Equation (14).

The variance can be calculated as follows:

$$\text{Var}(g(Y_b, R_b)) = E\big[\{g(Y_b, R_b) - E[g(Y_b, R_b)]\}^2\big] \tag{17}$$

Considering that $E[g(Y_b, R_b)]$ is being squared in the expression above, we use first order Taylor series expansion to get the value of $E[g(Y_b, R_b)]$ and substitute it in Equation (17).

$$E[g(Y_b, R_b)] = E\Big[(Y_b - \theta_1)\frac{\partial g}{\partial Y_b} + (R_b - \theta_2)\frac{\partial g}{\partial R_b}\Big] + g(\theta_1, \theta_2) + O(j^{-1})$$

$$= \Big[(0)\frac{\partial g}{\partial Y_b} + (0)\frac{\partial g}{\partial R_b}\Big] + g(\theta_1, \theta_2) + O(j^{-1}) \approx g(\theta_1, \theta_2)$$

Substituting the value of $E[g(Y_b, R_b)]$ and using the first order Taylor series expansion of $g(Y_b, R_b)$ in (17), we get

$$\text{Var}(g(Y_b, R_b)) = E\Big[\{(Y_b - \theta_1)\frac{\partial g}{\partial Y_b} + (R_b - \theta_2)\frac{\partial g}{\partial R_b}\}^2\Big] + O(j^{-1})$$

$$\approx \text{Var}(Y_b)\Big(\frac{\partial g}{\partial Y_b}\Big)^2 + 2\text{Cov}(Y_b, R_b)\frac{\partial g}{\partial Y_b}\frac{\partial g}{\partial R_b} + \text{Var}(R_b)\Big(\frac{\partial g}{\partial R_b}\Big)^2 \tag{18}$$

Evaluating the partial derivatives of g as required in the equation above, we get

$$\frac{\partial g(Y_b, R_b)}{\partial Y_b}\Big|_{\substack{Y_b=\theta_1 \\ R_b=\theta_2}} = \frac{1}{\theta_2}$$

$$\frac{\partial g(Y_b, R_b)}{\partial R_b}\Big|_{\substack{Y_b=\theta_1 \\ R_b=\theta_2}} = -\frac{\theta_1}{\theta_2^2}$$

Putting these values in Equation (18) and using $\theta_1 = E[Y_b]$ and $\theta_2 = E[R_b]$, we get Equation (15). □

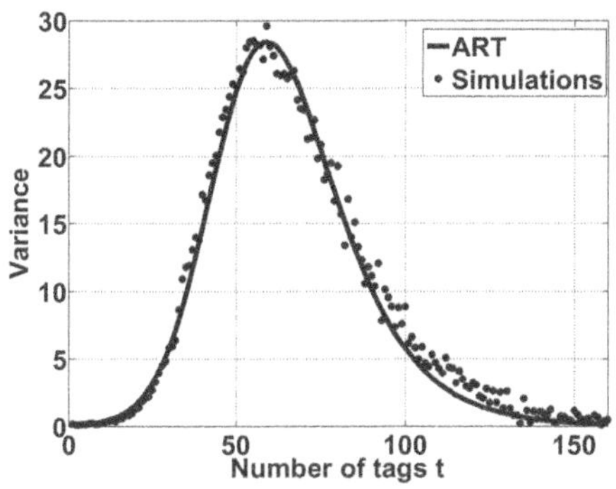

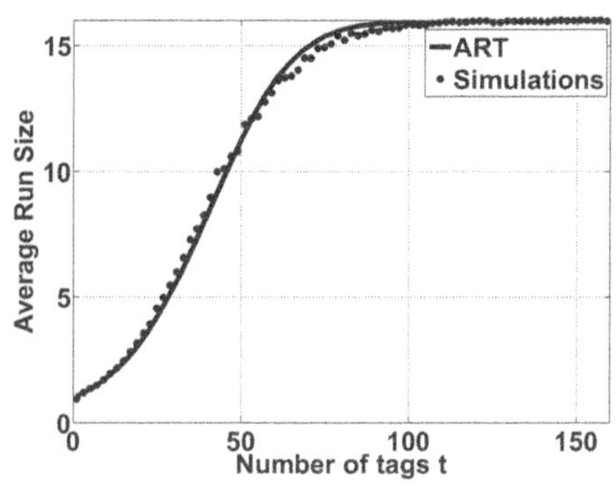

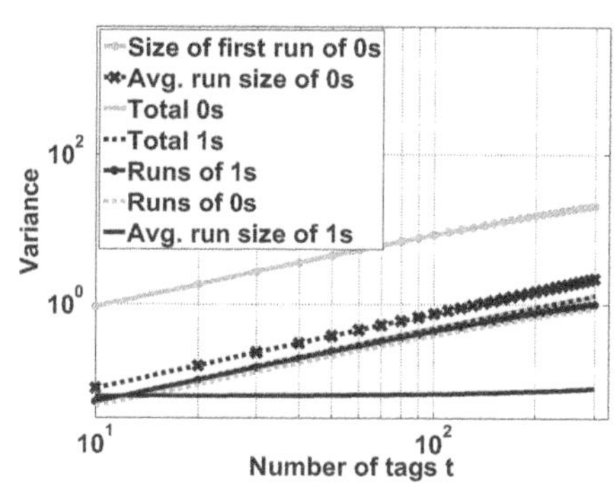

Figure 2: Variances of ART estimator

Figure 3: Expectation of ART estimator

Figure 4: Variance of different estimators vs. t

Figure 2 contains a line plotting the variance of X_1 using Equation (15), where $f = 16$ and $p = 1$. This figure also contains many dots where each dot represents the variance of X_1 that we obtained through 100 times of simulation for each tag population size. This figure shows that for the variance of X_1, simulation results match with the variance calculated from Equation (15). Figure 3 contains a line plotting the expectation of X_1 using Equation (14), where $f = 16$ and $p = 1$, and many dots where each dot represents the average value of X_1 that we obtained through 100 times of simulation for each tag population. This figure shows that the expected value calculated from Equation (14) tracks the simulation results very well.

4.2 Analytical Comparison of Estimators

Although both X_0 and X_1 can be used to estimate the tag population size, we choose X_1 for ART because the tag population size estimation calculated from X_1 has smaller variance compared to X_0 as we show below. It is worth noting that X_0 and X_1 are not equivalent estimators. The average run size of 0s cannot be inferred from the average run size of 1s, and vice versa. For example, 1100011 and 1100110 have the same average run size of 1s, but they have different average run size of 0s. Fundamentally, X_0 and X_1 are not equivalent estimators because for any slot, the probability of it being 0 (which means no tag chooses this slot) and that of it being 1 (which means one or more tags choose this slot) are different.

Next, we show that the ART estimator, namely the average run size of 1s, has less variance than many other framed slotted Aloha based estimators, namely (1) the size of the first run of 0s (used by FNEB [6]), (2) the average run size of 0s, (3) the total number of 0s (used by UPE [7] and EZB [8]), (4) the total number of 1s, (5) the total number of runs of 0s, and (6) the total number of runs of 1s. The higher the variance of an estimator is, the more number of rounds n are needed to improve reliability, and more rounds means more estimation time. Figure 4 shows the analytical plots of the variances of the ART estimator and the above 6 other estimators with frame size $f = 16$ versus tag population sizes. This figure shows that *the variance of ART estimator is significantly lower than all other estimators.* Runs of 1s and runs of 0s have smaller variance compared to ART for very small tag population sizes. This observation, however, is insignificant because both these quantities are non-monotonic functions of tag population size and therefore, cannot be used alone for estimation. The variances of these estimators are calculated as follows. The variance of the total number of 0s and 1s can be calculated using Equation (5). The variance of the size of the first run can be calculated using Equation (3) by setting $i = 1$. The variance of the number of runs of 0s and that of 1s can be calculated using Equation (7). We emphasize that plots in Figure 4 are not based on experimental results, instead, they are based on analytical formulas.

4.3 Unbounded Tag Population Size

For fixed values of required reliability α, frame size f, and persistence probability p, Theorem 3 calculates the upper bound t_M on the number of tags that ART can estimate; that is, for tag population sizes larger than t_M, all the slots in each frame are expected to be 1 and thus ART cannot make an estimate because the tag population size can be infinitely large.

THEOREM 3. *For given required reliability $\alpha \in [0, 1)$, frame size $f > 1$, persistence probability $p > 0$, the maximum number of tags t_M that ART can estimate is:*

$$t_M = \frac{\log\{1 - (1-\alpha)^{\frac{1}{f}}\}}{\log\{1 - \frac{p}{f}\}}$$

PROOF. ART fails if and only if all the slots are 1. To achieve the required reliability α, the failure probability of ART $P\{Y_1 = f\} = q_1^f$ has to be less than $1 - \alpha$. Using the value of q_1 calculated in Lemma 1, we have

$$P\{Y_1 = f\} = q_1^f = \left[1 - \left(1 - \frac{p}{f}\right)^t\right]^f < 1 - \alpha$$

Thus, by taking the log, we have

$$\log\left\{1 - (1-\alpha)^{\frac{1}{f}}\right\} < t \log\left\{1 - \frac{p}{f}\right\}$$

As $\log\left\{1 - \frac{p}{f}\right\} < 0$, dividing both sides by $\log\left\{1 - \frac{p}{f}\right\}$ changes the direction of inequality and results in:

$$t < \frac{\log\{1 - (1-\alpha)^{\frac{1}{f}}\}}{\log\{1 - \frac{p}{f}\}} = t_M \quad \square$$

From this theorem, we observe that as $p/f \to 0$ we have $t_M \to \infty$. Figure 5 shows the plot of t_M for increasing values of p and 4 different values of f with required reliability $\alpha = 99\%$. In theory, for any given required reliability α, we can increase the estimation scope of ART to any value by decreasing the value of p/f. In practice, however, p/f has

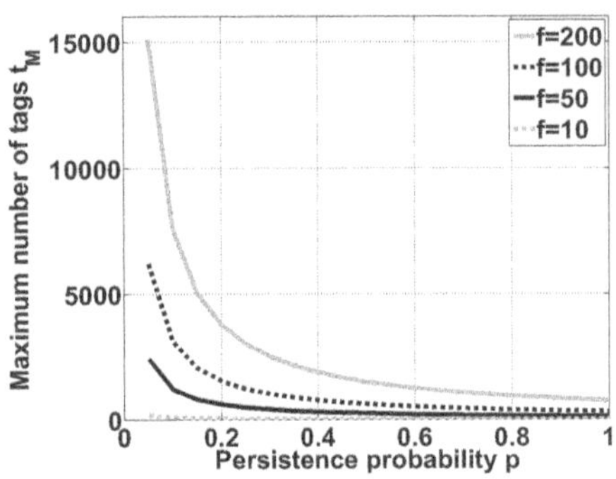

Figure 5: t_M vs. p and f

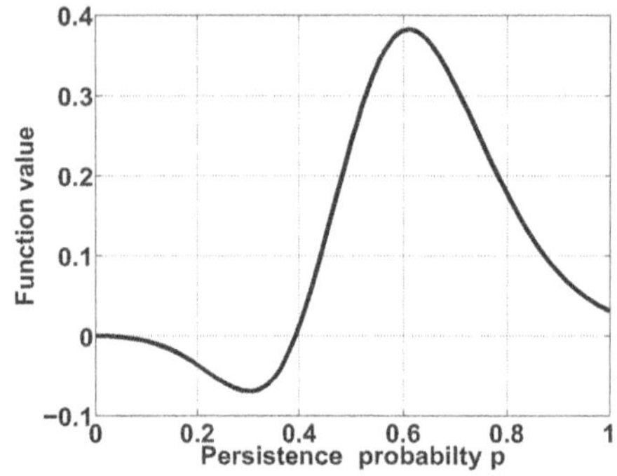

Figure 6: Eq. (19) as func. of p

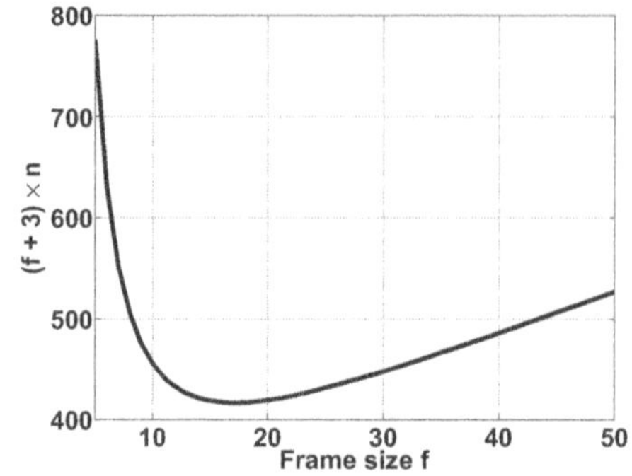

Figure 7: $(f+3)\times n$ vs. f

a minimum value of $1/2^{32}$ due to the 32-bit limitation of the hash functions used by passive tags as specified in the C1G2 standard. Recall that in ART, the reader announces a virtual frame size of f/p (although terminates the frame after the first f slots) and each tag uses the result of a hash function h to select a slot in the range $[1, f/p]$. The number of bits to encode the hash function result is specified to be 32 in the C1G2 standard. Thus, the maximum value of f/p is 2^{32}. Therefore, the maximum number of tags that ART can estimate based on the C1G2 standard is:

$$t_{M_{C1G2}} = \frac{\log\{1-(1-\alpha)^{\frac{1}{f}}\}}{\log\{1-\frac{1}{2^{32}}\}}$$

Here $t_{M_{C1G2}}$ is large enough for all practical applications. For example, with $f = 512$ and $\alpha = 90\%$, $t_{M_{C1G2}}$ is 2.3221e+10, which is over 23 billion; with $f = 512$ and $\alpha = 99\%$, $t_{M_{C1G2}}$ is 2.0254e+10, which is over 20 billion.

5. ART — PARAMETER OPTIMIZATION

To minimize estimation time while achieving required reliability, next, we optimize the three ART parameters: frame size f, persistence probability p, and number of rounds n.

5.1 Optimizing Persistence Probability p

Theorem 4 gives the condition that p needs to satisfy so that the actual reliability is equal to its lower bound, which is the required reliability α. We first prove this theorem and then show how to calculate the optimal value for p.

THEOREM 4. *Given confidence interval β, tag population size t, and frame size f, denoting $E[X_1]$ by $\mu\{t\}$, the optimal persistence probability p_{op} satisfies the following equation.*

$$2\mu\{t\} - \mu\{(1-\beta)t\} - \mu\{(1+\beta)t\} = 0 \qquad (19)$$

PROOF. To find the optimal value p_{op} for p, we first find the conditions that p_{op} needs to satisfy so that the actual reliability $P\{|\tilde{t}-t| \le \beta t\}$ is equal to its lower bound, which is the required reliability α.

$$P\{(1-\beta)t \le \tilde{t} \le (1+\beta)t\} = \alpha \qquad (20)$$

Denoting the observed average value of X_1 from the n frames by $\tilde{X_1}$, $E[X_1]$ by $\mu\{t\}$, and $\text{Var}(X_1)$ by $\sigma^2\{t\}$, we have $\tilde{t} = \mu^{-1}\{\tilde{X_1}\}$. Using $\mu^{-1}\{\tilde{X_1}\}$ to substitute $\tilde{t}$ in (20), we get

$$P\left\{(1-\beta)t \le \mu^{-1}\{\tilde{X_1}\} \le (1+\beta)t\right\}$$
$$= P\left\{\mu\{(1-\beta)t\} \le \tilde{X_1} \le \mu\{(1+\beta)t\}\right\} = \alpha \qquad (21)$$

Based on the fact that the variance of a random variable is reduced by n times if the same experiment is repeated n times, by running n rounds and getting n frames, the variance of X_1 becomes $\frac{\sigma^2\{t\}}{n}$ and the standard deviation of X_1 becomes $\frac{\sigma\{t\}}{\sqrt{n}}$. Let Z denote $\frac{\tilde{X_1}-\mu\{t\}}{\sigma\{t\}/\sqrt{n}}$. Thus, (21) becomes

$$P\left\{\frac{\mu\{(1-\beta)t\}-\mu\{t\}}{\frac{\sigma\{t\}}{\sqrt{n}}} \le Z \le \frac{\mu\{(1+\beta)t\}-\mu\{t\}}{\frac{\sigma\{t\}}{\sqrt{n}}}\right\} = \alpha \qquad (22)$$

By the central limit theorem, Z approximates a standard normal random variable. The area under the standard normal curve gives the success probability, which is the required reliability in our context. As our confidence interval requirement is symmetric on both the upper and lower sides of the population size, we can represent the required reliability α in terms of a constant k as follows:

$$P\{-k \le Z \le k\} = \alpha \qquad (23)$$

From Equations (22) and (23), we get

$$\frac{\mu\{(1-\beta)t\}-\mu\{t\}}{\frac{\sigma\{t\}}{\sqrt{n}}} = -k, \qquad \frac{\mu\{(1+\beta)t\}-\mu\{t\}}{\frac{\sigma\{t\}}{\sqrt{n}}} = k \qquad (24)$$

As the absolute values of the right hand size (R.H.S.) of both equations above are k, we get

$$2\mu\{t\} - \mu\{(1-\beta)t\} - \mu\{(1+\beta)t\} = 0 \quad \square \qquad (25)$$

Next, we show how to calculate the optimal value for p using Theorem 4, which shows that p_{op} only depends on confidence interval β, tag population size t, and frame size f. For now, we first assume that we know an upper bound on the tag population size denoted by t_m. Later we give a method to obtain t_m automatically. Second, we assume that we know the optimal frame size f_{op}. Later we give a method to calculate f_{op}. Replacing t by t_m and f by f_{op}, left hand side (L.H.S) of Equation (19) in Theorem 4 becomes a well behaved function of p as shown in Figure 6. The numerical solution of this equation gives the optimal value p_{op}.

5.2 Minimizing Number of Rounds n

Using optimal persistence probability p_{op}, the two equations in (24) hold. From them, we get

$$\left(\frac{k\sigma\{t\}}{\mu\{(1+\beta)t\}-\mu\{t\}}\right)^2 = n = \left(\frac{-k\sigma\{t\}}{\mu\{(1-\beta)t\}-\mu\{t\}}\right)^2 \quad (26)$$

Let Φ be the cumulative distribution function of a standard normal distribution and erf{.} be the standard error function, we get

$$P\{-k \leq Z \leq k\} = \Phi(k) - \Phi(-k) = \operatorname{erf}\left\{\frac{k}{\sqrt{2}}\right\} \quad (27)$$

From Equations (23) and (27), we get

$$k = \sqrt{2}\operatorname{erf}^{-1}\{\alpha\} \quad (28)$$

From Equations (26) and (28), we get

$$\left(\frac{\sqrt{2}\operatorname{erf}^{-1}\{\alpha\}\times\sigma\{t\}}{\mu\{(1+\beta)t\}-\mu\{t\}}\right)^2 = n = \left(\frac{-\sqrt{2}\operatorname{erf}^{-1}\{\alpha\}\times\sigma\{t\}}{\mu\{(1-\beta)t\}-\mu\{t\}}\right)^2 \quad (29)$$

Based on this equation, with β, α, $p = p_{op}$, $t = t_m$, and $f = f_{op}$, we can calculate the minimal value n_{op} for n.

5.3 Optimizing Frame Size f

Our goal of optimizing ART parameters, namely p, n, and f, is to minimize the total estimation time, which is $(f+3)\times n_{op}$. Because β, α, and t_m are known and p_{op} is a function of f, n_{op} is essentially a function of f. Thus, $(f+3)\times n_{op}$ is a function of f. Next, we show how to find the optimal frame size f_{op} so that $(f+3)\times n_{op}$ is minimized.

Function $(f+3)\times n_{op}$ is a convex function of f as seen from Figure 7. This means that an optimal frame size f_{op} exists and can be obtained by differentiating $(f+3)\times n_{op}$ with respect to f as shown in the following equation:

$$\frac{d}{df}\{(f+3)\times n_{op}\} = 0 \quad (30)$$

As both expressions for n given in Equation (29) have the same value when $p = p_{op}$, either of them can be used to calculate the value of n_{op}. By substituting n_{op} in Equation (30) by the L.H.S of the n_{op} expression in Equation (29), we get

$$\Big[\mu\{(1+\beta)t_m\} - \mu\{t_m\}\Big]\Big[\sigma\{t_m\} + 2\bar{f}\frac{\partial\sigma\{t_m\}}{\partial f}\Big] \\ -2\bar{f}\sigma\{t_m\}\Big[\frac{\partial\mu\{(1+\beta)t_m\}}{\partial f} - \frac{\partial\mu\{t_m\}}{\partial f}\Big] = 0 \quad (31)$$

where $\frac{\partial\mu\{.\}}{\partial f}$ and $\frac{\partial\sigma\{.\}}{\partial f}$ are obtained through the differentiation of expressions for $E[X_b]$ and $\mathrm{Var}(X_b)$ in Equations (14) and (15), respectively.

To obtain of f_{op} from Equation (31), we need the value of p_{op}, while to obtain the value of p_{op} from Equation (19), we need the value of f_{op}. Therefore, we have two simultaneous equations, (19) and (31), with two unknowns, f_{op} and p_{op}. These equations can be numerically solved simultaneously using Levenberg-Marquardt Algorithm to obtain the values of f_{op} and p_{op}.

Note that the estimation scheme will still work if we do not use $f = f_{op}$, but in that case the value of n will be such that the $(f+3)\times n$ will not be minimum. If we make f arbitrarily small, theoretically, it will still be possible to obtain the estimate accurately; however, in this case, n can be prohibitively large making it practically impossible to obtain the estimate as per the accuracy requirements.

5.4 Observably Constant Estimation Time

There are three inputs to ART: confidence interval β, required reliability α, and a population of t tags where t is unknown. The total estimation time of ART is $(f_{op}+3)\times n_{op}$, which from our experiments we observe to be dependent only on β and α, *i.e.*, $(f_{op}+3)\times n_{op}$ is constant with respect to (w.r.t.) t. Because $(f_{op}+3)\times n_{op}$ is highly complex due to complex components such as $\mathrm{Cov}(Y_b, R_b)$ as expressed in Equation (9), we have not yet formally proven that $(f_{op}+3)\times n_{op}$ is independent of t, although we hypothesize that this is mathematically true. Intuitively, the larger t is, the smaller p_{op} is. Although t plays an important role in computing p_{op}, n_{op}, and f_{op} individually, in formula $(f_{op}+3)\times n_{op}$ the impact of t gets canceled out. From Equations (29) and (31), we observe that the value of n_{op} depends upon α, β, μ, σ and value of f_{op} depends upon β, μ, σ. Here $(f_{op}+3)\times n_{op}$ is independent of t because α and β are given values and μ and σ are functions of q_1, if q_1 is independent of t. Indeed q_1 is intuitively independent of t because $q_1 \propto 1/t$ and p_{op} obtained from Equation (19) decreases as the value of t_m increases, as shown in Figure 8. A constant value of q_1 means that the probability of any slot in a frame being non-empty is the same, which in turn implies that average run size in a frame will also be the same regardless of t. Figure 9 plots the value of q_1 w.r.t. t using Equations (19) and (1). We observe that the value of q_1 stays perfectly constant w.r.t. t. Figure 10 shows that the total estimation time of ART stays constant w.r.t. t for given β and α.

5.5 Obtaining Population Upper Bound t_m

So far we have assumed the knowledge of an upper bound t_m on tag population size t. In some applications, t_m is readily available. For example, the number of TVs in a TV warehouse can be reasonably estimated by the warehouse size and minimum TV size. However, this may not be available for some applications where the tag population size changes in large magnitude. We next present a fast scheme to obtain t_m based on Flajolet and Martin's probabilistic counting algorithm used in databases [5]. Before running ART, the reader uses this scheme to obtain t_m. In this scheme, the reader keeps issuing single-slot frames, where the persistence probability p follows a geometric distribution starting from $p = 1$ (*i.e.*, $p = \frac{1}{2^{i-1}}$ in the i-th frame), until the reader gets an empty slot. Suppose the empty slot occurred in the i-th frame, then $t_m = 1.2897 \times 2^{i-2}$ is an upper bound on t [5, 14].

5.6 ART with Multiple Readers

We next discuss how to obtain t_m and $\tilde{t}$ using multiple readers whose covered regions may overlap. To obtain t_m using multiple readers, we can let each reader obtain the t_m value on its own and then sum them up as the final overall t_m because of two reasons. First, our requirement on t_m is only a rough upper bound estimate. Second, deployment of multiple readers in practice often requires site surveys to ensure minimal overlapping between readers. To use multiple readers to obtain $\tilde{t}$ more precisely, we adapt the approach

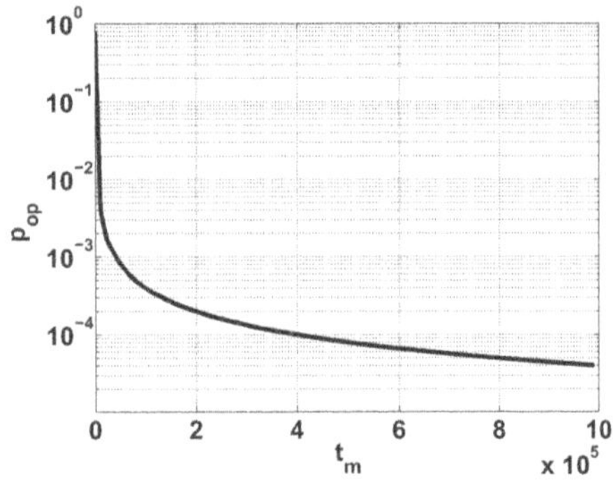

Figure 8: p_{op} **vs.** t_m

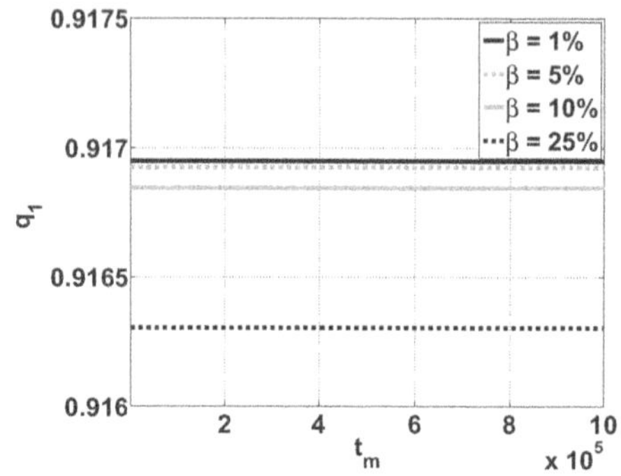

Figure 9: q_1 **vs.** t_m

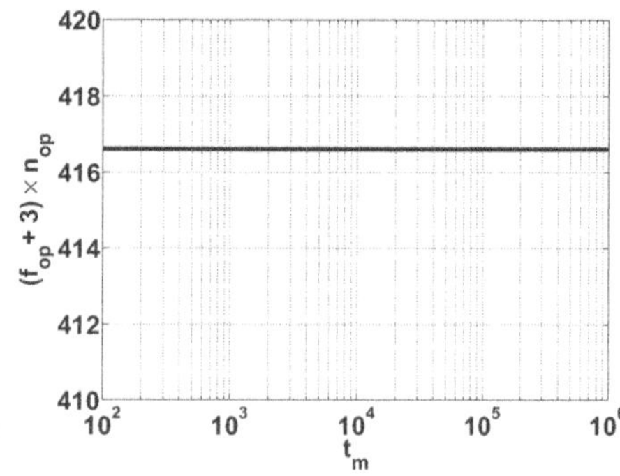

Figure 10: $(f_{op}{+}3){\times}n_{op}$**vs.**$t_m$

proposed by Kodialam *et al.* in [8], which uses a central controller for all readers. ART parameters β, α, t_m, f_{op}, p_{op}, and n_{op} have the same value across all readers. When a reader transmits seed R_i in its i-th frame, it does not generate R_i on its own, rather it uses the i-th seed R_i issued by the central controller. That is, each reader generates the same sequence of n_{op} seeds. In the i-th frames from different readers, because all readers use the same seed R_i, the slot number that a given tag chooses is the same (*i.e.*, $h(f, R_i, ID)$) in the frame of each reader covering this tag. Once a reader has completed its frame, it sends the frame to the central controller. The controller applies the logical OR on all the i-th frames from all readers, and gets the same i-th frame as if using a single reader. ART uses the n_{op} frames computed by logical OR to estimate the population size.

5.7 ART - Pseudocode

Algorithm 1 shows the pseudocode of ART.

Algorithm 1: EstimateRFIDTagPopulation(α, β, n_r)

Input: (1) Required reliability α
(2) Required confidence interval β
(3) Number of readers n_r
Output: Estimated tag population size $\tilde{t}$

$t_m :=$ **GetMaximumTags**(n_r)
Solve (19) and (31) to get f_{op} and p_{op}.
Evaluate k using (28).
Evaluate n_{op} by using $\alpha, \beta, k, f_{op}, p_{op}$, and t_m in (29).
for $i := 1$ *to* n_{op} **do**
    Provide all readers with f_{op}/p_{op} and a random seed R_i.
    Run Aloha on readers and gather all readers' frames.
    Perform slot wise OR on all frames to obtain one frame.
    Obtain $\tilde{X}_1(i)$, the average run size of 1s in this frame.
$\tilde{X}_{avg} \leftarrow \sum_{i=1}^{n_{op}} \tilde{X}_1(i)/n_{op}$
Use $E[X_1] := \tilde{X}_{avg}$ and solve (14) to obtain an estimate $\tilde{t}$ of t.
return $\tilde{t}$
GetMaximumTags(n_r)
$f := 1$
for $j := 1$ *to* n_r **do**
    $i := 1$
    $p_i := 1$
    repeat
        Provide reader j with f/p_i and a random seed R_i.
        Run Aloha on reader j and get the response.
        if *slot is not empty* **then**
            $p_{i+1} := p_i/2$
            $i := i + 1$
    until *slot is empty*
    $t_{m,j} := 1.2897 \times 2^{i-2}$
$t_m := \sum_{j=1}^{n_r} t_{m,j}$
return t_m

6. PERFORMANCE EVALUATION

We numerically evaluated in Matlab our ART scheme as well as four prior RFID estimation schemes: UPE [7], EZB [8], FNEB [6], and MLE [10]. We did not evaluate the other estimation scheme LoF [14] because it is non-compliant with C1G2. In terms of implementation complexity, the number of lines of code required to implement ART were almost the same as all four prior protocols. To ensure compliance with the C1G2 standard, in all our simulations, each tag picks up exactly one slot at the start of frame as soon as the reader broadcasts the frame size.

Next, we first conduct a side-by-side comparison on estimation time between ART and the four prior schemes. Then, we conduct experiments to show that ART indeed achieves the required reliability.

6.1 Estimation Time

The results in Figures 11, 12, and 13 show that *the estimation time of ART is significantly smaller than all prior schemes*. Note that in Figures 12 and 13, the plots for FNEB are out of the range of the vertical axes, and the plots of UPE and EZB are almost overlapping.

We make three main observations from Figures 11 (a), (b), and (c), which show the estimation time needed by each scheme with population sizes of up to one million tags for different configurations of α and confidence interval β. First, we observe that ART is faster than all four prior schemes in all these configurations. For $\alpha = 99.9\%$ and $\beta = 0.1\%$, ART is 7 times faster than the fastest prior estimation schemes, which are UPE [7] and EZB [8]. For $\alpha = 99\%$ and $\beta = 1\%$, ART is 1.96 times faster than UPE and EZB. For $\alpha = 95\%$ and $\beta = 5\%$, ART is 1.68 times faster than UPE and EZB. Second, we observe that ART, UPE, EZB, and MLE perform estimation in constant time, which attributes to the use of persistence probabilities. Third, we observe that FNEB, whose estimator is the size of the first run of 0s, is the slowest. This concurs with our analytical analysis in Figure 4, where we show that FNEB has the largest variance. The larger the variance of an estimator, the more the rounds of execution needed to achieve the required reliability, and therefore the longer the estimation time.

We make three main observations from Figures 12 (a), (b), and (c), which show the estimation time for 5000 tags required by each scheme with the required reliability α varying from 90% to 99.9% for different configurations of confidence interval β. First, we observe that ART is faster than all four prior estimation schemes in all these configurations. Second, the difference between the estimation time of ART

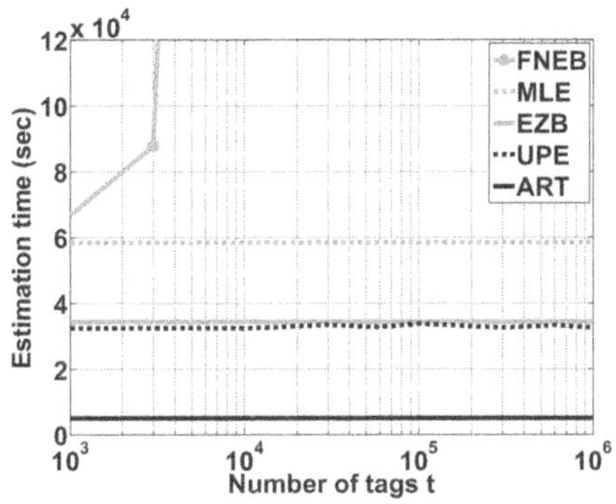

(a) $\alpha = 99.9\%, \beta = 0.1\%$

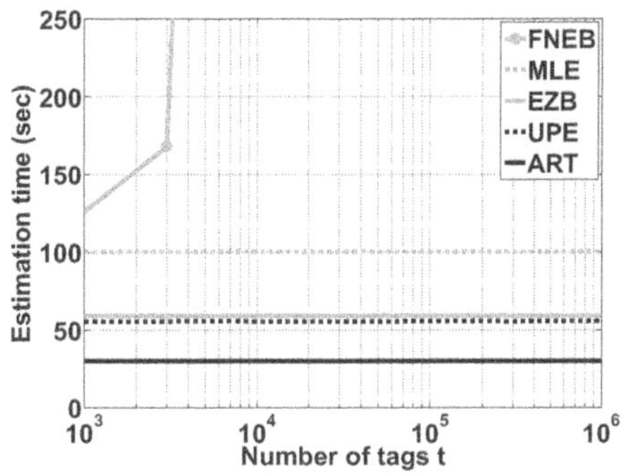

(b) $\alpha = 99\%, \beta = 1\%$

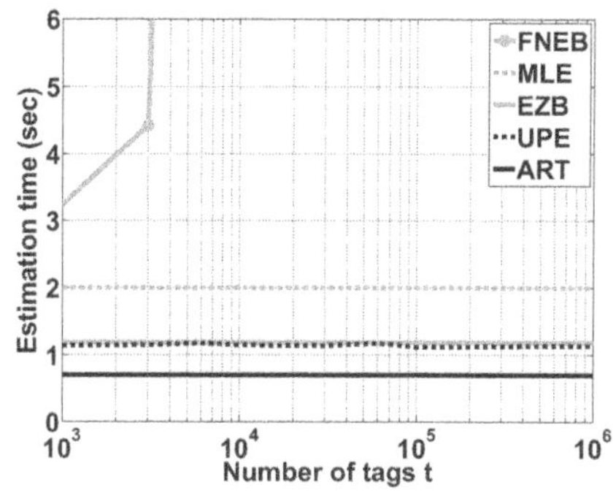

(c) $\alpha = 95\%, \beta = 5\%$

Figure 11: Time vs. t

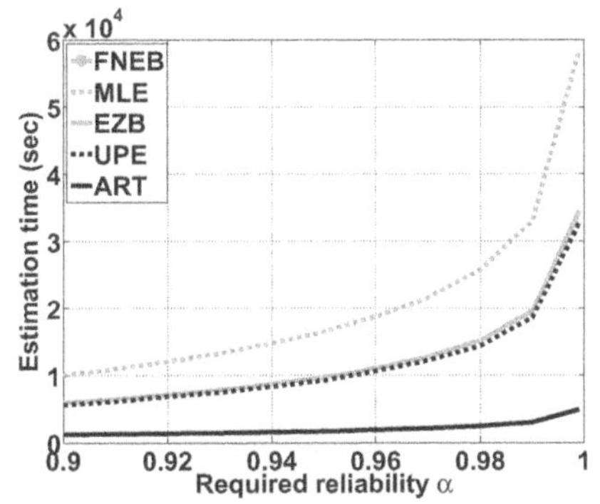

(a) $\beta = 0.1\%$

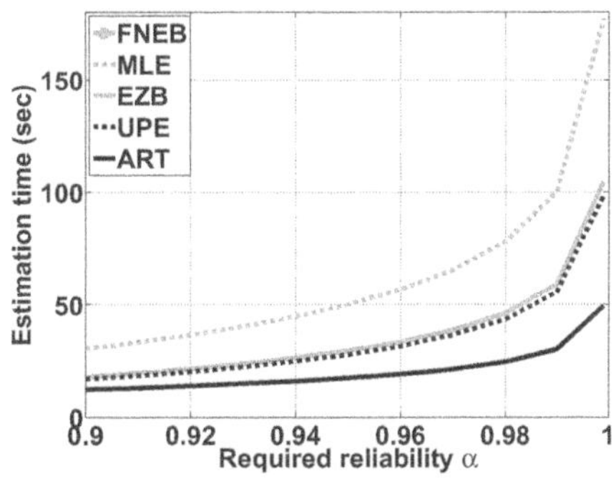

(b) $\beta = 1\%$

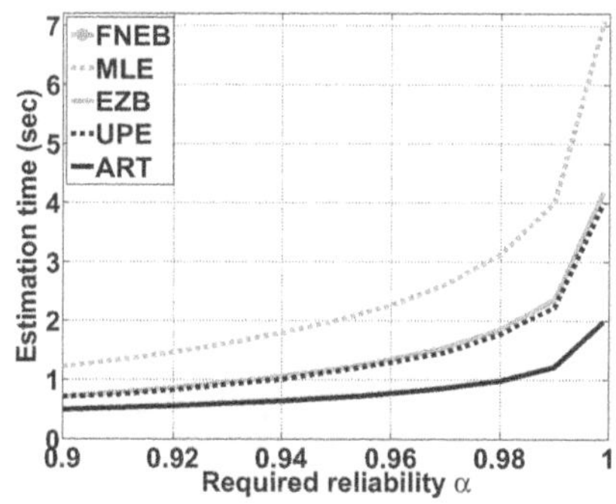

(c) $\beta = 5\%$

Figure 12: Time vs. α

and those of prior schemes increases as the required reliability increases. For example, for $\beta = 5\%$ and $\alpha = 95\%$, ART is 1.68 times faster than UPE and EZB while for $\beta = 0.1\%$ and $\alpha = 99.9\%$, it is 7 times faster. This shows that ART becomes more and more advantageous over existing schemes when the required reliability increases. Third, for all schemes, the estimation time increases as the required reliability increases because more number of rounds are needed to achieve the required reliability. We further observe that ART's estimation time increases at the lowest rate as the required reliability increases because its estimator has the smallest variance.

We make three main observations from Figures 13 (a), (b), and (c), which show the estimation time for 5000 tags required by each scheme with the confidence interval β varying from 0.1% to 10% for different configurations of α. First, we observe that ART is faster than all estimation schemes in all these configurations. Second, for all schemes, the estimation time decreases as the confidence interval increases because lesser number of rounds are needed to achieve the required reliability.

6.2 Actual Reliability

The subfigures in Figure 14 show the actual reliability of ART versus the number of tags for different configurations of required reliability α and confidence interval β. We observe that *ART always achieves the required reliability.*

7. CONCLUSIONS AND FUTURE WORK

The key technical novelty of this paper is in proposing the new estimator, the average run size of 1s, for estimating RFID tag population size. Using analytical plots, we show that our estimator has much smaller variance compared to other estimators including those used in prior work. It is this smaller variance that makes our scheme faster than the previous ones. In future work, we will work on mathematically proving that our estimator has smaller variance compared to other estimators. The key technical depth of this paper is in the mathematical development of the estimation theory using this estimator. Our experimental results show that our scheme ART is significantly faster than all prior RFID estimation schemes. Furthermore, both the analytical and experimental results show that the estimation of ART is independent of tag population size. In future work, we will work on mathematically proving this independence.

8. REFERENCES

[1] C. Bordenave, D. McDonald, and A. Proutiere. Performance of random medium access control, an asymptotic approach. In *Proc. of Int. Conf. on Measurements and Modeling of Computer Systems SIGMETRICS*, 2008.

[2] J.-R. Cha and I. Jae-Hyun Kim. Novel anti-collision algorithms for fast object identification in rfid system. In *Proc. of Int. Conf. on Parallel and Distributed Systems*, 2005.

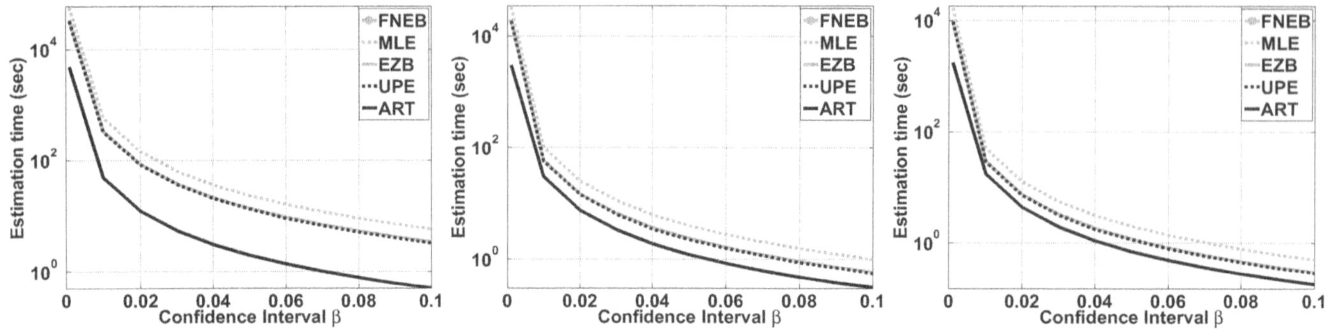

(a) $\alpha = 99.9\%$ (b) $\alpha = 99\%$ (c) $\alpha = 95\%$

Figure 13: Time vs. β

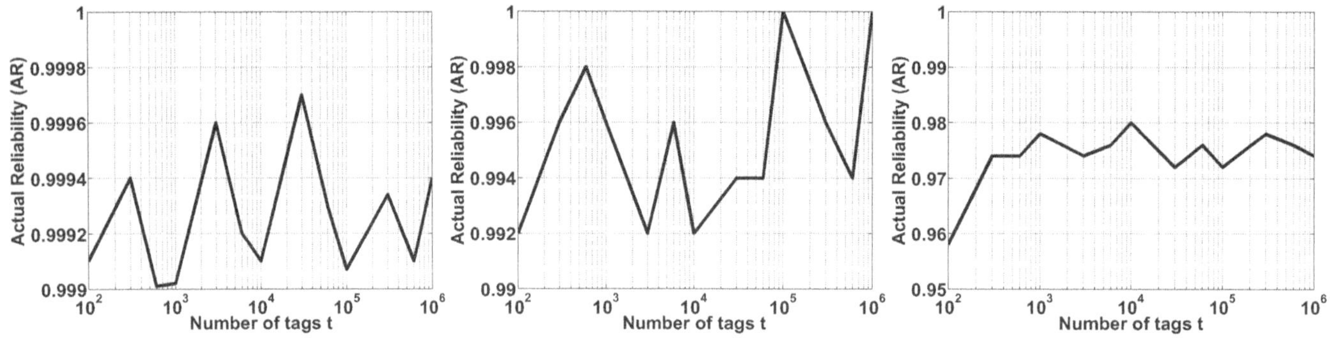

(a) $\alpha = 99.9\%, \beta = 0.1\%$ (b) $\alpha = 99\%, \beta = 1\%$ (c) $\alpha = 95\%, \beta = 5\%$

Figure 14: Actual reliability

[3] EPCGlobal Inc. *Radio-Frequency Identity Protocols Class-1 Generation-2 UHF RFID Protocol for Communications at 860 MHz–960 MHz*, 1.2.0 edition, 2008.

[4] K. Finkenzeller. *RFID Handbook: Fundamentals and Applications in Contactless Smart Cards, Radio Frequency Identification and Near-Field Communication*. Wiley, 2010.

[5] P. Flajolet and G. N. Martin. Probabilistic counting algorithms for data base applications. *Journal of Computer and System Sciences*, 31(2):182–209, 1985.

[6] H. Han, B. Sheng, C. C. Tan, Q. Li, W. Mao, and S. Lu. Counting RFID tags efficiently and anonymously. In *Proc. IEEE INFOCOM*, 2010.

[7] M. Kodialam and T. Nandagopal. Fast and reliable estimation schemes in RFID systems. In *Proc. 12th MobiCom*, pages 322–333, 2006.

[8] M. Kodialam, T. Nandagopal, and W. C. Lau. Anonymous tracking using RFID tags. In *Proc. IEEE INFOCOM*, 2007.

[9] C. H. Lee and C. W. Chung. Efficient storage scheme and query processing for supply chain management using RFID. In *Proc. ACM SIGMOD*, pages 291–302, 2008.

[10] T. Li, S. Wu, S. Chen, and M. Yang. Energy efficient algorithms for the RFID estimation problem. In *Proc. IEEE INFOCOM*, 2010.

[11] B. Nath, F. Reynolds, and R. Want. RFID technology and applications. *IEEE Pervasive Computing*, 5:22–24, 2006.

[12] A. Nemmaluri, M. D. Corner, and P. Shenoy. Sherlock: Automatically locating objects for humans. In *Proc. MobiSys*, pages 187–198, 2008.

[13] L. M. Ni, Y. Liu, Y. C. Lau, and A. P. Patil. Landmarc: Indoor location sensing using active RFID. *Wireless networks*, 10:701–710, 2004.

[14] C. Qian, H. Ngan, and Y. Liu. Cardinality estimation for large-scale RFID systems. In *Proc. 6th IEEE PerCom*, pages 30–39, 2008.

[15] M. Roberti. A 5-cent breakthrough. *RFID Journal*, 5(6), 2006.

[16] P. Semiconductors. *SL2 ICS11 I.Code UID Smart Label IC Functional Specification Datasheet www.advanide.com/datasheets/sl2ics11.pdf*, 2004.

[17] V. Shah-Mansouri and V. W. Wong. Cardinality estimation in RFID systems with multiple readers. In *Proc. IEEE GLOBECOM*, 2009.

[18] H. Vogt. Efficient object identification with passive RFID tags. *Pervasive Computing*, 2414:98–113, 2002.

[19] C. Wang, H. Wu, and N.-F. Tzeng. RFID-based 3-d positioning schemes. In *Proc. IEEE INFOCOM*, pages 1235–1243, 2007.

[20] B. Zhen, M. Kobayashi, and M. Shimizu. Framed ALOHA for multiple RFID objects identification. *IEICE Transactions on Communications*, 88:991–999, 2005.

Temporal Reachability Graphs*

John Whitbeck[†‡]
john.whitbeck@lip6.fr

Marcelo Dias de Amorim[†]
marcelo.amorim@lip6.fr

Vania Conan[‡]
vania.conan@fr.thalesgroup.com

Jean-Loup Guillaume[†]
jean-loup.guillaume@lip6.fr

[†]UPMC Sorbonne Universités
4, place Jussieu
75005 Paris
France

[‡]Thales Communications & Security
160 Boulevard de Valmy
92700 Colombes
France

ABSTRACT

While a natural fit for modeling and understanding mobile networks, time-varying graphs remain poorly understood. Indeed, many of the usual concepts of static graphs have no obvious counterpart in time-varying ones. In this paper, we introduce the notion of *temporal reachability graphs*. A *(τ, δ)-reachability graph* is a time-varying directed graph derived from an existing connectivity graph. An edge exists from one node to another in the reachability graph at time t if there exists a journey (i.e., a spatiotemporal path) in the connectivity graph from the first node to the second, leaving after t, with a positive edge traversal time τ, and arriving within a maximum delay δ. We make three contributions. First, we develop the theoretical framework around temporal reachability graphs. Second, we harness our theoretical findings to propose an algorithm for their efficient computation. Finally, we demonstrate the analytic power of the temporal reachability graph concept by applying it to synthetic and real-life datasets. On top of defining clear upper bounds on communication capabilities, reachability graphs highlight asymmetric communication opportunities and offloading potential.

Categories and Subject Descriptors

C.2.1 [**Computer-Communication Networks**]: Network Design and Architecture—*Store and forward networks*; F.2.2 [**Analysis of Algorithms and Problem Complexity**]: Nonnumerical Algorithms and Problems; G.2.2 [**Discrete Mathematics**]: Graph Theory

*An extended technical report containing the mathematical proofs is available online from `http://www-npa.lip6.fr/~whitbeck/`.

MobiCom'12, August 22–26, 2012, Istanbul, Turkey.

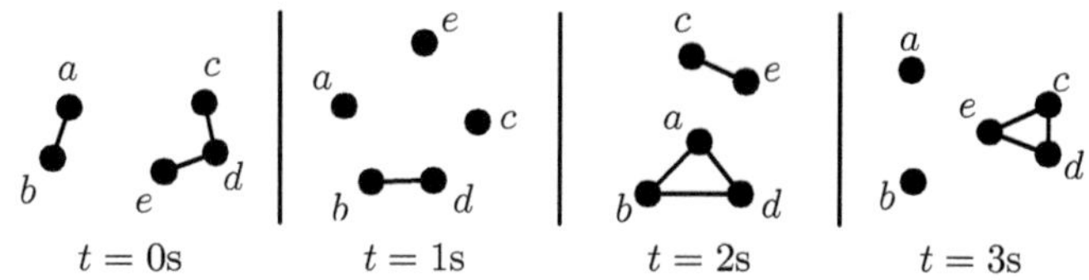

Figure 1: Snapshots of a time-varying graph with five nodes.

General Terms

Algorithms, Theory, Performance

Keywords

Time-varying graphs, Reachability, Opportunistic networks, Communication performance bounds

1. INTRODUCTION

In time-varying graphs (TVG), vertices and edges appear and disappear as a function of time. Alternatively called temporal networks [15] or evolving graphs [29], time-varying graphs have emerged over the past few years as a key model for a variety of complex systems [10, 13, 16, 26, 27]. Time-varying graphs can also serve as a solid theoretical framework for investigating fundamental properties of mobile networks [3, 9]. In Fig. 1, we show an example of a time-varying graph representing a five-node mobile network.

Many of the typical concepts of static graphs such as paths, distance, diameter, or node degree have no obvious counterpart in time-varying graphs. Theorems that are true on static graphs may not hold in time-varying ones [15] and dynamic equivalents of well-known problems on static graphs, such as finding strongly connected components, turn out to be intractable [2]. Naturally, it is always possible to track the evolution of static metrics such as node degree or clustering on snapshots of the time-varying graph at certain time intervals [6] but, while instructive, this approach often fails to capture the correlations between successive snapshots [14]. Recently a number of concepts specific to time-varying graphs have been studied such as *journeys* [29], *temporal diameter* [5], or *reachability time* [14].

In the context of opportunistic mobile networking, much attention has been focused on aggregate inter-contact time

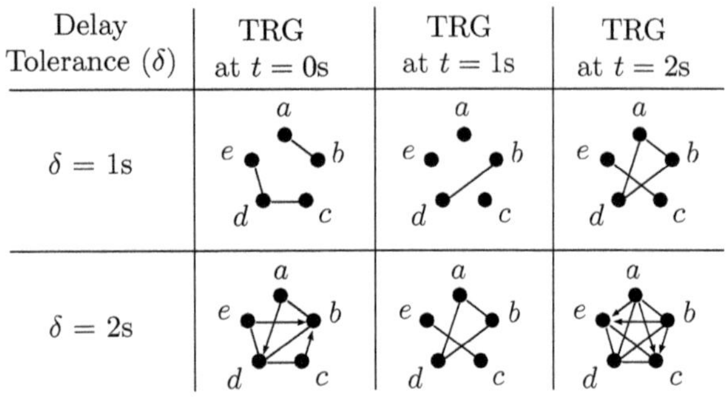

Figure 2: Temporal reachability graphs obtained from the time-varying graph of Fig. 1 for δ=1s and 2s. Arrows indicate uni-directional communication opportunisties, while their absence indicates bi-directional ones. We show snapshots for different times t. We assume in this example that one-hop communications require τ=1s.

distributions. These have been found to fit a power-law (with an α parameter smaller than 1) followed by an exponential cutoff in a number of real-life datasets [4]. Should this power-law also hold for pairwise inter-contact times, then the expected values of these would be infinite, a pessimistic result indeed. However, further work has shown that aggregate power-law distributions can emerge from a diversity of pairwise inter-contact laws (including exponential) [7, 22]. Moreover, all information on multi-hop connectivity is ignored. Inter-contact time measurements do not therefore suffice to characterize connectivity in mobile networks.

Temporal reachability graphs (TRG), our contribution, on the other hand, offer an immediate view of communication possibilities in a dynamic network. Given a time-varying graph $\mathcal{G}$, an arc exists from vertex a to vertex b at time t in its derived *(τ, δ)-reachability graph* if a journey exists in $\mathcal{G}$ from a to b leaving a after time t and arriving at b before $t+\delta$, given that each single-hop communication takes time τ. Here, τ is the edge traversal time and δ is the upper bound on journey times in $\mathcal{G}$. By definition, temporal reachability are *directed* time-varying graphs. For example, in the context of studying information dissemination in delay-tolerant networks (DTN) [11], τ would be the one-hop transmission delay and δ is the maximum tolerated delay.

In Fig. 2, we show the instances of the temporal-reachability graph extracted from the opportunistic mobile networks of Fig. 1 at different times (assuming that the graph does not change between the snapshots shown in Fig. 1). We consider in this example that the link traversal time is τ=1s. When δ = 1s, the only possible journeys in $\mathcal{G}$ at $t \in \{1s, 2s, 3s\}$ are the direct links between nodes – communications tolerate a delay of 1s, which is the time to traverse one single link. When δ = 2s, several more journeys appear. Consider for instance the reachability graph at t = 0s. There is a directed link between a and d as a can wait one time unit for the link (b, d) to appear ($a \rightsquigarrow b$ and $b \rightsquigarrow d$); thus, a two-hop journey $a \rightsquigarrow d$ through b becomes possible.

We make three main contributions:

Temporal reachability graphs. We formalize the concept of reachability graphs. In particular, under conditions that hold for all empirical datasets, we prove an additive property on the delay of reachability graphs. Roughly speaking, knowledge of the reachability graphs for delays δ and μ is enough to derive the reachability graph for delay $\delta + \mu$.

Table 1: Notations used in this paper

Notation	Meaning
N	Number of vertices
$\mathcal{G}$	A time-varying graph (TVG)
τ	Edge traversal time of a TVG
η	Time step of an η-regular TVG
$\mathcal{R}_\delta$	Reachability graph with maximum delay δ derived from a TVG $\mathcal{G}$
$\|\mathcal{R}_\delta(t)\|$	Number of arcs in $\mathcal{R}_\delta$ at time t
$\mathcal{U}_\delta$, $\mathcal{L}_\delta$	Upper and lower bounds of a reachability graph $\mathcal{R}_\delta$. $\mathcal{L}_\delta \subseteq \mathcal{R}_\delta \subseteq \mathcal{U}_\delta$.
$\mathcal{D}_\delta$	A time-varying out-dominating set of a reachability graph $\mathcal{R}_\delta$
$\mathsf{sym}(\mathcal{R}_\delta)$	Symmetric subset of $\mathcal{R}_\delta$: $(u, v) \in \mathsf{sym}(\mathcal{R}_\delta)(t) \Leftrightarrow (v, u) \in \mathsf{sym}(\mathcal{R}_\delta)(t)$
$\mathsf{asym}(\mathcal{R}_\delta)$	Asymmetric subset of $\mathcal{R}_\delta$: $(u, v) \in \mathsf{asym}(\mathcal{R}_\delta)(t) \Leftrightarrow (v, u) \notin \mathsf{asym}(\mathcal{R}_\delta)(t)$

Algorithm for efficient computation of TRG. We translate this additive property into an efficient single-pass streaming algorithm [1] to calculate entire families of reachability graphs.

Insights into real-world mobility traces. Once calculated, reachability graphs yield many original insights on the temporal structure of the time-varying graphs they derive from. In this paper, we calculate the reachability graphs of several synthetic and real-life high-resolution connectivity traces [18, 20, 23, 25]. These highlight the concepts of *temporal connectivity*, *temporal asymmetry*, and *temporal dominating set*. From the point of view of opportunistic communications, reachability graphs immediately provide, at all times, the maximum delivery ratio and the minimum set of users required to broadcast a message to the entire network, thereby offering a complete view of opportunistic communication capabilities and offloading potential.

The rest of the paper is structured as follows. Sections 2 and 3 introduce the concepts of time-varying and reachability graphs. Their theoretical properties justify their computation by a streaming algorithm detailed in Section 4. Section 5 then analyzes the reachability graphs of synthetic and real-life connectivity graphs. Related work is presented in Section 6. Finally, Section 7 concludes.

2. TIME-VARYING GRAPHS

Time-varying graphs are a very useful high-level abstraction for studying connectivity over time in mobile network. Terms, notations, and definitions around time-varying graphs vary considerably, but recently, Casteigts et al. have proposed a unified framework for TVGs and, wherever applicable, we will use their definitions and notations [3].

Definition 1. (**Time-varying graph**) Let V be a set of vertices, and $E \subseteq V \times V$ the set of possible edges between vertices in V. Events occur over a time span $\mathcal{T} \subseteq \mathbb{T}$, where $\mathbb{T}$ is the temporal domain ($\mathbb{N}$ for discrete-time systems or $\mathbb{R}+$ for continuous-time ones). In the general case, a TVG is a tuple $\mathcal{G} = (V, E, \mathcal{T}, \rho, \zeta)$ where $\rho : E \times \mathcal{T} \rightarrow \{0, 1\}$, called *presence function*, indicates whether a given edge exists at a given time and $\zeta : E \times \mathcal{T} \rightarrow \mathbb{T}$, called *latency function*, indicates the time it takes to cross a given edge if starting at a given date (the latency of an edge could vary in time).

In this paper, we will mostly consider continuous-time TVGs (i.e., $\mathbb{T} = \mathbb{R}+$). Furthermore, we will always assume a *constant* ζ function such that $\forall (e,t) \in E \times \mathcal{T}, \zeta(e,t) = \tau$ where $\tau \geq 0$ is our uniform edge traversal time. We also note $\mathcal{G}(t) \subseteq E$ the set of edges in the snapshot of $\mathcal{G}$ at time t. Hence $e \in \mathcal{G}(t) \Leftrightarrow \rho(e,t) = 1$. Finally, given an edge $e = (u,v) \in E$, we define $\mathsf{from}(e) = u$ and $\mathsf{to}(e) = v$.

Definition 2. (**Inclusion**) Let $\mathcal{G}$ and $\mathcal{G}'$ be two TVGs that differ only by their presence functions ρ and ρ'. We write $\mathcal{G} \subseteq \mathcal{G}'$ if and only if $\forall t \in \mathcal{T}, \forall e \in E, \rho(e,t) \leq \rho'(e,t)$, or equivalently if and only if $\forall t \in \mathcal{T}, \mathcal{G}(t) \subseteq \mathcal{G}'(t)$.

Definition 3. (**Union**) Let $\mathcal{G}$, $\mathcal{H}$, and $\mathcal{I}$ be three TVGs that differ only by their presence functions. We write $\mathcal{G} = \mathcal{H} \cup \mathcal{I}$ if and only if $\forall t \in \mathcal{T}, \mathcal{G}(t) = \mathcal{H}(t) \cup \mathcal{I}(t)$.

Definition 4. (**Journey**) A *journey* in $\mathcal{G}$ is a sequence of couples $\mathcal{J} = \{(e_1,t_1),(e_2,t_2),\ldots,(e_k,t_k)\}$ such that $\forall i < k$: (i) $\mathsf{to}(e_i) = \mathsf{from}(e_{i+1})$; (ii) $\forall t$ s.t. $t_i \leq t < t_i + \tau, \rho(e_i,t) = 1$; and (iii) $t_{i+1} \geq t_i + \tau$. Here, $|\mathcal{J}| = k$ is this journey's *topological length* (i.e., the number of hops). Furthermore, $\mathsf{departure}(\mathcal{J})$ and $\mathsf{arrival}(\mathcal{J})$ denote the starting date t_1 and the last date $t_k + \tau$, respectively. Finally $\delta_{\mathcal{J}} = \mathsf{arrival}(\mathcal{J}) - \mathsf{departure}(\mathcal{J})$ is the journey's *temporal length*. A journey may represent, for example, the sequence of hops that a message follows through an opportunistic network.

3. TEMPORAL REACHABILITY GRAPHS

Now that we have the necessary background, let us formally define the notion of *temporal reachability graphs* (or simply *reachability graphs* in the remainder of this paper, for the sake of readability). Furthermore, we write $\mathbb{N}^*$ for $\mathbb{N} \setminus \{0\}$. To the best of our knowledge, the definitions and results presented in this section are completely novel.

3.1 Reachability Graphs

Definition 5. (**Reachability graph**) For $\delta \in \mathbb{R}+$, let $\mathcal{R}_\delta$ be the reachability graph with maximum delay δ derived from $\mathcal{G}$ (with edge traversal time τ). Formally, $\forall t$, $(u,v) \in \mathcal{R}_\delta(t)$ if and only if $u \neq v$ and there exists at time t in $\mathcal{G}$ a journey $\mathcal{J}$ from u to v such that $\mathsf{departure}(\mathcal{J}) \geq t$ and $\mathsf{arrival}(\mathcal{J}) \leq t + \delta$.

Note that if $\delta < \tau$ then $\mathcal{R}_\delta$ is empty, as even one-hop journeys do not have enough time to arrive before the maximum delay.

PROPOSITION 1 (**Growth**). *Let $\mathcal{R}_\delta$ and $\mathcal{R}_\mu$ be two reachability graphs of $\mathcal{G}$. Then $\delta \leq \mu \implies \mathcal{R}_\delta \subseteq \mathcal{R}_\mu$.*

This follows naturally from the definition of a reachability graph. Note that the reverse is not true. For example let $\mathcal{G}$ be a TVG with edge traversal time $\tau > 0$. Its derived reachability graphs $\mathcal{R}_{\tau/2}$ and $\mathcal{R}_{\tau/3}$ are both empty as their maximum delays are smaller than the time it takes to cross one edge. Therefore, as per definition 2, $\mathcal{R}_{\tau/2} \subseteq \mathcal{R}_{\tau/3}$ even though $\frac{\tau}{2} > \frac{\tau}{3}$.

Definition 6. (**Composition**) Let $\mathcal{R}_\delta$ and $\mathcal{R}_\mu$ be two reachability graphs of $\mathcal{G}$. We define their *composition* $\mathcal{R}_\delta \otimes R_\mu$ as the TVG such that, at all times t,

$$(u,v) \in (\mathcal{R}_\delta \otimes \mathcal{R}_\mu)(t) \Leftrightarrow \begin{cases} (u,v) \in \mathcal{R}_\delta(t), \text{ or} \\ (u,v) \in \mathcal{R}_\mu(t+\delta), \text{ or} \\ \exists w \in V, (u,w) \in \mathcal{R}_\delta(t) \text{ and} \\ \quad (w,v) \in \mathcal{R}_\mu(t+\delta) \end{cases}$$

THEOREM 1 (**Decomposition**). *Let $\mathcal{G}$ be a TVG with edge traversal time τ and $\bar{\mathcal{R}} = \{\mathcal{R}_\delta\}_{\delta \geq 0}$ the set of all its reachability graphs. Let $\delta \geq 0$ and $\mu \geq \tau$. Then:*

$$\mathcal{R}_{\delta+\mu} = \bigcup_{0 \leq \epsilon < 1} \mathcal{R}_{\delta+\epsilon\tau} \otimes \mathcal{R}_{\mu-\epsilon\tau}.$$

While not yet in a calculable form, this theorem is at the heart of our approach. Intuitively, it states that if all the reachability graphs with delays close to δ and μ are known, then one can calculate the reachability graph with delay $\delta+\mu$. A full formal proof is provided in the appendix (Section **??**), but intuitively, it relies on the following simple idea. Any arc in $\mathcal{R}_{\delta+\mu}$ corresponds to a journey in $\mathcal{G}$ starting after time t and arriving before $t+\delta+\mu$. Either it can be neatly divided into a (possibly empty) journey in $\mathcal{R}_\delta$ starting after t and arriving before $t+\delta$ and another (possibly empty) journey in $\mathcal{R}_\mu$ starting after $t+\delta$ and arriving before $t+\delta+\mu$, or it starts crossing an edge before $t+\delta$ but after $t+\delta-\tau$ that straddles both time intervals. In the latter case, since the edge traversal time is τ, one can incorporate the straddling edge into a journey arriving before $t+\delta+\epsilon\tau$ and the rest into another journey leaving after $t+\delta+\epsilon\tau$ of temporal length less than $\mu-\epsilon\tau$.

3.2 Regular reachability graphs

Real-life datasets all have a *maximum resolution* (e.g., a second or a millisecond) that corresponds to the precision with which they were measured. While it is tempting to map their time domain $\mathcal{T}$ to $\mathbb{N}$, in reality one must account for 0-second edge durations. For example, if the edge traversal time τ is null, then even ephemeral edges that last 0 seconds may be part of a journey. In the more common situation where τ is strictly positive, one may still encounter 0-duration arcs in a reachability graph. For example, consider an edge (u,v) that is present for only one second. If τ and δ are both also equal to one second, then a one-hop journey from u to v using that edge only exists at precisely the instant t that the edge appears. This, in turn, corresponds to two 0-second arcs in $\mathcal{R}_\delta(t)$: (u,v) and (v,u). Note that both of these arcs will become one-second-long arcs in the reachability graph when $\delta = 2s$. These observations lead to the definition of *regular time-varying graphs*.

Definition 7. (**Regular TVG**) A TVG $\mathcal{G}$ is an η-regular TVG if there exists $\eta > 0$ such that $\forall k \in \mathbb{N}, k\eta < t_1 \leq t_2 < (k+1)\eta \implies \mathcal{G}(t_1) = \mathcal{G}(t_2) \subseteq \mathcal{G}(k\eta)$. Here, η is called $\mathcal{G}$'s resolution. The time interval $[k\eta,(k+1)\eta[$ is the kth epoch of $\mathcal{G}$ with starting time $k\eta$ and ending time $(k+1)\eta$.

Without loss of generality, this definition assumes that the first epoch starts at $t = 0$. Intuitively, an η-regular TVG is one whose instantaneous graph topology cannot change arbitrarily quickly, as, during each epoch, the graph topology remains constant. Ephemeral 0-second edges or arcs may exist at the start of an epoch, but the TVG then remains constant until the start of the next epoch. Not only are regular TVGs a natural fit for real-life traces, but, as we will see, they also represent a class of TVGs whose reachability graphs are calculable.

THEOREM 2 (**Regular reachability graphs**). *Let $\mathcal{G}$ be an η-regular TVG whose edge traversal time is $\tau \in \eta\mathbb{N}^*$. For $\delta \in \eta\mathbb{N}$, let $\mathcal{R}_\delta$ be a reachability graph of $\mathcal{G}$. Then $\mathcal{R}_\delta$ is an η-regular TVG and $\forall k \in \mathbb{N}, k\eta < t < (k+1)\eta \implies \mathcal{R}_\delta(t) \subseteq \mathcal{R}_\delta(k\eta) \cap \mathcal{R}_\delta((k+1)\eta)$.*

This theorem ensures that, as long as δ is a "multiple" of the resolution η, $\mathcal{R}_\delta$ carries on the η-regularity of $\mathcal{G}$. The full formal proof is a little technical (Section **??** in the appendix), but it relies on the following idea. Let us consider a one-hop journey in an η-regular TVG $\mathcal{G}$ leaving in the middle of an epoch at time t. Let k be the integer such that $k\eta < t < (k+1)\eta$. It arrives in the next epoch at time $(k+1)\eta < t+\tau < (k+2)\eta$ (the theorem assumes that $\tau \geq \eta$). Because $\mathcal{G}$ is η-regular, the departure time of this one-journey can be nudged forwards or backwards as long as it remains with the same epoch. Hence all one-hop journeys with a departure time t' such that $k\eta \leq t' \leq (k+1)\eta$ are also valid journeys in $\mathcal{G}$. This "nudging" can be extended, though not trivially, to multi-hop journeys which proves the theorem.

Furthermore, the $\tau \geq \eta$ hypothesis guarantees that an arc present in a reachability graph during an epoch is present not only at the start of the epoch like in any regular graphs both also at the start of the next epoch. This property will be leveraged later in this section. However $\mathcal{R}_\delta(t) \neq \mathcal{R}_\delta(k\eta) \cap \mathcal{R}_\delta((k+1)\eta)$. An example of such a situation is discussed in Section 3.3.

THEOREM 3 (**Sampling**). *Let $\mathcal{G}$ be an η-regular TVG whose edge traversal time is $\tau = n\eta$ with $n \in \mathbb{N}^*$. Let $\bar{\mathcal{R}} = \{\mathcal{R}_{i\eta}\}_{i\in\mathbb{N}}$ be the set of all its derived η-regular reachability graphs. For $(d,m) \in \mathbb{N}\times\mathbb{N}$ such that $m \geq n$, we have, $\forall a \in \mathbb{N}$:*

$$\mathcal{R}_{(d+m)\eta}(a\eta) = \bigcup_{0\leq k<n} \left(\mathcal{R}_{(d+k)\eta} \otimes \mathcal{R}_{(m-k)\eta}\right)(a\eta).$$

This theorem is the adaptation of Theorem 1 to regular TVGs. The formal proof is in the appendix (Section **??**) but, as previously, it relies on the idea that within an epoch, there exists a little freedom to "nudge" journey departure and arrival times backwards or forwards. In a sense, if both the minimum departure time and maximal arrival time of a journey fall exactly at the start of a time epoch, then this journey can be divided into two sub-journeys whose departure and arrival times also map exactly to the start of epochs. Under this form, the value of a reachability graph at the start of each epoch can be *exactly* calculated from the values, at the start of each epoch, of the proper set of τ pairs of reachability graphs.

Unfortunately, this exact formula does not extend to the time spent strictly within an epoch. In a way, the state of the reachability graph during epochs is more important than its state at their starting times. Indeed, the time spent at *exactly* the start of epochs is infinitesimally small. Therefore any metric averaged over time (e.g. average density) will depend only on the states strictly within epochs. In the next section, we propose an upper and a lower bound on the reachability graph during epochs. As we will later see in Section 5, these upper and lower bounds are in fact nearly always equal and therefore achieve an excellent approximation of the real reachability graph.

3.3 Upper and lower bounds

We define for every regular reachability graph $\mathcal{R}_\delta$ an upper and lower bound, i.e., two TVGs $\mathcal{U}_\delta$ and $\mathcal{L}_\delta$ such that $\mathcal{L}_\delta \subseteq \mathcal{R}_\delta \subseteq \mathcal{U}_\delta$. Both of these bounds are equal to the reachability graph at the start of each epoch but only differ slightly during the epoch. The upper bound is a straightforward application of Theorem 2.

PROPOSITION 2 (**Upper Epoch approximation**). *Let $\mathcal{G}$ be an η-regular TVG whose edge traversal time is $\tau \in \eta\mathbb{N}^*$. For $\delta \in \eta\mathbb{N}$, let $\mathcal{R}_\delta$ be a reachability graph of $\mathcal{G}$. We define $\mathcal{R}_\delta$'s* upper approximation, *the TVG $\mathcal{U}_\delta$, as follows ($\forall a\eta \leq t < (a+1)\eta$ with $a \in \mathbb{N}$):*

- *if $t = a\eta$, then $\mathcal{U}_\delta(a\eta) = \mathcal{R}_\delta(a\eta)$;*
- *if $t > a\eta$, then $\mathcal{U}_\delta(t) = \mathcal{R}_\delta(a\eta) \cap \mathcal{R}_\delta((a+1)\eta)$.*

As defined, $\forall\delta, \mathcal{R}_\delta \subseteq \mathcal{U}_\delta$.

The lower bound is a little more complicated and involves a modified version of the composition operator (Definition 6).

Definition 8. (**Approximate composition**) Let $\mathcal{R}_\delta$ and $\mathcal{R}_\mu$ be two reachability graphs of an η-regular TVG $\mathcal{G}$ with δ and μ in $\eta\mathbb{N}^*$. Let $\mathcal{L}_\delta$ and $\mathcal{L}_\mu$ be two TVGs such that $\mathcal{L}_\delta \subseteq \mathcal{R}_\delta$ and $\mathcal{L}_\mu \subseteq \mathcal{R}_\mu$. For all times $a\eta \leq t < (a+1)\eta$ with $a \in \mathbb{N}$, we define their *approximate composition* $\mathcal{L}_\delta \odot \mathcal{L}_\mu$ as follows:

- if $t = a\eta$, then $(\mathcal{L}_\delta \odot \mathcal{L}_\mu)(a\eta) = (\mathcal{R}_\delta \otimes \mathcal{R}_\mu)(a\eta)$;
- if $t > a\eta$, then

$$(u,v) \in (\mathcal{L}_\delta \odot \mathcal{L}_\mu)(t) \Leftrightarrow \begin{cases} (u,v) \in \mathcal{R}_\delta((a+1)\eta), \text{ or} \\ (u,v) \in \mathcal{R}_\mu(a\eta+\delta), \text{ or} \\ \exists w \in V, (u,w) \in \mathcal{L}_\delta(t) \text{ and} \\ \quad (w,v) \in \mathcal{L}_\mu(t+\delta) \end{cases}$$

By definition, at the start of each epoch the approximate composition is equal to the regular composition of the reachability graphs. During an epoch, the condition $(u,v) \in \mathcal{R}_\delta((a+1)\eta)$ is *easier* to meet than the more intuitive $(u,v) \in \mathcal{L}_\delta(t)$. Indeed, with Theorem 2, $\mathcal{L}_\delta(t) \subseteq \mathcal{R}_\delta(t) \subseteq \mathcal{R}_\delta((a+1)\eta)$, and this composition operator will catch the arcs (including the 0-second ones) occurring at the start of the next epoch. The same is true for the $(u,v) \in \mathcal{R}_\mu(a\eta+\delta)$ condition for arcs ending at the start of the current (δ-shifted) epoch. Combined, these more inclusive conditions make for a *tighter* lower bound.

PROPOSITION 3 (**Lower Epoch approximation**). *Let $\mathcal{G}$ be an η-regular TVG whose edge traversal time is $\tau = n\eta$ with $n \in \mathbb{N}^*$. Let $\bar{\mathcal{R}} = \{\mathcal{R}_\delta\}_{\delta\in\eta\mathbb{N}}$ be the set of all its derived reachability graphs. We recursively define $\mathcal{R}_\delta$'s* lower approximation *(i.e., the TVG $\mathcal{L}_\delta$) as follows. For $d < 2n$, $\mathcal{L}_{d\eta} = \mathcal{R}_{d\eta}$ and for all (d,m) in $\mathbb{N}\times\mathbb{N}$ such that $d \geq n$ and $m \geq n$,*

$$\mathcal{L}_{(d+m)\eta} = \bigcup_{0\leq k<n} \mathcal{L}_{(d+k)\eta} \odot \mathcal{L}_{(m-k)\eta}.$$

As defined, $\forall d \in \mathbb{N}, \mathcal{L}_{d\eta} \subseteq \mathcal{R}_{d\eta}$.

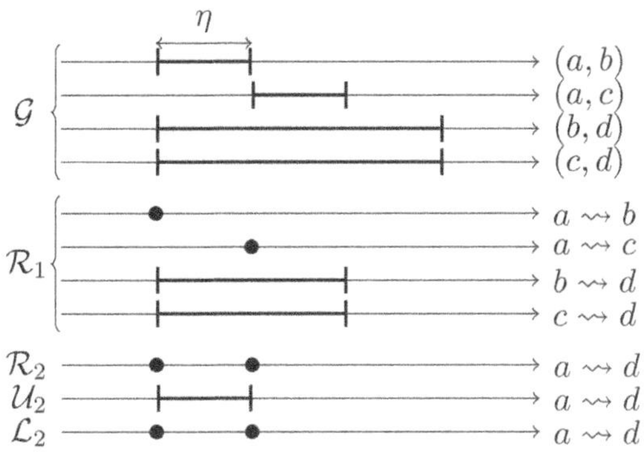

Figure 3: Upper epoch approximation with $\tau = \eta = 1$. Arrows point in the direction of time. Intervals denote the presence of an edge or arc during that time. Dots are intervals reduced to a single point in time. In this example the lower bound $\mathcal{L}_2$ is equal to the reachability graph $\mathcal{R}_2$ but the upper bound $\mathcal{U}_2$ is not.

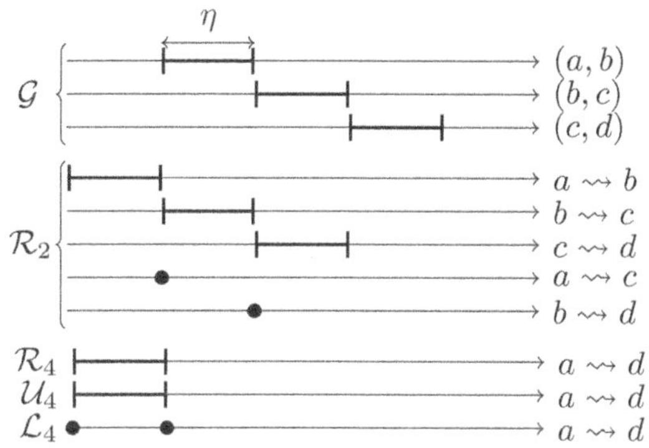

Figure 4: Lower epoch approximation with $\tau = \eta = 1$. Arrows point in the direction of time. Intervals denote the presence of an edge or arc during that time. Dots are intervals reduced to a single point in time. In this example the upper bound $\mathcal{U}_4$ is equal to the reachability graph $\mathcal{R}_4$ but the lower bound $\mathcal{L}_4$ is not.

The proof, detailed in Section **??** of the appendix, is fairly straightforward and involves setting the value of ϵ in Theorem 1 as a function of η, t, k, and n.

This recursive definition means that the exact value of $\mathcal{L}_\delta$ during a time epoch *depends on the sequence of compositions that was used to calculate it.* For example, in the simple situation where $\tau = \eta = 1$, the lower bounds $\mathcal{L}_4 = \mathcal{L}_2 \odot \mathcal{L}_2$ and $\mathcal{L}'_4 = \mathcal{L}_3 \odot \mathcal{L}_1$ may be different. Proposition 3 only guarantees that they are both included in the real reachability graph $\mathcal{R}_4$. However, as we will see, this is not a concern because this lower bound is tight regardless of how it is calculated.

Fig. 3 is an example of a situation with four vertices where the upper bound has an arc during an epoch that is not in the reachability graph. Here $\tau = \eta = 1$s. $\mathcal{R}_1$ is exactly derived from $\mathcal{G}$ by shortening the end time of each edge by τ. For example, this leads to an ephemeral arc from a to b and a two-second-long arc from b to d (down from a three-second-long edge (b,d) in $\mathcal{G}$). Obviously, not all arcs are represented on Fig. 3. In $\mathcal{R}_2$, there are two ephemeral arcs from a to d, corresponding to the $a \leadsto b \leadsto d$ journey, and the $a \leadsto c \leadsto d$ journey an epoch later. In between those instants, it is both too late to use the first journey and too early to use the second. Since this arc exists at the start of two successive epochs, $\mathcal{U}_2$ errs in considering that it exists during the entire epoch. In this example, the lower bound $\mathcal{L}_2$ matches the reachability graph.

Conversely, Fig. 4 is an example of a situation with three vertices where the lower bound misses an arc during an epoch that is present in the reachability graph. Here again $\tau = \eta = 1$. The one-second-long edges in $\mathcal{G}$ all become ephemeral 0-second arcs in $\mathcal{R}_1$ (not shown on the figure), that in turn become one-second-long arcs in $\mathcal{R}_2$. Furthermore, $\mathcal{R}_2$ also contains two ephemeral arcs (a,c) and (b,d) that correspond to the $a \leadsto b \leadsto c$ and $b \leadsto c \leadsto d$ journeys respectively. Here $\mathcal{R}_2 = \mathcal{L}_2 = \mathcal{U}_2$. Finally, $\mathcal{R}_4$ contains an arc (a,d) that corresponds to the existence of the three-hop journey $a \leadsto b \leadsto c \leadsto d$. $\mathcal{L}_4$ correctly identifies this journey at two instants k and $k+1$. At time k, it composes arc (a,b) from $\mathcal{R}_2(k)$ with the ephemeral arc (b,d) in $\mathcal{R}_2(k+2)$. At time $k+1$ it composes the ephemeral arc (a,c) in $\mathcal{R}_2(k+1)$ with the arc (c,d) in $\mathcal{R}_2(k+3)$. However, for $k < t < k+1$, $\mathcal{L}_4$ has no way of finding the journey from a to b. Note that in this case, $\mathcal{U}_4$ is equal to $\mathcal{R}_4$.

The example of Fig. 4 also helps us understand the absence of error propagation when calculating successive lower approximations from previous lower approximations. In this example, thanks to the conditions in Definition 8 that refer to the starting times of epochs, $\mathcal{L}_8$ will "bridge" over the missing (a,d) arc in $\mathcal{L}_4$. However it may miss composed arcs that use the (a,d) arc in $\mathcal{R}_4$. These missed arcs will be, in-turn, "bridged" in, for instance, a $\mathcal{L}_{16}$ approximation. Roughly speaking, errors during epochs do not propagate beyond two compositions. Combined with the fact that the lower approximation matches the reachability graph at the start of each epoch, this leads, as we will verify empirically, to a tight lower approximation or the reachability graph.

3.4 A few case studies

In this section, we examine how the theoretical results above apply to some simple situations.

Zero edge traversal time. The reachability graphs of η-regular TVGs with zero edge traversal time ($\tau = 0$) are very easy to calculate. In particular $(u,v) \in \mathcal{R}_0(t)$ if and only if u and v are in the same connected component at time t. Similarly, $\mathcal{R}_\eta(t)$ can be calculated from the connected components at time t and $t+\eta$. Furthermore, Theorem 1 becomes: $\mathcal{R}_{\delta+\mu} = \mathcal{R}_\delta \otimes \mathcal{R}_\mu$. This relation can then be applied repeatedly (e.g., in a binary-exponentiation) to obtain any $\mathcal{R}_\delta$ with $\delta \in \eta\mathbb{N}^*$. Furthermore, the reachability graphs thus calculated are *exact*. However, in this specific case where the edge traversal time is null, more efficient algorithms for calculating reachability graphs exist. For example the algorithm proposed by Chaintreau et al. for computing delay-optimal paths could be easily adapted to this purpose [5].

Several hops per epoch. While all of the results on regular reachability graphs in Sections 3.2 and 3.3 assume that the edge traversal time τ is in $\eta\mathbb{N}^*$, they can be simply adapted to the case where $\tau = \eta/n$ with $n \in \mathbb{N}^*$, i.e., when up to n edges may be crossed during a single epoch. Indeed, in this case, any η-regular graph would also be τ-regular, and we can apply all of our results as if $\tau = \eta$. For example, Proposition 3 is reduced to $\mathcal{L}_{(d+m)\eta} = \mathcal{L}_{d\eta} \odot \mathcal{L}_{m\eta}$.

Unit delay. For $\tau > 0$, the reachability graph $\mathcal{R}_\tau$ is trivially calculable from $\mathcal{G}$. Indeed, for any edge (u, v) in $\mathcal{G}$ that appears at time t_1 and disappears at time $t_2 \geq t_1 + \tau$, the arcs (u, v) and (v, u) appear in $\mathcal{R}_\tau$ at time t_1 and disappear at time $t_2 - \tau$. This derivation is simple but essential for bootstrapping iterations of lower bound compositions (e.g., repeated applications of Proposition 3).

4. EFFICIENT COMPUTATION OF REACHABILITY GRAPHS

4.1 Families of reachability graphs

Another interesting property of lower approximations of reachability graphs is that an upper approximation may be derived from it. Indeed, since it contains the exact values of the reachability graph at the start of each epoch, it is trivial to calculate their intersection during each epoch. While the method described in this section focuses on the efficient computation of the lower approximations of reachability graphs, it simultaneously computes the upper approximations.

At a high level, the algorithm presented in this section is a *binary exponentiation* on families of lower bounds of reachability graphs using a special additive operator. To simplify notations, we will consider in this section that $\eta = 1$ and $\tau \in \mathbb{N}^*$.

Definition 9. (**Lower bound family**) Let $\mathcal{G}$ be a 1-regular TVG whose edge traversal time is $\tau \in \mathbb{N}^*$. For $d \in \tau\mathbb{N}$, let $\mathcal{L}_d$ be the lower approximation of a reachability graph $\mathcal{R}_d$ of $\mathcal{G}$. For $d \geq \tau$, we define $\bar{\mathcal{L}}_d$, the *family* of $\mathcal{L}_d$ such that $\bar{\mathcal{L}}_d = \{\mathcal{L}_{d+i}\}_{-\tau<i<\tau}$.

PROPOSITION 4 (**Family additivity**). *Let $\mathcal{G}$ be a 1-regular TVG whose edge traversal time is $\tau \in \mathbb{N}^*$. Let $\mathcal{R}_d$ and $\mathcal{R}_m$ be two reachability graphs of $\mathcal{G}$ such that $d \in \tau\mathbb{N}^*$ and $m \in \tau\mathbb{N}^*$, and $\bar{\mathcal{L}}_d$ and $\bar{\mathcal{L}}_m$ the respective families of their lower approximations. We define $\bar{\mathcal{L}}_d \oplus \bar{\mathcal{L}}_m$ the set of TVGs such that*

$$\bar{\mathcal{L}}_d \oplus \bar{\mathcal{L}}_m = \left\{\bigcup_{0\leq k<\tau} \mathcal{L}_{d+k} \odot \mathcal{L}_{m+i-k}\right\}_{0\leq i<\tau} \cup \left\{\bigcup_{0\leq k<\tau} \mathcal{L}_{d+i+k} \odot \mathcal{L}_{m-k}\right\}_{-\tau<i<0}$$

Then (i) any element $\mathcal{L}_{d+m+i} \in \bar{\mathcal{L}}_d \oplus \bar{\mathcal{L}}_m$ with $-\tau < i < \tau$ is a lower approximation of $\mathcal{R}_{d+m+i}$, and (ii) any TVG in $\bar{\mathcal{L}}_{d+m}$ can be calculated from τ compositions of pairs of TVGs in $\bar{\mathcal{L}}_d \times \bar{\mathcal{L}}_m$.

PROOF. Consider any lower approximation $\mathcal{L}_{d+m+i} \in \bar{\mathcal{L}}_{d+m}$. If $-\tau < i < 0$, $\mathcal{L}_{d+m+i} = \mathcal{L}_{(d+i)+m} = \bigcup_{0\leq k<\tau} \mathcal{L}_{d+i+k} \odot \mathcal{L}_{m-k}$ (Theorem 3). If $0 \leq i < \tau$, $\mathcal{L}_{d+m+i} = \mathcal{L}_{d+(m+i)} = \bigcup_{0\leq k<\tau} \mathcal{L}_{d+k} \odot \mathcal{L}_{m+i-k}$ (Theorem 3). □

We now have our self-sufficient elements, the lower bound families, and an additive operation $\oplus$ between them. For illustration, Fig. 5 details the process of adding $\bar{\mathcal{L}}_4$ and $\bar{\mathcal{L}}_8$ families to obtain an $\bar{\mathcal{L}}_4 \oplus \bar{\mathcal{L}}_8 = \bar{\mathcal{L}}_{12}$ family. The $\mathcal{L}_3$, $\mathcal{L}_4$, $\mathcal{L}_5$, $\mathcal{L}_7$, $\mathcal{L}_8$, and $\mathcal{L}_9$ lower bounds are combined as inputs for three applications of Proposition 3 that yield $\mathcal{L}_{11}$, $\mathcal{L}_{12}$, and $\mathcal{L}_{13}$. Viewed as a black box, this operation combines the $\bar{\mathcal{L}}_4$ and $\bar{\mathcal{L}}_8$ families into an $\bar{\mathcal{L}}_{12}$ family.

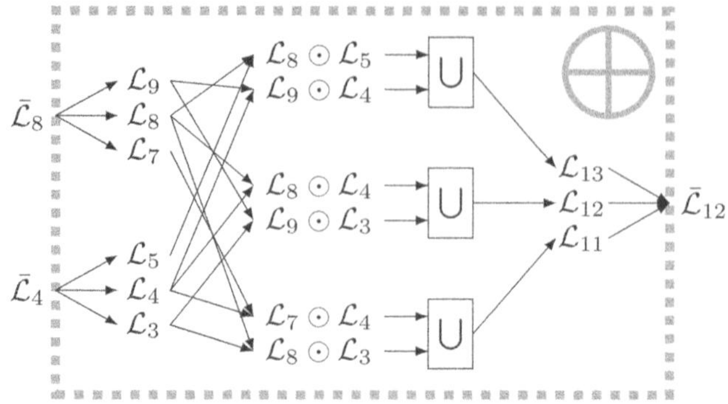

Figure 5: **Adding $\bar{\mathcal{L}}_4$ and $\bar{\mathcal{L}}_8$ families to obtain an $\bar{\mathcal{L}}_4 \oplus \bar{\mathcal{L}}_8 = \bar{\mathcal{L}}_{12}$ family. In this example $\tau = 2$ and $\eta = 1$.**

The inner workings of the $\oplus$ operator are embarrassingly parallel. Indeed, each application of Proposition 3 can be run completely independently of the others. This opens the way for highly distributed implementations, whose speed will be determined by that of the composition of τ pairs of lower bounds of reachability graphs. Accordingly, the next section proposes an efficient algorithm for this composition operation.

4.2 Composing reachability graphs

A time-varying graph $\mathcal{G}$ may be stored as a time-indexed sequence of edge UP and DOWN events. For example, if at time t an event $\{(u, v), \mathsf{DOWN}\}$ occurs, then the edge (u, v) disappears at time t in $\mathcal{G}(t)$. Such a representation is well suited for algorithms that sequentially examine all states of the TVG.

We present a streaming algorithm for composing τ reachability graph lower bounds as in Proposition 3. Streaming algorithms are well suited to TVGs as their memory requirements do not depend on the duration of the trace but only on the number of vertices [1]. Memory is indeed a limited resource, as reachability graphs can become fully connected cliques and data structures such as adjacency matrices cannot therefore be considered sparse. Furthermore, in our case, a streaming algorithm facilitates a parallel implementation of the $\oplus$ operator as one process can read the input families, duplicate their events, and dispatch these to various workers, each calculating one composition operation.

Algorithm 1 reads its input from τ pairs of event streams $\{\mathcal{L}_{d+k}, \mathcal{L}_{m-k}\}$ and writes $\mathcal{L}_{d+m} = \bigcup_{0\leq k<\tau} \mathcal{L}_{d+k} \odot \mathcal{L}_{m-k}$ to its output stream. It makes use of an arc counter that tracks how many of the conditions in the definitions of $\mathcal{L}_{d+k} \odot \mathcal{L}_{m-k}$ (see Definition 8) are verified by each arc. When a previously down link fulfills one of these conditions, its counter is initialized to 1 and an UP event is written to the output stream (Line 5). When this counter goes to 0 the arc is removed from $\mathcal{L}_{d+m}$ and a DOWN event is written to the output stream (Line 7). Note that an arc may be brought up and down at the same time i if its counter goes to 0 right after it appears (i.e., an ephemeral arc). In this case, the arc triggers both lines 5 and 7.

In more detail, this algorithm directly maps to the three conditions in the definition of the approximate composition $\odot$ (Definition 8):

Condition 1. $(u, v) \in \mathcal{R}_{d+k}(i + 1)$. This condition is

ALGORITHM 1: Lower bound composition

Require: $\{\mathcal{L}_{d+k}, \mathcal{L}_{m-k}\}$ `// τ pairs of input streams`
Ensure: $\mathcal{L}_{d+m}$ `// the output stream`
Local: $\{adj_{d+k}, adj_{m-k}\}$, `// τ pairs of adjacency matrices`
Local: *counter*[] `// arc counter`
Local: $delayed_k$[] `// τ lists of delayed down events`

1 **for** $i \leftarrow 0$ **to** T **do**
2   **for** $k \leftarrow 0$ **to** $\tau - 1$ **do**
3   **forall the** $\{a, \mathsf{UP}\} \in \mathcal{L}_{d+k}[i]$ **do** *counter*$[a]$++
4   **forall the** *new arcs a* **do**
5   append $\{a, \mathsf{UP}\}$ to $\mathcal{L}_{d+m}[i-1]$
6   **forall the** *arcs a s.t. counter*$[a] = 0$ **do**
7   append $\{a, \mathsf{DOWN}\}$ to $\mathcal{L}_{d+m}[i-1]$
8   **for** $k \leftarrow 0$ **to** $\tau - 1$ **do**
9   **forall the** $\{(u,v), \mathsf{UP}\} \in \mathcal{L}_{d+k}[i]$ **do**
10   add (u,v) to adj_{d+k}
11   **forall the** $(v,w) \in adj_{m-k}$ **do** *counter*$[(u,w)]$++
12   **forall the** $\{(u,v), \mathsf{UP}\} \in \mathcal{L}_{m-k}[i+d+k]$ **do**
13   *counter*$[(u,v)]$++
14   add (u,v) to adj_{m-k}
15   **forall the** $(w,u) \in adj_{d+k}$ **do** *counter*$[(w,v)]$++
16   **forall the** $a \in delayed_k$ **do** *counter*$[k]$−−
17     $delayed_k \leftarrow \emptyset$
18   **forall the** $\{(u,v), \mathsf{DOWN}\} \in \mathcal{L}_{m-k}[i+d+k]$ **do**
19   add (u,v) to $delayed_k$
20   remove (u,v) from adj_{m-k}
21   **forall the** $(w,u) \in adj_{d+k}$ **do** *counter*$[(w,v)]$−−
22   **forall the** $\{(u,v), \mathsf{DOWN}\} \in \mathcal{L}_{d+k}[i]$ **do**
23   *counter*$[(u,v)]$−−
24   remove (u,v) from adj_{d+k}
25   **forall the** $(v,w) \in adj_{m-k}$ **do** *counter*$[(u,w)]$−−

handled on line 3, *before* making a decision on bringing arcs up or down at the *previous* epoch. Indeed, the output stream at time i can only be written to after reading the input streams up to time $i+1$.

Condition 2. $(u,v) \in \mathcal{R}_{m-k}(i+d+k)$. This condition means that down events in $\mathcal{L}_{m-k}$ must be delayed for one epoch before lowering an arc's counter. This accounts for the local $delayed_k$ lists that are processed on lines 16 to 19.

Condition 3. $\exists w$, $(u,w) \in \mathcal{L}_{d+k}(t)$ and $(w,v) \in \mathcal{L}_{m-k}(t+d)$. This condition is checked by maintaining for each k two adjacency lists adj_{d+k} and adj_{m-k} and checking upon UP/DOWN events whether the end of an arc in adj_{d+k} corresponds to the origin of an arc in adj_{m-k} (lines 10-11, 14-15, 20-21, and 24-25).

Algorithm 1 is then used as a building block to implement the $\oplus$ addition of lower bound families. Starting from $\mathcal{L}_\tau$'s family, for any $n \in \mathbb{N}^*$, $\mathcal{L}_{n\tau}$ is obtained in $\log(n)$ applications of the $\oplus$ operation using a binary exponentiation process.

An implementation of Algorithm 1, fully integrated into a binary exponentiation algorithm over lower bound families, is available as a part of our dynamic trace library (DiTL [28]). This package also contains the code for transforming a lower bound TVG into an upper bound TVG, as well as the time-varying dominating set computation used in Section 5.

4.3 Complexity analysis

The worst-case memory requirements for this one-pass streaming algorithm are straightforward. Its local memory contains a non-sparse arc counter that requires up to $O(N^2)$ space. Furthermore, for each $0 \le k < \tau$, it maintains two non-sparse adjacency matrices and an arc event list thereby requiring $O(N^2)$ space. Adding everything together yields a worst-case space complexity of $O(\tau N^2)$ that is *independent* of the duration of the trace.

Before examining the worst-case time complexity, a word must be said about the implementation of the adjacency matrices (the arc counter is backed by an adjacency matrix). In our implementation, these are backed by per-vertex hash tables. Insertion and removal are therefore constant time operations but this approach may not scale to TVGs with much greater number of vertices than those considered in this paper. In this analysis we will consider that insertion and removal cost $O(\log(N))$ (e.g., by using binary trees).

At each epoch, arcs are brought up and/or down. For each $0 \le k < \tau$, we note M_{d+k} and M_{m-k} the number of UP and DOWN events in the entire TVGs $\mathcal{L}_{d+k}$ and $\mathcal{L}_{m-k}$, respectively. Let $M = \max_k\{\max\{M_{d+k}, M_{m-k}\}\}$ be the maximum number of events in all the involved TVGs. In the worst case scenario where all possible arcs are updated at each epoch, we have $M = O(TN^2)$ events, where T is the number of epochs. Each processed event costs a modification of the arc counter ($O(\log(N))$), an adjacency matrix ($O(\log(N))$), and up to N counters for the composed paths ($O(N\log(N))$). All of these must be performed for the τ pairs of input TRGs. Adding everything together yields a worst-case time complexity of $O(\tau MN\log(N))$, or $O(\tau TN^3\log(N))$ in terms of N exclusively. In practice however $M \ll TN^2$, so the first formulation is more accurate.

5. APPLICATIONS

Having formalized reachability graphs and detailed a method for efficiently computing upper and lower approximations of them in the previous sections, we now study the reachability graphs of several synthetic and real-life traces for a variety of edge traversal times τ and maximum delays δ. In particular, we show how the properties of these reachability graphs place bounds on communication capabilities and highlight the asymmetric nature of dynamic networks.

5.1 Datasets and metrics

While we have calculated reachability graphs on many publicly available datasets, in this paper we present results based on two real-life contact traces selected for their short beaconing periods:

Table 2: Dataset characteristics. N is the number of vertices, M the number of edge UP/DOWN events, T the duration, P the beaconing period, and η the time resolution.

Name	N	M	T	P	η
Rollernet	62	$99k$	$3h$	$15s$	$1s$
Stanford	782	$704k$	$8h$	$20s$	$20s$
Random Waypoint	50	$46k$	$8h$	$1s$	$1s$
Community	50	$824k$	$8h$	$1s$	$1s$

Stanford [23]. As part of an epidemiology study, this trace captures face-to-face contacts among all students, teachers, and staff in a US high school between 7 a.m. and 4 p.m. The 782 ZigBee motes (TelosB Crossbow) sent beacons every 20 seconds (sending times are synchronized in the published trace).

Table 3: **Metrics on η-regular reachability graphs of duration T. $\mathcal{X}_k$ is the value of $\mathcal{X}$ during the kth epoch, i.e., $\mathcal{X}_k = \mathcal{X}\left((k+\frac{1}{2})\eta\right)$.**

Metric name	Definition
Avg. dominating set size	$\frac{\eta}{TN}\sum_{k=0}^{T/\eta} \lVert\mathcal{D}_k\rVert$
Avg. density	$\frac{\eta}{TN(N-1)}\sum_{k=0}^{T/\eta} \lVert\mathcal{R}_k\rVert$
Avg. asymmetry	$\frac{\eta}{T}\sum_{k=0}^{T/\eta} \frac{\lVert asym(\mathcal{R}_k)\rVert}{\lVert asym(\mathcal{R}_k)\rVert + \lVert sym(\mathcal{R}_k)\rVert/2}$

Rollernet [25]. Opportunistic sighting of Bluetooth devices by groups of rollerbladers carrying Intel iMotes during a roller tour. The 62 iMotes performed neighborhood scans every 15 seconds.

For comparison purposes, we also provide results based on two well understood synthetic mobility models:

Random-Waypoint [18]. We simulated 50 nodes with speeds between 3 and 7 m/s from the stationary state in a $1,000 \times 500\text{m}^2$ rectangle sending beacons every second with a 20 m transmission range.

Community Model [19]. We simulated 50 nodes using the same parameters as above with 8 communities. In the community model, nodes with stronger social ties are more likely to be in geographic proximity.

The characteristics of these four η-regular traces are summarized in Table 2.

For each reachability graph, we compute its time-varying dominating set (TVDS), i.e., a mutable set such that, at all times, there exists an incoming arc to any vertex in the TVG from a member of the TVDS. Thus, at any given time t, the TVDS is a traditional dominating set on the directed graph at time t. We note this TVDS $\mathcal{D}_\delta$ if it is derived from a reachability graph $\mathcal{R}_\delta$.

Even on static graphs, calculating a minimal dominating set is a classic NP-complete problem. Here we adapt the well-known greedy algorithm for choosing multipoint relays for broadcasting in a wireless network to the time-varying context [17]. The size of the dominating set calculated by this algorithm is within a factor $\log(N)$ of the optimal, where N is the number of nodes. Our TVDS calculation algorithm reacts to arc UP/DOWN events as follows.

1. If, after the arc event is processed, the previous dominating set is no longer a dominating set, then proceed to step 2. Otherwise do nothing.
2. Iteratively build the new dominating set, starting from an empty set, by greedily adding the node with the greatest outgoing degree to nodes not yet covered by the new dominating set. If two nodes have equal outgoing degrees, pick the one that is in the previous dominating set.

The time-varying dominating set thus calculated attempts to be both reasonably stable over time and close the optimal.

The rest of this section focuses on the following metrics. Their formulas and average values are found in Table 3.

- **Density.** The ratio of the number of arcs in a reachability graph at time t over the total number of possible arcs.

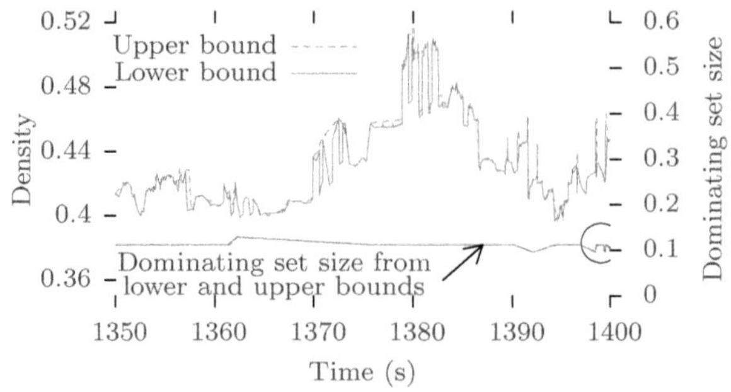

Figure 6: The upper and lower bounds give nearly identical values of density and dominating set size. Example taken from the Rollernet trace with τ=5s and δ=1min.

- **Dominating set size**. The normalized number of vertices in the dominating set at time t.
- **Asymmetry**. Here we no longer consider directed arcs but undirected pairs of vertices. At time t, the asymmetry is the ratio of asymmetric pairs among pairs that have at least one edge between them.

5.2 Algorithm accuracy

The approximation algorithm detailed in Section 4 is *extremely accurate*. Indeed, thanks to Theorem 3 and Definition 8, the approximation calculates the *exact* value for all times $t \in \eta\mathbb{N}$. Furthermore, the upper and lower bounds are nearly identical during epochs ($t \notin \eta\mathbb{N}$). Fig. 6 plots the density over time for both the upper and lower bounds of the Rollernet reachability graph for τ = 5s and δ = 1min. The plot has been zoomed in to show only a small span of the y-axis for 30 seconds. The upper and lower bounds on density are nearly equal at all times. So are the values of the dominating set size computed from the upper and lower approximations that only disagree in the circled area on Fig. 6. These observations hold for all the reachability graphs computed in this paper. Indeed, looking at the values of average density and average dominating set of the over 5,000 pairs of upper/lower bound TVGs calculated in this paper, the maximum disagreement is 8.10^{-3} for the former and 4.10^{-2} for the latter. In practice, the difference is smaller than the width of the line in plots. Therefore, in the rest of this paper, we only plot the values based on the lower approximation.

5.3 Revealing temporal structural properties

The dynamics of reachability graphs highlight temporal structural properties of the original connectivity trace that are not otherwise accessible. Fig. 7 shows snapshots of a subset of the reachability graph in the high school network captured in the Stanford dataset. When the delay is 20 minutes, the classroom structure of the trace is clearly visible. In the original contact trace, edges within a classroom are unstable and lead to merges and splits of small connected components. In the reachability graph, a classroom is a stable complete subgraph with one teacher. A group of interacting teachers are also visible at the top of Fig. 7a. When some teachers and students later change classrooms, they create strong asymmetries in higher-delay reachability graphs such as the 40-minute one depicted in Fig. 7b.

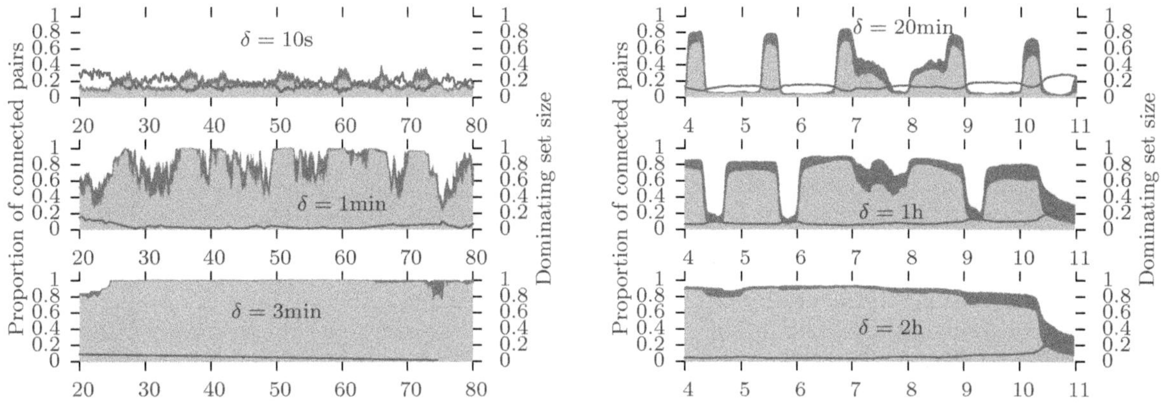

(a) Rollernet ($\tau = 5$s)

(b) Stanford ($\tau = 1$s)

Figure 9: Proportion of connected pairs of vertices over time for two real-life datasets. Connected pairs are divided into symmetric pairs (yellow) and asymmetric pairs (dark red).

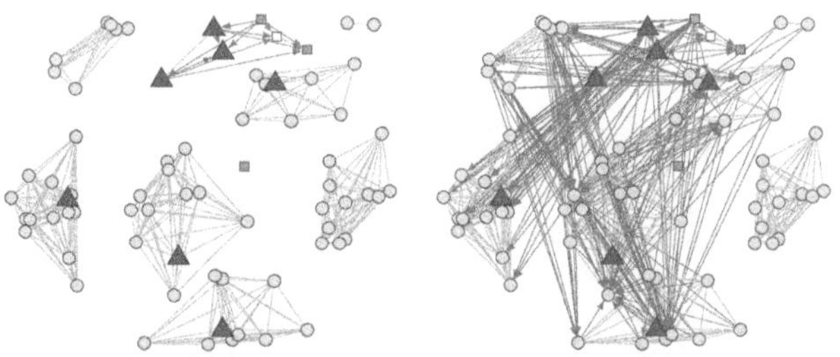

(a) $\delta = 20$min

(b) $\delta = 40$min

Figure 7: Subset of Stanford's 20 and 40-minutes reachability graphs during morning classes ($\tau = 20s$). Dark blue triangles are teachers; circles are students. The dark red arrows represent asymmetric arcs. The classroom structure is clearly visible.

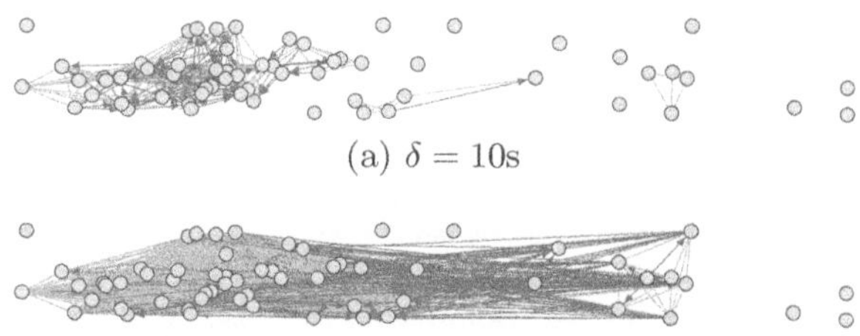

(a) $\delta = 10$s

(b) $\delta = 60$s

Figure 8: Snapshots of rollernet's 10 and 60-seconds reachability graphs after 20 minutes ($\tau = 5s$). The rollerblading tour is moving from left to right. The dark red arrows represent asymmetric arcs. The asymmetry is caused by the acceleration phase in the accordion phenomenon [25].

Fig. 8 shows snapshots of the reachability graph captured in the Rollernet dataset. These correspond to an acceleration phase, where the head of the tour (right on Fig. 8) pulls ahead of the rest of the rollerbladers. Due to the accordion phenomenon, the tail of the tour does not react immediately [25]. This completely prevents short delay communications between these two groups ($\delta = 10$s on Fig. 8a). With a longer delay (e.g., $\delta = 60$s on Fig. 8b), *backward* communications towards the rear of the tour become possible (in particular through organizers who stop on the side of the road and let the tour pass them), thereby creating strong asymmetry in the reachability graph.

A more systematic study of dynamic properties over time is shown on Fig. 9. It plots both the proportion of connected pairs of vertices (left axis) and the size of the dominating set (right axis). The pairs of vertices are further divided into symmetric pairs (yellow) and asymmetric pairs (red). At a given time, if the red histogram reaches 1 then a journey exists in at least one direction between all pairs of nodes. If the yellow histogram reaches 1, then a journey exist in both direction between all pairs of nodes.

For a 1-minute delay, the Rollernet reachability graph alternates, sometimes very rapidly, between fully connected states and highly asymmetric partially connected states (Fig. 9a). Therefore, any opportunistic communication system aiming for latencies under a minute will be strongly impacted by the accordion phenomenon. However, if a communication system can tolerate up to three minute delays, then it should be possible to smooth out the dynamic mobility. Indeed, the 3-minute reachability graph is fully connected for the entire duration of the trace, and the size of its dominating set is almost always equal to one.

The Stanford trace alternates static phases in classroom (the valleys on Fig. 9b) and dynamic phases moving between classrooms or around the food court (the peaks on Fig 9b). The progressive "shrinking" of the valleys illustrates how reachability graphs "grow backwards" with increasing delays. Indeed, if there exist a journey from a to b within 20 minutes at time t, then there exists a journey from a to b within 40 minutes at time t minus 20 minutes. Of course, after the increase in delays exceeds the width of the valleys, the reachability graph eventually reaches it maximum density. Note that in this case, an incompressible amount of asymmetry subsists throughout the trace.

Due to space constraints, we have only shown, in this

section, results for specific values of τ but, as a general rule, the smaller the value of τ, the faster the TRG becomes a clique. When analysing a network with a specific application in mind, the value of τ can be set to a realistic value (e.g. message size over bit-rate), and the plots in this section give an immediate visual understanding of the communication possibilites. The next section examines quantitatively the importance the delay (δ) and the edge traversal time (τ) parameters.

5.4 Bounds on communication capabilities

Reachability graphs give straightforward bounds on communication capabilities. Indeed, the density at a given moment is exactly equal to the *maximum delivery ratio expectancy* of a perfect opportunistic routing protocol whose delay-tolerance is equal to δ and whose message size over bit-rate ratio is equal to τ. Furthermore, the size of the dominating set indicates the achievable *offload ratio* in a scenario where the opportunistic network is assisting an infrastructure (e.g., 3G) for disseminating content to all nodes in the network [12].

Given real-world system requirements, i.e., wireless bit-rate estimates, messages sizes, target delivery ratio, and delay-tolerance, reachability graphs provide an immediate answer to the following question: can an opportunistic network support this service? If not, can it effectively supplement an infrastructure in an offloading scheme?

Fig. 10 plots the average density against the delay tolerance for all datasets and for increasing edge traversal times, while Fig. 11 plots the average dominating set size. Rollernet and Random Waypoint share similar characteristics. Both are very sensitive to increasing edge traversal times. When these are close to one second, near 100% density is achievable with a couple of minutes of delay tolerance (Figs. 10a, 10c, 11a, and 11c). In this case, they can support pure opportunistic communications. However, when edge traversal times are longer, tight delay constraints are impossible unless as part of an offloading scenario (e.g., τ = 10s and δ = 60s for Rollernet, Fig 10a). For τ = 20s, Random Waypoint cannot even provide offloading for reasonable delays (Fig. 11c).

Similarly, Stanford and Community share similar features. Regardless of the delay-tolerance and message size, no pure opportunistic routing protocol can provide anything near 100% delivery ratio [1] (Figs. 10b and 10d). Despite this, they are both good offloading scenarios as the size of their dominating sets is consistently below 20% of the total number of nodes, thereby offering potential offload ratios of around 80%. Indeed, in order to disseminate content to the entire network, pushing one copy per classroom in the Stanford case, or one copy per community in the Community case, plus copies to single nodes is an obvious strategy. However, for larger values of τ (e.g., 20), Community is no longer able to offload content with reasonable delays (Fig. 11d).

5.5 Asymmetry

Asymmetric communications are a fundamental aspect of opportunistic networks. Too strong focus on inter-contact times may lead to overlook asymmetry, but reachability graphs provide a natural way of studying and quantifying it.

Fig. 12 plots for all datasets the average asymmetry against the maximum delay for increasing values of the edge traversal time. The asymmetry for the $\mathcal{R}_\tau$ graph directly derived from the original contact trace is always 0 as it only contains symmetric arcs (see Section 3.4). Asymmetry in Random Waypoint follows a regular pattern: a bell-shaped curve of constant max value, whose width and center increase with τ (Fig. 12c). In this scenario, depending on the delay constraint, up to 80% on average of connected pairs of vertices can only communicate in one direction. This is due to nodes traveling long straight distances creating asymmetric reachability to nodes they meet from nodes they had met earlier. As the maximum delay increases, return journeys using different intermediate nodes appear, the reachability becomes complete, and the asymmetry returns to 0.

As previously, Rollernet and Random Waypoint show similar behavior. However, for Rollernet the maximum asymmetry also increases with τ (Fig. 12a). Indeed, as fewer edges are traversable, journeys become less likely between the front and the rear of the rollerblading tour. When such a journey is possible, it creates a longer-lasting asymmetry. In the Stanford trace, asymmetry follows the same increasing bell shape as in Rollernet, but never returns to 0 (Fig. 12b). In fact, this incompressible *minimum* asymmetry also increases with τ.

6. RELATED WORK

Adaptations of traditional static graph distance metrics and algorithms to time-varying graphs have yielded many different concepts. For example, Orda *et al.* propose a shortest path algorithm for TVGs based on different waiting policies (unrestricted, forbidden, and source waiting) [21]. Our work corresponds to the *unrestricted* policy, in which a message may wait for an unlimited amount of time anywhere along its path through the TVG. Bui Xuan *et al.* have proposed efficient algorithms for calculating shortest (in number of hops), fastest (in path traversal time), and foremost (i.e., earliest arrival) paths in TVGs [29]. All these algorithms are designed to compute the shortest paths to all destination from a source and a fixed starting time. Our algorithms, in contrast, computes the reachibility graph by estimating the shortest paths for all possible starting time.

Several approaches to reachability in time-varying graphs exists. For strictly positive edge traversal times, a simple heuristic consists in dividing time into successive slots of length τ and keeping only edges that are persistently present during each slot [8]. This provides a good lower-bound approximation for small values of τ (i.e., less than η) whereas our approach can handle arbitrary edge traversal times. From a given starting time t, reachability among all pairs of nodes can be calculated by iterating over all edge UP/DOWN events [24]. This calculation can then be repeated for a sample of starting times [14]. This approach yields static reachability graphs for a discrete sequence of starting times, whereas the temporal reachability graphs defined in this paper calculate reachability in continuous time.

Chaintreau *et al.*, in their work on the diameter of opportunistic networks, calculate a *Last Departure / Earliest Arrival* structure for each pair of nodes [5]. This structure can tell for any pair of nodes (A, B), at any time t, when is the earliest arrival for a message leaving A for B at time t. A reachability graph could be easily derived from these structures, but would unfortunately only cover the $\tau = 0$ case. Our approach is more general since it can also handle non-zero edge traversal times.

[1]The $\tau = 0$ results for Stanford are an artifact of the 20-second resolution.

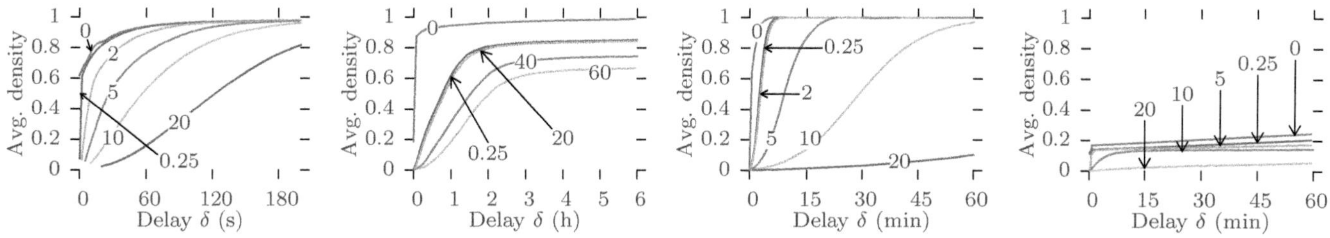

(a) Rollernet (b) Stanford (c) Random Waypoint (d) Community

Figure 10: Average density vs. maximum delay δ for different edge traversal times τ (in seconds).

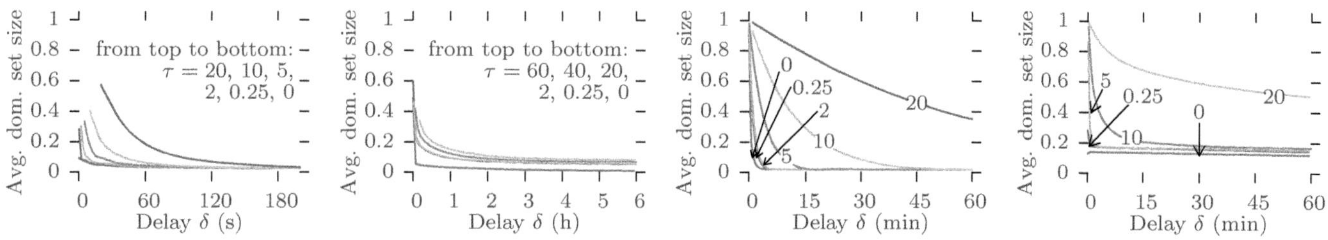

(a) Rollernet (b) Stanford (c) Random Waypoint (d) Community

Figure 11: Average dominating set size vs. maximum delay δ for different values of τ (in seconds).

7. CONCLUSION AND FURTHER WORK

In this paper, we introduced the notion of *temporal reachability graphs.* Given an edge traversal time and a maximum journey delay, temporal reachability graphs capture the temporal connectivity of the time-varying graphs they derive from. After formalizing the concept, we proved that regular reachability graphs, which encompass all experimental datasets, can be composed to compute reachability graphs of higher maximum delay with very high accuracy. Furthermore, we proposed a scalable highly-parallel streaming algorithm for their efficient computation. By applying this algorithm to synthetic and real-life contact traces, we showed how reachability graphs provide fresh new insights on temporal connectivity in time-varying graphs. In particular, they yield an immediate and intuitive picture of the communication capabilities and offloading potential of opportunistic networks.

This work on reachability graphs will be pursued in several directions. Firstly, our results could be extended to the more general case where the edge traversal time τ is not constant but may take values among multiples of a TVG's resolution η. Secondly, as seen in this paper, reachability graphs seem to reveal community structures that are not immediately apparent in the contact traces. They could therefore lead to new approaches for the difficult problem of community detection in time-varying graphs. Finally, much work remains to be done on the statistical analysis and modeling of reachability graphs. For example, what are the correlations between symmetric and asymmetric arcs from a given node? How do degree and inter-arc time distributions evolve with edge traversal time and delay? If reachability graphs turn out to be easier to model than their underlying time-varying graphs, could they be used as a first step for realistic synthetic connectivity models?

8. REFERENCES

[1] N. Alon, Y. Matias, and M. Szegedy. The space complexity of approximating the frequency moments. In *Proc. ACM STOC*, 1996.

[2] S. Bhadra and A. Ferreira. Computing multicast trees in dynamic networks using evolving graphs. Technical Report RR-4531, INRIA, 2002.

[3] A. Casteigts, P. Flocchini, W. Quattrociocchi, and N. Santoro. Time-varying graphs and dynamic networks. In *Ad-hoc, Mobile, and Wireless Networks*, volume 6811 of *Lecture Notes in Computer Science*, pages 346–359. Springer Berlin / Heidelberg, 2011.

[4] A. Chaintreau, P. Hui, J. Crowcroft, C. Diot, R. Gass, and J. Scott. Impact of human mobility on opportunistic forwarding algorithms. *IEEE Transactions on Mobile Computing*, 6(6):606 –620, 2007.

[5] A. Chaintreau, A. Mtibaa, L. Massoulie, and C. Diot. The diameter of opportunistic mobile networks. In *Proc. ACM CoNEXT*, 2007.

[6] A. Clauset and N. Eagle. Persistence and Periodicity in a Dynamic Proximity Network. In *Proc. DIMACS Workshop on Computational Methods for Dynamic Interaction Network*, 2007.

[7] V. Conan, J. Leguay, and T. Friedman. Characterizing pairwise inter-contact patterns in delay tolerant networks. In *Proc. ACM Autonomics*, 2007.

[8] E. M. Daly and M. Haahr. Social network analysis for information flow in disconnected delay-tolerant manets. *IEEE Transactions on Mobile Computing*, 8(5):606–621, May 2009.

[9] F. De Pellegrini, D. Miorandi, I. Carreras, and

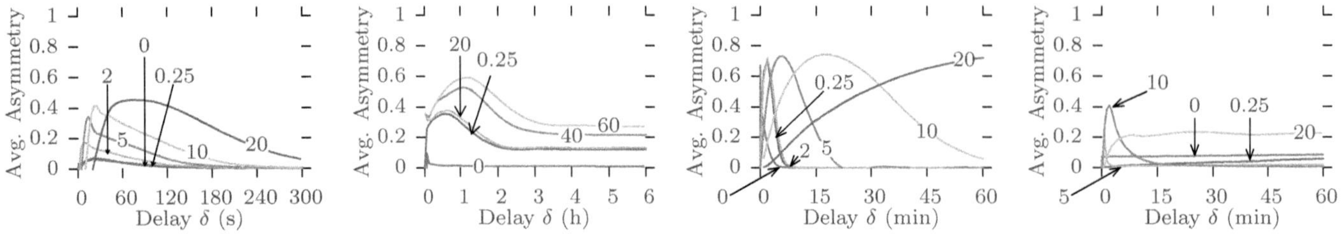

(a) Rollernet (b) Stanford (c) Random Waypoint (d) Community

Figure 12: Average asymmetry vs. maximum delay δ for different values of τ (in seconds).

I. Chlamtac. A graph-based model for disconnected ad hoc networks. In *Proc. IEEE INFOCOM*, 2007.

[10] S. N. Dorogovtsev and J. F. F. Mendes. Evolution of networks. *Advances In Physics*, 51(1079-1187):4, 2002.

[11] K. Fall. A delay-tolerant network architecture for challenged internets. In *Proc. ACM SIGCOMM*, 2003.

[12] B. Han, P. Hui, V. Kumar, M. Marathe, J. Shao, and A. Srinivasan. Mobile data offloading through opportunistic communications and social participation. *IEEE Transactions on Mobile Computing*, 11(5):821 – 834, 2011.

[13] F. Harary and G. Gupta. Dynamic graph models. *Mathematical and Computer Modelling*, 25(7):79–88, 2005.

[14] P. Holme. Network reachability of real-world contact sequences. *Phys. Rev. E*, 71:046119, Apr 2005.

[15] D. Kempe, J. Kleinberg, and A. Kumar. Connectivity and inference problems for temporal networks. In *Proc. ACM STOC*, 2000.

[16] V. Kostakos. Temporal graphs. *Physica A Statistical Mechanics and its Applications*, 388:1007–1023, 2009.

[17] A. Laouiti, A. Qayyum, and L. Viennot. Multipoint relaying: An efficient technique for flooding in mobile wireless networks. In *Proc. IEEE HICCS*, 2001.

[18] J.-Y. Le Boudec and M. Vojnovic. Perfect simulation and stationarity of a class of mobility models. In *Proc. IEEE Infocom*, 2005.

[19] M. Musolesi, S. Hailes, and C. Mascolo. Adaptive routing for intermittently connected mobile ad hoc networks. In *Proc. IEEE WoWMoM*, 2005.

[20] M. Musolesi and C. Mascolo. A community based mobility model for ad hoc network research. In *Proc. ACM REALMAN*, 2006.

[21] A. Orda and R. Rom. Shortest-path and minimum-delay algorithms in networks with time-dependent edge-length. *J. ACM*, 37:607–625, 1990.

[22] A. Passarella and M. Conti. Characterising aggregate inter-contact times in heterogeneous opportunistic networks. In *Proc. IFIP NETWORKING*, 2011.

[23] M. Salathé, M. Kazandjieva, J. W. Lee, P. Levis, M. W. Feldman, and J. H. Jones. A high-resolution human contact network for infectious disease transmission. *PNAS*, 107(51):22020–22025, 2010.

[24] J. Tang, M. Musolesi, C. Mascolo, and V. Latora. Characterising temporal distance and reachability in mobile and online social networks. *SIGCOMM Computer Communications Review*, 40(1):118–124, Jan. 2010.

[25] P.-U. Tournoux, J. Leguay, F. Benbadis, V. Conan, M. Amorim, and J. Whitbeck. Density aware routing in highly dynamic dtns: the rollernet case. *IEEE Transactions on Mobile Computing*, 10(12):1755–1768, 2011.

[26] A. Wagner. The yeast protein interaction network evolves rapidly and contains few redundant duplicate genes. *Molecular Biology and Evolution*, 18(7):1283–1292, 2001.

[27] D. J. Watts and S. H. Strogatz. Collective dynamics of small-world networks. *Nature*, 393(6684):440–442, 1998.

[28] J. Whitbeck. DynamIc Trace Library. http://github.com/neush/ditl.

[29] B. B. Xuan, A. Ferreira, and A. Jarry. Computing shortest, fastest, and foremost journeys in dynamic networks. *International Journal of Foundations of Computer Science*, 14(2):267–285, 2003.

A Simulation Tool for Automated Platooning in Mixed Highway Scenarios

Michele Segata[*†], Falko Dressler[*], Renato Lo Cigno[†], Mario Gerla[‡]
[*]University of Innsbruck, Austria, [†]University of Trento, Italy, [‡]University of California, Los Angeles
{segata,dressler}@ccs-labs.org, locigno@disi.unitn.it, gerla@cs.ucla.edu

ABSTRACT

Automated platooning is one of the most challenging fields in the domain of Intelligent Transportation Systems (ITS). Conceptually, platooning means creating clusters of vehicles which closely follow each other autonomously without action of the driver, neither for accelerating, nor for braking. This leads to several important benefits from substantially improved road throughput to increased safety. The control of such platoons depends on two components: First, radar is typically to be used to control the distance between the vehicles, and secondly, Inter-Vehicle Communication (IVC) helps managing the entire platoon allowing cars to join or to leave the group whenever necessary. Platooning systems have been mostly investigated in controlled environments such as dedicated highways with centralized management. However, platooning-enabled cars will be deployed gradually and might have to travel on highways together with other non-automated vehicles. We developed a combined traffic and network simulator for studying strategies and protocols needed for managing platoons in such mixed scenarios. We show the models needed and present first results using a simple IVC-based platoon management as a proof of concept.

Categories and Subject Descriptors

C.2.1 [**Computer-Communication Networks**]: Network Architecture and Design - Wireless Communication; I.6.5 [**Simulation and Modeling**]: Model Development

General Terms

Design, Verification

Keywords

Platooning, vehicular networks, cooperative adaptive cruise control, simulation

1. INTRODUCTION

In Intelligent Transportation Systems (ITS) platooning, as a mean of building an Intelligent Vehicle Highway Systems (IVHS) has always been a major challenge, because it encompasses several research fields from traffic management to control theory, and from vehicle dynamics to information technology [1,7]. The main objective is to reduce road

MobiCom'12, August 22–26, 2012, Istanbul, Turkey.
ACM 978-1-4503-1159-5/12/08.

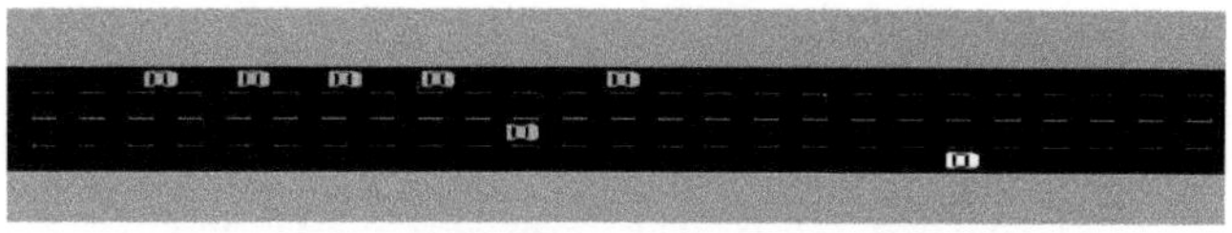

Figure 1: Screenshot of the simulator.

congestion and to increase traffic safety, as well as reducing CO_2 emissions thanks to the tight distance between cars. Radar-based techniques are used for Adaptive Cruise Control (ACC), i.e., for maintaining a safety distance from the vehicle in front. On the other hand, Inter-Vehicle Communication (IVC) is the basis for creating, managing, and organizing the platoons.

It is especially the IVC technique that need further investigation. So far, the ITS community mainly concentrated on fully automated and dedicated highways [1,4,7]. Only recently, the interest has moved to autonomous platooning, i.e., infrastructure-less operation on common roads together with human-driven vehicles [6]. We address one of the key issues in the development process of such systems, namely the integrated evaluation and performance assessment. Implementing and testing platoon management protocols in real environments without proper guarantees is not only expensive but possibly dangerous. We extended the IVC simulation toolkit Veins [8] for assessing the effectiveness of platooning management algorithms and protocols. In this paper, we introduce the resulting simulator and discuss the models we implemented using a simple IVC-based platoon management as a proof of concept. Figure 1 shows a snapshot of the simulator indicating an existing platoon and one car just leaving to take an exit.

Platooning has been studied since the 80's as a mean of increasing the throughput of the streets. PATH in California was one of the first pioneering projects [7]. Later on, Auto21 CDS [4] focused on the technologies needed for smooth merging and splitting. However, they considered dedicated highways for platooning-enabled cars only. Moreover, platoons were to be managed by a centralized center controlling form and composition of the platoons, a model that is still used in more recent studies [3]. One of the few projects considering fully autonomous platoons traveling on the road together with "common" drivers is SARTRE [6]. Here, the leader of the platoon is assumed to be a specially skilled driver, while all other vehicles are free to join and leave the platoon. Unfortunately, no information is available on how these maneuvers are to be performed or what IVC protocols are needed [1]. A Cooperative Adaptive Cruise

```
onInit:;
  setController(DRIVER);
onLeaveInRamp:;
  moveToLeftmostLane();
  setCCSpeed(150km/h);
  setController(ACC);
  sniffBeacons();
onStopSniffing:;
  if no beacons then
  |  startPlatoon();
  |  startBeaconing();
  else
  |  platoonId ← closestInFront();
  |  sendJoinRequest(platoonId);
  end
onJoinReply:;
  setCCSpeed(180km/h);
  setController(CACC);
onReplyTimeout:;
  startPlatoon();
  startBeaconing();
```

Algorithm 1: Vehicle start-up protocol

```
onDistanceToEnd < 1500:;
  sendLeaveNotification();
  leaveLeftMostLane();
  setController(DRIVER);
onLeaderLeaveNotification:;
  sendCurrentPosAndSpeed();
onChangeLeader(newLeader):;
  if me == newLeader then
  |  becomeLeader();
  |  startBeaconing();
  |  sendChangeAck();
  else
  |  updateLeader(newLeader);
  end
```

Algorithm 2: Follower life-cycle protocol

```
onDistanceToEnd < 1000:;
  if no followers then
  |  stopBeaconing();
  |  leaveLeftMostLane();
  |  setController(DRIVER);
  else
  |  Every 1 second:;
  |    sendLeaderLeaveNotification();
  end
onPosAndSpeedReceived:;
  saveFollowerData();
  if all data received then
  |  newLeader ← closestFollower();
  |  sendChangeLeader(newLeader);
  |  stopSendingLeaveNotifications();
  end
onChangeAck:;
  stopBeaconing();
  leaveLeftMostLane();
  setController(DRIVER);
```

Algorithm 3: Leader life-cycle protocol

Control (CACC) controller has been implemented in [2]. Yet, no ACC nor actuation lags have been considered and, again, only fully automated highways are supported.

With the emerging capabilities of IVC, there is now again a growing interest in autonomous platooning using CACC. Obviously, tool chains for testing systems and protocols and assessing their performance are strongly needed. We present a novel simulation toolkit allowing to study such platoon management systems, capturing both the mobility and the inter-vehicle communication issues. To the best of our knowledge, this is the first toolkit for investigating automated platoons in combination with other cars on a single highway.

2. SIMULATION ENVIRONMENT

We implemented the simulation tool based on the Veins simulator [8], which in turn uses OMNeT++[1] for network simulation and SUMO[2] for road traffic simulation. The main modifications have been made to SUMO, implementing a new car following model, which is able to either behave as a human driver or as an automated vehicle using ACC or CACC.

As a reference, we used the controllers detailed in [5]. Due to the lack of space, we only briefly described their properties. The simple Cruise Control (CC) follows the control equation $\ddot{x}_{des} = -k_p(\dot{x} - \dot{x}_{des})$, where $\dot{x}$ and $\dot{x}_{des}$ are the current and the desired speed, respectively, k_p is a design constant, and $\ddot{x}_{des}$ is the acceleration that should be applied. Whereas CC only controls the speed, ACC instead uses a radar in order to continuously measure the distance to the car in front. ACC uses $\ddot{x}_{i\_des} = -1/T\,(\dot{\varepsilon}_i + \lambda\delta_i)$ where $\dot{\varepsilon}_i$ is the speed relative to the vehicle in front, δ_i is the spacing error, so the difference to the desired gap distance, λ is a design constant, and T is the time headway, i.e., the desired distance in seconds to the vehicle in front (usually $T > 1\,\mathrm{s}$).

ACC is not suitable for tight car following in the order of 5 m to 10 m). Thus, a CACC controller is needed for platooning. CACC, takes into account the distance to the car in front using radar as well as information received from the platoon leader via IVC. The latter periodically broadcasts its current speed and acceleration. This way, all the followers can travel really close to each other without the risk of collisions. CACC control uses

$$\ddot{x}_{i\_des} = \alpha_1\ddot{x}_{i-1} + \alpha_2\ddot{x}_l + \alpha_3\dot{\varepsilon}_i + \alpha_4\left(\dot{x}_i - \dot{x}_l\right) + \alpha_5\varepsilon_i$$

where $\ddot{x}_{i-1}$ and $\ddot{x}_l$ are the accelerations of the vehicle in front and the leader, respectively, $\dot{\varepsilon}_i$ is the relative speed to the vehicle in front, $\dot{x}_i$ and $\dot{x}_l$ are the speeds of the car and of the leader, and ε_i is the spacing error. The α_i are design constants.

The desired acceleration $\ddot{x}_{i\_des}$ computed by the controllers cannot be applied immediately, because, due to the dynamics of the vehicle, there will be a certain actuation lag. As described in [5], this lag can be modeled as a first order lag, e.g., using a first order low pass filter with a time constant in the order of 0.5 s. The actual applied acceleration will be $\ddot{x}_t = \alpha \cdot \ddot{x}_{des} + (1-\alpha)\cdot\ddot{x}_{t-1}$ where $\ddot{x}_{t-1}$ is the acceleration at the previous time instant and $\alpha = 0.1666$, assuming a time constant of 0.5 s and a sampling time of 0.1 s.

We implemented all the controllers within SUMO and extended the TraCI interface used by Veins to perform operations like *switchOnACC*, *setCACCLeaderData*, *changeToLane(x)*, etc. These commands simplify the implementation of an application-level protocol in the Veins environment to manage the platoon and to evaluate its performance using metrics like traffic improvement or robustness to communications failures.

3. A SIMPLE PLATOONING PROTOCOL

As a proof of concept, we developed a simple platoon management protocol. We considered a highway scenario with five in/exit ramps in which the leftmost lane is reserved for platoons, similar to carpool lanes in the U.S. We configured SUMO to generate cars with different routes, different desired speeds and different capabilities. Human driven vehicles just follow their route following the human-driver model. Platooning-enabled cars (30 % of the all cars) instead follow the rules according to Algorithms 1, 2, and 3.

[1]`http://www.omnetpp.org/`

[2]`http://sumo.sourceforge.net`

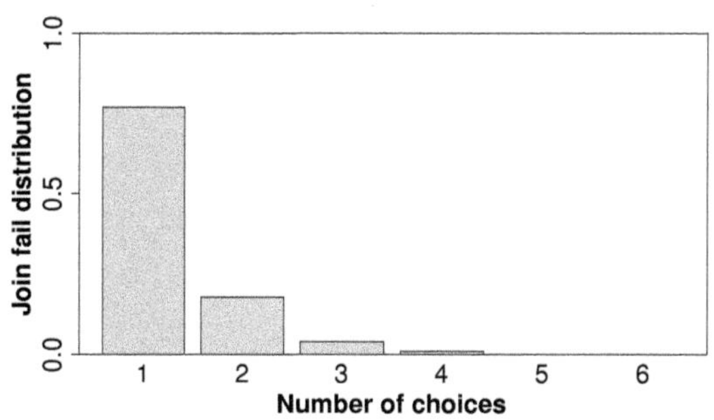

(a) Distribution of choices upon failure

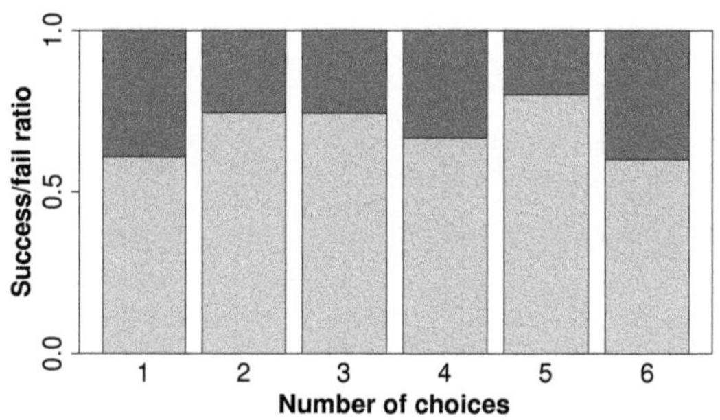

(b) Success/failure ratio

Figure 2: Distribution of join failure as a function of the number of available platoons (a) and comparison between successfully (lightblue) and failed (darkred) joins (b).

A platooning-enabled car entering the highway simply moves to the reserved lane and switches ACC on. The duty of steering the car in and out of the reserved lane is left to the driver. Moreover, we consider all CACC enabled cars to drive at the same speed of 150km/h while in the platooning lane. Deciding what is the best thing to do when a faster platoon reaches a slower one is an open problem which has to be addressed in future. If it receives a beacon message from platoons in front within 5 s, it asks the closest one for permission to join. If that fails, it simply starts its own platoon, otherwise it switches to CACC and accelerates to join the platoon. For leaving the platoon after getting close to its destination, a car notifies the platoon leader, exits the reserved lane, and gives the control back to the driver. If the leader exits, it elects a the new leader (the one following closest), and sends a *changeLeaderNotification*.

In order to illustrate the developed simulator, a screenshot of the simulator indicating a platoon of six cars (red) together with a second car leaving the platoon and a human driven car (yellow) is shown in Figure 1.

For the analysis of the protocol we plotted three graphs. The first two, depicted in Figure 2, show, as function of the number of platoons in front found during the discovery phase, how the join failures are distributed, and the ratio between successful and failed joins. The first one tells us that in the majority of the cases, when a failure happens, the car has no other choices (at least in front of it). In some rare cases, the cars were aware of two or three platoons, meaning that the protocol could be improved by trying to join another platoon. The second plot emphasizes the latter statement, since the ratio of success/failures is independent from the number of choices, i.e., probably due to communication failures.

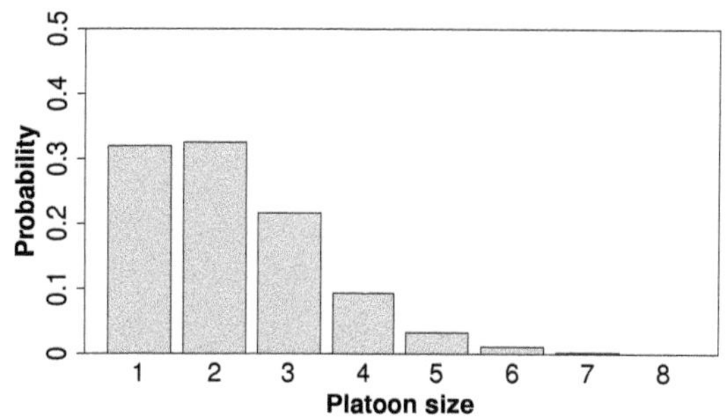

Figure 3: Distribution of platoon sizes computed from duration.

The fact that we are able to determine the existence of up to six platoons in the vicinity of a car, means that there may be several small platoons, thus the efficiency of the protocol might not yet be optimal. This is clearly depicted in Figure 3, where the distribution of platoon sizes is shown. We can see that the majority of the platoons have a size of one or two cars for the 60 % of the time. Only in some rare cases we have platoons of five to eight cars.

These early results and simple protocol implemented, show the power of the developed toolkit. They also suggest that reliable communications are fundamental for platooning efficiency. Moreover, it is clear that efficient and safe platooning needs means for merging, splitting and managing platoons, for instance by complementing the IVC communications with Road Side Units that enable inter-platoon communications even if these are out of direct communication range.

4. REFERENCES

[1] L. Baskar, B. De Schutter, J. Hellendoorn, and Z. Papp. Traffic control and intelligent vehicle highway systems: a survey. *IET Intelligent Transport Systems*, 5(1):38–52, 2011.

[2] P. Fernandes and U. Nunes. Platooning of autonomous vehicles with intervehicle communications in SUMO traffic simulator. In *IEEE ITSC 2010*, pages 1313–1318, Madeira Island, Portugal, September 2010.

[3] P. Fernandes and U. Nunes. Algorithms for management of a multi-platooning system of IVC-enabled autonomous vehicles, with high traffic capacity. In *IEEE ITSC 2011*, pages 1935–1941, Washington, D.C., October 2011.

[4] S. Hallé, B. Chaib-draa, and J. Laumonier. Car Platoons Simulated As A Multiagent System. In *4th Workshop on Agent-Based Simulation*, pages 57–63, Montpellier, France, April 2003.

[5] R. Rajamani. *Vehicle dynamics and control.* Springer, 2011.

[6] T. Robinson, E. Chan, and E. Coelingh. Operating Platoons On Public Motorways: An Introduction To The SARTRE Platooning Programme. In *ITS 2010*, Busan, Korea, October 2010.

[7] S. Shladover. PATH at 20 - History and Major Milestones. In *IEEE ITSC 2006*, pages 22–29, September 2006.

[8] C. Sommer, R. German, and F. Dressler. Bidirectionally Coupled Network and Road Traffic Simulation for Improved IVC Analysis. *IEEE Transactions on Mobile Computing*, 10(1):3–15, January 2011.

CloudMAC - Torwards Software Defined WLANs

Jonathan Vestin, Peter Dely, Andreas Kassler
Computer Science Department
Karlstad University
Karlstad, Sweden
impclaw@gmail.com, peter.dely@kau.se, andreas.kassler@ieee.org

Nico Bayer, Hans Einsiedler, Christoph Peylo
Telekom Innovation Laboratories
Berlin, Germany
{first.last}@telekom.de

Categories and Subject Descriptors

C.2 [**Computer Systems Organization**]: Computer Communication Networks

Keywords

OpenFlow, 802.11, Wireless

1. INTRODUCTION

Wireless Local Area Networks (WLANs) operated by large companies or universities can consist of hundreds or even thousands of Access Points (APs). In such networks, administrators typically do not manage individual APs, but use enterprise WLAN management systems. Those systems hide the complexity of managing heterogeneous networks with different hardware and software platforms. In addition, protocols such as CAPWAP [5] and LWAPP [2] allow vendor independent configuration of APs.

Besides the size of some WLANs, the handling complexity of APs themselves is a challenge. Network virtualization, roaming support, quality of service and energy saving lead to more and more complex APs. For example, the number of software packages included in the the current release of popular open source AP firmware OpenWRT (Backfire) by more than 800% over the past five years. For the widespread Cisco 500 series APs the operating system image size increased by more than 150% from 2007 to 2011.

Moving complexity away from networking devices and exposing data flow management functions via standardized interfaces and high level programming primitives is one of the core ideas of Software Defined Networking (SDN). OpenFlow [10], as one instance SDN, has recently gained a lot of attention. For example, Google has replaced all its inter-datacenter backbone routers with simple OpenFlow-enabled devices, that are controlled from applications run on standard servers [8]. Thereby the operational expenditure can be decreased and network flexibility can be increased.

To get similar benefits in WLANs, we introduce CloudMAC, an alternative management architecture in which APs just for forward MAC frames. Other functions, such as processing MAC frames, are implemented on standard servers that are operated in data centers and can be provided via cloud computing infrastructure. The main contributions of this paper are an implementation of this architecture, a performance evaluation and a list of applications that are enabled through CloudMAC.

MobiCom'12, August 22–26, 2012, Istanbul, Turkey.
ACM 978-1-4503-1159-5/12/08.

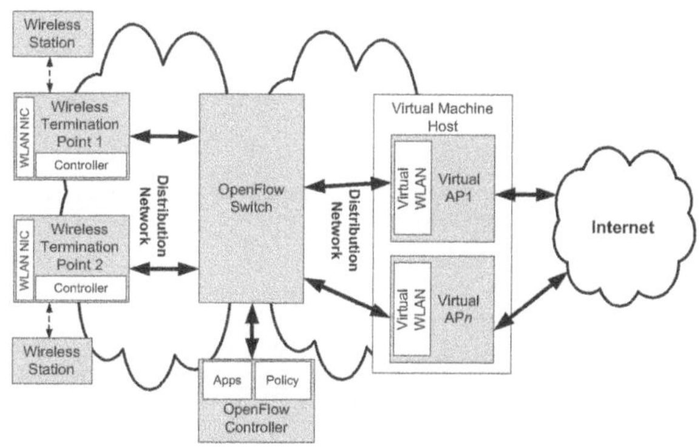

Figure 1: Architecture of a CloudMAC network

The remainder of the paper is organized as follows: In §2 we present the architecture and implementation of CloudMAC. We evaluate the performance in §3 and outline benefits and future applications in §4. Related work is reviewed in §5. The paper is concluded in §6.

2. ARCHITECTURE AND IMPLEMENTATION

2.1 Overview

Figure 1 shows a CloudMAC network, which consists of Virtual APs (VAPs), Wireless Termination Points (WTPs), an OpenFlow switch, an OpenFlow controller and tunnels to connect the entities.

WTPs are slim APs with WLAN cards operated in monitor mode (allows to send and receive raw MAC frames). The cards can send and receive MAC frames, but do not generate frames by themselves. Only ACK frames, which have very tight time constraints are generated by the cards. Each card includes a register of MAC addresses which it generates ACKs for. Other MAC frames are generated and received by VAPs. VAPs are operating system instances run on a virtualization host, such as Xen or VSphere Center. Each VAP has one or several virtual WLAN cards. A virtual WLAN card is driver, that appears to the operating system and user space applications like a normal physical WLAN card. All standard WLAN management tools can be used to set parameters of this virtual WLAN cards.

To send a packet from a VAP to a station, the virtual WLAN card adds a radio-tap and an IEEE 802.11 header and encrypts the frame (if WPA or WEP is used). The radio-tap header informs the physical card which modulation/coding scheme and transmission power to use. The frame is sent via layer 2 tunnels and an OpenFlow switch from the VAP to the physical card of a WTP. The WTP then transmits the frame to the station. In the opposite direction, from a station to the VAP, the process is done in the reverse order: the STA sends a frame, the WTP ACKs upon reception. The frame is forwarded to the VAP via a tunnel, where the virtual WLAN card receives it and hands it over to the networking stack of the operating system.

The OpenFlow switch and controller take a central role in the CloudMAC architecture. The switch contains a switch table, which specifies what frames to forward to which WTP or VAP. The controller runs applications that configure the switch table using the OpenFlow protocol. The forwarding table represents the binding between a VAPs (more specifically, the virtual WLAN cards) and WTPs. Hence, by reconfiguring the switch table, one AP can easily be moved from one WTP to another. As each physical card on a WTP can be bound to multiple virtual WLAN cards (traffic can be distinguished by the BSSID and MAC addresses), CloudMAC inherently supports network virtualization, in which multiple logical networks run on the same physical hardware.

VAPs run access point management software, for example to generate beacon frames, or respond to Association or Authentication MAC frames. As the virtual WLAN card appears like a real card, standard software such as hostapd can be used. One VAP can have many virtual WLAN cards, which are connected to physical cards on different WTPs. In the extreme case, one enterprise WLAN is only one VAP.

2.2 Control Commands

Besides forwarding MAC frames, CloudMAC also allows a fine grained control of configuration commands. Configuration commands are used to configure the WLAN card and typically issued by a user space application in a virtual AP. For example, if an application requests the virtual WLAN card to change its channel, the request is intercepted in the virtual WLAN card driver. The driver then sends a configuration command packet to the OpenFlow switch. The switch forwards it to the OpenFlow controller. If the configuration command is permitted according to a user-configurable policy, the OpenFlow controller forwards the command to a control application residing on the WTP. The control application executes the command and returns the execution status to the OpenFlow controller and the VAP.

2.3 Implementation

We have implemented CloudMAC on the KAUMesh testbed [3]. The WTPs are Cambria GW2358-4 embedded devices and use OpenWRT Backfire. The VAPs are Debian 6.0 VMs on a VSphere Center installation. The VAPs run hostapd 0.6. We use OpenVSwitch 1.3.0 as OpenFlow switch, which runs in a VM on the same VSphere installation. Th switch controlled by custom made Python applications. The virtual WLAN card driver is based on the mac80211_hwsim driver, which was modified to allow MAC frame injection.

3. PERFORMANCE EVALUATION

Potential performance degradations may result from distributing MAC processing as done in CloudMAC. Hence, we compared the performance of CloudMAC with a normal WLAN AP (using the same hardware).

3.1 Microbenchmarks

With ping and iperf we measured the round trip time (RTT) and the throughput between a station and the (V)AP. Figure 2 shows the ECDF of the round trip time and the throughput for different TCP segment sizes. When CloudMAC is used instead of a simple AP, the mean RTT increases from 0.85 ms to 2.35 ms. This increase is due to the additional processing at the OpenFlow switch and due to delay added by the tunnels. However, our experiments showed that time critical MAC frames like association response messages are delivered fast enough to allow standard clients (tested with Windows XP/Linux/MacOS X) to use CloudMAC VAPs.

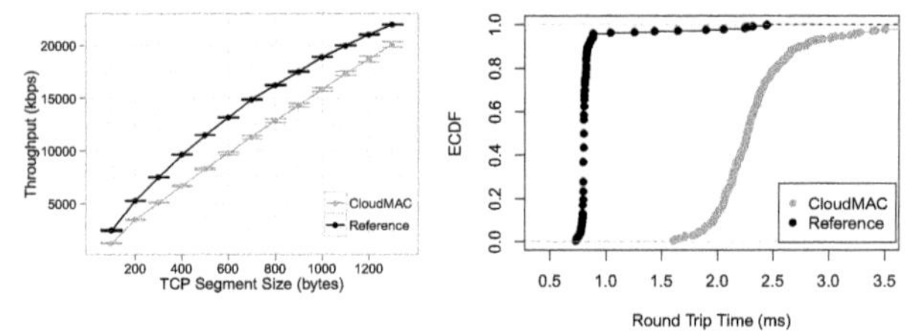

Figure 2: TCP throughput and Ping round trip time (error bars are standard deviation)

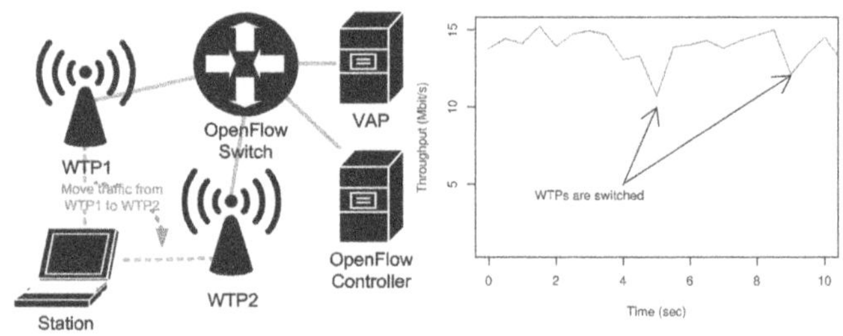

Figure 3: Scenario and throughput while switching WTPs (after 5 and 9 seconds)

Similarly, the TCP throughput is decreased slightly due to the additional components added by CloudMAC. For large TCP segments the performance decreases by approx. 8.5%. This performance decrease is due to the tunnels, which in our current implementation run in user space and therefore require context switching. In future work we plan to investigate the possibility of using kernel-level tunnels, which should result in improved performance.

3.2 Seamless AP Switch-Off

As demo application we implemented a seamless AP switch-off system. The idea behind that application is to save energy by switching of WTPs that cover the same area, if the network load is low. In traditional WLANs such a system would require stations to re-associate to a new AP, which leads a considerable service interruption. In CloudMAC the association state is kept in the VAP, so that moving traffic from one WTP to another only requires changes the forwarding rules in the OpenFlow switch and updating the register of MAC addresses that ACKs are generated for.

We evaluated if CloudMAC causes any noticable interruption when moving traffic from one WTP to another in the scenario depicted in Figure 3. The station downloads traffic from the VAP. After 5 and 9 seconds, the OpenFlow rules and MAC registers are changed, so that the traffic is moved from WTP1 to WTP2 and back again. As Figure 3 shows, there are small losses of throughput during moving. Moving traffic might cause some packets already buffered in the WLAN card to be lost or reordered. Nevertheless, the loss is so small that it is negligible in practice.

4. CLOUDMAC BENEFITS AND FUTURE APPLICATIONS

CloudMAC has several benefits when compared to traditional WLANs:

Thin APs: CloudMAC WTPs only need a basic WLAN driver and a small control application. This reduces software complexity, chances for software bugs and allows for a slimmer hardware. Furthermore, WTPs do not keep association and authentication state information and can therefore easily be replaced during operation (see also §3.2).

Simplified administration: For example, to add support for a new link layer encryption scheme or remove bugs in an existing one, only the VAPs (which might be only one for a whole network) need to be updated. This significantly simplifies administration.

Simple deployment of new applications: Below we will outline several applications, that can be implemented hardware and vendor independent in high level programming language on an OpenFlow controller such as NOX [6].

Integration with OpenFlow networks: OpenFlow is likely to supersede many of today's network management protocols. CloudMAC is fully integrated into the OpenFlow architecture and can utilize OpenFlow infrastructure, such as switches and controller.

CloudMAC enables a range of new applications such as:

On-demand AP: In today's virtualized WLANs one AP might broadcast the SSID for dozens of networks. As each SSID requires one beacon frame, an AP might broadcast hundreds of beacon frames per second and thereby reduce the available capacity for data transfers. CloudMAC enables the following scenario: By default, the OpenFlow switch does not forward beacon frames from a VAP to the connected WTP. When a new user arrives and sends a probe requests (that sometimes includes the SSID of the desired network), an application on the OpenFlow controller inspects the probe request and dynamically enables the beacon transmission. The new user now receives the beacon and can connect to the network. If the probe request includes no SSID, historical usage data and the user's MAC address can be used to identify the desired network. Thereby the number of beacons on the wireless medium can be reduced.

Downlink Scheduling: All traffic between the virtual APs and the WTPs pass through the OpenFlow switch. The switch hence can be used for downlink scheduling either by simple rate shaping as provided by OpenFlow or by time division. For time division scheduling, the OpenFlow controller instructs the switch to only forward the packet of one WTP, while putting the packets of interfering WTPs in a queue on the switch. After one time slot, the switch rules are changed, so that packets of another WTP are released from the queue and forwarded.

Dynamic Spectrum Use: In CloudMAC, one virtual WLAN card can be connected to several physical WLAN cards. This allows the following application: one WTP contains several physical cards, operated with the same MAC address, but on different channels. The physical cards periodically monitor the channel utilization. If a station is currently using a channel with high external interference, the OpenFlow controller creates a IEEE802.11h action frame to instruct the station to switch to a less used channel. The station does not experience any interruption, as it can continue to communicate on the new channel with another physical card of the same WTP. Since the station is associated with the virtual AP (and not the WTP), no re-association is required. This procedure does not require any modification on the client, as long as it supports IEEE 802.11h (which is mandated by IEEE 802.11a/n).

5. RELATED WORK

Related work falls into three main areas:

Split MAC: With a split MAC, some MAC frames are generated in a central server, others on the local WLAN card. CAPWAP [5] and LWAPP [2] implement such a split MAC. In comparison to CloudMAC, they only generate frames for association and authentication, whereas CloudMAC generates all MAC frames, except for ACKs. This makes CloudMAC WTPs simpler and more flexible, as they do not contain any state information of associated STA.

AP virtualization: Previously, [7] and [1] proposed to run an OS hypervisor on APs and thereby provide virtual APs. However, such full AP virtualization requires powerful APs (often also x86-based APs), which are not typical. Besides full AP virtualization, virtual WLAN cards are now standard in Linux. Several Virtual WLAN cards can be operated on top of one physical card. This approach does not allow for fine-grained access control on the virtual WLAN cards (for example by different users).

OpenFlow wireless: OpenFlow has been applied in the context of wireless networks in [9], [11], [4]. Those works focus on IP-layer management and on authentication and QoS-provisioning through OpenFlow.

6. CONCLUSIONS

We have presented a new WLAN management architecture called CloudMAC. CloudMAC implements an OpenFlow-controlled split MAC. The performance evaluation has shown that CloudMAC's performance is similar to normal WLANs and meets the timing requirements to interwork with standard IEEE 802.11 stations. CloudMAC enables a range of new applications, which we plan to implement in future.

7. REFERENCES

[1] O. Braham and G. Pujolle. Virtual wireless network urbanization. In *Proc. of NOF 2011*, pages 31 –34, nov. 2011.

[2] P. Calhoun, R. Suri, N. Cam Winget, M. Williams, S. Hares, B. O'Hara, and S. Kelly. Lightweight Access Point Protocol. RFC 5412 (Historic), February 2010.

[3] Peter Dely and Andreas Kassler. KAUMesh Demo. In *Proc. of 9th Scandinavian Workshop on Wireless Ad-hoc & Sensor Networks*, 2009.

[4] Peter Dely, Andreas Kassler, and Nico Bayer. OpenFlow for Wireless Mesh Networks. In *Proc. of WiMAN 2011*, jul 2011.

[5] S. Govindan, H. Cheng, ZH. Yao, WH. Zhou, and L. Yang. Objectives for Control and Provisioning of Wireless Access Points (CAPWAP). RFC 4564 (Informational), July 2006.

[6] Natasha Gude, Teemu Koponen, Justin Pettit, Ben Pfaff, Martín Casado, Nick McKeown, and Scott Shenker. Nox: towards an operating system for networks. *SIGCOMM Comput. Commun. Rev.*, 38:105–110, July 2008.

[7] Tsuyoshi Hamaguchi, Takuya Komata, Takahiro Nagai, and Hiroshi Shigeno. A framework of better deployment for wlan access point using virtualization technique. In *Proc. of WAINA 2010*, pages 968–973, Washington, DC, USA, 2010.

[8] Urs Hoelzle. Openflow @ google. Youtube online video, URL: http://www.youtube.com/watch?v=VLHJUfgxEO4, May 2012.

[9] R. Mortier, T. Rodden, T. Lodge, D. McAuley, C. Rotsos, A.W. Moore, A. Koliousis, and J. Sventek. Control and understanding: Owning your home network. In *Proc. of COMSNETS 2012*, pages 1 –10, jan. 2012.

[10] The OpenFlow Consortium. Openflow switch specification 1.1, 2012.

[11] Yiannis Yiakoumis, Kok-Kiong Yap, Sachin Katti, Guru Parulkar, and Nick McKeown. Slicing home networks. In *Proc. of HomeNets '11*, pages 1–6, 2011.

A Unified Approach to Identifying and Healing Vulnerabilities in x86 Machine Code

Kirill Kononenko
Secure Software Engineering Group
EC SPRIDE / TU Darmstadt
Mornewegstraße 30
64293 Darmstadt
kirill.kononenko@ec-spride.de

ABSTRACT

The security of software systems is threatened by a wide range of attack vectors, such as buffer overflows, insecure information flow, and side channels, which can leak private information, e.g., by monitoring a program's execution time. Even if programmers manage to avoid such vulnerabilities in a program's source code or bytecode, new vulnerabilities can arise as compilers generate machine code from those representations.

We propose a virtual execution environment for x86 machine code that combines information from compositional, static, and dynamic program analysis to identify vulnerabilities and timing channels, and uses code transformations to prevent those from being exploited. To achieve an appropriate level of performance as well as combine analysis results, our approach stores summary information in the form of conditional rules that can be shared among analyses.

Categories and Subject Descriptors

D.4.6 [**Programming Languages**]: Security and Protection – *access controls, invasive software, security kernels, verification.*

General Terms

Design, Performance, Security, Verification.

Keywords

Cryptography, cyber security, cloud computing, timing channels.

1. INTRODUCTION

In modern software systems, vulnerabilities can arise from many sources. They may occur through incorrect or compromised source code or though compromised tools in the programmer's tool chain, such as compilers or runtime libraries. An antivirus program can detect such vulnerabilities only if they match a known signature. If programs generate executable code at runtime, vulnerabilities in such code will be missed. They can be discovered only while the application is running. Another kind of vulnerability relates to timing channels. A timing channel declassifies some part of classified information for another process by modulating its own use of system resources. To solve this kind of problem, we introduce a virtual execution environment that disassembles code and performs a security analysis.

The Vx32 [1] virtual extension environment is an application-level virtual machine implemented as an ordinary user-mode library and designed to run the native x86 code. Applications can link with and use Vx32 in order to create a safe OS-independent execution environment in which to run mistrusted plug-ins or other extensions written in any language that complies with the x86 code. The machine code for Vx32 is compiled with the GNU Compiler Collection for a special architecture, which is based on x86. First, it is parsed and compiled to an intermediate form. This intermediate form represents a sound approximation of the program behavior and can easily be optimized. Although it cannot express the class of computable functions, it is suitable for domain-specific optimizations. This intermediate form represents variants of deterministic behavior of programs. Simplifying this intermediate representation and use of equivalent transformations allows us to optimize for a specific domain. Algorithms for these transformations are described in a special programming language.

MobiCom'12, August 22–26, 2012, Istanbul, Turkey.
ACM 978-1-4503-1159-5/12/08.

2. Approach and Uniqueness

Our analysis is based on the IFDS/IDE framework [2]. Elements in the analysis are theorems in Primal logic [3]. For this fragment of logic we have formulated an efficient algorithm of derivability based on that of Gurevich-Neeman [3]. If information from any classified function propagates towards any function that may declassify this information, this indicates a potential leak of this information. A deterministic pass and a non-deterministic pass of abstract interpretation allow us to perform dynamic verification efficiently in the most commonly used cases.

We represent the task of dynamic binary translation as a problem of analysis of streams of information at run-time in a specific area of code. A deterministic pass optimizes all paths of execution of the program. We then use a non-deterministic pass that represents the program properties as a stochastic process. As a practical instance of this problem, Figure 1 shows an example of accesses to variables that appear in code; on a primary level (compilation time), this requires that the detection of a target as well as the tracking of that target (with a certain level of security and stability) takes place on a secondary level at runtime.

For example, the non-deterministic part of the optimization collects statistics, such as the number of times each block was executed, branch frequencies, and information of past program behavior that can be used for better optimization. These statistics are used for improved value speculation, branch prediction, and static method lookup. Furthermore, the algorithm chooses between concurrent variants of optimization and finds the most beneficial variant. Finally, the goal is achieved in another thread by a computation of the expected results of these concurrent optimizations without distortion of the operating system. Figure 1 shows an example of both a static and a dynamic trace on a lattice for analysis of live variables. Each element of the lattice has a specific intensity of accesses to variables.

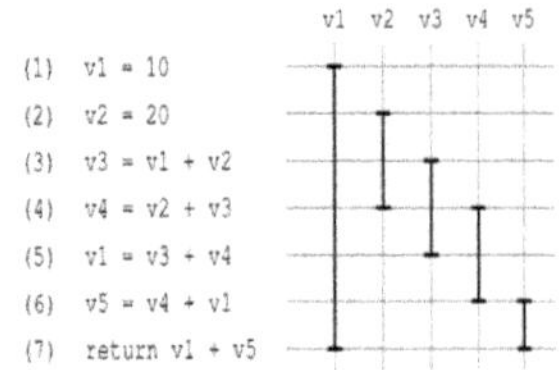

Figure 1 a. Example of a static trace

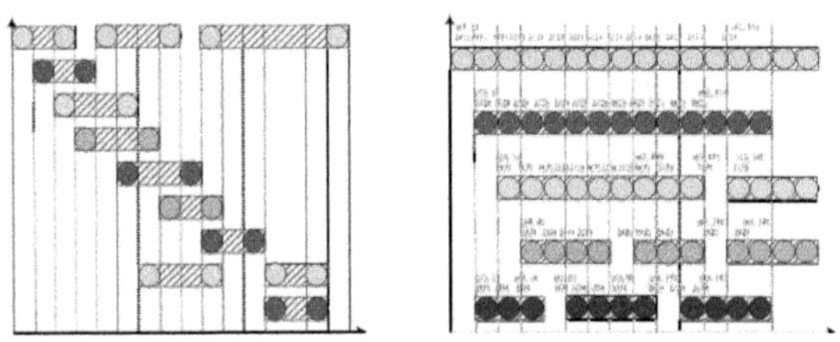

Figure 1 b. Example of a dynamic trace

Figure 2 shows the architecture of an extension of Vx32. The passes are performed in parallel, on different cores, for example. The non-deterministic pass is performed on good test samples, which an expert found very representative, as well as with sampling during execution of the program.

Our extension environment is useful for real-time embedded operating systems and virtual machine runtimes such as the Dalvik and .NET Micro frameworks. In fact, we found that a mid-level intermediate form is not required, as we can represent all properties of the bytecode with executable machine code. This executable machine code can then be parsed and optimized into another version.

There are a few limitations that we consider insignificant. For example, the program is compiled through the use of the GNU Compiler Collection or a compatible compiler. Other significant limitations include the restriction of self-patching and a finite control-flow graph. In addition, any non-deterministic behavior indicates viruses and introduces additional difficulties in debugging of software.

3. RESULTS AND CONTRIBUTIONS

Detection of memory leaks, CHROOT vulnerability, and Trojans is done using the rules of authorization in Primal logic. Detection of stack overflow is done with transformation of the prologue of the function. Additionally, an enhanced Vx32 version makes the code more deterministic from the point of view of the behavior of flags of instructions, accesses of memory, and synchronization of threads.

Instructions that may contain potential timing-channels can be detected statically. For this we use annotations of instructions that classify or declassify information. The weight of a leak is equal to the size of information declassified by the leak.

Timing channels can be fixed with register allocation. Below are the equations used in the optimization of register allocation. The exact solution can be found through the use of a third-party library. Registers are organized in such a way as to minimize the entropy of the timing channel.

$$v(i,1)\,x(1) + v(i,2)\,x(2) + \cdots + v(i,m)\,x(m) \le R(i),\ i \in (1,\ldots,N).$$

$$min(\Sigma^{m}_{j=1} x(j) f(j)) \ .$$

Each equation describes the elements of the lattice from figure 1.

R(i) is the number of available registers for the instruction with an ordinal number *i*, *f(j)* is a random variable, *x(i)* are binary values, *N* is a cardinal number of instructions, and *m* is a cardinal number of variables. Tensor *V* consists of elements *v(i, j)* and is a result of live variable analysis and dead-code elimination.

Vx32 solves these equations with the help of an external utility (a theorem prover). By default we use an approximate solution, which finds a solution in a linear time.

The use of this model for just-in-time compilation yields a significant increase in performance of bytecode execution. For example, use of this model for analysis of MS-CIL and Ruby improves performance of execution of applications in benchmarks.

The results of our research may be useful for the development of just-in-time compilers, interpreters, and compilers with a mixed execution mode of programs for virtual machines such as CLI, Java, Python, Perl, Ruby, LLVM, and Parrot. Further research directions might include the fields of dynamic compilation and dynamic verification. For example, there is the possibility of creating new domain-specific register allocation and code generation algorithms. These algorithms can then be optimized for applications in various domains, for example, in compression and security of binary code.

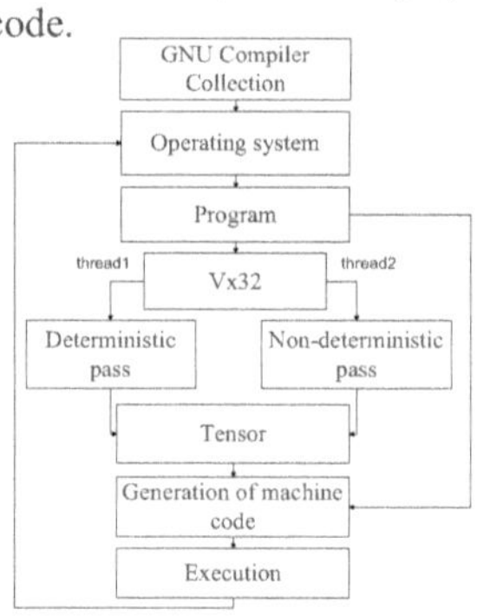

Figure 2. Architecture of extension of Vx32

4. ACKNOWLEDGMENTS

This work was supported by the German Federal Ministry of Education and Research (BMBF) in EC SPRIDE and by the Horst Görtz Foundation Hessian and the LOEWE excellence initiative in CASED.

5. REFERENCES

[1] Vx32 // http://pdos.csail.mit.edu/~baford/vm/.

[2] Mooly Sagiv, Thomas Reps, and Susan Horwitz. 1996. Precise interprocedural dataflow analysis with applications to constant propagation. *Theor. Comput. Sci.* 167, 1-2 (October 1996), 131-170. DOI= http://dx.doi.org/10.1016/0304-3975(96)00072-2.

[3] Distributed Knowledge Authorization Language // http://dkal.codeplex.com.

An Energy-Efficient Architecture for Multi-Hop Communication Between Rovers and Satellites in Extra-Terrestrial Surfaces

Daniel Irwin
Under-grad Student Member, ACM
School of Electronic Engineering
Dublin City University, Ireland
+353-17007645
daniel.irwin2@mail.dcu.ie

Hrishikesh Venkataraman
Member, ACM
School of Electronic Engineering
Dublin City University, Ireland
+353-17006401
hrishikesh@eeng.dcu.ie

Gabriel-Miro Muntean
Performance Eng Laboratory
School of Electronic Engineering
Dublin City University, Ireland
+353-17007648
munteang@eeng.dcu.ie

Abstract

Over the past three decades, several man-made vehicles have being sent into space to explore the extra-terrestrial bodies. As the search for water and other useful substances in the extra-terrestrial surfaces increases, this exploration activity is set to dramatically increase over the next decade (2020); with NASA planning to explore the surface of Mars, Moon and other planets and satellites. In this regard, it is imperative to build an extremely energy-efficient communication system that will cover a large area in the range of hundreds of kilometres which unfortunately is absolutely not possible today. A two-hop communication mechanism that has been well researched in the literature is insufficient to cover the extremely large distance of extra-terrestrial surface. In this paper, a novel three-hop cluster-based hierarchical communication architecture is proposed which could be easily extended to higher number of hops. In particular, it allows the individual rovers to move large distances for collecting data and at the same time provide an extremely energy-efficient mechanism for continuously exploring the surfaces. The simulation results show that in a Martian surface, the proposed three-hop design results in a higher capacity as compared to a single-hop or a cluster-based two-hop design even when the rovers move 100 km.

Categories and Subject Descriptors:
G.4 **[MATLAB]**
C.0 **[Systems Application Architecture]**

Keywords: Extra-terrestrial surface, Mars, Multihop, Rovers.

1. INTRODUCTION

Developing a communication mechanism for exploring surfaces of several hundreds of kilometres in extra-terrestrial surfaces is an extremely complicated operation.

MobiCom'12, August 22–26, 2012, Istanbul, Turkey.
ACM 978-1-4503-1159-5/12/08.

This is because; a lot of vehicle's energy was used solely to send data. This in-turn resulted in large battery requirement for the rovers which increases the overall load of the satellite sent to the extra-terrestrial surface.Hence, it is important that a communication system be built with low energy consumption in mind. A state-of-the-art mechanism for rover design is to have it solar powered, i.e., a rover that works by taking energy from the Sun. However, in these systems, the amount of energy consumed at any instant of time is nearly same as that of the energy produced during the same time period. Hence, these rovers work in the day time and 'sleep' in the night [1]. This is clearly not an efficient mechanism in terms of both capacity and signal monitoring. Also, the maximum possible travel distance of a rover is only 1 km [1] which is a great hindrance especially when the rovers are supposed to cover the entire surface of planet like Mars.

A multi-hop communication system has been proposed to carry out these exploratory works. However, as can be gathered from the extensive literature [2, 3, 4], most of the focus on multi-hop communication is geared towards two-hop communication. This is sufficient for cellular and wireless networks designed for establishing wireless communication in urban and densely populated areas in Earth. However, extra-terrestrial exploration has totally different requirements; ranging from distances, monitoring of signals, battery limitations, abundance of frequency resource, etc. Hence, it is imperative that an efficient architecture is developed that could form the basis for establishing communication mechanism in extra-terrestrial surfaces.

In this paper a three-hop cluster-based communication system is proposed that could be easily extended to higher number of hops. This design is developed, primarily for use in extra-terrestrial bodies like Moon, Mars, etc. Further, the communication mechanism would be investigated for two different surfaces; particularly for Earth and Mars. The difference in the propagation model of Mars is due to the weather, atmosphere composition and terrain. Mars's weather may be similar than the Earth's weather, but is still very complex [5]. Similarly, Moon has different set of variables to Mars and the Earth. For instance, Moon has no atmosphere. However the terrain, the cosmic radiation and the heat from the sun will all have an effect on how the communication system will work. It would be interesting to observe how the novel proposed architecture could be used for long-range communication and its efficiency across different surfaces and different extra-terrestrial bodies.

This paper is organized as follows. Section 2 describes the proposed mechanism for exploring Mars' surface using rovers and an efficient cluster-based architecture for realizing the same. Section 3 explains the simulation model, particularly the scheduling and communication mechanism; and importantly the propagation model in the surface of Mars and describes some initial results while Section 4 provides a brief conclusion of the work.

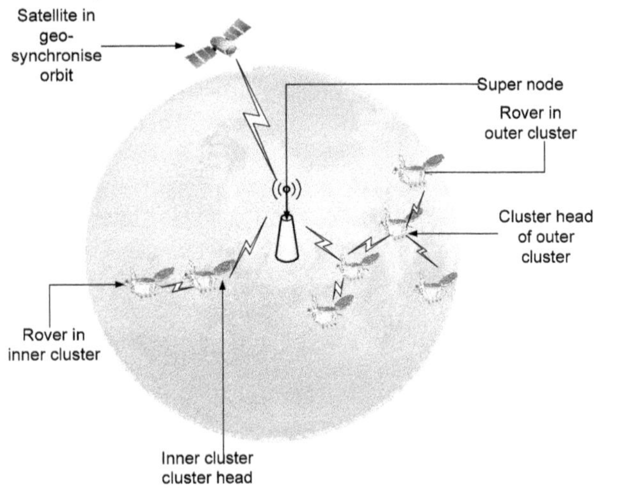

Fig. 1 Rovers used to explore extra-terrestrial surface (Mars)

2. COMMUNICATION SYSTEM

2.1 Architecture

At present, the Mars Odyssey is the current means of communication between the surface vehicles and Earth. It takes approximately 16 minutes to go from horizon to horizon and the surface vehicles send data to the satellite for 10 minutes [6]. Keeping this in mind, a cluster-based hierarchical multi-hop communication system is designed in this work; with the initial focus on three-hop model. This could be then further expanded to a four/higher hop system. The communication system consists of a central super node which would be the main source of communication between the rovers and the satellite. The satellite would be in geosynchronous orbit located directly above the super node. This way, the flow of data between the super node and geosynchronous orbit would remain constant.

In the proposed architecture, as shown in Fig. 1, the central super node is surrounded by several vehicles/ rovers. These vehicles are divided into groups/clusters. Hence, each cluster will contain several vehicles and one vehicle which will act as a cluster head as a means of communication between the cluster and the super node. The cluster head of each cluster will be placed at a minimum distance of 10 km (around 50 km) from the super node. All the vehicles/rovers in the cluster would communicate its information to the cluster-head which in turn would communicate the information of all the vehicles/rovers to the central super-node. As shown in Fig. 2, the proposed three-hop communication system has two sets of clusters – *inner-cluster* and *outer-cluster*. These inner clusters (marked by orange colour in Fig. 2) are surrounded by outer clusters. Each cluster in the inner set of clusters has the same structure; with all the vehicles in the cluster sending their information to a vehicle acting as a cluster head. However the vehicle that receives the information in the outer clusters does not communicate with the super node. These vehicles send their information to the cluster heads in the inner clusters. Through the inner cluster-heads, they communicate with the orbiting satellite.

Importantly, the communication mechanism would comprise of time division multiplexing (TDM) wherein, each vehicle in the cluster is provided a fraction of time period to communicate with the cluster-head. Further, since the cluster-head node communicates the information of all vehicles within the cluster to the super-node, the super-node to cluster-head communication is provided with M time slots, where M is the number of vehicles within the cluster. A detailed explanation of TDM for multi-hop communication is provided in [4].

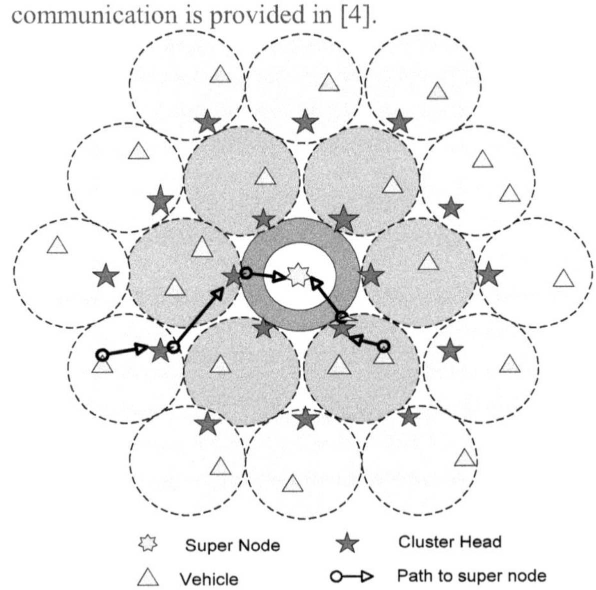

Fig. 2 Proposed architecture of three-hop communication system

2.2 Advantages of Cluster-based Model

The cluster-based architecture for three-hop model is considerably complex than a two-hop communication scenario. However, such three-hop architecture has two major benefits for exploration in extra-terrestrial surfaces.

Energy Recharge: The continuous exploration activity by the rovers would in general result in the battery of these rovers dying quite frequently. However, in case of the cluster-based design, if the battery of the vehicle where to reach a critical level, the communication of all the vehicles would not have to be stopped. The particular vehicle would stop its operation, handover its functionalities to some other vehicle/rover and then drive itself closer to the super node to charge. Vehicles will be only able to charge when it is close to the super node. The area where a vehicle is given permission to charge its self is shown as a white circle in the centre of the centre circles. If one of the vehicles that need to charge their battery is a cluster-head, other rovers/vehicles in the cluster would take over the responsibility of being the cluster-head and then communicate with the super-node.

Energy Consumption/ Large Coverage: The proposed cluster-based architecture for exploring large area surfaces is scalable beyond three; and even four hop communication system. With the increase in the number of hops, the communication complexity increases while at the same time, the overall coverage area also increases. With increasing number of hops, the energy levels needed for a surface vehicle to send a signal to a super node on the surface of the extra-terrestrial body is much smaller than the energy levels needed to send signals form surface vehicles to orbiting satellites

3. SIMULATION MODEL AND RESULTS

Most propagation models in the literature have been developed for use on Earth. When developing a propagation model for use on Mars, the differences in the environment of Mars compared to Earth need to be taken into account. This difference in environment is provided in [7]. In the simulations, an Irregular Terrain Model (ITM), also known as the Longley Rice Model is selected and used. ITM is described to fit the Mars surface very closely [7] and hence, is used in this paper. In the initial stage, the system was simulated for a free-space environment and for a rural environment of Earth based on Okumara-Hata model. Though it is a simple scenario, the overall surface of Mars would also be devoid of any building and other man made effects. Further, using the Longley Rice model, the capacity results in the surface of Mars is computed for a cluster-based three-hop design and compared against a single-hop and a cluster-based two-hop model. The system is simulated using Matlab with a transmit power of 1W and a path-loss exponent of *two*. Fig. 3 shows the improvement in the system capacity in the surface of Mars. It can be observed from Fig. 3 that the system capacity of a three-hop cluster-based system is 8% higher than a two-hop model and up to 23% higher than a traditional single-hop communication system. Notably, irrespective of the number of hops between the end-to-end link, the capacity results obtained for surface of Mars are higher than obtained traditionally in the surface of Earth. This is because; there is no man-made interference in the extra-terrestrial surface yet. Finally, the poster presentation during the conference would also compare the results obtained for surface of Mars with that obtained in Earth.

4. CONCLUSION

A hierarchical multi-hop cluster-based communication system is developed that could be used mainly for exploring extraterrestrial bodies such as Mars, Moon, etc. In particular, a three-hop the cluster-based model is proposed whereby the rovers could continuously sense the environment and send the signals to the geo-synchronize orbit through the central super-node. Such a design not only results in potential improvement in the energy efficiency but also provide a much larger increase in the coverage area – the two main factors in the design of communication system for any extra-terrestrial exploration.

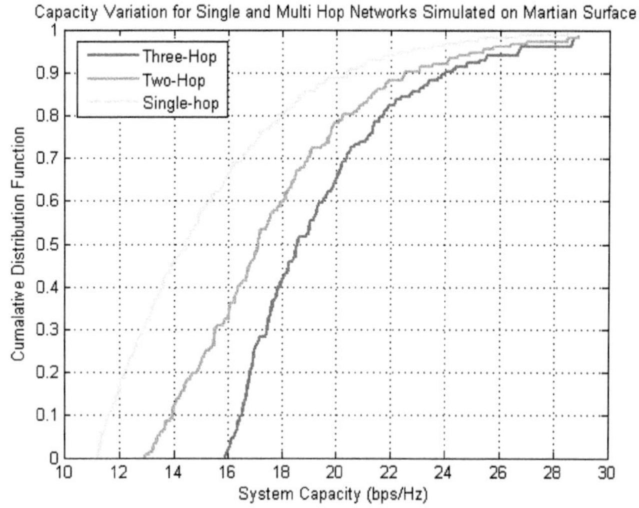

Fig. 3 System capacity variation of a single-hop and multi-hop network at Martian surface

References

[1] "Digital Learning Network Mars Rover", Glenn Research Centre, NASA, http://dlnrover.grc.nasa.gov/mer-briefing.html - Last accessed on 7th June 2012.

[2] H. Venkataraman, A. Chowdhury, A. Sikora and G.M. Muntean, "*COARSE – A Cluster-based Quality Oriented Adaptive Radio Resource Allocation Scheme*", IET Communications, Vol. 6, No. 1, pp. 46-54, January 2012.

[3] H. Venkataraman, S. Sinanovic and H. Haas, "*Cluster-Based Design for Two-hop Cellular Networks*", International Journal of Communications, Network and System Sciences (IJCNS), Vol. 1 No. 4, pp. 370-385, November 2008.

[4] H. Venkataraman and G.M. Muntean, "*Dynamic Time Slot Partitioning for Multimedia Transmission in Two-Hop Cellular Networks*", IEEE Transaction on Mobile Computing, Vol. 10, No. 4, pp. 532-543, April 2011.

[5] J. Annis, "*The Atmosphere of Mars and Optical Communication*", TDA Progress Report 42-91, pp. 124-132, July-September 1987.

[6] Mars Exploration Rovers, Jet Propulsion Laboratory, NASA, http://marsrovers.jpl.nasa.gov/technology/bb_telecommunications.html - Last accessed on 7th June 2012.

[7] V. Chukkala, P. DeLeon , S. Horan, V. Velusamy, "*Radio Frequency Channel Modeling for Proximity Networks on the Martian Surface*", Elsevier Computer Networks, Vol. 47, pp. 751–763, 2005.

Low-Cost Interferer Detection and Classification using TelosB Sensor Motes

Bastian Bloessl, Stefan Joerer, Fabian Mauroner and Falko Dressler
Institute of Computer Science, University of Innsbruck, Austria
{bloessl,joerer,dressler}@ccs-labs.org

ABSTRACT

Inter-protocol interference is one of the most critical issues in wireless communication. For example, this becomes extremely problematic in environments where robustness and real-time communication need to be considered, e.g., in industrial automation or health care applications. Recently, possible approaches for interference mitigation have been described in the literature assuming that the interferer is known in advance. We contribute to this line of research with a framework for interferer detection and classification. Essentially, we use a simple IEEE 802.15.4 transceiver as for example used on the TelosB sensor motes to scan the 2.4 GHz ISM band. This band is used by different technologies including Bluetooth, WiFi, and cordless phones. The key challenge is the accurate timing of the scanning of the frequency band. The presented framework supports flexible descriptions of such scan jobs allowing to adapt to the detectors requirements, depending on the interfering protocols.

Categories and Subject Descriptors

C.2.1 [**Computer-Communication Networks**]: Network Architecture and Design - Wireless Communication; C.4.3 [**Performance of Systems**]: Measurement Techniques

General Terms

Experimentation, Measurement

Keywords

Interference, ISM band, interferer detection, interferer classification

1. INTRODUCTION

We investigate the possibilities for low-cost interferer detection and classification based on a rather cheap and IEEE 802.15.4 standard conform transceiver chip integrated with the TelosB sensor mote platform. Inter-protocol interference is becoming a predominant problem for many wireless networks including WiFi access to the Internet. This is because the unlicensed ISM bands, especially the 2.4 GHz band, is a source of increasing interference among the different protocols. For example, WiFi, ZigBee-based sensor networks, Bluetooth devices, and dedicated application specific protocols have been developed. Studies have shown that insufficient knowledge of interfering signals may lead to substantial performance degradation [7,9]. However, current protocols, e.g., IEEE 802.11 or IEEE 802.15.4, do not take inter-protocol interference into account. This reduces the robustness, which is critical if reliable communication is required, e.g., in industrial automation environments [3]. Adaptive channel switching strategies can be realized assuming better knowledge about the interferer [5].

In 2009, Chowdhury and Akyildiz studied the problem of interferer detection and formulated a general methodology [4]. In particular, the authors used sensor nodes to collect spectral data which was used for offline interferer identification. Most of the commercial spectrum analyzers, e.g., AirMagnet Spectrum XT, Agilent Spectrum Analyzer or Bandspeed AirMaestro employ specialized hardware. They are expensive and cannot be easily integrated into an interference-aware protocol design. However, tools such as Wi-Spy and Ubertooth are more affordable, but provide limited capabilities. Based on commodity hardware, Airshark [8] uses WiFi cards to identify different interference sources.

We present an extended version of Spectrum Analysis Framework for Interferer Classification (SaFIC) [2], a sensor mote based spectrum analysis framework for the 2.4 GHz ISM band. Our framework allows to measure the Received Signal Strength Indicator (RSSI) in a predefined spectrum and visually displays the signal strengths and their corresponding frequencies in real-time. It is implemented on typical sensor mote hardware using a Chipcon CC2420 transceiver chip that is widely used in many application domains including health care, building automation, and industry automation. Our extended architecture supports fine grained scan jobs with very precise timing control to accurately detect multiple interfering networks at the same time. For improved scan performance, the jobs can be organized that only frequencies under investigation need to be evaluated. For rapid prototyping of detector modules, we support a flexible language to define complex scan jobs.

2. FRAMEWORK ARCHITECTURE

An overview of the framework is depicted in Figure 1. It consists of a host application and a customized firmware that is running on a TelosB mote [2,6]. The TelosB uses the very common Chipcon CC2420 transceiver chip. The transceiver mainly determines the measurement speed and resolution and hence defines the limitations of our framework.

MobiCom'12, August 22–26, 2012, Istanbul, Turkey.
ACM 978-1-4503-1159-5/12/08.

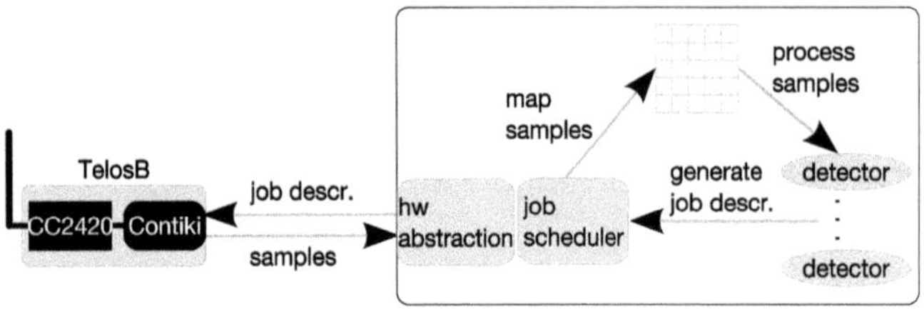

Figure 1: Interferer detection architecture

2.1 TelosB Firmware

The firmware is based on the Contiki operation system. It receives and parses the scan job descriptions which define how measurements should be recorded. The format and the capabilities of these jobs are described in Section 2.3. Before a measurement is started, all watchdogs and interrupts of the mote need to be disabled. This ensures that the measurements are not delayed or interrupted by any other event that may occur. Thus, the measurement process becomes deterministic and the results can be interpreted as the relative timing between the measurements is known. When the job is completed the watchdogs and interrupts are enabled again and the recorded samples are transmitted to the host computer for further processing.

2.2 Host Component

The host part of our framework is shown on the right hand side of Figure 1 which in essence consists of the job scheduler, multiple detection modules and a global matrix with measurement results. The job scheduler queries to detection modules for jobs in a round-robin fashion and transfers them to the sensor mote. When a sampling job has been completed by the sensor mote, the job scheduler receives the measurements and stores them in the matrix. Thereafter, the detection modules process the updated samples. Additionally, the matrix is visualized live in order to support an immediate, intuitive understanding of the issued scan jobs and the detection algorithms. Thus, the limitations of the sensing device can be investigated quickly and new findings can be incorporated in the development of detector modules.

2.3 Job Description Language

We extended the possibilities to describe jobs substantially. While the initial version supported only the most basic description of a single measurement, we can now nearly arbitrary combine scans, delays, and loops. This enhanced description of jobs can be formally specified in Backus-Naur Form as

$\langle job \rangle$ ::= $\langle command \rangle$ `'end'`

$\langle command \rangle$ ::= $\langle scan \rangle$ | $\langle loop \rangle$ | $\langle delay \rangle$
| $\langle command \rangle$

$\langle scan \rangle$ ::= `'scan' START END STEP DWELLS SWEEPS`

$\langle loop \rangle$ ::= `'loops' REP` $\langle command \rangle$ `'loope'`

$\langle delay \rangle$ ::= `'delay' DELAY-TIME` .

The *measurements* define where samples are to be recorded in the frequency band. In addition to the `START` and `END` of the spectrum also a `STEP` width can be specified. Due to the limitations of the transceiver, we can only measure frequencies that are multiples of 1 MHz and in the range from

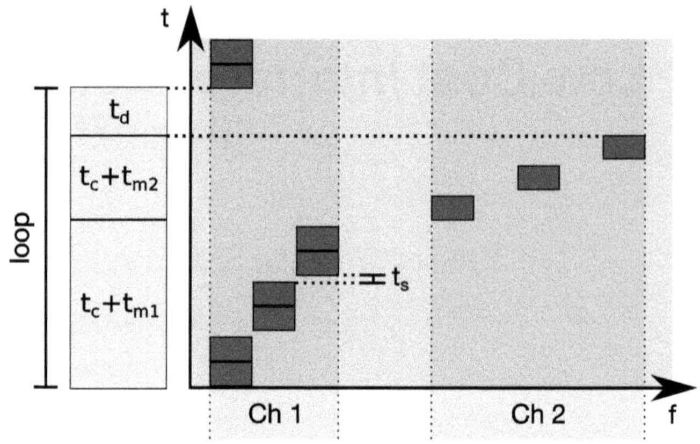

Figure 2: An interleaved measurement

2.4 GHz to 2.5 GHz. The `DWELL` and `SWEEP` parameters have the same semantic as in [2] and define how many samples are taken on a certain frequency and how often we hop through the spectrum. The arguments of *loop* and *delay* (`REP` and `DELAY-TIME`) specify how often the *commands* inside the loop are repeated and the duration to wait respectively.

3. FRAMEWORK CAPABILITIES

SaFIC offers the following capabilities for developing interferer detection and classification modules: **Wide range of sampling patterns**, i.e., the scan parameters offer extensive possibilities for sampling the entire frequency range; **Interleaving of scans**, i.e., the possibility to have two independent scans within one job giving the user the possibility to sample multiple channels nearly in parallel; **Delays**, i.e., support for delay commands between single measurements, which might be used to check for specific timings during protocol classification.

An example job is depicted in Listing 1 and Figure 2. It uses two specific scans to measure the signal spectrum of two frequency ranges *Ch 1* and *Ch 2*. The first scan uses two `DWELLS` in one `SWEEP`, whereas the second scan uses one `DWELL` in a single `SWEEP`, thus, representing two different sampling strategies. The `STEP` in the second scan defines a frequency stepping of 2 MHz. Finally, Listing 1 introduces a delay (t_d) after the two scans have been recorded.

```
loops 2
    scan f1 f2 1 2 1      // Ch1
    scan f3 f4 2 1 1      // Ch2
    delay t_d
loope
```

Listing 1: Example job

3.1 Timing of the Measurements on TelosB

In order to interpret the measured RSSI values, it is crucial that the relative timing between the samples is known. As we stated in Section 2.1 interrupts are disabled during the measurement process and, thus, its behavior becomes deterministic. Based on timing measurements and analysis of the assembler output, we created a model for our job descriptions. All measurements were made with the help of an oscilloscope that we connected to an I/O pin of the mote. According to our findings every *command* adds a

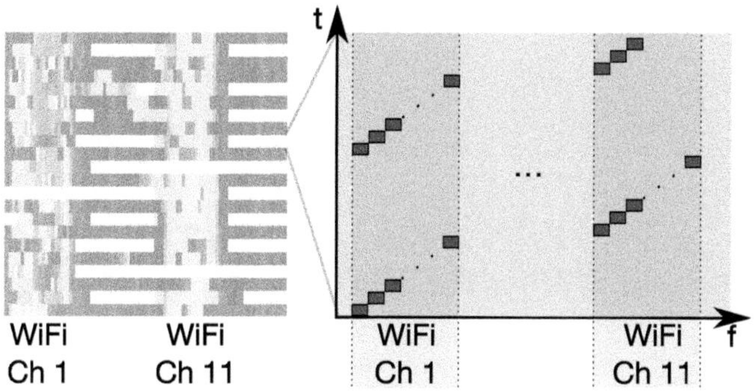

Figure 3: Example WiFi detection

constant delay $t_c = 7\,\mu s$. For a *scan* we figured out two relevant timings, while one covers a single measurement which takes a time $t_r = 80\,\mu s$, switching the frequency adds a delay $t_s = 300\,\mu s$. Apart from additional instructions that are required to tune to another frequency, t_s also includes the time to generate the first RSSI value. According to the standard, each RSSI measurement should be averaged over eight symbols [1], which correspond to $128\,\mu s$. Thus, with the number of sampled frequencies n_f, the total time for a *scan* t_{scan} can be calculated as

$$n_f = \lfloor (END - START)/STEPS \rfloor$$
$$t_{scan} = n_f \cdot SWEEP \cdot DWELL \cdot t_r + n_f \cdot SWEEP \cdot t_s + t_c.$$

With this formula the timings t_{m1} and t_{m2} from Figure 2 can be determined. The delay can be specified in the range from $7\,\mu s$ to 50 ms. Currently it is implemented as a busy-waiting loop that needs three CPU cycles per iteration. Hence, the delay resolution is $0.77\,\mu s$ which is determined by the clock speed of the microcontroller that runs at 3.9 MHz.

3.2 WiFi Detection

To demonstrate the features of our framework we implemented a simple detection module for WiFi. Figure 3 shows a screenshot of the visualization while the WiFi detector is running in our office. It can be seen that two separate WiFi networks have been recognized. To accomplish this, the detector takes rough sampling data from the entire 2.4 GHz spectrum and matches it with the spectral masks that are defined in the standard. Based on this matching a threshold is used to decide where WiFi networks may be present. Then, the detector focuses on the two parts of the spectrum where these networks operate and samples these candidate channels in an interleaved manner as shown on the right side of Figure 3. With this approach, the module is able to detect individual channels as long no other protocol interferes. However, differentiating multiple interference sources that operate in the same frequency range remains as open challenge.

4. CONCLUSION AND FUTURE WORK

We extended our interferer detection framework SaFIC to support flexible description of measurement jobs. For this reason, we defined a job description language which offers arbitrary combinations of loop, delay, and multiple scans within one measurement job.

The main advantage of our new framework is to support interleaved measurements. As a proof of concept, we demonstrated a WiFi detector that is now able to identify two WiFi networks operating on different channels more accurately. Furthermore, we improved the precision of the timings of each measurement job.

In future work, we aim to develop optimized detector modules that, for example, make use of the delay command to schedule local preprocessing tasks on the sensor mote. This would allow to transfer most of the detector logic to the mote as well.

5. REFERENCES

[1] Wireless Medium Access Control (MAC) and Physical Layer (PHY) Specifications for Low Rate Wireless Personal Area Networks (WPANs). IEEE Standard 802.15.4-2006, IEEE, 2006.

[2] B. Bloessl, S. Joerer, N. Nordin, C. Sommer, and F. Dressler. SaFIC: A Spectrum Analysis Framework for Interferer Classification in the 2.4 GHz Band. In *IEEE INFOCOM 2012, Demo Session*, Orlando, FL, March 2012. IEEE.

[3] F. Chen, R. German, and F. Dressler. Towards IEEE 802.15.4e: A Study of Performance Aspects. In *IEEE PERCOM 2010, IQ2S Workshop*, pages 68–73, Mannheim, Germany, March 2010.

[4] K. R. Chowdhury and I. F. Akyildiz. Interferer Classification, Channel Selection and Transmission Adaptation for Wireless Sensor Networks. In *IEEE ICC 2009*, pages 1–5, Dresden, Germany, June 2009.

[5] R. Gummadi, D. Wetherall, B. Greenstein, and S. Seshan. Understanding and Mitigating the Impact of RF Interference on 802.11 Networks. In *ACM SIGCOMM 2007*, pages 385–396, Kyoto, Japan, August 2007.

[6] J. Polastre, R. Szewczyk, and D. Culler. Telos: Enabling Ultra-Low Power Wireless Research. In *ACM/IEEE IPSN 2005*, pages 364–369, Los Angeles, CA, April 2005.

[7] S. Shin, H. Park, and W. Kwon. Mutual Interference Analysis of IEEE 802.15.4 and IEEE 802.11b. *Elsevier Computer Networks*, 51(12):3338–3353, 2007.

[8] R. Shravan, P. Ashish, and B. Suman. Airshark: Detecting Non-WiFi RF Devices Using Commodity WiFi Hardware. In *ACM SIGCOMM 2011*, pages 2–4, Berlin, Germany, November 2011.

[9] A. Sikora and V. Groza. Coexistence of IEEE 802.15.4 with Other Systems in the 2.4 GHz-ISM-Band. In *IEEE IMTC 2005*, pages 1786–1791, Ottawa, Canada, May 2005.

Autonomous and Distributed Recruitment and Data Collection Framework for Opportunistic Sensing

Güliz Seray Tuncay Giacomo Benincasa Ahmed Helmy
Department of Computer and Information Science and Engineering
University of Florida
Gainesville, FL 32611
{gstuncay, giacomo, helmy}@cise.ufl.edu

ABSTRACT

People-centric sensing is a novel apporach that exploits the sensing capabilities offered by smartphones and the mobility of users to sense large scale areas without requiring the deployment of sensors in-situ. Given the ubiquitous nature of smartphones, people-centric sensing is a viable and efficient solution for crowdsourcing data. In this work, we propose a fully distributed, opportunistic sensing framework that involves two main components which both work in an ad hoc fashion: Recruitment and Data Collection. We analyzed the feasibility of our distributed approach for both components through preliminary simulations. The results show that our recruitment method is able to select 66% of the nodes that are appropriate for the sensing activity and 88% of the messages sent by these selected nodes reach the sink by using our data collection method.

Categories and Subject Descriptors

C.2.4 [**Computer-Communication Networks**]: Distributed Systems—*distributed applications*; C.2.2 [**Network Protocols**]: Routing Protocols

General Terms

Design, Experimentation, Algorithms

Keywords

Crowdsourcing, DTN, Mobile Computing, Opportunistic Sensing, People-centric sensing

1. INTRODUCTION

Recent advances in sensing capabilities, storage capacity, and computation power of smartphones–along with the increasing pervasiveness of their adoption–led to the development of people-centric applications that exploit the capabilities of these devices to crowdsource data collection about the users' surroundings. People-centric sensing has therefore become a viable approach for large-scale urban sensing activities that do not require the deployment of sensors in-situ. Lane et al. discern two classes of people-centric sensing activities: participatory and opportunistic sensing [1]. Participatory sensing requires the participants to *consciously opt to meet the application requests* [1]; therefore, it is necessary that the participants are engaged in the sensing activity, or that the participants are rewarded to achieve better involvement. Conversely, in opportunistic sensing, minimal or no interaction with the participants is required; in fact, opportunistic sensing applications may run in the background and opportunistically collect data with no need to interact with the participant. High participation is crucial for the success of people-centric activities. Opportunistic sensing may be a suitable and inexpensive way to achieve high participation; however, current approaches for people centric-sensing rely on centralized registries to recruit possible participants [2]. Moreover, the collected data is uploaded from the participants to remote servers by using cellular networks. These approaches may lead to privacy and economical concerns that may work as a disincentive for participation. This paper describes a fully distributed and privacy-preserving framework for opportunistic sensing. The framework includes a recruitment component and a data collection component, both of which work in an ad hoc fashion and do not rely on the presence of an infrastructure. The only exceptions are the nodes that are designated as data sinks for collecting the sensed data, and the nodes designated as recruiters for the data collection campaign. The number of these nodes is expected to be orders of magnitude less than the number of sensing nodes. The framework consists of two main components: *recruitment* and *data collection*, both of which rely on previous work on profile-cast [3].

MobiCom'12, August 22–26, 2012, Istanbul, Turkey.
ACM 978-1-4503-1159-5/12/08.

2. PROFILE-CAST OVERVIEW

Profile-cast [3], is a behavior-oriented protocol that aims to send messages to the nodes matching a certain target *behavioral profile*, rather than explicit IDs, in opportunistic networks. Each node can autonomously construct its profile by monitoring the places it visits and the time spent at those locations. Based on this information, the node can build an association matrix that represents the percentage of time the node spent at various locations. Singular value decomposition (SVD) is then applied to this association matrix to obtain the dominant set of eigen-behavior vectors that represent the behavioral profile of the user. A target profile is a summarized mobility preference vector which is represented using the same format as that of the user behavioral profile. The similarity of two behavioral profiles is determined by computing the cosine similarity between their eigen-behavior vectors. In profile-cast, information is routed in two phases: *gradient ascend* and *group spread*. During the gradient ascend phase, messages are diffused following an ascending gradient of similarity between the behavioral profile of the node and the target profile. No replication is allowed; the message is transfered from a node to another with higher similarity to the target profile. When the message reaches a node with a higher similarity to the target profile than a certain threshold th_{sim}, the protocol switches to group spread mode. Group spread mode allows replication, and messages are copied from the holder to an encountered node with a higher similarity than th_{sim}. This double-phase of operation allows profile-cast to achieve a fair trade-off between overhead and delivery ratio.

3. THE OVERALL FRAMEWORK

3.1 Scenario

Profile-cast can be used for both recruitment and data collection. In the context of recruitment, specifically-designated nodes can be deployed prior to the start of a sensing activity and can disseminate recruitment messages that advertise the upcoming sensing activity. Since only the nodes that are likely to appear in the sensing area during the sensing time need to be recruited, the target profile of the recruitment messages will contain the locations that are included in the sensing area. For data collection, the sensing nodes need to send the collected data to the deployed data sinks. In this case, the messages can be sent using a target profile that describes the behavioral profile of the sinks.

3.2 Recruitment

The aim of the recruitment phase is to notify the nodes of an upcoming sensing activity. We exploit the stability of the user behaviors and based on the mobility history information, we recruit only the nodes that are likely to be in the sensing area when the sensing activity is taking place. In order to efficiently achieve a distributed recruitment, we propose the deployment of one or more recruiting nodes that visit the sensing area prior to the start of the sensing activity and then disseminate recruitment messages. Profile-cast can be used to perform the dissemination, and a profile which includes the locations in the sensing area can be used as the target profile. However, unlike profile-cast, where the aim is to reach nodes matching a certain target profile, for recruitment it is enough to reach the nodes that match only a *part* of the target profile. In other words, if the sensing area is large and consists of multiple locations, nodes that visit only few of the sensing locations should also be recruited. It is therefore necessary to change the similarity metric to be able to evaluate the similarity between single locations and the behavioral profile of the nodes. Finally, since the recruiting nodes are deployed in the sensing area, they can be configured to operate only in group spread mode; thus, they hold a copy of the message after replicating it to other nodes.

3.3 Data Collection

Once data is collected, it needs to be returned to the organizer of the sensing activity. We propose the use of data sinks for the collection of user data and suggest that the participating nodes opportunistically exploit ad hoc encounters to reach data sinks that are temporarily deployed in the sensed area. The data sinks are assumed to be temporary and mobile; moreover, they are not expected to be available for the whole duration of the sensing activity. In many cases, it is plausible that the locations visited by the data sinks are known or explicitly selected prior to the start of the sensing activity. This information can be included in the recruitment message and therefore can be used as the target profile to reach the sinks. While in profile-cast, the aim is to reach each nodes matching the target profile; in the context of data collection, the aim is to reach *any* of the nodes that match the target profile. For this reason, along with the fact that the sensing nodes are expected to greatly outnumber the number of data sinks and that they may produce a large number of messages, it is critical to limit the data replication by having the nodes run solely in gradient ascend mode. While this may significantly increase the delivery time, in many sensing applications it is not an issue.

4. EVALUATION AND RESULTS

In order to evaluate our approach, we developed a trace-driven simulation using USC Wi-Fi traces, in a way analogous to what is described in [3]: users are assumed to be in communication range when they are associated with the same access point at the same time. For the opportunistic sensing activity, we designed a test scenario in which the sensing activity is scheduled on a part of the USC campus. Since opportunistic sensing works best when there is high participation in the sensing activity, the sensing activity was scheduled to take place in the area including the seven most-visited locations. A single node, designated both as a recruiter and a data sink, is deployed one week before the start of the sensing activity, and continuously moves along a seven waypoint predefined path, that is assumed to be included in the recruitment message. The sink spends five minutes at each waypoint, before moving to the next. This behavior is continuously repeated along the duration of the sensing (from 8 am to 8 pm).

4.1 Recruitment

In our recruitment scenario, the recruiter node is configured to send a recruitment message every hour until the start of the sensing activity. The similarity threshold, th_{sim}, used in the recruitment simulation, is 0.8. A node is recruited if it receives the recruitment message before the sensing activity starts and is similar to the target profile in the recruitment message by at least 80%. We analyzed the traces and learned that the number of nodes that appear in the sensing area during the sensing activity, and hence should be recruited, is 376. The number of nodes which received the sink's recruitment messages is shown in Figure 1. 247 nodes are recruited for the sensing activity, which corresponds to 66% of the nodes that are eligible for taking part in the opportunistic sensing activity. It is worth highlighting that all 247 recruited nodes were among the 376 nodes that were eligible to take part in the sensing activity. Figure 1 also shows the overhead in terms of the number of nodes that are reached by the recruitment message. The reason for this inefficiency is that the behavioral profile is based on historical information and it is therefore possible that the behavior of the nodes is not always consistent with their mobility.

Figure 2 shows the compound number of recruited nodes as a function of the recruitment start time. It can be observed that the number of recruited nodes converges very quickly.

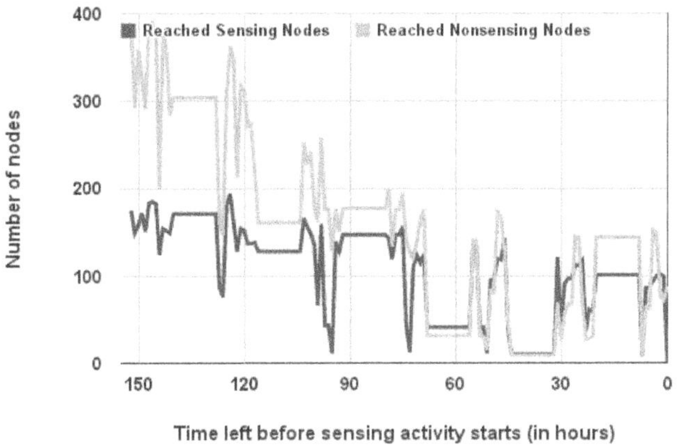

Figure 1: Number of nodes that received the recruitment message (recruited and not recruited) and number of nodes recruited (progressively) versus time

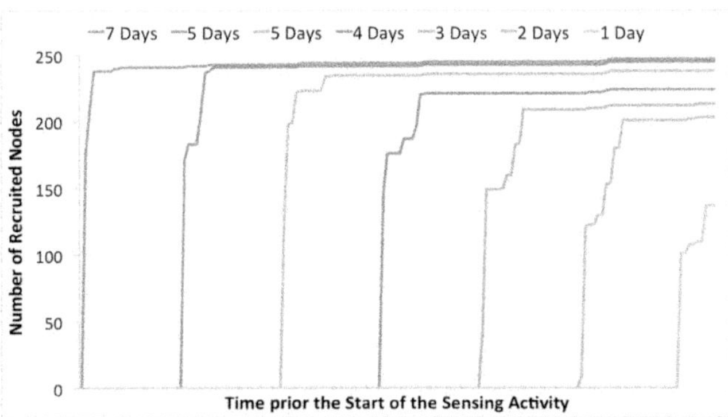

Figure 2: Number of nodes recruited (progressively) versus time

4.2 Data Collection

The nodes send their collected data every hour when they are in the sensing area. In the first scenario, we assumed ideal recruitment, meaning that all the nodes that are in the sensing area are recruited and take part in the sensing activity. Each of these nodes send one message to the sink per day. 2419 out of 2879 messages sent during the sensing activity are able to reach the sink node; that is, the delivery ratio of our method is 84%. In the second scenario, only the nodes that are selected by the recruitment algorithm send messages to the sink node. 1718 out of 1952 messages reached the sink, meaning that the delivery ratio of the proposed method is 88%. Figure 3 shows the messages sent by the nodes in the areas that are visited by the sink and the messages that are delivered. For each location, the first bar shows the number of delivered and undelivered messages sent from this location for the ideal recruitment case, while the second bar represents the case in which only the nodes selected by the recruitment phase send messages to the sink. It can be observed that a significant portion of the sent messages are delivered for both cases and the results for the second case are slightly better.

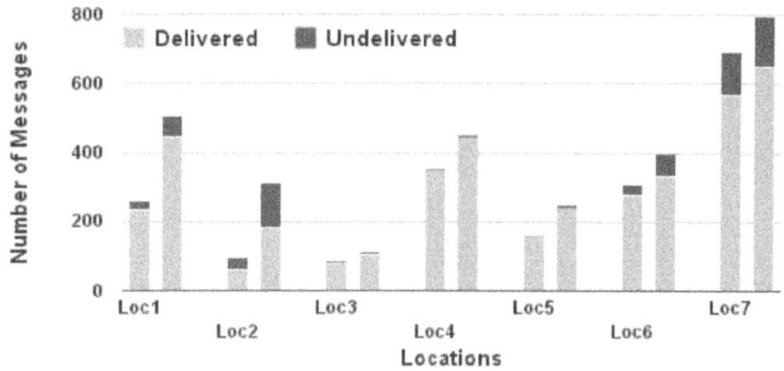

Figure 3: Delivery status of the messages from the locations visited by sink for 1)ideal recruitment case and for 2)only selected nodes.

5. CONCLUSION AND FUTURE WORK

In this work, we proposed a framework for autonomous and distributed recruitment and data collection in opportunistic sensing. For participant recruitment, we plan to investigate the addition of the progress review phase [2] into the framework and introduce online recruitment for selecting additional participants while the sensing is already taking place. For data collection, we plan to analyze the efficiency of appropriate unicast routing protocols and compare them with our proposed method. Additionally, we plan to consider recruiting sinks from the pool of users based on their mobility histories, by taking into account metrics such as behavior stability and variety of encounters. Furthermore, we plan to develop an opportunistic sensing application for smartphones in the future.

6. REFERENCES

[1] N. D. Lane, S. B. Eisenman, M. Musolesi, E. Miluzzo, and A. T. Campbell. Urban sensing systems: opportunistic or participatory? In *Proceedings of the 9th workshop on Mobile computing systems and applications*, HotMobile '08, NY, USA, 2008. ACM.

[2] S. Reddy, D. Estrin, and M. Srivastava. Recruitment Framework for Participatory Sensing Data Collections. In *International Conference on Pervasive Computing*, 2010.

[3] W. Hsu, D. Dutta, and A. Helmy. CSI: A paradigm for behavior-oriented profile-cast services in mobile networks. *Ad Hoc Networks*, 2011.

Protocols for Highly-Dynamic Airborne Networks

Egemen K. Çetinkaya, Justin P. Rohrer, Abdul Jabbar, Mohammed J.F. Alenazi,
Dongsheng Zhang, Dan S. Broyles, Kamakshi Sirisha Pathapati, Hemanth Narra,
Kevin Peters, Santosh Ajith Gogi, and James P.G. Sterbenz
Information and Telecommunication Technology Center
Department of Electrical Engineering and Computer Science
The University of Kansas, Lawrence, KS 66045, USA
{ekc, rohrej, jabbar, malenazi, dzhang, dbroyl01, kamipks, hemanth, kevjay, santoshag, jpgs}
@ittc.ku.edu

ABSTRACT

End-to-end communication in highly-dynamic airborne networks is challenging due to the presence of highly mobile nodes and the inherent nature of wireless communication channels. Domain-specific protocols are required that can address these challenges and enable reliable transmission of data in this environment. We develop the ANTP (airborne network and transport protocols) suite that operates in this highly-dynamic environment while utilising cross-layer optimisations between the physical, MAC, network, and transport layers. We show how each component in the ANTP suite outperforms the traditional TCP/IP and MANET protocols through simulation using ns-3. Having verified these protocols through simulation and analysis, the next step towards deployment of the ANTP suite is developing a cross-platform implementation of the protocols. Towards this end we present an architecture for the protocol stack to be implemented in the Python programming language.

Categories and Subject Descriptors

C.2.1 [**Computer-Communication Networks**]: Network Architecture and Design—*Network topology, Wireless communication*; C.2.2 [**Computer-Communication Networks**]: Network Protocols—*Routing protocols*

General Terms

Algorithms, Design, Experimentation, Performance

Keywords

Airborne mobile wireless network, DTN (disruption-tolerant network), MANET, ns-3 simulation, Python implementation

1. INTRODUCTION

Highly-dynamic airborne networks pose unique challenges to end-to-end data transmission. Mobility poses a great challenge since the airborne nodes can travel at speeds as high as Mach 3.5. In addition, the network is bandwidth-constrained due to limited spectrum. Intermittent connectivity is also a challenge, which is caused by the extremely short contact duration between any two nodes [11, 10, 9].

MobiCom'12, August 22–26, 2012, Istanbul, Turkey.
ACM 978-1-4503-1159-5/12/08.

A typical airborne tactical network as depicted in Figure 1 consists of three types of nodes: airborne nodes (AN), ground stations (GS), and relay nodes (RN). The airborne nodes contain a variety of data collection devices. The GSs are located on the ground (stationary or portable) and typically have a much higher transmission range than that of an AN. The GS also houses a gateway (GW) that connects the airborne network to several terminals that may run control applications for various devices on the AN. Furthermore, the GSs can be interconnected to do soft-handoffs from one to another while tracking an AN. The RNs are dedicated airborne nodes to improve the connectivity of the network [10].

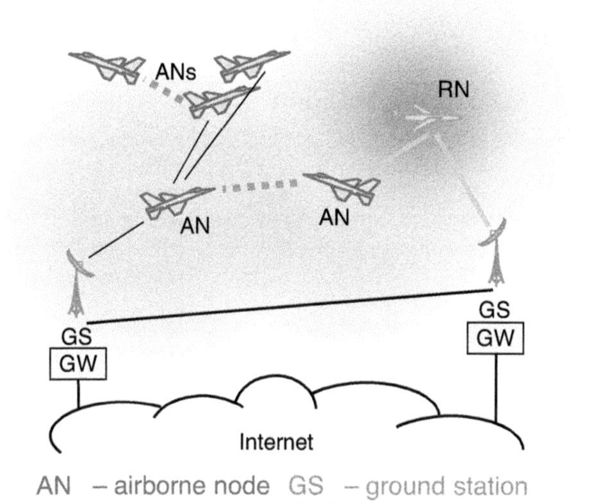

Figure 1: Dynamic airborne environment

The current TCP/IP-based Internet architecture is not designed to function in this environment. We have developed the ANTP suite (shown in Figure 2) that is optimised for the highly-dynamic airborne environment, while maintaining edge-to-edge compatibility with the legacy Internet architecture. These protocols include: *AeroTP* – a TCP-friendly transport protocol introduced in [12] with multiple reliability and QoS modes, *AeroNP* – an IP-compatible network protocol (addressing and forwarding) introduced in [4], and *AeroRP* – a routing protocol introduced in [4] and further evaluated in [7, 8, 5], which exploits location information to mitigate the short contact times of high-velocity airborne nodes. Both the source and destination for data transmitted may be native Aero-protocol devices or TCP/IP-based systems, however the IP protocol stack is not suitable for use within the airborne network itself. To overcome this challenge without requiring a total redesign of all sensors, peripherals, applications, and workstations, we introduce the Aero Gateway (AeroGW) [3]. This protocol suite is designed to perform well in an environment in which rapidly-changing topology prevents global routing convergence, as well as those in which long-lasting stable end-to-end paths do not exist.

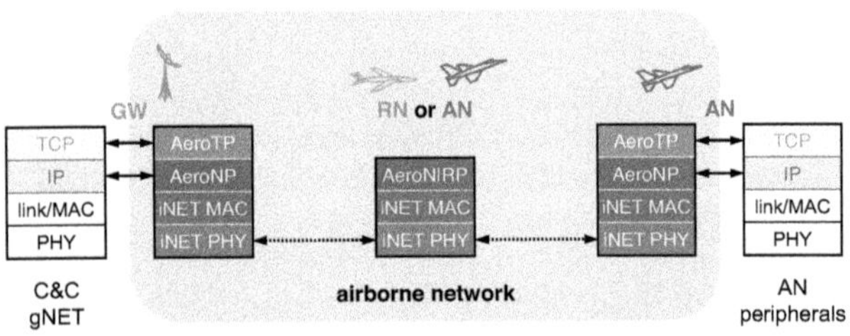

Figure 2: Airborne network protocols

2. SIMULATION MODELLING

In this section, we present the results of simulations conducted with the ns-3 simulator [1] to compare the performance of AeroTP with other transport protocols. We also compare the performance of AeroRP with the traditional MANET routing protocols.

2.1 AeroTP Simulation

AeroTP is designed to be TCP-friendly to allow seamless splicing with conventional TCP at the network edge in the GS and on the AN. AeroTP has several operational modes that support different service classes: reliable, nearly-reliable, quasi-reliable, best-effort connection, and best-effort datagram. The first of these is fully TCP compatible, the last fully UDP compatible, and the others TCP friendly with reliability semantics matching the needs of the mission. We compare AeroTP modes with TCP and UDP protocols. The selective-repeat ARQ algorithm is used to provide reliable edge-to-edge connection between nodes for the reliable mode, and FEC is used for the quasi-reliable mode of the AeroTP protocol [6].

Over the course of the simulation, both TCP and AeroTP are able to deliver the full 1 MB of data transmitted for low error rates (3.5×10^{-5}), but above that TCP performance drops rapidly while AeroTP is still able to deliver nearly all the data at the highest error rates as shown in Figure 3. In the same plot we see that UDP looses a percentage of the data due to corruption as the BER increases, and that the AeroTP quasi-reliable mode losses a much smaller percentage.

2.2 AeroRP Simulation

Both reactive and proactive routing protocols fail to operate in partially connected networks since a complete path may not exist at all time. Determining the next-hop is based on a metric called time to intercept (TTI) that is calculated based on inter-node distance, transmission range, and speed

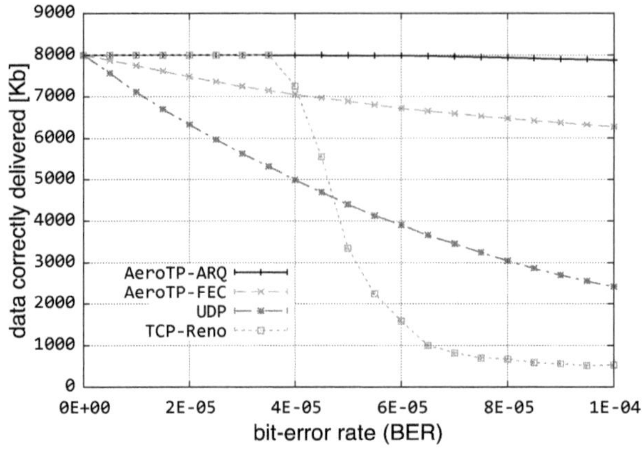

Figure 3: Cumulative goodput

component [7]. AeroRP is tested in the ferrying, buffer, and drop modes as well as in both beacon and beaconless promiscuous mode that were discussed [8, 7, 5].

Figure 4 shows the average PDR (packet delivery ratio) as the number of nodes are increased. The node density of the network affects all of the routing protocols with AeroRP ferrying packets in beaconless promiscuous mode performing the best. The PDR for all AeroRP modes increases as the number of nodes increase with the exception of a slight performance degradation as the number of nodes approaches 90% and higher. This suggests that as the number of nodes increase, AeroRP is able to make more intelligent decisions on how to move the data packets towards the destination whereas the MANET routing protocols are relying on non-geographic based links to move the packet to the destination.

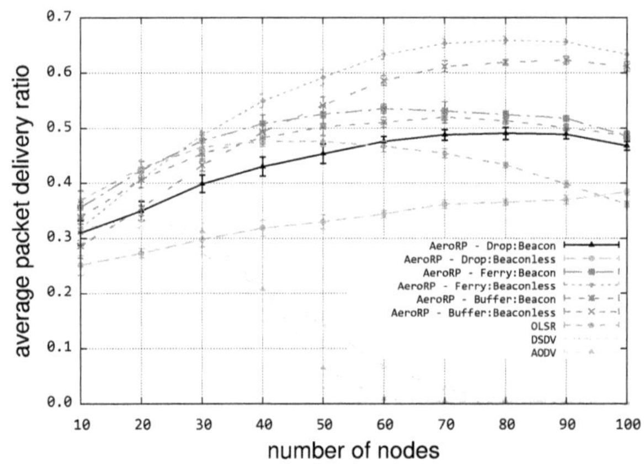

Figure 4: Effect of node density on PDR

3. IMPLEMENTATION ARCHITECTURE

The implementation architecture is designed to provide several features: maintainability, reliability, and data analysis accessibility. To achieve maintainability, the system is fully designed based on the object oriented programming (OOP) approach. This approach tries to eliminate the dependency between data structures, which gives us the option to upgrade the components with fewer errors. For reliability, the system employs try-catch error handling to avoid any I/O errors during runtime. For performance analysis, the system provides a shared logging system that can aggregate the logs in a single web server. Based on these considerations, the system is divided into several components as shown in Figure 5. AeroNP provides the interface between the higher layers of the ANTP protocol suite and airborne network [2]. The system architecture is implemented in Python in two phases. First, we implement each protocol separately and test their functionality on PlanetLab testbed. Finally, the implementation is deployed to embedded processors on radio-controlled aircraft and ground vehicles.

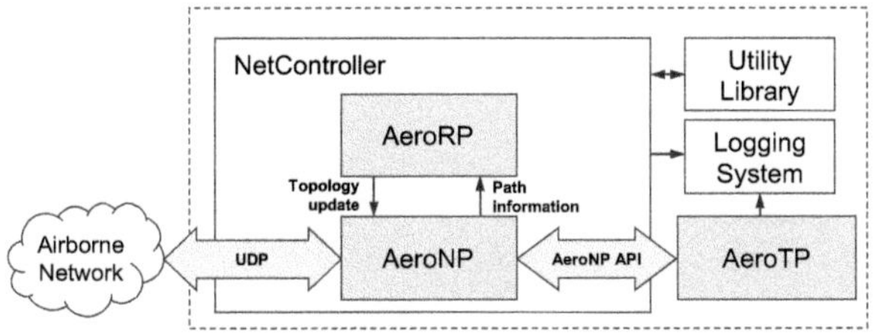

Figure 5: System architecture

4. CONCLUSIONS AND FUTURE WORK

The existing TCP/IP protocols are not well suited for applications in highly-dynamic airborne networks. We have developed domain-specific ANTP suite to leverage cross-layer information in optimising end-to-end performance in this challenging environment. We performed simulations showing significant improvements in end-to-end data delivery when using AeroTP instead of TCP. By predicting when links will be available based on trajectory information, as well as actively listening for nearby nodes, AeroRP can send data opportunistically towards its destination. We showed the prototype implementation design and architecture. For future work, we will deploy our Python implementation on embedded devices on radio-controlled aircraft and ground vehicles.

Acknowledgments

This abstract is a highly abbreviated version and contains text from [12, 4, 11, 10, 9, 3, 8, 7, 5, 6, 2]. The authors would like to thank the Test Resource Management Center (TRMC) Test and Evaluation/Science and Technology (T&E/S&T) Program for their support. This work was funded in part by the T&E/S&T Program through the Army PEO STRI Contracting Office, contract number W900KK-09-C-0019 for AeroNP and AeroTP: Aeronautical Network and Transport Protocols for iNET (ANTP). The Executing Agent and Program Manager work out of the AFFTC. This work was also funded in part by the International Foundation for Telemetering.

5. REFERENCES

[1] The ns-3 network simulator. `http://www.nsnam.org`, July 2009.

[2] M. Alenazi, S. A. Gogi, D. Zhang, E. K. Çetinkaya, J. P. Rohrer, and J. P. G. Sterbenz. ANTP Protocol Suite Software Implementation Architecture in Python. In *Proceedings of the International Telemetering Conference (ITC)*, Las Vegas, NV, October 2011.

[3] E. K. Çetinkaya and J. P. G. Sterbenz. Aeronautical Gateways: Supporting TCP/IP-based Devices and Applications over Modern Telemetry Networks. In *Proceedings of the International Telemetering Conference (ITC)*, Las Vegas, NV, October 2009.

[4] A. Jabbar, E. Perrins, and J. P. G. Sterbenz. A cross-layered protocol architecture for highly-dynamic multihop airborne telemetry networks. In *Proceedings of the International Telemetering Conference (ITC)*, San Diego, CA, October 2008.

[5] H. Narra, E. K. Çetinkaya, and J. P. Sterbenz. Performance Analysis of AeroRP with Ground Station Advertisements. In *Proceedings of the ACM MobiHoc Workshop on Airborne Networks and Communications*, pages 43–47, Hilton Head Island, SC, June 2012.

[6] K. S. Pathapati, T. A. N. Nguyen, J. P. Rohrer, and J. P. Sterbenz. Performance Analysis of the AeroTP Transport Protocol for Highly-Dynamic Airborne Telemetry Networks. In *Proceedings of the International Telemetering Conference (ITC)*, Las Vegas, NV, October 2011.

[7] K. Peters, A. Jabbar, E. K. Çetinkaya, and J. P. Sterbenz. A geographical routing protocol for highly-dynamic aeronautical networks. In *Proceedings of the IEEE Wireless Communications and Networking Conference (WCNC)*, pages 492–497, Cancun, Mexico, March 2011.

[8] J. P. Rohrer, E. K. Çetinkaya, H. Narra, D. Broyles, K. Peters, and J. P. G. Sterbenz. AeroRP Performance in Highly-Dynamic Airborne Networks using 3D Gauss-Markov Mobility Model. In *Proceedings of the IEEE Military Communications Conference (MILCOM)*, pages 834–841, Baltimore, MD, November 2011.

[9] J. P. Rohrer, A. Jabbar, E. K. Çetinkaya, E. Perrins, and J. P. Sterbenz. Highly-dynamic cross-layered aeronautical network architecture. *IEEE Transactions on Aerospace and Electronic Systems*, 47(4):2742–2765, October 2011.

[10] J. P. Rohrer, A. Jabbar, E. K. Çetinkaya, and J. P. Sterbenz. Airborne telemetry networks: Challenges and solutions in the ANTP suite. In *Proceedings of the IEEE Military Communications Conference (MILCOM)*, pages 74–79, San Jose, CA, November 2010.

[11] J. P. Rohrer, A. Jabbar, E. Perrins, and J. P. G. Sterbenz. Cross-layer architectural framework for highly-mobile multihop airborne telemetry networks. In *Proceedings of the IEEE Military Communications Conference (MILCOM)*, pages 1–9, San Diego, CA, November 2008.

[12] J. P. Rohrer, E. Perrins, and J. P. G. Sterbenz. End-to-end disruption-tolerant transport protocol issues and design for airborne telemetry networks. In *Proceedings of the International Telemetering Conference (ITC)*, San Diego, CA, October 2008.

LoadingZones: Leveraging Street Parking to Enable Vehicular Internet Access

Riccardo Crepaldi, Ryan Beavers, Braden Ehrat, Jason Sze, Matt Jaeger, Steven Biersteker, Robin Kravets
Dept of Computer Science, University of Illinois
Urbana-Champaign, IL, US
{rcrepal2,beavers1,ehrat1,sze1,jaeger3,bierste1,rhk}@illinois.edu

Categories and Subject Descriptors

C.2.1 [**COMMUNICATION NETWORKS**]: Network Architecture and Design; C.2.4 [**COMMUNICATION NETWORKS**]: Distributed Systems

Keywords

VANET, Opportunistic, Internet Access, Ad-Hoc

Introduction

Wherever we are, we assume that we will be able to access the network to read our email, visit a website or download our favorite music, and to do this we often rely on the cellular infrastructure, especially when we are on the streets, walking or driving. With our mobile devices in our hands, we perceive network access as an ubiquitous service. However, we often overlook that the place where many people spend the majority of their mobile time, the car, is a lot more than a means of transportation. Not only are cars an extremely advanced mechanical tool, they are also equipped with cutting-edge communication and sensing technology. For the first time, users are not carrying their mobile computing device with them but are being carried by the device itself, and at ever increasing speeds. This change encompasses more than a simple size difference: it relaxes some of the constraints of mobile networks, and defines a set of new ones.

When we drive, our expectations about the connectivity we can achieve are very different compared to when we are inside a building. While ubiquitous connectivity has come to users in cars in the form of 3G and 4G networks, such connectivity is extremely unpredictable and has limited reliability and bandwidth. Essentially, although cellular networks at first glance offer an attractive backhaul medium, the reality is that the load that mobile users currently put on it already pushes the limits of the infrastructure. The sheer demand of today's 3G users, who put all of their data over the 3G networks, overloads the limited bandwidth and so adversely affects the original design goals. If more delay-tolerant communication, such as sensed data, sending and retrieving emails and other delay-tolerant personal communications, prefetching web content for off-line browsing, were supported by other means, the 3G network could operate far more effectively. We have lost the potential benefits of more efficient data transfer in a world where data availability has exploded, including hundreds of sensor readings per minute in each car, the need for constantly faster acquisition of traffic information that must be delivered to navigation devices within the cars, and targeted multimedia content, such as advertisements or information that users store in the cloud expecting to have ubiquitous access.

MobiCom'12, August 22–26, 2012, Istanbul, Turkey.
ACM 978-1-4503-1159-5/12/08.

To enable the full potential of cars as mobile devices, we must be able to provide a network that insures predictability, reliability and high capacity. However, it is important to note that each of these requirements, despite all being necessary, may not always be needed at the same time. A predictable and reliable connection is required to enable time-sensitive operations such as credit card purchases, while a high bandwidth connection is necessary to transfer car-generated and other large chunks of data that can tolerate some delays and unpredictability. 3G networks were designed to satisfy part of this picture: predictability and reliability. When not overloaded, the cellular network is the best medium for timely responses. One step taken to relieve it of the unnecessary load is that of extending the in-building model of mobile handoffs to cars on the street [1, 2], allowing these cars to connect to any in-building access point (AP). While this seems to be a natural extension to mobile communications, a number of problems arise in a vehicular setting that have only partially been identified and addressed in the past. There are three major obstacles to such an approach:

Delay. Setting up an 802.11 link requires several steps: scanning available channels for AP beacons, associating to an AP, which often requires authentication, and finally running a DHCP request to configure the network parameters.

Quality. The connection between a vehicle and an indoor AP is impaired by several factors: the signal has to pass through thick walls, the AP location is not designed to cover the street, and contention happens with regular indoor users.

Duration. The coverage range of an AP depends on obstacles and transmit power. The former are inevitable, the latter limited by regulation. A car driving on the street passes through the area in which the indoor AP's signal is strong enough in only a few seconds.

Combined, these three factors have a deadly effect on the data transfer potential between cars and in-building APs. However, tackled individually, it is possible to provide the desired predictability, reliability and high capacity. While researchers have tackled the connection problem [1, 3], the only real solution to the other two problems requires bringing the APs closer to the cars. Researchers and companies have tried to deploy outdoor APs [4] and have failed simply due to the high cost of deployment. While we believe these outdoor AP approaches have merit, we take a novel approach to this problem that leverages inter-vehicle communications and enables individual cars to play the role of a relay to an AP.

Our solution to this problem of vehicle connectivity is LoadingZones, a communication system that uses parked cars as relay agents between moving vehicles and APs. The key to LoadingZones is a divide-and-conquer approach. By separating a single connection into two, moving vehicle to parked car and parked car

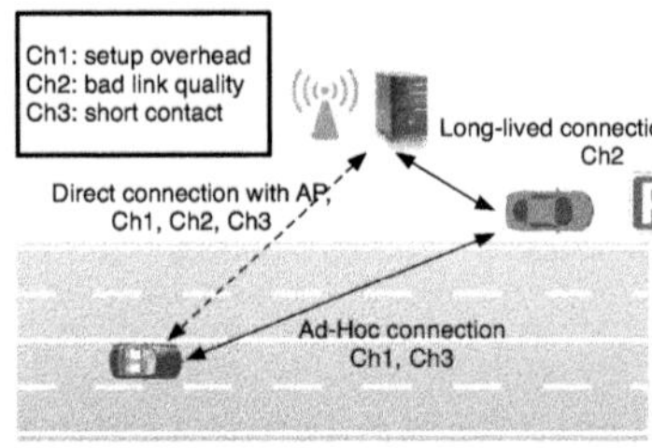

Figure 1: The two-hop approach of LoadingZones separates the challenges that make a direct link to an AP almost unusable.

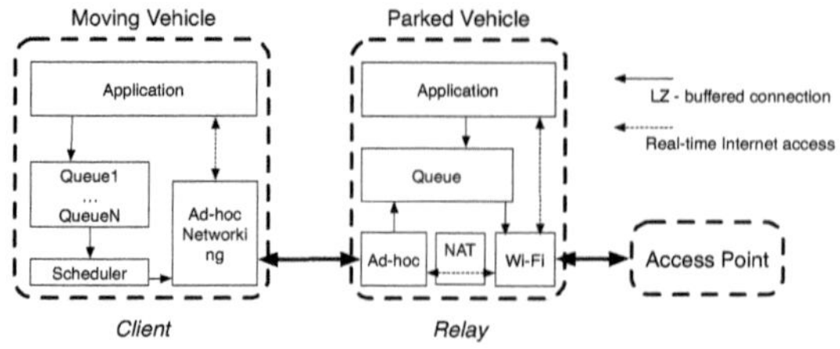

Figure 2: LoadingZones multi-layered architecture

to AP, as shown in Figure 1, LoadingZones isolates the detrimental effects of slow connection times and poor throughput to the car. The role of parked cars has been recognized as very useful in maintaining the DTN fabric of vehicular networks [5], but none of the existing solutions identified the primary role that parked vehicles can have in alleviating most of the issues that limit Internet connectivity for VANETs. Despite being stationary, they see both the vehicular network and the Internet from a vantage point, and can act as bridges between them. When parked in front of a building equipped with Wi-Fi, a vehicle can connect with it for a long time. The still long setup delay can be amortized over a much longer connection that may last several minutes if not many hours. The channel between the parked car and the AP still suffers from low quality, but the steady nature of the link helps alleviating this issue as well. The long life of this link, despite the poor quality and the setup overhead enables uploading and downloading large volumes of data to and from the Internet. A moving car consequently changes its search target. Not anymore an available AP, but parked car instead, that acts as a relay. Such a dedicated connection to the relay vehicle can be designed to significantly reduce setup delay. The channel quality between the moving car and the parked car is also better, and the contact duration is longer between two vehicles, due to a better position compared to that of an indoor AP, and lack of obstacles.

LoadingZones (LZ) is a system that can work as a companion to the cellular infrastructure with a dual benefit. It is a system that can support the increasing demand of an emerging scenario such as VANETS, and at the same time allows a cheap alternative to subscriber services for *delay-tolerant* data. This, in turn, releases the load on the current infrastructure, potentially providing a better service for everyone. Our work constitutes an initial effort to understand their potential and their limitations using a real implementation. In terms of throughput, our experimental results show that LoadingZones enables the transfer of several megabytes of data during a brief contact between a moving and a parked car. When the parked car is used as a relay, LoadingZones enables an increase in throughput of more than 100% compared to a direct connection to indoor APs even when the channel setup overhead is not considered. In the end, LoadingZones improves throughput, reduces user cost and results in a less loaded cellular infrastructure, ready to better serve users that require real-time access.

LoadingZones

LoadingZones (LZ) provides Internet connectivity to moving cars in short bursts as they pass by cooperating parked cars. Since applications may have diverse requirements, LZ's design enables two communication paradigms: *interactive* communication that supports short-lived web browsing or small data transfers such as sending and receiving email, and *bulk* communication for transferring larger data, although downloads are only available when prefetching content at a relay is supported. The application determines the specific style of communication, but LZ hides the complexity of supporting the distinct communication styles from the application.

System Design. The key design feature of LZ is a clear distinction between the role of a moving vehicle and that of a parked one. When a moving car has an *interactive* communication request, it initiates the search for a parked car to connect to. The parked vehicle acts as a relay to the Internet for the moving vehicle for as long as that connection lasts. Since *bulk* communication does not require any responses, LZ includes a local queuing system that resides locally on each vehicle to store the bulk data for uploading (or downloading in the case of prefetched data at a parked car). In the queuing system, a queue manager establishes a new queue for each application with bulk data. Applications can then continuously enqueue their data and continue their normal operations. This time when a connection between a moving and a parked vehicle happens, the moving car starts sending data to the parked car. The transmission order is determined by the *scheduler*, which can implement specific policies to guarantee fairness across applications. Since the channel between the vehicles generally has higher throughput than the channel between the parked car and the indoor AP, the *relay* does not forward the data immediately, but caches it in its own queue first. In this way, for the duration of the connection, the Internet connection is free for servicing *interactive* requests. A similar procedure happens in the opposite direction, with prefetched data such as map updates or local advertisements.

Implementation. We implemented a prototype of LZ using the Illinois Vehicular Project (IVP) [6] In IVP, devices are being installed in University of Illinois service vehicles that normally operate within the campus boundaries. The university campus, as with many other campuses worldwide, has good Wi-Fi coverage that includes all university-owned buildings, and extends to a large portion of the outdoor campus area.

Connectivity is automated by a number of scripts running in the background. First, the GPS receiver measures location and current speed to determine whether the vehicle is parked or moving. In a parked vehicle, a background process tries to locate the campus network and associate with it. Once associated, the parked vehicle broadcasts a beacon over the car-to-car interface. Upon establishing a connection with a moving car, the *interactive* connectivity is supported using NAT, IP forwarding and the `iptables` MASQUERADE policy. For *bulk* communication, the local queuing system simply binds to a local socket that applications can connect to and send messages to at any time. Applications must be specifically programmed to send messages to the queue. Beyond the local queue, however, applications do not have to implement any particular intelligence for sending packets through the network. They simply specify during enqueue the recipient's address and port.

The queue and message passing system is built to handle both small messages (e.g. sensor readings) and large messages (e.g., images and videos). For fairness, an independent queue is created for

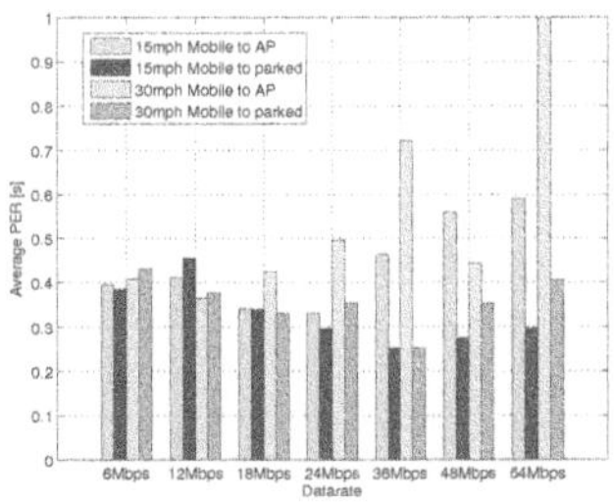

Figure 3: Average PER during a contact.

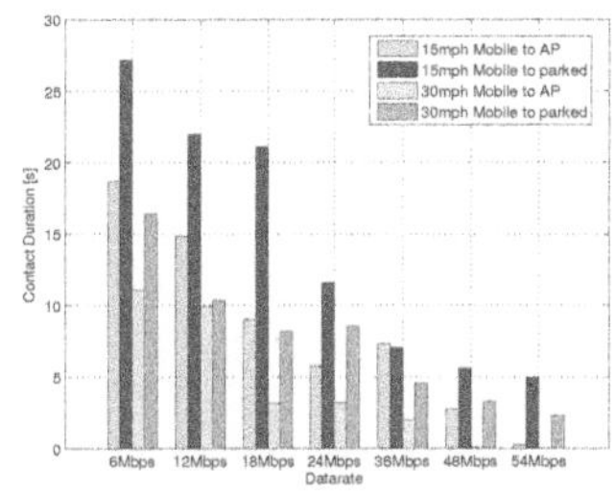

Figure 4: Link duration to an AP Vs a link between vehicles.

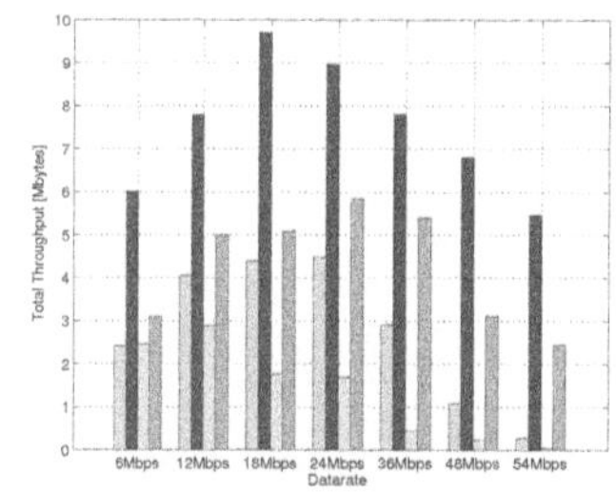

Figure 5: Total traffic delivered per contact.

data from each individual application. When a car-to-car connection is available, a scheduler is triggered and starts pulling packets from the queues according to its policy (the default is Round Robin for our prototype). The messages are then transferred and stored in the queue of the parked vehicle, where another scheduler dequeues and transmits them to the AP once the network is idle (i.e., the *interactive* applications do not require any bandwidth). Upon the messages' arrival at the destination, the data is simply passed to the specified port.

Evaluation

We compared LZ's performance to using a direct connection to an indoor AP. In both cases, we ran experiments with the moving car driving at 15 and 30mph to evaluate the impact of speed on system performance. In high mobility scenarios, automatic rate adaptation algorithms are known to perform poorly due to slow reactivity, which keeps them in a less than optimal state all of the time. For this reason, we turned off rate adaptation.

Delay. Connection setup time accounts for three operations: scanning time, association and authentication, and IP setup. The impact of security is substantial due to the additional overhead required. Our experiments show that the use of WPA2 accounts for more than 33% of the total connection time. Our experiments show that the average time of a successful connection setup, including association, authentication and DHCP, is not significantly influenced by mobility. Regardles of whether the car was parked or moving, the average time is $8.8s$, with a variance of less than $2s$. However, these values are computed only for the successful connections, ignoring timeouts. In our experiments, only 54% of the connection attempts worked on the first try. This is mainly due to the poor channel quality. 18% succeeded after the first timeout, 11% after 2 timeouts, and the remaining required even more retries. The key observation here is that parked vehicles have enough time to eventually recover from these timeouts. A moving vehicle, on the other hand, has a limited time to connect to to an AP before leaving its coverage range. As a result, for vehicles moving at $15mph$, the success rate was approximately 90% and decreased to 57.14% when for 30mph.

Quality. To measure the quality, and next duration, of a connection, we isolated the effect of association by establishing an ad hoc link between two IVP devices. We then placed one of them in a car, and drove it at different speeds. We measured the packet reception rate at the other device placing it inside a building first, to emulate the channel to an infrastructure AP, as in the single-hop approach. We then repeated our experiments with the second device outdoors, in a parking spot on the street shoulder, to emulate the channel with a parked vehicle. We measured the packet error rate (PER) for each link from the moment the first message was received to the moment the last one reached the destination. We define a contact as the interval between the moment the success rate is higher than 0.3 for the first time, until it reaches the same value for the last time before the connection fades out completely.

At low data rates, the robustness of the modulation scheme and coding dominates the effect of mobility or AP placement, and all configurations have a similar PER (see Figure 3). However, as data rate increases, more aggressive modulation schemes are required, and the robustness of coding is sacrificed to improve the bandwidth. In these configurations, speed has a severe impact on the channel quality. This impact is higher for connections to an indoor AP.

Duration. Channel quality considered alone is not representative of contact performance, since it does not indicate how long the contact was. As expected, faster speeds result in shorter contacts, and the line-of-sight channel with a parked vehicle enables contacts that can last more than twice as long as those with an indoor AP at the same data rate (see Figure 4).

Throughput. As always, throughput must be the deciding metric. To understand the total amount of data that can ideally be transferred using the LZ as compared to the single-hop approach, we evaluate the aggregate effects of quality and duration (see Figure 5). The higher throughput to the parked vehicles is a consequence of comparable or lower PER and longer contact durations. Our experimental results show that LZ enables opportunistic data delivery of large volumes of data over a single contact, and unleashes the potential role of VANETs as an ubiquitous opportunistic mobile network. LoadingZones is a first effort in understanding the potential of cooperation between vehicles to expand the range of cheap, high bandwidth links that are so far confined to indoor environments.

1. REFERENCES

[1] B. Hull, V. Bychkovsky, Y. Zhang, K. Chen, M. Goraczko, A. Miu, E. Shih, H. Balakrishnan, and S. Madden, "CarTel: a distributed mobile sensor computing system," *ACM SenSys*, 2006.

[2] Z. Zheng, P. Sinha, and S. Kumar, "Alpha Coverage: Bounding the Interconnection Gap for Vehicular Internet Access," *IEEE INFOCOM*, 2009.

[3] J. Eriksson, H. Balakrishnan, and S. Madden, "Cabernet: vehicular content delivery using wifi," 2008.

[4] R. Frenkiel, B. Badrinath, J. Borres, and R. Yates, "The infostations challenge: balancing cost and ubiquity in delivering wireless data," *IEEE PerCom*, 2000.

[5] N. Liu, M. Liu, W. Lou, G. Chen, and J. Cao, "PVA in VANETs: Stopped cars are not silent," *IEEE INFOCOM*, 2011.

[6] R. Crepaldi, R. Beavers, B. Ehrat, and R. Kravets, "Illinois vehicular project, live data sampling and opportunistic internet connectivity," *ACM MobiOpp*, 2012.

Physarum Optimization: A New Heuristic Algorithm to Minimal Exposure Problem

Yuning Song, Liang Liu, Huadong Ma
Beijing Key Lab of Intelligent Telecomm.
Software and Multimedia
Beijing University of Posts and Telecomm., China
{herufsong, liangliu82}@gmail.com,
mhd@bupt.edu.cn

Athanasios V. Vasilakos
Dept. of Computer and Telecommunication
Engineering
University of Western Macedonia, Greece
vasilako@ath.forthnet.gr

ABSTRACT

Nature is an invaluable source of inspiration for long-standing computational problems. Genetic algorithms and ant colony optimization are the well-known successes. Drawing the inspirations from these successful experience, we exploit the computing model in a single-celled organism – physarum polycephalum, and further design a new heuristic algorithm, Physarum Optimization, to the minimal exposure problem of wireless sensor networks. More importantly, our proposed physarum optimization algorithm can be also applied to the general Steiner tree problem.

Categories and Subject Descriptors: I.0 [Computing Methodologies]: GENERAL

General Terms: Algorithm, Design.

Keywords: Bio-inspired computing, Physarum Optimization, minimal exposure problem, Steiner problem.

1. INSPIRATION & INTRODUCTION

Minimal Exposure Problem (MEP), which corresponds to the quality of coverage, is a fundamental problem in wireless sensor networks [1], [2]. As shown in Fig. 1(a), the goal of MEP is to find a road-network among the Points of Interest (PoIs) such that a target moving along the paths of the road-network has the least observability. Finding such the worst-coverage road-network is of significance because additional sensors could be deployed along these paths to increase the quality of coverage.

For an arbitrary point $\mathcal{P}$ in the monitoring filed, the *sensing intensity* $I(\mathcal{P})$ is defined as the sum of all sensor sensitivities at $\mathcal{P}$. For a road-network R, the exposure $E_x(R)$ is the sum of $I(\mathcal{P}), \forall \mathcal{P} \in R$. Given several PoIs in the monitoring filed, our goal is to determine a road-network which connects PoIs with the minimal exposure. To construct such a road-network, some extra junction points may be introduced (see Fig. 1(a)), which correspond to the Steiner Points in the Steiner problem. We discretize the monitoring filed into a fine grid such that the target moves only along the grid (see Fig. 1(b)). The weight of each edge is set to the sensitivity along the edge. Then, we transform the MEP to the minimum Steiner tree problem in weighted graphs [5] which is a classical NP-hard problem.

MobiCom'12, August 22 26, 2012, Istanbul, Turkey.
ACM 978-1-4503-1159-5/12/08.

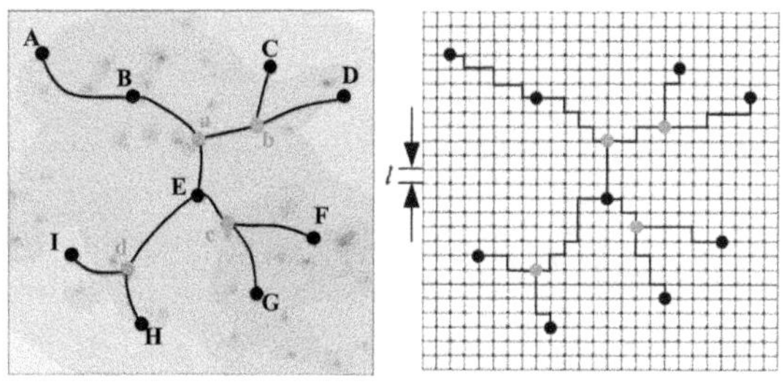

(a) (b)

Figure 1: (a) The minimal exposure road-network in a sensing field, where darker regions have greater exposure values. (b) The approximation of MEP in a weighted grid.

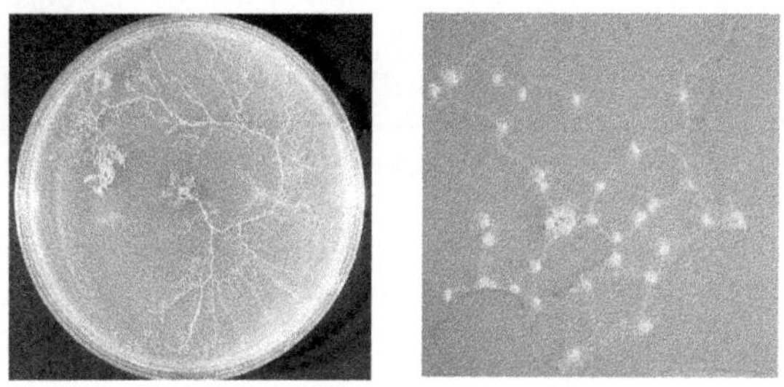

(a) (b)

Figure 2: (a) Photograph of the physarum's tubular network. (b) Physarum for Tokyo rail system design (This figure comes from [4]).

On the other hand, looking at nature as model, measure, and mentor, could give us inspirations for problem solving, especially for some NP-hard problems. The *physarum* is a single-celled tubular organism which has survived for almost 1 billion years. Its body contains a tubular network through which nutrients, signals, and body mass are transported (see Fig. 2(a)). When Food Sources (FSs) are distributed patchily around it, the physarum forages for them and finally forms a tubular network, which connects FSs while avoiding potential dangerous. In addition, because the physarum is photophobic, when the organism is illuminated in an inhomogeneous light field, the tubes connecting the FSs react to the illumination inhomogeneity. Researchers from Hokkaido University applied the physarum to Tokyo rail system design [4]. Fig 2(b) illustrates an approximate optimal tubular network.

The characteristics of the physarum presented above bring us inspirations for some minimization problems such as the shortest path problem and the *Steiner tree problem.* If we regard the illumination and tube length as the exposure of the paths, and FSs as PoIs, then the physarum seems to be able to achieve an approximate optimal solution for MEP. In this paper, we exploit the cellular computing model in the physarum to find the minimal exposure road-network among multiple PoIs. By advancing the physarum model, we propose a practical heuristic algorithm, called Physarum Optimization Algorithm (POA), and further present the parallel implementation of POA.

2. MODELS & METHODS

As we introduced in Section 1, the physarum is a single-celled amoeba like organism. Suppose the initial shape of a physarum is represented by a randomly meshed lattice, in which the edge and the vertex correspond to the plasmodial tube and the junction between tubes, respectively. Let $V = \{v_1, v_2, \cdots, v_n\}$ denote the set of all vertices and E denote the set of all edges. There are two special types of vertices, the physarum's core c and the set of FSs T. Let s be the number of FSs. Let $e_{i,j}$ denote the edge between v_i and v_j. Let the edge length be l and the weight of $e_{i,j}$ be $w_{i,j}$. Let p_i and p_j be the pressures at v_i and v_j, respectively. Assuming it is a Poiseuille flow, the flux through $e_{i,j}$ from v_i to v_j is:

$$q_{i,j} = \frac{\pi r_{i,j}^4 (p_i - p_j)}{8\xi l_{i,j}} = \frac{D_{i,j}}{l_{i,j}} (p_i - p_j), \quad (1)$$

where ξ is the viscosity of the fluid, and $D_{i,j}$ is a measure of the conductivity of the tube.

The flux sources T drive the flow through the network, so that for each $v_i \in T$, $\sum_j q_{i,j} = \Phi$, where Φ is the flux flowing into v_i. By considering the conservation law of flux, the vertex c is the sink with a corresponding withdrawal of $\sum_j q_{c,j} = -s\Phi$. Also, the inflow and outflow at any other junction vertex must balance. Then, the network Poisson equation for the vertex pressures, derived from the above equations, is given by:

$$\sum_j \frac{D_{i,j}}{l_{i,j}} (p_i - p_j) = \begin{cases} \Phi, & \text{for } v_i \in T; \\ -s\Phi, & \text{for } v_i = c; \\ 0, & \text{for } v_i \neq c, v_i \notin T. \end{cases} \quad (2)$$

By setting the pressure of physarum's core $p_c = 0$ as the basic pressure level, all p_i's can be determined by solving above equation system, and each $q_{i,j}$ is also obtained.

To accommodate the adaptive behavior of the tubes, all corresponding conductivities $D_{i,j}$'s change in time according to the following equation:

$$\frac{\mathrm{d}D_{i,j}}{\mathrm{d}t} = f(|q_{i,j}|) - \alpha D_{i,j}. \quad (3)$$

The first term on the right side of Eq. (3) describes the expansion of tubes in response to the flux. The functional form $f(|q_{i,j}|)$ is generally given by $f(|q_{i,j}|) = |q_{i,j}|$ for the sake of simplicity. The second term on the right side of Eq. (3) represents the rate of tube constriction, so that in the absence of flow the tubes will gradually disappear. The parameter α expresses how rapidly the thickness decreases in response to the risk (exposure). Because the total amount of fluid in the tube network must be conserved, the paths of the road-network appear to compete with each other. This competition continues for a period of time, and the minimum risk road-network is finally selected.

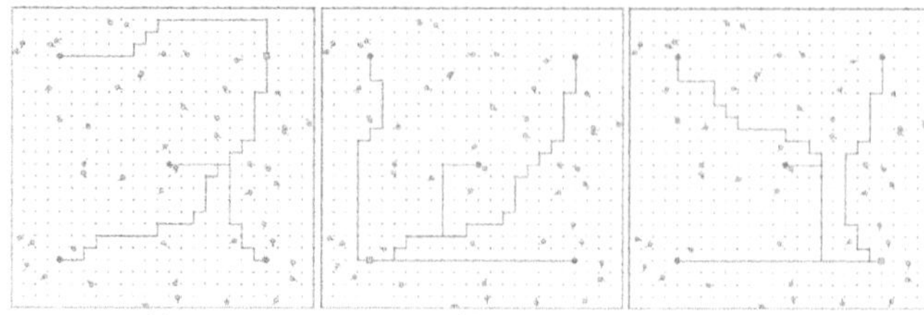

Figure 3: Road-networks with different centers.

For an arbitrary vertex v_i in the grid, let N_i denote the set of neighboring vertexes, and E_i denote the set of adjacent edges of v_i. Let $[variable]^t$ denote the value of this variable at time t and $[variable]^0$ denote the initial value. From Eq. (4), the equations of initial pressures for all vertices are:

$$\begin{cases} \sum_{j\in N_i} p_i^0 - p_j^0 = \frac{\Phi l}{D^0}, & \text{for } v_i \in T; \\ \sum_{j\in N_i} p_j^0 = -\frac{s\Phi l}{D^0}, & \text{for } v_i = c; \\ \sum_{j\in N_i} p_i^0 - p_j^0 = 0, & \text{for } v_i \neq c, v_i \notin T. \end{cases} \quad (4)$$

By solving these linear equations, we can get the initial pressure p_i^0 of each vertex. Because the values of initial pressures are independent of the edge weights. Therefore, the initial pressures can be regards as inputs of our POA.

For a given edge $e_{i,j}$, the weight $w_{i,j}$ expresses how rapidly the conductivity $D_{i,j}$ decreases. The main characteristic of the physarum model described above is the *positive feedback* between the conductivity of each tube and the flux through it. Therefore, we utilize an *iterative process* to express the feedback-driven evolution of the physarum.

Considering the numerical scheme used in the model simulation, Eq. (2) yields a linear equation system with sparse symmetric matrix which is numerically solved by standard *Incomplete Cholesky Conjugate Gradient* (ICCG) method. At each time stamp ICCG runs in $O(|V|^2)$, where $|V|$ denotes the number of the vertices. From simulations for the physarum model, we observed that the variation of pressure is continuous and relatively small between two iterations. Moreover, the evolution process cares about the trend of evolution and the final state more than the specific values of intermediate states. Thus, we substitute $p_j^{t+1} = p_j^t$ to derive an approximate expression of p_i^{t+1} as follows:

$$p_i^{t+1} \approx \begin{cases} \frac{\Phi l + \sum_{j\in N_i} D_{i,j}^{t+1} p_j^t}{\sum_{j\in N_i} D_{i,j}^{t+1}}, & \text{for } v_i \in T; \\ 0, & \text{for } v_i = c; \\ \frac{\sum_{j\in N_i} D_{i,j}^{t+1} p_j^t}{\sum_{j\in N_i} D_{i,j}^{t+1}}, & \text{for } v_i \neq c, v_i \notin T. \end{cases} \quad (5)$$

For each iteration, we use Eq. (5) to calculate the pressure p_i and use Eq. (2) to get the pressure p_i'. When t is relatively small, both p_i and p_i' decrease swiftly. Experiment results show that all vertex pressures calculated by Eq. (5) are close to the corresponding pressures obtained by solving Eq. (2) after enough iterations.

Following the process presented above, POA could be applied to MEP if we regard the sensing field as a exposure-weighted grid and the road-network as the Steiner tree with a center PoI. However, extensive simulations show that the road-network varies when different centers are given, as shown in Fig. 3, where the green square represents the center PoI and red circles represent intrusion PoIs. Please note that the fluxes around the physarum's core are greater than, even several times of, those around FSs. The nonequivalence of the fluxes results in the asymmetry between the PoIs. Here, we discretize each iteration by $s+1$ time steps corresponding to $s+1$ PoIs. For the k-th time step, let the k-th PoI be the center PoI of the centered Steiner system. We associate a data structure to each junction node, called *pressure list* P_i, that memorizes the pressure states of $s+1$ centered Steiner systems at last iteration. Similarly, each path maintains a *flux list* $Q_{i,j}$ which memorizes the flux states of $s+1$ centered Steiner system at this iteration. The POA for the Steiner tree problem is described by Algorithm 1.

Algorithm 1 POA for the Steiner tree problem

for $t = 1$ to M **do**
  for each centered Steiner system **do**
    for each vertex v_i **do**
      for each neighboring vertex v_j **do**
        Judge whether $e_{i,j} \in E_i$ will be deleted.
        Update $p_{k,i}$.
        $q_{k,i,j} \leftarrow D_{i,j}(p_{k,i} - p_{k,j})/l$
      end for
    end for
  end for
  for each vertex v_i **do**
    Update $D_{i,j}$ according to $\sum_k q_{k,i,j}$.
  end for
end for

Please note that the computational complexity does not include the step of calculating the edge weights and the initialization of the pressures. The complexity of POA is $O(M|V|)$ where M is the number of iterations. Moreover, from POA described above, we have that each vertex v_i updates $D_{i,j}$'s and p_i according to the pressures of neighboring vertices. Thus, in one iteration each vertex can update in parallel. The parallel POA runs in $O(M)$ when each vertex runs a process. For an arbitrary vertex v_i, the parallel implementation of POA is described by Algorithm 2.

Algorithm 2 Parallel implementation of POA

for $t = 1$ to M **do**
  for each centered Steiner system **do**
    for each neighboring node v_j **do**
      Judge whether $e_{i,j} \in E_i$ will be deleted.
      Update $p_{k,i}$.
      $q_{k,i,j} \leftarrow D_{i,j}(p_{k,i} - p_{k,j})/l$
    end for
  end for
  Update $D_{i,j}$ according to $\sum_k q_{k,i,j}$.
  Communicate with each neighboring node to update the state.
end for

3. EXPERIMENTS & EVALUATIONS

To demonstrate our proposed algorithm for MEP, we implement a simulation platform by VC++. We compare the solutions achieved by POA for the Steiner tree problem with

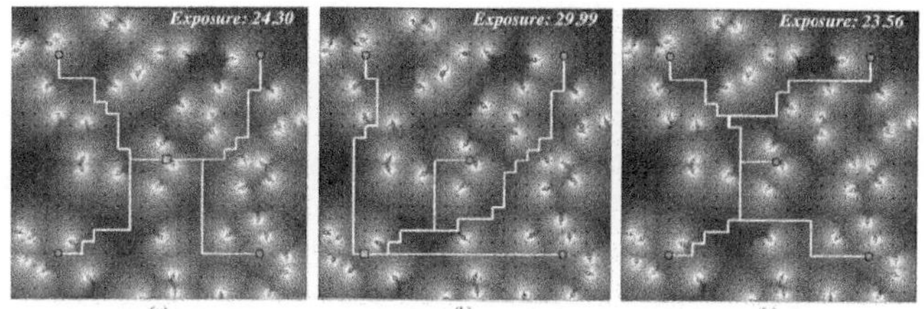

(a) (b) (c)

Figure 4: The comparison of the exposure values for MEP with five PoIs.

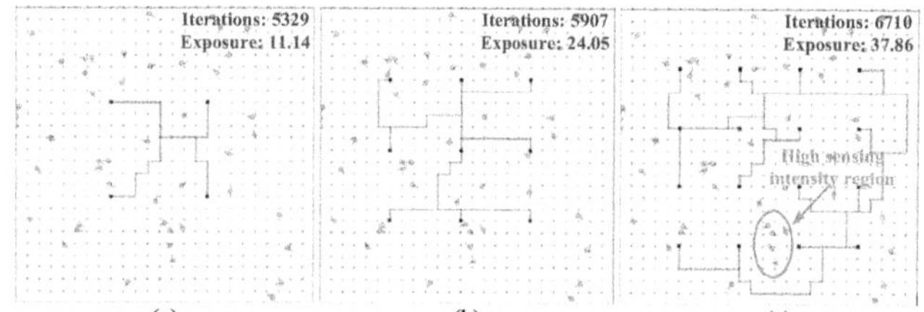

(a) (b) (c)

Figure 5: Examples of the POA for the Steiner tree problem with multiple PoIs.

a center PoI and that without center PoIs. We simulate on a 25 × 25 lattices with 50 sensors deployed randomly, shown in Fig. 4. Five PoIs are deployed in the sensing region. Fig. 4(a) and (b) are the solutions to the centered Steiner system. The green square represents the center PoI and the red circles represent the intrusion PoIs. Fig. 4(c) is the solution to the Steiner system with symmetric PoIs, which means that the intruders traverse among the PoIs. We can observe that the exposure of the solution to the Steiner tree problem without center PoIs is smaller than that to the Steiner tree problem with the center PoI.

By using some case studies for the Steiner tree problem with multiple PoIs, we qualitatively validate the feasibility of POA for the Steiner tree problem. We also simulate on a grid of 25×25 lattices and deploy 50 sensors. Fig. 5 illustrate the achieved road-network with 4, 9, and 16 PoIs deployed in the sensing region, respectively.

4. ACKNOWLEDGEMENT

The research reported in this paper was supported in part by the National Basic Research Program of China (973 Program) under Grant No. 2011CB302701; the National Natural Science Foundation of China under Grant No. 60833009, No. 61003280, and No. 61133015; the National Science Funds for Distinguished Young Scientists under Grant No. 60925010.

5. REFERENCES

[1] S. Megerian, F. Koushanfar, G. Qu and M. Potkonjak, "Exposure in Wireless Sensor Networks." *in Proc. of MobiCom*, 2001.

[2] L. Liu, X. Zhang, and H. Ma, "Minimal Exposure Path Algorithms for Directional Sensor Networks," in *Proc. of GLOBECOM*, 2009.

[3] T. Nakagaki, M. Iima, T. Ueda, Y. Nishiura, T. Saigusa, A. Tero, R. Kobayashi, and K. Showalter, "Minimum-risk path finding by an adaptive amoebal network," *Phys. Rev. Lett.*, 99, 2007.

[4] A. Tero, S. Takagi, T. Saigusa, K. Ito, D. Bebber, M. Fricker, K. Yumiki, R. Kobayashi, and T. Nakagaki, "Rules for biologically inspired adaptive network design," *Science*, 327: 439-442, 2010.

[5] F. K. Hwang, D. S. Richards, and P. Winter, "The Steiner Tree Problem," North-Holland, Amsterdam, 1992.

Modelling Wireless Challenges

Dongsheng Zhang, Santosh Ajith Gogi, Dan S. Broyles,
Egemen K. Çetinkaya, and James P.G. Sterbenz
Information and Telecommunication Technology Center
Department of Electrical Engineering and Computer Science
The University of Kansas, Lawrence, KS 66045, USA
{dzhang, santoshag, dbroyl01, ekc, jpgs}@ittc.ku.edu

ABSTRACT

A thorough understanding of the network behaviour when exposed to challenges is of paramount importance to construct a resilient MANET (mobile ad hoc network). However, modelling mobile and wireless networks as well as challenges against them is non-trivial due to dynamic and intermittent connectivity caused by channel fading and mobility of the nodes. We treat MANETs as time-varying graphs (TVGs) represented as a weighted adjacency matrix, in which the weights denote the link availability. We present how centrality-based attacks could affect network performance for different routing protocols. Furthermore, we model propagation loss models that represent realistic area-based challenges in wireless networks.

Categories and Subject Descriptors

C.2.1 [**Computer-Communication Networks**]: Network Architecture and Design—*Network topology, Wireless communication*; G.2.2 [**Discrete Mathematics**]: Graph Theory—*Graph labeling*; I.6.5 [**Simulation and Modeling**]: Model Development—*Modeling methodologies*

General Terms

Algorithms, Performance, Reliability, Security

Keywords

MANET routing, ns-3 simulation, time varying graphs, mobile wireless network challenges, resilience, survivability

1. INTRODUCTION

The performance of mobile and wireless networks suffer due to dynamic and intermittent connectivity resulting from channel fading and mobility of the nodes. Furthermore, some MANET (mobile ad hoc network) environments suffer from the constraint of limited energy and unpredictable propagation delays due to distance or episodic connectivity [8]. Hence, it is complex to model these networks as well as the challenges against them. In order to construct a resilient wireless network, we need to understand network behaviour in the face of various challenges [7]. Furthermore, understanding the network behaviour under perturbation can help devise Future Internet architectures [4].

MobiCom'12, August 22–26, 2012, Istanbul, Turkey.
ACM 978-1-4503-1159-5/12/08.

We model MANETs as TVGs (time-varying graphs) and pairwise node interactions are aggregated within a certain time window. The network can be represented as a weighted adjacency matrix, in which the weights refer to the link availability. We utilise centrality metrics of weighted graphs to measure the significance of a node. Attacks targeted toward nodes with high significance could degrade network performance severely. As opposed to node and link failures that affect single or multiple elements, area-based challenges could affect numerous network components. Natural phenomena that are geographically correlated might impact quite large areas [4]. In this extended abstract, we present our ongoing efforts to model attack- and area-based challenges that disrupts communication in mobile and wireless networks.

2. TIME-VARYING GRAPHS

A TVG is defined as $\mathcal{G} = (V, E, \mathcal{T}, \rho, \zeta)$, where the definitions of V and E is the same as in static graphs except that $V(G)$ and $E(G)$ vary over time [2]. Since it is used to describe dynamic systems, the relation between nodes change with time; $\mathcal{T} \subseteq \mathbb{T}$ is called the *lifetime* of the system; ρ: $E \times \mathcal{T} \to \{0, 1\}$, is called the presence function that indicates the availability of a specified edge at a given time; ζ: $E \times \mathcal{T} \to \mathbb{T}$, is called the latency function that indicates the time needed to traverse a certain edge E. Since information propagates at a speed that is close to velocity of light and is far higher than the speed of mobile nodes, latency function ζ is negligible for the cases investigated here.

The *footprint* of a TVG $\mathcal{G}$ from t_1 to t_2 can be represented as a static graph $G^{[t_1,t_2)} = (V, E^{[t_1,t_2)})$ such that $\forall e \in E, e \in E^{[t_1,t_2)} \Leftrightarrow \exists t \in [t_1, t_2), \rho(e,t) = 1$ [2]. Fundamentally, the footprint denotes an aggregation of node interactions within a certain time window $[t_1, t_2)$. Thus, we can have a static graph for each time interval. The time interval between two instants t_i and t_j can be denoted as $\tau_{i,j} = [t_i, t_j) \subseteq \mathcal{T}$. The link availability during interval $\tau_{i,j}$ between pairwise nodes can be represented as the ratio of $\tau_{\text{up}} \subseteq \tau_{i,j}$ to the time window length $\tau_{i,j}$, where τ_{up} is the time two nodes are within the transmission range of each other. Then we can have availability matrices of all time windows, in which each element denotes the link availability of certain pair of nodes with a value ranging from 0 to 1. Based on different ranges of time windows, we can obtain the availability matrix of different granularities. Atemporal metrics of the static graph are applied on the availability matrix. Since the matrix is aggregated over time, the atemporal metrics become less accurate as the time window increases.

3. MODELLING CHALLENGES

We model three types of network challenges: non-malicious, malicious, and area-based. Non-malicious challenges can simply be modelled as failures of randomly selected nodes. For malicious attacks, the purpose is to model attacking specific nodes with certain characteristics to maximise overall network degradation. For area-based challenges, we exploit moving impairments of varying size and shapes to model certain large-scale disasters that impact a wide area. We use the ns-3 version 3.13 as our simulation tool [1]. We have developed a new propagation loss model, `MovingPropagationLossModel` in ns-3, which includes a mobility model parameter and range of influence to manipulate the communication in a wireless network [4, 3].

3.1 Malicious Attacks

In a MANET environment, all the nodes are mobile and the pairwise node connectivity is dynamic. The evolution of the network can be described as a sequence of static graphs. We aggregate all the interactions between nodes given a time range into a static weighted graph, where the link weights represent link availability between node pairs. Next, we calculate three atemporal metrics (degree, betweenness, and closeness centrality) of the weighted graphs [6]. We employ them as the node significance indicators and model attacks toward the most critical nodes. Aggregation of node activities of different time window sizes impact the accuracy of using centrality metrics as significance indicators, since time range affects granularities of the aggregation.

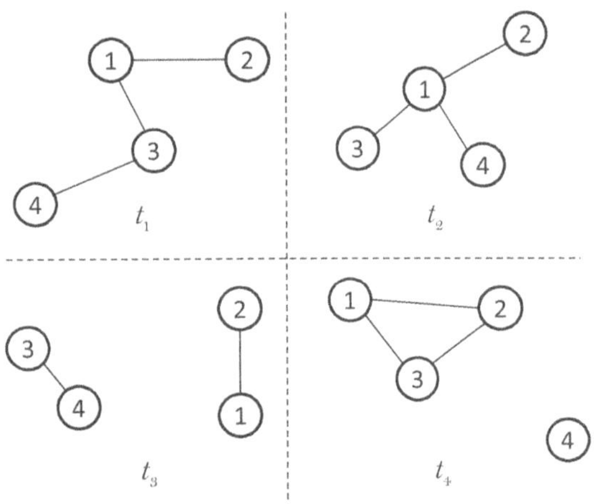

Figure 1: MANET topology at four time steps

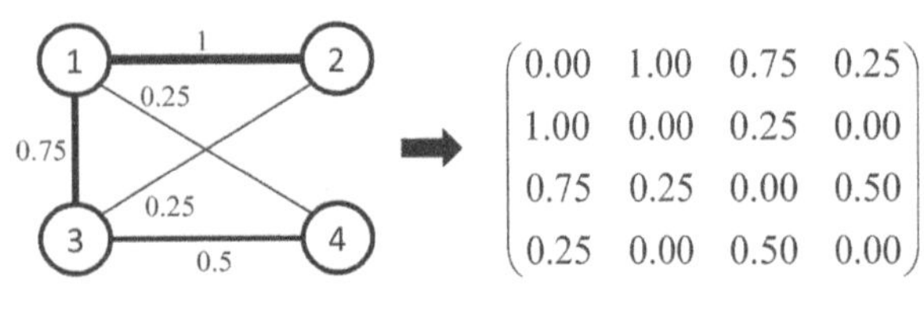

Figure 2: Pairwise link availability in a matrix

An adjacency matrix representing the instantaneous topology can be obtained in each time step. We sum up the matrices for each time step within the time window and the link availability of any pair of nodes can be calculated as the number of 1s divided by the total number of time steps during that time window. Therefore, node interactions for each time window are aggregated into a static graph, based on which centrality metrics can be calculated. Figure 1 presents MANET topologies at four consecutive time steps and Figure 2 shows the aggregation of MANETs over time and its representation as an adjacency matrix. By feeding centrality information into ns-3, we can obtain simulation results of attacks according to different metrics.

3.2 Area-based Challenges

The challenge specification for area-based challenges is a polygon with user-specified behaviour and a circle centered at a user-specified coördinates with radius r as in [3]. Both of these propagation loss models determine the wireless channels that are encompassed by the defined shape and do not allow transmission over that channel during the challenge interval. These models can behave dynamically by moving or scaling (expanding or contracting) over time.

4. SIMULATION ANALYSIS

The simulation consists of two major parts. In the malicious attack model, we assume the channel strength depends only on distance so as to better concentrate on pure topological properties. In the area-based challenge scenario, a couple of realistic models are introduced to simulate large-area radio channel failures. PDR (packet delivery ratio) is used to measure the network performance under challenges.

4.1 Malicious Attacks

We select a scenario with node number of 20 and node speeds given by uniform random variable in the interval of [5, 10] m/s for our studies of network behaviour under malicious attacks using the Gauss-Markov mobility model.

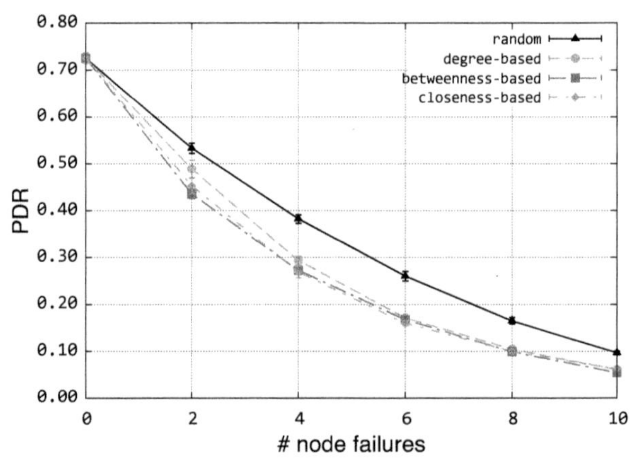

Figure 3: Random and malicious attacks

The difference between random attacks and centrality-based attacks for the AODV routing protocol with window sizes of 10 s are presented in Figure 3. The overall PDR degrades with an increased number of simultaneous node attacks. The maximum difference between random and centrality-based attacks are approximately 0.1 for AODV. PDR decreases resulting from the attacks based on three centrality metrics are close to each other generally, even though they play different roles in the network. The minor differences between attacks based on different centrality metrics can be examined by studying special cases in the future. In a *highly-connected* network environment, the difference between network behaviour under random attack and centrality-based attacks is negligible due to the greedy routing algorithm. However, high network connectivity usually comes with high cost and might not be a representative of most network deployments.

4.2 Area-based Challenges

We measure the network performance during a simulated rainstorm [5], which is modelled as an 8-sided polygon shown in Figure 4. The topology consists of 16 stationary nodes in a square mesh structure with link distance between each pair of nodes being 1000 m. Each node is both the CBR traffic source and sink.

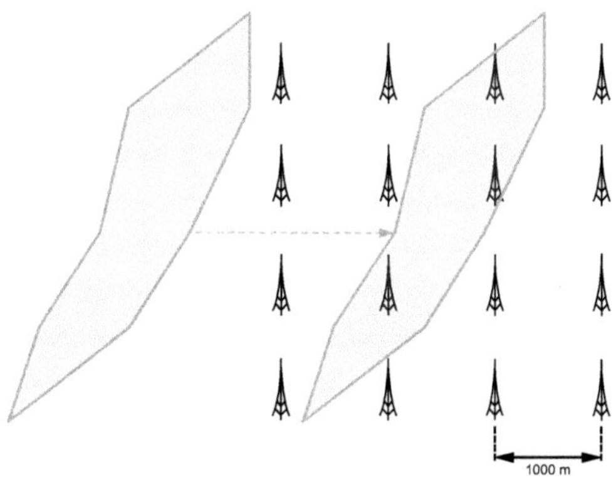

Figure 4: Moving polygon challenge scenario

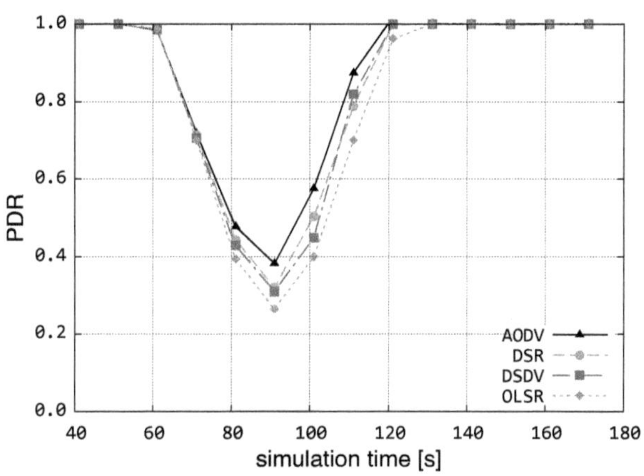

Figure 5: PDR for moving polygon

As the challenge moves across the network, it experiences loss due to the effect of the storm as shown in Figure 5. During the challenge scenario, well-known MANET routing protocols behave similarly, with the AODV routing protocol performing slightly better. The severe degradation due to the large-scale effect of weather disruption can be observed from 82 to 86 s as the network is partitioned. The backbone network experiences maximum degradation of service by approximately 75% during this period. As the rainstorm moves away, the routing reconverges to provide full services.

5. CONCLUSIONS AND FUTURE WORK

We modelled time-varying MANETs as a link availability matrix by aggregating evolving graphs into a static graph. Three centrality metrics exploited as node significance indicators are more accurate within a relatively short time window. For large-scale challenges, we simulated a rainstorm using a moving polygon and network performance severely degrades due to multiple channel failures. We will investigate combined centrality metrics that might provide a more precise indication of node significance than single metric as part of our future work, and examine the impact of mobility and node density.

Acknowledgments

This research was supported in part by NSF FIND (Future Internet Design) Program under grant CNS-0626918 (Postmodern Internet Architecture), by NSF grant CNS-1050226 (Multilayer Network Resilience Analysis and Experimentation on GENI), and by the EU FP7 FIRE Programme ResumeNet project (grant agreement no. 224619).

6. REFERENCES

[1] The ns-3 network simulator. `http://www.nsnam.org`, July 2009.

[2] A. Casteigts, P. Flocchini, W. Quattrociocchi, and N. Santoro. Time-varying graphs and dynamic networks. In *ADHOC-NOW*, volume 6811 of *LNCS*, pages 346–359. Springer, Paderborn, 2011.

[3] E. K. Çetinkaya, D. Broyles, A. Dandekar, S. Srinivasan, and J. P. G. Sterbenz. A comprehensive framework to simulate network attacks and challenges. In *Proceedings of the 2nd IEEE/IFIP International Workshop on Reliable Networks Design and Modeling (RNDM)*, pages 538–544, Moscow, October 2010.

[4] E. K. Çetinkaya, D. Broyles, A. Dandekar, S. Srinivasan, and J. P. G. Sterbenz. Modelling Communication Network Challenges for Future Internet Resilience, Survivability, and Disruption Tolerance: A Simulation-Based Approach. *Springer Telecommunication Systems*, pages 1–16, 2011. Published online: 21 September 2011.

[5] A. Jabbar, J. P. Rohrer, A. Oberthaler, E. K. Çetinkaya, V. Frost, and J. P. G. Sterbenz. Performance comparison of weather disruption-tolerant cross-layer routing algorithms. In *IEEE INFOCOM*, pages 1143–1151, Rio de Janeiro, April 2009.

[6] T. Opsahl, F. Agneessens, and J. Skvoretz. Node centrality in weighted networks: Generalizing degree and shortest paths. *Social Networks*, 32(3):245–251, 2010.

[7] J. P. G. Sterbenz, D. Hutchison, E. K. Çetinkaya, A. Jabbar, J. P. Rohrer, M. Schöller, and P. Smith. Resilience and survivability in communication networks: Strategies, principles, and survey of disciplines. *Computer Networks*, 54(8):1245–1265, 2010.

[8] J. P. G. Sterbenz, R. Krishnan, R. R. Hain, A. W. Jackson, D. Levin, R. Ramanathan, and J. Zao. Survivable mobile wireless networks: issues, challenges, and research directions. In *Proceedings of the 3rd ACM workshop on Wireless Security (WiSE)*, pages 31–40, Atlanta, GA, 2002.

Proximity Enhanced Mobile D2D Video Streaming

Marcel Großmann
Faculty of Information Systems and Applied Computer Science
Otto-Friedrich University
Bamberg, Germany
marcel.grossmann@uni-bamberg.de

ABSTRACT

Mobile video consumption establishes as a massively growing market. To guarantee an acceptable quality of experience to the end user, current studies try to improve data transmissions on air interfaces. However, the bottleneck in cellular mobile networks is the traversal of the infrastructure. A paradigm to circumvent this fact and to increase the system's capacity is given by device-to-device communication in combination with proximity detection. Those networks promise scalability and performance improvements in utilizing scarce resources. This principle is easy to integrate into an existing mobile peer-to-peer video streaming system. In this study, we present measurements on application level to determine the performance that a smart device contributes to different wireless networks. The results outline that content uploading from smart devices is negligible in cellular networks, but shows acceptable performance in WiFi networks. This evaluation is the foundation to build a proximity enhanced video streaming system for mobile devices.

Categories and Subject Descriptors

C.2.1 [**Computer-Communication Networks**]: Network Architecture and Design—*Wireless communication; Network topology*; C.4 [**Performance of Systems**]: Measurement techniques; Performance attributes

General Terms

Design, Measurement, Performance, Reliability

Keywords

Cellular Network, Device-to-Device Communication, Mobile Video Streaming, Peer-to-Peer Network

1. INTRODUCTION

Among a few other scientific studies on mobile peer-to-peer (P2P) video streaming, Eittenberger *et al.* [1] present a prototype for a mobile P2P video streaming application running on the Android operating system (OS). The bottleneck for the video streaming of the framework and similar P2P networks is their current restriction to rely on the existing infrastructure. Of course, the connections to a P2P network through an assigned IP address, which mobile devices can establish, are twofold: On the one hand, a mobile handheld can connect via an IEEE 802.11 wireless network, if the smart device provides the corresponding hardware. This infrastructure is used for short distances, dedicated to an area with access points, and offers a limited mobility to the smart device. On the other hand, cellular air interfaces are providing another entry point to the Internet that is used to access a P2P network. Additionally, a cellular network over relatively narrow air interfaces suffers from hand-overs between the base stations (BS), which are handled by radio network controllers (RNC), and areas that are not covered by a BS. Hua *et al.* [3] propose a multicast approach with adhoc-transmissions. In their study the video is encoded into different layers, one base layer and multiple enhancement layers. Thus, devices with a higher distance from the BS receive less enhancement layers. The proposed adhoc multicast extension allows the reception of layers with a higher quality farer away, if there are a few more peers between the BS and the receiver. Since most smart devices provide the ability to open an IEEE 802.11 access point, a proximity based approach for P2P networks is possible in such a way that device-to-device (D2D) connections are utilized for data dissemination. However, this paradigm should be analyzed further to consider the battery drainage of mobile devices. Obviously, it is higher if a mobile device opens two connections over different interfaces concurrently. Of course, proximity based network establishment is only useful if a crowd of peers, which are in near distance to each other participate in a P2P network. In this context realtime video streaming is a good example, when a lot of users request the same content at the same time. The results generated in this scenario are fundamental to develop applications for current and new mobile devices that are capable for standards proposed by the 3GPP Long Term Evolution (LTE) system. Upcoming approaches to use D2D communication on the physical layer are, for example, FlashLinQ [4] by Qualcomm or the LTE advanced approach presented by Fodor *et al.* [2] that describe methods to increase the spectrum and energy efficiency of traditional cellular networks.

2. ENHANCING VIDEO STREAMING IN NEXT GENERATION NETWORKS

The architecture of the core network for mobile communication is designed hierarchically. The data traffic stream of each mobile traverses, e.g. in UMTS, along the core network from the UMTS Terrestrial Radio Access Network through the packet-switched domain and back. At the Gateway

MobiCom'12, August 22–26, 2012, Istanbul, Turkey.
ACM 978-1-4503-1159-5/12/08.

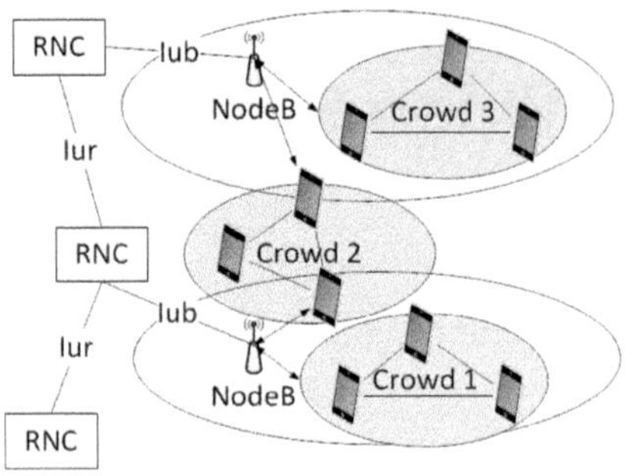

Figure 1: A blueprint for D2D communication in a P2P overlay network. The communication in each crowd is established via a mobile network and D2D connections. Each device accesses the P2P overlay either via the mobile network or via a direct connection to another device.

GPRS Support Node (GGSN) the mobile hosts get assigned to an IP address. Therefore, the GGSN is both the interface to the Internet and to other mobile devices in the mobile domain, making it also the point in the core network where all packet traffic is concentrated. This hierarchical, very centralized topology is in strong contrast to the flat, mesh-like overlay network topologies of most P2P systems. For a P2P video streaming system this hierarchy is the bottleneck of efficient participation, as shown in the measurements of Section 3. However, since a smart device is able to communicate with other devices through an ad-hoc WiFi network, a D2D approach can be used to improve a mobile P2P video streaming system as shown in Figure 1. *NodeB* is the common notation for the BS and *Iub* interfaces connect them to the RNC, while *Iur* interfaces interconnect the RNCs that access the Internet through GGSNs. These interfaces and RNCs are responsible for the hand-over between the BSs. Without D2D data transmissions must traverse the infrastructure especially when hand-overs occur. For example, the devices of *Crowd 2* are currently in a hand-over process, where two handhelds are connected to different BSs and one has no connection at all. However, the data transmission is performed by the D2D paradigm, such that the effect of a hand-over is alleviated as long as one device has access to the Internet. For the D2D connection establishment the signal strength of each peer is an important parameter in the context of real-time streaming over an unreliable transmission protocol (UDP). The stronger the signal, the lower the error ratio of the transfer, which means less packet retransmissions, lower energy consumption and higher throughput. According to the geo-location of a device determined by a Global Position System (GPS) sensor, the mobility or moving speed must be estimated to find peer crowds that could participate in a D2D communication paradigm. An identification of crowds is represented by a feature vector that describes, for example, the position of the crowd, the acceleration and direction of the movement and a threshold that limits how many devices can connect to a specific crowd. If a new device tries to connect, it compares its own feature vector with the vectors of all crowds and connects to the top match. Additionally, the utilization of a device that can be described by concurrently running transmission threads and the battery level are considered in the feature vector. Here, smart algorithms must be applied to create an incentive for devices that have a relatively high amount of energy

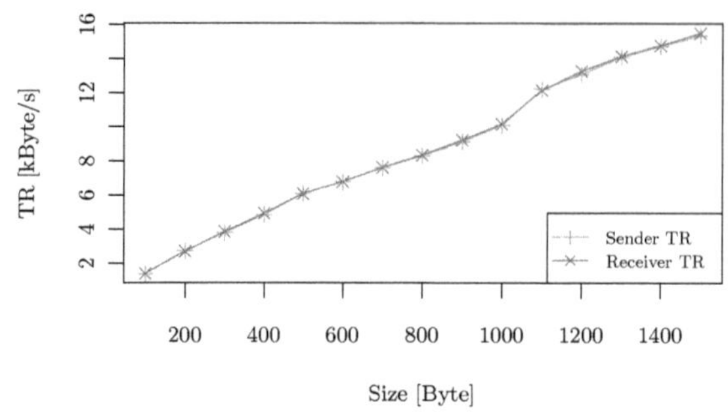

Figure 2: The median of the instantaneous throughputs of a smart device for TCP packets in uplink direction for HSUPA. The packet size is adjusted from 50 bytes to the maximal payload for TCP. The median values are calculated on the sender side (red) and on the receiver side (blue).

to offer it voluntarily to other devices and support the D2D enhanced video streaming network in this way.

3. PERFORMANCE OF SMART DEVICES IN WIRELESS NETWORKS

Current mobile networks, which are defined in the 3GPP releases, promise increasing throughput rates. Today, 3G and its High Speed Packet Access (HSPA) extension is the most commonly distributed standard. Nevertheless, the next generation mobile Internet, LTE, is currently expanding and also defines D2D connections at the physical layer. Even if the new standard promises higher data rates, it is feasible to evaluate devices participating in current mobile networks. We conducted first measurements to gain insights into the performance of smart devices and the way they could participate in P2P networks. Therefore, we developed the measurement framework *Houston* for the Android OS to evaluate the network performance of a smart device on application level. Houston generates instantaneous throughput measurements for one sender-receiver pair over an available network interface. It transmits uniquely identifiable packets with randomly generated payload. To collect exact results for one smart device, it is essential that one participant of this pair is a standard PC; otherwise, the performance of the less powerful device influences the results and the evaluation. The packet size, additional inter-packet delay and the number of sent packets in a sequence are adjustable for the measurements. Houston calculates the throughput by packet timestamps that are produced on two levels: On the one hand, timestamps are taken from the Java program and on the other hand from the underlying Linux Kernel via a packet capture trace. Finally, Houston generates a basic statistical evaluation of the gathered data.

The following results are gained by the analysis of the performed measurements at application level of the Android OS. Regarding the directions, uplink denotes the data transmission from the smart device to the server and downlink vice versa. The instantaneous throughput is calculated with the inter-arrival time (Δt) of two consecutive, equal-sized packets by $TR\ [\frac{\text{Byte}}{s}] = \frac{\text{packetsize [Byte]}}{\Delta t\ [s]}$. For a sequence of

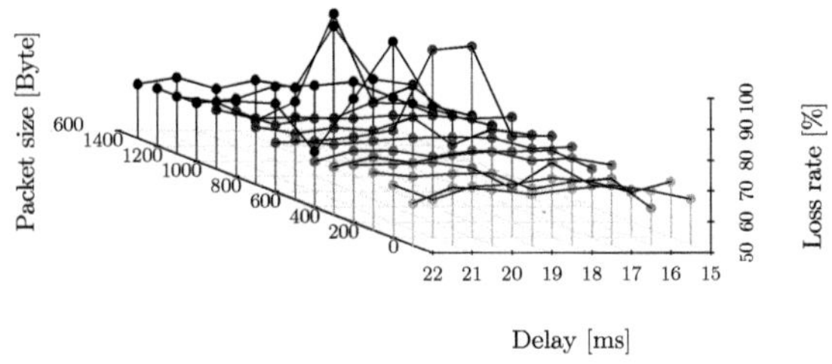

Figure 3: Loss rates for packet transmission in uplink direction in a cellular HSUPA network. The measurements were performed for packet sizes from 100 to 1,400 bytes with an inter-packet delay from 15 to 22 ms for a sequence of 2,000 packets.

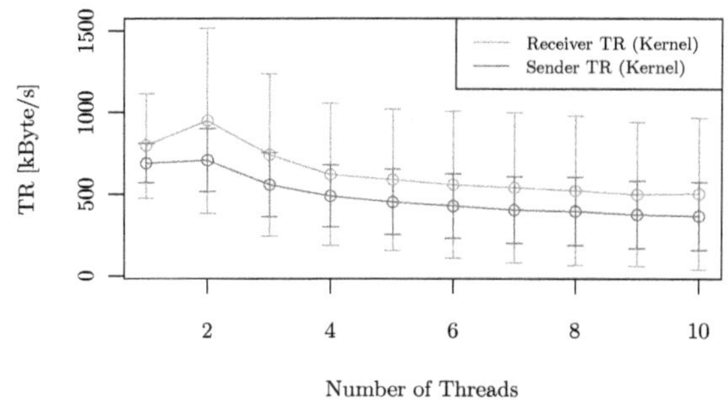

Figure 4: Throughput per thread of up to ten concurrently running ones. Each thread is sending UDP packets with maximal payload in uplink direction over WiFi with no additional inter-packet delay. The centers are the mean values of the calculated throughput, while the bars display the standard deviation.

inter-arrival times the mean value ($\overline{\Delta t}$) denotes the throughput of our measurements at application level. In addition, Houston calculates the loss rate $LR\ [\%] = \frac{n-n_A}{n} \cdot 100$, where n denotes the number of packets sent and n_A the number of packets that arrived. The test conditions for the following cellular experiments with TCP and UDP were identical. We performed stationary tests at one place with the same smartphone and HSPA connection.To evaluate reliable data transmission, we performed measurements on Houston with TCP packets. When using a reliable transmission protocol, the overall performance of the throughput in a HSUPA cellular network in uplink direction is a bottleneck regarding acceptable P2P participation. As shown in Figure 2 the maximal instantaneous throughput is about 0.12 $^{Mbit}/_{s}$, if packets with the maximal payload are sent. This may be caused by the reliability mechanism of TCP. Due to eliminate packet loss the mechanisms recovers lost packets and restricts the throughput to avoid the losses. However, for real-time video streaming the unreliable transmission with UDP packets is more interesting. First measurements show a throughput of 0.53 $^{Mbit}/_{s}$. In contrast, the average percentage of loss rates is about 65% with a few outliers appearing randomly, as shown in Figure 3. Even a higher inter-packet delay does not result in lower loss rates, since the test run was conducted with a delay range from 15 to 22 ms. By using multiple threads with UDP transmissions, D2D is a possible solution for real-time video streaming in wireless networks. The first test was conducted with a dual core smartphone, where one to ten threads were sending randomly generated payload via UDP. The blue curve in Figure 4 is the instantaneous throughput calculated on the smartphone, while the red one denotes the calculation on the receiving PC. The throughput per thread achieves about 4$^{Mbit}/_{s}$ over an IEEE 802.11 WiFi connection.

In summary, TCP shows a weak performance in uplink direction in a cellular network and UDP is not loss resilient enough. In contrast to traversing the cellular infrastructure, the evaluation for WiFi reveals that D2D communication is able to support a P2P streaming framework as proposed in Section 2. Of course, this support is restricted to a crowd, but in uplink direction over a cellular network a device can hardly participate.

4. CONCLUSION

The results of Houston provide insights into application level performance of smart devices. By this knowledge it is possible to develop a D2D video streaming system that respects the bottlenecks of wireless networks and their participating devices. The insights of Houston and part of the code can be reused to generate attractive algorithms that support proximity enhanced video streaming. Application level measurements provide the ability to adjust the upload throughput of a device to satisfy both the capacity of the device and the network's requirements. In future, battery drainage must be considered to estimate how effective a device can participate. If all features are represented in a vector, similarity tests can reveal to which crowd a specific peer connects through a D2D paradigm. The D2D approach is a local rebirth of P2P to establish a fast, reliable mobile video streaming.

5. REFERENCES

[1] P. Eittenberger, M. Herbst, and U. Krieger. RapidStream: P2P Streaming on Android. In *19th International Packet Video Workshop (PV 2012), Munich, Germany, 10-11 May 2012*, 2012.

[2] G. Fodor, E. Dahlman, G. Mildh, S. Parkvall, N. Reider, G. Miklós, and Z. Turányi. Design aspects of network assisted device-to-device communications. *IEEE Communications Magazine*, 50(3):170–177, 2012.

[3] S. Hua, Y. Guo, Y. Liu, H. Liu, and S. S. Panwar. Scalable Video Multicast in Hybrid 3G/Ad-hoc Networks. In *Global Telecommunications Conference, 2009. GLOBECOM 2009. IEEE*, pages 1–6, 2009.

[4] X. Wu, S. Tavildar, S. Shakkottai, T. Richardson, J. Li, R. Laroia, and A. Jovicic. FlashlinQ: a synchronous distributed scheduler for peer-to-peer ad hoc networks. In *Communication, Control, and Computing (Allerton), 2010 48th Annual Allerton Conference on*, pages 514–521, 2010.

An Efficient Way to Track Peers in Mobile P2P Network

Saumay Pushp
Department of Computer Science
KAIST, Daejeon, 305-701
Republic of Korea
saumay@nclab.kaist.ac.kr

Tae Hun Cho
Division of Web Science Technology
KAIST, Daejeon, 305-701
Republic of Korea
taehuncho@kaist.ac.kr

Jongwon Han
Department of Computer Science
KAIST, Daejeon, 305-701
Republic of Korea
addnull@cs.kaist.ac.kr

Dongman Lee
Department of Computer Science
KAIST, Daejeon, 305-701
Republic of Korea
dlee@cs.kaist.ac.kr

Junehwa Song
Department of Computer Science
KAIST, Daejeon, 305-701
Republic of Korea
junesong@nclab.kaist.ac.kr

Sunghee Choi
Department of Computer Science
KAIST, Daejeon, 305-701
Republic of Korea
sunghee@kaist.ac.kr

ABSTRACT

In this paper, we propose a new class of message routing scheme for a distributed network which can scale up to the needs of a very large number of mobile users. We leverage the work in the GeoKad paper which uses the Distributed Geographic Table, allowing node retrieval from resources that are close to any given region. The paper addresses the problem regarding efficient message passing and peer misses. We propose a new routing scheme that addresses this kind of limitation. In fact, the problem can be solved in two ways, either by preserving the state of peer and further use a prediction algorithm or by better message routing. The proposed scheme is based on swarm intelligence (SI) and its attribute of exploit and explore. We leverage the statistical properties of time-varying network connectivity for opportunistic message forwarding. Considering the present 3G or Wi-Fi environment this approach can also help in optimizing energy as it can prevent query over flooding which is mostly encountered in such network scenarios.

Categories and Subject Descriptors

C.2.1 [**Computer-Communication Networks**]: Network Architecture and Design – Distributed networks; C.4 [**Performance of Systems**]: Performance attributes

General Terms

Performance, Design, Theory.

Keywords

P2P, Mobile Computing, Swarm Intelligence.

1. INTRODUCTION

Location-based search in P2P networks usually re-use existing protocols to provide good discovery results [5]. In recent years tracking and localization devices have become of widespread use and are increasingly applied both to industrial and social services. Moreover, ubiquitous and pervasive space has seen a large opportunity in sharing resources and contexts due to increase in the number of mobile devices [8]; as a result, discovering potential users under a heterogeneous environment generally becomes challenging. The situation becomes more severe when there are multitude platform devices and dynamic resources. Additionally, the mobile situation is still based on low range personal area network (PAN), therefore, it's always difficult for a peer to know there proximate peer, which happens to be sporadic. Considering an example of a large mall where hundred thousand customers visit every day. People show different and diverse moving patterns. In such a case, to figure out a potential proximate peer for sharing any kind of resource (e.g. sharing advertising information for a restaurant with a certain group that feels hungry after hanging around for a while) requires understanding the customer mobility. In such situations, to figure out the *optimal* is always cumbersome. Collective intelligence in form of swarm intelligence [7] provides a platform to mitigate such problems inherently.

In this paper, we figured out an efficient way to track reasonable peers in intermittent P2P networks in case of mobile devices. Our approach will solve the problem when there is an increase in the number of peers (inability to pass global knowledge in a given time period result into peer misses). In addition to connectivity, in general it is not guaranteed that two mobile nodes will ever meet each other. Therefore, even if infinite delay is allowed some nodes may never be able to deliver messages, especially in the case of fewer nodes. Applications can sometimes obtain network information dynamically or measure link performance with respect to particular peers, but even when this is an option, it takes time. The application cannot always start out with an optimal arrangement of peers, thus risking at least temporary poor performance and excessive cross-domain traffic. Providing more information in peer selection can improve P2P performance and lower ISP costs.

The mobile P2P network needs to address the following issues. First, the system should provide a network aware discovery scheme for mobile hosts to find new resource providing peers that share files in their currently resident wireless network. Second, the system should provide a *resource discovery* control policy for mobile hosts to obtain fresh status of peers, i.e., *peer's join and leave*, in a mobile network. The approach which we are considering solves the above two problems, better peer updation (message passing time) as well as reduced peer misses. The main objective of our work is to enable continuous resource discovery and better peer management for mobile users in mobile networks.

2. RELATED WORKS

So far, location-based searching in P2P networks is mainly approached by re-using existing structured overlays that are used to provide efficient one dimensional lookups. Related works [1]

MobiCom'12, August 22–26, 2012, Istanbul, Turkey.
ACM 978-1-4503-1159-5/12/08.

[6] have shown methodologies on tracking peers in mobile environment. The Geokad [1] uses the technique of geographic based table to update the peer information in such a dynamic environment. The problems they faced were peer misses and inefficient message passing due to the poor peer updating (message passing time) latency. Mobile P2P networks are typically used for connecting nodes via largely ad-hoc connections. In this paper, we have used the concept of swarm intelligence (an AntNet [4] concept) to propose a distributed P2P service. We believe that the swarm intelligence is one way we can counter the problem of efficient message passing (countering latency).

3. THE MODEL

This section gives an overview of the proposed system. We exploit the situation described in the paper and apply a state preserving look up concept. We use the concept of super node (which behaves as a server node). To reduce the overhead the routing state will be preserved as in position and velocity. Since any node can behave as a super node, accounting its stability we assume some time period. The whole idea is to provide a better interconnection between the super peer P_i and all other neighboring peers P_u where $u \in (1, \ldots. n)$. Not to forget, the DGT information will be stored in a table τ of size $C \times n$, where C is the size of controlled vocabulary for distance and n is the number of peers moving with a velocity v (each peer maintain this table). The information is based on the latency recorded from different peers. We assume that at start up, all table entries are initialized with the same small value τ_{init}. Our approach is different from the the existing GeoKad approach as the later uses an iterative based concept while we incorporate a query based concept.

In this model we consider a TTL (time to live) parameter T_{max} to prevent infinite query passing. If a query is received by a peer P_i that stores documents satisfying through query, it creates a backward response. Hence the total time period for querying peers will be $2 \times T_{max}$. Our routing strategy is based upon the Adaptive transition rule [3] which consists of two strategies which actually complement each other. Based on probability w_s each query A^Q decides if it applies the *exploiting* or the *exploring* strategy. In exploiting strategy the link to the best neighbor peer P_j with the highest quality is selected by applying the following equation:

$$j = argmax_{u \in U \wedge u \notin S(A^Q)}[\tau_{cu}] \qquad (1)$$

Where U is the set of neighbor peers of P_i, and S $[A^Q]$ is the set of peer already visited by A^Q.

The exploring strategy encourages the query to discover further new paths. In case A^Q utilizes the exploring strategy, the transition rule shown in (2) and (3) is applied for each neighbor peer P_j. The first step is to derive a *good-ness value* p_j for each neighbor peer P_j not already visited. In the second step adapted version of *roulette wheel selection* is applied for selecting peers [3]. This strategy allows to more than one peer to be selected to account for the fact there are multiple possible destination peers which contains answers for the query. To ensure that at least one peer is selected the model boils to the exploiting based strategy in case of exploring strategy does not select any peer.

$$p_j = \frac{\tau_{cj}}{\sum_{u \in U \wedge u \notin S(A^Q)} \tau_{cu}} \qquad (2)$$

$$GOTO_j = \begin{cases} 1, & if\ q \leq p_j \wedge j \in U \wedge j \notin S(A^Q) \\ 0, & else \end{cases} \qquad (3)$$

Where q is a random value, $q \in [0,1]$, and $\sum p_j = 1$. In case of $GOTO_j = 1$, A^Q sends a clone of it to peer P_j.

If an query arrives at a certain peer P^D storing appropriate storage documents D, it generates an response and supplies with D and with a copy of stack that contains all visited peers S (A^Q). The response calculates the sum of all entries in S (A^Q) to get the total number of hops T_D for the path from the querying peer P^Q. The strength of the path depends upon the goodness of the path. The goodness of the path is found by the comparing the number of documents found and the length of the path to optimal values. As shown in the below equation, parameter w_d weights the influence of document quantities and path lengths.

$$\tau_{cj} \leftarrow \tau_{cj} + Z, \qquad (4)$$

$$\text{Where} \quad Z = w_d . \frac{|D|}{D_{opt}} + (1 - w_d) . \frac{T_{max}}{2.T_D} \qquad (5)$$

Then strength of the path is governed by the function:

$$\tau_{cu} \leftarrow (1 - \rho)\tau_{cu} \qquad (6)$$

$$\text{Where } \rho \in (0,1]$$

System consists of two distinctive components one is the super node which is chosen as the most stable node and the other comprise of all flash mobile nodes, we call them clients. Our system specifies the query-response message exchange between a client and a super peer. Since we are leveraging Swarm Intelligence mobility of peers is guaranteed. The client would query the other client's information about candidate peers from the super peer. Then the client could connect to one peer or set of peers (remember that we based our concept on exploit and explore) is chosen by client based on the information, and then client exchange some data with choosing peer without any action of super node as P2P system. Generally this kind of system assumes a super node for a certain time period based upon the strength of the edges which can be referred as the amount of pheromone if we take AntNet [4] as an example.

The *super node* (considered to be a stable node, for a given time interval) has responsibility to manage the overall network topology; a probabilistic overlay network – that indicates the most promising path and can consequently create ratings for any combinations of resource provider and resource consumer location. Each super node makes a list called *bucket list* (derived from the GeoKad paper) connected to client peers. Bucket list is constructed based on network latency. Each super node peer can be the center of k different concentric circles, each having a different radius and corresponding to a list that contains the client peers.

The *client* is kind of mobile peer and knows where the super node from the query passing concept (exploring one). When client enters the network, client peer should sequentially invocate to super node which is close to get all candidate's data that is in proximity. In nutshell, we don't need to make a whole *look up* structure which generally is done in case of GeoKad as well as we provide efficient query based approach which will prevent over flooding of queries and hence provide energy efficiency in building such a system.

4. IMPLEMENTATION AND COMPARISON

We designed our own simulation environment for implementing our proposed system using Matlab. We considered a test condition which can support both inner and outer ambiance. Our test condition include the fact that the peer join also depends upon latency, we also added an extra attribute to the existing concept so that same peer in the same distance range can also consider latency Figure 1 shows the bucket error index comparison. As compared to GeoKad like system (figure 1(a)) we achieved reduction in the number of missing nodes. This metric is important to evaluate accuracy of system.

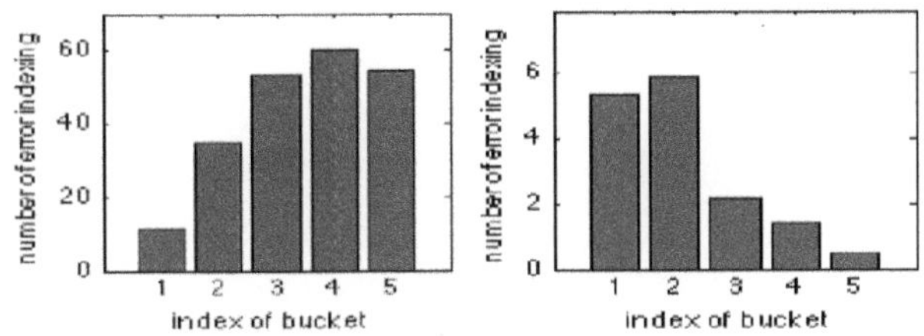

(a) Geokad (b) From our Model

Figure 1. Bucket Error Index.

The average message rate is the number of initialization message required to join the system from new peer. Figure 2 shows the result of the number of initialization messages. From the figure, GeoKad like system should obtain geographic information on from all nodes in the initialization, so the number of messages is higher than our proposed methodology which is requested by peer to get candidate peers.

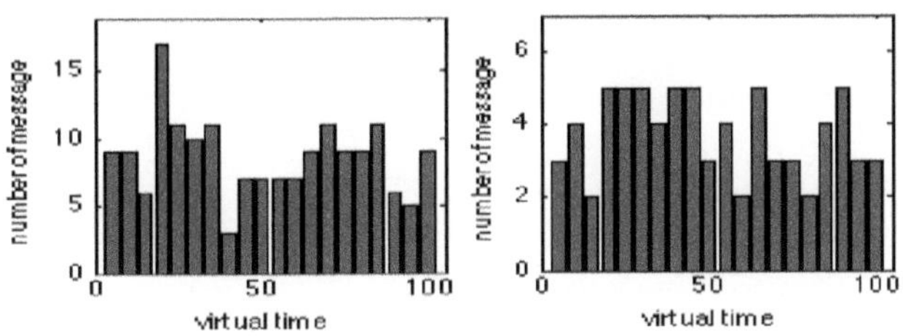

(a) GeoKad (b) From Our Method

Figure 2. Message Rate.

Figure 3 shows the data transmission rate that our methodology can determine which has higher rate than the original system.

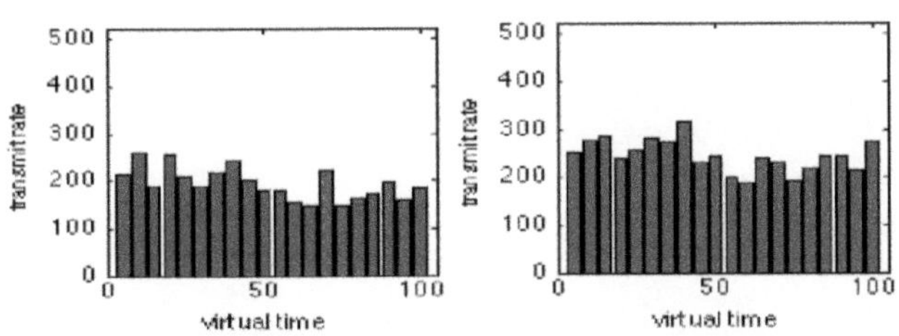

(a) Geokad (b) From our Model

Figure 3. Data Transmission Rate.

5. CONCLUSION AND FUTURE WORK

We leveraged GeoKad and proposed a new way to route messages in this kind of system to achieve better message passing, which further could lead to less peer misses. We didn't take into consideration about the intermittency due to latency in this work (as the original GeoKad doesn't take it into consideration, in fact, to our knowledge this is first of its kind research to introduce latency resulting to intermittency) and planning to adopt it using *eta-strategy* [2] in case of vehicular peer to peer network using the present GeoKad protocol. Our system specifies the query-response message exchange between *a client* and *a super peer*. The *super peer* stores a list of connected *client* peers to manage this overall network topology. This allows us to find available peers that have higher network latency. Through computer simulation, we have evaluated the performance of our proposed location-awareness mobile system and leveraged GeoKad kind of service. Simulation results have shown that our proposed system achieves lower bucket error index and message rate with higher data transmission range than GeoKad type system.

6. ACKNOWLEDGEMENT

This Research was supported by the KCC (Korea Communications Commission), Korea, under the R&D program supervised by the KCA (Korea Communication Agency) (KCA-2012-11911-05005).

7. REFERENCES

[1] Marco, P., Michele, A., and Francesco, Z. 2010. GeoKad A P2P Distributed Localization Protocol. In *IEEE PERCOM Workshop*, Mannheim, 2010, 800 - 803.

[2] Guntsch, M. and Middendorf, M. 2001. Pheromone Modification Strategies for Ant Algorithms Applied to Dynamics TSP. In *Lecture Notes in Computer Science*, 213–222. Springer, 2001.

[3] Dorigo, M. and Gambardella, L. M. 1997. Ant Colony System: A Cooperative Learning Approach to the Travelling Sales Man's Problem. In *IEEE Transaction on Evolutionary Computation*, VOL. 1, NO. 1, 53–66, April. 1997.

[4] Caro, G. D. and Dorigo, M. 2007. AntNet: Distributed Stigmergy Control for Communications Networks. In *Journal of Artificial Intelligence Research,* VOL. 9, 317–365, 1998.

[5] Zhou, S., Gregory, R. G., and Steenkiste, P. 2003. *Location-based Node IDs: Enabling Explicit Localities in DHTs*. Technical Report CMU-CS-03-171. Carnegie Mellon University, 2003.

[6] Kovacevic, A., Liebau, N., and Steinmetz, R. 2007. Globase.KOM-A P2P Overlay for fully Retrievable Location-based Search. In *Seventh IEEE International Conference on Peer-to-Peer Computing,* Ireland, 2007, 87-96.

[7] Swarm Intelligence: Is the Group Really Smarter? http://peopletriggers.wordpress.com/2011/11/14/swarm-intelligence-is-the-group-really-smarter/

[8] Lee, Y., Ju, Y., Min, C., Kang, S., Hwang, I., and Song, J. 2012. CoMon: Cooperative Ambience Monitoring Platform with Continuity and Benefits Awareness. In *MobiSys*, 2012. 43-56.

Poster: A Marine Experiment of a Long Distance Communication Sensor Network -MAD-SS-

Norihisa Segawa, Jun Sawamoto
Iwate Prefectural University
Takizawa, Iwate, Japan
mad-ss@dais.soft.iwate-pu.ac.jp

Masato Yazawa, Haruo Tamaki
Mathematical Assist Design Laboratory,
Ohta, Gunma, Japan
{yazawa,mad}@mail.wind.ne.jp

ABSTRACT
We conducted long distance radio propagation experiments at 1-10mW/145MHz, to realize a low-power long-distance communication for wildlife research and disaster prevention telemetry [1]. In this paper, we describe that we succeeded in long distance communications, from a ferryboat to the top of Mt. Asugiyama (elevation: 501m, distance: 15km) in Kure, Hiroshima, Japan, in the verification test of our method using 3μW radio power. We found out that our method has sufficient capability to achieve such a long distance communication at 10BPS/10mW in battery cell operation on a marine, if we use SSB mode and the SNR in SSB bandwidth is better than -10dB.

Categories and Subject Descriptors
B.4.1 [Data Communications Devises]: Transmitters

Keywords: MAD-SS, Direct Sequence of Spread Spectrum communication (DS-SS), Sensor Network, Narrow Band Spread Spectrum

1. INTRODUCTION
Japan is located on the circum-Pacific volcanic zone, mainly comprised of mountains, and almost every people lives in the mountains and coastal area. On March 11, 2011, the Japanese coast was attacked by huge tsunami after the M9 earthquake (Fig.1). Such a tsunami attack is a common phenomenon from the perspective of 1000-year time span. March 11th's disaster could be attributed to Japanese themselves, because they had been living without the recognition that they were under such a fragile condition, which is subject to tsunami attack.

Figure 1. Tsunami Attack in Iwate, Japan

Disaster and communication have really deep relations with each other. It is said that we live in an advanced information communication society, so we tend to think it is easy to establish communications even when a disaster happens. However, in March 11th's disaster, the public telecommunication networks alongside the Japanese coast were totally devastated. Therefore, disaster communication systems have to be designed and installed without relying on public telecommunication networks.

The objectives of our technology presenting hereafter is to achieve "long distance communication at low power" for disaster communication and environmental telemetry. In disaster monitoring, sending only 1-bit per several seconds is well enough for its objective, thus it is not required to transmit large amount of information at a time. Regarding a tide gauge for tsunami warning, it is required to realize 5-10km radio communications from coastal areas to inland for a long time, as well as to endure the tsunami attack, only with built-in batteries instead of using solar cells or commercial power supply.

In this paper, we describe the construction of a long distance communication sensor network node and the marine experiment.

MobiCom'12, August 22–26, 2012, Istanbul, Turkey.
ACM 978-1-4503-1159-5/12/08.

2. Analyzing the Existing Technology High Sensitive Reception by Narrow Band Communication
If high-speed communication is not required, the bandwidth can be narrowed to get higher sensitivity. The narrower the bandwidth becomes such as AM (6kHz), SSB (3kHz), and CW (300Hz), the more the sensitivity improves and a long distance communications becomes possible. SSB or CW is used for long distance communications in ham radio. If the bandwidth is narrowed to 1/100 of SSB (i.e. 30Hz), the sensitivity becomes 100 times (i.e. 20dB) better than that of SSB. However, another problem arises when the bandwidth is narrowed.

(1) Higher stability of frequency is required
If the bandwidth is 30Hz at 145MHz, the frequency stability should be about ±10%(±3Hz) of the center frequency, which corresponds to the accuracy of 3Hz /145MHz = 0.02ppm. On the other hand, the accuracy of TCXO is 1ppm, namely, 50 times worse than required accuracy.

(2) Can't handle the interference
If there are interfering waves from electronics devices around the radio stations, those stations can't execute radio communication properly. If there is someone around the transmitter, the operational frequency could be changed, but such an alteration is impossible for unmanned telemetry equipment.

3. MAD-SS
As a communication technology that has both ultra-high sensitivity and interference rejection capability, we can take advantage of Spread Spectrum communication (hereafter, refer to it as "SS" for short). Therefore, we decided to address the realization of low-power long-distance communication by SS technology.

In order to perform communication using the SS, it is necessary to solve the following problems.

1) **Long Duration for Synchronization** [3]
At the receiver, the received signal is multiplied by a spreading code (this is used at the transmitter for spreading process) to restore the original carrier wave. If the start point of the spreading code which is used at the transmitter is unknown, the expected spreading signal is shifted on a chip-by-chip basis, the level of the carrier wave is verified, and then the properties of the received signal is inspected for synchronizing the spread code between the transmitter and the receiver. This is called "Sliding Correlation

method"("Sliding method" for short). In sliding method, the required time before detecting the starting point of a spreading code is (cycle time of spreading code)×(number of chips contained in received signal) on average. Here, "chip" means the minimum unit, which comprise of a spreading code. If the cycle time of a spread code is 1 second, 1023 chips corresponds to 1023 seconds, so the receiver can't receive a signal until 17 minutes passes after starting a receiving procedure. This means that the receiver can't receive a signal unless the transmitter sends the signal for 17 minutes. Low-power communication can't be achieved unless signals are transmitted only when it is actually required. And such a large delay makes it impossible to apply SS technology to warning systems (because "Warning for the Dead" is useless). Thus SS had been believed to be inappropriate for intermittent communications at low data rate.

2) Frequency Stability

In SS communication, a certain amount of spreading gain is earned by using a narrow band filter (this filter is implemented to detect the carrier wave), after the de-spreading process. Because the spreading gain is a ratio between two bandwidth (i.e. the bandwidth of a communication channel and the bandwidth of a narrow band filter), the frequency should be accurate to synchronize the received signal, if the spreading code which is used in the process has a high spreading gain. The frequency can't be determined unless the spreading code is synchronized. In equipment whose frequency accuracy is unassured, two-dimensional synchronization is required to adjust the deviation of spreading codes as well as the deviation of frequency.

3) Complicated Receiver

The SS transmitter can be designed in a simple structure. On the other hand, designing the SS receiver to be operated with battery cells is difficult, because the receiver requires considerable power although how deliberately we design it. However, if we give up full-duplex mode, the sensor installed outside can be operated with a simple transmitter. This is the very situation that disaster communications and environmental telemetries are required.

We have developed a Spread Spectrum with High Speed Synchronization Method to solve this problem [3], and we implemented as a MAD-SS. Mad-SS is based on direct sequence of spread spectrum communication (DS-SS). In our practical example, we were able to establish radio communications at 10BPS, 145MHz SSB bandwidth, SNR = -10dB[4]. This means that our method can ensure radio communications if the receiving power is not less than 1/10 of noise power. Practical receiving sensitivities were; - 125dBm in urban area, -135dBm in suburb, and -145dBm in a laboratory. Receiving power is quite influential on receiving success rate. The receiving success rate drastically changes from 0% to 90s% even though the receiving power changes only several dBs.

4. MAD-SS Algorithm

The problem of sliding method is that confirming procedure of a carrier wave and sliding procedure of a spread code is combined into one loop process. We found out a breakthrough by separating those two procedures into estimating procedure of a sync position and confirming procedure of a carrier wave.

1) Estimate the Sync Position (Frequency insensitive method)

In this step, the transition points of a BPSK signal (hereafter, refer to it as "toggle points") are extracted and these toggle points compose a time series sequence. Before estimating the sync position, a 1-cycle sequence of an expected spreading code (which is also comprised of toggle points) is prepared. And the time series sequence mentioned above is correlated with this pre-configured spreading code to estimate the shift amount of the spreading code. A signal of 2-chip length is enough for extracting the toggle points. In a spreading code which has 1024 chips and its cycle is 1 second, 2 chips corresponds to 2msec. And if deviation of 0.5 cycle is acceptable, allowable deviation is equivalent to 0.5cycle/2msec = 250Hz. ±250Hz at 145MHz corresponds to ±1.7ppm, so it can be implemented with TCXO.

2) Confirm the Carrier Wave

The validity of the received signal is assured, if a proper carrier wave is detected as a result of de-spreading the received signal by applying the estimated shift-amount of the spreading code. An interference signal can be regarded as a noise when it is de-spreaded, because its spectrum is scattered widely as a result of de-spreading. Using our method, synchronization can be established within 2 seconds if the cycle of the spreading code is 1 second.

3) Data Transmission by MAD-SS's Shift Method

Now we know how to achieve high-speed synchronization, and then go on to the transmission procedure of information. Generally, 1bit of information is sent by one or some spreading code in SS communication. GPS uses 20 spreading codes to send 1bit(i.e. 0.05bit per one spreading code). Our high-speed synchronization method allows certain amount of sensitivity degradation, so we chose a method that can send information as fast as possible, with the minimum difference of sensitivity between synchronization and data transmission. Thus to send information as fast as possible, we utilize the position information of spreading code in time domain, harnessing the advantages of high-speed synchronization method. We refer to this method (which utilizes the position information in time domain) as "Shift Domain Keying"(SDK for short).

In our experiment, a spreading code of 1024 chips is used, therefore, considering its shift position resolution, up to 10bit data can be transmitted at a time. If 8bit data is sent, 2 bits of the data are wasted because its shift position is specified based on 4-chip unit. The transmitter uses a special spreading code called "FLAG" for synchronizing, and uses another spreading code for data transmission. Regarding the spreading code for data transmission, the starting point could be anywhere among the 1024 chips. Therefore, say the transmission data has 8bit(here we denote the data as TxD), data transmission can be started at 4×TxD away from the top of the code (the spreading code for data transmission), and go back to the top of the code when the shift position reaches the end. That is, all the 1024 chips are sent by starting somewhere in the code of 1024 chips and coming back to the start point.

On the other hand, the receiver can know the synchronization position by FLAGs, so it can get 4×TxD as shift amount information whenever it receives a DATA block, and by this shift amount information, the receiver can restore the transmitted information.

Example of Synchronization

At first spreading codes are not synchronized between the transmitter and the receiver, thus 2 spreading codes are sent consecutively from the transmitter. On the receiver side, extracting a certain length of signal (whose length corresponds to that of 1- cycle spreading code) out of those consecutive 2 spreading codes, always earns a signal which is comprised of each part of the consecutive signal (e.g. 50% out of the first signal and 50% out of the latter one). Therefore, degradation of sensitivity will not occur in spite of rapid synchronization, and synchronization will always be completed within the period equivalent to 2 spreading codes (if enough SNR for receiving is guaranteed).

Use FLAG1 FLAG2 DATA1 DATA2 codes as a data to be transmitted. Consecutive 2 FLAGs are sent. The receiver decodes the signal with 2 buffers for each spreading code (Fig. 2).

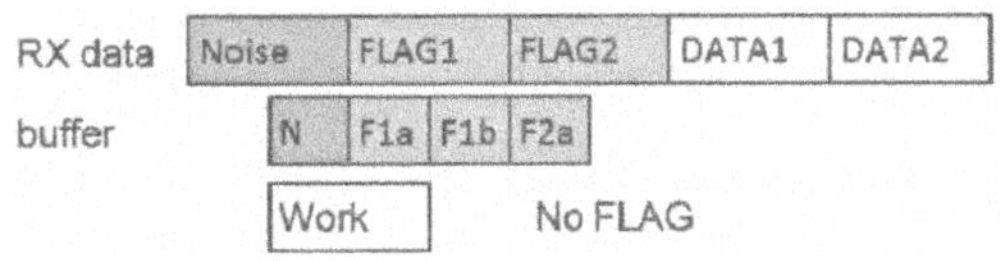

Figure 2. Receiver decodes the signal with 2 buffers for each spreading code

The spreading codes are yet to be synchronized between the transmitter and the receiver, so when the receiver has completed to receive FLAG1, only F1a is stored in the buffer. Synchronization can't be established at this moment (sometimes establishes, though)(Fig. 3).

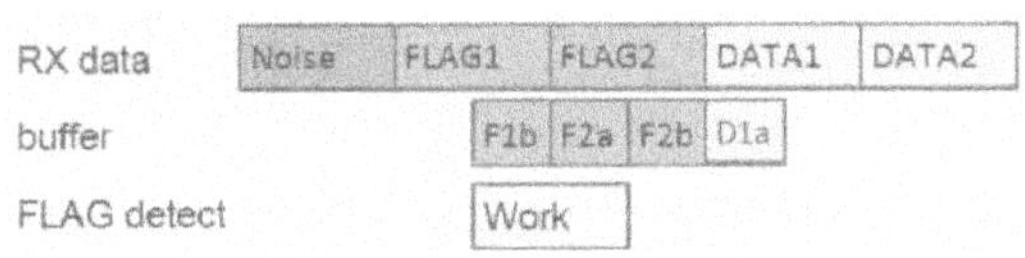

Figure 3. Receiver detects FLAG

Then after FLAG2 has been received, the latter part of FLAG1 (F1b) and the first part of FLAG2 (F2a) are stored and processed in the buffer, results in establishing synchronization with the maximum sensitivity.

Processing the shift amount as a position from the top of the buffer (position of "Work" as shown above), we can see that synchronization between transmitter and receiver has been established as for DATA1 and following data. In our experiment, the period of one spreading code is 0.7 seconds, so it can synchronize within 1.4 seconds.

5. Field Test on a marine

We conducted an experiment for a warning system in Kure, Hiroshima, Japan. The system will be operated in VHF band with battery cells, and expected to be quite effective although the communication speed is low. Setting up a receive station on the top of a mountain (Mt. Asugiyama; elevation 501m, antenna height 6m); we conducted an experiment to send signals from the ferryboat which goes to and come from mainland (10m above sea level: Fig.4). The condition was 3μW (-25dBm), 144.45MHz, and transmit every 20 seconds. The receiving probability was 50% at 15 km away from main land (Fig.5). In this experiment, vertical dipole antennas were used both at the transmit/receive station.

By these experiments, it is estimated that a signal could be sent by MAD-SS at 10mW (3333 times relative to 3 μ W) from the coast to a 100km distance. In addition, it is estimated that if a warning signal is sent by MAD-SS at 10mW from the coast to a 200km distance that is 3000m high, the system can notify the earthquake warning 30 seconds before the seismic wave reaches the mainland.

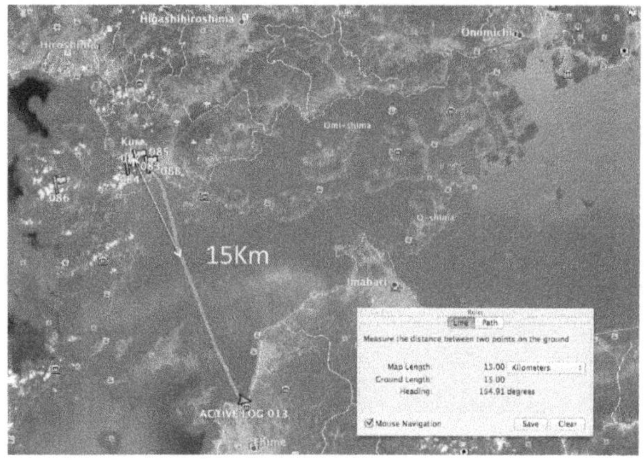

Figure 4. The ferry route from Main Island

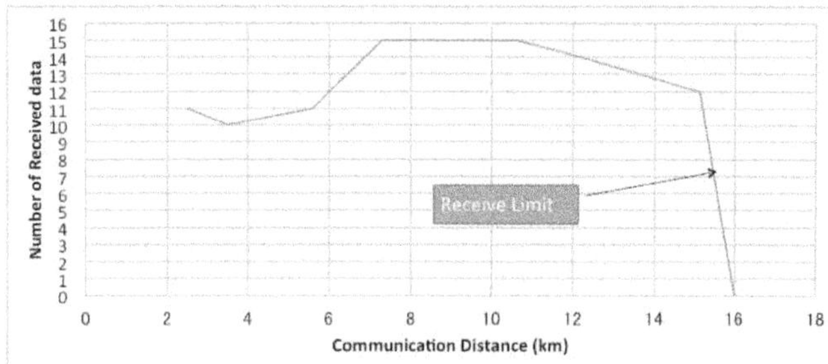

Figure 5. Relation between Distance and Number of Received Data of every 5 minutes (MAX=15)

For example, the distance is 181km from Siberut Island off the Coast of Sumatra Island to Mt. Singgalang (2867m) on Sumatra Island. Siberut Island is located on the potential epicenter of a massive earthquake believed to happen in the near future. Satellite communications are available, but satellite dish and other facilities would be out of service in case of a central earthquake. A 10mW transmitter would be able to run for 2 years with only built-in batteries, while sending keep-alive signals every 15 minutes and a warning signal in case of an earthquake.

6. Conclusion

We describe a construction of a long distance sensor network. We show the idea of Spread Spectrum with High Speed Synchronization Method, and an implementation of the MAD-SS. Now we are making efforts to realize environmental telemetry systems, as well as disaster prevention communication systems for Pacific nations.

Acknowledgements

A part of this work was supported by MEXT/JSPS Grant-in-Aid for Scientific Research(C) (#24500086)

References

[1] N. Segawa, J. Sawamoto, K. Asakawa, M. Kashiwada, A. Azuma, H.Takahashi, C. Nishi, K. Tokita, K. Fujinaga, M. Yazawa, H. Tamaki , Construction of Position Tracking System of Birds using MAD-SS Sensor Network, 2nd International Workshop on Mobile Sensing, http://research.microsoft.com/en-us/um/beijing/events/ms_ipsn12/posters.html (Apr. 2012)

[2] Robert C. Dixon, Spread Spectrum Systems with Commercial Applications ISBN-10: 0471593427 (3rd edition 1994)

[3] http://patentscope.wipo.int/search/en/WO2004070967

[4] Mad-SS technology, http://www.madlabo.com/mad/product/ss/English/20090500esecEng.pdf

Field Tests and Indoor Emulation of Distributed Autonomous Multi-Hop Vehicle-to-Vehicle Communications over TV White Space

Onur Altintas, Yutaka Ihara, Haris Kremo, Hideaki Tanaka
TOYOTA InfoTechnology Center
Tokyo, Japan
onur@jp.toyota-itc.com

Masaaki Ohtake, Takeo Fujii
The University of Electro-Communications
Tokyo, Japan
fujii@awcc.uec.ac.jp

Chikara Yoshimura, Keisuke Ando, Kazuya Tsukamoto, Masato Tsuru, Yuji Oie
Kyushu Institute of Technology
Fukuoka, Japan
tsukamoto@cse.kyutech.ac.jp

ABSTRACT
Vehicular networking has significant potential to enable diverse range of applications, including safety and convenience. As the number of vehicles and applications using the specially designated wireless spectrum grow, one can expect substantial increase in the bandwidth requirements. In this paper and demonstration, we advocate that dynamic spectrum access techniques facilitating the utilization of white spaces for vehicles will be the first step towards solving the expected spectrum shortage. In the demonstration, we will introduce field tests of a multi-hop inter-vehicle communication system with distributed and autonomous TV white space channel selection, performed in Japan from January to February 2012. Furthermore, in addition to the field test videos, we will show an indoor emulation of the entire system used during the field tests in the vehicles.

Categories and Subject Descriptors
C.2.1 [**Computer-Communication Networks**]: Network Architecture and Design – *distributed networks, wireless communication.*

General Terms
Design, Experimentation, Measurement.

Keywords
Vehicular communications, TV white space, multi-hop, spectrum sensing, dynamic spectrum access.

1. INTRODUCTION
Vehicular networking is envisioned to enable various applications involving vehicles, drivers, passengers, pedestrians, and the traffic flow in general. These applications have tremendous potential to improve situational awareness of the driver, to increase safety and convenience of transportation systems, as well as to improve road traffic efficiency and to reduce greenhouse gas emissions [1, 2]. Although we are yet to understand the spectrum requirements of future vehicular networking applications, as the number of vehicles and applications using the specially designated wireless spectrum grow, one can expect substantial increase in the bandwidth requirements. Eventually, we can expect that such applications might suffer from spectrum scarcity, as has been already experienced by the other mobile wireless communications sectors. One solution for accommodating this growth of demand for wireless spectrum is the dynamic spectrum access (DSA) paradigm where unlicensed devices temporarily borrow licensed but spatially and/or temporally unused spectrum (a.k.a. white spaces). Simultaneously, DSA must ensure that rights of the incumbent spectrum license holders are respected. One method of achieving DSA functionality is cognitive radio, which employs autonomous communications techniques [3, 4]. In this paper and demonstration we speculate on the use of DSA techniques facilitating the utilization of white spaces in vehicular networks, in addition to the already assigned spectrum bands for Intelligent Transportation Systems (ITS).

In previous work [5], we have developed a distributed and autonomous dynamic spectrum coordination method tailored for vehicular environments where vehicles coordinate to agree on a control channel, to subsequently setup data channels, and from there to further exchange information on spatial and temporal spectrum variations. Vehicles in this method make use of each other's temporal and spatial proximity relationships to autonomously agree on control and data channels. Furthermore, in [6], we investigated the spatial dependencies in selecting such an appropriate vacant channel for multi-hop vehicle-to-vehicle (V2V) communications by taking into account several factors, such as the distance between the vehicles, channel bit rate, vehicle velocities, statistical information of channel utilization and propagation range of candidate channels. In [7], we looked into the problem of detecting spatial and temporal spectrum holes (a.k.a. spectrum sensing) in vehicular environments. Finally, in [8], we reported and demonstrated the world's first field tests of vehicle-to-vehicle communications over TV white space between two moving vehicles.

This work is a combination and extension of the previous work cited above, with the distributed and autonomous control and data channel selection algorithms tailored for a group of vehicles. In the demonstrations, we will first show videos of proof-of-concept field tests of a multi-hop inter-vehicle communication system with distributed and autonomous TV white space detection and selection. The tests were performed in southwest Japan from January to February 2012. Next, we will demonstrate an indoor emulation of the exact same system employed in the vehicles during the field tests.

MobiCom'12, August 22–26, 2012, Istanbul, Turkey.
ACM 978-1-4503-1159-5/12/08.

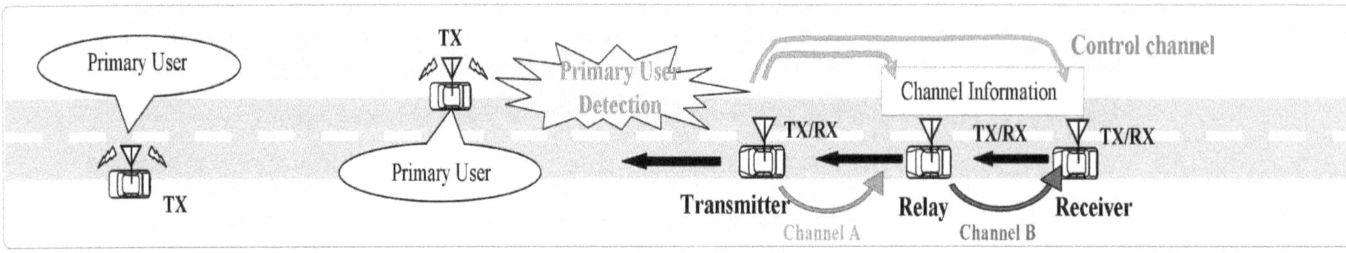

Figure 1. Field tests layout.

2. DEMONSTRATIONS

2.1 Field Tests of V2V Multi-Hop Communications over TV White Space

Broadcast television spectrum is one of the possible candidates for dynamic spectrum access in vehicular environments due to its relatively static channel utilization by the incumbents. In this part of the demonstration we will show a video of actual multi-hop vehicle-to-vehicle communications field tests conducted in Japan using TV white space (Figure 1). In our experiments, three moving vehicles autonomously establish a multi-hop wireless network entirely over the TV white space. Vehicles do not rely on pre-assigned channels nor use the spectrum bands allocated for ITS. For doing so, vehicles first agree on control and data channels autonomously, while also observing the channel occupancy via spectrum sensing. Note that the vehicles select different data channels on each hop; i.e., the first hop does not share the same data channel with the second hop. If a primary system (TV broadcast) is detected in any of the borrowed channels, vehicles switch to another vacant channel without disrupting the ongoing multi-hop session. They collectively and cooperatively detect the primary system appearances by exchanging messages over the control channel. In the experiments, we employed two types of spectrum sensing – one for vacancy detection over the complete set of channels, and the other for the detection of the probabilistic appearance of a primary user signal on the currently borrowed channel.

Figure 2. Field test scene with primary user emulator vehicle shown on the left.

All of the functions described above are implemented onto the general purpose software defined radio platform USRP N210 and integrated with a circuitry that captures and sends real time vehicle speed and sudden break scene picture information from the first vehicle to the following vehicle(s). Figure 2 shows a snapshot from the demo video of the field tests showing the transmitting, relay and receiving vehicles (right side of the picture), as well as a vehicle emulating incumbent (primary) user appearances (left side of the picture) on public roads. Figure 3 depicts a snapshot of the information display inside the third (rear) vehicle, while similarly Figure 4 shows the instantaneous vacant and used channel status of all vehicles with the center frequencies indicated next to the channel "names". Experiments are conducted over 8 different channels ranging from 412 MHz to 491 MHz, each with 1 MHz bandwidth. Duration and location-limited experimental spectrum licenses for the above bands have been obtained from the Japanese Ministry of Internal Affairs and Communications. Modulation schemes used are GMSK and QPSK and the maximum transmission power is 40 mW.

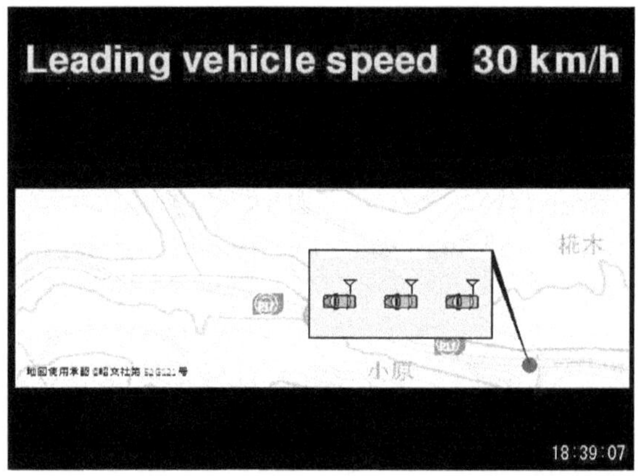

Figure 3. Information display inside the rear vehicle.

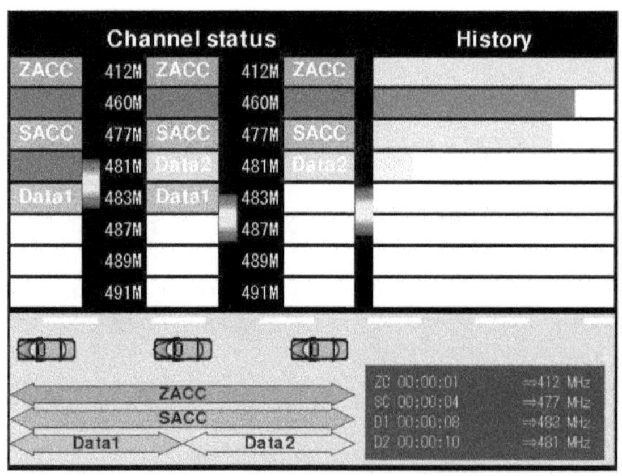

Figure 4. Channel usage status display inside the rear vehicle.

2.2 Indoor Emulation

In this part of the demonstrations, we will show an indoor emulation of the V2V TV white space multi-hop communications by using the exact same equipment employed during the field tests inside the vehicles. For this, we ported the entire system so as to emulate the operation over RF cables appropriately attenuated to avoid RF leakage onto the TV bands. Furthermore, we ported control channel radio exchanges into a leakage-free shielded tent with four UHF band antennas. Since there is no vehicle movement in this setup, we emulated the speed and break information via a game console controlling the steering wheel and gas/break pedal operation. The game console packs and sends the speed and break data to the transmitter radio of the first "vehicle" which in turn gets relayed over the RF cables onto following two vehicles. Another tweak for the indoor emulation is to mock the approach of the vehicles to a primary user propagation area. As opposed to two continuously transmitting primary users in the field experiments, we turn on and off a radio on any of the available frequencies to emulate the movement of vehicles into and out of the primary user area. Figure 6 shows part of the indoor setup with software defined radios and the active GUI display.

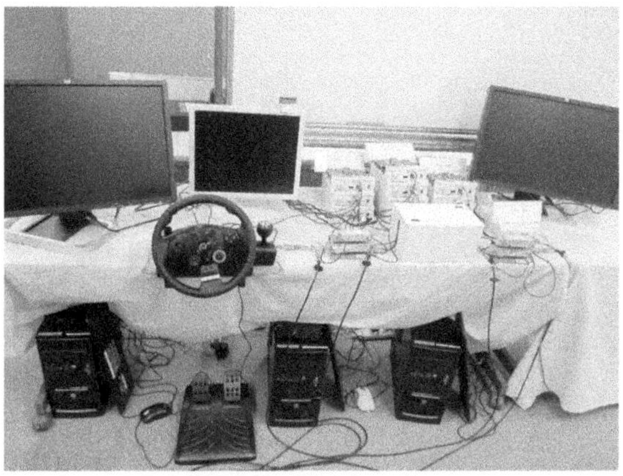

Figure 5. Indoor emulation setup.

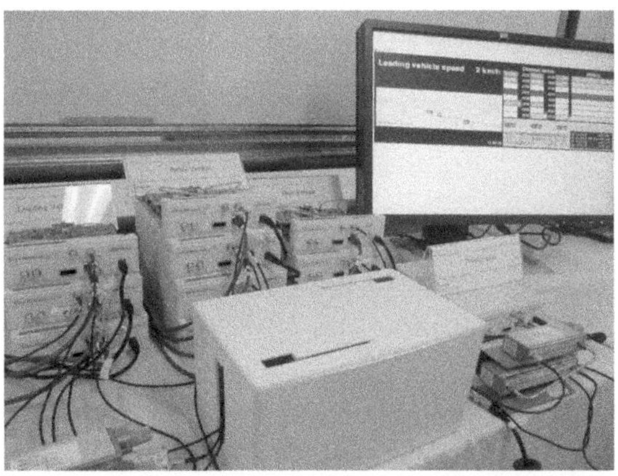

Figure 6. Indoor emulation with active GUI.

3. SUMMARY

Vehicular networking is expected to become ubiquitous in the future resulting in a significant increase in the bandwidth required by various applications. Anticipating the need for more spectrum, we have developed and tested vehicular dynamic spectrum access techniques. This paper and demonstration presented proof-of-concept field tests of distributed and autonomous multi-hop V2V communications over TV white space as well as an indoor emulation of it. Subsequent papers will describe details of the algorithms developed, as well as the performance evaluation results and lessons learned in the field experiments.

4. ACKNOWLEDGMENTS

Our thanks to Kyushu Island Alliance of ICT (KIAI) for their help with the local arrangements, and to Misato Town of Miyazaki Prefecture for their cooperation during the field tests..

5. REFERENCES

[1] P. Papadimitratos, A. de La Fortelle, K. Evenssen, and R. Bringnolol, "Vehicular communication systems: Enabling technologies, applications, and future outlook on intelligent transportation," *IEEE Communications Magazine*, vol. 47, 84–95, 2009.

[2] G. Karagiannis, O. Altintas, E. Ekici, G. Heijenk, B. Jarupan, K. Lin and T. Weil, "Vehicular Networking: A Survey and Tutorial on Requirements, Architectures, Challenges, Standards and Solutions," *IEEE Communications Surveys & Tutorials*, vol. 13, no. 4, 584-616, 2011.

[3] J. Mitola III, "Cognitive radio for flexible mobile multimedia communications", in *Proc. IEEE International Workshop on Mobile Multimedia Communications*, vol. 1, 3–10, San Diego, CA, USA, Nov. 1999.

[4] S. Haykin, "Cognitive radio: Brain-empowered wireless communications," *IEEE Journal on Selected Areas in Communications*, vol. 23, pp. 201–220, Feb. 2005.

[5] K. Tsukamoto, S. Matsuoka, O. Altintas, M. Tsuru and Y. Oie, "Distributed channel coordination in cognitive wireless vehicle-to-vehicle communications," in *Proc. International Conf. on Wireless Access in Vehicular Environment (WAVE 2008)*, Dec. 2008, Dearborn, MI, USA.

[6] K. Tsukamoto, Y. Omori, O. Altintas, M. Tsuru and Y. Oie, "On spatially-aware channel selection in dynamic spectrum access multi-hop inter-vehicle communications," in *Proc. IEEE VTC 2009-Fall*, Sept. 2009, Anchorage, AK, USA.

[7] H. Kremo, R. Vuyyuru and O. Altintas, "Spectrum Sensing in the Vehicular Environment: An Overview of the Requirements," *SDR'12-WInnComm-Europe*, June 2012, Brussels, Belgium.

[8] O. Altintas, M. Nishibori, O. Takuro, Y. Ihara, M. Saito, R. Vuyyuru, C. Yoshimura, Y. Fujii, K. Nishida, K. Tsukamoto, M. Tsuru, Y. Oie, A. Al Abbasi, M. Ohtake, M. Ohta, T. Fujii, S. Chen, S. Pagadarai and A.M. Wyglinski, "Demonstration of Vehicle to Vehicle Communications over TV White Space," *Proc. IEEE WiVeC 2011*, Sept. 2011, San Francisco, CA, USA.

A Bundle Protocol Implementation for Android Devices

Johannes Morgenroth Sebastian Schildt Lars Wolf

Institute for Operating Systems and Computer Networks
Technische Universitt Braunschweig
Braunschweig, Germany
morgenroth|schildt|wolf@ibr.cs.tu-bs.de

ABSTRACT

In this demo we present IBR-DTN for Android: IBR-DTN is a fully featured RFC5050 [5] compliant Bundle Protocol implementation that can run on un-rooted Android devices starting from Android Version 2.3 (Gingerbread). IBR-DTN for Android supports all features of the IBR-DTN version for PCs and embedded systems. It is available in the Google Play Store for free.

In addition to the protocol stack we provide two simple real world applications: a text messaging system and a push-to-talk application. They can serve as an example how to build DTN applications for mobile phones as both applications as well as the protocol implementation itself are open sourced.

Categories and Subject Descriptors

H.4.3 [**Information System Applications**]: Communications Applications; C.2.2 [**Computer-Communication Networks**]: Network Protocols; K.8.0 [**Personal Computing**]: General—*Games*

Keywords

Bundle Protocol, DTN, Smartphones, Mobile Phones, Android

1. INTRODUCTION

Today it is accepted within the networking community that continuous end-to-end connectivity will not be available all the time, not even in wired networks. DTNs replace the end-to-end semantics of common protocols such as IP with a hop-by-hop store-and-forward architecture [3]. Mobile nodes carry messages ("bundles") and forward them to other nodes they might meet later.

The standardized Bundle Protocol [5] specifies how to realize a DTN network. Several well-known implementations exist: DTN2[1] is the reference implementation of the Bundle Protocol by the Delay Tolerant Networking Research Group (DTNRG). It provides a flexible framework for DTN related experiments. IBR-DTN [4] is a lightweight, modular and highly portable Bundle Protocol implementation which has been designed to run on embedded systems as well as on big servers. The Interplanetary Overlay Network (ION)[2] is a Bundle Protocol implementation from the JPL that focusses on deterministic scenarios from Interplanetary Networking and can not only run on Linux but also supports the VxWorks RTOS.

Recently there has been some research interest into realizing DTN like networks with smartphones. These endeavours have been termed Pocket Switched Networks (PSN) [2]. The Haggle project[3] offers a version for their data centric opportunistic communication middleware for Android and IOS smartphones. For RFC 5050 based DTNs the Bytewalla project [1] offers a Java implementation for Android that is basically a reimplementation of the DTN2 C++ code in Java.

In this demo, we present IBR-DTN for Android. IBR-DTN is an actively developed efficient Bundle Protocol daemon with a special focus on embedded devices. We added an Android service abstraction layer and developed some DTN Apps for Android. The resulting implementation is available in the Google Play Store and can run on all un-rooted Android devices which support at least version 2.3 of the Android operating system.

2. ARCHITECTURE

IBR-DTN for Android is not a reimplemented or cut-down version of a Bundle Protocol implementation. Instead it runs the full-featured IBR-DTN daemon including security support and extended routing schemes. This gives all the flexibility and capabilities of a full IBR-DTN installation. It is not necessary to root the phone in order to use IBR-DTN. The daemon process that implements the Bundle Protocol is written in C++ and compiled for the device's target architecture. It runs in the background and is managed by an Android service. Currently the IBR-DTN Android distribution available in the Google Play Store includes support for Android devices with ARM and x86 processors.

Since Android applications have different needs as applications on standard computers, we have designed a intermediate layer between the applications and the daemon process. As shown in figure 1, this layer translates IBR-DTN's normal TCP based API into the Android concept of asynchronous event-based communication which notifies or wakes up applications when needed.

Installed DTN applications register to a *DTN Service* which is provided by the IBR-DTN application package. The *Registration* includes the own application identifier and one or

[1] `http://sourceforge.net/projects/dtn/`

[2] `https://ion.ocp.ohiou.edu/`

[3] `http://www.haggleproject.org/`

MobiCom'12, August 22–26, 2012, Istanbul, Turkey.
ACM 978-1-4503-1159-5/12/08.

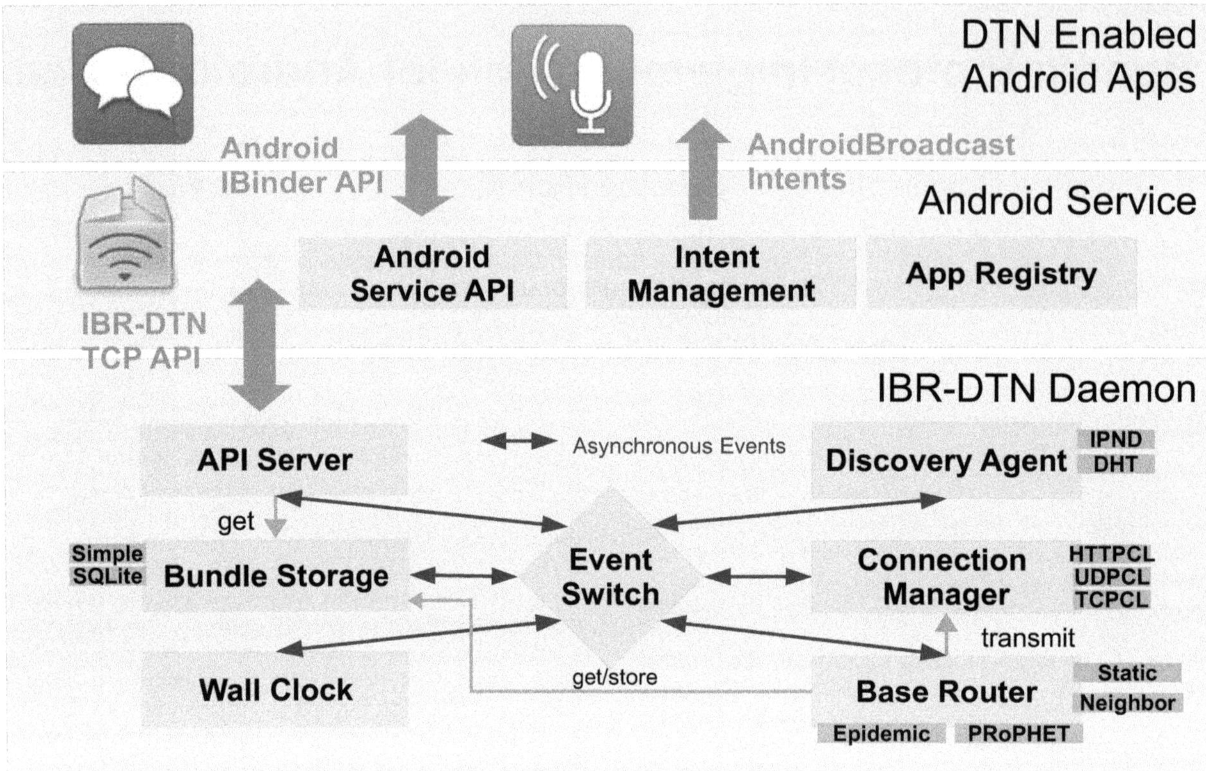

Figure 1: IBR-DTN Architcture for Android

more Endpoint Identifier (EID) the applications is interested in. Once the registration process is finished, the application binds to the *DTN Service* and initiates a *Session*. This *Session* information will be persisted by the *DTN Service* even after the application decides to terminate itself to save resources on the phone.

If the daemon receives a new bundle for a registered application, the *DTN Service* will send an Intent to the previously registered application. This Intent will wake-up the application if it is not running. When receiving the intent, the application will receive and process the bundle.

Besides the API, the *DTN service* also provides a frontend (figure 2(a)) to configure and manage the underlying daemon. Some of the management functionality is available through the service API as well. This includes notifications about discovered neighbours and daemon states such as “daemon running” or “daemon stopped”.

3. APPLICATIONS

As the Bundle Protocol ecosystem matures and more and more research work is done in the area of PSNs using the Bundle Protocol to implement PSN applications is a logical choice as it offers the needed opportunistic communication paradigm. Furthermore you can immediately profit from easy integration with backend servers that already support the Bundle Protocol.

With IBR-DTN for Android we contribute a framework that allows to build DTN applications on mobile devices with ease. Together with the IBR-DTN package, we published two applications in the Android Play Store. Both are available as open-source and provide an introduction to application developers.

3.1 Whisper

Whisper (figure 2(b)) is a DTN based messaging application. It supports presence sensing by using DTN group messages and allows the sending of text messages to other users. A user can set a nickname for the chat and can choose the validity of sent messages up to two weeks. Messages can be received even when the *Whisper* application is not running. The IBR-DTN daemon uses an Android Intent to notify an inactive *Whisper* application of incoming messages.

3.2 Talkie

Talkie (figure 2(c)) is a DTN based push-to-talk application. When *Talkie* is activated using the speak button or by the proximity sensor of the mobile phone, sound will be recorded and upon deactivation of the recording function it will be sent to the Talkie group using IBR-DTN. A receiving *Talkie* instance will either only add received messages to the GUI or play them automatically.

4. DEMONSTRATION

The demonstration of IBR-DTN will be a hands-on-event. We will show some prepared smartphones with the application presented in section 3. Since IBR-DTN can run on any Android device starting from Android version 2.3 and because the software is available in the Google Play Store we invite visitors to install IBR-DTN and the application on their private devices. As un-rooted Android devices do not support Ad-Hoc WiFi, we will bring some WiFi APs, turn

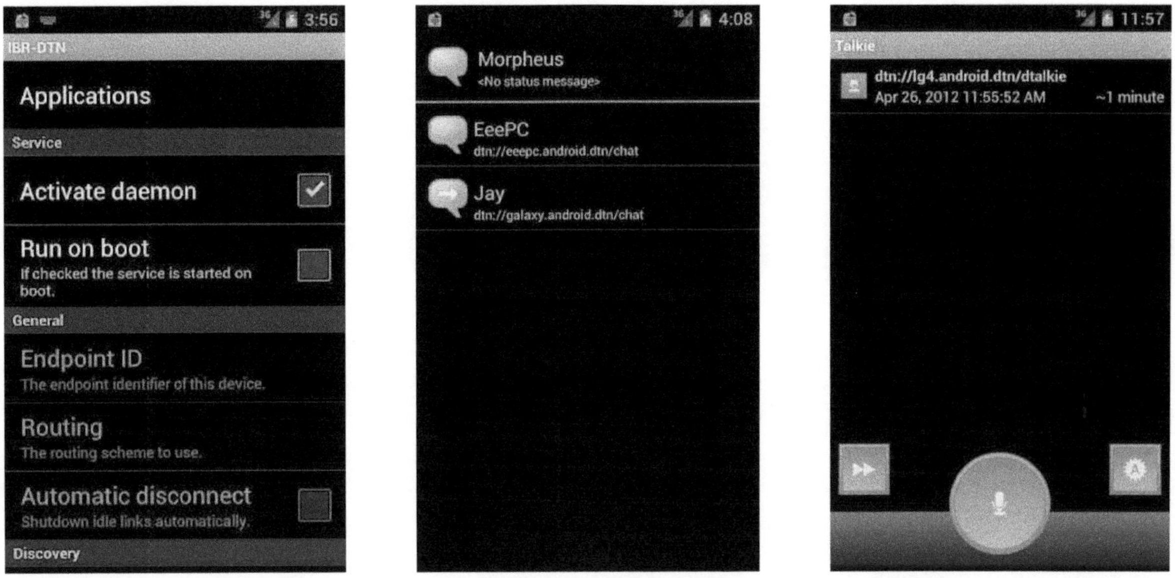

(a) Daemon Configuration (b) Whisper application (c) Talkie application

Figure 2: IBR-DTN service and application GUI

them into DTN nodes with the IBR-DTN software and deploy them on the conference campus to connect the smartphones to each other. Depending on the configuration it might also be possible to use the conference WiFi.

With this everybody can use the the system just like a real Walkie Talkie system, although depending on the number of interested users some radio discipline might be required. We hope this demonstration will motivate some attendees to realize their own Bundle Protocol based applications for Android using IBR-DTN.

5. CONCLUSIONS

We presented IBR-DTN for Android, a fully featured, no-compromise Bundle Protocol implementation for Android. IBR-DTN can run on un-rooted smartphones and is compatible with other Bundle Protocol implementations such as DTN2, ION and the standard IBR-DTN version. IBR-DTN for Android can be used to develop novel applications in the area of PSNs or DTN research in general. IBR-DTN and demo applications are available for download in the Google Play Store. Additionally we provide full sources for IBR-DTN, the Android extensions as well as the example applications on the project website[4].

Acknowledgments

This work has been supported by the NTH School for IT Ecosystems.

[4]http://www.ibr.cs.tu-bs.de/projects/ibr-dtn

6. REFERENCES

[1] S. A. Abdullah Azfar, Jiong Jiang, Li Shan, Maria Jose Peroza Marval, Rerngvit Yanggratoke. ByteWalla : Delay Tolerant Networks on Android phones. Technical report, KTH Telecommunication Systems Laboratory, 2010.

[2] C. Caini, P. Cornice, R. Firrincieli, M. Livini, and D. Lacamera. DTN meets smartphones: Future prospects and tests. *Wireless Pervasive Computing (ISWPC), 2010 5th IEEE International Symposium on DOI - 10.1109/ISWPC.2010.5483772*, 2010.

[3] V. Cerf, S. Burleigh, A. Hooke, L. Torgerson, R. Durst, K. Scott, K. Fall, and H. Weiss. Delay-Tolerant Networking Architecture. RFC 4838 (Informational), Apr. 2007.

[4] S. Schildt, J. Morgenroth, W.-B. Pöttner, and L. Wolf. IBR-DTN: A lightweight, modular and highly portable Bundle Protocol implementation. *Electronic Communications of the EASST*, 37:1–11, Jan. 2011.

[5] K. Scott and S. Burleigh. Bundle Protocol Specification. RFC 5050 (Experimental), Nov. 2007.

An Implementation of Secure Two-Party Computation for Smartphones with Application to Privacy-Preserving Interest-Cast

Gianpiero Costantino, Fabio Martinelli, Paolo Santi, Dario Amoruso
IIT-CNR, Pisa, Italy
Email: name.surname@iit.cnr.it

ABSTRACT

For this demo, we present an implementation of the FairPlay framework for secure two-party function computation on Android smartphones, which we call MobileFairPlay. MobileFairPlay allows high-level programming of several secure two-party protocols, including protocols for the Millionaire problem, set intersection, etc. All these functions are useful in the context of mobile social networks and opportunistic networks, where parties are often requested to exchange sensitive information (list of contacts, interest profiles, etc.) to optimise network operation.

Categories and Subject Descriptors

D.0 [**GENERAL**]; D.4.4 [**Communications Management**]: Network communication; J.4 [**SOCIAL AND BEHAVIORAL SCIENCES**]: Sociology

General Terms

Performance, Security

Keywords

Opportunist Networks, Android, Secure two-party computation, Bluetooth

1. INTRODUCTION

We introduce for the first time a *feasible* implementation of a cryptographic framework for Secure Multi-Party computation (the FairPlay framework proposed in [1]) on the Android mobile platform. Secure Multi-Party computation is a general framework that can be used for privacy-preserving computation of a variety of functions.

Our application was developed with the aim to: 1) find people in the user's (Alice) neighbourhood through a Bluetooth scan operation, 2) connect to another user (Bob) and discover whether Bob and Alice have similar interest profiles without disclosing sensitive information, and 3) share messages between Alice's and Bob's devices only if their profiles are similar.

MobiCom'12, August 22–26, 2012, Istanbul, Turkey.
ACM 978-1-4503-1159-5/12/08.

2. FAIRPLAY PROJECT

FairPlay [1] is a framework for secure two-party and multi-party [2] computation that allows users to write and run secure functions. In particular, a user writes high-level procedures that are compiled by FairPlay into optimised boolean circuits. Since in mobile environments, and especially in OppNets, interactions are mostly pair-wise, in the following we focus on the two-party version of FairPlay, which is sufficient to our purposes.

3. MOBILEFAIRPLAY

Two main issues have to be addressed in porting FairPlay to a mobile environment such as Android. First, the Java object computed as outcome of the FairPlay compilation phase has to be made compatible with the JavaVM used in the mobile phone, which is different from the standard one. In particular, Android phones use the DalvikVM. The first step in our porting process has then been translating a Java object as produced by FairPlay into a `.dex` file executable on the DalvikVM. Second, FairPlay uses TCP/IP for communication between parties, which is not suitable for setting up and operating a direct radio communication between two smartphones. Instead, we used the Bluetooth interface for communications between the two parties. In particular, when Alice wants to communicate with Bob, she hooks to the Bluetooth socket, and then sends a Bluetooth request to Bob. Finally, the communication between the two parties can start once Bob has accepted Alice's connection request.

4. INTEREST-CASTING IN OPPNETS

In this section, we briefly describe the principles underlying interest-casting, and the specific interest-casting application we have implemented using MobileFairPlay.

Interest-casting has been recently proposed as a novel forwarding strategy in OppNets [3]. Here, each user in an opportunistic network is characterized by an *interest profile*, which is used to drive the information dissemination process within the network. User interests can be modeled as an m-dimensional vector in a common m-dimensional *interest space*, where the number m of interest is typically much smaller than the number n of nodes in the network. More formally, the *interest profile* of user A is defined as:

$$I_A = (a_1, \ldots, a_m) ,$$

where $a_i \in [1, max]$ is an integer representing A's interest in the i-th topic of the interest space. Note that interests are expressed as integers in the range $[1, max]$, with 1 representing no interest and max (an arbitrary integer > 0) representing maximum interest[1].

Let S be a user denoted as the message *source*. According to the definition of interest-casting, the message M generated by S (which can be thought of as a piece of information node S wants to share within the network) should be delivered to all nodes in the set $\mathcal{D}(S, \gamma)$, where

$$\mathcal{D}(S, \gamma) = \{U \in \mathcal{N} | sim(U, S) \geq \gamma\} \ ,$$

where $sim(U, S)$ is a similarity metric used to express similarity between a node U and S's interest profiles, with relatively higher similarity values representing relatively more similar interests, and γ is the *relevance threshold* (set by S). Set $\mathcal{D}(S, \gamma)$ is called the set of *relevant destinations*, and in principle it is not known to node S. Instead, set $\mathcal{D}(S, \gamma)$ is implicitly defined by S's interest profile, and by the relevance threshold γ.

In the interest-cast implementation presented in the following, we consider the vector-component-wise (vcw) similarity metric defined in [4], which we recall here. Let $S = (s_1, \ldots, s_m)$ and $U = (u_1, \ldots, u_m)$ be the interest profiles of users S and U, respectively. We have:

$$vcw(U, S, \lambda) = \begin{cases} 1 & \text{if } \forall i \in \{1, \ldots, m\},\ |u_i - s_i| \leq \lambda \\ 0 & \text{otherwise} \end{cases} \ ,$$

where $\lambda \in [0, max]$ is an integer parameter used to narrow/widen the scope of the interest-cast. More specifically, by setting $\gamma = 1$, we have that $\mathcal{D}(S, 1)$ corresponds to the set of all nodes in the network if $\lambda = max$, while $\mathcal{D}(S, 1) = \{S\}$ if $\lambda = 0$.

5. INTEREST-CASTING WITH MOBILEFAIRPLAY

In the developed application, a user can:

1. set up his own profile regarding different topics;
2. start a new connection with another user and checking if they have similar interests;
3. wait for incoming connections.

5.1 User profile

When the application is run for the first time, the preference window is shown to the user, see Fig. 1. He/she must insert a value for each topic in the window. The possible values that can be inserted are between 1 and 100, where the lowest value means no interest, and the highest value, maximum interest for a topic. Finally, the user sets the value of λ by means of a slide-bar, and accepts values between 0 and 10.

5.2 Privacy-preserving hand-shaking

This part represents the main section of our App. We consider the case in which Alice is carrying her smartphone, and starts to discovers people around her. Alice decides to connect to another device, which is owned by Bob. Our app manages both the discovery and connection phase with the Bluetooth interface, hence Alice can covers a range of ten-twenty meters around her. Thus, while Bob is waiting for an incoming connection, Alice tries to connect with Bob. After that both devices have been paired, the App works as described in Fig. 2.

[1]The notion of interest profile can be straightforwardly extended to represent also information about a user's habits, such as living in a certain neighborhood, working in a certain place, and so on. For details, see [3].

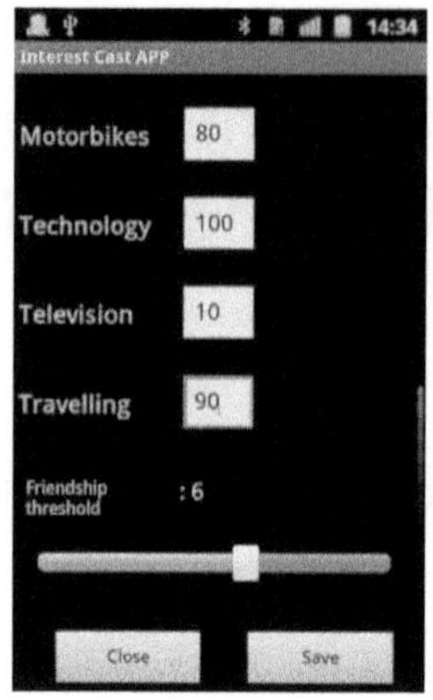

Figure 1: User's profile window

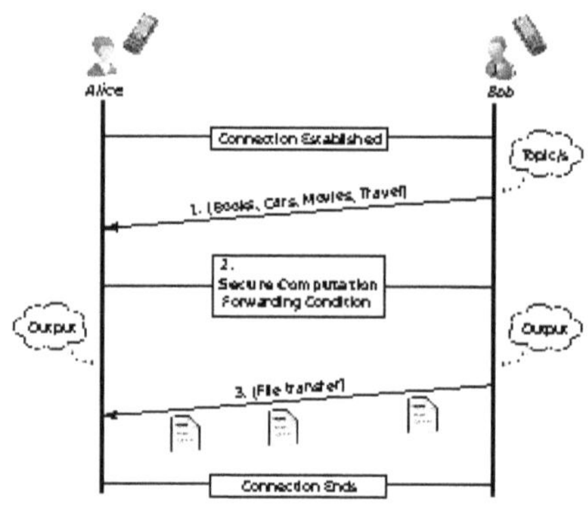

Figure 2: Protocol steps

5.2.1 *Topic/s selection*

When Alice and Bob have been established a new connection, Bob, who received the connection, randomly selects different topics to verify the forwarding condition. As soon as Alice receives the packet containing the topics selected by Bob, they start verifying the forwarding condition in a secure manner.

5.2.2 *Secure Computation of the forwarding condition*

When the Bluetooth connection is established, Bob and Alice are ready to run the secure function. Bob is the first that runs the function and waits for Alice. She uses the Bluetooth socket that has just established to connect with Bob. At this point, they start to run the secure function according to the secure steps implemented in MobileFairPlay. During this execution, both Bob and Alice use their own value of the selected topic, extracted from the interest profile, to compute vcw_e. However, these values are not sent to the other participant in plain, but they are encoded in the garbled Boolean circuits exchanged through MobileFairPlay.

Smartphone	CPU	RAM	Bluetooth Ver.	Android O.S.
Samsung Galaxy S2	Dual-core 1228 MHz	1 GB	3.0	2.3.6
Samsung Galaxy S-Plus	Single-core 1443 MHz	512 MB	3.0	2.3.5
Samsung Galaxy S	Single-core 1024 MHz	512 MB	3.0	2.3.3
Lg Optimus Dual	Dual-core 1024 MHz	512 MB	2.1	2.3.4
Htc Desire	Single-core 1024 MHz	576 MB	2.1	2.2.3

Table 1: Smartphones used for testing the interest-cast application

Smartphone	One Topic (ms)
Samsung Galaxy S2	382
Samsung Galaxy S-Plus	446
Samsung Galaxy S	492
Lg Optimus Dual	449
Htc Desire	489

Table 2: Compilation time required for one topic function

This way, at the end of the hand-shaking phase Alice and Bob only knows the result of jointly computing vcw_e, without knowing the specific interest values of the other party.

5.2.3 Files Transfer

Once the hand-shaking phase has established that Alice and Bob have similar interests, Bob sends his files to her. Our application is developed to exchange files of any kind and any extension; in fact, raw bytes are exchanged during this phase, allowing users to transfer files of arbitrary format. In our tests, we have successfully transferred text files (`.txt`), pdf files, image files (`.jpg`), etc.

6. COMPUTATIONAL - TIME EVALUATION

Here, we present the results —in terms of execution-time— of two main studies, regarding 1) compilation of the SFDL code, and 2) running of the secure function (hand-shaking phase).

Table 1 reports the specification of the models considered in our evaluation.

6.1 Compiling the secure function

The current version of the APP required that when the compilation phase is run for the first time. In Table 2, the time that each device requires to compile the secure function is shown.

6.2 Running the secure function (hand-shaking)

In this section we evaluate the time that a user, for instance Bob, needs to run the secure function, i.e., the duration of the hand-shaking phase.

Fig. 3 shows the time needed to run the secure function in the case of one topic comparison. This time includes the time needed to exchange the garbled boolean circuits through the Bluetooth interface, and to compute the output of vcw_e. The results are obtained considering the case in which the role of Alice is kept constant. In fact, while Bob is run four times with four different smartphones, Alice is always run on the Samsung Galaxy S-Plus.

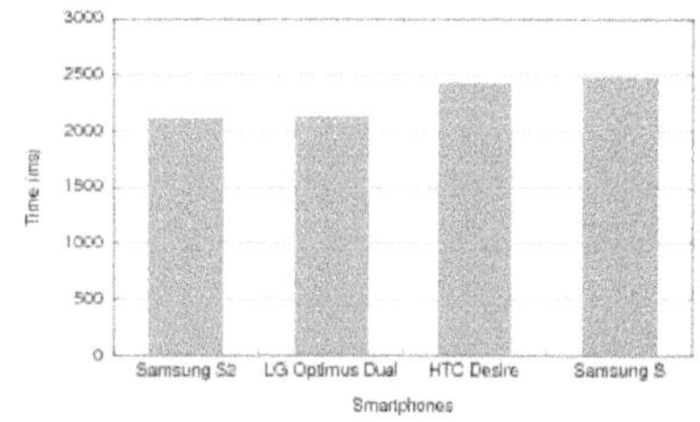

Figure 3: One topic: Bob running time of the secure function

7. OTHER DEMO INFO

7.1 Equipment to be used

For the Demo we have to use at least two smartphones. We may take with us two or three smartphones.

7.2 Space needed and setup time required

A desktop to keep our smartphones. Set up time required is 15 minutes.

7.3 Additional facilities needed

We need the power for our smartphones. Internet is optional.

Acknowledgment

Work partially supported by the EU project FP7-257930 Aniketos: *Ensuring Trustworthiness and Security in Service Composition* and by the EU project FP7-256980 Nessos: *Network of Excellence on Engineering Secure Future Internet Software Services and Systems.*

The work of P. Santi was partially supported by MIUR, program PRIN, Project COGENT.

8. REFERENCES

[1] D. Malkhi, N. Nisan, B. Pinkas, and Y. Sella, "Fairplay;a secure two-party computation system," in *Proc. of the 13th conf. on USENIX Security Symposium*, Berkeley, CA, USA, 2004, pp. 20–20.

[2] A. Ben-David, N. Nisan, and B. Pinkas, "Fairplaymp: a system for secure multi-party computation," in *Proc. of the 15th ACM conf. on Comp. and communications security*, ser. CCS '08. ACM, 2008, pp. 257–266.

[3] A. Mei, G. Morabito, P. Santi, and J. Stefa, "Social-aware stateless forwarding in pocket switched networks," in *IEEE Infocom*, 2011.

[4] G. Costantino, F. Martinelli, and P. Santi, "Privacy-preserving interest-casting in opportunistic networks," in *IEEE Wireless Communication and Net.Conf. (WCNC)*, 2012.

On-demand Content-centric Wireless Networking

Hanno Wirtz, David Martin, Benjamin Grap, Klaus Wehrle
RWTH Aachen University
Chair of Communication and Distributed Systems
{wirtz, wehrle}@comsys.rwth-aachen.de
{martin, grap}@rwth-aachen.de

ABSTRACT

Typical scenarios in the city or on campus, show a high proliferation of wireless communication devices such as smartphones, laptops or netbooks. Wireless 802.11 networks between these devices allow for a spontaneous exchange of content or provision of services without the need for infrastructure-based services. However, the above mentioned proliferation of devices and therefore large number of networks hinders users in identifying and selecting the network that serves a specific request.

We propose an approach to client-driven content-centric wireless networking in which the user specifically signals his request for a user, content item or service via 802.11 management frames. Upon reception of these frames, wireless devices that serve this request establish a dedicated wireless network on-demand. Our approach seamlessly integrates into the 802.11 association process and therefore provides support for unmodified wireless devices. Furthermore, by leveraging the wireless broadcast medium, we achieve pervasive service and peer discovery without the overhead of iterating through existing networks or running a traditional service discovery protocol.

Categories and Subject Descriptors

C.2.1 [**Computer Communication Networks**]: Network Architecture and Design—*Wireless Communication*

Keywords

Pervasive Communication, Opportunistic Networking, Content-centric, Wireless Communication

1. INTRODUCTION

The ongoing proliferation of mobile wireless devices [1] results in an abundance of possible encounters between highly capable, mobile devices in every-day scenarios, e.g., on a campus, in a company or in public places. Further scenarios in which users benefit from direct device-to-device communication instead of Internet-based communication are remote or disaster regions without network infrastructure, spontaneous data exchange, e.g., at a conference, as well as roaming and emergency scenarios. The capabilities of, e.g., netbooks and smartphones, in terms of computational power, storage and wireless communication thereby allow the provision of

MobiCom'12, August 22–26, 2012, Istanbul, Turkey.
ACM 978-1-4503-1159-5/12/08.

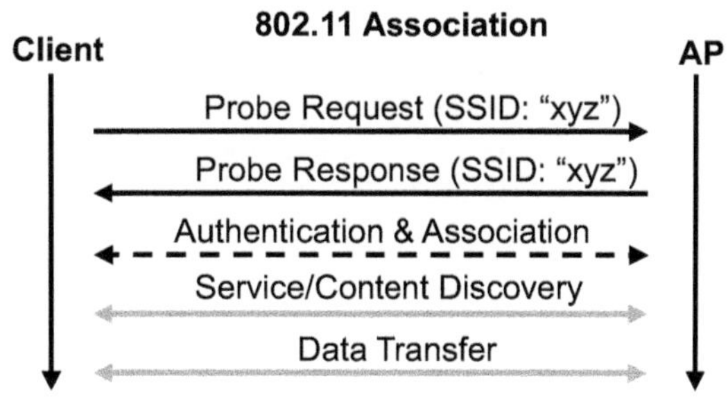

Figure 1: 802.11 association. Clients have to choose a network and associate to it before discovering whether it serves the request.

communication, sophisticated services and content directly between the mobile devices.

Applications that envision direct communication between mobile devices range from plain file downloads over specific content exchange between two or multiple peers [2, 3] to network-wide opportunistic content sharing via sequential unilateral transmissions [4]. However, the number of possible communication partners also poses a problem in spontaneous communication, namely how to find the device that serves a given request for content or a specific service or is operated by a specific user.

Imagine the following scenarios: *1)* While at a conference, a participant wants to obtain the minutes of the day's presentations and discussions, *2)* A visitor in a foreign city wants to obtain a map of the city in electronic form while being mobile, *3)* In a crowded space, a user wants to find out if a specific colleague of his is in the vicinity, i.e., in communication range. In all cases, it is impractical to ask all people around but, given the proliferation of mobile wireless devices, the user could download the requested content from another user in the vicinity or contact his colleague wirelessly.

To establish a network between the requesting user's device and the target device, the 802.11 standard defines two mechanisms. While network creation in the 802.11 ad-hoc mode, despite being designed for exactly this use case, is not supported by most current smartphones [5–7], access point (AP) functionality for infrastructure mode networks is a built-in commodity in current mobile devices. However, a user typically has no means of identifying the 802.11 networks that serve his request. This is because, even un-

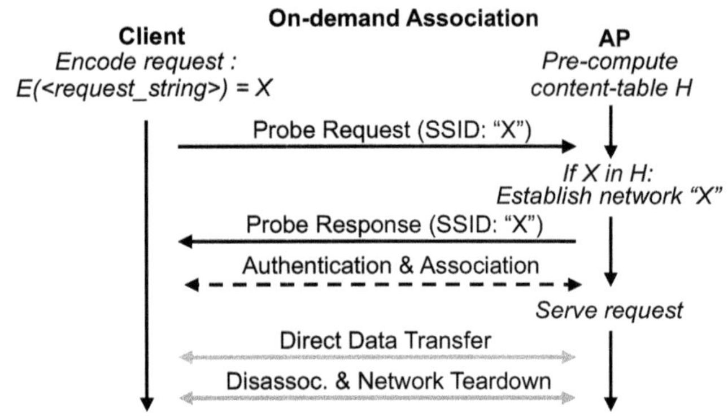

Figure 2: Client-driven, content-centric on-demand wireless network association. Clients signal their request encoded in the SSID of probe request frames. Potential AP devices check whether they serve the specific request and, if positive, establish the respective network solely for this request.

der the assumption that all nearby users, i.e., their mobile devices, provide a network, the network identifier (SSID) typically does not reflect the available content or services in the network. Furthermore, 802.11 requires devices to partake in a *specific* network prior to further communication. A mobile user thus *first* has to associate to a network (cf. Figure 1) and only as a *second* step is able to query devices in the network for the desired content, services or communication partners, e.g., using service discovery protocols such as SLP [8] or UPnP/SSDP [9].

Existing approaches indicate the semantics of the wireless network [10, 11] to facilitate the network choice of clients, e.g., in 802.11 beacon frames. However, these approaches are not able to simultaneously reflect the current user and available services and content due to the very constrained size of 802.11 frames in contrast to the number and variance in files and service descriptors. Furthermore, it is infeasible in such *AP-driven* approaches to correctly predict the upcoming requests of clients, especially with regard to local or dynamic content.

We thus propose an approach to *client-driven* content-centric wireless networking in which the requesting client device specifically requests a user, content item or service via 802.11 management frames. Wireless devices in the vicinity then establish an on-demand wireless network specifically for this request if they provide this content or service or are operated by the requested user. Client-side requests trade the necessity for the AP to signal all available content or services for a single message by the client and avoid consecutive associations to networks that do no serve the request. This efficiently enables arbitrary requests via the broadcast nature of Wi-Fi in spite of an abundance of users and potential AP devices as well as an even higher number and variance of content and services. We detail our approach in Section 2 and outline our demonstration proposal in Section 3.

2. ON-DEMAND CONTENT-CENTRIC WIRELESS NETWORKING

The goal of our approach is to let clients express their request for a user, content, or service *actively towards the wireless network environment.* We furthermore want devices in the vicinity to be able to effortlessly observe these requests and eventually react to them. To this end, we integrate the signaling of client requests and the response of devices into the 802.11 association process.

Figure 1 shows the original 802.11 process to associate with a (known) wireless network. We modify this process to accommodate client-side requests and the establishment of wireless networks triggered by these requests. Figure 2 shows the modified *on-demand* association process with a wireless network that serves the request. We provide a client-side application and extend AP functionality to support client-driven, content-centric wireless networking as follows.

Client-side request encoding.

A user that searches for a specific person, content item or service encodes the search string, i.e., the item's name or the person's name or public key, by first sanitizing the input string and then hashing the value to the corresponding key. Input sanitizing serves to ensure that, for example, `"minutes_mobicom"`, `"minutes mobicom"` and `"minutes-mobicom"` are encoded to the same key. Hashing serves to reduce the size of the input string and to provide a consistent encoding scheme at the client and the AP. In our first example, the participant searching for the presentation minutes would thus encode the input string `"presentation_minutes"` to the key `"X"` (cf. Figure 2, Client side).

To query the devices in the vicinity for the request, the client then actively sends a probe request for a network with the SSID `"X"` (cf. Figure 2). This is analogous to joining a known or hidden 802.11 network and is a built-in functionality of current operating systems. By probing for this network, the user queries for a *content-centric* network establishment, in contrast to network-centric approaches in which the user first has to connect to an existing network. However, APs in 802.11 traditionally provide a permanent network with a fixed SSID, respond only to probe requests for this SSID and are unable to react to information transported in, e.g., the SSID field.

On-demand wireless network establishment.

To allow for client-driven networking, we thus modify the AP functionality of wireless devices. In contrast to the current provision of pre-configured, fixed networks in terms of their SSID, wireless devices in our approach do not permanently provide a network. Instead, they monitor the incoming probe requests and check the request encoded in the SSID against a look-up table H that holds the encoded keys of the requests that this devices wants to serve. Construction of the look-up table follows the same encoding scheme as on the client side to derive identical keys for matching search strings. This is a one-time effort similar to generating and maintaining the content database in [4] and can be done in the background.

In case the request, i.e., the corresponding key, is listed in the look-up table, the AP device establishes an 802.11 infrastructure mode network *on-demand* for this request. To this end, we modify the functionality of the *hostapd* daemon which provides AP functionality on Linux-, Android-, and OpenWrt-based devices, among others. To signal the presence of this network to the client, we send a probe response frame that triggers the original 802.11 association process at the client. If the look-up table does not contain the key, the AP device ignores the probe request.

As the AP device only provides this network in response to the specific request, the client is able to associate directly to the network that serves his request. By integrating our approach into the original 802.11 association process we gain compatibility to legacy devices and do not require an additional discovery protocol. The on-demand nature of the network also frees the user from performing additional service discovery in the resulting network. Therefore, the AP device may serve the request based on its semantics, e.g., a file request triggers the provision of this file via an HTTP server at a predefined IP address while a person request triggers a chat or localization application. Once the client request is served, e.g., the requested content is transmitted, the AP device tears down the network as it only served for this specific request.

3. DEMONSTRATION

In our demonstration, we will showcase the integration of our approach into the 802.11 association process and its practical usage and benefits. We specifically want to introduce the community to our approach and hope to get feedback on the design and possible further development of the approach.

To this end, we will provide a number of AP devices that provide services and content. The AP devices run the aforementioned modified version of *hostapd* to monitor the incoming probe requests and to establish 802.11 networks on-demand. In our approach, AP devices are mobile wireless devices such as netbooks or smartphones (instead of Wi-Fi routers). In the demonstration, we use netbooks as APs for the better presentation capabilities compared to the small displays and cumbersome input of smartphones. Using netbooks instead of smartphones further allows the use of network virtualization. AP devices are thus able to establish multiple AP interfaces concurrently. A single request thus does not block the device for further requests.

We further provide a number of pre-installed client devices that run a small application that encodes a given search string and eventually associates with a provided network. As client devices, we use Android-based smartphones to show the compatibility of our approach to current operating systems. To further highlight the practicality and ease of use, conference participants will be able to download the client-side Android application and partake in the demonstration with their own devices[1]. Client devices are able to query for arbitrary content, services or persons, demonstrating the flexibility of our approach. However, the AP devices will only respond and create networks for the example content, services and persons they serve.

While not discussed in detail in this summary, our demonstration will highlight further important design aspects next to the aforementioned network and connection establishment. Examples are security aspects, e.g., denial of service protection and communication confidentiality, and protocol aspects, such as client behavior in case multiple AP devices respond to the same request. The demonstration further makes a distinction between *legacy* clients, i.e., devices that can not be modified except for the installation of a user-space client-side application, and *native* clients. Native clients allow a substitution of the installed 802.11 functionality with regard to the handling of, e.g., probe response frames that contain additional information in 802.11 Information Elements such as IP assignments for the client.

Setup and Requirements.

For the demonstration, we require a table to put the AP devices and access to a power plug for the netbook devices. We would benefit from a monitor to better present the functionality running on the netbooks, but this is very optional. To demonstrate the idea, design and scope of our approach, we would like to additionally present a poster.

[1]We do not provide the AP-side application as it modifies the AP functionality and accesses data on the device.

4. REFERENCES

[1] International Business Times. Report: Samsung Global Number 1 in 2012 Q1 Smartphone and Feature Phone Shipments. [Online] http://au.ibtimes.com/articles/336621/20120503/report-samsung-global-number-1-2012-q1.htm.

[2] Sewook Jung, Uichin Lee, Alexander Chang, Dae-Ki Cho, and Mario Gerla. Bluetorrent: Cooperative content sharing for bluetooth users. *Pervasive Computing and Communications, IEEE International Conference on*, 0:47–56, 2007.

[3] NearVerse. Lokast. [Online] http://www.lokast.com.

[4] Ólafur R. Helgason, Emre A. Yavuz, Sylvia T. Kouyoumdjieva, Ljubica Pajevic, and Gunnar Karlsson. A mobile peer-to-peer system for opportunistic content-centric networking. In *Proceedings of the second ACM SIGCOMM workshop on Networking, systems, and applications on mobile handhelds*, MobiHeld '10, pages 21–26, 2010.

[5] Google. Ad-Hoc Support in Android. [Online] http://code.google.com/p/android/issues/detail?id=82.

[6] MSDN Developer Forums. Wireless issues. [Online] http://forums.create.msdn.com/forums/p/64562/394857.aspx.

[7] iOS Developer Library. Document Transfer Strategies. [Online] http://developer.apple.com/library/ios/technotes/tn2152/_index.html.

[8] E. Guttman, C. Perkins, J. Veizades, and M. Day. Service Location Protocol, Version 2. RFC 2608 (Proposed Standard), June 1999. Updated by RFC 3224.

[9] UPnP Forum. Upnp device architecture, white paper. http://www.upnp.org/specs/arch/UPnP-arch-DeviceArchitecture-v1.1.pdf.

[10] Ranveer Chandra, Jitendra Padhye, Lenin Ravindranath, and Alec Wolman. Beacon-stuffing: Wi-fi without associations. In *Proceedings of the Eighth IEEE Workshop on Mobile Computing Systems and Applications*, HOTMOBILE '07, 2007.

[11] Yui-Wah Lee. Network selection and discovery of service information in public wlan hotspots. In *Proceedings of the 2nd ACM international workshop on Wireless mobile applications and services on WLAN hotspots*, WMASH '04.

Using SOS Message Propagation to Estimate the Location of Immobilized Persons

Noriyuki Suzuki, Jane Louie Fresco Zamora, Shigeru Kashihara, Suguru Yamaguchi
Nara Institute of Science and Technology
8916-5 Takayama Ikoma, Nara, Japan
{noriyuki-s, louie-z, shigeru, suguru}@is.aist-nara.ac.jp

ABSTRACT

This demonstration presents SOSCast, an application that will assist the search for immobilized persons in a disaster area. When a catastrophic disaster such as an earthquake or tsunami occurs, people may be immobilized because of an injury or become trapped in buildings and debris. However, in such a situation, since conventional communication services are likely to be severely damaged, it would be difficult for immobilized persons to ask for rescue using their mobile phones. As a solution, we present SOSCast, which propagates SOS messages through direct communication between smartphones most especially inside the disaster area. By collecting these SOS messages, rescuers can estimate the locations of people who need help using the application. In this demonstration, we will show our prototype implementation of the SOSCast for SOS message processing and its map function to estimate the location of immobilized persons.

Categories and Subject Descriptors

J.0 [**Computer Applications**]: GENERAL

General Terms

Design

Keywords

Disaster network, Location estimation, Smartphone, Store-Carry-Forward, Bluetooth

1. INTRODUCTION

Immediately after a catastrophic disaster such as an earthquake or a tsunami, a prompt rescue operation is critical to save many lives in the disaster area. It is particularly important to rescue as many disaster victims as possible within "the Golden 72 hours", which is the first 72 hours right after a disaster occurs, because the survival rate rapidly decreases when the duration has passed. It is also essential to collect information such as the current locations and statuses of disaster victims for a successful rescue operation, even if the information may not be 100 % accurate [1].

However, due to disrupted communication networks, disaster victims may not be able to easily ask for help. In a large disaster such as the Great East Japan Earthquake, communication services were disrupted because base stations were severely damaged and have ceased operation with the loss of electric power supplies [2]. For this case, not only disaster victims experienced difficulties in asking for help using their mobile phones, but rescue teams also had problems in collecting disaster information and communicating to each other.

So far, many researchers have been studying various means of communicating between smartphones directly via wireless communication media (e.g., WiFi and Bluetooth) within the affected area [3, 4, 5] as solutions for this particular problem in emergency communications. They generally focus on applications using Store-Carry-Forward routing in Delay/Disruption Tolerant Networks [4, 5], where such a direct communication between smartphones is expected to be utilized in such a situation.

In the demonstration, we present a prototype application called SOSCast that enables searching for immobilized persons using smartphones. An immobilized person with an installed SOSCast application in his or her smartphone can transmit via Bluetooth an SOS message that includes significant information, such as their name, state, and location. Upon capturing packets from smartphones carried by people passing by the area, the smartphone of an immobilized person will establish a communication with the smartphones of the pedestrians passing by and subsequently transmit the SOS message. The SOS message will then be propagated to other people in the same manner, disseminating the information to people that may be able to identify the immobilized person. In addition, every smartphone will be capable of collecting the information received from every immobilized person who has sent the SOS message. Non-rescuers who have gathered the collected information may then forward it to rescuers who are capable of rescuing immobilized persons. However, it is difficult to provide GPS information for devices located indoors. Then, to estimate the location of an immobilized person, we also utilize GPS location information from other smartphones which connected with the smartphone of the immobilized person. In the demonstration, we will show the following features:

- Creation, collection, and propagation of SOS messages via Bluetooth
- Display of an immobilized person's location utilizing locations of other smartphones on the smartphone

MobiCom'12, August 22–26, 2012, Istanbul, Turkey.
ACM 978-1-4503-1159-5/12/08.

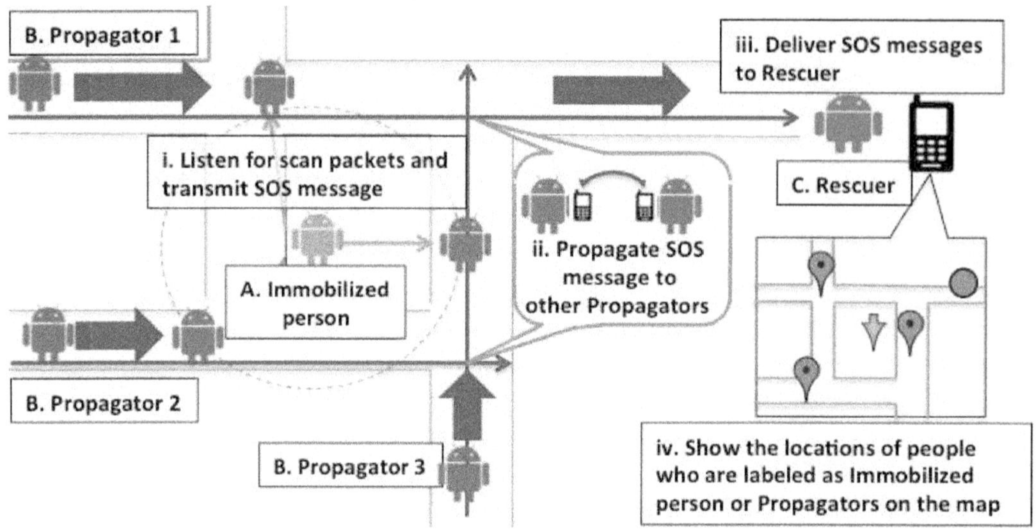

Figure 1: Overview of SOSCast

2. SOSCAST OVERVIEW

The SOSCast application is presumed to be implemented in a disaster area cut off from the conventional communication services. That is, smartphones with the installed SOSCast application can directly communicate with each other. Figure 1 illustrates how the SOSCast works. The smartphone of an immobilized person (who is presumed to be trapped in a collapsed building or is severely injured) will start to listen for scan packets broadcasted by rescuers and civilians passing by the area (see i of Figure 1). When a scan packet has been captured, the smartphone of the immobilized person communicates with the smartphone that sent the scan packet, i.e., propagator. The immobilized person then transmits an SOS message including his or her name, current condition, and GPS information to the propagator. After the propagator receives the SOS message, a corresponding message that includes the location where the propagator has received the SOS message will be sent back to the immobilized person. Hence, people that have smartphones with the installed SOSCast can automatically collect SOS messages as they approach the vicinity of the immobilized person.

Thereafter, the SOS messages will be continuously forwarded from one propagator (see ii of Fig. 1) to another until rescuers receive them. When one propagator meets another propagator and a rescuer, they will be sharing collected SOS messages with one another (see ii and iii of Figure 1). All SOS messages will be accounted for while propagating the messages promptly, such that all persons involved in the exchange of information will have a similar number SOS messages by then.

Although a person may have received an SOS message, there is a possibility that the immobilized person may not be found as he or she is not visible within the vicinity of the reported location. It is also difficult to specify the location of the immobilized person from only one SOS message since there is no guarantee that the GPS information is always accurate. As a solution, rescuers will first need to collect as much SOS messages as possible and estimate the location from the collected information, including that of the immobilized person and propagators. Utilizing not only the GPS information of the immobilized person but also of the propagators provide additional information for a precise estimation of the location of the immobilized person (see iv of Fig. 1). Note, however, that the difference between a propagator and rescuer is whether the person is capable of searching for the immobilized person or not; in the SOSCast application, both have the same function.

3. SOSCAST DEMONSTRATION

In this section, we specify the contents of our demonstration. In the demonstration, we show the two following functions, (1) creation, collection, and propagation of SOS messages, and (2) display of the locations for an immobilized person and propagators on a smartphone. These functions are implemented in an Android application, i.e., SOSCast.

We would like to demonstrate how SOSCast operates during an actual use of the application in a disaster-like situation. SOSCast has two modes, i.e., one is for an immobilized person and the other is for a propagator. Figure 2 illustrates the interface of how SOS messages will be created by an immobilized person. The immobilized person first has to fill in his or her name and describe their current state by choosing from the provided list in the right figure. If none of the descriptions in the list are satisfied, he or she can manually write the comments in "3. Other comments".

After the SOS message has been created, it will be transmitted to the propagator's device right after the immobilized person receives the scan packets from the propagator. Shown in Figure 3(a) is a screenshot of the propagators or rescuers' device when they receive SOS messages from an immobilized person. The description provides information of the immobilized person, which includes his or her name, situation, generation time of the message, and remaining battery power. Moreover, since propagators exchange SOS messages they have received as described in Section 2, the SOS messages are actually shared among several smartphones. Figure 3(b) indicates the locations of the immobilized person and the propagators that made contact with the immobilized person. By having the GPS information, we can easily dis-

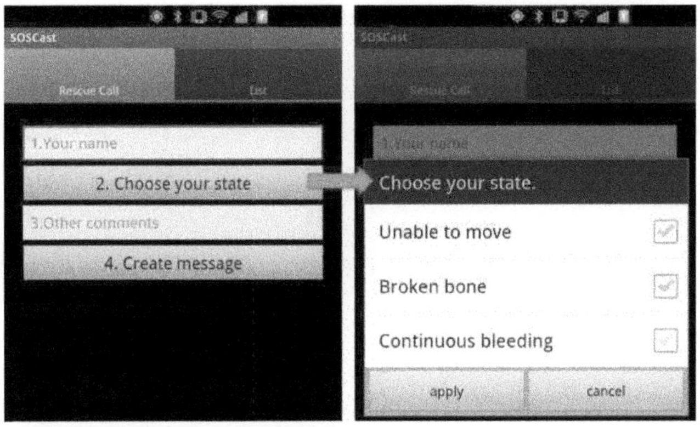

Figure 2: SOS message creation by an immobilized person

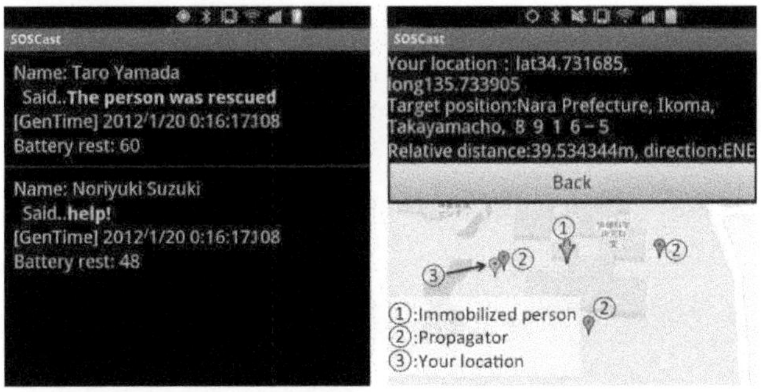

(a)SOS message view (b)Map view

Figure 3: Propagator's display

tinguish the location of an immobilized person on the map. However, to obtain more accurate information, they need to continue to obtain the locations of the immobilized person based on the propagators' location information considering that it may be incorrect or is not obtained.

4. CONCLUSION

This demonstration presented SOSCast, an application capable of estimating the location of immobilized persons by propagating SOS messages. In SOSCast, propagators can collect SOS messages of immobilized persons and distribute these collected messages to other propagators. Rescuers can then estimate the location of the immobilized person from the obtained SOS messages. In a disaster area where conventional communication services are impaired, SOSCast has potential for assisting the work of rescuers as it can enable direct communication between smartphones.

5. ACKNOWLEDGEMENT

We would like to thank Prof. Kazutoshi Fujikawa, Mr. Toru Tsuda, Mr. Toshifumi Saito, and Mr. Christopher Michael Yap, of NAIST for supporting the work. This work was supported in part by IT3 project of NAIST and Space-Time Engineering Japan, Inc.

6. REFERENCES

[1] S. Underwood. Improving disaster management. *Commun. ACM*, 53:18–20, Feb. 2010.

[2] Ministry of Internal Affairs and Communications (Japan). WHITE PAPER Information and Communications in Japan 2011. http://www.soumu.go.jp/johotsusintokei/whitepaper/eng/WP2011/2011-index.html.

[3] H.-C. Jang *et al.* Rescue information system for earthquake disasters based on MANET emergency communication platform. In *Proc. of the 2009 International Conference on Wireless Communications and Mobile Computing(IWCMC '09)*, pages 623–627, Jun. 2009.

[4] X. Wu *et al.* Emergency message dissemination system for smartphones during natural disasters. In *Proc. of 11th International Conference on ITS Telecommunications (ITST2011)*, pages 258–263, Aug. 2011.

[5] Y. Teranishi, S. Shimojo. MONAC: SNS message dissemination over smartphone-based DTN and cloud. In *Proc. of 2011 IEEE International Conference on Peer-to-Peer Computing (IEEE P2P 2011)*, pages 158–159, Sept. 2011.

Demo Abstract: MARVEL: Multiple Antenna based Relative Vehicle Localizer

Dong Li[†], Tarun Bansal[†], Zhixue Lu[†] and Prasun Sinha

Department of Computer Science and Engineering
The Ohio State University
Columbus, OH 43210
{lido, bansal, luz, prasun}@cse.ohio-state.edu
†Co-primary authors

ABSTRACT

Access to relati e location o near y e icles on t e local roa s or on t e ree ays is se l or ro i ing critical alerts to t e ri ers t ere y en ancing t eir ri ing e erience as ell as re cing t e c ances o acci ents e ro le o eter ining t e relati e location o t o e icles can e ro en into t o s aller s ro le s: i elati e lane locali ation ere a e icle eter ines i t e ot er e icle is in le t lane sa e lane or rig t lane it res ect to it an ii elati e ront-rear locali ation ere it nee s to e eter ine ic o t e t o e icles is a ea o t e ot er on t e roa n t is e o e s o a no el antenna i ersity ase sol tion MA t at sol es t e ro le o relati e lane locali ation

Categories and Subject Descriptors

Computer-Communication Networks : istri te Syste s *Distributed Applications* **Computer Applications** : o ters in t er Syste s *Consumer Products* **Information Systems** : n or ation Syste s A lications *Miscellaneous*

General Terms

esign eri entation Meas re ent Per or ance

Keywords

S art one ireless anging ocation lassi cation

1. INTRODUCTION

e se o PS e i e s art ones as een increasing ra i ly B t t e PS e ices on t e ones o not a e s cient acc racy to locali e t e e icles to t e lane-le el A aila ility o tra c in or ation at t e icro-le el or lane-le el gran larity is se l or lti le a lications t at a e t e otential to not only re ce t e c ances o acci ents t also en ance t e ri ing e erience So e a lications are as ollo s: i Alerting t e ri ers o co ing o stacles or ot oles t at are in t e sa e lane as t e e icle an r-t er g i ing t e ri er to o e to t e a ro riate lane to a oi t e ii Alerting t e ri ers i t ere is a e icle in t e lin one or i t is e icle is tailgating anot er e i-cle t ere y re cing t e c ances o collision iii Alerting t e ri er t at t e e icle a ea is slo ing o n i t e t o e icles are in t e sa e lane i etecting t e lane-le el location o slo o ing lnera le e icles an eter-ining t e i erences in s ee s o i erent lanes to assist in tra c lanning

Alt o g PS tec nology is i ely se or e ic lar locali ation ario s actors s c as signal lti at n-no n elays e to ionos ere an tro os ere error in t e cloc s o PS e ices an inacc racies in t e locations o satellites re ce its acc racy e ice an act rers s c as ar in re ort t e a erage PS acc racy to e e-ters e en or e ices e i e it ne er AAS i e Area A g entation Syste an PS i erential PS tec nology ic is not eno g or relati e locali ation o er icles n o r o n e eri ents See ig re e o-ser e t at error in PS rea ings e cee in o t e cases s PS isco tes t e *relative location* o t e ot er e icle in o t e cases Si ilarly ra ar laser an aco stic ase sensors an ca eras are so e ot er co on e ices t at a e also een se or locali ation o e icles o e er t ose sensors are s ally li ite in range to line o sig t i c lt to install es ecially on e isting e icles an e i it a tra eo et een acc racy an cost Si ilarly sing ca eras or e icle etection an lane recognition is ig ly s sce ti le to errors e to ari-o s actors s c as: i Ba lig t con itions e g nig t ti e s n glare ea lig t glare s a o s ro near y il ings ii ro er eat er con itions e g sno rain an iii S rro n ing noise e g a e lane ar s e icles ar e on roa si e roa si e cras arriers trees store ronts etc

n o r recent or e ro ose a no el lti-antenna i ersity ase sol tion calle MA t at eter ines t e relati e location o t o e icles a ong t e si ossi le ossi ilities See ig re MA co rises o t o co -onents or e ery e icle e rst co onent is a s art-one t at co l also e t e ersonal s art one o t e ri er o t e e icle e secon co onent co rises o o r ireless ra ios locate at ario s ositions on t e lat-eral si es o t e e icle e s art one or t e MA a lication r nning on it is se to co nicate it ot er s art ones in ot er e icles as ell as to is lay alert es-sages e o r ireless ra ios co nicate it ra ios locate on ot er e icles as ell as it t e s art one locate in t e sa e e icle an el in eter ining t e rel-ati e location o e icles

MobiCom'12, August 22–26, 2012, Istanbul, Turkey.
ACM 978-1-4503-1159-5/12/08.

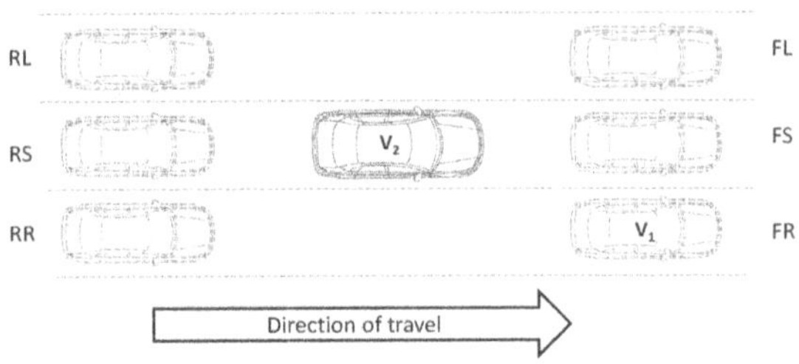

Figure 1: Vehicle V_1 can be in six different relative positions with respect to V_2: Front-Left, Front-Right, Front-Same, Front-Right, Rear-Left, Rear-Same and Rear-Right. Note that Front-Left, Front-Right, Rear-Left, and Rear-Right also include the cases that V_2 is immediately to the left or right of V_1.

e asic i ea o MA is s o n in ig re ac e icle is o nte it t o ireless ra ios on eit er si e o t e e icle ser e t at en t e t o cars are in t e sa e lane t en lin s AD an BC are ro g ly sy etrical s t e at loss al es o t ese t o lin s o l e si ilar o e er en t e cars are in i erent lanes t e lin s are not sy etrical S eci cally en V_1 is in t e le t lane o V_2 lin BC as lo at loss co are to at loss on lin AD since BC is a irect line o sig t at it no o stacles ile AD asses t ro g o ies o e icles or alls an t e engine co art ent o one o t e e icles So it is ossi le t at t e relati e signal strengt o t ese t o lin s can e se to isting is t e t ree scenarios t ere y sol ing t e relati e lane locali ation ro le n t e sa e set a ing t o ore ra ios to eac e icle si ilarly ro i es s it ore in or ation t at can e tili e to sol e relati e ront-rear locali ation ro le

n o r sc e e t e s art one transitions et een t ree i erent ases see ig re n t e onitoring ase t e s art one onitors t e accelero eter rea ings nce t e accelero eter rea ings cross a certain t res ol in icating a ossi le lane c ange e ent o r algorit o es to ne t ase n t e Beacon P ase one o t e t o ones irects its o r ireless ra ios to sen a re eter ine n - er o eacons ile t e ra ios on t e ot er e icle listen or t e eacons t ere y esti ating t e at loss et een t e airs o ireless ra ios A ter t e at loss al es o t e ossi le lin s are o taine y a s art one ro t e associate ireless ra ios in t at e icle t e algorit o es to t e t ir ase ere it eter ines t e relati e location o t e t o e icles

e i enti e t e ro le o n ing anot er e icle s o-sition a ong t e si ossi le ositions as a s er ise clas-si cation ro le an a resse it sing ac ine learning algorit s By syste atically e loring t e lace ent o t e ra ios at i erent ositions o t e e icle See ig re e o n t at t e est acc racy is ac ie e en sing t e o el create y lacing o r ra ios a o e t e o r eels o a car ig re s o s o r o t o any ra io osition co inations e a e trie it t o cars

e also e al ate t e classi cation acc racy on i erent e icle ty es ario s roa ty es an tra c con itions s-ing t e collecte ata set e create an S M classi cation

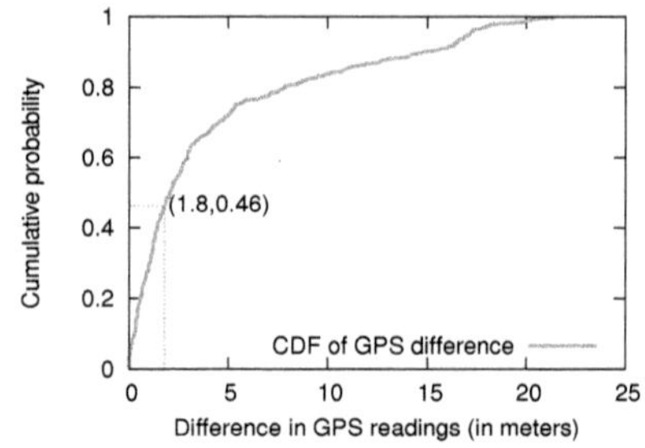

Figure 2: Comparison of GPS trace of two smartphones located in the same car: Cumulative distribution function of GPS error (in meters).

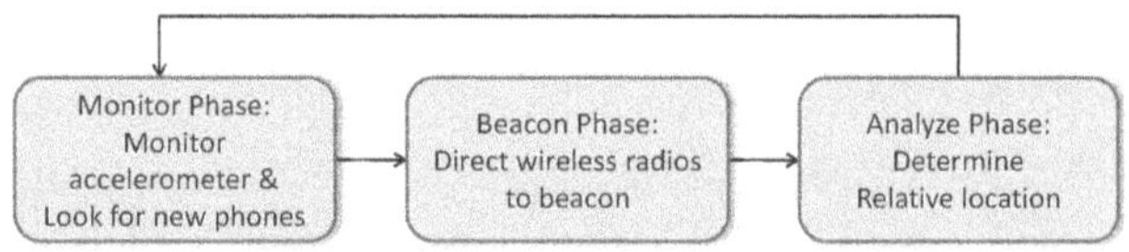

Figure 4: Transitions made by the smartphone among the three phases.

Table 1: Test results of the classifier trained by mixing all the driving data.

Testing data set	Accuracy
Mi e ata	
o se ans ocal Mo erate tra c	
o se ans ree ay Mo erate tra c	
o e an S ocal Mo erate tra c	
o e an S ree ay Mo erate ra c	
o e an S ocal ea y ra c	
o e an S ree ay ea y ra c	
o se ans on a ree ay it c r es	

o el it ig re iction acc racy e res lts are re-sente in a le

r e eri ents s o t at MA eter ines rela-ti e location o t o e icles it acc racy en teste on ata sets it i erent tra c con itions an i erent

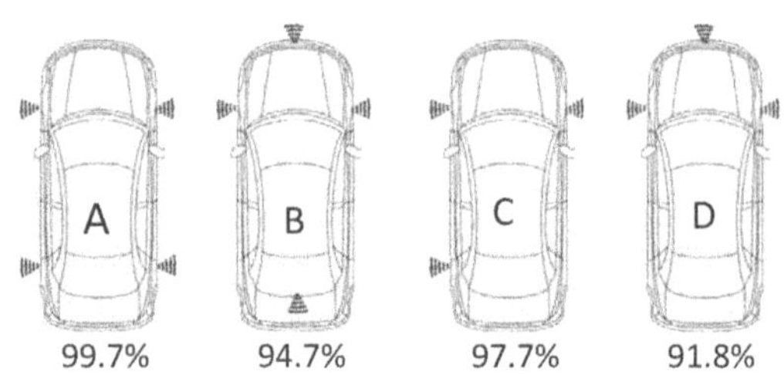

Figure 5: Some combinations of placement of the wireless radios and the corresponding relative lane localization accuracy. A and B have four radios attached while C and D have 3 radios. The first classifier from configuration A has the highest prediction accuracy.

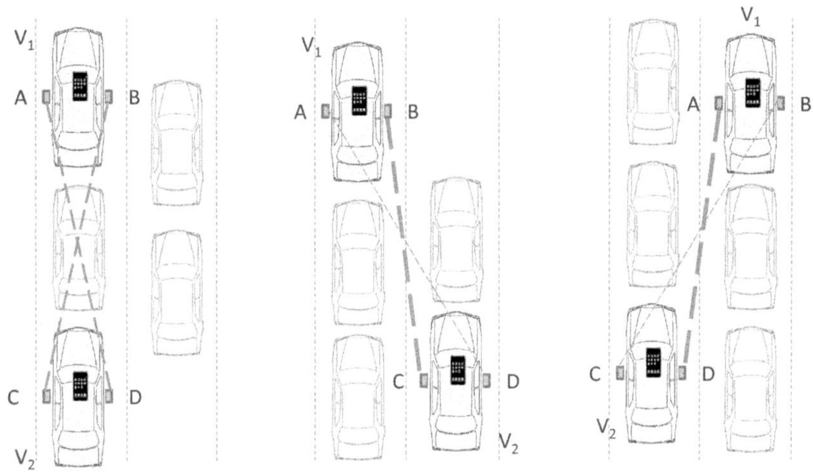

(a) Both vehicles in same lane (b) Front vehicle in left lane (c) Front vehicle in right lane

Figure 3: RSSI based Relative Lane Localization with two radios: A, B, C, and D are radios mounted on vehicles V_1 and V_2. The expected RSSI for two links AD and BC are shown for three different cases with thicker lines representing links with higher RSSI value.(a) When V_1 and V_2 are in the same lane, then the path loss for links AD and BC is almost symmetrical (Regardless of whether a vehicle is present between V_1 and V_2). (b) When V_1 is to the left of V_2, then BC is a direct line of sight link while AD passes through bodies of 2 vehicles (including one vehicle's heavy machinery compartment). (c) When V_1 is to the right of V_2, link AD is stronger but BC is weaker. Nearby vehicles may affect the signal strengths to a certain degree due to multipath.

roa ty es e elie e t at any ri ing a isory a lica- tions s c as ot ole a oi ance s en ra e alert an lin s ot arning can e e elo e sing MA

2. DEMONSTRATION

n t is e o e i le ent t e Beacon P ase an t e Analy e P ase o MA sing lti le toy- e icles t at can e o e on lti le lanes ra n on a car oar ac toy e icle ill a e t o elosB otes attac e on eac si e A la to connecte it a *central* elosB ote ill e se to gat er real-ti e res lts ro t e e icles or t e r ose o is ali ation A art ro s o ing a iniat ri e roo - o -conce t or MA n er nor al con itions t ro g t e e o e inten to s o t at MA also or s n- er ario s n a ora le con itions s c as: i Presence o lti le e icles in t e neig or oo t at inter ere it t e ireless signals trans itte et een ra ios ii Presence o r an o tacles s c as so n alls an store ront ne t to roa ays an iii en e icles a e i erent ysical ro les

n t e e o e i le ent t ese scenarios y e loying lti le toy e icles it i erent ysical i ensions on t e car oar trac r an o stacles are o ele sing t e iniat re alls on t e si es o t e car oar trac t o l e ossi le or e o artici ants to o e t e cars aro n an o ser e t e er or ance o o r lane-locali ation algo- rit Mo ing a toy e icle ill c ange its relati e location it res ect to ot er e icles t re y c anging t e signal strengt al es o t e lin s a ong t e elosB otes is c ange o l e con eye y t e otes on t e e icles to t e ote connecte to t e la to A la to - ase a lica- tion ill a to atically s o t is c ange in t e relati e lane locations A art ro o ing t e cars an ally e ill also e r nning a contin o s i eo on one o t e ot er la to s t at ill s o t e ollo ing: i o t e ireless ra ios are installe on act al e icles an ii A i eo s o ing t at a ter o ing t e toy cars MA is a le to ate t e relati e osition o t e t o e icles

3. ACKNOWLEDGEMENTS

is aterial is ase on or artially s orte y t e ational Science o n ation n er rants S- an S-

4. REFERENCES

aleo ayt eon Syste s aleo co

ella a ian an eng e icle etection an rac ing sing Aco stic an i eo Sensors n *Proc. IEEE Int'l Conf. Acoustic, Speech, and Signal Processing*

ar in at is PS tt : gar in co a o t PS in e t l

ei es an agel o ar s Acti e Mac ine- ision-Base ri er Assistance or r an Areas *Int'l J. Computer Vision* : –

i an Bansal an an Sin a P MA : M lti le Antenna ase elati e e icle ocali er n *Proc. of ACM Mobicom*

S Par i S ang an eon A o el Signal Processing ec ni e or e icle etection a ar *IEEE MTT-S Int'l Microwave Symp. Digest*

ang or e an A S e a ar-Base etection an rac ing o Mo ing ects ro a ro n e icle at ig S ee s n *Proc. of IEEE Intelligent Vehicle Symp.*

Demo: Clock Calibration Using Fluorescent Lighting

Zhenjiang Li[1], Cheng Li[1], Wenwei Chen[1], Jingyao Dai[1], Mo Li[1], Xiang-yang Li[2], Yunhao Liu[3]
[1]Nanyang Technological University, Singapore
[2]Illinois Institute of Technology, USA
[3]Tsinghua University, China
{lzjiang, cli6, chen0746, jydai, limo}@ntu.edu.sg, xli@cs.iit.edu, yunhao@greenorbs.com

ABSTRACT

In this demo, we propose a novel clock calibration approach called FLIGHT, which leverages the fact that the fluorescent light intensity changes with a stable period that equals half of the alternating current's. By tuning to the light emitted from indoor fluorescent lamps, FLIGHT can intelligently extract the light period information and achieve network wide time calibration by referring to such a common time reference. We address a series of practical challenges and implement FLIGHT in TelosB motes. In this demonstration, we will show that by taking advantage of the stability of the AC frequency, the detected light intensity, even from different lamps, exhibits a consistent and stable period. FLIGHT can achieve tightly synchronized time with low energy consumption. In addition, since FLIGHT is independent to the network message exchange, time synchronization can be retained even when the network is temporarily disconnected. Such characteristics particularly suit various mobility-enabled scenarios.

Categories and Subject Descriptors

C.2.2 [**Computer Communication Networks**]: Network Protocols; C.2.4 [**Computer Communication Networks**]: Distributed Systems

General Terms

Design, Performance

Keywords

Clock calibration, Fluorescent lighting, Energy efficiency

1. INTRODUCTION

Maintaining a common notion of time is one fundamental service in many distributed networking systems. A variety of applications depend on the availability of tightly synchronized time cross network nodes. Retaining a common time in the network, however, faces substantial challenges. Due to the low-cost design, CMOS crystal oscillators serve as the most privileging signal source to generate on-board clocks. The frequency of an oscillator is not stable and it fluctuates with the surrounding environment. Thus, clocks on different nodes need to be precisely calibrated such that consistent time can be maintained. Due to the inherent uncertainty, no matter how accurately clocks are initially calibrated, they will ultimately tick towards divergency.

MobiCom'12, August 22–26, 2012, Istanbul, Turkey.
ACM 978-1-4503-1159-5/12/08.

Great efforts have been made in the past decade to address above issues. Proposed solutions mainly rely on heavy intercommunications between nodes to exchange their local time references [1]. The calibration error, however, gets accumulated hop by hop exponentially. Even that, those solutions lead to excessively high communication overhead and significant power drain in the system. Against the problems, an emerging type of solutions have been recently proposed, which make use of certain external signal sources as common time references. Typical examples are using power lines [2], FM radio [3], Wi-Fi [4]. Although such newly emerging solutions can dramatically reduce the communication cost, they also utilize radios for the clock calibration at the expense of higher power consumption. Even if clock calibration can be performed less frequently, the energy consumption of *each* calibration could remain high. Besides, most of those solutions need specific hardware components to generate or capture periodical signals, which introduce extra cost and design complexity.

In this study, we propose a new clock synchronization approach that allows network nodes to receive a globally available clock reference by tuning to the fluorescent lighting. Our approach is based on the following observations and facts. Alternating Current (AC) is a periodical signal with a frequency of 50 or $60Hz$ and the frequency has adequate stability of $5 \cdot 10^{-5}$ measured by the Allan variance. Powered by AC, the fluorescent light intensity thus changes with a stable period that equals half of the alternating current's Commonly available in most indoor environments like universities, airports, hospitals, and supermarkets, the fluorescent light provides a universal period reference. On the other hand, the light sensor or camera is widely embedded in commodity wireless platforms, e.g., sensor nodes, smart phones and notebooks. This offers a great opportunity for a plenty of indoor applications to maintain common time by referring to the light emitted from fluorescent lamps.

We propose the Fluorescent LIGHTing (FLIGHT) in this demo and the goals of the demo can be summarized as the following four aspects.

- FLIGHT does not require any extra hardware support. By sampling the light sensor or camera with a sufficiently high rate, accurate period information can be detected for clock calibration. Hence, our solution is widely available on most existing commodity platforms.

- Taking advantage of the stability of the AC frequency, the detected light intensity, even from different lamps, exhibits a consistent and stable period, which ensures high synchronization accuracy.

- Compared with prior radio operation based solutions, sam-

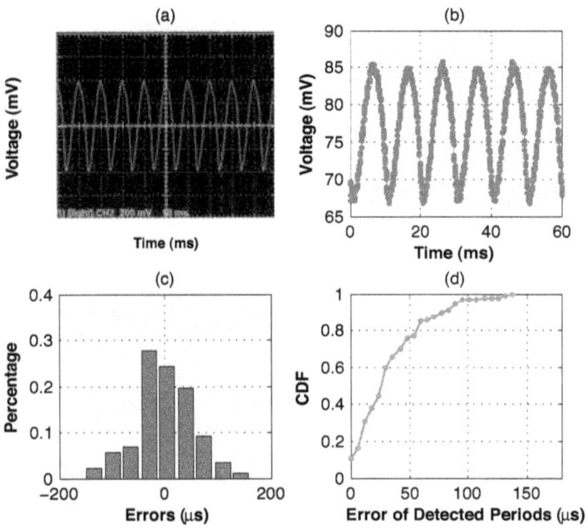

Figure 1: Single-lamp experiment in the laboratory

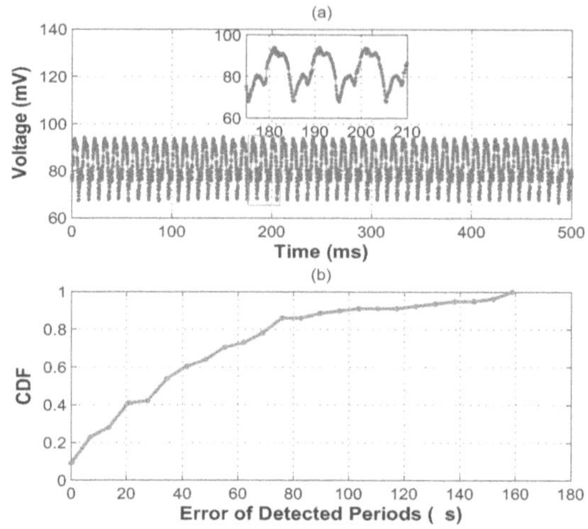

Figure 2: Multi-lamp experiment in the laboratory

pling the light sensor consumes substantially less energy, which provides us a lightweight solution.

- Since our approach is independent to the network message exchange, time synchronization can be retained even when the network is temporarily disconnected. When individual devices are moving, they can still obtain desired periodical patterns for clock calibration. Such characteristics particularly suit various mobility-enabled scenarios.

The rest of this demo will be organized as follows. In Section 2, we will introduce the concept of clock calibration, the mechanism of fluorescent lighting, the design challenges, and the system architecture of FLIGHT. The facilities requirements of the demo is presented in Section 3.

2. SYSTEM OVERVIEW

2.1 Clock calibration

A *native clock* is the clock driven by a node's built-in crystal oscillator directly. We denote the native clock as $c_n(t_0+t)$, referring to the measured time duration from an initial time t_0 until t by the native clock. Without loss of generality, we assume t_0 to be zero; hence the native clock can be written as $c_n(t)$ for short. The periodical pattern of the fluorescent light intensity can be treated as a *global reference clock*, which is denoted as $c_g(t)$. In our system, the AC frequency is $50Hz$; thus, the frequency of the global reference clock f_g is $100Hz$ and the time unit of the global reference clock is $1/(100Hz) = 10ms$. In addition, each node maintains a *logic clock*, written as $c_l(t)$. The logic clock is used by upper-layer applications and it advances as follows:

$$c_l(t) \quad = \quad c_l(0) + \int_0^t r(\tau)d\tau, \qquad (1)$$

where $r(\tau)$ is the instant rate of the native clock at time τ. The goal of FLIGHT is to ensure **logic clocks** consistent among different nodes.

2.2 Mechanism of fluorescent lighting

The fluorescent lamp is a gas discharge lamp using electricity to excite mercury vapor. When the lamp is turned on, the electric power heats up the cathode to emit electrons. Emitted photons are absorbed by electrons in the atoms of the interior fluorescent coating of lamp, leading to physical reactions with emission of visible light. After being heated up, the gas conductivity inside the lamp rapidly rises, allowing the alternating current to flow through and continuously emit light. From the principle of fluorescent lighting, the emitted light is expected to exhibit a periodical pattern since its AC power source is a periodical signal. As the period of AC is adequately stable and phase-coherent, we expect that the light generated by fluorescent lamps can be used as a periodic reference for clock calibration.

We have conducted some preliminary experiments to validate the initial idea of FLIGHT in Figure 1 [5]. As a benchmark, we plot the instant voltage value from two frontend pins of the light sensor using an oscilloscope. We can observe that the reading from oscilloscope is quite stable and exhibits a regular fluctuation with a period of $10ms$ (depicted in Figure 1(a)). In practice, sensor nodes need to sample the light sensors. After sampling, we see that the sampling results exhibit the same periodical property as depicted in Figure 1(b), which suggests that the obtained light intensity pattern can be used for clock calibration. Figure 1(c) and (d) show summarize the statistics of the detected period lengths from the trace.

2.3 Design challenges

There exist plenty of design challenges for FLIGHT.

First, a node needs to extract the light period by sampling and further generates signals with the same period to calibrate clocks. There may exist multiple peaks in each light period as we have shown in Figure 2 [5]. The light intensity can further fluctuate when we consider node mobility (movement, rotation, etc.). We find that simple heuristics may not precisely extract periods. Existing digital processing techniques, however, introduce unacceptable computation burdens. Thus, it is not trivial to design lightweight approach for nodes to extract accurate period information from fluorescent lighting.

Second, in order to precisely calibration the clock, nodes need to select an appropriate calibration window size. By intuition, if the window size is too small, the sampling error and jitter cannot be effectively canceled out. On the other hand, if the window size is too large, the calibration will suffer a long latency to finish, which may disturb the processing of MCU for other tasks. Both scenarios will deteriorate the overall system performance. The interval

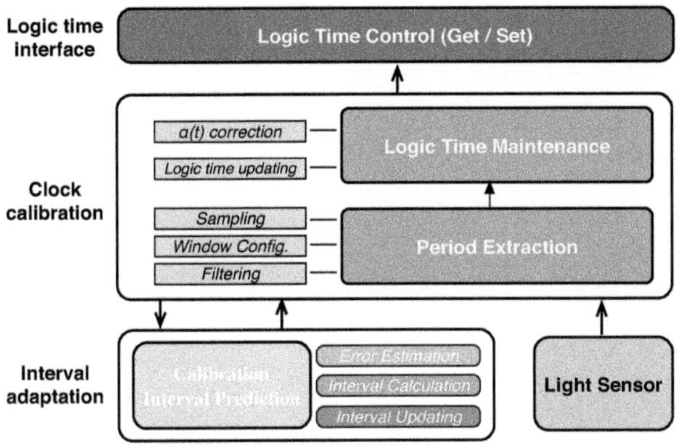

Figure 3: Illustration of the FLIGHT architecture

length between two consecutive calibrations should be appropriate as well. As we configure the clock calibration in a periodical work mode, we need to guarantee that the clock errors should not be significantly accumulated during each calibration interval. Optimizing such calibration settings needs in-depth investigation in mathematics and practices.

Third, several practical issues must be addressed in the system implementation, e.g., although nodes can launch the sampling operation with a sufficiently high rate, e.g. $32KHz$, the standard sampling module for the light sensor may not be able to respond in time. For instance, by default, the sampling module in TelosB motes needs around $1ms$ to sample the light sensor once. No matter how fast nodes trigger the module, the effective sampling rate will not become higher as long as the built-in module is adopted directly. Even we can derive a proper sampling window size to guarantee the calibration accuracy, the limited buffer size in most commodity wireless devices prevents to hold sufficient samples during the calibration. As a result, we need to break all such technical barriers to approach a good calibration performance in FLIGHT.

2.4 System architecture and implementation

Figure 3 depicts the system architecture. There exist three major components in FLIGHT: clock calibration, logic time interface, and interval adaptation. Based on the readings from the light sensor, the period generation module in the calibration component is launched to produce the global reference clock for later calibration. The output of the period generation module acts as the input of the logic time maintenance module, which is responsible for updating the frequency ratio and correcting the accumulated drift between two calibrations. The newly obtained logic time can be used by upper layer applications. Meanwhile, the output from the logic time maintenance module can be adopted to determine the interval length for the next calibration.

We implement FLIGHT in TinyOS 2.1x on the TelosB platform. The kernel of FLIGHT includes 600 lines of NesC code that was complied to 2242 bytes of RAM and 16114 byes of ROM. As a result, FLIGHT only occupies 23% of the total buffer and the remaining space is available for other applications. We use the built-in light sensor on the TelosB mote in our experiments. The light sensor utilizes an ADC module to measure the light intensity and records it as a corresponding voltage value in the register. The standard TinyOS packages the ADC module and relies on the MSP430 Timer to control the ADC sampling. The maximum sampling rate of the MSP430 Timer is around $1KHz$, which is not fast enough for FLIGHT. To overcome this issue, we write a driver to directly access the register of the ADC module. By so doing, the effective sampling rate can break the $1KHz$ barrier and achieves up to $83KHz$ in FLIGHT.

3. FACILITY REQUIREMENTS

Our facility requirements can be listed as follows.

- **Equipment requirement**: Several TelosB sensor motes are needed, which will be synchronized by FLIGHT in the exhibition. In addition, one or two notebook computers will also be used to collect the information sent back from sensor motes and display the results through the GUI interface.
- **Space requirement**: There is no special requirements for the exhibition space. A normal desk or table will work. On the desk or table, notebooks and sensor motes will be placed.
- **Power and Internet requirement**: The power plug is needed for the notebooks. Sensor motes will be powered by the battery. In the demo, the Wi-Fi connection might be required.

4. OTHER INFORMATION

We acknowledge the support from NTU Nanyang Assistant Professorship (NAP) grant M4080738.020 and Microsoft research grant FY12-RES-THEME-001. The research of Xiang-Yang Li is partially sûpported by NSF CNS-0832120 and NSF CNS-1035894. The research of Yunhao Liu is supported by the NSFC Distinguished Young Scholars Program under grant No. 61125202 and NSFC Major Program 61190110. This demo is related to our recent work [5].

5. REFERENCES

[1] M. Maroti, B. Kusy, G. Simon, and A. Ledeczi, "The flooding time synchronization protocol," in *Proc. of ACM Sensys*, 2004.

[2] A. Rowe, V. Gupta, and R. Rajkumar, "Low-power clock synchronization using electromagnetic energy radiating from ac power lines," in *In Proc. of ACM Sensys*, 2009.

[3] L. Li, G. Xing, L. Sun, W. Huangfu, R. Zhou, and H. Zhu, "Exploiting fm radio data system for adaptive clock calibration in sensor networks," in *In Proc. of ACM Mobisys*, 2011.

[4] T. Hao, R. Zhou, G. Xing, and M. Mutka, "Wizsync: Exploiting wi-fi infrastructure for clock synchronization in wireless sensor networks," in *In Proc. of IEEE RTSS*, 2011.

[5] Z. Li, W. Chen, C. Li, M. Li, X. Li, and Y. Liu, "Flight: Clock calibration using fluorescent lighting," in *Proc. of ACM Mobicom*, 2012.

A Wideband Compressed Spectrum Sensing Platform for Dynamic Spectrum Access Networks

Qiang Liu
Institute of Computing Technology,
Chinese Academy of Sciences
Beijing, China
+86 10 62600766
liuqiang@ict.ac.cn

Ze Zhao
Institute of Computing Technology,
Chinese Academy of Sciences
Beijing, China
+86 10 62600723
zhaoze@ict.ac.cn

Li Cui
Institute of Computing Technology,
Chinese Academy of Sciences
Beijing, China
+86 10 62600742
lcui@ict.ac.cn

ABSTRACT

Dynamic spectrum access (DSA) networks can significantly improve the networks performance and efficiency of spectrum utilization. To harness this capability, spectrum sensing is a fundamental problem for DSA networks. This demonstration shows our wideband compressed spectrum sensing (WCSS) platform. This platform supports the controllable spectrum environment, high performance and low cost wideband spectrum sensing based USRP2, powerful and flexible compressed sensing computing platform based MATLAB. With this platform, we can research the wideband compressed spectrum sensing algorithm by actual and controllable spectrum environment.

Categories and Subject Descriptors

C.2.1 [**Computer-Communications Networks**]: Network Architecture and Design – *Wireless communication*; C.4 [**Performance of Systems**]: *Design studies, Measurement Techniques.*

General Terms

Design, Measurement, Verification, Experimentation.

Keywords: Wideband Spectrum Sensing, Compressed Sensing, Platform, Dynamic Spectrum Access.

1. INTRODUCTION

Accurate and timely sensing of a wide spectrum band in a highly dynamic environment is the main challenge for DSA networks [1], [2]. Fortunately, if a large part of the spectrum is unoccupied, that is, the signal is frequency- domain sparse, we can use the recently developed compressed sensing to reduce the sampling rate by a large margin [3], [4]. A powerful and flexible platform is the wideband compressed spectrum sensing research and application foundation. In [5], the scholars use a spectrum analyzer sampling live spectrum signal over the air. MATLAB is a very powerful tool to study compressed sensing, and a lot of literatures utilize MATLAB to implement the compressed sensing algorithms [6], [7]. In [2], the compressed sensing recovery performance is evaluated in MATLAB by using simulated data and USRP measured data for continuous waves and pulses, and the randomly spaced temporal sampling compressed sensing solution on USRP captured data. Therefore a powerful and flexible platform for wideband compressed spectrum sensing study is important.

MobiCom'12, August 22–26, 2012, Istanbul, Turkey.
ACM 978-1-4503-1159-5/12/08.

2. MOTIVATION

Since primary users have priority in using the licensed spectrum, when secondary users coexist with primary users, they have to perform timely wideband monitoring of the spectrum to be used. However the spectrum sensing for wideband is a serious burden to be endured by second users. Compressed sensing can remarkably decrease the hardware requirement for wideband spectrum sensing. Therefore wideband compressed spectrum sensing is a promising method. Meanwhile compressed sensing studies have brought plenty of algorithm programs, and most of them are based on MATLAB, such as SparseLab [8], l1-Magic [9], GPSR [10], SGPL1 [11] and so on. Thus constructing the WCSS algorithms based on MATLAB will be very convenient.

In terms of the spectrum sensing platform, if we can make a spectrum environment which we expect, it will lend itself to the spectrum sensing study. In our WCSS platform, we use two USRP2 software defined radio devices, one as a transmitter and the other as the spectrum sensing receiver. For implementing a controllable spectrum environment, we connect these two USRP2 with a coaxial cable, and the transmitter USRP2 can generate many kinds of signal including noise. Hence the repeatable experiments can be carried out with the same wireless environment by our platform.

For wideband spectrum sensing, the front-end is required to attain wide band spectrum information, USRP2 owns the wideband spectrum sensing ability.

3. DESIGN AND IMPLEMENTATION

Considering the requirements of WCSS platform, there are two main parts, one is the transmission part and the other is reception part. The USRP2s of two parts are connected with a coaxial cable. The System architecture of platform is shown as Figure 1.

The transmission program bases on Simulink. Simulink is integrated with MATLAB, providing immediate access to an extensive range of tools that is used to develop algorithms, analyze and visualize simulations. With the various signal source of Simulink, the transmission part can provide many spectrum environments including additional noise. MATLAB and Simulink connect to the USRP software-defined radios from Ettus Research LLC to provide a radio-in-the-loop design and modeling environment. With the support package, Communications System Toolbox, and a USRP radio, researchers can design and verify practical SDR systems starting with Release 2011a of MATLAB and Simulink. At the same time GNU Radio open-source software is not required to use this support package [12].

We impose MathWorks company support for the USRP2, using the Communications System Toolbox, which supports the USRP2 Hardware Driver to provide a real-time connection to the USRP family of radios directly from Simulink.

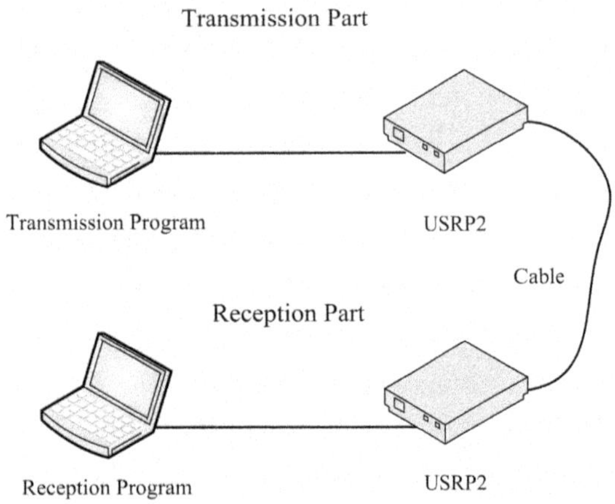

Figure 1. System architecture of the platform.

The USRP2 enables researchers to rapidly design and implement powerful, flexible software radio systems [13]. High speed and high precision ADCs and DACs allow for wider band signals while at the same time increasing dynamic range. The USRP2 owns two 100 MS/s 14-bit analog to digital converters, two 400 MS/s 16-bit digital to analog converters, so we can utilize the USRP2 to achieve the wideband spectrum data by four high-speed analog to digital converters.

Similar with the transmission part, the reception part is composed of a Laptop computer with Simulink and the USRP2. To fulfill the compressed sensing, we use the SparseLab package. SparseLab is a MATLAB software package designed to find sparse solutions to systems of linear equations, particularly underdetermined systems. SparseLab includes a number of examples to demonstrate compressed sensing algorithms. Since there are SparseLab functions which call WaveLab functions, so we download and install the WaveLab first from website [14]. We base on the SparseLab program to build the Simulink block.

The Block diagram of transmission program is shown as Figure 2. We mix a sine wave source and a random source to generate baseband signal. Then the baseband signal is inputted to the USRP2 device. With the SDRu Transmitter block, we can set the center frequency, gain, interpolation and so on. This block accepts a column vector input signal from Simulink and transmits signal and control data to the USRP2 board.

The Block diagram of reception program is shown as Figure 3. SDRu Receiver block receives the RF signal from the USRP2 device. The SDRu Receiver block enables communication with a USRP2 board on the same Ethernet subnetwork. This block is a Simulink source that receives data from a USRP board and outputs a column vector signal of fixed length. With the SDRu Receiver block, we can set the center frequency, gain, decimation, sample time, output data type, frame length and so on. The compressed spectrum block fulfills the compressed spectrum sensing.

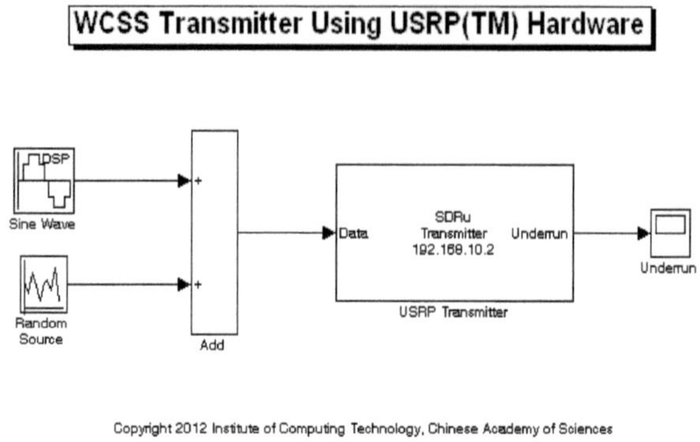

Figure 2. Block diagram of transmission program.

The block diagram of compressed spectrum sensing is shown as Figure 4. We first transform the IQ samples from the SDRu Receiver block to amplitude data. Then compressed sensing starts with taking a weighted linear combination of samples (compressive measurements) in a basis different from the basis in which the signal is known to be sparse. The signal is reconstructed and we use FFT to conclude if the primary user is active.

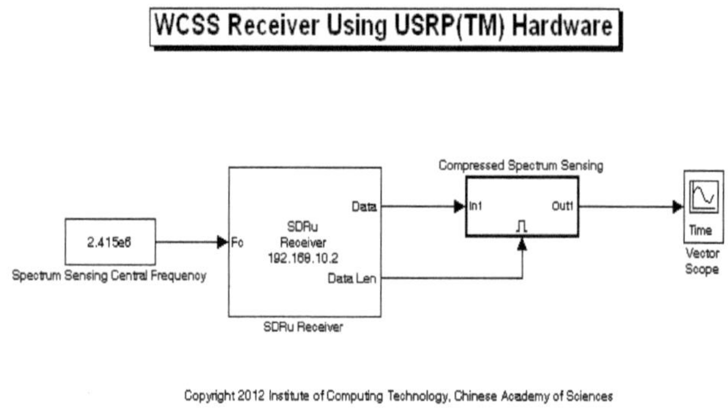

Figure 3. Block diagram of reception program.

According to many compressed sensing program of the SparseLab package, we need to create a matrix in each time step to do certain operations. Nevertheless, the matrix dimensions depend on a parameter that changes from one time step to another. That is, matrix dimensions change in each time step. However when we put the program into Simulink block, the matrix dimensions cannot be changed during runtime. Our solution is that if we know the maximum value of this parameter we can create a matrix of that dimension and then operate only on the size specified by the parameter.

4. GOALS OF THE DEMONSTRATION

Our demo will show the implementation of the wideband compressed spectrum sensing platform.

To demonstrate the capabilities of the WCSS platform, the different compressed sensing methods under different wireless environments is setup and evaluated in this section as Figure 5. We will show three compressed sensing methods: BP, OMP and StOMP.

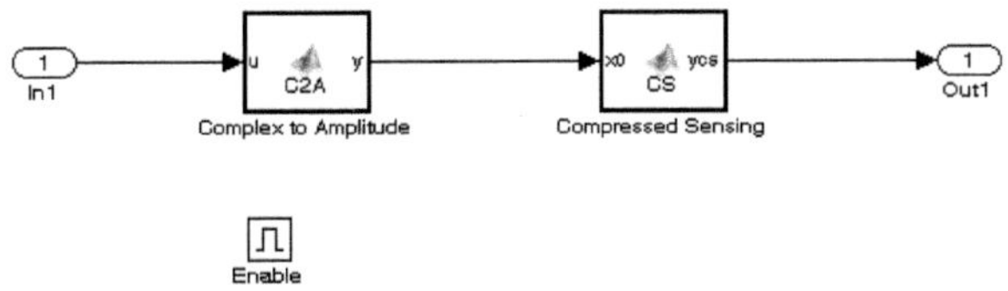

Figure 4. Block diagram of compressed spectrum sensing.

With signals of different SNR, we will present the probability of detection in various SNR sensing environments when the false alarm probability is fixed.

Compressed sample rate is a key parameter for WCSS, so we will show the average spectrum sensing performance of the BP, OMP and StOMP algorithms when different compressed sample rates are applied.

Figure 5. Demonstration scenario.

5. EQUIPMENT TO BE USED

The equipment that will be used in the demo includes: 2 Laptop computers and 2 USRP2 devices.

6. SPACE NEEDED-SETUP TIME

A sample setup is presented in Figure 5. The space needed is a 120 cm x 60 cm table space. We need 10 minutes to setup our demo.

7. ADDITIONAL FACILITIES NEEDED

To setup our demo we will need one connection to a power outlet with ground. We will provide our own extension cord to provide 4 plugs for our setup. Note that our demo will not need any wired or wireless connection to any network or Internet.

8. ACKNOWLEDGMENTS

The authors thank Prof. D. Donoho et al. for sharing the SparseLab software package online. This work is supported by the National Basic Research Program of China (973 Program) under Grant No. 2011CB302803, the "Strategic Priority Research Program" of the Chinese Academy of Sciences, Grant No. XDA060307000, and the IoT Development Project of MIIT and MoF under title "Research & Development of IoT Application Middleware and Its Industrialization".

9. REFERENCES

[1] Wang, B., and Liu, K. J. R. 2011. Advances in Cognitive Radio Networks: A Survey. *IEEE Journal of Selected Topics in Signal Processing*. 5, 1 (Feb. 2011), 5-23.

[2] Sakarya, F. A., Nagel, G. S., Tran, L. H., and Molnar, J. A. 2011. Wideband Compressed Sensing for Cognitive Radios. In *Proceedings of the 2011 Military Communications Conference* (Baltimore, Maryland, USA, November 07 - 10, 2011), MILCOM 2011. IEEE, Piscataway, NJ, 31-36. DOI= http://dx.doi.org/10.1109/MILCOM.2011.6127684.

[3] Candes, E. J., Romberg, J., and Tao, T. 2006. Robust uncertainty principles: exact signal reconstruction from highly incomplete frequency information. *IEEE Transactions on Information Theory*. 52, 2 (Feb. 2006), 489-509.

[4] Donoho, D. L. 2006. Compressed sensing. *IEEE Transactions on Information Theory*. 52, 4 (Apr. 2006), 1289-1306.

[5] Laska, J. N., Bradley, W. F., Rondeau, T. W., Nolan, K. E., and Vigoday, B. 2011. Compressive sensing for dynamic spectrum access networks: Techniques and tradeoffs. In *Proceeding of 2011 IEEE International Symposium on Dynamic Spectrum Access Networks* (Aachen, Germany, May 03 – 06, 2011), DySPAN 2011. IEEE, Piscataway, NJ, 156-163. DOI= http://dx.doi.org/10.1109/DYSPAN.2011.5936202

[6] Ji, S., Xue, Y., and Carin, L. 2008. Bayesian Compressive Sensing. *IEEE Transactions on Signal Processing*. 56, 6 (Jun. 2008), 2346-2356.

[7] Maleki, A., and Donoho, D. L. 2010. Optimally Tuned Iterative Reconstruction Algorithms for Compressed Sensing. *IEEE Journal of Selected Topics in Signal Processing*. 4, 2 (Apr. 2010), 330-341.

[8] SparseLab: Sparse solutions to systems of linear equations, particularly underdetermined systems. http://sparselab.stanford.edu/

[9] l1-Magic: A collection of MATLAB routines for solving the convex optimization programs central to compressive sampling. http://www.l1-magic.org/

[10] GPSR: Gradient projection for sparse reconstruction. http://www.lx.it.pt/~mtf/GPSR/

[11] SGPL1: A solver for large-scale sparse reconstruction. http://www.cs.ubc.ca/labs/scl/spgl1/

[12] USRP hardware support from MATLAB and Simulink. http://www.mathworks.cn/discovery/sdr/usrp.html;jsessionid=629363ec2cc3c2efc644be51a5b2

[13] Ettus Research LLC. http://www.ettus.com/

[14] WaveLab: a variety of algorithms related to wavelet analysis. http://www-stat.stanford.edu/~wavelab/

Author Index

www.ingramcontent.com/pod-product-compliance
Ingram Content Group UK Ltd.
Pitfield, Milton Keynes, MK11 3LW, UK
UKHW051413290926
15471UKWH00009B/1590